A Collection of Technical Papers

35TH INTERSOCIETY ENERGY CONVERSION ENGINEERING CONFERENCE & EXHIBIT (IECEC) VOLUME 1

LAS VEGAS, NEVADA
24 JULY – 28 JULY 2000

IEEE Catalog Number: 00CH37022
ISBN: 1-56347-375-5 (Softbound)
ISBN: 0-7803-5707-8 (Casebound)
ISBN: 0-7803-5708-6 (Microfiche)

Printed in the U.S.A.

A Collection of the
35th Intersociety Energy Conversion Engineering Conference & Exhibit
Technical Papers
24 July – 28 July 2000
Las Vegas, Nevada

Table of Contents

AIAA-2000-2800

EFFICIENCY AND REGULATION OF COMMERCIAL LOW POWER DC/DC CONVERTER MODULES AT LOW TEMPERATURES

Malik E. Elbuluk
Electrical Engineering Dept.
University of Akron
Akron, OH 44325-3904

Scott Gerber & Ahmad Hammoud
Dynacs Engineering, Inc.
NASA Glenn Group
Cleveland, OH 44135

Richard L. Patterson
NASA Glenn Research Center
MS 301-5
Cleveland, OH 44135

ABSTRACT

DC/DC converters that are capable of operating at cryogenic temperatures are anticipated to play an important role in the power systems of future NASA deep space missions. Design of these converters to survive cryogenic temperatures will improve the power system performance, and reduce development and launch costs.

At the NASA Glenn Research Center Low Temperature Electronics Laboratory, several commercial off-the-shelf dc/dc converter modules were evaluated for their low temperature performance. Various parameters were investigated as a function of temperature, in the range of 20 °C to -190 °C. Data pertaining to the efficiency and voltage regulation of the tested converters is presented and discussed.

BACKGROUND

Electrical power components and systems for future NASA space missions, such as outer planetary exploration and deep space probes, must operate reliably and efficiently in very low temperature environments. For example, inter-planetary probe launched to explore the rings of Saturn would experience a temperature of about -183 °C. Table 1 shows the operational temperatures for an unheated spacecraft in the vicinity of each of the outer planets. These spacecraft include deep space probes, planetary orbiters and landers, and surface exploratory instrumentation. Presently, spacecraft operating in these regions utilize Radioisotope Heating Units (RHUs) to maintain an operating temperature for the on-board electronics of approximately 20 °C [1]. RHUs, which are always on, require associated containment structures and thermal systems such as shutters to maintain the 20 °C over the course of the entire mission. However, if the electronics were capable of operating at the temperature of the mission environment, the RHUs and their associated structures and thermal systems could be eliminated, reducing system size and weight and thereby reducing system development and launch costs, and improving reliability and lifetime [2].

Mission	Temperature °C
Mars	-20 to –120
Jupiter	-151
Saturn	-183
Uranus	-209
Neptune	-222
Pluto	-229

Table I. Typical operational temperatures for an unheated spacecraft.

In addition to deep space applications, low temperature electronics have potential uses in terrestrial applications that include magnetic levitation transportation systems, medical diagnostics, cryogenic instrumentation, and super-conducting magnetic energy storage systems. The utilization of power electronics designed for and operated at low temperature is expected to result in more efficient systems than room temperature systems. This improvement results from better electronic, electrical, and thermal properties of materials at low temperatures. In particular, the performance of certain semiconductor devices improves with decreasing temperature down to liquid nitrogen temperature (-196 °C). An example is the power MOSFET which has lower conduction losses at low temperature due to the reduction in the drain-to-source resistance $R_{DS}(on)$ resulting from increased carrier mobility.

The Low Temperature Electronics Program at the NASA Glenn Research Center focuses on the research and development of electronic components, circuits and systems suitable for applications where cold temperatures are expected (i.e. deep space). This program has emphasized screening and characterization of candidate components for the development of low temperature electronic circuits [2,3,4]. As part of the circuit development, a number of dc/dc converters has been designed or modified to operate from room temperature to -196 °C using commercially-available components such as CMOS-type devices and MOSFET switches. These converters ranged in output power from 5 W to 1 kW with switching frequencies of 50 kHz to 200 kHz. Pulse-width modulation technique was implemented in most of these converters with open- as well as closed-loop control. The topologies included buck, boost, multi-resonant, push-pull and full-bridge configuration [1,2,5-9].

AEROSPACE POWER SYSTEMS

Most of aerospace power systems are DC-based. Therefore, dc/dc power converters are required to provide the outputs needed for different loads. Such outputs range from 1.5 V to 15 V at various power levels. Recently, there has been a tremendous progress in the design of high power density dc/dc converters. Converters that operate at power densities of 50% or more greater than the available standard conventional converter designs have been developed. This increase in power density is achieved using new designs, advanced devices and components, and packaging techniques. For example, the newly developed synchronous rectifier-based dc/dc converters modules with multi-layer thick film hybrid packaging provide more usable output power without the use of a heat sink than do the conventional, schottky diode based converters with a heat sink and thick-film single layer packaging. However, all of the existing dc/dc converter systems are specified to operate above a minimum temperature of -40 °C or -55 °C. As a result, it was the goal of this work to investigate the performance of these high power density converters and to determine their tolerance to temperatures below their minimum specified temperature.

In this work, six commercial-off-the-shelf modular, low power DC/DC converters, with specifications that might fit the requirements of specific future space missions, have been selected for investigation. Some of the specifications of these converters, which ranged in electrical power from 8 W to 13 W and output voltage from 3.3 V to 5 V, are listed in Table II. The converters were characterized in terms of their performance as a function of temperature in the range of 20 °C to -190 °C. The experimental procedures along with the experimental data obtained on the investigated converters are presented and discussed.

Module	Input (V)	Output (V)	Power (W)	Temp Range (°C)
1	9 -36	3.3	10	-40 — +60
2	36-75	5.0	10	-25 — +70
3	36-72	3.3	10	-40 — +85
4	18-36	3.3	8.4	-40 — +70
5	18-36	3.3	13	-40 — +85
6	9-36	3.3	10	-40 — +85

Table II. Converter Module Specifications.

LOW TEMPERATURE TEST SET-UP

The converters listed in Table II were characterized as a function of temperature from 20 °C to -190 °C in terms of output regulation, efficiency, and input and output current distortions. At a given temperature, these properties were obtained at various input voltages and at different load levels; from no-load to full-load conditions. The tests were performed as a function of temperature using a Sun Systems environmental chamber utilizing liquid nitrogen as the coolant. A temperature rate of change of 10 °C/min was used throughout the experiment. The modular converters were tested separately at the following temperatures: 20; 0; -20; -40; -60; -80; -100; -120; -140; -160; -180; and -190 °C. At every test temperature, the device under test was allowed to soak at that temperature for a period of 30 minutes before any measurements were made. After the last measurement was taken at the lowest temperature, the converters were allowed to stabilize to room temperature and then the measurements were repeated at room temperature to determine the effect of one thermal cycle on the converters.

RESULTS AND DISCUSSIONS

During the investigations, data was generated for both steady and dynamic states. In this paper, only data pertaining to the steady state efficiency and voltage regulation of the tested converters are presented and discussed.

Figures 1-6 show the output voltage and efficiency of the modules versus temperature for four conditions of input voltage and output load levels. These conditions include: minimum input voltage under light and heavy loads, and maximum input voltage under light and heavy loads. In Figures 1(a) through 6(a), the output voltage variations are shown. An offset in the output voltage at light and heavy loads can be attributed to the resistive drop (≅ 70 mΩ) in the wiring from the output terminals of the module to the electronic load where the output voltage was actually measured.

Figure 1(a) shows the output voltage for module #1. At a given load, the output voltage maintains a steady value from room temperature to -120 °C. For temperatures beyond -120 °C, the converter begins to show loss in regulation. For example, the output voltage increases slightly when the input voltage is 36V but decreases drastically when the input voltage is 12V. As expected, the output voltage drops slightly when the load is increased. The effect of temperature on the efficiency of converter module #1 under different input voltage and load conditions is shown in Figure 1(b). In general, the efficiency drops as the temperature is lowered with the heavy load condition having a higher efficiency than that of a light load. For the same loading, the efficiency is higher as the input voltage is decreased. For a given input voltage, the converter has lower efficiency when the load level is low.

Figure 2(a) shows the effect of low temperature on the output voltage of module #2. It can be seen that the converter maintains a regulated voltage up to -100 °C. The converter loses regulation between -100 °C and -180 °C, after which the converter ceased to function. The effect of temperature on the efficiency of module #2 is shown in Figure 2(b). Similar to the voltage regulation, the efficiency of the converter drops rapidly for temperatures below -100 °C.

In Figure 3(a), module #3 shows reasonable performance to -20 °C, but shows complete loss of voltage regulation for temperatures below -80 °C. It did, however, continue to operate with no regulation down to -190 °C. This module displayed similar behavior in its efficiency with temperature change as depicted in Figure 3(b).

The output voltage of module #4 does not exhibit any dependence on either the input voltage or the test temperature at low loads as shown in Figure 4(a). At heavy loads, it does however decrease slightly upon lowering the test temperature regardless of the level of the input voltage. In general, the efficiency of this converter exhibits a slight decrease with decreasing temperature. This reduction becomes apparent at temperatures below -60 °C, as shown in Figure 4(b).

In Figure 5(a), module #5 shows excellent output voltage regulation with temperatures down to -120 °C. The only exception is at -120 °C where at minimum input voltage and light load the output voltage increases to over 4 volts. In addition, this module ceased to operate for temperatures below -120 °C, but regained operation once its temperature rose above -120 °C. The efficiency of this module (Figure 5(b)) held relatively steady values at heavy load to -60 °C and then dropped off as temperature decreased down to -120 °C.

In Figure 6(a), module #6 shows relatively good output regulation down to -120 °C. Beyond that temperature, the output voltage seems to slightly increase as the temperature is decreased further. Its efficiency (Figure 3(b)), however, exhibits a gradual decrease as temperature is decreased. Module #6 ceased to operate for temperatures below -180 °C, but regained operation once its temperature rose above -180 °C.

CONCLUSIONS

Several commercial-off-the-shelf, low power, dc/dc converter modules were evaluated as a function of temperature in the range of 20 °C to -190 °C. Data pertaining to the output voltage regulation and efficiency of the tested converters were presented and discussed.

Test results obtained on the modules have shown that they operated as expected within the manufacturer's specified temperature range as well as with reasonably good performance down to temperatures between -80 °C and -100 °C. For temperatures below -100 °C, performance was either out of range, erratic, or non-existent.

In all cases, the temperature range for which these modules were designed and specified does not include the severe temperature range for which they were subjected to in this investigation. Additional testing taking into account long-term evaluation and thermal cycling may reveal the potential for extending the operational temperature range and/or improving their performance at these very low temperatures through component screening and/or modification to the module design.

ACKNOWLEDGEMENTS

This work has been supported by a grant from the Low Temperature Electronics Group at the NASA Glenn Research Center to the University of Akron and under the NASA Contract NAS3-98008 (Dynacs Engineering , Inc.).

REFERENCES

1. Gerber, S.S., Patterson, R.L., Ray, B. and Stell, C., "Performance of a Spacecraft DC-DC Converter Breadboard Modified for Low Temperature Operation," IECEC 96, Vol.1, 1996, pp. 592-598.

2. Patterson, R.L., Dickman, J.E., Hammoud, A. and Gerber S.S., "Low Temperature Power Electronics Program," NASA EEE Links, Electronic Packaging and Space Parts News, Vol. 4, No. 1, January 1998.

3. Patterson, R.L., Hammoud, A. and Gerber, S.S., "Evaluation of Capacitors at Cryogenic Temperatures for Space Applications," IEEE International Conference on Electrical Insulation, Washington DC, June 7-10, 1998.

4. Hammoud, A., Gerber, S.S, Patterson, R.L. and MacDonald, T., "Performance of Surface-Mount Ceramic and Solid Tantalum Capacitors for Cryogenic Applications," IEEE Conference Phenomena, Atlanta, GA, Oct. 25-28, 1998.

5. Ray, B., Gerber, S.S., Patterson, R.L. and Myers, I.T., "Power Control Electronics for Cryogenic Instrumentation," Advances in Inst. and Control, Vol. 50, Part 1, Int. Soc. for Measurement and Control, 1995, pp. 131-139.

6. Ray, B., Gerber, S., Patterson, R. and Myers, I., "77K Operation of a Multi-Resonant Power Converter," IEEE PESC'95 Record, Vol. 1, pp. 55-60.

7. Ray, B., Gerber, S.S. and Patterson, R.L., "Low Temperature Performance of a Full-Bridge DC-DC Converter," IECEC '96, Vol. 1, 1996, pp. 553-559.

8. Ray, B., Gerber, S.S., Patterson, R.L. and Myers, I.T., "Liquid Nitrogen Temperature Operation of a Switching Power Converter," Symp. on Low Temp. Electronics and High Temperature Superconductivity, Vol. 9, 1995, pp. 345-352.

9. Gerber, S.S. Miller, T., Patterson, R.L. and Hammoud, A., "Performance of a Closed-Loop Controlled High Voltage DC/DC Converter at Cryogenic Temperature," IECEC '98, Vol. 1, 1998.

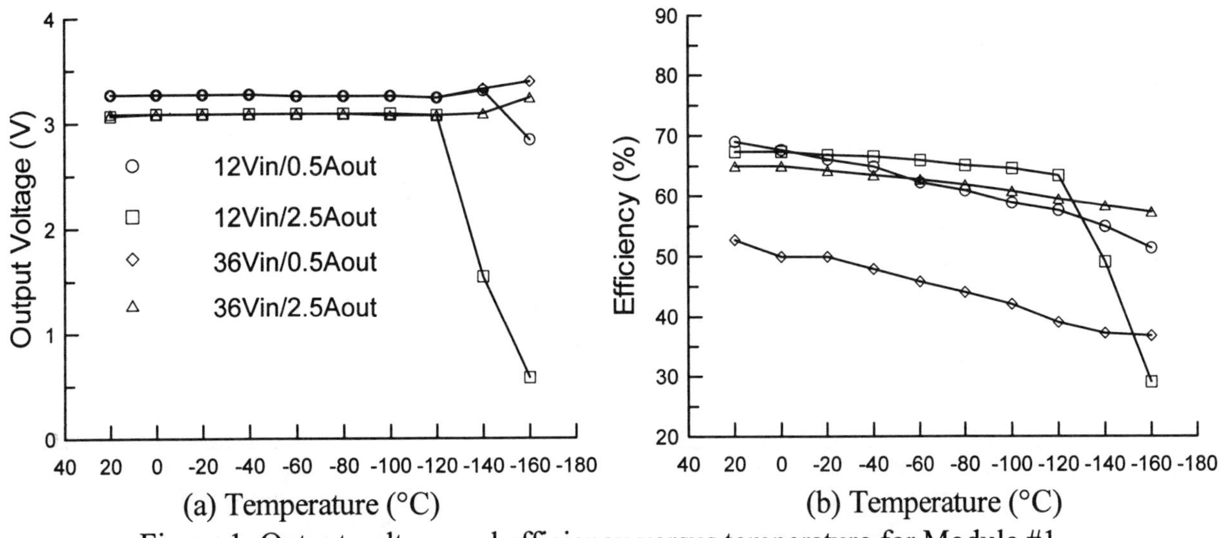

Figure 1. Output voltage and efficiency versus temperature for Module #1.

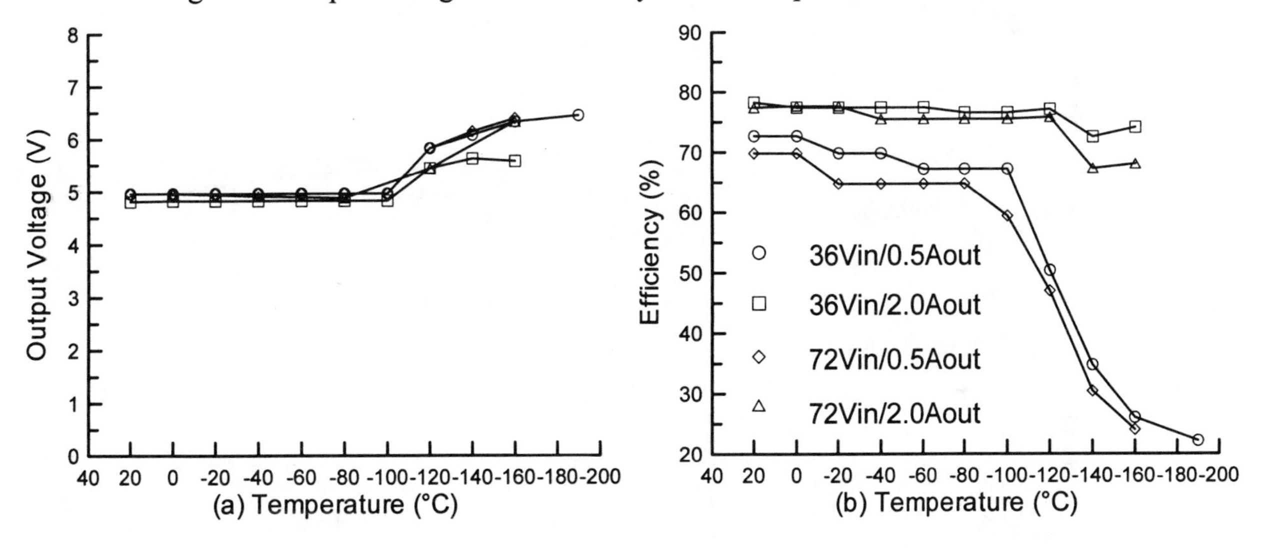

Figure 2. Output voltage and efficiency versus temperature for Module #2.

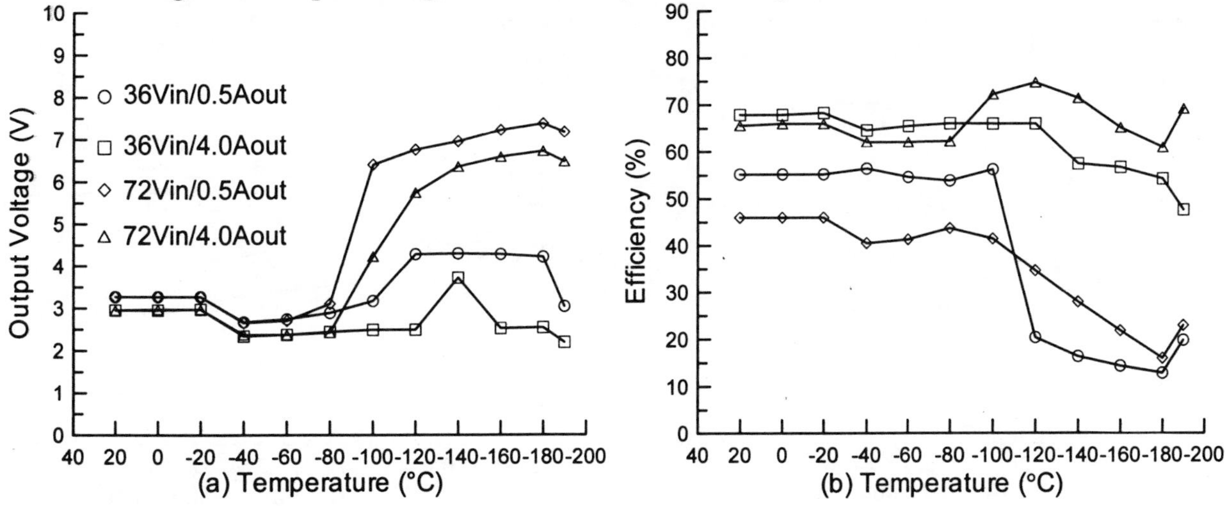

Figure 3. Output voltage and efficiency versus temperature for Module #3.

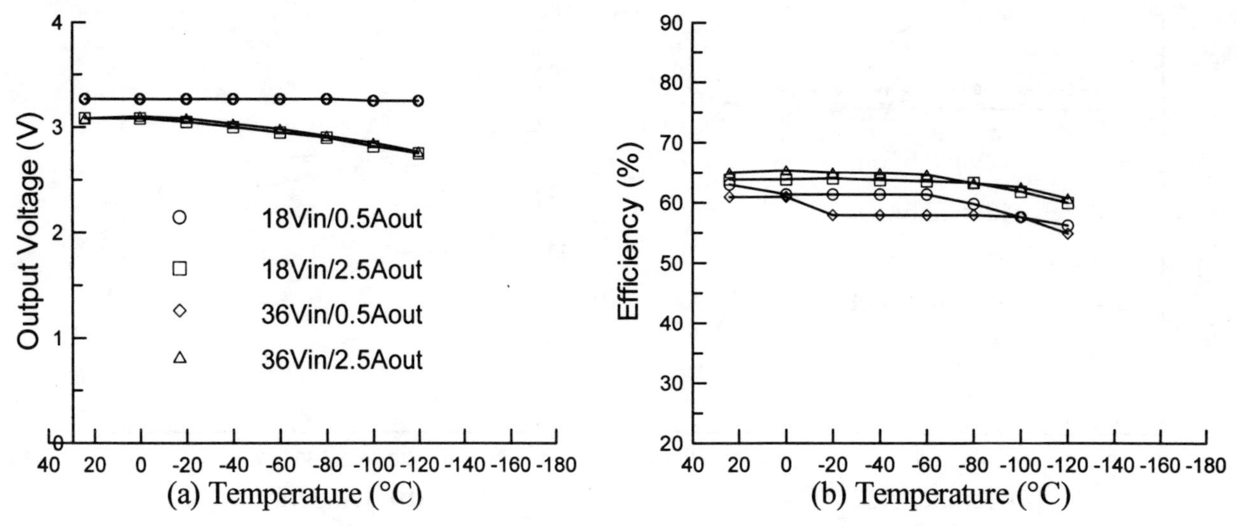

Figure 4. Output voltage and efficiency versus temperature for Module #4.

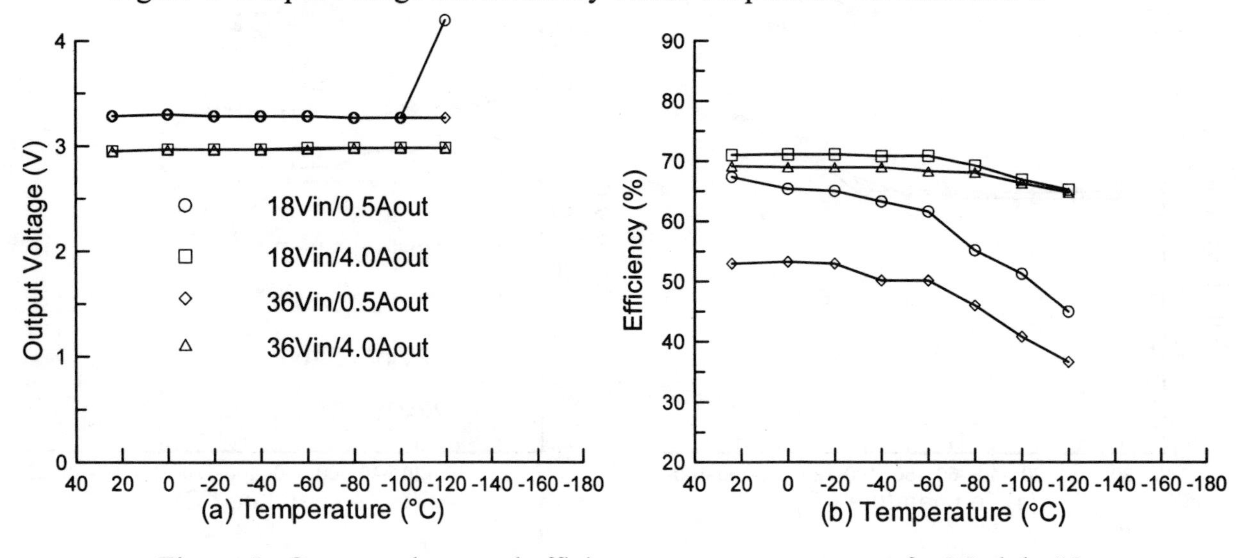

Figure 5. Output voltage and efficiency versus temperature for Module #5.

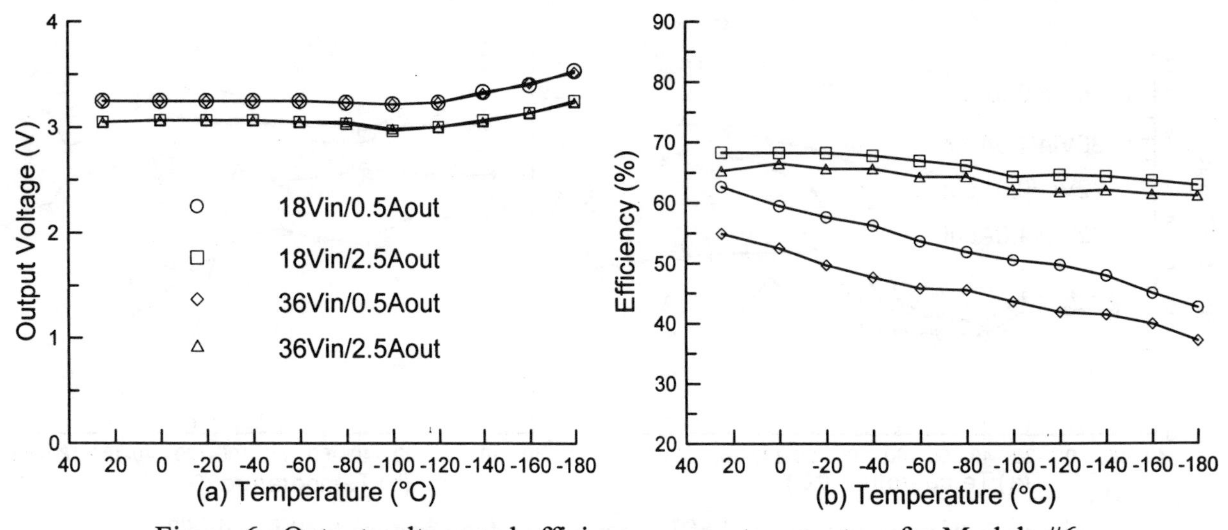

Figure 6. Output voltage and efficiency versus temperature for Module #6.

AIAA-2000-2801

A NOVEL START SYSTEM
FOR AN AIRCRAFT AUXILIARY POWER UNIT

Cristian Anghel
Honeywell Engines & Systems Toronto
3333 Unity Drive, Mississauga, Ontario, Canada, L5L 3S6
Cristian.Anghel@Honeywell.com

ABSTRACT

This paper describes the design and the main technical characteristics of a new Start System for an Aircraft Auxiliary Power Unit (APU). By using the latest improvements in power electronics and microelectronics this system eliminates the conventional DC starter by driving the generator installed on the APU as a motor to achieve the start.

The developed Start System eliminates one battery, dedicated for APU starting, the DC Starter and its Clutch Assembly, while providing a faster start, assisting the APU acceleration up to 8,400 rpm (vs. 6,000 rpm provided by the DC Starter).

This new system extends the APU life by maintaining lower engine temperatures during start. The battery voltage control feature, that maintains this voltage at 18/20 VDC during start and the option of using AC power for start, to save battery power is another advantage over the classical DC.

This new Start System is installed in over 650 commercial airplanes and the operation in revenue service has proved its high reliability. Presently its MTBF is 600 % higher than the Time Between Overhauls (TBO) for the DC Starter, while assuring an increased performance and safety in operation.

I. INTRODUCTION

A typical Electrical Power System of a commercial twin-jet aircraft has two 400 Hz, 3 phase, 115/200 VAC buses, each supplied by main engine driven AC generators. These are the primary electrical power sources. The DC power is obtained from the AC power, employing Transformer Rectifier Units, and from battery.

An APU driven generator is installed to supply secondary electric power, on the ground or in flight to the Generator AC Buses, when power from the main generator is not available.

The APU Generator (400 Hz, 3 phase, 115/200 VAC) output is connected to one or both Generator AC Buses via the APU Power Breaker (APB). The external power source is connected to one or both Generator AC Buses when the aircraft is on the ground.

The APU also provides pneumatic power for the Environmental Control System and Main Engine Starting. Figure 1 presents the new APU Start System.

The Electronic Control Unit (ECU) governs the operation of the APU by providing auto-start sequencing, self-monitoring and auto-shutdown based on airplane demands to provide pneumatic air and electrical power to the airplane.

The APU Start System developed by Honeywell comprises the following main components: the Start Power Unit (SPU), the Start Converter Unit (SCU) and the brushless synchronous APU Generator used in system as Starter/Generator (S/G).

The SPU and SCU convert airplane electrical power, either AC or DC, into accelerating power supplied to the S/G. The S/G converts this power to torque to rotate the APU. Following a successful APU start, the S/G supplies 3-phase AC power to the aircraft power distribution system.

The APU Generator Control Unit (AGCU) is responsible for monitoring the APU generator's output power quality and enabling the operation of the Voltage Regulator (VR) installed in the SCU.

The Power Distribution Panel (PDP) houses the APB and the S/G feeder current transformers. The APB channels 3-phase AC power in the Generate Mode by interfacing the S/G output with the aircraft power distribution system. The Current Transformers (CTs) sense the generator output current.

This new system has two distinct operating modes: Start Mode and Generate Mode.

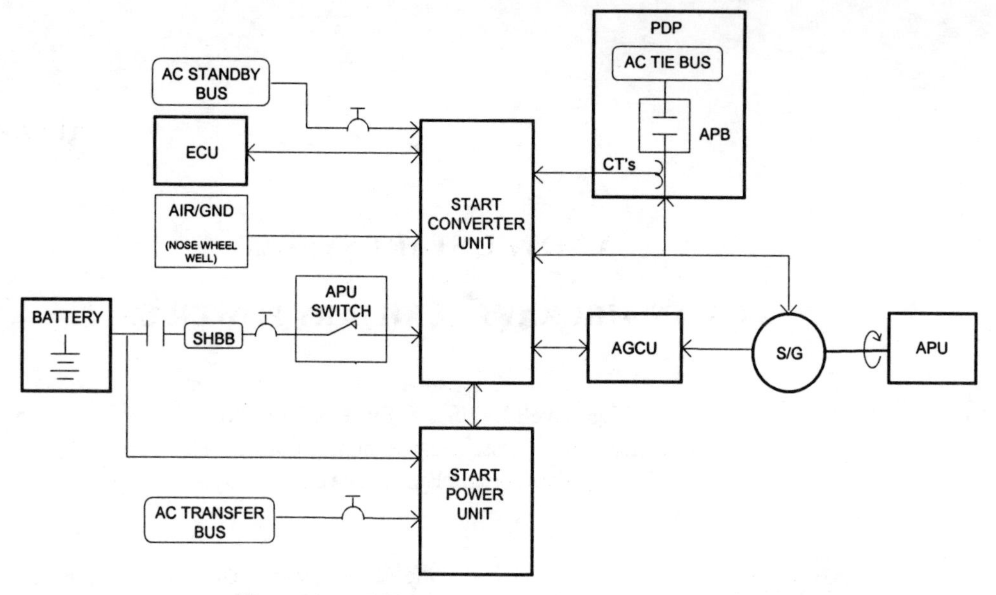

Figure 1 – APU Start System Block Diagram

Section II of this paper contains the APU Start System description and operation. Section III presents the experimental results and Section IV is designated for conclusions.

II. APU START SYSTEM DESCRIPTION AND OPERATION

A. Starter/Generator

The S/G contains three generators in one case. The simplified schematic of this generator is presented in Figure 2. The Generate Mode of operation is described first.

A Permanent Magnet Generator (PMG) supplies power for voltage regulation. An Exciter Generator supplies field excitation power for the main AC generator. The Main AC Generator supplies AC three phase power for the electrical system. The AC output of the PMG is used to provide exciter field power controlled by the Voltage Regulator. The VR rectifies and conditions this output and sends it back to the exciter generator field winding. The voltage magnitude at the Point Of Regulation (POR) is compared to a set value to control the exciter field winding current.

The DC current in the exciter field winding produces an electromagnetic field in the rotor exciter armature which induces an AC voltage (in the exciter's armature). A three phase, full-wave diode bridge rectifies the AC output of the exciter generator into DC power which is supplied to the main AC generator rotor to produce an electromagnetic field.

This DC field induces three phase AC voltage in the main stator windings.

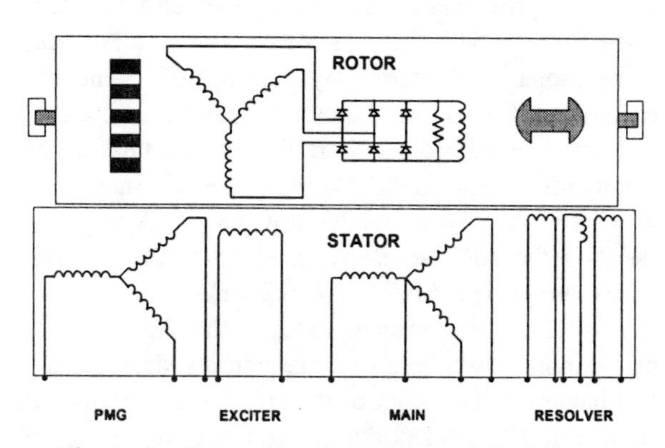

Figure 2 – Starter/Generator Schematic Diagram

During Start Mode the stator excitation winding is supplied with single phase, constant frequency 400 Hz, AC voltage that produces, through a transformer effect, an electromagnetic field in the exciter's armature at zero rotational speed.

The SCU supplies the main winding of the stator with a balanced set of currents, controlled in function of the rotor position, such that maximum torque is achieved. This position is provided to the SCU by the resolver installed in the S/G. In this mode the PMG power is not used.

The electromagnetic torque [1] produced by a synchronous machine in the ***d-q*** reference frame is defined by:

$$T_e = \frac{3}{2} \cdot \frac{P}{2} \cdot \left(\lambda_{ds} \cdot i_{qs} - \lambda_{qs} \cdot i_{ds}\right) \quad (1)$$

λ are flux linkages, i are currents, P is the number of poles and X_{md} is the d axis magnetizing inductance.

This Start System is using a control scheme that impresses the stator current vector along with the q axis, eliminating the i_{ds} component of the current, and equation (1) is reduced to:

$$T_e = \frac{3}{2} \cdot \frac{P}{2} \cdot X_{md} \cdot i_{df} \cdot i_{qs} \qquad (2)$$

The two controllable terms dictating the torque produced by the S/G become in this case the stator and the field current, making the synchronous machine analogous to a DC machine.

B. **Start Power Unit**

The SPU converts aircraft electric power, either AC or DC, to steady state DC voltage for the DC link used by the SCU. The SCU controls the SPU's conversion mode, selecting AC power conversion as a priority.
If AC is not available then DC battery power conversion is selected.

The AC/DC Converter converts 3-phase AC power into a 270 VDC output. The diode bridge is supplied through a 3 phase, step-up (1:1.15), isolation transformer. The output of this bridge is utilized by the SCU as APU start power.
The AC/DC section includes relays, soft start circuit, snubber and input EMI filter (not shown in the block diagram).
The converter supplies the DC link with 270 VDC power when commanded by the SCU.

C. **Start Converter Unit**

The SCU converts DC power from the SPU to provide conditioned 3-phase stator power and single phase field excitation to the S/G. In this mode the S/G operates as a motor for starting the APU. The SCU also contains the S/G's Voltage Regulator. When the APU reaches 95 % of rated speed, the VR in the SCU is commanded to regulate the S/G output by

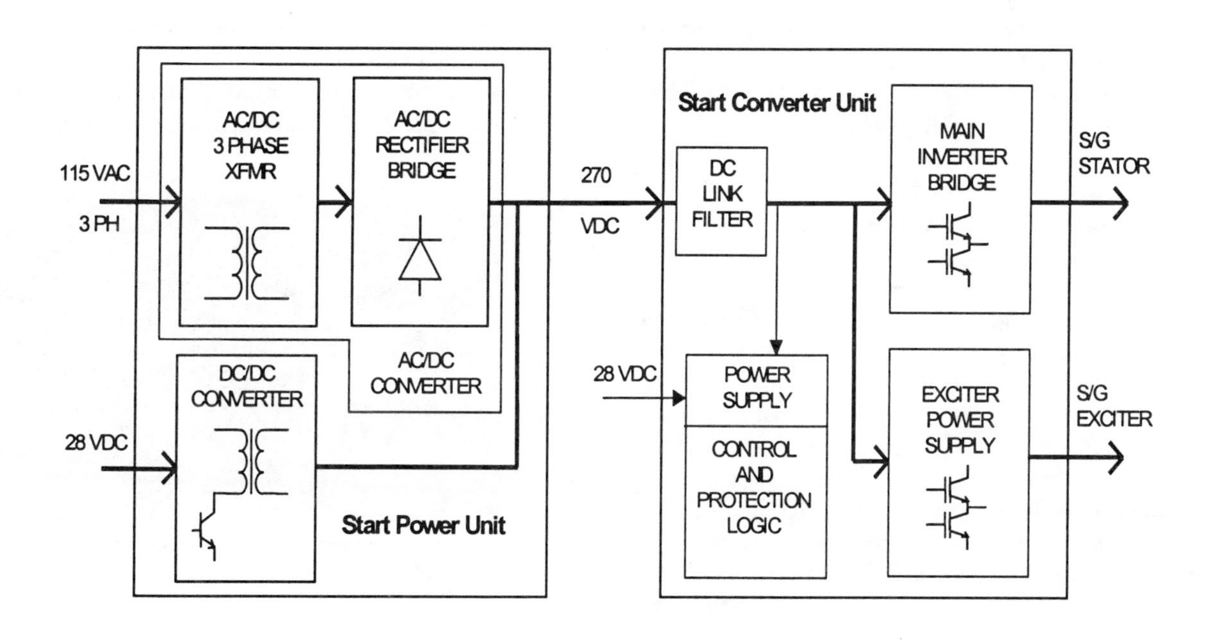

Figure 3 – SPU and SCU Functional Block Diagram

Figure 3 shows the SPU block diagram. The functional groups of the SPU are:

1. DC/DC Converter
2. AC/DC Converter

The DC/DC Converter converts 28 VDC battery power into a 300 VDC output. This output is utilized by the SCU. This converter provides internal health monitoring signals to the SCU.

controlling the field excitation.

Figure 3 shows the SCU functional block diagram. The SCU's functional groups applicable to Start Mode are:

1. DC/AC Inverter
2. Exciter Power Supply (EXPS)
3. Control Power Supply (CPS)
4. Control and Protection Logic (CPL)

The DC/AC Inverter converts the DC voltage from the DC Link into PWM regulated 3-phase AC output power to accelerate the SG. During the start mode the inverter varies the 3-phase output voltage, current, and frequency as a function of S/G shaft position and speed. The inverter provides a current schedule as a function of speed to the S/G to produce the torque required to start the APU.

The inverter control includes battery voltage control circuitry. This circuitry prevents the battery voltage from falling below 18 V on ground and 20 V in air.

The gate drivers of the inverter provide over-current protection for the switching devices. Also the inverter control provides over-current and unbalance protection for the output current. In case of either over-current/unbalance condition the inverter flags the condition to the CPL. The inverter monitors the DC link current and sends this information to the CPL.

The CPL contains the reference voltages and logic to control the sequencing of the SCU and SPU operation. The CPL enables or disables functional modules, relays, and cooling fan depending on the operating state. The CPL provides an enable or disable signal to the following items:

- DC/AC Inverter
- Exciter Power Supply
- DC/DC Converter (SPU)
- AC/DC Converter (SPU)
- SPU and SCU Relays

This module supplies excitation power to the resolver located in the generator. The resolver provides generator angular position and speed information to the inverter. CPL also provides a signal to the Exciter Power Supply to regulate the exciter field current.

The CPL receives command signals from other Line Replaceable Units (LRUs) in the APU system. The functional modules relay information back to the CPL functional group. Data related to the CPL include temperature sensors readings, failure flags, Built-In-Test-Equipment (BITE) flags and status flags. BITE provides fault monitoring and fault isolation to the LRU level: SCU, SPU or S/G failure.

The Exciter Power Supply (EXPS) converts DC link power into a single phase, fixed frequency, variable current/voltage output. The output of the EXPS is used to excite the S/G during the Start Mode. The EXPS receives a control signal from the CPL to control the output current supplied to the exciter field. The EXPS provides two fault flags to the CPL: a flag for internal failures of the EXPS and a flag for excitation wiring or S/G winding failure.

The Control Power Supply (CPS) provides control power to various modules in the SCU. It converts 28 VDC from the aircraft Switched Hot Battery Bus (SHBB) and 270-300 VDC Link voltage into various levels of analog, digital, and switched control voltages. The CPS also supplies the power to energize the SPU and SCU relays.

The Voltage Regulator (not shown in the block diagram) provides APU S/G field excitation control. It converts 3-phase AC Permanent Magnet Generator voltage to DC voltage for the VR control power and the S/G exciter field. The VR provides voltage regulation of the 3-phase S/G output. The regulator provides S/G current limiting during POR short circuits.

The VR provides BITE data to the CPL for fault isolation of the Voltage Regulator, S/G Shorted Rotating Diode (SRD) and the S/G, for selected fault conditions. These faults are flagged to the CPL. The Voltage Regulator provides a signal representing PMG frequency to the AGCU. The VR receives the VR Enable, OV and UV signals from the AGCU via SCU conditioning circuits.

Figure 4 presents the actual hardware, SCU and SPU, presently in aircraft revenue service.

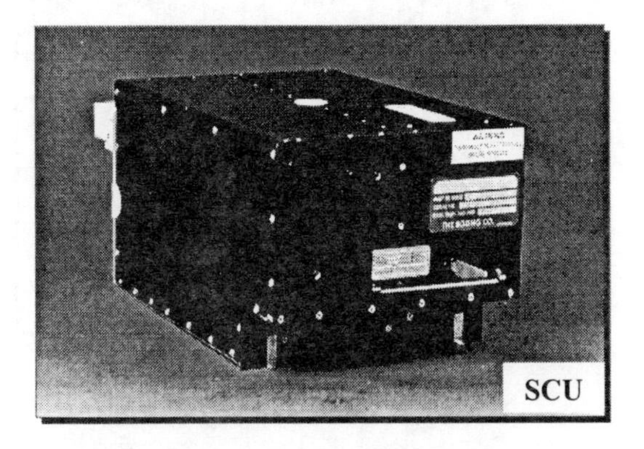

Figure 4

III. TEST RESULTS

The verification and validation of the new Start System required extensive testing during the development of the SPU and SCU. The testing was performed in the Honeywell Power Lab in Toronto.

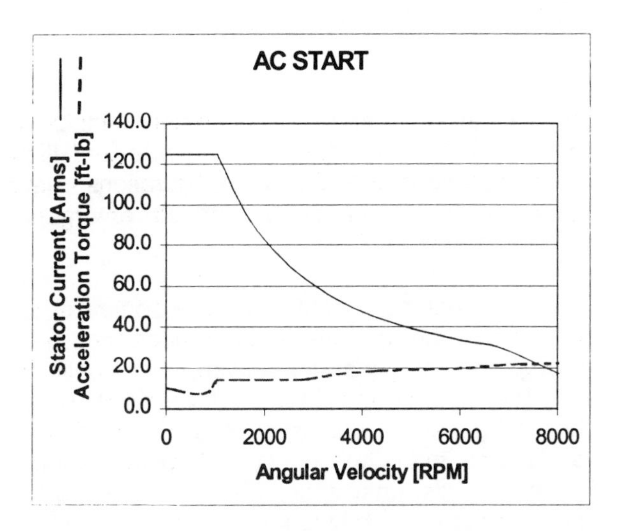

Figure 5

Figure 5 shows the two key parameters contributing to the start: the S/G stator phase currents and the difference between the S/G torque and the APU drag torque vs. angular speed. This torque difference is named on the plot Accelerating Torque.

It is important to note that when the main stator winding is supplied with 125 Arms maximum, less than half of rated current in Generate Mode at rated power, the S/G torque value exceeds substantially the APU drag torque (i.e. positive Accelerating Torque), resulting in a fast APU start.

IV. CONCLUSIONS

The SPU/SCU Start System is a high performance motor controller that can be easily adapted for any 3-phase AC machine (induction, permanent magnet) used in aerospace applications requiring precise torque/speed control.

This Start System presents important advantages over the conventional DC Starter:

- eliminates one battery, dedicated for APU starting, the DC Starter and its Clutch Assembly
- provides a faster start, assisting the APU acceleration up to 8,400 rpm (vs. 6,000 rpm provided by the DC Starter)
- extends the APU life by maintaining lower engine temperatures during start
- maintains the battery voltage at 18/20 VDC during start (vs. DC Starter, that causes at start a dip in battery voltage)
- uses AC power for start, to save battery power and extend the battery life (non-existent with a conventional starter)
- efficiency of the new system during start is higher than that of the conventional system (75% vs. 60%)
- eliminates significant wire weight due to battery cables between avionics bay and APU

The voltage regulation function provides, under steady-state conditions, a POR voltage of 115±2.0 Vrms. The voltage regulation is maintained within 2 % throughout the range of the continuous load (0-90 KVA) and within 2.6 % under 200 % overload conditions.

This new Start System is installed in over 650 commercial airplanes and the operation in revenue service has proved its high reliability. Presently its MTBF exceeds the required/predicted numbers.

The MTBF is 600 % higher than the Time Between Overhauls for the DC Starter, while assuring an increased performance and safety in operation.

The solid-state unit with modular cabling and component packaging decreases substantially the maintenance cost. The field data show that this system provides very good performances for all operating conditions encountered in the aircraft.

ACKNOWLEDGMENTS

The author wishes to thank to Mr. Rocco Divito of Honeywell for sharing some of the design experience of the APU Start System and for the suggestions provided during the review of this paper.

REFERENCES

1. P. Krause, *Analysis of Electric Machinery,* p 227, IEEE PRESS, New York (1995)

AIAA 2000-2802

Recent Developments in Aircraft Emergency Power

Mike Koerner
Honeywell Engines & Systems
Torrance, CA

ABSTRACT

Aircraft have become increasingly dependent on uninterrupted electric and hydraulic power for flight control. Aircraft have also become increasingly dependent on shaft power to assist with engine starting in the event of a flameout. The use of turbofan engines, fly-by-wire flight control systems, and less stable aircraft configurations has increased the need for emergency power systems. Several configurations of aircraft emergency power systems are currently in use. These include ram air turbines, bleed air driven power units and propellant power systems. The propellant power systems include stand-alone emergency power systems, emergency engine start systems, and integrated auxiliary and emergency power systems. Two fuels are currently used in emergency power systems: H-70 monopropellant and jet fuel and air bipropellant. H-70 systems are small and lightweight and have been used for many years. However, H-70 is toxic and requires special handling. Jet fuel and air are less toxic and easier to handle. Jet fuel and air bipropellant systems have been used in several recent emergency power applications. Four specific aircraft emergency power systems are described: the F-16 emergency power unit; the U-2 emergency start system; The F-2 emergency power system; and the F-22, integrated auxiliary and emergency power system.

INTRODUCTION

Aircraft emergency power systems provide backup electric, hydraulic or shaft power in the event an aircraft's primary systems become inoperative, or in the event an aircraft's main engines flameout. To minimize the duration of power interruption resulting from such events, emergency power systems startup quickly and operate at any aircraft attitude, altitude and airspeed.

Recent trends in aircraft development - including the use of turbofan engines, computerized flight controls and unstable aircraft configurations - have increased the demands on aircraft emergency power systems.

Turbofan engines, in which a large portion of the air flowing through the engine does not pass through the gas generator section, offer significant increases in performance and efficiency. However, the gas generator spools of these engines have less inertia than the spools of turbojet engines. Since it is this inner spool which must be tied to the accessory gearbox to facilitate engine starting, the reduction in inertia means that the accessories, including aircraft hydraulic pumps and electric generators, decelerate more quickly when the engine flames out. Furthermore, the inner spool on turbofan engines does not windmill as efficiently as a turbojet engine. As a result, after an engine flameout less energy can be extracted from the aircraft's airspeed to assist with engine starting or to continue to motor the accessories.

Computerized flight control systems, in which there is no mechanical link between the pilot's inputs and control surface deflections, offer benefits in aircraft performance and safety; and when integrated with aircraft offensive and defensive systems, weapon lethality and aircraft survivability also improve. However, computerized flight control systems have little tolerance for interruptions in electrical or hydraulic power.

Unstable (and relaxed stability) aircraft configurations, in which continuous control adjustments are necessary to maintain steady flight, offer reduced drag, reduced radar cross-sections and enhanced maneuverability. However unstable and relaxed stability aircraft need uninterrupted electric and hydraulic power to maintain controlled flight.

Thus current trends in modern aircraft development - including the use of turbofan engines, computerized flight controls and unstable aircraft configurations - increase the need for emergency power systems which can quickly supply electric and hydraulic power, under any flight condition, in the event the aircraft's primary electric or hydraulic systems become inoperative, or if the aircraft's main engines flameout.

SYSTEM CONFIGURATIONS

There are several configurations of aircraft emergency power systems currently in use. These include ram air turbines, engine bleed air driven systems and propellant power systems.

Ram air turbines extract energy from the aircraft's airspeed. Ram air turbines are commonly used on commercial aircraft. An example is the ram air turbine that Honeywell developed for the Boeing 777, Figure 1. Honeywell has also supplied ram air turbines for other applications such as power systems on external pods.

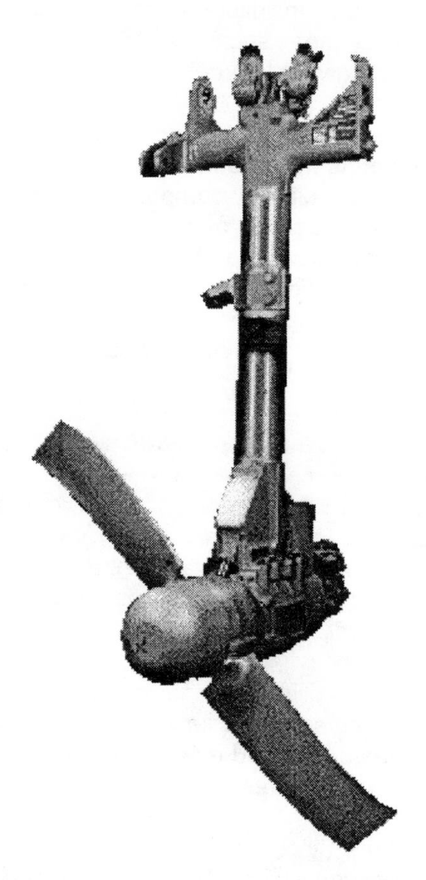

Figure 1. 777 Ram Air Turbine

Bleed air driven emergency power systems utilize engine bleed air to produce emergency power. An example is the emergency power system on the Lockheed Martin F-117 Nighthawk, Figure 2. The F-117 emergency power system operates on bleed air from either the aircraft's main engine or auxiliary power unit. The auxiliary power unit operates continuously during flight allowing the emergency power system to startup quickly in the event of an emergency. The F-117's auxiliary power unit and emergency power system were both developed by Honeywell.

Figure 2. F-117 Turbine Power Module

Propellant power systems utilize the chemical energy from combustion or decomposition reactions generate power. They are generally self-contained systems in that they are not dependent on aircraft airspeed or on the availability of engine bleed air. Propellant power systems include stand-alone emergency power systems, emergency engine start systems, and integrated auxiliary and emergency power systems.

Stand-alone emergency power systems typically include a propellant supply system, a reaction chamber, a turbine wheel, a gearbox and a gearbox mounted hydraulic pump and electric generator. The propellant supply system stores the propellant until needed then delivers the required quantity to the reaction chamber. The reaction chamber converts the propellant into hot gas and delivers the hot gas to the turbine. The turbine converts the hot gas into torque on the turbine shaft. The gearbox reduces the turbine shaft speed to the pump and generator operating speeds. The pump and generator convert the gearbox output shaft torque into electric and hydraulic power.

Honeywell has developed stand-alone emergency power systems for the Lockheed Martin F-16 and for the Japanese F-2 Fighter. Honeywell is currently developing a stand-alone emergency power system for the Korean KTX-2 trainer. Honeywell has also provided stand-alone emergency power systems for a number of developmental and flight test aircraft including the Israeli Lavi, F-16XL, X-29, B-2, X-31, YF-23, British EAP/EFA, French Rafael, and JIST.

Emergency engine start systems operate in much the same manner as stand-alone emergency power units: a propellant supply system delivers propellant to a reaction chamber, the reaction chamber delivers hot gases to a turbine, and the turbine delivers torque to a gearbox. However, instead of the gearbox being used to drive a generator and pump, it is instead used to start the aircraft's main engine.

Emergency engine start systems provide engine start assist in-flight, if the aircraft's main engine has flamed out, or they may be used to provide a rapid scramble start or alert launch capability on the ground.

Honeywell developed an emergency start system for the re-engined U-2 aircraft. Honeywell has also has provided emergency start systems for a number of flight test or developmental aircraft including the F-14B, F-20, Singapore A-4, and X-31.

Integrated auxiliary and emergency power systems combine the functions of both systems with a single shared gearbox and single set of shared accessories. The emergency power portion of integrated systems is much the same as a stand-alone emergency power system with a propellant supply system, reaction chamber, turbine wheel, gearbox and accessories. An auxiliary power unit is also attached to the gearbox and either the auxiliary power unit or the emergency power system can drive the accessories.

Auxiliary power units offer significant advantages including autonomous aircraft ground checkout, maintenance, weapon loading, and main engine starting; and in-flight main engine start assists. For aircraft with an auxiliary power unit the shared gearbox and accessories of an integrated auxiliary and emergency power system offer advantages in cost, weight, reliability and maintainability over separate, independent systems.

For emergency power, integrated auxiliary and emergency power systems combine the advantages of air-breathing gas turbine engines with those of self-contained propellant power systems. An auxiliary power unit can operate continuously for long durations. An emergency power system can start quickly and can operate at any aircraft altitude, attitude and airspeed.

Honeywell has developed integrated auxiliary and emergency power systems for the Taiwanese Indigenous Defensive Fighter and for the U.S. Air Force F-22 Raptor.

EMERGENCY POWER FUELS

Although many different propellants have been considered for aircraft emergency power systems, only two are currently in use: H-70 monopropellant and jet fuel and air bipropellant.

H-70 monopropellant is a blend of 70% hydrazine and 30% water. Hydrazine (N_2H_4) is a clear liquid with about the same density as water. It is flammable, explosive and a strong reducing agent. In high concentrations hydrazine is used as a rocket propellant. In dilute form it is used in industrial processes such as the removal of oxidation from power plant boilers.

In aircraft emergency power systems, H-70 monopropellant is injected into a bed of Shell 405 catalyst. The hydrazine decomposes into nitrogen and ammonia. A portion of the ammonia further dissociates into nitrogen and hydrogen. Heat is a product of the reaction. The water is included only to reduce the decomposition gas temperature and to depress the freezing point of the fuel.

As a propellant hydrazine offers a high specific impulse and the simplicity of a monopropellant system, but it is also toxic and a suspected carcinogen. H-70 monopropellant, like any hydrazine derivative, requires special handling: specially trained technicians with protective garments and special breathing apparatus (Figure 3), remote storage and handling facilities, contamination monitors and special logistics.

In an effort to eliminate the special handling requirements associated with hydrazine systems, Honeywell has conducted extensive investigations of hydrazine alternatives. Among the alternatives investigated are hydrogen peroxide monopropellant, alcohol and oxygen bipropellant, jet fuel and oxygen bipropellant, and jet fuel and air bipropellant systems.

Figure 3. Loading an F-16 Fuel Tank with H-70 Monopropellant

Each of these fuels has unique characteristics. Many factors must be considered in selecting an optimum fuel for a given application. The jet fuel and air bipropellant combination is particularly suited to aircraft emergency power applications.

Jet fuel and air are both readily available, have low toxicity and are relatively easy to handle. With the addition of an on-board air compressor and a connection to the aircraft fuel supply, jet fuel and air systems can be automatically reserviced after use. Operating the combustor in a fuel-rich mode minimizes air storage penalty. Honeywell has developed a jet fuel and air combustor which burns clean in both fuel-rich and lean-burning modes.

Honeywell has recently developed emergency power systems that use jet fuel and air for the Taiwanese Indigenous Defensive Fighter, the Japanese F-2 fighter and the U.S. Air Force F-22. Jet fuel and air is also used on the emergency power system Honeywell is currently developing for the Korean KTX-2 aircraft.

THE F-16 EMERGENCY POWER UNIT

The Lockheed Martin F-16 is a single-engine, fly-by-wire aircraft. The F-16 has an H-70 monopropellant fueled, stand-alone emergency power unit that was developed by Honeywell. The system provides up to 55 output shaft horsepower for driving an electric generator and hydraulic pump for up to 10 minutes in the propellant power mode of operation. It can also operate on engine bleed for a period of up to five hours. The F-16 emergency power unit starts within three seconds and can operate anywhere in the aircraft's flight envelope.

The F-16 emergency power unit, Figure 4, consists of a nitrogen bottle, a nitrogen valve, a fuel tank, a turbine power unit, a heat exchanger and an electronic controller. The nitrogen bottle contains gas that is used to expel fuel from the fuel tank. The nitrogen bottle is a lightweight Kevlar composite wound bottle. The nitrogen valve initiates the flow of nitrogen and regulates the pressure in the fuel tank. The fuel tank contains the H-70 monopropellant. The tank is a piston-type, positive expulsion tank that is sealed with a burst disc. When pressurized with nitrogen the burst disc ruptures and the fuel is expelled from the tank.

The turbine power unit includes a fuel control valve, a decomposition chamber, a turbine wheel, a bleed air valve, and a gearbox. The fuel control valve opens and closes as necessary to maintain turbine speed. The decomposition chamber contains a bed of shell 405 catalyst that converts the H-70 propellant into hot gas. The turbine is an axial impulse wheel which converts the hot gas, or engine bleed air, into torque. The bleed air valve controls the flow of engine bleed air to the turbine. The gearbox reduces the turbine shaft speed and provides mounting for a hydraulic pump and electric generator. The heat exchanger, which is immersed in an aircraft fuel tank, cools the gearbox lubrication oil. The electronic controller provides control signals to the valves in response to aircraft and speed sensor inputs.

THE U-2 EMERGENCY START SYSTEM

The Lockheed Martin U-2 is a single-engine reconnaissance aircraft that was recently re-engined. The re-engined U-2 has a single-start, H-70 monopropellant emergency engine start system developed by Honeywell. The system is intended for use in the event of an in-flight engine flameout. The U-2 system can start the engine from stopped in about 15 seconds.

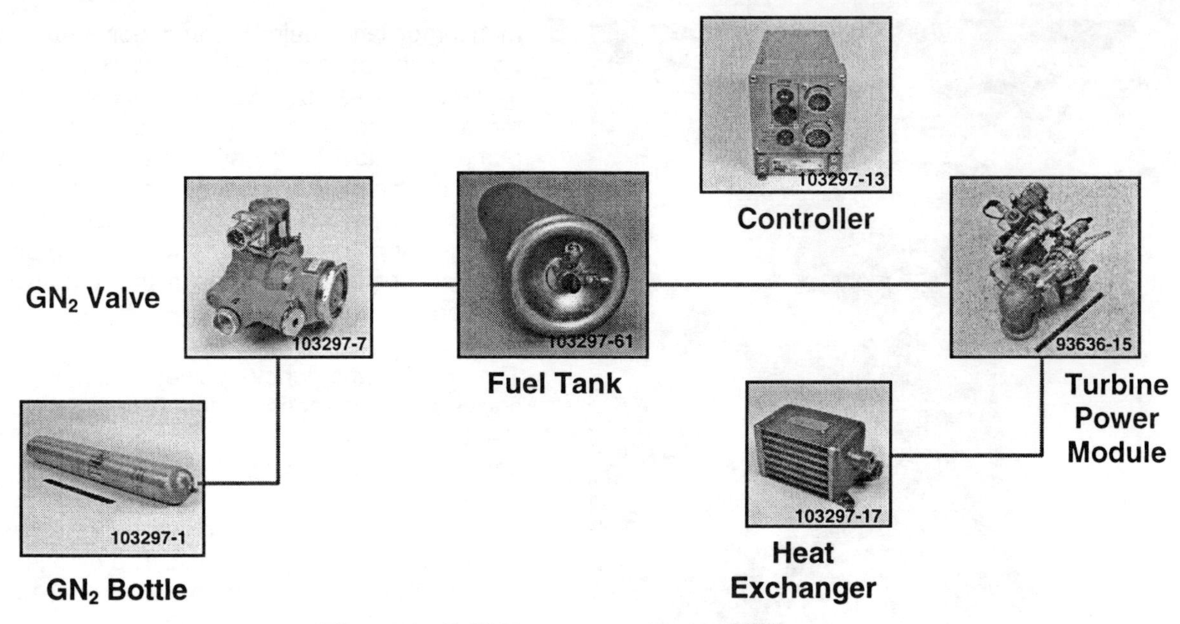

Figure 4. F-16 Emergency Power Unit

The U-2 start system includes a nitrogen bottle, a nitrogen valve, a fuel tank, a fuel valve, a gas generator, an air turbine starter and a controller. The nitrogen bottle contains gas used to expel fuel from the fuel tank. The nitrogen valve initiates the flow of the gas and regulates the pressure in the tank. The fuel tank contains the H-70 monopropellant. The tank is a free-surface expulsion tank that is sealed with burst discs. When pressurized with nitrogen the burst discs rupture and the fuel is expelled from the tank. The fuel valve gradually opens to initiate starter operation and shuts when the starter reaches cutout speed. The gas generator contains a bed of shell 405 catalyst that converts the H-70 propellant into hot gas. The air turbine starter has a reaction-bladed axial-flow turbine which converts the hot gas (or ground cart bleed air during normal operations) into torque. It also has the integral gearing required to reduce the turbine shaft speed to a speed compatible with the airframe mounted auxiliary drive gearbox on which it mounts. The controller energizes the valves in response to an aircraft initiation command and de-energizes them when the starter cutout speed is reached.

THE F-2 EMERGENCY POWER SYSTEM

The Japanese F-2 is a single-engine, fly-by-wire aircraft. It has a jet fuel and air stand-alone emergency power system developed by Honeywell. The F-2 emergency power system provides up to 55 output shaft horsepower to drive an electric generator and hydraulic pump for up to five minutes in the propellant power mode of operation. It can also operate on engine bleed for a period of up to five hours. It starts within three seconds and can operate anywhere in the aircraft's flight envelope.

The F-2 emergency power system, Figure 5, consists of two air bottles, a pressure regulating and shutoff valve, a fuel tank, an air control valve, a fuel control valve, a turbine power unit, a heat exchanger and an electronic controller. The two air bottles contain the air used for combustion and the air used to expel the fuel from the fuel tank. The bottles are high-pressure, lightweight, carbon fiber wound composite bottles. The pressure regulating and shutoff valve initiates the flow of air and regulates the air and fuel supply pressure. The fuel tank contains aircraft jet fuel. The tank is a piston-type, positive expulsion tank. The air control valve modulates the flow of air into the combustor as required to maintain turbine speed. The fuel control valve modulates the fuel to maintain the combustion temperature.

The turbine power unit, Figure 6, includes the combustor, a turbine wheel, a bleed air valve, and a gearbox. The combustor converts the fuel and air into hot gas. The turbine is an axial impulse wheel that converts the hot gas, or engine bleed air, into torque. The bleed air valve controls the flow of engine bleed air to the turbine. The gearbox reduces the turbine shaft speed and provides mounting for a hydraulic pump and an electric generator.

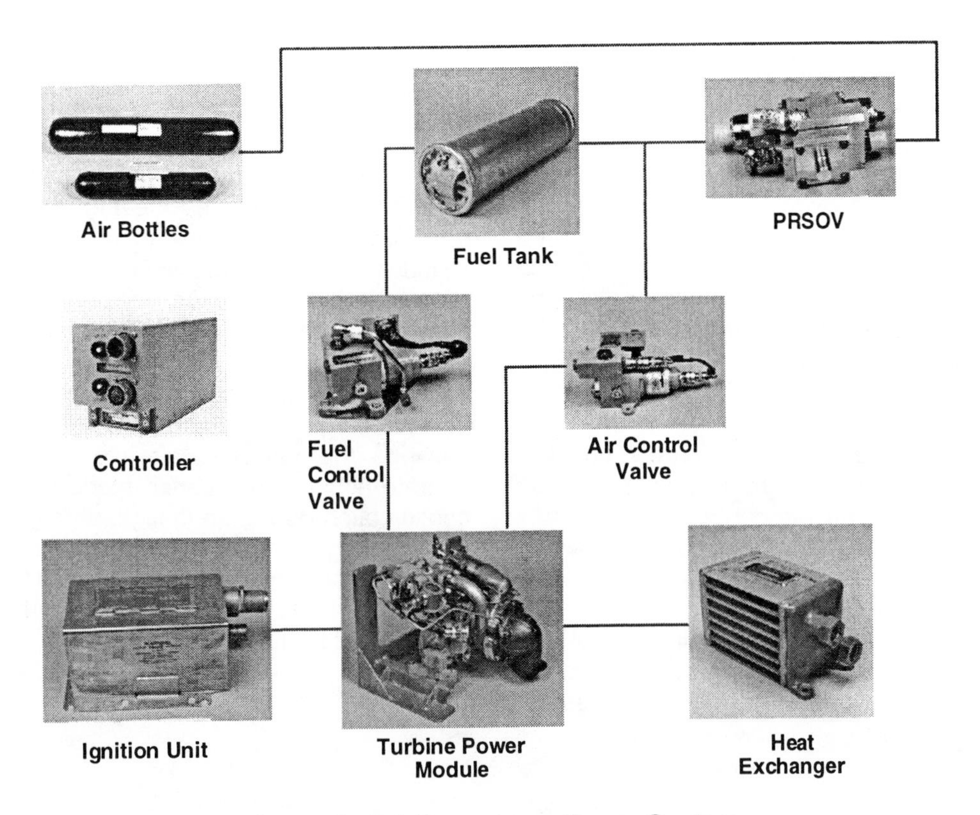

Figure 5. F-2 Emergency Power System

Figure 6. F-2 Turbine Power Unit

The heat exchanger, which is immersed in an aircraft fuel tank, cools the gearbox lubrication oil. The electronic controller provides control signals to the valves in response to aircraft signals, gearbox speed sensors and a combustion temperature sensor.

THE F-22 APGS

The U.S. Air Force F-22 is a twin-engine, fly-by-wire fighter that features integrated flight and weapon control systems. The F-22 has a jet fuel and air integrated auxiliary and emergency power system developed by Honeywell. This system is referred to as the F-22 auxiliary power generating subsystem (APGS).

On the ground the F-22 APGS provides electric and hydraulic power for aircraft checkout, maintenance and weapon loading; and bleed air for the environmental control system and for aircraft main engine starting.

In flight the F-22 APGS provides emergency electric and hydraulic power within five seconds and for up to 90 seconds, anywhere with in the aircraft envelope. Within the auxiliary power unit operating envelope it can supply electric and hydraulic power indefinitely, and can also provide bleed air for the environmental control system and for in-flight engine start assists.

The main elements of the F-22 APGS are the auxiliary power unit, a stored energy system, and a gearbox. The auxiliary power unit is an air-breathing gas turbine engine that provides shaft power to the gearbox and provides bleed air to other aircraft systems. The stored energy system is a jet fuel and air system that provides shaft power to the gearbox. The stored energy system is used both to drive the accessories in flight and to start the auxiliary power unit, whether in flight or on the ground. The gearbox is a two-speed gearbox that allows either the auxiliary power unit or the stored energy system to drive the accessories at rated speed while also allowing the stored energy system to drive the auxiliary power unit at half speed.

The F-22 stored energy system, Figure 7, consists of two air bottles, a pressure regulating and shutoff valve, a fuel tank, an air control valve, a fuel control valve, a turbine power module, and an air recharge system.

The two air bottles contain the air used in the combustor and the air used to expel fuel from the fuel tank. The bottles are high-pressure, lightweight, carbon fiber wound composite bottles. The pressure regulating and shutoff valve initiates the flow of air and regulates the air and fuel supply pressure. The fuel tank, Figure 8, contains aircraft jet fuel. The tank is a bladder-type, positive expulsion tank.

The air control valve modulates the flow of air to the combustor to maintain turbine speed. The fuel control valve modulates the fuel to maintain the combustion temperature. The fuel valve also includes provisions allowing the fuel tank to be automatically refilled from the aircraft fuel supply after each stored energy system operation.

The turbine power module, Figure 9, includes the combustor and a turbine wheel. The combustor converts the fuel and air into hot gas. It can operate either in a fuel-rich mode to minimize air consumption during an in-flight emergency, or in a lean-burning mode to eliminate carbon accumulation during APU ground starts. The axial impulse turbine wheel converts the hot gas into output shaft torque.

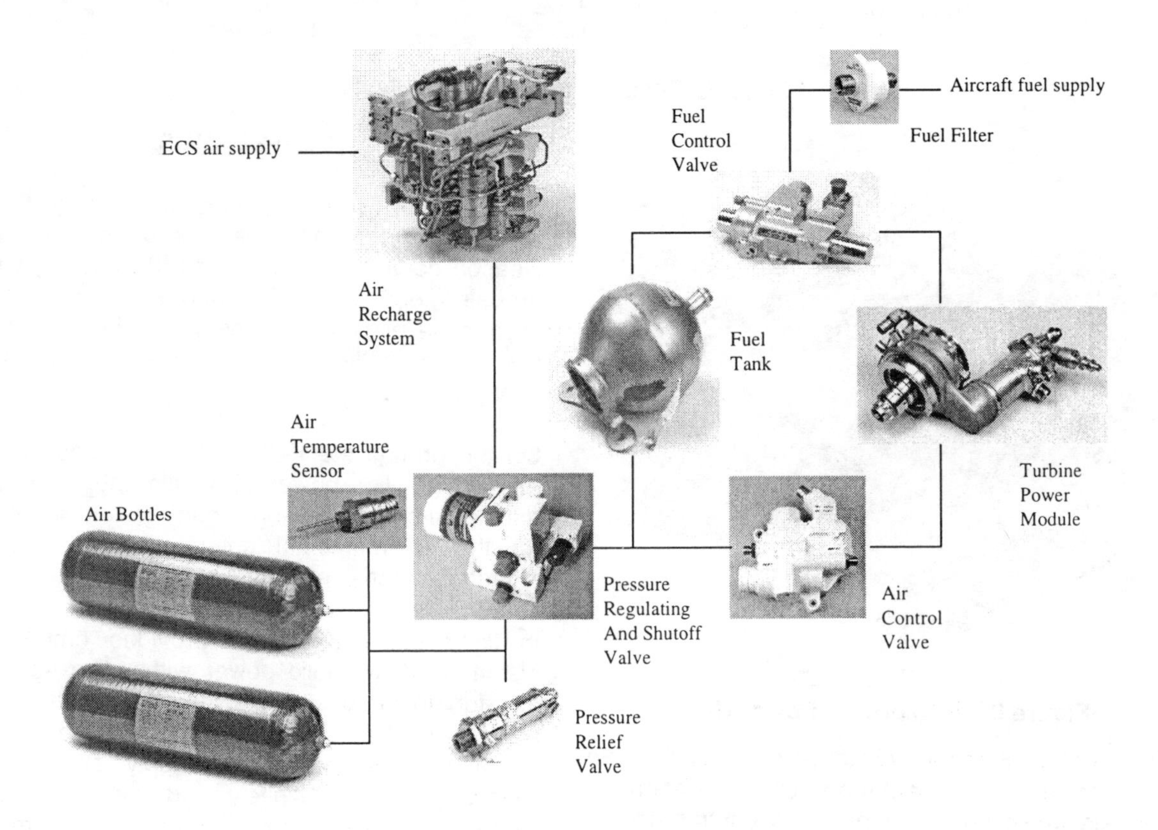

Figure 7. F-22 Stored Energy System

Figure 8. F-22 Fuel Tank

Figure 9. F-22 Turbine Power Module

The air recharge system, Figure 10, automatically recharges the air bottles after each stored energy system operation. The air recharge system is a hydraulic motor driven, 5000-psi, 4-stage piston compressor with an integral water removal system.

Figure 10. F-22 Air Recharge System

SUMMARY

Emergency power is increasingly needed for modern aircraft. Several emergency power system configurations are currently in use including ram air turbines, bleed air driven emergency power systems, and propellant power systems. Among the propellant power systems are stand alone emergency power systems like the F-16 and F-2 systems, emergency engine start systems like the U-2 system, and integrated auxiliary and emergency power systems like the F-22 system. There are also two different fuel systems currently in use: H-70 monopropellant like the F-16 and U-2 systems, and jet fuel and air like the F-2 and F-22 systems. The best emergency power system configuration and fuel combination for a given aircraft depends on the unique requirements of that application.

AIAA-2000-2803

FUZZY CURRENT-MODE CONTROL AND STABILITY ANALYSIS

George Kopasakis
National Aeronautics and Space Administration
Glenn Research Center
Cleveland, Ohio 44135

Abstract

In this paper a current-mode control (CMC) methodology is developed for a buck converter by using a fuzzy logic controller. Conventional CMC methodologies are based on lead-lag compensation with voltage and inductor current feedback. In this paper the converter lead-lag compensation will be substituted with a fuzzy controller. A small-signal model of the fuzzy controller will also be developed in order to examine the stability properties of this buck converter control system. The paper develops an analytical approach, introducing fuzzy control into the area of CMC.

1. Introduction

Fuzzy control has emerged as one of the most active and promising control areas, especially because it can control highly nonlinear, time-variant, and ill-defined systems. The work of Mamdani and his colleagues on fuzzy control[1–2] was motivated by Zadeh's work on the theory of fuzzy sets,[3–4] and its application to linguistics and systems analysis. Layne later modified work by Procyk, Mamdani, and others on the linguistic self-organizing controller to what it is now, Fuzzy Model Reference Learning Control (FMRLC).[5]

Existing CMC techniques[6–8] primarily employ analog lead-lag compensation to shape the closed-loop gain of the converter (magnitude and phase) in order to achieve certain design criteria, such as stability margins, low output impedance, low audio susceptibility, and response time characteristics like overshoot, settling time, and zero steady-state error. In this paper fuzzy control will be used, instead, to control a 6-kW converter with constant-power or negative-resistance loading. The rest of the design procedure, such as the insertion of the control-stabilizing ramp signal, the inductor current feedback and the design of the slopes of the feedback signals, will follow the same procedures as the traditional CMC design. The existing CMC methodology offers a comprehensive and methodical approach to meeting design criteria in both the frequency and time domains. The approach introduced here offers a design alternative for converter control by incorporating modern nonlinear control design approaches, such as fuzzy control, while still being able to use powerful frequency domain analysis techniques.

Initially, in this paper, the FMRLC approach will be employed to control the converter. Later the control law learned by using FMRLC will be employed using instead a conventional fuzzy controller to control the process. The FMRLC control technique could be used as the final control. However, because the FMRLC exhibits nonstationary nonlinearities, it is not amenable to small-signal, frequency domain analysis. A small-signal model of the converter will be constructed initially to help understand how to develop a heuristic, inverse fuzzy model of the converter, which is needed to employ the FMRLC technique. Later the small-signal model of the converter will be used to help analyze the stability properties of the control system. Sinusoidal sweeps will be applied to the fuzzy controller to construct frequency domain responses by employing fast Fourier transforms (FFT's). The small-signal transfer functions of the fuzzy controller will be constructed by using pole-zero approximation of the frequency responses.

2. Converter Model

2.1 Large-Signal, Switch-Mode Model

Figure 1 shows the model of a switch-mode, 6-kW buck converter regulator. The power stage equations for the large-signal nonlinear model are

$$\frac{di_l}{dt} = \frac{1}{L_f}\left(v_g - v_o\right) \tag{1}$$

$$\frac{dv_c}{dt} = \frac{i_c}{C_f} \tag{2}$$

$$\frac{di_c}{dt} = \frac{1}{L_c}\left(v_o - i_c R_c - v_c\right) \tag{3}$$

$$i_o = i_l - i_c \tag{4}$$

$$v_o = P_o / i_o \tag{5}$$

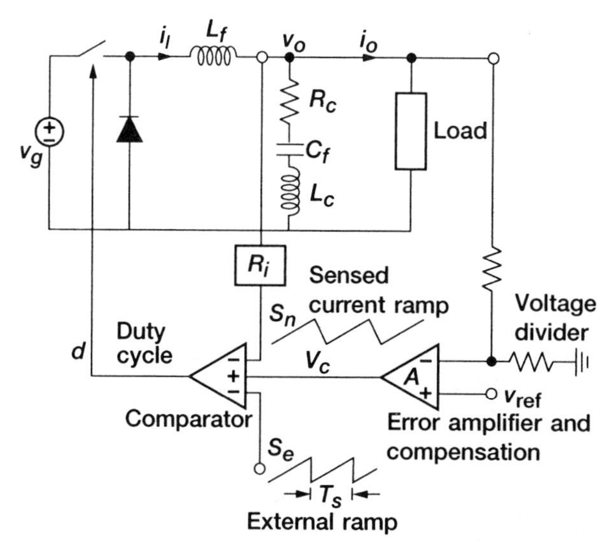

Figure 1.—Switch-mode, 6-kW buck converter regulator.

where R_c and L_c are the filter capacitor effective series resistance and inductance, respectively, and P_o is the load power. The nonlinear switch-mode model has two operating modes per conducting cycle: mode 1, when the switch is closed and Eqs. (1) to (5) are active; and mode 2, when the switch is open ($v_g = 0$) and the fly-back diode is conducting.

2.2 Small-Signal, Power Stage Model in Continuous Conduction Mode

Figure 2 shows the small-signal continuous conduction mode (CCM) perturbation model of the buck converter power stage. This model is derived from performing state-space average[8] of the nonlinear switch-mode model above, to come up with the large-signal average model, and then perturbing this model to calculate the small-signal perturbation model:

$$\hat{i}_l = \frac{1}{sL_f}\left(V_g\hat{d} - D\hat{v}_o - V_o\hat{d}\right) \tag{6}$$

$$\hat{v}_o = \left(I_l\hat{d} + D\hat{i}_l\right)Z(s) \tag{7}$$

$$Z(s) = \frac{R_l\left(C_f L_c s^2 + R_c C_f s + 1\right)}{C_f L_c s^2 + R_l C_f s + 1}, \quad \text{for } R_c \langle\langle R_l \tag{8}$$

Substituting Eq. (6) into Eq. (7) and after some simplifications:

$$\frac{\hat{v}_o}{\hat{d}} = \frac{V_g - V_o}{D}\frac{P_n(s)}{P_d(s)} \tag{9}$$

where

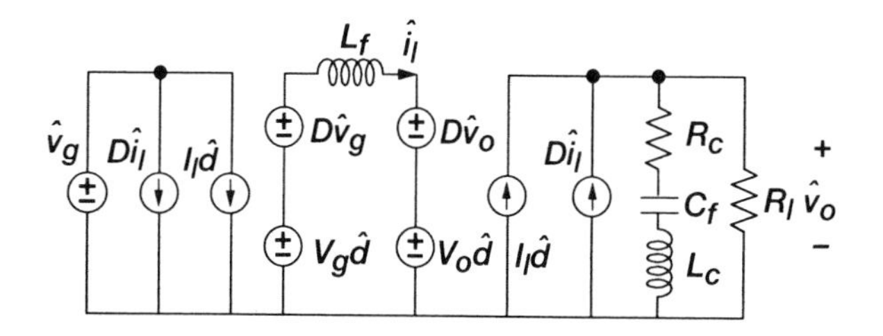

Figure 2.—Small-signal CCM perturbation model of buck converter power stage.

$$P_n(s) = \frac{L_f L_c C_f I_l}{A}s^3 + \frac{L_f C_f I_l R_c}{A}s^2 + \frac{L_f I_l}{A}s + 1$$

$$P_d(s) = \frac{L_f L_c C_f}{B}s^3 + \frac{R_l L_f C_f}{B}s^2 + \frac{L_f}{B}s + 1$$

$A = D(V_g - V_o)$, $B = D^2R_l$, and D is the duty cycle. Similarly, substituting Eq. (7) into Eq. (6) and after some simplifications:

$$\frac{\hat{i}_l}{\hat{d}} = \frac{E}{B}\frac{C_f L_c s^2 + \left[\left(V_g - V_o\right)R_l C_f / E\right]s + 1}{P_d(s)} \tag{10}$$

where $E = V_g - V_o - DI_l$ and D, V_g, V_o, and I_l, in Eqs. (6) to (10) represent average quantities over a conduction cycle of the duty cycle, input voltage, output voltage, and inductor current, respectively.

The corresponding lower case variables with the hat represent the perturbation quantities of these variables. Equations (9) and (10) represent the small-signal transfer functions of the duty cycle to the output voltage and the inductor current, respectively. Other transfer functions necessary for a complete CCM model of the converter power stage or the discontinuous conduction mode (DCM) model will not be derived here as this is beyond the scope of this paper. Figures 3 and 4 show frequency responses of Eqs. (9) and (10), respectively, at 100% constant-power load ($-R_l$) with $V_g = 160$ volts, $V_o = 120$ volts, $L_f = 80$ µH, $C_f = 320$ µF, $R_c = 20$ mΩ, $L_c = 0.5$ µH.

Figure 5 shows the small-signal, control-to-output model[8] in CCM. From Fig. 5 with $G_v = \hat{v}_o / \hat{d}$ and $G_i = \hat{i}_l / \hat{d}$,

$$\hat{d} = F_m\left[K_r\hat{v}_o + F_c\hat{v}_c - R_i H_e(s)G_i(s)\hat{d}\right] \tag{11}$$

$$\hat{v}_o = G_v\hat{d} \tag{12}$$

Solving for $\hat{d}$ in Eq. (11) and substituting into Eq. (12) to solve for $\hat{v}_o / \hat{v}_c$ gives

$$\frac{\hat{v}_o}{\hat{v}_c} = \frac{F_c F_m G_v(s)}{1 + F_m R_i H_e(s)G_i(s) - F_m K_r G_v(s)} \tag{13}$$

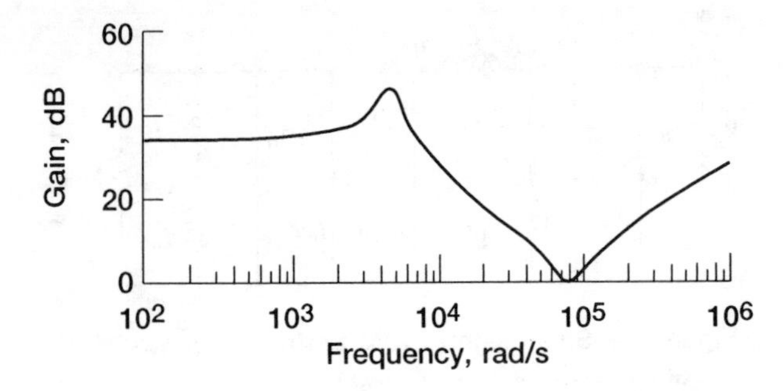

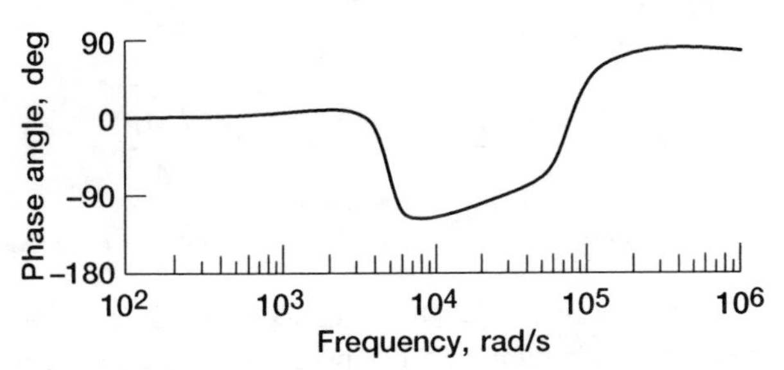

Figure 3.—Frequency response of $\hat{v}_o/\hat{d}$.

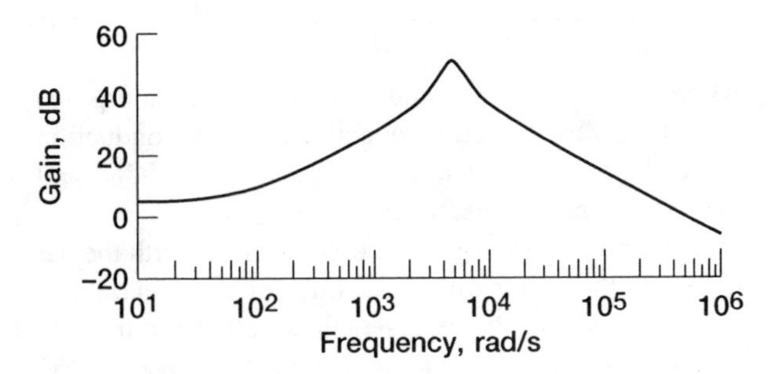

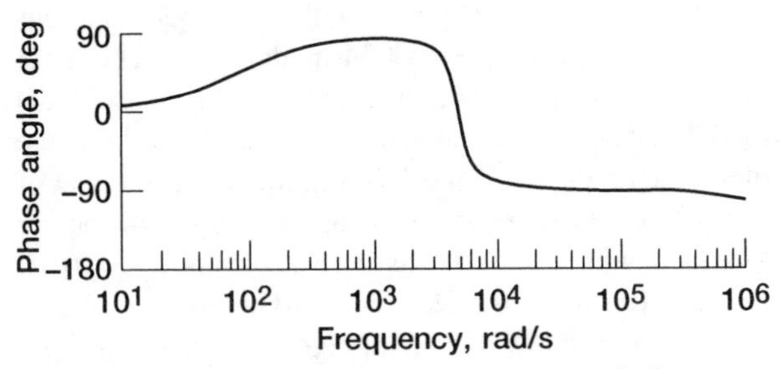

Figure 4.—Frequency response of $\hat{i}_l/\hat{d}$.

where, for the buck converter case,[7]

$$H_e = 1 + \frac{s}{\omega_n Q_z} + \frac{s^2}{\omega_n^2}, K_r = \frac{T_s R_i}{2L_f}, Q_z = \frac{-2}{\pi},$$

$$\omega_n = \frac{\pi}{T_s},\ F_m = \frac{1}{(S_n + S_e)T_s},\ F_c = 1$$

T_s is the conduction period in seconds, S_n is the rising slope of the inductor current feedback, and S_e is the slope of the stabilizing sawtooth signal. Figure 6 shows a frequency response of Eq. (13) at 100% load with a switching frequency of 40 kHz, R_i = 0.02, S_e = 80 000, and S_n = 11 000.

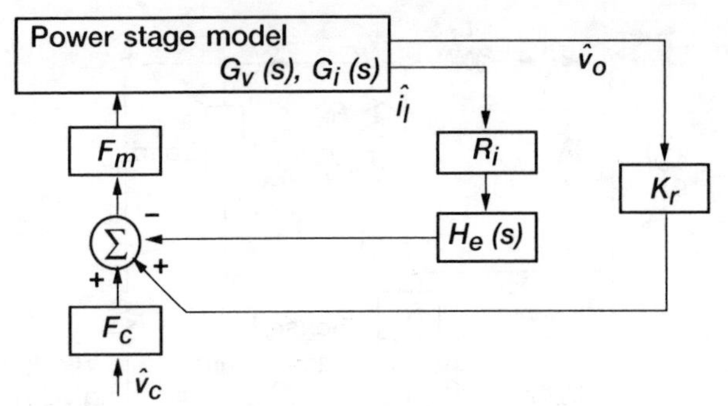

Figure 5.—Small-signal, control-to-output model in CCM.

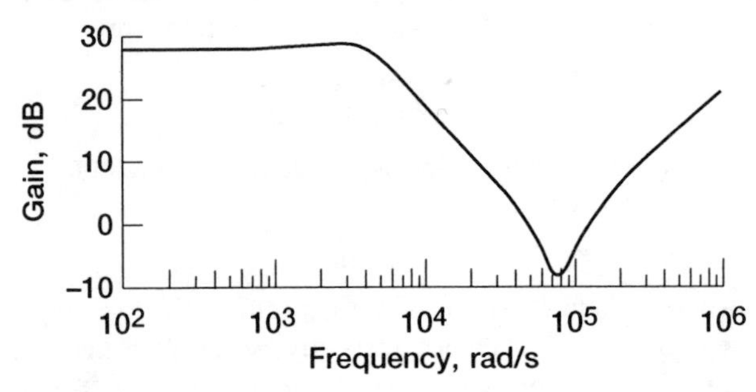

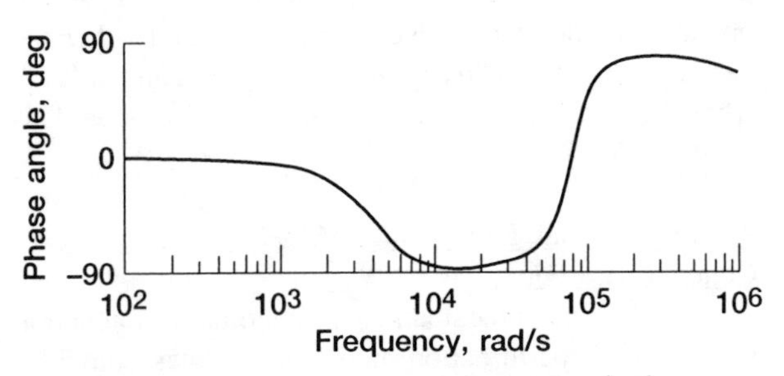

Figure 6.—Frequency response of $\hat{v}_o/\hat{v}_c$.

3. Fuzzy Current-Mode Control

3.1 FMRLC Current-Mode Control

Figure 7 shows the basic FMRLC structure[5] with the addition of a pole at the origin placed at the output of the fuzzy controller. The proposed fuzzy controller will be replacing the traditional lead-lag compensation. The gains of the FMRLC controller are tuned on the basis of information gained from an open-loop step response of the process and guidelines generated by the author.[9–10] For each fuzzy input 11 evenly distributed triangular membership functions are chosen, resulting in 121 rules. Each membership function has a base width of 0.4. For the FMRLC the inverse-model knowledge base of the process needs to be developed. This knowledge base does not necessarily need to exhibit high accuracy; however, it needs to exhibit the right output control directionality. The

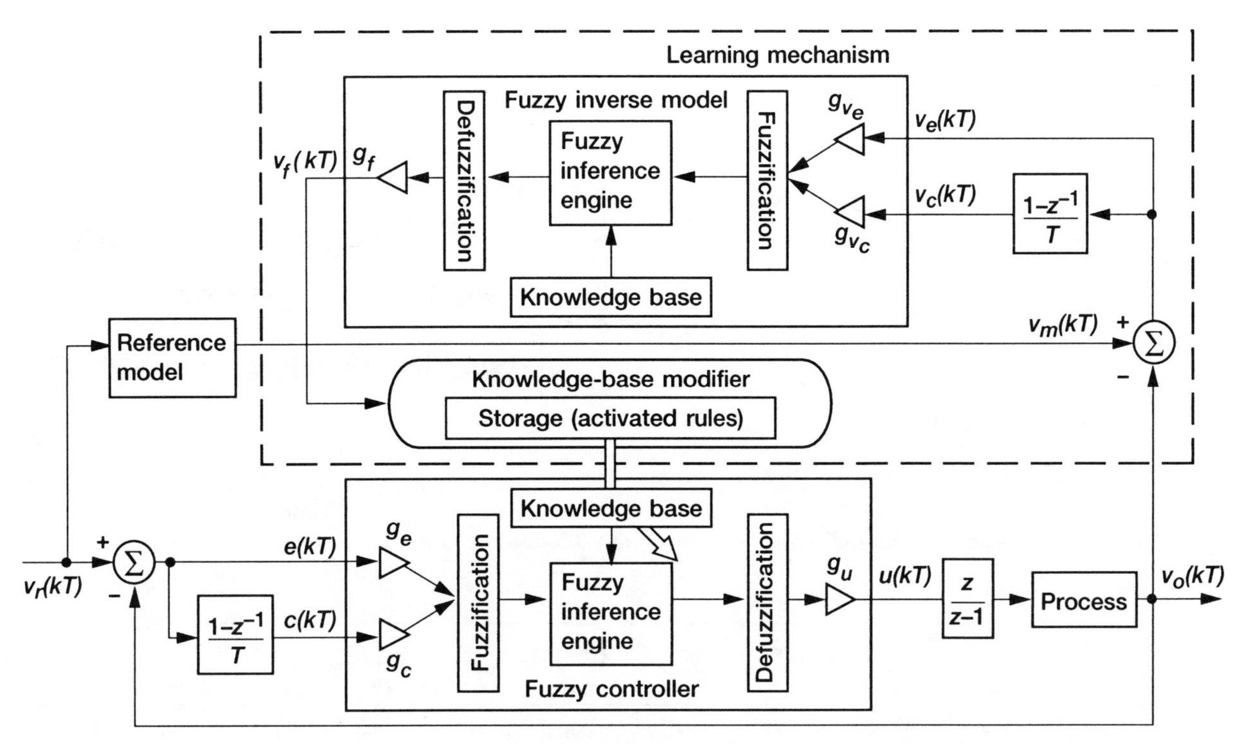

Figure 7.—FMRLC structure.

$P_i^{j,k}$	V_c^k										
V_e^j	−5	−4	−3	−2	−1	0	+1	+2	+3	+4	+5
−5	−1.0	−0.9	−0.8	−0.7	−0.6	−0.5	−0.4	−0.3	−0.2	−0.1	+0.0
−4	−0.9	−0.8	−0.7	−0.6	−0.5	−0.4	−0.3	−0.2	−0.1	+0.0	+0.0
−3	−0.8	−0.7	−0.6	−0.5	−0.4	−0.3	−0.2	−0.1	+0.0	+0.0	+0.0
−2	−0.7	−0.6	−0.5	−0.4	−0.3	−0.2	−0.1	+0.0	+0.0	+0.0	+0.0
−1	−0.6	−0.5	−0.4	−0.3	−0.2	−0.1	+0.0	+0.0	+0.0	+0.0	+0.0
0	−0.0	−0.0	−0.0	−0.0	−0.0	+0.0	+0.0	+0.0	+0.0	+0.0	+0.0
+1	−0.0	−0.0	−0.0	−0.0	+0.0	+0.1	+0.2	+0.3	+0.4	+0.5	+0.6
+2	−0.0	−0.0	−0.0	+0.0	+0.1	+0.2	+0.3	+0.4	+0.5	+0.6	+0.7
+3	−0.0	−0.0	+0.0	+0.1	+0.2	+0.3	+0.4	+0.5	+0.6	+0.7	+0.8
+4	−0.0	+0.0	+0.1	+0.2	+0.3	+0.4	+0.5	+0.6	+0.7	+0.8	+0.9
+5	+0.0	+0.1	+0.2	+0.3	+0.4	+0.5	+0.6	+0.7	+0.8	+0.9	+1.0

Figure 8.—Inverse fuzzy model rule base, also showing zeroed-out elements to eliminate sensitivity to undamped inductor current response.

inputs to the FMRLC (Fig. 7) are the output voltage error and the output voltage error derivative.

The error derivative is proportional to the output voltage derivative V_o, where the derivative of V_o is approximately proportional to the inductor current I_l, from Eq. (2) $V_o \cong V_c$, and Eq. (4) for I_o is approximately constant over a conduction cycle. From the frequency response of $\hat{i}_l / \hat{d}$ (Fig. 4), it is evident that the response of the inductor current to the duty cycle is highly undamped. This result was expected, since very little damping was used in terms of the R_c value chosen, purposely, to make the control design more challenging. Therefore, in constructing this knowledge base the elements associated with the error derivative that can cause a sign reversal of the control output variable are zeroed out (Fig. 8). Simulations, not shown here, showed that the time response of the closed-loop system will stall out before the output voltage reaches its reference and that the output voltage will oscillate at that point if these elements of the fuzzy model have other than zero values.

Figure 9 shows the closed-loop step response of the FMRLC system. The reference model of the controller has been modified to automatically reset every time the output voltage recovers and starts moving toward the set point, in

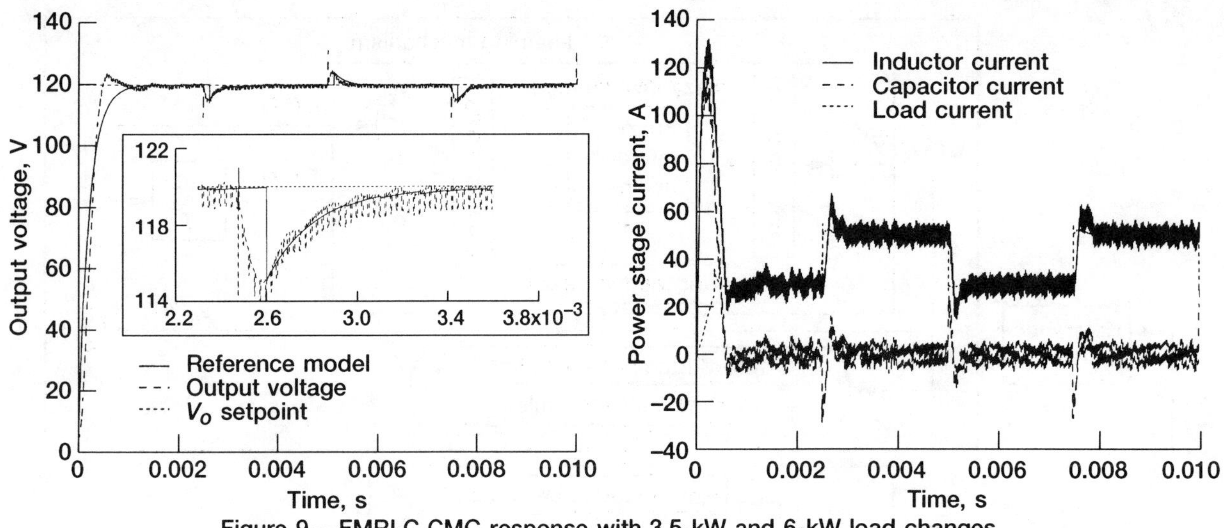

Figure 9.—FMRLC CMC response with 3.5-kW and 6-kW load changes.

$P_i^{j,k}$		C^k −5	−4	−3	−2	−1	0	+1	+2	+3	+4	+5
E^j	−5	0.000	0.000	0.000	0.000	0.000	0.000	0.000	0.000	0.000	0.000	0.000
	−4	0.000	0.000	0.000	0.000	0.000	0.000	0.000	0.000	0.000	0.000	0.000
	−3	0.000	0.000	0.000	−0.535	−1.000	−1.000	−1.000	−0.392	0.000	0.000	[−0.200] 0
	−2	−0.200	0.000	−1.000	−1.000	−1.000	−1.000	−1.000	[−1.000] 0	0.000	0.000	[−1.000] 0
	−1	−1.000	−1.000	−1.000	−1.000	−1.000	[−0.330] −.36	[0.352] .35	[−1.000] 0	[−0.038] 0	0.000	[−1.000] 0
	0	−1.000	−1.000	−1.000	[−1.000] −5	[−1.000] −.1	[−0.311] 0	[0.677] .1	[1.000] .5	1.000	[0.054] 1	[−1.000] 1
	+1	−1.000	−1.000	[−1.000] 0	−0.316	[−1.000] −.35	[−0.171] .36	[0.677] 1	1.000	1.000	1.000	−0.167
	+2	[−1.000] 0	[−0.400] 0	[−0.354] 0	0.046	[1.000] 0	1.000	1.000	1.000	1.000	1.000	1.000
	+3	[0.200] 0	0.000	0.000	[0.131] 0	[0.389] 1	1.000	1.000	1.000	1.000	1.000	1.000
	+4	[1.000] 0	[0.400] 0	0.700	0.431	0.584	1.000	0.985	1.000	0.762	0.272	1.000
	+5	[1.000] 0	[0.400] 0	0.814	0.414	0.357	0.870	0.255	0.000	0.000	0.000	1.000

Figure 10.—Learned knowledge base and modified elements (in boxes) used for fuzzy control knowledge base.

order to achieve a damped response. The reference model chosen for this controller is a first-order type with the transfer function

$$G_r = \frac{\omega_r}{s + \omega_r} \tag{14}$$

The value of ω_r was chosen to approximate the natural frequency of the converter to an open-loop step response. The gains of the FMRLC per the tuning guidelines[9–10] have been selected with the following values:

$$\left[g_e g_c g_{v_e} g_{v_c} g_u g_f k_i \omega_r\right] = [0.01\ \ 2\times10^{-7}\ \ 4.3\times10^{-4}\ \ 8\times10^{-4}\ \ 120\ 120\ 3000\ 4714]$$

The FMRLC controller knowledge base was initialized with zeros as its elements, indicating no knowledge of how to control the process initially. The resulting controller knowledge base from the learned control law, which was derived from the step response simulation (Fig. 9), is shown in Fig. 10. This figure also shows, in boxes, the elements modified and the modified values, outside the boxes, to be used as the knowledge base for the straight fuzzy controller (i.e., with the learning mechanism disabled). The modifications are primarily in the region associated with sign reversal of the controlled variable because of the undamped inductor current response. Some values other than zero are allowed near the center region of the knowledge base, where the tendency for sign reversal is not too drastic. This allowance was made to achieve a relatively faster response at the region of the knowledge base that becomes active near steady-state operation of the process.

As mentioned before, the FMRLC controller can be used for the final control of the process, in fact it shows very good response to step load changes. However, if small-signal analysis is desired, the learning controller is not suitable because the controller nonlinearity is nonstationary.

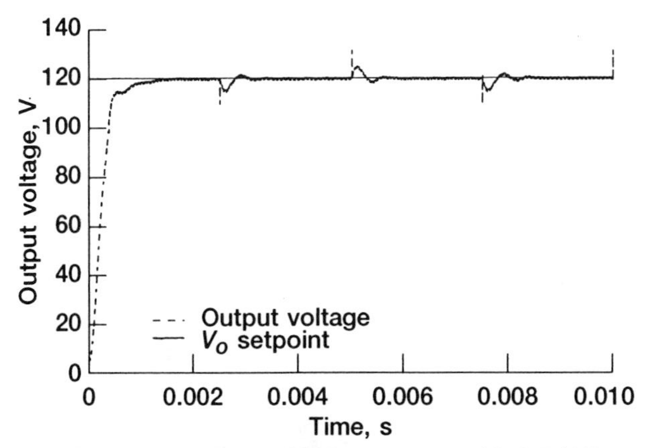

Figure 11.—Fuzzy CMC response with 3.5 kW and 6 kW load changes.

3.2 Fuzzy Current-Mode Control

The conventional fuzzy controller has been implemented here for CMC by disabling the learning mechanism in Fig. 7. The control knowledge base learned from applying the FMRLC CMC (Fig. 10 with indicated modifications) was used for the fuzzy CMC controller. Figure 11 shows step responses of the conventional fuzzy CMC system.

4. Stability Analysis of Fuzzy Current-Mode Control

Various nonlinear stability analysis methodologies could be applied for analyzing fuzzy control systems. Some are the Lyapunov's indirect method, the circle criterion approach, and describing function analysis (DFA).[11] The stability analysis covered in this section is based on the DFA approach supplemented with Bode analysis.

4.1 Describing Function Analysis

First, some background will be presented in computing the describing function. For an input $e(t) = C\sin(\omega t)$ to the nonlinear fuzzy controller there will be a periodic output $u(t)$. Expanding $u(t)$ into a Fourier series results in

$$u(t) = \frac{a_o}{2} + \sum_{n=1}^{\infty}\left[a_n \cos(n\omega t) + b_n \sin(n\omega t)\right] \quad (15)$$

The Fourier coefficients a_i and b_i are functions of C and ω and are determined by

$$a_o = \frac{1}{\pi}\int_{-\pi}^{\pi} u(t)d(\omega t) \quad (16)$$

$$a_n = \frac{1}{\pi}\int_{-\pi}^{\pi} u(t)\cos(n\omega t)d(\omega t) \quad (17)$$

$$b_n = \frac{1}{\pi}\int_{-\pi}^{\pi} u(t)\sin(n\omega t)d(\omega t) \quad (18)$$

Assuming that the fundamental component $u_1(t)$ is dominant (i.e., a_o and the higher order harmonics can be neglected), then

$$\begin{aligned} u(t) \approx u_1(t) &= a_1\cos(\omega t) + b_1 \sin(\omega t) \\ &= M(C,\omega)\sin[\omega t + \phi(C,\omega)] \end{aligned} \quad (19)$$

where

$$M(C,\omega) = \sqrt{a_1^2 + b_1^2}, \phi(C,\omega) = \arctan\left(\frac{a_1}{b_1}\right)$$

Equation (19) can be written in complex form as

$$u_1 = M(C,\omega)e^{j[\omega t + \phi(C,\omega)]} = \left(b_1 + ja_1\right)e^{j\omega t} \quad (20)$$

The describing function of the nonlinear fuzzy controller is defined to be the complex ratio of the fundamental component representing the controller nonlinearity to the input sinusoid

$$\begin{aligned} N(C,\omega) &= \frac{u_1}{C\sin(\omega t)} = \frac{M(C,\omega)e^{j[\omega t + \phi(C,\omega)]}}{Ce^{j\omega t}} \\ &= \frac{1}{C}(b_1 + ja_1) \end{aligned} \quad (21)$$

Therefore, the nonlinear controller can be treated as if it were a linear element in a closed-loop control system consisting of the controller $N(C,\omega)$ and the process $G(j\omega)$ represented by Eq. (13). The harmonic balance equation of this closed-loop control system is

$$G(j\omega)N(C,\omega) + 1 = 0 \quad (22)$$

This equation can also be written as

$$G(j\omega) = -\frac{1}{N(C,\omega)} \quad (23)$$

If the closed-loop system has any limit cycles, its amplitude and frequency can be predicted by solving Eq. (23). If there are no solutions, there are no harmonic oscillations. However, solving Eq. (23) can be quite complex, especially for higher order systems, and it is best to solve this equation graphically by plotting $-1/N(C,\omega)$ against $G(j\omega)$ and finding its intersection points. For an oscillation to exist the intersection must occur at the same frequency ω. The values C and ω, at this intersection point, are the amplitude and frequency of the harmonic oscillation. If points near the intersection and along the increasing C side of the curve $-1/N(C,\omega)$ are not encircled by the curve $G(j\omega)$, the corresponding limit cycle is stable. Otherwise, the limit cycle is unstable.

For experimental evaluation of the fuzzy controller describing function, first the fuzzy controller will be excited with sinusoidal frequency sweeps of different amplitudes to compute the control output $u(t)$. The input sinusoidal sweep function is constructed here as

$$e(t) = C\sin\left(2\pi\alpha f_s t\right) \quad (24)$$

where

$$\alpha = 1 + \frac{\left(f_s - f_f\right)}{f_s}\frac{t}{t_e}$$

and f_s, f_f, and t_e are the starting frequency of the sinusoidal sweep, the final frequency, and the simulation time, respectively. The minimum sweep frequency is chosen to be at least a decade below the expected crossover frequency of the closed-loop gain, and the maximum sweep frequency is chosen to be equal to the subharmonic frequency (i.e., half the switching frequency). After the control outputs are computed for different values of C, FFT's are used to compute transfer function estimates of Eq. (21) as

$$N(C,\omega) = \frac{P_{eu}(C,\omega)}{P_{ee}(C,\omega)} \qquad (25)$$

where P_{eu} and P_{ee} are the cross spectrum of e and u and the power spectrum of e, respectively.

Figure 12 shows the transfer function estimates for different amplitude sweeps. MATLAB was used to compute these transfer functions in Eq. (25). These transfer functions were approximated with poles and zeros by drawing the corresponding straight-line asymptotes. A slope transition of the transfer function indicates first- or second-order zeros or poles depending on the direction and order of the transition. The damping of the second-order responses is a measure of the sharpness of the response transition. Occasionally, final adjustments need to be performed to match the phase responses of Eq. (25), not shown here. On this basis the transfer function approximations for the responses of Eq. (25), displayed in Fig. 12, are of the form

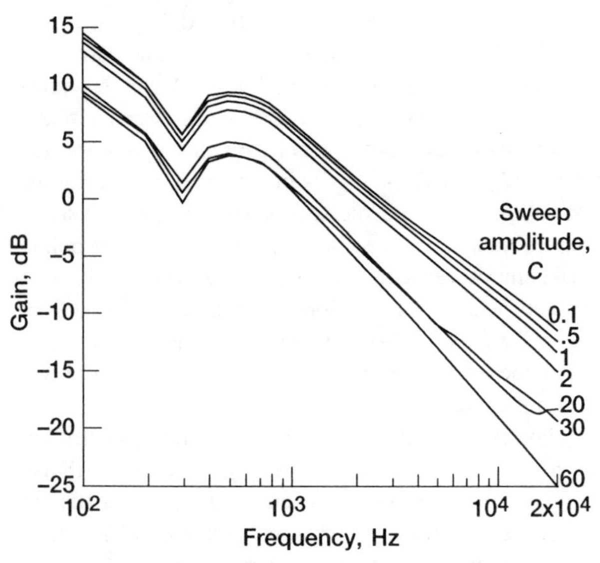

Figure 12.—Transfer functions of fuzzy controller constructed using sine sweeps with different amplitudes and FFT's.

$$N_a(C,s) = K\frac{P_{an}(C,s)}{P_{ad}(C,s)} \qquad (26)$$

where $P_{an}(C,s)$ and $P_{ad}(C,s)$ are the approximation polynomials of the numerator and denominator, respectively. All the polynomials, as a function of the sweep amplitude C, are found to have the same general characteristic as

$$P_{an}(s) = \left(\frac{s^2}{\omega_{z1}^2} + \frac{2\zeta_{z1}s}{\omega_{z1}} + 1\right)\left(\frac{s}{\omega_{z2}} + 1\right)\left(\frac{s}{\omega_{z3}} + 1\right)\left(\frac{s}{\omega_{z4}} + 1\right)$$

$$P_{ad}(s) = s\left(\frac{s^2}{\omega_{p1}^2} + \frac{2\zeta_{p1}s}{\omega_{p1}} + 1\right)\left(\frac{s^2}{\omega_{p2}^2} + \frac{2\zeta_{p2}s}{\omega_{p2}} + 1\right)\left(\frac{s}{\omega_{p3}} + 1\right)$$

Figure 13 shows a frequency response based on FFT's (Eq. (25)) and its approximation (Eq. (26)) for C = 2.0. The approximations have less accuracy at low frequencies. However, the low-frequency response will have little influence approximately one frequency decade above, where the main interest of the response lies (i.e., at the neighborhood of the crossover frequency), as will be seen later. For the transfer function approximations shown in Fig. 14 corresponding to Eq. (26) the parameters of the transfer functions are

$$\left[C\,K f_{z1} f_{z2} f_{z3} f_{z4} f_{p1} f_{p2} f_{p3} \zeta_{z1} \zeta_{p1} \zeta_{p2}\right] =$$

$$\begin{bmatrix} 0.1 & 3000 & 325 & 450 & 720 & 1770 & 375 & 850 & 6000 & .29 & .28 & .57 \\ 2.0 & 3000 & 325 & 450 & 720 & 1770 & 375 & 850 & 4900 & .29 & .32 & .60 \\ 20. & 2000 & 325 & 450 & 720 & 1770 & 375 & 850 & 3000 & .25 & .29 & .60 \\ 60. & 1500 & 325 & 450 & 720 & 1770 & 375 & 850 & 3000 & .29 & .26 & .54 \end{bmatrix}$$

As discussed before, Eq. (23) is solved graphically by plotting its Nyquist, Fig. 15, of $G(j\omega)$ represented by Eq. (13) and $-1/N(C,\omega)$, where $N(C,\omega)$ is represented by

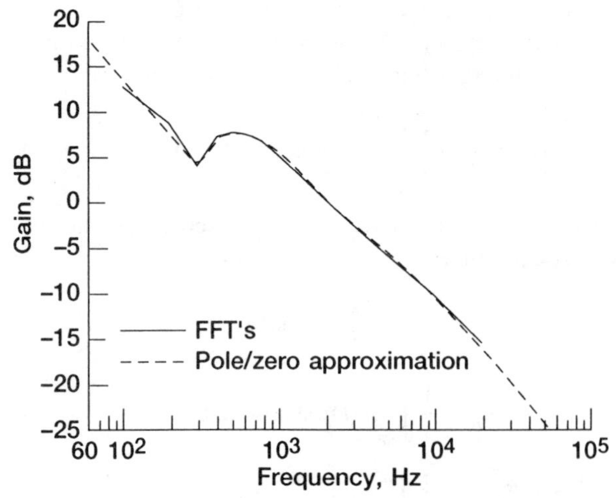

Figure 13.—Fuzzy controller transfer function and its pole-zero transfer function approximation.

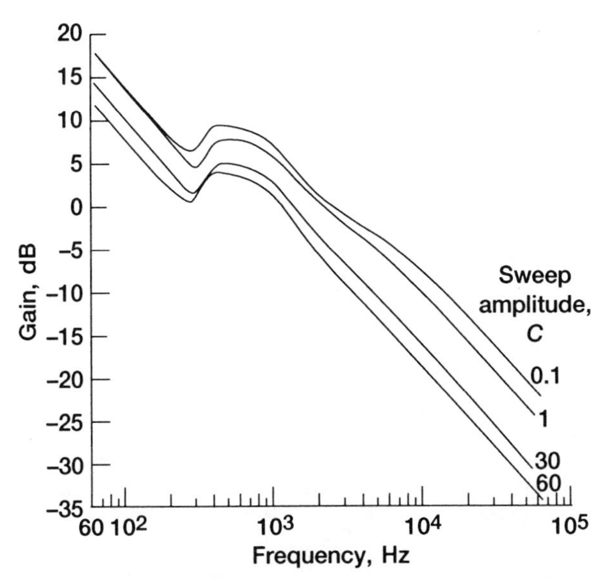

Figure 14.—Approximated transfer functions of fuzzy controller by applying sinusoidal sweeps of different amplitudes.

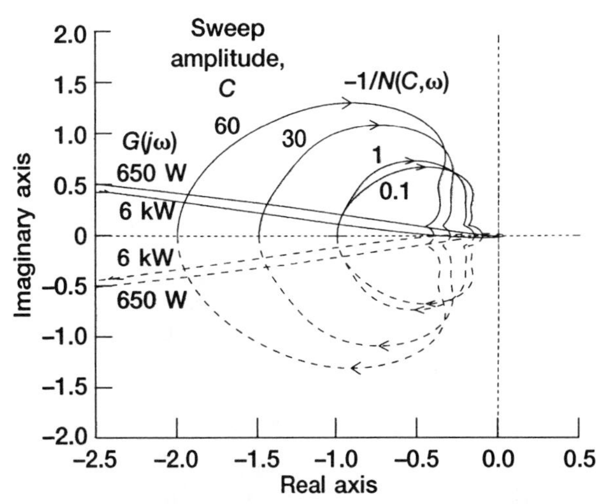

Figure 15.—Nyquist plot of $G(j\omega)$ at 6 kW and 650 W versus $-1/N(C,\omega)$.

Eq. (26). The $G(j\omega)$ of Eq. (13) is plotted at two different operating points of the converter power stage (i.e., at 100% load and at approximately 10% load, which represents the borderline between CCM and DCM). The fact that the plot $G(j\omega)$ passes very near the -1 point on the Nyquist plot is an indication that the open-loop gain of the power stage is nearly unstable without proper compensation. Besides the power level operating points, the input line voltage constitutes operating point changes that will also influence $G(j\omega)$. If desirable, an analysis for changes in line voltages can also be carried out in the exact same way, but this analysis is beyond the scope of this paper.

As shown in Fig. 15 there are several intersection points. However, a closer examination of the complex vectors of these transfer functions as a function of ω reveals that none of these intersections occurs at the same ω, which is the condition for predicting sustained harmonic oscillations or instability. The closest these transfer functions came to intersecting for the same ω was for the $C = 60$ plot, as was expected from the Bode analysis that follows. The time domain response shown in Fig. 11 confirms the prediction of the describing function analysis (i.e., the absence of sustained system limit cycles).

4.2 Bode Analysis

4.2.1 Closed-loop gain.—Bode analysis allows for a convenient way to carry out direct quantifiable measurements of the system's stability margins. In a closed-loop control system consisting of a controller with a transfer function $N(s)$ and a process with a transfer function $G(s)$, the closed-loop gain of the system is

$$L_{cl}(s) = G(s)N(s) \tag{27}$$

The overall input-to-output transfer function of the closed-loop system will have the characteristic equation of $G(s)N(s) + 1 = 0$, the same as Eq. (23). From this characteristic equation it is observed that when $L_{cl} = -1$, or in phasor terms $L_{cl} = 0$ dB $\langle 180°$, the characteristic equation and therefore the overall closed-loop transfer function will have a zero denominator. Therefore, the system will be unstable for $L_{cl} = -1$. The difference, between the actual phase angle of L_{cl} and 180° at the point where L_{cl} crosses 0 dB, is defined as the system stability phase margin. Similarly, the gain margin of the system is defined to be the difference between 0 dB and the actual magnitude of L_{cl} when the phase of L_{cl} crosses 180°. When the system phase and/or gain margins are relatively small, the system will normally exhibit an oscillatory response, similar to an underdamped response with increased sensitivity.

Figures 16 and 17 show the closed-loop gain responses, based on the converter transfer function (Eq. (13)) and the fuzzy controller transfer function (Eq. (26)), at 100% power and at the borderline between CCM and DCM, respectively. In both Figs. 16 and 17 the converter has less phase margin for larger values of C. For instance, for $C = 60$ the phase margins are ~9° and ~5°, respectively, for loads of 6 kW and 650 W. At low values of C the corresponding phase margins for high and low power levels are ~36° and ~20°, respectively. Theoretically, the gain margins are infinite in all cases, since the phases do not cross the 180° line. However, there is another measure of stability, conditional stability, which in brief is a measure of how close the phase approaches 180° while the magnitude approaches 0 dB. Not shown here, the phase margins can be improved by adjusting the values of the membership functions (Fig. 10 at the center region of the knowledge base) to move them closer to the origin. This

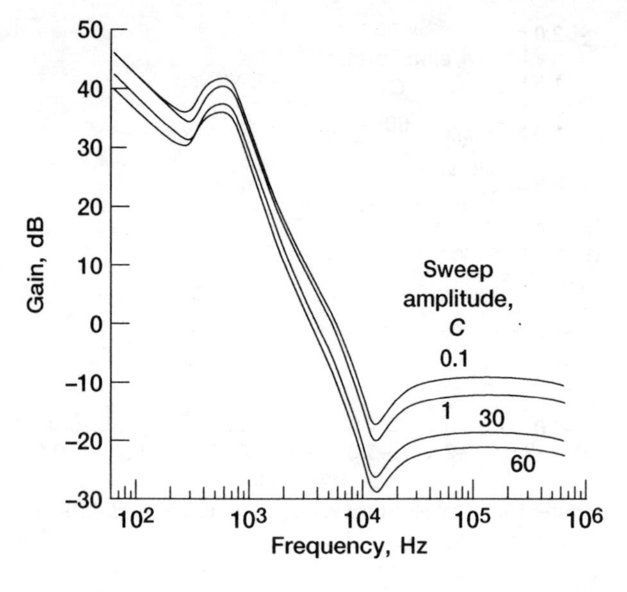

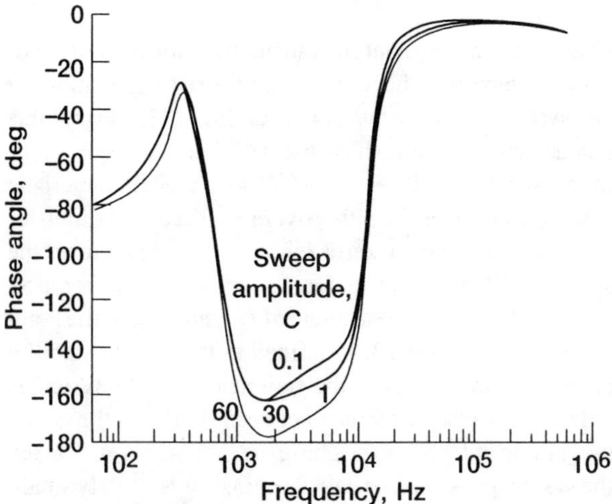

Figure 16.—Closed-loop gain and phase angle of buck converter with fuzzy CMC control at 6 kW.

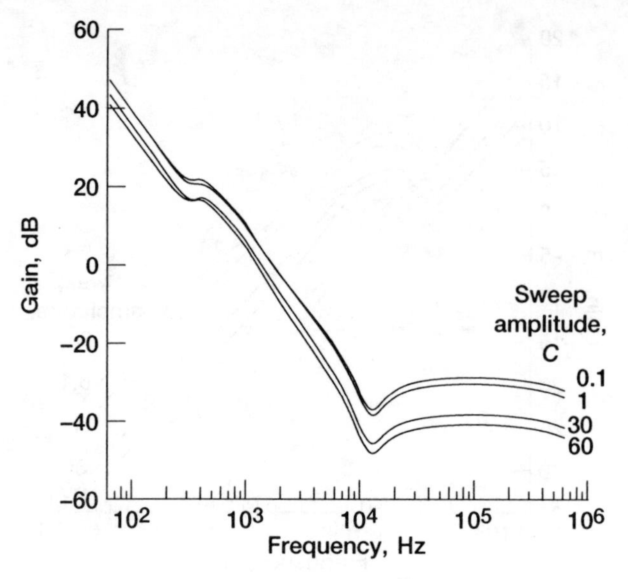

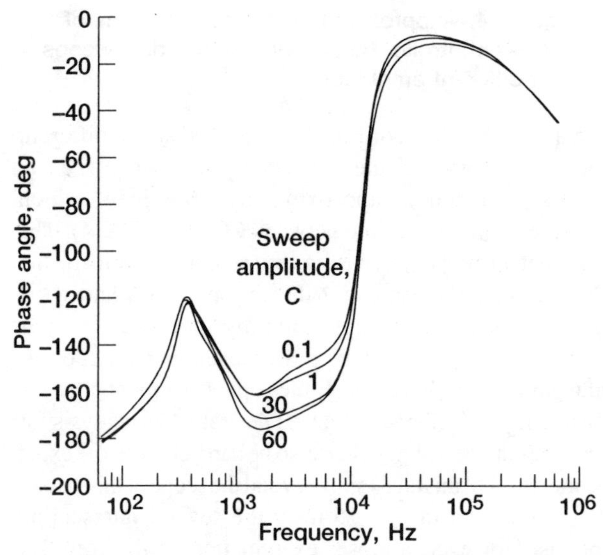

Figure 17.—Closed-loop gain and phase angle of buck converter with fuzzy CMC control at borderline of CCM and DCM (~650 W).

will make the response more damped, but slower, with a little more overshoot. Of course, improving the filter design, which was purposely underdamped, is another way to improve the stability margins.

4.2.2 Output impedance.—Another important measure of system performance is the converter output impedance. A source supplying a single or an aggregate load, as a system, will be subjected to additional stability criteria at the source-load interface as

$$G_t(s) = G_s(s)G_l(s) \Big/ \left[1 + \frac{Z_o(s)}{Z_l(s)}\right] \tag{28}$$

where $G_t(s)$, $G_s(s)$, and $G_l(s)$ are the input-to-output transfer functions of the overall system, the source, and the load, respectively; $Z_o(s)$ is the source output impedance; and $Z_l(s)$ is the load input impedance. For the power converter in Fig. 1 its open-loop output impedance, which is the source impedance of the output filter, including the inductor effective series resistance R_{in}, is

$$Z_P(s) = R_{in}\frac{\left(C_f L_c s^2 + R_c C_f s + 1\right)\left(\frac{L_f}{R_{in}} s + 1\right)}{C_f L_f s^2 + C_f R_c s + 1} \tag{29}$$

for $L_f \gg L_c$ and $R_c \gg R_{in}$. The closed-loop output impedance is

$$Z_o(s) = \frac{Z_p(s)}{1 + L_{cl}(s)} \tag{30}$$

where $Z_p(s)$ is the open-loop output impedance (i.e., $Z_p = \frac{\hat{v}_o}{\hat{i}_o}\Big|_{\hat{d}=0}$).

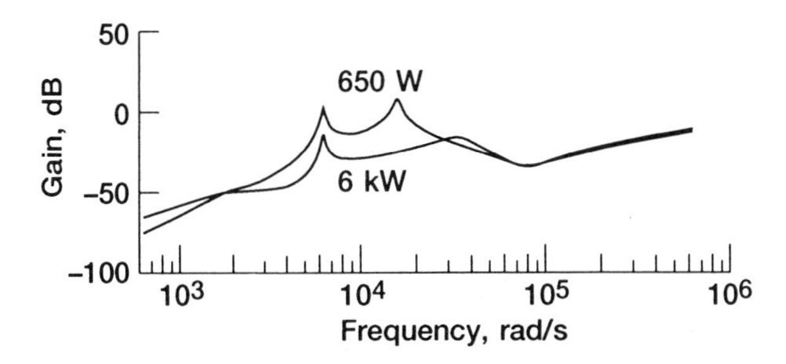

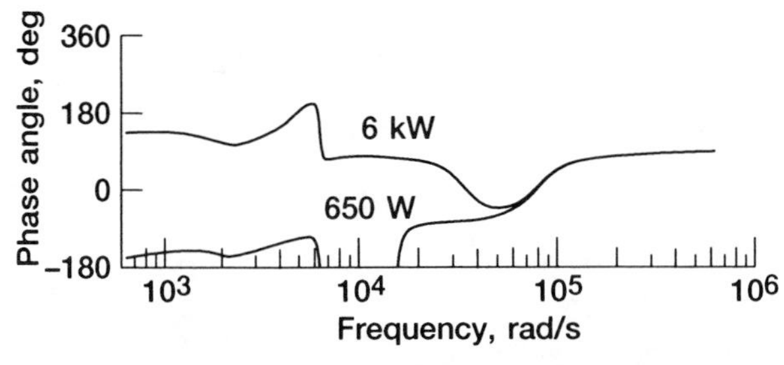

Figure 18.—Closed-loop output impedance of buck converter with fuzzy CMC control at C = 1.0.

Figure 18 shows the closed-loop output impedance of the converter at power levels of 6 kW and 650 W and for $C = 1$, with a parasitic value of R_{in}. The phase of the impedances in Fig. 18 needed to be unramped. A complete analysis will include the output impedance at different values of C and an actual load impedance, to study the stability margins of the system. For such an analysis the phase margin, measured at the intersecting points of the two impedances and based on Eq. (28), will be computed as the addition of the load impedance phase minus the source impedance phase subtracted from 180°. The output impedance can also be improved by redesigning the output filter.

5. Concluding Remarks

In this paper fuzzy control was applied to traditional current-mode control (CMC) by replacing the typical lead-lag compensation design. It was demonstrated that, with proper design of the Fuzzy Model Reference Learning Control (FMRLC), very good time responses to load step changes can be obtained even with a highly undamped power converter design. The control knowledge base learned with the FMRLC was used to develop a straight fuzzy controller for the converter in order to conduct small-signal analysis. The paper demonstrated the ability to conduct small-signal analysis of the converter with fuzzy control by developing a small-signal model of the fuzzy controller using sinusoidal sweeps and fast Fourier transforms. The small-signal analyses conducted by employing the describing function and Bode analysis showed the feasibility of applying fuzzy control to CMC, while still employing powerful traditional analysis tools.

References

1. Mamdani, E.H., "Applications of Fuzzy Algorithms for Simple Dynamic Plant," Proceedings of IEEE, vol. 121, no. 12, pp. 1585–1588, 1974.
2. Mamdani, E.H., and Procyk, T., "A Linguistic Self-Organizing Process Controller," Automatica, vol. 15, no. 1, pp. 15–30, 1979.
3. Zadeh, L.A., "Fuzzy Algorithm," Informat. Control, vol. 12, pp. 94–102, 1968.
4. Zadeh, L.A., "Outline of a New Approach to the Analysis of Complex Systems and Decision Process," IEEE Trans. Man Cybern., vol. SMC–3, pp. 28–44, 1973.
5. Layne, J.R., and Passino, K.M., "Fuzzy Model Reference Learning Control," Proceedings of IEEE Conference on Control Applications, pp. 696–691, Sept. 1992.
6. Middlebrook, R.D., and Cuk, S., "A Generalized Uniform Approach To Modeling Switching-Converter Power Stages," Proceedings of IEEE Power Electronics Specialist Conference, June 1976.
7. Middlebrook, R.D., "Topics in Multi-Loop Regulators and Current-Mode Programming," Proceedings of IEEE Power Electronics Specialist Conference, June 1985.
8. Ridley, R.B., "A New Small-Signal Model for Current-Mode Control," Proceedings of Power Conversion and Intelligent Motion Control Conference, October 1989.
9. Kopasakis, G., "Adaptive Performance Seeking Control Using Fuzzy Model Reference Learning Control and Positive Gradient Control," AIAA, July 1997.
10. Kopasakis, G., "Nonlinear Performance Seeking Control Using Fuzzy Model Reference Learning Control and the Method of Steepest Descent," AIAA, July 1997.
11. Passino, K.M., and Yurkovich, S. "Fuzzy Control," Addison-Wesley Longman, Inc., 1998.

AIAA-2000-2804

DESIGN AND TECHNOLOGY OF COMPACT HIGH-POWER CONVERTERS[1]

Krishna Shenai
Electrical Engineering and Computer Science Department
1120 SEO, M/C 154, 851 South Morgan Street
The University of Illinois at Chicago
Chicago, IL 60607
Tel: (312) 996-2633; Fax: (312) 996-0763; E-mail: shenai@eecs.uic.edu

and

Philip G. Neudeck[+] and Gene Schwarze[++]
NASA Glenn Research Center
Lewis Field, 21000 Brookpark Road
Cleveland, OH 44135-3191
[+]Tel: (216) 433-8902; Fax: (216) 433-8643; E-mail: neudeck@grc.nasa.gov
[++]Tel: (216) 433-6117; Fax: (216) 433-8311;
E-mail: Gene.E.Schwarze@grc.nasa.gov

ABSTRACT

New material technologies such as Silicon Carbide (SiC) are promising in the development of compact high-power converters for next-generation power electronics applications. This paper presents an optimized converter design approach that takes into consideration non-linear interactions among various converter components, source and load. It is shown that with the development of high-temperature, high-power SiC power module technology, magnetic components and capacitors become important technology challenges, and cannot be ignored. A 50% improvement in power density is calculated for a 100V-2kV, 7kW SiC DC-DC power converter operating at 150°C compared to a silicon power converter. The SiC power converter can be operated at junction temperatures in excess of 300°C (as compared to 150°C for a silicon power converter) with reasonable efficiency that potentially leads to a significant reduction in thermal management.

I. INTRODUCTION

New millennium spacecraft will require efficient, compact and reliable power management and control concepts for significant reduction in the spacecraft bus mass, volume and cost [1]. Commercial applications of the same technology include electric utility [2], motor control [3], automotive electronics [4] and power supplies [5]. State of the art silicon high-power, high-voltage DC-DC converters are bulky, heavy, inefficient, and require extensive thermal management and cooling. For example, a 100V-2kV, 7kW silicon power converter switching at 50 kHz has 85% efficiency and a power density of less than 4W/cm^3. It can be operated reliably up to a junction temperature of only 150°C, thus requiring extensive thermal management and cooling. A key requirement to develop a compact, light-weight, high power DC-DC converter is to increase the switching frequency. However, switching losses in a silicon high-power switch become excessively large above 50 kHz [6].

Silicon Carbide (SiC) high power switches are more efficient than silicon devices at much higher switching frequencies and operating temperatures [7,8]. Our calculations show that a100V-2kV, 7 kW SiC power converter switching at 500 kHz has an efficiency of 89% at 150°C and a power density of 7W/cm^3. This result is more than a 50% increase in converter power density as compared to a silicon converter with identical ratings. Increasing the power density results in a reduction in size, weight and cost of the power converter. The SiC power converter is also capable of efficient operation (85% at 300°C) at elevated junction temperatures. This results in a reduction in mass/volume and complexity of the cooling system. However, high-temperature, high-power converter operation requires efficient, high-frequency (> 500kHz) high-temperature (300°C) circuit components (switches, diodes, control electronics, transformers, inductors and capacitors) [9] and power module packaging technologies. Presently, SiC suffers from a high density of defects [10] that may impact long-term device reliability, especially when stressed under extreme dynamic switching conditions in a high power converter [11].

This paper presents the results obtained from a "top-down" design approach to develop high power-density DC-DC power converters. It takes into account circuit-level non-linear interactions and develops components that lead to an optimized overall power system for performance and reliability. Component ratings were determined by the circuit topology. Finite-element (FE) simulations were used to design SiC power switches and the switch design was performed in a realistic manner by taking into account practical technology limits. For example, maximum voltage and current stresses in the switch were determined from circuit calculations. The breakdown voltage and on-state current dictate the internal cell design, edge termination design, drift region doping density and thickness, and device width for the power switch.

First-order calculations were performed to design high-temperature, high-frequency components for 300°C operation of an 85% efficient, 100V-2kV, 7kW SiC DC-DC power converter. The required component specifications are robust SiC diode (2.5kV/3.5A, 130V/20A) and JFET (130V/20A) modules; high temperature low-loss capacitors, inductors, transformers and Silicon-On-Insulator (SOI) control electronics; and, high-temperature power packaging. This design results in more than a 50% increase in power density for the SiC power converter as compared to its silicon counterpart when operated at 150°C. The converter power density is largely limited by the magnetic components, especially that of the power transformer. The upper limit of the transformer's power density is determined by either the minimum allowable transformer efficiency or maximum allowable temperature rise. The temperature rise and efficiency are not independent of each other, because the temperature rise is determined by the transformer's power losses (heat), and thus, the efficiency.

II. DESIGN PROCEDURE AND RESULTS

Our design calculations are based on a "top-down" approach. From application requirements, we select the circuit switching topology and control strategy. Simple circuit calculations yield component ratings. Component designs are performed utilizing available material and device data and are based on extensive computer simulations.

1. Converter Design: The high power DC-DC converter performance calculations presented are for the basic asymmetrical bridge topology shown in Fig. 1 [12]. This DC-DC converter operates in the Zero-Voltage Switching (ZVS) mode by utilizing the resonance between transformer leakage inductance and output capacitance of the switch, without penalties of extra voltage or current stress on the power semiconductor switches. This topology was used to increase the efficiency at high switching frequencies required to reduce the size of the magnetic components and capacitors.

A current-doubling circuit [13] is used on the transformer secondary side. As a result, the transformer secondary winding is rated for one half the load current and does not require a center-tapped connection.

For the 100V-2kV, 7kW converter, component power densities and efficiencies were first estimated as a function of frequency. These estimates were then used to calculate the converter efficiency and power density as a function of frequency and the results are plotted in Fig. 2. Because of junction temperature limitations, silicon converter calculations are shown only at 150°C. Note that the switch power losses limit the silicon power converter operation to 50 kHz whereas SiC power converter can be efficiently operated at 500 kHz. High-frequency operation results in over 50% increase in the power density for the SiC power converter as compared with the silicon converter. A reduction in the mass/volume and complexity of the cooling system can be obtained by operating the SiC power converter at 300°C. For high-temperature operation, however, efficient magnetic components and capacitors are needed.

2. SiC Device Design: Our device calculations are based on two-dimensional (2-D) finite-element (FE) simulations [14]. From the converter circuit calculations, switch ratings were obtained. These are SiC Schottky diodes (2.5kV/3.5A and 130V/20A) and JFET's (130V/20A). The design of low-voltage high-current SiC JFET's on the transformer primary follows the approach previously reported [15].

The rectifiers on the transformer secondary are required to support high voltage and conduct moderate amount of current. Performance of large area, high voltage SiC devices is degraded by the presence of material defects, and poor edge-termination structures [16]. Improved edge-termination designs are needed to raise the breakdown voltage to near-ideal values without the penalty of leakage current or die area. The calculated power loss data for the secondary-side rectifiers is plotted in Fig. 3 as a function of frequency for a junction temperature of 150°C. For comparison, power loss calculated for silicon PiN diodes is also shown. The calculated silicon power loss closely corresponds to the measured data available for commercial devices. It is seen that above 50 kHz, silicon rectifiers becomes excessively inefficient. SiC Schottky diodes show excellent performance even at a switching frequency of 500 kHz. In fact, these devices can be efficiently operated at switching frequencies in excess of 1 MHz.

3. Magnetic Design : The development of high temperature, high frequency, high power density, and high efficiency magnetic components (transformers and inductors) is of paramount importance in the realization of high power density DC-DC power converters. In order to decrease the mass/volume of magnetic components, and thus increase the converter power density, requires an increase in switching frequency. For power transformers [17] a frequency is approached beyond which no appreciable gain in transformer power density is realized. Our designs of magnetic components (transformer and inductor) are based on the available materials data and follow a first-order simple design approach. We have developed an extensive core loss database for crystalline and amorphous soft magnetic materials for frequencies up to 50 kHz and temperatures to 300°C; limited core loss data up to 1MHz at room temperature is also available.

To choose the transformer core size the area product, W_aA_e, can be derived from Faraday's law. Based on this value the *EI* ferrite core was chosen from the manufacturer's data sheets [18]. TSF-7099 material was chosen for the transformer core, which has a minimum core loss at 100°C. We used this value in all our calculations. After choosing the proper core size, the transformer core loss was calculated at each frequency and flux density from the core loss vs. frequency plots in the manufacturer's catalog. These calculations were used to estimate the number of primary and secondary turns. The size of the winding wires was chosen considering the skin effect and current density. The transformer efficiency was calculated after determining the core and wire size at the operating frequency. The transformer was designed to have an efficiency in the range of 96%-98%.

Fig. 4 shows typical results obtained for a transformer design where the power density was calculated for the *EI* ferrite core. The results show less than a linear reduction in transformer size with increasing frequency. It should be noted that the core loss increases with temperature. Also, the Curie temperature of the ferrite core is low, and hence, ferrite cores are generally not used above 150°C. Improved trade-off analysis and magnetic component design will require core loss data up to 1MHz at temperatures up to 300°C. We are presently performing such analysis using high Curie temperature magnetic materials.

The output filter component values were calculated as follows [19]. The filter inductance (L_f) value was calculated from the output voltage, switching frequency and for a load current variation of 3.5A assuming that the converter operates at the edge of continuous conduction. The filter capacitor (C_f) values plotted in Fig. 4 were calculated for an output voltage ripple of 5% and for a switching duty cycle of 50%. At higher operating temperatures, the leakage current in the capacitor becomes too high. High-temperature capacitors capable of reliable operation up to 300°C are required for the SiC power converter.

III. CONCLUSIONS AND DISCUSSIONS

In this paper, we present a design approach to develop compact, high power DC-DC converters based on emerging SiC material and device technologies. For comparison, the performance of a silicon power converter is also calculated that utilizes commercially available state of the art silicon power switches. The design calculations are based on a simplified model for the magnetic components and can be refined using experimental magnetic core loss data at higher temperatures and frequencies.

It is shown that a 100V-2kV, 7kW silicon DC-DC power converter can be operated at a switching frequency of only 50 kHz with 85% power conversion efficiency at 150°C. A SiC power converter with identical ratings can be operated at 500 kHz with an efficiency of 89%. This amounts to over a 50% increase in the converter power density. The switching frequency (and hence, the converter power density) is limited by the performance of the magnetic components (transformers and inductors).

Junction temperature limitations and poor efficiency constrain silicon power converters to below 150°C operation. Our calculations suggest that a SiC power converter can be operated at 300°C with 85% efficiency provided high-temperature magnetic components and capacitors become available. High temperature converter operation results in a dramatic reduction in the mass/volume and complexity of the cooling system.

REFERENCES

1. J. A. Hamley, G. I. H. Cardwell, J. McDowell, and M. Matranga, "The design and performance characteristics of the NSTAR PPU and DCIU," AIAA Paper 98-3938, 1998.

2. Narain G. Hingorani, "Introducing Custom Power," *IEEE Spectrum*, June 1995, pp. 41-48.

3. Austin Bonnett "Analysis of AC induction Motor Transients Caused by PWM Inverters," *Proceedings of PQA '93/Pecon IV*, 1993.

4. Speicial Issue, *Proceedings of the IEEE*, August 1994.

5. Deepak Divan, "Low-Stress Switching for Efficiency," *IEEE Spectrum,* December 1996, pp. 33-39.

6. K. Shenai, " Technology Trends in High-Frequency Power Semiconductor Discrete Devices and Integrated Circuits," in *Technical Papers of the Fourth Int. High Frequency Power Conversion*, Intertec Communications Inc., Ventura, CA, pp. 1-23, May 1989.

7. K. Shenai, R. S. Scott, and B. J. Baliga, "Optimum Semiconductors for High-Power Electronics, " *IEEE Trans. Electron Devices*, vol. 36, pp. 1811-1823, Sept. 1989.

8. K. Shenai, "Potential Impact of Emerging Semiconductor Technologies on Advanced Power Electronic Systems, " *IEEE Electron Device Lett.*, vol. 11, no. 11, pp. 520-522, Nov. 1990.

9. K. Shenai, "Application-Specific Power Electronic Modules (ASPEMs)-Technologies and Challenges," *Proc. of PowerSystems 1996*, Las Vegas, Nevada, September 1996.

10. Neudeck, W. Huang, and M. Dudley, "Study of bulk and elementary screw dislocation assisted reverse breakdown in low-voltage (<250 V) 4H-SiC p+n junction diodes – Parts I & II," IEEE Trans. Electron Devices, vol. 46, no. 3, pp. 478-492, 1999.

11. N. Keskar, M. Trivedi, and K. Shenai, "Device Reliability and Robust Power Converter Development," in *Proc. European Symp. Reliability of Electron Devices, Failure Physics and Analysis (ESREF)*, 1999, pp. 1121-1130.

12. P. Imbertson and N. Mohan, "Asymmetrical Duty Cycle Permits Zero Switching Loss in PWM Circuits with no Conduction Loss Penalty," *IEEE Trans. Power Electronics*, vol. 29, no. 1, pp.121-125, January 1993.

13. K. O'meara, "A New Output Rectifier Configuration Optimized for High-Frequency Operation," in *Technical Papers of the Fourth Int. High Frequency Power Conversion*, Intertec Communications Inc., Ventura, CA, 1991, pp. 219-225.

14. K. Shenai, "Mixed-mode circuit simulation: An emerging CAD tool for the design and optimization of power semiconductor devices and circuits," *Proceedings of IEEE 4th PELS Workshop on Computers in Power Electronics*, Aug. 7-10, 1994, IEEE, Cat # 94TH0705-4, pp. 1-5.

15. K. Shenai, " Optimally Scaled Low-Voltage Vertical Power DMOSFET's for High-Frequency Power Switching Applications, " IEEE Trans. Electron Devices, vol. 37, no. 4, pp. 1141-1153, Apr. 1990.

16. P. G. Neudeck, "SiC Technology," in *The VLSI Handbook*, Edited by W. K. Chen, CRC Press/IEEE Press, 2000.

17. G. E. Schwarze, "Development of high frequency low weight power magnetics for aerospace power systems," Proc. 19th Intersociety Energy Conversion Engineering Conference, San Franscisco, CA, August 19-24, 1984, pp. 196-204.

18. TSC International Catalog, 1997.

19. N. Mohan, P. Undeland, and W. Robbins, *Power Electronics Devices, Circuits and Converters*, Academic Press, 2nd Ed., 1992.

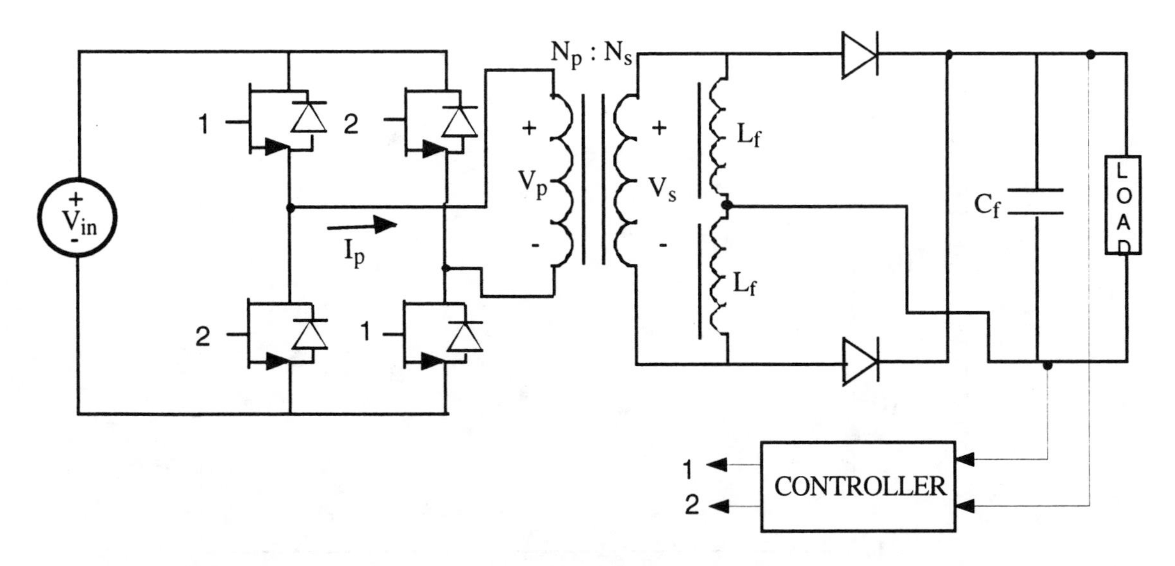

Fig. 1 A circuit schematic of a full-bridge ZVS PWM DC-DC power converter with a current doubler in the output circuit.

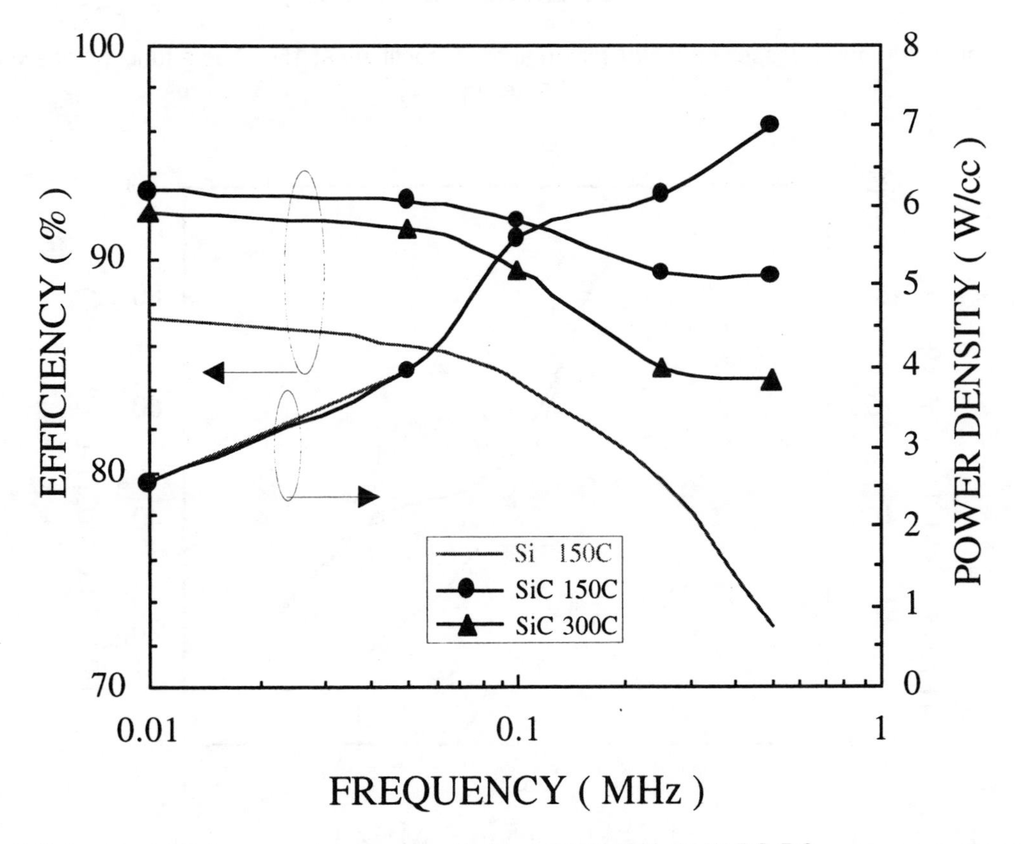

Fig. 2 Calculated efficiency and power density for a 100V-2kV, 7 kW DC-DC power converter as a function of switching frequency.

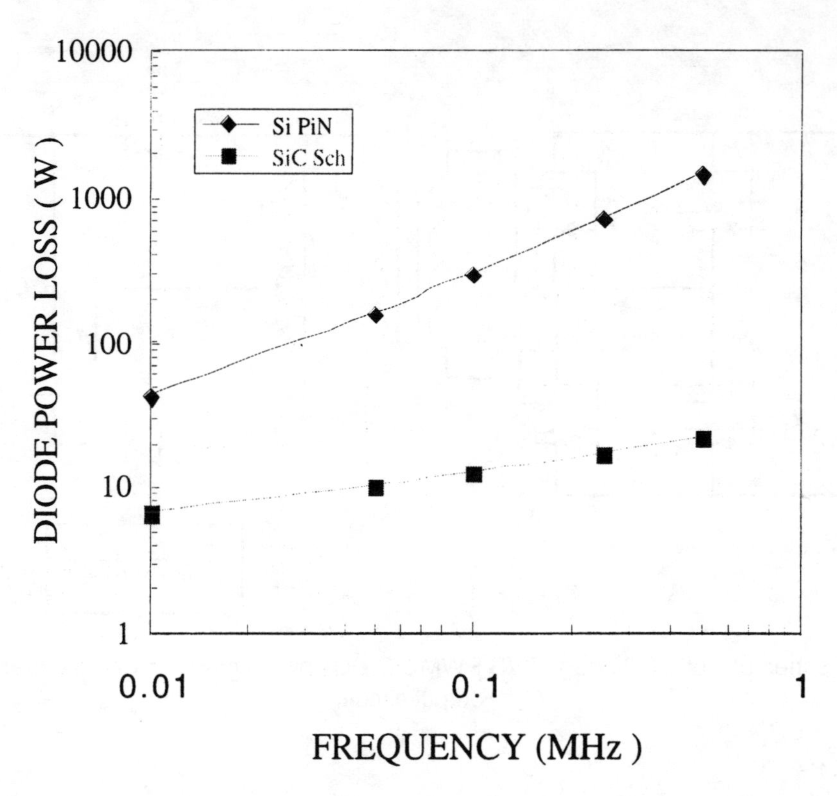

Fig. 3 Calculated power losses for 2.5kV silicon and SiC rectifiers at 150°C as a function of switching frequency.

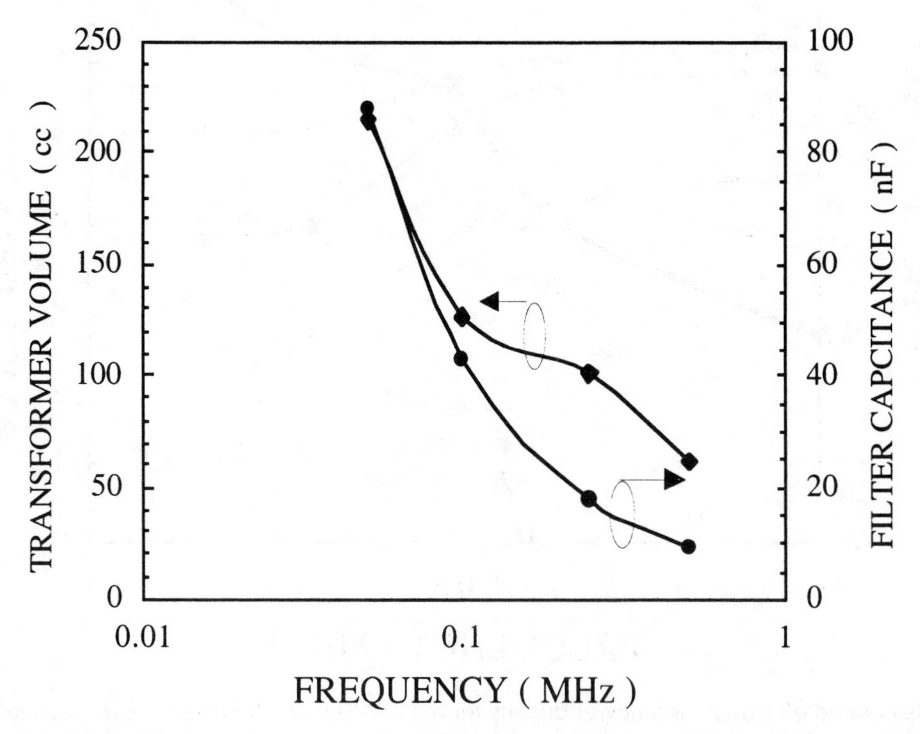

Fig. 4 Calculated transformer volume and output filter capacitance as a function of frequency in a 100V-2kV, 7kW DC-DC power converter.

PERFORMANCE EVALUATION OF SILICON CARBIDE DEVICES IN POWER CONVERTERS[1]

Krishna Shenai
Electrical Engineering and Computer Science Department
1120 SEO, M/C 154, 851 South Morgan Street
The University of Illinois at Chicago
Chicago, IL 60607
TEL: (312) 996-2633; FAX: (312) 996-0763; E-mail: shenai@eecs.uic.edu

and

Philip G. Neudeck
NASA Glenn Research Center
Lewis Field, 21000 Brookpark Road
Cleveland, OH 44135-3191
TEL: (216) 433-8902; FAX: (216) 433-8643; E-mail: neudeck@grc.nasa.gov

ABSTRACT

Commercial Silicon Carbide (SiC) Schottky and PiN diodes, and JFET's were characterized for static conduction and dynamic switching performances in hard- and soft-switching buck converters. The results are compared with the measured data obtained from similarly rated silicon and GaAs devices. It is shown that for low-voltage rectifiers the only real advantage of SiC over silicon and GaAs is the potential for high-temperature operation. A significant improvement in reverse recovery performance was measured for high-voltage SiC PiN diodes compared to silicon PiN diodes. Low-voltage SiC JFET's were found to perform poorly in power converters compared to similarly rated silicon power MOSFET's. It is shown that further technology optimization could make SiC JFET's attractive candidates for high-power high-temperature high-frequency electronics switching.

I. INTRODUCTION

Silicon-based power converter technology is relatively mature and further major advances are difficult. For example, silicon power switching device performance has reached nearly the material limit [1] and further circuit topology advances are difficult to achieve [2]. The trend is toward custom power electronics [3] where device technology is optimized for specific end application taking into consideration all of the non-linear circuit level interactions, layout and interconnect parasitic effects, and thermal management.

In many cost-sensitive power electronics applications such as automotive electronics and computer/telecommunication power supplies, product cost critically hinges on manufacturing yield and long-term reliability. For example, 1% - 2% field failures in the first couple of years of field operation are typical in such applications [4]. In many cases, redundant units are used to avoid hazards and catastrophic failures. However, this approach is costly and leads to non-optimal systems. In military, space and some key commercial systems, power converter reliability is of paramount importance.

Soft switching is gaining acceptance although large-scale adaptation of soft-switching techniques is unlikely in the near future [5]. Soft switching reduces component stresses and results in increased power conversion efficiency [6]. However, power-switching technology specifically optimized for soft switching is currently not available in the commercial market [7]. In high-performance AC-DC and DC-DC power converters, soft-switching transition of the main switch is employed to reduce power losses and improve switch reliability. In addition, rapid turn-off of power switches is carried out to increase the switching frequency and obtain performance advantages in a highly competitive market. Fast switching leads to high dynamic stresses (such as high *di/*dt and dv*/dt*), and power semiconductor devices slowly degrade with time and eventually fail in the field under stressful repetitive switching conditions [4].

Further improvements in strategic power electronics systems are therefore critically dependent on fundamental advances in solid-state switching technology [8]. Silicon Carbide (SiC) is widely recognized as a material of choice for future high-power and high-temperature electronic applications because of its superior electrical breakdown performance, high thermal conductivity, and improved ruggedness compared to silicon [9]. From the theoretical calculations, improvement in power-handling capability by more than three orders of magnitude has been predicted for SiC power devices compared to silicon devices. However, commercialization of SiC technology is hampered by a high density of material defects [10]. Although significant advances in SiC material growth technology have been reported in reducing micropipes, other types of defects such as the screw dislocations exist in large quantities [11,12]. It is not clear how these defects manifest on device characteristics particularly when stressed under dynamic switching conditions typical of a power converter.

This paper presents the results obtained from an experimental study of commercial SiC diodes and JFET's. Low-voltage and high-voltage SiC Schottky and PiN rectifiers and low-voltage SiC JFET's were extensively characterized for static conduction and dynamic switching performances in power converters. The SiC devices studied in this work were fragile, and hence, low-level electrical and thermal stresses were applied to avoid their destruction. Devices were first characterized in detail for static I-V, C-V and breakdown performance up to 100°C. The same devices were then tested in hard- and soft-switching buck converters. The results are compared with GaAs Schottky diodes, and silicon PiN diodes and MOSFET's with similar ratings. Low-voltage SiC diodes performed comparably with silicon and GaAs devices. High-voltage SiC PiN diodes were found to be far superior in terms of reverse recovery performance as compared to silicon counterparts of similar ratings. Low-voltage SiC JFET's show promise for efficient power conversion, especially at elevated junction temperatures as compared to silicon power MOSFET's, provided significant technology improvements are made.

II. EXPERIMENTAL PROCEDURE AND RESULTS

The rectifiers studied included 250V/0.1A Schottky diodes and 3kV/0.25A and 3kV/1A PiN diodes. JFET's were rated at 100V/0.5A. For the buried gate JFET structure, the source and drain contacts are made to the top surface and the gate electrode is at the bottom. No special edge termination was used. Consequently, the reverse-biased breakdown voltage measured was only a fraction of the ideal breakdown voltage determined from drift region doping density and thickness [13]. The residual material defects may have also caused breakdown degradation [14]. Special care was used in wafer handling and measurements since these devices were extremely fragile. The applied stress levels, both electrical and thermal, were intentionally kept low during static as well as dynamic testing to avoid device destruction.

Device dimensions were obtained from SEM measurements and the doping levels and layer thickness were extracted from C-V measurement [15] and two-dimensional (2-D) finite-element (FE) simulations [16].

The circuit schematics used for hard-switching and Zero Voltage Switching (ZVS) buck converters are shown in Figs. 1(a) and 1(b), respectively [17]. The Device Under Test (DUT) is either the SiC diode or JFET. The diode provides a path for inductor to freewheel when the main switch is turned-off. The ZVS conditions across the DUT are established by the resonance between the inductor L_r and capacitor C_r. The gate control electronics was constructed using silicon commercial chips and the entire power converter was mounted on a Printed Circuit Board (PCB) specially designed to minimize the layout parasitic effects.

Rectifier Results: Forward and reverse I-V and C-V characteristics were measured over the temperature range of 23°C to 100°C. Figs. 2(a) and 2(b) show typical variation of SiC Schottky diode ideality factor n and barrier height ϕ_B with temperature, respectively. The device is modeled as two devices in parallel -- a defect diode corresponding to bulk defects and a normal diode that is defect-free. The defect diode causes excessive leakage at low voltages in the forward bias region. It also results in premature breakdown in the reverse bias region [10]. For both diodes, the ideality factor n follows the standard T_0 anomaly [18]. The barrier height ϕ_B extracted from I-V and C-V analyses for the defect-free diode merge at higher temperatures suggesting that the interfacial contamination layer becomes nearly transparent for charge transport.

Extensive diode reverse recovery measurements were performed with varying on-state current I_{ON}, turn-off di/dt, bus voltage V_{BUS}, and temperature. Typical reverse recovery waveforms are shown in Fig. 3(a) for 250V/0.1A SiC Schottky diode and in Fig. 3(b) for 3kV/1A SiC PiN diode at room temperature. The extended reverse recovery in Fig. 3(a) is an artifact of the measurement and is caused by large value of the gate resistance R_G used for the control switch to reduce circuit ringing. Table 1 lists the reverse recovery performance parameters extracted for 250V/0.1A SiC Schottky diodes where I_{rr} and t_{rr} are reverse recovery peak current and time, respectively. A comparison of these results with similarly rated GaAs Schottky diodes suggests nearly the same performance for SiC and GaAs devices [19,20]. However, for high-voltage SiC PiN diodes, the results shown in Fig. 3(b) suggest a dramatic improvement in reverse recovery performance as compared to silicon PiN rectifiers. This is because of reduced minority carrier stored charge in the drift region and faster charge removal during the reverse recovery process. Similar tradeoff at higher junction temperatures shows further advantage of SiC over silicon because of reduced temperature dependence of minority carrier lifetimes in SiC. It must be noted that the rating of the SiC PiN diode (3kV/1A) is not exactly the same as that of the silicon PiN diode (1.5kV/10A) studied. However, the tradeoffs in reverse recovery performance and trends identified are expected to be valid.

The performance of 3kV/0.25A SiC PiN diodes in the buck converters is summarized in Figs. 4(a) and 4(b). The oscillations in the voltage waveforms resulted from circuit parasitic effects. The effect of diode reverse recovery can be seen in the hard-switching converter current waveform shown in Fig. 4(a) (for the time period of 6 μs to 8 μs). Tables 2(a) and 2(b) summarize the converter switching conditions as well as the turn-off energy loss E_{off} calculated from the measured circuit waveforms for 3kV/0.25A and 3kV/1A SiC PiN diodes. The turn-off energy was calculated by integrating the instantaneous power over the turn-off time period [2]. It is seen that the turn-off energy loss increases with temperature, consistent with the measured reverse recovery performance. No appreciable change in turn-off energy was measured for hard-switching and ZVS converters.

JFET Results: A number of 50V/0.5A SiC JFET's were measured in the package form for I-V and C-V characteristics. At T=23°C, the measured gate pinch-off voltage was -8V. A large output conductance was measured and is attributed to a combination of surface and bulk states within the SiC material. The non-planar buried gate structure may have also enhanced surface conduction. More detailed investigation is currently underway to conclusively identify the cause of this effect. These devices also had large leakage current in the voltage-blocking state and degraded breakdown performance.

The switching performance was first evaluated in a test circuit [17] and the turn-off energy E_{off} was extracted from measured transient voltage and current waveforms during turn-off. The results are plotted in Fig. 5 for hard-switching and ZVS for two values of snubber capacitor. It can be seen that ZVS results in up to a factor of 4x reduction in switching energy loss as compared to hard-switching.

The same devices were used to construct the buck converters shown in Figs. 1(a) and 1(b). For comparison, 70V/2A silicon power MOSFET converters were also characterized under identical circuit switching conditions [20]. The energy loss calculations were made from the measured switching waveforms and the results are shown in Tables 3(a) and 3(b). The converter was operated at 25 kHz at T=23°C with a 50% duty cycle. The energy is calculated by integrating the instantaneous power over the specified time period [2]. The on-state conduction energy loss is labeled as "conduction loss" and the off-state energy loss is labeled as "off-state loss." The switching loss is the sum of turn-on and turn-off energy losses. It can be seen that ZVS results in more than 2x improvement in total energy loss for SiC JFET's. The tradeoff is slightly smaller for silicon power MOSFET's. For both devices, total switching energy loss is reduced at higher temperatures although the conduction and off-state energy losses increase dramatically with temperature, especially for SiC JFET's. The off-state energy loss is a significant portion of the overall energy loss for SiC converters at elevated temperatures. This energy loss can be nearly eliminated by reducing the off-state leakage current. When the off-state power loss is minimized, SiC JFET's appear to be superior for power conversion at elevated temperatures compared to silicon power MOSFET's. A further reduction in SiC JFET conduction power loss is possible by optimizing the material growth, device processing and device design.

III. CONCLUSIONS AND DISCUSSIONS

This work presents experimental static conduction and switching performances of SiC power rectifiers and JFET's in hard- and soft-switching power converters. For comparison, similarly rated silicon and GaAs power devices were also characterized under identical terminal conditions. The switching stresses were intentionally kept low to avoid destroying the SiC devices that were found to be very fragile.

It is shown that low-voltage SiC rectifiers do not have any performance advantage over silicon or GaAs rectifiers except for potential high-temperature operation. SiC rectifiers show non-ideal I-V characteristics. The excess current at low voltages in the forward-biased region is attributed to bulk defects. Bulk defects also cause increased leakage in the reverse biased regime. High-voltage SiC PiN diodes show a dramatic improvement in reverse recovery performance as compared to silicon PiN diodes. This improved reverse recovery characteristics results in nearly an order of magnitude lower switching losses, especially in hard-switched buck converters operated above elevated junction temperatures of 100°C.

The results clearly suggest that the commercial SiC JFET's used in this work show a high output conductance and poor reverse voltage-blocking characteristics. These devices also have a high channel pinch-off voltage (-8V). The channel pinch-off voltage can be tailored to application requirements by controlling the channel layer thickness and doping density. The breakdown characteristics can be improved using proper edge termination structures. A large output conductance may have been caused by a combination of surface and bulk states within the SiC material. The buried gate structure may have also resulted in enhanced surface conduction. This phenomenon is presently under investigation.

Our results show that for low-voltage power switching applications, SiC JFET's do not offer performance advantages as compared to well-established silicon power MOSFET technology. More work is needed to evaluate high-voltage high-current SiC JFET devices in power converters and a direct comparison with silicon power switching devices.

The reliability of the power switch is of paramount importance to minimize field-failures of power converters. Because of the fragile nature of SiC devices available today, we have intentionally applied low-level electrical and thermal stresses under static conduction and dynamic switching conditions. The reliability of SiC devices, in general, needs to be critically investigated in the application circuit to evaluate potential hazards and prevent catastrophic failures.

REFERENCES

1. K. Shenai, " Technology Trends in High-Frequency Power Semiconductor Discrete Devices and Integrated Circuits, " *in Technical Papers of the Fourth Int. High Frequency Power Conversion*, Intertec Communications Inc., Ventura, CA, pp. 1-23, May 1989.

2. N. Mohan, P. Undeland, and W. Robbins, *Power Electronics Devices, Circuits and Converters*, Academic Press, 2nd Ed., 1992.

3. S. Abedinpour and K. Shenai, "Power Electronics Technologies for the New Millennium," *IEEE Third international Caracas Conference on Devices, Circuits, and Systems (ICCDCS2000),* IEEE Catalog Number: 00TH8474C, pp. P 111-1 -P 111-9.

4. N. Keskar, M. Trivedi, and K. Shenai, "Device Reliability and Robust Power Converter Development," in *Proc. European Symp. Reliability of Electron Devices, Failure Physics and Analysis (ESREF)*, 1999, pp. 1121-1130.

5. Deepak Divan, "Low-Stress Switching for Efficiency," *IEEE Spectrum,* December 1996, pp. 33-39.

6. M. Trivedi and K. Shenai, "Internal Dynamics of IGBT Under Zero-Voltage and Zero-Current Switching Conditions," *IEEE Trans. Electron Devices*, vol. 46, no. 6, pp. 1274-1282, June 1999.

7. S. Pendharkar and K. Shenai, "Zero Voltage Switching Behavior of Punch-Through and Non Punch-Through IGBTs," *IEEE Trans. Electron Devices,* vol. 45, no. 8, pp. 1826-1835, August 1998.

8. K. Shenai, " Potential Impact of Emerging Semiconductor Technologies on Advanced Power Electronic Systems, " *IEEE Electron Device Lett.*, vol. 11, no. 11, pp. 520-522, Nov. 1990.

9. K. Shenai, R. S. Scott, and B. J. Baliga, " Optimum Semiconductors for High-Power Electronics, " *IEEE Trans. Electron Devices*, vol. 36, pp. 1811-1823, Sept. 1989.

10. P. G. Neudeck, "SiC Technology," in *The VLSI Handbook*, Edited by W. K. Chen, CRC Press/IEEE Press, 2000.

11. P. G. Neudeck, W. Huang, and M. Dudley, "Study of Bulk and Elementary Screw Dislocation Assisted Reverse Breakdown in Low-Voltage (< 250 V) 4H-SiC P^+N Junction Diodes - Part I: DC Properties," *IEEE Trans. Electron Devices*, vol. 46, no. 3, pp. 478-484, 1999.

12. P. G. Neudeck and C. Fazi, "Study of Bulk and Elementary Screw Dislocation Assisted Reverse Breakdown in Low-Voltage (< 250 V) 4H-SiC P^+N Junction Diodes - Part 2: Dynamic Pulse-Breakdown Properties," *IEEE Trans. Electron Devices*, vol. 46, no. 3, pp. 485-492, 1999.

13. M. Trivedi and K. Shenai, "Performance Evaluation of High-Power Wide Band-Gap Semiconductor Rectifiers," *J. Applied Physics*, vol. 85, no. 9, pp. 6889-6897, May 1999.

14. P. G. Neudeck, "SiC Technology," in *The VLSI Handbook*, Edited by W. K. Chen, CRC Press/IEEE Press, 2000.

15. K. Shenai and R.W. Dutton, " Channel-Buffer (Substrate) Interface Phenomena in GaAs MESFET's Fabricated by MBE, " *IEEE Trans. Electron Devices*, vol. ED-35, no. 5, pp. 590-603, May1988.

16. K. Shenai, "Mixed-mode circuit simulation: An emerging CAD tool for the design and optimization of power semiconductor devices and circuits," *Proceedings of IEEE 4th PELS Workshop on Computers in Power Electronics*, August 7-10, 1994, IEEE Cat # 94TH0705-4, pp. 1-5.

17. S. Pendharkar, C. Winterhalter, M. Trivedi, H. Li, A. Kurnia, D. Divan, and K. Shenai, "Test circuits for verification of power device models," in *Proc. 1995 Digest of IEEE Industrial Applications Society (IAS) Annual Meeting*, Orlando, FL, October 8-12, 1995.

18. E. H. Rhoderick, *Metal-Semiconductor Contacts*, Oxford University Press, Oxford, UK, 1978.

19. S. Pendharkar, C. Winterhalter, and K. Shenai, "Modeling and Characterization of the Reverse Recovery of a High-Power GaAs Schottky Diode," IEEE Transactions on Electron Devices, Vol. 43, No. 5, May, 1996.

20. S. Pendharkar and K. Shenai, "Performance Evaluation of High-power GaAs Schottky and Silicon P-i-N Rectifiers in Hard- and Soft-Switching Applications," *IEEE Trans. Power Electronics*, vol. 13, no. 3, pp. 441-451, May 1998.

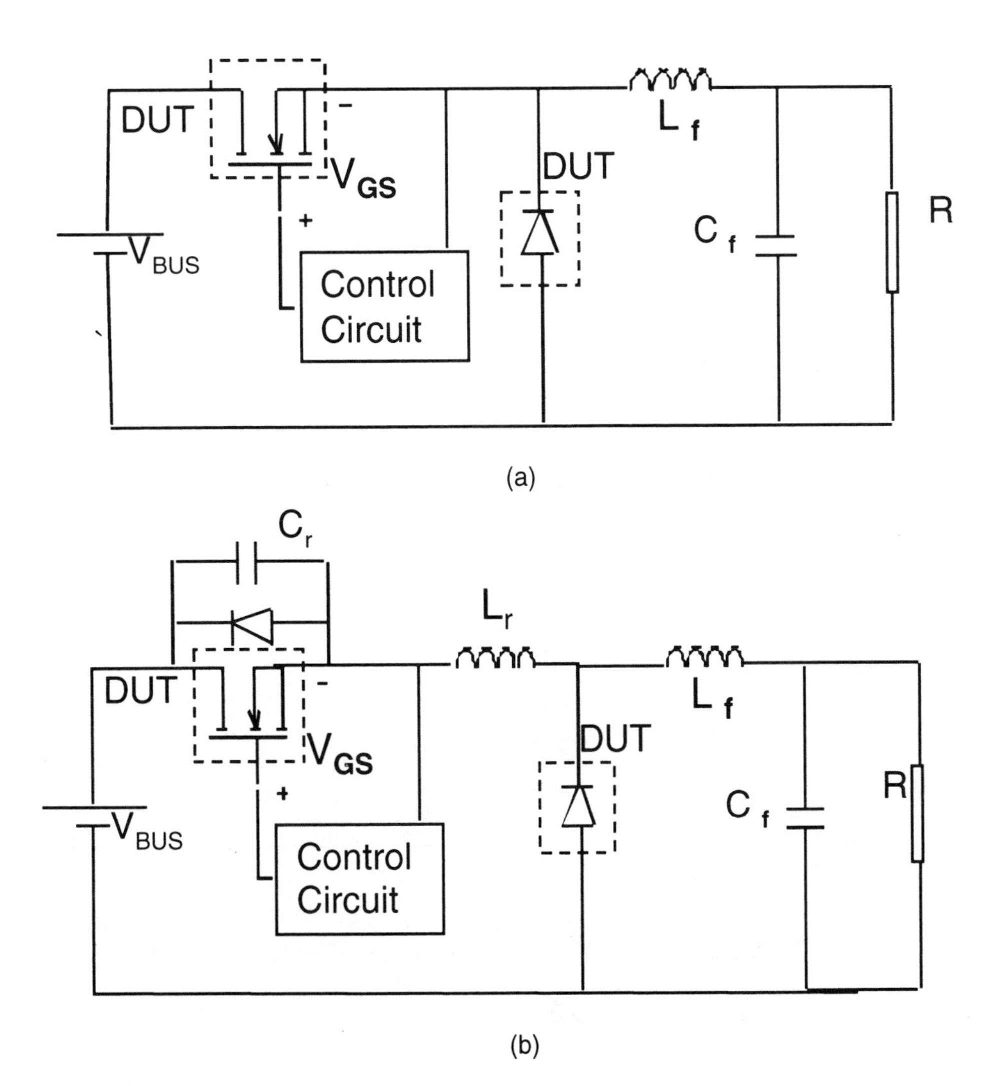

Fig. 1 DC-DC buck converter circuits for (a) hard-switching and (b) Zero Voltage Switching (ZVS).

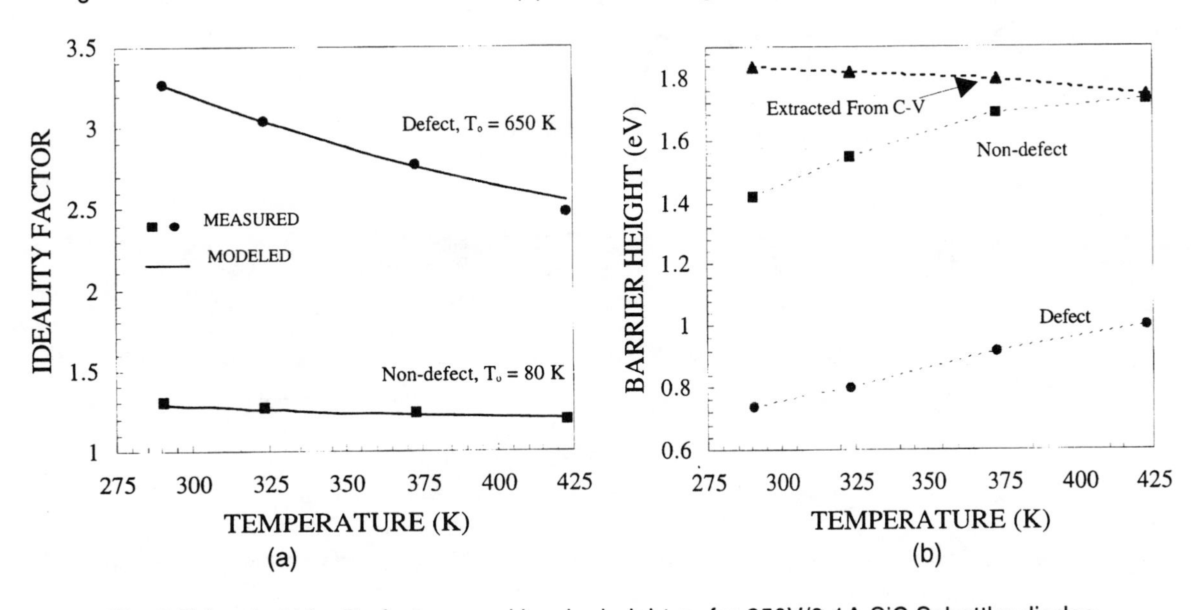

Fig. 2 Extracted ideality factor n and barrier height $ø_B$ for 250V/0.1A SiC Schottky diodes.

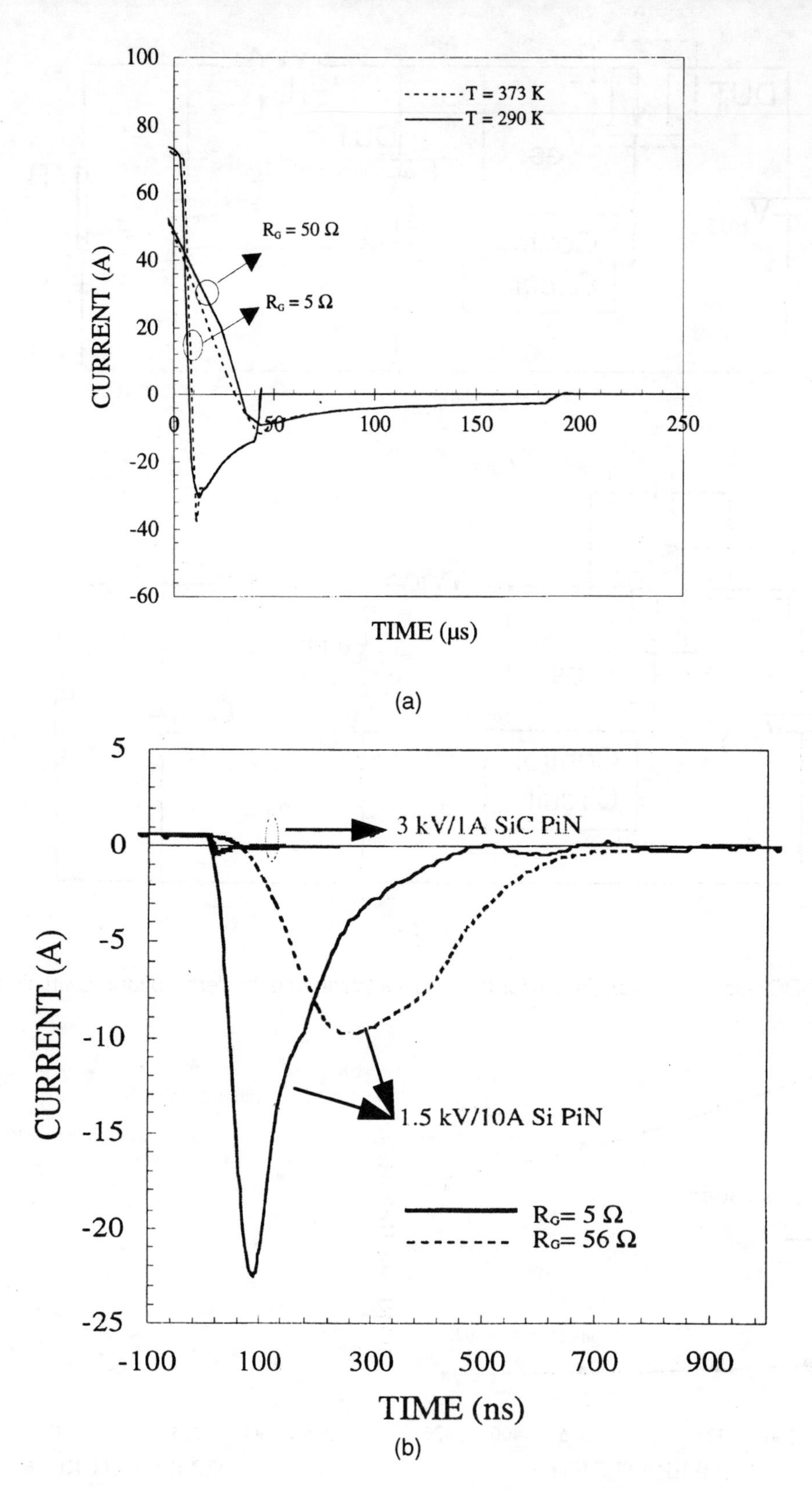

Fig. 3 Measured reverse recovery current waveforms for (a) 250V/0.1A SiC Schottky diode, and (b) 3kV/1A SiC PiN diode at T=23°C as a function of gate resistance R_G.

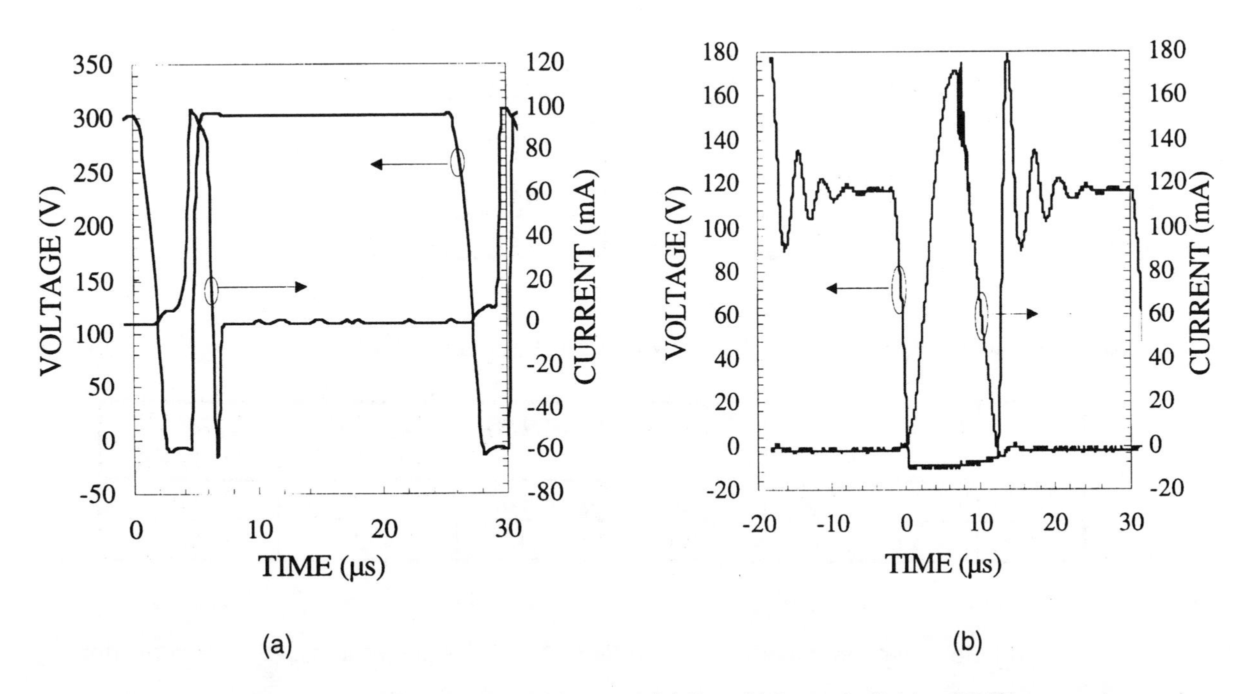

Fig. 4 Measured waveforms for (a) hard-switching and (b) Zero Voltage Switching (ZVS) converters using 3kV/0.25A SiC PiN diodes at T=23°C.

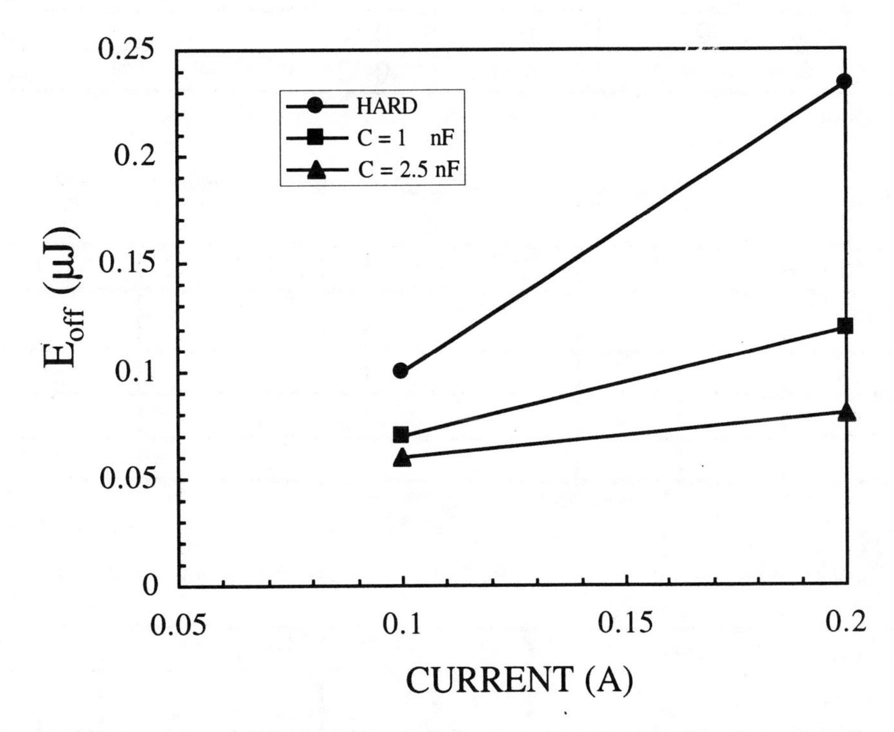

Fig. 5 Turn-off energy of 50V/0.5A SiC JFET at T=23°C in a special ZVS test circuit [17] as a function of snubber capacitance.

Table 1. Extracted reverse recovery parameters for 250V/0.1A SiC Schottky diodes for V_{BUS} = 100V and I_{ON} = 60 mA.

R_G (Ω)	T (K)	I_{rr} (mA)	t_{rr} (ns)
5	290	45	41
50	290	11	130
5	373	44	38
50	373	12	110

Table 2(a) Measured turn-off energy E_{off} for 3kV/1A SiC PiN diodes in a hard-switching buck converter.

Device	V_{BUS} (V)	I_{ON} (A)	E_{OFF} (μJ) @ 290 K	E_{OFF} (μJ) @ 373 K
SiC PiN 3 kV/0.25 A	150	0.16	10	12
SiC PiN 3 kV/1 A	300	0.42	52	67

Table 2(b) Measured turn-off energy E_{off} for 3kV/1A SiC PiN diodes in a ZVS buck converter.

Device	V_{BUS} (V)	I_{ON} (A)	E_{OFF} (μJ) @ 290 K	E_{OFF} (μJ) @ 373 K
SiC PiN 3 kV/0.25 A	150	0.16	10	12
SiC PiN 3 kV/1 A	300	0.42	51	68

Table 3(a). Measured energy losses for SiC JFET converters at 25 kHz with 50% duty cycle.

Energy Loss (μJ)	HARD 23°C	ZVS 23°C	HARD 100°C	ZVS 100°C
Conduction Loss	3.7	3.8	3.4	3.7
Switching Loss	10	0.4	9	0.3
Off-state Loss	7	6	12	5.8
Total Loss	20.7	10.2	24.4	9.8

Table 3(b). Measured energy losses for silicon MOSFET converters at 25 kHz with 50% duty cycle.

Energy Loss (μJ)	HARD 23°C	ZVS 23°C	HARD 100°C	ZVS 100°C
Conduction Loss	2.8	2.9	2.8	2.8
Switching Loss	4.4	2.2	4	1.8
Off-state Loss	0.2	0.3	0.6	0.3
Total Loss	7.4	5.4	7.4	4.9

AIAA-2000-2807

SOLAR POWER SYSTEM OPTIONS FOR THE RADIATION AND TECHNOLOGY DEMONSTRATION SPACECRAFT*

Thomas W. Kerslake, Francis M. Haraburda, John P. Riehl
NASA Glenn Research Center, Cleveland, OH 44135

ABSTRACT

The Radiation and Technology Demonstration (RTD) Mission has the primary objective of demonstrating high-power (10 kilowatts) electric thruster technologies in Earth orbit. This paper discusses the conceptual design of the RTD spacecraft photovoltaic (PV) power system and mission performance analyses. These power system studies assessed multiple options for PV arrays, battery technologies and bus voltage levels. To quantify performance attributes of these power system options, a dedicated Fortran code was developed to predict power system performance and estimate system mass. The low-thrust mission trajectory was analyzed and important Earth orbital environments were modeled. Baseline power system design options are recommended on the basis of performance, mass and risk/complexity. Important findings from parametric studies are discussed and the resulting impacts to the spacecraft design and cost.

INTRODUCTION

Various electric propulsion systems have been flying on spacecraft since the 1960's[1]. Because of their low mass and high specific impulse (Isp), electric propulsion systems are being baselined on an ever increasing number of missions. To keep mission trip times manageable, higher thrust levels are desirable with an attendant increase in spacecraft power level. Power levels in the 10's of kilowatts (kW) [2,3,4], 100's of kW[5] and even 1000's of kW[6,7,8] have been proposed to operate electric thruster systems.

To enhance the flight readiness of high power electric propulsion systems, the Radiation and Technology Demonstration (RTD) Mission[9] is under joint study by three NASA Centers: Johnson Space Center, Goddard Space Flight Center and Glenn Research Center. This Earth-orbiting mission, that may launch on the Space Shuttle within the next 5 years, has the primary objective of demonstrating high-power (10 kW) electric thruster technologies. Secondary scientific objectives include: better characterization of Earth's Van Allen trapped radiation belts, measurement of shielding effectiveness for human protection from trapped radiation and galactic cosmic radiation, measurement of radiation effects on advanced solar cells, demonstration of radiation tolerant microelectronics and measurement of the interactions between the spacecraft and its ambient environment.

The 1500-kilogram (wet) RTD spacecraft, shown in Figure 1, consists of a spacecraft bus, the Hall thruster system[3,4] on top, the VAriable Specific Impulse Magnetoplasma Rocket (VASIMR) thruster system[10] on the bottom and a microsatellite stowage/deployment system located in the spacecraft midsection. The spacecraft bus includes a solar Electric Power System (EPS) that is dominated by two deployable, rectangular photovoltaic array wings with single-axis Solar Array Drive Assemblies (SADAs) for solar tracking.

This paper will discuss the RTD mission trajectory analysis, the conceptual design of the RTD spacecraft photovoltaic (PV) EPS and predicted EPS mission performance for the various power system options. Baseline solar EPS design options are recommended on the basis of performance, mass and risk/complexity. Important findings from parametric studies are discussed and the resulting impacts to the spacecraft design and cost.

MISSION TRAJECTORY ANALYSIS

The two mission scenarios presented herein represent the latest iteration in the design process whereby spacecraft mass and mission duration are traded for scientific and research objectives. During the proposed mission duration of about 270 days, the RTD spacecraft spirals outward from the Shuttle-deployed, circular low Earth orbit. During the planar spiral out at an orbit inclination of 28.45° or 51.6°, four microsatellites will be deployed to provide simultaneous radiation measurements of various Van Allen belt locations. By the phased operation of the 10 kW Hall thruster and 10 kW VASIMR, the RTD spacecraft will attain an orbit radius greater than 5 Earth radii. The Hall thruster, because of its lower specific impulse (Isp), produces relatively short transfer times while requiring more propellant whereas the VASIMR thruster, with its very high Isp, produces very long transfer times and uses very little propellant. To control mission cost, the overall mission must be performed in no more than one year. All of these competing needs have produced a mission that is at best a compromise. Mission requirements and thruster system performance for RTD are summarized in Table 1.

Trajectory analysis for this kind of mission is initially rather straight forward. Using Edelbaum's expression for optimal minimum time transfer between circular orbits, one can compute the velocity increment, ΔV, to transfer from an initial circular orbit with velocity V_1 to a final circular orbit with velocity V_2 while performing a plane change of Δi:

$$\Delta V = \{ V_1^2 + V_2^2 - 2 V_1 V_2 \cos(\pi/2 \, \Delta i) \}^{1/2} .$$

When used with the "Rocket" equation,

$$m_f / m_0 = \exp(-\Delta V / g I_{sp})$$

one has a means for estimating propellant consumption, $m_0 - m_f$ (kg), and thrusting time (seconds) because thrust, T (Newtons), Isp (seconds), thruster efficiency, η, and input power level, P_0, (W) are related by:

$$T = 2 \eta P_0 / (g I_{sp}) ,$$

while the

$$\text{Transfer Time} = (m_0 - m_f) / \text{mass flow rate}.$$

These expressions demonstrate the inherent synergism between the spacecraft trajectory, electric propulsion performance and power system performance.

To estimate the impact of shadowing and launch date on the performance, the computer program SEPSPOT[11] was used. This program uses orbital averaging and solves for the minimum time trajectory between two closed conic orbits. This combination of resources, namely spread sheets which incorporate the above expressions and the SEPSPOT program, permits the determination of preliminary mission durations and altitudes at which to switch from one thruster to another. This allows one to simulate a mission in much more detail and fidelity. The results presented herein were derived from trajectories generated using the Glenn Research Center's high fidelity integrator SNAP which incorporates an eighth order Runge-Kutta integrator, the AE8 and AP8 trapped radiation models[12], and the JPL DE403 Ephemeris file of the solar system. Earth shadow crossings were precisely modeled and an eighth order earth gravity model was used along with solar and lunar gravitational effects. The vehicle was assumed to steer with tangential steering, but no attempt was made to remove eccentricity that accumulates because of discontinuous thrusting when crossing into and out of shadow regions. It was further assumed that the EPS provided a constant 10 kW of power to the thrusters during orbit sun times. The microsat masses were jettisoned at the prescribed altitudes at perigee.

Spacecraft Flight Mode

The nominal flight mode has the RTD spacecraft thruster-axis in the orbit plane and tangent to the orbit ellipse. The solar array wings are perpendicular to the orbit plane and rotate about a single axis. This nominal flight mode is used for absolute solar beta angles, $|\beta|$, less than 25°, where β is the angle between the orbit plane and Earth-Sun line. During this flight mode, solar array cosine pointing losses are limited to 9%.

For $|\beta|$ greater than 25°, the spacecraft switches to a so-called "yaw-steering" flight mode utilized by several other spacecraft[13-16]. For the RTD spacecraft, the flight mode is more properly called "roll-steering" as the spacecraft rolls about the thruster axis (maintained in plane and tangent to the orbit ellipse). This type of spacecraft steering was studied for a solar electric transfer vehicle mission[17]. Using momentum wheels, the spacecraft is rolled through an angle $\leq 2(90° - |\beta|)$ in one-half orbit period. The combined spacecraft roll angle and single axis SADA rotation enables sun-tracking array pointing. The implications of this flight mode for spacecraft attitude control system (ACS) design will be discussed later in the paper.

EPS DESIGN OPTIONS

Photovoltaic Array

Nine different PV array design options were evaluated and are briefly described in Table 2. Options 1 and 2 are typical of deployable rigid-panel designs with cascade (or multi-junction) GaInP2/GaAs/Ge solar cells[18] used on communication satellite buses, such as the A2100[19] and HS702[20]. Panels are constructed of composite face sheets bonded to a 1-inch thick aluminum honeycomb core. The HS702 arrays also deploy side reflectors to achieve about 2X solar concentration on the solar cells. Options 3, 4 and 5 are mast-deployable, flexible composite panel designs using multi-junction GaInP2/GaAs/Ge solar cells, high efficiency silicon cells[21] or three-junction amorphous SiGe cells[22] on stainless steel panels. Options 6 and 7 use multi-junction GaInP2/GaAs/Ge solar cells operating under 7.5X solar illumination afforded by rigid refractive linear concentrators on deployable rigid panels or flexible refractive linear concentrators on mast-deployable flexible panels[23]. The former array type is successfully flying on the Deep Space 1 Mission spacecraft[24]. Options 8 and 9 employ a multi-gore array design that deploys circumferentially to form a quasi-circular, array geometry[25]. The gore material is a low-mass open-weave fabric. The multi-junction GaInP2/GaAs/Ge solar cells or the high efficiency silicon cells populate the gore sections. Array wings similar to option 9 have been built for an up-coming Mars Lander mission[26]. In all options, each solar array wing is mounted to a single-axis SADA with slip ring power transfer.

Energy Storage

Both nickel-hydrogen (NiH2) individual pressure vessel and prismatic lithium ion battery cell technologies were considered to fulfill energy storage requirements. Cell properties, such as charge/discharge voltage limits, dimensions, mass, and operating temperature range, were obtained from typical values found in commercial battery product data sheets (see Table 3). For this mission, a modest 1000-cycle cell life was required. Cells were series-connected and housed in an aluminum containment box to afford environmental protection and enhance thermal control.

PMAD Architecture, Loads and Power Requirements

End-of-mission power system requirements included 10 kW of sun time power delivered to the electric thruster Power Processing Unit (PPU) input and 0.4 kW delivered to spacecraft loads continuously through sun and eclipse times. The maximum eclipse time for this mission was 1.14 hrs. This translated to an energy storage requirement of 0.46 kW-hrs exclusive of system losses. Power system reliability and fault-tolerance requirements have not been yet specified. As such, the Power Management And Distribution (PMAD) architecture does not include design features to address these operational requirements.

The direct energy transfer, direct current (DC) power distribution architecture, as shown in Figure 2, provides sun time regulated primary power for the operating thruster and for battery charging. Sun time and eclipse time regulated secondary power at 28 VDC is provided to the spacecraft loads. The power generation of each solar array wing (discussed in detail below) is controlled by a sequentially-shunted, pulse-width-modulated, Array Regulator Unit (ARU). Approximately 5.5 kW is transferred from each solar array wing through the ARU and is paralleled at the input of the Main Distribution Panel (MDP). The MDP contains all of the associated fault detection and isolation hardware for the individual power feeds to the vehicle loads, the capability to isolate solar arrays, the required power supply and control processor. Primary distribution bus set point voltages of 120 VDC, 50 VDC and 28 VDC were assessed. These voltage levels were chosen to allow use of existing components (120-V space station, 28-V typical satellites) or to match the operating voltage, 50-V, of the VASIMR radio frequency plasma generators[10].

Grounding of the system negative return will be made in the MDP. The switchgear used in the MDP will be similar to those used in Space Station hardware, which uses a 120 VDC secondary architecture. The MDP power supply will receive input power from the MDP power bus and from the vehicle battery for the eclipse portions of the orbit. Isolation between the two feeds will be provided for.

RTD vehicle and experiment load power will be provided using a proven design similar to that used for the Microwave Anisotropy Probe mission[27]. The Power System Enclosure (PSE) will provide power conditioning, switching, SADA control, fault detection and isolation for vehicle subsystems and instrumentation. It will accept power from the MDP to charge the battery and to power loads through a DC-to-DC converter unit.

Each of the vehicle's two thrusters is shown with an accompanying PPU. The PPU will be used to condition the input power to the voltage/current requirements of the various loads associated with the thruster. The PPU will consist of converters with the ability to isolate failed components of the thruster and isolate load effects from the vehicle power system. Thruster startup and maintenance power will be provided via 28 VDC secondary power. The actual power requirements for these loads and their number have yet to be determined.

PMAD cable sizes were selected based on the assumed number of vehicle loads, run lengths, number of parallel conductors (two-hot, two-two ground), operating temperature (100°C). The sizes shown in the power architecture (Figure 2) were selected based on the assumed load power requirements. Those shown were rated using the Mil-W-22759D specification for cables in hard vacuum and current derating due to bundling.

ANALYSIS & COMPUTATIONAL METHODS

To assess the relative merits of these PV array and PMAD design options, a dedicated Fortran code was developed to predict power system performance and estimate system mass. EPS component design and mission information are read in via data input files. Mission data, provided at 15-minute intervals throughout the 270-day mission, include spacecraft position/velocity, Sun/Earth angles, orbit sun/eclipse indicator, insolation strength and local radiation fluences. This information is used to calculate environmental heating rates, solar cell equivalent radiation dose[28,29] and solar cell micrometeoroid / orbital debris damage area[30-34].

EPS performance analysis is performed in a time-stepping, load-driven fashion. Based on load demand and setpoint voltages, PMAD system currents and voltages are calculated for the current time step. Component and cable losses are calculated based on input resistances, diode voltage drops and converter efficiency (if present). PV array string current is iteratively determined to satisfy solar cell and ARU voltage and current constraints. The number of string series-connected solar cells is determined iteratively such that cell operating voltage becomes no larger than maximum power voltage throughout the mission. The number of strings per wing is also determined iteratively such that minimum number of shunted strings is < 3. This minimizes array area while ensuring that the ARU can maintain sun time voltage regulation throughout the mission.

Solar cell electrical performance is modeled using a single exponential current-voltage (IV) function that is adjusted for operating temperature, illumination intensity, PV array sun pointing error and flatness, coverslide transmittance, environmental degradation and cell mismatch. Solar cell IV operating point and temperature are iteratively determined. Cell operating temperature is calculated using a lump-mass, transient thermal model accounting for environmental heating/cooling, electrical power extraction and interconnect wiring ohmic heating. Array area is calculated as the total solar cell area divided by a packing factor, 0.85. A wing length-to-width ratio of 6 was selected to obtain a width of approximately 2-m that corresponds to the approximate length of the spacecraft microsat storage section (see Figure 1).

The battery is sized based on the input design characteristics of Table 3. These inputs are used to calculate the number of cells, cell capacity, design Depth-Of-Discharge (DOD), design charge/discharge rates and trickle charge rate. For mission analysis, the battery charge and discharge rates are determined based on the required load, battery charge and discharge efficiencies and the orbit sun and eclipse time. For orbits with little or no eclipse period, the battery charge current is set to a trickle charge value (C/50).

The temperature of several EPS components is also calculated using a simplified, lumped-mass transient model. Calculated values of component power dissipation and environmental heating are used to determine operating temperature and the required thermal control heating or cooling to satisfy operating temperature limits.

Mass Estimation

PV array mass was comprised of panels, structures/mechanisms/miscellaneous, power harness and the SADA. Panel masses were calculated for each array option based on panel layer material thicknesses, densities and areal fractions (see Table 2). Array structure and mechanism masses were estimated to be 10% of the panel mass (for rigid concepts), 0.6 kg/m2 for deployable mast concepts and 0.3 kg/m2 for multi-gore concepts. Power harness mass was based on a commercial flat ribbon design sized for 3% voltage drop. The SADA mass was based on a commercial product. The battery mass was calculated to be 1.1 times the cell mass to account for cell interconnects, by-pass diodes, cell voltage control (Li-ion only), cell heaters, the containment box and connectors. PMAD component masses were based on scaling ISS EPS component masses and power levels: ARU (2 kg/kW), MDP (2 kg/kW), and PSE (9 kg/kW). Cable conductor masses, from Mil-W-22759D, were multiplied by 1.1 to account for insulation mass. PMAD component and battery thermal control were achieved via thermal control coatings (no mass assumed) and electric resistance heaters (included in component masses).

MISSION PERFORMANCE RESULTS

Selected mission and EPS performance analysis results are discussed in this section for an initial baseline EPS design and orbit defined as follows:

- Option 1 solar array (rigid panels, cascade GaAs cells)
- NiH2 Battery
- 120 VDC PMAD Primary Voltage
- 51.6° orbit inclination

Figure 3 shows the spacecraft orbit altitude throughout the 270-day mission. Lack of smoothness in this curve was due to a small buildup of orbit eccentricity (about 5%) that resulted from only thrusting during orbit sunlight periods. Solar array wing power is shown in Figure 4. The top curve shows the full wing power with no strings shunted while the bottom curve shows power delivered by the array through the ARU to the MDP. The difference between the two curves was the power shunted in the ARU. Most of the power degradation occurred between mission days 50 and 100 while the spacecraft was passing through the trapped radiation proton belts. As shown in Figure 5, during this period of time, most of the solar cell current capability was reduced due to radiation degradation. Other current reducing factors are also shown on Figure 5. Similar behavior was seen with solar cell voltage degradation, i.e. the largest degradation mechanism was radiation damage.

The ratio of solar cell operating voltage, Vop, to maximum power voltage, Vmp, is displayed in Figure 6. At the mission start, the ratio ran at about 0.9 since the number of series solar cells must be oversized for degradation. After receiving the bulk of radiation damage, the voltage ratio approached 1.0. Thereafter, it decreases slightly as array temperatures cooled off with increasing spacecraft altitudes and solar cell voltage capability was improved.

EPS component temperatures are given in Figure 7. The solar array operated at suntime temperatures of about 47°C at the mission start to 35°C at mission end. Array eclipse temperatures reached -120°C 90-days into the mission at orbit altitudes of about 7000 km. Thereafter, the spacecraft did not encounter eclipse periods. PMAD components remained within a modest temperature envelope: 0°C to 75°C at the mission start and 5°C to 10°C at mission end.

PARAMETRIC STUDY RESULTS

Parametric studies were performed to assess EPS performance and mass versus several system design options and mission operation options. The primary objectives of these studies were to minimize spacecraft mass and to quantify impacts of various mission operation scenarios. The parametric trade space is shown in Table 4. The initial baseline EPS design and orbit were defined in the previous section. These items were used in the following parametric studies unless noted otherwise.

Coverslide Thickness

The first parametric study conducted was EPS mass versus solar cell coverslide thickness. Coverslide thickness affects many properties including radiation shielding, transmittance, solar cell operating efficiency, solar array panel areal mass and thermal capacitance and solar cell operating temperature. Thus, an iterative EPS sizing and performance analysis must be performed for each case of coverslide thickness to ensure power requirements are met. The normalized results from EPS sizing analyses are shown in Figure 8.

EPS mass (and solar array mass) was minimized with a 10-mil coverslide thickness. Greater thicknesses decreased the solar cell Damage Equivalent Normally Incident (DENI) mission fluence of 1-MeV electrons, decreased array area and increased cell beginning-of-life (BOL) operating efficiency, i.e. the cell operating voltage was closer to the maximum power point voltage. However, these benefits came at the expense of greater EPS mass. Thus, a 10-mil coverslide thickness was baselined.

As an aside, the optimum coverslide thickness for GaAs cells mounted to a flexible substrate, with lower backside shielding than that of rigid panels, would be thicker. Also, if silicon cells were assumed, the effective transmittance loss of thicker coverslides would be reduced since most transmission losses are in the blue region away from the peak spectral response of silicon cells. This effect would lead to greater optimum coverslide thicknesses for silicon cells compared to GaAs cells.

Photovoltaic Array Technology

EPS sizing analyses were performed for the baseline design and mission operation conditions for each of the nine solar array options. Results are shown in Table 5. The baseline EPS had a mass of 258 kg comprised of a 167 kg array, 36 kg NiH2 battery and a 55 kg PMAD system. The solar array had two wings, each with an area of 24.9 m2.

The lowest EPS mass, 180 kg, and lowest solar array mass, 89 kg, were provided by option 7 (flexible panel, flexible concentrator, GaAs cells). A power system mass savings 78 kg over the baseline design was achieved. Option 2 (rigid panel with side reflectors, GaAs cells) and option 8 (flexible gore, GaAs cells) were also strong low-mass contenders with EPS masses of 188 kg and 195 kg, respectively.

The highest EPS masses, 298 kg and 313 kg, were obtained with options 4 and 9, respectively, which both use high efficiency silicon cells. During the mission, these solar cells received a very high radiation dose ($4x10^{15}$ e/cm2). This dictated that the array had to be oversized considerably to make-up for radiation degradation losses. For this reason, crystalline silicon cells are not a good choice for the RTD mission.

The smallest array wing area, 23.5 m2, was obtained with option 6 (rigid panel, rigid concentrator, GaAs cells). The option had the highest areal mass and provided excellent radiation shielding for the solar cells, i.e. dose of only $2x10^{14}$ e/cm2. As such, cell radiation losses were small and array oversizing was minimal. The option 2 array (rigid panel, side reflectors, GaAs cells) had the smallest solar cell panel area. However, when including the area of the side reflectors, the wing frontal area increased to 29.5 m2 which was the largest of the array options using GaAs cells.

The option 5 array (amorphous SiGe thin film cells on stainless steel) did not provide a design solution for this mission. The reason for this was twofold: high radiation degradation and high operating temperature. Both of these factors lowered the cell operating efficiency to below 3% at 150-days into the 270-day mission. At this point, stable solar cell string operation could not be obtained. At the expense of added mass, the thin film panel could be encapsulated with a fluoropolymer to provide radiation shielding and improve surface emittance to lower cell operating temperature. In the longer term, alternative thin film technologies, such as Cu(In,Ga)Se2 or Cu(In,Ga)S2 on a polymer substrate, promise higher stable conversion efficiencies, greater radiation tolerance and lower areal mass compared to three-junction amorphous-SiGe thin film cells on a stainless steel substrate. An excellent assessment of the benefits of thin film photovoltaic arrays was reported by Hoffman[35].

Battery Technology

Preliminary battery designs were developed based on NiH2 and Li ion cell technologies. Design results are shown in Table 6. The Li ion cell option provided a 23 kg battery mass savings over the NiH2 cell option in addition to a considerable reduction in battery volume. For both battery technologies, preliminary values of heater power and cooling load were negligible to maintain operating temperatures within design limits. Assuming cell capacity loss is minimal during 1000 cycles of operation at 50% DOD, the Li ion battery technology is clearly preferred over the NiH2 technology from a performance standpoint.

PMAD Primary Voltage

EPS design sizing and performance were analyzed for primary PMAD voltage set points of 120-V, 50-V and 28-V. Results are provided in Table 7. The most obvious effect of reducing primary voltage was a large increase (64%) in power system mass, i.e. from 258 kg at 120-V to 424 kg at 28-V. For most system components, mass increased with decreasing voltage due to higher operating currents, larger voltage drops and larger physical size. In the array, size and mass increased for panels, structure, power harness and the SADA. For PMAD components, the bulk of the mass increase was from the ARU that would require a 3X to 4X increase in the number of shunt channels. PMAD cable mass increased as the number of parallel conductors was increased to satisfy increased derated current requirements. For example, the PPU input current increased from 84 Amp at 120-V to 382 Amp at 28-V. Higher PPU currents would also increase the PPU mass not addressed in this paper (tallied with the propulsion system budget).

The only exception to the trend of increasing mass was the battery. Battery mass decreased for the lower voltage cases since it was possible to better match the required cell capacity with commercially available 10 A-hr, 20 A-hr, etc., capacity cells. For example, the 120-V system required a 6 A-hr cell for which a 10 A-hr standard cell sized was specified.

With decreasing PMAD voltage, the required solar array wing area also increased (27%) from 24.9 m2 at 120-V to 31.7 m2 at 28-V. The 28-V system had increased voltage losses in the power harness, SADA, PMAD components and cabling. This dictated that array strings operate at a higher voltage relative to the setpoint voltage level. To satisfy the power requirements, many more array wing strings were required, i.e. from 138 strings at 120-V to 624 strings at 28-V. With this many strings, panel solar cell lay-down pattern and string-power harness integration would become increasingly complex.

To minimize system mass, array area and array wiring complexity, a 120-V power system is a clear winner. In the future, even better mass performance will be possible using a high-voltage EPS (400-V for Hall thrusters) and "direct-drive" electric thruster operation: that is, using a less complex PPU without a step-up voltage converter[36].

Orbit Inclination

EPS sizing and performance were analyzed for planar spiraling mission trajectories with orbital inclinations of 51.6° and 28.45°. Results showed that orbit inclination in the range of 28.45° to 51.6° had little impact on EPS design and performance. At 28.45°, the EPS mass was about 5 kg more than that at 51.6° primary due to an increase in array mass. A slightly larger array was needed for the 28.45° inclination orbit since the solar cell radiation dose was slightly larger, i.e. more mission time was spent passing through the proton belts. Also, the solar cell operating temperature was slightly higher due to a larger Earth view factor and the attendant higher array backside albedo and infrared heating fluxes.

Impacts to ACS

In addition to the high degree of synergism between the EPS and electric propulsion systems, there is a strong synergism between the spacecraft EPS and ACS. Specifically, the EPS solar array wings contribute to the bulk of spacecraft inertia and control the magnitude of environmental disturbance torques and the required roll torques to achieve the "roll steering" flight mode.

To quantify the consequence of array selection on the ACS design, spacecraft maximum disturbance torques[37] and roll steering torques were estimated. Contributing disturbances included gravity gradient, aerodynamic drag, thruster misalignment, solar pressure and magnetic in order of magnitude. Orbital momentum accumulation was also estimated. Results are shown in Table 8 for the most demanding part of the mission at a BOL orbit altitude of 400 km.

Predicted roll steering torques were dominant over disturbance torques and consistent with those of Jenkin[17]. Combined torques were within the torque capability of commercial reaction wheels. Momentum accumulation would be managed via periodic hydrazine thruster firings. Assuming 50 N-m-sec wheels, wheel desaturation would be needed, in the worst case, about 30 times daily at BOL, 12 times daily after mission day 20 (1000 km altitude) and 9 times daily after mission day 50 (5000 km altitude). Over the mission, wheel desaturation would require about 68 kg of hydrazine propellant. Since mission momentum accumulation was dominated by electric thruster misalignment torque, considerable hydrazine mass savings could be achieved using a Hall thruster gimbal system. At the expense of 55 kg of added mass, reaction wheel momentum could be doubled thereby reducing the required frequency of momentum dumps by a factor of 2. Aside from solar array options 4 and 9 that utilize silicon cells (and thus, have large areas and large inertias), there is not a clearly preferred option to reduce ACS mass or complexity.

Spacecraft Benefits from EPS Mass Savings

The RTD spacecraft has a wet mass to dry mass ratio of 1.4. Thus, for every 1 kg saved in spacecraft dry mass, 1.4 kg is saved in spacecraft wet mass. Based on these trade studies, the best possible EPS mass reduction would be to transition from the baseline design to solar array option 7 using a Li ion battery. This would provide mass savings of 101 kg for the EPS and 141 kg for the spacecraft including 40 kg (14%) savings for Hall thruster xenon propellant and 7 kg (14%) savings for VASIMR hydrogen propellant.

Given the RTD spacecraft is launched by the Space Shuttle, spacecraft mass is not a critical design driver. Instead, spacecraft cost is probably the most important hardware design factor. EPS mass savings were mostly obtained by using an advanced technology solar array. Thus, cost savings associated with reduced array size are not likely to outweigh the multi-million dollar development and qualification costs required by an advanced, first-unit design. Also, a 14% propellant savings will not appreciably affect the propellant storage and delivery system design or cost aside from a modest savings in xenon propellant procurement cost.

Therefore, after this preliminary assessment, it appears that a commercially available EPS design would be the best option from a mission/spacecraft cost standpoint. However, the cost-benefit calculus may dramatically change in favor of a low-mass, high-technology EPS for spacecraft launched with expendable vehicles on high-energy trajectories (inter-planetary missions). Here, mass savings can be a critical cost factor and many times, a mission-enabling factor.

Final Baseline Options

At both ends of the spectrum, two final baseline EPS options should be considered:

(1) For lowest cost and lowest risk at moderate mass, a final baseline EPS is the initial baseline EPS: that is, an option 1 rigid panel solar array with cascade GaAs cells, a NiH2 battery and a 120-V PMAD system.

(2) For lowest mass, moderate risk, higher cost yet high technology demonstration value, a final baseline EPS design employs an option 7 flexible array with flexible concentrators, cascade GaAs cells, a Li ion battery and a 120-V PMAD system.

CONCLUDING REMARKS

Several conceptual EPS designs were specified and analyzed with the intent of fulfilling RTD spacecraft and mission requirements while minimizing mass and cost. The spacecraft mission trajectory and the nature of EPS performance through a mission were described. The mass and performance benefits of various power system technology choices was quantified through parametric studies. From these, we concluded that solar cells should be glassed with 10-mil thick coverslides, Li ion battery cell technology is preferred over NiH2 technology and a 120-V PMAD offers substantial mass/size savings over lower voltage systems. The impacts of EPS sizing on spacecraft ACS was examined and found not to be discriminatory. And finally, the benefits of spacecraft mass savings in the context of the RTD mission were assessed with the fundamental finding that low cost components are favored over low mass components. In the end, a low-cost, moderate mass EPS option and a low-mass, moderate cost EPS option were recommended.

ACKNOWLEDGEMENTS

The authors would like to acknowledge Mr. Leon Gefert of NASA Glenn Research Center for his contributions to the trajectory analysis, Mr. Scott Benson, NASA Glenn RTD Project Manager for his valuable technical suggestions and Mr. H. James Fincannon of NASA Glenn Research Center for his valuable contributions to the spacecraft flight mode and attitude control assessment.

REFERENCES

1. Sovey, James S., et al., "A Synopsis for Ion Propulsion Development in the United States: SERT I to Deep Space I," NASA/TM-1999-209439, Oct 1999.
2. Oleson, Steve, "Advanced Electric Propulsion For RLV-Launched Geosynchronous Spacecraft," NASA/TM-1999-209646, Dec 1999.
3. Jankovsky, Robert S., et al., "Preliminary Evaluation of a 10 kW Hall Thruster," NASA/TM-1999-209075, Apr 1999.
4. Jacobson, D. and Jankovsky, R., "Performance Evaluation of a 50 kW Hall Thruster," AIAA 99-0457, 37th AIAA Aerospace Sciences Meeting and Exhibit, Reno, NV, Jan 11-14, 1999.
5. Oleson, Steve, "Advanced Electric Propulsion For Space Solar Power Satellites," NASA/TM-1999-209307, Aug 1999.
6. Kerslake, Thomas W., and Gefert, Leon P., "Solar Power System Analyses for Electric Propulsion Missions," 34th Intersociety Energy Conversion Engineering Conference, SAE99-01-2449, Vancouver, British Columbia, Canada, Aug 1-5, 1999. (see also NASA TM-1999-209289).
7. Gefert, Leon P., Hack, Kurt J., and Kerslake, Thomas W., "Options for the Human Exploration of Mars using Solar Electric Propulsion," AIP Conference Proceedings, No. 458, STAIF-99, January 1999, p. 1275-1280.
8. Dudzinski, Leonard A., Hack, Kurt J., Gefert, Leon P. and Kerslake, Thomas W., " Design of a Solar Electric Propulsion Transfer Vehicle for a Non-Nuclear Human Mars Exploration Architecture," 26th International Electric Propulsion Conference, paper IEPC-99-181, Kitakyushu, Japan, October 17-21, 1999.
9. Kerslake, Thomas W. and Benson, Scott W., "Power System Options for the Radiation & Technology Demonstration (RTD)," Research & Technology 1999, NASA/TM-2000-209639, March 2000, p. 163-164.
10. Baity, F. W., et al., "Design of RF Systems for the RTD Mission VASIMR," ORNL/CP-103576, Apr 12, 1999.
11. Sackett, Lester L., Malchow, Harvey L., and Edelbaum, Theodore N., "Solar Electric Geocentric Transfer with Attitude Constraints: Analysis" , NASA CR-134927, Aug 01, 1975.
12. Jordan, Carolyn E., " NASA radiation belt models AP-8 and AE-8," Report AD-A223660, Sep 30, 1989.
13. Chetty, P.R.K., et al., "TOPEX Electrical Power System," 26th Intersociety Energy Conversion Engineering Conference, August 4-9, 1991.
14. Hosken, Robert W., and Wertz, James, R., " Microcosm Autonomous Navigation System On-Orbit Operation," http://www.smad.com/ analysis/mans1.html
15. Anonymous, http://www-projet.cst.cnes.fr:8060/ JASON/MissionRequirements.html
16. Robertson, Brent, et al., "TRMM On Orbit Attitude Control System Performance," AAS-99-073, Jan 01, 1999.
17. Jenkin, Alan B., "Attitude Maneuvers of a Solar-Powered Electric Orbital Transfer Vehicle," AAS PAPER 91-481, Jan 01, 1992.
18. Anonymous, http://www.spectrolab.com/
19. Ehsani, M. and Salim, A. "Flawless IN-orbit Performance of Lockheed Martins' Premier A2100 Electrical Power Subsystem for Communications Satellites," AIAA-2000-2809, 35th Intersociety Energy Conversion Engineering Conference, Las Vegas, NV, Jul 24-28, 2000.
20. Anonymous, http://www.hughespace.com/ factsheets/702/702.html
21. Suzuki, Akio, "High-Efficiency Silicon Space Solar Cells," Solar Energy Materials and Solar Cells, Vol. 50, 1998, p. 289-303.
22. Jang, Jeffrey, et. al., "Recent Progress in Amorphous Silicon Alloy Leading to 13% Stable Cell Efficiency," 26th IEE PVSC, Anaheim, CA, Sep 29- Oct 3, 1997.
23. O'neill, Mark, ENTECH Inc., personal communication, March 2000.
24. Murphy, David M. and Allen, Douglas M., "SCARLET Development, Fabrication and Testing for the Deep Space 1 Spacecraft," 32nd Intersociety Energy Conversion Engineering Conference Proceedings, paper no. 97539, Aug 1997.
25. Jones, P. Alan, et al., "A high specific power solar array for low to mid-power spacecraft," Proceedings of the 12th Space Photovoltaic Research and Technology Conference (SPRAT 12), May 01, 1993, p. 177-187.
26. Plaut, J. J. and Spencer, D. A., "Mission Plan for the Mars Surveyor 2001 Orbiter and Lander," Workshop on Mars 2001: Integrated Science in Preparation for Sample Return and Human Exploration, Jan 01, 1999, pp. 83.
27. Castell, Karen and Wingard, Robert, "Recent Advances in Power System Design at GSFC," 34th Intersociety Energy Conversion Engineering Conference, SAE99-01-2534, Vancouver, British Columbia, Canada, Aug 1-5, 1999.
28. Tada, H. Y. et al., "Solar Cell Radiation Handbook," NASA-CR-169662, Nov 01, 1982.
29. Anspaugh, B. E., "GaAs Solar Cell Radiation Handbook," NASA-CR-203421, Jul 01, 1996.
30. Cour-Palais, B. G., "Meteoroid Environment Model -1969 (Near Earth to Lunar Surface)," NASA SP 8013, 1969.
31. Kessler, D. J., et al., "Orbital Debris Environment for Spacecraft Designed to Operate in Low Earth Orbit," NASA TM 100471, April 1989.

32. Kessler, D. J., et al., "A Computer-Based Orbital Debris Model for Spacecraft Designs and Observations in Low-Earth Orbit," NASA TM 104825, Nov 1, 1996.
33. Eichelberger, R. J., and Kineke, J. H., Jr., Hypervelocity Impact," SPRINGER-VERLAG, Jan 1, 1967, p. 659-692.
34. Myre, Craig A., "Hypervelocity Particle Impact Testing of Solar Array Coupons," Preliminary Information Report #259, NASA Lewis Research Center, May 30, 1991.
35. Hoffman, David J., et al., "Thin-Film Photovoltaic Solar Array Parametric Assessment," AIAA-2000-2919, 35th Intersociety Energy Conversion Engineering Conference, Las Vegas, NV, Jul 24-28, 2000.
36. Hamley, John A., et al., "Hall Thruster Direct Drive Demonstration," paper AIAA97-2787, 33rd AIAA/ASME/SAE/ASEE Joint Propulsion Conference and Exhibit, Seattle Washington, Jul 6-9, 1997.
37. Wertz, James R. and Larson, Wiley J., Space Mission Analysis and Design, 3rd Ed., Microcosm Press, Torrance, CA and Kluwer Academic Publishers, Boston, MA, Sec. 11-1, 1999.

Parameter	Requirement
Orbit Type Inclination(°) Initial Altitude (km) Final Altitude (km)	Near Circular 28.45 or 51.6 400 ~5 Earth Radii
Initial Mass (kg)	1500
Transfer Time (yrs)	≤ 1
Microsats Number Mass (kg) Drop-off Altitudes (km)	 4 25 each 2000, 12000, 22000, 32000
Thruster System Power Input (kW)	 10
Hall Thruster Isp (sec) η (PPU & thruster) Propellant	 2100 0.54 xenon
VASIMR Thruster Isp (sec) η (PPU & thruster) Propellant	 10000 0.5 hydrogen
Operating Thruster Type Versus Mission Altitude 400 to 9000 km 9000 to 20000 km 20000 km & above	 Hall VASIMR Hall

Table 1. Mission Requirements & Vehicle Definition

Option Description	Cell η 1-Sun, AM0 28°C (%)	Calculated Panel Mass* (kg/m2)	Estimated Mechanism & Structure Mass† (kg/m2)
1. Rigid panel, cascade GaAs cells	25	2.74	10% of panel mass
2. Rigid panel, side reflectors, cascade GaAs cells	25	2.88	10% of panel mass
3. Flexible panel, cascade GaAs cells	25	1.64	0.6
4. Flexible panel, high η, 4-mil Si cells	17	1.16	0.6
5. Flexible 5-mil stainless steel panel, amorphous SiGe cells	9 (stable)	1.06	0.6
6. Rigid panel, rigid linear concentrators, cascade GaAs cells	25	2.64	10% of panel mass
7. Flexible panel, flexible linear concentrators, cascade GaAs cells	25	0.88	0.6
8. Flexible gore, cascade GaAs cells	25	1.43	0.3
9. Flexible gore, high η, 4-mil Si cells	17	0.96	0.3

Table 2. PV Array Technology Options

* -10-mil coverslides, GaAs cells with 5.5-mil Ge substrates

† - Power harness and SADA masses calculated separately and included in PV Array wing mass total.

Design Feature	NiH2	Li Ion
Maximum Cell Voltage (V)	1.5	4.1
Minimum Eclipse Time Bus Voltage (% of Sun Time Voltage Setpoint)	75	75
# of Cycles	1000	1000
Recharge Ratio	1.1	1.0
# of Failed (Open-Circuited) Cells	1	1
By-pass Diode Voltage Drop (V)	0.7	0.7
Battery Round Trip Energy Efficiency (%)	79	85
Design Temperature Limits (°C) Minimum Maximum	 0 20	 -5 30

Table 3. Battery Design Inputs

Parameter	Options
Solar Cell Ceria-Doped Coverslide Thickness	4-mil through 32-mil
PV Array Technology	See Table 2.
Battery Technology	NiH2, Li Ion
PMAD DC Voltage	28-V, 50-V, 120-V
Mission Orbit Inclination	28.45°, 51.6°

Table 4. Parametric Study Trade Space

Option	EPS Mass (kg)	Array* Mass (kg)	Wing Area (m2)	Specific Power (W/kg) Panel	Array	EPS	In-Service, Cell BOL η (%)	Typical Cell BOL Temp. (°C)	log_{10}DENI Fluence (# 1-MeV e/cm2) Current	Voltage
1	258	167	24.9	104	80	40	21.7	45	14.62	14.79
2	188	99	13.8 29.5†	198	133	55	19.8	105	14.60	14.77
3	220	130	25.3	175	104	47	21.7	45	14.71	14.88
4	313	218	56.4	165	68	33	10.9	60	15.47	15.82
5	N/S	N/S	N/S	108	29	11	4.3	110	>15.33	>16.06
6	280	191	23.5	98	68	37	22.2	65 - cell 20 - lens	14.18	14.36
7	180	89	24.7	330	150	58	21.7	70 - cell 25 - lens	14.55	14.72
8	195	105	25.3	199	124	53	21.7	45	14.79	14.95
9	298	199	70.8	200	87	35	10.0	65	16.03	16.47

* - Array is comprised of two wings

† - Array plus side reflector area, N/S - No Solution

Table 5. Effects of Solar Array Technology Option

Design Feature	NiH2	Li Ion
# Series Cells	81	30
Cell Capacity (Amp-hrs)	10	10
Design Depth-of-Discharge (%)	75	50
Design Charge Rate	C/3.0	C/3.3
Design Discharge Rate	C/2.1	C/2.1
Trickle Charge Rate	C/50	C/50
Battery Mass (kg)	35.6	12.5
Battery Volume (m3)	0.0537	0.0069
Maximum Heater Power (W)	9	0
Maximum Cooling Load (W)	2	2

Table 6. Battery Design Results

Design Characteristic	120-VDC	50-VDC	28-VDC
EPS Masses (kg)			
Solar Array	167.2	198.6	253.2
Battery	35.6	26.2	33.5
PMAD Boxes	51.5	72.4	116.4
PMAD Cabling	4.1	10.6	20.7
Total	258.4	307.8	423.8
PPU Current (Amp)	84	205	382
Array Wing Area (m2)	24.9	26.6	31.7
# Solar Cells per String	64	28	18
# Strings per Wing	138	336	624
Battery Cell Capacity (Amp-hrs)	10	20	50

Table 7. Design Impacts of PMAD Voltage

Option	Max Spacecraft Mass Moment of Inertia (kg-m2)	Max Sum Environ. Torque (N-m)	Max Roll Steering Torque (N-m)	Max Sum Environ. Momentum per Orbit (N-m-sec)	Max Roll Steering Momentum per Orbit (N-m-sec)	# Momentum Dumps per Day: BOL at 400 km alt.	# Momentum Dumps per Day: EOL at 35000 km alt.
1	13289	0.0312	0.1811	81.24	21.73	32	9
2	5685	0.0139	0.0775	50.50	9.30	19	9
3	10727	0.0259	0.1462	73.49	17.54	28	9
4	32977	0.0755	0.4494	155.39	53.92	65	9
5	N/S	N/S	N/S	N/S	N/S	N/S	N/S
6	14482	0.0327	0.1973	79.32	23.68	32	9
7	7685	0.0202	0.1047	67.67	12.57	25	9
8	4997	0.0151	0.0681	63.34	8.17	22	9
9	18258	0.0503	0.2488	144.02	29.86	54	9

Table 8. EPS Design Impacts to ACS

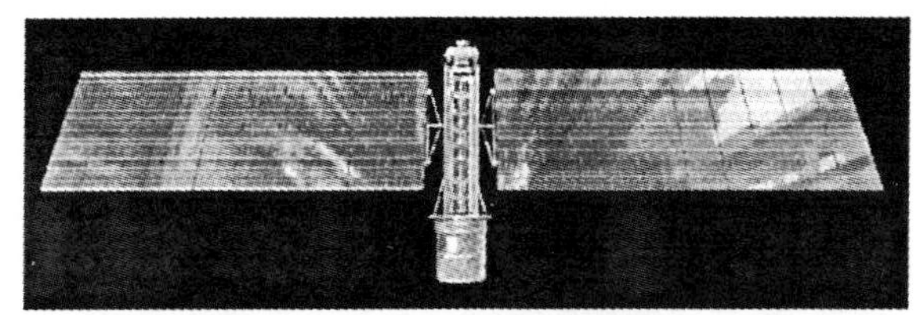

Figure 1. Conceptual RTD Spacecraft

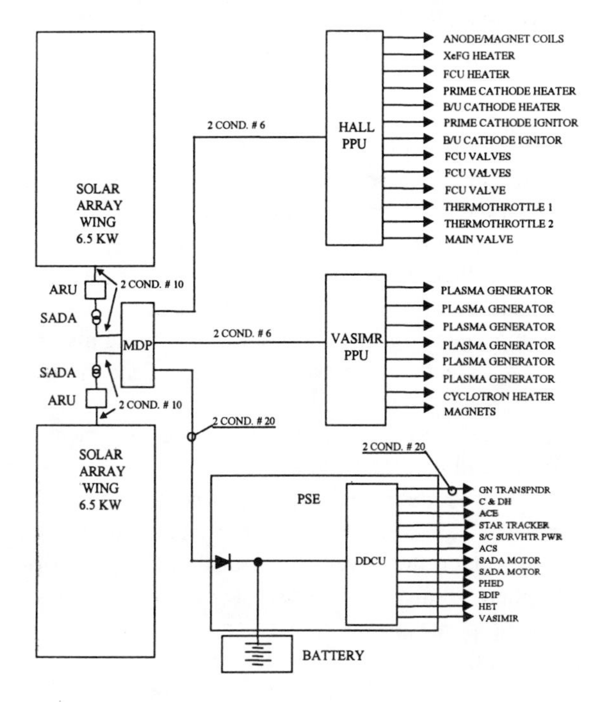

Figure 2. PMAD Architecture

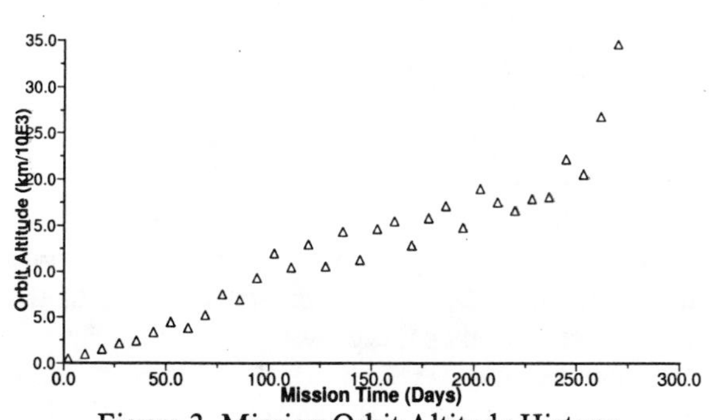

Figure 3. Mission Orbit Altitude History

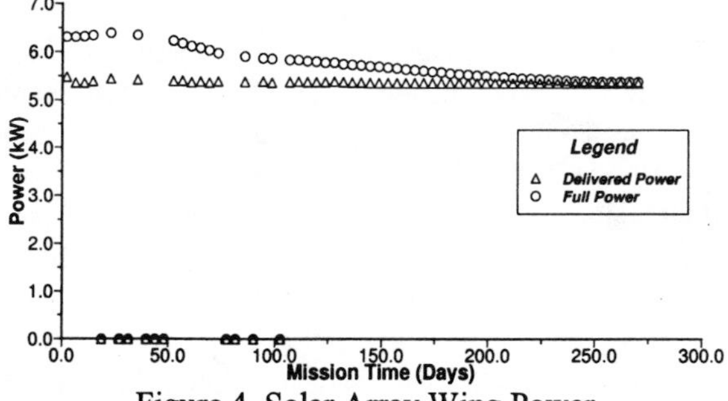

Figure 4. Solar Array Wing Power

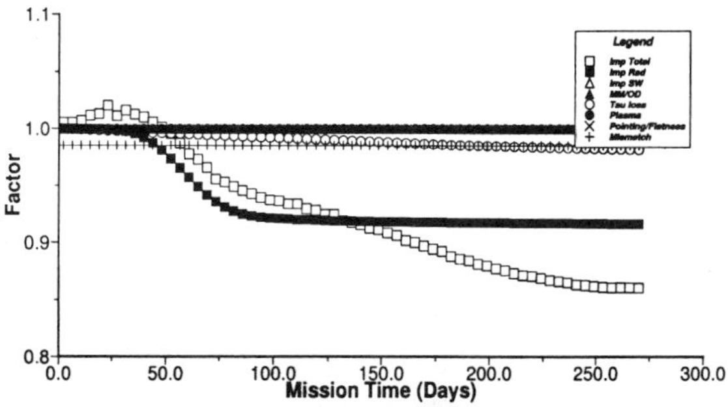

Figure 5. Solar Cell Current Degradation Factors

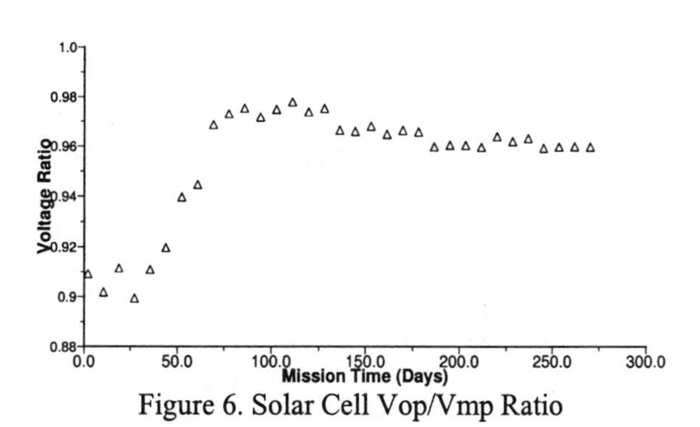

Figure 6. Solar Cell Vop/Vmp Ratio

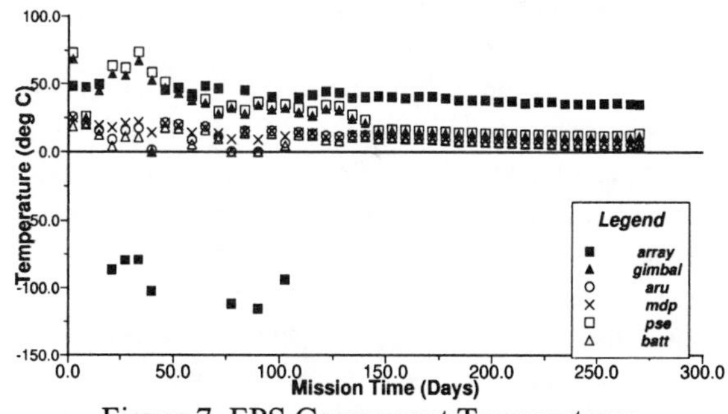

Figure 7. EPS Component Temperatures

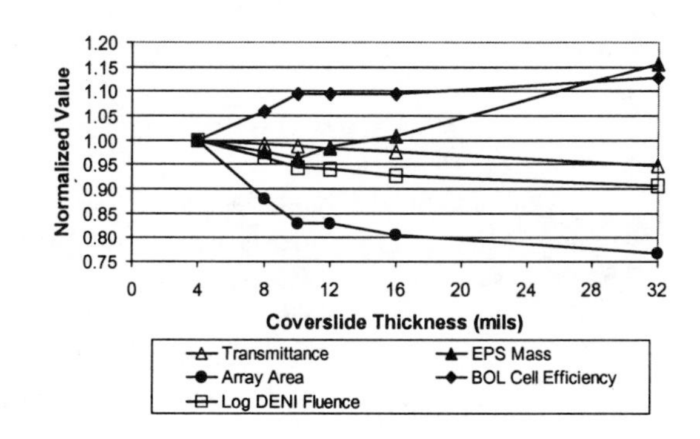

Figure 8. Effects of Solar Cell Coverslide Thickness

INTEGRATED THIN-FILM SOLAR POWER SYSTEM

R.P. Raffaelle[a], J.D. Harris[b], D. Hehemann[b], D. Scheiman[c], G. Rybicki[d], and A.F. Hepp[d]

[a]*Rochester Institute of Technology, Rochester, NY 14623*
[b]*Kent State University, Kent, OH 44242*
[c]*Ohio Aerospace Institute, Brookpark, OH 44142*
[d]*NASA Glenn Research Center, Cleveland, OH 44135*

ABSTRACT

The need for small and lightweight modular power systems is growing rapidly as the space science community continues to move toward smaller and less costly spacecraft (e.g., nanosatellites). Thus the use of lightweight thin-film photovoltaic solar cell arrays for power generation is an attractive possibility. Thin-film lithium ion energy storage with its large power densities and long cycling lifetimes should also prove a valuable resource. We have been developing a thin-film device capable of both solar energy conversion and storage. This device combines a thin-film lithium polymer battery with a thin-film solar cell. In a typical satellite application, the solar cell would be used to provide power for the spacecraft and charge the battery during the illuminated portion of the orbit. The battery would then provide the necessary "stay-alive" power for the satellite when in eclipse.

The deposition techniques we have been using to deposit the component films of this device are inherently scalable. Sectioning the solar cells, varying the series and parallel combinations of the batteries, and adjusting the overall size of the device can be used to fulfill a wide variety of power requirements. We will discuss the device design and the electrical properties of the component films. The output of the thin-film solar cell under an AM0 illumination and the cycling characteristics of the thin-film battery under a low-earth orbit or LEO (e.g., 55 min. illumination, 35 min. eclipse) timing sequence will be presented.

INTRODUCTION

One of the major limitations in the design of new micro and nano-satellites is power generation and storage systems.[1-2] These systems restrict the scientific payloads and objectives due to their size and weight, deployment systems, and lifetime. They also pose significant hurdles when it comes to attitudinal control, spacecraft ballast and systems integration. These limitations can be addressed through the use of modular power supplies capable of both solar power generation and storage. It is desirable that these devices be small, lightweight, mechanically flexible, conformable, and compatible with microelectronics. These devices should also be able to meet a variety of voltage and power requirements. These specifications can be met by combining a thin-film photovoltaic solar array with a thin-film lithium ion battery.

The development of self-contained power supplies that are suitable for space applications depends upon the development of thin-film batteries and photovoltaic cells. These devices need to have the appropriate power density and lifetime (five to ten years), and a means for processing that is compatible with IC (integrated circuit) or MCM (multi-chip module) manufacturing technology.[3]

Lithium Ion Batteries

Lithium ion batteries have received considerable attention since the commercialization of the first Li-ion cell by Sony in 1991. These cells are based on the concept of intercalation or ability of Li ions to enter a framework structure without transforming or altering the crystal structure of the host material. This concept was a subject of intense experiments and research in the early and mid 1970's.[4] The concept was demonstrated largely in host materials such as TiS_2,[5] the demonstration of similar intercalation behavior of Li in high voltage $LiCoO_2$ in the early 1980's by Mizushima and Goodenough led to the development of the commercial Sony Li-ion battery.[6] Since then there have been a host of transition metal oxides developed as cathode materials for Li-ion batteries.[7] Similarly, intercalation of Li in ordered and disordered carbon led to the usage of carbon as an alternative anode material to metallic Li.[8] The rechargeable electrode materials area experienced intense research activity mainly for the development of bulk battery packs for consumer electronic applications.[9] Research activity

mainly for the development of bulk battery packs for consumer electronic applications.[9] Research activity into extension of similar ideas for thin-film Li-ion batteries or micro-batteries has in comparison been minimal. However, the interest in thin-film batteries is growing due to the fact that they are ideal for providing energy to low power microelectronic systems.[10]

Thin-Film Photovoltaic Solar Cells

The ability to produce large area high efficiency thin-film photovoltaic solar arrays for space is still a tremendous challenge to the photovoltaics community. The potential benefits of weight reduction, radiation resistance, mechanical flexibility, low-cost substrates, and inexpensive processing techniques remain a strong driving force. Thin-film $CuInSe_2$ based cells have achieved efficiencies that are approaching the level necessary to justify their use in space and they are already being produced for terrestrial application.[11] Thin-film amorphous silicon dual-junction and triple-junction cells are now achieving reproducible efficiency in large area roll-to-roll processing.[12]

EXPERIMENTAL PROCEDURE

Lithium Thin-Film Batteries

The lithium thin-film batteries were prepared in the procedure outlined in a previous paper.[13] The batteries are comprised of two sheets of Kapton™ that serve as the outer layers. The electrical contacts to the anode and cathode materials were made by magnetron sputtering 1.0 μm of molybdenum on the inner surfaces of the Kapton™. The battery cathode consisted of (0.75 g) lithium cobaltate ($LiCoO_2$), (0.05 g) carbon black, and (0.20 g) polyvinylchloride (PVC), and 10 ml of tetrahydrofuran (THF). The cathode solution was stirred for 1 hour and then ultrasonicated for 1 hour. The thin-film cathode material was then directly applied to the Mo coated Kapton™. The electrolyte was prepared by soaking Whatman® 1 filter paper in a solution of (6.52 g) ethylene carbonate (EC), (1.38 mL) propylene carbonate (PC), (0.089 g) polyethylene oxide (PEO), (1.22 g) lithium hexaflurophosphate ($LiPF_6$), (10 mL) tetrahydrofuran (THF). The filter was soaked for 5 minutes in the solution, removed and allowed to dry for 20 min. The anode was prepared from lithium foil that was rolled to a nominal thickness of 0.2 mm. The edges of the battery were hermetically sealed using a thin bead of epoxy (see Figure 1). Battery assembly was performed under an argon atmosphere in a Vacuum Atmospheres glove box. The final battery was 0.45 mm thick and had a mass of 0.33 g.

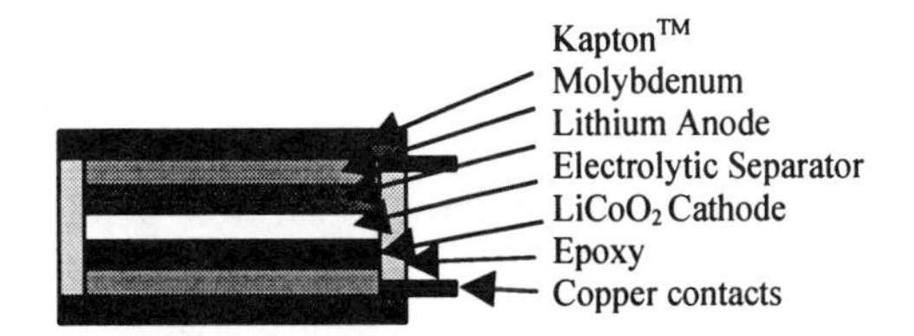

Figure 1. Battery schematic.

Solar Cell Calibration

Solar cells are solid-state devices that generate electricity when exposed to sunlight (or light energy of sufficient wavelength and intensity), this is known as the photovoltaic effect. They are large area diodes. These cells produce current proportional to the light intensity and voltage relative to the semiconductor bandgap. They can be connected in series and parallel strings (arrays) like batteries and generate a wide range of voltages and currents.

Sunlight in space has a different spectrum and intensity than that on earth. Since the sunlight does not pass through the earth's atmosphere it is referred to as the Air Mass Zero Spectrum or AM0. The AM0 spectrum has a total intensity 136.8 mW/cm^2. Since satellite manufacturing and launch costs are very high, accurate prediction and calibration of array performance is critical, and solar arrays must also be precisely sized for the application to avoid excess mass and power. Predicting solar power in space requires an AM0 calibration facility which consists of an AM0 calibrated solar cell and a close-match light source. The AM0 calibration is accomplished using standard solar cells that have been tested in or near AM0 via high altitude balloons or aircraft. A calibrated standard cell must be made for each type of solar cell. These calibrated solar cells are used to adjust the light source intensity to match the performance of the cell in AM0. Once the light source is adjusted, the AM0 performance of the experimental array can be measured.

Characterization of an array is done by measuring the solar cell current output as a function of voltage or I-V curve. Critical parameters of the IV curve are the Maximum Power point (P_{max}), which is the point of optimum efficiency of the solar cell, the open circuit voltage (V_{oc}) at 0 A, and short circuit current (I_{sc}) at 0 V. The ratio of $P_{max}/(V_{oc} \times I_{sc})$ is known as the fill factor (FF), this is a gauge of the cell quality, and higher numbers indicate higher quality. The efficiency is based on the P_{max} of the cell over the AM0 intensity per unit area.

The solar cell output was connected to a battery through a simple circuit that was designed to control constant charging and discharging currents. The battery voltage was monitored using an ordinary analog to digital board interfaced to a personal computer. The control circuit schematic is shown in Figure 2. All the tests on the system were performed at room temperature and no accommodations were made for the likely LEO temperature variations.

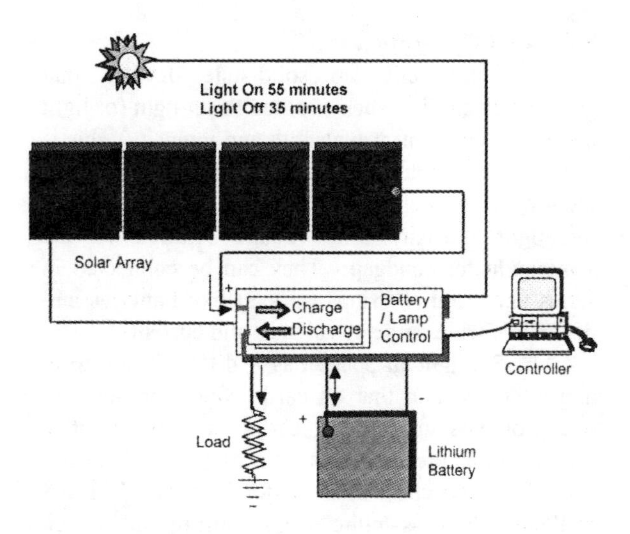

Figure 2. Test bench schematic.

RESULTS AND DISCUSSION

LEO Simulation

An un-optimized amorphous Si thin-film solar array (Iowa Thin-Film Technologies, 2/92) was used to charge the battery. The active array area was 49.9 cm^2 and was comprised of 4 tandem cells connected in series, with each cell yielding approximately 1.4 V under AM0 illumination (see Figure 3). This array was selected because the voltage had to exceed the open circuit voltage of the thin-film battery (V_{oc} = 3.6 V), due to the voltage drops in the associated test circuitry, while providing a charging current of 100 μA. These parameters are well below the available power of the array. This array was also selected because it is a thin-film solar array that complements the lightweight/high energy density of the thin-film battery. The mass of the array was 1.17 g and its thickness was 0.05 mm.

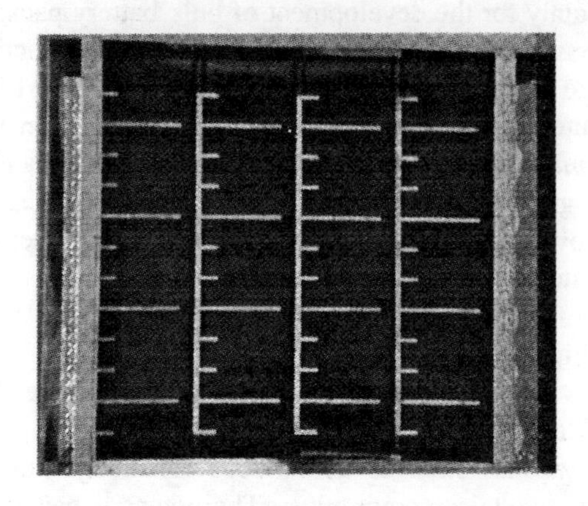

Figure 3. Photograph of the thin-film a–Si array used in the power supply.

The current versus voltage behavior of the array is shown in Figure 4. The maximum power point corresponded to a current of 45.35 mA and a voltage of 3.75 V, yielding a power of 170 mW. Thus, the array provided 150 W/kg. The fill factor was 42.5 giving an AM0 efficiency of 2.49 %. The combined weight of the integrated power supply (battery + solar cell) was under 1.5 g and its total thickness was under 0.5 mm.

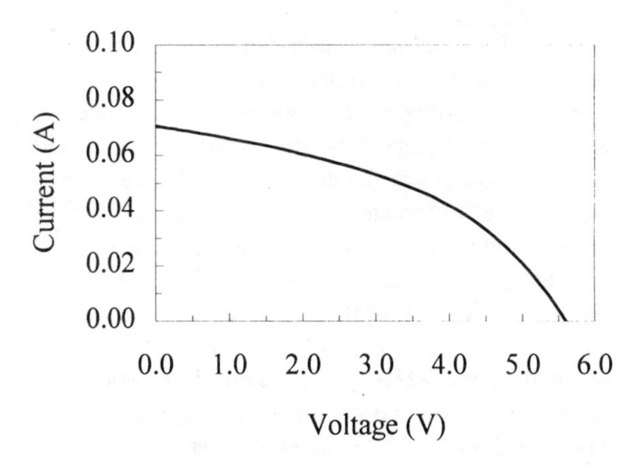

Figure 4. I-V photoresponse of a-Si thin-film solar cell under AM0 illumination.

The amorphous Si array was connected to the charging circuit of the battery and illuminated with a General Electric Quartzline® ELH lamp. The light source was adjusted to get the same performance as shown in Figure 4. Once the light intensity was adjusted, the array was illuminated using a Low Earth Orbit cycling sequence (i.e., 55 minutes ON and 35 minutes OFF). During testing the array was operating at nearly the V_{oc} of the array.

Therefore, the battery would be charged for 55 minutes and discharged for 35 minutes. Discharge current was adjusted so that the battery was not depleted at the end of an orbit. Once the LEO cycle was programmed into the computer, the battery discharge and charge voltages were recorded every 5 seconds throughout the test.

The thin-film lithium battery used in the power supply is shown in Figure 5. The discharge and charge characteristics of the battery were found by charging the battery at a constant current density of 0.5 mA/cm^2 until it reached a voltage of 4.2 V, at which point a variable load was applied to the battery in order to maintain a constant discharging current of 0.5 mA/cm^2. The battery was then discharged to a voltage of 3.0V at which the constant current charge was again initiated (see Figure 6).

Figure 5. The thin-film lithium battery used in the power supply.

The output of the thin-film a-Si solar cell was connected to the battery through the control circuit. The solar cell was then illuminated using the simulated AM0 source and the system was cycled using a LEO timing sequence. The battery was charged at a constant current density of 0.1 mA/cm^2 and the discharge current was a constant 0.3 mA/cm^2.

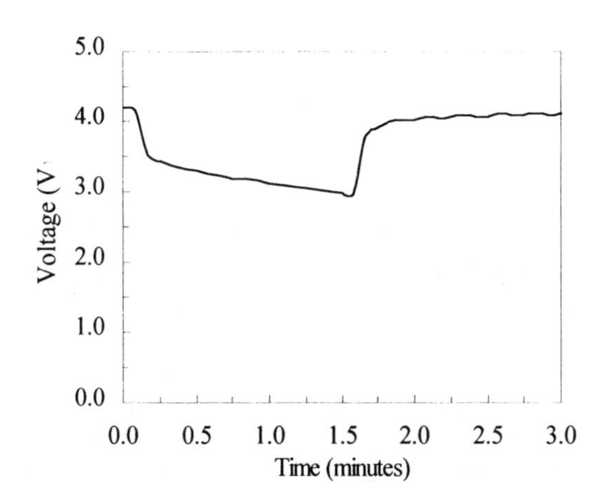

Figure 6. A single charge/discharge cycle of the thin-film battery between 4.2 V and 3.0 V using a constant charging and discharging current density of 0.5 mA/cm^2.

These current densities were based upon the charging and discharging times of the LEO cycle and the charging discharging characteristics shown in Figure 2. These values were chosen so that the battery would not be overcharged or be driven to complete discharge. The first ten charging/discharging curves are shown in Figure 7. The system was cycled for 40 hours and showed little or no loss of battery capacity in this limited test. Other batteries of this type have been cycled up to 20,000 times.[13]

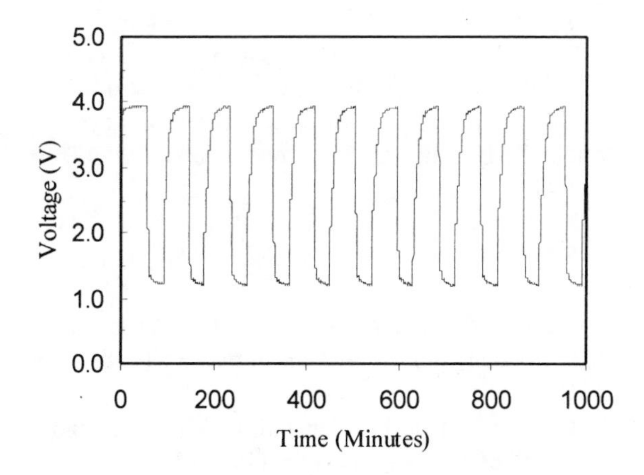

Figure 7. LEO orbit cycling simulation of the integrated power supply using AM0 illumination.

CONCLUSIONS

We have produced a robust thin-film integrated solid state battery and photovoltaic power supply that, while extremely small, is capable of delivering sufficient power during an entire LEO orbit for a multi-chip module (MCM) or microelectromechanical system (MEMS) designed for space applications. The thin-film lithium battery in this supply has the ability to be discharged and recharged over thousands of cycles. The performance of this battery with its simple and reliable manufacture method coupled with the advantages of thin-film photovoltaic solar power should provide suitable energy storage for a wide variety of space electronic and power delivery systems.

ACKNOWLEDGEMENTS

The authors wish to thank Prashant Kumta of Carnegie Mellon University and Harry Shaw of the NASA Goddard Space Flight Center for their advice, assistance and encouragement pertaining to this work. The authors would also like to acknowledge the support of the NASA cooperative agreements NCC3-710, NCC3-563, and the NASA Glenn Research Center Directors Discretionary Fund.

REFERENCES

1. B.B. Roberts and D. Bland, *NASA TM-4075-VOL-2* (1988).
2. A.B. Chmielewski, A. Das, C. Cassapakis, D. Allen, W.J. Schafer, J. Sercel, F. Deligiannis, M. Piszcor, P.A. Jones, D. Barnett, S. Rawal, T. Reddy, 31st *Inter. Energy Conv. Eng. Conf.*, **4** (1996) 2193.
3. J.B. Bates, G.R. Gruzalski, N.J. Dudney, C.F. Luck, X.-H. Yu, and S.D. Jones, *Solid State Tech.*, July (1993) 59.
4. K. Brandt, *Solid State Ionics* **69** (1994) 173-183.
5. S.D. Jones and J.R. Aridge, *Solid State Ionics* **86** (1996) 1291-1294.
6. K. Mizushima and P.C. Jones, P.J. Wiseman and J.B. Goodenough, *Solid State Ionics* **34** (1981) 171-174.
7. D. M. Schleich, P. Fragnaud, R. Marchand and T. Brousse, *ECS Proceedings* **96**, (1996) 71-75.
8. T. Brousse, R. Retoux, U. Herterich, and D.M. Schleich, *J. Electrochemical Society* **145**, (1998) 1-4.
9. R.J. Jasinski, *High Energy Batteries* (Plenum Press, New York, 1967).
10. B. Scrosati, *Nature* **373** (1995) 557.
11. J.R. Tuttle, J.S. Ward, A. Duda, T.A. Berens, M.A. Contreras, . Ramanathan, A.L Tennant, J. Keane, E.D. Cole, K. Emory, R. Noufi, *Proc. Mater. Res. Symp.* **426** (1996) 143.
12. S. Guha, J. Yang, A. Banerjee, P. Nath, J. Call, T. Glatfelter, F.J. Boelens, and G. Oomenn 34th *IECEC*, 2553 (1999).
13. R. P. Raffaelle, J.D. Harris, D. Hehemann, G. Rybici, D. Scheiman, and A.F. Hepp, In press, *J. of Power Sources* (2000).

AIAA 2000-2810

SPACECRAFT CHARGING EFFECTS ON ANODIZED ALUMINUM SURFACES

G.B. Hillard*
NASA Glenn Research Center
Cleveland, OH 44135

Boris Vayner†
Ohio Aerospace Institute
Cleveland, Ohio 44135

Introduction

Because of a number of key properties, there is increased interest in using anodized aluminum on external spacecraft surfaces. Originally used by the space station program because aluminum is cheap and easy to anodize, it attracted attention because its optical properties, especially absorptivity (α) and emissivity (ε), can be tailored to produce a surface having excellent thermal control properties. Furthermore, since it is an oxide and is almost impervious to atomic oxygen (AO) it is ideal for use in Low Earth Orbit (LEO) where AO attack is a major design issue.

In addition to these traditional uses of the material, the recent introduction of high-voltage space power systems has caused designers to take note of the excellent insulating properties of this material. As a result, both laboratory work as well as limited space testing has been performed. We will show a typical sample from high voltage arc testing as well as a flight sample from the Solar Array module Plasma Interactions Experiment (SAMPIE). We believe our findings to apply generally to this material as we assess its value as a high voltage insulator in space.

Background

Anodized aluminum became a high-priority issue in the late 1980s when Space Station Freedom (SSF) designers specified vast amounts of the material (more than 2000 m^2). Virtually all exposed surfaces, including the main truss structure as well as all hab modules, were to be made from this material. The power system for SSF was originally designed to use 20Khz alternating current (ac). Plasma interactions of high-frequency ac systems are poorly understood but arcing is believed to be much less severe than for dc.

When the power system was changed to a 160-volt direct current (dc) system in late 1989, it was realized immediately that serious interactions between high voltage surfaces and the ambient plasma were probable. The issues were studied over the next two years by the Space Station Freedom Electrical Grounding Tiger team (hereafter called the "Tiger Team") involving several dozen experts from the government, industry and academia. Among their conclusions were that the space station main truss would "float" as much as 140 volts negative with respect to space.[1]

The ideal solution to such a problem would be a complete redesign of the solar arrays to eliminate current collection. The technology necessary for such an effort was in its infancy at

* Senior member, AIAA
† Member, AIAA

the time and such a course would have meant significant program delays. The solution selected was to add a plasma contactor to SSF to actively maintain the spacecraft "ground" potential at all points to within 40V of the plasma. All of these considerations continued as SSF was redesigned and became the International Space Station (ISS). With the addition of the Plasma Contactor, the problem is under control.

During this effort there were many questions about the behavior of materials and systems if a suitable plasma contactor could not be designed, failed or was not ready in early stages of ISS operation. It was also clear that similar considerations would arise in future programs, as high voltage power systems become more common. The suitability of anodized aluminum as a high voltage insulator continued, therefore, to be of interest.

Among the implications for high voltage on SSF external surfaces was that the anodized aluminum was unlikely to stand off such potentials. Dielectric breakdown and arcing to space has the potential to destroy the coating in time as well as to produce significant electromagnetic interference. Exactly how the coating would behave became an area of research and the subject of extensive numerous ground testing.

Ground Tests

Testing of anodized aluminum concentrated on material made to SSF, and later ISS, specifications. For all tests samples the alloy was type 2219 (Federal Specification QQ-A-250/30). Nominal 0.5 mil anodization was done by sulfuric acid to specifications STP 0554-0101 and Mil-A-8625. Details of these experiments are part of the Tiger Team records.

Figure 1 is a photograph of a typical anodized aluminum plate used in plasma chamber testing. This particular sample was subjected to a wide range of applied voltage, up to 600V dc, and most of the visible damage occurred at higher voltages. Figure 2 is a closer look at several of the arc sites.

Of interest to us here is the fact that the ground test program was beset from the beginning by an inability to reliably reproduce experimental results. In one case, for example, two identical samples (from the same batch) were tested and found to break down at -300V and -700V. In another case, the original sample arced at –120V while a new one made to spec broke down at –250V. A number of samples arced with less than -100 volts and, in the worst case, samples arced at as little as –55V. When these tests were concluded, it was clear that ISS was safe with the plasma contactor on board and no surfaces exceeding 40V. The suitability of the material for high voltage surfaces was, however, still unclear.

SAMPIE

SAMPIE was flown on STS-62 in March 1994. The experiment has been documented extensively elsewhere including the behavior of the anodized aluminum sample.[2] The sample, which withstood voltages to –220V, was unremarkable in its flight behavior and gave only one more (unreproducible) data point. It is shown in figure 3 on its mounting plate prior to flight.

Post-flight Analysis

After the flight, a review of all results to date failed to shed light on the widely variable arcing threshold. It was decided to sacrifice the sample to allow SEM analysis of its surfaces. The original interest was in details of the various arc pits and of ejected material around them as several theories of arc initiation differed in their predictions. Much more interesting was a close look at the coating in cross section.

In figure 4, the lower white portion is solid metal while the dark black upper part is open air. The gray layer is the coating and, as can be seen, is highly non-uniform. Nominally 0.5 mil thick, there are numerous gaps and voids, places where the actual thickness is less than one tenth of nominal thickness. While we show only this one picture, it is not exceptional but typical.

Clearly this high degree of non-uniformity is the reason for the material's erratic behavior under test. This material was never designed to be an electrical insulator and nothing in its production specs covers coating uniformity.

Conclusion

We conclude that anodized aluminum, while a superb thermal control surface, is not well suited in the role of electrical insulator. Limited in-house efforts at NASA Glenn have suggested that it may be possible to significantly improve uniformity.[3] These efforts may eventually point the way to a coating process that would meet a uniformity specification for the material and allows its widespread use on high voltage spacecraft.

Because of various constraints and priorities, no further research efforts are underway at this time.

References

[1] Clouber, D., and Tye, C., Space Station Freedom Program (SSFP) Electrical Grounding Tiger Team Preliminary Integrated Assessment Report, Under NASA Contract: NAS 9-18200/B, McDonnell Douglas Space Systems Company, Space Station Division, Houston, TX, 1991.

[2] G.B. Hillard and D.C. Ferguson, *Measured Rate of Arcing From an Anodized Sample on the SAMPIE Flight Experiment*, 33rd Aerospace Sciences Meeting & exhibit, AIAA 95-0487

[3] David B. Snyder, NASA Glenn Research Center, private communication.

Figures

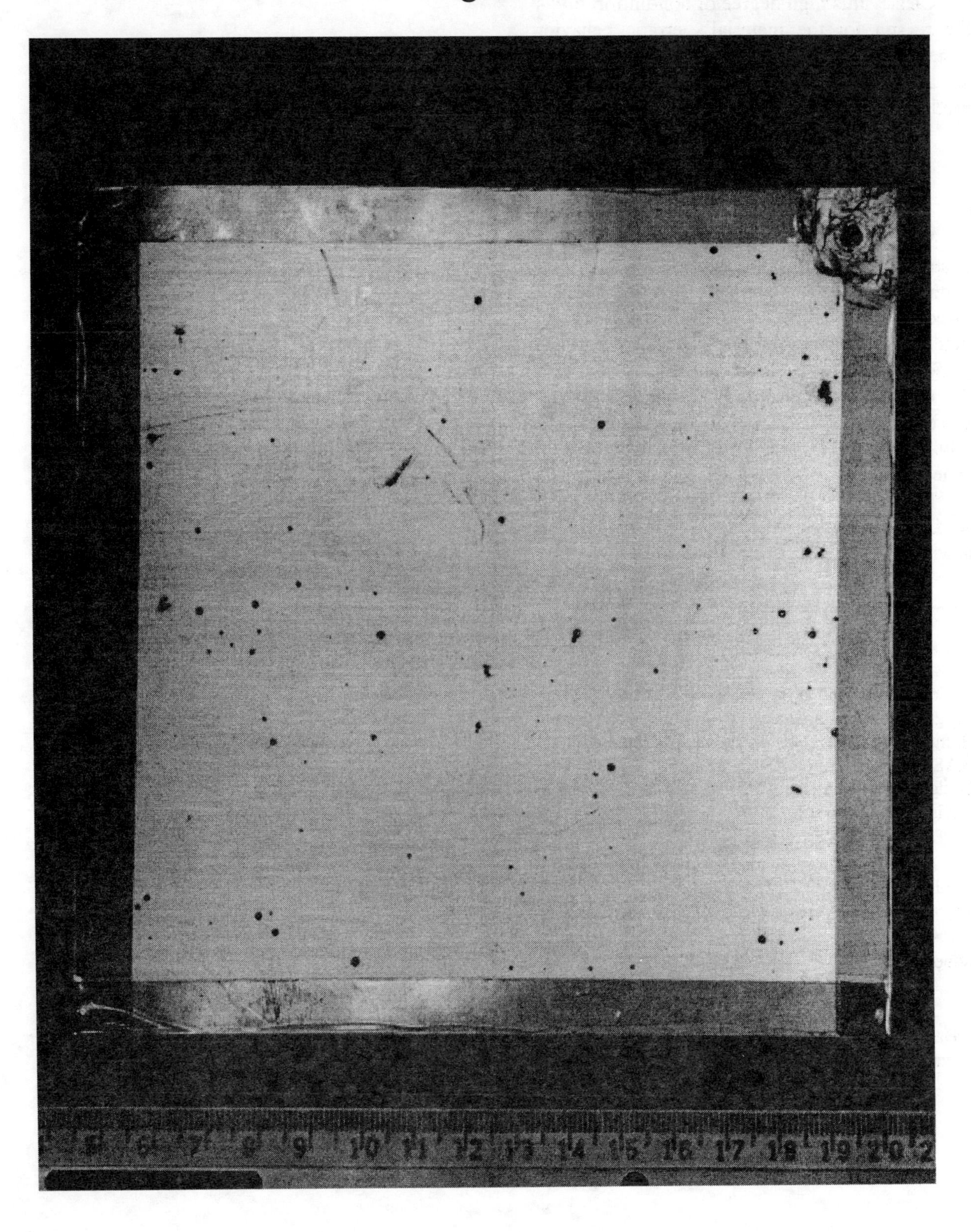

Figure 1: Typical damage from arcing

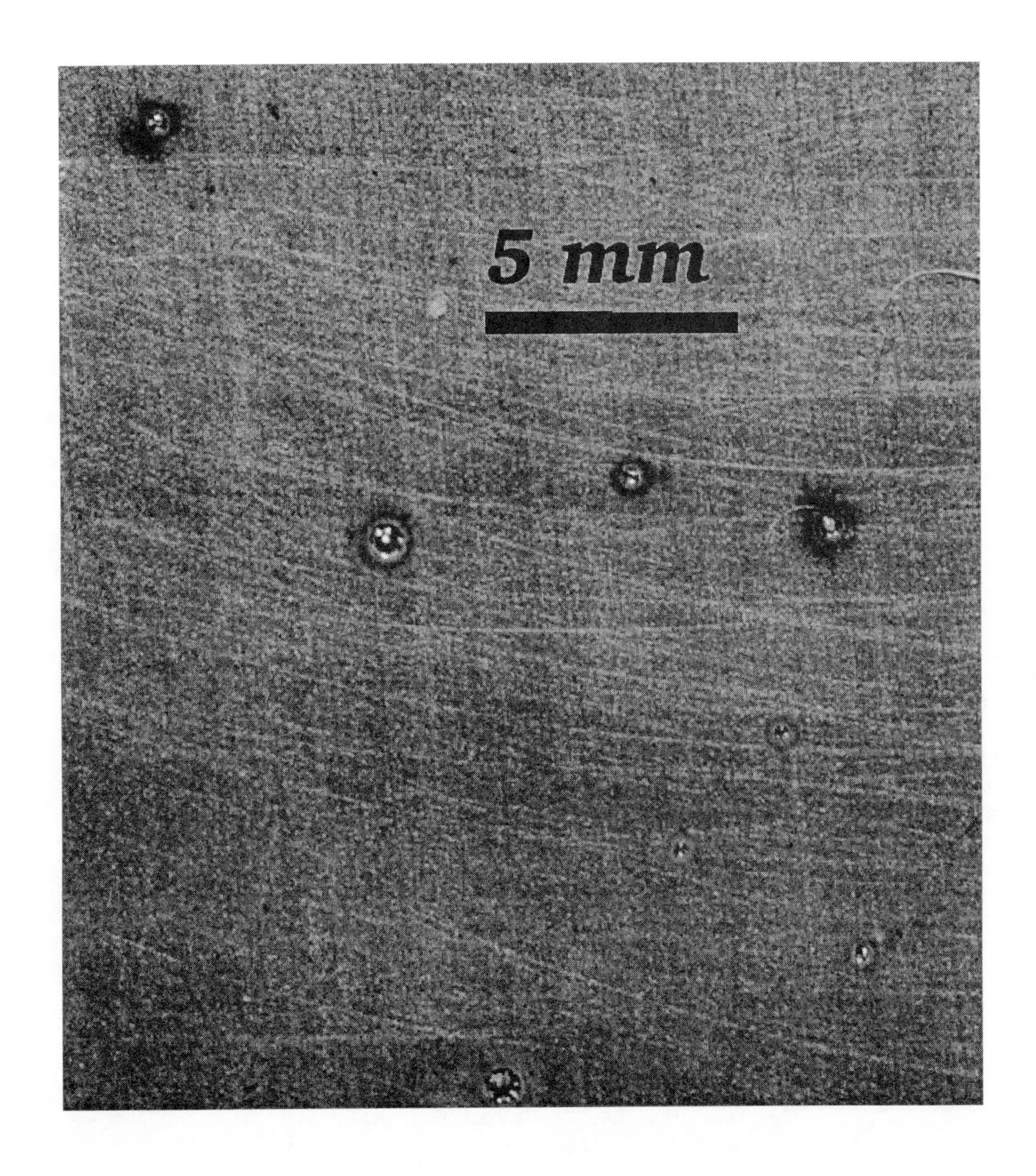

Figure 2: Close up of arc damage

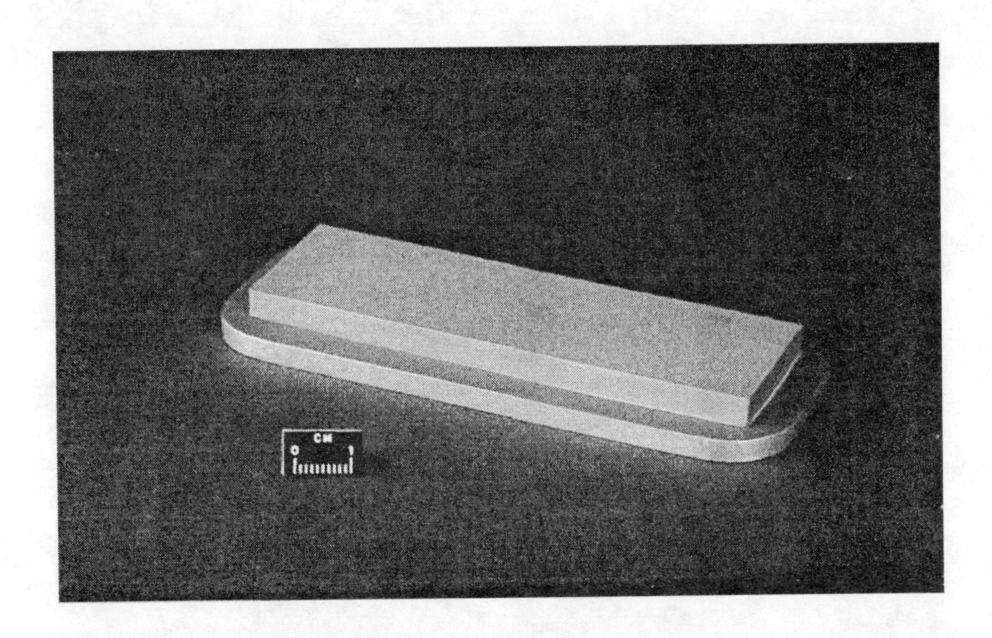

Figure 3: Photo of SAMPIE flight sample

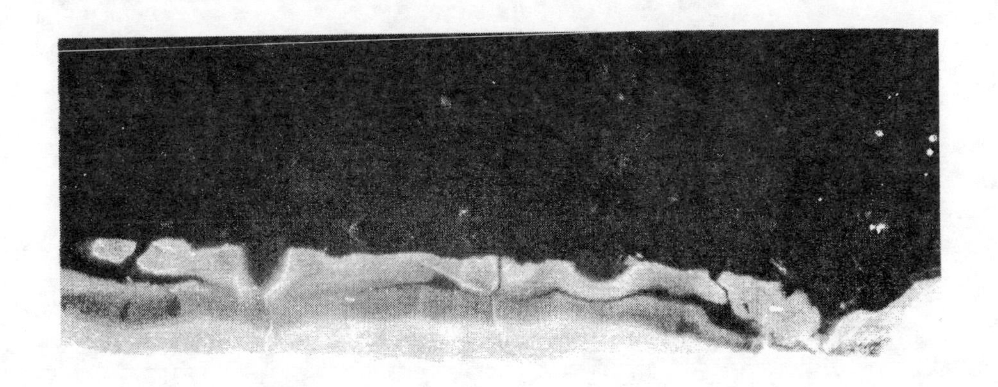

Figure 4: SEM photo of coating in cross-section. Scale: 1 cm = 10μm (≈ .39 mil)

AIAA-2000-2811

REAL GAS EFFECTS IN STIRLING ENGINES

Gianfranco Angelino - Dipartimento di Energetica, Politecnico di Milano, Milano, Italy
Costante Invernizzi - Dipartimento di Ingegneria Meccanica, Università di Brescia, Brescia, Italy

Abstract
Real gas effects in the working fluid critical region is shown to be a powerful tool to increase the energy conversion efficiency of Stirling cycles, mainly at low top temperatures. Organic working substances are shown to represent the only practical choice to optimize real gas effects for applications at given minimum cycle temperatures. The moderate thermal stability of organic compounds limits the cycle top temperature to 350-450 °C and, consequently, the achievable top efficiency. However the good intrinsic thermodynamics of real gas cycles allows conversion efficiencies in excess of 25% for power only and in excess of 20% for combined heat and power plants. Adequate working fluids, both as pure substances and as mixture are shown to be practically available. An optimization procedure is illustrated allowing to select the best cycle performance for a given cycle top pressure and temperature.

Introduction

The ideal Stirling Engine cycle, equivalent, in terms of thermodynamic efficiency, to the Carnot cycle, is performed resorting to a perfect gas: only under this assumption regeneration can be reversible and the cycle does not exhibit any internal loss. Many permanent gases such as helium, hydrogen, nitrogen, air, carbon dioxide in the working conditions involved in practical Stirling engines behave substantially as perfect gases and, being thermally stable, do not represent a limiting factor from the point of view of achievable operational temperatures. In principle real gas effects implying a difference in heat capacity between the high and the low isochores are detrimental to efficiency. However, when the actual engine operation is considered, the effect of the nature of the working fluid on losses could determine an overall picture which favours the non-ideal behaviour of the active medium. Assuming that such benefits are a practical possibility the question arises whether non ideal fluids are available for operation in actual engines.

In this paper only non-idealities related to molecular interaction in the vicinity of the saturation curve will be considered. Aiming to evaluate the effect of strong deviations from the perfect gas behaviour it was decided to include in the conversion cycle a portion of the fluid critical region, on the gas side, thus excluding any change of phase. Strong real gas effects vanishing rapidly as temperature is increased above the critical, only the coldest portion of the cycle will experience a definite departure from usual ideal gas assumption. In any case the cycle considered will be formed by two isotherms and two isochores. Since the level of the lower isotherm is determined by practical reasons (the need of rejecting waste heat to a specified ambient or the needs of the heat user in a cogeneration system) the critical temperature of the fluid must be carefully selected for each heat rejection temperature. Only organic fluids offer a variety of critical temperatures suitable to meeting the aforesaid requirements. Although thermodynamically attractive organic fluids exhibit a limited thermal stability. There is experimental evidence that safe operating temperatures around 400 °C can be achieved. [**1-3**]. A careful selection of both fluid and containing materials could made a top temperature of 450 °C attainable.

Such thermal levels are more typical of Rankine (steam or vapour) rather than of gas cycles.

Such a constrain, which is not experienced by inert gas engines, represents an objective limitation to the performance of organic fluid cycles but, conversely, relaxes high temperature material problems.

Basic thermodynamics

With reference to Figure 1 relating to HFC-23 (CHF_3) let us consider two ideal Stirling cycles (i.e. cycles bounded by two reversible isotherms connecting the fluid with external sources and two loss-free isochores between which a regenerative heat transfer process is performed) in the one at high entropy the working fluid behaves like a perfect gas (cycle B), in the one at low entropy strong real gas effects are present in the compression process which takes place in the fluid critical region (cycle A). Reduced rather than ordinary variables are used for convenience.

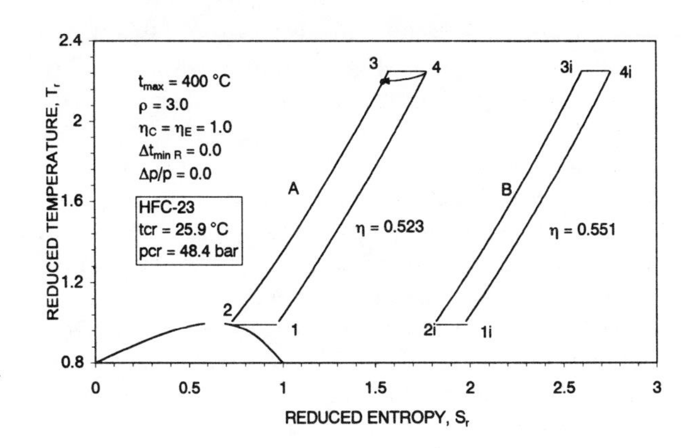

Figure 1 - Configuration of real gas (cycle A) and ideal gas (cycle B) Stirling cycles using ideal components in T_r-S_r plane.

For both cycles volume pressure ratio, maximum and minimum temperatures are the same: $\rho = 3.0$, $T_{r\,min} = 1.01$, $T_{r\,max} = 2.25$). Two main differences are visible in the T_r-S_r plane: the area of cycle A is considerably larger and the regeneration process for the same cycle is not reversible in that the average heat capacity along the low volume isochore is larger than that along the large volume isochore (temperature t_6 instead of t_3 is reached at the end of the regenerative pre-heating). Consequently specific work in cycle A is much larger than in cycle B (68 kJ/kg against 48.4 kJ/kg at t_{min} = 29 °C, t_{max} = 400 °C), but efficiency is somewhat lower (52.3% against 55.1%) owing to heat transfer irreversibility losses mainly within the regenerator. The mechanism which generates the increase in specific work is explained in Figure 2 in which the real gas cycle in the p_r-v_r plane is compared with a theoretical ideal gas cycle having the same limiting temperatures and volumes.

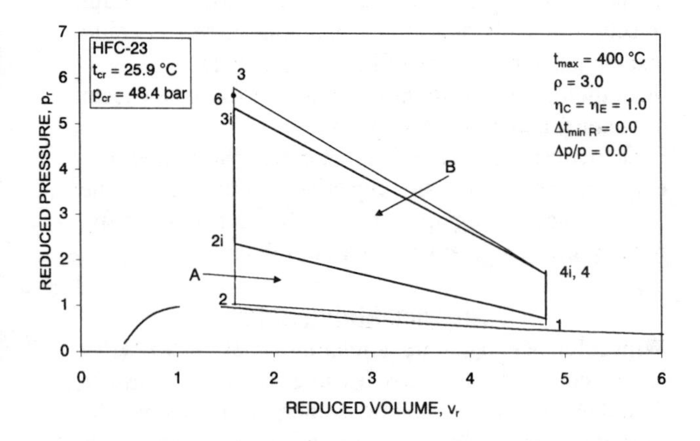

Figure 2 - Real gas (A) and ideal gas (B) Stirling cycles, relying on ideal components, in the p_r-v_r plane.

In the real gas compression process pressure increases only slightly from point 1 to point 2 (from 29.0 bar to 49.8 bar) while under the perfect gas assumption the pressure rise is much steeper (from point 1i at 38.6 bar to point 2i at 115.6 bar). Consequently compression work is much larger in the latter case (39.4 kJ/kg against 24.1 kJ/kg). Real gas expansion work from 3 to 4 is only slightly larger than that of the ideal gas expanding from 3i to 4i (92.2 kJ/kg against 87.8 kJ/kg). Aiming to achieve a more realistic picture of cycles performance a method was developed to take into account the main losses which are typical of a real engine. No attempt was made to describe the detailed operating mechanism of actual reciprocating engines. What we shall define as "real" cycles are in fact still a schematic representation of a theoretical model which includes a number of thermal and fluid-dynamic losses. The following modifications were introduced in the ideal cycles of Figure 1 **[4]**: a) compression and expansion are not isothermal, but involve a certain temperature change. Temperature rise in compression and temperature drop in expansion are assumed to be half that of isentropic changes; i.e. compression takes place between 1 and 7 (or 1i and 7i), Figure 3, and expansion between 3 and 8 (or 3i and 8i).

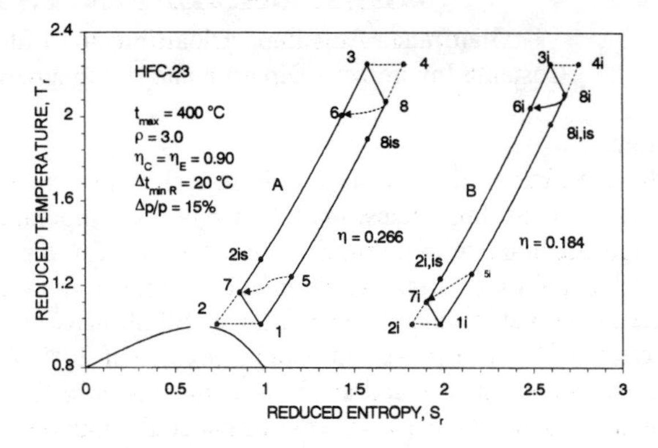

Figure 3 - Configuration of real gas (cycle A) and ideal gas (cycle B) Stirling Cycles in the Tr-Sr plane under assumption of standard internal losses.

Furthermore a compression efficiency η_C is introduced, defined as the ratio of reversible to actual work along the same temperature path and an expansion efficiency η_E is considered similarly defined as actual to reversible work along the same temperature path; b) in the regenerative heat exchange process a finite minimum temperature difference is assumed from cold to hot gas, $\Delta t_{min\ R}$, no matter whether it takes place at the cold or at the hot regenerator end; c) pressure losses are taken into account by concentrating a given $\Delta p/p$ in the low pressure side of regenerator which typically accounts for the most severe loss. Using as working fluid CHF_3 at t_{max} = 400 °C, t_{min} = 29 °C, assuming ρ = 3, $\eta_C = \eta_E = 0.9$, $\Delta t_{min\ R}$ = 20 °C, $\Delta p/p$ = 15% two cycles were computed, one at very low pressure (perfect gas) and a second one at $p_{r1} = 0.6$ (p_1 = 29 bar, real gas). The above component efficiencies were selected in such a way that if applied to usual Stirling engines predict a reasonable overall performance. Resulting conversion efficiency is 18.4% for perfect gas and 26.6% for real gas. The reasons of the better performance of the real gas cycle is explained in Figure 4a and 4b which give the loss analysis according to the entropy production method. The starting input is in both cases 55.1 exergy units corresponding to a Carnot cycle output using 100 units at 400 °C with an ambient at 29 °C: thus the various losses detract from the Carnot cycle performance. **[5]**

Regeneration, owing to both driving temperature differences and pressure drop, is the main reason of loss but is much more severe in the perfect gas cycle (15.8 against 10.0 units). Other surplus losses are found in the compression process (6.2 against 4.9 units, depending on both fluid-dynamic and heat transfer irreversibilities) in the expansion process (7.4 against 7.6 units) and in gas pre-cooling (5.8 against 4.3 units). Only in compressed gas pre-heating after regeneration real gas loss is slightly larger (1.7 against 1.5 units).

The main reason for the greater work losses in the perfect gas cycle is traceable in the fact that similar irreversible processes (for example the transfer of similar amounts of heat under the same Δt), which produce similar absolute work losses are responsible for a greater relative penalty in perfect gas cycles owing to their smaller net work output.

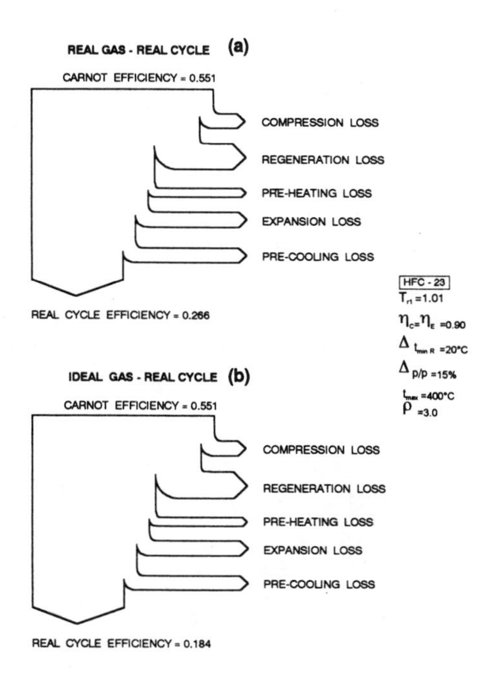

Figure 4 - Losses due to entropy production in real (Figure 4a) and ideal gas (Figure 4b) Stirling cycles. Starting energy input is the exergy of 100 heat unuts at the top temperature of 400 °C with an ambient at 29 °C.

Although in a reduced measure with respect to perfect gas cycles, the performance of real gas Stirling cycles is rather sensitive to fluid-dynamic and thermal irreversibilities.

The influence of expansion and compression efficiency, of pressure drops and of regenerator minimum temperature difference on cycle efficiency is illustrated in Figure 5 a, b and c. In a typical cycle with $\rho = 3$ changing both η_C and η_E from 0.85 to 0.95 produces an efficiency gain of about 35%; a similar increase is obtained by reducing pressure drop $\Delta p/p$ from 25 to 5%.

$\Delta t_{min\ R}$ is only slightly less effective in controlling the overall cycle performance: 25% efficiency gain when it is reduced from 30 to 5 °C.

Working fluids

The main requirements for a working fluid to be suitable for the proposed application are technical and environmental acceptability (i.e. the fluid must not be corrosive or toxic or in some other way dangerous), thermodynamic suitability, basically referable to its critical temperature and thermal stability at temperatures above at least 300 or 350 °C. Typical working fluids for organic power cycles (straight chain and aromatic hydrocarbons, fluoro-substituted hydrocarbons, siloxanes) implying a high temperature evaporation process have critical temperatures by far too high (for example t_{cr} = 197 °C for n-pentane, 319 °C for toluene while values of interest for this application are in the range 30 to 70 °C). Chlorine containing fluids besides being ozone depleting do not exhibit an adequate thermal stability.

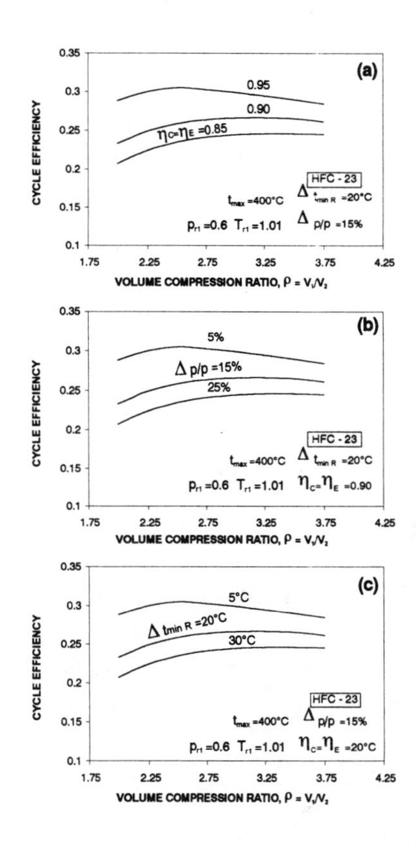

Figure 5 - Effects of components performance in real gas Stirling cycles: influence of compression and expansion efficiencies (Figure 5a); of pressure drops (Figure 5b) and of regenerator minimum temperature difference (Figure 5c).

Some of the new zero ozone depletion potential compounds developed as working fluids for the refrigeration industry have an excellent thermal stability at temperature around 400 °C and suitable critical temperatures. In particular HFC-23 (CHF_3) was experimentally found stable in static tests at about 450 °C **[6]**. Its critical temperature of 26 °C is marginal in that it implies a heat rejection process approximately at temperatures in the range 70-30 °C which could not be easy to perform in a hot environment. A second stable fluid available as a refrigerant is HFC-125 (C_2HF_5) with a critical temperature of 66 °C and typical heat rejection temperatures in the range 70-110 °, which are suitable for cogeneration systems but tend to damage excessively the thermodynamics of power-only cycles.

An effective way of obtaining a working fluid with an optimized critical temperature relying on the limited number of available substances is the use of mixtures. In the case of HFC-23 and HFC-125 [7-9]for instance a binary mixture of varying composition exhibit critical temperatures regularly passing from that of the first fluid to that of the second one (Figure 6). For each specific application requesting heat rejection at an appropriate temperature, an optimum mixture can be selected. Although up to now there is not experimental evidence that mixing does not affect thermal stability, it seems likely that, provided the fluids are of the same family, as in the quoted case, the thermal stability of the mixture is similar to that of the least stable compound. Another potential advantage of mixing is that of making practically usable fluids with unsuited critical temperatures (either too low or too high).

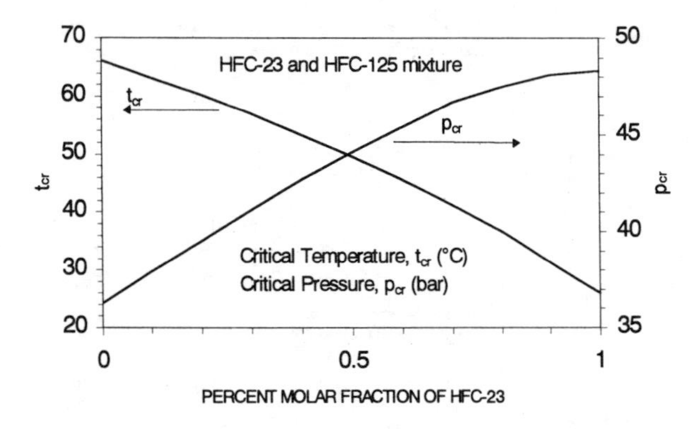

Figure 6 - Effectiveness of fluid mixing in producing a varying critical temperature and pressure.

Although the thermodynamics of mixture is much more complex than that of pure substances, in our case, in which all processes are performed in the gas-phase region and no phase changes are involved, mixture behaviour is similar to that of pure fluids, provided reliable data on critical constants can be obtained through appropriate experimental or computational techniques **[10]**. On the whole, accepting an operational limit of 350 to 450 °C for top temperature it seems that adequate working fluids for the proposed application are available.

Thermodynamic optimization

The main variables which determine the cycle performance for a given fluid are: starting pressure and temperature of the compression process (p_1, t_1 or p_{r1}, T_{r1}), volume compression ratio ρ and top temperature t_3. Top cycle pressure p_3 depends on the previous variables and is significant from a structural point of view. In order to rely on a strong real gas effect, compression starting point 1 must be located in the critical region of the fluid. What does this means in quantitative terms is shown in Figure 7 giving cycle efficiency as a function of p_{r1}, T_{r1} for ρ = 2 and 3 for HFC-23 at T_{max}/T_{min} = 2.23 (approximately t_{max} = 400 °C). Perfect gas cycle efficiency is evaluated at p_{r1} around zero.

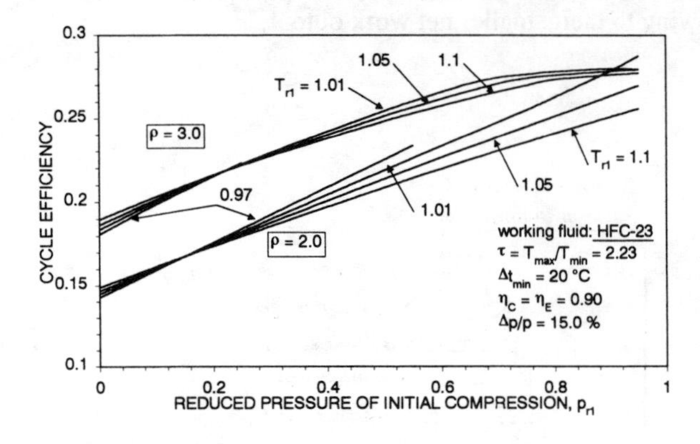

Figure 7 - Influence of compression starting temperature and pressure on real gas cycle efficiency for ρ = 2 and 3.

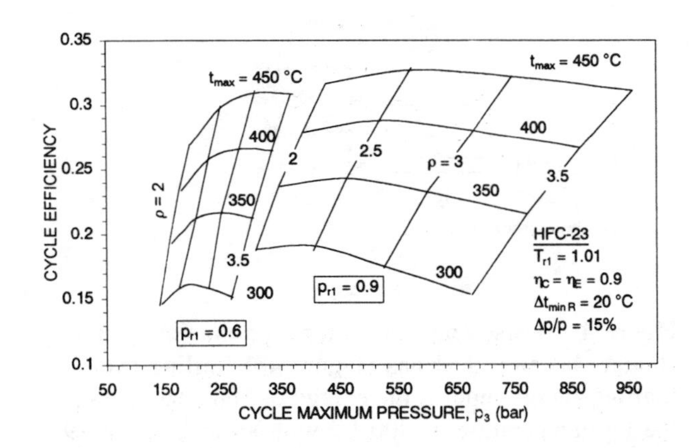

Figure 8 - Real gas cycle efficiency as a function of the maximum pressure for two values of p_{r1} in the case of HFC-23.

The positive real gas effect increases gradually from p_{r1} = 0.0 to p_{r1} around 1.0 (higher values were not considered to avoid excessive top pressure): at T_{r1} = 1.01 and ρ = 3 cycle efficiency is about 18% for perfect gas, 24% at p_{r1} = 0.4 and almost 28% at p_{r1} = 0.9.

The efficiency gain decreases only slightly at higher T_{r1} (on the average one point between p_{r1} = 0.4 and p_{r1} = 0.9 changing from T_{r1} = 1.01 to T_{r1} = 1.1). This means that in the case of HFC-23 there is a 30 °C region of temperature above the critical point where the real gas effect is fully active.

In practical applications maximum cycle pressure is likely to be one of the main limiting factors. Consequently performance graphs as functions of top pressure p_3 are particularly meaningful. In Figure 8 relating to HFC-23 the cycle efficiency is reported as a function of p_3 for temperatures in the range 300-450 °C, ρ in the range 2-3.5 and for two values of p_{r1} (0.6 and 0.9). The inspection of Figure 8 suggests the following comments: 1) at the highest temperature (i.e.

450 °C) the best cycle efficiencies at $p_{r1} = 0.6$ or $p_{r1} = 0.9$ are similar and around 30-32%, however smaller ρ are required at $p_r = 0.9$; 2) at low temperature (i.e. 300 °C) cycles at $p_{r1} = 0.9$ are superior; 3) top pressures at $p_{r1} = 0.9$ are very high, in excess of 300 bar for ρ = 2 even at the lowest temperatures. In order to limit p_3 also p_{r1} must be limited.

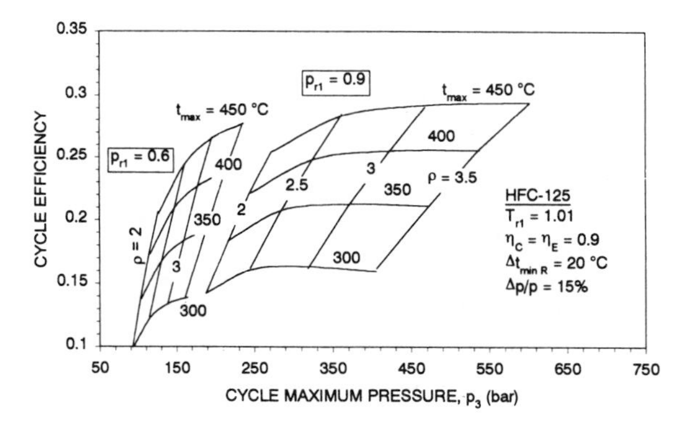

Figure 9 - Real gas cycle efficiency as a function of maximum pressure for two values of p_{r1} in the case of HFC-125.

A similar diagram, but relating to HFC-125 is given in Figure 9. The superiority of cycles at $p_{r1} = 0.6$ at low p_3 is even more evident. The higher minimum temperature (69.6 °C) reduces somewhat all cycle efficiencies. At t_{max} = 400 °C, p_{max} = 200 bar, η is about 23%.However waste heat is now available at temperatures at which it is fully usable in a cogeneration system. The largest fraction of waste heat is transferred to the ambient between t_5 and t_1 (Figure 3) in the pre-cooling process, the remainder in the compression process between t_1 and t_7.

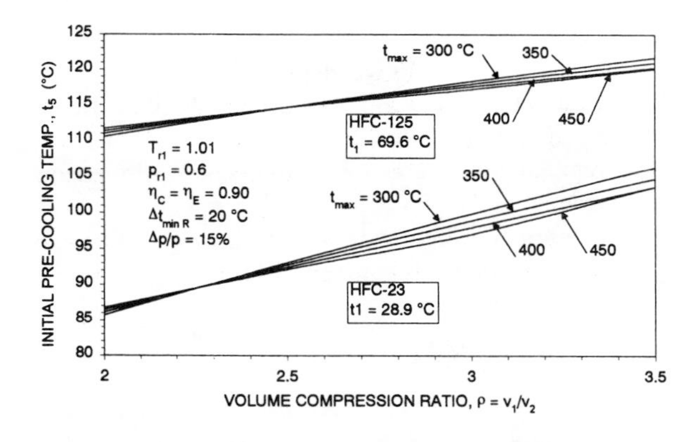

Figure 10 - Initial temperature of the pre-cooling process as a funtion of the volume compression ratio for HFC-23 and HFC-125.

To figure out which temperatures are involved in a heat and power system or in the waste heat rejection process Figure 10 reports t_5 for both HFC-125 and HFC-23 as a function of ρ at different t_{max}. With reference to HFC-125 at ρ = 3.0 for example a large portion of heat is available between 118 and 70 °C. In the case of HFC-23 t_5 is important for heat rejection technical implications rather than for cogeneration (waste heat temperatures are only marginally useful). At ρ = 3, t_{max} = 400 °C for instance pre-cooling heat must be dumped in the ambient between 98 and 29 °C at a mean temperature that simplifies heat transfer problems even if the cooling medium is air. In the previous examples the influence of the nature of the working fluid on cycle performance is not evident, being concealed by the more important effect of t_{cr}. In order to isolate this aspect of the question in Figure 11 cycle efficiency is reported as a function of ρ at a fixed τ for a number of fluids of increasing molecular complexity (CO_2, CHF_3, SF_6 and C_2HF_5).

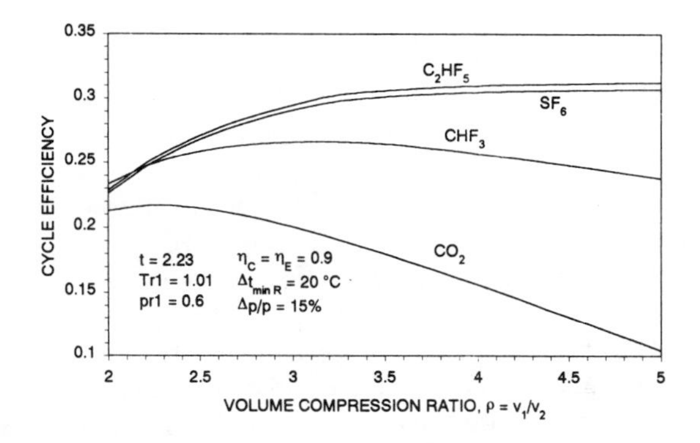

Figure 11 - Influence of working fluid molecular structure on cycle efficiency.

As a general rule fluids with a relatively simple molecule exhibit the optimum efficiency at low ρ (2.4 for CO_2, 3 for CHF_3) while fluids with more complex molecules require higher compression ratios (4 to 5). Furthermore for standard losses levels the best cycle efficiency increases with the molecular complexity (0.22 for CO_2, 0.26 for CHF_3, 0.30 for SF_6 and C_2HF_5).

Examples of potential applications

Some examples of possible cycle configurations of practical interest will be considered in more details. Resorting to HFC-23 as working fluid, limiting top pressure to 230 bar, with standard levels of losses ($\eta_C = \eta_E = 0.9$; Δp/p = 0.15; $\Delta t_{min R}$ = 20 °C) at a minimum temperature of 29 °C, a maximum temperature of 400 °C and ρ = 2.5, cycle efficiency turns out to be 25.9% (perfect gas efficiency 17.3%). The most important fraction of waste heat is transferred to the ambient between 93 and 29 °C, the remaining portion is rejected during compression between 67 and 29 °C. These temperatures could allow a limited operation of the conversion cycle in a cogenerative mode. Assuming an improved set of losses ($\eta_C = \eta_E = 0.92$; Δp/p = 0.10; $\Delta t_{min R}$ = 15 °C) cycle efficiency rises to 31.7% with the same t_7 and a slightly reduced t_5 (82 °C instead of 93). If, in addition, temperature is increased to 430 °C efficiency improves to 34.2%. On the contrary with standard losses but with tmax = 350 °C

efficiency drops to 21.3% with a maximum pressure of 207 bar.
With reference to a combined heat and power plant HFC-125 with its critical temperature of 66.25 °C is selected. At a standard level of losses, $\rho = 3$, $t_{max} = 400$ °C maximum pressure is 179 bar and efficiency is 22.7% (13.4% perfect gas efficiency). Somewhat more than half of the waste heat is transferred to the ambient at temperatures between 118 and 69 °C (minimum cycle temperature), the remainder between 96 and 69 °C during compression. Assuming the previously quoted improved set of losses efficiency increases to 27.5% (t_5 = 111 °C). If, in addition, temperature is raised to 430 °C cycle efficiency reaches 29.9%.
All the data reported relate to a minimum reduced temperature of cycle of 1.01. As previously illustrated efficiency is only slightly penalized for T_{r1} up to 1.1. This means that minimum cycle temperature could increase up to 56 °C for HFC-23 and to 100 °C for HFC-125 with a small efficiency penalty provided the same τ is maintained. Such flexibility could be of use either as a design tool or for off-design operation (ambient temperature higher than assumed, increased temperature request for cogenerated heat).
The use of mixtures, however, is even more effective in allowing the tailoring of t_{min}. Each composition determines, above its critical temperature, an interval for the minimum temperature variation of about 0.1 T_{cr}. Since temperatures involved in organic real gas Stirling cycles are below 400-450 °C, carbon steel seems an appropriate building material even for high temperature components. However the large pressures needed call for demanding mechanical solutions.

Conclusions

The results of the analysis carried out in the previous sections justify the following conclusions:

1) real gas effects in the critical region during the compression process, although theoretically detrimental in ideal cycles, give important efficiency gains in real Stirling cycles mainly owing to a diminished compression work;
2) in order to make the critical region accessible to the compression process, the working fluid critical temperature must have a well defined value: only organic fluids offer a wide selection of critical temperatures for different technical applications;
3) owing to the limited thermal stability of organic substances only cycles at a top temperature in the range of 350-450 °C are feasible: consequently real gas Stirling cycles have an operating temperature range more similar to that of vapour Rankine cycles than to that of gas cycles;
4) the good energy performance of real gas cycles is obtained at the cost of a high maximum cycle pressure, in the range of at least 150-200 bar;
5) both power only and combined heat and power cycles can be obtained with commercially available working fluids;
6) two basic procedures can be followed to obtain an about optimum real gas effect at a given minimum temperature: the mixing of fluids with different critical temperatures and the variation of the starting compression temperature above the critical point witin an interval of about 0.1 T_{cr};
7) the moderate thermal stability of organic substances if on one side limits the absolute cycle performance on the other has a positive influence on material problems: ordinary carbon steel seems an adequate choice even for the hottest engine parts.

Nomenclature

p	pressure, bar
S	specific entropy (J/kg K)
S_r	reduced entropy (= $(S-S_{sat\ liq\ Tr=0.8})/\Delta S_{evap\ Tr=0.8}$)
T	absolute temperature, K
t	temperature, °C
v	volume, m^3/kg
Δp	cycle pressure loss, bar
Δt	temperature difference, °C
η	efficiency
ρ	volume compression ratio
τ	ratio of maximum to minimum cycle temperature

$()_{1\div 8}$	points of real gas cycle (Figure 3, cycle A)
$()_{1i\div 8i}$	points of ideal gas cycle (Figure 3, cycle B)
$()_{is}$	isentropic conditions
$()_{max}$	maximum conditions
$()_{min}$	minimum conditions
$()_C$	compressor
$()_E$	expansion
$()_{sat}$	working fluid in saturated conditions
$()_{liq}$	liquid
$()_{cr}$	identifies physical parameters at the critical point of the working fluid
$()_r$	reduced conditions ($T_r = T/T_{cr}$; $p_r = p/p_{cr}$; $v_r = v/v_{cr}$)

References

1 Miller D. R., and alii, "Optimum working fluids for automotive Rankine engines", APTD-1565, prepared for U.S. Environmental Protection Agency, 1973.

2 Angelino G., Invernizzi C., Macchi E., "Organic working fluid optimization for space power cycles", in Modern research topics in aerospace propulsion, Springer-Verlag, 1991, 297-326.

3 Calderazzi L., Colonna P., "Thermal stability of R-134a, R-141b, R-13I1, R-7146, R-125 associated with stainless steel as a containing material", Int. J. Refrig., 1997, 20, 6, 381-389.

4 Angelino G., Invernizzi C., "Potential Performance of Real Gas Stirling Cycle Heat Pumps", Int. J., Refrig., Vol. 19, No. 6, 390-399.

5 Angelino G., Invernizzi C., "Macchina Stirling a rendimento migliorato", in Italian, patent No. 01284448, 1998.

6 Galas E., Vasconi A., "Experimental and numerical investigation on organic working fluids

for Brayton cycles", graduation thesis, in Italian, Politecnico di Milano, 1998.

7 McLinden M. O., "Thermodynamic properties of CFC alternatives: A survey of the available data", Int. J. of Refrig., 1990, Vol. 13, May, pp. 149-162.

8 Reid R. C., Prausnitz J. M., Poling B. E., "The Properties of Gases & Liquids", Fourth Edition, McGraw-Hill International Editions, 1988.

9 AA VV, "1997 ASHRAE HANDBOOK - Fundamentals", Chapter 19, American Society of Heating, Refrigerating and Air-Conditioning Engineers, Inc.

10 Fromm M., "Calculation of Mixture Critical Points as a Tool to Optimize Thermodynamic Power Cycles", Studienarbeit, 1996 (Politecnico di Milano - Stuttgart University).

AIAA-2000-2812

UPDATE ON THE EVALUATION OF DIFFERENT CORRELATIONS FOR THE FLOW FRICTION FACTOR AND HEAT TRANSFER OF STIRLING ENGINE REGENERATORS

Bernd Thomas, Deborah Pittman
Reutlingen University, University of Applied Sciences
Alteburgstr. 150, 72762 Reutlingen, Germany
Phone: +49 7121 271-362
Fax: +49 7121 271-530
Email: bernd.thomas@fh-reutlingen.de

Abstract

This paper is meant to update the overview of different correlations for the friction factor and heat transfer of Stirling engine regenerators given in references[1,2] by the Oscillating-Flow Regenerator Test Rig data collected by Gedeon and Wood[3]. For that purpose the Gedeon/Wood data was recalculated using the general applicable set of equations for Reynolds number, friction factor, pressure drop, Nusselt number and heat transfer area derived in references[1,2]. Based on these general definitions, the correlations can be shown in one plot for the flow friction factor and one plot for the Nusselt number, giving a good impression of their characteristics and the deviations among them, which are discussed in the body of this paper.

Introduction

The regenerator is the most crucial part when designing Stirling engines or other regenerative machines. For that reason, various experimental data dealing with fluid flow friction and heat transfer in regenerator matrices have been published. Two papers have already been presented to summarize and evaluate different correlations for flow friction factor[1] and heat transfer[2] of Stirling engine regenerators. Because the first two papers only cover a selection of correlations published, i.e. by Tong/London[4], Blass[5], Miyabe et.al.[6], Tanaka[7] et.al., ILK Dresden[8] and Kühl[8], this paper is an update, including the Oscillating-Flow Regenerator Test Rig data presented by Gedeon and Wood[3]. This data, collected under contract of NASA Glenn Research Center (formerly NASA Lewis), is a significant contribution, since the flow friction and heat transfer data was collected under oscillating flow conditions applying a wide range of parameters including Reynolds number, Valensi number, flush ratio, wire diameter, mesh width and porosity. In addition, Gedeon and Wood used at least two working gases - helium and nitrogen - in order to view the effects of the gas properties.

Gedeon and Wood derived correlations for the flow friction factor and Nusselt number for both wire screens and metal felts. The geometric constraints and the governing equations of these correlations are summarized in Table 1. The following paragraphs discuss in greater detail how these correlations were recalculated using a general applicable set of equations for Reynolds number, friction factor, pressure drop, Nusselt number and heat transfer area[1,2]. This is necessary to directly compare the friction factor and Nusselt number data to the correlations investigated previously. Since Gedeon and Wood distinguish between wire screens and metal felts (also termed random wire material), the correlations are separated in the same way, yielding the following wire screen and metal felt correlations:

Wire screen: Gedeon/Wood, Tong/London, Blass (friction factor only), Miyabe, Tanaka

Metal Felts: Gedeon/Wood, ILK Dresden, Kühl

It should be noted, that the metal felt correlations given by ILK Dresden and Kühl are based on the same set of experimental data collected by ILK Dresden.

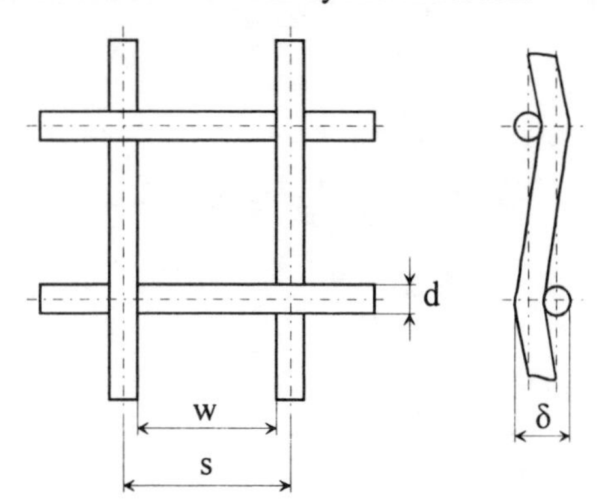

Figure 1 Geometric parameters of wire screen

Table 1: Geometric constraints and governing equations of friction factor and heat transfer correlations given by Gedeon and Wood[3]

	Friction factor correlation			**Nusselt number correlation**	
Matrix material	Wire screens	Metal felts	Matrix material	Wire screens	Metal felts
Number of samples	3	5	Number of samples	3	5
Wire diameter range	0.0533 to 0.094 mm	0.0127 to 0.0508 mm	Wire diameter range	0.0533 to 0.094 mm	0.0127 to 0.0508 mm
Porosity range	0.6232 to 0.781	0.688 to 0.8405	Porosity range	0.6232 to 0.781	0.688 to 0.8405
Re_{max} number range	1.04 to 3400	0.79 to 1400	Re_{max} number range	1.04 to 3400	0.79 to 1400
Valensi number range	0.0052 to 21	0.0021 to 5.6	Valensi number range	0.0048 to 16	0.0037 to 3.3
Range of tidal amplitude ratio (= Flush ratio)	0.028 to 2.2	0.043 to 2.6	Range of tidal amplitude ratio (= Flush ratio)	0.17 to 3.0	0.17 to 3.8
Equation for porosity	$\varepsilon = 1 - \frac{V_{mat}}{V_{tot}}$	$\varepsilon = 1 - \frac{V_{mat}}{V_{tot}}$	Equation for porosity	$\varepsilon = 1 - \frac{V_{mat}}{V_{tot}}$	$\varepsilon = 1 - \frac{V_{mat}}{V_{tot}}$
Equation for flow velocity	$u = \frac{u_0}{\varepsilon}$	$u = \frac{u_0}{\varepsilon}$	Equation for flow velocity	$u = \frac{u_0}{\varepsilon}$	$u = \frac{u_0}{\varepsilon}$
Equation for Reynolds number	$Re = \frac{d \varepsilon u}{(1-\varepsilon)\nu}$	$Re = \frac{d \varepsilon u}{(1-\varepsilon)\nu}$	Equation for Reynolds number	$Re = \frac{d \varepsilon u}{(1-\varepsilon)\nu}$	$Re = \frac{d \varepsilon u}{(1-\varepsilon)\nu}$
Equation for pressure drop	$\Delta p = c_F \frac{L(1-\varepsilon)}{d\varepsilon} \frac{\rho}{2} u^2$	$\Delta p = c_F \frac{L(1-\varepsilon)}{d\varepsilon} \frac{\rho}{2} u^2$	Equation for specific heat transfer area	$\varphi = \frac{A}{V_{tot}} = \frac{4}{d}(1-\varepsilon)$	$\varphi = \frac{A}{V_{tot}} = \frac{4}{d}(1-\varepsilon)$
Friction factor correlation	$c_F = \frac{129}{Re} + \frac{2.91}{Re^{0.103}}$	$c_F = \frac{192}{Re} + \frac{4.53}{Re^{0.067}}$	Equation for Nusselt number	$Nu = \frac{\alpha d \varepsilon}{\lambda(1-\varepsilon)}$	$Nu = \frac{\alpha d \varepsilon}{\lambda(1-\varepsilon)}$
			Nusselt number correlation	$Nu = \left(1 + 0.99 (Re \cdot Pr)^{0.66}\right) \cdot \varepsilon^{1.79}$	$Nu = \left(1 + 1.16 (Re \cdot Pr)^{0.66}\right) \cdot \varepsilon^{2.61}$
			Overall thermal regenerator loss	$\frac{\dot{Q}_{Reg.-loss}}{\lambda A_{free} \Delta T} = \frac{0.194 (Re_{max} \cdot Pr)^{1.30}}{\varepsilon^{1.81}}$	$\frac{\dot{Q}_{Reg.-loss}}{\lambda A_{free} \Delta T} = \frac{0.253 (Re_{max} \cdot Pr)^{1.24}}{\varepsilon^{2.67}}$

Experiments

The Oscillating-Flow Regenerator Test Rig was specially designed to analyze regenerator pressure drop and heat transfer for oscillating flow. For that purpose a motor-driven piston-cylinder assembly was installed to push gas back and forth through the regenerator sample. The friction factor was calculated from the piston pV-power, which was measured directly. To measure heat transfer, the regenerator sample was inserted between a heater and cooler in order to impose an axial temperature gradient along the regenerator. The net axial energy flux, caused by the thermal imperfections in the regenerator, was calculated using the measured heat rejection in the cooler (minus static conduction).

Thus, in contrast to the other correlations evaluated, the data given by Gedeon and Wood is based on the energy losses such as piston pV-dissipation and net axial energy flux along the regenerator. All other correlations are based on pressure drop and temperature measurements. This is an important difference, which must be kept in mind when comparing the data sets.

Furthermore, Gedeon and Wood varied the flush ratio over a wide range in order to include its effect in the correlations (see Table 1). The flush ratio indicates how much gas in terms of mass flow amplitude is pushed through the regenerator compared to the mass of gas in the void volume of the regenerator. For example, a flush ratio of 1 means that exactly every gas particle in the regenerator is exchanged one to one by the incoming mass flow. However, the friction factor and Nusselt number correlations derived by Gedeon and Wood are not dramatically affected by the flush ratio, which indicates that its effect is of minor importance.

Figures 2 and 3 show the ranges of wire diameter and porosity investigated by the different authors. It is important to keep in mind, that Kühl used the

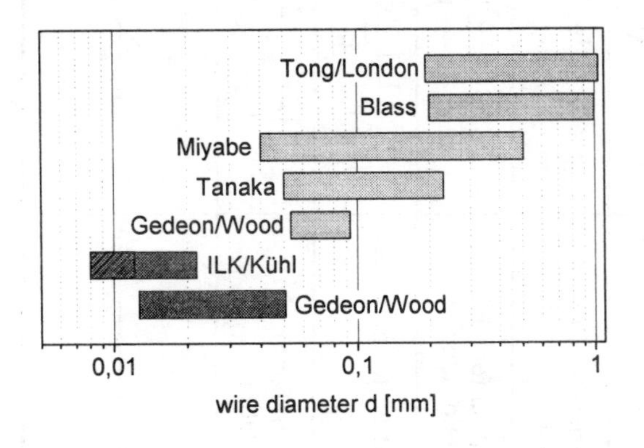

Figure 2 Ranges of wire diameter investigated
(shaded areas: pressure drop experiments only)

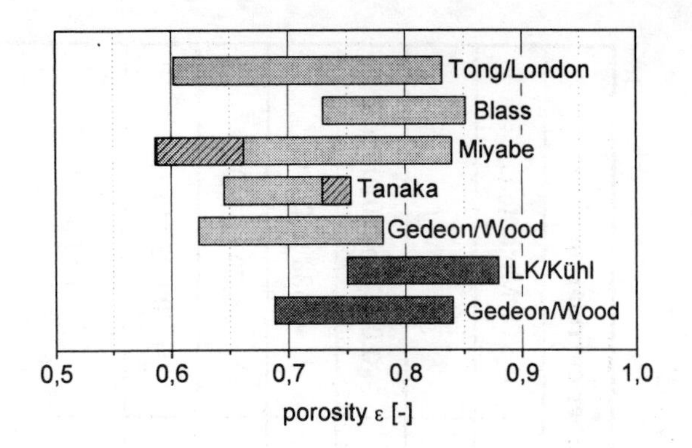

Figure 3 Ranges of porosity investigated
(shaded areas: pressure drop experiments only)

experimental data collected by ILK Dresden. The light gray bars indicate the wire screen data and the dark gray bars refer to metal felts.

The small range of wire diameter Gedeon and Wood applied to their wire screen experiments does not affect the significance of their correlations, since the effect of wire diameter on friction factor and Nusselt number is fully defined by the hydraulic diameter. It is more important to apply a wide range of porosity especially when heat transfer is affected, because porosity shows an additional effect on heat transfer area as the investigations carried out so far reveal[2].

Friction factor correlations

The analysis in reference[1] indicated that the open mesh width w (see Figure 1) is appropriate to serve as characteristic length for the evaluation of the friction factor in the regenerator. Consequently, Reynolds number and flow velocity in the matrix are based on this parameter:

$$Re = \frac{w \cdot u}{\nu} \qquad u = u_0 \frac{s^2}{w^2} \tag{1), (2}$$

For metal felts, an equivalent open mesh width can be derived from wire diameter and porosity. Treating the felts as cross rod matrices and setting each layer thickness δ to 2d yields:

$$w = d\left(\frac{1}{\sqrt{\sqrt{0.25 + 4(1-\varepsilon)^2/\pi} - 0.5}} - 1\right) \tag{3}$$

The pressure drop in the matrix is caused by two physical factors: form drag and skin friction. Therefore, their effect on the resulting equations for pressure drop and friction factor can be seen in the following:

$$\Delta p = c_F n \frac{\rho}{2} u^2 \qquad c_F = C_{fd} + \frac{C_{sf}}{Re} \qquad (4), (5)$$

The pressure drop is proportional to the number of "flow resistors" n in the matrix, given by the number of screens or by the number of layers L/2d, if the matrix is made of metal felts.

It has been shown that this set of equations (1-5) is generally applicable to pressure drop data of regenerator matrices regardless of whether wire screens or metal felts are used[1]. Therefore, the correlations given by Gedeon and Wood were recalculated in order to fit their data to the proposed equations. This was necessary to compare the different correlations. Although the resulting data must be compared separately for wire screens and for metal felts, the friction factors of the two types of matrices can be compared easily, since the diagrams are plotted using the same scale.

Wire screens

Figure 4 shows the friction factors according to the equations given above as a function of Reynolds number for wire screens. It can be seen that Gedeon and Wood predict the highest friction factors. In the region of small Reynolds numbers, the friction factors differ by a factor of approximately 2. For high Reynolds numbers, the data given by Tanaka and Blass approaches the Gedeon and Wood correlation. Miyabe predicts the smallest friction factors over the entire range of Reynolds numbers.

The line graph in Figure 4 shows the results of the friction factor correlation according to equation (5) for Gedeon and Wood's data. The two correlation constants C_{fd} and C_{sf} are given below:

Gedeon/Wood

$C_{sf} = 68.556 \qquad C_{fd} = 0.5274$

In order to complete the analysis presented in this paper, the correlation constants for the other data sets are repeated from reference[1]:[†]

Tong/London

$C_{sf} = 44.710 \qquad C_{fd} = 0.3243$

Blass

$C_{sf} = 47.245 \qquad C_{fd} = 0.4892$

Miyabe

$C_{sf} = 33.603 \qquad C_{fd} = 0.3370$

Tanaka

$C_{sf} = 40.7413 \qquad C_{fd} = 0.5315$

[†] The correlation constants differ slightly from the values given in reference[1]. This is due to an improved fitting function based on the least square fit.

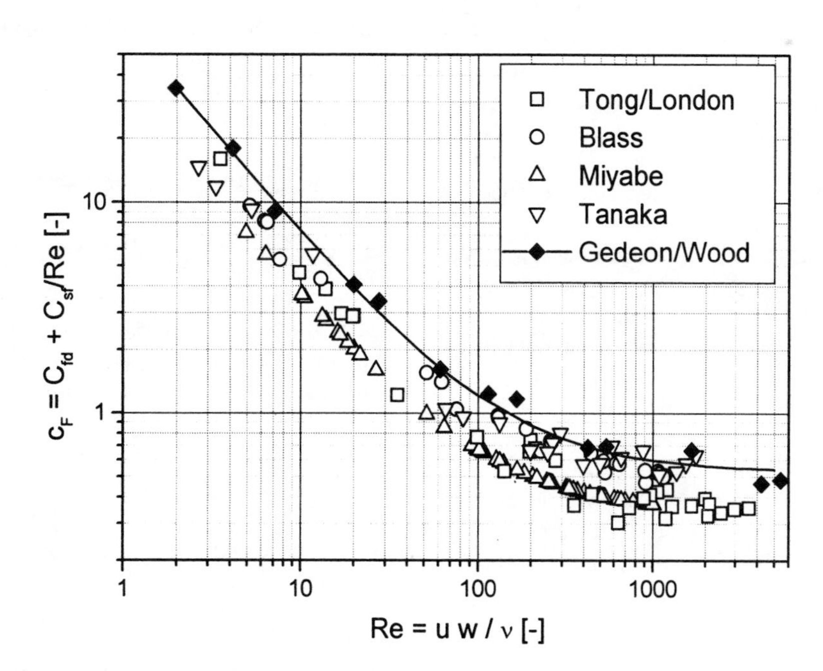

Figure 4 Friction factor according to the general applicable set of equations vs. Reynolds number for wire screens

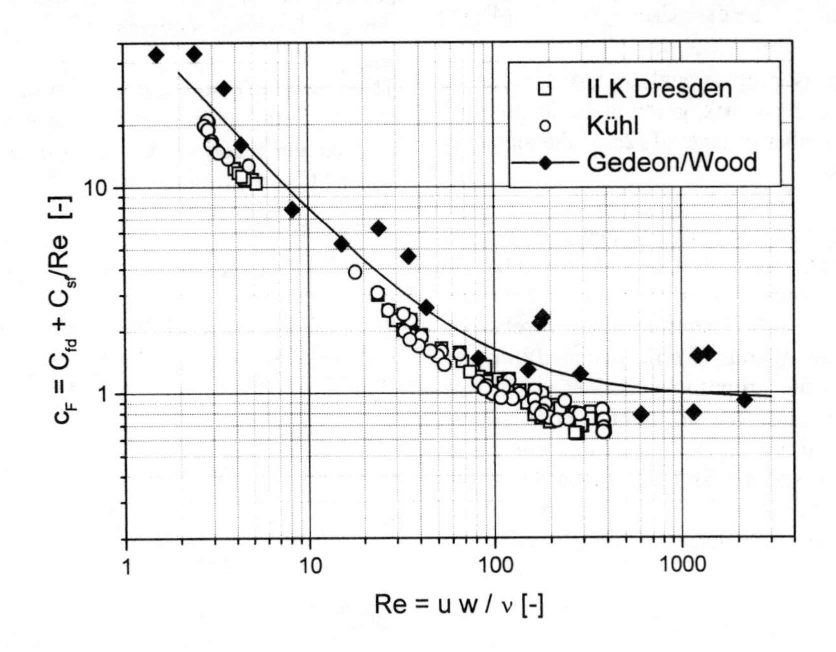

Figure 5 Friction factor according to the general applicable set of equations vs. Reynolds number for metal felts

Metal felts

The friction factors for metal felts are presented in Figure 5. There is a marked scattering of the recalculated Gedeon/Wood friction factors due mainly to the broad dispersion of the original data. Gedeon and Wood obtained a maximum pressure drop error of 27%, using their metal felt friction factor correlation as given in Table 1. The maximum pressure drop error for wire screens is much smaller (10%). Looking at porosity and wire diameter, the scattering of the data offers no obvious trend, which would reveal an additional effect not covered by the correlation constants. Gedeon and Wood tried to include an additional porosity term in their correlation, but its effect was minor since the pressure drop error decreased only marginally. Due to the fact that all other data sets are well correlated, more experimental friction factor data for metal felts in an oscillating flow must be analyzed in order to prove the scattering of the Gedeon/Wood data and to derive an appropriate correlation.

However, the friction factor correlation according to equation (5) can be applied to the Gedeon/Wood data for metal felts and the correlation constants C_{fd} and C_{sf} follow below:

Gedeon/Wood

$C_{sf} = 70.035$ $C_{fd} = 0.9307$

Additionally, the constants for the data sets of ILK Dresden and Kühl are repeated:

ILK Dresden

$C_{sf} = 49.465$ $C_{fd} = 0.5726$

Kühl

$C_{sf} = 50.379$ $C_{fd} = 0.5747$

Figure 5 shows higher friction factors when using the correlation given by Gedeon and Wood. In comparison the ILK Dresden data is approximately 1.5 times smaller.

Discussion of the friction factor results

It is obvious that the correlations for the Gedeon/Wood data predict the highest friction factors for both wire screens and metal felts. This behavior can be attributed to the oscillating flow, even though the Tanaka data, which was also obtained under oscillating flow conditions, does not support this statement for small Reynolds numbers (see Figure 4 and constants C_{sf}). However, it has been outlined before that Gedeon and Wood measured the pV-power dissipated by the fluid flow in the regenerator instead of the pressure drop itself. This may be another reason for the higher friction factor results. Evidently, the recalculation of the friction factor from the dissipated pV-power causes greater errors than using the steady flow pressure drop data. This may explain the larger scatter in Gedeon and Wood's friction factor results compared to others. For that reason, the steady flow friction factor results should be more accurate, even though Gedeon and

Wood's correlations directly refer to the term of greatest interest in a simulation code - the power dissipation in the regenerator.

Finally, the friction factor results for wire screens and metal felts were compared. Looking at the correlation constants and Figures 4 and 5, which are plotted using the same scale, it can be seen that the friction factors of metal felts are greater than those for wire screens, especially for higher Reynolds numbers. This yields a higher form drag for metal felts than for wire screens, which can be related to its random structure compared to the regular structure of wire screens. Therefore wire screens should be preferred for minimizing regenerator pressure drop.

Nusselt number correlations

The general applicable set of equations for the analysis of the regenerator heat transfer was derived in reference[2]. It was outlined that the hydraulic diameter is appropriate to define Reynolds and Nusselt numbers as follows:

$$d_{hy} = d \cdot \frac{\varepsilon}{1-\varepsilon} \qquad u = \frac{u_0}{\varepsilon} \tag{6), (7}$$

$$Re = \frac{u \cdot d_{hy}}{\upsilon} \qquad Nu = \frac{\alpha \cdot d_{hy}}{\lambda} \tag{8), (9}$$

Using these equations and neglecting the effect of Prandtl number, which can always be approximated as 0.7 for the gases used as working fluids in Stirling machines, the Nusselt number can be written as a function of Reynolds number in the following three-parameter form:

$$Nu = C_1 + C_2 Re^{C_3} \tag{10}$$

When analyzing heat transfer, the heat transfer area must be specified as well. For regenerator matrices it was found that the net heat transfer area is smaller than the gross surface area of the wires, because the contact areas between the wires are not active for heat transfer[2]. The reduction can be attributed to porosity, and the specific net heat transfer area is given by:

$$\varphi_{net} = \varphi\left(1 - C_4(1-\varepsilon)\right) \tag{11}$$

For practical reasons it is convenient to refer the Nusselt number to the gross surface area, which can be easily determined. Thus, defining Nu_{tot} as

$$Nu_{tot} = \frac{Nu \cdot \varphi_{net}}{\varphi} \tag{12}$$

yields a "total" Nusselt number, which includes the additional effect of porosity:

$$Nu_{tot} = \left(C_1 + C_2 Re^{C_3}\right) \cdot \left(1 - C_4(1-\varepsilon)\right) \tag{13}$$

Gedeon and Wood also observed the additional effect of porosity and attributed it to their Nusselt number equation,

$$Nu = \left(1 + a \cdot (Re \cdot Pr)^b\right) \cdot \varepsilon^c \tag{14}$$

which is based on the gross specific heat transfer area as well (see Table 1). Also, Gedeon and Wood used the same equations for hydraulic diameter, Reynolds number and Nusselt number as given in equations (6-9). Thus, their correlations can be directly transferred to the form of equation (13). Using Pr = 0.7, the correlations constants a, b, c become C_1, C_2, C_3, C_4 as follows:

Gedeon/Wood (wire screens)

$$Nu_{tot} = \left(1.010 + 0.790 \cdot Re^{0.662}\right) \cdot \left(1 - 0.845 \cdot (1-\varepsilon)\right)$$

Gedeon/Wood (metal felts)

$$Nu_{tot} = \left(0.972 + 0.892 \cdot Re^{0.656}\right) \cdot \left(1 - 1.404 \cdot (1-\varepsilon)\right)$$

Wire screens

The correlations resulting from the recalculation of the other wire screen data sets are repeated below from reference[2]:

Tong/London‡

$$Nu_{tot} = \left(0.384 + 1.075 \cdot Re^{0.578}\right) \cdot \left(1 - 1.356 \cdot (1-\varepsilon)\right)$$

Miyabe

$$Nu_{tot} = \left(0.036 + 1.150 \cdot Re^{0.562}\right) \cdot \left(1 - 1.333 \cdot (1-\varepsilon)\right)$$

Tanaka

$$Nu_{tot} = \left(-0.001 + 0.330 \cdot Re^{0.670}\right) \cdot \left(1 + 0.0002 \cdot (1-\varepsilon)\right)$$

It can be seen that the correlation for the data given by Tanaka does not exhibit the additional effect of porosity. This was discussed in reference[2] and two potential reasons for this are the oscillating flow and the small range of porosity investigated experimentally (see Figure 3). Since the oscillating flow data given by Gedeon and Wood does show the additional effect of porosity, the first possible reason can be omitted. Thus, the small range of porosity investigated is the most likely reason for the absence of an additional effect on the Nusselt number correlation given by Tanaka. This should be kept in mind when using Tanaka's correlation.

‡ The correlation constants differ slightly from the values given in reference[2]. This is due to an improved fitting function based on the least square fit.

Figure 6 shows the Nusselt number according to equation (13) as a function of Reynolds number. In order to omit the additional effect of porosity, its value is fixed at 0.75 with respect to the experimental ranges covered by the authors (see Figure 3).

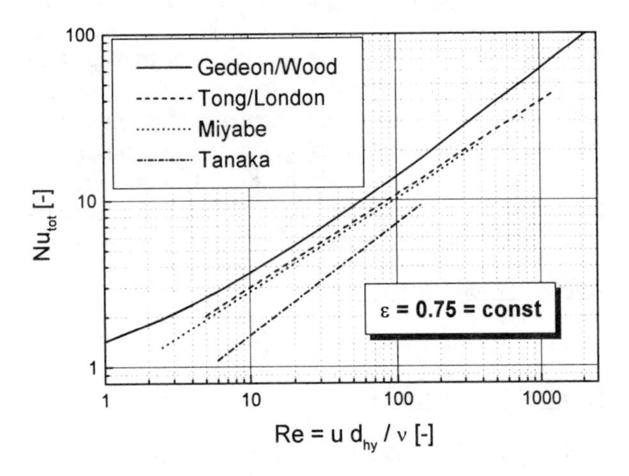

Figure 6 Nu_{tot}-correlations based on equation (13) versus Reynolds number for wire screens and a porosity of 0.75

It can be seen that the Gedeon/Wood correlation predicts the highest Nusselt numbers. The deviations between the data given by Tong/London and Miyabe are small, and the Nusselt numbers calculated from Tanaka's correlation are considerably smaller than the others. For a Reynolds number of 10, the factor between highest and lowest Nusselt number is approximately 2.5. The order of the correlations can also be observed for porosities other than 0.75, even though Tanaka's correlation does not show the additional effect of porosity as outlined above.

Metal felts

The Nusselt number correlations for metal felts are plotted in Figure 7 as a function of Reynolds number. Again, the porosity is fixed at a value of 0.75 with respect to the experimental ranges covered by the authors.

The correlations for the data given by ILK Dresden and Kühl are given follows[2]:

ILK Dresden

$$Nu_{tot} = \left(0.098 + 0.627 \cdot Re^{0.702}\right) \cdot \left(1 - 2.010 \cdot (1 - \varepsilon)\right)$$

Kühl

$$Nu_{tot} = \left(1.190 + 0.410 \cdot Re^{0.779}\right) \cdot \left(1 + 2.154 \cdot (1 - \varepsilon)\right)$$

Figure 7 shows that the highest Nusselt numbers occur when using the correlation given by Gedeon and Wood. Their prediction exceeds the values calculated from the ILK Dresden or Kühl correlation for small Reynolds numbers by a factor of approximately 2. These deviations, however, decrease with increasing Reynolds numbers . This result is only slightly dependent upon porosity, which was also observed for wire screens.

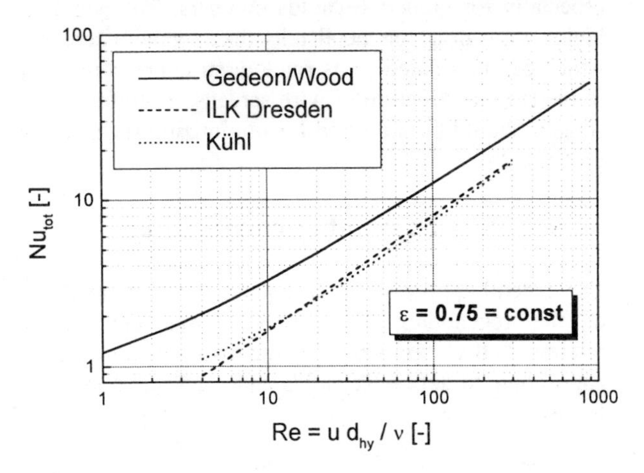

Figure 7 Nu_{tot}-correlations based on equation (13) versus Reynolds number for metal felts and a porosity of 0.75

When comparing the heat transfer between the two matrix materials, it can be seen from Figures 6 and 7 that wire screens show a slightly better heat transfer than metal felts.

Discussion of the Nusselt number results

When comparing the oscillating flow data to the steady flow correlations, the different experimental methods must be addressed. As outlined above, Gedeon and Wood measured the net axial heat flux along the regenerator as its overall thermal loss. From this loss they calculated the fraction caused by the limited heat transfer resulting in their Nusselt number correlation. The other part of the regenerator loss was attributed to an enhanced axial conduction. An analysis of the enhanced axial conduction loss, which is also correlated by Gedeon and Wood[3], revealed that it can be reduced to the temperature swing loss in combination with the loss caused by axial flow dispersion[9], speaking in terms of 2nd order regenerator loss inventory.

Looking at the steady flow experiments, the heat transfer is measured by applying a temperature step function at the regenerator entrance and examining the temperature signal at the exit as a function of time. From this function, the number of transfer units (NTU) of the regenerator is calculated, which directly leads to the heat transfer coefficient and the Nusselt number. When analyzing this experimental method, it is obvious that losses other than the heat transfer loss are also present, such as the temperature swing loss and the dispersion loss. Consequently, these losses are incorporated into the Nusselt number, since no separation of the different loss mechanisms is applied. Using another experimental method, Tanaka also did not separate the different regenerator losses. He attributed the total loss, which was determined experimentally from instantaneous gas temperature measurements in an oscillating flow, to heat transfer. Thus, the Nusselt numbers resulting from the steady flow experiments as well as from Tanaka's oscillating flow experiments tend to be smaller than those given by Gedeon and Wood, who applied a method for separating the different losses.

This fact can be proved by plotting the "effective" Nusselt number Nu_e for the Gedeon/Wood data, which was calculated and correlated by Gedeon and Wood assuming the regenerator loss from their experiments can be totally attributed to the heat transfer loss[3]. After rearranging these correlations with respect to the definition of the total Nusselt number (see equation 13), the effective Nusselt numbers are given as follows:

Gedeon/Wood (wire screens)

$$Nu_{e,tot} = \left(1.010 + 0.503 \cdot Re^{0.720}\right) \cdot \left(1 - 0.847 \cdot (1-\varepsilon)\right)$$

Gedeon/Wood (metal felts)

$$Nu_{e,tot} = \left(0.963 + 0.352 \cdot Re^{0.794}\right) \cdot \left(1 - 1.485 \cdot (1-\varepsilon)\right)$$

In order to demonstrate the impact of the effective Nusselt number, Figures 6 and 7 were replotted using the correlations for $Nu_{e,tot}$ instead of Nu_{tot} for Gedeon and Wood's data (see Figures 8 and 9).

It can be seen that for both wire screens and metal felts the effective Nusselt numbers given by Gedeon and Wood are smaller than their Nusselt numbers according to Table 1, resulting in a better correspondence to the other correlations. Especially for the Reynolds number range of 10 to 100, the Nusselt numbers given by Gedeon/Wood, Tong/London and Miyabe for wire screens and Gedeon/Wood, ILK Dresden and Kühl for metal felts show very small deviations. Only Tanaka's data does not fit the other correlations, which may be caused by the absence of the additional porosity effect, as outlined above.

This analysis revealed that one has to be very careful when using any Nusselt number correlation for the evaluation of regenerator heat transfer, since other losses

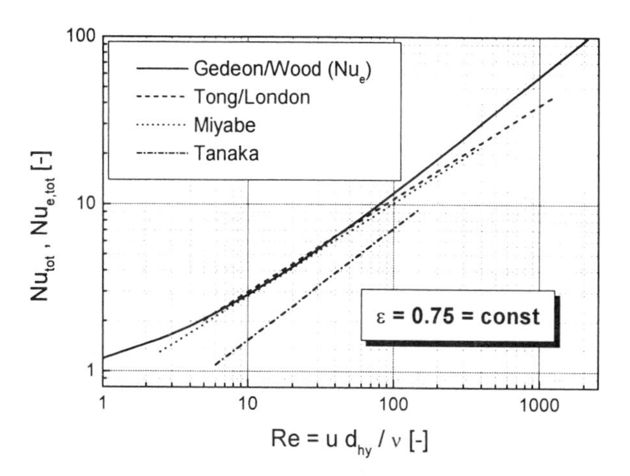

Figure 8 Nu_{tot}-correlations based on equation (13) versus Reynolds number for wire screens and a porosity of 0.75. (Gedeon and Wood data represented by effective Nusselt number $Nu_{e,tot}$)

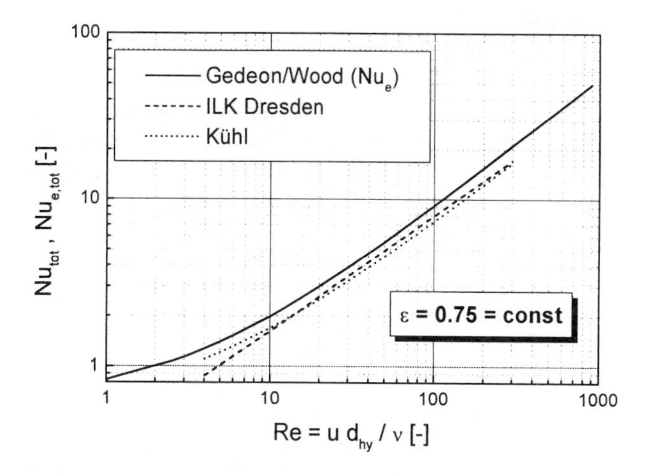

Figure 9 Nu_{tot}-correlations based on equation (13) versus Reynolds number for metal felts and a porosity of 0.75. (Gedeon and Wood data represented by effective Nusselt number $Nu_{e,tot}$)

may be incorporated. This must be considered when applying a Nusselt number correlation to a simulation code, because depending on the correlation used, losses such as flow dispersion and temperature swing loss may or may not be calculated additionally.

For that reason, it seems best to implement the correlation for the overall thermal regenerator loss, as given by Gedeon and Wood (see Table 1), into the simulation code. This directly links the term required for the simulation to the experimental data. Any Nusselt number correlation can be used afterward in order to decide which fraction of the overall loss is caused by the imperfect heat transfer in the regenerator matrix.

Conclusion

The analysis in this paper was meant to present an overview of different friction factor and Nusselt number correlations for the regenerator of Stirling machines including the Oscillating-Flow Regenerator Test Rig data collected by Gedeon and Wood. The discussion was divided for wire screen matrices and metal felts. The resulting plots show that the correlations derived by Gedeon and Wood predict the highest friction factors as well as the highest Nusselt numbers for both matrix materials. However, due to the different experimental method Gedeon and Wood applied in contrast to the steady flow experiments, it is questionable if the results are totally comparable.

If friction factor is considered, Gedeon and Wood experimentally determined the term of interest for the simulation - power dissipation in the regenerator caused by pressure drop. However, the recalculation of this term in order to achieve a correlation for the friction factor does cause errors, which can be omitted when using any correlation based on direct pressure drop measurements. Gedeon and Wood also concluded that the impact of the oscillating flow on the friction factor in the regenerator matrix is not meaningful for Valensi numbers below 20 [3].

When looking at the heat transfer data, the impact of other thermal losses, such as temperature swing and dispersion loss, must be considered in order to make sure that they are not incorporated in the Nusselt number. For that reason it seems most appropriate to use the correlation for the overall thermal loss given by Gedeon and Wood, which is a direct result of their experimental data.

Nomenclature

A	(gross) heat transfer area
A_{net}	net heat transfer area
A_{free}	free flow area
c_F	friction factor
C_{sf}, C_{fd}, $C_1..C_4$	correlation constants
d	wire diameter
d_{hy}	hydraulic diameter
L	total length of matrix
n	number of screens (or layers)
p	pressure
Δp	pressure drop
$\dot{Q}_{Reg.-loss}$	overall thermal regenerator loss
s	pitch or mesh width (see Figure 1)
T	temperature
ΔT	regenerator axial gas temperature difference
u	flow velocity in matrix
u_0	empty tube velocity
V_{mat}	volume of matrix material
V_{tot}	total volume of regenerator
w	open mesh width (see Figure 1)
α	heat transfer coefficient
δ	screen thickness
ε	porosity of matrix
φ	specific (gross) heat transfer area ($=A/V_{tot}$)
φ_{net}	specific net heat transfer area ($=A_{net}/V_{tot}$)
λ	thermal conductivity of fluid
ν	kinematic viscosity of fluid
ρ	density of fluid
Nu	Nusselt number
Nu_e	"effective" Nusselt number; given by Gedeon and Wood assuming the heat transfer loss equals the total regenerator loss
Nu_{tot}, $Nu_{e,tot}$	Nusselt number based on the gross heat transfer area A or φ, respectively
Pr	Prandtl number
Re	Reynolds number

References

1 Thomas, B.: „Evaluation of 6 different correlations for the flow friction factor of Stirling engine regenerators“, Proc. 34th IECEC, Vancouver, 1999

2 Thomas, B., Bolleber, F.: „Evaluation of 5 different correlations for the heat transfer in Stirling engine regenerators“, Proc. Europ. Stirling Forum 2000, Osnabrück, 2000

3 Gedeon, D., Wood, J.G.: „Oscillating-Flow Regenerator Test Rig: Hardware and Theory With Derived Correlations for Screens and Felts“, NASA Contractor Report 198442, 1996

4 Tong, L.S., London, A.L.: „Heat-Transfer and Flow-Friction Characteristics of Woven-Screen and Crossed-Rod Matrixes“, Transactions of the ASME, 10, S. 1558-1570, 1958

5 Blass, E.: „Geometrische und strömungstechnische Untersuchungen an Drahtgeweben“, Chemie-Ing.-Techn., 36. Jahrg., Nr. 7, 1964

6 Miyabe, H., Takahashi, S., Hamaguchi, K.: „An Approach to the Design of Stirling Engine Regenerator Matrix using Packs of Wire Gauze“, Proc. 17th IECEC, New York, 1982

7 Tanaka, M., Yamashita, I., Chisaka, F. „Flow and Heat Transfer Characteristics of the Stirling Engine Regenerator in an Oscillating Flow“, JSME Int. Journal, Series 2, Vol. 33, No. 2, 1990

8 Kühl, H.-D., Schulz, S., Walther, C.: „Theoretical Models and Correlations for the Flow Friction and Heat Transfer Characteristics of Random Wire Regenerator Materials“, Proc. 33rd IECEC, paper No. 207, Colorado Springs, 1998

9 Kühl, H.-D., Schulz, S.: „A 2nd order Regenerator Model including Flow Dispersion and Bypass Losses“, Proc. 31th IECEC, Washington D.C., 1996

PERFORMANCE CHARACTERISTICS OF A GAS ENGINE DRIVEN STIRLING HEAT PUMP

Sumio Yagyu, Ichiro Fujishima, Yuji Fukuyama, Tomoyuki Morikawa & Norio Obata
Kubota Corporation, Hyogo, Japan

John Corey
Clever Fellows Innovation Consortium, Inc., New York, USA

Naotsugu Isshiki & Isao Satoh
Tokyo Institute of Technology, Tokyo, Japan

ABSTRACT

This paper describes recent results in a project at KUBOTA to develop an efficient CFC-free multi-functional heat supply system. A heat pump in the system is a gas engine driven Stirling heat pump. The heat pump is mainly driven by engine shaft power and is partially assisted by thermal power from the engine exhaust heat. By proportioning two energy sources to match heat balance of the driving engine, this heat-assisted Stirling heat pump can be supplied with the maximum share of the original energy fueling the engine and can be operated at the most efficient point. We have developed a system heat pump composed of 6 cylinders, the doubled E-3 prototype. This prototype uses helium gas as a working gas and is constructed as two sets of three-cylinder machines, each a combination of two Stirling sub-systems (one a power producer and one a heat pump). Design and performance simulations of the prototype are presented in conjunction with the driving engine characteristics. This heat supply system is expected to produce cooling and heating water at high COP. Developing the system will provide a CFC-free thermal utilization system technology that satisfies both wide heat demands and various fuel systems.

INTRODUCTION

Kubota is developing an efficient CFC-free multi-functional heat supply system.[1] The system consists of a multi-fuel gas internal combustion engine and a heat-assisted Stirling heat pump utilizing both shaft power and thermal power. The heat pump is mainly driven by engine shaft power and is partially assisted by thermal power from the engine exhaust heat.

This compares with other CFC-free heat pumps using helium in a gas cycle, including Stirling and Vuilleumier (VM) machines. They utilize either shaft power or thermal power. In this heat-assisted Stirling heat pump, by proportioning the two energy sources to match heat balance of the driving engine, the heat pump can be supplied with the maximum share of the original energy fueling the engine and can be operated at the most efficient point.

The heat-assisted Stirling heat pump is essentially composed of a set of three cylinders. Otomo et al. performed experiments on a similar three-cylinder machine, mostly for validation of the vector method of cycle analysis, using the Schmidt assumption.[2] Yagyu et al. presented a third-order method for analysis and optimization of multi-cylinder regenerative machines and applied it to a three-cylinder heat-assisted Stirling heat pump case.[3] The three-cylinder machine has been modeled as comprising two merged Stirling sub-systems, one a power producer and one a heat pump, using the Sage simulation code.[4] The performance predictions made in the Sage simulation code were validated by test results of the third-generation three-cylinder machine (C-3)[5] and the fourth-generation three-cylinder machine (D-3).[6] These machines achieved sufficient COP on an indicated basis, however, the actual COP reduced by the large mechanical loss. In this paper, we present the design and performance simulations for a system heat pump composed of 6 cylinders, the doubled E-3 prototype, which is symmetrically composed of two sets of 3-cylinder machine (E-3). The doubled E-3 prototype has reduced pressure loss in each cycle to raise the indicated COP and improved mechanical loss to raise the actual COP. This prototype is expected to achieve final project performance goals. The performance is discussed in conjunction with the driving engine heat balance.

OVERVIEW OF HEAT SUPPLY SYSTEM

Utilizing gaseous fuel such as methanol reformed gas (MRG), natural gas, and hydrogen, the system produces four thermal output levels (263 K, 280 K, 318 K, and 353 K) for air-conditioning, hot water supply, and refrigeration, all in a single system. The system specifications are shown in Table 1. Figure 1 shows a schematic of the system configuration. The heat pump includes a specialized H-cylinder, which is heated by the exhaust gas of the driving engine and phased to produce added mechanical power. This arrangement recaptures some of the waste heat in the exhaust in order to reduce shaft power needed for driving the heat pump. By phase shifting the piston in the H-cylinder to match the engine heat balance, the heat pump can be supplied with the maximum share of the original energy fueling the engine.

Table 1: Specifications of Heat Supply System

Working Gas	Helium Gas
Fuel Gas	$CO+2H_2$ Methanol Reformed Gas Natural Gas, and Hydrogen
Rated Output (Total: 15 kW)	Cooling (5 kW @280 K, Also Refrigeration @263 K) Heating (8 kW @318 K) Hot Water (2 kW @353 K)
Total COP	2.42 (HHV Basis of MRG)

SYSTEM HEAT PUMP

Three-cylinder prototypes, C-3 and D-3, were built and have been tested against performance analyses and component tests. Validated by the results, a 6-cylinder machine was planned as the complete system heat pump, the doubled E-3 prototype.

HEAT FLOW

Figure 2 shows a heat flow diagram of the system heat pump. This heat pump is a heat-assisted Stirling heat pump and the cycle can be considered as two 2-cylinder Stirling sub-systems: one between high and medium temperature (H-M), and one between medium and low temperature (M-C). The M-C sub-system acts as a heat pump, while the H-M sub-system serves as a power producer which assists with extra shaft power. The downstream heat of the H-M sub-system, which is waste heat of the sub-system, is also utilized as additional heating output. Taking both shaft power and thermal power, the heat pump produces cooling/refrigeration and heating water.

MACHINE DESIGN

Figure 3 shows a system heat pump composed of 6 cylinders, the doubled E-3 prototype. Table 2 shows its specifications. This prototype utilizes both shaft power and thermal power and produce 280 K cold water and 318 K hot water. Simulating an engine driven machine, the tests were conducted using shaft power supplied by a motor and thermal power supplied by hot air.

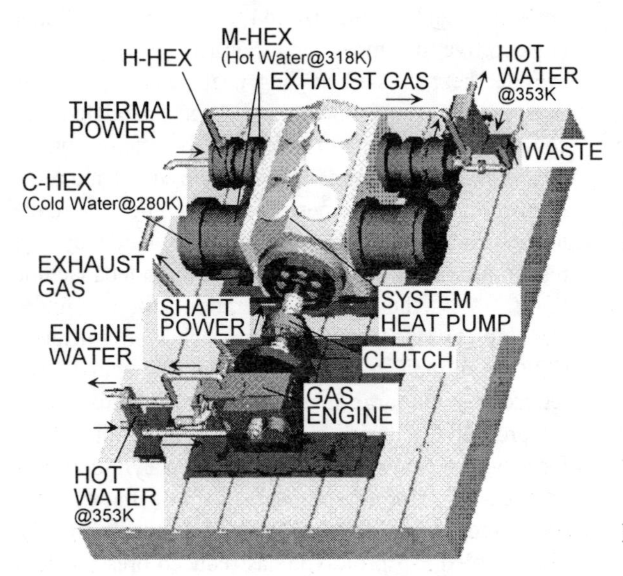

Figure 1: Schematic of System Configuration

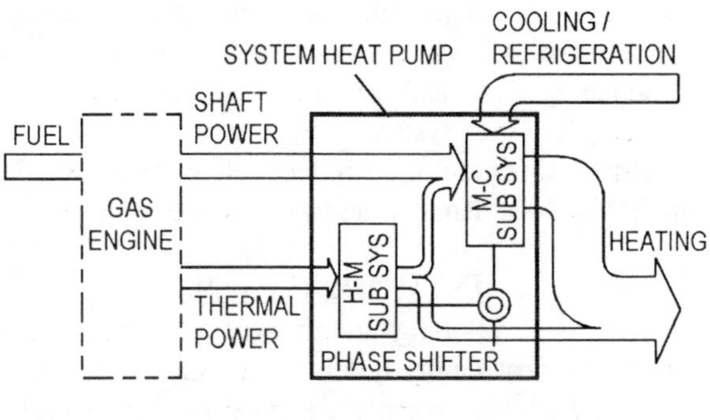

Figure 2: Heat Flow Diagram

Table 2: Doubled E-3 Specifications

Rated Output: (Temp.)	Cooling: (280K) 5.0 kW
	Heating: (318K) 8.0 kW
Cylinder: (Phase) Bore x Stroke	C-Cylinder: (67deg.) 110 mm x 52 mm
	M- Cylinder: (0deg.) 100 mm x 58 mm
	H- Cylinder: (Variable Phase) 96 mm x 20 mm

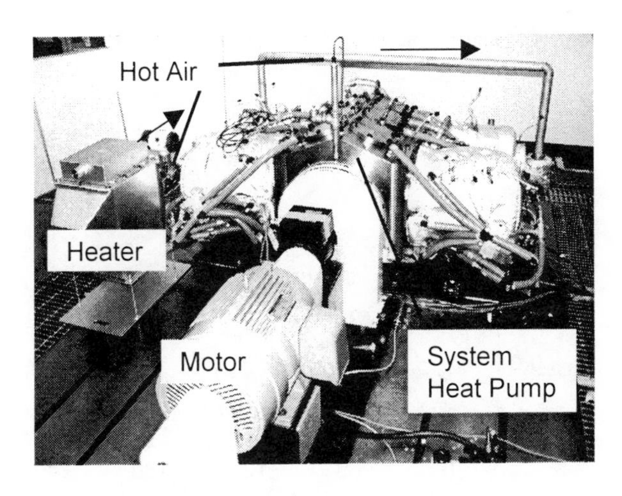

Figure 3: Doubled E-3 Prototype Machine

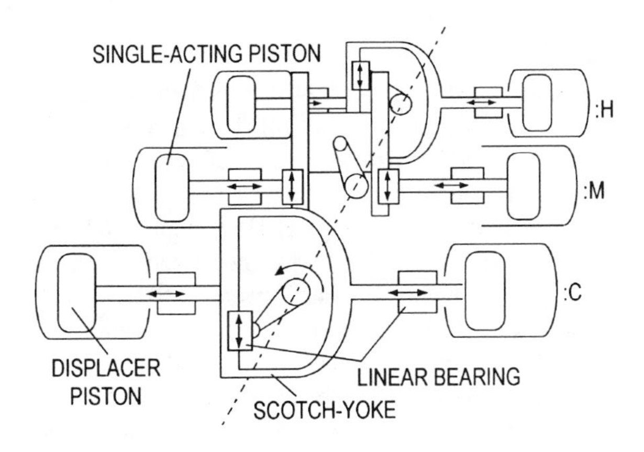

Figure 4: Cylinder Arrangement

The prototype was a kinematic type and the crank case was pressurized. Helium as a working gas was sealed at the ends of crank shafts by mechanical seals. This prototype, which was constructed with two sets of 3-cylinder machines, was symmetric with respect to the main shaft, and composed of 2X2 displacers and 1X2 single-acting pistons. The M-C sub-system was constructed as a Stirling cycle, whose cylinders were set in 67 deg phase difference. The phase of H-piston could be arbitrary set by using a phase shifter located between crank shafts. The H-cylinder in each set was staged in series, so the hot temperatures of the two 3-cylinder sets will differ. This has been found by analysis to give a higher COP than parallel, equal-temperature hot side connections. With careful consideration for minimizing pressure loss in a cycle, an annular type regenerator of stacked stainless steel was installed between opposing Bayonet heat exchanges[7] and was connected to displacer top and bottom spaces. The Bayonet heat exchangers were employed for cold and hot water outputs. The hot heat exchanger was a U-shape heat exchanger optimized for engine exhaust heat recovery. As shown in Figure 4, horizontal opposing cylinder arrangement was adopted in the prototype to ease the mechanical forces on the phase shifter and bearings, then to reduce mechanical loss in the multi-cylinder machine. Also linear movement of opposing pistons was planned by utilizing the Scotch-yoke mechanism with linear bearings. The linear bearings were arranged for sustaining piston side thrust and sliding yoke-crank connection. Since the piston linear movement was secured, clearance seals instead of piston rings as the gas sealing system, were employed at displacer pistons and gas leakage was minimized with low friction.

PERFORMANCE SIMULATION

The system heat pump was designed by simulating its thermodynamic performance using the Sage modeling code (from Gedeon Associates). Figure 5 shows a half analysis model of the heat pump (E-3). This E-3 model was generated by combining two Stirling (2-cylinder) models after each was separately optimized. The H-M part was merged with the M-C sub-system using a common M-cylinder, which requires that the two sub-cycles have the same mean pressure and pressure waves in their M-cylinders. This simulation method of division-optimize-merge, which was detailed and discussed for earlier prototype D-3 case, is useful in applying the greater experience and familiarity with simple Stirlings to this more complex device, especially during parametric variation for performance optimization.

In the three-cylinder machine, the performance of the M-C cycle dominates overall performance of the

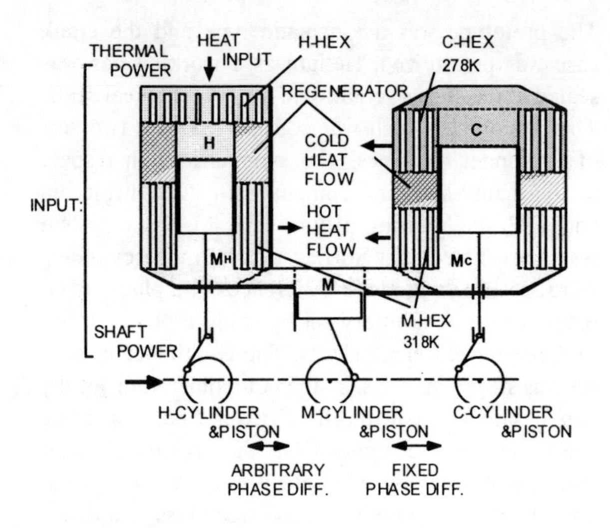

Figure 5: Simulation Model (E-3)

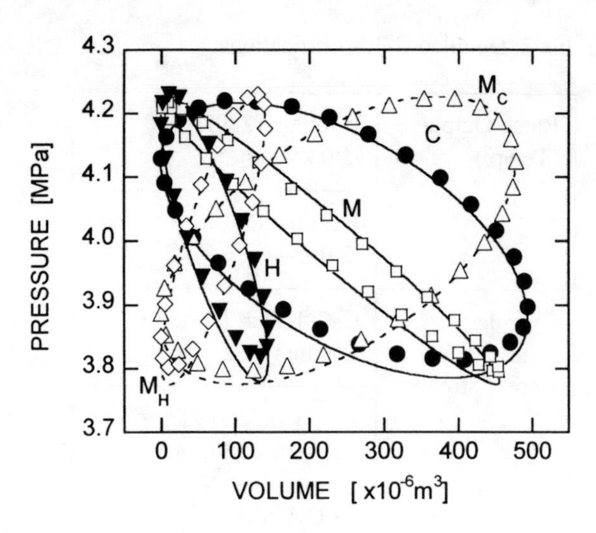

Figure 6: PV diagram

machine, particularly when matched to the engine system. Therefore the M-C cycle was first optimized, without prior constraints on complex pressure, to produce the desired cooling capacity at maximum COP. Then the first stage H-M cycle was optimized, but subject to a complex pressure constraint that forces its mid-temperature pressure to match those in the M-C mid-temperature cylinder. The H-M cycle is also sized by the ratio of heat available (in the engine exhaust) to the net work supplied by the engine to drive the heat pump. Finally, the second stage H-M cycle was optimized, subject to the pressure constraints, but also the mechanically-imposed same stroke and phase as the first stage H-M. Because the second stage H-M contributes less power, even ideally, than the first stage, the COP penalty for mechanical constraint is less on the second stage.

Heat exchanger temperatures were set for an engine driven machine for air-conditioning application, that is, 278 K at C-HEX wall, 318 K at M-HEX wall, and temperatures available at H-HEX walls individually. Considering the low temperature difference between C and M temperatures, the prototype was designed to have a low pressure ratio. The pressure amplitude was planned to be about 5% of the mean pressure.

COMPARISON OF TEST & SIMULATION

The performance tests were conducted at a mean pressure of 4 MPa (helium), a speed of 1,000 rpm, and with H-piston phase around 50 degrees (relative to M-piston). Each inlet water stream was controlled by a cooling or heating device. Temperature was regulated at the outlet to 280 K on the cold side and 318 K on the hot (mid-T) side.

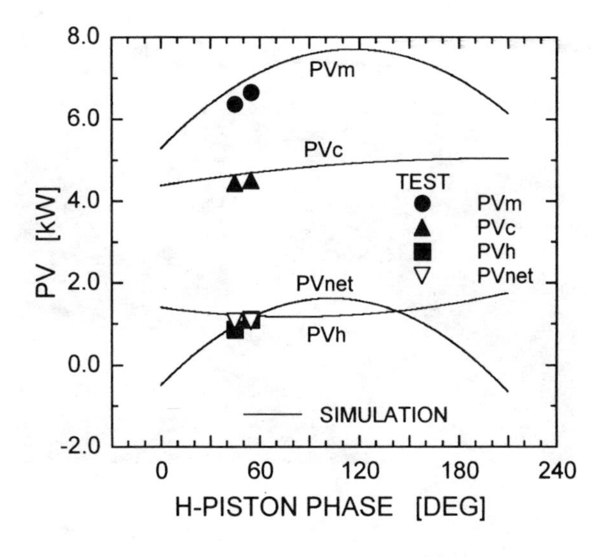

Figure 7: PV Characteristics vs H-Piston Phase

The hot air stream for H-HEX which simulated engine exhaust gas was controlled by a AC electric power device and resistance heaters. Temperature was regulated to the equivalent temperature of the engine driven case at the first stage H-HEX inlet to 677 K. Shaft power was supplied by the motor and measured by a torque meter. Encoders were fitted to each crankshaft and the difference in crank angle was precisely measured. RTDs were used for water temperature measuring and TCs were used for gas temperature measuring. Dynamic gas pressures were measured by quartz pressure transducers. The test results have been compared to corresponding Sage model results.

PV diagrams of C-cylinder, three M-cylinders (M, Mc, and Mh), and H-cylinder at the H-phase of 45 degrees are shown in Figure 6. The indicated work at each cylinder had a good oval shape and the pressure ratio was around 1.1 as planned. The simulation results were agreed well with test results in 5-10 % accuracy. This indicates that the fundamental physics and geometry of the doubled E-3 prototype are well modeled and can be used to predict the performance of the real machine. It also enables the design of the next-generation system heat pump.

Figure 7 shows primary characteristics of the prototype on a PV basis. Changing the H-piston phase, the tests were performed at some phases at a mean pressure of 4 MPa and a speed of 1,000 rpm. The independent parameter, H-phase, was defined as the phase difference between H-piston and M-piston. The cooling indicated work PVc, the heating indicated work PVm, the indicated work of thermal power PVh, and the indicated shaft power PVnet are plotted there. The simulation results are shown by continuous curves in the same figure. PVnet is summation of PVs at all cylinders and it means the net shaft power. The indicated work from thermal power input had the maximum value around the phase of 100 degrees. At that phase, the shaft power was most assisted by the thermal power. COPind-c was defined as the ratio of PVc to PVnet and the maximum value was around the same point. As indicated, the performances of this heat-assisted Stirling heat pump have strong H-phase dependency. Utilizing this feature, the heat pump can be controlled in both capacities and heat input ratio (a ratio of thermal power to shaft power).

DISCUSSION & APPLICATION

EFFECT OF PRESSURE LOSS

Heat transfer characteristics of heat exchangers in Stirling type machines are essentially worse than those of evaporators or condensers in conventional CFC heat pumps. Therefore a smaller and more efficient heat exchanger is needed for the machines. However in heat exchangers, the better the heat transfer characteristics are, the worse the pressure loss characteristics are. These characteristics are trade-off. Since the pressure ratio is relatively small in the prototype, pressure loss in heat exchangers is vital and critical factor to define the COP.

The prototype machine employed the Bayonet heat exchangers. A schematic configuration of the Bayonet HEX is shown in Figure 8. The Bayonet HEX consisting of inner tubes and outer tubes has a bundle of annular passages with spiral fins as an extended surfaces to improve heat transfer on the working gas side. The working gas flows back and forth, in an oscillatory motion, in annular passages between tubes. The heat exchanger performances were carefully examined. Since the size and shape of the tubes and their fins can be easily changed, this configuration provides good design flexibility and it also provides wide choices for setting various heat transfer conditions.

The prototype was planned that a roughness of Bayonet heat exchanger was 0.157. The roughness is a ratio of fin height (h) to hydraulic diameter (Dh) in an annular gas passage. Since the roughness is easily changed by machining fins or substituting parts, the pressure loss was tested by some combinations of elements with almost the same hydraulic diameter in steady gas flow. Figure 9 shows pressure loss characteristic of the Bayonet heat exchanges in the prototype. In the figure, Fmult is a multiplication factor for the loss factor in annual passage without fins, so that it will be 1 in the smooth tube case. Fins greatly affect Fmult and that tendency would be accelerated in oscillating flow. Figure 10 shows measured pressure wave in the M-HEX of the prototype, and the Sage simulation results at some roughness cases shown by continuous curves. The component-tested value of Fmult was adopted, and generally agrees with test results. The COP and the indicated work PVc characteristics by the simulations are shown in Figure 11 conjunction with roughness. In the figure, the pressure loss characteristic in M-HEX is also indicated as a reference and the loss dP is calculated based on averaged absolute pressure swing in a cycle. The COPind-c increased by a pressure loss decrease in heat exchanger and greatly affected by the roughness. It indicates that optimal roughness would be 0.12, which corresponds to just smaller fin height than the current design. In Stirling type machines especially working in low temperature difference, pressure loss distribution or design should be more carefully considered than when we design the high temperature difference machine such as Stirling engines.

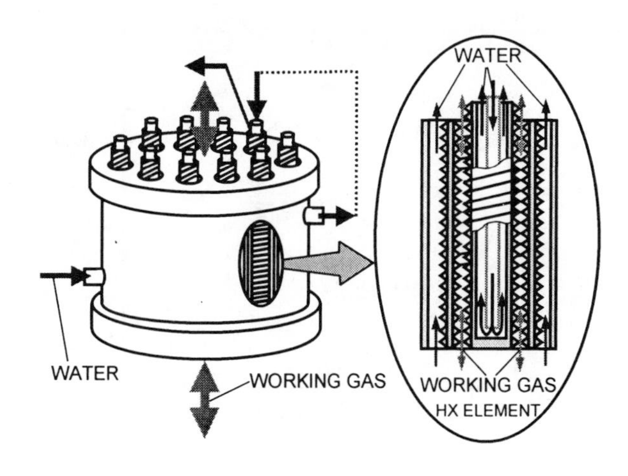

Figure 8: Schematic Configuration of Bayonet HEX

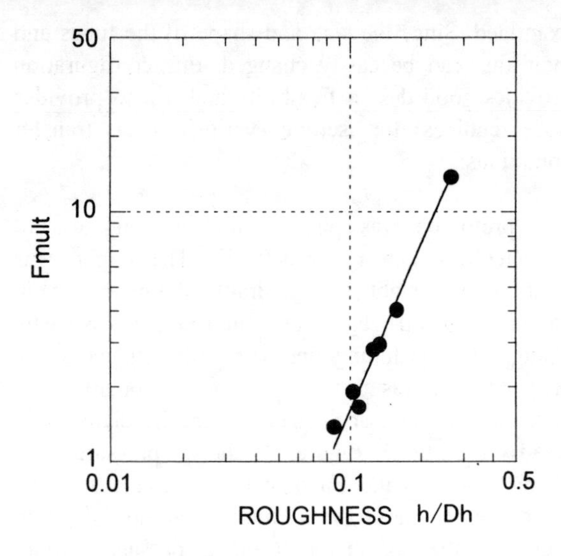

Figure 9: Pressure Loss Factor in Bayonet HEX

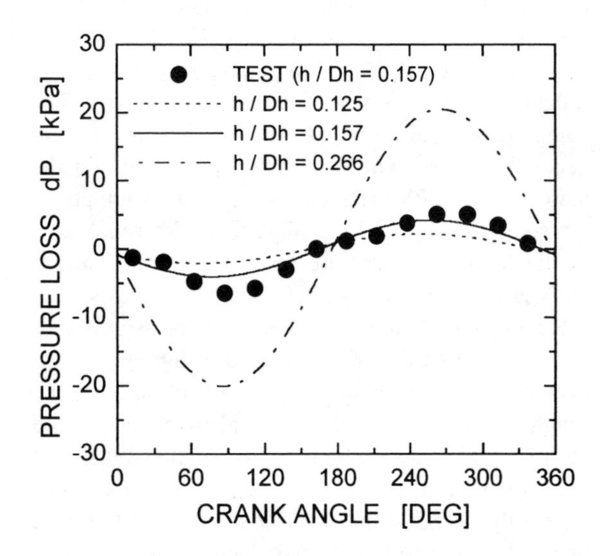

Figure 10: Pressure Swing in Bayonet HEX

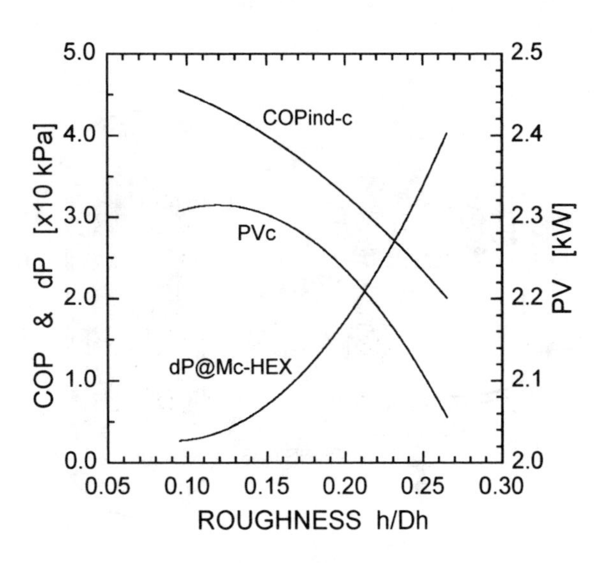

Figure 11: HEX Characteristics vs Roughness

ENGINE DRIVEN CASE

The system heat pump should be operated in conjunction with the driving engine characteristics. The gas engine was newly developed for the system heat pump in the project. Table 3 shows the driving gas engine heat balance with methanol reformed gas. The thermal efficiency is 32 % of the originally fueled gas HHV, which corresponds to shaft power. This is very good value in such a small gas engine around 2 kW output. The exhaust gas heat is 37 % of the originally fueled gas HHV, which corresponds to thermal power. This heat is essential to the heat-assisted heat pump case. The engine water heat is also used for supplying 80 degC hot water.

We have not conducted the complete system tests of the engine driven case. Based on the test results of the prototype, the estimated performances of the system are shown in Table 4. Since the mean pressure of working gas in the prototype has not reached to the rated value, the total output is not sufficient and it is around 76 % of the rated total output of 15 kW. However the heat cycle characteristics on an indicated basis have attained very reasonable level. The maximum achieved COPind-c of the prototype machine was more than 4 at the specified engine heat balance.

Table 3: Typical Engine Heat Balance (HHV basis)

Thermal Efficiency	32%
Exhaust Gas	37%
Engine Water	22%
Miscellaneous Loss	9%

Table 4: Estimated System Performances

Input	6.0kW (HHV)
Output	Cooling 3.1kW@280K Heating 6.2kW@318K Hot Water 2.1kW@353K
COPind-i (PVi／PVnet)	i:c 4.2 i:m 6.0
COPnet-i (Qi／PVnet)	i:c 3.0 i:m 5.8
Total COP (Output／Input)	1.9 (HHV)

Still there is low mechanical efficiency and the actual COP which included mechanical efficiency becomes lower than the targeted value. In the next step, by better matching the driving engine characteristics, the system heat pump should be pressurized up to 5 MPa and be improved in mechanical efficiency. Then we expect that the system would provide higher performance than the conventional CFC-free heat pumps.

CONCLUSIONS

Kubota is developing a gas engine driven CFC-free heat supply system. From testing the system heat pump (the doubled E-3 prototype machine), the following results are known:

1. A system heat pump, the heat-assisted Stirling heat pump, was successfully designed with 3rd-order simulations merged Stirling sub-cycles in Sage, in conjunction with the driving engine characteristics. The maximum achieved COPind-c of the prototype machine was more than 4 at the specified engine heat balance. The system performance is expected to be high COP in CFC-free systems.

2. The heat pump performance was greatly affected by pressure loss and mechanical loss. In such a machine working in small temperature difference, since the pressure swing of working gas was relatively small, PV work in each cylinder was directly and severely affected by pressure loss, especially in heat exchangers. Also, since the net driving power was relatively small comparing with mechanical loss, improved mechanical efficiency is a major opportunity to raise the COP.

3. Investigating further the characteristics of both pressure loss and mechanical loss, an efficient gas engine driven CFC-free heat supply system should attain the target system total COP of 2.42.

ACKNOWLEDGMENT

This project is a part of the New Sunshine Project promoted by the Ministry of International Trade and Industry (MITI). NEDO is the implementing organization of this project and ECC is a contract to NEDO. The authors would like to thank the ECC and NEDO for their support in this project.

REFERENCES

[1] ECC, 1995, "Multi-Temperature Heat Supply System", *Element Technology Development Program at Eco-Energy City Project*, pp 32-33

[2] Ohtomo M., Isshiki N., and Watanabe H., 1996, "Studies on Three-Temperature, Three-Cylinder Stirling Cycle Machine", *Proceedings of 31st IECEC*, Vol. 2, pp 1238-1242

[3] Yagyu S., Fujishima I., Corey J., and Isshiki N., 1996, "An Analytical Approach to Multi-Cylinder Regenerative Machines with Application to 3-Cylinder Heat-Aided Stirling Heat Pump", *Proceeding of 31st IECEC*, Vol. 2, pp 1201-1205

[4] Gedeon D., 1994, "Sage: Object Oriented Software for Stirling Machine design", *Proceeding of 29th IECEC*, Vol. 4, pp 1902-1907

[5] Yagyu S., Fujishima I., Corey J., Isshiki N., and Satoh, I., 1997, "Design, Simulation, and Test Results of a Heat-Assisted Three-Cylinder Stirling Heat Pump (C-3)", *Proceeding of 32nd IECEC*, Vol. 2, pp 1033-1038

[6] Yagyu S., Fujishima I., Fukuyama Y., Morikawa T., Corey J., Isshiki N., and Satoh, I., 1999, "Performance Modeling of 5th Generation Heat-Assisted Stirling Heat Pump", *Proceedings of 34th IECEC*, Paper No. 1999-01-2700

[7] Yagyu S., Fukuyama Y., Morikawa T., Corey J., Isshiki N., and Satoh, I., 1998, "Bayonet Heat Exchangers in Heat-Assisted Stirling Heat Pump", *Proceeding of 33rd IECEC*, Paper No. 080

AIAA-2000-2814

AN EXPERIMENTAL STUDY OF A 3-KW STIRLING ENGINE†

Noboru Kagawa
Associate Professor, Department of Mechanical Systems Engineering, National Defense Academy
Yokosuka, Japan

ABSTRACT

A 3-kW Stirling engine, NS03T which is a two-piston type with a unique V-shape cylinder arrangement, was developed by Toshiba. By adopting the cylinder arrangement and a well-designed heat exchanger system, a 50% indicated efficiency was obtained. In this paper, the specifications and modifications of the 1984-year NS03T engine, which is the first NS03T engine with the higher efficiency and performance, are presented. The higher indicated efficiency of the NS03T engine is presented by using reliable experimental data and analyzed data by a mathematical model, SETMA. Also, the regenerator loss is analyzed with a proposed method, which is able to derive reasonable regenerator loss from engine operating data. The results show that the design of the heat exchanger is important for high efficiency engines.

INTODUCTION

Related to recent substantial environmental and energy problems, Stirling cycle technology has attracted attention. With an aid of a heat recovery mechanism between two isochoric processes of Stirling cycle, the cycle efficiency becomes higher. The regeneration mechanism, so-called regenerator, plays a very important role. If the regenerator works without any loss, the required heat input, which should be supplied at the heater and/or the expansion cylinder, becomes less. For realization of Stirling engine with higher regenerator efficiency, however, it is essential to design and develop a suitable regenerator for each machine.

A 3-kW Stirling engine, NS03T was developed by Toshiba from 1982 to 1988[1]. The NS03T engine is a two-piston type with a unique V-shape cylinder arrangement. The engine has a 50 % indicated efficiency and a 31 % thermal efficiency. The NS03T was reassembled to study its high performance in detail at the National Defense Academy. There are very limited engines, which have such higher indicated efficiency. It is interested to analyze the engine performance based on reliable experimental data. Also, it is useful to arrange such technical information for design and development of new Stirling engines.

In this paper, the indicated efficiency of the NS03T engine is analyzed by using reliable experimental data measured officially at the Mechanical Engineering Laboratory in Tsukuba. The regenerator loss is rearranged with a proposed method, which is able to derive the reasonable regenerator loss from the engine operating data. In addition, the paper will try to specify the specifications of the NS03T engine.

SPECIFICATIONS OF NS03T

Table 1 shows the design parameters of a 3-kW two-piston type Stirling engine, NS03T. The New Energy Development Organization under a contract funded it from 1982 to 1988. The final model was constructed with lighter and compact components, which were resulted in by several improvements. Figure 1 represents the schematic view of the engine of the final type, the 1987-year model. The engine is a deformed in-line design with two pistons and cylinders mounted on a pressurized crankcase. The cylinders are arranged in unsymmetrical `V` form at an angle of 60 degrees.

Table 1 Design parameters of NS03T engine

Main fuel	Natural gas
Working fluid	Helium
Mean pressure	3 – 6 MPa
Max. expansion space temp.	975 ± 50 K
Compression space temps	< 323 K (water cooling)
Engine speed	500 – 1500 rpm
Max. shaft power	> 3 kW
Max. thermal efficiency	32 %
NOz	<150 ppm
Noise level	<60 dB(A)
Mass	<75 kg
Height	<800 mm

The target engine of this study is a 1984 model engine which shows in Figure 2. The initial specifications and performance of the engine were reported[2]. After several improvements, the engine performance was improved dramatically[3]. Table 2 shows detailed information about the 1984 model engine.

Combustion systems

The optimized combustion chamber and a set of a fuel nozzle, a swirler, and a burner-throat realized the shorter flame length of about 150 mm. It has uniform temperature distribution of the combustion gas whose temperature goes up to 1773 K. The preheater is a counter-flow type fresh air heater surrounding the combustor to reduce heat loss. The matrix consists of six thin stainless cylinders and one hundred eighty tubes. The cylinders are arranged concentrically and the tubes are set between the cylinders to enlarge the heating surface. The combustion gas flowing in the opposite direction in the tubes heats fresh air flowing through each annual duct between the cylinder walls. It

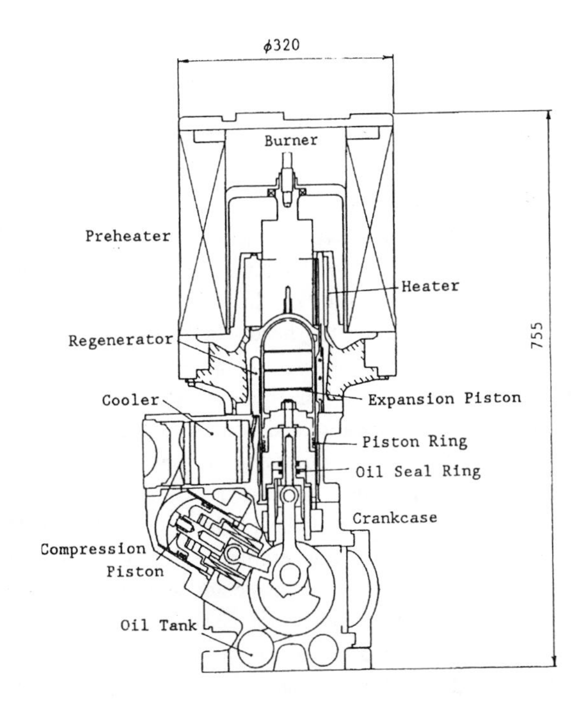

Figure 1 1987-year model engine

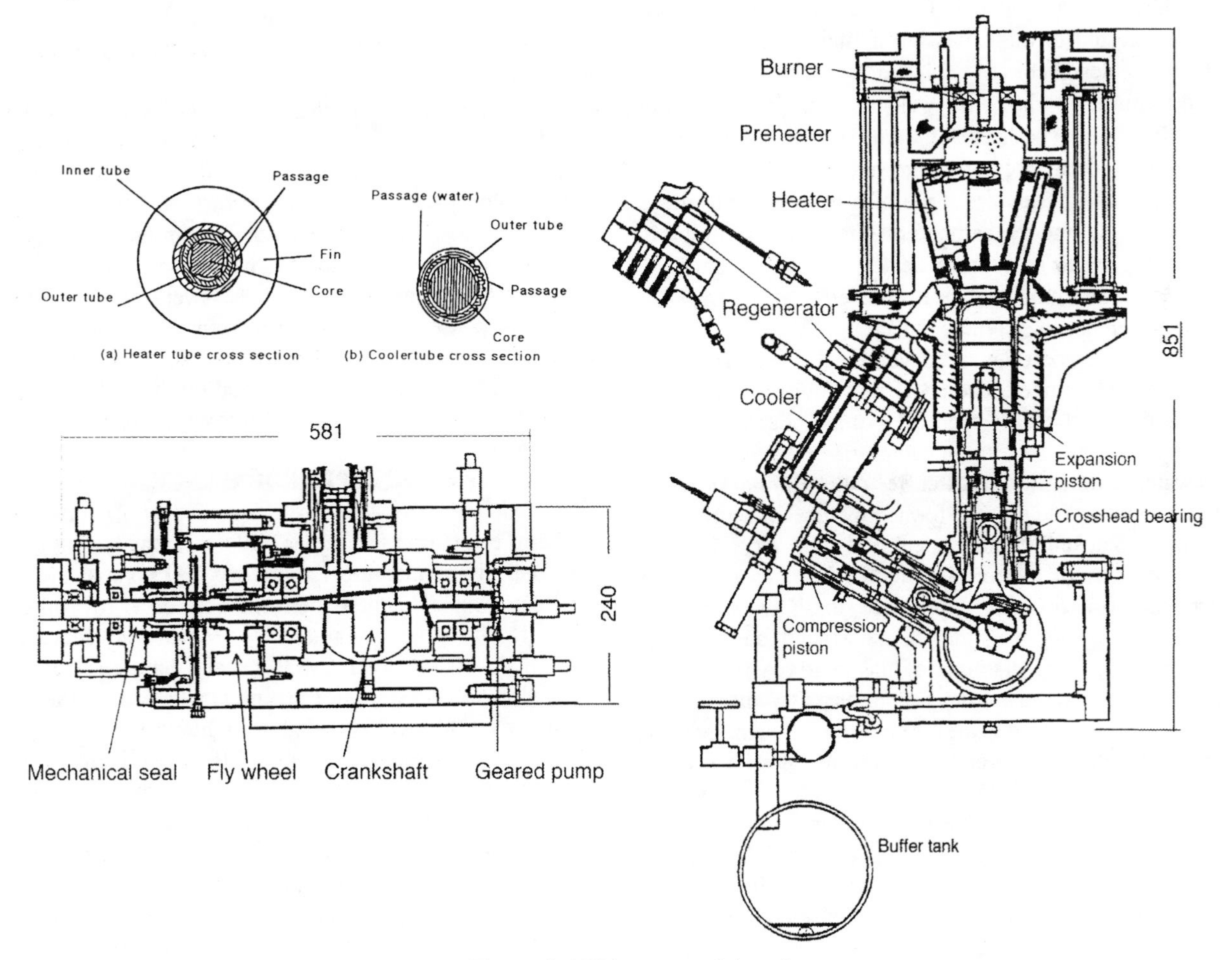

Figure 2 1984-year model engine

has a heating surface of 1.3 m^2. The pressure drop of the combustion system is about 138 mmAq which includes 93 mmAq of the flow control valve.

The convective heat-transfer would be remarkably influenced by tangential and turbulent flow developed by the swirler of the combustor, which is located above the heater head. In addition, thermal radiation from high temperature materials surrounding the heater increases the heat transfer. Also, uniformities of heater wall temperature and combustion gas velocity distribution which passes between the fins of the heater, are required to increase the heat transfer. The distance between the nozzle and the heater head was adjusted to obtaine the minimum temperature difference between the top and the bottom of the heater with 140 K. The layers of wire mesh screens, one sheet of #35-20 and seven sheets of #35-50, and a 1 mm thickness radiation plate located around the heater tube enhance both the radiation and convective heat transfer.

Heat exchanger system

The heater is constructed of twelve heater tubes, which are connected to the expansion cylinder concentrically. It was designed to achieve uniformity of the temperature of the tube wall and the flow of the working fluid. The heater is a double tube composed of the outer tube with external fins and the inner tube with the inner core. They are made of a heat-resistant alloy, Hastelloy-X. The temperature difference of each tube is less than 10 K at average heater wall temperature of 1067 K. The amount of heat transfer is achieved 7760 watts when net shaft power is 3 kW.

The cooler is a conventional shell and tube-type heat exchanger. The cooler has fourteen tubes and is similar to those of the heater tubes. Each tube has longitudinal grooves in the outer surface to discard heat to water with higher heat transfer ratio. Helium flows in the 0.75 mm wide passage formed by tube and core.

Stacked stainless steel 304 wire mesh is used as a regenerator matrix. The matrix is made up of 560 screens of #150 and #200 wire mesh. Screens are packed by using a 3 mm diameter bolt and formed cylindrically. The diameter of the matrix with 54 mm length is the same as the piston diameter of 70 mm. Three wheel shaped plates with 1 mm thickness were inserted in the matrix and two were put at both ends to decrease side leakage loss.

For the regenerator design concept, the thermal efficiency defined as temperature difference of one way divided by maximum temperature difference at the regenerator ends, was more considered than the pressure drop. As the result, the measured thermal efficiency was around 86 % for high temperature side and 94% for low temperature side. The pressure drop is approximately 0.3 MPa at 6 MPa mean engine pressure and 1000 rpm engine speed.

Driving system

Four-stage low-friction loss and low-wear piston rings are used for the NS03T engine. Three-piece type, which consists of two plane rings, an inner ring and a steel tension ring, is used for each piston. The material of the plane and inner rings is polyimide in which carbon fiber is added to reduce wear and increase heat resistance of the rings. Molybdenum disulphide was added to the piston ring material of the compression piston. The friction factor of the assorted rings, measured by a friction-measuring instrument, is less than 0.1.

A low-leakage and low-friction loss mechanical seal is also developed for the shaft seal device. The mechanical seal is set in a room where lubricant is filled with an enforced lubrication system. The material of mechanical seal, which is set at the crankcase end, is carbon. The other rotating part attached to the crankshaft is made of hard tungsten steel. The mechanical loss of the seal is about 140 W at 6 MPa and 1000 rpm.

The crank mechanism with cross head is adopted. The crosshead plays a role to endure the side force loading to the piston. Both expansion and compression pistons are connected to the crosshead by rods. The crossheads slide in the linear ball bearings. The lubricant is transferred with a controlled oil pressure by a geared pump whose swept volume is 1 cm^3. A phase difference between expansion and compression pistons is selected at an angle of 100 degrees. The crankshaft is supported on the ball bearing set. A flywheel, made of heavy-alloy is attached to the crankshaft.

ENGINE PERFORMANCE

The engine performance was measured at the official midterm evaluation test under NEDO. It was held at the Mechanical Engineering Laboratory in Tsukuba, 1984. At that time, detailed performance data for other Moonlight-program engines by Mitsubishi Electric (NS03M), Aishin Seiki (NS30A), and Sanyo (NS30S) were also measured. Major testing results of the NS03T engine were already reported in a literature[2]. Table 3 shows the more detailed results of the NS03T engine at the maximum power and efficiency conditions. The definitions of the powers, heats, losses, and efficiencies are shown in Table 4. Even though the engine was developed after only two and half years from the start of the project, it has enough performance as a practical engine. It is noteworthy that the indicated efficiency of 48.8 % is very high. The associated targets, like mass,

Table 2 Engine specifications

1. Engine	
Type	Two piston
Swept volume	
Expansion	192 cm^3
Compression	173 cm^3
Volume Phase angle	100 degrees
Compression ratio	1.63
Crankcase	Pressurized
2. Combustor	
Type	Swirled diffusion burner
Combustion capacity	4.5 – 13 kW
Turndown ratio	1:3
Air excess ratio	0.8 – 2.0
Combustor volume	1000 cm^3
Max. load ration	13 MW/m^3
3. Preheater	
Type	Shell and tube
Heat transfer area	
Combustion gas side	1.3 m^2
Fresh air side	1.3 m^2
4. Piston	
Bore	70 mm
Stroke	
Expansion	50 mm
Compression	45 mm
Rod diameter	20 mm
Connecting rod length	108 mm
5. Heater	
Type	Bayonet tube with fin
Heater tube	
Outer diameter	16 mm
Length	152 mm
Number	12
Material	Hasteroy-X
Heat transfer area	
Working fluid side	0.09 m^2
Combustion gas side	0.63 m^2
6. Regenerator	
Type	Canned
Number	1
Matrix	
Material	Stainless steel 304
Mesh	#40-150+#40-200
Outer diameter	70 mm
Sheet number	210+250
Length	54 mm
Void ratio	0.65
7. Cooler	
Type	Double tube w/inner fin
Cooler tube	
Outer diameter	13 mm
Length	150.5 mm
Number	14
Material	Stainless steel 304
Heat transfer area	
Working fluid side	0.10 m^2
Combustion gas side	0.14 m^2
8. Seal	
Piston seal	
Type	3 piece type piston ring
Ring height	2.0 mm
Stage number	4
Main material	Polyimide18/Polyimide W
Rod seal	
Type	Scraper type
Stage number	2
Main material	VS83
Shaft seal	
Type	Mechanical seal
Balance ratio	0.60
Seal surface wide	2.0 mm
Material	Carbon/Super hard steel
Helium storage volume	
Crank room	7700 cm^3
Buffer tank	10000 cm^3
9. Output mechanism	
Type	60 degree V-crank
Lubrication system	Integrated geared pump
Fly wheel inertia	0.9 kg m

dimension, exhaust gas characteristics were realized until the final evaluation test after another three years[1].

As same as other engines, the initial performance of the 1984-year model engine was not so good. As shown in Figure 3, the performance was improved by many experimental procedures with an aid of a mathematical analysis. All of the improvements were done during six months. The modifications of the heat exchanger system influenced significantly the engine performance, especially, the optimization of the cooler and the regenerator. About the other modifications and optimization including the heater and the cooler will be reported at the next paper. In this paper, the optimization of the regenerator is presented.

REGENERATOR LOSS OF 3-KW STIRLING ENGINE

In order to optimize the regenerator, losses, which have complicated relations each other, must be clarified and be reduced. To regenerate heat during the short period of time when the working gas passes through, a good regenerator must possess a large heat transfer surface and should allow low temperature difference between incoming and outgoing working gas. However, a regenerator with good heat transfer characteristics and high heat capacity is likely to cause too much pressure drop of the working gas flow. Desirable matrix is with high porosity, high heat capacity, and low flow impedance.

Table 3 Engine performance

1. Maximum output conditions	
Heater tube ave. temp.	1062 K
Gas temp. in exp. space	959 K
Mean cycle pressure	6.1 MPa
Engine speed	1250 rpm
Cooling water temp.	298 K
Cooling water flow rate	20×10^3 cm^3/min.
Output power	3.46 kW
Thermal efficiency η	27.3 %
Boiler efficiency η_b	81.4 %
Preheater efficiency η_{pre}	57.0 %
Inter conversion efficiency η_{int}	49.9 %
Indicated efficiency η_i	46.5 %
Mechanical efficiency η_{mech}	72.2 %
2.Maximum efficiency conditions	
Heater tube ave. temp.	1071 K
Gas temp. in exp. space	971 K
Mean cycle pressure	6.1 MPa
Engine speed	900 rpm
Cooling water temp.	298 K
Cooling water flow rate	20×10^3 cm^3/min.
OUtput power	3.00 kW
Thermal efficiency η	31.1 %
Burner efficiency η_b	80.7 %
Preheater efficiency η_{pre}	57.1 %
Inter conversion efficiency η_{int}	52.7 %
Indicated efficiency η_i	48.8 %
Mechanical efficiency η_{mech}	78.9 %
3. Partial load characteristics	
Pressure range	3-6.1 MPa
Speed range	500-1400 rpm
Power variable ratio	1:4.13
Turndown ratio	1:3.69
4. Cooling temperature dependence	
Cooling inlet water temp.	283-323 K
Decrement of shaft power	12 %
Decrement of thermal efficiency.	9 %
Exhaust gas	
NOx	290 ppm
HC	0.9 ppm
CO	185 ppm
Mass	270 kg
Noise level	App. 72dB(A)
Temperature distribution on heater tube (at 6 measuring points)	±10 K

Table 4 Definitions of powers and efficiencies

1.Power	
Expansion work (power)	$W_e = n\oint P_e dV_e$
Compression work (power)	$W_c = n\oint P_c dV_c$
Indicated work (power)	$W_i = W_e - W_c$
Output power	W_s
Mechanical loss	$W_{mech} = W_i - W_s$
2.Quantity of heat	
Heat input	Q_{in}
Exhaust gas heat at preheater inlet (quantity)	$Q_{ex,in}$
Preheater recovered heat (quantity)	Q_{pre}
Cooling heat in cooler (quantity)	Q_c
Cooling heat in exp. cyl. water jacket (quantity)	Q_{ec}
Cooling heat in comp. cyl. water jacket (quantity)	Q_{cc}
Regenerator loss	$Q_{rlos} = Q_c + Q_{ec} - W_c$
Effective heat input	$Q_{eff} = Q_{rlos} + W_e$
3.Efficiency	
Thermal efficiency	$\eta = W_s / Q_{in} = \eta_b \eta_i \eta_{mech}$
Burner efficiency	$\eta_b = Q_h / Q_{in}$
Preheater efficiency	$\eta_{pre} = Q_{pre} / Q_{ex.in}$
Internal conversion efficiency	$\eta_{int} = W_i / W_e$
Indicated efficiency	$\eta_i = W_i / Q_{eff}$
Mechanical efficiency	$\eta_{mech} = W_s / W_i$

There are always considerable losses, which can not be avoided: heat transfer, pressure drop, heat conduction, side leakage, matrix friction, and duct shape losses. Their interrelation becomes more complicated by the fact that the flow in the regenerator is no uniform flow and is reciprocating with phase angle of 90 to 100 degrees. For the purpose of providing appropriate design information for regenerator, many papers have been published and reported empirical information about regenerator and matrix[4-6]. However, such information was normally arranged based on experimental data, which were obtained by using element test stand not equipping whole heat exchanger system and cylinder set. Only with such information, in some cases, it is difficult to evaluate the actual performance of a regenerator installed in an actual engine[7].

To predict engine and cooler performance, a Stirling Engine Thermodynamic and Mechanical Analysis, SETMA has been developed. The SETMA divides working space into five controlled volumes and

Table 5 Specifications of tested regenerator

	100S	100	150	150/200
Dimension (mm)	70-56	70-60	70-60	70-60
Mesh No.	#40-100	#40-100	#40-150	#40-150+#40-200
Sheet num.	227	240	428	210+250
Wire diameter (mm)	0.1	0.1	0.07	0.07,0.05
Pitch (mm)	0.254	0.254	0.169	0.169,0.127
Matrix surface (mm^3)	$2.56x10^9$	$2.76x10^9$	$4.18x10^9$	$5.52x10^9$
Equiv. flow area (mm^2)	$1.41x10^6$	$1.41x10^6$	$1.32x10^6$	$1.41x10^6$
Dead volume (mm^3)	$1.29x10^5$	$1.38x10^5$	$1.35x10^5$	$1.38x10^5$

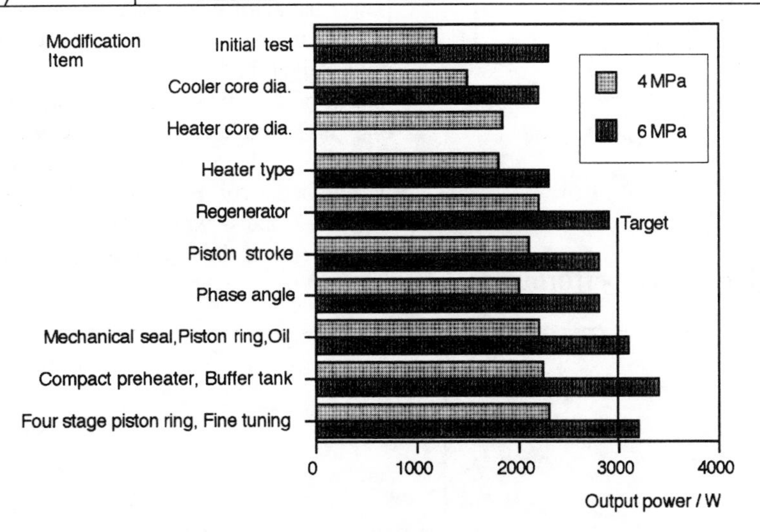

(a) Output power

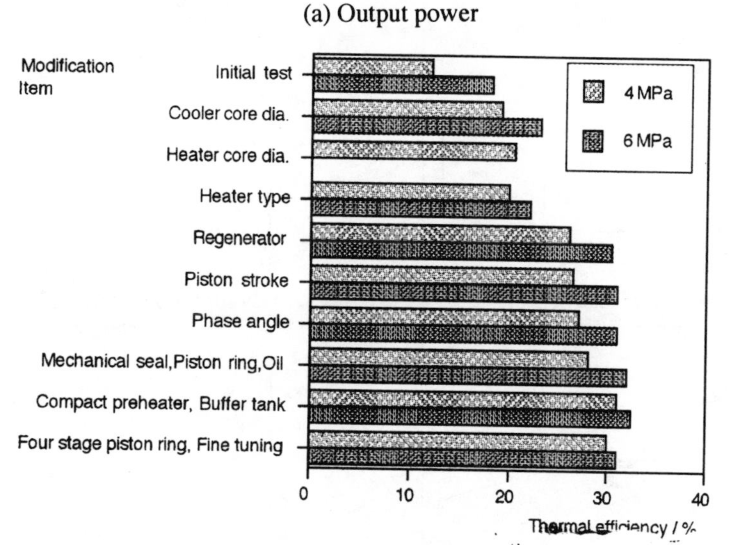

(b) Thermal efficiency

Figure 3 Improved engine performance and modification items

can analyze engine and cooler performance with enough accuracy for industrial usages[3,8]. It is a useful analysis to design Stirling cycle machines and to determine their dimensions. Figure 4 shows the SETMA analysis result of indicated efficiency of the NS03T engine. It is a function with the regenerator length and the diameter at 4.1 MPa, 1000 rpm with 1070 K heater temperature and 398 K cooler temperature. The result shows that the highest efficiency is obtained with 60mm length and 70 mm diameter. The calculated regenerator loss (reheat loss) is 480 W.

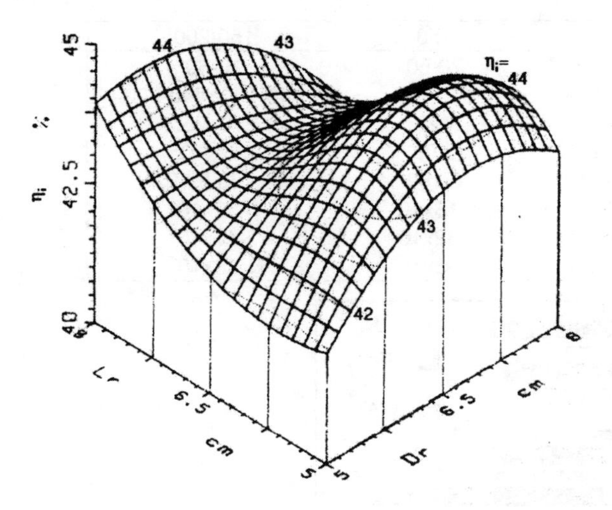

Figure 4 Calculated indicated performance

Such a reliable mathematical model is very useful to design and develop engine, however, experimental process is required to determine the final dimension. Figures 5–8 show measuring results about one of the parameter tests of the 1984 engine to determine the final dimension and the matrix of the regenerator. Four matrix sets shown in Table 5, were prepared for the test. Figure 5 shows the indicated power behavior when installing each matrix. It indicates that the matrix of #100 with short body and #150+#200 with longer body promise to play a good role as the matrix of the 1984 engine. As shown in Figure 6, #100 one is suitable for lower engine speed operation concerning the indicated efficiency. The mixed type will be better for using at higher engine speed region. In Figures 7 and 8 which show average gas temperatures at each end of the regenerator, it is clear that the mixed

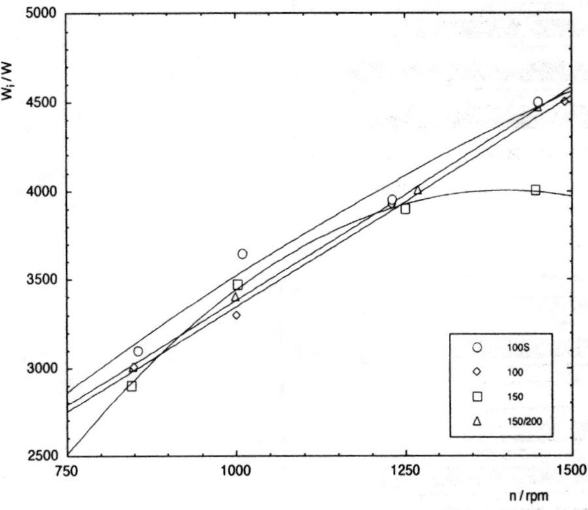

Figure 5 Indicated power of matrix sets

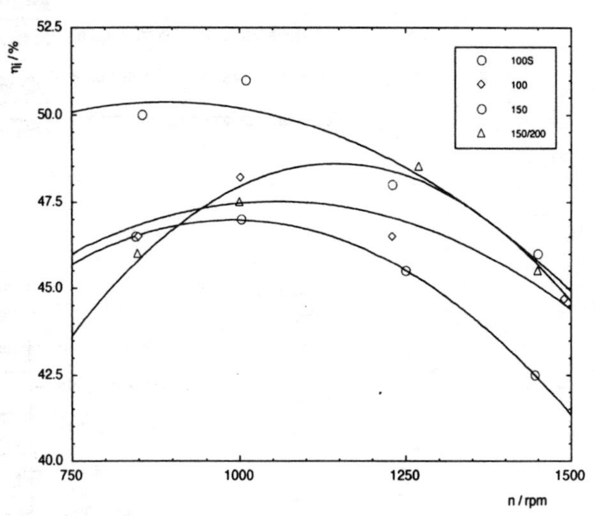

Figure 6 Indicated efficiency of matrix sets

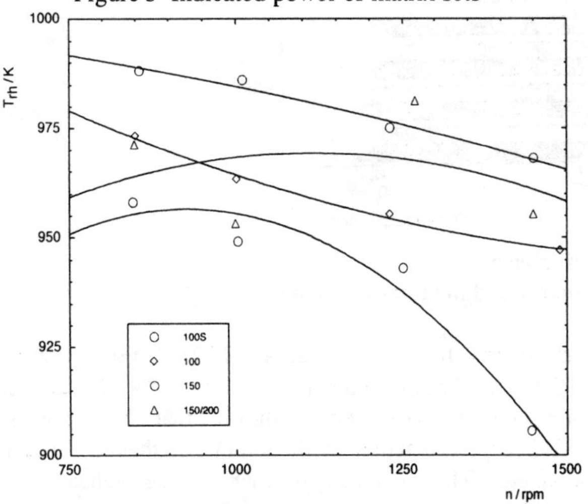

Figure 7 Gas temperature of high temperature side

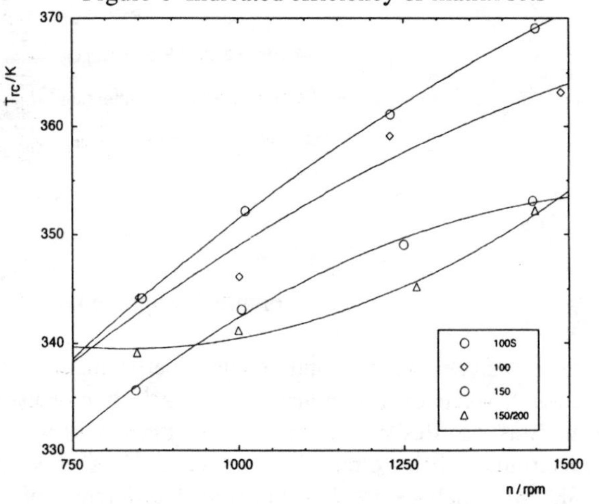

Figure 8 Gas temperature of low temperature side

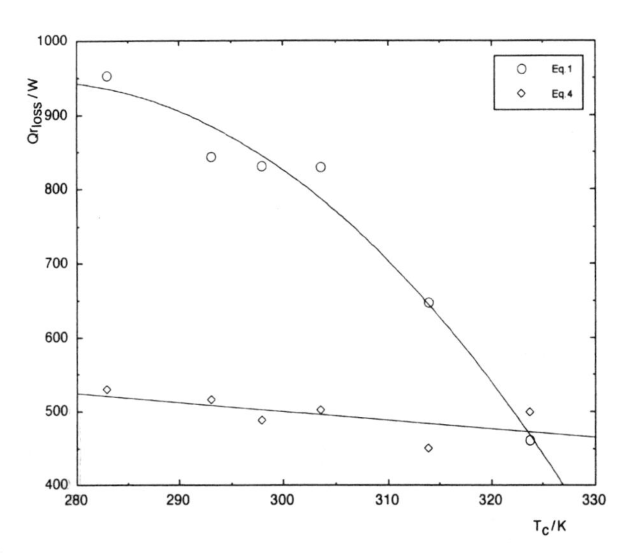

Figure 9 Regenerator loss versus cooling temperature

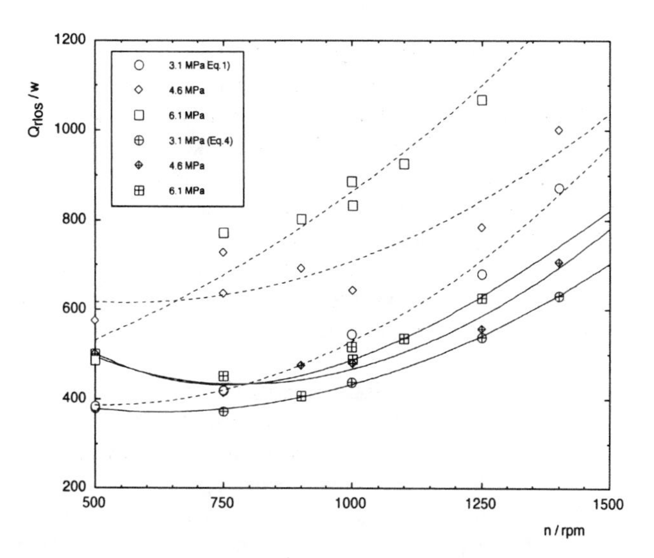

Figure 10 Regenerator loss of NS03T

type has a higher efficiency. From the results and the SETMA analysis results, the mixed one was selected for the 1984-year model engine. And the engine has a higher performance with the aid of the regenerator and matrix as show in Figure 3.

EVALUATION OF REGENERATOR LOSS

The regenerator performance is difficult to evaluate in actual engine. Eq. 1 is commonly referred to derive regenerator loss.

$$Q_{rlos} = Q_c + Q_{ec} - |W_c| \qquad (1)$$

However, the regenerator loss calculated from Eq.1 shifted with operating conditions. For example, Figure 9 shows the temperature dependence of the regenerator loss during the midterm evaluation test of the 1984-year model engine. Even though the temperature efficiency of the regenerator was kept at 86 % and 94 % with changing the cooling water temperatures, the regenerator loss is changed considerably. The reason seems that the heat flows in the engine were influenced by the circumstance, such as temperatures of water and air. Proposing a model of the heat flows is defined by using the following heat transfer fundamentals[8].

Heat conduction from element in engine:

$$Q_{conduction} = \lambda A \frac{\Delta T}{l} \qquad (2)$$

Free convection from engine to air:

$$Q_{heat\ transfer} = hS\Delta T \qquad (3)$$

where λ [W/m K] and h [W/m^2 K] represent thermal conductivity and heat transfer coefficients, A, l, S, and ΔT represent cross-area, surface area, distance (thickness), and temperature differences, respectively.

A proposed method to evaluate regenerator loss from actual engine data is

$$Q_{rlos} = Q_c + Q_{cc} + Q_{air} - Q_{cond} \qquad (4)$$

where, Q_{air} is heat transfer from the cooler to air and, Q_{cond} is heat leakage from the compression cylinder to the cooler. Figures 9 and 10 include the regenerator loss of the engine based on Eq.4. It shows more reasonable behavior and it is similar to one of the SETMA analysis results. The reheat loss is 500 W at 6 MPa an 1000 rpm. The calculated thermal conduction loss along the regenerator wall is 130 W. One through the matrix is 18W. At 500 rpm, the heat transfer of the matrix is decreased by lower flow rate. Furthermore, the amount of reheat becomes less so that the thermal conduction losses are dominated. It is cleared from that the regenerator efficiencies derived from gas temperatures remarkably decreased at lower engine speed.

CONCLUSIONS

In this paper, the specifications of the 1984-year NS03T engine were presented. The higher indicated efficiency of the NS03T engine was analyzed by using the reliable experimental data and the mathematical model, SETMA. The regenerator loss was analyzed with a proposed method, which is able to derive the reasonable regenerator loss from the engine operating data. And the result showed that the design of the heat exchanger is important, especially, the low-temperature parts, cooler and regenerator, for the high efficiency engines.

The NS03T was reassembled to study its high performance in detail at National Defense Academy. The engine performance and the regenerator loss are currently being measured.

ACKNOWLEDGEMENTS

The author would like to thank Iwao Yamashita of the Mechanical Engineering Laboratory (currently professor of Tokyo Denki University) for presenting the important data of the midterm evaluation test. And the author also thanks Moriyoshi Sakamoto (currently president of Tokyo Technical Junior College), Shigemi Nagatomo, Tsuyoshi Komakine, and Ichiro Hongo of Toshiba for their technical activities to develop the NS03T engine.

REFERENCES

1. Kagawa, N. et al., Development of a 3 kW Stirling Engine for a Residential Heat Pump System. Proc 4th ICSE. 1988. Tokyo: JSME, pp.1-6.
2. Komakine, T. et al., Performance of A 3 kW Prototype Stirling Engine for a Gas-fired Residential Heat Pump. SAE Paper No. 859142, 1985.
3. Kagawa, N. et al., Performance Analysis and Improvement of a 3 kW Stirling Engine. Proc. 21st IECEC. 1986. San Diego: ACS No. 869109.
4. Kays, W.M. and London, A.L., Compact Heat Exchangers. 1964: McGraw Hill.
5. Hamaguchi, K. et al., Improvement on Regenerator Matrix Properties by the Combined Mesh Wire Gauze. Proc. 4th ICSE. 1988. Tokyo: JSME, pp.387-392.
6. Organ, A.J., Thermodynamics & Gas Dynamics of the Stirling Cycle Machine. 1992:Cambridge Univ. Press.
7. Thomas, B., Evaluation of 6 Different Correlations for the Flow Friction Factor of Stirling Engine Regenerators. SAE Paper No. 1999-01-2456, 1999.
8. Kagawa, N. Analysis Method for Stirling Engines and Coolers. JSME Int. Journal, 1997: Series B, 41(3): pp.632-637.
9. Kagawa, N. and Ohyama, T., Regenerator Loss of 3-kW Stirling Engine. Proc. 9th ISEC. 1999, Pilanesburg: pp.201-206.

THEORETICAL INVESTIGATIONS ON THE STIRLING ENGINE WORKING PROCESS

K. Makhkamov
Physical –Technical Institute
Uzbek Academy of Sciences
700048, 2-b Mavlyanova, Tashkent, Uzbekistan
E-mail: khamid@amsta.leeds.ac.uk
Fax: +998 3712 354291

D.B. Ingham
School of Mathematics
The University of Leeds
LS2 9JT Leeds, UK
E-mail: amt6dbi@amsta.leeds.ac.uk
Fax: +44 113 242 9925

Abstract

The use of CFD models significantly extends the capabilities for the detailed analysis of the complex heat transfer and gas dynamic processes which occur in the internal gas circuit of a Stirling Engine. In this paper some restrictions on the implementation of a two-dimensional CFD model for the description of the working process of the Stirling Engine with a complex geometry for its heat exchangers are discussed. A mathematical model for a 2-dimensional simulation of a "V"-type Stirling Engine is described and a simplified axis-symmetric calculation scheme for the Engine has been assumed using the standard κ-ε turbulence model with a deforming mesh to describe the piston movements for the analysis of the Engine's working process. Gas temperature and pressure distributions and velocity fields in the internal circuit of the Engine have been obtained and pressure-volume diagrams for the working spaces of the machine have been calculated. Comparison of the numerical results obtained from the two-dimensional simulation of the Engine's working process with those computed with the use of second-order mathematical models shows that there are considerable differences. In particular, an analysis of the data obtained shows that the changing of the gas temperature in the compression space depends on the location in the cylinder and it may differ substantially from being harmonic in time. At present work on the experimental validation of the computational results is underway and it is planned to use the developed approach in the design process of new Stirling Engines in the Laboratory for Stirling Engines at the Physical-Technical Institute in Tashkent.

1. Introduction

One of the main components of the research activities of the Laboratory for Stirling Engines, Physical-Technical Institute in Tashkent, Uzbekistan, in the development of Solar and Micro Combined Heat and Power (MCHP) Co-generation installations on the basis of Stirling Engines, is to create computational tools which allow a more accurate mathematical simulation of the Stirling Engine's working process and an estimation of the performance of the machines.

Until recently the Laboratory for Stirling Engines in Tashkent has mainly used a second-order type of mathematical model for the description of the Stirling Engine working process to evaluate its performance. The second-order mathematical models consider a one-dimensional scheme for the Engine and assumes a uniform fluid velocity, temperature and pressure profiles at the planes which divide the internal gas circuit of the Engine into a limited number of cells (usually 5-30). Our experience is that in order to improve the accuracy in the estimation of the engine's performance then it is necessary to gain a much greater insight into the heat transfer processes which occur in the internal gas circuit of the Stirling Engine with the use of CFD.

In this paper mathematical simulation methods, which are in use in the Laboratory, are demonstrated in application to the particular Engine's design. Figure 1 shows a schematic diagram of the 1-kW Stirling Engine which has been installed on the Solar Power Unit and it has the α-configuration. The main future of the design is that a cylindrical cavity type annular heat receiver (3) is used for the solar Engine and its external surface (1) is for the input of the heat flux due to solar insolation from the Concentrator.

2. Computer Simulation with the Use of a Second-Order Mathematical Model

The second-order type of mathematical models (MM) for computer simulation of Stirling cycle machines has been in the use for a long period of time and it has been described in a wide variety of scientific papers and books, see for example [1-4]. In accordance to this model, the Engine's internal volume is divided into the following five chambers (see Figure 2): the expansion space, the heater, the regenerator, the cooler and the compression space and the chambers are seen as open thermodynamic systems with a variable gas mass. The major assumptions of the MM are that the temperatures of each chamber wall and the regenerator matrix are constant and the Engine's regenerator operates as an ideal regenerator, i.e. the efficiency of the regenerator is 100%.

The main equations for the MM are the energy conservation equations, the equations of mass conservation which may be written for each chamber and the van der Waals gas state equation which is written for the whole internal volume of the Engine. Expressions for the determination of the current temperature of the working fluid in the chambers can be found by solving the energy equations. This set of equations are integrated, using the Euler method, and then the determination of the temperature change due to the pressure change, gas mixing and heat transfer are subsequently carried out at every time step. The solution of the governing differential equations depends on the initial conditions and so the results for a steady state regime are obtained by reaching a heat balance over the cycle. In so doing it was assumed that the heat transfer in the expansion and compression spaces between the gas and the chamber walls is zero over the whole cycle. After obtaining a heat balance, the correctness of the mean temperature of the regenerator matrix is checked. The mean temperature of the regenerator matrix depends on the matrix temperature profile along the regenerator axis and in its turn the temperature profile of the matrix depends on the working process parameters. If it is necessary, the value of the mean temperature of the regenerator matrix is refined by the use of a simple iterative method. The value of the indicated cycle work is determined as the difference between the area of the PV -diagrams of the expansion and the compression spaces, and when determining this value, the pressure drop in the channels of the heater, the regenerator and the cooler are taken into account. Results of the simulations of the Engine's working process are typical of those obtained by the use of the second level MM.

The temperature of the gas as a function of time in the chambers relative to the wall temperature of the chambers is shown in Figure 3. To improve the accuracy of the prediction of the gas temperature as a function of time in the regenerator it is necessary to use both a much more complex calculation scheme and detailed mathematical model to describe the heat transfer processes which occur in the regenerator. Figure 4 shows the computed pressure-volume diagrams for the engine and the indicated cycle work (L_i) and the indicated power (N_i) are 160J and almost 3190W, respectively.

3. Two-dimensional CFD model of the engine's working process

The use of two-dimensional CFD models can significantly extend the capabilities for the more detailed analysis of the complex heat transfer and gas dynamical processes which occur in the internal gas circuit. Therefore, at present, specialists of the Laboratory for Stirling Engines in Tashkent and the Computational Fluid Dynamics Centre of the University of Leeds (UK) are developing a two-dimensional model for the working process of the Stirling Engine with the use of the standard κ-ε turbulent model for compressible flow.

This type of CFD model can be used for obtaining a two-dimensional description of the working process with certain restrictions because, in fact, the heat input into the working gas inside the engine, and the heat output from the working gas into the cooling system, are such that laminar, transient and turbulent regimes of the gas flow occur in each location but in various instances of time in the engine cycle. Further, all these regimes take place at the same instance but in various locations along the channels of the engine. An additional restriction is that an "equivalent" axi-symmetrical calculation scheme should be used to describe the geometry of the engine in order to satisfactorily estimate the performance of the engine.

An axi-symmetric simplified calculation scheme is employed for the internal circuit of the 1-kW Stirling Engine and the physical dimensions of the elements of this scheme are shown in Figure 5 for the instance when the volume of the expansion space is a maximum and the piston in the compression space is at its middle position (crankshaft angle ϕ=0). The side lines P_e and P_c represent the surfaces of the moving pistons in the expansion and compression spaces of the Engine, respectively. The expansion and compression spaces are also bounded by the lines W_e and W_c, respectively, and the lines H_e and H_c are the top of the "hot" and the "cold" cylinders, respectively. The channel of the heater, which is bounded by the lines HW_1-HW_4, is connected to the expansion space by a special channel, which is bounded by the lines W_1 and W_2. The cylindrical regenerator, with the side wall RW_1, the top RW_2 and the bottom RW_3, is connected to the channel of the cooler which is bounded by the lines CW_1-CW_4. In its turn, the cooler is connected to the compression space by a tube W_3.

The control volume technique for a moving structured grid [5] has been used for solving the governing equations of the standard κ-ε turbulent model for compressible flow, which describes the gas dynamical and heat transfer processes which occur in the internal circuit of the machine. In the computations we assume that the temperature of the sides W_e, H_e, W_1, W_2 and HW_2 is 1000K, the temperature of the sides HW_4 and RW_1 is 800K, and the sides CW_1-CW_4, W_3, H_c and W_c have a temperature of 333K. Additionally, it was assumed that there is zero heat flux on the sides HW_1, HW_3, RW_3 and on the surfaces of the pistons.

Figure 6 shows the typical computational grid employed for the internal gas circuit of the Stirling

Engine and Figures 7-11 present some of the obtained computational results.

The fluid velocity fields in the expansion space, the regenerator and compression space are shown in Figure 7 for the moments when ϕ is 0, 90, 180 and 270 degrees, respectively.

Figures 8 shows the temperature distribution in the expansion space, the regenerator, the compression space and in one of the sections of the heater of the Engine for the moment when $\phi=0$.

Figures 9 (a) and (b) show the temperature of the working gas as a function of time in the middle plane of the expansion space and regenerator for different values of their radii. It can be observed that the computed curves are similar to those obtained when using the second-order MM. Figure 9(c) presents the average temperature in different planes along the axis of the compression space. These results show that the character of the change of the working gas temperature is quite different from being harmonic in time and this may be due to the complex interaction and mixing between the working gas in the "cold" cylinder and a jet in the inlet of the cylinder.

High velocities of the working gas and the complex character of its motion also causes a non uniform pressure distribution in the compression space, see Figure 10.

Figure 11 presents the pressure-volume diagrams for the 1-kW Stirling Engine. Pressure curves P_E and P_c are the average pressures of the working gas on surfaces of the "hot" and "cold" pistons, respectively. The computed value of the indicated power is 1682W and this value is significantly less than the value of 3190W obtained with the use of the second-order MM.

In general, it may be observed from the presented computational results that the fluid flow in the channels of the Engine has a very complex character and the parameters of the gas are inhomogeneous. Clearly, the inclusion of this effect into the analysis of the working process enables a more accurate prediction of the Engine's performance.

4. Outlook

At present further theoretical investigations on the working process of Stirling Engines and Combustion Chambers for MCHP systems with the use of CFD are ongoing.

It is planned to build and conduct experimental tests on the MCHP installations at the Laboratory for Stirling Engines in Tashkent in order to experimentally validate the numerical results which have been obtained. During the tests measurements of the Engine's and Combustion Chamber's working process parameters, such as instant and average temperatures and pressures of the gas, speed of the Engine, heat losses and output power, will be performed with the use of a PC based data acquisition system.

Acknowledgements

This research work was partially supported by a NATO Science for Peace Grant (SfP- 972296) and an EU INCO-Copernicus Grant (Contract ERB IC15-CT98-0501).

References

1. **Finkelstein, T.**, 1965, "Simulation of a Regenerative Reciprocating Machine on an Analogue Computer", *Proceedings of the SAE International Congress on Automotive Engineering*, 11-15 January, Detroit, Paper 949F.

2. **Tew, R., Jeffries, K. and Miao, D.A.**, 1978, "A Stirling Engine Computer Model for Performance Calculations", *US DOE/NASA Report TM-78884*.

3. **Urieli, I. and Berchowitz, D. M.**, 1984, "Stirling Cycle Engine Analysis", *Adam Hilger*, Bristol.

4. **Makhkamov, Kh. and Ingham, D.B.**, 1999, Analysis of the Working Process and Mechanical Losses in a Stirling Engine for a Solar Power Unit, *ASME Journal of Solar Energy Engineering,* Vol. 121, No 2, 121-127.

5. **Hirsch, C.**, 1990, "Numerical computation of internal and external flows", Vol.1-2, *Wiley,* New-York.

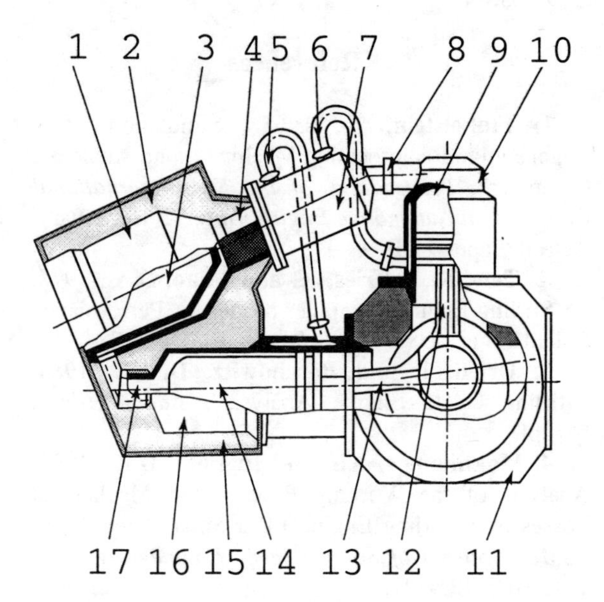

Figure 1. A 1-kW Solar Stirling Engine.

1-The external surface of the heat receiver; 2,15-thermoinsolation; 3-the cavity of the heat receiver; 4-the regenerator; 5,6-the tubes of the cooling system; 7- the cooler; 8-the joint between the cooler and the "cold" cylinder; 9-the piston of the compression space; 10-the cylinder of the compression space; 11-the engine's crank-case; 12,13-connecting roads; 14-the piston of the expansion space; 16-the cylinder of the expansion space; 17-the joint between the "hot" cylinder and the heater.

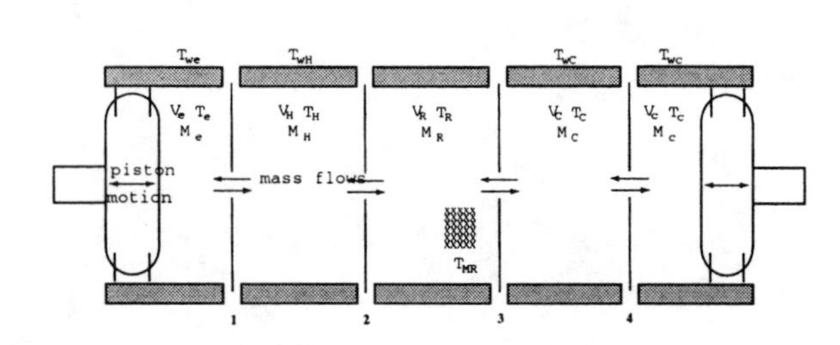

Figure 2. The 5-chamber calculation scheme for the Stirling Engine.

1, 2, 3, 4 - separating planes between the chambers.

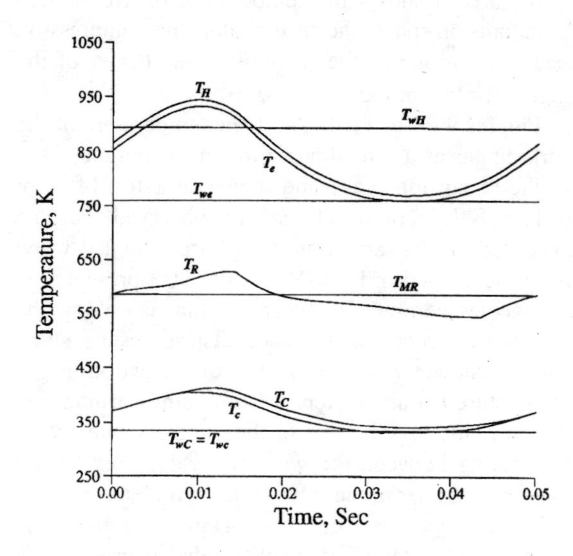

Fig. 3. The temperature as a functions of time.

T_e, T_H, T_R, T_C and T_c are the temperature of the gas in the expansion space, the heater, the regenerator, the cooler and in the compression space, respectively.

T_{we}, T_{wH}, T_{wC} and T_{wc} are the temperature of the walls of the expansion space, the heater, the cooler and of the compression space, respectively.

T_{MR} is the temperature of the regenerator matrix.

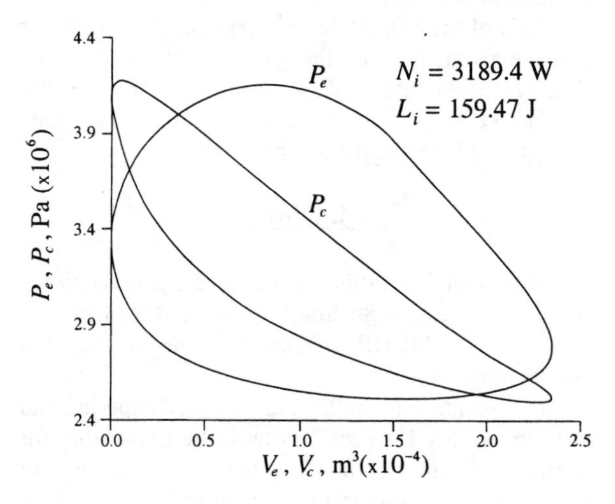

Fig. 4. Pressure-volume diagrams.

P_e, P_C are the pressure of the gas in the expansion and compression space, respectively.

L_i, N_i are the indicated cyclic work and indicated power, respectively.

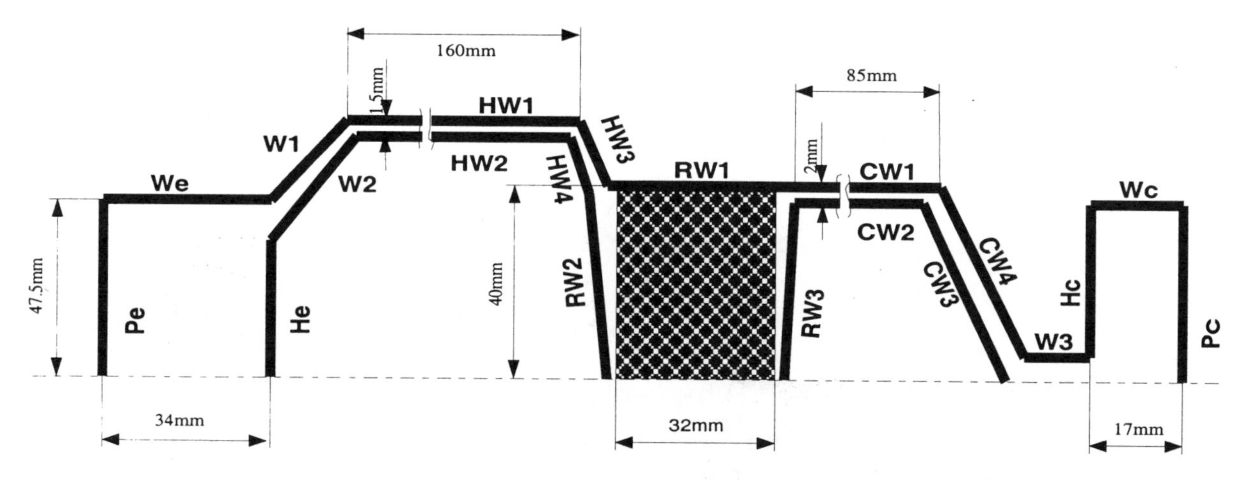

Figure 5. Schematic calculation scheme of the 1-kW Stirling Engine.

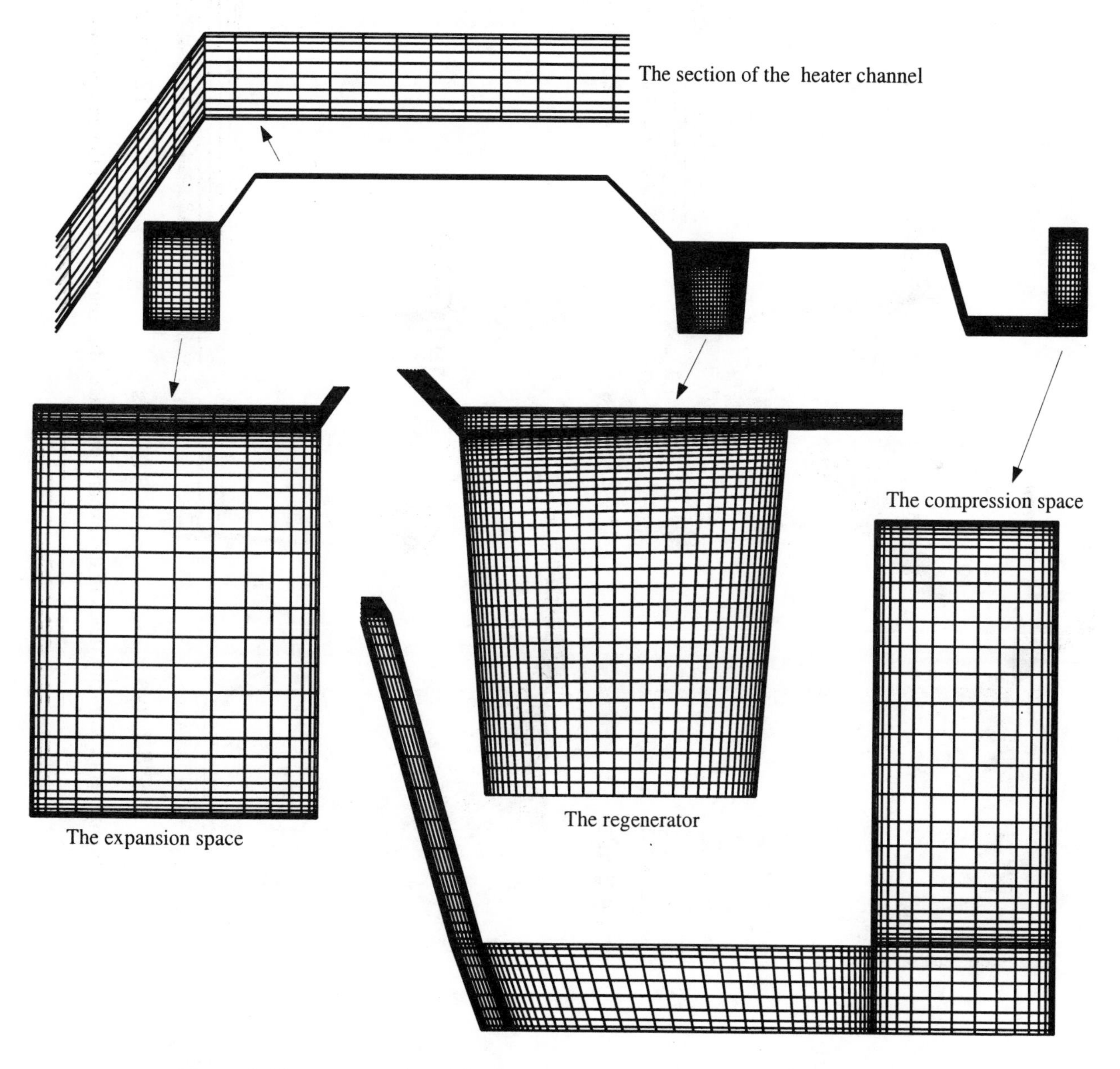

Figure 6. Computational grid for the internal gas circuit of the 1-kW Stirling Engine.

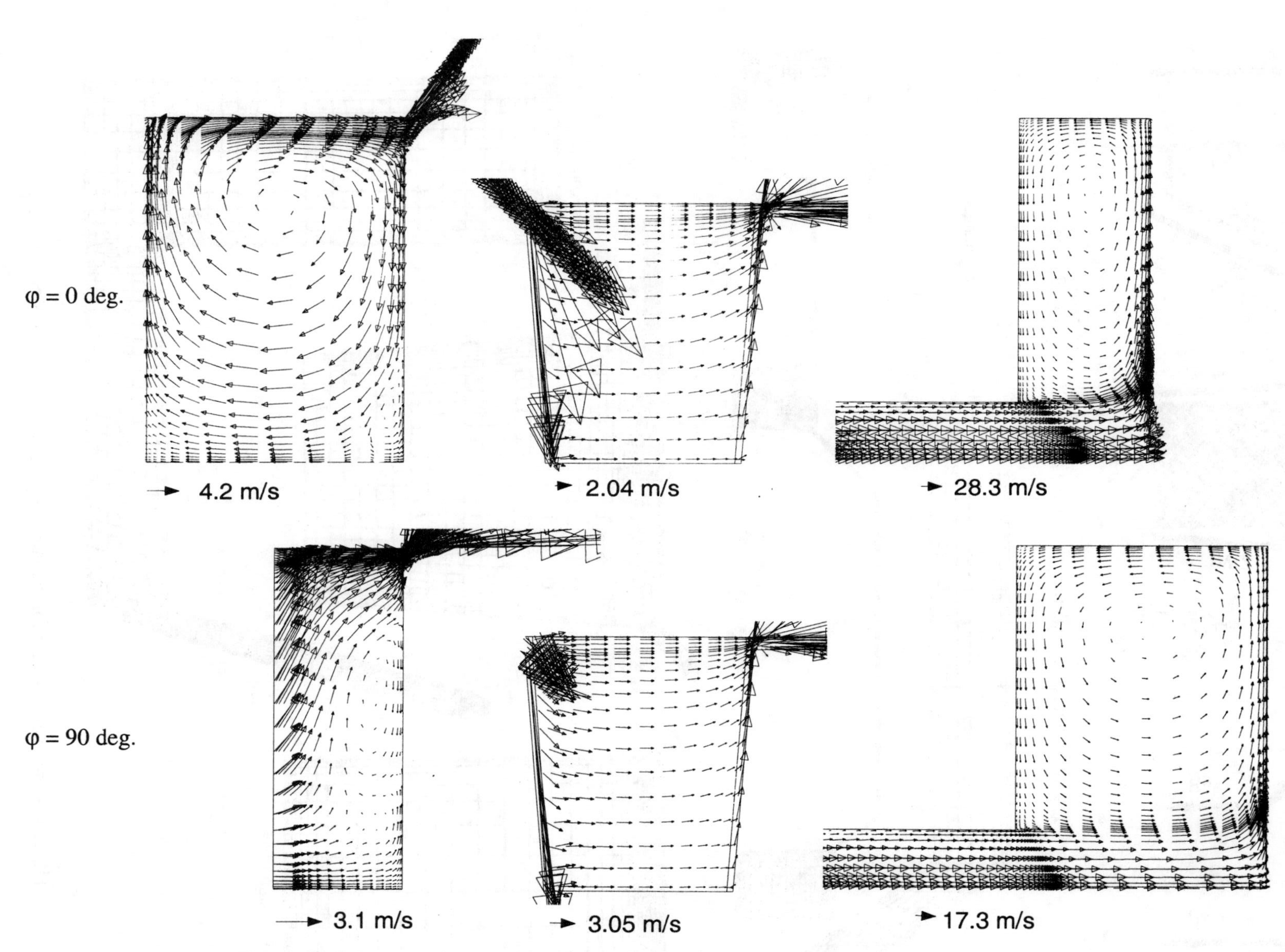

Figure 7(a). Velocity fields in the expansion space, the regenerator and the compression space.

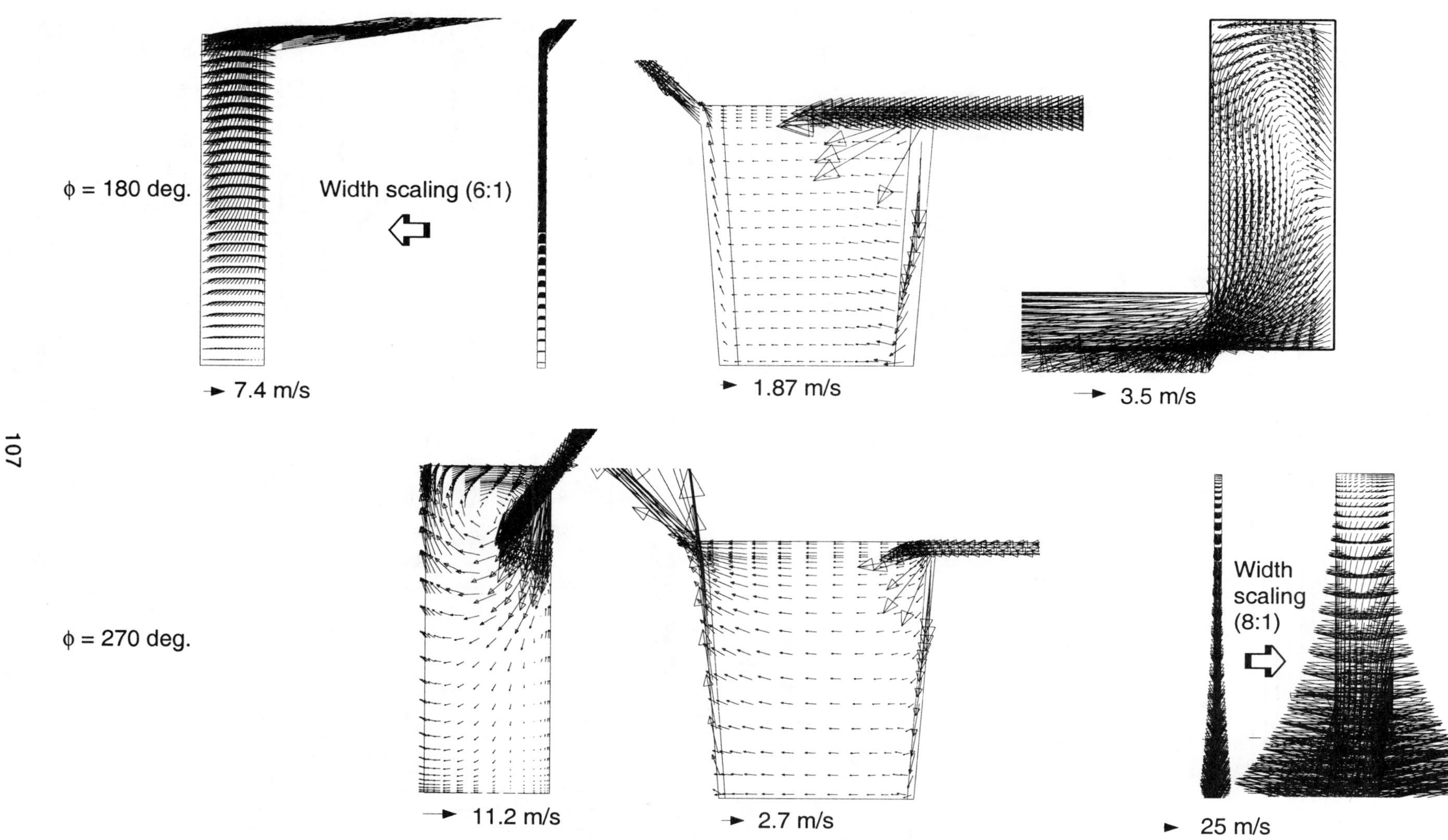

Figure 7(b). Velocity fields in the expansion space, the regenerator and the compression space.

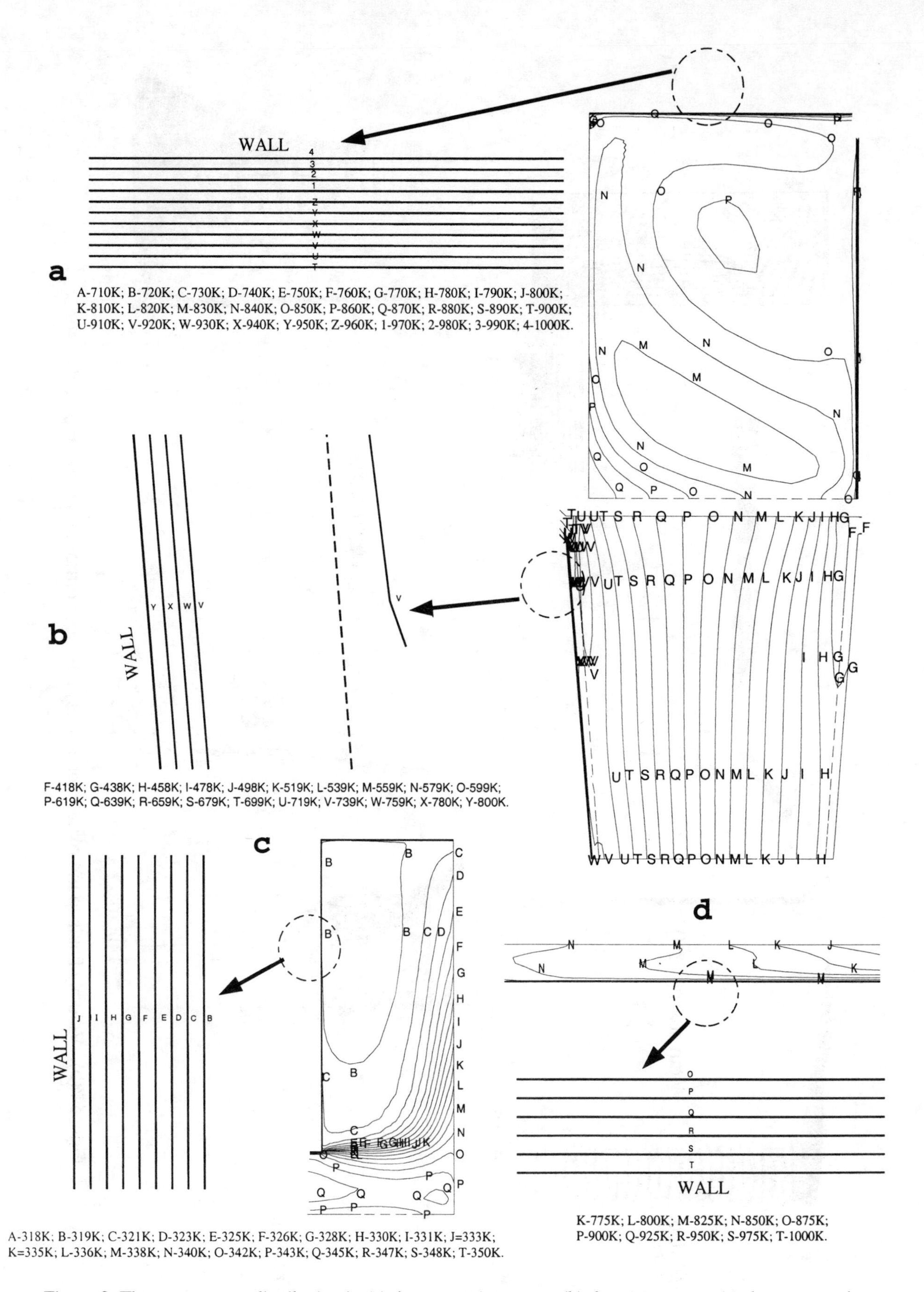

Figure 8. The temperature distribution in (a) the expansion space, (b) the regenerator, (c) the compression space and (d) the section of the heater for the instant $\phi = 0$ degree.

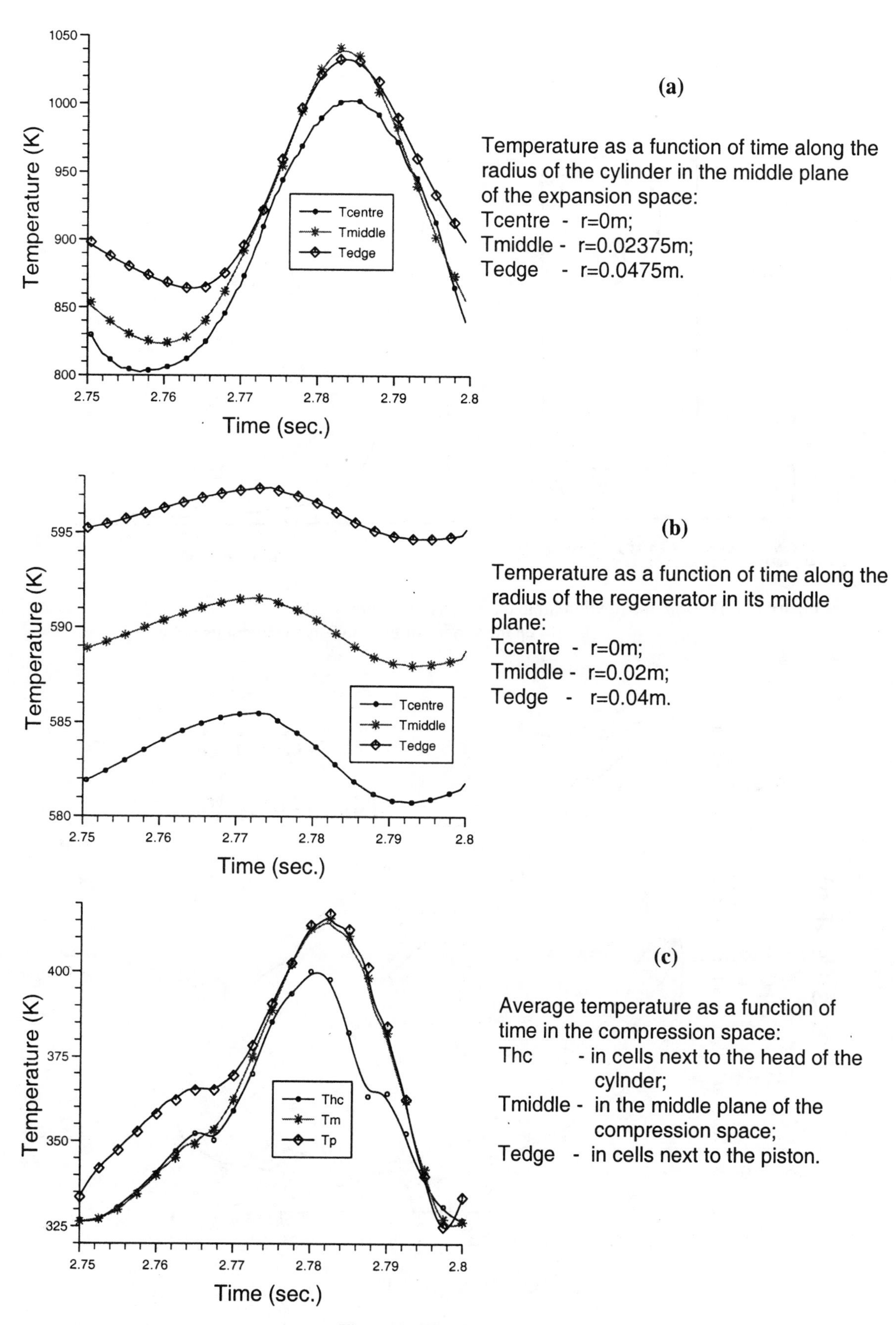

Figure 9. Temperature as a function of time.

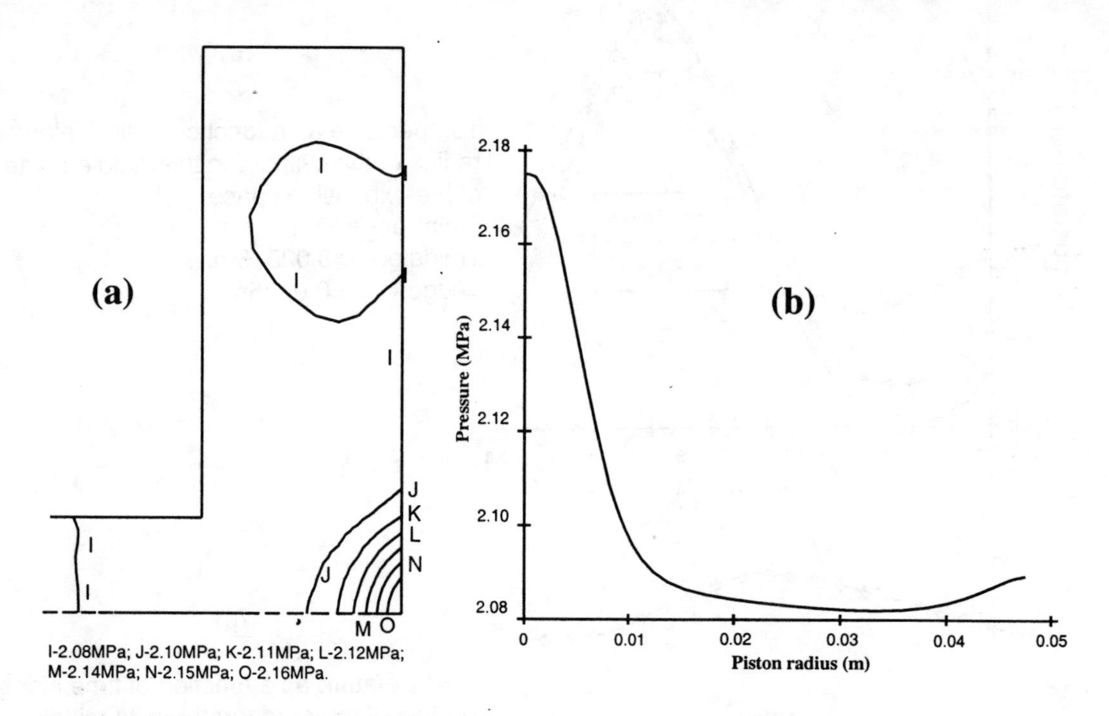

Figure 10. Pressure distribution (a) in the expansion space;
(b) on the surface of the "hot" piston along its radius.

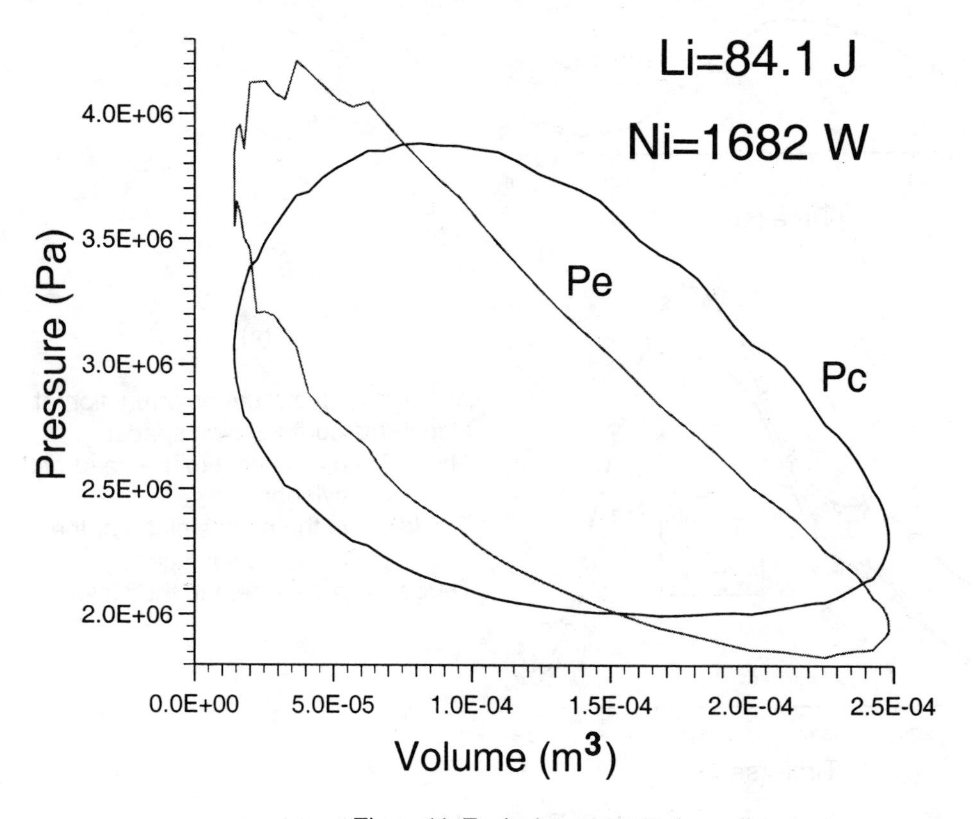

Figure 11. Engine's pressure-volume diagrams.

Report on the Developments of Steam Super Stirling Engine

Naostugu ISSHIKI
Tokyo Inst. of Tech, Kodo, Setagaya, Tokyo, 1560052, Japan

Hiroshi KOJIMA
Maeta Techno-Research, Inc., Sakatashi, Ymagata-ken, 9988611, Japan

Izumi USHIYAMA,
Ashikaga Inst. of Tech, Ashikagashi, 3268558, Japan

Seita ISSHIKI,
Fukushima National College of Tech. Iwakishi, 9708034, Japan

ABSTRACT

"Steam Super Stirling Engine (SSSE)" is a hybrid engine of Rankins cycle steam engine and steam Stirling engine.

This engine has power piston and displaced piston. and when the power piston comes near its top dead center, saturated steam of 0.2~1.0 Mpa is injected into the cold side of the displaced cylinder, and the steam expands into power cylinder at constant high temperature after passing regenerator and heater, as Stirling engine. Many small remodeled atmospheric engines were tasted. Then a small double expanding SSSE was installed on an electric vehicle for generations of electricity. Each engine above was successful, and below points was known.

(1) Power increases up to several times of atmospheric Stirling engines.
(2) This engine needs no Helium gas.
(3) Nearly no cooling system and pressurized crankcase are needed. (4) Theoretical thermal efficiency is expected to be middle of both cycles.

Now three kW double expansions SSSE is being made.

INTRODUCTION

This " Steam Super Stirling Engines(SSSE)" is a quite new engine which is a hybrid of classic steam engine (steam expander) and steam Stirling engine combining both good points.

The main engine uses steam from any steam generator as its working fluid and it has power pistons and displacer pistons, heat regenerators, and steam heaters just like general Stirling engines. By any type valves, steam is injected and ejected from this engine same as steam expander, but hot steam heat higher than 500 ℃ is regenerated by the regenerator.

So it works along combined Stirling and Rankins cycle, and this engine is very light weight and cheap and easy to manufacture, because it dose not need high pressure Helium and cooling system if part of steam is injected and rejected to outside as exhaust.

First many small model conventional hot air Stirling engines were reformed to this engine type with various valves for injection and rejection of compressed air instead of steam. They worked very well with several times higher powers, when heated.

Then first small steam Super Stirling engine of this kind called "Mini SSSE" which is a double expansion engine with 26mm and 38mm diameter power pistons.

This Mini SSSE engine worked well with 3 at a steam from rice husk boiler, and then it worked well as a prime mover of a small manned Electric Vehicle combined with one small model boiler.

Now new complete prototype of this engine of 3KW target named "3KW-SSSE" is under construction.

It will work with 0.9Mpa saturated steam, through two stages expansion cylinder each has heater and regenerator. The first results will be reported at this conference.

This engines is very light weight with open crankcase and no need of Helium and cooling system except condenser if needed, and it fits to all kinds of heat sources including biomass and solar etc.

PRINCIPLE OF SSSE

For this engine, there are two cases, one is single expansion with low-pressure steam, and the other is two or more stages expansion by high pressure steam. At Fig.1 the illustration of basic principle of double expansion SSSE case with single boiler and two burners is shown.

There are two stages Stirling engine type steam expanders. PP1 and PP2 are power pistons each in power cylinders.

PP2 is about two or three times bigger than PP1.

DP1 and DP2 are displacer pistons containing wire mesh heat regenerators RG1 and RG2 in each inside same as Stirling engines.

PP1, DP1,RG1 makes the first stage, and PP2, DP2, RG2 makes the second stage, and H1 and H2 are each stage's heaters heated by two burners just same as conventional Stirling engines. The two stages have 180° phase difference.

B1 is a boiler, and its steam is furnished to the engine through a horizontal rotary valve RV, and it sends steam to each stage one by one with appropriate timing.

Steam is injected first into the bottom of DP1 cylinder and steam expands in PP1 at high temperature after passing RG1 and H1.

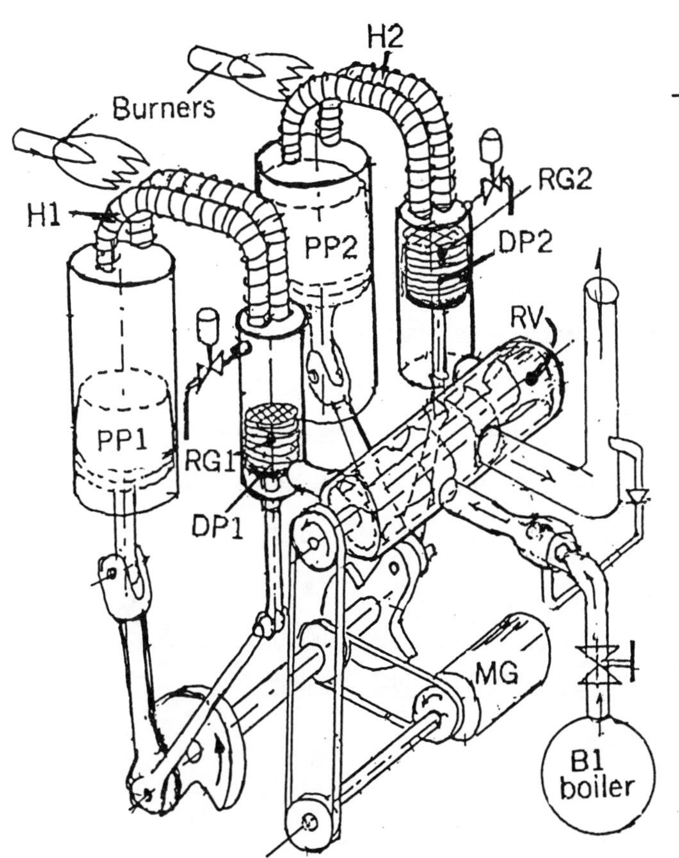

Fig.1 Principle of Double Expansion SSSE

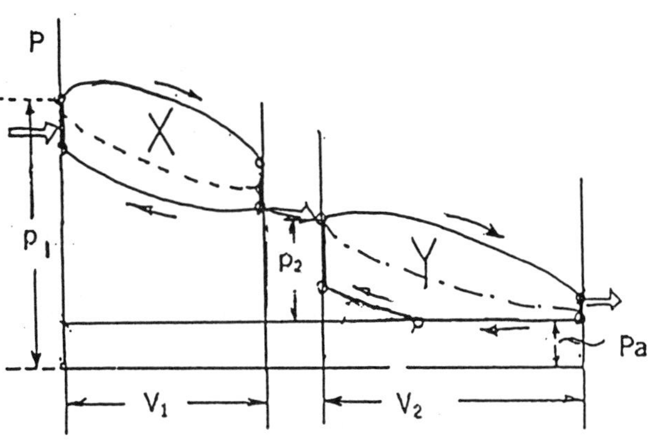

Fig.2 P~V Diagram of Double Expansion SSSE

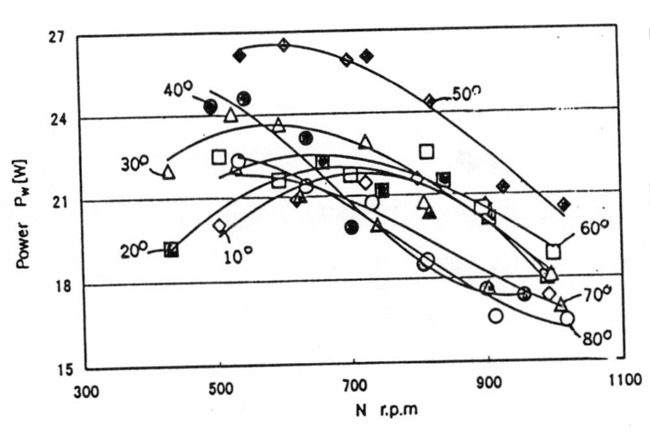

Fig. 4 An Example of Power Output of SE-45 with various Phase angle of Valve Timing, at 0.2Mpa injection Air Pressure and 333 ℃ Heater Temperature.

Timing Disc
Carbon Piston Valve
Air Input
Power cylinder
Water Cooler
Air Tube
Gas Heater
RG
Displacer Cylinder

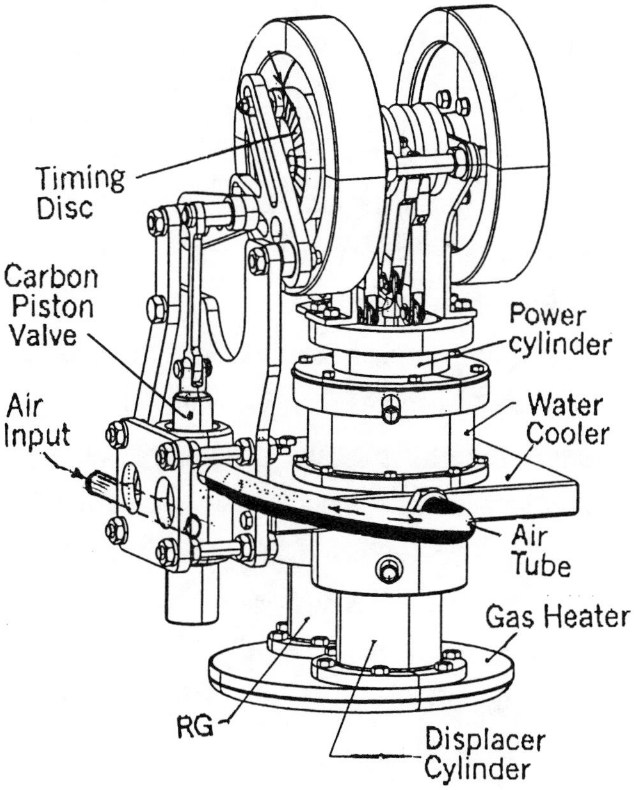

Fig. 3 SE-45 Hot Air Stirling Engine Equipped with Piston Valve for Air Injection

On the return of steam, it gets outside at lower temperature after passing RG1.

Then steam is sent to the second stage at lower pressure and works same cycle as the first one.

Exhaust steam from the second stage will be expanded again by any bottoming turbine, and or sent to the condenser, which are not shown here.

At Fig.2 an idealized p~v diagram of this engine is shown,

p_1 is the input boiler pressure, and starting pressure of PP1.

p_2 is the input of the second PP2 stage, and V_1 and V_2 are the volume of each power pistons. pa is the back pressure. X and Y are the p~v diagrams of both power cylinders.

As seen there, the cycle of this engine is the hybrid of Stirling and Rankine Cycle. So, the thermal efficiency of this engine is just between these of both cycles.

Actually, by rough calculations, thermal efficiency becomes higher with higher temperature and pressure of working steam, while smaller consumption of steam per KW, is attained. This means the ratio of Stirling cycle to Rankine cycle is hoped to be bigger to get higher efficiency. So at the end of the steam line, a choke valve is hoped to be set to choke the ejected steam, that may results higher efficiency and lower power output and steam consumption. if needed.

Actually thermal efficiency of this combined cycle is hoped to be between 20 ~ 30 %, which will be proved in future. If this engine is used for co-generation of power and heat, the total efficiency will reach 80%.

Actually, flow resistance at various parts and leakage of valves give biggest loss today, which will be much improved in future.

Also, any valves with wider steam inlet area are needed for speed up.

EXPERIMENTS BY SMALL MODEL STIRLING ENGINES

As already reported in (1), (2), first experiments of this engine were done with various conventional atmospheric models Stirling engines made by the authors for long time.

Compressed air was used instead of steam, and rotary valves and or piston valves of many types were tried for comparison.

A representative experiments with SE-45, which is Sanden Companies model Stirling engines designed by Isshiki will be shown here. As seen at Fig.3 a complicated piston type reciprocating carbon valve is installed to it, and into the cold side of the engine, compressed cold air of 0.2 Mpa is injected and ejected timely to make expected Stirling-Rankine cycle of this SSSE, and heater is heated by gas burner.

One of the results of the measured power with various valves timing of this engine is shown at Fig.4.

As seen there, the power of SE-45 reached near 27 watt, with 333℃ heater temperature, and 0.2Mpa air pressure. This power is about four times of without air injection, but its speed went down.

By other several atmospheric conventional Stirling engine models have shown very similar results with power increase nearly proportional to Stirling heater temperature and injected air pressure lower than 0.25Mpa.

MINI SSSE AND ITS EXPERIMENTS

First small double expansion steam engine is manufactured in 1999, named Mini SSSE, its section figure is shown at Fig.5.

It has a rotary valve made with brass and directly driven by the main shaft.

Although it is double expansion engine, but its second expansion cylinder has no DP, RG and Heater, so it is not yet the complete double expanding SSSE, as Fig.1.

First, the rotating parts of the valve rotor was made by carbon, but with steam, steam leaks very much by the difference of thermal expansion between carbon and metal cylinder, so the rotor was changed to brass same to the valve cylinder.

The output of this engine was about 10 to 12 watt at 0.2 Mpa (gauge) steam at 250℃ heater temperature with 400 rpm .

First, this engine was set to a steam generator specially designed on a biomass rice chaff burner for the demonstration of multi fuel capability of this engine at Maeta Company.

At Fig.6 the figures, of the rice chaff steam generator and the position of SSSE engines are shown.

There were two engines were tried, first, this mini SSSE was tested by the generated steam from rice chaff boiler and solidified chaff tips for heater, and other engine modified from conventional swash plate two-cylinder Stirling engine was tested as seen in the figure. Both engines worked well by steam, and it was a good demonstration of this kind of engine, for biomass utilization.

At Fig.7, this Mini SSSE installed on a manned small EV (electric vehicle) is shown, the SSSE is supplied steam from a small marine type three drums model boiler at its left, and reject steam at car end.

The engine is heated from downside by a gas burner, and it drives motor generator at right of the car by chain and it charges electric battery set below the seat. Solar car type directs driving motor at rear wheel drive the car at 6 km/h at maximum. On Fig.8, this Mini SSSE EV car is running at the 3 rd Stirling Techno Rally race held at Nov. '99.

About two-hour generation and charging of electricity by SSSE runs the car for half-hour same as conventional EV cars with battery.

The EV car experience has shown that

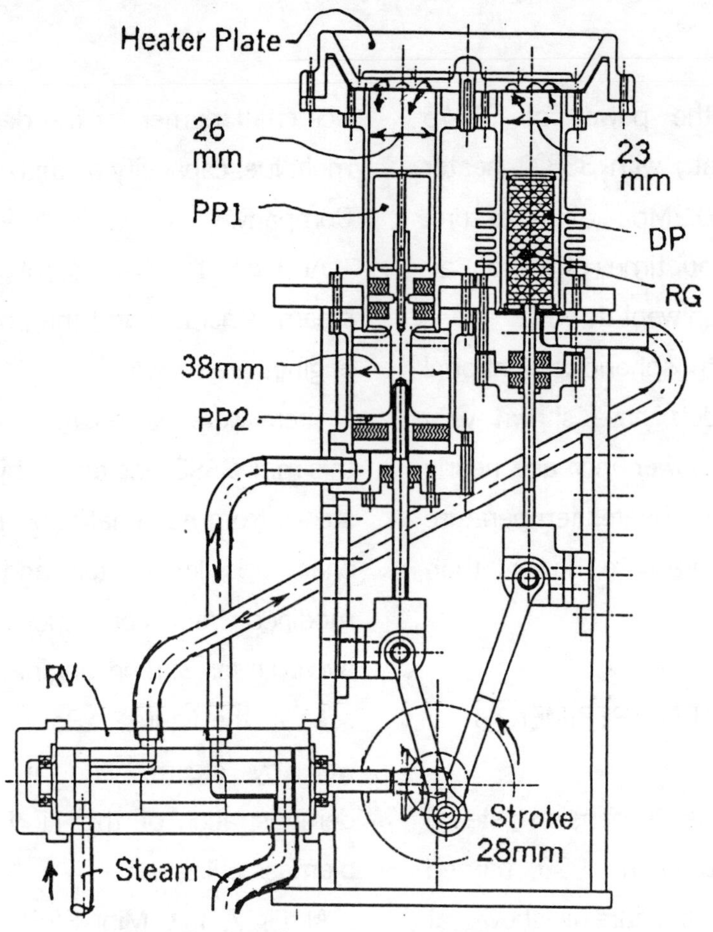

Fig. 5 Section of "MINI SSSE"

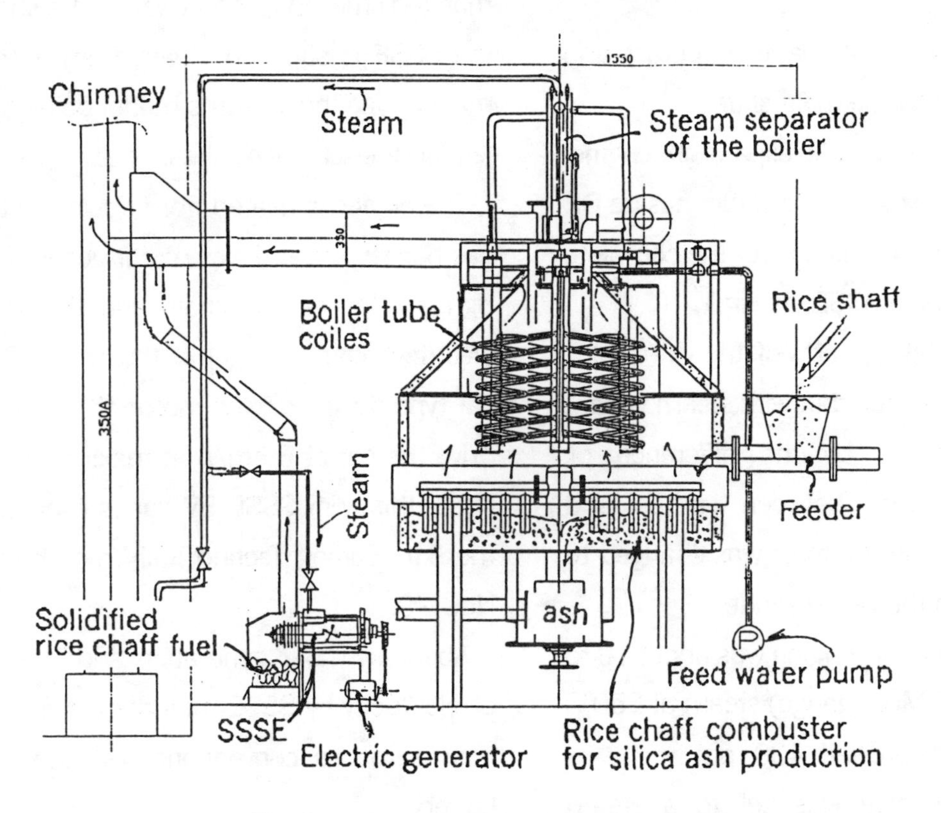

Fig.6 Rice Chaff Furnace and Boiler sends Steam to SSSEs.

SSSE engine fits very well to the EV hybrid car, because constant running of the engine and boiler are quite suitable. By this experiments it is known that any kind of Stirling engine fits to EV cars very well.

PROJECT OF 3KW SSSE

Now, for one more higher step to the practical application, "3KW SSSE" project is going on, and the design of engine itself is nearly completed.

This has just same principle as Fig. 1, and it has two-stage expansion type cylinders with two sets of PP, DP; H and RG. Steam is distributed by one long cylindrical rotary valve on the back of the engine.

The diameters of PP and DP are 92mm and 40mm, and their stroke are 60mm and 92mm for first and second stage, with 840 rpm designed speed.

The head heaters are bayonet types made of double tube, and each heater has nine bayonets.

The pressure of injected steam is 0.9 Mpa (gauge) and it should be heated up to 550℃ at the heaters. The consumption of the steam is planned less than 24 Kg/h, if leakage is small. Other type of valve than rotary valve will be tested for better efficiency and less leakage and to get best timing.

At Fig.9, the general view of the proposed total SSSE system is shown. At the furnace, biomass wood tips, solidified rice chaff and plastic tips, or their mixture will be burnt by natural convection of a tall chimney. The style of the boiler of this time is a mono tubes one through type steam generator with one steam and water separator. It has a circular coil tube heated by rotating flue gas as cyclone, at the back of engine, and the boiler furnishes steam to the main engine, also over generated steam than the engine capacity will be used for co-generation. The maximum thermal efficiency of co-generation wills reaches 85%.

CONCLUSION

As shown here, through many experiments with modified conventional model Stirling engine, power increase very much by air or steam injection from the bottom of DP at appropriate timing, has been known very well.

Also heating of each heater effects very well.

Then, small size mini SSSE was made and successfully tested by steam from rice chaff boiler. Then it was used to the main electricity generator engine of a small manned EV car, and it has proved that this engine fits very well to EV hybrid car.

Then 3KW SSSE prototype engine system is under construction now for up stage practical development.

These experiments and experiences have shown the following many good points of

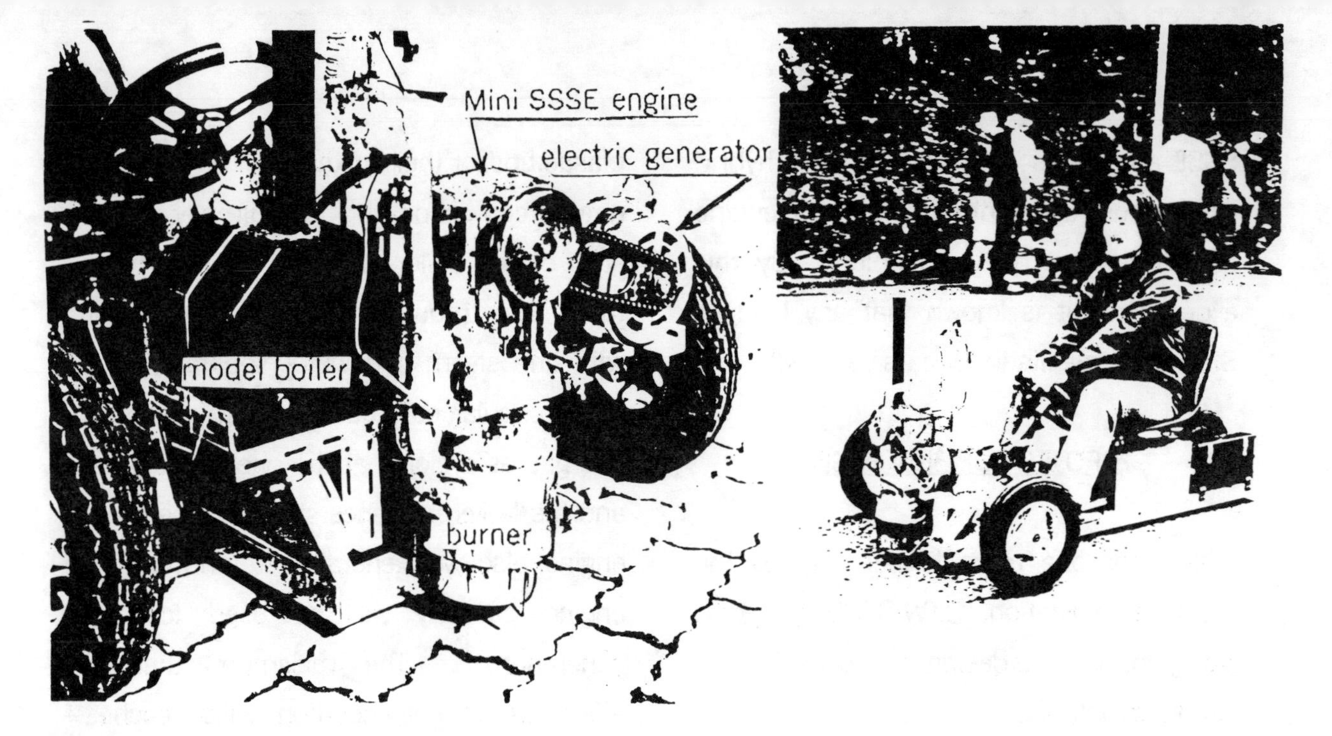

Fig. 7 Head of the Mini SSSE·EV Car, the engine is at the center, and it is heated from below by a gas burner, a model boiler for steam supply and an electric generator at left and right.

Fig.8 The Mini SSSE·EV is running.

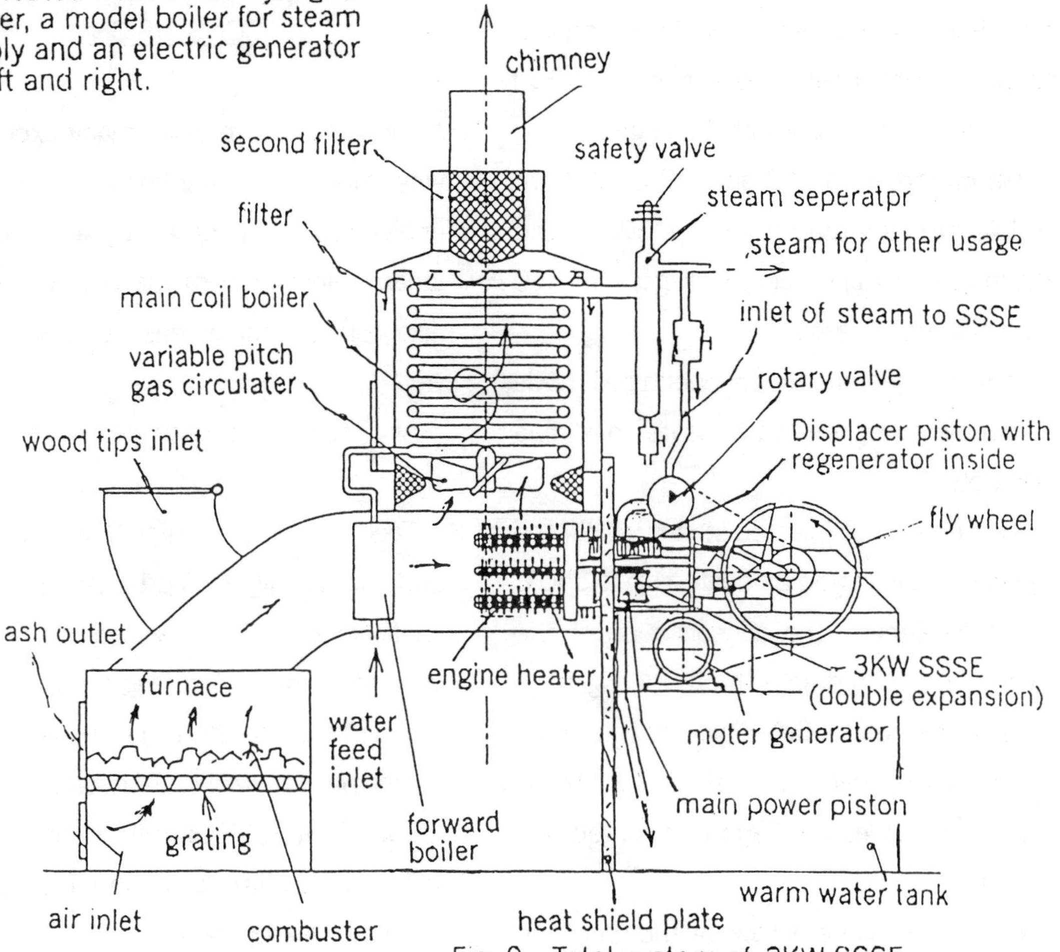

Fig. 9 Total system of 3KW SSSE cogeneration project, set to a biomass, woodtips, solidified rice chaffs, and or plastic tips furnace.

this SSSE Systems.

(1) It uses steam, so high pressure, and expensive Helium which leaks easily is not needed.

(2) It needs no coolers and no highly pressurized crankcase, so its weight is very light.

(3) Cost of the engine is far cheaper than conventional high pressure Helium Stirling engines, even if low-pressure conventional boiler system is included.

(4) Any kind of heat source, as biomass, solar, city or industrial waste incinerator heat, and hot heat from fuel batteries of high temperature can be used for the heater of the engine and or boilers.

(5) By experiments, it is known that SSSE and Stirling engine fit to hybrid EV car very well.

The new type engine, SSSE which is the combination of Stirling and Rankine cycle engines, should be widely researched in the world, of 21 st century for the use of renewable energy sources.

Especially this should be studied and utilized at developing countries where multi natural heat sources as solar and biomass are abundant.

ACKNOWLEDGEMENTS

The authors acknowledge very much helpful assistance and discussions from the president Mt. Naomi MAETA of MAETA C.T. company and Mr. Hiroshi KASHIYAMA and Mr. Shigetou OKANO of KASHIYAMA Industrial Company, and experimental efforts of Mr. WU Chungmin of Ashikaga Institute of Technology, and others.

REFERENCE

(1) Naotsugu ISSHIKI and five others "Introduction of the Super Stirling Engines" 9 th Int. Stirling Engine Conf. ISEC-99009 South Africa (1999).

(2). Naotsugu ISSHIKI and three others "Proposal and Basic Experiments of Super Stirling Engines" 34 th IECEC 01-2502, Vancouver (1999).

AIAA-2000-2818

QUICK CHARGE BATTERY WITH INTERNAL THERMAL MANAGEMENT

Ronald J. Parise
Parise Research Technologies
101 Wendover Road
Suffield, CT 06078

ABSTRACT

Thermoelectric generators (TEGs) are unique solid state components that can be used as a cooling device when supplied with electric power. These coolers can be built in many configurations for unusual applications where parasitic thermal energy must be managed. Such an application is in secondary (rechargeable) batteries where thermal build-up extends recharge time and reduces battery life. These maladies increase system operating costs and add inconvenience to the operation of the battery-powered device.

In particular, battery usage in electric vehicles has been especially disappointing since recharge time needed between limited vehicle travel distances (approximately 160 km/100 mi) are not acceptable to consumers. Vehicle down-time has discouraged usage of this pollution-free means of transportation.

Higher charge rates would result in lessening vehicle recharge time. However, these charge rates are undesirable due to the thermal build-up in the battery. The bulk and mass of a typical electric vehicle battery, usually a lead-acid battery, does not lend itself to being cooled quickly, strictly from external surfaces. Therefore a means of removing the thermal energy at the point of production, internal to the battery, can be accomplished using thermoelectric coolers in the battery case, in cell partitions and/or between positive-negative plate pairs where the heat is produced.

Although the lead-acid battery is considered in this study, any other battery type with thermal management problems can be considered, including fuel cell applications. A preliminary study utilizing this unique thermal management scheme inside the battery is presented. The simple model investigates the parameters that influence parasitic heat build-up in a lead-acid battery.

INTRODUCTION

There are several sources of heat generation in secondary batteries that result in thermal build-up. Some are due to particular battery types or designs. However, there are two main causes inherent to all batteries. One is electrical I^2R losses during discharge or recharging. Quick recharge times requiring high charging rates can result in very large losses. Typically, with the resulting thermal build-up in the battery, charging must be interrupted to allow the heat to dissipate, extending recharge time.

The second source of thermal energy in electrochemical batteries is the various reactions that take place during both discharge and charging. For example, in lead-acid batteries the chemical reaction that produces the electromotive force, lead and lead dioxide with sulfuric acid, causes an exothermic heating of the battery. During charging, this is an endothermic reaction that actually cools the battery. However, during typical recharge rates, the I^2R losses more than offset this cooling, resulting in a net increase in thermal energy. And for very high charging rates, the cooling effect becomes insignificant.

Thus the metal grid plates are the prime source of heat production, both at their surface where the chemical reactions take place and internally where the current flows and the electrical resistance causes the heat generation. This is the primary location and source of heat in the battery. And this is why the heat is so difficult to remove. The bulk and configuration of large batteries do not lend themselves to easy heat removal.

The subsequent heat build-up produces an obvious temperature increase in the battery. This elevated operating temperature can cause physical damage and/or electrolyte changes that will adversely affect the operation of the battery. Also, the electrical resistance will increase, causing the performance of the battery to deteriorate further.

This heat generation takes place most internal to the battery where heat transfer augmentation is almost impossible to apply without significantly altering the design of the battery. However, if cooling can be provided at the point of this heat generation, improved recharge times and battery performance will result.

The application of thermoelectric coolers at the source of waste heat production will aid in battery longevity and increase practical usage of the battery, especially in the automotive industry.[1] The coolers can be configured to be positioned in any or all of several locations internal to the battery for the best results of heat removal.

First, the TEG coolers can be installed between the positive and negative grid plates of the cell pairs. Construction has to be such that the electrolyte can still flow easily between the plates during both discharge and recharge of the battery. In this location, the coolers are in good thermal contact with the prime source of heat production, the metal plates.

Second, the TEG coolers can be positioned between positive/negative plate pairs. This reduces the influence of electrolyte flow between the positive and negative plates. This is not as integral to the origination of the heat generation, but well within contact of that main location.

Third, the TEG coolers may be placed in cell partitions or battery case walls. In this location, there is no effect on electrolyte flow between plates. The battery still enjoys the cooling effects of close proximity to the heat source.

Figure 1 illustrates the thermoelectric cooler between a pair of positive/negative grid plate pairs. The cold junctions of the TEG cooler are in intimate thermal contact with the electrolyte and the surfaces of the grid plates through the cold junction plates. The cold junction plate is a good thermally conducting square or rectangular ring that allows the free flow of electrolyte between the grid pairs.

The warm junctions of the TEGs are on the exterior of the battery. Cascading of elements can facilitate the hotter junctions being outside the core of the battery where heat transfer augmentation can be utilized to remove the waste heat.

Under certain operating conditions, the interior of the battery may require heating. In this application, current flow can be reversed in the TEG to control the battery internal temperature. Obviously control circuitry would be needed to maintain a specified temperature.

A two-phased approach will be used to investigate the effectiveness of the TEG cooler inside the battery. The basic model developed here will include only the heat generated by I^2R losses. The chemical reactions that take place in a lead-acid battery can be of the same order of magnitude as the I^2R losses,[2] but will not be considered in this simple model. This study will focus initially on the overall battery-cooler concept.

The second phase will include chemical reactions that influence the overall thermal energy in the battery, geometric considerations and battery types other than lead-acid. This advanced model, considered for future research, will be used to optimize the TEG cooling effect for locations in the battery, battery type and TEG cooler configuration.

MODELING APPROACH

A two-dimensional, steady-state thermal model with a known, uniform heat generation will be developed to determine the temperature gradient in the region between the two grid plate stacks. The model takes advantage of symmetry for a square or rectangular battery design. Figure 2 shows the section to be modeled for the battery plates. The TEG cooler is located between two positive/negative plate pairs in intimate thermal contact with the two plates, situated in the electrolyte. Although in actuality the heat is generated in the metal plates, the model will consider the bulk of the generation to occur in the adjacent electrolyte section, around which the cooling plate is located.

Shown in Figure 3 is the configuration used for the thermal analysis. The heat removal by the TEG cooler is depicted by the two adjacent boundaries with the constant heat flux. The cooling plate can be configured to maintain a constant heat flux along these two sides. The primary concern is the maximum temperature that the battery will achieve normally during charging. This temperature would be at the centerline of the battery, at x=0 and y=0 in the model.

This initial study shows the effect of the TEG cooler with a one-dimensional model of the battery cooler. The analysis does present the solution to the two-dimensional case, but only the results to the 1-D case are presented here.

The 1-D model is used for the cooling of a typical lead-acid battery with a TEG cooler. This configuration is shown in Figure 4(A). The material of the battery in thermal contact with the cold junction plate is the electrolyte between the cell grid plates. The full width of the battery is typically 16cm. With two cooling plates in the battery, the distance from the centerline to the cold junction plate is L=4cm. The maximum allowable internal temperature of the battery is 350K.

The TEG cooler is modeled as a single junction with bismuth telluride. The elements are 0.1cm x 0.5cm and 4cm long. The hot junction of the TEG cooler is in the ambient at T_{amb}=300K. The internal resistance[2] of the battery is assumed to be 0.0025ohms and the charging current varies from 0 to 2.0amps.

The maximum internal temperature of the battery is considered for three cases, two with no TEG cooling and the third with the TEG cooler. In the first two cases, the external temperature of the battery is maintained at 305K and 310K, respectively. This is analagous to the battery being cooled by natural or forced convection when the ambient temperature is 300K and there are no TEG coolers. In the third case, the TEG cooler is utilized to maintain the cold junction plate at a temperature T_c, hence, the heat is removed by the TEG cooler. Figure 4(B) shows the parameters used for the model.

The maximum temperature that the battery will experience is at x=0, the furthest point from the cooling surface. This is the centerline location of the cold junction plate, or the axis of symmetry. In the model, this is depicted by T_{CL}.

EQUATION DEVELOPMENT

HEAT CONDUCTION MODEL

The two-dimensional Fourier Law for heat conduction[3] in the battery with internal heat generation, q''', is:

$$\partial^2T/\partial x^2 + \partial^2T/\partial y^2 + q'''/k_a = 0, \quad (1)$$

where T is the local temperature at (x,y) and k_a is the thermal conductivity of the battery electrolyte. The generation term is

$q''' = I_{batt\ chrg}^2 \times R_{batt}/Volume$,

where $I_{batt\ chrg}$ is the charging amperage to the battery, R_{batt} is the internal resistance of the battery, and Volume is the dimensional volume of the generation region. For the 2-D case, Volume=A_{batt}, that is, the area of the 2-D battery per unit depth.

The centerline for the region of cooling in the battery is the axes of symmetry. The boundary conditions at the centerline are:

at x,y = 0:

$$\partial T/\partial x|_{x=0} = 0$$
$$\partial T/\partial y|_{y=0} = 0 \quad (2)$$

Internal to the battery and where the cold junctions of the TEG cooler are in thermal contact with the cooling plate, the boundary is maintained at a constant temperature T_c. Therefore the boundary conditions are:

at $x = l$, y = L:

$$T(l) = T_c$$
$$T(L) = T_c. \quad (3)$$

The solution to the two-dimensional heat conduction equation[3] is shown in Appendix A. Equations (A1) - (A3) provide the temperature distribution throughout the battery region with the cold junction plate maintained at a constant temperature T_c. The maximum temperature in the battery, at x,y=0, can be calculated from Equation (A4). Equations (A5) and (A6) relate the heat flux that must be removed at the boundaries, Q_x and Q_y, by the TEG cooler.

ONE-DIMENSIONAL MODEL

In this initial study, the 1-D model is used. The solution for one-dimensional heat flow in the battery is:

$$T(x)=(q'''L^2/2k_a)[1-(x/L)^2]+T_c \quad (4)$$

where T_c is the temperature of the boundary at x=L. With the TEG cooler, T_c is the temperature of the cold junction plate. When no TEG cooler is present, T_c is the applied temperature at this boundary.

The boundary conditions for the 1-D case are the same as the 2-D case. That is, the boundary at x=0 is an axis of symmetry and the cold junction plate is at T_c.

For the charging of the battery in the 1-D case:

$$q''' = I_{batt\ chrg}^2 \times R_{batt}/L_{batt} \quad (5)$$

This is the heat addition per length of battery due to charging.

THERMOELECTRIC EQUATIONS

The Seebeck cooling effect of the TEG junction is also influenced by the heat generation in the TEG elements and the conduction of heat from the hot to the cold junctions.[4] Therefore the heat removed at the cold junction is:

$$q_c = \alpha T_c I_{cool} - 1/2\, I_{cool}^2 R_{cool} - \kappa_c \Delta T, \quad (6)$$

where

$$R_{cool} = (\rho_n l_n/A_n) + (\rho_p l_p/A_p)$$

and

$$\kappa_c = (\lambda_n A_n/l_n) + (\lambda_p A_p/l_p).$$

The physical properties of the p- and n-type TEG materials are the combined Seebeck coefficients of the two materials, $\alpha=|\alpha_p|+|\alpha_n|$; the thermal conductivities, λ_p, λ_n; and the thermal resistivities, ρ_p, ρ_n. The geometry of the TEG elements are the lengths of the elements l_p, l_n and the areas A_p and A_n.

Equation (4) is used to calculate the heat rate, dT/dx at x=L, due to heat generation in the battery, to determine q_c. The value of I_{cool} that will maximize the heat removed,

q_c, by the TEG cooler is

$$I_{coolmax} = \alpha T_c / R_{cool}. \quad (7)$$

Therefore, with the required q_c known and $I_{coolmax}$ chosen to maximize the heat removal, T_c can be calculated. With T_c known, $T(0)=T_{CL}$ is then calculated from Equation (4).

RESULTS

The maximum internal temperature of the battery due to heat generation is shown in Figure 5. The temperatures T_{CL}, at x=0 or the centerline of the cold junction plate, are shown with and without the benefit of the TEG cooler. The surface temperatures of 305K and 310K correspond to the temperature of the cold junction plate in the model, T_c. For the two cases where there is only convection cooling at the battery surface (no TEGs), the internal temperature exceeds the maximum allowable temperature of 350K. But when the TEG cooler is utilized, the maximum internal battery temperature remains well within the acceptable limits. The figure also shows the relative temperature of the TEG cold junction plate, T_c, increasing only slightly over the full range of charging rates.

When the charging rate for the battery is increased from 0.4 to 0.8, the battery with the TEG cooler still maintains an internal temperature well below what is acceptable. The battery with only convection cooling at the surface is no longer able to meet this requirement.

DISCUSSION

For the cases where the internal maximum temperature exceeds the allowable limit, the charging rate would have to be reduced. This would result in an increased recharge time of the battery. Otherwise the longevity and performance of the battery would be reduced over time if operated at this elevated temperature.

However, utilizing the TEG cooler indicates that the charging rate is well within the acceptable limits for the internal temperature of the battery and in fact the charging rate can be increased further to hasten the recharge time. For the example cited, battery recharge time can be reduced 40% by providing internal cooling to the battery.

CONCLUSIONS

The addition of the TEG cooler increases the charging rate that the lead-acid battery can endure. This will decrease the amount of time that the battery is out of service for recharge. Or, by better monitoring and controlling the internal temperature of the battery, the longevity and performance of the battery may be improved greatly. Therefore the effectiveness of internal cooling for lead-acid batteries can show promise in many applications where recharge cycle time must be reduced for the usefulness of the battery powered device to be practical.

REFERENCES

1. Parise, Ronald J., Quick Charge Battery, IECEC98, Colorado Springs, CO, 1998, Paper No. IECEC-98-I-136.

2. Crompton, T. R., Battery Reference Book, 2nd Edition, SAE International, Warrendale, Pennsylvania, USA, 1997.

3. Arpaci, V. S., Conduction Heat Transfer, Addison-Wesley, Reading, MA, 1966.

4. Angrist, Stanley W., Direct Energy Conversion, Fourth Edition, Allyn and Bacon, Inc., Boston, Massachusetts, 1982.

Appendix A
2-D Temperature Equations

The temperature distribution in a quadrant of the battery of width 2L in the x-direction and height of $2l$ in the y-direction is:

$$\frac{T(\eta,\xi) - T_c}{q'''L^2/k_a} = \frac{1}{2}(1-\eta^2) - 2\sum_{n=0}^{\infty}\frac{(-1)^n}{\lambda_n^3}\,\frac{\cosh(\lambda_n\xi)}{\cosh(\lambda_n\alpha)}\,\cos(\lambda_n\eta), \quad \text{(A1)}$$

where

$$\alpha = \frac{l}{L}, \quad \eta = \frac{x}{L}, \quad \xi = \frac{y}{L}, \quad \text{(A2)}$$

and

$$\lambda_n = (n + \frac{1}{2})\,\pi. \quad \text{(A3)}$$

The maximum temperature in the battery, T_{CL}, at $(\eta,\xi) = (0,0)$:

$$\frac{T_{CL} - T_0}{q'''L^2/k_a} = \frac{1}{2} - 2\sum_{n=0}^{\infty}\frac{(-1)^n}{\lambda_n^3\cosh(\lambda_n\alpha)} \quad \text{(A4)}$$

The heat flow rate through one quadrant of the battery (adjacent sides) is:

$$\frac{Q_x}{q'''L^2/k_a} = \alpha - 2\sum_{n=0}^{\infty}\frac{(-1)^n}{\lambda_n^3}\,\tanh(\lambda_n\alpha)\,\sin(\lambda_n), \quad \text{(A5)}$$

$$\frac{Q_y}{q'''L^2/k_a} = 2\sum_{n=0}^{\infty}\frac{(-1)^n}{\lambda_n^3}\,\tanh(\lambda_n\alpha)\,\sin(\lambda_n)\,. \quad \text{(A6)}$$

Figure 1: Cooler Between Plate Pairs.

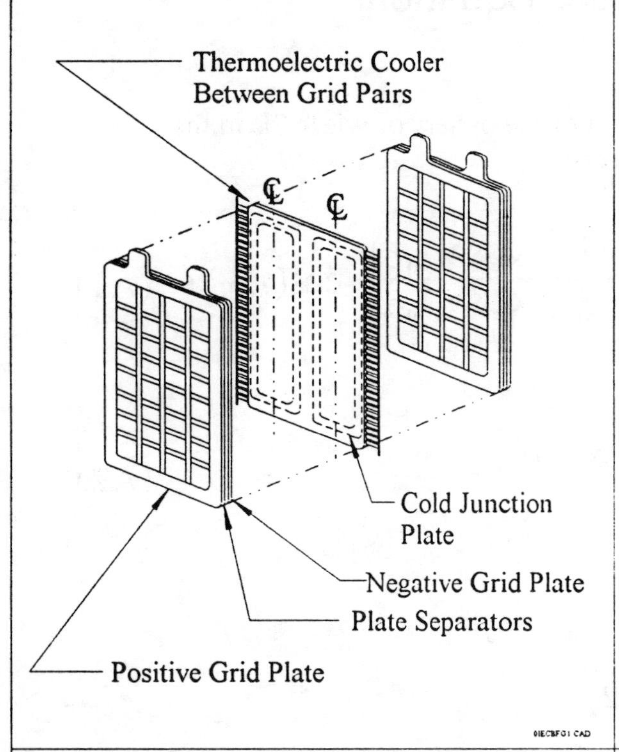

Figure 2: Cooler Location Between Grid Pairs.

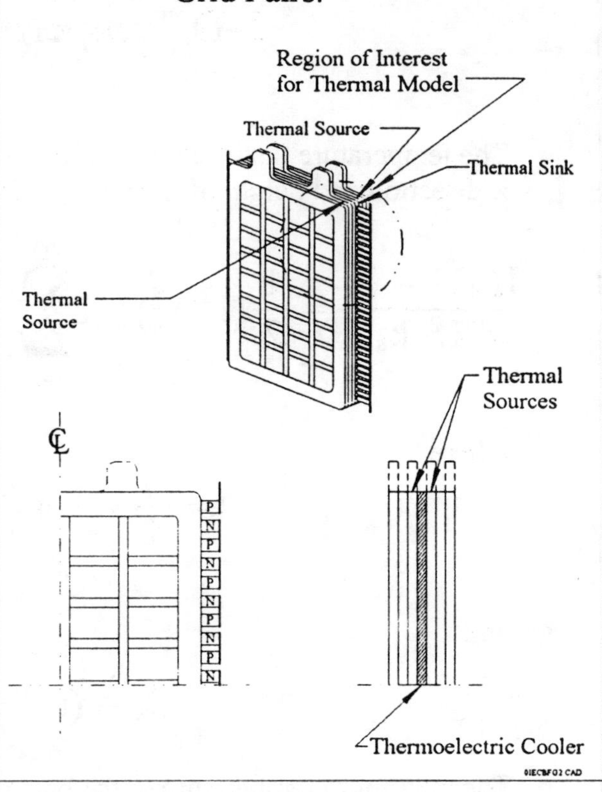

Figure 3: Basic Model Configuration for Thermal Analysis

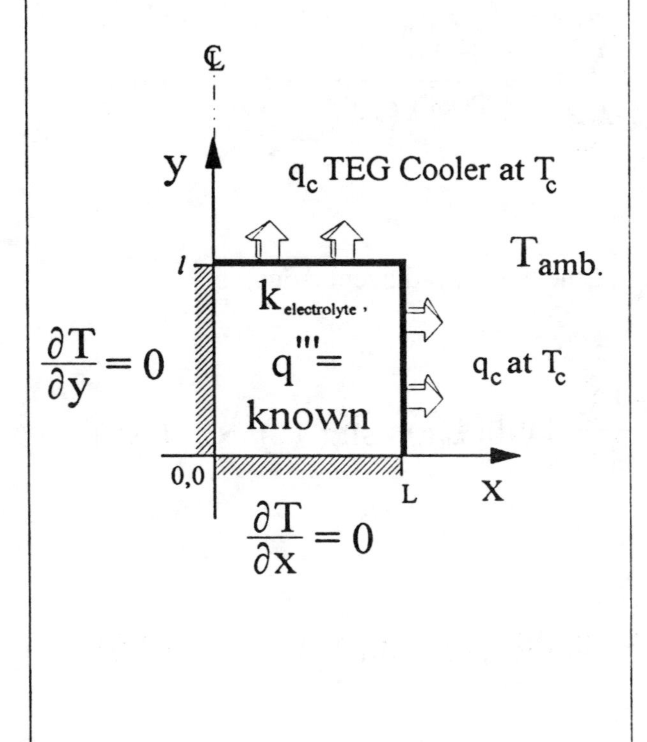

Figure 4: Basic Configuration for the 1-D Model.

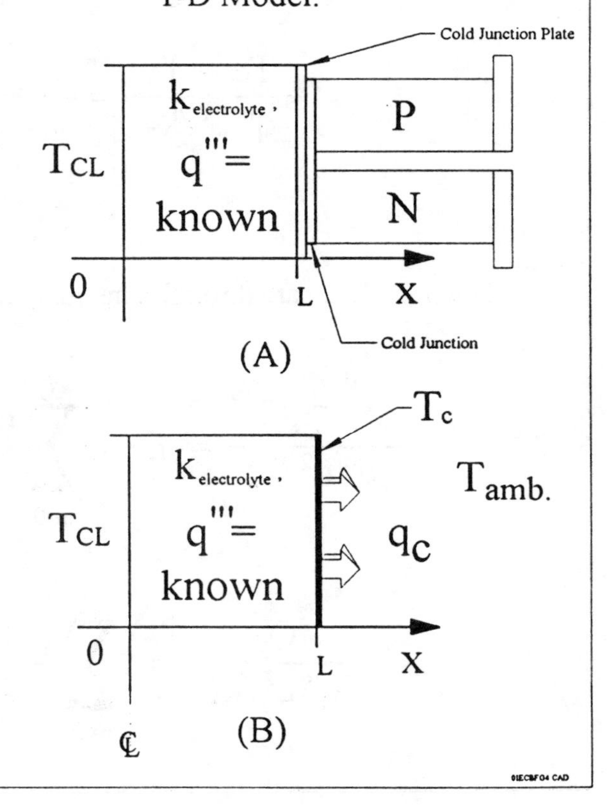

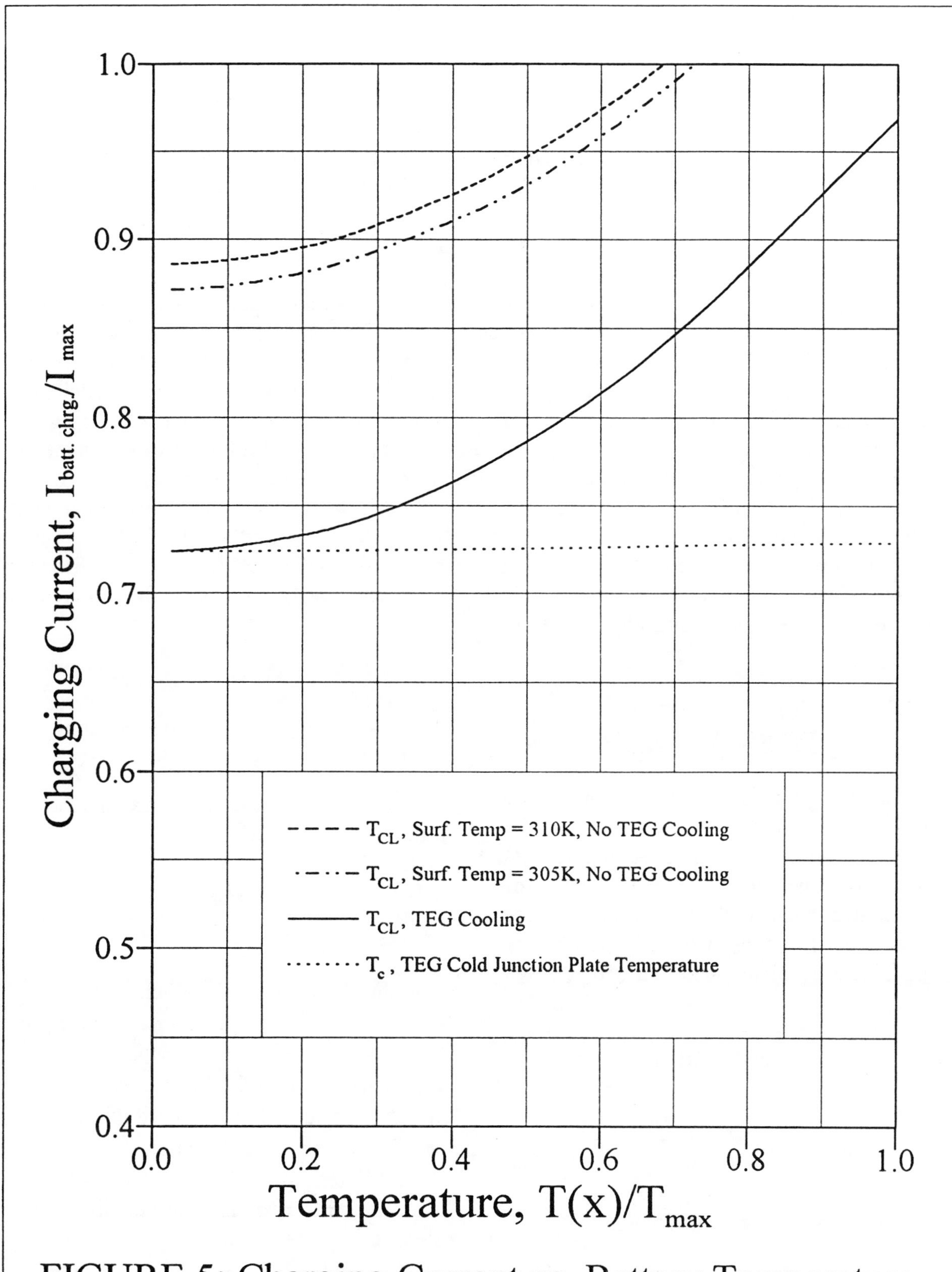

FIGURE 5: Charging Current vs. Battery Temperature

EFFECT OF A LIMIT TO THE FIGURE-OF-MERIT ON THERMOELECTRIC GENERATION

D.M. Rowe, Gao Min, V. Kuznetsov and A. Kaliazin
University of Wales Cardiff, School of Engineering
Division of Electronic Engineering, The Parade, PO Box 689
Cardiff. CF2 3TF, United Kingdom

ABSTRACT

Evidence is presented which indicates that the dimensionless figure-of-merit (ZT) of a thermoelectric material has a limit of around 2. In this paper the consequences of the existence of this limit in material performance, on the development of thermoelectric technology is discussed. It is concluded that in specialised applications where efficiency, device compactness and fuel costs are major considerations, the existence of this low material performance limit may well have a capping effect on the planning of future advanced material development programmes. However, in applications where the cost of fuel is low or essentially free, as in waste heat recovery, the economic factor (cost per watt) rather than efficiency, is the overriding consideration. In such applications a limit to ZT will have only a marginal impact on the competitiveness of thermoelectrics as an environmentally friendly source of electrical power. Alternative strategies for effectively increasing the conversion efficiency of thermoelements are discussed.

INTRODUCTION

A thermocouple is a unique heat engine which employs the electron gas as the working fluid. Like all heat engines, its mode of operation obeys the laws of thermodynamics and application of these laws enables an expression to be derived for the efficiency of its operation and is given by;

The conversion efficiency =

$$\frac{\textit{Electrical energy supplied to the load}}{\textit{Heat supplied}}$$

with the Max conversion efficiency =

$$\frac{T_H - T_C}{T_H}\frac{\sqrt{(1+Z\overline{T})}-1}{\sqrt{(1+Z\overline{T})}+(T_C/T_H)}$$

$$\overline{T} = \frac{T_H + T_C}{2} \quad \text{T = absolute temperature}$$

When the branch resistances are matched to minimise heat absorption

$$Z = \frac{\alpha^2\sigma}{\lambda}$$

Z is the figure-of-merit which is expressed in its dimensionless form ZT with α the Seebeck efficiency, σ the electrical conductivity and λ the thermal conductivity which consists of lattice contribution λ_L and an electronic contribution λ_e. In thermoelectric materials λ_L is ~ 50-75% λ. The quantity $\alpha^2\sigma$ is referred to as the electrical power factor. In a generating system improvements in efficiency can, of course, be achieved by reducing unwanted heat losses and ensuring that all the available heat flux passes through the active components of the generator, i.e. the modules. Evidently once the temperature regime of operation is established the efficiency of a generating module can only be increased by improving the figure-of-merit. Improvements in efficiency has many accompanying benefits such as a reduction in the fuel inventory, an important factor when fuel is expensive as in isotopic powered systems, improvement in power to weight ratio, which is important in space applications. Consequently, during the past forty years or so, there has been concerted research to improve the figure-of-merit and hence the efficiency of thermoelectric materials. Effort has focused on reducing the lattice component of the thermal conductivity without attracting an unwanted degradation in electrical properties. This route was taken primarily because there is a reasonable understanding of thermal conductivity mechanisms and a theoretical framework available which serves to provide some material guidelines. However, it has proved to be extremely difficult to achieve a significant reduction in lattice thermal conductivity. Historically, silicon germanium alloy has received the most attention and the efforts to reduce its lattice thermal conductivity over more than thirty years has enjoyed marginal, if any, success. The various mechanisms that have been employed during this period to increase phonon scattering together with figure-of-merit data are collected in Figure 1. The increase in the figure-of-merit of silicon germanium alloys achieved in the early 90's was due to an improvement in electrical properties which accompanied high temperature annealing sequences.

In spite of a lack of success, research continues to focus on reducing the lattice thermal conductivity and in recent years these efforts received a boost following the formulation of the concept of a PGE crystal (phonon glass-electronic crystal) approach. The crystal structure of these materials is comprised of weakly bound atoms or molecules that rattle within an atomic cage

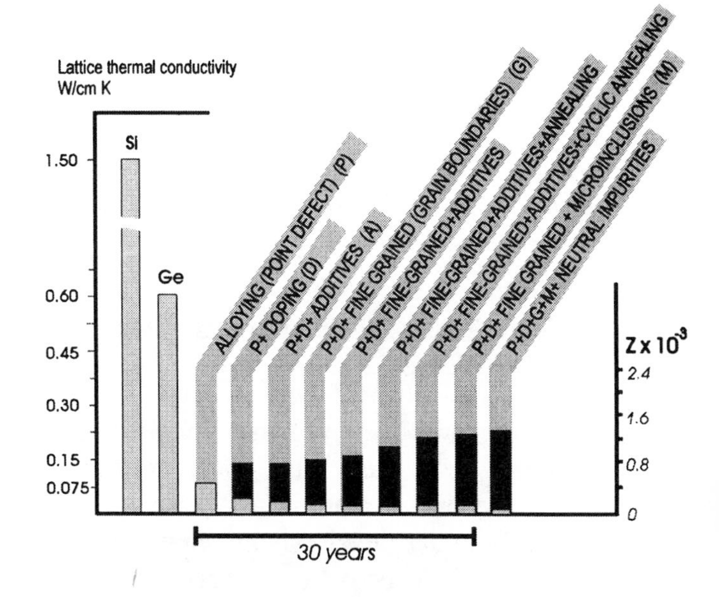

Figure 1. Historical catalogue of the effect of various phonon scattering mechanisms on reducing the lattice thermal conductivity and increasing the figure-of-merit of silicon germanium alloys.

and conduct heat like a glass, but conduct electricity like a crystal.[1] Following the implementation of these guidelines a number of promising 'new' materials have been developed, referred to as 'inclusion materials' among which are filled skutterudites[2,3] and clathrates.[4,5] Considerable interest has also been shown in the thermal transport properties in low dimensional structures with the lattice thermal conductivity of superlattices differing appreciably from their bulk counterparts due to phonon scattering effects resulting from the materials reduced dimensions and the presence of interfaces[6]

In this paper an analysis is presented which indicates that the figure-of-merit cannot be significantly increased by reducing the lattice thermal conductivity alone and that parallel attention must be paid to increasing the electrical power factor. Based on the concept of a minimum thermal conductivity[1] and an analysis of currently known materials, it is concluded that an upper limit of around 2 appears to exists for the dimensionless figure-of-merit and that exceeding this barrier will require the development of materials with unconventional transport properties. The sensitivity of thermoelectric applications due to this limit of the figure-of-merit are considered and strategies for increasing the performance of semiconductor thermoelements discussed.

Limit to the dimensionless thermoelectric figure-of-merit

In Figure 2 is displayed the dimensionless figure-of-merit of state-of-the-art and thermoelectric materials which are currently under development. State-of-the-art thermoelectric materials fall conveniently into three groups depending upon their temperature range of operation. Materials based on bismuth telluride, which were originally developed for refrigeration, can be operated from below ambient up to around 450K while materials based on lead telluride and silicon germanium, which were originally developed for space applications, can operated up to 750K and 1300K respectively. Evidently the dimensionless figure-of-merit is around unity.

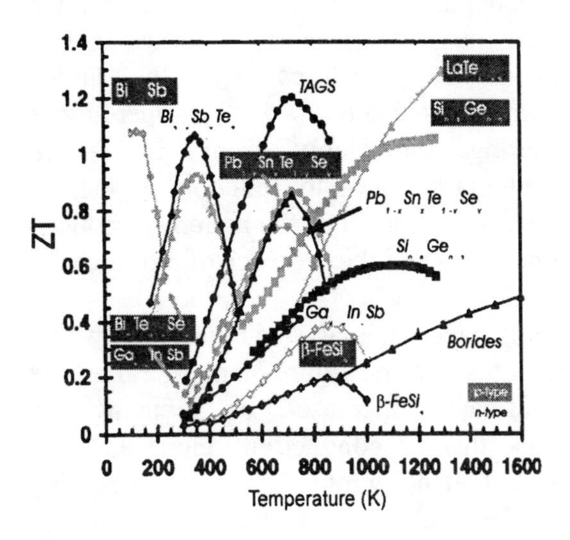

Figure 2. ZT vs temperature for thermoelectric 'generating' materials

Although it has been almost universally accepted that thermodynamics does not place an upper limit on the figure-of-merit it does appear that a bound does exist and some reservations have been expressed in the literature.[7,8] A more recent analysis in which thermoelectric behaviour of an electron gas was considered in terms of an ideal gas analogue indicated that thermodynamics may place an upper limit to the dimensionless figure-of-merit[9].

A recent analysis of available thermoelectric data provides phenomenological evidence that a ZT of the order of two appears to be the limit achievable by reducing the lattice thermal conductivity to its minimum value[10]. This can be explained by the fact that since the thermal conductivity has a lattice and an electronic component. Consequently, a reduction in the former by introducing phonon scattering agencies serves to increase the relative importance of the latter. This is particularly so in heavily doped materials employed in thermoelectric conversion. In Figure 3 is displayed the

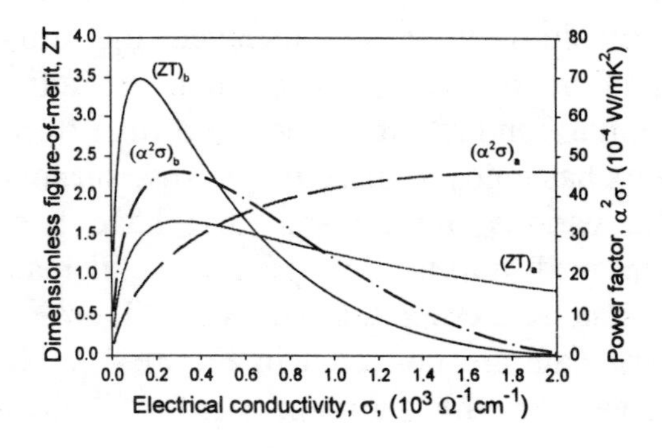

Figure 3. ZT as a function of electrical conductivity for different $\alpha^2\sigma$-σ profile

calculated $(ZT)_{max}$ as a function of σ at a minimum lattice thermal conductivity λ_l=0.25 W/mK. The results were calculated based on two different $\alpha^2\sigma$-σ profiles with identical values of $(\alpha^2\sigma)_{max}$. The dashed line represents a realistic $\alpha^2\sigma$-σ profile based on Bi_2Te_3 alloy data. The dash-dot line represents an imaginary $\alpha^2\sigma$-σ profile obtained by shifting σ_p to a small value while $(\alpha^2\sigma)_{max}$ remains unchanged in order to achieve ZT=3.5. Although both 'materials' possess identical values of λ_{min} and $(\alpha^2\sigma)_{max}$ the maximum value of $(ZT)_b$ is more than double that of $(ZT)_a$. Evidently any change in the ratio $(\alpha^2\sigma_p/\sigma_p)$ has a profound effect on ZT and it is apparent that without an improvement in the electrical properties ZT will be limited to less than 2 at room temperature.

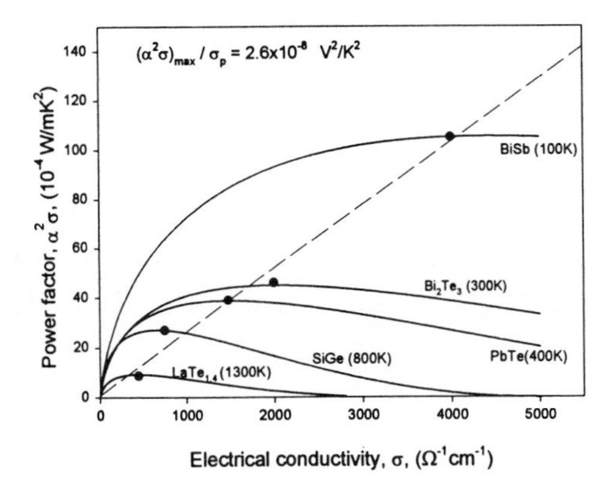

Figure 4. Electrical power factors as a function of electrical conductivity for a number of established thermoelectric materials.

However, it has been discovered that the ratio appears to be a constant and in Figure 4 is displayed $\alpha^2\sigma$ as a function of σ for a number of established thermoelectric materials. Although these materials exhibits a very similar $(ZT)_{max}$, the magnitude of $(\alpha^2\sigma)_{max}$ and σ_p varies significantly for different materials. It has been discovered that the ratio of the maximum electrical power factor to optimum electrical conductivity, $(\alpha^2\sigma)_{max}/\sigma_p$, appears constant as indicated by the dashed line in the figure. The value of this constant obtained using linear regression is found to be 2.6×10^{-8} V^2/K^2, which is very close to the Lorenz number. No known thermoelectric materials are located above this border line to a significant degree and this result, when combined with the concept of a minimum thermal conductivity, gives rise to an apparent upper limit to ZT. In Figure 4 is plotted the ZT values of established thermoelectric material together with the maximum figure-of-merit based on the above analysis. Evidently the maximum of ZT is around 2 at room temperature.

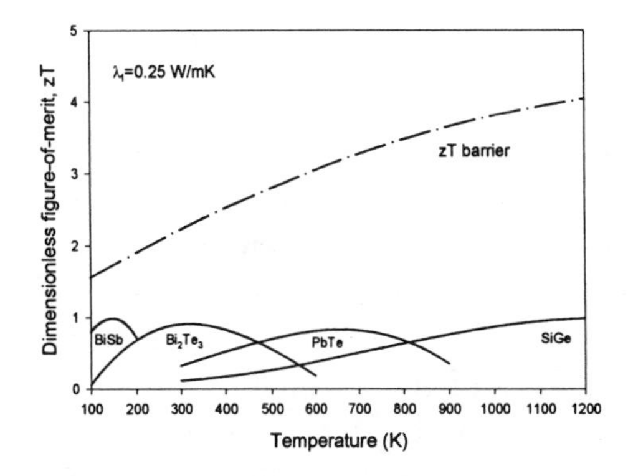

Figure 5. Dimensionless figure-of-merit as a function of temperature for established thermoelectric materials together with the ZT barrier.

APPLICATION'S SENSITIVITY TO ZT

Application sensitivity to the figure-of-merit

The use of thermoelectric generators to date has been almost exclusive to applications where the attractive features of this method of generating electrical power outweighs its relatively low efficiency (typically less than 5%). Historically research has focussed on improving the thermoelectric generators conversion efficiency by optimising the generators design through developments in engineering and in attempting to increase the figure-of-merit and hence the efficiency of the semiconductor thermoelements through material research. From a material scientists perspective improvement in efficiency is synonymous with improving the thermoelectric figure-of-merit. Evidently an improvement in Z would be accompanied by a reduction in the fuel inventory and an increase in specific power, an important consideration in isotopic powered generators where fuel cost is extremely expensive. In these specialised applications the existence of an upper bound to the dimensionless figure-of-merit is clearly a capping factor, although there remains room to almost double its value from the present unity. However, the existence of such a limit, will have only a marginal impact on the growing volume of thermoelectric generating applications which utilise cheap or essentially free heat sources of energy. In commercial applications of thermoelectric generation the viability of the technology depends on the cost per watt of electricity produced. Which can be conveniently discussed in terms of an economic factor E.F. (defined as £/kWh). The economic factor depends both on the efficiency and the electrical power output with the former reflecting the funning costs and the latter the capital costs.[11]

Both conversion efficiency and power output depend on thermoelement length. In order to obtain high conversion efficiency, the thermoelements should be long while they should be short for a large power output. The capital cost is mainly related to the power output per unit area while the running cost is dependent upon conversion efficiency. Consequently, optimisation of the thermoelement to achieve a minimum total cost will be guided by the economic factor. In Figure 6 is displayed the economic factor as a function of thermoelement length for different thermal energy (fuel) costs based on data for commercially available Peltier modules. It is apparent that as the cost of input heat energy decreases so does the required thermoelement length and in the extreme

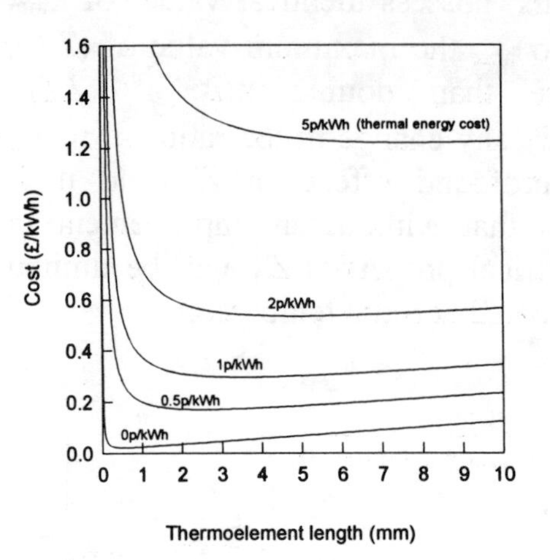

Figure 6. Economic factor as a function of thermoelement length for different thermal energy costs.

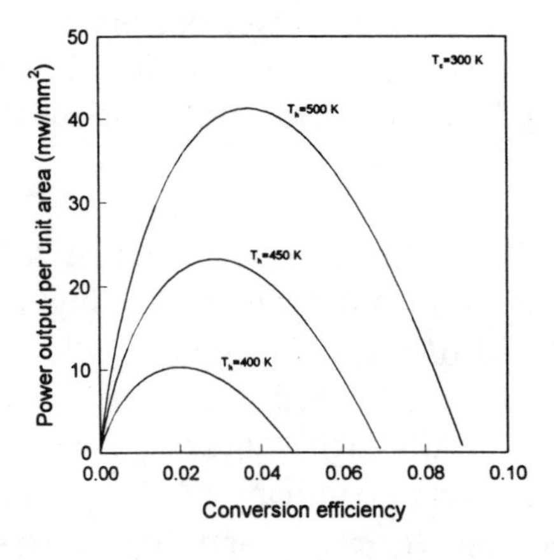

Figure 7. Power output per unit area vs conversion efficiency for different temperature differences.

case when the input heat energy is free the thermoelement length should be optimised to obtain maximum power output. In Figure 7 is displayed the power output per unit area vs conversion efficiency for different temperature differences. It is apparent that the maximum output power per unit area is obtained at an efficiency which is approximately half the value of the attainable maximum conversion efficiency.

It is concluded that when heat input is cheap the figure-of-merit is not a major consideration. Indeed it is desirable not to utilise present available maximum in order to achieve economic competitiveness.

ALTERNATIVE STRATEGIES TO IMPROVE THERMOELECTRIC PERFORMANCE,

As indicated in a previous section in specialised applications achieving an improved conversion efficiency by increasing the figure-of-merit remains the major objective. However, little, if any, progress has been made during the past forty years in attempting to increase the figure-of-merit of thermoelectric semiconductors and based on the content of this paper, there is little evidence that a breakthrough will be achieved in the near future. An alternative strategy for improving the conversion efficiency of a thermoelectric generator is to increase the conversion efficiency of the thermoelements. It is apparent from Figure 1 that the thermoelectric figure-of-merit varies with temperature and reaches a maximum at a precise value. Consequently, in a thermoelement which consists of a single material, only a specific location along length operates at its full potential. Clearly, matching the figure-of-merit along the length of the material to the temperature it is subjected to under operating conditions would in principle result in a substantial increase in the 'effective' thermoelectric figure-of-merit and consequent improvement in the generator's conversion efficiency. Two strategies currently being explored at Cardiff are segmented and functionally graded thermoelements. Evidently the overall performance of the thermocouple would be increased by connecting in tandem two or more materials, whose figure-of-merit are maximised at different temperatures. The arrangement shown in Figure 8 is known as segmentation with the overall figure-of-merit estimated to be 12%.

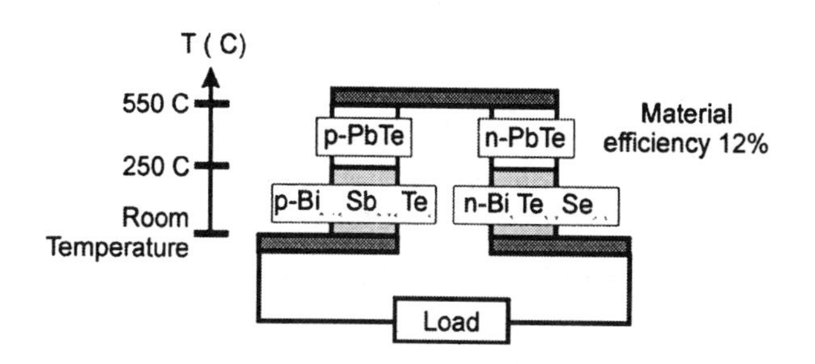

Figure 8. Segmented thermoelement

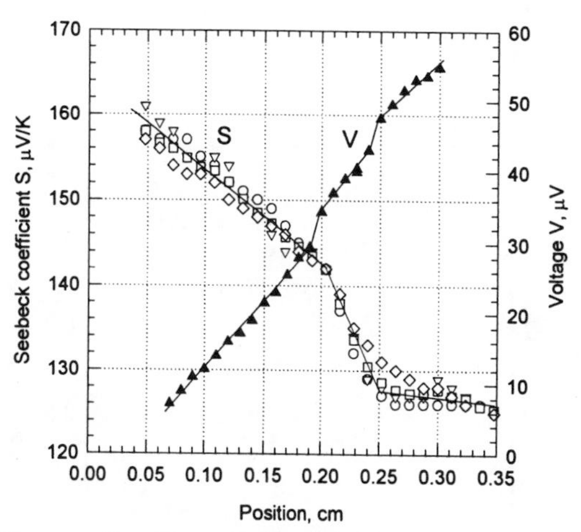

Figure 9. Experimental electrical data on segmented thermoelement

In Figure 9 is displayed the Seebeck coefficient and electrical resistivity distribution along the length of a segmented n-type Bi_2Te_3 solid solution. The junction between the two materials is located at 0.2cm. A standard n,p-Bi_2Te_3 based material has an efficiency of 8% when operated between 223-423K, compared to an efficiency of 9.5% for a segmented thermoelement operating over the same temperature difference.

A problem with successfully implementing segmentation is the decrease in carrier mobility at the interface boundary and hence increase in electrical resistivity. In some materials this problem can be avoided by adjusting the carrier transport properties in an attempt to match the maximum figure-of-merit to the temperature along the thermoelement. In practice this can be achieved by carefully controlling the alloy composition and carrier concentrations. In applications where the temperature range is too large to be covered by single materials, several functionally graded materials can be segmented together.

In Figure 10 is displayed the preliminary results of measurement made on a functionally graded thermoelement based on bismuth telluride technology. The conversion efficiency of a functionally graded thermocouple to be in excess of 10% when operated between 223K and 423K.

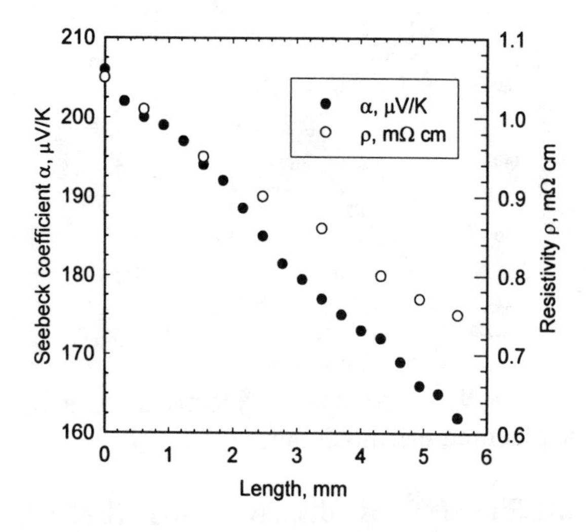

Figure 10. Distribution of Seebeck coefficient and electrical resistivity along a crystal of functionally graded p-type Bi_2Te_3 based solid solution.

CONCLUSION AND DISCUSSION

Based on the lack of progress in improving the thermoelectric figure-of-merit during the past 40 years or so and the results of a phenomenological analysis of current thermoelectric materials reported in this paper, it appears that an upper limit to this parameter of around 2 exists. Exceeding this bound will require the development of materials whose lattice thermal conductivity approaches the theoretical minimum and with hitherto unobserved electrical properties and in particular ones which $\alpha^2\sigma/\sigma$ ratio exceeds 2.6 x 10^{-8} V^2/K^2.. Although the existence of this limit will have considerable impact of attempt to develop materials with substantially improved performance, it will have only marginal effect on applications which utilise cheap or waste heat energy inputs. In these situations it is the electrical power factor rather than the figure-of-merit which is of major importance. Finally improvements in conversion efficiency of up to 10% can be achieved when operating between 223-423K by employing segmented and functionally graded materials.

REFERENCES

1. Slack, G.A. Chapter 34, CRC Handbook of Thermoelectrics. CRC Press, London. Ed. D.M. Rowe, p. 617, 1995.
2. Morelli D.T. and Meisner G.P., J. Appl. Phys. 77, 3777, 1995
3. Rowe D.M, Kuznetsov V.L, Kuznetsova L.A, Proc. 17th Int. Conf on Thermoelectrics, Nagoya, Japan, 323-325, 1998
4. Nolas N.S, Cohn J.P, Slack G.A, and Schiyman S.B, Appl. Phys. Letts. Vol. 73, No. 2, p. 178-180, 1998.
5. Kuznetsov V.L, Kuznetsova L.A, Kaliazin A, Rowe D.M. Accepted for publication in J. Appl. Phys. Vol. 87, No. 11, 1st June 2000.
6. Chen G, Phys Rev B 57, (23) p. 14958, 1998
7. Ure Jr. R.W, Energy Conversion, 12, 1972
8. Mahan, G.D. J. Appl. Phys 65. 4, p. 1578, 1989.
9. Vining C.B. Proc MRS Symposium 478, San Francisco, p 3, 1997
10. Min G. and Rowe D.M., Joun Mat. Science Letts 18, p 1305, 1999
11. Rowe D.M. and Min G. Journal of Power Sources 73, p. 193-198, 1998

AIAA-2000-2821

Thermoelectric power measurements between 300-1200 K using high temperature Peltier modules

H. Scherrer, M. M' Jahed, M. Riffel, C. Roche, S. Scherrer
Laboratoire de Physique des Materiaux
Ecole des Mines
Parc de Saurupt
F-54042 Nancy cedex
France
Fax 33 383584161
hubert.scherrer@mines.u-nancy.fr

Abstract

A measurement technique for the thermoelectric power of thermoelectric materials between 300 and 1200 K is proposed and described. The temperature regulation uses the Peltier effect or the Joule effect provided by thermoelectric modules. The Peltier modules are built using $FeSi_2$ material, allowing for a precise and fast temperature control. The technique of fabrication of the modules as well as the measurement set-up are described and discussed. Some thermoelectric power experimental data are shown.

Contact author :

Professor H. Scherrer
Laboratoire de Physique des Materiaux
Ecole des Mines
Parc de Saurupt
F-54042 Nancy cedex
France
Fax 33 383584161
hubert.scherrer@mines.u-nancy.fr

Introduction :

An apparatus has been designed which allows measurement of Seebeck coefficient of semi-conducting thermoelectric materials from room temperature up to 1200K .We have used a $FeSi_2$ high temperature Peltier module to adjust the temperature and to create the temperature gradient along the sample .We have also described the technique for building the $FeSi_2$ module.

I – Apparatus design :

The sample at a given temperature, we have to establish a temperature gradient to its extremities and to measure the generated voltage .Figure 1 is a schematic diagram of an isothermal heater housing, sample holder, thermocouple instrumentation .The whole assembly is placed inside a vacuum chamber connected to a diffusion pump system. The sample is hold between two copper disks in which chromel/alumel thermocouples and copper wires are inserted for temperature and voltage measurements .They are placed between two copper blocks ,the upper block by its weight has decreased the electrical and thermal contacts at the interfaces sample – copper disks. Also the surfaces of the sample and of the copper disks are very well polished. A thermoelectric module is constituted of 8 p-n couples of $FeSi_2$. A direct current in the module can modify very quickly and locally the temperature of the sample. In a reversible and reliable way the temperature can be decreased or increased by Peltier or Joule effects .The average temperature of the system is maintained by a monitored heater housing .

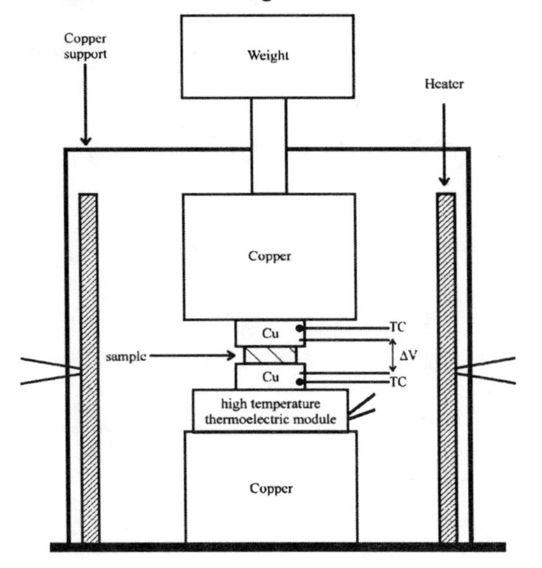

Figure 1: Schematic diagram of the apparatus

II – Experimental measurements :

The measurements can be done by a dynamic or static way .In the first way ,at a temperature T we create a temperature gradient with the thermoelectric module .The corresponding voltages for the temperature gradients (ΔT = 0.1-1K) has been recorded .The temperature gradients were very weak and we could consider that during the recording the Seebeck coefficient was constant .Thus the curve representing ΔV versus ΔT was a straight line and its slope gave the thermoelectric power of the sample . The residual voltage due to the contact resistances has only shifted the origin of the line .We have performed again the same experiment with an inverse temperature gradient .The average of the two measurements has given a reliable value of the thermoelectric power.

In the static way ,at a temperature T we have canceled the residual temperature ΔT_R on the sample by operating the Peltier module .Thus we could measure the residual voltages of the electrical and thermal contacts and create an absolute temperature gradient . As in the dynamic way the temperature gradients were very weak and we could consider the thermoelectric power of the sample as constant during the experience .After the stabilization of the temperature gradient, the ΔV voltage generated at which we have subtracted the residual voltage divided by ΔT has given the thermoelectric power .We have done the same experiment with an inverse temperature gradient and the average of the two measurements.

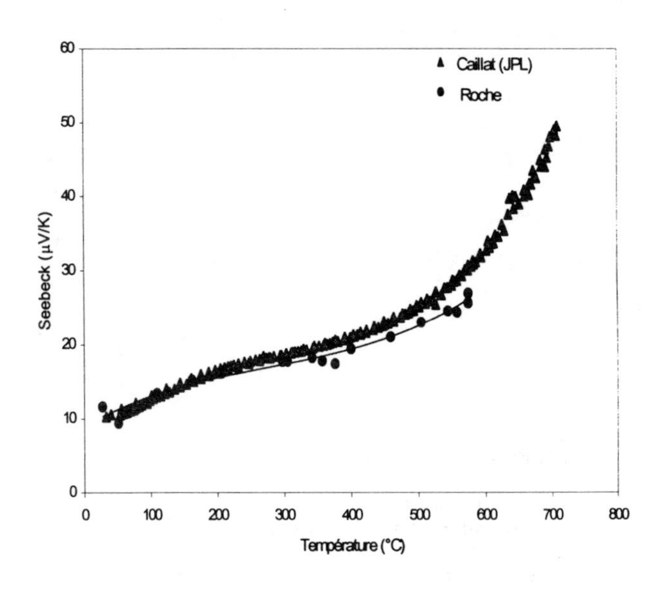

Figure2:Thermoelectric power versus the temperature

We have taking into account the different sources of errors in minimizing :

- the electrical resistances of contacts
- the thermoelectric power of measurement wires
- the generation of thermoelectric couples
- the voltages due to the position and the standardization of the thermocouples

Therefore, we think that we have measured the thermoelectric power inside ± 1μV/K of errors between room temperature up to 1200K .We have displayed on the figure 2 our results on a sample (Chevrel phase) elaborated in our laboratory and also characterized at Jet Propulsion Laboratory .The values of the thermoelectric power were very close.

III – $FeSi_2$ thermoelectric modules :

Since more than 30 years, $FeSi_2$ is well-known as a thermoelectric material[1]. It has the advantages of being cheap, non-toxic, and easy to handle. $FeSi_2$ thermoelectric devices were made of Co-doped n-legs and Mn- or Al-doped p-legs. The junctions that work at high temperatures were realized by a direct sintered contact [2-5] or by a soldered bridge with high electrical conductivity [6-8]. In the past all proposed devices work in the generating mode. i.e. a thermal flux is partially converted into an electrical power by the use of the Seebeck effect [1,9,10]. In this paper we propose to use a $FeSi_2$ device as a thermal pump which operates at high temperatures. Thus, it is the first time that $FeSi_2$ is applied in a small temperature gradient of only a few degree centigrade and in the Peltier effect's mode.

The starting materials in our production route were commercial powders. They are produced by gas atomization and have a spherical particle size with a mean diameter of 30 μm. The composition of the n-type material is $Fe_{0,95}\,Co_{0,05}\,Si_2$ and of the p-type one $FeSi_{1,95}\,Al_{0,05}$ Due to the rapid quenching in the gas atomization process both types of powders are in the metastable metallic high temperature a-phase. Thus, the phase transformation into the stable semi-conducting β-phase must be performed during the further production steps. The consolidation method we use is hot uniaxial pressing. The $FeSi_2$ powder is filled into a graphite die and pressed between two graphite stamps. To protect the die and the stamps from a reaction with the $FeSi_2$ and to allow an easier taking out of the pressed ingot the graphite parts are covered with a thin graphite foil. The hot uniaxial pressing is performed in high vacuum at a temperature of about 900 °C and a pressure of about 50 MPa. Thus after 15 minutes pressing time the ingots have reached at least 95 % of the theoretical density. The graphite die we used has an inner diameter of 20 mm. Each hot pressing's powder amount is 40 g leading to a pressed cylinder with a height of about 26.5 mm. From such a cylinder the upper and the lower 1.5 mm are cut off to be sure that no graphite has diffused into the material which is used for the production. From the inner 23.5 mm, four disks each with a height of about 5. mm are cut (see figures 3 and 4).

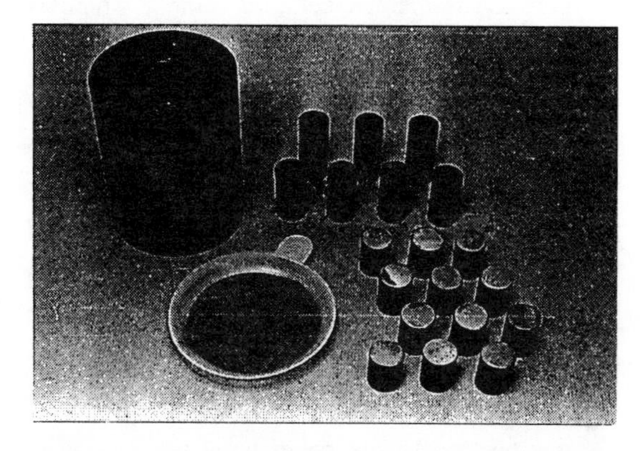

Figure 3: Powder, pressed samples and disks

Figure 4 : Cutting of the disks

In the following step these disks are machined ,so each of them has exactly the same height. To achieve this we first polished one face of the disks with emery paper (grid 800). Additionally, one surface of an alumina plate with the dimension 100 * 100 mm^2 is made plain in a milling machine. On this plain surface 24 disks. i.e. 12 of the n type and 12 of the p-type material. are glued with a tar-like glue. As the alumina plate rests in exactly the same position with respect to the milling machine the disks can be surfaced to a constant height with a variation of less than 0,03 mm . In our case we have chosen exactly 5 mm for the height of the disks.

Each of the disks is then cut into four cubes with the dimension 5*5 mm^3. It is important to mark the milled surface because only in this direction the cubes have the same height with a variation of less than 0,03 mm.
At this point of the production route the thermoelectric legs are annealed in high vacuum at 800°C. After the annealing the legs are in the semi-conducting ß-phase.
Subsequently , 6 p-type legs and 6 n-type legs are alternatively glued together in a T-shaped form . As the glue a commercial ceramic cement which is electrically insulating is used.
The glued devices are then machined in a way similar to the one of the disks:

- One face of each device is polished with emery paper (grid 800).
- The sample holder of an automatic polishing machine is polished to have a plain surface.
- The polished faces of 6-12 devices are glued onto the plain surface of the sample holder . As the glue we use wax.
- The second face of the devices is polished to the final height of 4,9 mm.

As the electrical isolation material we have used alumina plates which are on one side covered with a tungsten metallization . On the other side we have soldered small pieces of a silver titanium foil and two nickel wires.
The silver titanium alloy we have used, is a foil of a commercial solder called Degussa CB2 .Pieces with the size 4,5 * 9,5 mm^2 are cut from the foil and arranged on the alumina plates in the chosen form .The foils arrangement differ for the "top" alumina plate and for the "base" alumina plate (see figure 5) In the "first" and the 'last" silver titanium piece of the base plate, two nickel wires are embedded. After polishing with emery paper (grid 800) and cleaning with acetone, this arrangement is soldered in high vacuum at a temperature of about 1000 °C .

Figure 5 : Metallized alumina plates and glued and polished thermoelectric legs.

The silver titanium foil has been used as the electrical contact between two neighboring thermoelectric legs and at the same time as the mechanical contact between the $FeSi_2$ and the alumina plates. The nickel wires are the electrical connection to a source,etc. In a subsequent annealing process the connection between the silvcr-titanium-foil and the prepared $FeSi_2$ devices is realized. The base plate, the $FeSi_2$ and the top plate are polished with emery paper (grid 800), cleaned with acetone ,and then stacked one onto the other. This arrangement is slowly heated up to a temperature which is between the eutectic temperature of the silicon silver binary system (830 °C) [11] and the melting temperature of the used silver titanium alloy. The connection is then formed by the eutectic reaction but the solder is not liquid enough to lead to electrical short circuits between neighboring silver titanium foils. Thus, small thermoelectric modules with 8 legs are produced .

IV - Conclusion

A measurement technique for the thermoelectric power of thermoelectric materials between 300 and 1200 K is proposed and described. The temperature regulation uses the Peltier effect or the Joule effect provided by thermoelectric modules.We developed a production process for a $FeSi_2$ thermoelectric device that works in small temperature gradients between room temperature up to 1200K. Besides the known advantages of $FeSi_2$ it has metallic surfaces thus allowing easy matching.

References :

[1] R. Ware and D. NcNeill. in Proc.IEE. 111. 178 (1964)

[2] T. Nakajima M. Suzuki. .and J. Ohta. in Proc. l6th IECEC, 2013 (1981)

[3] K. Ue:mura Y. Mori. T. Imai, I. Nishida. S. Horie. and M. Kawaguchi. in Proc. 8th ICT, 151 (1989)

[4] U. Stöhrer. R. Voggesberger. G. Wagner and U. Birkholz. in Proc. 8th ICT, 130 (1989)

[5] U. Stöhrer. R. Voggesberger. G. Wagner and U. Birkholz. Energy Conv. Mgmt. 30 (2), 143 (1990)

[6] J. Hesse. Z. Angew. Phys. 28. 133 (1972).

[7] E. Gross. M. Riffel. and U. Stöhrer, J. Mater. Res. 10 (1) , 34 (1995)

[8] M. Riffel. E. Gross. U. Stöhrer. J. Mater. Sci: Mater. in Electronics 6. 182 (1995)

[9] Komatsu Electronics Inc.: Thermoelectrics and Electronic Equipement. Information on Komatsu $FeSi_2$ thermogenerators, Kyoto, Japan (1983)

[10] E. Gross. U. Stöhrer. R. Voggesberger. U. Birkholz. G. Druden. in Proc. 9th ICT (1990)

[11] M. Hensen. Constitution of Binary Alloys. McGraw Hill Book Company (1958)

AIAA-2000-2822

ENERGY FROM DEEP SPACE
THE NIGHTTIME SOLAR CELL™
ELECTRICAL ENERGY PRODUCTION

Ronald J. Parise
Parise Research Technologies
101 Wendover Road
Suffield, CT 06078

G. F. Jones
Depart. of Mech. Engrg.
Villanova University
Villanova, PA 19085

ABSTRACT

The primary objective of the Nighttime Solar Cell™ is to produce electric power at night. The lack of energy production when there is no incident solar energy is a major drawback to photovoltaic cells. Nighttime utilization of the new device produces electrical energy using a thermoelectric generator (TEG) operating in the temperature differential that exists between deep space at an effective temperature of 4K and the surrounding ambient temperature (nominally at 300K). Thus the ambient or surroundings of the device are the source of thermal energy while deep space provides a thermal sink. The cold junction of the TEG is insulated from the surroundings by a vacuum cell, improving its overall effectiveness.

This research is an on-going effort to develop a clean, reliable, safe, inexpensive, alternate source of electric power using deep space. The model discussed herein investigates the many design parameters that influence electrical power production including semiconductor configuration, cold junction plate area and depth of vacuum required in the cell for acceptable performance.

The "hot" junction can be supplied energy from a low grade thermal stream, previously considered too low a temperature source to recover the energy economically, greatly improving electrical energy production. Cell performance using this low grade thermal energy is presented.

Specific design configurations can be used for electrical energy production both day and night, regardless of weather conditions. For example, when only a thermoelectric generator is utilized, shielding from direct sunlight can produce power from deep space without photovoltaic cells. Nighttime Solar Cell™ performance with this configuration is discussed.

Finally, solar thermal energy can heat the daytime hot junction, utilizing the ambient as the thermal sink. This power-producing configuration is also investigated, showing great promise for daytime operation of the cell.

The thermal model will be utilized for parameter selection in the design and subsequent building of a prototype Nighttime Solar Cell™.

INTRODUCTION

The original function of the Nighttime Solar Cell™ is to produce electric power both day and night in a terrestrial application.[1] This solid state device, operating in a vacuum, utilizes a combination of photovoltaic cells for daytime operation and thermoelectric generators for nighttime operation. The photovoltaic-thermoelectric device operates in a vacuum, called a Vacuum Pod, to isolate the components from the ambient temperature, improving electric power production capability.

However, this investigation focuses on energy production using TEGs only. This means of operation can be employed during the day with shielding from direct sunlight, or, in a reverse mode of operation for the TEG module, by direct heating from the sun.

Consider the amount of energy that is available from nighttime operation of the cell. A blackbody at 300K, typically the ambient temperature at the surface of the earth, can radiate 450 W/m^2. (This is about one-half the energy available at the surface of the earth due to solar energy.) Utilizing deep space as a thermal sink having an effective temperature of 4K,[2] a temperature differential can be created between the surface of the earth and this thermal sink in a device with TEGs to produce electrical power.

The energy spectrum between 8μm and 13μm is nearly transparent under all atmospheric conditions for radiating energy to deep space, with smaller windows occurring in a few other bands. This represents approximately 40% of the total energy radiated at 300K.

Therefore, depending on the efficiency of the TEGs that are selected or developed, upwards of 180 W/m^2 of energy can be utilized. In dry, arid climates, more energy would be available, improving the operation of the cell significantly. The Nighttime Solar Cell™ is independent of the tilt of the earth, time of day, weather, etc., which are all problems that plague solar cells.

In reality, temperature differentials are on the order of 40K to 100K, and TEG module efficiencies are around 4% to 7%. Hence, during nighttime operation, cells can produce about 4.5 W/m^2 to 5.0 W/m^2 with today's technology, under all atmospheric conditions. Therefore this means of energy production can be utilized in many applications where small, reliable energy requirements are needed for remote monitoring sites, sensors, etc.

The operation of the Nighttime Solar Cell™ is currently at the mercy of available materials that operate in this temperature range. The development of new materials, an ongoing research effort worldwide, will vastly improve the operation of the device. Heat transfer studies of the system have shown new designs that will improve these numbers even without new material breakthroughs. This research effort reflects those advances.

THERMAL MODEL

The thermal model determines the temperature differential between the hot and cold junctions of the TEGs when using deep space as a thermal sink. Figure 1 shows the physical configuration and parameters utilized in the model. The thermal source supplies energy to the module at T_∞. The energy travels through the TEG elements (the p-n material), through the cold junctions into the cold junction plate (CJP).

Figure 1 shows the height of the TEG elements to be L, and the height of the distance from the CJP to the window to be L_a. The TEG cold junctions are made of copper and the CJP is made of aluminum. The thermal losses in both these components are three orders of magnitude less than in the TEG elements or the space above the CJP when air is present. Therefore these losses will be neglected in the model.

The CJP is 20cm^2 and the aperture opening is slightly larger to avoid physical or thermal interference. The thickness of the CJP is 5mm to minimize the fin effect from the TEG module being smaller.

The surface of the CJP is assumed to be a gray, diffuse surface with $\epsilon_c = \alpha_c$ at temperature T_c. The surface facing the nighttime sky is treated to have an emissivity ϵ_c of 0.96. The CJP then radiates thermal energy to deep space through the window covering the aperture of the vacuum cell.

The input temperature to the module, T_∞, can also be used to portray the temperature of a low grade thermal waste stream for the addition of thermal energy to the pod, incrementally improving the performance of the Vacuum Pod.

Deep space is modeled as a black body at temperature T_s = 4K. Initial model development will have a three-band radiation capability for energy transmission through the window. The two wavelengths that separate the three bands are cutoff wavelengths λ_l and λ_u. The radiation surface properties, for absorptivity (α), reflectivity (ρ), and transmissivity (τ), are given subscripts 1, 2, and 3 to designate their values in each of the three bands. In any band, emissivity and absorptivity are assumed to be equal, $\epsilon = \alpha$. Initial model results will use a single band between $8\mu m$ and $13\mu m$ where approximately one-third of the total radiative energy spectrum is transmitted. This is well within the spectral capability of zinc selenide (ZnSe), the material chosen for the window. The view factor between the CJP and deep space is assumed to be one.

The exterior of the window is exposed to the ambient temperature, T_∞, through a specified heat transfer coefficient, h_W. Temperature variations through the window are neglected.

Bismuth telluride is the material selected for the TEG module based on the expected operating temperature range of the pod. The specific design configuration of the TEG elements and module are based on standard sizes available from industry. The CJP and Vacuum Pod size are based on the size availability of the ZnSe window (nominal size: 2" x 2").

The model is utilized to show the effects of heating the CJP during the day by incident solar energy. In this mode of operation for the Vacuum Pod, the current in the TEG module will be reversed. The heating at the CJP is input as a thermal flux at the surface. Radiation shape factors, other input from the surroundings, and any cooling effects that may occur due to deep space are neglected. No additional cooling is used at the now cold junction of the module; only the ambient temperature of the air is the thermal sink.

Finally, the model is used to determine the effect of not having a full vacuum in the cell. That is, there will now be a slight trace of air in the Vacuum Pod. With this air, the only effect that will be considered in the model is the gap between the CJP and the ZnSe window, referred to as the air gap.

Except at extreme pressures, the thermal conductivity is a function of temperature. When the pressure is very high, or when the pressure is reduced to a level where the gas is at rarified levels and the mean free-molecular path between collisions is on the order of a physical dimension of the space (the gap between the CJP and the window in this case) will the conductivity be affected. That is, at low pressures, the conductivity becomes a function of pressure.

When not fully evacuated (a full vacuum would be on the order of 10^{-5} torr) the geometry of the pod can influence the physics of the thermal model with the entrapped air by both thermal conduction and the buoyancy effects of free convection. Eaton and Blum[3] have shown that pressures on the order of 1 to 25 torr in an enclosed area negate the effects of natural convection. For nighttime operation of the cell, the cooler surface is lower than the warmer surface and only conduction heat transfer is present; there is no natural convection.

However, depending on the pressure in the pod during daytime operation, air can produce buoyancy effects when the sun provides the thermal energy to the module. Therefore the analysis considers the operation of the cell when the pressure is increased from a full vacuum to 25 torr.

The thermoelectric properties of the Seebeck coefficients (α_n,α_p), thermal conductivity (λ_n,λ_p), and the electrical resistivity (ρ_n,ρ_p) are assumed to be constant. The length of the thermoelectric elements in the direction of heat flow is L.

One mode of daytime operation the TEG-only module in the Vacuum Pod could assume would be utilizing the ambient as the thermal source and deep space as the thermal sink, similar to the nighttime operation. However, shielding from direct sunlight would be required to prevent solar heating of the CJP. The thermal model will predict this mode of operation, assuming no incident solar energy strikes the CJP. This means of operation of the system is advantageous for 24-hour energy production where no storage is needed or wanted.

Therefore the same mode of operation can be used day or night using the thermal model with deep space as the thermal sink.

EQUATION DEVELOPMENT

The thermal model has been developed in a previous study[4] and the results will be summarized here.

RADIATION MODEL

The net radiative heat flux on the CJP comes from three sources: the fraction of energy from the night sky that is transmitted through the window, emission from the window, and the fraction of the radiosity, J_c, that is reflected from the window. Thus

$$q_c = (1 - \rho_w)J_c - \tau_w \sigma T_s^4 - \epsilon_w \sigma T_w^4, \quad (1)$$

where the ρ_w, τ_w, ϵ_w refer to the radiative properties of reflectivity, transmissivity and emissivity of the window and T_w is the temperature.

An energy balance on the window accounts for the convective heat transfer rate at the external surface of the window and the net radiative energy it absorbs. Therefore

$$h_w(T_\infty - T_w) = 2\epsilon_w \sigma T_w^4 - \epsilon_w \sigma T_s^4 - \epsilon_w \sigma T_w^4 + h_a(T_a - T_w), \quad (2)$$

where h_w is the convective heat transfer coefficient at the surface of the glass. When the effect of air is present, h_a is the thermal resistance of the air gap in the pod of height L_a and thermal conductivity $k_a(P)$, and T_a is the temperature of the air.

All radiative properties in eqns. (1) and (2) are written as the sum of three contributions from their respective bands. For example, ϵ_w in the term $\epsilon_w \sigma T_s^4$ in eqn. (3) may be written as

$$\epsilon_w = \epsilon_1 F(0, \lambda_l T_s) + \epsilon_2 F(\lambda_l T_s, \lambda_u T_s) + \epsilon_3 F(\lambda_u T_s, \infty), \quad (3)$$

where $F(x,y)$ is the blackbody emissive power fraction over the band of λT defined by the values of the first (x) and second (y) arguments in $F(x,y)$.[5] In eqn. (1), the radiation properties of J_c are based on the band model at temperature T_c.

HEAT CONDUCTION MODEL

A steady-state, quasi one-dimensional heat conduction model with internal energy generation is used.[4] One boundary condition is the radiative heat flux (q_c) at the CJP, the second is the convection heat transfer at the hot junction plate.

The area for heat conduction in the individual thermoelectric elements is A_e. The area ratio, A_r, is equal to A_w/A_e. This is the area parameter used in the development of the conduction model, where A_r is greater than 1.

From this information, the temperature distribution in the thermoelectric elements may be written as

$$T_t(\eta) = -\phi\eta^2/2 - Bi\,\eta T_\infty + (\phi/Bi + T_\infty - q_c/h_b)(1 + Bi\,\eta), \quad (4)$$

where ϕ is the energy generation parameter, $q'''L^2/\lambda_n$, where q''' is the rate of energy generation per unit volume; Bi is the Biot number, $A_r h_b L/\lambda_n$; and η is the dimensionless local coordinate, x/L.

The energy generation term is related to the Seebeck coefficients and one of the junction temperatures[4] as

$$\phi = \{A_r q_c L/[(T_c(\alpha_p - \alpha_n))]\}^2 \rho_p/\lambda_p, \quad (5)$$

where ρ_p and λ_p or ρ_n and λ_n, properties of the TEG elements, may be used, respectively. The cold and hot plate temperatures, T_c and T_h, are

$$T_c = T_t(\eta = 1), \text{ and } T_h = T_t(\eta = 0). \quad (6)$$

This system of equations defines the thermal model for determining the temperature differential between the hot and cold plate junctions of the TEG module.

THERMOELECTRIC EQUATIONS

The thermoelectric generator (TEG) equations will be selected to maximize the thermal efficiency[6,7] of the module based on the semiconductor material properties and the geometry of the module. Therefore the maximum value for the figure of merit is

$$Z = \frac{(|\alpha_p| + |\alpha_n|)^2}{[(\rho_n\lambda_n)^{1/2} + (\rho_p\lambda_p)^{1/2}]^2}, \quad (7)$$

where α_p, α_n are the respective Seebeck coefficients, ρ_n, ρ_p are the electrical resistivities and λ_n, λ_p are the thermal conductivities of the materials.

Utilizing the figure of merit, the calculation for the current output of the TEG module is based on optimizing the internal and external resistances of the system.[7] Therefore the equation for the current produced by the module to maximize the thermal efficiency is

$$I_{out} = \frac{(|\alpha_p| + |\alpha_n|)(T_h - T_c)}{R\,[x + 1]}, \quad (8)$$

where
$R = (\rho_n l_n/A_n) + (\rho_p l_p/A_p)$, and

$x = [1 + Z(T_h+T_c)/2]^{1/2}$,

A_n is the area of n-type material, A_p is the area of p-type material, l_n and l_p are the lengths of the elements, and T_h and T_c are the temperatures of the hot and cold junctions, respectively.

The open circuit voltage for the thermoelectric generator is

$$V_{oc} = (|\alpha_p| + |\alpha_n|)[T_h - T_c]. \quad (9)$$

The selection of the TEG module will be based on utilizing off-the-shelf or near-off-the-shelf materials. That is, a minimum of modifications to existing tooling will be sought. Therefore, both n-type and p-type elements are chosen with the same cross-sectional area and length. Typical assembly techniques of copper junctions, ceramic endfaces, etc., are used.

RESULTS

The full thermal model is not used to determine the parameters required for the prototype. The model is simplified in three ways: the hot junction plate is maintained at a constant temperature T_h; the fin efficiency of the CJP is considered unity; and all radiative interactions occur between the black sky and the surface of the CJP only.

The TEG elements utilized are p- and n-doped bismuth telluride with a 1mm x 1mm square cross-section with a length of 25mm. These were element sizes available from suppliers. For this geometry of the elements, the model is used to optimize the number of junction pairs, based on the power output of the cell.

Figure 2 shows the power output of the cell as a function of the number of elements or element pairs. Forty-four elements correspond to a module having 22 TEG junctions, the maximum energy output developed for this geometry. With this design, the cell will produce approximately 7mW of electrical power. Note the power output is maximized when the thermal conductance ($A_e\lambda_{n or p}/L$) and the electrical conductance ($A_e\rho_{n or p}/L$) are optimized for the chosen geometry. Although not shown, this corresponds to about 0.42 volts.

The temperature of the CJP varies from 208.6K with four junction pairs to 281.5K with 100 junction pairs. For the maximum electrical energy production with 22 TEG junctions, the CJP temperature is 252K. This corresponds to a 48K temperature differential across the TEG elements.

In Figure 3, increasing the temperature of the hot junction plate by utilizing low grade thermal waste heat can affect the operation of the 22-junction module significantly. For the example shown, only a 10-degree temperature rise in the thermal source will result in a 22% increase in the power output of the module. Therefore, with a low temperature source only slightly above the ambient available to drive the pod, a considerable increase in the electrical output can be achieved.

Figure 4 shows the power output of the 22-junction module when the CJP is heated by the sun. This daytime operation of the cell shows promise for higher electrical energy production during daylight hours, compared to nighttime energy production. However, this analysis does not take into account sun angle, cloud cover, shape factors, etc. Therefore this should only be considered a first attempt at demonstrating the feasibility of a possibly promising system.

Figure 5 illustrates the effect on the performance of the 22-junction cell with air at 25 torr in the Vacuum Pod. With an air gap of 1.2cm, well proportioned to the physical dimensions of the pod, the output of the cell is reduced by about 2% compared to the full vacuum. The 1.2cm gap is between the CJP and the aperture window cover.

This moderate reduction in energy output may show that the cost advantages for the lower vacuum in the pod will more than offset the slight loss of electrical power.

DISCUSSION

The 22 TEG junction module design is selected for the prototype. This provides both convenience for the selection of a standard module and utility in the optimum power range of the device. The performance of the Vacuum Pod with solar heating of the CJP and utilizing low grade waste thermal heat shows the practicality of such a power producing device. Without the benefit of a full vacuum in the pod, the cell still provides sufficient electrical energy production for many applications, the decrease in power being slight. The advantages of a less costly cell will far outweigh the reduction in energy production.

CONCLUSIONS

The thermal model shows the performance of the Vacuum Pod to be satisfactory under several operating conditions and modes of operation, while providing valuable parametric guidelines for the design of the prototype Nighttime Solar Cell™. The electrical power output of the cell, nominally sized at 6cm x 6cm x 3cm, will produce 7mW of power at night. Four cells connected in series, a 12cm x 12cm panel, will produce about 1.6 volts. This corresponds to a single D-sized battery, with an almost infinite life.

Cell performance can also be improved significantly utilizing previously unusable, low grade thermal waste heat. Daytime operation without the use of solar cells can also be achieved successfully. Therefore this new mode of electric power production may be the next source of clean, reliable, safe and inexpensive energy.

CONTACT

For information regarding the operation of the Nighttime Solar Cell™, contact Ronald J. Parise at PARISE RESEARCH TECHNOLOGIES, Suffield, Connecticut 06078.

REFERENCES

1. Parise, Ronald J., Nighttime Solar Cell™, IECEC98, Colorado Springs, CO, 1998, Paper No. IECEC-98-133.

2. Bliss, Raymond W., Jr., Atmospheric Radiation Near the Surface of the Ground: A Summary for Engineers, Solar Energy, Vol. 5, p103, (1961).

3. Eaton, C.B. and Blum, H.A., The Use of Moderate Vacuum Environments as a Means of Increasing the Collection Efficiencies and Operating Temperatures of Flat-Plate Solar Collectors, Solar Energy, Vol. 17, p151, (1975).

4. Parise, R.J., Jones, G.F., Strayer, B., Prototype Nighttime Solar Cell™, Electrical Energy Production from the Night Sky, IECEC99, Vancouver, British Columbia, Canada, 1999, Paper No. 1999-01-2566.

5. Dunkle, R.V., Thermal Radiation Tables and Applications, Transaction of the ASME, Paper No. 53-A-220, 1953.

6. Culp, Archie W., Jr., Principles of Energy Conversion, McGraw-Hill Book Company, New York, NY, 1979.

7. Angrist, Stanley W., Direct Energy Conversion, Fourth Edition, Allyn and Bacon, Inc., Boston, Massachusetts, 1982.

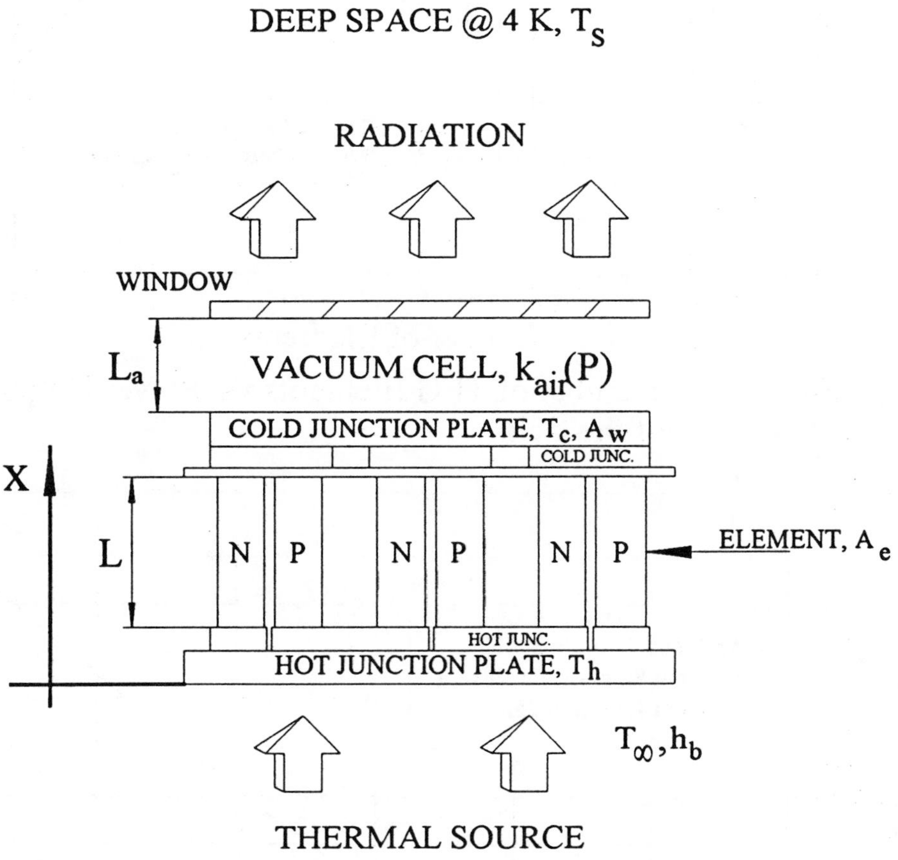

DEEP SPACE @ 4 K, T_S

RADIATION

WINDOW

L_a

VACUUM CELL, $k_{air}(P)$

COLD JUNCTION PLATE, T_c, A_w

COLD JUNC.

X

L

N P N P N P

ELEMENT, A_e

HOT JUNC.

HOT JUNCTION PLATE, T_h

T_∞, h_b

THERMAL SOURCE

FIGURE 1: Model Configuration.

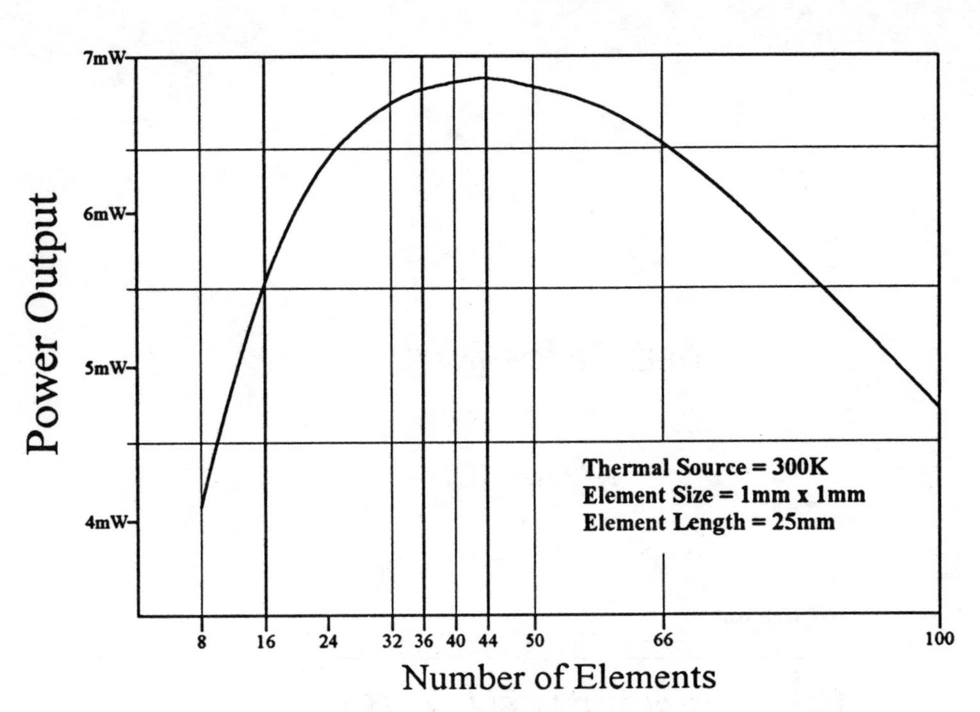

FIGURE 2: Number of TEG Elements vs. Power Output of the Cell.

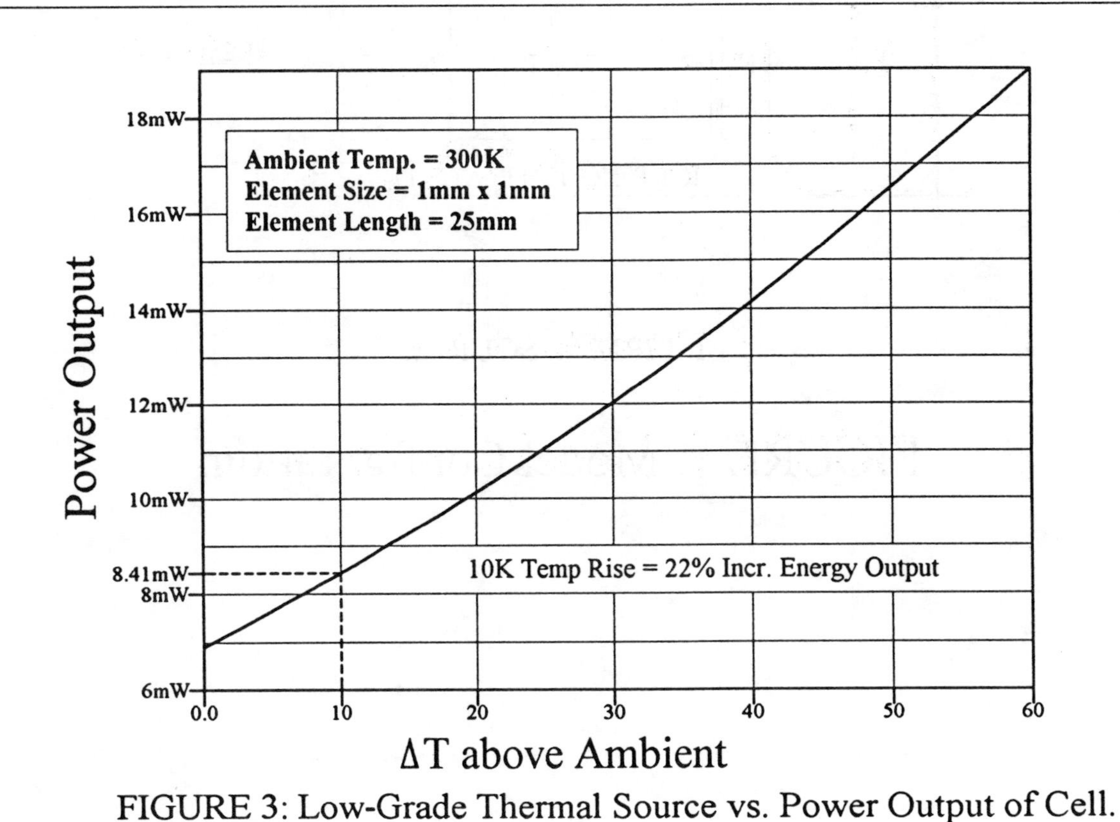

FIGURE 3: Low-Grade Thermal Source vs. Power Output of Cell.

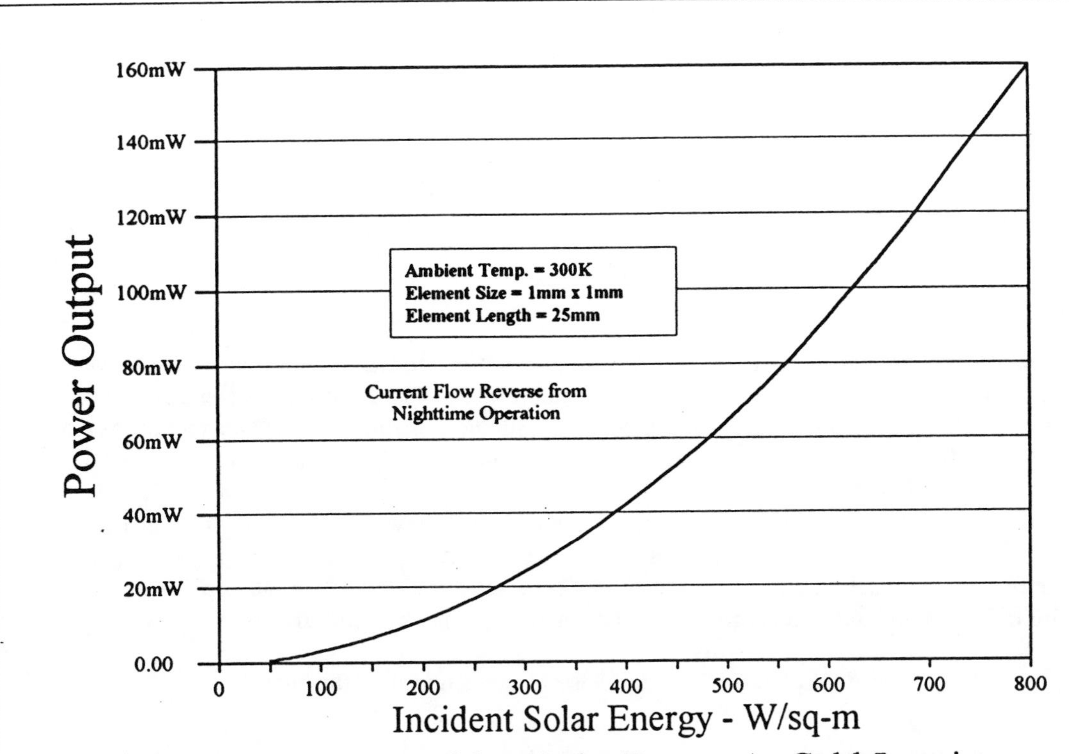

FIGURE 4: Incident Solar Energy on Cold Junction Plate vs. Power Output.

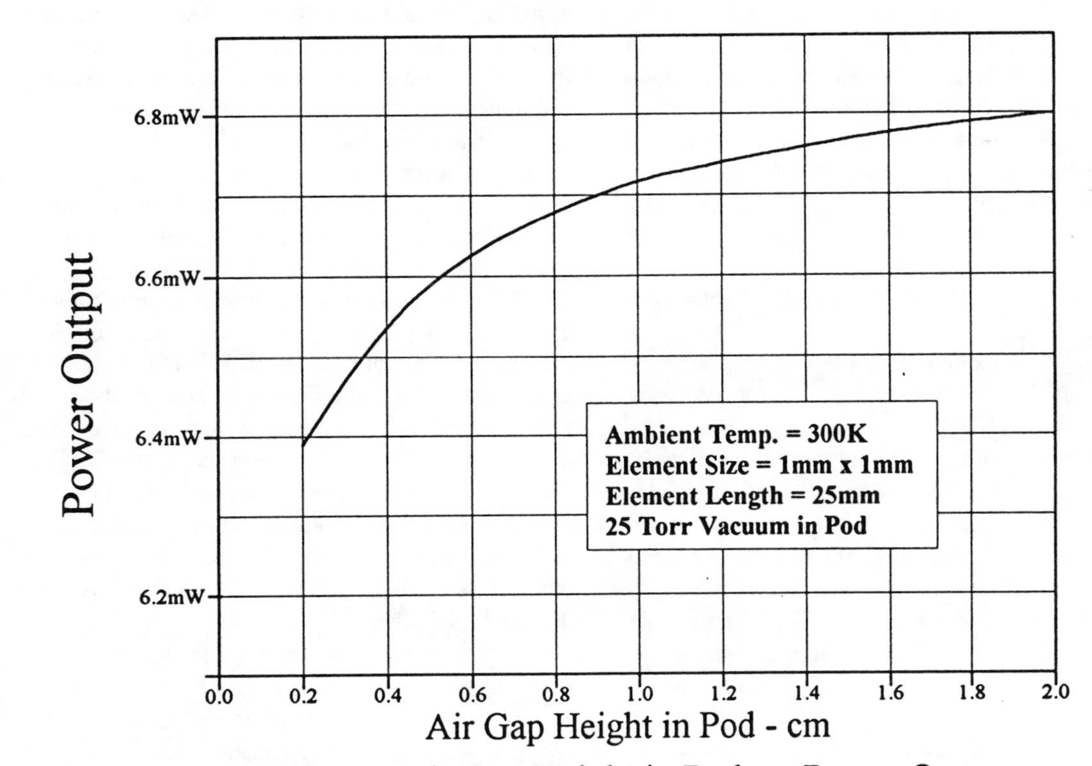

FIGURE 5: Air Gap Height in Pod vs. Power Output.

AIAA-2000-2823

A FLAT PLATE SOLAR COLLECTOR WITH DUAL CHANNELS FOR AIR FLOWS

Sadasuke Ito*, Minoru Kashima** and Naokatu Miura#
Kanagawa Institute of Technology, Atugi, 243-0292, Japan
Tel: 81-462-91-3091, Fax: 81-462-42-6806, E-mail: ito@sd.kanagawa-it.ac.jp

ABSTRACT

An analysis was made on the thermal performance of a flat-plate solar collector in which the air flows both in upper and under channels of the absorber plate. The results were compared with experimental results. The analytical results agreed well with the experimental results in the case that the ratio of the flow rate of the upper channel to the total flow rate, β_2, was in the range between 0 and 0.5. It was found that generally the flow in the upper channel increased the thermal efficiency. For example, the increase in thermal efficiency for turbulent flow was 14% at β_2=0.2 when the inlet air temperature was the same as the ambient air temperature. For laminar flow, larger increase of the efficiency, as much as 30 %, was obtained. The effect of the total flow rate and β_2 on the collector efficiency and temperature rise of the flow in each channel was also discussed.

Key words : Solar Air Collector, Solar House, Collector Efficiency

INTRODUCTION

The global environmental problem is becoming serious, and the utilization of clean energy is more and more desired. The goal of the utilization of solar heat in 2010 in Japan is 4.5 Mm^3 in equivalent oil. In order to achieve goal, 4 times the present utilization quantity will be required. Though solar heaters and comparatively small-scale solar hot water systems are fairly popular, further popularization of these systems and the development of large solar thermal applications such as solar houses are desired. The number of air heated solar houses has increased steadily in the past ten years in Japan due to advances in the technology of the construction of solar systems and the high evaluation of demand side in amenity and in safety. Further development of air heated solar houses can be expected.

Black painted iron sheets are usually used as the absorber plates of collectors on the roof of a solar house. The air taken in from the space under the eaves of the roof flows though the channels between the absorber plates and the inside bottom plates on the roof boards. Glass plates cover all of the absorber plates or part of them near the top of the roof to reduce heat loss from the absorber plates to the ambient air.

Generally, there is no forced flow of air between the absorber plate and the glass cover. It was reported by Komano et al[1] that a better characteristics of collecting heat with a flow in the channel between the absorber plate and the glass cover than without a flow was obtained in experiments. However, the effects of flow rate, inlet air temperature and weather conditions on the collector efficiency were not examined in these studies.

Ong[2] made an analysis on the thermal performance of a solar collector with two channels. Heat balance equations at a local point on the glass cover, absorber plate, inside bottom plate were derived and each flow was solved mathematically by employing a matrix inversion technique. The calculated results of the distributions of temperatures of the air streams, absorber plate, bottom plate, and efficiency at the collector were compared with experimental results[3]. The agreement was slightly less satisfactory. The author concluded that the measurement of temperatures of the air streams might not be accurate. Parker et al[4]. solved the problem in the case that the flow rates in the two channels were the same each other.

In this paper, an analysis is made on the performance of a collector with a flow in the upper channel on the absorber plate as well as in the lower channel. The heat balance equations are the same as the ones given by Ong[2]. The temperature of the glass cover, absorber plate, bottom plate and air streams at an arbitrary location are easily obtained numerically

* Professor, Department of System Design Engineering
** Graduate Student
# Research Associate

by iteration. The results of the analysis were compared with experimental results to examine the validity of the analysis. Then, the effect of flow in the upper channel on the collector efficiency was examined analytically.

NOMENCLATURE

A = collector area
a. = thermal diffusivity
c_p = specific heat
D = hydraulic diameter
g = acceleration of gravity
h = heat transfer coefficient
I = solar radiation intensity on collector
K = thermal conductivity
L = length of collector area
m = mass flow rate
Nu = Nusselt number
Pr = Prandtl number
Q = collected heat
q = collected heat per unit area
Ra = Raylaigh number
Re = Reynolds number
S = solar heat collected by absorber plate
T = temperature
Δt = difference between mixing temperatures of flows at exits and inlets of channels
U = overall heat transfer coefficient
V = wind velocity
W = width of collector area
X = distance from inlet
Δx = increment of distance x
α = absorptivity
β = ratio of the flow rate of upper flow to that of lower flow
η = collector efficiency
τ = transmissibility
ν = kinematic viscosity
ξ = volume coefficient of expansion

Subscripts
a = ambient air
b = inside bottom plate
c = glass cover
in = inlet
out = outlet
p = absorber plate
r = radiation
s = sky
t = total
th = analysis
w = wind
x = distance from inlet
1 = lower channel
2 = upper channel

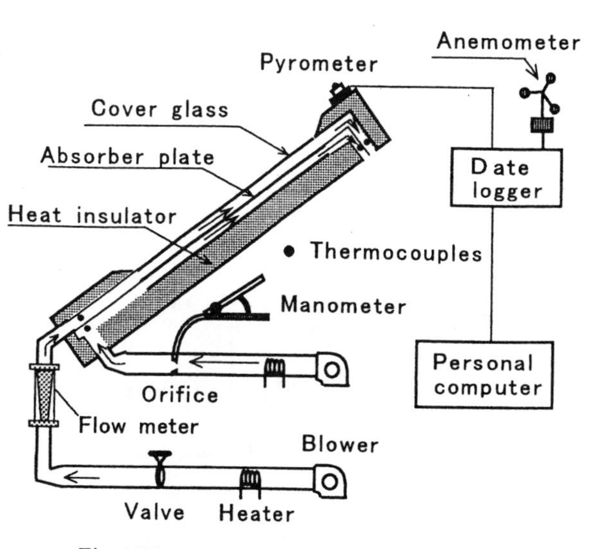

Fig. 1 Diagram of experimental apparatus.

TEST EQUIPMENT AND EXPERIMENTAL METHOD

The experimental apparatus is shown in Fig. 1. The glass cover of the collector was 5mm thick and the absorber plate was a black painted copper plate 2mm thick. The spacing between the glass cover and the absorber plate was 25mm. The spacing between the absorber plate and the inside bottom plate was also 25mm. The width of the channel was 460mm. The ambient air was taken into the upper and lower channels in the collector. The length and area of the sun light penetration part of the glass plate was 2.5m and 1.15m², respectively. The collector was installed at 30°tilt angle facing to the south direction. Heaters adjusted the temperatures of the air at the inlets of the channels of the collector. Both temperatures were maintained the same during experiments. The flow rates of the air in the upper and lower channels were measured by an orifice and float type flow meter, respectively. The solar radiation intensity on the collector, wind velocity, temperatures of the flows at the inlets and outlets of the channels and ambient air were measured by a pyrometer, a three cup anemometer and thermocouples, respectively, every 5 minutes. The heat collected and collector efficiency at a certain time were obtained from the mean values of the data taken in 30 minutes intervals. When the flow rate in the upper channel was zero, the flow rate in the lower channel was chosen to be either 0.0137kg/s (m_1/A=0.012kg·m^{-2}·s^{-1}) or 0.0069kg/s (m_1/A=0.006 kg·m^{-2}·s^{-1}). The corresponding Reynolds number of the flow at the flow rate of 0.0137kg/s was about 2800. This flow was in turbulent or transition zone. The Reynolds number of the flow at the flow rate of the 0.0069kg/s was about 1400. The flow was in the region of a laminar flow.

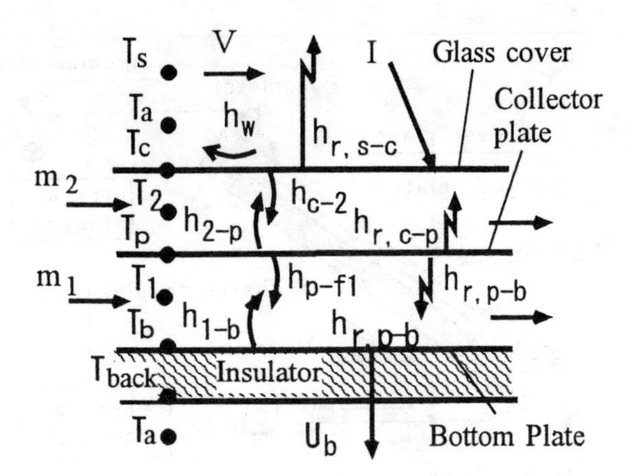

Fig. 2 Model of heat transfer in collector.

When the air was run in the upper channel, the total flow rate was kept constant at 0.0137kg/s.

ANALYSIS

The model proposed by Hottel and Whillier is representative of the analysis on the thermal performance of flat-plate collectors. This model is explained in detail in Duffie and Beckman[6]. In the model, it is assumed that the phenomenon is steady and that the heat transfer from the absorber plate to the ambient air over the glass cover and to the collector back is one dimensional. Also, the absorption of solar heat by the glass cover is assumed to be able to be negligible. Under the assumption of steady state, heat loss from the collector in the upper direction is proportional to the difference between the temperature of the absorber plate and the ambient air temperature.

For the collector without a flow between the absorber plate and glass cover (m_2=0), the present analysis uses the same assumptions as those in the model proposed by Hottel and Whillier[5], except that the glass cover absorbs solar heat. When there is a flow in the upper channel (m_2>0), coexistence of natural and forced convection is considered in the heat transfer.

For m_2>0, the equations of energy balance on the glass cover, absorber plate and inside base plate are given by Eqs. (1)- (3), respectively.

$$I\alpha_c+h_{r,s\text{-}c}(t_s-t_c)+h_w(t_a-t_c)+h_{c\text{-}2}(t_2-t_c)+h_{r,c\text{-}p}(t_p-t_c)=0 \quad (1)$$
$$S+h_{r,c\text{-}p}(t_c-t_p)+h_{2\text{-}p}(t_2-t_p)+h_{p\text{-}1}(t_1-t_p)+h_{r,p\text{-}b}(t_b-t_p)=0 \quad (2)$$
$$h_{r,p\text{-}b}(t_p-t_b)+h_{1\text{-}b}(t_1-t_b)+U_b(t_a-t_b)=0 \quad (3)$$

where S is the solar heat collected by the absorber plate, h is heat transfer coefficient, α is the absorptivity of the absorber plate, t is the temperature, U_b is the overall heat transfer coefficient between the inside bottom plate and the ambient air, the subscripts 1 and 2 refer to the flows in the lower and upper channels respectively, a is the ambient, b is the inside bottom plate, c is the glass cover, p is the absorber plate, r is radiation, and s is the sky. The heat transfer coefficients with and without subscript r are radiative and connective heat transfer coefficients, respectively. Denoting the transmissibility of the glass cover by τ_c, S is made to be 1.02 $\tau_c \alpha_p$I. The temperature of the glass cover t_c at an arbitrary location x can be obtained by solving Eqs. (1)- (3), simultaneously.

Finding the temperatures t_c, t_p and t_b from the above equations arithmetically or mathematically is rather complicated. However, these temperatures can be easily found by calculation as follows. From Eq. (1), the temperature of the glass cover is expressed as a function of the temperature of the absorber plate as shown by $t_c=f_1(t_p)$. Similarly, from Eqs. (1) -(3), the equations $t_p=f_2(t_c,t_b)$ and $t_b=f_3(t_p)$ can be obtained. Assuming some values for t_c, t_p and t_b in the beginning of calculation, new t_c, t_p and t_b values are obtained from the above equations. By repeating the calculations, each temperature reaches its true steady state temperature.

The heat collected by the lower flow per unit area q_1 at location x and the heat collected by the upper flow q_2 is be given by Eqs. (4) , (5), respectively.

$$q_1=h_{p\text{-}1}(t_p-t_1)+h_{1\text{-}b}(t_b-t_1) \quad (4)$$
$$q_2=h_{c\text{-}2}(t_c-t_2)+h_{2\text{-}p}(t_p-t_2) \quad (5)$$

At location x+Δx, the temperature of the lower flow $t_{1,x+\Delta x}$ is given by the following equation.

$$t_{1,x+\Delta x}=t_{1,x}+q_1W\Delta x/(m_1c_p) \quad (6)$$

The heat collected by the lower flow Q_1 and by the upper flow Q_2 is obtained by Eqs. (7), (8), respectively. The total heat collected by the collector Q_t and the collector efficiency η are given by Eqs. (8), (9) respectively.

$$Q_1=m_1c_p(t_{1,out}-t_{1,in}) \quad (7)$$
$$Q_2=m_2c_p(t_{2,out}-t_{2,in}) \quad (8)$$
$$Q_t=Q_1+Q_2 \quad (9)$$
$$\eta=Q_t/(A\cdot I) \quad (10)$$

In order to obtain Q_1, Q_2, Q_t and η by calculations under given conditions of the flows at the inlets and of the weather, each heat transfer coefficient should be known.

The heat transfer coefficient between the glass cover and the ambient air h_w is obtained by Eq. (11) as a function of the wind velocity V [m/s].

$$h_w=5.7+3.8V \quad [W/(m^2\cdot K)] \quad (11)$$

For the flow in the laminar region in the upper channel, natural and forced convection would coexist. Then, Eq. (12) is used for the flow. The effect of the entrance region on the mean Nusselt number of turbulent flow in a tube was considered by Kays[7]. In the present case, the factor $1+6D_2/L$, which is 1.11, is multiplied in eq.(12) to the Nusselt number given by Mori et al[8] in order to consider the entrance region of the channel.

$$Nu_2 = h_{2\text{-}p}D_2/k_2 = \{0.61(Re_2Ra_2)^{1/5}+1.10\}\ (1+6\ D_2/L) \quad (12)$$

where Nu_2 is the Nusselt number, D_2 is the hydraulic diameter, k_2 is the thermal conductivity of the air, Re_2 is the Reynolds number, and Ra_2 is the Rayleigh number given by Eq. (13).

$$Ra_2 = g\,\xi_2(dt_2/dx)(D_2/2)^4/(a_2\,\nu_2) \quad (13)$$

where g is the acceleration of gravity, ξ_2 is the volume coefficient of expansion, ν_2 is the kinematic viscosity, and a_2 is the thermal diffusivity.

Equation (14) is recommended for a laminar flow in Ref. 5. For a turbulent flow, the equation without $1+6D_1/L$ in Eq. (15) is recommended in Ref. 6. Considering the effect of the entrance region, Eq. (15) is used in the present analysis.

$$Nu_1 = 4.9 + \frac{0.0606(Re_1Pr_1D_1/L)^{1.2}}{1+0.0909(Re_1Pr_1D_1/L)^{0.7}Pr_1^{0.17}}$$

$$Re_1 \leqq 2300 \quad (14)$$

$$Nu_1 = 0.0158Re_1^{0.8}(1+6\ D_1/L) \quad Re_1 > 2300 \quad (15)$$

In Eqs. (14), (15), D_1 denotes the hydraulic diameter. The spacing between the absorber plate and inside bottom plate is taken as the representative length.

For $m_2=0$, heat is transferred from the absorber plate to the glass cover by natural convection. The amount of the heat transferred is given by the heat transfer coefficient times the difference of the temperature of the absorber plate and temperature of the glass cover. The heat transfer coefficient used in this case is obtained from the Nusselt number given by Holland et al[9]. However, the equations of energy balance at the glass cover and the absorber plate are slightly different from Eqs. (1), (2). The heat collected in the upper channel Q_2 is zero. The other equations necessary for the analysis are the same as those for $m_2>0$.

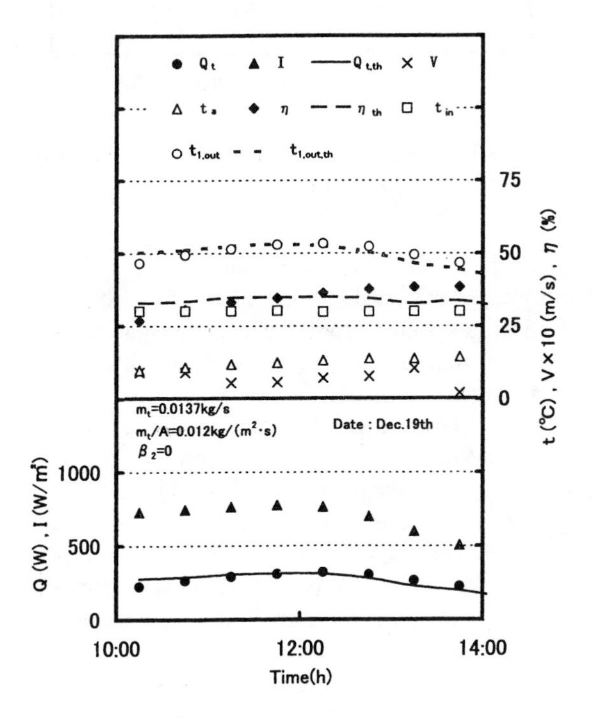

Fig. 3 Experimental and analytical results at $\beta_2=0$ and $m_t/A=0.012kg/(m^2 \cdot s)$.

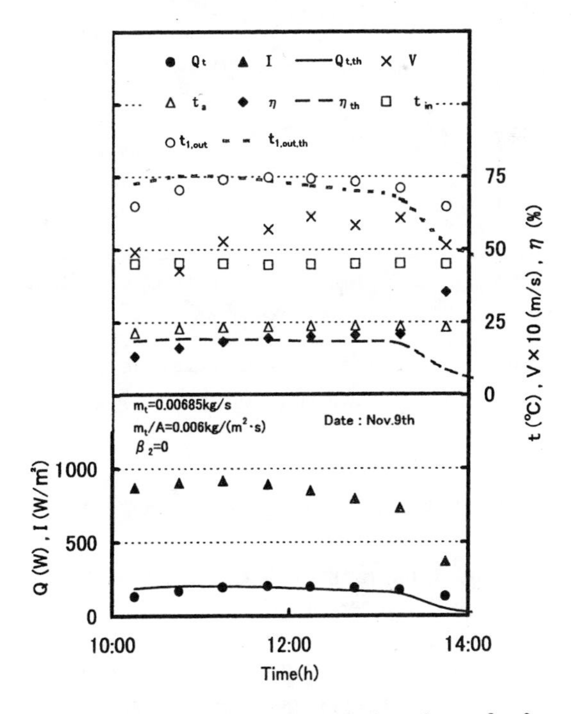

Fig. 4 Experimental and analytical results at $\beta_2=0$ and $m_t/A=0.006kg/(m^2 \cdot s)$.

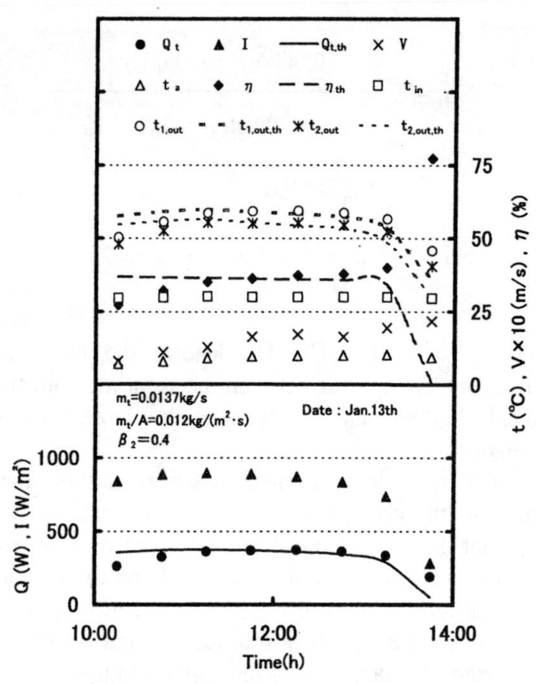

Fig. 5 Experimental and analytical results at β_2=0.4 and m_t/A=0.012kg/($m^2 \cdot$s).

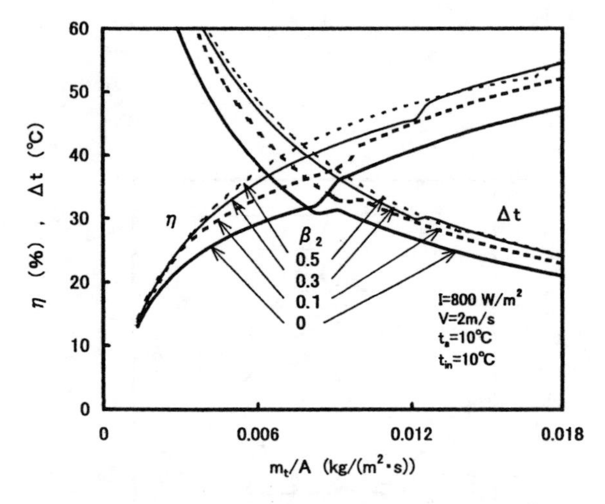

Fig. 6 The effect of total mass flow rate on the collector efficiency and temperature increase of the air.

RESULTS AND DISCUSSIONS

In Fig. 3, the results of the experiment and the analysis on December 19 are shown. The air flowed only the lower channel at the flow rate of 0.0137 kg/s (m_1/A=0.012kg$\cdot m^{-2} \cdot s^{-1}$). The collector efficiency at 11:45 a.m. was 34.5 % in the experiment and 34.8 % in the analysis. Right after the beginning of the experiment and after 1p.m., when the solar radiation

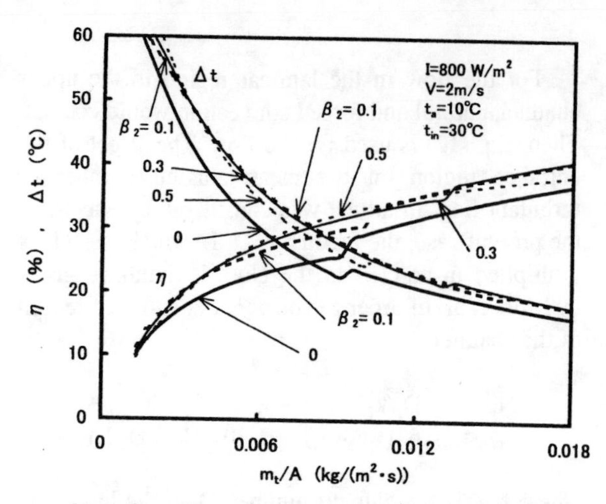

Fig.7 The effect of total mass flow rate on the collector efficiency and temperature increase of the air.

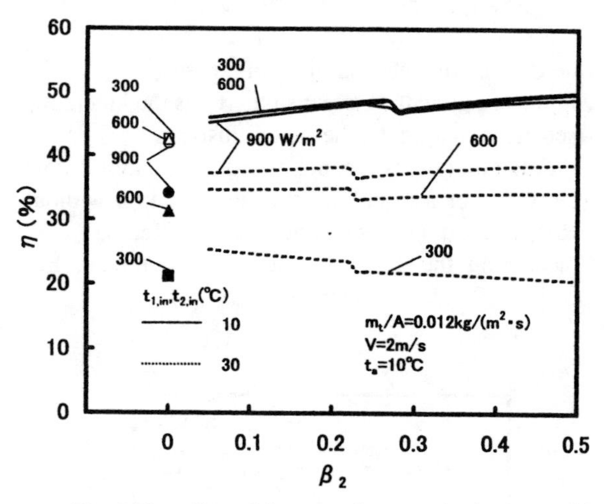

Fig. 8 The effect of the mass flow rate in the upper side on the collector efficiency.

decreased rapidly, the collector efficiency exit temperature, collected heat, which were obtained analytically, differed from the experimental results but agreed well at around noon when the insulation was stabilized.

Fig. 4 shows experimental and the analytical results on November 9 when the flow rate m_1 was chosen to be 0.0069kg/s (m_1/A = 0.006 kg$\cdot m^{-2} \cdot s^{-1}$). The collector efficiency at 11:45 a.m. was 19.6 % in the experiment and 18.9 % in the analysis. From these results, it was shown that good analytical results were obtained for both laminar and turbulent flows in the lower channel.

Fig. 5 shows the results on January 13 when the ratio of the flow rate in the upper channel to the total flow rate, β_2, was 0.4. The Reynolds number of the lower flow was 1600. The collector efficiency at

11:45 a.m. was 36.4 % in the experiment and 36.3 % in the analysis. Under the equal weather condition, the collector efficiency by the analysis became 32.7 % when the same total amount of the air flows through only the lower channel. This shows that the collector efficiency in the case with the upper flow increased 11 % compared to the collector efficiency in the case without the upper flow.

In Fig. 6, the relation between m_t/A and the collector efficiency is shown. Also the relation between m_t/A and the difference between the mixing temperatures of the flows at the exits and the inlets of the upper and the lower channels, Δt, is shown. The conditions in the calculations are that I=800W/m^2, t_a=10 ℃, V=2m/s, $t_{1,in}$= $t_{2,in}$=10 ℃. It is shown from Fig. 6 that the collector efficiency increases and Δt decreases for any β when the total flow rate increases. It is also shown from Fig. 6 that the collector efficiency and the temperature rise Δt increase by letting a part of the total amount of the ambient air pass through the upper channel. When m_t/A=0.012 kg·m^{-2}·s^{-1}, the collector efficiency is 40.9 % at β_2=0. At β_2=0.2, the collector efficiency is 46.5 %, which is 14 % higher than that at β_2=0. If the flow is laminar, an additional increase in the efficiency of about 28 % for β_2=0.5 is obtained. The same collector efficiency of 40.9 % is obtained at m_t/A=0.0096 kg·m^{-2}·s^{-1} for β_2=0.2. For this condition, the temperature rise Δt changes from 27 ℃ at β_2=0 to 34 ℃ at β_2=0.2. The total flow rate at β_2=0.2 becomes 80 % of that at β_2=0. The flow rate of the lower flow at β_2=0.2 becomes 64 % of that at β_2=0. This decreases the pressure loss of the flow in the collector. As a result, the capacity of the fan to draw in air at β_2=0.2 is smaller than that at β_2=0.

Fig. 7 shows the analytical results of the case that the inlet temperatures are 20 ℃ higher than the temperature of the ambient air and I=800W/m^2. When the insulation is high, like in this case, the effect of the upper flow on η and Δt is similar to the case that the inlet air temperatures are the same as the ambient air temperature.

The change of collector efficiency with the flow ratio β is shown in Fig. 8. If the solar radiation intensity I is more than 600 W/m^2, the collector efficiency is higher with upper flow than without it, even when the inlet air temperature is 20 ℃ higher than the ambient air temperature. At I=300 W/m^2, the collector efficiency decreases with an increase of β_2. The reason why the collector efficiency decreases with an increase of β_2 (at β_2=0.3) for $t_{1,in}$ = $t_{2,in}$ = 10 ℃ is that the mode of the flow changes from turbulent flow to laminar flow decreasing heat transfer between the absorber plate and the lower flow.

CONCLUSIONS

The analytical results of the thermal performance of the collector with the upper flow, in addition to the lower flow, agreed well with experimental results. Generally, the collector efficiency was higher with the upper flow than without it for the same total flow rate. The effect of the upper flow on raising the collector efficiency decreased when the inlet air temperature became high. Generally, the total flow rate can be smaller with some upper flow than without it to get the same collector efficiency. The temperature of the air at the exit of the collector increased with the upper flow and reduced the pressure loss in the collectors and ducts in the solar heating system. The power of the fan necessary to push or draw the air was also reduced.

REFERENCES

[1]Komano, S., Wada, H., and Ebara, Y., "Study on Heat Collector of the Solar System Utilizing Outdoor Air," Proceedings of JSES/JWEA Joint Conference, 149-152 (1996) (in Japanese).

[2]Ong, K. S., "Thermal Performance of Solar Air Heaters : Mathematical Model and Solution Procedure," Solar energy, Vol.55, No.2, 93-109 (1995).

[3]Ong, K. S., "Thermal Performance of Solar Air Heaters−Experimental Correlation, " Solar Energy, Vol.55, No.3, 209-220 (1995).

[4]Parker, B. F., Lindley, M. R., Colliver, D. G., and Murphy, W. E., "Thermal Performance of Three Solar Air Heaters," Solar Energy, Vol.51, No.6, 467-479 (1993).

[5]Hottel, H. C., and Whillier, A., "Evaluation of Flat-Plate Solar-Collector Performance," Trans. of Conference on the Use of Solar Energy, University of Arizona, 1958, Vol II, 74-104 (1958).

[6]Duffie, J. A., and Beckman, W. A., Solar Engineering of Thermal Processes, John Wiley & Sons, New York, 197-249 (1980).

[7]Kays, W. M., Convective Heat and Mass Transfer, Tata McGraw-Hill Publishing Co., New Delhi 194-196 (1975).

[8]Mori, Y., Futagami, K., Tokuda, S., and Nakamura, A., "Forced Convective Heat Transfer in Horizontal Tube (1st Report, Experimental Study on the Effect of Buoyancy)," Transactions of the Japan Society of Mechanical Engineers, Vol.30, No.219, 1378-1385(1964) (in Japanese).

[9]Hollands, K. G. T., Unny, T. E., Raithby ,G. D., and Konicek, L., "Free Convective Heat Transfer Across Inclined Air Layers, "Trans. ASME, J. of Heat and Mass Transfer, 189-193 (1976).

AIAA-2000-2824

AN EXPERIMENTAL STUDY OF HEAT TRANSFER IN CURVED PIPE WITH PERIODICALLY VARYING CURVATURE FOR APPLICATION IN SOLAR COLLECTORS AND HEAT EXCHANGERS

Ru Yang and Fan Pin Chiang
Department of Mechanical Engineering
National Sun Yat-Sen University
Kaohsiung, Taiwan

ABSTRACT

Experiments were conducted for a varying curvature curved pipe within a double-pipe heat exchanger with water as the working medium. The heat transfer coefficients were obtained using the Wilson plot method. The effect of the Dean, Prandtl, Reynolds number and curvature ratio on the average heat transfer coefficients and the friction factors are presented. A higher Dean number results in a higher heat transfer rate. It is found that the heat transfer rate may be increased by up to 100%, as compared with a straight pipe, while the friction coefficient increased by less than 40%. Therefore, it is promising to use S-shaped pipes instead of straight pipes for the enhancement of a solar collector performance.

INTRODUCTION

Curved pipe curvature induces a secondary flow across the main stream that may enhance heat transfer rate significantly [1-6]. Therefore, it should receive much attention in the heat transfer enhancement applications. However, literature for the study of periodically varying-curvature curved pipes (e.g. wavy curved pipe) is very little [7-9]. Although the perturbation models of [7-9] are available, the models are restricted to very small amplitude of wavy pipe that limits its application drastically. In addition, the study of [7] only deals with the flow solution while heat transfer is very important in applications. This study is motivated to provide experimental data for curved pipe applications, especially for solar collectors.

EXPERIMENTAL SYSTEM

In this study, experiments are made for measurements of the heat transfer and the pressure drop of flow inside a wavy curved pipe with axial function $y=a\cdot\sin(\kappa x)$. The geometry of the pipe is illustrated in Fig 1. The curved pipe is placed inside a larger circular pipe with cold or hot fluid flowing in between to form a double-pipe heat exchanger.

Figure 2 is the schematic of the experimental system. The test section, also illustrated in Fig.3, consists of a curved pipe made of 9.52mm O.D. copper tube and a 56mm O.D. PVC pipe insulated externally. The length of the test section is 1 m. The upper loop of Fig. 2 is for the supply of cold water at designed flow rate and temperature to the test section flowing through the space between the curved pipe and the PVC pipe. Temperature is controlled by a constant temperature bath, and the flow rate is controlled with a constant mass flow rate pump (Fasco Dizark).

The lower loop in Fig. 2 is the loop for hot water flowing inside the curved pipe. Temperature is controlled by the adjustment of a heater, and flow rate is controlled by the adjustment of a valve and a centrifugal pump. The pressure drop across the curved pipe is measured with a differential pressure gauge. All temperatures are measured with RTD thermometers. Two flow loops are interchangeable such that both heating and cooling of the flow inside the curved pipe can be tested.

ANALYSIS

The pipe curvature at any location is

$$\kappa_c = \kappa^2 a \sin \kappa x / (1 + \kappa^2 a^2 \cos^2 \kappa x)^{1.5} \quad (1)$$

Heat transfer to the cold water in the test section is

$$\dot{Q}_c = \dot{m}_c \cdot C_{p,c} \cdot (T_{c,o} - T_{c,i}) \quad (2)$$

where $\dot{m}_c$ is the flow rate of cold water and $C_{p,c}$ is the specific heat of the cold water. $T_{c,i}$ and $T_{c,o}$ are the inlet and the outlet temperatures of the cold fluid.

Similarly, heat transfer from hot water is

$$\dot{Q}_h = \dot{m}_h \cdot C_{p,h} \cdot (T_{h,o} - T_{h,i}) \quad (3)$$

where $\dot{m}_h$ is the flow rate of hot water and $C_{p,h}$ is the specific heat of the hot water. $T_{h,i}$ and $T_{h,o}$ are the inlet and the outlet temperatures of the cold fluid. The energy balance for the test section gives $\dot{Q}_c = -\dot{Q}_h$. This provides a check for data accuracy. Averaged heat transfer rate is defined as

$$\dot{Q}_{av} = \frac{1}{2}\left(|\dot{Q}_h| + |\dot{Q}_c|\right) \quad (4)$$

the overall heat transfer coefficient U_0 is defined as:

$$\dot{Q}_{av} \equiv U_0 A_0 \Delta T_{lm} \quad (5)$$

where A_0 is total heat transfer area, ΔT_{lm} is the LMTD (log mean temperature difference) defined as

$$\Delta T_{lm}(\text{LMTD}) \equiv \frac{\Delta T_a - \Delta T_b}{\ln(\frac{\Delta T_a}{\Delta T_b})} \quad (6)$$

where $\Delta T_a \equiv T_{h,i} - T_{c,o}$ (7)

$\Delta T_b \equiv T_{h,o} - T_{c,i}$ (8)

therefore, the overall heat transfer resistance can be expressed by

$$\frac{1}{U_o A_o} = \frac{\Delta T_{lm}}{\dot{Q}_{av}} \quad (9)$$

The overall resistance consists of three resistances in series. They are outer surface convection resistance, pipe wall conduction resistance and inner surface resistance. That is

$$\frac{1}{U_o A_o} = \frac{1}{h_o A_o} + R_{wall} + \frac{1}{h_i A_i} \quad (10)$$

where h_0 and h_i are the outer surface and inner surface convective heat transfer coefficients, respectively.

Parameters that affects the problem including flow parameter of the Reynolds number and geometric parameters of a and κ that determine the curvature of a curved pipe.

The inner surface heat transfer coefficient is cast into the form

$$h_i = C_i \text{Re}_i^{\,m} \frac{k_i}{D_i} \quad (11)$$

where m and C_i are constants. The outer surface heat transfer coefficient can be expressed in the form

$$h_o = C_o \text{Re}_o^{\,0.8} \frac{k_o}{D_o} \quad (12)$$

Substitution of equations (11) and (12) into Eq. (10) results in:

$$\frac{1}{U_o A_o} - R_{wall} = \frac{1}{(C_i \text{Re}_i^{\,m} \frac{k_i}{D_i})A_i} + \frac{1}{(C_o \text{Re}_o^{\,0.8} \frac{k_o}{D_o})A_o}$$

(13)

The value of $1/U_0A_0$ can be evaluated by Eq. (9), and R_{wall} can be neglected. C_i, m and C_0 are the coefficients to be determined. Equation (13) can be expressed by

$$\frac{\text{Re}_o^{\,0.8}\frac{k_o}{D_o}}{U_o} = \frac{1}{C_i}\frac{\text{Re}_o^{\,0.8}\frac{k_o}{D_o}A_o}{\text{Re}_i^{\,m}\frac{k_i}{D_i}A_i} + \frac{1}{C_o} \quad (14)$$

Reform the equation into a linear form

$$Y = AX + B \quad (15)$$

where

$$Y = \frac{\text{Re}_o^{\,0.8}\frac{k_o}{D_o}}{U_o}$$

$$X = \frac{\text{Re}_o^{\,0.8}\frac{k_o}{D_o}A_o}{\text{Re}_i^{\,m}\frac{k_i}{D_i}A_i}$$

$$A = \frac{1}{C_i}; \quad B = \frac{1}{C_o}$$

If the value of m is assumed, experimental data are fit into Eq. (15) to determine C_i and C_0. This is the Wilson plot method [10]. However, in order to determine m, modified Wilson plot method [10] is employed. Equation (13) is reformed into

$$C_i \text{Re}_i^{\,m} = 1/[(\frac{1}{U_o} - \frac{1}{h_o})\frac{\frac{k_i}{D_i}A_i}{A_o}] \quad (16)$$

which is then expressed by

$$Y_2 = DX_2 + E \quad (17)$$

where $$Y_2 = \ln[1/\{(\frac{1}{U_o} - \frac{1}{h_o})\frac{\frac{k_i}{D_i}A_i}{A_o}\}]$$

$$X_2 = \ln(Re_i)$$

$$D = m, \quad \text{and} \quad E = \ln C_i$$

Experimental data are fit into Eq. (17) to determine m and C_i. The obtained new value of m is then compared with the assumed value of m for Eq. (15). If they are not equal, the value of m for Eq. (15) is adjusted and the procedure is repeated until m, C_i and C_0 are converged. Heat transfer coefficient h_i is then obtained.

Dimensionless parameters that affect the flow and heat transfer of the problem including the wavy curved pipe amplitude a', wavy number κ', the flow Reynolds number and the Prandtl number.

RESULTS AND DISCUSSIONS

Figures 4 and 5 illustrate the straightness of the Wilson plot for varies a and κ, respectively. The good straight lines confirm that the method is appropriate.

It is shown in Fig.6 that heat transfer coefficient increases with increasing κ. The reason is that the larger the κ, the larger the curvature as indicated in Eq. (1). A larger curvature corresponds to a stronger secondary flow. Therefore, the heat transfer rate becomes higher. The effect of a on the heat transfer rate shown in Fig. 7 is similar to the effect of κ since an increased a also corresponds to an increased curvature as indicated in Eq. (1). The value of κ or a equals to zero corresponds to zero pipe curvature, that is the pipe is straight. In Figs. 6 and 7, it is clearly shown that curved pipe has higher heat transfer coefficient than that of a straight pipe. Especially for Re around 3000, the heat transfer rate is increased by 100%.

As expected, pressure drop is higher in a curved pipe than in a straight pipe. Figures 8 and 9 show that friction coefficient increases with increasing curvature parameters of κ and a, respectively. This is owing to the higher strength of the secondary flow corresponding to the higher curvature. It is shown that, in laminar region, f equals the theoretical value of 64/Re for a smooth pipe (a=0 or κ=0). In turbulent region, data for smooth pipes are fit well into Blasius' [11] correlation

$$f \cong 0.316 Re^{-\frac{1}{4}}, \qquad Re < 2 \times 10^4 \qquad (21)$$

$$f \cong 0.184 Re^{-\frac{1}{5}}, \qquad Re \geq 2 \times 10^4 \qquad (22)$$

Dimensionless pump power can be expressed by fRe^3 [12]. Figures 10 and 11 show the effects of k and a, respectively, on the heat transfer coefficient for various pump power assumptions. It can be seen that curved pipe gives net benefit (up to 100% increase) in heat transfer rate for any fixed pump power assumption rate.

Data can also be presented in terms of the dimensionless curvature parameters of the Dean number (De) [2] and the curvature ratio,δ. Note that since the curvature is periodic varying, averaged curvature is employed. The results for the heat transfer coefficient are illustrated in Figs. 12 and 13 for cooling and heating respectively. It is shown that for cooling process, the curvature ratio effect is not profound, and the heat transfer rate is almost depending on the Dean number only. Figures 14 and 15 show that the friction coefficient as function of the Dean number and the curvature ratio.

The results within the current experimental data range can be correlated into the following forms.

For turbulent flow (Re>2000):

$$Nu = 2.87 De^{0.4} \delta^{-0.203} Pr^{0.386} \qquad R^2 = 0.85$$

$$f = 1.69 De^{-0.159} \delta^{0.488} \qquad R^2 = 0.95$$

$$2.1 \times 10^6 \leq De \leq 5.5 \times 10^7, \; 0.050 < \delta < 0.096, \; 4.0 < Pr < 5.2$$

For laminar flow (Re<2000):

$$Nu = 0.185 De^{0.325} \delta^{-0.157} Pr^{0.234} \qquad R^2 = 0.89$$

$$f = 739 De^{-0..507} \delta^{0.988} \qquad R^2 = 0.87$$

$$2.5 \times 10^4 \leq De \leq 6 \times 10^5, \; 0.050 < \delta < 0.096, \; 3.9 < Pr < 4.5$$

CONCLUSIONS

The present study provides data and correlations for heat transfer and friction of the flow in wavy curved pipes. The effect of the Dean, Prandtl, Reynolds number and curvature ratio on the average heat transfer coefficients and the friction factors are presented. A higher Dean number results in a higher heat transfer rate. It is found that the heat transfer rate may be increased by up to 100%, as compared with a straight pipe, while the friction coefficient increased by less than 40%. Therefore, it is promising to use S-shaped pipes instead of straight pipes for the enhancement of a heat exchanger or a solar collector performance. Flow ranges of low Reynolds number (Re<600) was not tested due to incompatible experimental setup. Revising the experimental system is under progress and the study will be made in the near future.

NOMENCLATURE

A	heat transfer area (m^2)
a	amplitude of wavy pipe (cm)
a'	dimensionless a, a/r
C_p	constant pressure specific heat (kJ/kg°C)
D	pipe diameter (mm)
De	Dean number, $Re^2 \delta$
f	friction coefficient, $(D_i/L)(2\Delta p/\rho v^2)$
h	heat transfer coefficient (W/m^2 °C)
k	conductivity (W/m°C)
L	length of test section (cm)
$\dot{m}$	mass flow rate (kg/s)
Nu	Nusselt number, hD/k
P	pressure (kpa)
Pr	Prandtl number, $C_p\mu/k$
$\dot{Q}$	heat transfer rate (W)
r	pipe radius (cm)
Re	Reynolds number, $\rho vD/\mu$
R_{wall}	pipe wall thermal resistance $(W\ °C)^{-1}$
T	temperature (°C)
U	overall heat transfer coefficient (W/m^2 °C)
v	velocity (m/s)
x	x-coordinate
y	y-coordinate

Greek symbols:

ΔP	pressure deference (N/m^2)
ΔT_{lm}	log mean temperature difference
δ	curvature ratio, $r\bar{\kappa}_c$
κ	wave number of wavy pipe, $2\pi/L\ (cm)^{-1}$
κ'	dimensionless κ, $r\kappa$
κ_c	curvature of curved pipe
$\bar{\kappa}_c$	average curvature of wavy curved pipe
μ	dynamic viscosity (Ns/m^2)
ρ	density (kg/m^3)

Subscrips:

av	average quantities
c	cold
h	hot
i	inner
o	outer
s	quantities associated with the straight pipe

ACKNOWLEDGEMENT

This study was sponsored by the National Science Council, Taiwan, ROC under the contract NSC88-2212-E-110-020.

REFERENCES

1. Eustic, J. 1911 "Experiments of Streamline Motion in Curved Pipes" Proc. R. Soc. London Ser. A85, pp119-131.
2. Dean, W. R. 1927 "Note on the Motion of Fluid in a Curved Pipe ",Philos. Mag. Ser. 7, Vol. 5, 208-223.
3. Berger, S. A., Talbot, L. and Tao, L. S. 1983 "Flow in Curved Pipe" Annual Reviews in Fluid Mech., Vol. 15, 461-512.
4. Yang, Ru. and Chang, S. F. 1993 "Numerical Study of Fully Developed Laminar Flow and Heat Transfer in a Curved Pipe With Arbitrary Curvature Ratio", Int. J. Heat Fluid Flow.
5. Yang, Ru. and Chang, S. F. 1994 "Combined Free and Forced Convection for Developed Flow in Curved Pipe with Finite Curvature Ratio", Int. J. Heat Fluid Flow, Vol.15, No.6, pp470-476.
6. Yang, R., Chang, S. F., and Wu, W. 1995 "A Numerical Study of Flow and Heat Transfer in a Curved Pipe with Periodically Varying Curvature," ASME paper 95-WA/HT-9.
7. Murata, S., Miyaka, Y. and Inaba, T. 1976." laminar Flow in a Curved Pipe with Varying Curvature" J. of Fluid Mech., Vol. 73, pp735-752.
8. Itmoto, y., Nagata, M. and Yamamoto, F. 1986. "Steady Laminar Flow of Viscoelastic Fluid in a Curved Pipe of Circular Cross-section with Varying Curvature" J. of Non-Newtonian Fluid Mech., Vol. 22,pp101-114.
9. Gopalan, N. P. 1985 "Laminar Flow of a Suspension in a Curved Pipe With Varying Curvature", Int. J. Engery, Vol.23, pp621-632.
10. Shah, R. K.1990. "Assessment of Modified Wilson Plot Techniques for Obtaining Heat Exchanger Design Data." Proc. 9th Int. Heat Transfer Conf., 5, 51-56.
11. Blasius, H. 1913 "Das Ahnlichkeitsgesetz bei Reibungsvorgangen in Flussigkeiten", Forschungs-arbeiten des Ver. Deutsch. Ing., No. 131 . quoted in :Webb, R. L. , 1987, Handbook of single-Phase Heat Transfer, Chapter 4, S. Kakac, R. K. Shah, and W. Aung, Eds., John Wiley & Sons, New York .
12. Snyder, B., Li, K. T. and Wirtz, R. A. 1993 "Heat Transfer Enhancement in a Serpentine Channel," Int. J. Heat Mass Transfer, Vol. 36 ,No. 12, pp. 2965-2976.

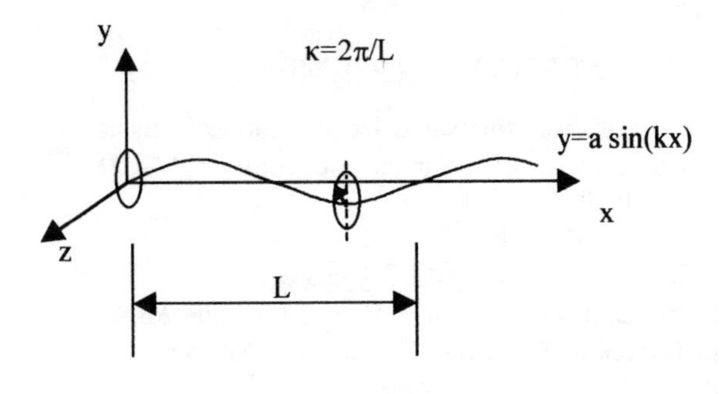

Fig. 1 Geometric shape of wavy curved pipe.

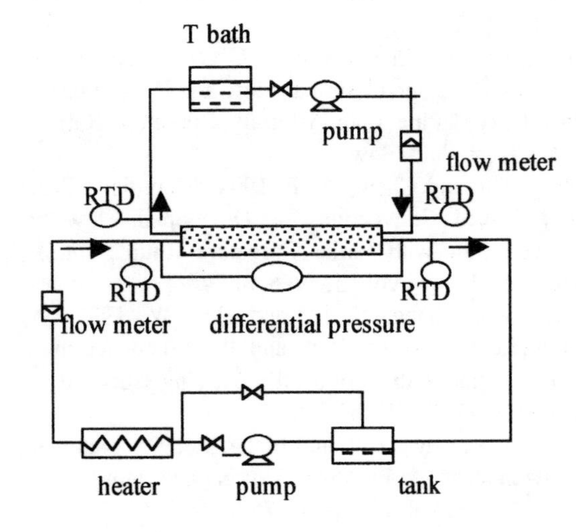

Fig. 2 Schematic of experimental system.

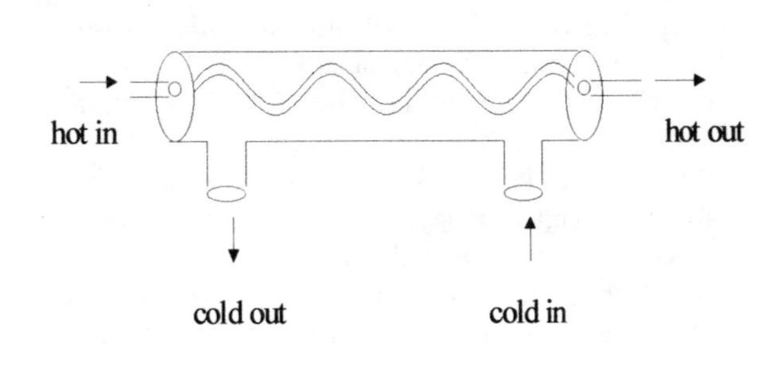

Fig. 3 schematic of test section.

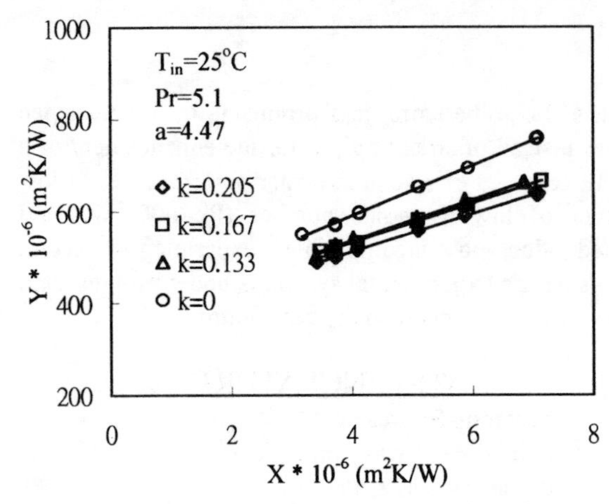

Fig. 4 Wilson plot for various k.

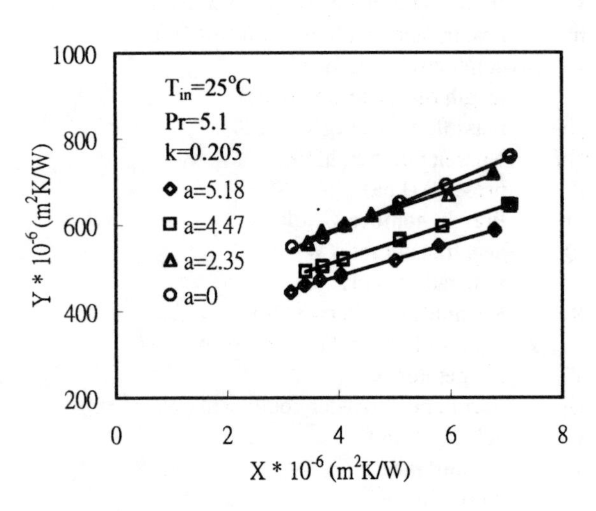

Fig. 5 Wilson plot for various a.

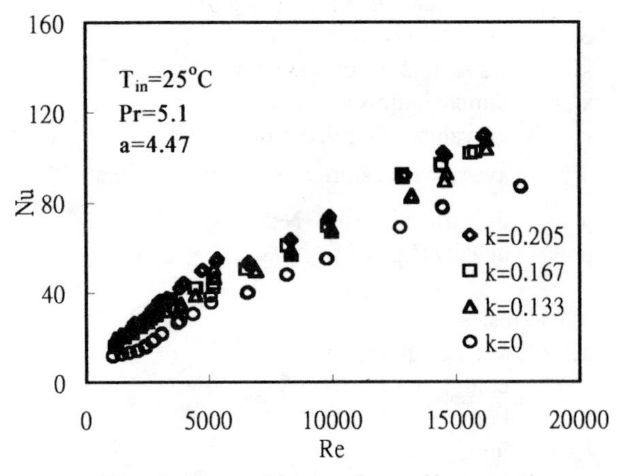

Fig. 6 Reynolds number effect on the Nu for various k for heating process

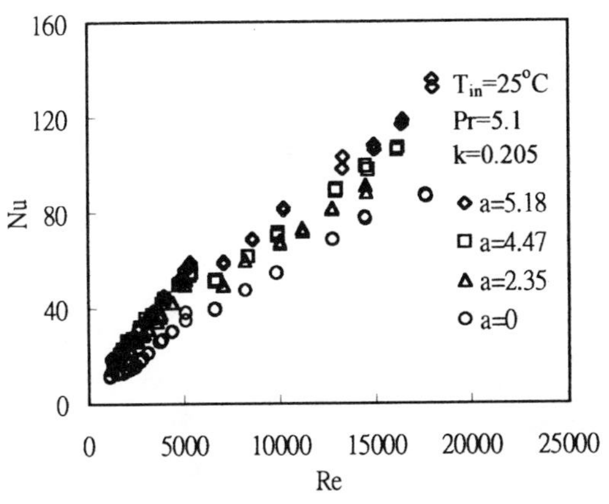

Fig. 7 Reynolds number effect on Nu for various a for heating process.

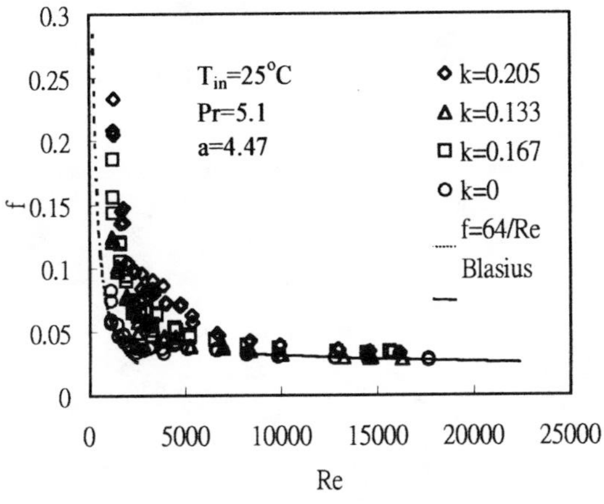

Fig. 8 Reynolds number effect on f for various k for heating process.

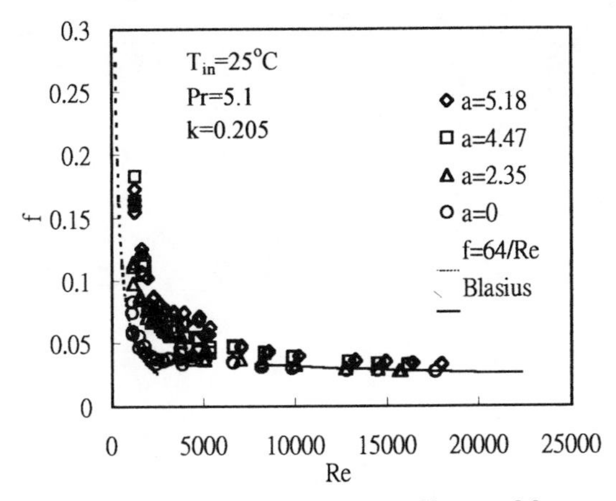

Fig. 9 Reynolds number effect on f for various k for heating process.

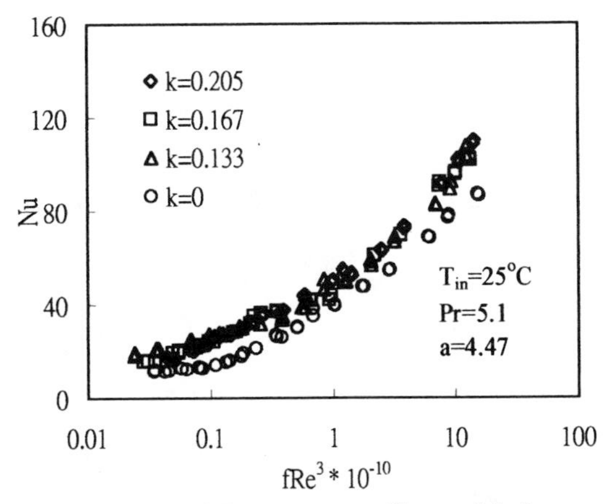

Fig. 10 Pump power effect on Nu for various k for heating process.

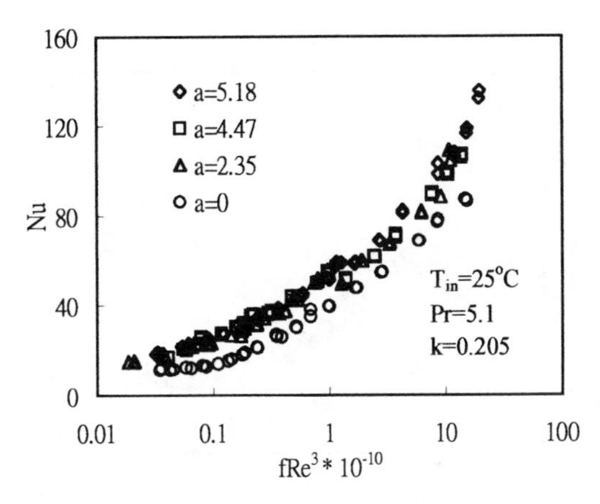

Fig. 11 Pump power effect on Nu for various a for heating process.

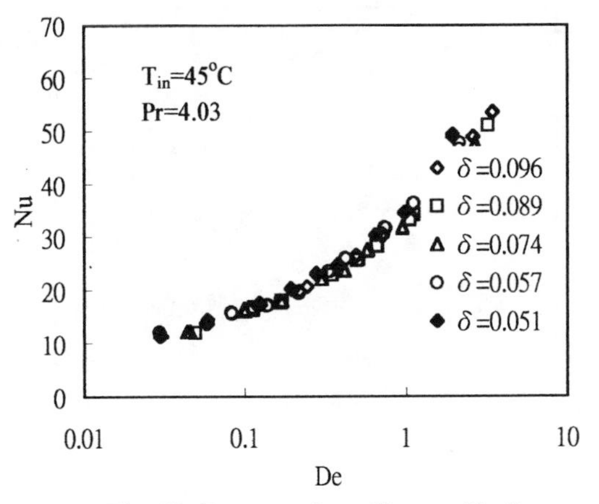

Fig. 12 Dean number effect on Nu for various δ for cooling process.

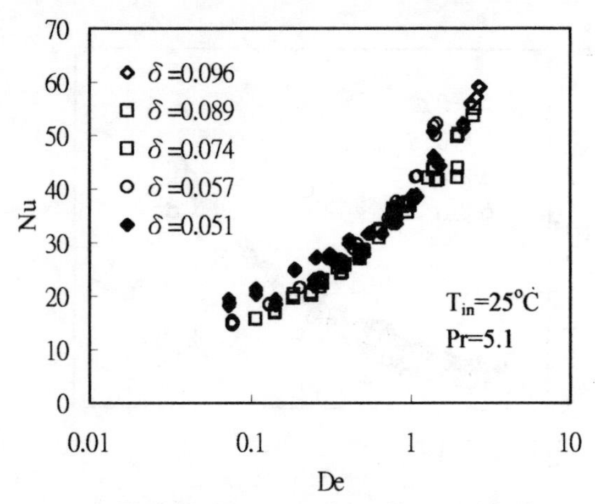

Fig. 13 Dean number effect on Nu for various δ for heating process.

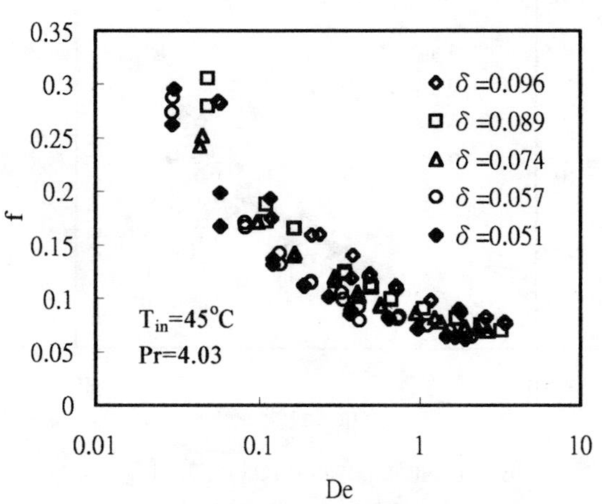

Fig. 14 Dean number effect on f for various δ for cooling process.

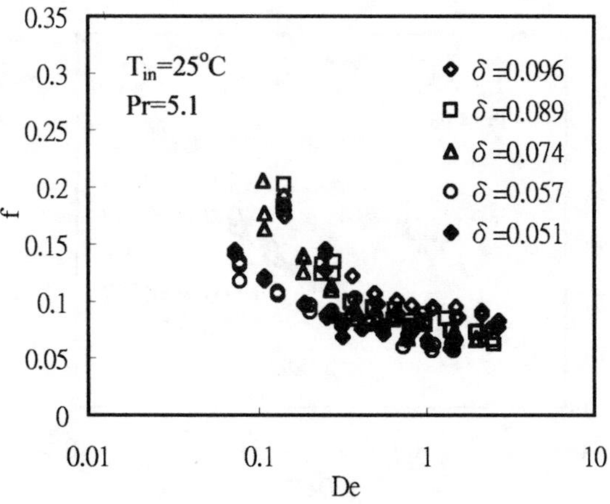

Fig. 15 Dean number effect on f for various δ for heating process.

EFFECT OF FLAT REFLECTORS ON THE PERFORMANCE OF PHOTOVOLTAIC MODULES

Dr. A. Md. Aziz-ul Huq*
Professor, Mechanical Engineering Department
Bangladesh University of Engineering and Technology
Dhaka 1000, Bangladesh

Mohammad Irteza Hossain
Post Graduate Student, KTH
Stockholm, Sweden

Muhammad Mustafizur Rahman
Associate Professor, Department of Mechanical Engineering
University of South Florida
Tampa, Florida 33620, U.S.A.

ABSTRACT

To convert solar energy directly into electrical energy, photovoltaic modules are widely being used. Different techniques such as retrofitting of reflectors and concentrators are adopted to increase the amount of radiation flux falling on the module and the output of the module thereby increases. The present work investigates the effect of flat reflecting materials such as flat mirrors, shinny aluminum, and white tiles on the performance of a PV module. The performance was compared with an identical module without reflector; both studied under the same environmental conditions. It was found that the performance depends on the sky condition and has no advantage on a rainy or cloudy day. The reflector fitted module output based on Maximum Power Point showed a significant increase, and there was also an increase in temperature of the module.

NOMENCLATURE

b	PV module length (m)
c	PV module width (m)
F_{r-c}	View factor from reflector to collector
I	Current (A)
P	Power (W)
r	Reflector width (m)
V	Voltage (V)
ψ	Angle

* Corresponding Author

INTRODUCTION

In many remote and isolated areas far from the grid line and in many island areas, a photovoltaic system is a very promising option. High initial installation costs of a complete PV system in many cases becomes a deterrent for such options, overriding the advantages. In addition to this, there are two major handicaps of a PV system. Solar energy is very much seasonal and energy is not available at night. An electrical storage system like a battery is required to maintain power supply during such periods when solar energy is not available or not enough to meet the demand. Moreover, solar energy flux is dilute and rarely exceeds $1 kW/m^2$. The use of some form of reflector or concentrator will increase the intensity of solar radiation flux falling on the PV module. As cell output is also dependent on cell temperature, this intensified flux will increase the output depending on the cell temperature. Any form of retrofit involves cost, and all these analyses enter into the final choice of options.

A number of studies, such as those by Stacey and McCormick[1], Nann[2], Peters and Karlsson[3], Gordon[4] have been done to study the effect of concentrators on the performance of PV modules. Mills et al.[5] performed a relative comparison of certain truncated symmetrical and asymmetrical fixed reflector designs for solar collectors. Bollentin and Wilk[6] presented an analytical model to determine solar irradiation on flat collectors augmented with planar reflectors. Kumar et al.[7] analytically studied the general case of a collector with four reflectors. An alternative to glass mirrors coated with an aluminum or silver reflecting material is a silvered polymer reflector developed by Schissel et al[8]. The silver acrylic film has the advantages of being low cost and lightweight. The focus of these studies was to investigate reflectors for large systems with potential commercial application.

Haque[9] studied the performance of a small PV (40 watt peak) module using flat mirrors as reflectors and found that on typical days, module output increased around 11-13% with a module temperature rise of 2-10 °C. The present work extended this work and checked the effect of other reflecting materials such as flat shinny aluminum, flat tiles and also flat mirrors. All of these are spectral in nature. The findings were compared with an identical module having no reflector placed side by side so as to expose them to similar environmental conditions.

EXPERIMENTAL SET –UP

The set-up was installed on the roof of a six-story building on the campus of Bangladesh University of Engineering and Technology (longitude 90.0° E and latitude 23.5° N). A structure was specially designed to mount the reflectors on all four sides of the PV module. The angle between each reflector and the PV module could be varied independently so as to change the shape factor. The experimental set-up is shown in Fig.1.

For the experiment, 40-watt TATA BP solar modules were used. During the experiment, solar radiation was measured, module temperature was measured by thermocouples, and wind speed and psychrometric data were recorded. Loads on the PV modules were controlled by rheostats, and corresponding volts and amperes were recorded by digital meters. The data was used for plotting the corresponding I-V and Power-Voltage curves.

Three types of reflecting materials were used during this experimental investigation. These were flat mirrors, shinny aluminum, and white tiles. All of these are spectral in nature. PV modules were placed facing south at the latitude angle, 23.5° N.

The view factor is an important parameter that gives the effect of reflectors on the PV panel, and it is defined as how much of the reflector could be seen by the PV panel. The view factor between the module and the reflector varied throughout the season and also over the day, depending on the position of the sun in the sky. The theoretical value of the view factor for a particular day and for a particular setting was calculated using the following relation given by Duffie and Beckman[10].

$$F_{R-PV} = (c + r - s)/ 2r$$

where, $s = [c^2 + r^2 - 2 c*r \cos\psi]^{1/2}$

Table 1 gives the angles subtended by the reflector with respect to the module. In determining these angles, the shading effect of different reflectors at different times of the day was considered. The angle between the module and the reflector is schematically shown in Fig. 2. This paper presents experimental results involving all four reflectors, and the performances were compared with respect to the maximum power point; the percentage increase was calculated by using the following relationship:

Percentage increase of output

$$= \{(P_{reflector} - P)/P\} *100$$

RESULTS AND DISCUSSION

The tests were performed by putting reflectors on all sides of the PV module (i.e. top, bottom, left, and right). A number of tests were performed at different times of the day. For each test, the load current and voltage readings of the module fitted with the reflector and for the module without the reflector were recorded; then, the power output at each loading condition was calculated by multiplying the current and voltage readings. Analyses were done on the basis of maximum power point.

Figure 3 shows the performance of mirrored reflectors in a clear sky environment. It can be noticed that the output power increased by 33 % at 15:15 hrs. The output power dropped to nearly 10% at 15:25 hours based on the maximum power point. The temperature of the panel fitted with four reflectors increased by 6-10 °C compared to the module without reflectors.

For Shinny Aluminum reflectors, the output power increased by 30% (Fig. 4) at 15:40 hours; at other times, it was around 15% (based on maximum power point). The temperature rise with shinny aluminum sheet reflectors was around 7 °C more compared to the module without reflector.

For White Tile reflectors, the power from the PV module increased by 16.12 % (Fig. 5) and was around 8% on other days. The temperature of the PV module having White Tile reflectors was increased by 2-3 °C compared to module without reflector.

Figure 6 shows the effect of cloud cover on the performance of the two modules. There is practically no difference between the two outputs. During tests with mirrored reflectors or with tiles, a complete cloudy condition was not present, so similar results could not be obtained. Similar experiments were done with pairs of two reflectors and a single reflector. All showed some degree of augmentation compared to the module without any reflector depending on the time of the day and the sky condition.

CONCLUSIONS

From the above findings, it is clear that the use of reflectors augments energy gain per meter square of module area. The percentage gain is very much dependent on the sky conditions. The experiment was done for a limited period of time. A more extensive study is required with different reflecting materials, including the diffuse type (such as white paint) and under different sky conditions, to identify some inexpensive reflective materials that can be economically viable. The optimum angles of the reflectors with the module for different months in the year is needed to design systems for best performance.

ACKNOWLEDGEMENT

The authors gratefully acknowledge the financial support provided by the Department of Energy Technology, Royal Institute of Technology, Stockholm, Sweden.

REFERENCES

1. Stacey, R.W. and Mccormick, P.G., "Effect of Concentration on the Performance of Flat Plate Photovoltaic Modules," *Solar Energy*, Vol. 33, 1984, pp. 565-569.

2. Nann, S., "Potential for Tracking Photovoltaic Systems and V-Troughs in Moderate Climates," *Solar Energy*, Vol. 45, No. 6, 1990, pp. 385-393.

3. Peters, B. and Karlsson, B., "External Reflectors for Large Solar Collector Arrays, Simulation Model and Experimental Results," *Solar Energy*, Vol. 51, 1993, pp. 327-337.

4. Gordon, J., "A 100-Sun Liner Photovoltaic Solar Concentrator Design from Inexpensive Commercial Components," *Solar Energy*, Vol. 57, No.4, 1996, pp. 301-305.

5. Mills, D.R., Monger, A., and Morrison, G.L., "Comparison of Fixed Asymmetrical and Symmetrical Reflectors for Evacuated Tube Solar Receivers," *Solar Energy*, Vol. 53, 1994, pp. 91-104.

6. Bollentin, J.W. and Wilk, R.D., "Modeling of Solar Irradiation on Flat Plate Collectors Augmented with Planar Reflectors," *Solar Energy*, Vol. 55, 1995, pp. 343-354.

7. Kumar, R., Kaushik, S.C., and Garg, H.P., "Analytical Study of Collector Solar-Gain Enhancement by Multiple Reflectors, *Energy*, Vol. 20, 1995, pp. 511-522.

8. Schissel, P., Neidlinger, H., and Czanderna, A., "Silvered Polymer Reflectors," SERI/TP-255-2670, Solar Energy Research Institute, Golden, Colorado, 1985.

9. Haque, M.J., "Study of the Effect of Reflectors on the Performance of PV Module", M.S. Thesis, KTH, Sweden, 1998.

10. Duffie, J.A. and Beckman, W.A., *Solar Engineering of Thermal Processes*, Second Edition, Wiley- Interscience, 1991.

Table 1 Angles between Reflector and Module

Reflector	Mirror ψ	Shinny reflector ψ	White tiles ψ
Top	133.5	123.5	123.5
Bottom	128.5	128.5	128.5
Two sides	152	152	152

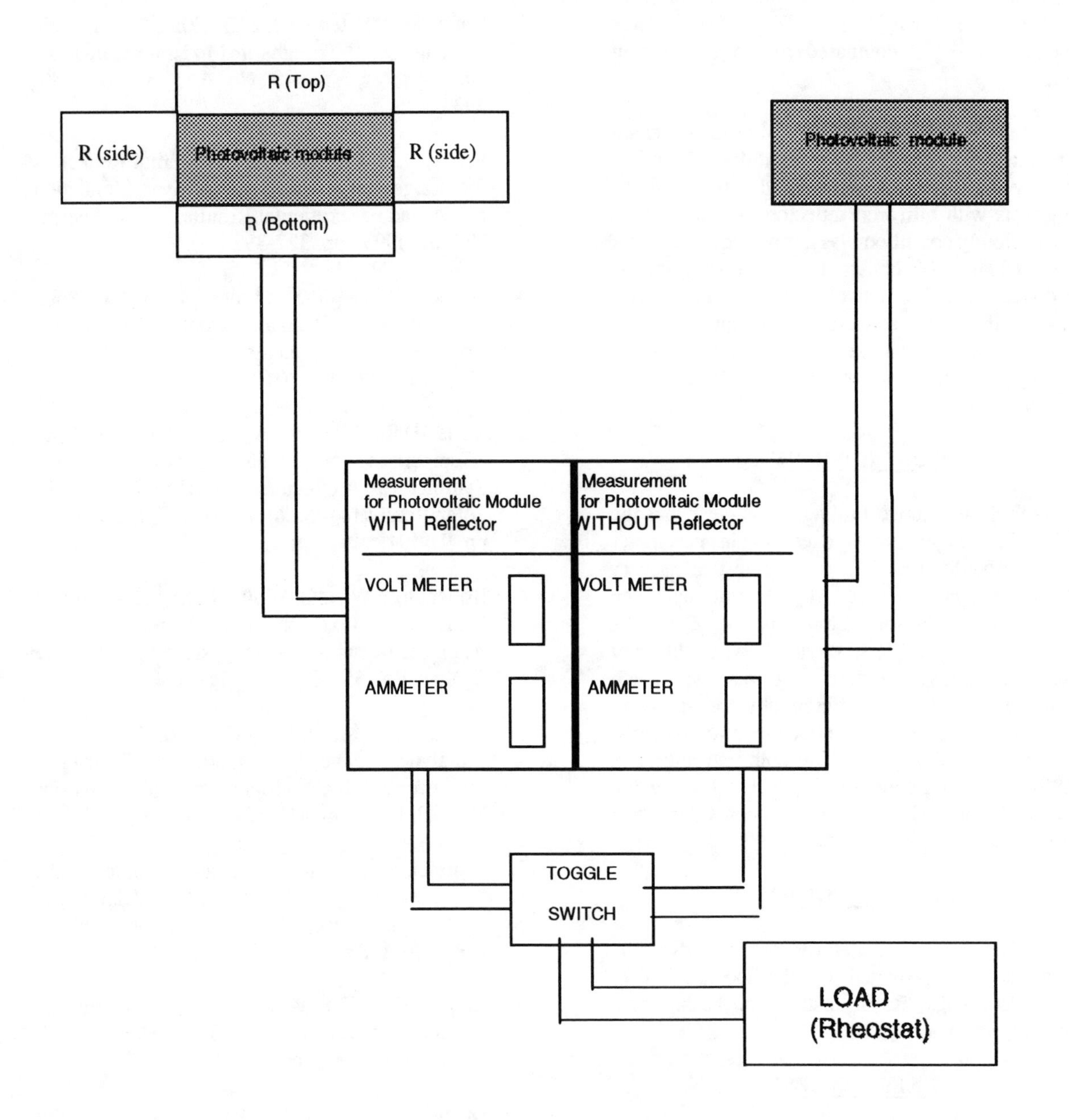

Fig. 1 Schematic diagram of the set-up

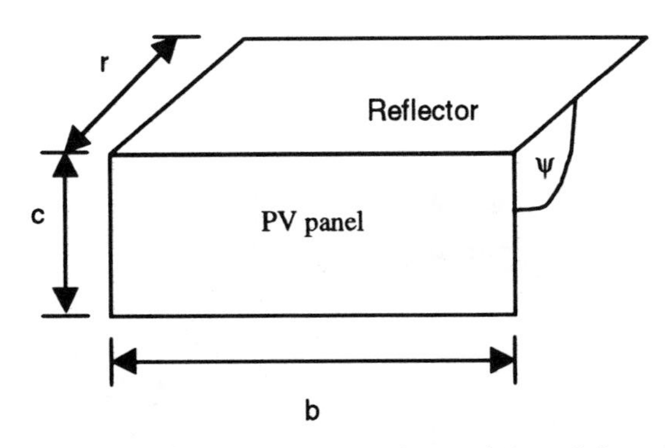

Fig. 2 Angle between the module and the reflector

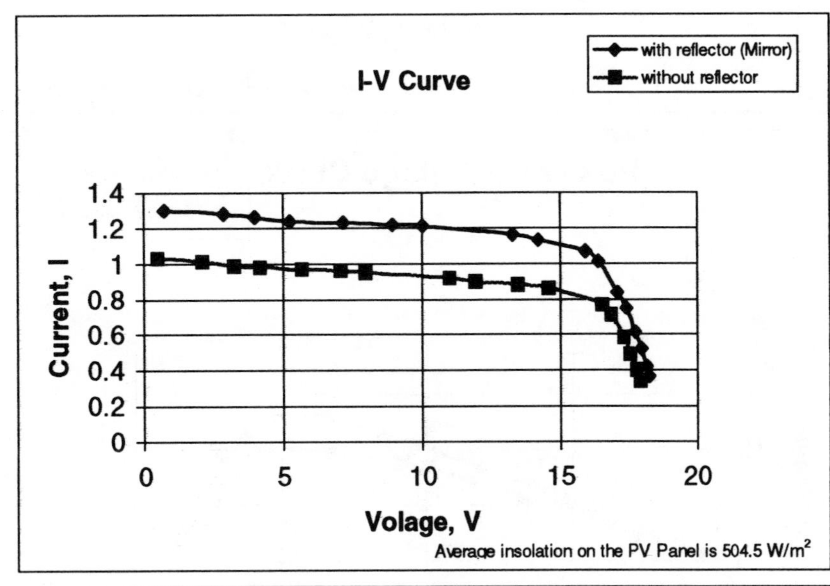

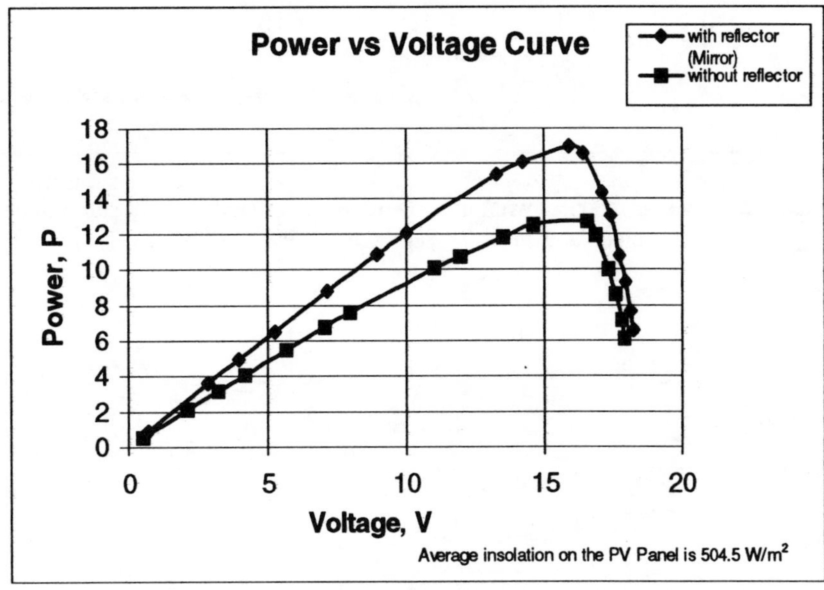

Fig. 3 I-V and P-V curve for modules without and with reflector (four mirror reflectors, clear sky), 3rd August 1999, 15:15 hours

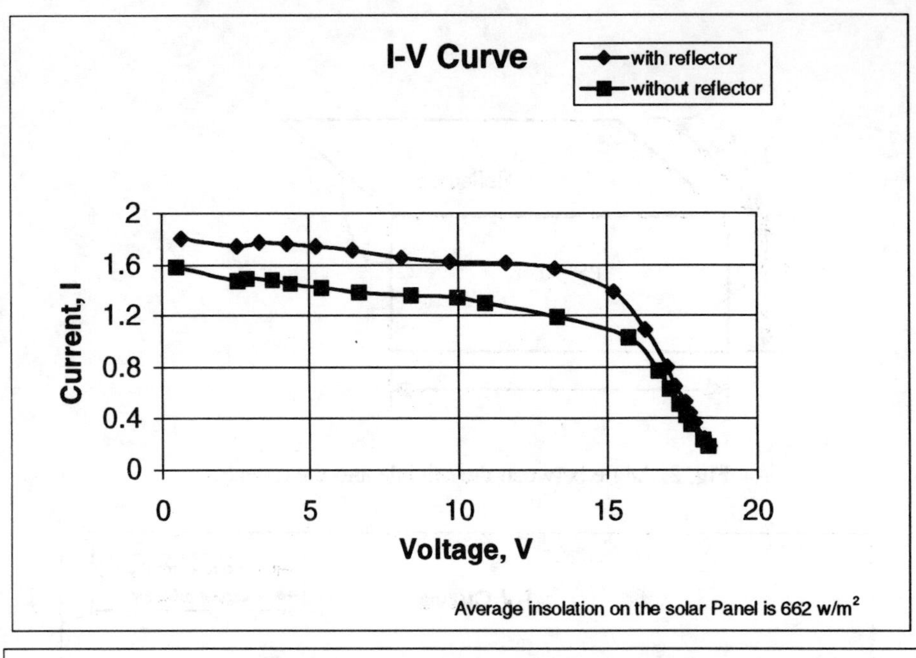

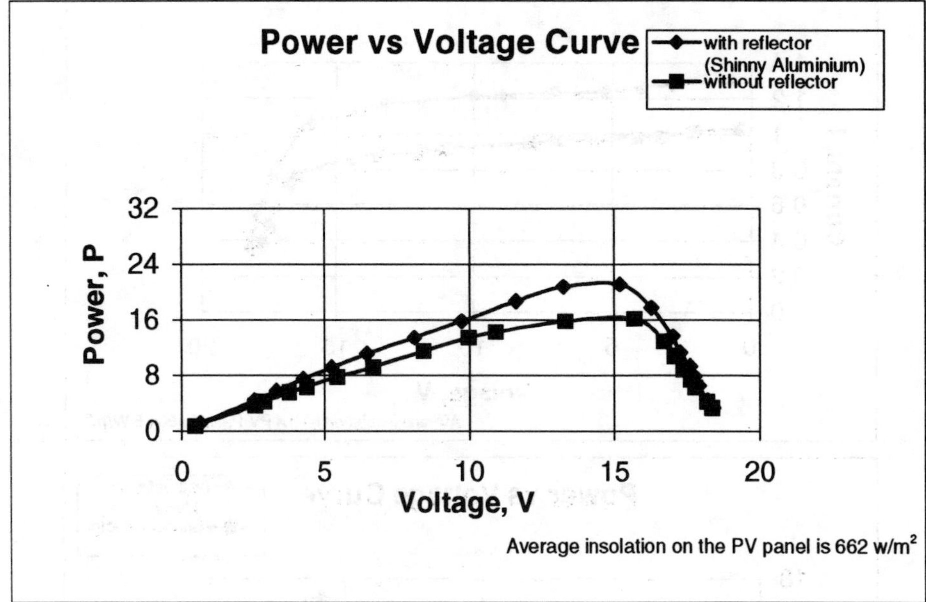

Fig. 4 I-V and P-V curve for modules without and with reflector (four shinny aluminum reflectors, white cloud and sun), 7th September 1999, 15:40 hours

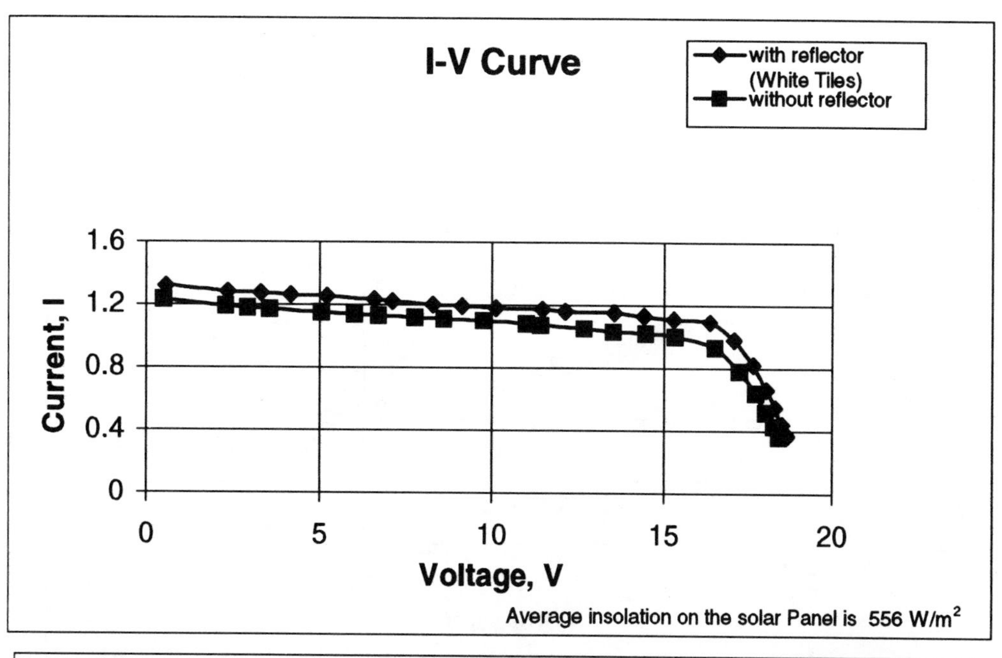

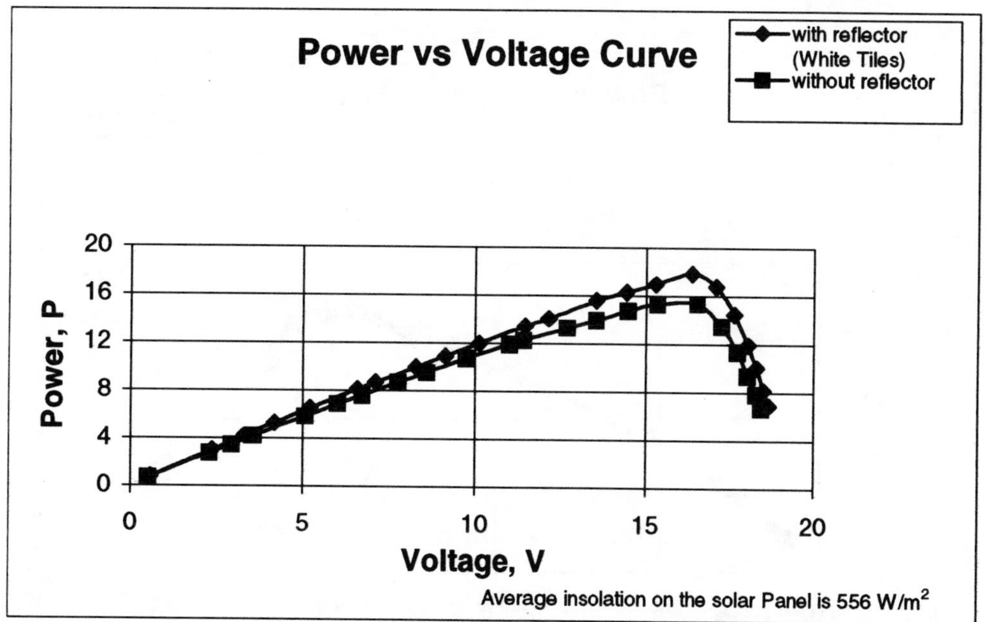

Fig. 5 I-V and P-V curve for modules without and with reflector (four white tile reflectors, white cloud and sun), September 1999, 15:30 hours

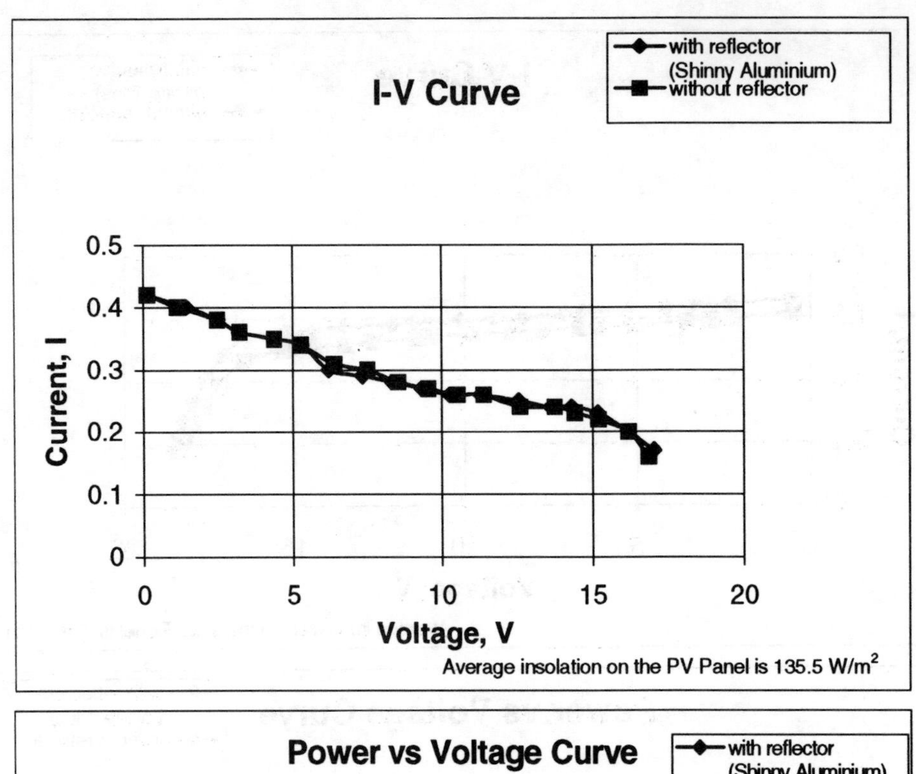

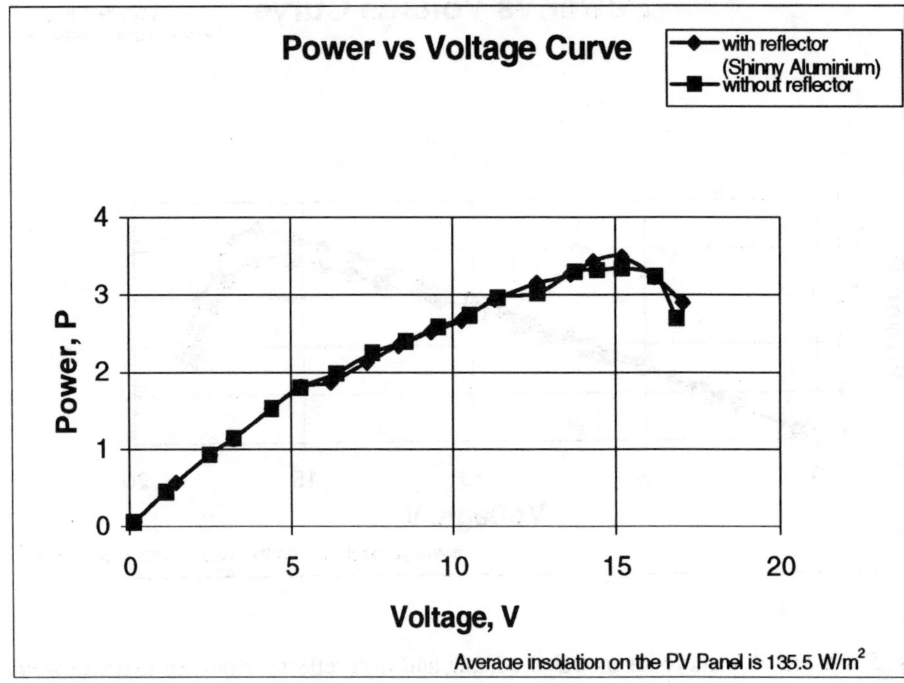

Fig. 6 I-V and P-V curve for modules without and with reflector (four shinny aluminum reflectors, very cloudy day), 24th August 1999, 10:45 hours

AIAA-2000-2827

Dual Use Power Management System for Space Radar

Michael Brand
Raytheon Electronic Systems
El Segundo, CA 90245
310 334-0010
mbrand@west.raytheon.com

Rene Thibodeaux
Air Force Research Laboratory
Wright-Patterson Air Force Base
Dayton, OH 45433
937-255-6016
rene.thibodeaux@wpafb.af.mil

Gregory Fronista
Air Force Research Laboratory
Wright-Patterson Air Force Base
Dayton, OH 45433
937-255-6235
gregory.fronista@wpafb.af.mil

Jim Wilson
Raytheon Electronic Systems
Dallas, TX 75243
972 344-4815
jsw@raytheon.com

Steve Levin
Raytheon Electronic Systems
P.O. Box 801 MS8019
McKinney, TX 75070
972 952-3707
SteveL@raytheon.com

ABSTRACT

*Space Based Radar (SBR) applications impose challenging weight, volume, efficiency, reliability, environmental, performance and cost requirements on the radar power subsystem (RPS). To meet SBR requirements, the Direct Point of Use (DPU) architecture was chosen. A 90 % efficient, 100 watt power converter will be demonstrated in a sub-panel that simulates a portion of the radar antenna. The power converter will improve military radar, communications equipment, and man-portable electronics. To achieve high-density packaging, inductors and transformers will be embedded in a conventional printed wiring board. Power electronics is a highly competitive $26 billion industry largely driven by the automotive industry and portable electronics. Development of this converter will benefit both the defense and commercial market.

INTRODUCTION

Space based radar activities are concentrating on cost effective solutions for on-demand surveillance, track, and ID of space-borne, air-borne, and ground targets. The performance of these missions requires large, high power radar systems. Such systems present significant design challenges for the radar's power and thermal management subsystems. SBR antenna arrays are

Table 1. Four Major Radar Power System Architectures.

Architecture	*Power Distribution*	*Reliability*	*Thermal Load*	*Energy Storage*	*Regulation*	*Technology*
Centralized	Single Converter: Heavy Bus Bars	Single Point Failure	Concentrated	Large Storage at POU	Poor Due to Load Distance	Standard
Semi-distributed	Multi Converters: Heavy Bus Bars	Single Point Failure	Moderate	Large Storage at POU	Moderate at Loads	State-of-the-Art
Highly Distributed	Many Converters: Low Bus Weight, High Voltage on Array Backplane	High Reliab: Redundance, Graceful Degradation	Distributed with Low Loads	Minimal Storage at POU	Excellent at Loads	Needs: Dev. of Adv. Pkging and Component Technologies
Direct POU	Direct Prime Power: Low Bus Weight, High Voltage on Array Backplane	High Reliab: Redundancy, High Degree of Graceful Degradation	Fully Distributed	Minimal Storage at POU	Excellent at Loads	Needs: Dev. of Adv. Pkging and Semiconductor Technologies

large, occupying hundreds of square meters and consisting of thousands of low voltage T/R modules. Projected power requirements for these active arrays exceed 35 kW, requiring the distribution of thousands of amperes of bias current across the array. The distribution of high currents can adversely impact power subsystem efficiency, reliability and weight. Thermal systems must also be in place to effectively manage the heat load generated by both the power converters and the T/R modules. Finally, the antenna and the power subsystem mechanical structures must withstand the thermal and orbital stress expected during a 20 to 30 year mission. Development of a power distribution architecture which minimizes the power distribution weight while maximizing power conversion efficiency is the foundation for the realization of a preferred space vehicle power subsystem. Therefore, the RPS and the antenna mechanical structure must be optimized across the total spacecraft/payload system in trades that include power generation, battery technology, heat load, orbital parameters and antenna size.

POWER MANAGEMENT SUBSYSTEM

This program will develop the power conversion and thermal subsystems to support space radar and potential space lasers. Prototype circuits, devices, and thermal technologies will be fabricated and integrated with a demonstration array sub-panel. To meet SBR requirements, a highly distributed architecture was chosen, as the most likely to meet space radar needs. The distributed approach demands the development of advanced semiconductors, packaging approaches and circuit topologies, will result in a 90% efficient, 100 Watt power converter by the end of the proposed

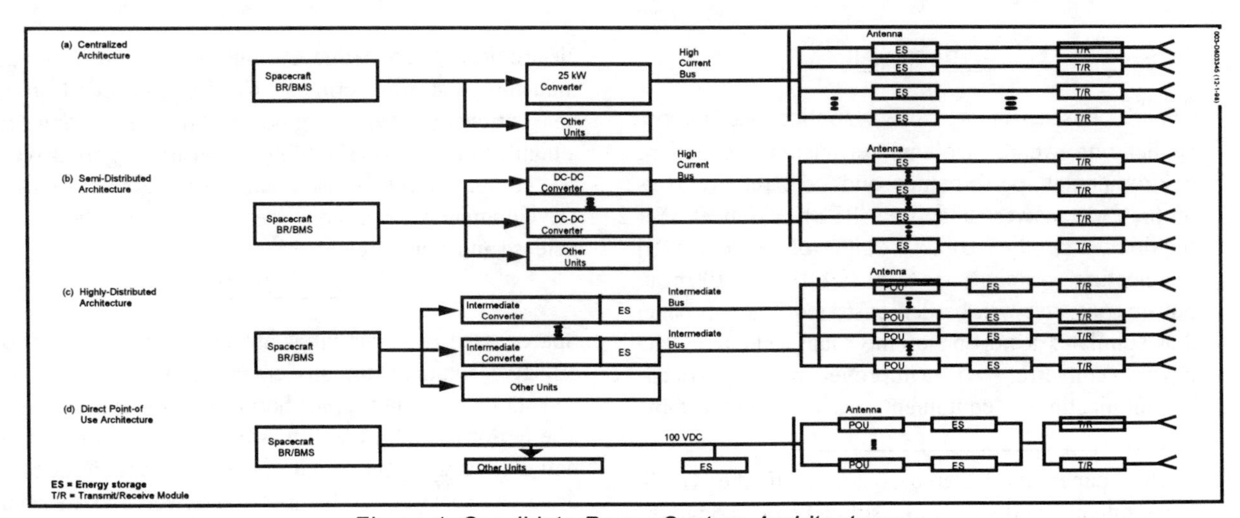

Figure 1. Candidate Power System Architectures

TABLE 2. Summary Of Power Converter Characteristics Supporting Power Subsystem Concept.

Parameters	*Existing (SOA)*	*Program Goal*	*SBR Objective*
Power Efficiency	79%	90%	>90%
Reliability/MTBF/Useful life	N/A	220 kHrs	>220 kHrs >20 years
Radiation hardness	10 Mrad	1 Mrad	1 Mrad
Power density	75 $W/in.^3$	150 $W/in.^3$	Very high (>300 $W/in.^3$)
Power-to-weight	1600 W/lb	3200 W/lb	Very high (>7 $KW/in.^3$)
Distribution bus voltage	50V	100V	240V
Transient response	<0.2μs	<2μs	<2μs
Output regulation for ±10% input line voltage variation and disturbance	±1%	±1%	±1%
Cost of Converter			
Military Use		\$2.00/W	
Commercial Use		\$0.20/W	

program. The power converter will use advanced semiconductors in its inverter and rectifier subsections. The advanced semiconductor is capable of high switching speeds up to 100 MHz at high voltages between 200 and 700 volts. Low loss rectifier diodes developed from this program will enable the power converter to develop low voltage outputs with high efficiency. The power requirements of the SBR will necessitate bus voltages higher than 100 volts, so the transistor must accommodate input bus voltages of 240 to 270 Volts.

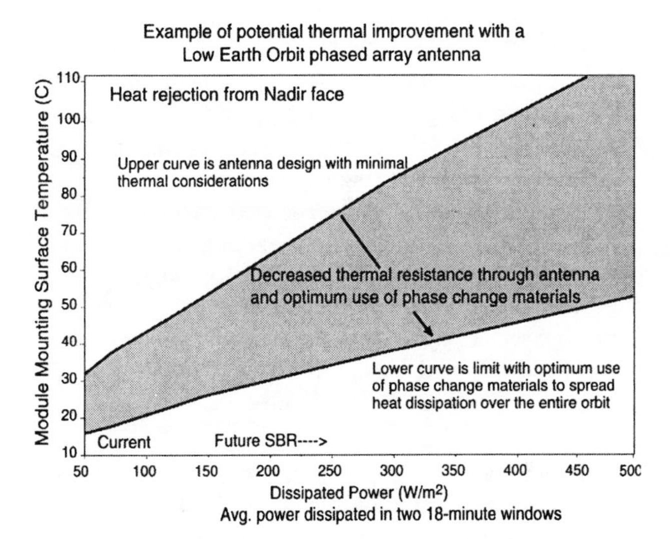

Figure 2 - Trade Study

The preferred approach to achieving high-density packaging is to embed inductors, transformers, and high density interconnects into a conventional Printed Wiring Board (PWB). These technologies have pushed power densities, input voltage, and cost by leveraging both commercial and other military applications. Raytheon has built and demonstrated power converter efficiencies up to 79% and expect to achieve 90% during this program through the use of advanced transistors, diodes, and embedded low loss components. The military cost-per-watt assumes production in the thousands while reducing the radiation hard requirements. Similarly, the commercial cost assumes a production run of one million with reduced radiation hardening and reliability objectives, and random production quality testing.

THERMAL MANAGEMENT SUBSYSTEM

The SBR thermal management system must maintain an acceptable thermal environment for both operating and non-operating conditions, as well as prevent excessive temperature gradients both through and across the antenna. The study will focus on a range of dissipated power levels and multiple imaging windows per orbit. The primary focus is on SBR time-averaged power dissipation levels in the 200 to 600 W/m^2 range. If time permits, the study will address reduced levels consistent with current efforts (~75 W/m^2) and increased levels on the order of 1 to 2 kW/m^2 for future applications. The large area of the antenna allows its use as a thermal radiator. This concept is viable because the dissipated power from the modules is highly distributed. The more concentrated heat dissipation from the other electronics (such as the power converters) is about 20% or less of the total

TABLE 3. Market For Power Supplies (Global) And Power Semiconductors (North American) 2001 Through 2003

Products	*Consumer Electronics*	*Automotive*	*Defense*	*Computers*	*Telecommunications*	*Industrial Electronics*
Power Supplies	$230M	$80M	$523M	$900M	$660M	$270M
Power Semi-conductors	$325M	$298M	$135M	$182M	$266M	$475M

dissipation and can be controlled with local heat spreaders and/or phase change materials. This concept is the simplest thermal design and produces the lowest cost. Using the antenna as a thermal radiator also significantly reduces or eliminates the need for heaters for non-operating conditions, if the antenna is earth facing in a low earth orbit.

A preliminary trade study, shown in Figure 2, compared peak X-band module mounting temperature versus array dissipation while using the antenna as a thermal radiator. The array was powered for two 18-minute imaging windows per low earth orbit. The power dissipation on the x-axis is the average power dissipated during the imaging window. The upper limit of the shaded area indicates the peak module mounting temperatures without the use of any phase change material and typical circuit board construction for the antenna. The lower temperature limit of the shaded area gives an absolute temperature limit assuming optimum design using phase change materials. A typical module temperature rise is about 45°C and would be added to the peak module mounting surface temperature.

The representative SBR power dissipation at 200 W/m^2 would have peak mounting-surface temperatures in the range 25° to 60°C (junction temperatures between 70° to 105°C). While the 70°C junction temperature would be acceptable, significant thermal design using phase change materials would be required to come close to this limit. The first objective of the proposed study will be to establish the limits of acceptable performance (power dissipation and imaging window duration) of this concept. This study will include methods of increasing the heat transfer through the structure for higher power levels, such as through-thickness heat pipes and high thermal-conductivity materials like graphite.

The options for heat removal above the limit for using the antenna as a heat radiator include adding louvered heat radiators, heat pipes, and capillary pump loop systems. Module-packaging alternatives will be evaluated to decrease the module internal and interface temperature rises, since decreasing these temperature rises becomes important with higher power modules.

SYSTEM INTEGRATION

An essential aspect of the design is the mechanical integration of the power and thermal control systems with the rest of the antenna. If these systems are combined into "multi-purpose" structures, there is the potential for a large amount of weight savings. The parasitic weight of the structure can be reduced if DC power is distributed through aluminum structural members or solar cells and batteries are distributed directly on the back of the antenna. Combining DC power distribution with heat pipe structures or thermal insulation materials potentially offers similar benefits. Trade studies will be performed to investigate innovative concepts for reducing weight.

COMMERCIAL APPLICATIONS

The global power processing market has expanded steadily over the past 20 years and will attain annual sales of $26 billion in 2003. Within this market, the United States produces $9.7B (37.4%), Europe $6.4B (24.7%) and Pacific Rim $8.1B (31.1%). The industry trend is toward distributed power systems that are smaller in size, and higher in power density and efficiency. To meet the associated challenges, most commercial manufacturers are looking toward higher switching speeds and more exotic packaging. The resultant technologies will benefit both commercial and military sectors by offering dramatic improvements in power density and cost. At power densities of 150 $W/in.^3$, cost of $0.20/W is projected.

Projected demand for power supplies in 2001 for the computer, telecommunications, defense, industrial/ consumer electronics, and automotive markets exceeds 27 million units with an annualized growth rate of 11%. This growth reflects demonstrated consumer trends toward increased electronic convenience in daily life as well as the growing military need for greater operational capability at lower costs. For example, the advanced semiconductor-based power converters will provide more power in a fraction of the space and for less cost per watt. The lower cost is estimated using the advanced semiconductors. Projected commercial power converter cost is $0.20 per watt compared with $0.75 per watt using the most cost-effective current technologies. Similarly, advanced military converters available today cost about $6 per watt compared with $2 per watt projected for the most compact power converters planned for development on this program. Based on data obtained from studies [3,4] and assuming a conservative 10% initial penetration into the markets identified in Table 2, the advanced semiconductor-based power converter demand represents a substantial market opportunity of over $315 million.

REFERENCES

[1] Raytheon Proposal for the Air Force BAA 98-023, "Power Management, Distribution, Conditioning, and Thermal Management Technologies for Space Based Radar."

[2] Ferrara, *Satellite Power Processing Development*, Technical Report AFWAL-TR-87-2056, October 1987.

[3] Frost and Sullivan, "World Switching Power Supply Markets," 1997.

[4] Frost and Sullivan, "North American Power Semiconductor Markets," 1998.

Michael Brand is an engineer in the Power Electronics Department of Raytheon Electronic Systems, El Segundo, California. He has 18 years of experience in power electronics including low and high voltage power supplies and radar transmitters. Currently he manages a research program to develop high density a point-of-use power converter. He has a B. S. in engineering from UCLA (1981), a M. S. in engineering management from USC (1983), and an MSEE in integrated circuits from USC (1985).

Rene Thibodeaux is an electrical engineer in power systems and power electronics. He has worked in the development of power technology for aerospace systems for the U. S. Air Force since 1985. He works in the Power Division, Propulsion Directorate of the Air Force Research Laboratory, Dayton, Ohio. He has a B. S. in physics from the University of Southwestern Louisiana (1979) and a MSEE from Louisiana State University (1983).

Jim Wilson is a mechanical engineer in thermal design and analysis for electronics cooling. He has worked in the thermal design group for Raytheon since 1985 and has focused on microwave systems since 1990. He has a BSME from Texas Tech (1984), MSME from Stanford (1985) and a Ph.D. in ME from SMU (1997).

Steve Levin is a mechanical engineer in antenna design and packaging. He has worked in the Antenna/Non-metallics Technology Center for Raytheon since 1997 and for Texas Instruments since 1982. His work has focused on space based antennas since 1996. He has a BSME from Rensselaer Polytechnic Institute (1982) and an MSME from Southern Methodist University (1985).

Gregory Fronista is the Space Power Manager in the Power Division, Propulsion Directorate of the Air Force Research Laboratory, Dayton, Ohio. Mr. Fronista joined the Laboratory in 1989, serving as a technical specialist in the field of aerospace electrical power systems technology with emphasis in power electronics, controls of electric machines and motor drives.

AIAA-2000-2828

COMPARISON of 1200V SILICON CARBIDE SCHOTTKY DIODES AND SILICON POWER DIODES

H.-R. Chang, R. N. Gupta, C. Winterhalter, and E. Hanna

Rockwell Science Center, 1049 Camino Dos Rios, Thousand Oaks, CA 91360
Phone: (805) 373-4769
FAX: (805) 373-4869
Email: hrchang@rsc.rockwell.com

Abstract - This paper describes the design and characterization of 1200V silicon carbide (SiC) Schottky diodes using a p-n junction barrier to control the leakage current. A low forward voltage drop of 1.9V at 150A/cm^2 and a low leakage current of 300 μA/cm^2 are demonstrated. We compare the performance of SiC Schottky diodes with the 1200V silicon (Si) PiN diodes fabricated and optimized for military and commercial motor drive applications. The power loss and related voltage and current stress of the SiC Schottky diodes and Si PiN diodes are evaluated using a half-bridge test circuit. The static and dynamic characterization of 1200V SiC Schottky diodes and Si PIN diodes is performed at 25°C and 150°C. The SiC Schottky diode exhibits a 30% reduction in the total power loss as compared to the state-of-the-art Si PiN diodes, mainly due to a negligible reverse recovery charge in the SiC Schottky diode, which is 20X lower than state-of-the-art Si power diode. Furthermore, the SiC Schottky diode contributes to lowering the overall voltage and current stress by a factor of 6X on the switches (in this case IGBTs are used) due to the superior reverse recovery performance of the SiC Schottky diode.

1. INTRODUCTION

Since the early 1990's the silicon (Si) IGBT has been the power switch of choice in industrial applications. Recent developments in Si trench (the 4th generation) IGBT technology have further enhanced IGBT performance by increasing the device switching speed and reducing the total system power losses [1,2]. The progress in the improvement of IGBTs to blocking capabilities > 1200V makes further development of the freewheeling diode(FWD) important in order to take full advantage of the excellent switching capabilities of the IGBTs without using a snubber circuit. Fast recovery P^+-*i*-N rectifiers using carrier lifetime control by gold or platinum doping, or by electron irradiation have been extensively used in high voltage power electronic applications due to their good blocking capability. However, these diodes exhibit large peak reverse recovery current and snappy reverse recovery behavior, resulting in large voltage overshoots. Various concepts have been proposed to improve the diode characteristics including low forward voltage drop, low peak reverse recovery current I_{RP} for high di/dt values, low reverse recovery charge Q_{rr} and last but not least a soft recovery behavior for a wide current range [3-5].

To complement the improved performance achieved in the IGBTs, advanced FWD technologies have been developed in order to fully utilize the performance advantages of the IGBT. There are a couple of recent publications on advanced high voltage Si power diode technology [6,7]. These technologies reduce the component voltage and current stress, Electro-Magnetic Interference (EMI), and the overall system losses. However fundamental limitations in Si diode technology are rapidly being approached. Further performance improvements can be obtained with the use of wide-bandgap materials such as silicon carbide (SiC). High voltage SiC Schottky diodes have shown promising static and dynamic characteristics for breakdown voltages up to several kV [8]. The fast switching speed and low forward voltage drop of SiC Schottky diodes can significantly reduce the total power loss and component stress as compared to conventional Si P^+-*i*-N diodes.

This paper reports the design and performance of 1200V SiC Schottky diodes using p-n junction barriers. The static and dynamic characterization of the optimized SiC Schottky diodes is performed at 25°C and 150°C using a double pulse half-bridge test circuit.

2. DEVICE STRUCTURE

Silicon p-n junction power diodes have been the diode of choice in medium voltage motor drive applications. Recent progress has been made in Si power diode technology to improve the diode performance for high power switching applications [6,7]. All state-of-the-art high power Si diodes use minority carrier injection to modulate the drift region resistance to achieve low on-state voltage. This results in low conduction losses but substantially increases the switching losses due to a large amount of stored charge. Conversely, Schottky diodes are majority carrier devices with fast switching speed and negligible switching loss. However, high voltage Si Schottky diodes are limited in practical use by high on-state voltages and excessive leakage currents that result from Schottky barrier lowering [9] and large thermionic currents flowing through the Schottky barrier. The use of SiC, a material with a 10X increase in the breakdown field, a 3X larger bandgap, and a 3X higher thermal conductivity as compared to Si, can result in Schottky diodes with superior forward conducting, reverse blocking and switching performance at high temperatures. These attributes make the SiC Schottky diode ideal for use as a FWD in motor drive applications.

A cross-section of the SiC Schottky diode evaluated in this study is shown in Fig. 1. The diode design integrates a *p-n* junction grid under the conventional metal-semiconductor junction of the Schottky diode. The *p-n* junction grid is designed to allow current to flow from the Schottky metal to the drift region between the junction grid regions during the forward-biased condition (see Fig.2a). During the reverse bias, the junction depletion layers expand and create a potential barrier under the Schottky interface after the adjacent junctions intersect (see Fig. 2b). The junction-induced potential barrier is used to protect the Schottky barrier from the applied reverse voltage. This prevents Schottky-barrier lowering with increasing reverse bias and eliminates the soft breakdown characteristics of conventional Schottky diodes. This concept has been used to increase the blocking voltage of Schottky diodes [10,11]. This SiC diode structure permits majority carrier operation without minority carrier injection from the p-n junctions for blocking voltages below 3000 volts. The SiC Schottky diode described here are designed to achieve a breakdown voltage of 1200V. It uses a field ring termination at the device periphery for high blocking voltage.

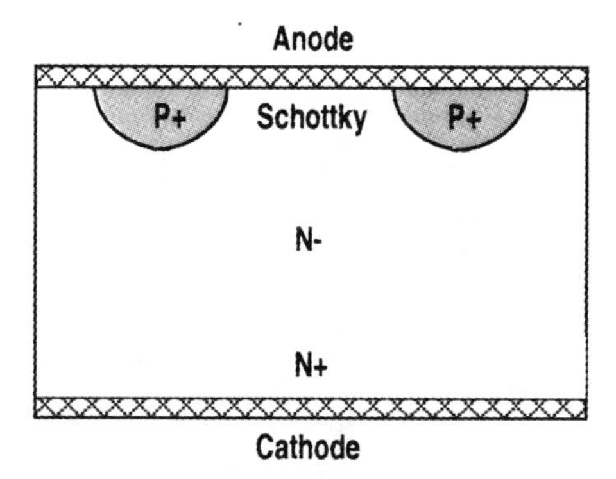

Fig. 1. Cross-section of the SiC Schottky diode with the pn junction barrier to reduce reverse leakage current.

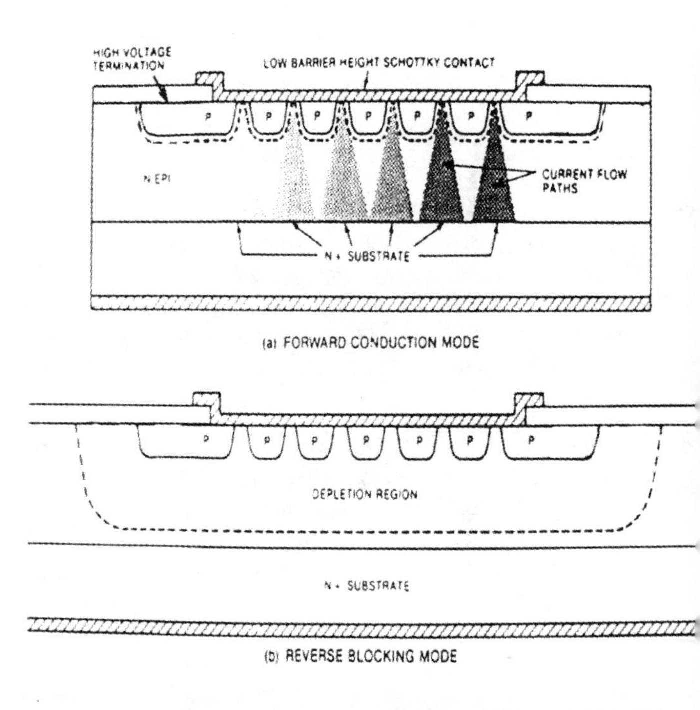

Figure 2. (a) The current flow pattern in the forward conduction mode and (b) the depletion region in the reverse-blocking mode exhibited in the *p-n* junction barrier controlled SiC Shcottky diode.

3. EXPERIMENTAL RESULTS

3.1 Static Characteristics

Fig. 3 shows the reverse blocking characteristics of the *p-n* junction barrier controlled SiC Schottky diodes with an active area of 1.1mm^2 at 25°C. The leakage current for the SiC Schottky diodes is approximately 300μA/cm^2 at 1200V. This value is larger than the leakage current of the conventional fast recovery Si P$^+$-i-N diode but is still within a reasonable range.

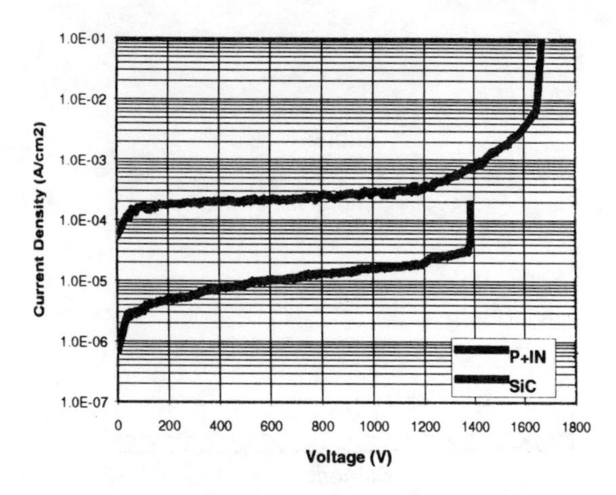

Fig. 3. The reverse blocking characteristics of a *p-n* junction barrier controlled SiC Schottky diode and a Si P^+-i-N diode at 25°C.

A comparison of the forward conduction characteristics for 1200V Si P^+-i-N, advanced Si TOPS [7] and *p-n* junction barrier controlled SiC Schottky diodes at 25°C and 150°C is shown in Fig.4. At a current level of 150A/cm^2 and 25°C, the *p-n* junction barrier controlled SiC Schottky diode has a forward voltage drop of 1.74V while the Si TOPS and Si P^+-i-N diodes have forward voltages of 2.14V and 2.48V, respectively. Even at a high current density such as 400A/cm^2, the SiC Schottky diode has a forward drop of 2.44V; approximately equal to that of the Si P^+-i-N diode at 150A/cm^2.

At 150°C and 150A/cm^2, the forward voltage drop of the SiC Schottky diode increases to 2.15V while the forward voltage drop of the Si TOPS and P^+-i-N diodes reduce to 1.73V and 1.98V, respectively. These forward voltages are reasonable for operation at high temperatures. However, at 400A/cm^2 and 150°C the SiC Schottky diode forward voltage drop increases to nearly 4V and is significantly large for use in motor drive applications with switching frequencies below 20kHz-30kHz.

The difference in the temperature coefficients of the three diodes is due to the different device structures, doping levels, and material properties. The Si power diodes use epitaxial structures and have a negative temperature coefficient while the SiC Schottky diode is a unipolar device with a positive temperature coefficient. Although the positive temperature coefficient increases the conduction losses in the SiC Schottky diode at high temperatures, this characteristic aids in the symmetric current sharing capability when paralleling individual die for higher current capability.

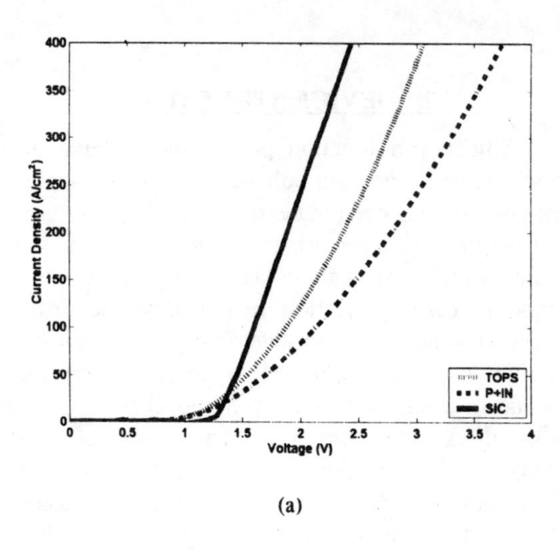

(a)

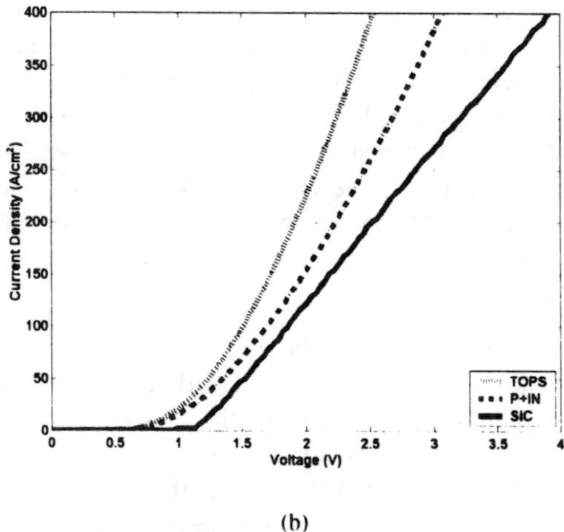

(b)

Fig. 4. FWD forward I-V Characteristics at (a) 25°C and (b) 150°C.

3.2 Switching Characteristics

The reverse recovery switching performance of the SiC Schottky and the Si power diode are evaluated in a half-bridge test circuit using IGBTs as the switches (shown in Fig. 5).

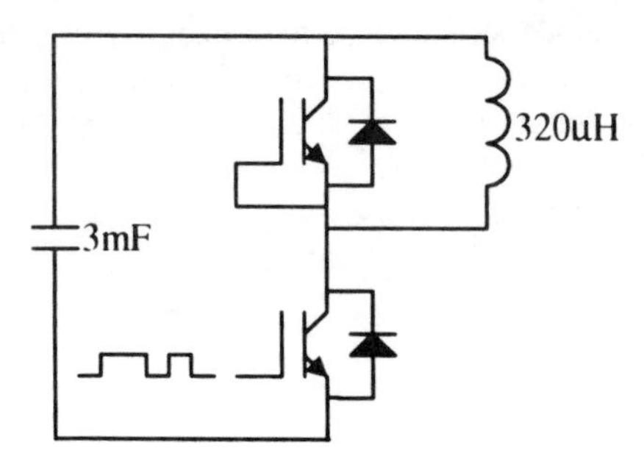

Fig. 5. Half-bridge circuit used to characterize the SiC Schottky and Si power diodes.

The reverse recovery switching characteristics of the three diodes were measured at 25°C and 150°C with a di/dt of 2,000 (A/cm^2)/μs and a reverse bias of 300 volts. The results are shown in Fig. 6. At 25°C the SiC Schottky diode has a significantly shorter switching time, t_{rr}, along with a lower peak reverse current, I_{RP}, and a smaller stored charge, Q_{rr} as compared with either of the two Si diodes. Furthermore, as the temperature increases from 25°C to 150°C, there is no significant change in the SiC Schottky diode reverse recovery current. It is very clear at 150°C that the SiC Schottky diode has a much smaller stored charge, Q_{rr}, approximately 20X smaller than the Si diodes. This ultimately results in a reduction in the diode switching loss, E_{diode}, by a factor of 12 and allows the SiC diode to be used at much higher switching frequencies. A summary of the normalized diode reverse recovery characteristics at 150°C is shown in Table 1.

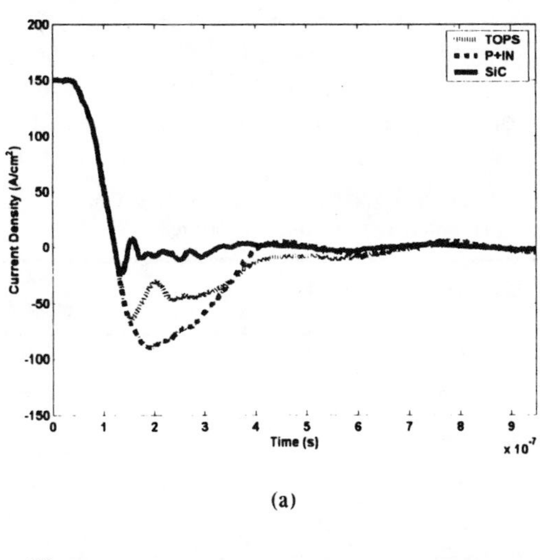

(a)

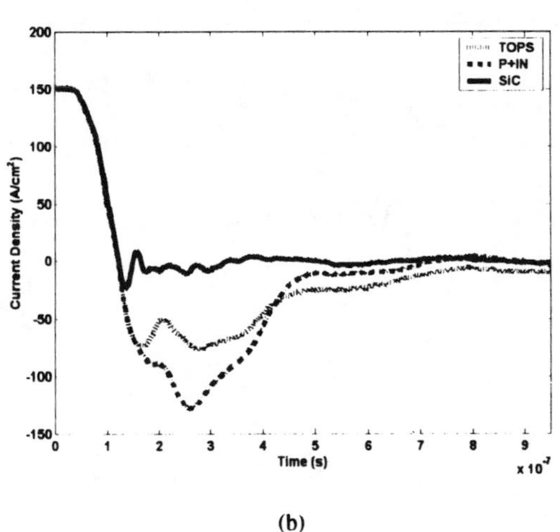

(b)

Fig. 6. Reverse recovery waveforms of SiC Schottky and Si power didoes at (a) 25°C and (b) 150°C at 2000A/μs/cm^2.

TABLE 1
NORMALIZED DIODE REVERSE RECOVERY CHARACTERISTICS AT 150A/cm^2 AND 150°C.

Parameter	Si P$^+$-i-N	Si TOPS	SiC Schottky
Q_{rr} (C/cm^2)	1.00	0.92	0.05
I_{RP} (A/cm^2)	1.00	0.60	0.18
E_{rr} (J/cm^2)	1.00	2.05	0.09
t_{rr} (s)	1.00	1.69	0.08

It is also important to note that even at high current densities (400A/cm^2) the reverse recovery charge in the SiC Schottky diodes does not increase significantly. This indicates that the forward voltage drop is not sufficiently high at this current density to cause excessive minority carrier injection [9]. The normalized reverse recovery current waveforms are shown in Fig. 7 for SiC Schottky diodes operating at current densities of 100A/cm^2, 200A/cm^2, and 400A/cm^2.

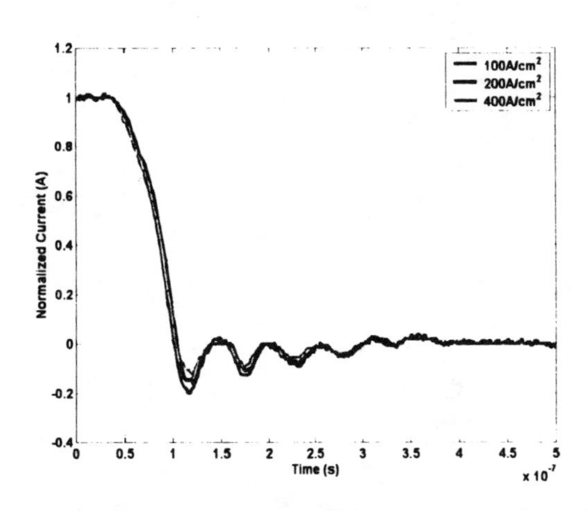

Fig. 7. Normalized SiC Schottky diode reverse recovery currents at 100A/cm^2, 200A/cm^2, and 400A/cm^2 at 150°C.

4. TOTAL POWER LOSSES

To determine the overall performance of the Si and SiC diodes, the total power loss versus switching frequency is measured for the IGBT with the Si P$^+$-i-N, TOPS, and SiC Schottky diodes. The approach used to evaluate the devices is based on the total losses, namely the turn-on and turn-off losses in the IGBT and the reverse recovery loss in each of the diodes. This methodology evaluates the effect of the diodes on the entire system; in full consideration of all device losses in the system. It should be mentioned that the power loss distribution is affected by the device characteristics and also the test circuit parasitic inductance and capacitance. This will later become apparent when the IGBT turn-on waveforms are examined.

Fig. 8 shows the total power loss density (W/cm^2) versus switching frequency at 150A/cm^2 in the diodes at 25°C and 150°C for sinusoidal modulation. At 25°C the SiC Schottky diode performance is far superior to that of either the P$^+$-i-N or the TOPS diodes (see Fig. 8a). However, at 150°C, the forward voltage drop of the SiC Schottky diode substantially increases, resulting in higher losses as compared to the Si diodes at switching frequencies below 6 kHz. From Fig. 8 it is clear that the advantages of the SiC Schottky diode become greater at higher switching frequencies. For higher frequency applications (i.e. power supplies), the current density of the SiC Schottky diodes can be significantly increased (up to 400A/cm^2) without paying the penalty of high forward voltage drop because of its negligible stored charge and switching losses.

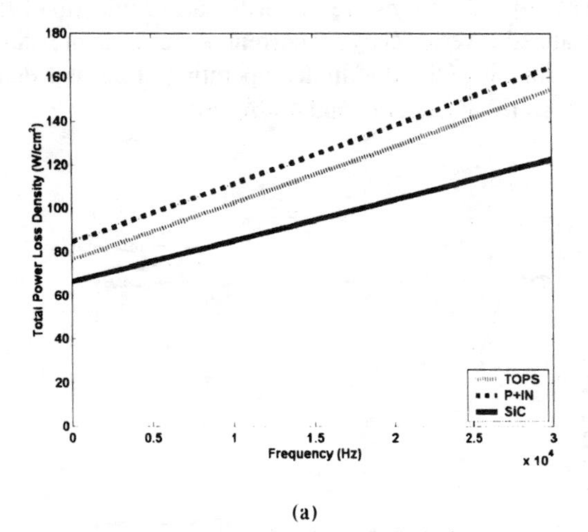

(a)

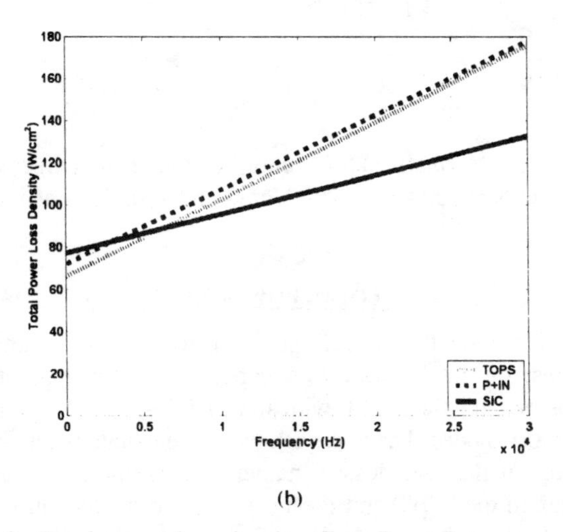

(b)

Fig. 8. Total power loss density (E_{ON}+E_{OFF}+E_{rr}) versus switching frequency at 150A/cm^2 : (a) 25°C and (b)150°C.

A summary of the normalized IGBT and diode energy losses is shown in Table 2. It is interesting to note that the reverse recovery loss, E_{rr}, of the SiC Schottky diode is approximately one-tenth smaller than of the Si P$^+$-i-N diode. This is inconsistent with the turn-on energy loss, E_{ON}, for the SiC Schottky and Si P$^+$-i-N diodes. The turn-on energy loss of the IGBT with the SiC diode is only one-half that with the Si P$^+$-i-N diode. This is due to the effect of the parasitic capacitance across the upper diode of the half-bridge test circuit shown in Fig. 5. The parasitic capacitance is formed when using a planar bus structure that successfully reduces the parasitic inductance in the dc bus but significantly increases the parasitic capacitance across each of the devices in the half-bridge circuit configuration. To verify this, Fig. 9 shows the IGBT turn-on current waveform along with the inverted and shifted SiC Schottky diode reverse current. It is clear that the IGBT current is not simply the sum of the load and reverse recovery currents. Capacitive current also flows through the IGBT during the switching transition and significantly contributes to turn-on losses in the IGBT. These results illustrates the importance of circuit layout when considering both the total device loss and the loss distribution among each of the devices in the circuit.

Table 2

NORMALIZED IGBT AND DIODE STATIC AND DYNAMIC PARAMETERS AT 150A/cm^2 AND 150°C

Parameters	Si P+IN	Si TOPS	SiC Schottky
V_F (V)	1.00	0.87	1.15
E_{OFF} (J/cm^2)	1.00	1.00	0.95
E_{ON} (J/cm^2)	1.00	0.76	0.47
E_{rr} (J/cm^2)	1.00	2.05	0.09
E_{TOTAL} (J/cm^2)	1.00	1.04	0.52

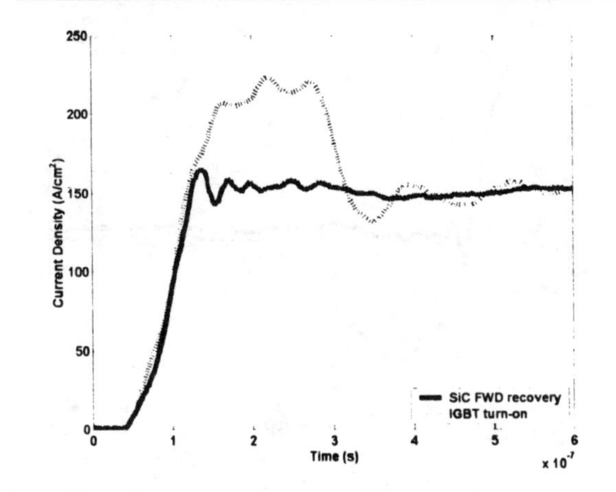

Fig. 9. IGBT turn-on current and SiC Schottky diode reverse recovery current (inverted and shifted) showing the effect of parasitic capacitive current.

In determining the overall performance of the devices the voltage and current stress on the IGBT and diode must also be considered. In Table 3 a summary of the normalized stress parameters are shown for the IGBT, P^+-i-N, TOPS, and SiC Schottky diodes. The peak voltage, V_{PEAK}, at IGBT turn-off is similar for the Si diodes and 10% lower for the SiC Schottky diode. The peak reverse recovery current, I_{RP}, of the SiC Schottky diode is about 6 times smaller than that of the P^+-i-N. This attribute reduces the stress on the diode and also reduces the peak current, I_{PEAK}, through the IGBT by 30%.

TABLE 3
NORMALIZED TRENCH IGBT AND DIODE VOLTAGE AND CURRENT STRESS PARAMETERS AT 150A/cm^2 AND 150°C

Parameter	Si P+IN	Si TOPS	SiC Schottky
V_{PEAK} (V)	1.00	1.00	0.89
I_{PEAK} (V)	1.00	0.89	0.69
I_{RP} (A)	1.00	0.60	0.18

5. CONCLUSION

To fully utilize the performance enhancements of the IGBT, an advanced SiC Schottky diode is proposed. The SiC Schottky diode demonstrated a 30% decrease in total power loss over existing Si P^+-i-N technology. The majority of the loss reduction was due to the superior switching characteristics of the SiC Schottky diode with a 20X reduction in the reverse recovery charge at 150°C. Furthermore, the SiC Schottky diode contributes to lowering the voltage and current stress on the IGBTs and diodes in the switching power circuits. This results in an improvement in the overall performance of the power electronic circuit and also an increase the volumetric power density of the entire system.

ACKNOWLEDGMENT

This project is supported by Rockwell Automation, DARPA/ONR contract number N00014-99-C-0376 and DUS&T Air Force contract number F33615-00-2-2005.

REFERENCES

[1] E. Motto, J. Donlon, H. Takahashi, M. Tabata, and H. Iwamoto, *IEEE-IAS Conference Record*, pp. 811-816 (1998).

[2] M. Hierholzer, T. Laska, M. Loddenkotter, M. Minzer, F. Pfirsch, C. Schaffer, and T. Schmidt, *IEEE-IAS Conference Record*, pp. 1787-1792 (1999).

[3] M. Naito, H. Matsuzaki and T. Ogawa, *IEEE trans. Electron Devices*, vol. 23, no. 8 (1976).

[4] M. Mori, Y. Yasuda, N. Sakurai, and Y. Sugawara, *Proc. IEEE*, pp. 113-117 (1991)

[5] B.J. Baliga and H.-R. Chang, *IEDM Technical Digest*, pp. 658-661(1987).

[6] A. Porst, F. Auerbach, H. Brunner, G. Deboy, F. Hille, *ISPSD Proceedings*, pp. 213-216 (1997).

[7] H.-R. Chang, C. Winterhalter, R. Gupta and K. Humphrey, *IEEE-IAS Conference Record*, pp. 353-358 (1999).

[8] R. Raghunathan, D. Alok, and B.J. Baliga, *IEEE Electron Devices Letters*, vol. 16, no.6, pp. 226-227 (1995).

[9] S. M. Sze, *Physics of Semiconductor Devices*, 2nd Edition, 1981.

[10] B. M. Wilamowski, *Solid-State Electron.*, vol. 26, pp. 491(1983).

[11] H.-R. Chang and B.J. Baliga, *Solid-State Electron.*, vol. 29, pp. 359 (1986).

AIAA-2000-2829

DESIGN OF DUAL USE, HIGH EFFICIENCY, 4H-SIC SCHOTTKY AND MPS DIODES[1]

Clarence Severt
Air Force Research Lab, Power Div.
AFRL/PRPE, 1950 5th St
WPAFB, OH 45433

A. Agarwal, R. Singh, S-H. Ryu and J. W. Palmour
Cree, Inc.,
4600 Silicon Dr.,
Durham NC 27703.

Ph: 937-255-6235, Fax: 937-656-4781, E-mail: severt@wl.wpafb.af.mil

ABSTRACT

The prime benefit of the SiC Schottky diode lies in its ability to switch fast (<50 ns) and with almost no reverse recovery charge. The incorporation of a SiC Schottky-type rectifier in typical power-electronic systems, such as motor drive circuits and switching power supplies, will virtually eliminate the switching losses. In the forward bias, Ti Schottky diode provides a forward current density of 100 A/cm^2 at a forward drop of 1.15 V, at room temperature. However, the low barrier height of Ti Schottky diode results in a very high room temperature leakage current density of 300 μA/cm^2 at 500 V. Furthermore, the leakage current becomes unacceptable at temperatures higher than 100°C. A 4H-SiC Merged PiN Schottky (MPS) diode uses interdigitated p$^+$ regions between Schottky contacts to limit the electric field at the Schottky interface during the off-state operation of the device. It has a low on-state voltage drop and fast switching of a Schottky diode and offers a low off-state leakage current like the PiN diode. The on-state and off-state performance was optimized by analyzing the effect of adjacent p$^+$ region width and spacing by 2D device simulations.

INTRODUCTION

Although there are about 170 known polytypes of SiC, only two (4H-SiC and 6H-SiC) are available commercially. 4H-SiC is preferred over 6H-SiC for most electronics applications because it has a higher and more isotropic electron mobility than 6H-SiC. Table 1 compares some key electronic properties of 4H-SiC to Si and GaAs.

The higher breakdown electric field strength of SiC enables the potential use of SiC Schottky diodes in 600-1200 V range. The power electronic systems operating in the above voltage range currently utilize silicon PIN diodes which tend to store large amounts of minority carrier charge in the forward-biased state. The stored charge has to be removed by recombination before the diode can be turned off. This causes long storage and turn-off times.

Table 1: Key electronic properties of Si, GaAs, and 4H-SiC

Property	Silicon	GaAs	4H-SiC
E_g (eV)	1.12	1.5	3.25
μ_n (cm^2/Vs)	1400	9200	800
μ_p (cm^2/Vs)	450	400	140
n_i (cm^{-3}) at 300 K	$1.5x10^{10}$	$2.1x10^6$	$5x10^{-9}$
v_{nsat} (x10^{7}cm/s)	1.0	1.0	2.0
E_{crit} (MV/cm)	0.25	0.4	3.0

The prime benefits of the SiC Schottky diode lie in its ability to switch fast (<50 ns) and with almost no reverse recovery charge and the high junction temperature operation. The GaAs Schottky diodes can be made in this voltage range but have limitations with regards to the high junction temperature operation. The comparable Silicon P-i-N diodes (Si Schottky diodes are not viable in the 600 V range because of their large on-state voltage drops) have a reverse recovery charge of 400-700 nC and take at least 200 ns to turn-off. This places a tremendous burden on other switching elements in the system in terms of the required forward safe operating area and the switching losses incurred. The incorporation of a SiC Schottky diode capable of operating at a junction temperature of 350°C in the system will not only virtually eliminate the switching losses **(thus greatly increasing the efficiency)** but will also reduce the cooling requirements by at least a factor of 3 due to the higher junction temperature. Additional reduction of weight and volume of magnetic components in certain systems such as a switching power supply can be achieved by the increase in switching frequency allowed by the SiC Schottky diode.

The weight and volume reduction of power conditioning systems is particularly attractive for airborne and space applications where there is neither the volume- nor weight-budget to accommodate the cooling systems that are normally required for power electronic systems. For airborne applications, the associated savings for every pound of weight reduction is typically quoted in excess of $1 Million over the life cycle of a system across a fleet. In addition, this component will find applications across the board on many military platforms requiring efficient, compact, and light power conditioning systems. Almost all the major military systems such as Space Based Laser (SBL); Orbital Transfer Vehicle (OTV); Space Based Radar (SBR); Electro-mechanical Actuators and Integral Starter-Generator on More Electric Aircraft (MEA); Fire Control X-band Phased Array Radars; L and S-band Surveillance Radars; Electronic Warfare and Electronic Countermeasure (EW/ECM); land, air and sea based Communications; Missiles; Combat Hybrid Electric Vehicle etc. have a Common Power Supply Module or some kind of power conversion sub-system. All these systems, and especially the airborne systems would greatly benefit from the SiC Schottky diode.

DEVICE DESIGN AND RESULTS

Ti and Pt Schottky Diodes

It is important to achieve a good quality Schottky interface for low on-state voltage Schottky diodes. The interface quality of Schottky diodes can be measured by the value of the ideality factor on these diodes. Using test wafers with the aim of improving the interface quality, Ti Schottky diodes with an ideality factor as low as 1.07 have been achieved as shown in Figure 1. These 600 V diodes had an on-state voltage drop of only 1.15 V at a forward current density of 100 A/cm^2 at room temperature. The forward drop decreases up to 200°C due to increase in emission over the Schottky metal to SiC barrier. At 300°C, the forward drop increases due to an increase in the resistance of SiC. The reverse J-V characteristics for the Ti Schottky diodes, at room temperature, are shown in Fig. 2. The leakage current increases rapidly with reverse bias (due to the low barrier height) and reaches 300 μA/cm^2 at 500 V reverse bias. The leakage current data for higher temperatures is not shown as the leakage current increases rapidly with temperature rendering the device useless at temperatures in excess of 100°C. Thus, the use of a low barrier Schottky metal such as Ti, results in low forward drop and high reverse leakage current.

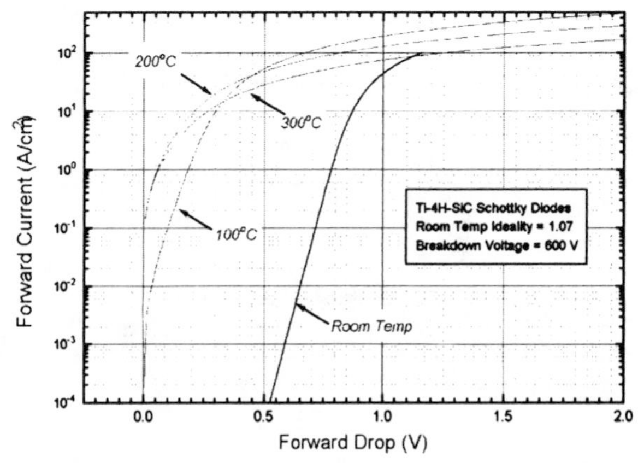

Fig. 1 Forward J-V characteristics of Ti Schottky diodes at different temperatures.

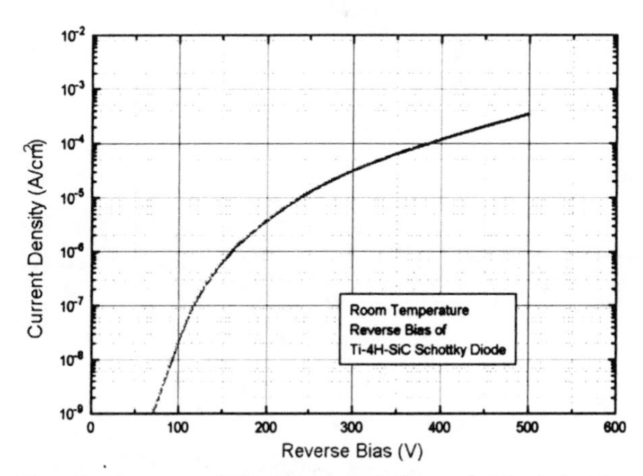

Fig. 2 Reverse J-V characteristics of Ti Schottky diodes.

Pt has a higher work-function compared to Ti and therefore, Pt Schottky diodes on 4H-SiC result in a higher barrier height. Fig. 3 shows the forward J-V characteristics of a 700 V, Pt Schottky diode on 4H-SiC. These diodes show an on-state voltage drop of

1.88 V at a forward current density of 100 A/cm^2 at room temperature with an ideality factor of 1.13. The forward drop decreases to 1.5 V at 300°. The reverse J-V characteristics are shown in Fig. 4. The reverse leakage current density of 10^{-5} A/cm^2 is obtained at 500 V reverse bias at room temperature. It increases to 10^{-4} A/cm^2 at 300°C, which is quite acceptable. The Pt Schottky has a higher barrier height as compared to Ti Schottky. This leads to lower forward and reverse currents in the Pt Schottky diode. While the lower reverse current is desirable, the lower forward current leads to higher forward drop. On the other hand, Ti Schottky diodes have an acceptable forward drop but the reverse leakage current can be very high – especially at elevated temperatures. As explained in the next section, the use of Merged P-i-N Schottky (MPS) diode can rectify this problem by means of embedding P-i-N diodes within Ti Schottky diodes.

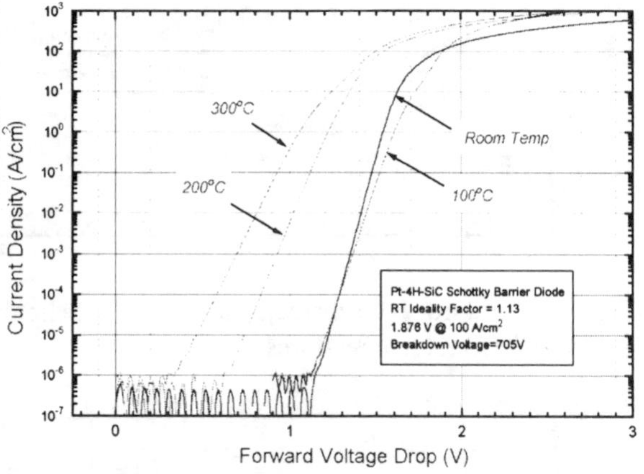

Fig. 3 Forward J-V characteristics of Pt Schottky diodes at different temperatures.

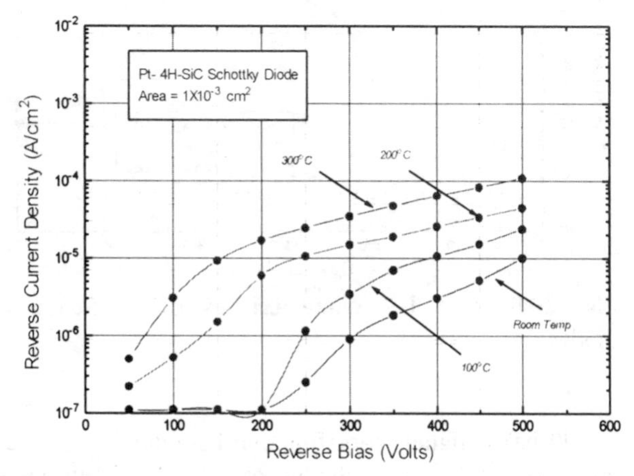

Fig. 4 Reverse J-V characteristics of Pt Schottky diodes at different temperatures.

Merged P-i-N Schottky (MPS) Diode

A diode structure that achieves the low on-state voltage drop of a Schottky diode while having good high temperature, high voltage rating is the merged P-i-N Schottky (MPS) diode.[1-3] A MPS diode consists of interdigitated Schottky and p$^+$ areas that pinch off the reverse bias leakage current from the Schottky areas of the diode. It behaves like a Schottky diode in the on-state and a P-i-N diode in the off-state. This allows the fabrication of high voltage, low leakage and high temperature diodes. A cross-section of a MPS diode operating in the forward bias is shown in Figure 5. For on-state voltages <3 V, only the Schottky regions of the diode conduct. The on-state voltage drop of the device is determined by the metal-SiC barrier height, resistance of the drift region, and the relative area of the Schottky vs. the p$^+$ implanted regions. Also, a near unity ideality factor Schottky contact ensures a low on-state voltage drop at high current densities.

The presence of p$^+$ implanted regions give a very high surge current, dV/dt, and dI/dt rating to this device. The reverse bias operation of an MPS diode is shown in Figure 6. As the reverse bias increases, the depletion regions from adjacent p$^+$ implanted regions pinch-off the leakage current arising from the Schottky contacts of the device. The leakage current in the Schottky regions occurs due to Schottky barrier lowering at the metal-N$^-$ junction. The presence of p$^+$ implanted regions reduces the electric field at the metal-SiC junction because of two-dimensional charge sharing. This property is especially useful in reducing the leakage current when the diode is operating at elevated temperatures.

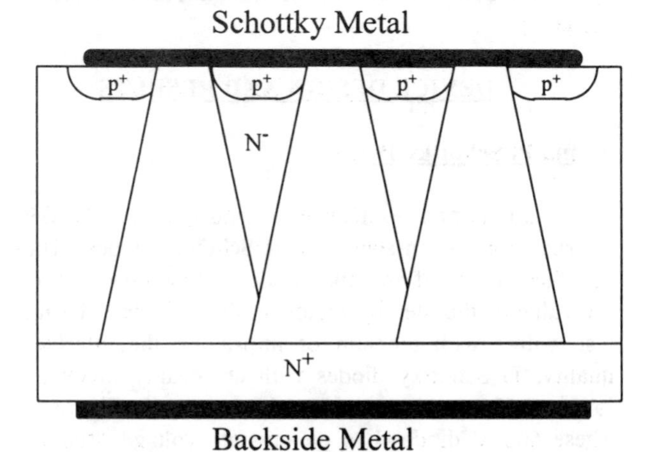

Fig. 5 MPS diode under on-state operation.

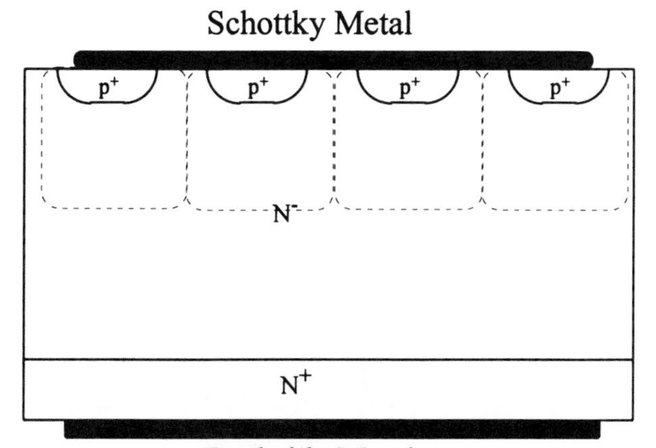

Fig. 6 MPS diode under reverse bias operation.

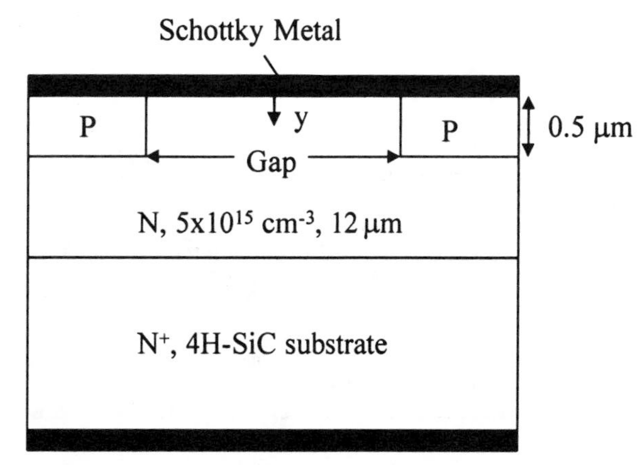

Fig. 7 The MPS diode structure used for 2D device simulation.

The effect of grid spacing on the electric field in the Schottky region under reverse bias has been simulated using a 2D device simulator. For this study, we have selected a 600 V application. The simulated structure is shown in Fig. 7. The drift-layer doping and thickness have been selected as $5x10^{15}$ cm^{-3} and 12 μm, respectively. The 0.5 μm thick p grid has been doped at $2x10^{17}$ cm^{-3}. The electron mobility in the n-type drift layer has been assumed to be 800 $cm^2/V{\bullet}s$. The Schottky metal was assumed to have an ohmic contact to the p-grid and was grounded. To carry out reverse voltage simulations, a positive voltage (600 V) was applied to the n^+ substrate. The plots of electric field in the y-direction along the center-line of the structure are shown in Fig. 8, for different gap dimensions. As expected, the smallest gap dimension of 1 μm results in the lowest electric field (0.86 MV/cm) and the maximum gap dimension of 6 μm results in the highest electric field (1.3 MV/cm), at the surface. The reduction in electric field at the surface is important in reducing the field emission current under reverse bias at elevated temperatures.

The simulated forward J-V characteristics for various values of the gap, are shown in Fig. 9. It is evident that although the gap of 1 μm provides maximum protection from electric field, it also results in the lowest forward current density. Comparing Figures 8 and 9, we can conclude that a gap of 2 μm is close to an optimum value. The designed dimension should be somewhat higher (closer to 3 μm) to take into account the process bias and straggle of the Aluminum implant.

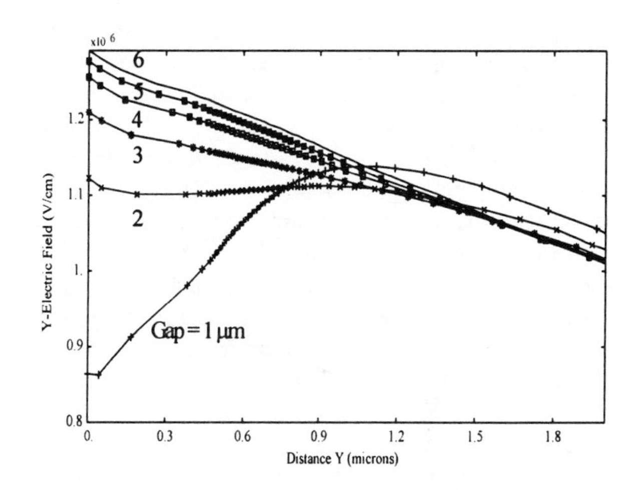

Fig. 8 The variation of the electric field in the y-direction along the center line of the cell, shown in Fig. 7, for different values of the gap (1 to 6 μm).

DISCUSSION AND CONCLUSIONS

The MPS diodes described in this paper are presently being fabricated at Cree, Inc. The detailed measurements including switching transients and temperature dependence of forward and reverse I-V characteristics will be presented at the conference. We will also describe the use of different edge termination structures which are required to attain high breakdown voltages in these structures.

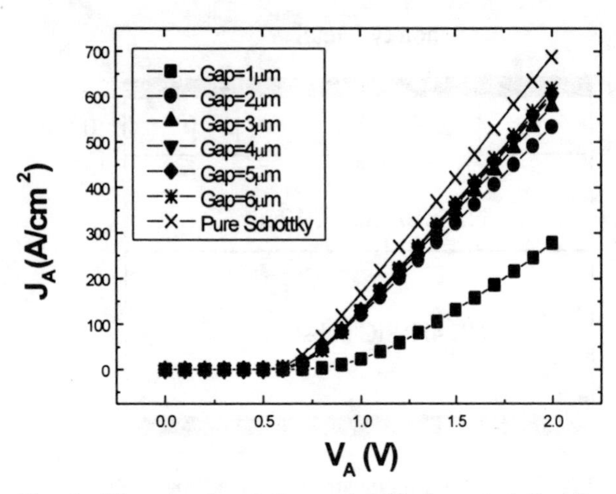

Fig. 9 The simulated forward J-V characteristics for various values of the gap (1 to 6 μm).

In summary, the prime benefit of the SiC Schottky diode lies in its ability to switch fast (<50 ns) and with almost no reverse recovery charge. The incorporation of a SiC Schottky-type rectifier in typical power-electronic systems, such as motor drive circuits and switching power supplies, will virtually eliminate the switching losses. The pure Schottky diodes with Ti as a Schottky metal tend to have low forward drop but high reverse leakage current. On the other hand, the Pt Schottky diodes have a higher forward drop and low reverse leakage currents. The MPS diodes using interdigitated p^+ regions and Ti as a Schottky barrier metal, combines the low forward drop with low reverse leakage currents. The on-state and off-state performance of these MPS diodes was optimized by analyzing the effect of adjacent p^+ region spacing by 2D device simulations. It is concluded that a p^+ region spacing of 2 μm is optimum.

REFERENCES

[1] M. Wilmowski, Solid-State Electronics, Vol. 26, p. 491, 1983.

[2] Y. Sugawara, K. Asano and R. Saito, presented at the International Conference on Silicon Carbide and Related Materials (ICSCRM'99), North Carolina, U.S.A., October 10-15, 1999.

[3] K. Tone, J. H. Zhao, M. Weiner, and M. Pan, presented at the International Conference on Silicon Carbide and Related Materials (ICSCRM'99), North Carolina, U.S.A., October 10-15, 1999.

DUAL-USE DEVELOPMENT OF HIGH DENSITY DC/DC CONVERTERS†

Jian Sun, Ph.D.
Rockwell Collins, 400 Collins Road NE, Cedar Rapids, IA 52498
Phone: (319) 295-5929; E-mail: jsun@collins.rockwell.com

Vivek Mehrotra, Ph.D. and Jun He, PhD.
Rockwell Science Center, 1049 Camino Dos Rios, Thousand Oaks, CA 91358-0085
Phone: (805) 373-4484; E-mail: vmehrotra@rsc.rockwell.com

ABSTRACT

Rockwell Collins and Rockwell Science Center have teamed with the Office of Naval Research (ONR) to develop high density, high efficiency, programmable dc/dc converters for both military and commercial applications. This three-year program aims at the development of two 100W dc/dc converter modules that run off a 50V dc bus and provide 3.3V and 1.5V outputs, respectively, at 90-95% efficiency, and with a power density at or higher than 100W/in³. This paper provides an overview and a detailed analysis of the program goals and then discusses the technical challenges to achieve these goals. Key technology development efforts are also discussed. Up-to-date achievements in the areas of the evaluation and selection of power devices, circuit design, and magnetics analysis are outlined.

INTRODUCTION

The demand for increasing speed and integration of processors and other ICs has led to continuously decreasing supply voltages. It is estimated that by 2003, the supply voltages required by CMOS compatible devices and ICs will drop below 2 V (Fig. 1) [1]. By lowering the supply voltages the chip speed has been increased and the power consumption lowered. But the total power consumption in a system has actually increased due to higher functionality, integration and larger number of chips per system. The problem is further complicated by the requirement for power management that turns on and off the processor at high clock rates, resulting in changing load currents at very high slew rates. All these requirements together have created an enormous design constraint on power supplies for next-generation electronic systems [2].

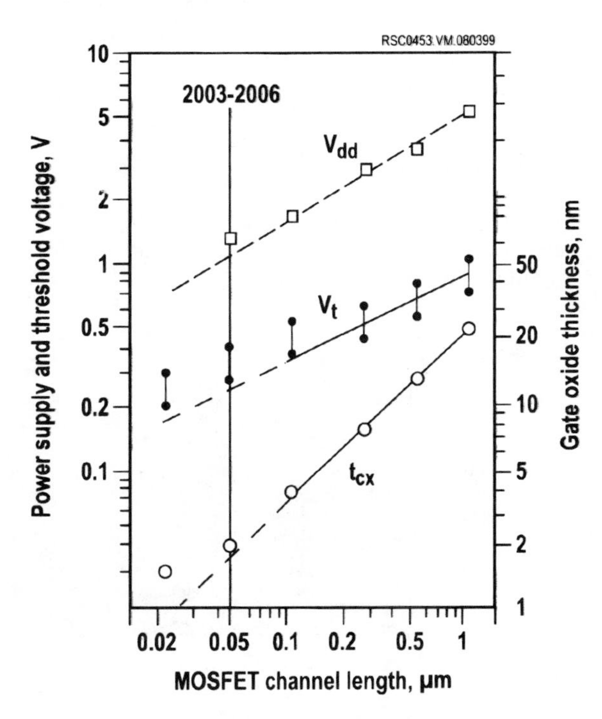

Figure 1. Change of CMOS feature sizes and supply voltage in the future.

As the output voltage requirement dropped from 5 V to 3.3 V and now even lower (1-2 V), the distributed power system architecture emerged as the only viable choice. This has placed additional constraints on the power supply in terms of thermal management and EMI since the power supply has to be co-located with the load. Additional requirements that have emerged on the power supply for the distributed power system architecture are increased efficiency, higher power density, increased reliability and lower cost. Higher power density mandates increased efficiency due to the need to eliminate large heat sinks.

† This effort is partially supported by the Office of Naval Research (ONR) under the Dual-Use Science and Technology program (Contract No. N00014-99-3-0006).

In response to these challenges in future electronic systems, the Office of Naval Research (ONR) and Rockwell have teamed to develop the next generation of dc-dc converters for both commercial and military markets. The project aims at the development of two 100W dc/dc converter modules that provide 3.3V and 1.5V outputs, respectively, and achieve 90-95% efficiency and a power density of 100W/in^3. Other goals include low cost (< $1/W) and high reliability (> 200,000 h Mean-Time-To-Failure, MTBF). This paper gives an overview of the program goals, the design challenges and technology development strategies for achieving these goals. A summary of up-to-date progress is also outlined.

EFFICIENCY AND POWER DENSITY ISSUES

A loss analysis for the 3.3 V output and the 1.5 V output module has been carried out. The results for the 1.5V output module are illustrated in Fig. 2.

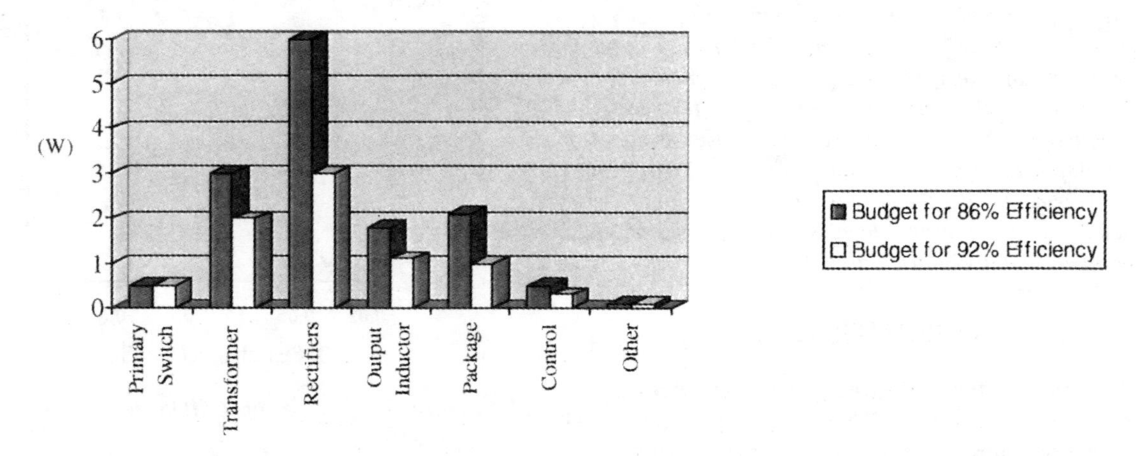

Figure 2. Loss distribution from all sources in a 1.5 V converter and that required for achieving 86-92% efficiency.

The analysis reveals that, with current technologies, only 80.5% efficiency can be achieved for 3.3 V output, and the efficiency drops to 68% if the output voltage is reduced to 1.5 V. Major barriers to efficiency improvement are

- primary switching losses,
- transformer losses at high current and high operation frequency (1 MHz),
- conduction losses in the output rectifiers
- conduction losses due to interconnection and packaging.

The footprint budget shown in Fig. 3 has been established on a comparative basis as necessary to achieve the proposed 100 W/in^3 power density goal of the project. The total maximum footprint is taken to be 3.2 in^2 assuming a 1 in^3 total volume and 0.3 inch maximum height.

The loss and footprint budgets presented here define the maximum size and power dissipation for each of the key components: primary switches, rectifiers, transformer and inductor, output capacitors, EMI filter, and control circuitry. The next section describes the technologies to be developed and applied for each of these components as well as the overall system packaging and integration.

In order to achieve the proposed efficiency and power density goals, a three-step technology development and demonstration plan has been proposed:

- *90% efficiency at 3.3 V output.* This will be achieved by using incrementally improved silicon devices and packaging techniques, soft-switching for the primary switch, and optimized transformer design.
- *95% efficiency at 3.3 V output and 86% efficiency at 1.5 V output.* Significant improvements in packaging and interconnect technology, improvement in semiconductor device performance such as low voltage trench Si MOSFETs with improved cell designs, and loss reduction in magnetics are key to this goal.
- *92% efficiency at 1.5 V output.* This will require new, ultra fast devices with very low on-state resistance, such as GaAs vertical FETs developed by Navy SPAWAR and Texas Instruments under a prior effort, further loss reduction in the transformer and output inductor, and a radical approach to packaging and interconnects.

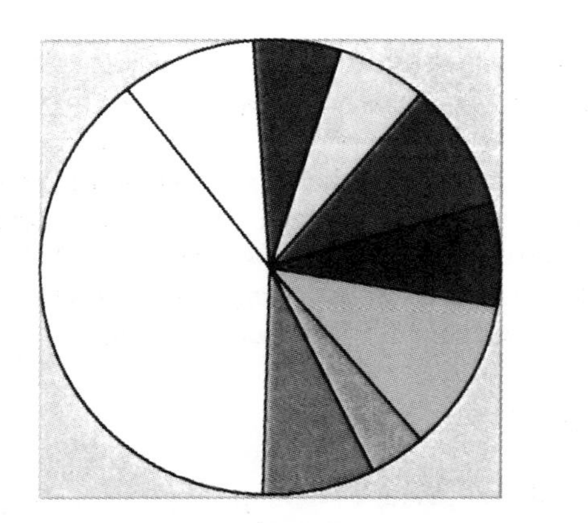

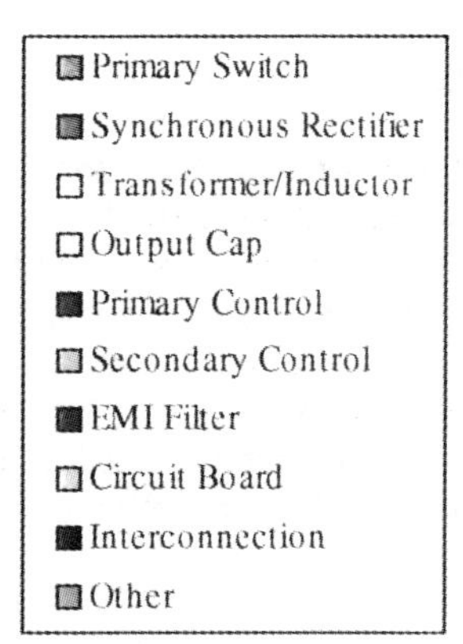

Figure 3. Footprint budget for the proposed modules assuming a 3.2 in^2 area and 0.3 in height.

TECHNOLOGY DEVELOPMENT PLAN

The development of the above power supplies for both commercial and military markets will require development of various technologies, which are outlined below.

Topologies

Since many dc/dc converter topologies have been developed and several of them are suited for the targeted power of 100 W and input voltage of 50 V, this task has been concentrated on comparing and modifying existing topologies rather than attempting to invent new ones. Both forward and half-bridge topologies are being evaluated. Forward topologies are simpler than half-bridge, but suffer from several potential drawbacks for this specific application. These drawbacks include the need for transformer resetting, limited maximum duty ratio, lower throughput power of the transformer, and higher switching and conduction losses of the primary switch because of the higher voltage stress, which is estimated to be in the range of 100-150 V for the 50 V input. On the other hand, a half-bridge converter requires a gate transformer for driving the high-side switch, and leads to increased size and cost. The overall comparison and trade-off takes into account:

- device characteristics (conduction loss vs. switching loss of various MOSFET designs)
- zero-voltage switching techniques for reducing turn-on losses of primary switches
- transformer and integrated magnetics design for different topologies
- packaging and interconnect approaches
- dynamics and control requirements

Switching and conduction losses of semiconductor devices constitute more than 50% of the total loss of a converter, and are therefore a major barrier to achieving high efficiency and high power density. In the present program, zero-voltage-switching techniques will be applied for the primary MOSFET to minimize its turn-on loss. Low-voltage MOSFETs or GaAs VFETs [3] with low on-state resistance will be used to replace power diodes for secondary rectification (synchronous rectifier) to minimize the conduction loss. Furthermore, in order to reduce the size and interconnection losses, integrated magnetic structures that allow both the transformer and inductor to be built into a single structure are being optimized for the chosen topology. Depending on the main converter topology, different approaches to integrating the transformer and inductor are being investigated. The possibility of integrating secondary rectifiers into the magnetic structure to further reduce interconnections and the associated conduction losses will be necessary.

The two control strategies that need to be addressed are the control of the primary switch for fast transient response, and the control of the synchronous rectifiers for high efficiency and reliable operation.

To meet the load transient regulation requirement (0.1 V maximum output voltage deviation under 50% load step), the main control loop must be carefully designed. Traditional linear design methods based on averaging and linearization do not address large-signal transients and are therefore inadequate in the present design. Nonlinear large-signal models and design methods are being used to ensure superior large-signal transient response under load current step change.

The synchronous rectifiers may be directly driven by the secondary voltage of the transformer (self-driven), eliminating the need for external gating signals. However, the secondary voltage, especially at 1.5 V output, may not be high enough to drive the synchronous rectifiers. A self-driven scheme also doesn't allow exact timing between the high and low side rectifiers and is therefore not optimal from an efficiency standpoint. External gate drive signals for the synchronous rectifiers offer better control. The timing between the drive signals must be adjusted in accordance with device characteristics and optimized for high efficiency. The goal is to minimize the delay between the gate signals so that the conduction of the body diode can be avoided.

Although integrated ICs are available for primary switch control, discrete components, i.e. passive resistors and capacitors and active devices, are still required for compensation, timing, housekeeping of power, and protection. These, together with the control circuitry for synchronous rectifiers, will consume a significant portion of the circuit board area and are one of the limiting factors for achieving high power density.

Devices for Synchronous Rectification

The full load current of the module will be 30 A at 3.3 V output and 67 A at 1.5 V output. Hence each milli-ohm of resistance in the secondary-side current path will translate to 0.918 W of conduction loss at 3.3 V output and more than 4.4 W loss at 1.5 V output! The on-state resistance of the synchronous rectifiers is therefore most critical for achieving the efficiency goal. At the present time, the drain-to-source resistance $R_{DS(on)}$ of commercially available high current, low voltage MOSFETs is in the range of 2-10 mΩ. For example, Motorola MTP75N03HDL, which is rated for 25V and 75A, has an on-resistance of 9 mΩ. This translates to more than 8.1 W conduction loss at 3.3 V output and 40 W conduction loss at 1.5 V output, which is unacceptable for the efficiency goals. Some newer MOSFET devices from ST Microelectronics have an on-resistance of about 2-mΩ and represent the best available choice for synchronous rectification.

The gate capacitance associated with the secondary side devices also has a major effect on the switching loss characteristics since this capacitance must be charged at each turn-on event. Typically, large area Si devices are required to bring down the on-state resistance to reasonable values, and this tends to increase the device capacitance. A resonant gate drive may enable lowering of the net switching loss by recycling some of the gate energy.

In a prior Navy program, Texas Instruments developed GaAs Vertical FETs [4] for synchronous rectification, in which both the on-resistance and device capacitance can be lowered to achieve very high efficiencies. An on-state resistance of about 1.5 mΩ at 10 A and a blocking voltage of 10 V were demonstrated. However, seven to eight parallel devices would be needed to meet the current requirement, which is not practical from commercial and cost viewpoints. A GaAs based switch with a higher current rating is necessary to achieve 95% efficiency for 1.5 V output converters.

Magnetics Design

With increasing switching frequencies, the size of the required magnetic element decreases with a concomitant increase in loss owing to both proximity and skin effects in the winding (copper losses) and losses in the core. At frequencies of about 1 MHz, ferrite is the material of choice from the loss and cost viewpoints. The winding loss is dependent on the winding geometry and thus can be optimized. The objective is to develop a planar magnetic structure that incorporates a transformer and inductor(s) into a single component while lowering the losses. At least a 25% reduction in copper loss and a 50% reduction in the core temperature rise compared to the state of the art magnetic structures, while restricting the height to less than 0.25", are necessary to achieve the program goals.

Figure 4 shows the variation of box volume of a transformer as a function of frequency for different

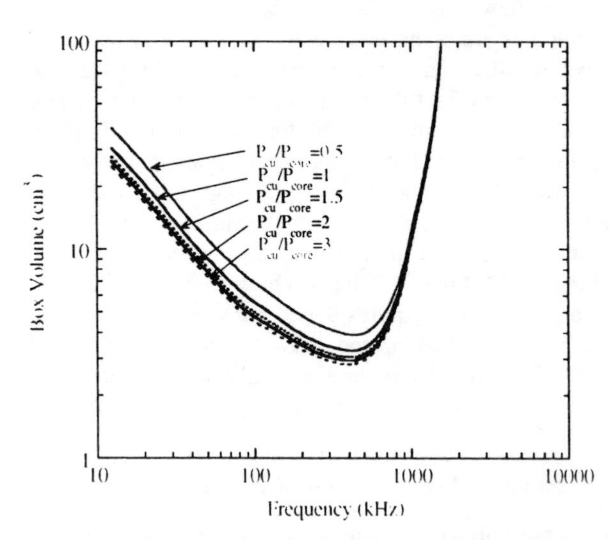

Figure 4. Plot showing the variation of the box volume of the transformer at different copper to core loss ratios (P_{cu}/P_{core}) as a function of frequency. The box volume starts to increase beyond a critical frequency (500 kHz) due to the high thermal resistance from the core to the surface in traditional designs. Improved packaging and winding designs are necessary to lower the box volume beyond 500 kHz.

ratios of copper to core power loss. A fixed 30°C rise in temperature from core to surface has been assumed and a 3F3 ferrite material with an EI core has been used in these calculations. The box volume decreases with increasing frequency but then tends to increase. This is due to decreasing surface area of the transformer, and the size must be increased in order to maintain a 30°C temperature gradient. The high core-to-surface thermal resistance gives rise to a thermally limited transformer size and will affect its reliability if pushed beyond the calculated guideline. This situation can be alleviated with a marked decrease in the size of the magnetics by improving the core to surface thermal resistance. Furthermore, in traditional magnetic components, the core temperature near the gap is much higher than the surface temperature due to poor conduction through the windings, thus limiting the size of the magnetics. Since the bobbin structure in traditional transformers does not allow a good contact with the substrate, the heat generated is removed primarily through the pins soldered to the board.

Planar magnetics are necessary to improve the contact with the substrate and to provide for an efficient removal of heat. Another approach possible with the magnetic structures is to provide thermally conductive but electrically insulating inserts. Yet another approach is to use copper traces attached to high thermal conductivity AlN and Al_2O_3 substrates. Inserting flat heat pipes may further reduce the core-to-surface thermal resistance of the magnetic structure. Heat pipes are capable of carrying high thermal loads without appreciable thermal resistance, and their widespread use in laptop computers has made them cost-effective. The core to surface thermal gradient must be reduced by improved packaging so that the magnetics size is not thermally limited. In addition, an integrated magnetics structure in which both the transformer and filter inductor are combined into a single component is required to achieve the power density goals.

Reliability and Packaging

Wire bonds and solder joints mainly limit the reliability of conventional power device modules and converters. Wire bond failures are known to be dictated by the mechanical stress that is induced as a result of the differences in thermal expansion mismatch between the materials comprising the wire and the underlying device/substrate assembly. Of special concern is that this stress is repeatedly generated and removed during service, since a number of progressive damage mechanisms, such as metal fatigue, are known to occur under these conditions. Therefore, the fatigue life of wire bonds is measured in the number of temperature or power cycles necessary to induce failure. Among all wire bonds in the assembly, the stitch bonds on the device itself are expected to have the minimum MTBF by virtue of their larger CTE mismatch (Al versus GaAs or Si) and shorter loop length. The predicted cycles to failure as a function of peak junction temperature assuming a loop length of 1.5 mm and an ambient temperature of 20°C are shown in figure 5.

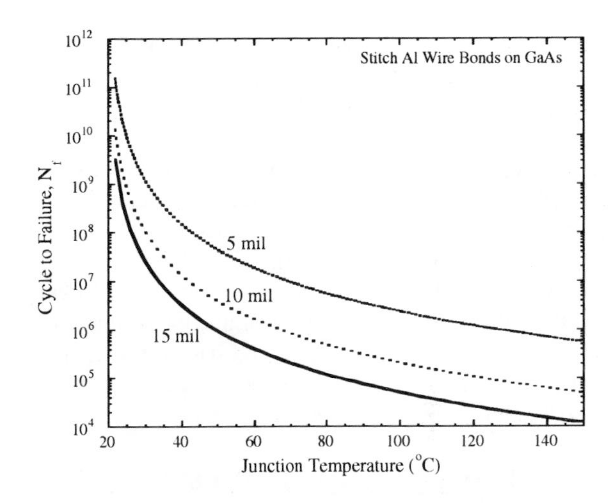

Figure 5. An estimate of the reliability of stitch Al wire bonds on GaAs at different wire diameters.

In order to achieve an order of magnitude increase in MTBF over the present commercial dc-dc converters, either a 40-50°C reduction of peak junction temperature or a switch to smaller diameter wirebonds is necessary. A significant reduction in peak junction temperature requires dramatic improvements of packaging and cooling technologies and is unrealistic to realize within the cost and time constraints of the proposed effort. Meanwhile, the usage of small diameter wire bonds will increase the power dissipation at full load by as much as 2 W, as shown in the Table 2. In addition to the reliability and power dissipation issues, the wire bonds introduce parasitic inductance and limit the di/dt response. In addition, both parasitic inductance and capacitance contribute to EMI.

Therefore, a flip chip approach is proposed to replace the Al wire bonds as the chip level interconnect. In this approach, the bare semiconductor chip with Al contact pads can be converted into proper solder bumping structures, such as Ni/Cu or Cr/Cu/Au, and flipped upside down and bonded directly to the substrate using continuous solder joints. Metal bus bars or DBC substrates can be soldered to the backside as the drain interconnection. This not only allows minimum power dissipation in interconnects and reduces the parasitic

Table 1. Interconnection Losses Using Al Wire Bonds and Flip-Chip Approach

		Power Loss in Interconnection at Full Load			
		Al Wire Bonds (loop length 15 mm)			Flip Chip
Wire Diameter		5 mil	10 mil	15 mil	
Outputs	1.5 V	2.16 W	0.54 W	0.24 W	0.03 W
	3.3 V	0.98 W	0.25 W	0.11 W	0.01 W

inductance and EMI, but also provides a significant advantage over wire bonded systems in terms of thermal management and system reliability.

The replacement of Al wire bonds by solder joints thus provides a solution to the dilemma associated with improving reliability while reducing the power dissipation in interconnects. Although the solder joints exhibit similar fatigue fracture under thermal or power cycling, their reliability can be dramatically improved by incorporating new substrate and baseplate materials with CTE matched to that of the semiconductor and also by choosing a solder with high yield stress.

PROGRESS AND CURRENT STATUS

Topology Selection

A detailed comparison of different dc/dc converter topologies that are suitable for the targeted power level of 100 W has been conducted. A half-bridge converter with current-doubler rectification circuit [5] seems to be the best choice for achieving the efficiency and power density goals. The overall converter topology is shown in Fig. 6.

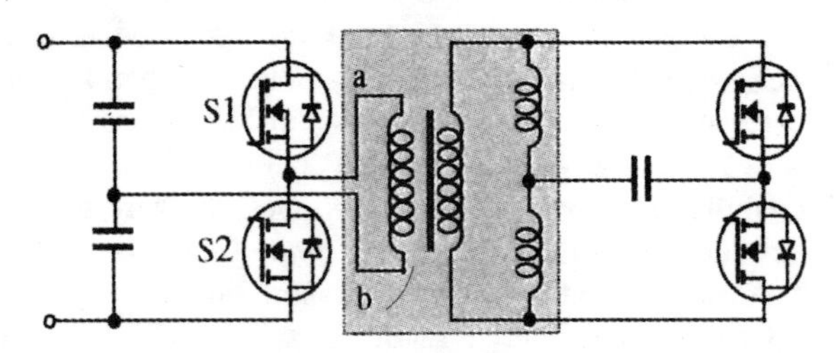

Figure 6. Half-bridge converter with current-doubler rectifier.

The half-bridge arrangement on the primary side of the transformer has the advantage of low voltage stress on primary switches due to clamping to the input voltage as compared to single-ended topologies such as flyback and forward converters. In addition, the half-bridge offers a simpler gate drive compared to the full-bridge topology. The current-doubler rectifier on the secondary side halves the current stress of the magnetic components compared to conventional center-tapped approach, significantly lowering the conduction losses, which is most critical for low-voltage, high-current applications. It also reduces the conduction loss of the synchronous rectifiers when the converter operates in the PWM control mode.

A disadvantage of the current-doubler rectifier is that it requires two inductors. The size of the magnetic components including the transformer and the two inductors, and the losses due to interconnections among them, would significantly impact the converter size and efficiency if discrete magnetic components were to be used. A solution to this problem is to use integrated magnetics that provide the function of both inductors and the transformer using a single magnetic structure. An integrated magnetic structure that serves this function is depicted in Fig. 7 [6]. A brief summary of the loss analysis and some preliminary design results are presented later in this section.

Control of Half-Bridge Converter

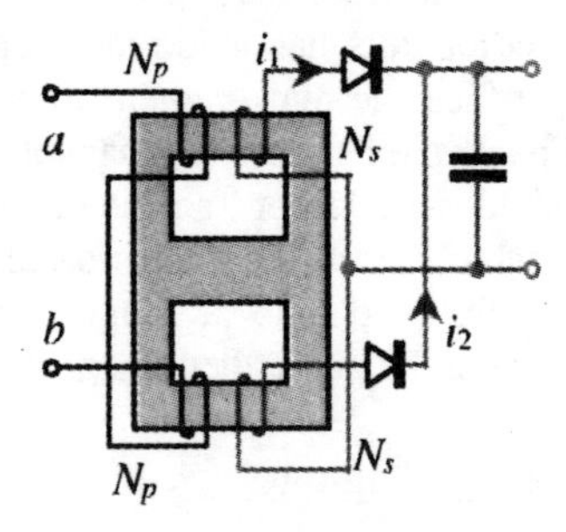

Figure 7. Integrated magnetics that provides the function of the transformer and the two inductors for use in the selected topology.

The half-bridge converter shown in Fig. 6 can be controlled in two different modes: PWM and asymmetric control [7]. Fig. 8 shows the gate signals for the primary switched under each control mode. The PWM control provides an off-time period in which both synchronous rectifiers are on, resulting in reduced rms current for synchronous rectifiers. Therefore, the conduction loss of synchronous rectifiers is lower in the

PWM control. On the other hand, asymmetric control allows the primary switches to be turned on under zero drain-to-source voltages, eliminating the turn-on losses.

Other factors that are considered when comparing the two control methods include the efficiency of the transformer as well as the transient characteristics. Asymmetric control requires higher turns ratio of the transformer, which could lead to reduced efficiency and more complicated winding and interconnect designs.

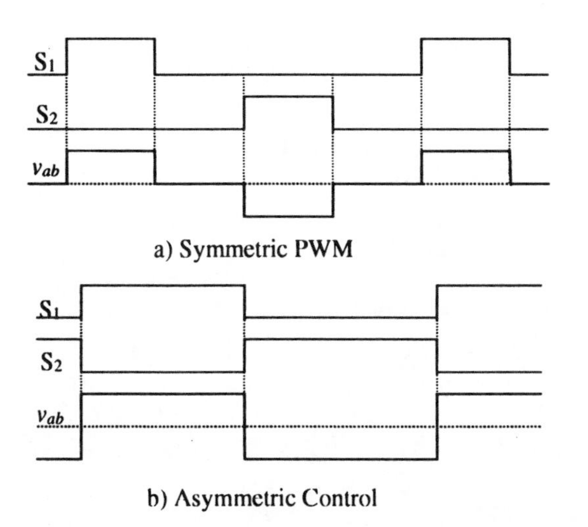

Figure 8. Primary switch gate signals for half-bridge converter under a) PWM and b) asymmetric control. v_{ab} is the resulting voltage across the transformer primary winding.

Averaged models of the converter under both PWM and asymmetric control have been derived to predict the transient responses. It was found that the two inductors and the magnetizing inductance of the transformer in a current-doubler arrangement form an inductor loop that may have a circulating current, which is not detectable from outside the loop. The circulating current would offset the transformer operating point and increase the core loss, especially during large-signal transient. The total inductance of the loop together with the parasitic resistance forms a low-frequency pole that affects the transient response of inductor currents. This pole has to be taken into account when current-mode control is to be used in either PWM or asymmetric control.

Small-signal analysis of the averaged models under PWM control also indicates that the overall model of the half-bridge converter is a second-order system when neglecting the imbalance in the transformer, while asymmetric control changes the converter to a fifth-order system. Therefore, loop design under asymmetric control is more complicated than with PWM control. This is an extremely important characteristic especially when wide control bandwidth is required.

As is clear from the above discussion, PWM and asymmetric controls have both advantages and disadvantages. We plan to design and develop prototypes with both control methods and evaluate their merits and shortcomings based on actual measurements.

Magnetics Analysis and Design

The integrated magnetics design shown in fig. 7 was simulated using finite element analysis method with trapezoidal voltage input. A turns ratio of 2:1 has been used with the voltage excitation at the primary winding from a half-bridge circuit with a duty ratio of 0.6 and at an operation frequency of 500 KHz. The magnitude of the input voltage to the primary winding during the conduction period is thus 25 V. Figure 9 shows the simulated current waveform in the three windings.

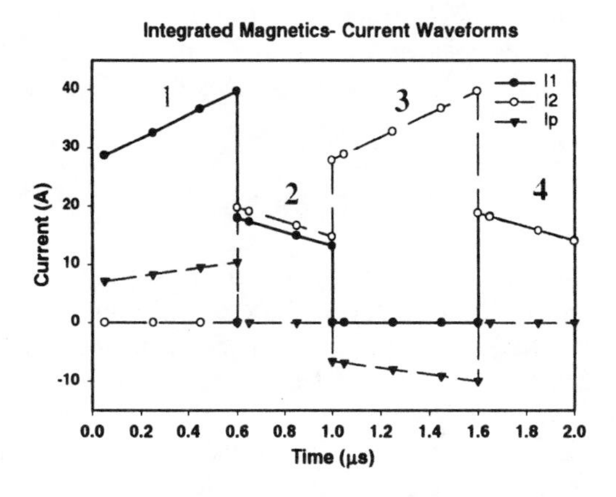

Figure 9. Simulation showing the input current (Ip) and currents through the secondary windings (I1 and I2) as identified in figure 7. The magnetics structure is a transformer in phases 1 and 3 and an inductor in phases 2 and 4.

An EI22 planar core with a center gap of 0.2 mm has been used in the calculation. The flux density distribution is shown in figure 10 during phase 1, where the windings on the left side of the core perform the transformer action.

The magnetic structure design shown in figure 10 is dominated by ac losses in the windings (copper loss). The center air-gap is required to avoid core saturation due to the large dc component of the flux in the center leg, although the flux ripple is smaller compared to the outer legs. During the inductive phases (phases 2 and 4), the primary windings are in the field of the secondary, but it is estimated that this loss contribution is relatively small, with a major reduction in overall interconnect losses. The total loss has been reduced to about 2.5 W, with optimization of winding geometry, interleaving, copper thickness optimization and winding keep-away regions due to the fringing field created by the center air-gap.

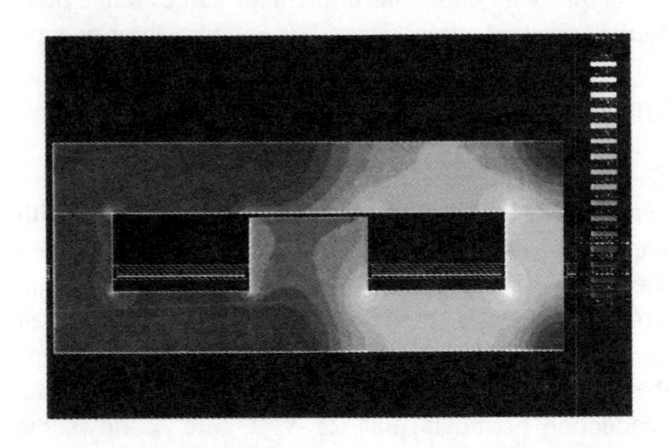

Figure 10. Flux density in the planar EI22 core with interleaved S-P-S structure. The two secondary windings are connected in parallel with two turns of primary interleaved between the secondary windings.

CONCLUSIONS

The challenges in achieving a high density and high efficiency dc/dc converters at 3.3V and 1.5V outputs at 100 W were outlined. An integrated magnetics structure combined with a current doubler rectifier is shown to be an attractive approach for minimizing losses and reducing the size of the converters. In addition, the control strategy for synchronous rectification is shown to have a major impact on the efficiency. Finally, a GaAs power switch is necessary for achieving 95% efficiency at 1.5V output.

REFERENCES

[1] Y. Taur, "The incredible shrinking transistors," in *IEEE Spectrum* July 1999, pp. 25-29.

[2] R. Huljak et al., "Where are power supplies headed," in *Proceedings of 2000 IEEE Applied Power Electronics Conference*, pp. 10-17.

[3] V. Niemela and W. Bowman, "Comparison of GaAs and silicon synchronous rectifiers in a 3.3V out, 50W dc-dc converter," in *Records of 1996 IEEE Power Electronics Specialists Conference*, pp. 861-867.

[4] "Microelectronics power supply," *Final report*, submitted by TI to NCCOSC/NRAD, 1997.

[5] C. Peng and O. Seiersen, "A new efficient high frequency rectifier circuit," *in Proceedings of 1991 High Frequency Power Conversion Conference*, pp. 236-243.

[6] P. Xu et al., "A novel integrated current doubler rectifier," in *Proceedings of 1999 VPEC Annual Seminar*, pp. 159-164.

[7] T. Ninomiya et al., "Static and dynamic analysis of zero-voltage-switched half-bridge converter with PWM control," in *Records of 1991 IEEE Power Electronics Specialists Conference*, pp. 230-237.

AIAA-2000-2832

DUAL USE APPLICATION OF CONTINUOS TRANSFER STUBBS (CTS) TECHNOLOGY

Todd Kastle and Jacqueline Toussaint

Electronic Engineers

Air Force Research Laboratory

Sensor Directorate

Wright-Patterson AFB

ABSTRACT

Military and civil radar and communication systems utilize phased array antennas that require hundreds or in some cases, thousands of transmit/receive (T/R) modules. T/R modules, even when low cost, drive the total RF system cost to an undesirable level. This results in fewer systems being purchased. This paper will review the application of continuous transverse stub (CTS) antenna technology to reduce the number of T/R modules required and thus the overall RF system cost. The paper will review military and civil applications and their respective cost savings.

INTRODUCTION

In surveying today's available "off-the-shelf" antenna technology, electronic scanning at X-band or other frequencies is a challenging proposition, particularly within the cost constraints typical of both the commercial and military marketplaces.

The realization of efficient, low-cost, planar array apertures at X-band frequencies is a difficult task. Nevertheless, the potential system benefits and capabilities associated with the agility and versatility of electronic (compared with mechanical) scanning realized in a low-profile planar aperture is significant in both military and commercial markets. Therefore, there is a need to develop an affordable electronically scanned antenna (ESA) technology that can support both military and commercial products.

The proposed military and commercial applications have many common X-band ESA requirements and resulting benefits. Desirable attributes include high efficiency (low dissipative losses/high power handling); scalability to large (high gain) aperture sizes; lightweight, thin (planar) mechanical profile; high accuracy/precision beam pointing; wide field-of-view (scan volume); reconfigurable shaped-beam capability; fast beam repositioning rates; wide instantaneous and tunable bandwidths; good sidelobe control; and low life cycle cost. CTS ESA technology offers innovative performance, packaging, cost reduction, and reliability solutions for antennas in advanced military radar systems, commercial communications systems and automotive radar systems. The development of an ESA at X-band frequencies is viable with the CTS architecture.

The CTS ESA architecture provides several advantages when compared with conventional ESA implementations. These include

lightweight and simpler packaging, wider instantaneous operating bandwidth, and more affordable recurring cost.

The CTS array is a new microwave coupling/radiating element that has been patented by Raytheon Systems Company.

The CTS element, illustrated in Figure 1, is used as the radiating element for a micro electromechanical system (MEMS) CTS ESA antenna application.

Incident parallel plate waveguide modes, launched via a primary line feed of arbitrary configuration, have associated with them longitudinal electric current components interrupted by the presence of a continuous stub, thereby exciting a longitudinal, z-directed

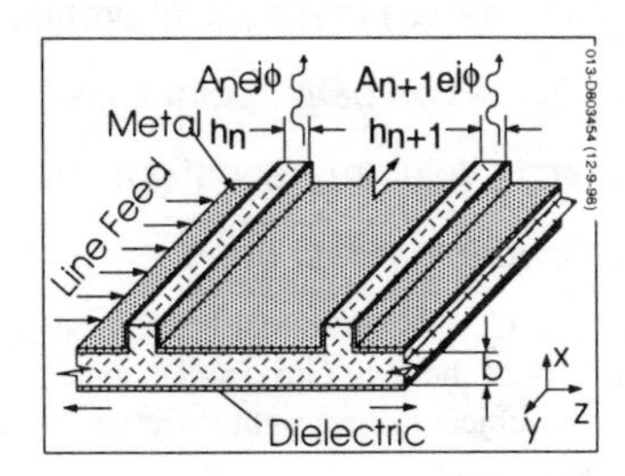

Figure 1. CTS Concept

displacement current across the stub/parallel plate interface. This induced displacement current in turn excites equivalent EM waves traveling through the stub in the x-direction to its terminus and radiates into free space.

One of the primary advantages of the CTS antenna is its simple design. The antenna consists of a dielectric (a plastic such as rexolite or polypropylene) that is machined or extruded to the shape shown in Figure 1. This dielectric is then metal plated to form the final CTS antenna. Thus the CTS lends itself to high volume plastic extrusion and metal plating processes (common in automotive manufacturing operations) thereby enabling low-cost production. These nonscanning antennas are now being manufactured in the range of <$10 to $200 depending on the application, which demonstrates the manufacturability of CTS for our intended applications.

To date, several CTS nonscanning antennas have been built including a Ku-band antenna for reception of a Direct Broadcast Satellite signal, a 38 GHz point-to-point (PTP) cellular communications antenna, a 77 GHz collision avoidance radar antenna for automotive applications, and most recently a 94 GHz CTS antenna and nonscanning line feed. In addition, electronically scanned CTS antennas have been designed and built within the 6–18 GHz and Ka-bands.

W-Band CTS

Some of the new millimeter-wave radar systems have demanding packaging requirements, which require planar array antenna architectures. The most popular planar antenna architecture being proposed for advanced radar applications is the fully populated two-dimensional (2-D) array. Active planar arrays consists of discrete radiating elements each backed by their own phase shifter or transmit/receive module and are interconnected with large and often multiple, multi-port feeding networks. At W-band, a fully populated array would require over 4000 elements. While some feasibility has been demonstrated [1] for a linear array implementation, the module cost and mechanical complexity make such architecture at W-band (millimeter-wave) frequencies prohibitive. In an attempt to reduce cost and increase manufacturability, antenna research efforts at the Raytheon Systems Company have focused on CTS based alternatives to the fully populated planar architecture. Described below is one application of CTS planar antenna technology to an arrayed W-band antenna. The resultant antenna array aperture size is 4 in. x 4 in. The antenna is microscanned via ferrite phase shifters behind the four quadrants of the array. A dramatic reduction in parts count and fabrication complexity has been demonstrated with this W-band planar array. The next section summarizes some of the array design features and array performance. "thinned array" architecture was chosen among other alternatives to meet the goal of a planar array with some beam microscanning. In this approach, over 4000 active elements are replaced by four 2 in. × 2 in. square planar subarrays. The "thinned array" consists of four subarrays. Each subarray consists of a traveling-wave fed rexolite CTS array with 33 radiating stub elements. Metalization is sputtered silver. A 31 element subarray was first built to demonstrate feasibility. The gain for the 2 in. × 2 in. subarray was 31.5 dB at the design frequency of 94 GHz. This design success was followed by the design of the 33-element, single panel configuration with a slightly different geometry to allow the seamless arraying of the subarrays to avoid error sidelobes. A four-way power divider with WR-10 waveguide input and outputs was designed and built. An aluminum brazing process was developed with significant accuracy to implement the power divider design. Next, four, latching, ferrite phase shifters, were attached between the output ports of a four-way power divider and the input ports of each subarray. The phase shifters provide five- (5) bit resolution and are controlled via a controller card connected to a PC computer. Final phasing and some beam steering are achieved under computer command. The subarrays are supported by an aluminum mounting plate that allows for arranging the array in a two element high by two-element long configuration; shown in Figure 2.

Figure 2. Four-Panel W-Band Planar Array

PREDICTED AND MEASURED RESULTS

All performance objectives were achieved. The measured results match very closely with the predicted results. Performance highlights include measured gain of 34.5 dB versus predicted gain of 34.6 dB; a difference pattern null depth over frequency of –30 dB; E- and H-plane beam widths of 1.84 and 1.87 degrees, respectively. The array, excluding phase shifter loss, has an efficiency of 55 percent. It should be noted that the important performance characteristics of the antenna are maintained across the frequency range 93.5 GHz to 94.5 GHz.

CONCLUSIONS

CTS ESAs has been successfully designed and demonstrated at a variety of frequencies. The reduced number of T/R modules in the CTS ESA translates to a reduction in array, harnessing and Beam Steering Controller (BSC) complexity. This reduction in complexity results in significant cost savings when compared to conventional ESAs. Other significant advantages of the CTS ESA include low production cost, high design flexibility and fast cycle times from need to prototype.

REFERENCES

1. Edward, B.J., Helms, D.R., Webb, R.S., and Weinreb, S., "W-band Active Transmit and Receive Phased Array Antennas", 1995 IEEE MTT-S International Microwave Symposium Digest, pp. 1095–1098.

2. Bradley, T.W., Chen, W.W., Kastle, T.A., et al, "Development Of A Voltage Variable Dielectric (VVD), Electronic Scan Antenna", IEE Radar 97 Proceedings.

3. Lemons, A., Lewis, R., Milroy, W.W.,Robertson, R.S., Coppedge, S.B. and Kastle, T.A., "W-Band CTS Planar Array", 1999 IEEE MTT-S International Microwave Symposium Digest.

AIAA-2000-2833

EARTH OBSERVING SYSTEM (EOS) TERRA SPACECRAFT 120 VOLT POWER SUBSYSTEM[©]

Requirements, Development, and Implementation

Denney J. Keys, P.E.

Power Systems Branch, Electrical Systems Center

National Aeronautics and Space Administration (NASA)

Goddard Space Flight Center (GSFC)

Greenbelt, MD 20771

Phone: (301) 286-6202 Fax: (301) 286-1743

Email: denney.j.keys.1@gsfc.nasa.gov

ABSTRACT

Built by the Lockheed-Martin Corporation, the Earth Observing System (EOS) TERRA spacecraft represents the first orbiting application of a 120 Vdc high voltage spacecraft electrical power system implemented by the National Aeronautics and Space Administration (NASA) Goddard Space Flight Center (GSFC). The EOS TERRA spacecraft's launch provided a major contribution to the NASA Mission to Planet Earth program while incorporating many state of the art electrical power system technologies to achieve its mission goals. The EOS TERRA spacecraft was designed around five state-of-the-art scientific instrument packages designed to monitor key parameters associated with the earth's climate.

The development focus of the TERRA electrical power system (EPS) resulted from a need for high power distribution to the EOS TERRA spacecraft subsystems and instruments and minimizing mass and parasitic losses. Also important as a design goal of the EPS was maintaining tight regulation on voltage and achieving low conducted bus noise characteristics.

This paper outlines the major requirements for the EPS as well as the resulting hardware implementation approach adopted to meet the demands of the EOS TERRA low earth orbit mission. The selected orbit, based on scientific needs, to achieve the EOS TERRA mission goals is a sun-synchronous circular 98.2-degree inclination Low Earth Orbit (LEO) with a near circular average altitude of 705 kilometers. The nominal spacecraft orbit is approximately 99 minutes with an average eclipse period of about 34 minutes. The scientific goal of the selected orbit is to maintain a repeated 10:30 a.m. +/- 15 minute descending equatorial crossing which provides a fairly clear view of the earth's surface and relatively low cloud interference for the instrument observation measurements.

The major EOS TERRA EPS design requirements are single fault tolerant, average orbit power delivery of 2,530 watts with a defined minimum lifetime of five years (EOL). To meet these mission requirements, while minimizing mass and parasitic power losses, the EOS TERRA project relies on 36, 096 high efficiency Gallium Arsenide (GaAs) on Germanium solar cells adhered to a deployable flexible solar array designed to provide over 5,000 watts of power at EOL. To meet the eclipse power demands of the spacecraft, EOS TERRA selected an application of two 54-cell series connected Individual Pressure Vessel (IPV) Nickel-Hydrogen (NiH2) 50 Ampere-Hour batteries. All of the spacecraft observatory electrical power is controlled via the TERRA Power Distribution Unit (PDU) which is designed to provide main bus regulation of 120 Vdc +/-4% at all load interfaces through the implementation of majority voter control of both the spacecraft's solar array sequential shunt unit (SSU) and the two battery bi-directional charge and discharge regulators.

This paper will review the major electrical power system requirement drivers for the EOS TERRA mission as well as some of the challenges encountered during the development, testing, and implementation of the power system. In addition, spacecraft test and early on orbit performance results will also be covered.

SYSTEM REQUIREMENTS OVERVIEW

The EOS TERRA spacecraft utilizes a multi-tiered requirement documentation approach. Top level requirements are contained in four basic categories that were used to define a Contract End-Item (CEI) Specification. The documents are; (1) Performance Assurance Requirements, (2) Requirements Document, (3) Unique Instrument Interface Documents, and (4) General Instrument Interface Specification.

The CEI Specification was used to derive the Interface Control Requirements, Verification Requirements, Spacecraft Control Plans, and the General Interface Specification (GIS). All of these requirements were, in turn, flowed down to Major Assembly Specifications, Subsystem Specifications, Flight Software System Specifications, and Instrument Interface Control Documents (ICDs). The focus of this paper will address requirement overview of the EPS specifically with regard to the GIS requirements and the resulting spacecraft implementation architecture.

EPS DRIVING REQUIREMENTS

Major GIS Requirements

The EOS TERRA major power subsystem requirements are defined in a top level General Interface Specification (GIS) and subsequently flow down into an Electrical Power Subsystem (EPS) Specification. In addition to electrical requirements, the GIS also defines the mechanical, thermal, and environmental requirements that are common to most of the spacecraft bus elements. There are two basic power interfaces defined in the GIS, one for the "primary" power users and the other being for "secondary" power users. The major difference between the two is the voltage provided at the user load interface. The primary user voltage is defined as a 120 Volt direct current (Vdc) interface and the secondary user voltage is a 28 Vdc interface. The primary interface requirement for bus regulation is 120 Vdc with a tolerance of +/-4% at the load interface. As a result of this tight voltage regulation, the source impedance requirement is specified as a series combination of resistance less than or equal to 0.8 ohms and an inductance of 2.5 microhenries (with a resulting corner frequency of 51 kHz) over the frequency range of dc to 150 kHz. Under all abnormal conditions (e.g. fault conditions) the power system is required to maintain primary EPS voltages between 0 and 132 Vdc. The primary power system is also required to provide fusing to protect the EPS and spacecraft harnesses from all potential load or harness faults.

The secondary power voltage regulation requirement is 28 Vdc +/-2% at the load interface. The secondary power system has two specifications for impedance requirements depending on the length of the harness to the load interface. One impedance requirement is less than 0.4 ohms at dc, increasing at a 750 microhenry slope up to 300 Hz, then 1.5 ohms from 300 Hz to 100 kHz, then increasing at 0.7 microhenry slope from 100 kHz to 1 MHz, for harness lengths of 10 feet or less. For those users with harness lengths greater than 10 feet, the impedance specification is less than 0.9 ohms at dc increasing at a 1000 microhenry slope from 300 Hz to 100 kHz, then increasing at a 2 microhenry slope from 100 kHz to 1 MHz. Again, under abnormal or fault conditions, the secondary voltage requirement is to always be within 0 to 35 Vdc range.

The EPS is required to provide a single point ground reference for the spacecraft primary power system while all secondary power returns are required to be referenced either at the component chassis reference or at the power source chassis reference. The use of the power source chassis reference was normally the preferred means for secondary grounding. The maximum allowable resistance for all ground reference ties is 2.5 milliohms.

Major EPS Performance Specification Requirements

The GIS requirement flow down forms the basis for development of the major EPS performance requirements. In turn, the EPS performance specification requirements are the basis for development of individual component performance specifications, once the architecture is defined. For this paper, only the major subsystem performance specification requirements are presented.

The overall architecture for the EPS was derived initially on the top-level requirements and was refined through multiple trade studies in order to meet the necessary tolerances imposed. Given the nature of the spacecraft orbit and the power distribution needed, a Direct Energy Transfer (DET) system was the obvious choice. The details of the actual component implementation will be covered later in this paper.

The maximum average spacecraft load power requirement for the EOS TERRA spacecraft was defined at 2,530 Watts, excluding energy storage requirements, at a defined end of life (EOL) scenario of 5 years. The power system is also required to provide up to 3000 watts in peak load power for up to 15 minutes per orbit, all without violating the 2530-Watt orbit average power requirement.

In the event of a non-catastrophic loss of all spacecraft power on orbit, the EPS is also required to be capable of autonomously restarting whenever the bus voltage rises above 50 Vdc.

To meet the maximum average load distribution power requirement as well as the battery charge requirement after five years on orbit, the solar array power requirement was derived to be 5 kW at year 5, EOL worst day conditions. The energy storage requirements for the eclipse portion of the TERRA orbit were derived based on the defined maximum average orbit load power requirements and ultimately resulted in the selection of two 50 Ampere-hour Nickel-Hydrogen batteries to provide the necessary energy capacity.

Recognizing the requirement for 120 Vdc +/-4% at the load interface, the EPS derived the requirement for the main bus regulation to be defined as 120 Vdc +4/-2%, in order to allow for worst case voltage drop conditions in spacecraft harnesses. The voltage ripple requirement imposed on the EPS was 280 millivolts peak-to-peak (100 millivolts rms) in order to assure low noise characteristics for the instrument reference planes. The peak-to-peak ripple requirement represents a value of only 0.23% of the nominal primary bus voltage. Under transient load conditions, the EPS is required to maintain regulation within a maximum of 3 volts overshoot or undershoot and is required to recover to within 10% of it's final value within 10 milliseconds for a primary load step of 10 amperes (1200 Watts).

To meet the requirement for autonomous start whenever the bus voltage rises above 50 Vdc, the EPS derived requirements that it must operate within specification down to 50 Vdc. Also, as a result of the GIS imposed impedance requirement, the EPS derived a requirement for source impedance, as measured across the main bus capacitor, of less than 200 milliohms for frequencies between 0.1 Hz to 100 kHz. This impedance requirement was derived based on subtracting expected maximum impedance characteristics of worst case harness lengths to load interfaces with appropriate margin included.

In addition to the derived performance requirements, conducted and radiated noise characteristics were also imposed on the EPS. Although they will not be covered in this paper, the details of the imposed requirements can be found in the EOS TERRA Electromagnetic Compatibility (EMC) Control Plan.*

Using normal spacecraft program development mass allocation techniques, the EPS was given a mass allocation based, ultimately, on the total lift capability of the baseline launch vehicle. After the Atlas IIAS was selected as the launch vehicle, the total allocated mass for the EOS TERRA EPS was defined at 1315 pounds, or about 15% of the vehicle lift capability. Of primary importance in the design phase was the requirement for the EPS to be single fault tolerant to any failure within the subsystem. The implications of this requirement resulted in all EPS components having at least internal redundancy to any single failure mechanism. Some aspects of the redundancy implementation will also be covered in discussions concerning the components later in this paper.

The last major top level requirements are the maximum orbital eclipse period and expected beta angles. The eclipse is defined as 34.84 minutes, and is consistent with the worst (longest) case for the selected spacecraft orbit altitude and inclination. The beta angle variation is defined as being between 13.5 and 30.8 degrees, while the EPS has a derived requirement to also operate at a beta angle of 0 degrees when in a sun pointing safe hold mode.

These requirements represent the major top level imposed requirements from which the TERRA EPS design was generated with the hardware implementation being covered next.

HARDWARE IMPLEMENTATION

Architecture Selection

Given the LEO orbit selected for the EOS TERRA spacecraft and the relatively high power requirements necessary to support the instrument packages, the project initially performed a trade study to evaluate high voltage power distribution versus the more traditional (heritage) 28 Vdc bus architecture. Of particular importance for TERRA was the mass trade for spacecraft harnesses and the resulting parasitic I^2R losses. The high power distribution around a large spacecraft structure (TERRA spacecraft dimensions are diameter of 3.5 m, length of 6.8 m, and weight of 5190 kg), especially when considering the single fault tolerance requirement, becomes very unwieldy at 28 Vdc. For just the distributed load power requirements, the 28 Vdc architecture would be required to distribute approximately 90 amperes of current and when considering the expected maximum

* Electromagnetic Compatibility (EMC) Control Plan for the EOS-AM Spacecraft (SEP-106), PN20005869B, July 7, 1994, Lockheed-Martin Corporation.

current flow from the solar array, the total current flow becomes almost 180 amperes.

Selection of higher distribution voltages result in proportionately lower overall electrical current flow and, in this case, at 120 Vdc the current flow becomes a much more manageable 42 amperes from the solar array. Also, given the initial association of TERRA with the International Space Station program, where 120 Vdc was ultimately selected as the baseline for power distribution, the options considered for TERRA were ultimately limited to 28 Vdc and 120 Vdc in hopes of utilizing space qualified parts that were either already available or in development for 120 Vdc. One additional benefit perceived by NASA/GSFC in this trade study evaluation is the development of a second "standard" voltage (in addition to 28 Vdc) for any large future high-powered spacecraft power systems developed at or for NASA/GSFC.

NASA GSFC engineers worked closely with Lockheed-Martin Corporation and Virginia Polytechnic Institute (specifically the Virginia Power Electronics Center or VPEC) to develop a viable spacecraft EPS architecture utilizing 120 Vdc. A significant portion of the development took place while the EOS TERRA project was associated with the International Space Station program. The results of the efforts at VPEC culminated in a 120 Vdc test bed development and verification.[†] Based on these successful results and the need to maintain highly regulated power distribution characteristics, minimize I^2R losses, minimize harness mass, and lower overall launch cost considerations, the base line selection of 120 Vdc as the preferred power system distribution architecture for TERRA was made.

Major Subsystem Trade Studies

During the preliminary design review process of the EOS TERRA power subsystem, there were many trades studies that were undertaken to optimize cost, schedule, and mass. The primary trade study, as discussed earlier, focused on 120 Vdc versus 28 Vdc distribution of power with 120 Vdc ultimately being selected on the basis that it would save a large amount in both mass and distribution losses. The estimated mass savings for the EOS TERRA spacecraft using 120 Vdc versus 28 Vdc was determined to be approximately 900 pounds.[‡]

Another early trade study involved whether to use fuses for circuit protection or resettable type circuit breakers. Again, the trade focused mainly on mass and ultimately was decided in favor of the reliable and lighter solid state fuses made by Mepcopal, Inc.

From a power generation perspective, the solar array was required to provide 5+ kW power to the spacecraft at EOL and obviously could be derived using flexible or rigid panels with either Silicon or Gallium Arsenide (GaAs) on Germanium solar cells. From a launch vehicle perspective, the ability to package the array within the launch vehicle fairing dynamic envelope leaned heavily in favor of a flexible array that could be stowed in a relatively small volume. TRW, Inc. was nearing completion of a Jet Propulsion Laboratories (JPL) funded research program to develop a lightweight flexible array concept named Advanced Photovoltaic Solar Array Program (APSA) that held promise for increasing the power to mass ratio for larger arrays. The results of the development effort are contained in the APSA Final Technical Report[§] and ultimately also heavily influenced a decision in favor of selecting the TRW APSA based design for TERRA using higher efficiency GaAs solar cells.

As with all power system selection processes, the fundamental options for the power system architecture focused on Peak Power Tracker versus Direct Energy Transfer systems. Given the high power requirements and the need for tight regulation on the power system bus, the DET architecture was the clear favorite in the trade, as can be expected for the given requirements.

Energy storage requirements were also driven by a desire to maximize energy storage capability while minimizing mass. To this end, the primary options available for TERRA were Nickel Cadmium (NiCd) versus the more recently developed Nickel Hydrogen (NiH2) battery cells. Even though the cost was somewhat higher for the NiH2 cells, the overall ability to maximize the energy storage density for a given weight led the decision process to select NiH2 cells manufactured by Eagle-Picher Technologies.

Since the desire was to tightly regulate the bus, the batteries were going to be isolated from the bus via some form of charge/discharge regulation and therefore it was not imperative that the number of cells closely mimic the subsystem operating voltage. Based on a variety of factors during the evaluation process, as well as the availability of NiH2 IPV ampere-hour ratings at the time of the study, two batteries, each consisting of 54 50-Ampere-hour Individual Pressure Vessel (IPV) cells connected in series were selected for TERRA. The selection of this combination

[†] EOS Satellite Power System Testbed, Dr. Dan Sable, VPEC, Virginia Polytechnic Institute, Blacksburg, VA presented at the High Voltage Spacecraft Power Technology Workshop, May 4-5, 1993.

[‡] 120 Vdc and 28 Vdc Power Distribution Trade Off White Paper, R. Stone and B. Beaman, NASA GSFC Internal Memorandum, February 12, 1992,

[§] Advanced Photovoltaic Solar Array Program, Final Technical Report (CDRL 012), November 15, 1994, TRW Space & Technology Division, TRW Report No. 51760-6006-UT-00

resulted in a nominal on-orbit useable battery voltage range of approximately 60 to 90 Vdc.

Major EPS Components

With the experience of testing performed at VPEC, the block diagram architecture took shape in the form shown in Figure 1. The basic major

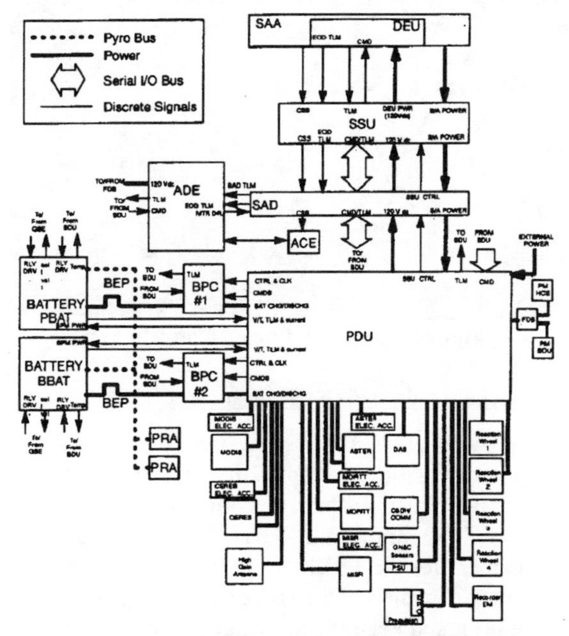

Figure 1 EOS TERRA EPS Block Diagram

components comprising the EPS were; Sequential Shunt Unit (SSU), Power Distribution Unit (PDU), Battery Power Conditioners (BPC), Two Nickel-Hydrogen Batteries, and One Solar Array utilizing GaAs solar cells. For primary to secondary power conversion, Electronic Power Conditioners (EPC) were proposed and adopted which allowed conversion to 28 Vdc at, or very near, the load interface.

Deployable Flexible GaAs Solar Array

All of the power needed to operate the EOS TERRA spacecraft is generated by the solar array. The TERRA solar array is a flexible blanket design utilizing GaAs on Germanium solar cells attached to a Germanium coated Kapton substrate with graphite fiber reinforced plastic (GFRP) reinforcement. The entire substrate was vapor deposited with a 1000-Angstrom thick Germanium coating to provide ESD charge dissipation capability.

The solar array contains 36,096 solar cells, each cell measuring 2.4 cm x 4 cm x .014 cm. The solar cells are inter-connected together to form a total of 192 solar cell strings, each consisting of 188 series connected cells. Eight strings are wired in parallel to form 24 circuits. All 24 circuits are brought to the SSU via a slip ring assembly within the solar array drive assembly. The beginning-of-life (BOL) design characteristics attempted to achieve a total power generation of approximately 8 kW of power at standard temperature conditions (STC). Completed array flash test results at TRW verified this design capability.

A picture of the fully deployed TERRA solar array is shown in Figure 2. Full solar array deployments were only performed at the manufacturer's facility (TRW) on a specially made deployment fixture.

Figure 2. EOS TERRA Solar Array

Power Distribution Unit (PDU)

The PDU is the main bus and contains the electronics used to maintain the 120 Vdc bus regulation. The PDU regulates the bus voltage via current control of both the SSU and the BPCs, depending on whether the spacecraft is in sunlight or orbital eclipse. Control is achieved using a majority voter control logic with 3 voter circuits. The PDU also houses all of the main bus switches for non-essential loads (such as instruments and High Gain Antenna) and can be used in the event load shedding is necessary or if there is a desire to switch from A to B-side operation.

A picture of the flight PDU while in test is shown in Figure 3.

Figure 3. Power Distribution Unit

Sequential Shunt Unit (SSU)

The SSU actively controls the amount of current supplied to the TERRA PDU from the 24 available solar array circuits. The current provided by each circuit is controlled via pulse width modulation (PWM) and can be up to approximately 2.5 amperes of current at about 130 Volts for each circuit. All unneeded bus current generated by the solar array is shunted back to the array via the SSU. The input feeds from the solar array to the PDU are each protected by a 10-amp fuse that primarily protects the bus from a solar array to PDU harness short.

A picture of one of the 6-shunt modules that make up the SSU is shown in Figure 4. Each SSU shunt module contains 4 shunt circuits.

Figure 4. One of 6 SSU Modules

Battery Power Conditioners (BPC)

The two BPCs provide the means to buck regulate charge current to the spacecraft batteries during the sun lit portion of the orbit as well as providing a boost regulation capability from the batteries to the main bus during the eclipse portion of the orbit. Again, the BPCs utilize a PWM boost regulator design approach which will vary the electrical current provided to the main bus, while in orbit eclipse, proportionally based on bus load demand. The discharge current control signal is provided from the PDU via a MIL STD-1553 data bus.

The BPCs are each designed for four-channel operation, in parallel, with a 4 for 3 redundant architecture. Each channel of the BPC can provide up to 5.5 amperes charge current at battery voltage and provide up to 30 amperes output under fuse clearing conditions (limited ultimately by battery fuse sizing).

The PDU, BPCs and one of the two spacecraft batteries are contained within a Power Equipment Module (PEM). The equipment module approach was implemented throughout the TERRA spacecraft in an effort to minimize the integration complexity.

A picture of one of the flight BPCs as installed in the Power Equipment Module is shown in Figure 5.

Figure 5. Battery Power Conditioner

Nickel-Hydrogen (NiH2) Batteries

The two NiH2 batteries were sized to provide all of the energy storage capability for the TERRA spacecraft. Each battery is comprised of 54 series connected 50 amp-hour NiH2 battery cells. The requirement for the TERRA mission was to meet all of the power demands of the spacecraft during the eclipse of the orbit and keep the maximum normal orbit depth of discharge (DOD) of the batteries to 30%, or less, of nameplate capacity. The DOD requirement assumes maximum spacecraft design load conditions and the loss of one battery cell in each battery.

The batteries were designed to be capable of monitoring individual cell voltages as well as individual cell temperatures through a Heater Control Electronics (HCE) component. These parameters were considered crucial during the design phase for such large series connected battery cells. The active heater control in the HCE utilized heater groups, each composed of 9 battery cells. The heater group temperatures are averaged in the spacecraft control computer and that value is used to provide the HCE with the temperature control signal. The HCE design temperature range for each battery is between –1 and –5 degrees C. The active thermal control pulse width modulated heaters will operate at 0% duty cycle at –1 C and at 100% duty cycle at –5 C.

An example of one of the two spacecraft flight batteries is shown in Figure 6 without its battery cover.

Figure 6. One TERRA 54 Cell NiH2 Battery

Electronic Power Conditioners (EPC)

The EPCs were developed in conjunction with the 120 Vdc bus architecture as a means to provide various low voltage regulated power to all spacecraft subsystems that required them. The EPCs were originally envisioned to be a common component capable of various load demands with standard secondary voltage levels (+/-5, +/-10, +/-15, +18, +28 Vdc). After taking stock of the wide power requirements of the various subsystem needs, the single common EPC power supply design approach was eliminated to more appropriately tailor the converters to their specific load requirements. This modified approach resulted in ten basic similar designs for EPCs with a total numerical count of 42 EPC converters needed for the various spacecraft subsystems.

From an EPS perspective, only one of the spacecraft's subsystem EPCs is considered as an integral part of the EPS and that is the Propulsion System EPCs. The Propulsion System EPCs were solely 120/28 V converters and were used to adapt heritage based propulsion components into the EOS TERRA spacecraft design.

A picture of a typical EPC in an open configuration is shown in Figure 7.

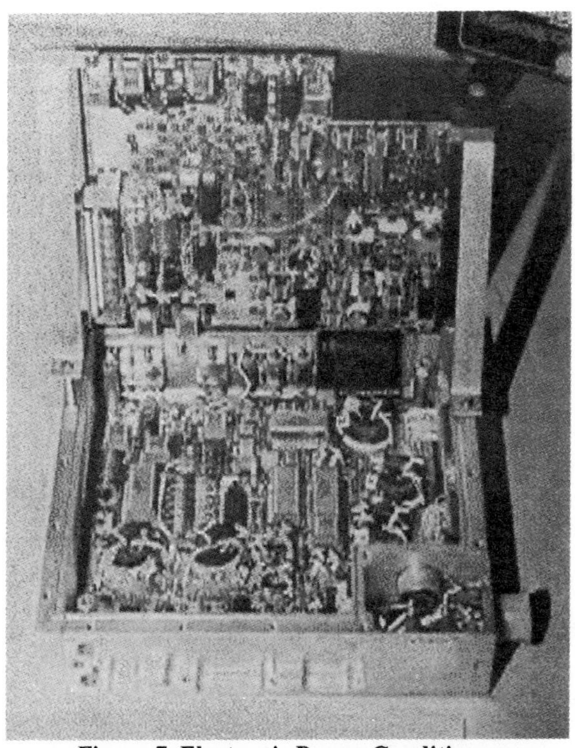

Figure 7. Electronic Power Conditioner

TEST PERFORMANCE RESULTS

The EOS TERRA spacecraft 120 Vdc power system operation concept was initially verified through the development and test effort at VPEC, and once those breadboard versions of the power system components were considered mature enough, a development effort using engineering models was undertaken by Lockheed-Martin Corporation. After comprehensive component level testing was completed the engineering models were placed in a test bed and operated over the required limits imposed on the EPS. Some minor differences were expected between the test bed and the actual flight system and were needed in order to conveniently, and cost effectively, perform the system level testing. The deviations from planned flight configuration were primarily the use of lead acid batteries in place of the NiH2 batteries and the use of only one half of the control electronics in the PDU (equivalent to using only one side of the redundant control electronics).

A solar array simulator was manufactured by ELGAR Corporation that accurately simulated up to 28 solar array circuits and, utilizing sophisticated software programming capability, precisely matched the actual GaAs cell and string characteristics. A photo of the EPS engineering test model (ETM) test bed configuration is shown in Figure 8. The operating performance of the ETM test bed confirmed well-behaved stable performance as well as low conducted noise characteristics.[¶]

Figure 8. EOS TERRA ETM Test Bed

Problems Encountered During Development/Fabrication

During the course of developing the flight EPS hardware, the project was faced with many challenges, from a technical perspective, associated with a new power system architecture, new solar array and battery technologies, as well as development and test schedules. Early in the acceptance testing of some of the flight EPCs, the TERRA project found evidence of via failures on EPC printed wiring boards (open circuit conditions as a result of trace to barrel fractures), which were ultimately traced to the board manufacturing processes. In all cases, the flight printed wire board coupons were evaluated for acceptance and all of the identified failed boards initially appeared to have acceptable coupons. A second round of coupon analysis was performed at which time several board lots were found to be suspect. Unfortunately, at this time at least two of the suspect flight boards were already fabricated and tested and were scheduled to be integrated with their next higher assemblies. To mitigate this potential problem, and minimize a potentially large schedule impact needed in order to re-fabricate the boards, a series of thermal cycle tests were performed on the completed boards that were specifically designed to surface the potential problem. In the case of the two suspect boards, there was no evidence of this problem during the testing and they were ultimately cleared for flight use.

A second challenge became apparent when the stability characteristics of some of the EPCs were found to be marginal at low load conditions, primarily as a result of the predicted design loads on the converters being significantly more than the actual loads. The approach taken to alleviate the performance problems at low load conditions was to implement pre-loads on those EPCs that were being significantly under utilized. While this provided somewhat of a quandary as to effectively lowering the converter efficiency of the EPCs, the trade off was, again, a significant slip in schedule while the designs were modified to accommodate the revised loads. In most cases, the pre-loads were not substantially altering the full load efficiency, although the largest power EPC (Propulsion EPC) ultimately required an externally switchable 28-Watt load resistor bank to meet the necessary performance desired at low load conditions.

A third problem that surfaced during fabrication and testing of the EPCs was a series of failures of Vishay (precision surface mount) resistors. This particular problem was ultimately traced to improper handling of the resistors during the parts kit process and once tighter control over the handling was implemented, the failure rate dropped to near zero.

One of the most challenging problems encountered during the TERRA EPS development was the identification of electrostatic discharge (ESD) induced solar array arcing as a potentially destructive phenomenon to the TERRA solar array design. This particular problem was identified as a result of multiple orbiting spacecraft solar array degradation failures. While the mechanism is not absolutely conclusive, the on-orbit array degradation failures were most likely linked to cover glass ESD surface charging which, in turn, resulted in sporadic discharge arcing that ultimately provided a condition that supported high to low voltage cell-to-cell string short circuit failures.[#] This particular phenomenon was identified as a potentially high risk item after the on-orbit anomalies because the EOS TERRA array was designed similar to those on orbit with the highest voltage solar cell in a string located directly next to the lowest voltage cell in that string. Substantial testing of the EOS solar array design was conducted in a specially designed vacuum chamber at NASA Glenn Research Center. The results of the testing indicated that the EOS TERRA solar array was most definitely susceptible to this form of potentially destructive

---

¶ EOS-AM EPS ETM S/S Test Report, Design Note EOS-DN-EPS-090, Lockheed-Martin Corporation, August 1, 1995,

Katz, Davis, and Snyder, "Mechanism for Spacecraft Charging Initiated Destruction of Solar Arrays in GEO", AIAA, 1998

arcing on orbit. As a result, several modifications** were made to the EOS solar array, most notably the addition of diodes in every solar array string intended to severely limit the available current for supporting the cell-to-cell arcing in the event the arcing phenomenon was indeed present on orbit.

Another particularly disturbing problem that required attention late in the TERRA test schedule was significant capacity fading noted on the workhorse NiH2 batteries. The original project plan was to install of the TERRA flight batteries during spacecraft environmental testing (just prior to spacecraft thermal vacuum testing) and leave the flight batteries installed on the spacecraft until TERRA launched. What appeared at the time to be a result of ambient handling, partial charge and discharge cycles, and potentially improper battery letdown procedures, a significant capacity fading was found to have occurred on the workhorse batteries. Further evaluation of available trend data, prior to shipping the spacecraft to the launch site, indicated this degradation was also beginning to appear on the flight batteries as well. Destructive physical analysis (DPA) performed on sample battery cells were not conclusive as to the cause of capacity fading, but significant blistering was noted on at least one cell. Ultimately, the battery letdown equipment being used was found to promote battery cell reversal while not accurately displaying the reversal effect during letdown. Once discovered, the procedures for battery conditioning and letdown were altered to minimize the possibility of cell reversal in the future and a battery test program was initiated to evaluate the long term effects of the degradation in on-orbit conditions.

During environmental testing, specifically in spacecraft thermal vacuum testing, another problem was identified regarding significant noise on the battery cell voltage telemetry. While never confirmed through disassembly of the flight batteries, the cause of this noise was ultimately believed to be a lack of shielding of the battery cell voltage monitoring wiring within the battery itself and appeared to only be a problem when the heater circuitry (pulse width modulated) was active in PWM mode. The noise was not normally present on telemetry signals of cell voltage in ambient temperature because the thermostat that controls the circuit operation was normally open (thermostat is designed to close when the temperature is approximately 7 degrees C).

In addition to the cell voltage noise, there was also a low level of noise observed on the temperature sense circuits of all of the battery cells. This temperature noise is not believed to be caused by the same phenomenon as the voltage noise but, is instead believed to be a result of noise induced within the heater control electronics itself. The noise associated with the temperature monitoring has been of more concern than the voltage monitoring because the temperature measurements are used by the flight computer in a control manner to ultimately determine the heater duty cycle. If the noise were to get significantly worse, the heater pulse width operation could be adversely affected.

The ultimate resolution of this problem was to fly the hardware "as is" and to monitor the noise levels and plan to turn off the active control if the temperature noise were ever to adversely affect the heater operation. When disabled, the method of maintaining thermal control on the batteries would then be equivalent to the spacecraft safe hold operation, which utilizes thermostatic control of 100% duty cycle heater power whenever the battery temperature groups reach about –10 degrees C. The upper thermostat will open and disable the heater power when the temperature reaches about 0 degrees C.

This particular operating mode was tested in spacecraft thermal vacuum testing and ultimately did provide an adequate backup means to alleviate the problem. Another means that was determined to be an acceptable back up was the reprogramming of the flight computer to operate in a full-on/full-off mode of 100% pulse heater power while disabling the software controlled PWM mode of heater operation. Basically, armed with these alternative means for resolving this potential thermal control problem was rationale enough to proceed with the hardware "as is."

Launch Performance

Launch of EOS TERRA occurred on December 20, 1999 with near flawless performance of all systems. A photo of the complete spacecraft during launch preparations at Vandenberg Air Force Base is shown in Figure 9. Soon after orbit was achieved, the deployment of the flexible solar array was initiated. The onboard software in the flight computer controlled the deployment and all was nominal until one of two micro switches failed a telemetry check to indicate separation of the solar array blanket box.

** Davis, Stillwell, Andiario, and Snyder, "EOS-AM Solar Array Arc Mitigation Design", IECEC, 1999.

Figure 9. EOS TERRA Spacecraft at VAFB

This particular failure mode (along with a long potential list of other failure scenarios) was handled automatically on board using back up deployment procedures. In this particular case, the on-board software was programmed to automatically switch to the back up side of the deployment electronics after a prescribed wait time. After the switchover the remainder of the deployment went as expected.

Although there was an investigation into this deployment failure, it is still unclear as to whether the micro switch itself failed to indicate separation or if the failure was in the deployment electronics telemetry.

On-Orbit Performance

For the roughly 5 months that EOS TERRA has been on orbit, there have been no significant anomalies experienced with the EPS. One EPS related anomaly identified during early TERRA orbit correction maneuvers was a significant plume impingement of the thrusters onto the solar array. When one of the first correction maneuvers was performed, the spacecraft began to roll unexpectedly and subsequently put the spacecraft into a safe hold. Analysis performed after the problem was encountered finally reached the conclusion that the plume field of the thrusters was impacting the solar array and, for that particular maneuver, the solar array happened to be in the ideal position for significant impingement. This analysis was subsequently verified through measured thruster firings with the array in prescribed positions and the resulting momentum changes predicted and verified.

Since there were a number of Delta-V burns yet scheduled to achieve final orbit position, the spacecraft operations team defined a method of stopping the solar array rotation in a position least impacted by the thruster burns with a resumption of array rotation immediately following the correction burns.

From a performance perspective, all EPS subsystem operations are performing as expected with only minor operational anomalies experienced to date.

SUMMARY

EOS TERRA represents the first successfully developed and launched high voltage 120 Vdc based spacecraft power system for NASA. Through significant development efforts and attention to many details associated with high voltage operation in vacuum, EOS TERRA stands as a model for future high power and high voltage power systems. The resulting savings in both parasitic power losses and harness weight coupled with experience on-orbit provides a proven basis for consideration of high voltage dc power systems in the future.

AIAA-2000-2834

THE POWER SUBSYSTEM FOR THE NEXT GENERATION GOES SATELLITE

William Krummann
Hughes Space and Communications
P.O. Box 92919
Los Angeles, CA 90009
Bldg. S24, M/S D507
310-416-5860
310-662-6879 (fax)
william.krummann@hsc.com

Abstract

Hughes Space and Communications was awarded a contract in January 1998 from NASA's Goddard Space Flight Center for the design, manufacture, integration and launch of two next generation Geostationary Operational Environmental Satellites (GOES) and options for two additional satellites. The satellites measure weather phenomena as the primary mission and X-rays and charged particles as the secondary missions. This paper describes the GOES Electrical Power Subsystem (EPS) requirements, architecture, hardware and energy balance analysis. As of this writing, the design of the EPS has completed the Critical Design Review phase and the hardware has entered the manufacturing phase.

Background

The GOES satellite is a three axis stabilized spacecraft located in geosynchronous orbit with a planned lifetime of 7 years. The first launch is scheduled in 2002. The primary mission is to provide accurate location of severe storms and other weather phenomena across the United States. The instruments (Imager and Sounder) stare at the earth and image clouds, monitor surface temperatures and sound the atmosphere for vertical temperature and water vapor distribution. The instruments to perform the primary mission are provided from the government and produced by ITT Industries. The secondary missions provide a search and rescue capability to detect distress signals from ships and planes and space environment monitoring to track the sun's X-rays for early detection of solar flares.

Additional instruments will also detect particle emissions such as solar protons and alpha particles, determine the local magnetic field and map lightning events.

Spacecraft Description

The GOES spacecraft is a three axis stabilized vehicle based upon the Hughes satellite design. An illustration of the GOES spacecraft in flight configuration is shown in Figure 1. The spacecraft utilizes a single solar array located on the –Y side to provide an unobstructed view for the instruments. The primary instruments are mounted on an optical bench located on the +Y nadir side. The secondary instruments are located on the solar array yoke assembly, the aft corner (-X, -Y) of the spacecraft and the nadir shelf. The magnetometer boom deploys from the -X aft side. The communications payload equipment is located on the internal side of the radiator panel on the -Y side of the spacecraft. The communications antennas are mounted on the nadir shelf.

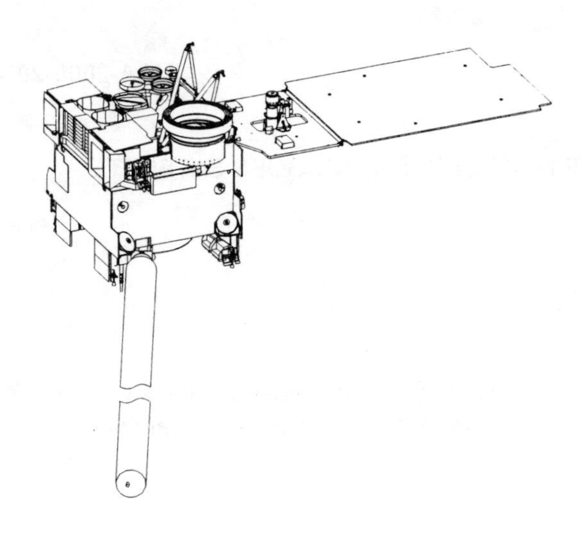

Figure 1. Spacecraft in Deployed Configuration

EPS Description and Requirements

The GOES EPS is a direct energy transfer system utilizing elements from several of the Hughes product lines. A block diagram of the EPS is shown in Figure 2. The top level requirements are shown in Table 1. The main bus voltage is regulated at 53.1+/- 0.25 volts under all sunlight and eclipse conditions by the Integrated Power Controller (IPC) electronics. Secondary bus voltages are provided for previously used bus components and instruments. During sunlight periods, bus voltage control (BVC) modules within the IPC match the amount of power supplied to the loads from the solar array. Excess solar array current is shunted, thereby maintaining a regulated bus voltage. During eclipse periods, the same BVC modules within the IPC control the regulated bus. Power is extracted from the nickel hydrogen battery to support the spacecraft loads during eclipse and peak power periods. The solar array is composed of a single main panel and yoke panel with dual junction solar cell circuits. Power distribution is achieved by the IPC and the bus and payload power distribution units (PDU), which provide fused and switched power to the spacecraft. All returns are brought back to the single point ground at the IPC to minimize chassis current on the spacecraft and reduce magnetic field generation. Load switches provide switching functions for bus and payload units.

Table 1. Top Level Requirements

Parameter	Value
Main Bus Voltage	53.1 +/1 0.25 V
Instrument Bus Voltage	42.0 +/- 0.5 V
Secondary Bus Voltage	29.9 +/- 0.5 V
Total Max Bus Load	1940 W
Orbit Location & Longitude	GEO, 74.5 to 135.5 Deg West
Nickel Hydrogen Battery	24 Cells, 123 Amp-Hr Capacity, 75% DOD Maximum
Dual Junction Solar Cells	117 Circuits, 106 Main Panel and 11 Yoke Panel
Fused Loads	All loads redundantly fused except primary instruments
Charge Efficiency	87% Minimum
Discharge Efficiency	91% Minimum
42V Secondary Bus Efficiency	89.5% Minimum
Data Bus Compatible	1553 Data Bus
Fault Tolerant	Include one failure of battery cell, solar circuit, power module/circuit
Single Point Ground	No current on chassis
Minimal Magnetic Dipole	Backwire of solar array, battery and power paths internal to IPC

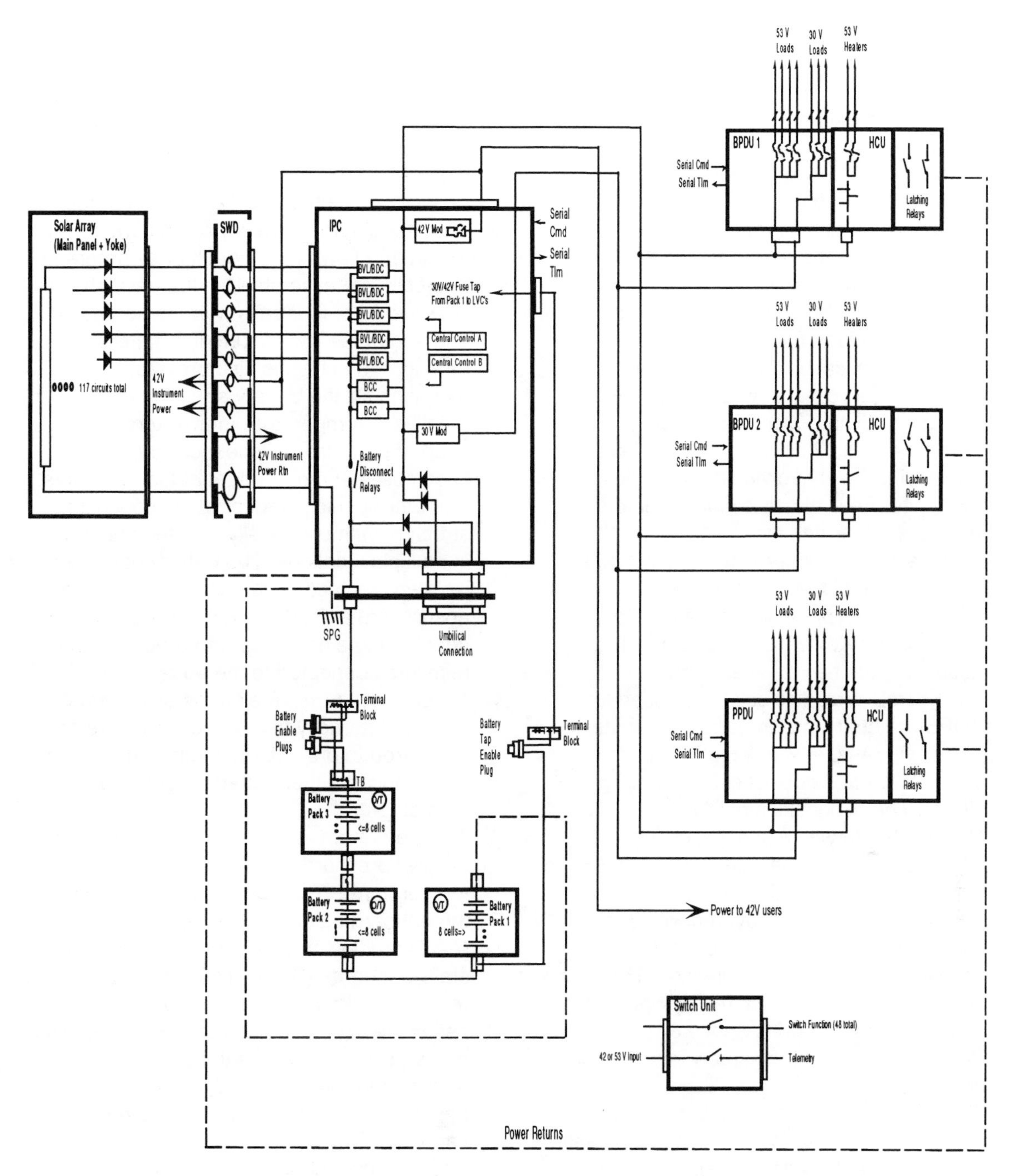

Figure 2. EPS Subsystem Block Diagram

EPS Hardware

Solar Array

The primary power source for the spacecraft is the solar array. The array consists of a single rigid main panel and composite yoke assembly rotating about the pitch (-Y) axis of the satellite. The main panel has dimensions of 274 x 382 x 2.54 cm and consists of 106 individual solar circuits. The yoke assembly, which connects the main panel to the solar wing drive, provides a location for the remaining 11 solar circuits as well as a mounting platform for various payload units. The individual circuits are joined together to provide 5 main groups which are connected to the solar wing drive and eventually to the IPC. There are 32 cells in series to develop the specified array voltage of 54.8 Vdc at end-of-life. Each circuit is diode isolated from each other and the bus to ensure that a single failure will not propagate to the other circuits or affect the performance of the EPS. The main panel substrate is composed of lightweight Kevlar facesheets and Tedlar insulation attached to an aluminum honeycomb core. The solar cells used for GOES are dual junction gallium arsenide ($GaInP_2$/GaAs/Ge) with integral bypass diodes to protect against cell shadowing effects. The dimensions of the cell are 3.97 x 6.91 x .014 cm with a nominal conversion efficiency of 21%. The cells are connected together using molybdenum welded interconnects and are covered with 3 mil ceria doped microsheet coverglass for ultraviolet and radiation protection. The cells and panels are manufactured by Spectrolab and supplied to Hughes Space & Communications for assembly into the solar array.

The solar array capability at end-of-life of 7 years is 2278 watts (autumnal equinox) and 2084 watts (summer solstice). The predictions do not include any solar circuit failures.

The main solar panel and yoke panel fold up against each other in the stowed launch configuration. The solar circuits on the main panel and yoke panel both face outward from the spacecraft spin axis to generate power while spinning during transfer orbit.

Battery

The EPS uses a single nickel hydrogen battery to store energy and supply load power during eclipse and peak power periods. The battery consists of 24 individual pressure vessel cells housed in three separate battery packs. The nameplate capacity of each cell is 123 ampere-hours. This is the same capacity cell as previously tested and flown on other Hughes programs. The cells are assembled in the pack chassis in two rows of four cells with the domes of one end of the cells facing the battery radiator. The chassis also provides mounting for the battery strain gauge amplifiers, temperature thermistors, overtemperature switches, cell bypass assemblies and redundant heater elements. The battery packs mount to the spacecraft structure with electrically isolated fasteners in multiple locations. The battery packs are isolated electrically from the spacecraft to ensure that a short in the cell insulation does not result in a power path to ground. Bleed resistors connected to the outer cover of the battery pack remove electrostatic charge buildup. Internally to the pack, the cells are backwired to provide magnetic cancellation and reduce the generated magnetic dipole moment.

Telemetry from each battery pack includes two cell pressures and two chassis temperatures. The telemetry is utilized by the onboard attitude control electronics (ACE) processor to determine the battery state-of-charge (SOC). The pressure is measured directly from strain gauge assemblies mounted on the cell domes. The pressure is adjusted for the actual temperature and the SOC is calculated based upon comparison to the nameplate capacity of the cell.

Each battery cell includes a bypass assembly to isolate the cell should a failure occur during discharge. In the event of an open circuit, the cell is bypassed and removed from the battery.

The nominal temperature range of the battery is –10 to +20 degrees C. The minimum temperature is maintained by redundant heaters mounted on the pack chassis. The heater set point is controlled by the spacecraft processor and is selectable. The maximum temperature occurs at the

end of the longest eclipse period and is determined by the heat dissipation and dissipative surface area.
The spacecraft integration and test sequence will utilize non-flight test batteries to perform all aspects of the test program. Flight batteries will be manufactured based upon the need date of each spacecraft.

Electronics
The spacecraft bus electronics consists of an Integrated Power Controller (IPC), which performs the functions of bus regulation, battery charging and voltage conversion in a single unit. The IPC is composed of the following assemblies: One Master Module (MM), five Bus Voltage Control (BVC) modules, two Battery Charge Control (BCC) modules and two secondary low voltage control (LVC) modules. The design of the IPC is modular such that several combinations of power modules are possible to meet specific spacecraft requirements. The IPC contains the primary and redundant control electronics to maintain the power bus at 53.1 +/- 0.25 Vdc under normal sunlight/eclipse operation. The control electronics has three operating modes; eclipse, sunlight and battery charging. During eclipse, the BVC modules receive a common control signal to allow boosting of the battery voltage to the regulated bus voltage and provide power to the loads. During sunlight, the BVC modules regulate the bus by controlling the amount of solar array power that is provided to the spacecraft. The BCC modules charge the battery with the excess solar array power. The MM contains the primary and redundant control electronics to maintain the main bus voltage at 53.1 +/- 0.25 Vdc. The bus voltage is compared to a temperature compensated reference voltage and a control signal is developed to operate the BVC modules and maintain the regulated bus voltage. The control signal has sufficient voltage separation to ensure that battery charging and discharging do not overlap. The MM also houses the battery connection relays, charge and discharge current sensors, solar array current sensor, umbilical and bypass diodes, command and telemetry functions and the single point ground.
The BVC modules house the bus voltage limiting (BVL) and battery discharge control (BDC) power stages to provide power to the spacecraft loads. The 5 BVC modules operate in parallel using the common control signal generated in the MM. The modules can also operate in a local control mode using a local voltage error signal should the common control signal not be available. The BVL stage of the BVC modules can pass solar array current, shunt current or actively regulate current thereby maintaining the regulated bus voltage. Only two modules will be in the sunlight regulation mode at any given time to minimize power dissipation. Each BVL stage of the BVC module receives power from a dedicated group of solar array circuits and modulates the input power to that required by the loads by means of a DC-DC switching power converter. The converter utilizes multiple N-channel MOSFET devices in parallel for current sharing purposes. If all of the power is required from a particular BVC module, the pulse width modulated (PWM) switch duty cycle is 0 and the power is passed through a set of parallel blocking diodes directly to the load. The BVC module also processes battery power when required for eclipse or peak power operation. The battery voltage is boosted to the main bus voltage by the BDC stage of the BVC utilizing a separate DC-DC converter. Each BVC module has its own housekeeping power supply which is commandable to turn the module on and off. The BDC stage has overvoltage protection and will automatically shut off if the module output voltage reaches 58 Vdc. The BVL stage will still operate even if the BDC stage has turned off.
The BCC provides for the charging of the battery during all mission phases. There are two electrical circuits per BCC module which provide a maximum charge current of 40 amperes. The charge rate is programmable in 255 increments. The BCC modules receive a control signal from the charge control electronics located in the MM and use excess solar array current to charge the battery. If the commanded amount of current is not available from the solar array, the BCC modules operate in a charge reduction mode and use the available solar array current to charge the battery. The BCC

modules inhibit charging in response to an overtemperature signal indicating the battery has reached a temperature of 35 degrees C. The signal turns the PWM switch off preventing the BCC from passing current to the battery. The BCC circuits are 4-for-3 redundant when providing the maximum charge current.
The secondary bus is provided by one 30 volt and one 42 volt LVC power module. Each module contains four low voltage converter circuits using a forward topology. The LVC modules are designed to provide the required steady state power for each bus and support peak load conditions. The spacecraft bus components use the 30 volt module while the instruments use the 42 volt module. Each converter circuit contains overvoltage protection and will turn off if the voltage exceeds a preset limit. The 42 volt LVC also provides fused outputs to the instrument loads. The LVC circuits are 4-for-3 redundant.

Power Distribution

The spacecraft power is distributed to the user loads, with the exception of payload instruments, through fuse networks located in the bus and payload power distribution units (PDU). Each user is provided with a pair of fuses to distribute either 53V or 30V power. The fuse utilized on the GOES program is a gold conductor printed on a ceramic substrate with plastic housing and copper leads. The leads provide for feedthrough mounting onto circuit boards, which utilizes less space than the axial type of fuse. The fuses are connected in parallel and are interconnected to assure that a single open fuse does not prevent power being distributed to the load. The fuses are derated 50% including temperature effects. The PDU's also contain 36 switch circuits for controlling power application to spacecraft heaters. The switches use two N channel MOSFET devices placed in series which are controlled by redundant serial command decoders. This arrangement allows for the failure of one device in the 'on' state and maintaining control of power to that particular heater. Implementing heater redundancy accommodates devices that fail 'off'. The PDU's also provide two mechanical latching relays for use as switches for loads which do not normally incorporate an on/off switch.

Analysis

The energy balance of the satellite is analyzed for each possible operating mode. Several tools are used to perform the calculations, including EXCEL spreadsheets and MATLAB models. Data such as solar array power generation as a function of degradation over time and battery watt-hour capability as a function of temperature and state-of-charge (SOC) are incorporated into the analysis to verify positive energy balance and battery depth-of-discharge for each spacecraft operating mode. The following table provides the power budget load requirements for spinning transfer orbit and on-orbit end-of-life conditions:

Operation	Bus Load (W)	Payload Load (W)
Spin Sunlight	393	110
Spin Eclipse	447	110
End-of-life Solstice	1171	711
End-of-life Equinox	1202	711
End-of-life Eclipse	1297	625

The battery SOC during transfer orbit is maintained between 25 and 55% under all conditions of orbit raising, changing orbit periods and varying loads.
The total battery recharge power at end-of-life equinox is calculated at 250 W including harness losses, a charge/discharge ratio of 1.15 and a BCC efficiency of 87%. The charge rate required to recharge the battery and provide time at the end of the orbit for any battery cooldown is 6.0 amperes (C/20.5 rate). The battery watt-hours required to support the load during the longest eclipse (72 minutes total) is 2601 W-Hrs. The equinox solar array power provided from the solar array at end-of-life, including one circuit failure and distribution losses, is 2209 W.

Conclusion

The design of the EPS for the next generation GOES satellite utilizes elements from several different Hughes Space & Communications product lines to meet the particular demands of this program. The power electronics are accommodated in a single unit and three separate units provide power distribution. The single solar array provides the sunlight power and the single battery supports the spacecraft during eclipse. All of the unit designs have passed the Critical Design Review phase. Manufacturing of the first and second flight units has begun and delivery of the units to the spacecraft will take place in the next few months.

THE PRELIMINARY DESIGN OF THE COMET NUCLEUS TOUR SPACECRAFT POWER SUBSYSTEM

P. Panneton, G. Dakermanji, D. Temkin, J. Jenkins
The Johns Hopkins University, Applied Physics Laboratory
11100 Johns Hopkins Road, Laurel, MD, 20723-6099, USA
Phone (240) 228-5649
Fax (240) 228-6556
E-mail: paul.panneton@jhuapl.edu

ABSTRACT

The power subsystem for the Johns Hopkins University, Applied Physics Laboratory, Comet Nucleus Tour (CONTOUR) mission spacecraft which will visit two comets is described. The spacecraft is octagonal shaped with body mounted solar panels. The power system is an unregulated bus direct energy transfer system with a "Super" NiCd battery and utilizes triple junction solar cells. The power control is performed by digital and linear sequential shunts. The sun distance, the temperature variations, the shadowing on the solar panels and the requirements on the power subsystem to support the spinning and the inertially stabilized modes of operation of the spacecraft make the solar array and power subsystem design a challenging task. The power system design uses heritage circuits and emphasizes reliability with low cost, weight, and power consumption.

SPACECRAFT OVERVIEW

The Johns Hopkins University Applied Physics Laboratory (APL) is developing the CONTOUR spacecraft for the NASA Discovery satellite program. CONTOUR will visit the comets Encke and Schwassmann-Wachmann 3 (SW3) to take high-resolution pictures, perform detailed analyses of their composition, and determine their precise orbits. It will navigate to the two comets over a period of four years by Earth gravity-assist flybys and propulsive burns from a solid rocket motor (SRM) and a liquid propulsion system (LPS). The spacecraft will be in spin-stabilized hibernation during most of the mission. However, during comet encounters and collected data playback, it will be three-axis stabilized. The required variations in: pitch attitude (0 to 140 degrees), spin rate (up to 60RPM), Sun-spacecraft distance (0.87 to 1.13AU), solar panel peak temperatures, shadows and solar cell keep-out zones from various spacecraft appendages, and load power, make the power system design a challenging task.

The spacecraft is an octagonal-shaped cylinder with triple-junction (TJ) solar cells body-mounted on all eight side panels and one aft panel. To reduce cost, extensive use is made of existing APL power system designs: The NEAR[1] Battery and Power System Electronics (PSE) and TIMED[2] Power Switching Unit (PSU) are to be rebuilt with minimal modifications for CONTOUR.

Figure 1 shows the mast and dish antennae, rocket engine modules (REM, which house the LPS thrusters), instrument and attitude sensor protrusions, and dust shield, which all cast shadows on the solar panels depending on the spacecraft solar attitude. Also shown are the many solar panel cutouts for: mounting hardware, test connector access, propulsion fill-valves, battery radiator, and instrument and attitude sensor ports. CONTOUR spins about the axis defined by the mast antenna, and pitch is defined as the Sun angle down from the mast. Maximum solar power is produced with a 45-degree pitch and with a roll angle close to that shown in the left view of Figure 1. The areas available for solar cells on the three leftmost side panels are three inches taller than those of the other side panels to maximize the solar power produced when pointed at the Sun during a comet encounter in three axis

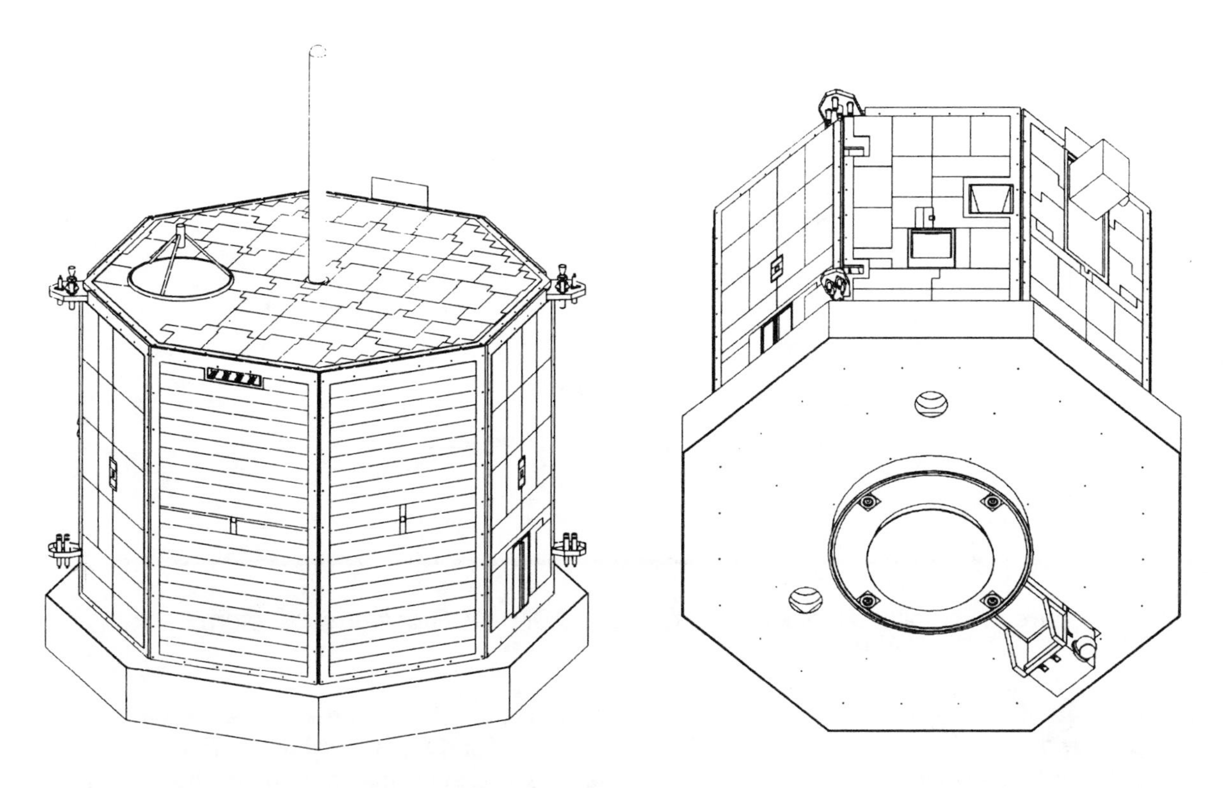

Figure 1. Views of solar array keep-out zones and spacecraft appendages.

stabilized mode. The three unpopulated inches at the bottom of the other side panels are used to provide launch pad access to pyrotechnic arming plug connectors, the solid rocket motor safe and arm device, and the battery enable plug, and also provide mounting area for the analog shunt resistors.

MISSION OVERVIEW

CONTOUR will be launched on a Delta-7425 Med-Lite in mid-2002. The launch vehicle will inject CONTOUR into a phasing orbit at 60RPM with the solar attitude shown on the right hand side of figure 1. The roll angle shown produces the minimum power for each revolution. As the spacecraft continues to roll, solar power variations produce a considerable disturbance for the shunt regulator, synchronous with the spin rate. The pitch angle of 140 degrees is sufficiently steep to cause the dust shield to shadow the solar strings on the bottom of the side panels. The spacecraft also cools down at this attitude, and the increasing heater power demand will eventually exceed the solar power being generated at certain roll angles. Therefore, a spin axis adjustment to a more favorable solar attitude (~110 degrees pitch) is required within the first several hours after launch. The thruster firing for this maneuver produces load transients that are also synchronous with the spin rate. Depending on the heater load or if a failure occurs affecting power margin, this combination of solar and load power disturbances may cause the power bus to have large cyclic fluctuations between the battery voltage, minimum of 22 V, and regulation voltage, up to 34 V.

The 60RPM-spin rate is not only required for precise phasing orbit injection via the Delta launch vehicle, but also for the SRM burn. After several weeks of phasing orbits, the pitch angle is adjusted to 118 degrees and the SRM is fired to put CONTOUR into a one-year Earth-return trajectory. Additional trajectory correction maneuvers may be necessary via the LPS. CONTOUR will then be spun-down to between 5 and 20RPM and monitored for stability until it is put into hibernation mode at a pitch angle of ~90 degrees (sun normal to the spin axis). CONTOUR will experience a half-hour solar eclipse as it swings by the Earth and heads toward the comet Encke. Four months before the closest approach to Encke, CONTOUR will be further spun down and transitioned into 3-

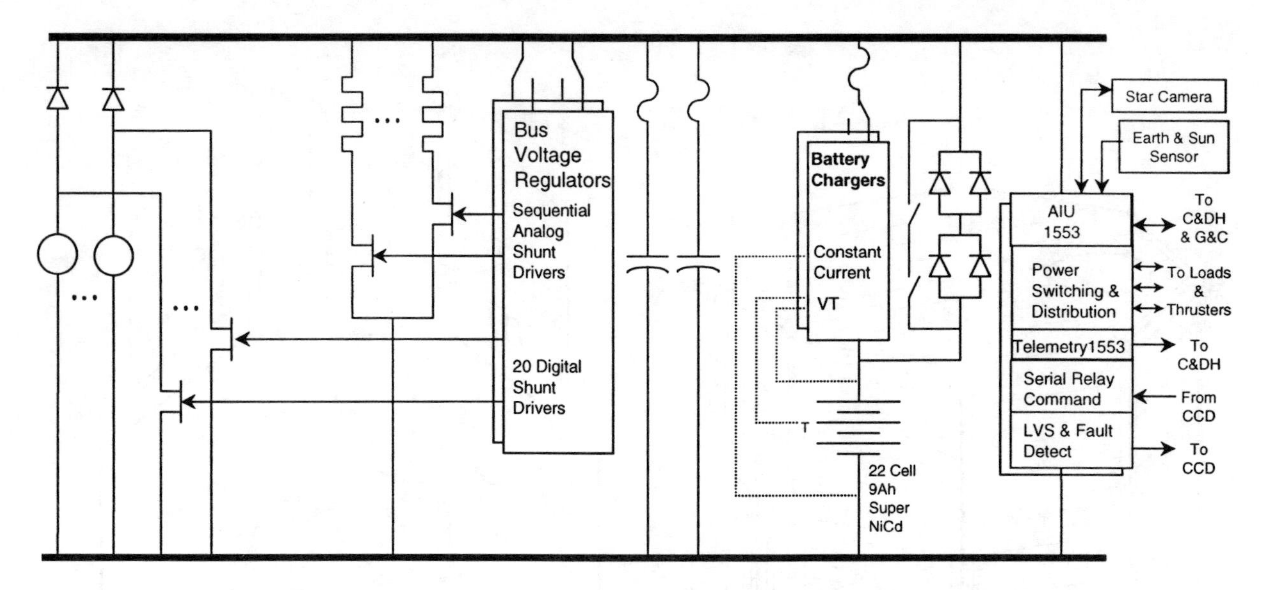

Figure 2. Simplified block diagram of the CONTOUR power subsystem

axis mode, where it will begin imaging the comet for optical navigation of the final approach. Several sequences of spin-less comet-pointed data collection followed by one revolution in twelve minutes Earth-pointed data playback will occur in the final weeks of preparation for each comet encounter. After the Encke encounter, CONTOUR will be spun-up and directed to swing by Earth for a trajectory toward comet SW3. After the SW3 encounter, there may be a third comet encounter, funding-permitted.

The total power requirement of the spacecraft during cruise, comet encounter and data playback are respectively: 250W, 450W, and 350W.

POWER SYSTEM DESCRIPTION

The power subsystem shown in the simplified block diagram in Figure 2 consists of the solar array, battery, power system electronics, power distribution and propulsion thruster switching. The power system is a Direct Energy Transfer (DET) system. It is single fault tolerant. To reduce the size of the analog shunt dissipators, the solar array (S/A) is divided into segments with a digital shunt transistor across each segment. The "coarse" S/A power control and bus voltage regulation are performed by sequentially turning ON/OFF the digital shunts in response to the available solar array power and S/C load demand. To minimize the noise and ripple generation on the power bus, the "fine" control is performed by a six stage linear sequential full dissipative shunt. When the power changes are such that the linear shunt range is exceeded, the digital shunts are activated by either turning ON or OFF to maintain bus regulation. The bus is regulated at 33.5 ±0.5 V when the solar array power is adequate to supply the load and battery charge power. The bus follows the battery voltage during the launch and earth fly-by phases and also during abnormal conditions when the solar array power capabilities are exceeded.

In the three axis stabilized comet encounter phases, the solar array panels on the sides toward the sun are sized to provide the higher load power required. During the spin mode, the spacecraft minimum power capability is determined by the minimum power level during a revolution caused by the cut-outs and the shadowing of the solar array strings.

To mitigate the effects of possible large cyclic fluctuations of the bus voltage between battery voltage and the bus regulation levels, relays have been placed across the battery discharge diodes to clamp the battery to the bus during the first several hours of the mission. This causes somewhat less solar power to be available due to the lower bus voltage, but more battery energy will be available to the loads. The battery will filter out the voltage variations until the spin axis is adjusted to more favorable sun angles and the relays opened. The charge current during this limited duration mode is C/20 rate utilizing a separate solar array controller for current limiting.

The battery is charged from redundant battery chargers. Battery charging is performed using a C/20 constant current charge, then tapering to a ground-selected temperature compensated voltage (V/T) limit, then commanding from the ground to C/100 trickle charge when full state of charge is reached. The V/T limit control is active simultaneously with the constant current and trickle charge control.

The selection of the NEAR spacecraft power subsystem design for CONTOUR was strongly influenced by the need to use heritage circuits to minimize power system electronics hardware development cost, time and risk.

SOLAR ARRAY

The solar attitudes required during 3-axis mode place varied thermal loads on the solar panels. The SW3 encounter (100 degrees pitch) will heat the long side panels to 100ºC, while comet pointing at Encke (12 degrees pitch) will heat the aft solar panel to 90ºC. Temperatures near the dust shield will be higher than near the aft panel. The varied maximum temperatures require that the number of cells in series will vary from panel-to-panel. Earth pointing at one revolution per twelve minutes at Encke (65 degrees pitch) will cause all the side panels to cycle 11,000 times from 0ºC to 40ºC, while Earth pointing at one revolution per twelve minutes at SW3 (93 degrees pitch) will cause all the side panels to cycle 11,000 times from 0ºC to 60ºC. The aft solar panel is expected to see only minor thermal cycling.

Triple junction cells are required to satisfy the necessary power margins over the CONTOUR mission. To reduce cost of the solar panels, large cells will be used. To minimize the effect of thermal cycling on these large multi-junction cells, the side panels have composite face sheets.

All solar panel substrates will consist of an aluminum (5056) honeycomb core. The 8 side panel cores will be sandwiched between front and rear graphite-epoxy face sheets. Face sheets, (200-micron), are comprised of 3 layers of M55J/Cyanate oriented at 0 and +/-60 degrees. The aft panel core will be sandwiched between front and rear aluminum face sheets. All solar cell side face sheets will be covered with 51-micron Kapton insulator co-cured with the facesheets. The rear face sheets will be painted with Aeroglaze Z306 black thermal control paint. All exposed core at panel edges will be coated with epoxy and covered by Kapton tape which overlaps the face sheets. The dimensions of a side panel are 148.5 x 82.5 cm and 1.28 cm thick.

Each solar cell string will have two blocking diodes: one connects to the main bus and the other connects to a digital shunt. A typical digital shunt circuit will consist of one string from each side panel plus two or three strings from the aft panel. If the strings selected for a circuit are not shadowed then the circuit current will stay reasonably smooth as the spacecraft spins. Some circuits, however, will have contributing strings from only a few of the larger panels. These circuits will have greater variation in the digital shunt current that becomes even greater at pitch angles that cause shadows to fall on the contributing strings. Careful sorting of the power output of each string as a function of pitch angle leads to an optimum circuit selection.

Figure 3 shows the worst case solar array current variation for 11, 12, and 13 digital shunts turned on. To minimize repetitive bus voltage disturbances during this sensitive maneuver, the analog shunt range will be sufficiently large to accommodate the worst case combination of load and solar power.

BATTERY

A 9-ampere-hour, 22-cell "Super" NiCd battery is required to support the launch, spacecraft deployment and earth flybys. It will also support emergency conditions during the mission. Figure 4 illustrates the battery assembly. The battery is mounted to the spacecraft from the top bracket, thermally isolated by fiberglass stand-offs. Heat is rejected to space from the bottom plate. The battery weight is approximately 11.5 kg and the size is 30x18x12 centimeters.

The electrolyte is 31% potassium hydroxide and contains a Hughes-proprietary additive. An advantage of the "Super" NiCd for this mission is that the manufacturer's recommended storage mode is to maintain it fully charged at cold

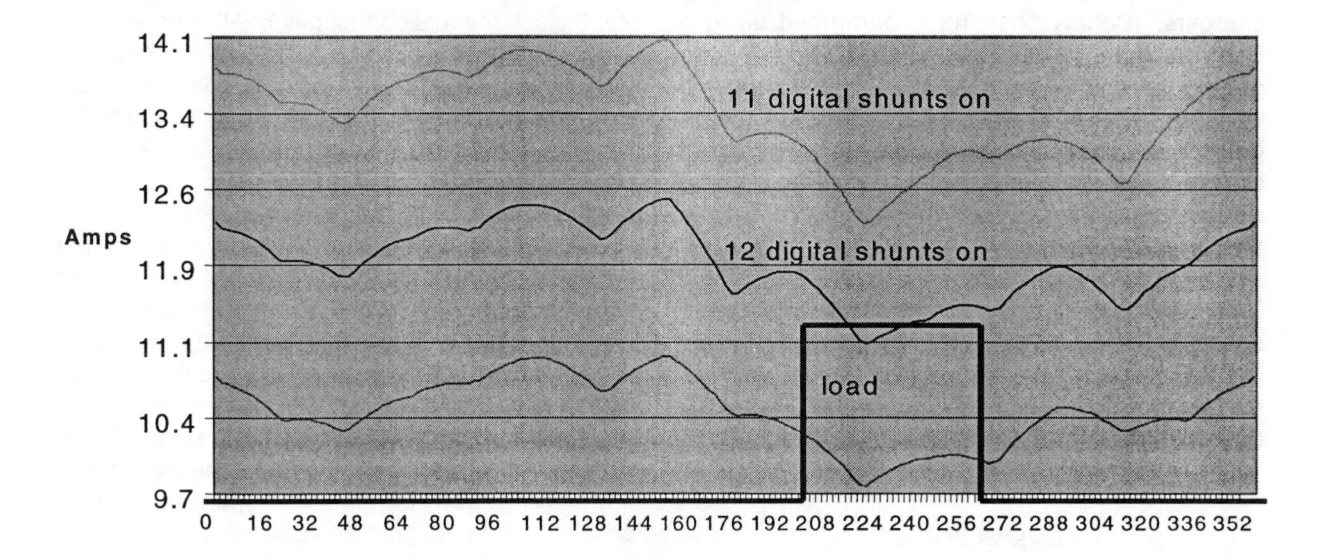

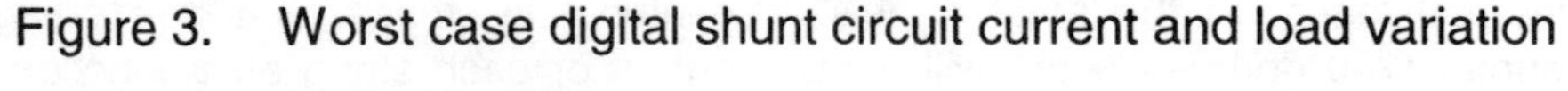

Figure 3. Worst case digital shunt circuit current and load variation

Figure 4. CONTOUR battery

temperature on trickle charge, which is the planned operational mode during most of the mission. The battery temperature is expected to remain between 0ºC and 10ºC, during the mission except during discharge and comet encounter phases. No battery reconditioning is planned. Tests at APL and experience on the NEAR spacecraft[3], which used the same battery, indicate that "Super" NiCd cells which have been at trickle charge for over 3 years, retain most of their capacity. The depth of discharge expected during launch is approximately 50%.

SHUNTS

To minimize the weight of the power subsystem, the analog shunt elements consist of thin film foil heater elements encapsulated in Kapton that are bonded to aluminum plates mounted in the available space at the bottom of solar panels which do not face the Sun in the comet encounter modes. The thin film metalization layout is designed to minimize the magnetic field generation.

POWER SYSTEM ELECTRONICS

The Power System Electronics (PSE) contains the circuits that regulate the power bus, control battery charging, provide shunt control of the excess solar power, and provide the autonomy, telemetry and command interfaces to the spacecraft. The battery charger, analog shunts, and the digital shunt drivers are redundant. There are 20 digital shunts. The power system design can tolerate the loss of one digital segment.

The power bus regulation is performed using digital and analog shunts. MOSFET switches across the solar array segments are sequentially turned ON/OFF depending on the available S/A power and the load demands. The fine control is performed by a six stage linear sequential dissipative shunt. The sequence in which the six segments will operate is derived from a voltage divider string. Only one of the linear shunt MOSFET transistors operates in its linear region at a time. The remaining transistors are either fully ON or fully OFF. Each shunt segment has its own amplifier and operates as a constant current source which is proportional to the control signal during the linear mode operation. This technique reduces the power dissipation in the analog shunt transistors compared to a single stage linear analog dissipative shunt design, to allow all the electronics to be placed in one box. When the error control signal is outside the range of the linear shunts, it generates a signal that drives an up/down counter. The counter outputs are activated sequentially and drive the digital shunt MOSFET transistors. The analog shunt resistors are sized such that five sections of the analog shunt can dissipate the maximum power generated by any one of the 20 solar array segments and are sized larger than the maximum variation of power from the solar arrays during a revolution of the spinning spacecraft.

Battery charging is performed using both constant current charge and V/T controls. The high current charge rate is set to C/20, the charge current tapers as the selected V/T limit is reached. When the battery is fully charged the constant charge rate is reduced by ground command to C/100 trickle charge rate. The C&DH computer will place the charge rate limit back to C/20 rate if a battery discharge is detected. Eight V/T levels are provided in the design which supports the battery even with one cell shorted. The cell V/T levels of are defined by the following:

- Separation between levels: 0.02 ± .002V
- Slope: -2.33 ± .2mV/ºC
- Highest level voltage at 0ºC: 1.5. ±0.02 V

The eight V/T levels are selected by binary switching of resistors in the voltage divider circuit and comparing these with a precision reference in the error amplifier. Latching relays are used to switch the resistors, thus any temporary power interruption or logic reset will not affect the settings of the V/T controller. The charger is designed to operate with a minimum of 0.5 volt across it to maintain full charge on the 22 cell battery at the coldest expected battery temperature. The spacecraft power bus and the battery are protected from catastrophic failures in the chargers by fuses both at the power bus and battery power interfaces.

CONTOUR is a three axis stabilized and spinning satellite, with spin rates up to 60 rpm. To insure that the high spin rate, pulse loads and shadowing changes which occur synchronized with the spin, and solar incidence variations do not adversely affect the power system response and bus ripple, the breadboard of the power system has been built and is being modified for the CONTOUR requirements. Analysis and breadboard testing are in progress to insure that the control loops are stable and that adequate bandwidth can be achieved to maintain low bus ripple during the load solar array power transient conditions.

POWER SWITCHING UNIT

The PSU contains the circuitry for the S/C pyrotechnic firing control, the power distribution relays, fuses, load current monitoring, and the power system internal relays. The commands to control the spacecraft power distribution relays and the power system relays are encoded by the two spacecraft up-link receiver Critical Command Decoder (CCD) and sent to the power System Command Decoders (SCD) as serial data. Each of the two command decoders in the PSU are dedicated to one of the CCDs.

Hardware features are incorporated into the design to prevent false commanding. The interface between the CCD and the SCD consists of 4 twisted pair lines: enable, gate, data, and clock. The SCD converts the serial command into a parallel command, which the relay cards inside the PSU can then execute. For the SCD to send the command to the relay cards, the second 8 bits must be inverted from the first, 16 clocks cycles must have been received with the data word, the enable must be present for the entire time of the command, and the gate signal must be present allowing the relay pulse driver to turn on the selected relay drivers.

Distribution relays have been selected which allow bipolar operation: relay set and reset activation can occur by driving the current in the relay coil bi-directional. This allows the primary and redundant command systems to independently control each relay through isolated coils. This technique has been used successfully on many APL developed spacecraft.

Telemetry of all relay status, load currents, and power system analog telemetry, which includes battery voltages and currents, solar array currents, and voltages and temperatures, are collected and digitized in a 12 bit A/D converter and sent to the C&DH through redundant MIL-STD-1553 buses. The PSU command and telemetry capabilities are:

- 128 Commands.
- 128 Analog telemetry.
- 160 Bilevel Telemetry.

In addition, the PSU contains the spacecraft propulsion thruster switches and the redundant propulsion Mil-Std-1553 interface circuits.

FAULT DETECTION AND CORRECTION

To protect the bus from failures which can cause an over-voltage condition on the bus, both bus regulators are ON during normal operation. The reference point of the primary controller is set slightly lower than the redundant controller reference, thus only one regulator is controlling the bus voltage. The two bus regulator ON/OFF relays are interlocked to eliminate the possibility of accidental removal of both regulators from the bus. The spacecraft C&DH processor performs the fault detection and correction of internal power subsystem faults.

The C&DH processor will also detect any potential S/A-battery lock-up condition. If a lock-up is detected, the C&DH will remove for about 10 minutes the non-critical loads and heater power which constitute a substantial amount of the S/C power especially during the cruise phase. This will allow the recovery from a lock-up condition.

CONCLUSION

This paper describes the preliminary design of the CONTOUR power subsystem. The power system uses heritage designs modified to meet the unique CONTOUR mission requirements while reducing the development schedule, cost and risks.

The breadboard of the power system electronics has been built and is being tested to verify the design modifications and system requirements.

REFERENCES

1. J.E. Jenkins, G.D. Dakermanji, M.H. Butler, P.U. Carlsson, "NEAR Power Subsystem Design and Early Mission Performance," *JHU/APL Technical Digest*, Vol 19, No 2, pp. 195-204, 1998. For more Information, visit the APL web site: http://www.jhuapl.edu/digest/issues/td1902/jenkins.pdf

2. G.D. Dakermanji, M.H. Butler, P.U. Carlsson, D.K. Temkin, "The Thermosphere, Ionosphere, Mesosphere Energetics and Dynamics Spacecraft Power System" *Proceedings of the 1997 Intersociety Energy Conversion Engineering Conference (IECEC97), Honolulu, Hawaii, Vol. 2, pp. 544-549.*

3. J.E. Jenkins, G. Dakermanji, "The Near Earth Asteroid Rendezvous Spacecraft Power System Flight Performance" *Proceedings of the 1999 Intersociety Energy Conversion Engineering Conference (IECEC99), Vancouver, Canada, 1999-01-2485.*

AIAA-2000-2837

THE X-RAY MULTI MIRROR (XMM-NEWTON) POWER SYSTEM

J. E. Haines, B. Jackson, L. Gerlach

European Space Technology Centre (ESTeC),
European Space Agency,
Keplerlaan 1, 2201 AG, Noordwijk, The Netherlands
Phone No. : INT + 31-71-565 3855, Fax No. : INT + 31-71-565 4994,
E-Mail : jhaines@estec.esa.nl

ABSTRACT

The XMM-Newton X-ray observing astronomy mission, which is a prime component of ESA scientific spacecraft programme for the new century, was launched by Ariane-504 from French Guiana on the 10th of December 1999. It was injected into a highly elliptical, 834 km by 113990 km orbit, having an inclination with respect to the equator of 40 degrees. The spacecraft was built in Europe by an industrial team led by Dornier Space Systems (DSS) of Friedrichshafen, Germany with the onboard instruments provided by scientific institutions. The technical paper describes the XMM-Newton power system topology, its equipment characteristics and the early in-flight operations. In addition two novel features related to the design of the Electrical Power System for XMM-Newton will be reported in the technical paper. These being the implementation of a surface conductive network for the cover glasses of the solar cells and the extensive use of telecommandable and resettable latching current limiters (LCLs) for payload failure protection and ON/OFF switching.

INTRODUCTION

In recent years, new astronomical pictures of the sky have shown the hot places within the universe. Wherever temperatures are elevated to millions or billions of degrees, X-rays pour out; they are however blocked from the view of terrestrial observers by the Earth's atmospheric gases. X-ray telescopes operating in space can however see these sources, which range from the surprisingly hot atmospheres of commonplace stars to the gravity-powered furnaces surrounding giant black holes in distant galaxies.

Immense clouds of hot gas pervade the galaxies and clusters of galaxies whilst a background glow of X-rays fills the sky and may be announcing the origin of these massive cosmic features.

The archetypal X-ray star is a destructive entity, a white dwarf or neutron star that sucks gas from a companion star and heats it by intense gravity. In some of these cases the X-ray source may even be a small black hole.

X-ray astronomers have spent an exciting quarter of a century finding these things out. The first successful X-ray satellite was NASA's UHURU mission that flew in 1970 and charted 339 cosmic sources. Since then a long succession of American, Russian, European and Japanese missions have made X-ray observations one of the strongest mediums of space astronomy. ROSAT a German-US-UK collaboration launched in 1990 pushed up the number of known X-ray sources to 120,000.

Fig. 1. XMM-Newton In-orbit Concept

The European XMM-NEWTON mission has been developed with the intention of raising the identification of these X-ray sources into the millions.

XMM-Newton which was launched in December of 1999 is the most sensitive X-ray satellite ever built. NASA's Chandra spacecraft that was launched a few months earlier in 1999, hunts for small, faint sources and is capable of seeing fine detail better than XMM-Newton. Where XMM-Newton excels and is complementary to Chandra, is in having telescopes of a design that provide it with a large X-ray collecting capability, this allowing it to surpass that of CHANDRA in the identification of independent X-ray sources.

XMM-Newton's capacity to pick out large numbers of objects is incidental to its main purpose which is to make X-ray astronomy a more exact science. The keyword in this respect is spectroscopy. The spreading of light into a spectrum is indispensable for astronomers working in the visible and ultraviolet radiation domains and thus by latching on to precise wavelengths emitted by atoms, spectroscopy can measure a source's composition, temperature, and motion.

XMM-Newton is being operated as an observatory for the use of the world's astronomers and they will exploit its capacity not merely to observe millions of objects but to conduct general spectroscopy on these sources and detailed spectroscopy on perhaps about 30,000 of them over a ten year period.

THE XMM-NEWTON SPACECRAFT

Overview

As depicted in figure 1, in order to be the world's most sensitive X-ray observatory, XMM-Newton is the largest scientific spacecraft ever built in Europe. Overall it is nearly 10 metres long and nearly 4 tonnes in mass, the 7 metre long tube structure which accommodates the required focal length of the X-ray telescopes, dominating the design.

Most of the spacecraft's mass is concentrated at one end, where the short but heavy X-ray multi-mirror modules are mounted in the Mirror Support Platform (MSP). Surrounding this assembly the spacecraft Service Module (SM) carries the solar powered electrical system and the data handling and telecommunications equipment.

Systems for controlling the spacecraft's orbit and attitude/orientation in space are also located within the Service Module. Coarse attitude control is achieved utilising hydrazine thrusters while fine pointing for astronomical observation purposes is accomplished using reaction wheels which provide the necessary torques to orientate this large spacecraft with precision. Star trackers are used to pick out visible guide stars close to the position of the desired X-ray target, with the attitude control system then maintaining the pointing steady to within a few thousandths of a degree for several hours on end.

Since the deployable solar panels, cannot be rotated about their attachment axes, the telescope axis of XMM-Newton must always remain more or less at right angles to the Sun. A sunshield is mounted on one side of the Service Module in order to keep solar rays out of the telescopes.

At a given time of year XMM-Newton's observations are confined to a swath 40 degrees wide, or about one third of the whole sky. As the seasons progress and the Earth together with the spacecraft orbit the Sun, the entire Universe will become accessible to XMM-Newton's gaze.

At the far end of the tube structure, the X-ray detectors straddle the foci of the telescopes and spectrographs that are accommodated within the cold locations of the Focal Plane Assembly (FPA). In contrast to the multi-mirror assemblies of the X-ray telescopes which remain close to +20 degrees C, the X-ray detectors will operate at -100 degrees C, this design necessitating the implementation of specific radiators that are screened from solar input, within the Focal Plane Assembly.

Payload Instruments

There are three scientific instruments aboard XMM-Newton spacecraft, these being the European Photon Imaging Camera (EPIC), the Reflection Grating Spectrometer (RGS) and the Optical/UV Monitor (OM).

The major payload function of establishing X-ray focussing is achieved by the three Mirror Assemblies housed within the Service Module, where each individual assembly contains a Mirror Module which is comprised of 58 concentric mirror shells. The mass associated with these complex assemblies and the supporting mirror platform account for just over 50% of the spacecraft dry mass.

European Photon Imaging Cameras (EPIC)

The mirror modules will send the image beam along the telescope tube to five cameras at the extremity of the spacecraft. At the prime focus of each of the telescopes, behind six-position filters, are three European Photon Imaging Cameras (EPIC). With silicon chips that can register extremely weak X-ray radiation, these advanced Charge-Coupled Device cameras (CCD) are capable of detecting rapid variations in intensity, down to a thousandth of a second and less.

Reflection Grating Spectrometer (RGS)

For a complementary analysis of the spectrum, a grating structure on two mirror modules reflects about half of the incoming rays to a secondary focus, with its own CCD camera.

This Reflection Grating Spectrometer (RGS) "fans out" the various wavelengths, thus indicating in more detail than EPIC the exact condition of individual elements, such as oxygen and iron.

Optical/UV Monitor (OM)

The third instrument aboard XMM-Newton is a conventional but very sensitive Optical/UV Monitor (OM) which can observe simultaneously the same regions as the X-ray telescopes, but in the ultraviolet and visible wavelengths. This gives astronomers complementary data about the X-ray sources. In orbit, this 30 cm telescope is as sensitive as a four metre instrument on the Earth's surface.

Module / Subsystem	before separ.	LEOP SA deployed.	Transfer Orbit	Sunlit Operational Orbits	Eclipse Orbits Operation	Eclipse Orbits Perigee		ESAM		DNEL	Peak Power during Operation	
					sun	sun	eclipse	sun	eclipse	eclipse	sun	eclipse
duration [h]	0.7		46.2	48	40	6.2	1.8	46.2	1.8	1.8	48	1.8
Focal Plane Assembly												
Instruments	0	12	0	256	256	256	0	223	0	0	409	12
Data Handling	10	10	10	10	10	10	10	10	10	10	10	10
Thermal Control	0	230	230	0	0	0	21	28	49	0	0	24
Power S/S	9	13	13	12	12	12	9	12	10	9	14	9
Telescope Tube												
Thermal Control	0	0	0	0	0	0	0	0	0	0	45	0
Mirror Support Platform												
Instruments (OM + RM)	0	0	0	66	66	57	5	66	5	0	88	10
Thermal Control	0	105	154	229	231	482	87	239	86	0	444	119
Mirror Thermal Control Unit	0	27	28	30	30	37	26	31	26	25	36	26
Service Module												
AOCS	0	141	141	128	128	156	156	91	91	49	726	726
Data Handling	30	30	30	30	30	30	30	30	30	30	30	30
Thermal Control	176	113	105	120	114	114	69	117	74	79	193	130
RCS	14	32	32	32	32	32	32	32	32	32	32	32
TT&C	16	42	42	42	42	42	42	42	42	42	42	42
S/C Harness (3% of User Power)	8	23	24	29	29	37	15	28	14	9	63	36
Power S/S	67	196	203	112	201	119	89	200	86	68	182	162
Spacecraft	330	974	1012	1096	1181	1384	591	1150	554	353	2314	1368

Table1. XMM-Newton Power Budget

System Level Power Budget

Detailed in Table 1 is the XMM-Newton system level power budget in watts for normal on-station operations in both full sun and eclipse orbits. In addition the table presents the spacecraft peak power requirements and the specific power demands associated with non nominal operations such as Emergency Sun Acquisition Mode (ESAM) and Disconnect Non Essential Loads (DNEL).

THE XMM-NEWTON POWER SYSTEM

The XMM-Newton Electrical Power System which is the technical responsibility of DASA-Dornier Satellite Systems in Germany, is depicted in figure 2.

The spacecraft power source consists of two identical and deployable wings, each wing consisting of three rigid panels and each panel supporting four independent yet identical electrical sections.

The energy storage element consists of two 24 Ah batteries, each battery comprised of 32 series connected nickel-cadmium cells.

Centralised power management of the 28 volt ± 2% power bus, is achieved by the Main Regulator Unit (MRU) which comprises a Sequential Switching Shunt Regulator (S3R) and redundant Charge/Discharge Regulators. These various regulators satisfy both the payload power demands and battery re-charge needs by actively controlling the available solar array power and battery energy.

Power distribution for the spacecraft electrical loads and heaters is achieved by the two Power Distribution Units, one housed within the Service Module (SM-PDU) and the other housed in the Focal Plane Assembly (FPA-PDU) at the remote end of the telescope tube.

For maintaining precision temperature differentials within the complex Mirror Assemblies (MA), the Mirror Support Platform (MSP) and the two Reflection Grating Assemblies (RGA), a dedicated Mirror Thermal Control Unit (MTCU) is utilised.

Finally the Pyrotechnic Release Unit (PRU) is used for the control and distribution of power to the pyrotechnic and other electro-mechanical release devices on the spacecraft.

Solar Array

The solar array has two wings with three rigid panels each, of 1.94 m x 1.81m, having a total area of 21 m² and a mass of 81.4 kg.

The Solar Array wings are body fixed and have a sun incidence angle variation around normal of up to 28°. In sunlight EOL (10 years), worst case including one section failed, the solar array is required to provide 1600 W at 30V at interface connectors.

During launch and early orbit phase, the 3 panels of each wing are folded and stowed to the spacecraft on four holddown points. Kevlar restraint cables keep the solar array in folded position until they are cut on command from the PRU

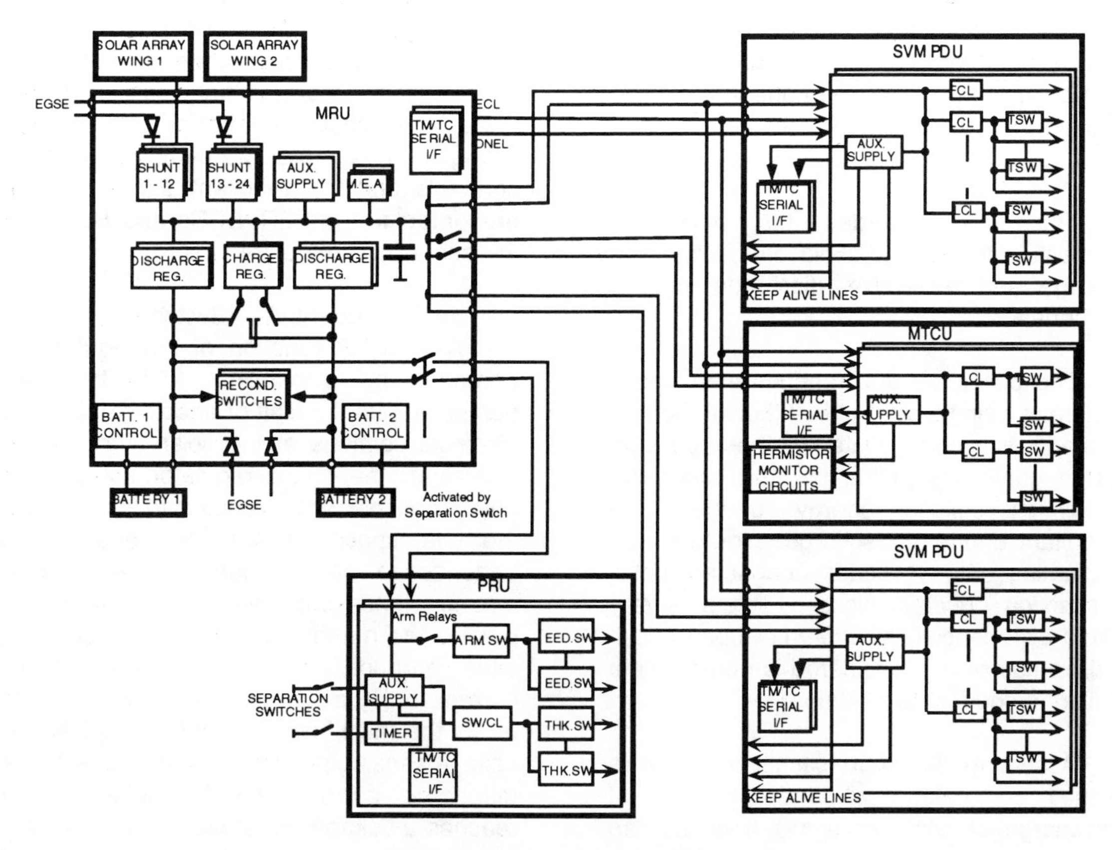

Fig.2. The XMM-Newton Power System

Each wing carries 12 power sections with four sections per panel and the solar cell strings are interconnected by blocking diodes and discrete panel wiring at the rear side of the panel. Power delivery is divided equally between the 24 solar array sections.

2 Ωcm Silicon BSR solar cells (210 μm) with a 300 μm CMX coverglass are mounted on CFRP panels. In order to prevent electrostatic discharge, the solar cell coverglasses are coated with a conductive ITO coating. As depicted in figure 3, the electrical grounding of this conductive layer is achieved by means of a silver patch bonding technology from cell corner to cell corner and ultimately via ground busbars to the solar array structure.

Battery

The XMM-Newton energy storage element consists of two 24 Ah batteries, each battery comprised of 32 series connected VOS 24 AMAA nickel-cadmium cells.

The XMM-Newton battery which was designed and built by Alcatel-SAFT of France, is arranged into two rows of cells, each row containing sixteen insulated, prismatic cells. The battery design is a construction where these two rows of cells are clamped between two machined aluminium end plates using a system of tie rods.

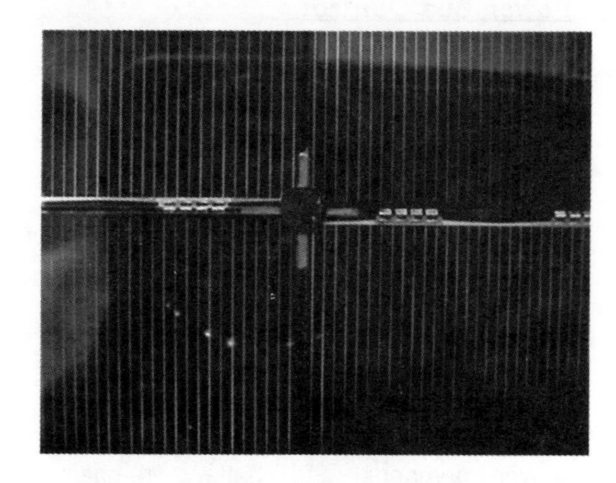

Fig. 3. ITO Grounding Concept

Interposed between each pair of cells is a thermal collector plate which enables heat conduction from the largest cell surface to the base of the pack. The battery pack also contains thermistors, heaters and appropriate wiring and connectors.

The MRU provides the necessary protection features to avoid both excessive charge and excessive discharge of the batteries

The possibility of a random short circuit failure of a single cell within either of the two batteries is accommodated by the imposition of adequate design margin. As a result the maximum eclipse energy budget and resultant depth of discharge is determined assuming only 31 series connected cells. Based on previous nickel-cadmium battery flight experiences, a battery cell open circuit failure has been considered a non-credible failure mode for the batteries.

Assuming the short circuit loss of one series connected cell and a depth of discharge of 64% for a 1.8 hour eclipse case, the overall energy of the two batteries equates to 650 watt-hours. In a spacecraft emergency case and a depth of discharge of 90% for a 1.8 hour eclipse the available energy is 760 watt-hours, whilst the worst case operating voltage of the battery is assumed to range from 31 volts to 51 volts.

The overall mass of the two XMM-Newton batteries amounts to 82 kg.

Power Bus Control

Utilising a direct energy transfer concept which is configured so as to provide a 28 volt fully regulated power bus, centralised power management is achieved by the Main Regulator Unit (MRU) which comprises a Sequential Switching Shunt Regulator (S3R), redundant Battery Charge Regulators (BCRs) and Battery Discharge Regulators (BDRs). This equipment which is designed and built by Alcatel-ETCA of Belgium, thus contains the necessary electronic circuits for satisfying both the XMM-Newton payload power demands and battery re-charge needs, by actively controlling the available solar array power and battery energy.

The Main Error Amplifier (MEA) used for regulating the XMM-Newton power bus is configured as a triple channel, analogue chain, with the outputs from each of these three independent channels. A majority voting circuit reliably combines these channels and the resultant signal is then presented to the S3R, BCRs and BDRs for the achieving the desired control.

The Sequential Switching Shunt Regulator (S3R) function of this equipment operates by controlling FET transistor dumps, to either inhibit or enable to bus, the electrical sections of the solar array. When operating, the shunt regulator selects the number of solar array sections to supply both the total spacecraft load and enable the batteries to be charged in the desired manner. Since this selection can rarely be precise the number of sections enabled will either produce an excess or deficiency of current to the main bus. If a current excess exists the main bus capacitor located at the voltage sensing point will charge up with the difference current until the MEA output reaches a voltage threshold where the next solar array section in the sequence will 'dump'. Since a deficiency of current will now exist, the bus capacitor will discharge into the load until the MEA output reaches a lower voltage threshold where the section previously dumped will again be enabled.

The shunt regulator concept is therefore basically a limit cycle oscillator which dependent upon load and array section current levels will operate at frequencies between DC (theoretically) and about 10 kHz.

Since on XMM-Newton twenty-four independent solar array sections are provided to the S3R, the loss of a single electrical section by any possible failure is fully accounted for in the power margin philosophy of the subsystem.

If the main bus voltage falls such that MEA output signal moves out of the shunt control range, then the Battery Discharge Regulators are automatically enabled and linear control of these circuits initiated by the loop. The four, 450 watt output power, BDR modules which are housed within the PCU, are connected in pairs to the two independent batteries of the spacecraft.

The two Battery Charge Regulators (BCRs) housed within the Power Control Unit are also operated under the centralised control of the MEA. In the case of these particular regulators their functioning band is located in between the shunt and BDR regimes and they are automatically initiated on entry into the sunlight phase.

The MRU also produces two system protection signals for load shedding reasons, these being the 'Eclipse' signal and the 'Disconnect Non Essential Load' signal. These two signals are routed to the power distribution equipments on the XMM-Newton spacecraft.

The mass of the XMM-Newton Main Regulator Unit (MRU) is 24 kg

Battery Management

The maximum charge rate that can be applied to the XMM-Newton batteries is 1.8 amps, this approximately equating to a C/13 rate for the battery. This charge current level can either be applied to a single battery, or equally divided between the two (0.9 amps into each).

For the XMM-Newton mission, end of battery charge management is achieved by use of redundant on-board ampere-hour monitoring with battery temperature acting as a back-up indicator. In addition total battery voltage, battery temperature and nickel-cadmium cell voltage information is processed by electronics within the MRU so as to allow spacecraft non essential load shedding and disconnection of an offending battery in case of anomalies.

Due to the application of nickel-cadmium battery cell technology for XMM-Newton, provision also exists for in-flight reconditioning of the energy storage system. The reconditioning process is conducted during permanent sunlight phases of the mission. This operation is achieved by using a 230Ω resistor.

Electrical Power Distribution

The XMM-Newton electrical power distribution is based on two independent Power Distribution Units, one for the Focal Plane Assembly (FPA-PDU) and one for the Service Module (SVM-PDU).

The two PDUs are manufactured by Terma in Denmark and although each of them are responsible for controlling different spacecraft loads, in general there is a high degree of similarity in the design of these two units.

Both PDUs are supplied with main bus power from the centrally regulated 28 volt main bus developed within the MRU. The primary function of both PDUs is to distribute and protect the spacecraft regulated bus via solid state current limited switches, the operation of these switches being controlled by serial loaded 16 bit commands and also via discrete commands in the case for the SVM-PDU.

The PDUs provide two types of current limiters, these being :

- Latching Current Limiters (LCLs)

These are devices which act as a telecommandable power switches and possess a current limited characteristic that will result in autonomous switch-off if the current limitation exceeds a defined duration.

- Foldback Current Limiters (LCLs)

These are permanently powered, (non digitally latching) devices which also possess a current limited characteristic and which in the event of output overload will exhibit a 'foldback' characteristic in order to minimise power dissipation.

At a secondary distribution level within the PDUs, distribution is also provided for a number of spacecraft heaters, this arranged as a group of simple non limiting transistor switches (TSW) which are protected by an upstream Latching Current Limiter.

In addition the PDUs provide four isolated and protected 'Keep Alive Lines' (KAL), these being arranged as two nominal and two redundant supplies. The 'Keep Alive Lines' are always present when the main bus is power active.

The SVM-PDU and FPA-PDU provide full redundancy since the architecture of these two units is divided between an 'A' and 'B' part, each part providing :

- O/P section of FCLs, LCLs and TSWs
- Serial TC/TM Interface
- Discrete Command Line Interfaces
- Discrete Telemetry Lines
- Internal auxiliary power supply
- Keep Alive Lines

Within the two PDUs, the LCLs and FCLs are independently configured such that no single point failure can lead to the loss of more than one output and any failure of one LCL/FCL will not propagate to other switches inside the assembly.

Two key signals produced within the MRU are interfaced with the PDUs these being the 'Eclipse' signal and 'Disconnect Non Essential Loads' signal.

The 'Eclipse' signal which indicates if the output current of the mid solar array panels has dropped below 6.7A, is used to autonomously ensure that all loads which are not required during eclipse are switched off by the PDUs. This function is thus a back up to ground commanding which in nominal circumstances would have already configured the spacecraft loads for eclipse operation.

The 'Disconnect Non Essential Loads' signal which is issued if one of the two spacecraft batteries is critically low during the eclipse period, is used to ensure that all equipments other than those required for spacecraft survival are switched off by the PDUs.

The mass of the XMM-Newton SVM-PDU is 16 kg whilst that of the FPA-PDU is 11 kg.

Mirror Thermal Control/Distribution

In addition to the two PDUs, one other power distribution unit exists within the electrical power system of XMM-Newton, this being the Mirror Temperature Control Unit (MTCU) which is responsible for temperature control of the Mirror Modules and Mirror Support Platform. The MTCU has been developed and manufactured by Patria Finivitec in Finland.

The MTCU has the task of processing the signals produced from 64 thermistors and in response to these measured temperatures, control appropriate heaters.

Similarly to the two PDUs, the MTCU power distribution function is performed by a combination of Latching Current Limiters and Transistor Switches. Each of the sixteen latching current limiters within the unit, (8 nominal on the 'A' side, 8 redundant on the 'B' side), control up to 3 transistor switches. The transistor switches provide the heater regulation function, by being duty cycled on and off in response to the appropriate temperature measurement.

Nominally both 'A' and 'B' parts of the MTCU are powered, although in the nominal mode of operation it is that the 'A' side that dominates the heater control function. The 'B' side however does monitor four critical temperatures and will initiate a changeover from the 'A' side to the 'B' side in the event that the temperature limits monitored by these four thermistors are violated. In this event the MTCU-B controller will :

- Switch off the 'A' side controller and LCLs
- Switch on the 'B' side LCLs

and

- Execute a default heater duty cycle on the 'B' side TSWs

Similar to the PDUs, the MTCU receives both the 'Eclipse' and 'Disconnect Essential Load Signal' from the MRU. In the first case the reaction is to switch off all heater power to the Mirror Modules whilst maintaining Mirror Support Platform heaters. In the second case the reaction is to switch off all heater loads.

The mass of the XMM-Newton MTCU is 16 kg.

Pyrotechnic Power Distribution

The Pyrotechnic Release Unit (PRU), which has been developed and manufactured by Sener of Spain, is comprised of two identical and redundant firing channels, each channel being independently powered from the one of the two spacecraft batteries.

The PRU is an internally redundant unit which contains excitation channels for both conventional pyrotechnic devices and for the 'thermal knives' which are responsible for the post launch, solar array deployment.

The circuit concept for the pyrotechnic chain is one which provides simple, resistor limited current drive to the ignitors, following an 'arming', 'ordnance selection, and 'fire' sequence of independent commands.

A similar sequence of commands is also necessary for the activating the solar array thermal knives; these being powered from the output of a DC/DC power supply in order to obtain a precise voltage level.

Following launch the thermal knives activation was performed autonomously whilst the spacecraft was out of contact but the ability also does exist to ground command the solar array deployment sequence.

The mass of the XMM-Newton Pyrotechnic Release Unit is 5.5 kg.

EPS FLIGHT PERFORMANCE

Following a nominal countdown the XMM-Newton spacecraft was launched on Ariane flight V119 at 14.32.07 UTC on the 10th of December 1999.

At the time of launch the MRU, SVM-PDU and FPS-PDU were all powered and the total main bus consumption was about 330 watts. The time from removal of ground support power to the batteries commencing charge in orbit was 57 minutes, which represented a discharged energy of 3.9 Ah from each battery.

The MRU successfully performed the powering up of the PRU at separation and at this point in time it was also confirmed that all relays within the MRU had maintained their pre-launch status.

The solar array deployment was performed as expected by the PRU exciting the nominal set of thermal knives. The cutting times of thermal knives 4 and 8 and the deployment times of the wings were the same as those recorded in the final deployment test at ESTEC.

Pictures showing the XMM-Newton solar arrays in their deployed state were recorded by the two experimental Visual Monitoring Cameras (VMCs) located on the Focal Plane Assembly of the spacecraft. These pictures were transmitted back to earth shortly following the solar array deployment sequence and one of them is shown in figure 4.

Fig.4. One of the XMM Solar Arrays

The solar array temperature sensors are located on the back of the solar array panels, these typically recording a temperature some 10 to 13 degrees C cooler than the front side. From this assumption this gave a solar cell temperature of approximately 44 deg. C for those strings powering the bus during the initial winter solstice period.

Following solar array deployment the MRU shunt regulators automatically started operation and battery charging began at the C/26 rate.

The PRU was activated as planned by the breaking of separation straps and the unit performed the automatic deployment sequences and also sent control signals to the PDUs for the attitude control and transmitter operation.

The pyro firings released the three Mirror Module doors, the Optical Monitor door and the Telescope Sun Shield (TSS). The TSS pyro was the most critical activation, as it required the simultaneous firing of two ignitors for the +Y and -Y hold-downs of the TSS.

The RFC-2 door was opened on the 25th of January and the RFC-1 door on the 6th of February. In both cases the nominal side of the PRU achieved these deployments. Since the RFC-1 door opening was the last activity to be performed by the PRU it was then switched off and will no longer be required during the mission.

To date the spacecraft electrical load has been varied between 300 watts and 1300 watts and the main bus voltage has maintained itself stable at 28.14 volts.

Following launch the battery temperatures decreased from 21 deg. C to -4.5 deg. C (+Z) and -2 deg. C (-Z). The on-board battery management was disabled at launch and re-enabled by ground command six hours after lift-off. At the beginning of March, battery re-conditioning was applied in preparation for the first eclipse season that commenced on the 6th of April.

Both PDUs were powered during the launch phase. After separation the SVM-PDU powered to AOCS units and TX-1 and all switched outputs performed as expected. The heater transistor switches were successfully configured into the post launch status by ground command.

The MTCU was switched on by ground command just three hours after launch and the unit started up as designed with the default duty cycling on chain A. Subsequent to this the heater duty cycles have been adjusted to take account of the attitude changes imposed during the Apogee raising manoeuvres and the opening of the mirror doors. Based on current measurements from the MTCU LCLs, it was confirmed that all heater mats were intact following the launch event.

In early February 2000 the first X-ray pictures were taken by the XMM-Newton spacecraft, two examples of which are shown in figure 5.

Both PDUs were powered during the launch phase. After separation the SVM-PDU powered to AOCS units and TX-1 and all switched outputs performed as expected. The heater transistor switches were successfully configured into the post launch status by ground command.

The MTCU was switched on by ground command just three hours after launch and the unit started up as designed with the default duty cycling on chain A. Subsequent to this the heater duty cycles have been adjusted to take account of the attitude changes imposed during the Apogee raising manoeuvres and the opening of the mirror doors. Based on current measurements from the MTCU LCLs, it was confirmed that all heater mats were intact following the launch event.

In early February 2000 the first X-ray pictures were taken by the XMM-Newton spacecraft, two examples of which are shown in figure 5.

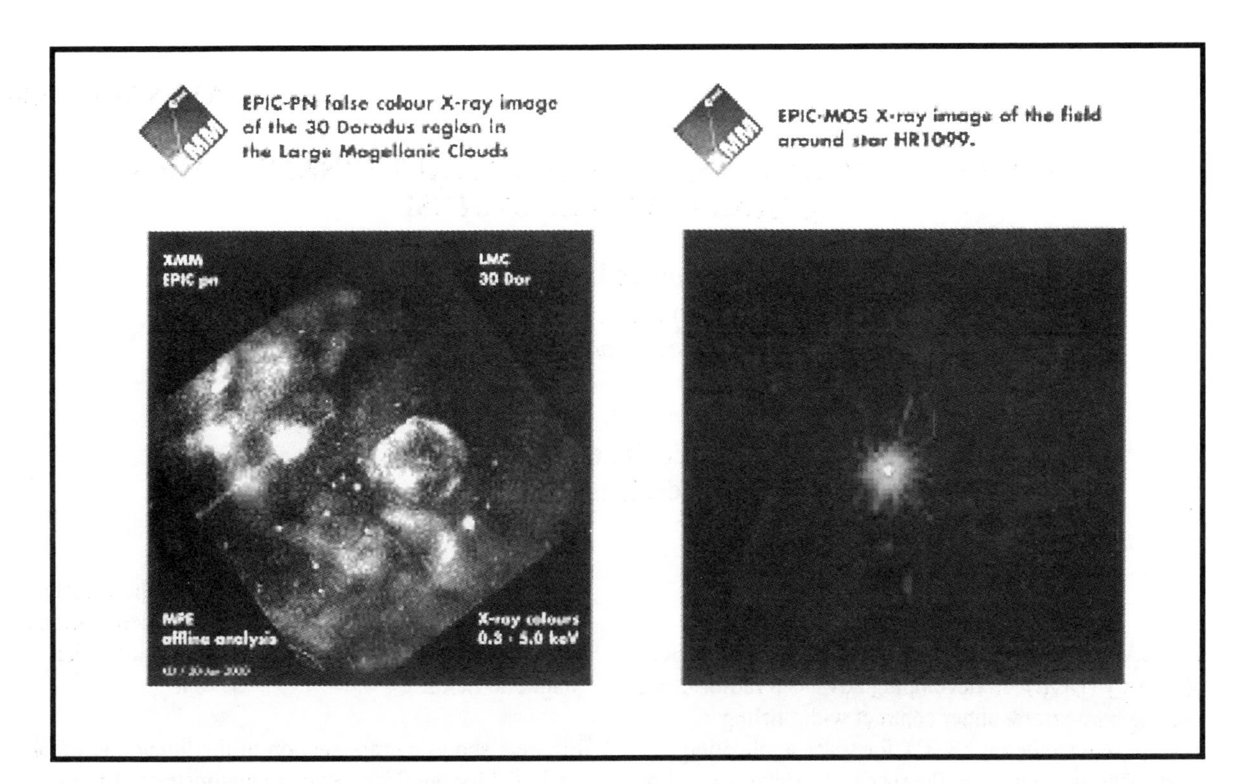

Fig.5. XMM-Newton X-ray Pictures - February 2000

CONCLUSIONS

The XMM-Newton spacecraft which was launched in December of 1999, has performed impressively to date, with a significant number of astronomical observations having already been made. In addition the Electrical Power Subsystem performed to expectations during the commissioning/verification phase of the mission and is continuing to operate nominally.

Additionally it is intended the Service Module design that was developed and manufactured for XMM-Newton programme will also be applied to the European Space Agency's INTEGRAL spacecraft, which is due for launch in 2001.

Thus a second flight of the Electrical Power System which has been described in this paper, is already a reality.

REFERENCES

World Wide Web Page : http://sci.esa.int/

ESA Brochure :
XMM : A leap forward for X-ray astronomy

XMM Project -Technical Documentation

The X-ray MultiMirror Mission Spacecraft Power Subsystem
T. Blancquaert. B. Jackson
Proc. 5th European Space Power Conference
Tarragona, Spain, 21-25 September 1999

AIAA 2000-2838

A 3-D MAGNETIC ANALYSIS OF A LINEAR ALTERNATOR FOR A STIRLING POWER SYSTEM

Steven M. Geng and Gene E. Schwarze
NASA John H. Glenn Research Center at Lewis Field
Cleveland, Ohio

Janis M. Niedra
Dynacs Engineering Co., Inc.
Cleveland, Ohio

ABSTRACT

The NASA Glenn Research Center and the Department of Energy (DOE) are developing advanced radioisotope Stirling convertors, under contract with Stirling Technology Company (STC), for space applications. Of critical importance to the successful development of the Stirling convertor for space power applications is the development of a lightweight and highly efficient linear alternator.

This paper presents a 3-D finite element method (FEM) approach for evaluating Stirling convertor linear alternators. Preliminary correlations with open-circuit voltage measurements provide an encouraging level of confidence in the model. Spatial plots of magnetic field strength (H) are presented in the region of the exciting permanent magnets. These plots identify regions of high H, where at elevated temperature and under electrical load, the potential to alter the magnetic moment of the magnets exists. This implies the need for further testing and analysis.

INTRODUCTION

Early linear alternator designs developed under the NASA Civil Space Technology Initiative (CSTI) generally used rare earth permanent magnets attached to a moving plunger. The harmonic motion of the magnets between the two gaps formed by the inner and outer stator laminations generated a sinusoidal flux in the stator. The sinusoidal flux then induced a voltage in the coils, which were wound inside the outer stator. The magnetic flux path through any of the stator laminations was coplanar with the axis of mover motion; therefore, the designs were somewhat symmetric about their axis of motion. In these early linear alternator designs, a 2-D, finite element method (FEM) analysis was found to adequately model the magnetic fields.

Figure 1 shows a cross-section of the linear alternator designed for the Space Power Demonstrator Engine (SPDE), which was developed in the 1980s by Mechanical Technology, Inc. (MTI) of Latham, New York. Reference 1 provides a discussion on the 2-D analyses that were performed. Reference 2 shows that the analytical open-circuit RMS voltage (i.e., theoretical value) of the SPDE linear alternator was within a few percentage points of the measured values over a wide frequency range.

Current linear alternator designs being developed for the Stirling radioisotope power systems for deep space missions feature stationary rare earth permanent magnets attached to the stator laminations. These designs obtain a sinusoidal, reversing magnetic flux perpendicular to the planes of the stator coils by varying the reluctance of the flux path, as seen by the individual magnets. One coil is wound around each of the stator legs. A sinusoidal voltage is induced in the coils as the moving plunger (or mover) oscillates adjacent to the permanent magnets. These designs are typically symmetric about the axis of motion to the quarter section and, as a result, a 2-D analysis can not accurately model the magnetic fields. In addition, the magnetic flux path is perpendicular to the axis of mover motion, which further complicates the issue. Linear alternators of this design are better modeled in 3-D.

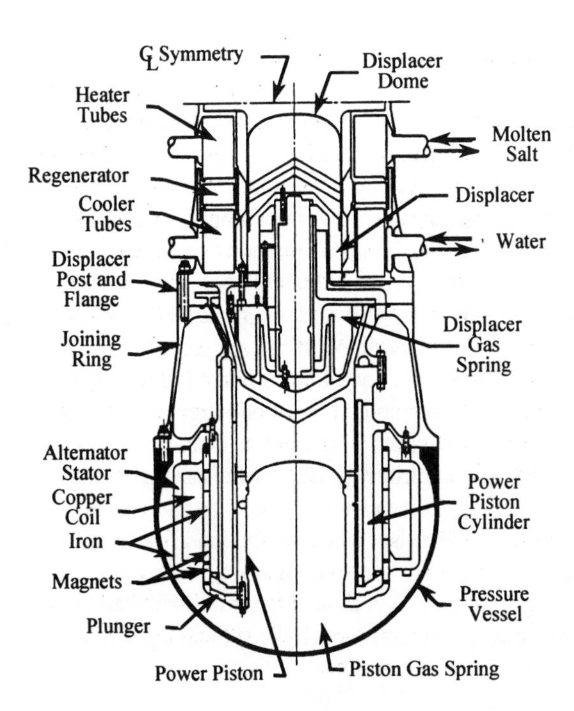

Figure 1 – Cross-section of Stirling Power Demonstrator Engine (SPDE)

Figure 2 shows a cross-section of the linear alternator designed for the 55W Technology Demonstration Convertor (TDC), which was developed in the 1990s by Stirling Technology Company (STC) of Kennewick, Washington, for the Department of Energy (DOE). A more detailed description of the TDC is given in References 3 and 4. The linear alternator model and results discussed in this paper are based on this STC 55W TDC linear alternator design.

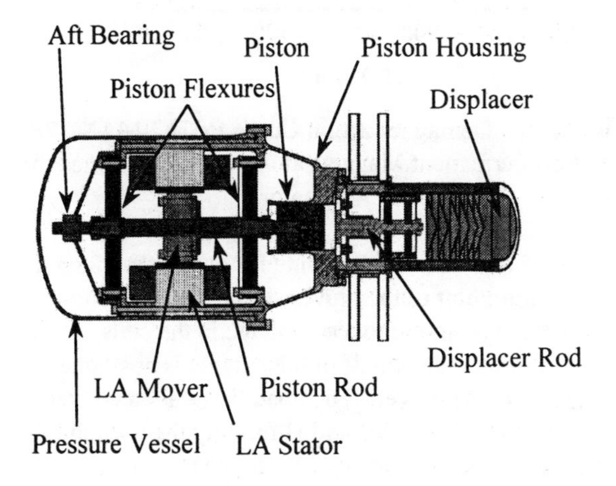

Figure 2 – Cross-section of STC's 55W Technology Demonstration Convertor (TDC)

OBJECTIVE

The objective of this work was to develop a numerical analysis model that could aid engineers in the design, development, evaluation, and understanding of advanced linear alternators.

LINEAR ALTERNATOR MAGNETOSTATIC MODEL

A three-dimensional magnetostatic model of STC's 55W linear alternator was developed using Ansoft's Maxwell 3-D finite element method (FEM) software (Ref. 5). The magnetostatic model consisted of a quarter section of the linear alternator and included the following components: 1) ¼ stack of Hyperco-50 mover laminations, 2) Neodymium-Iron-Boron half-magnets, 3) ¼ stack of Hyperco-50 stator laminations, and 4) copper half-coils. Since the actual hardware is symmetric to the quarter section about the axis of motion, only a quarter section was needed in the model.

Figure 3 shows the flux density (B) and field strength (H) data used in the model for the Hyperco-50 mover and stator laminations. The data shown in this plot was measured at the NASA Glenn Research Center. For illustrative purposes in the FEM model, a remanence B_r of 1.26 Tesla and a coercive force H_c of –931 kA/m were assumed. These values are consistent with the STC specifications for the magnets and are similar to the properties of UGIMAX 37B at 23°C.

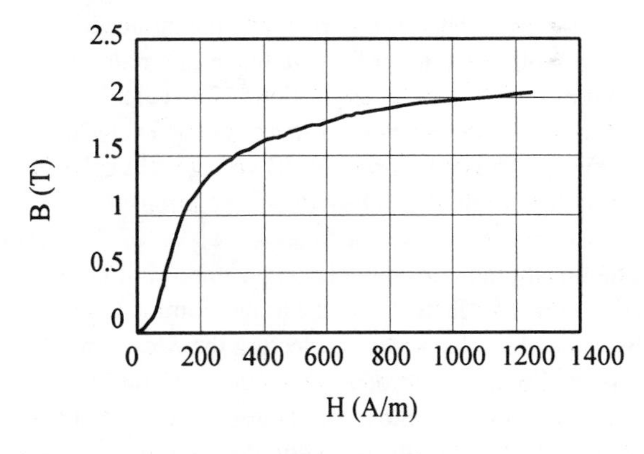

Figure 3 – BH Curve of Hyperco-50

The primary model parameter of interest for verification with experimental data is the magnetic field parallel to the axes of the coil windings. It is this value of magnetic flux that is used to calculate the open-circuit voltage. To reduce the numerical error associated with the FEM mesh that is used to calculate the magnetic flux, namely the mesh for the coil, stator, and coil/stator gap, 9 cut-planes were defined along the axis of the coil. The 9 cut-planes are shown in Figure 4. The model was set-up to calculate and average the magnetic flux normal to each of these cut-planes to obtain an overall average flux parallel to the axis of the coil.

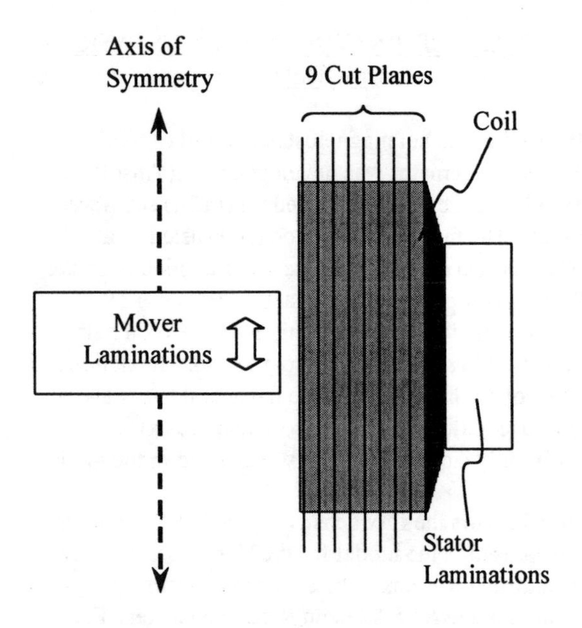

Figure 4 – Cut-Planes through Coil Used to Calculate Average B Perpendicular to Coil Plane

Magnetostatic flux solutions for a series of 13 different mover positions were generated. Faraday's Law of induction was then used to calculate the induced electromotive force (emf) in the linear alternator coil windings at the 13 mover positions. The calculated induced emf was then curve-fit, using the TableCurve software. The open-circuit RMS voltage was calculated using the amplitude of the curve-fit solution.

The linear alternator model being a static simulation, eddy current effects due to the mover motion could not be calculated. However, neglecting the eddy current effects should not introduce large errors to the static solutions, since the mover oscillates at a low frequency (≈80 Hz) and all of the magnetic flux conducting structures were laminated (except for the magnets).

RESULTS

Internal State of Magnets

Figure 5 shows the demagnetization curves estimated for the UGIMAX 37B magnets at 23°C and 75°C. The 23°C demagnetization curve was used in the magnetostatic analysis, since STC measures the linear alternator open-circuit voltage with the magnets at room temperature. Several load lines are shown in this figure with respect to one particular magnet. The load lines representing the mover positioned adjacent to this magnet and away from this magnet (i.e. at the ends of the mover stroke) were calculated based on the magnetic flux density B and magnetic field strength H averaged over the volume of the magnet. For the case where the mover is positioned adjacent to the magnet, the average load line is a fairly good representation of the internal operation of the magnet, since B and H are then somewhat uniform. But for the case where the mover is positioned away from the magnet, large variations in H exist. The average load line in this case may indicate a false margin of safety relative to demagnetization. Load lines for small, localized areas of the magnet may approach the critical knee where demagnetization can occur.

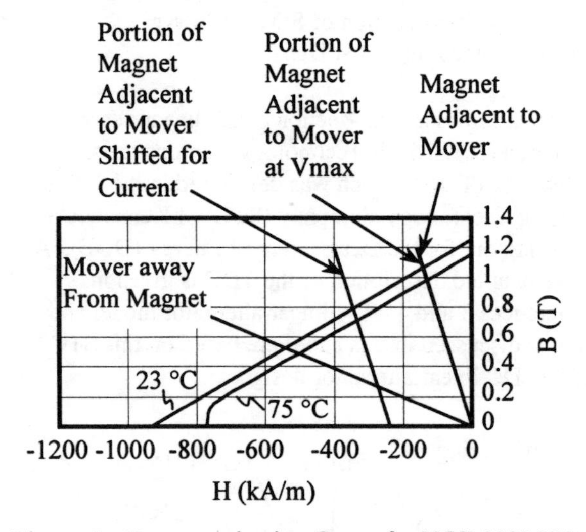

Figure 5 – Demagnetization Curve for UGIMAX 37B NeFeB Permanent Magnets along with Load Lines for the 55W TDC Linear Alternator.

Figure 5 also shows a load line for one of the magnets at the mid-point of the mover stroke, which corresponds with the maximum induced voltage in the coils. In this case, approximately half of the magnet is positioned adjacent to the mover. This load line was calculated based on the average B and H for only the portion of the magnet positioned adjacent to the mover. The reason for averaging over only this portion is that it is only this portion that is within the magnetic circuit at the mover mid-stroke position. This load line practically falls right on top of the magnet-adjacent-to-mover load line described earlier. The load line was then shifted by –238 kA/m (number of turns multiplied by the peak current divided by the thickness of the magnet and air gap) to account for the demagnetization effect caused by current flowing through the coil. Equations derived by Niedra (Ref. 6, Appendix A) were used to verify the load lines for the covered portions of the magnet and agreed well with the FEM model. Based on the load lines shown in Figure 5, it appears that the self-demagnetization effect of the uncovered magnets are more of a concern than the demagnetization effect caused by current flowing through the coils. A closer look at localized areas of the magnets is needed.

Figure 6 is a plot of magnetic field strength (H) vs. distance along one stator leg for the mover positioned at mid-stroke. This is the mover position for peak induced voltage in the coils. The plot shows three curves plotted along parallel lines which cut through the magnets: one represents a line along the inside surface (mover side of magnet), one represents a line along the center, and one represents a line along the outside surface (stator side of magnet) of the magnet pair. Each line is parallel to the mover motion (the z-direction) and lies in the plane of the magnet symmetry. This plot shows a localized maximum demagnetization field of roughly 670 kA/m on the inside surface of the uncovered magnets. The plot also shows a field spike of 680 kA/m on the inside surface of the magnets, where the two magnets meet. This spike is due to H_z, (the H in the z-direction), as predicted by Ampere's Law. The curves for the center and outside magnet surfaces in the covered regions of the magnets show a much smaller variation in the demagnetization fields. A further refinement of the FEM mesh in this region of the model should be explored to resolve irregularities of computation.

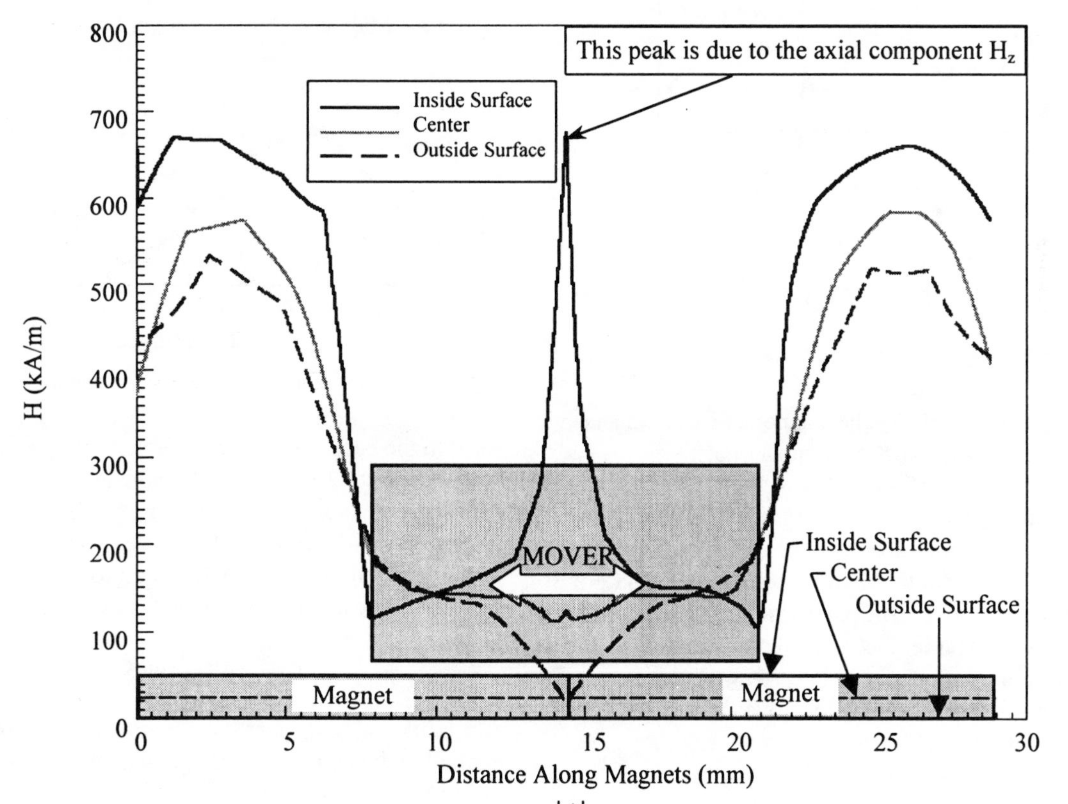

Figure 6 – Magnetic Field Strength $|\vec{H}|$ vs. Distance along Magnets at 23°C

Average Properties at a Higher Temperature

In actual Stirling convertor applications, the magnet temperatures may reach as high as 75°C. The model was used to generate the BH trace for the magnets over a cycle at the expected operating temperature. Figure 7 shows a plot of average B vs. average H, for the alternator magnets as the mover oscillates with a 6-mm amplitude. B and H are averaged over the volume of the magnet. This plot shows that the minimum average values of B and H are 0.58 Tesla and –503 kA/m, respectively. The critical H value for the Neodymium-Iron-Boron magnets assumed in the model is approximately –750 kA/m at 75°C. For H values ≤–750 kA/m, demagnetization may occur. Based on the large variations in the magnetic field strength seen in the magnets at the lower temperature, peak H values for localized portions of the magnets may exceed –750 kA/m at the higher temperature.

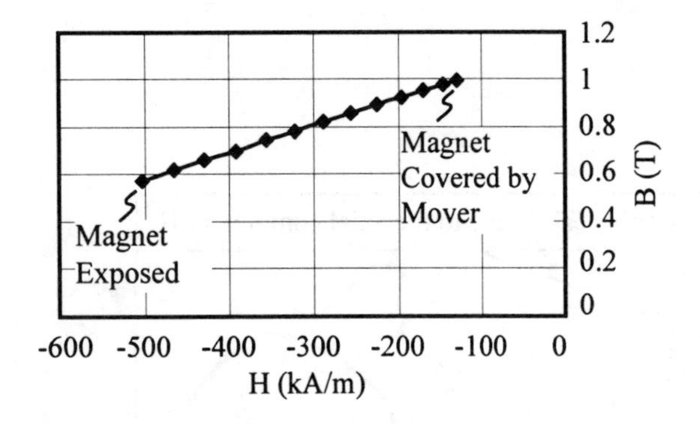

Figure 7 – Magnetic Flux Density vs. Field Strength Averaged over Magnet Volume at 75°C

Induced Voltage

Figure 8 shows a plot of the magnetic flux linking each alternator coil as a function of mover position. Although the peak flux of 1.52E-4 Webers occurs at the ends of the mover stroke, the largest change in magnetic flux with respect to time (and therefore the largest induced emf in the coils) occurs at the mid-point of the mover stroke. The plot shows that the flux is almost linear with respect to mover position. The fact that this plot is not perfectly linear may be due to minor variations in flux fringing with mover position or it may indicate minor inaccuracies associated with the numerical solution.

The magnetostatic flux solutions were considered converged, once the energy error between iterations (as defined by the Maxwell software) became less than 2%. The 2% convergence criterion was selected as a trade-off between accuracy and computation time.

Figure 9 shows a plot of the predicted alternator open-circuit voltage as a function of time. The diamond markers show the model predictions. The solid line is a sine-wave curve fit of the points. The curve fit was used to calculate the RMS open-circuit voltage for comparison with the measured value. The shape of the voltage curve should be nearly sinusoidal. The fact that the model predictions do not lie on a sinusoidal curve may once again be due to either flux fringing variations or numerical inaccuracies (c.f. Fig. 8).

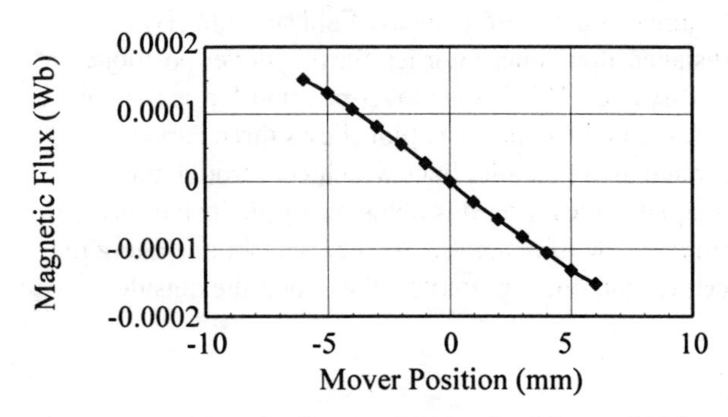

Figure 8 – Magnetic Flux vs. Mover Position at 23°C

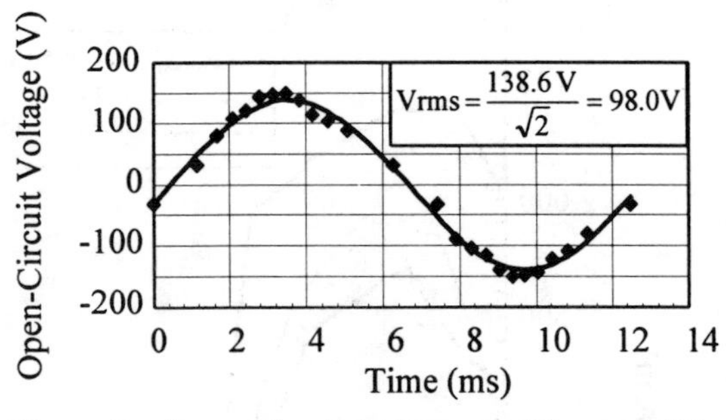

Figure 9 – Open-Circuit Voltage vs. Time at 23°C

As with the earlier SPDE open-circuit voltage comparisons to 2-D analytical predictions, the TDC open-circuit voltage correlates closely with the new preliminary 3-D analysis. As the model is developed further and validated with more comprehensive test data, it will be an increasingly valuable design tool.

CONCLUSIONS

A commercial magnetics computation tool has been applied to develop a valuable numerical analysis model of linear alternator designs for Stirling power convertors. This model can be used to explore design variations early in the development process. However, the model needs to be further validated through additional experimentation.

The results shown by the preliminary 3-D analysis indicate that a more comprehensive analysis and measurements are appropriate for the 55W TDC linear alternator design, especially since the magnet temperatures may reach 75°C in space power applications. The portion of the magnet with the highest potential for demagnetization appears to be on the surface of the magnet pair at the edge, where the magnets are in contact with each other and adjacent to the mover. The region near an advancing/receding edge of the mover is a location where fringing fields due to load current can reinforce the already high self-demagnetizing field in the area of a magnet not adjacent to the mover. The results presented here support the need for a detailed study of load current effects in these critical regions.

Special care is needed when selecting magnets because of the high, localized demagnetizing fields. Magnets with a higher resistance to demagnetization are being used in the TDC convertors currently being fabricated by STC for NASA Glenn.

ACKNOWLEDGEMENTS

The authors wish to acknowledge the Stirling Technology Company for providing hardware information and test data relative to the DOE/STC Technology Demonstration Convertor.

REFERENCES

1. Brown, A.T.,: "Space Power Demonstrator Engine, Phase I Final Report," NASA CR-179555, 1987.
2. Dochat, G.,: "SPDE/SPRE Final Summary Report," NASA CR-187086, 1993.
3. Thieme, L.G., Qiu, S., White, M.A.,: "Technology Development for a Stirling Radioisotope Power System," NASA TM-2000-209791, 2000.
4. White, M.A., Qiu, S., Augenblick, J.E.,: "Preliminary Test Results from a Free-Piston Stirling Engine Technology Demonstration Program to Support Advanced Radioisotope Space Power Applications," in Proceedings from STAIF–00, Space Technology and Applications International Forum–2000, Albuquerque, New Mexico, 2000.
5. Ansoft's Maxwell 3-D simulator, see World Wide Web page: www.ansoft.com
6. Niedra, J.M.,: "Lightweight Linear Alternators With and Without Capacitive Tuning," NASA CR-185273, 1993

AIAA-2000-2839

DYNAMIC CAPABILITY OF AN OPERATING STIRLING CONVERTOR

Thomas W. Goodnight*

William O. Hughes*
Associate Fellow AIAA

Mark E. McNelis*
Senior Member AIAA

ABSTRACT

The NASA John H. Glenn Research Center and the U.S. Department of Energy are currently developing a Stirling convertor for use as an advanced spacecraft power system for future NASA deep-space missions. NASA Headquarters has recently identified the Stirling technology generator for potential use as the spacecraft power system for two of NASA's new missions, the Europa Orbiter and the Solar Probe missions (planned for launch in 2006 and 2007 respectively).

As part of the development of this power system, a Stirling Technology Demonstration Convertor was vibration tested at NASA John H. Glenn Research Center to verify its survivability and capability of withstanding the harsh dynamic environment typically seen by the spacecraft when it is launched by an expendable launch vehicle.

The Technology Demonstration Convertor was fully operational (producing power) during the random vibration testing. The output power of the convertor and other convertor performance indicators were measured during the testing, and these results are discussed in this paper. Numerous accelerometers and force gauges also were used to provide information on the dynamic characteristics of the Technology Demonstration Convertor and as an indication of any possible damage due to the vibration. These measurements will also be discussed in this paper.

---

*Aerospace Engineer at NASA Glenn Research Center, Cleveland, Ohio, USA

The vibration testing of the Stirling Technology Demonstration Convertor was extremely successful. The Technology Demonstration Convertor survived all its vibration testing with no structural damage or functional performance degradation. As a result of this testing, the Stirling convertor's capability to withstand vibration has been demonstrated, enabling its usage in future spacecraft power systems.

INTRODUCTION

NASA John H. Glenn Research Center (GRC) and the U.S. Department of Energy (DOE) are currently developing a Stirling convertor for an advanced radioisotope power system that will provide spacecraft on-board electric power for NASA deep-space missions. Stirling Technology Company (STC) of Kennewick, WA is under contract to DOE to develop a radioisotope Stirling convertor. NASA GRC is providing technical consultation for this effort based on their expertise in Stirling technologies dating back to the mid-1970's.

In May 2000, NASA Headquarters identified the Stirling technology generator as an alternative space power system to Radioisotope Thermoelectric Generators (RTGs). The Stirling technology generator may potentially be used as the spacecraft power system for two new NASA deep-space missions. These are the Europa Orbiter and the Solar Probe missions, which are planned for launch in 2006 and 2007, respectively.

The Stirling system is an attractive alternative to the RTG system. Due to the Stirling system's efficiency (over 20 percent) just one-third the amount of Plutonium is required compared to the RTGs, thereby significantly reducing nuclear fuel cost.

Multiple Stirling units have demonstrated convertor power, efficiency and long life. Currently, one STC terrestrial radioisotope Stirling convertor is operating maintenance-free and without performance degradation after over 53,000 hours (6+ years). Such long lifetimes are required for deep-space missions to the outer planets where solar power is not an option.

Figure 1 illustrates the basic concepts of a Stirling convertor. Designing this convertor with long life and high reliability entails eliminating wear mechanisms and the proper selection and verification of materials. References 1 and 2 provide further insight into Stirling technology and current developmental efforts.

Before this convertor could be selected as a viable space-flight power system, a test program was needed to determine whether it could survive and function after exposure to the harsh launch vibration environment.

A spacecraft, and the components mounted to it, are exposed to both direct and indirect random vibration during its launch on an expendable launch vehicle. Random vibration from the vehicle's engines and rockets are directly transmitted through the vehicle to the spacecraft via its structural connection path. Usually however it is the indirect vibration that is the dominant source of random vibration to the spacecraft. This vibration is produced by the acoustic pressure waves produced during the vehicle's liftoff. These waves are transmitted through the vehicle's payload fairing protective layer and impinge on the spacecraft. It is typical for spacecraft to be exposed to overall sound pressure levels of 135 to 145 decibels (dB) during a liftoff. (This compares to a level of 100 dB that is produced by a modern jet fly-over at 1500 feet. On the decibel scale, every increase of 3 dB is a doubling of the mean square acoustic pressure.)

The high acoustic levels produced during the launch results in significant vibration to the spacecraft and its components. It is for this reason that the Stirling convertor underwent random vibration testing.

To address this issue a DOE/STC 55-We (electric watt) Stirling Technology Demonstration Convertor (TDC) was tested at NASA GRC's Structural Dynamics Laboratory (SDL). The TDC's heat source was electricity, not nuclear fuel. During this test program, the TDC was exposed to high random vibration levels and durations that simulated the maximum expected launch vibration.

VIBRATION TEST DESCRIPTION

On November 29 - December 2, 1999 vibration tests were performed on the Stirling TDC. The testing followed the requirements of the test plan (Reference 3). The Stirling TDC was supported in a specially made test fixture, as shown in Figure 2. This fixture was, in turn, attached to a slip table driven by a MB Dynamics Model C210 (28,000 pound-force) horizontal electrodynamic shaker. Testing was performed in both the lateral (X, perpendicular to the TDC's piston stroke) and axial (Y, direction of TDC's piston stroke) directions. The Stirling TDC was fully operational (producing power) during the vibration tests.

Two accelerometers located on the test fixture were used to control the vibration test input. Eleven other accelerometers, located on the fixture and TDC, were used to measure the response. The TDC was mounted in its test fixture by eight-force gauge measuring bolts.

Other non-vibration measurements were recorded during the testing to monitor the convertor's performance. These measurements included the input voltage and current at the heater end, the output voltage and current at the alternator end, the average output electric power, and the TDC's temperature and internal dynamic pressure.

Since little was known about the TDC's capability to withstand vibration, it was decided to incrementally test it to higher random vibration levels. Thus the TDC was initially tested to a modified workmanship (NASA-STD-7001) level of 6.8 Grms. This is a level of vibration that historically has been shown to detect hidden flaws or defects in materials, production or assembly for electronic and mechanical spaceflight hardware components.

Next the TDC was shaken to the higher Jet Propulsion Laboratory (JPL) vibration levels (-6 dB, -3 dB and 0 dB relative to the qualification level of 12.3 Grms). JPL's standard method is to test qualify its hardware to levels which are 3 dB greater than the maximum expected flight environment. Thus the "qualification - 3 dB" level represents a "flight" level. However, due to unknowns in both the launch vehicle and in the spacecraft mounting details, the JPL "flight" test levels should be representative of typical flight levels seen by a spacecraft's components during liftoff. The qualification test levels used for these tests are considered to provide margin above this typical "flight".

These test levels are shown in Figure 3. All random test durations were for one minute per axis, except for the 3 minutes per axis qualification tests.

To closely monitor the performance and structural

integrity of the TDC, low-level (0.25 g) sine sweep tests were performed before and after each major random vibration test. This provided a means of observing any structural changes caused by the random testing. Functional and electrical tests and visual inspections were also performed before and after each dynamic test (random or sine sweep) to monitor the TDC's functional performance and structural integrity.

VIBRATION TEST RESULTS

General

Figure 4 shows the test matrix and the order in which the tests were performed. The control of the test input was excellent for all tests.

The objectives of this test program were met and exceeded. The Stirling TDC was exposed to significant (greater than expected launch) vibration test levels and durations, and successfully passed post-test functional testing. The most severe test was to the JPL qualification test levels (12.3 Grms, 0.2 g^2/Hz from 50 to 250 Hz, 3 minutes/axis, in both the axial and lateral directions). As a comparison, the RTGs used for the Galileo/Ulysses/Cassini missions were test qualified to 7.7 Grms (laterally) and 6.1 Grms (axially).

The Stirling TDC survived all dynamic testing with no structural damage or functional performance degradation. The Stirling TDC operated at full-stroke and produced power during all its vibration testing. Dithering (reducing) the piston stroke was not necessary. The TDC produced full power at the end of each and every vibration test.

Power Performance

The power produced by the Stirling TDC did vary during testing. During the lateral random vibration testing, a slow steady degradation of output power was observed (on average) as a function of increasing vibration test input, but the power immediately returned to the full and stable level when each vibration test ended. This observed trend is illustrated in Figure 5. The mean and standard deviation (sigma) statistics were numerically calculated from periodically observed data taken from a wattmeter during each test. The "mean +/- 2 sigma" levels statistically represents the levels that the observed power should be recorded 95.45% of the time. This measure closely correlated with the actual range of the output power data from the wattmeter. The change in power during the lateral testing is believed due to the increased friction and leakage losses caused by the lateral vibration.

During the axial direction random vibration testing, the power output remained, on average, at full level, as shown in Figure 6. Instantaneous surges and dips were observed however, which were believed to be due to shaking the TDC in the direction of its piston stroke. These variations as measured by the standard deviation (sigma) calculation were larger in the axial direction than in the lateral direction. Once again, the power immediately returned to the full and stable level when each vibration test ended.

The output power plotted as a function of time is shown in Figures 7-9 for three different cases. Arbitrary time slices of 0.2 seconds are shown for illustration purposes, although the signal's frequency characteristics were consistent throughout the entire test duration. These instantaneous output power time histories were computed by multiplying the instantaneous output voltage by the instantaneous output current. The amplitudes are normalized and plotted on the same scale.

Figure 7 shows the "solo" production or the output power time history that is produced with no external vibration. This time plot shows a primary frequency of 158 Hz, or 2 times the piston stroke frequency of 79 Hz. It also shows a slight modulation frequency of 19 Hz.

Figure 8 shows the time history for the qualification test level in the lateral direction. Although the relative amplitude (compared to solo) is reduced, as expected from Figure 5, the frequency content of 158 Hz remains. Now the modulation is more pronounced and occurs at 8 to 9 Hz.

Likewise, Figure 9 shows the time history for the qualification - 6 dB test level in the axial direction. In this case the average amplitude is comparative to that of the solo condition, as expected from Figure 6. Once again the primary frequency of the output power is 158 Hz, with a modulating frequency of 8 to 9 Hz.

Frequency spectrums of the output power for these three tests were calculated using the entire available time histories. These output power spectral densities are shown in Figures 10 (solo), 11 (qualification in lateral direction) and 12 (qualification - 6 dB in axial direction). Note though that the amplitude is plotted on a log axis. Thus each major division is a factor of 10 on amplitude.

Figure 10 shows that 158 Hz dominates the solo output power, and to a much lesser extent the higher even harmonics of 79 Hz (316 Hz, 474 Hz, 632 Hz and 790 Hz). The presence of the modulating frequency of 19 Hz and the associated side-bands are also evident. Note that all these frequencies are very sharp and pure (sinusoidal) for the solo condition.

Figure 11 shows the frequency spectrum for the qualification test in the lateral direction. Once again note the dominating presence of 158 Hz and the other even harmonics of 79 Hz. A major difference however is the apparent spreading of the power into the "skirt" of each even harmonic. This spreading is attributable to the vibration excitation slightly distorting the pure piston stroke frequency. This observation should be tempered by remembering the magnitude is on a log axis. The small side-band frequencies, most notable at 158 Hz, seems to be the modulating effect of the 8 Hz frequency.

The frequency spectrum for the qualification - 6 dB test in the axial direction is shown in Figure 12. Similar conclusions on the effects of vibration can be made here as was done for Figure 11.

The output voltage, current and internal dynamic pressure all exhibited similar characteristics under the vibration excitation, peaking at the piston stroke frequency of 79 Hz. Figure 13 shows the frequency spectrum of the output voltage for the lateral direction tests for various test levels (solo, qualification - 6 dB, and qualification). This plot shows that the spreading of the signal (breadth of the skirt) correlates with the level of vibration excitation, as expected.

Figure 14 shows the output current for the qualification - 6 dB test level for the lateral and axial tests and for the solo condition. Here one sees that vibrating in the axial direction distorts the pure piston stroke even more than the lateral excitation does.

Structural Dynamic Response

The Stirling TDC structure was found to be dynamically well behaved and linear, with reasonable damping (due to its multi-layered design and inherent internal gas damping). The pre and post sine testing showed no significant changes in the dynamic characteristics of the TDC measurements, due to its exposure to the random environments.

The lateral direction sine sweep data shows a convertor casing structural resonance at 1030 Hz, with a mechanical Q (dynamic amplification) of 5 to 8. In the axial directions, the sine sweep data shows possible convertor casing/fixture structural resonances at 1492 Hz and 1730 Hz, with mechanical Q values of 2 to 5. In both directions, the structural resonances are high enough in frequency to avoid any dynamic coupling problems with the Stirling's piston stroke frequency of 79 Hz. The TDC responded rigidly to the random test input below 350 Hz.

Additionally details on the Stirling TDC's structural dynamic response during the vibration test may be found in References 4 and 5.

POST VIBRATION TESTING EVALUATION

Following the vibration testing at NASA GRC, the Stirling TDC was operated for an additional 10 million cycles (35 hours) to demonstrate its post-test life. On January 4, 2000 a team of STC, NASA GRC and Lockheed Martin personnel participated in the disassembly and inspection of the TDC unit at STC. This team concluded that there were no areas of concern attributable to the vibration testing (Reference 6). The bases for their conclusion were: (1) satisfactory alignment for all moving parts and properly retained clearances, (2) no broken parts found inside convertor and intact structural integrity, (3) all fasteners retained adequate tightness, and (4) all post-test electrical measurements were consistent with pre-vibration test values.

STRUCTURAL MOUNTING CONSIDERATIONS

The Stirling TDC's structural characteristics (refer to Figures 1 and 2) were utilized in achieving a successful vibration test. The TDC has two bulbous ends, one end larger than the other end. The larger end, the pressure vessel, houses the heavier mass of the linear alternator. The smaller end contains the thermally active working elements of the heater head assembly, including the displacer. A continuous conic section, called the piston housing, is centrally located between these two ends. As tested, the piston housing has two flanges. These two flanges are the only mounting planes for attachment for the test fixture (or for future spacecraft systems). For a flight system, heat will be rejected, via a radiator mount, from the system near the smaller flange. The linear alternator end cap is cantilevered from the larger flange. The center of gravity of the TDC is just outside the large flange on the linear alternator end.

The piston housing is central to the success of the TDC because it contains the piston, and the displacer rod is mounted from the smaller flange. Close clearance pressure seals are used on both the piston and displacer rod. The piston transmits the power generated from the heater head assembly to the linear alternator.

The two flanges at each end of the piston housing provided a choice of a large or small bolt circle for mounting the TDC to the test fixture. The rest of the TDC is supported from the piston housing. For ground operations, there is a water loop on the heater head assembly, adjacent to the piston housing, that limited access to the smaller ring. These water loop fittings were well supported during the vibration test to alleviate any vibration concern.

In fixture design, a rigid simulation of the interface is required. In other words, the impedance of the fixture as measured at the mounting face must be large as compared to the mounting flange compliance. For this test program, due to clearance issues and available mounting flanges, the fixture block enveloped the linear alternator (large bulbous end). This allowed sufficient surface area to seat the eight-instrumented force bolts on 0.25-inch spacers. These bolts were inserted through the large outer-diameter mounting flange and torqued directly to the aluminum fixture block.

For the purposes of system integration, previous Stirling system designs have shown mountings at the small flange. In this test, the water loop fittings made that flange inaccessible. Even if it was accessible however, the larger flange is still highly recommended for future mounting systems from a structural perspective. By mounting from the larger ring, the net results are that the critical clearances and pressure seals of the piston housing were lightly strained and the resulting overhung mass moments (produced in the simulated launch environment vibration) were minimized. Choosing the larger flange for the test mounting condition, in conjunction with the TDC's inherent damping, is thought to be a key contributor to the TDC's successful vibration test.

CONCLUSIONS

The Stirling convertor is a promising and viable alternative to the RTG for use as an electric on-board power source for future NASA deep-space missions. The vibration testing performed at NASA Glenn Research Center has proven that the Stirling convertor can survive and function after exposure to its expected launch random vibration environment. Both structural integrity and power performance were retained throughout and at the completion of the vibration testing. The structural dynamics of the convertor needs to be considered when integrating it to the rest of the spacecraft power system in order to minimize the vibration loads on critical internal components.

ACKNOWLEDGEMENTS

The authors acknowledge the support and expertise of Richard Shaltens, Jeff Schreiber, Lanny Thieme, Steve Geng, Lee Mason and Ken Mellott/NASA GRC, the NASA GRC SDL technical staff, Rich Furlong/DOE, Maury White and Ian Williford/STC, and Kurng Chang/JPL. A special note of thanks goes to the SDL's test engineer Sergey Samorezov/Zin Technologies for his expertise in producing and analyzing the test data from this test program.

REFERENCES

1. "Technology Demonstration of a Free-Piston Stirling Advanced Radioisotope Space Power System," M. A. White, S. Qiu, R. W. Olan and R. M. Erbeznik, *Proceedings for the Space Technology & Applications International Forum (STAIF-99)*, Albuquerque, New Mexico, 30 January-4 February 1999.
2. "Technology Development for a Stirling Radioisotope Power System," L. G. Thieme, S. Qiu, and M. A. White, NASA Technical Memorandum-2000-209791, March 2000.
3. "Vibration Test Plan to Characterize the TDC Stirling Engine for Future Technology Development," approved by R. Furlong (DOE), M. White (STC), J. Schreiber (NASA GRC), and T. Goodnight (NASA GRC), November 23, 1999.
4. "Technology Demonstrator Convertor (TDC) Stirling Engine, Prototype Unit, Vibration Test Report," S. Samorezov, NASA GRC Test Report SDL-TR 99-37, 24 January 2000.
5. "Vibration Testing of an Operating Stirling Convertor, " by W. O. Hughes, M. E. McNelis, T. W. Goodnight, Published in the Proceedings of the Seventh International Congress on Sound and Vibration, Garmisch-Partenkirchen, Germany, July 4-7, 2000.
6. "STC Convertor Disassembly and Inspection," by L. Mason, NASA GRC memorandum, January 19, 2000.

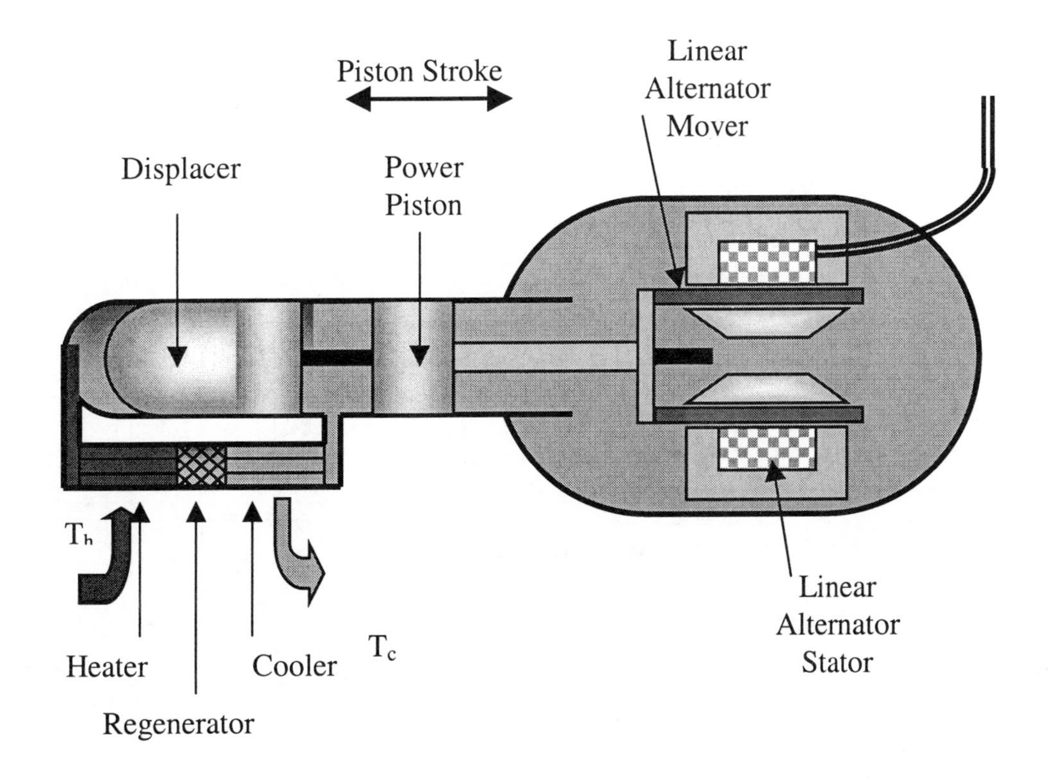

Figure 1. General Free-Piston Stirling Convertor Schematic

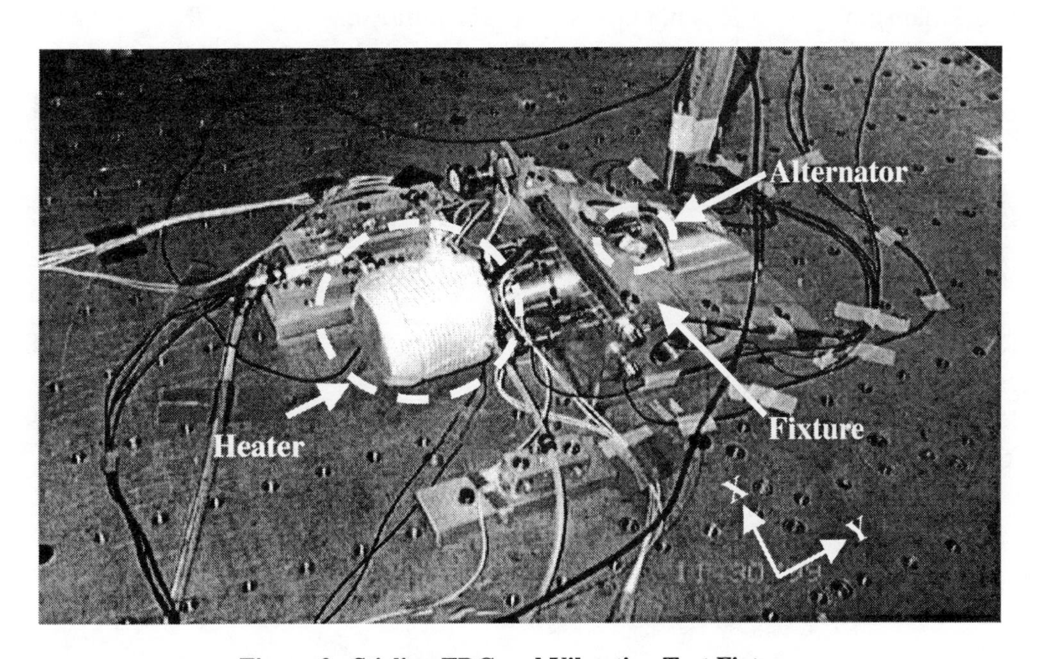

Figure 2. Stirling TDC and Vibration Test Fixture

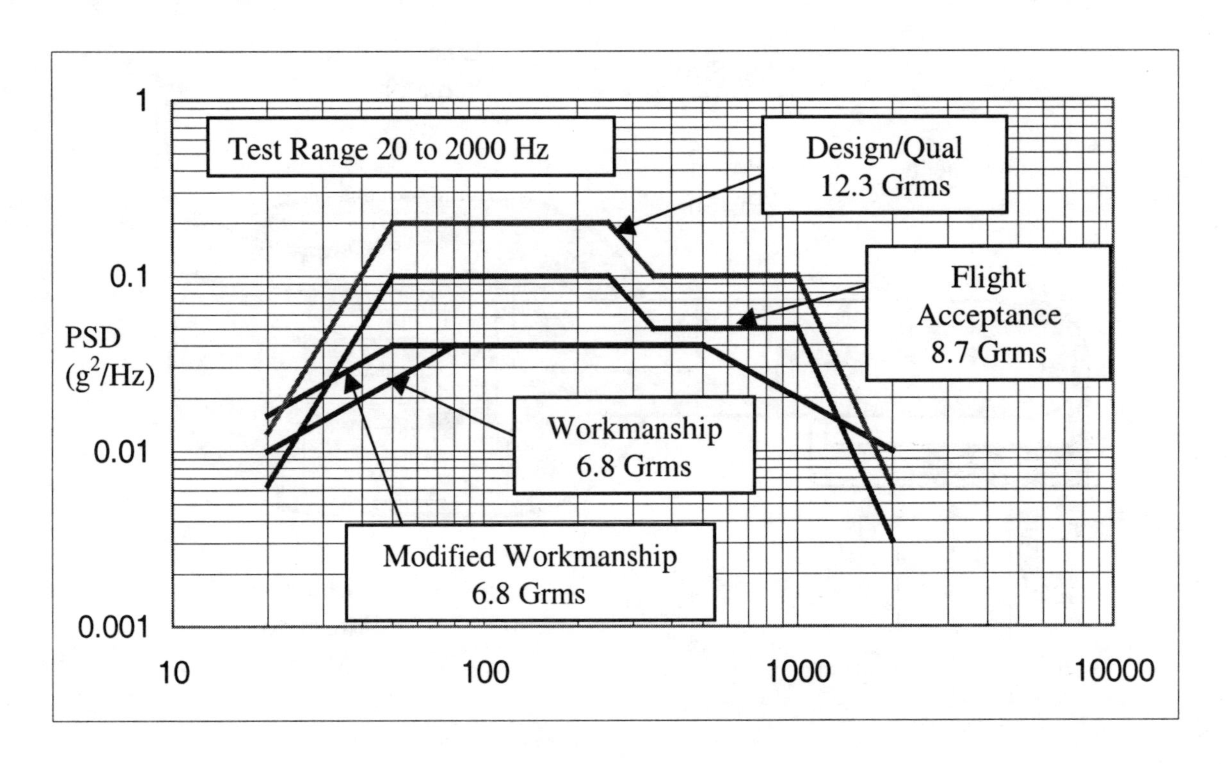

Figure 3. Stirling TDC Random Vibration Test Levels

Test Description	Test Magnitude	Test Duration	Axial * Direction	Lateral* Direction
Pre Sine Sweep	0.25 g	4 octaves/min	6	1
Workmanship Random Vibration	6.8 Grms	1 min/axes	7	2
Post Sine Sweep	0.25 g	4 octaves/min	8	3
Sine Sweep	0.375 g	4 octaves/min	9	4
Sine Sweep	0.50 g	4 octaves/min	10	5
Pre Sine Sweep	0.25 g	4 octaves/min	11	16
JPL Random Vibration Qualification - 6 dB	6.2 Grms	1 min/axes	12	17
JPL Random Vibration Qualification - 3 dB = Flight	8.7 Grms	1 min/axes	13	18
JPL Random Vibration Qualification	12.3 Grms	3 min/axes	14	19
Post Sine Sweep	0.25 g	4 octaves/min	15	20

Figure 4. Stirling TDC Vibration Test Matrix (* Order of testing; Random tests from 20 to 2000 Hz; Sine tests from 5 to 2000 Hz)

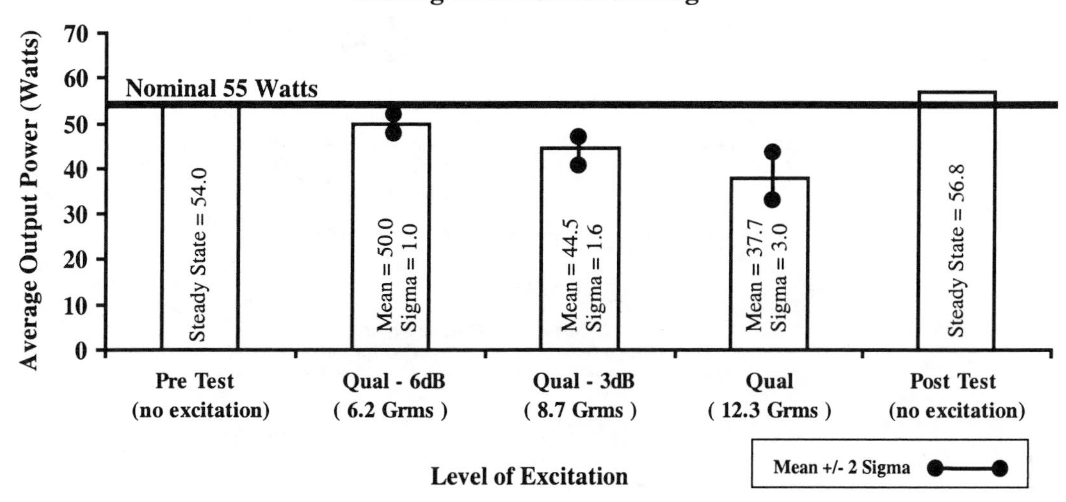

Figure 5. Comparison of Output Power versus Level of Input Vibration in Lateral Direction.

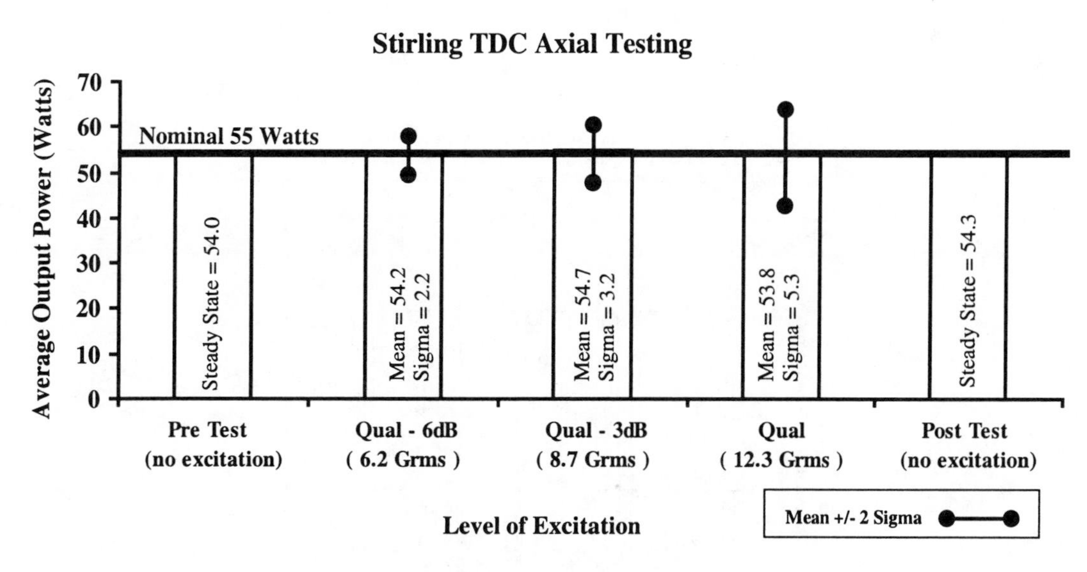

Figure 6. Comparison of Output Power versus Level of Input Vibration in Axial Direction.

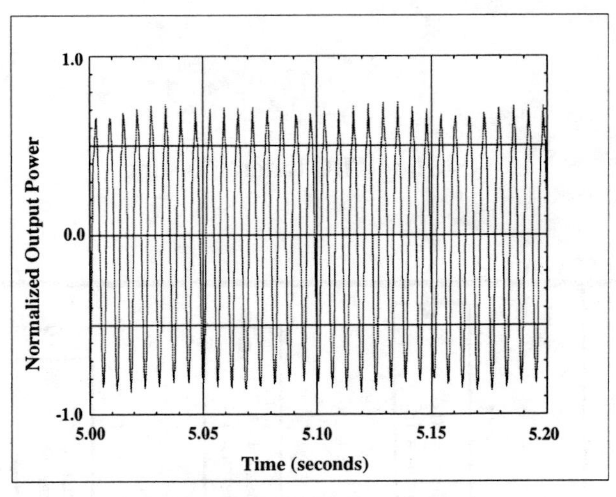

**Figure 7. Normalized Output Power
Time Plot for Solo TDC (no external vibration)**

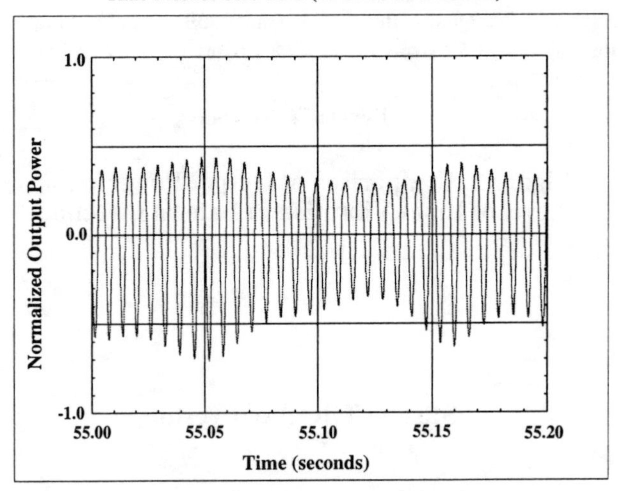

**Figure 8. Normalized Output Power
Time Plot for Qualification Test in Lateral Direction**

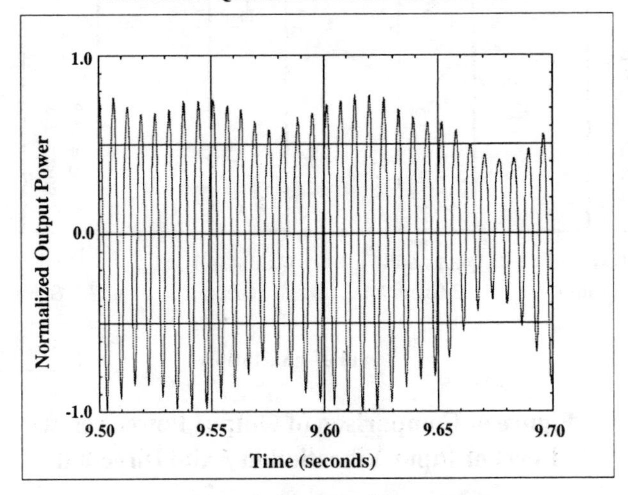

**Figure 9. Normalized Output Power
Time Plot for Qualification-6dB Test in Axial Direction**

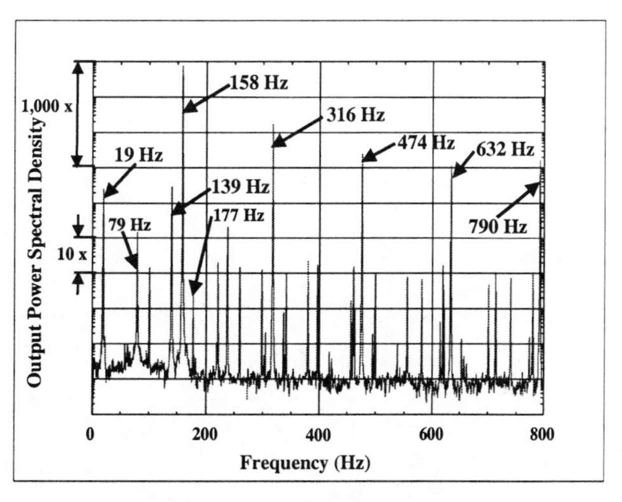

Figure 10. Frequency Spectrum of Output Power for Solo TDC (no external vibration)

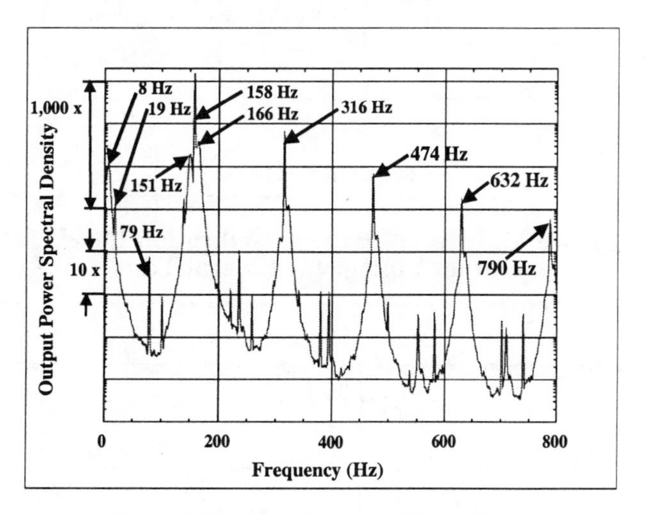

Figure 11. Frequency Spectrum of Output Power for Qualification Test in Lateral Direction

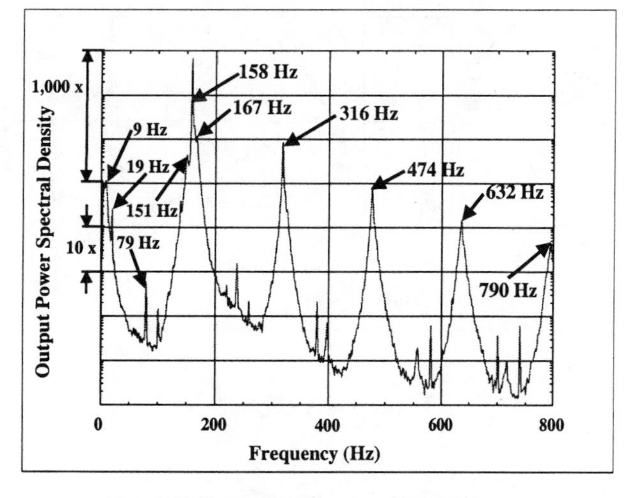

Figure 12. Frequency Spectrum of Output Power for Qualification-6dB Test in Axial Direction

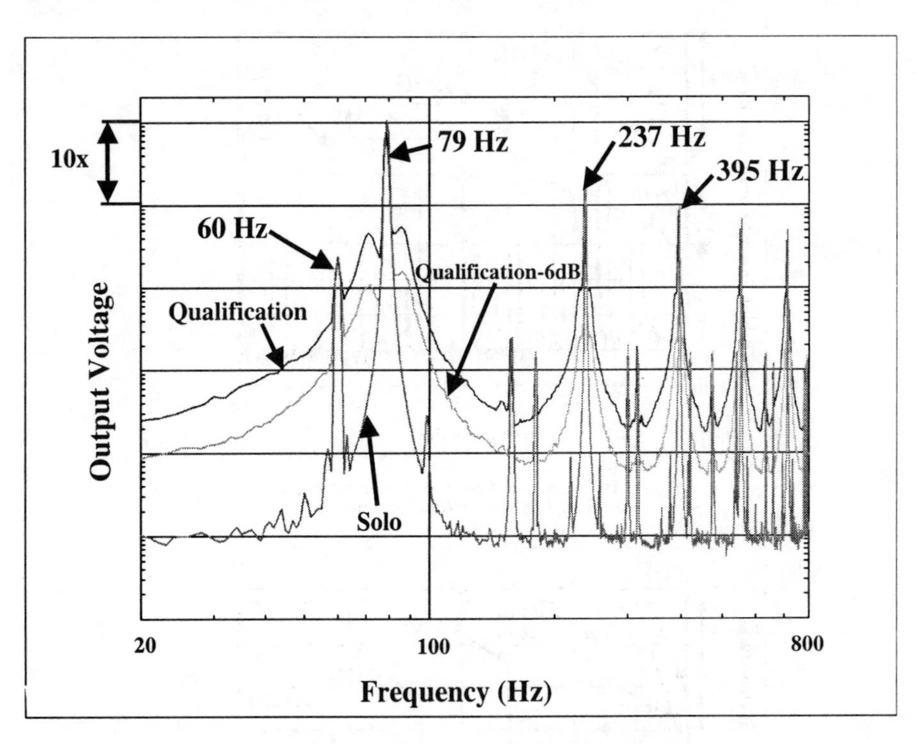

Figure 13. Effect of External Vibration Level on Output Voltage for Lateral Tests

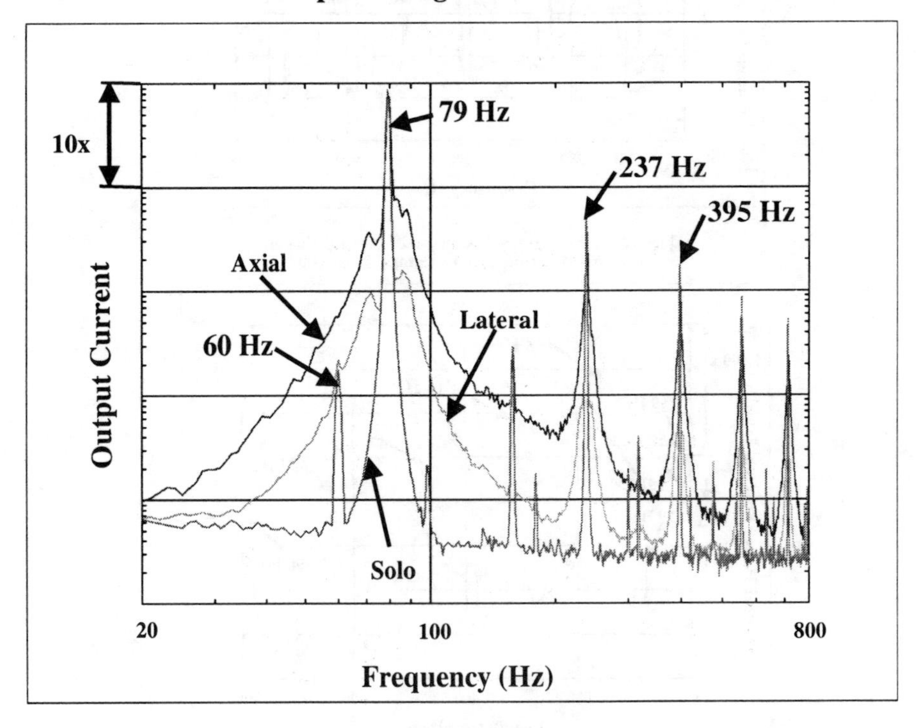

Figure 14. Effect of Test Direction on Output Current for Qualification-6dB Tests

AIAA-2000-2840

NASA GRC TECHNOLOGY DEVELOPMENT PROJECT FOR A STIRLING RADIOISOTOPE POWER SYSTEM

Lanny G. Thieme and Jeffrey G. Schreiber
NASA John H. Glenn Research Center at Lewis Field
Cleveland, Ohio

Abstract

NASA Glenn Research Center (GRC), the Department of Energy (DOE), and Stirling Technology Company (STC) are developing a Stirling convertor for an advanced radioisotope power system to provide spacecraft on-board electric power for NASA deep space missions. NASA GRC is conducting an in-house project to provide convertor, component, and materials testing and evaluation in support of the overall power system development. A first characterization of the DOE/STC 55-We Stirling Technology Demonstration Convertor (TDC) under the expected launch random vibration environment was recently completed in the NASA GRC Structural Dynamics Laboratory. Two TDCs also completed an initial EMI characterization at NASA GRC while being tested in a synchronized, opposed configuration. Materials testing is underway to support a life assessment of the heater head, and magnet characterization and aging tests have been initiated. Test facilities are now being established for an independent convertor performance verification and technology development. A preliminary FMEA, initial FEA for the linear alternator, ionizing radiation survivability assessment, and radiator parametric study have also been completed. This paper will discuss the status, plans, and results to date for these efforts.

Introduction

NASA Glenn Research Center (GRC) (formerly NASA Lewis Research Center), the Department of Energy (DOE), and the Stirling Technology Company (STC) of Kennewick, WA are developing a Stirling convertor for an advanced radioisotope power system to provide spacecraft on-board electric power for NASA deep space missions. Stirling is being evaluated as an alternative to replace Radioisotope Thermoelectric Generators (RTGs) with a high-efficiency power source and has recently been identified for potential use on the Europa Orbiter and Solar Probe missions now scheduled for launch in the 2006-2007 timeframe. The efficiency of the Stirling system, about 20-25% for this application, will reduce the necessary isotope inventory by a factor of 3 or more compared to RTGs.

STC has designed, fabricated, and completed first testing of the 55-We Technology Demonstration Convertor (TDC) under contract to DOE[1,2]. The TDC has been baselined by DOE for use in the upcoming missions. Two TDCs are now being tested by STC in a dynamically-balanced opposed arrangement, as shown in figure 1. Both design convertor power and efficiency have been demonstrated. Long life has also been demonstrated on a similar STC 10-We radioisotope terrestrial convertor, RG-10, that has been on life test at STC for over 54,000 hours (6.1 years) with no convertor maintenance and no degradation in performance. NASA GRC is providing technical consulting for this effort under an Interagency Agreement with DOE.

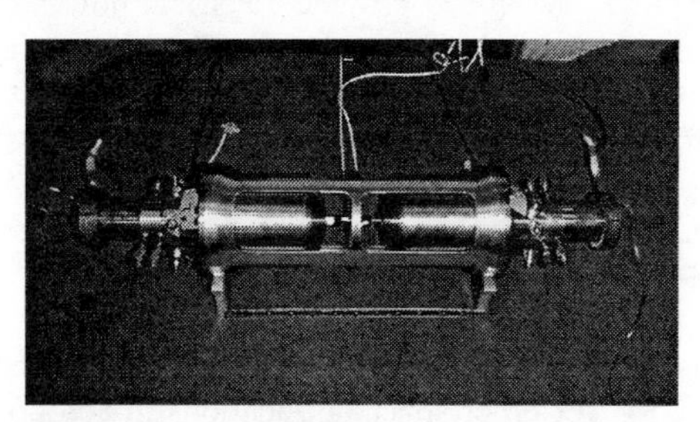

Figure 1. Two opposed 55-We TDC convertors on test (courtesy of STC).

As part of the overall Stirling radioisotope convertor development, NASA GRC is addressing key technology issues through the use of two NASA Phase II Small Business Innovation Research (SBIR) contracts with STC[3,4]. Under the first SBIR, STC demonstrated a synchronous connection of two thermodynamically independent Stirling convertors and a 40 to 50 fold reduction in vibrations compared to an unbalanced convertor. This connection method is now being used to connect the DOE/STC TDC convertors. The second SBIR contract is for the development of an Adaptive Vibration Reduction System that further reduces vibrations under normal operating conditions and will also add the ability to adapt to any changing convertor conditions over the course of a mission.

NASA GRC is also conducting an in-house project to provide convertor, component, and materials testing and evaluation in support of the overall Stirling radioisotope power system development. The project tasks build on NASA GRC expertise developed as part of previous Stirling research completed by NASA GRC over the last 25 years. Tasks include convertor performance verification, controls technology development, heater head structural life assessment, materials and joining evaluations, linear alternator finite element analyses (FEA), permanent magnet characterization and thermal aging tests, demonstration of convertor operation under launch and orbit transfer random vibration environments, electro-magnetic interference (EMI) characterization of the convertor, ionizing radiation survivability assessment, radiator parametric study, and failure modes and effects analysis (FMEA). This paper will discuss the status and results to date for this in-house project.

NASA GRC has been investigating Stirling radioisotope power systems for deep space missions since about 1990[5]. This work grew out of earlier Stirling efforts conducted for DOE for a Stirling automotive engine and for the NASA Civil Space Technology Initiative (CSTI) to develop Stirling for a nuclear power system to provide electrical power for a lunar or Mars base (part of the SP-100 program). NASA GRC also provided technical management for DOE for the Advanced Stirling Conversion System (ASCS) terrestrial dish Stirling project. Overall, NASA GRC has been developing Stirling technologies since the mid-1970's.

Systems using Stirling convertors are being analyzed by NASA GRC for other space applications in addition to Stirling radioisotope power for deep space missions. These include solar dynamic power systems for space-based radar[6] and as a deep space alternative to the radioisotope system[7,8], lunar/Mars bases and rovers, and a combined electrical power and cooling system for a Venus lander.

Convertor Characterization in Launch Environment

Following initial development of the TDC, the performance goals of electric power output and conversion efficiency had been demonstrated. Mass was projected for a flight version of the TDC that was acceptable for the integrated spacecraft power system. It was also shown that the TDC was based on technologies that were capable of leading to long life and reliability. Although these capabilities might be demonstrated analytically and in laboratory tests, it was unclear what the effect would be on the convertor when subject to the anticipated launch vibration. Specifications used during the initial design and development of the TDC had no requirement for surviving the random vibration of the launch environment.

Personnel representing DOE, GRC, the Jet Propulsion Laboratory (JPL), and STC worked jointly to develop a test plan to evaluate the effect that vibration would have on the operation and performance of the TDC. The vibration levels for the test are shown in figure 2. The standard workmanship test (NASA-STD-7001) was modified to move the knee of the vibration curve below the 80 Hz operating frequency of the TDC. Since little was known about the ability of the TDC to withstand vibration, it was decided that the test would proceed incrementally from low vibration levels to higher levels. It was also decided that the TDC would be tested with random vibrations imposed in the axial and lateral directions, and that it would be first tested at workmanship level in each direction before advancing to the flight and qualification level tests. Duration of each vibration test was one minute, with the exception of the qualification tests which lasted three minutes.

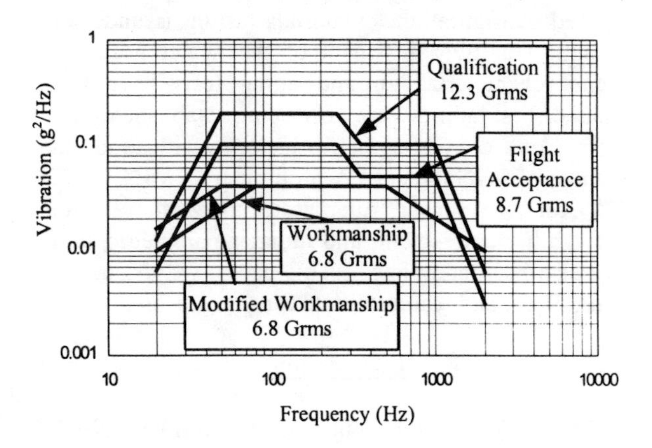

Figure 2. Random vibration test levels.

The TDC mounted in the vibration test fixture is shown in figures 3 and 4. The TDC was operated at full power and full stroke during each test. Power output was recorded as each test proceeded. Sine sweeps were performed before and after each major test to try to detect any changes in structural integrity that might occur. References [9,10] discuss the results of these tests in more detail.

The objectives of the test plan were met and exceeded. The TDC was exposed to significant vibration levels and durations, beyond those anticipated at launch, and continued to operate through all tests without problem.

The most severe test was the JPL qualification test at 12.3 Grms for three minutes in both the lateral and axial directions. As a comparison, the RTGs used for Galileo, Ulysses, and Cassini missions were tested to 7.7 Grms laterally and 6.1 Grms axially.

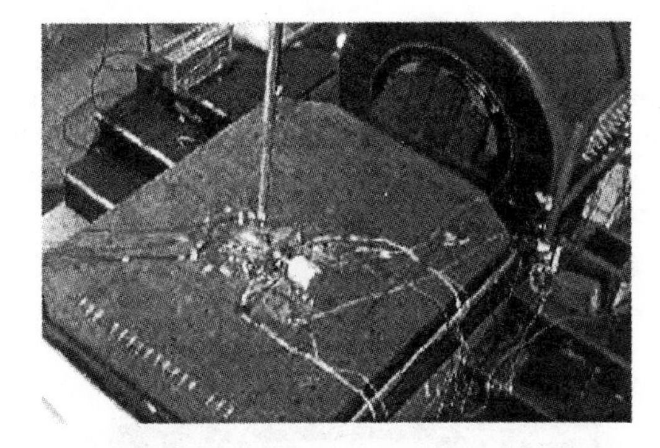

Figure 3. The TDC vibration test at the NASA GRC Structural Dynamics Lab.

Figure 4. The TDC mounted in the vibration test fixture.

The TDC was operated at full piston stroke and produced power through all of the tests. During the lateral vibration tests, the average power output did vary somewhat, with power output being reduced as the vibration level increased. At the completion of each lateral test, the operation returned to the pre-test condition. Table 1 summarizes the average power output of the TDC during the vibration tests. At the conclusion of the Flight/Qual test in the lateral direction, the TDC produced in excess of the 54-We starting condition due to residual heat that accumulated in the head during the vibration test. After a few minutes, the power output did return to the 54-We starting condition.

The reduced power output can be attributed to several factors. First, during lateral vibration tests, the non-contacting close clearance seals for the power piston, the displacer, and the displacer rod would be operating at time varying clearances. A piston operating coaxially in a bore produces the lowest leakage condition, therefore any deviation from coaxial operation would increase the leakage. Second, as the lateral vibration increased in intensity, the moving parts of the piston/linear alternator and the displacer might contact stationary support surfaces. It was anticipated that this would happen and appropriate materials and coatings were used on the surfaces that would come into contact. Contact would result in some friction that reduces the electric power output. Third, the friction acting on the moving parts could disrupt the tuned dynamics of the free-piston Stirling convertor and thus alter the thermodynamic cycle producing the power. The extent to which these three mechanisms existed, and the possibility of other loss mechanisms has not been studied.

	Workmanship		Flight/Qual	
	Lateral	Axial	Lateral	Axial
Pre-Test	54	54	54	54
-12 dB	53	54	54	55
-6 dB	53	54	50	54
-3 dB	51	54	45	55
0	48	55	38	54
Post-Test	54	54	57	54

Table 1. Average power output (watts) during Workmanship and Flight/Qual random vibration tests.

During the random vibration tests in the axial direction, there was no measurable change in the average power output. At the highest levels of axial vibration, there was a slight, irregular tapping sound coming from the convertor that sounded like metal-to-metal contact. In anticipation of possible over stroke during these tests, teflon bumpers were incorporated in the piston/linear alternator mover of the TDC to prevent damage. The displacer had no such bumper system and therefore was capable of metal-to-metal contact. No instrumentation measured this effect so it is unclear which component produced the sound.

It is worth noting that the TDC passed not just one, but a series of severe tests. The workmanship test was performed twice, once each in the lateral and axial directions. The flight/qualification test was then performed twice, once each in the lateral and axial directions. Before and after each major test, a sine sweep was performed to try to detect structural changes in the hardware; approximately 6 sine sweeps were performed. Thus, the total vibration exposure was far beyond what would be expected at launch. Also, the data from the sine sweeps indicated no structural change as a result of the random vibration tests.

Following successful completion of the vibration tests, the TDC was operated for over 35 hours at full power to accumulate more than 10^7 cycles on the hardware. Per the experience of STC, successful completion of 10^7 cycles is an indication that the flexures have the ability for essentially infinite life. This criteria is consistent with the views of the flexure-based Stirling cryocooler industry as an indication of long life. No anomalies were detected during the 35 hours of operation. The TDC was subsequently disassembled and inspected at STC with representatives present from Lockheed Martin Astronautics (LMA), Valley Forge, PA, and GRC. It was concluded that no wear or damage was found that could be attributed to the recently completed vibration tests.

Convertor EMI/EMC Characterization

For use as a power system on future NASA science missions, a Stirling radioisotope power system will be required to meet electromagnetic interference and electromagnetic compatibility (EMI/EMC) specifications. Similar to the situation with the vibration loads, there was no EMI/EMC specification applied to the TDC during the initial design.

A team of experts from JPL, GRC and LMA were convened for this assessment. The purpose was to characterize the radiated electromagnetic fields to determine whether the TDC is electromagnetically compatible with the NASA X2000 advanced technology requirements. The tests performed are outlined in table 2. If the measured levels were found to exceed the specifications, the assessment was also viewed as an initial opportunity to assess the levels that could be achieved in the future and what techniques this might entail. One of the main concerns was that the X2000 radiated emissions requirements used for this evaluation are up to 100 dB more stringent than the commonly used MIL-STD-461 requirements.

RE0 AC Magnetic Emissions, 50 Hz - 150 kHz, 13.3 cm loop antenna at 4 antenna positions from 12.5 cm to 1 m
RE04 AC Magnetics, 50 Hz - 150 kHz, larger loop antenna at 3 positions from 25 cm to 1 m
Search coil measurements, 25 cm and 1 m (3 axis)
RE02 Electric Field Emissions, 50 Hz - 150 kHz, and 14 kHz - 1 GHz at 1 m
CE01 test, measured at TDC interface for information only, not a spec
Characterized controller voltage and current waveforms
Magnetic Field Emissions test using partial mumetal shields over the alternators

Table 2. EMI/EMC tests performed on the TDC.

The EMI/EMC laboratory at GRC was selected as the site for the assessment. Two operational TDCs and the needed support equipment were installed in the EMI/EMC lab. Initial tests were run with the TDCs configured in the alternator-to-alternator configuration as shown in figure 1. Later in the test sequence, the configuration was changed so that the TDCs were tested in a hot-end to hot-end configuration as shown in figure 5. It should be noted that the TDCs were never optimized to reduce EMC, and further that there was no attempt to minimize loop area or clock the TDCs relative to one another for this particular test, nor to provide any shielding in the TDC.

Figure 5. A pair of TDCs on test at the GRC EMI/EMC Laboratory.

Based on the test data, the TDCs would meet the requirements for missions such as Europa Orbiter and Pluto Kuiper Express, but would exceed the Solar Probe magnetic field requirements (driven by the science package) by up to about 100 dB at the 80 Hz operating frequency. The electric field emissions exceeded the Solar Probe electric field requirements by about 60 dB, but further testing showed that these emissions were primarily being radiated by the interconnecting cable and were reduced by 30 dB with minimal effort. Harmonic content was largely a function of the current waveform which is expected to be near sinusoidal with few harmonics in the flight unit.

It was the conclusion that significant improvement in magnetic field emissions are available by optimizing wiring layout and providing counter-loops when necessary. It was also concluded that it is possible to use the TDC for science missions with plasma experiments; however, care would need to be taken to minimize current loops and magnetic shielding would also be required.

Heater Head Life Assessment

Heater head life is a critical element for achieving the 100,000+ hour life of the convertor. NASA GRC materials and structures personnel have developed an approach to characterize the long-term durability of the heater head using relatively short-term extrapolation methods[11]. A similar assessment was previously completed for a 12.5-kWe Stirling convertor as part of the NASA CSTI efforts for SP-100[12]. Life prediction models will be selected, and an independent thermal and structural FEA completed. The IN718 heater head material in the final process condition that will be used in the convertor will be characterized for creep (<20,000 hours), low cycle fatigue, thermal, and tensile data. Structural benchmark testing will also be done consisting of accelerated life tests on prototypical heater heads. Results of this testing and the IN718 material tests will be used to calibrate and validate the life prediction model. This calibrated model will then be used to project the heater head lifetime. Successful completion of these efforts is expected to give high confidence in the ability of the heater head to meet the operational life requirements.

To date, life prediction models have been identified, and the heater head FEA model completed. Creep testing has been initiated for the IN718 heater head material. A high temperature, high pressure test system has been designed and is now being assembled for the heater head accelerated life tests. This will apply both temperature gradients and pressure conditions based on those in the actual convertor. Elevated pressure and/or temperature conditions will be used to accelerate the testing.

A study was completed to investigate the possible depletion of chromium from the IN718 by the long-term, high-temperature exposure to the vacuum of space. Chromium depletion would decrease material creep resistance and strength. A model was set up and then characterized and verified by accelerated aging tests on IN718. This model was then used to predict the chromium loss after 100,000 hours at the 650°C hot-end temperature. The final results showed negligible chromium loss at these conditions.

An evaluation of the joining methods used in the convertor is also being conducted. Critical joints will be thermally aged and microstructurally investigated. This will include, in particular, the critical close-out hermetic seal. A test setup to evaluate the hermetic seal has been designed. The organic materials in the convertor are being reviewed for any compatibility and outgassing concerns; these efforts are also supporting the selection of radiation-hard organics (see section on Ionizing Radiation Survivability Assessment).

Linear Alternator Finite Element Analyses

Lightweight linear alternators with high efficiency are a key to obtaining the potential of the Stirling convertor for the radioisotope power system. NASA GRC has been developing finite element analyses (FEA) tools for performing various linear alternator analyses, including thermal and electromagnetic analyses and evaluating design configurations. A three-dimensional magnetostatic model of the TDC linear alternator has been developed using Ansoft's Maxwell 3D finite element method software[13].

A key thermal analysis was first completed. The cold-end of the convertor is expected to operate in the range of 80-120°C. The higher temperatures of this range could be marginal for the use of NdFeB magnets. However, it was recognized by LMA, one of the DOE contractors analyzing system designs, that the convertor pressure vessel will also act as a radiator. A model was developed by NASA GRC that included the linear alternator, pressure vessel, piston, and radiator. Boundary conditions and inputs included a 120°C convertor cold-end temperature, radiation sink temperature of -40°C, full power output of 55 We from the convertor with an 85% efficient linear alternator, a surface conductance between adjacent parts of 0.31 W/°C-cm^2, and an emissivity of 0.91 with an absorptivity/emissivity ratio of 1.0 for the radiation heat transfer. An average magnet temperature of 73°C was obtained from the analysis. This is expected to be conservative as the model did not include any convection heat transfer from the helium inside the convertor pressure vessel. This calculated magnet temperature is readily in the typical operating range of NdFeB magnets, and this calculation resolved concerns over the magnet selection. The calculated linear alternator temperature levels are also key inputs to the selection of radiation-hard organic materials for use in the alternator (see section on Ionizing Radiation Survivability Assessment).

Electromagnetic calculations of the TDC linear alternator open circuit voltage have also been recently completed and compared to the value measured by STC at room temperature[13]. The NASA GRC model calculated 98.0 Vrms at 23°C while STC measured 97.0 Vrms. Dynamic circuit simulation software capable of working with Maxwell 3D is expected soon; this will allow complete simulations under load.

This FEA tool should be useful for evaluating both linear alternator performance and design variations. A future task is planned to evaluate a lightweight linear alternator concept that has been previously tested by STC and had performance shortfalls. The FEA will be used to evaluate the alternator flux profiles and identify potential design modifications with STC to improve performance.

Permanent Magnet Aging Tests

NdFeB permanent magnets are used in the TDC linear alternators. Magnet choice is dependent on the final convertor cold-end temperature chosen based on system analyses (now expected to be between 80 and 120°C), magnet temperature calculations (preliminary calculations indicate that the magnet operating temperature is about 70-80°C – see previous section), and the safety margin necessary to prevent demagnetization. Characterization testing of appropriate NdFeB magnets will first be done using an existing test rig shown in figure 6 that was developed for testing SmCo magnets during the previous NASA CSTI research for SP-100[14]. Selected NdFeB magnets will then be put on short-term and long-term aging tests at operating temperature and in a demagnetizing field to quantify any potential magnet degradation with time and temperature. Such degradation could affect both the remanent magnetization and the demagnetization resistance.

Figure 6. Magnet characterization test rig.

The characterization testing of 9 NdFeB magnet types from 3 different manufacturers has been mostly completed. B-H curves for two different samples of each magnet type were measured over a range of temperatures from room temperature to 150°C. Comparisons between one of the better (for this application) NdFeB magnet types and one of the better Sm_2Co_{17} magnet types tested previously are shown in figures 7 and 8. Figure 7 compares the magnet remanence, a measure of magnet strength, over a range of temperatures. It can be seen that the crossover point in magnet strength between the NdFeB and SmCo is around 190°C. Figure 8 compares the intrinsic coercivity, a measure of demagnetization resistance, over a range of temperatures. At the expected magnet operating temperatures for the Stirling convertor, it can be seen from the two figures that the NdFeB magnets provide higher strength but lower demagnetization resistance.

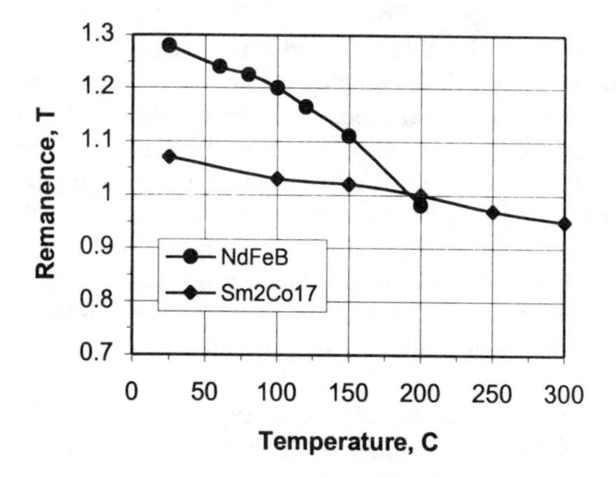

Figure 7. Comparison of magnet remanence vs. temperature for NdFeB and SmCo magnets.

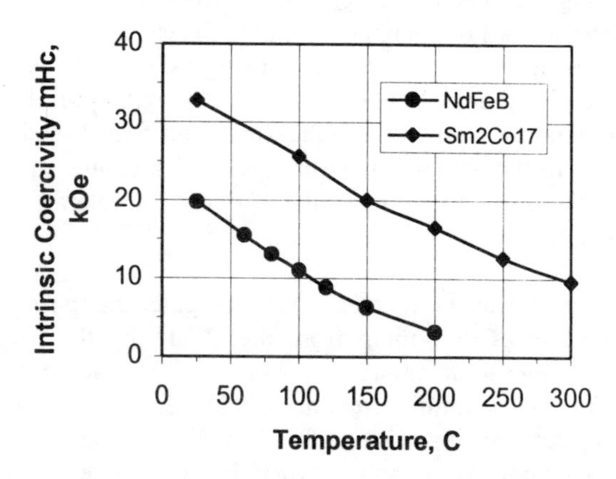

Figure 8. Comparison of magnet intrinsic coercivity vs. temperature for NdFeB and SmCo magnets.

Based on the NdFeB characterization tests, several NdFeB magnet types will be selected for short-term magnet aging tests. A magnet aging rig has been designed and fabricated by KJS Associates Inc., Indianapolis, IN and will soon be installed at NASA GRC. This test rig will allow up to 10 magnet samples to be tested at once in an inert gas. The inert gas will prevent oxidation effects from occurring over the test period and influencing the results. In actual operation, the magnets are located in the helium environment of the convertor. The samples will be maintained at a selected temperature and with a fixed DC demagnetizing field applied. This demagnetizing field will be chosen based on the maximum demagnetizing field that the magnets see in operation.

Preliminary short-term aging tests have been completed using the magnet characterization rig. Three NdFeB magnet types have been tested for about 100 hours at 120°C and in a 6-kOe demagnetizing field. No measurable change in magnet characteristics were found after these tests. Further short-term aging tests with the aging test rig may be run at higher temperatures. Either one or two NdFeB magnet types will then be chosen for the long-term aging tests. It is planned to run the long-term tests for a minimum of 12,000 hours. Magnet samples will be periodically removed from the test rig to quantify the rate of aging. These results will then be used to project magnet characteristics over the 6-15 year mission lifetimes.

In-House Testing of 350-We Convertors and 55-We TDCs

An in-house Stirling test facility is currently being established for testing both 350-We and 55-We convertors. Four 55-We TDCs are now being built by STC for NASA GRC. These will essentially be the same as the TDCs tested previously by STC. One main difference is that they will incorporate radiation-hard organic materials for the piston bearings and bumpers and in the linear alternator (see next section). The TDCs will be arranged in opposed pairs in a hot-end to hot-end configuration; this configuration was used during previous EMI/EMC testing at NASA GRC and is shown in figure 5.

The TDC convertors are expected to be used first for independent performance verification and further launch environment testing. They will also serve as a functional test bed for demonstrating the radiation-hard organics.

Figure 9. Two opposed 350-We convertors on test at STC.

Two 350-We convertors will be delivered to NASA GRC that were built under the NASA Phase II SBIR that developed the synchronous connection for two thermodynamically independent convertors[3,4]. These convertors are shown on test at STC in figure 9. STC has also used these convertors for developing the Adaptive Vibration Reduction System under a further NASA Phase II SBIR[3,4]. These 350-We convertors as well as the second pair of TDCs will be used by NASA GRC for controls technology development, including some further developments of the SBIR technologies.

Ionizing Radiation Survivability Assessment

Future missions for a Stirling radioisotope power system include a possible mission to the Jupiter moon Europa. The Europa radiation environment is harsh, and this ionizing radiation around Jupiter is much more severe than that produced by the GPHS radioisotope heat source.

As part of a DOE/NASA assessment of the Stirling technology, NASA GRC evaluated the radiation effects on the organic materials in the Stirling convertor[15]. Experts were also consulted from JPL, LMA, and General Electric, Schenectady, NY. The total expected ionizing dose inside the convertor was estimated at 4 MRAD considering the intrinsic shielding provided by the pressure vessel. Each organic material used in the Stirling convertor was identified. These included piston bearing coatings, piston bumpers, and alternator adhesives and electrical insulations. Each separate material was evaluated for acceptability, and new organic materials were recommended where necessary. Both STC and vendors of the organic materials were consulted as to the satisfactory functional use of the recommended materials. STC completed functional tests as required for some of the materials under consideration.

In each case, a satisfactory radiation-hard organic has been identified. These materials are now being incorporated into the TDC convertors that STC is currently building for NASA GRC. The testing of these convertors will verify the functionality for each of the modified organics.

Failure Modes and Effects Analysis

An independent reliability assessment of the TDC was performed at GRC. The assessment included two tasks. The first task was to develop a Failure Modes and Effects Analysis (FMEA). This analysis was intended to address the Stirling convertor, but not the radiator, controller, structural interface, and GPHS modules and housing, which were all viewed to be part of the system and beyond the envelope of the TDC. The second task was to develop a life and reliability model using all available test data that were deemed relevant. The ultimate purpose of this second task was to determine if a flight version of the TDC was capable of completing a 14-year mission.

An FMEA was performed for the TDC looking at the component level. This analysis included all phases of the TDC operational life starting when the fueled GPHS modules are loaded into the system through the completion of the mission. Following this, a top level FMEA was generated to determine if there were any failures in external systems that would cause a critical failure in the TDC.

Based on the results from the component level FMEA, a Critical Items List (CIL) was generated. A critical item was defined as an item whose failure would prevent the TDC from producing power at or near the full capacity. A failure that shuts down the TDC but allows a restart was not considered a critical failure. A total of 14 components were identified on the CIL, as shown in table 3.

1	Hermetic closure, pressure vessel
2	Hermetic closure, piston housing
3	Hermetic closure, electric feed throughs
4	Aft bearing support spider
5	Heater head assembly
6	Over stroke protection bumper
7	Over stroke protection bumper stops
8	Linear alternator stator permanent magnets
9	Linear alternator stator insulation
10	Linear alternator mover assembly
11	Displacer flexures
12	Piston flexures
13	Flexure stack fasteners
14	Linear alternator stator laminations

Table 3. TDC Critical Items List.

To evaluate the potential for long life, a life and reliability analysis was performed on each of the items identified on the CIL. A number of sources of information were used, including test data, materials data, and component design data. The conclusion was that there was sufficient relevant data and that all of the design issues were well understood and could be managed within the current quality practices of the aerospace industry.

Parametric Study of Radiator Concepts

Orbital Sciences Corporation[16,17], Germantown, MD, and LMA[18] have provided system analyses for DOE. As part of the Interagency Agreement with DOE, NASA GRC has contributed a parametric study of heat pipe and non-heat pipe radiator concepts using both state-of-the-art and advanced high thermal conductivity materials[19]. The heat pipe concepts were based on an earlier Orbital Sciences system concept shown in figure 10; see references [16,17] for later Orbital Sciences system designs. The non-heat pipe concept that was evaluated was developed by NASA GRC and is shown in figure 11. This concept uses solid disks and eliminates any concerns with heat pipe reliability, such as sensitivity to micrometeoroid damage.

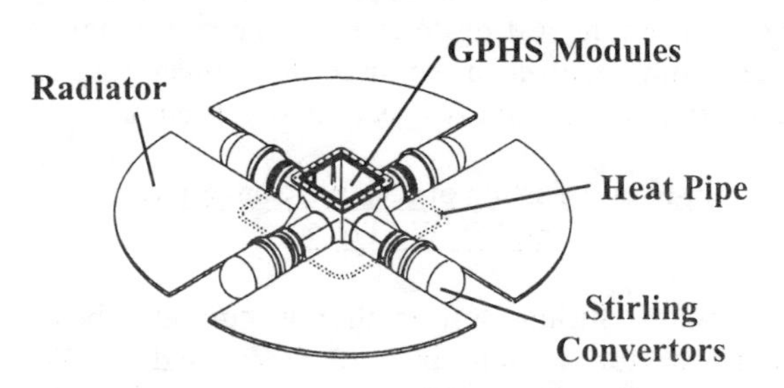

Figure 10. Orbital Sciences Corporation system concept[20].

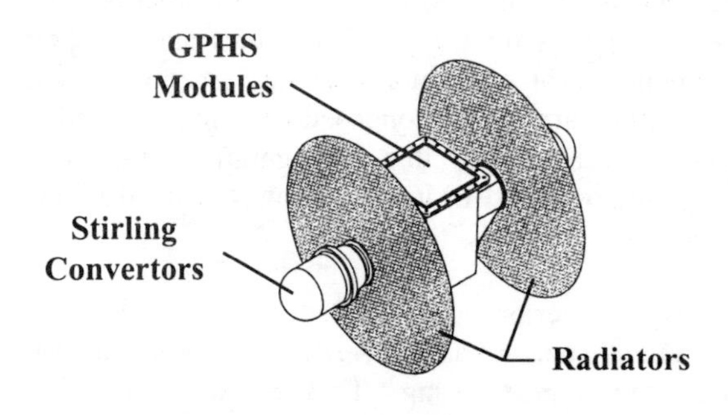

Figure 11. Non-heat pipe radiator concept.

For the heat pipe concepts, layouts using heat pipes located optimally for performance and radiator size were compared with layouts with the heat pipe fixed at the radial location of the convertor cooler (which was assumed to be easier to fabricate). The non-heat pipe radiators use thermal conduction through solid disks mounted perpendicular to the convertor axis of motion. Flat, tapered, and parabolic surfaces were considered for the solid disks. Both two- and four-convertor system designs were evaluated for both heat pipe and non-heat pipe concepts.

Materials considered for the radiator fins included aluminum, beryllium, carbon composite, and thermal pyrolitic graphite (TPG). TPG is an advanced high-thermal conductivity material and is not an available radiator technology at this time. The TPG must be encapsulated within another material to provide strength. TPG encapsulants considered in the study were aluminum, beryllium, and carbon composite. A Z-93 radiator surface coating with high emissivity and low absorptivity was assumed for the study.

These conceptual radiator evaluations were done with a dedicated radiator analysis tool, GPHRAD, recently developed at NASA GRC[21]. GPHRAD is a finite difference computational code developed for the analysis and design of circular sector radiators. It includes a novel subroutine to determine equilibrium space sink temperatures anywhere within the Solar System[22]. This takes into account the radiator surface characteristics, view factor to space, distance from the sun and any planets, and on-board heat generation.

For a nominal 105-We power system with two opposed Stirling convertors and a convertor cold-end temperature of 120°C, total radiator mass was found to be as low as 1.75 kg for a non-heat pipe concept and as low as 2.17 kg for a heat pipe concept. As the layouts went from the more complex (heat pipe radiators with heat pipes requiring a number of bends) to the simpler (simple disks and no heat pipes), the benefits of using the advanced high thermal conductivity TPG increased. However, designs with aluminum, beryllium, and carbon composites were also found to be attractive in terms of mass and thermal performance for each potential layout. The best non-heat pipe designs have outer radii that are 2/3 as large as the best designs using heat pipes; however, these best non-heat pipe designs require the use of TPG fins. In general, maximum temperature differences across the radiator were minimized by using optimally-located heat pipes and TPG material.

A view factor of 1.5 was assumed for these calculations. A sensitivity analysis to view factor showed that view factor significantly affects radiator size and mass and needs to be carefully calculated for a given system and spacecraft layout.

Concluding Remarks

The Department of Energy, NASA Glenn Research Center, and Stirling Technology Company are developing a Stirling convertor for a high-efficiency radioisotope power system to provide spacecraft on-board electric power for NASA deep space missions. Rapid progress has been made, and a Stirling power system has now been identified for potential use on the Europa Orbiter and Solar Probe missions.

A technology assessment was recently completed by DOE and NASA with a joint government/industry team to evaluate the technology readiness of STC's 55-We Technology Demonstration Convertor. The team concluded that no technical barriers exist and that the Stirling convertor is ready for the next step towards developing the Stirling technology into a radioisotope power system for future NASA deep space missions. A DOE procurement solicitation for a system integrator for this Stirling radioisotope power system has since been released.

NASA GRC is playing a key role in the overall development of the Stirling convertor for this application. As part of the NASA GRC in-house supporting technology project, four key tasks of the technology assessment were completed. These were the demonstration of the Stirling convertor operation under launch environments, convertor EMI/EMC characterization, survivability assessment of the convertor organic materials in a high-radiation environment, and failure modes and effects analysis. Further in-house tasks are providing life assessments for the heater head and permanent magnets, an independent performance verification for the convertor, a joining evaluation, and FEA analyses of the linear alternator. These NASA GRC efforts are moving along rapidly with a number of significant accomplishments in the first year of the project. In addition, Stirling technology expertise is provided to DOE as part of a Space Act Agreement. Under this Space Act Agreement, a parametric study of radiator concepts was also completed.

References

[1] White, M.A.; Qiu, S.; Olan, R.W.; and Erbeznik, R.M.: Technology Demonstration of a Free-Piston Stirling Advanced Radioisotope Space Power System, Proceedings of the Space Technology and Applications International Forum, 1999.

[2] White, M.A.; Qiu, S.; and Augenblick, J.E.: Preliminary Test Results from a Free-Piston Stirling Engine Technology Demonstration Program to Support Advanced Radioisotope Space Power Applications, Proceedings of the Space Technology and Applications International Forum, 2000.

[3] Thieme, L.G.; Qiu, S.; and White, M.A.: Technology Development for a Stirling Radioisotope Power System for Deep Space Missions, Proceedings of the 34th Intersociety Energy Conversion Engineering Conference - 1999, NASA/TM – 2000-209767, 2000.

[4] Thieme, L.G.; Qiu, S.; and White, M.A.: Technology Development for a Stirling Radioisotope Power System, Proceedings of the Space Technology and Applications International Forum - 2000, NASA TM-2000-209791, 2000.

[5] Bents, D.J.; Geng, S.M.; Schreiber, J.G.; Withrow, C.A.; Schmitz, P.C.; and McComas, T.J.: Design of Multihundredwatt DIPS for Robotic Space Missions, Proceedings of the 26th Intersociety Energy Conversion Engineering Conference, NASA TM-104401, 1991.

[6] Mason, L.S.: Technology Projections for Solar Dynamic Power, NASA/TM-1999-208851, 1999.

[7] Mason, L.S.: Solar Stirling for Deep Space Applications, Proceedings of the Space Technology and Applications International Forum – 2000, NASA/TM-1999-209656, 1999.

[8] Schreiber, J.G.: A Deep Space Power System Option Based on Synergistic Power Conversion Technologies, Proceedings of the Space Technology and Applications International Forum, 2000.

[9] Hughes, W.O.; McNelis, M.E.; and Goodnight, T.W.: Vibration Testing of an Operating Stirling Convertor, Proceedings of the Seventh International Congress on Sound and Vibration, Garmisch-Partenkirchen, Germany, 4-7 July 2000.

[10] Goodnight, T.W.; Hughes, W.O.; and McNelis, M.E.: Dynamic Capability of an Operating Stirling Convertor, to be published in the Proceedings of the 35th Intersociety Energy Conversion Engineering Conference, 2000.

[11] Bartolotta, P.; Bowman, R.; Krause, D.; and Halford, G.: Long-Term Durability Analysis of a 100,000-Hour Stirling Power Convertor Heater Head, to be published in the Proceedings of the 35th Intersociety Energy Conversion Engineering Conference, 2000.

[12] Abdul-Aziz, A., Bartolotta, P., Tong, M., and Allen, G.: An Experimental and Analytical Investigation of Stirling Space Power Convertor Heater Head, NASA TM-107013, 1995.

[13] Geng, S.M.; Schwarze, G.E.; and Niedra, J.M: A 3-D Magnetic Analysis of a Linear Alternator for a Stirling Power System, to be published in the Proceedings of the 35th Intersociety Energy Conversion Engineering Conference, 2000.

[14] Niedra, J.M.: Comparative M-H Characteristics of 1-5 and 2-17 Type Samarium-Cobalt Permanent Magnets to 300 C, NASA CR-194440, 1994.

[15] Golliher, E.L. and Pepper, S.V.: Organic Materials Ionizing Radiation Susceptibility for the Outer Planets/Solar Probe Radioisotope Power Source, to be published in the Proceedings of the 35th Intersociety Energy Conversion Engineering Conference, 2000.

[16] Or, C.; Kumar, V.; Carpenter, R.; and Schock, A.: Self-Supporting Radioisotope Generators with STC-55W Stirling Convertors, Proceedings of the Space Technology and Applications International Forum, 2000.

[17] Or, C.; Carpenter, R.; Schock, A.; and Kumar, V.: Performance of the Preferred Self-Supporting Radioisotope Power System with STC 55-W Stirling Convertors, Proceedings of the Space Technology and Applications International Forum, 2000.

[18] Cockfield, R.D.: Radioisotope Stirling Generator Concepts for Planetary Missions, to be published in the Proceedings of the 35th Intersociety Energy Conversion Engineering Conference, 2000.

[19] Juhasz, A.J.; Tew, R.C.; and Thieme, L.G.: Parametric Study of Radiator Concepts for a Stirling Radioisotope Power System Applicable to Deep Space Missions, NASA/TP-2000-209676, to be published in 2000.

[20] Schock, A.; Or, C.; and Kumar, V.: Radioisotope Power System Based on Improved Derivative of Existing Stirling Engine and Alternator, Proceedings of the Space Technology and Applications International Forum, 1999.

[21] Juhasz, A.J. and Thieme, L.G.: Design and Analysis Code for Radiators of Stirling Power Systems With General Purpose (GPHS) Heat Sources, Proceedings of the 34th Intersociety Energy Conversion Engineering Conference, 1999.

[22] Juhasz, A.J.: An Analysis and Procedure for Determining Space Environmental Sink Temperatures with Selected Computational Results, to be published in the Proceedings of the 35th Intersociety Energy Conversion Engineering Conference, Paper No. 2000-3010, 2000.

LONG-TERM DURABILITY ANALYSIS OF A 100,000+ HR STIRLING POWER CONVERTOR HEATER HEAD

Paul A. Bartolotta, Randy R. Bowman, David L. Krause, and Gary R. Halford
National Aeronautics and Space Administration
Glenn Research Center
Cleveland, Ohio 44135

Summary

DOE and NASA have identified Stirling Radioisotope Power Systems (SRPS) as a candidate power system for future deep space exploration missions. As a part of this effort, NASA has initiated a long-term durability project for critical hot section components of the Stirling power convertor to qualify flight hardware. This project will develop a life prediction methodology that utilizes short-term (t < 20,000 hr) test data to verify long-term (t > 100,000 hr) design life. The project consists of generating a materials database for the specific heat of alloy, evaluation of critical hermetic sealed joints, life model characterization, and model verification. This paper will describe the qualification methodology being developed and provide a status for this effort.

Introduction

A Stirling Radioisotope Power System (SRPS) has been chosen as a candidate power system for future NASA deep space exploratory missions. The Europa (one of Jupiter's moons) Orbiter and the Solar Probe are scheduled to launch by January 2006 and February 2007 respectively. DOE is responsible for the successful design and fabrication of the SRPS for these two missions. NASA is aiding DOE by conducting several projects that will reduce the design and fabrication risks of the SRPS.

Projected design life requirements of the SRPS for these missions are over 60,000 hr. Long-term durability of hot section components (i.e., heater heads) of power convertors is a prime area of concern. Since launch dates for both of these missions are within the 2006 and 2007 timeframe, conventional design approaches for long-term durability verification are not appropriate. Instead, a new innovative approach will be required. This new design and verification approach consists of generating a materials test database specific for the Stirling power convertor application, defining the appropriate definition of failure, developing a probabilistic design methodology, and verifying the critical flight hardware using benchmark tests.

Design reviews have identified the heater head of the Stirling power convertor as a critical component. The heater head is a high-temperature pressure vessel that transfers heat to the working medium of the convertor, which is typically helium. Efficient heater head designs result from a compromise between thin walls for increased heat transfer and thick walls for lower stresses thus improved creep resistance/durability. Existing long-term creep data (>50,000 hr) on thin specimens of the proposed heater head material, Inconel 718, for the operating conditions and long-term durability are limited. The proposed approach uses long-term test data generated from the same heat of material from which the flight hardware will be fabricated from and compares it to existing materials databases. Chrome depletion of heater head materials resulting from material temperatures as high as 650 °C while exposed to the vacuum of space is also a concern. With launch dates scheduled for early 2006, innovative life prediction methods and accelerated tests will be required to assure heater head lives of over 100,000 hr. The lifting methodology involves conducting a series of creep tests on thin specimens, characterizing/evaluating various prediction models, and analyzing the heater head structure. Final validation of the flight hardware design will be accomplished by calibrating/verifying the life models using benchmark tests on actual heater heads under prototypical operating conditions.

This paper will discuss the life prediction philosophy and design methodologies being utilized to increase the confidence level of achieving 100,000-hr design life for the convertor's heater head. Current supporting test data, benchmark test description, and structural analyses of the heater head will also be presented.

Stirling Power Convertor

The fundamental operation of a radioisotope Stirling Power convertor is a topic covered in several papers in the open literature[1,2] by NASA and the Stirling Technology Company (STC). This section identifies major components of the convertor and their probable damage modes caused by long-term exposure to high temperatures and mechanical

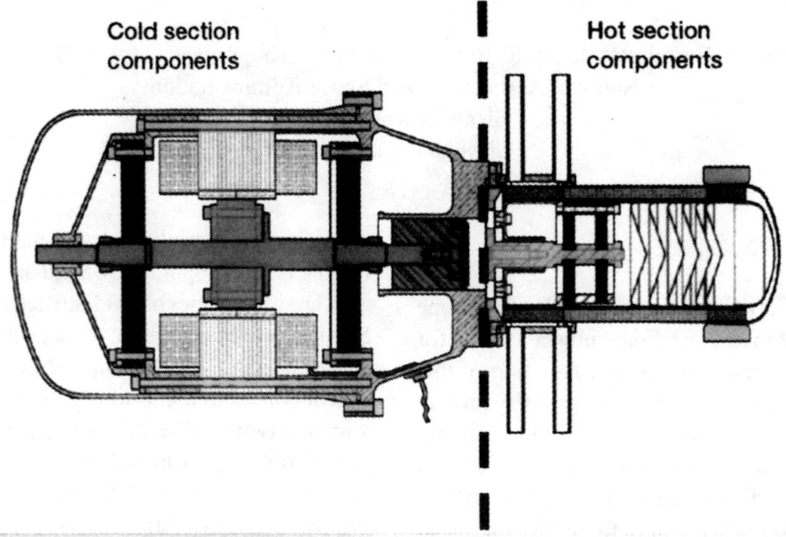

Figure 1.—Schematic of STC's Stirling power convertor showing cold and hot side components.

damage (fatigue, creep, abrasion). The radioisotope Stirling power convertor (Fig. 1) has two basic component sections (hot and cold). In the cold section, mechanical motion is converted into electrical power. Components that are located in this area consist of a linear alternator that produces the electrical power, a power piston that transfers the cycle energy, and the pressure vessel and alternator support structure. Nominal operating temperatures in this section are relatively cool ($T < 120$ °C). Therefore, thermal stability of the materials is not a prime concern. There are only three possible areas for mechanical damage in the cold section of the convertor and they are: (i) tensile rupture of the pressure vessel, (ii) high cycle fatigue (HCF) of the piston flexures, and (iii) abrasion of the power piston-to-sleeve or magnets-to-mover. The first and third damage mode can be eliminated through the use of good conventional design and fabrication methods. The only damage mode that needs to be addressed is the HCF of the piston flexures. However, due to years of test data generation and analysis of this flexure concept and materials by STC,[2] this damage mode is unlikely to occur.

The hot section (Fig. 1) is where heat energy from the radioisotope is converted into kinetic energy (and ultimately into electrical power in the cold section). The hot section of the convertor consists of a displacer that is used to move the working fluid from the convertor hot end to the cold end and vice versa, several heat exchangers to either add (or remove) heat energy from the working gas, and a heater head to contain the working gas and through which the heat energy is transferred from the radioisotope to the working gas. Due to their relatively high operating temperatures (up to 650 °C) and exposure to the vacuum of space, there are several thermal stability issues for the hot section components. Thermal aging of the hermetically sealed heater head joint for the convertor and Chromium loss in the heater head material are the two primary thermal stability issues that have been identified and are being addressed. As for mechanical damage modes, only one has been identified which is creep deformation of the heater head. The remainder of this paper describes risk mitigation on the aforementioned design and validation issues for the Stirling power convertor's heater head.

Heater Head Design/Validation Parameter

One of the most critical components of the Stirling power convertor is the heater head. Basic design considerations for the heater head are diametrically different. The heater head design must compromise between thin walls for optimal heat transfer properties and thick walls for increased structural creep resistance. Furthermore, a balance must be attained with the microstructure of the heater head material. First, the material needs to be creep resistant which requires large grains. For metals the larger grains will provide better creep resistance. However, the grains cannot be too large with respect to the thin walls of the heater head. In order to obtain a material's elastic, plastic, and creep properties, a minimum of 5 to 10 grains across the thin wall must be maintained.

Current heater head design practices are based on the only generally accepted, high-temperature structural design code in the United States' public domain, ASME Code Case N-47.[3] This code was developed for the design assessment of terrestrial-based nuclear power generation plants. The code addresses failure modes due to yielding, ratcheting, over-load, creep, creep-rupture, fatigue, and creep-fatigue interaction. N-47 is truly not a life prediction code, but rather a code for assessing, with a high degree of confidence, that a structure will last longer than some prescribed lifetime. The code is well suited for terrestrial applications that do not have limits on weight or mass. Large factors of safety are imposed on minimums from a large materials database to create design curves that are highly conservative. Furthermore, the code recognizes only a limited number of superalloys (under several limited conditions i.e., heat treatment, thickness) that are currently used in the nuclear pressure vessel and piping industry. None of the Code Case N-47 alloys are candidates for use in the hot section of the Stirling power convertor.

Generally, the essence of Code Case N-47 can be applied to the design philosophy of a Space Stirling power convertor and produce designs for launch-weight hardware. However, the basic methodology, minus the excessively large safety factors, could be employed if sufficient long-term data could be generated. Only short-term creep, creep-rupture, and thermal stability data can be generated on the heater head material within the remaining time before design details must be finalized to meet the mission schedule. Consequently, design curves for the 100,000+ hr time frame will have to be estimated using short-term data and extrapolation methods. These curves must be correlated with existing long-term curves of data that does not necessarily have the identical material conditions as the material for the flight hardware. For the subject NASA/DOE project, all test samples will come from the same heat of material as the flight hardware and will be heat-treated using an identical heat-treating schedule as the flight hardware.

Code Case N-47 is based on failure criterion that allows for gross deformations without the concern of how those deformations affect the operation of the structure. In other words, N-47 was developed for pressure vessels and pipes where 2 to 5 percent creep deformation can be allowed, as long as it does not cause a rupture, because the pressure vessel (or pipe) can still function with relatively large deformations. However, the heater head is closely coupled with the performance of the Stirling power convertor and therefore, a specific criteria of failure needs to be defined for the Stirling power convertor heater head. As the heater head is deformed due to long-term creep, the heater head volume for the working gas is increased and the displacer piston appendix gap is increased (increasing the loss in this region); both conditions will decrease the efficiency of the Stirling convertor. A more appropriate measure of failure and/or damage of a heater head should be based on these implications instead of those outlined in N-47.

Design and hardware validation for the NASA/DOE project will use a new failure criterion for the heater head based on the above premise and predictions from Stirling performance codes[4]. Present design criterion uses the time it takes to accumulate 1% creep in the heater head. Permanent creep deformation of the heater head could eventually increase the gap between the displacer piston and its sleeve. Whether and when creep of the heater head will cause increased appendix gap losses will be answered by the outlined NASA/DOE project. An additional design criterion for the heater head would be based on the time it takes the heater head to creep (deform) and significantly increase displacer appendix gap losses. For the NASA deep space missions, the time must be greater than 100,000+ hr for an adequate heater head design. This would be in addition to the conventional N-47 design criteria of time to rupture and time to 1% creep.

To estimate the degradation in the Stirling Convertor performance associated with 1% creep of the heater head, it is proposed to use Stirling performance codes in combination with a finite element analysis (FEA) of the heater head. In this method, FEA and appropriate creep models estimate the displacer appendix gap width for 1% creep of the heater head. This gap is then used for several performance code runs to estimate its contribution to a reduction in system power. Note: power reduction could be caused by a number of items such as isotope decay, magnet degradation, or an increase in the displacer appendix gap. All of these issues could contribute to the overall decrease in power and are being addressed in the NASA/DOE project. The specific distribution of each power loss mechanism and how they interact with each other is still an issue that needs to be determined.

NASA Heater Head Validation

To qualify the design and flight hardware for NASA deep space missions, NASA has developed a validation project for the critical hot section components. This project consists of (i) accelerated testing of IN718 to develop both creep and thermal stability data bases and characterize creep prediction models, (ii) thermal and mechanical finite element analysis (FEA) to verify designs and predict heater head life, and (iii) benchmark tests to characterize/verify the creep prediction model for the FEA and to evaluate the thermal stability of the heater head hermetic seals in the hot section. An outline and status of this project will be given for the remainder of this paper.

Inconel 718 (IN718) Long-Term Material Assessment Status

Due to the relatively volatility (high vapor pressure) of Chromium (Cr), long-term exposure at elevated temperatures in the vacuum of space poses a risk of Cr loss in the IN718 due to volatilization. Volatilization is especially important when the exposed surface area is high and the total volume of material is small, as is the case for the thin walls of the heater head. For these thin walls, even a small depletion layer can represent a large percentage of the total wall thickness. Bourgette demonstrated that, in general, the stability of metals alloyed with volatile elements (Chromium, Manganese, etc.) is only a concern for temperatures exceeding about 815 °C. While the heater head operating temperature is well below the 815 °C critical temperature for species loss, volatilization may still be an issue. This is because even minuscule evaporation rates could eventually result in appreciable Cr loss due to the extremely long times required for deep space missions.

Although vacuum level had little effect on the loss of Cr and Manganese (Mn),[5] the volatilization rate increased with temperature and decreased with accumulating time of exposure. Such behavior is consistent with diffusional processes, which bring the volatile species to the metal surface. Assuming that the loss of Cr is a diffusion-controlled process, then the Cr loss can be accurately predicted by solving Fick's diffusion law for the case of unidimensional diffusion in a semi-infinite medium. In this case, one half of the diffusion couple is the IN718 heater head with an initial Cr concentration of about 20 at % distributed uniformly through the thickness. The other half of the diffusion couple is the vacuum of space.

To verify the validity of Fick's equation to this problem, two 10×22×25 mm IN718 samples were placed in a vacuum furnace at 10^{-7} atm at 985 °C for 500 hr. The samples were weighed before and after exposure. The predicted total weight loss was within 90 percent of the actual measured loss for both samples. Also, the experimentally measured concentration profile of Cr as a function of distance from the surface was found to be very close to the predictions (Fig. 2). With the accuracy of the analytical solution verified, a prediction was then made

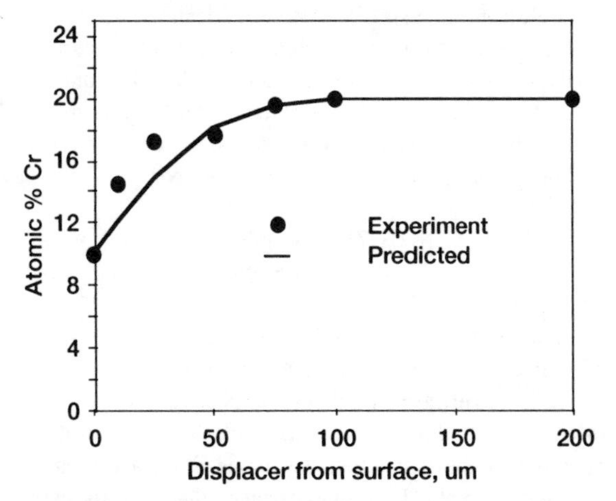

Figure 2.—Experimentally measured (circles) and analytically predicted (line) Cr concentrations as a function of distance from free surface after annealing in vacuum at 985 °C for 500 hours.

for the actual operating conditions of 650 °C and 100,000 hr. The predicted weight loss due to Cr evaporation was found to be only 0.0008 g/cm^2 with a surface depletion layer thickness of only about 10 μm. These predictions confirm the general consensus that at 650 °C Cr loss will not be a concern.

Accurately assessing the creep properties of IN718 for the heater head service conditions poses several challenges. The primary difficulty lies in the extremely long times involved – namely 100,000+ hr (~11.4 years). Not surprisingly, very little creep data exists for such long times. This dictates that the creep response will, by necessity, be extrapolated from shorter-time tests. Even when ignoring for the moment the accuracy or appropriateness of a particular predictive creep model, caution must be used if creep data from the open literature is used as the input to the model. Creep response is highly dependent on composition, heat treatment, and grain size of the material. Another complication arises due to the fairly thin wall thickness of the heater head. It is well known that the creep properties of superalloys are dependent on the specimen thickness-to-grain size ratio. Since it is unlikely that thin-section 11-year creep data will be available, it will be necessary to combine short and long-term data from the literature with short term-tests on the actual heater head material and rely on modeling to predict the viability of the material. It is for these reasons, that while the bulk of the creep database will be drawn from the literature, additional creep tests are being performed on samples with the identical thickness, heat treatment, and composition as the flight hardware.

An example of the creep data being generated is shown as a Larson-Miller Parameter (LMP) plot in Fig. 3. In Fig. 3, the rupture-time data are compiled from the literature on both thick (>0.25 in.) and thin (~0.025 in.) samples at various stresses and temperatures. Such plots suggests the possibility that creep tests can be performed in a practical time frame (at higher temperatures) under a wide variety of stress-temperature-size conditions and then extrapolate the results to the time-temperature-stress regime of interest. While a LMP plot is a fairly simplistic extrapolation technique, it is useful for quickly estimating long-term behavior. It is readily apparent from this plot that the creep lives for the thin (0.025 in.-thick literature data and 0.02 in.-thick heater head material) samples fall well below that predicted from the LMP graph for thick samples. Note that each three integers of the LMP correspond to a factor of 100 in time for the 650 °C, 100,000-hr design condition. Thus the challenge for accurate creep prediction is to not only accurately predict long-term lives based on short-term data, but also to account for the sample size effect as well.

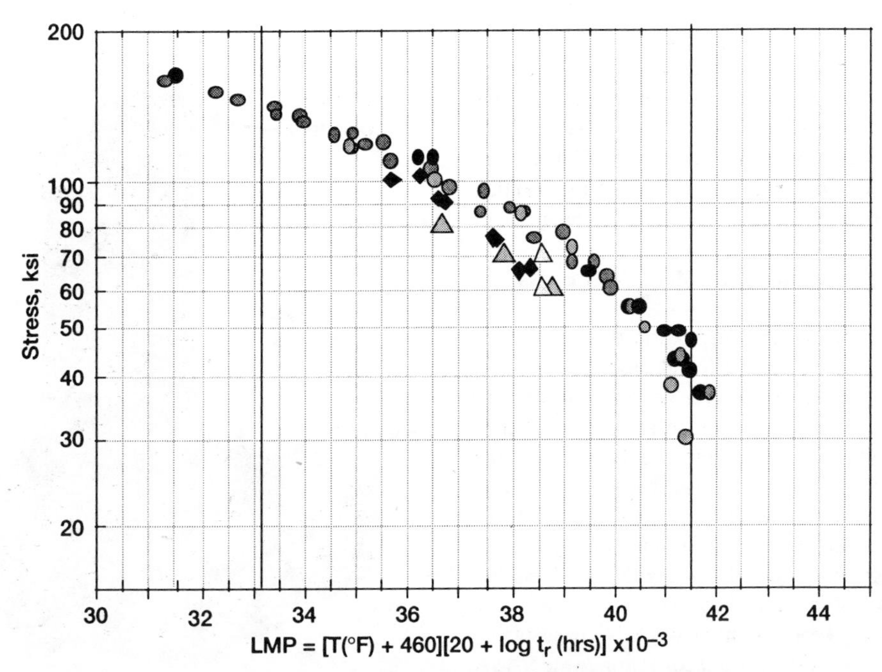

Figure 3.—Larson-Miller plot for IN718. Circles represent literature data generated on thick samples. Diamonds are literature data for thin sheet samples. Triangles are data generated in-house on samples in the heater-head configuration.

Finite Element Analysis of Heater Head

Several FEA will be conducted on the heater head. The first will be a thermal FEA where the temperature profile of the heater head will be based on the physics of the radioisotope Stirling power convertor. This FEA will take into account the heat flow between the radioisotope and the heater head, conduction losses, and convection losses. The outcome of this analysis will be an accurate temperature profile of the heater head and provide the input for the mechanical FEA, the output from which the life of this critical component will ultimately be predicted.

The IN718 creep data being generated in this project will be used to characterize a probabilistic creep model that will be used in the mechanical FEA. The mechanical FEA will be used to validate the design of the heater head for 100,000+ hr lifetimes by predicting the time it takes to increase the displacer gap to some maximum allowable size or 1% creep of the heater head (whichever comes first). An example FEA mesh and an exaggerated deformed output plot are shown in Fig. 4. A series of deformed shape plots can be produced for different time periods throughout the proposed mission life of the power convertor. This will provide information that can be input to the Stirling performance codes to predict power performance at those different time periods.

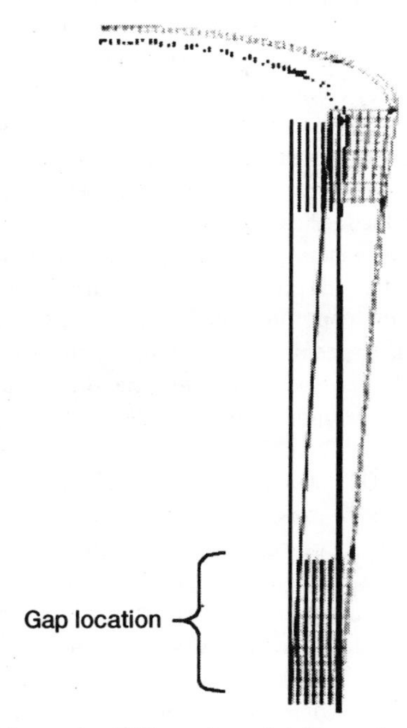

Figure 4.—FEA mesh and deformed output plot of a Stirling power convertor heater head showing the displacer appendix gap critical location.

The FEA predictions and creep models will be calibrated and verified through the use of a series of benchmark tests being developed by NASA. Several of these benchmark tests will be conducted for short durations (2 to 6 months) to calibrate the models. After the FEA prediction models have been calibrated a final long term test (>1 year) will be used to verify the FEA method that will be used, in turn, to qualify the flight hardware.

Benchmark Testing and Life Model Verification

Creep test data being generated in this project are from thin flat specimens of IN718 conducted under uniaxial test conditions. The IN718 heater head thin walls, like most pressure vessels, are under a stress state that is multiaxial. Typically, designers would account for multiaxial stress effects by either taking Von Mises (equivalent) stresses and compare them to uniaxial data or by taking a knockdown factor (based on past experience) in design allowables and design with the degraded properties. Both of these methods would produce a very conservative heater head design that would be unacceptable for the radioisotope Stirling power convertor. In the NASA project, uniaxial test data is being used for the initial model characterization and benchmark tests on heater heads will provide information for the final calibration of the FEA model.

Figure 5 illustrates the test method being developed to calibrate and verify the model. For these tests, purified argon is used as the working gas and is being controlled by a sophisticated computer control system (not shown in Fig. 5). The test system consists of an IN718 heater head (actual flight hardware type), a pressure manifold, a ceramic filler plug, five linear voltage displacement transducers (LVDT), induction heating system, and a computer control system. The induction heating system will provide the appropriate thermal profile to the heater head. Argon will flow from the controller through the manifold into the heater head. The ceramic volume fill plug is used to limit the amount of argon required for pressure cycling. The plug is slightly undersized from the heater head thus significantly reducing the fill volume of the heater head.

The LVDTs will measure the deformation of the heater head as the tests are performed. Four of the LVDT are placed in the hottest elevation around the circumference of the heater head to sense the hoop deformation. The fifth LVDT is placed at the top of the head to measure the axial deformation relative to the base of the heater head. These displacement measurements will be compared to the FEA predictions for each test condition to calibrate and verify the model.

Initially, test conditions of argon pressure and heater head temperatures will be chosen to produce a fixed amount of creep deformation corresponding to exposure

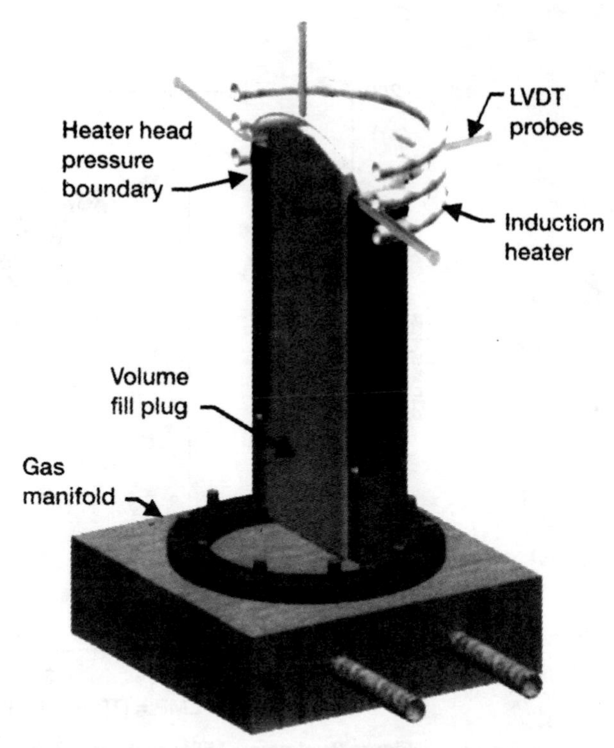

Figure 5.—Schematic of heater head benchmark test setup that will be used to calibrate and verify life prediction design method for 100,000+ hour mission life hardware.

lives of 2, 4, and 6 months. The test conditions are chosen to accelerate creep damage (higher pressures and temperatures) without inducing creep damage mechanisms that are different from those that the heat head will encounter during actual 100,000+ hr operations at a 650 °C maximum. Tests results from these short-term tests will be correlated with prediction results and the models will then be recalibrated.

The final long-term test will verify the prediction model by having the test conditions chosen by the model. Periodically throughout this test, the deformation will be monitored and compared to the predicted deformations. The life method will be valid if the measured strains (deformations) fall within a 10 percent error band. The final prediction of the verified life method will be for the actual 100,000+ hr design life of the flight hardware.

Concerns for the hermetic seal of the Stirling power convertor has prompted NASA to develop an innovative test method to verify the joint in the hot section. A Stirling power convertor pressure shell is fabricated using the stainless steel outer body in the cold section and the IN718 in the hot section of the convertor. The pressure shell is then charged with helium (He), which is the working gas for the deep space Stirling convertor. The pressure of the He is twice that of the normal operating pressure in flight applications. Likewise, the temperature of the hermetic joint location will be higher than normal at 500 °C (normal operating temperature at this location is ~100 °C).

The assembly will be installed into a leak detection chamber and He content will be monitored for a minimum of two years. The basic premise behind this test is that by heating the joint location up to 500 °C, the thermal aging of the joint will be accelerated. Likewise, doubling the working pressure of the He gas, the main driver for diffusion of the He gas is significantly increased. The failure definition for this test is if helium is detected at a certain parts per million (ppm) level. If failure in this accelerated test does not occur within the two-year period, the hermetic seal design and joining process will be verified. By passing this thermal aging test, there will be a high level of confidence in the design and fabrication processes of the hermetic seal.

Summary

Radioisotope Stirling Power Convertors are candidate power systems for deep space probes. Critical components have been identified and a design and validation philosophy has been defined. A new design and validation methodology has been created for the heater head that is based on system performance instead of structural damage such as creep rupture. FEA plays an important role in the process and will be used to calibrate life models and predict ultimate component life. A heater head material has been chosen and a preliminary materials database is being generated. Chromium loss due to the high operating temperatures and the vacuum of space has been shown to be insignificant for the heater head material. However, creep behaviors of thin specimens when compared to existing data have shown to be an area of concern. This does not represent an insurmountable issue but just an issue that needs to be resolved by using good engineering judgment and design methods.

A series of planned benchmark tests are in the process of being developed. These tests will aid in the calibration of life prediction methods and the eventual verification of those methods. A benchmark test for evaluating the hermetic joint has been defined and is presently being developed. This test will thermally age the joint and monitor for He gas leakage within a two-year period. The proposed plan will provide a significantly high level of confidence that the hot section components of a Radioisotope Stirling power convertor will survive its 100,000+ hr mission life for deep space probes.

Acknowledgments

The authors would like to acknowledge Mr. Roy Tew and Mr. Steve Geng for their HFAST results of the Stirling power convertor system.

References

1. Thieme, Qiu, and White, NASA TM-2000-209791, "Technology Development for a Stirling Radioisotope Power System," Space Technology and Applications International Forum 2000, AIP Conference Proceedings, American Institute of Physics, 2000.
2. White, Qiu, Olan, and Erbeznik, *"Technology Demonstration of a Free Piston Stirling Advanced Radioisotope Space Power System,"* Space Technology and Applications International Forum 1999, AIP Conference Proceedings 458, American Institute of Physics, 1999.
3. Anon. *Code Case N-47*, ASME, 1987.
4. Geng and Tew, NASA TM-1992-105549, *"Comparison of Glimps and HFAST Stirling Engine Code Predictions with Experimental Data,"* 27th IECEC, 1992
5. Bourgette, ORNL-3677, *"Evaporation of Iron-, Nickel-, and Cobalt-based Alloys at 760 to 980 °C in High Vacuums,"* Oak Ridge National Laboratory, 1964.

EXPERIMENTAL STUDY ON STIRLING ENGINE GENERATOR AND SOLAR RECEIVER SYSTEM FOR FUTURE SPACE APPLICATIONS

Takeshi HOSHINO[1], Hitoshi NAITO, Tsutomu FUJIHARA and Kunihisa EGUCHI
Space Energy Utilization Research Group, National Aerospace Laboratory,
7-44-1, Jindaiji-Higashi, Chofu, Tokyo 182-8522, Japan
E-mail: [1]hoshino@nal.go.jp

ABSTRACT

A fundamental study on solar Stirling power generation system has been performed as a part of the space solar power technology program in National Aerospace Laboratory (NAL). The research work involves both a solar receiver and a Stirling power generation technologies. The former work is focused on developing a high efficiency solar receiver for future space energy experiments on the Japan Experiment Module - Exposed Facility of the International Space Station. It is composed of a cavity receiver, thermal energy storage and sodium heat pipes, and aims at transporting solar heat to a bottoming system with the minimized heat loss. A ground test model of the solar receiver will be experimentally evaluated in NAL. In parallel, semi-free piston Stirling engine generators have been experimentally studied as one of candidate converters for future space power generation. Through a series of bench tests, the thermodynamic efficiency of 32% and system efficiency of 20% was obtained. Based on these achievements, the design work for an improved Stirling generator system is underway.

INTRODUCTION

The available primary energy sources in long-term space activity are solar energy or nuclear energy. However, it is impossible to utilize atomic energy for space use in Japan under the existing circumstances, and the energy source must depend on the solar energy. In present state, the solar energy is converted into electricity by solar batteries, and utilized only 20%, and approximately 80% of energy is often being disposed as unused energy. This is a waste of energy, and it is necessary to utilize the solar energy more effectively. The solar energy is possible to convert into high temperature and high density thermal energy by the appropriate concentration. This is a very useful energy source, and it is possible to effectively conjugate for various purposes such as material experiments, power generation and thermal propulsion.

In National Aerospace Laboratory, the basic research of solar receiver system and power generation system using the Stirling engine generator is carried out in order to obtain the basic data of the solar thermal application system in the space. Since 1997, the basic study of heat transport system composed of a cavity solar receiver, thermal energy storage and heat pipes have been performed as a part of "Ground Research Announcement for Space Environment Utilization Program" promoted by NASDA (National Space Development Agency of Japan) and Japan Space Forum[1)]. This study is aiming to establish the basic technology for the space experiments using solar thermal energy on the JEM-EF (Japanese Experiment Module - Exposed Facility) of the ISS (International Space Station). On the other hand, performance test and improvement of Stirling generators has been performed as part of the space solar power utilization program of NAL. Several types of semi-free piston Stirling generator have been fabricated and experimentally evaluated. These engines, called NALSEM (NAL Stirling Engine Model) series, are semi-free piston Stirling machines with a moving-magnet linear alternator. The latest engine is the NALSEM 500 designed to achieve a target thermodynamic efficiency of 30 % or more with an indicated power of 500 W[2,3]. The design performance has been available to date[4].

In this paper, the following points are described: design concept and test results of the receiver system, experiments and future research plan of the engine system.

RECEIVER SYSTEM

SYSTEM CONCEPT

Figure 1 shows a schematic of solar receiver system on the JEM. It is composed of a concentration mirror, a cavity heat absorber, and six modules of thermal energy storage (TES) and sodium heat pipes (HP). Solar light of approximately 1.6 kW is concentrated on a focal plane of the absorber aperture, and then thermal energy is transported to a bottoming system as a heat source via the TES/HP module behind the receiver.

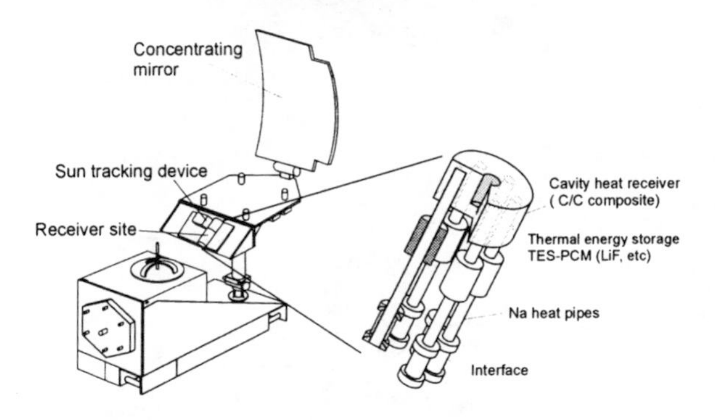

Fig.1 Schematic - solar heat receiver model.

TEST MODEL

An experimental model of receiver system has been fabricated, and the performance test is now underway. The experimental model is shown in figures 2. The experimental model is different from the whole model from restrictions of experimental facility. The number of heat pipe is three, a half of whole model. Therefore, the input heat quantity is also a half of the real system. The cavity receiver is made of three-dimensionally arranged C/C composite. It has high thermal conductivity and provides equalizing of temperature distribution in the receiver. Four modules of thermal energy storage are installed in each heat pipe. The LiF-CaF_2 is used as its medium. A cooler using helium gas is installed at the condenser part of each heat pipe in order to measure an amount of heat transport. The experimental model is tested in the vacuum chamber, and it is covered with the multilayer metal foil, which minimizes heat loss due to radiation.

Fig. 2 Outer view of ground test model.

TEST RESULTS

Though the receiver concept model is not designed for providing enough heat energy to NALSEM 500 generator system (the input power is smaller than the engine required power because of the mirror area limitation), it can be evaluated whether this system is adaptable to a Stirling engine system by setting appropriate conditions. Then, it was experimentally confirmed that this system was able to provide enough temperature for the engine operation during sunshine (40min) / sunshade (50min) cycle.

In the experiment, the halogen lamp was used for the heating, and an amount of heat transport was measured by a calorimeter of He-cooling jacket attached to a condenser part of each heat pipe. The pre-heater was connected to the cooler, in order to change the cooling gas temperatures from the room temperature to 873K. The cooling gas temperature and flow rate are correspondent to the gas temperature at the hot side of regenerator in the Stirling engine and engine operating speed, respectively. These conditions were adjusted to the NALSEM500 design conditions.

Figure 3 shows history of the vapor temperature at the heat pipe condenser (T_{Vc}) and each TES. It is found that the receiver system is able to continuously maintain a constant temperatures above the required level of 923K.

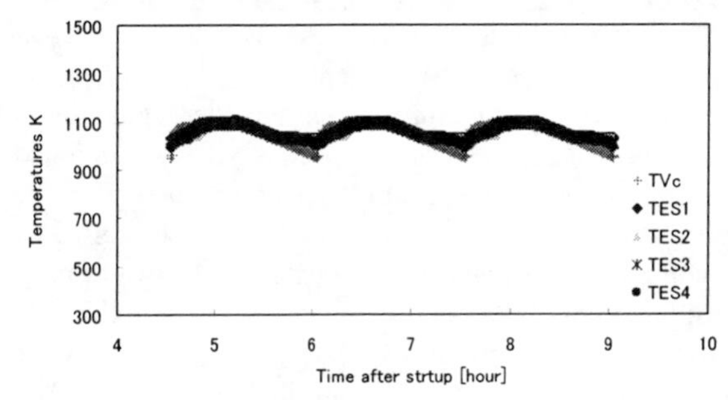

Fig. 3 Heat cycle test results.

The detailed performance evaluations are described in the reference 5.

STIRLING GENERATOR

DESCRIPTION OF NALSEM500

A photograph and a cross-section of NALSEM 500 are given in figure 4 and 5, respectively. Table 1 shows typical design specifications. This engine has gamma-type configuration and an integrated linear alternator.

The engine heater head has been developed for integration with solar receiver heat pipes. However, in our experiments, twelve electrical heaters are attached to the engine heater head, instead of the heat pipes. The regenerator has an annular flow configuration, and multi-layer stainless steel meshes are used for the regenerator matrix. The water cooler is a shell and tube type.

The alternator is a moving-magnet type in an annular configuration. The magnet material is Nd-F-B. To minimize helium leakage into bounce space, a clearance seal is adopted at the power piston. The design piston stroke is 20mm. The power piston is completely free, and the gas spring is used as spring device. Therefore, for centering and stable operations of the power piston, pressure control system is adopted. To keep centering, the mean cycle pressures in both of compression and bounce spaces are automatically controlled with one-way return flow valve, where the leaked gas flows back to the compression space, driven by a stepping motor.

The displacer is driven by a DC electric motor, not free piston. The engine operation speed ranges from 10 to 30 Hz by controlling displacer motor frequency. It is useful to make clear the performance characteristics, depending on engine speed. The displacer stroke is changeable using an eccentric disc, attached between a motor shaft and a yoke-cam. In our testing, the stroke is kept 20 mm constant.

EXPERIMENTS AND CONTROL SYSTEM

The pressure and temperature sensor are located at each working space. The displacer and power piston displacement are measured by an encoder signal and a laser displacement sensor, respectively. As mentioned before, the pressure control system is used for the piston centering. The return flow valve is controlled by a computer based on measured data of the power piston displacement. The control frequency is approximately two times per second, and the center location of the power piston stroke is controlled within ±0.1mm accuracy. In our experiment, this system operates only during unsteady state, in case of changing engine speed, temperature, and load. The generated electrical power is dissipated by an external load.

Fig. 4 Outer view of NALSEM 500 hardware.

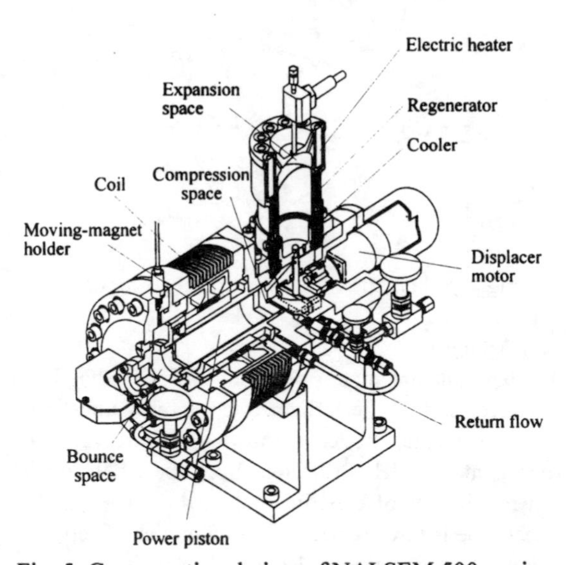

Fig. 5 Cross-sectional view of NALSEM 500 engine.

Table 1 Design specifications.

Initial charge pressure	3 MPa	
Expansion space temp.	873 K	
Coolant temp.	298 K	
Engine speeds	- 30 Hz	
Piston stroke	Displacer	20mm
	Power piston	20mm
Regenerator	Stainless steel wire mesh	
Linear alternator	Moving magnet	

PERFORMANCE EVALUATIONS

TEST CONDITIONS

The engine / alternator tests have been conducted for performance evaluation. Test conditions are specified in Table 2.

Table 2 Test conditions.

Operation parameters	Values
Initial charge pressure	3 MPa
Expansion space temperature	873 K
Coolant inlet temperature	298 K
Engine speed	10 to 30 Hz
External load resistance	0 to 1500 Ω

THERMODYNAMIC PERFORMANCE

The indicated power and efficiencies are shown in Fig. 6 and 7, respectively. The indicated efficiencies are based on the following definition.

$$\eta_i = \frac{W_i}{HI} \tag{1}$$

W_i: Indicated power

HI: Electrical heater power input

HI is all electric power that is added to the heater, and it contains the loss of heater lead and heat radiation and conduction to the circumference environment. (Therefore, this makes efficiency a few percent lower as compared with a value based on net heat input.)

The maximum values of indicated power (W_i) and efficiency (η_i) were 585W and 32%, respectively, at the conditions of 30Hz frequency, and 1000 Ω resistance. It was expected that more power and efficiency would be achieved at a higher resistance value from the data tendency, however, the data above 1000 Ω could not be obtained because of the piston stroke limitation. The stroke length of power piston was approximately 21mm at the maximum power conditions, and this value was larger than our design value of 20 mm (Fig. 8). As mentioned previously, this engine was designed to achieve 500W indicated power, and it was found that obtained power was greater than the design power. It was desirable to obtain greater power than design power, however, it caused a problem on the alternator operation. In other words, the alternator could not consume engine power. It will be discussed later.

SYSTEM PERFORMANCE

The alternator output power and system efficiencies are shown in Fig. 9 and 10, respectively. The system efficiency (η_s) is defined as follows.

$$\eta_s = \frac{W_G - W_{DM}}{HI} \tag{2}$$

W_G: Alternator power output

W_{DM}: Displacer motor driving-power

All of these values are in terms of electric power. The data when the displacer driving power is greater than the alternator output (negative system efficiencies) are not shown here. As seen from the plotted data, the maximum system efficiency and alternator power output (W_G) were 427W and 20%, respectively, under the conditions of 30Hz frequency, 1000 Ω resistance. The alternator efficiencies were around 80%.

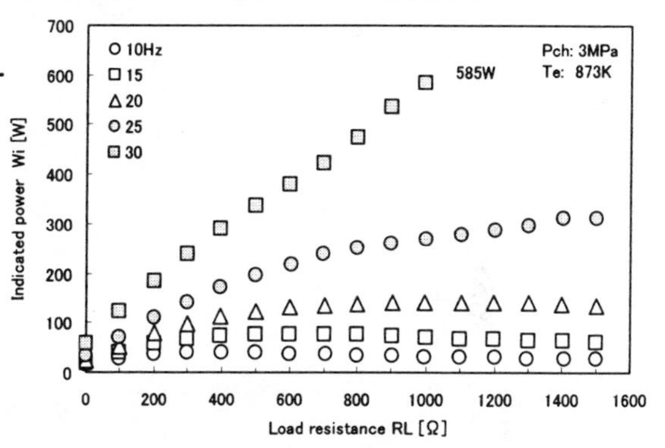

Fig. 6 Indicated power.

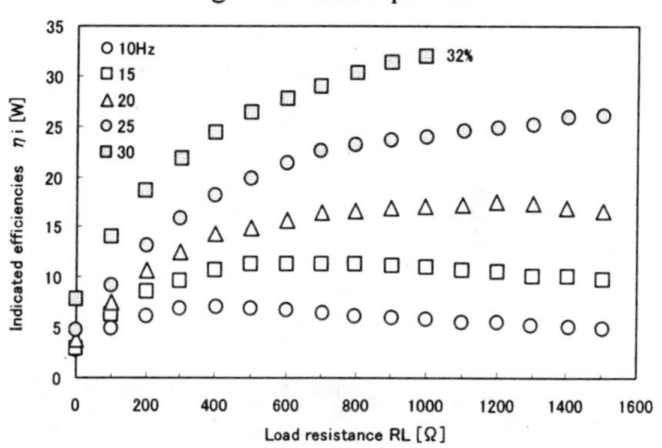

Fig. 7 Thermodynamic performance.

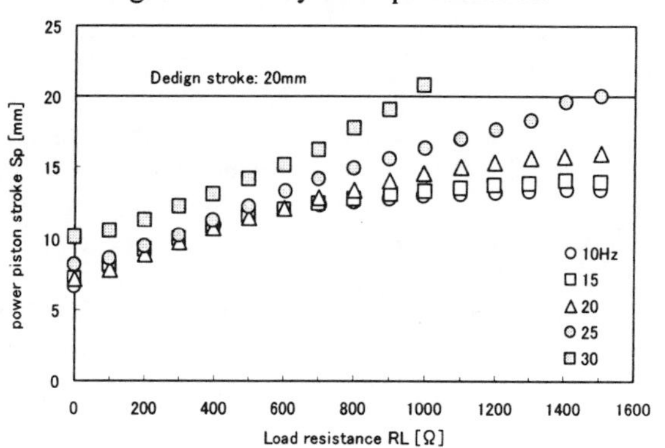

Fig. 8 Power piston stroke.

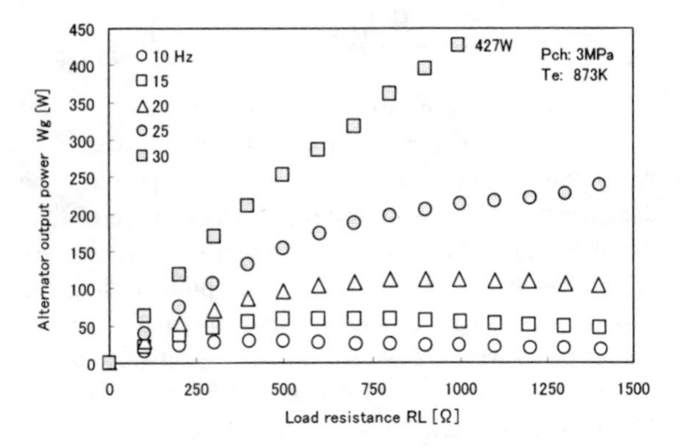

Fig. 9 Alternator output power.

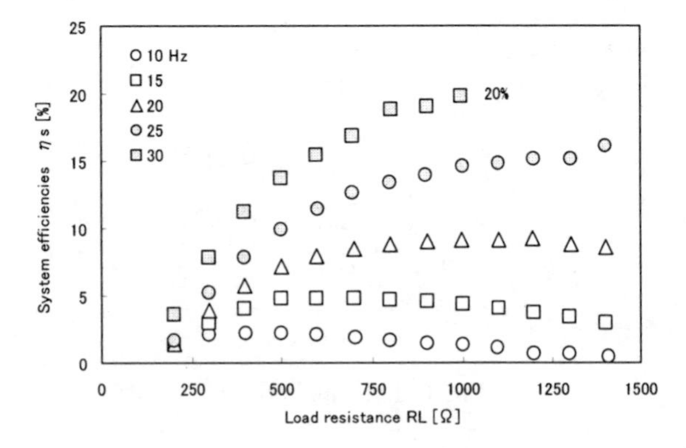

Fig. 10 System efficiencies.

DISCUSSIONS

The obtained system efficiency was not enough level. In order to enhance the system performance, the alternator efficiency should be improved. Though it is a necessary condition to increase alternator efficiency for system performance improvements, there is another important point.

Design operation frequency of this engine is 30Hz, and the piston stroke is 20mm. It is required that the alternator absorbs an output of the engine under this condition and converts it into the electric power.

The calculated result of input and output power of the alternator on operation frequency 30Hz and piston stroke 20mm is shown in figure 11. The calculation conditions are based on the NALSEM500 design specifications. The required constants, coil inductance, internal resistance, flux linkage depend on the piston position and etc., were experimentally measured.

W_G means electrical output power at the external load, and W_{G0} means all electrical power including the copper loss at the coil. W_P is consumption power of the alternator, and it includes alternator loss (e.g. eddy current loss at the iron core). Though the results could not be compared directly with experimental results because the stroke was assumed 20mm, it had good agreement with the experimental data at 1000 Ω. From this figure, it is found that both input and output power of the alternator show a maximum around 500 Ω and largest consumption power of this alternator is about 640W. To improve the efficiency, the alternator should be operated at higher load resistance than 500 Ω. Then, the maximum consumption power becomes lower than 640W.

On the other hand, output characteristics of the engine were similarly calculated on operation frequency 30Hz, piston stroke 20mm. The calculation method, though is not described here, it was based on the nodal analysis. The result shows that the piston power W_P reached approximately 650W at the condition of maximum efficiency in spite of 500W design power. The reason was that some engine modifications, which had been done, made engine output power increase.

If only alternator efficiency is improved, its consumption power is decreased and the alternator cannot absorb engine power. From them, it turns out that it is required to improve both alternator efficiency and capacity.

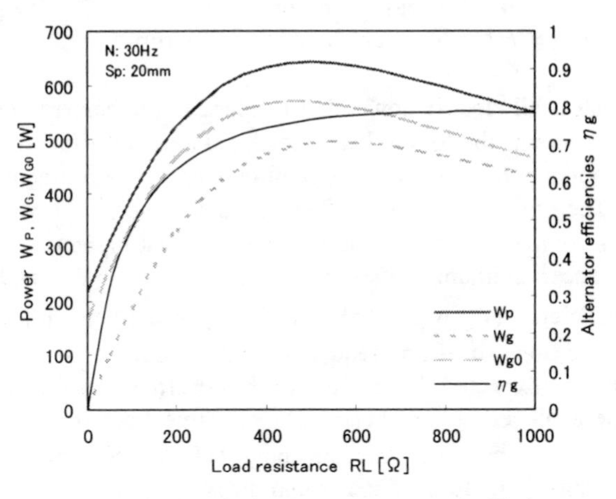

Fig 11 Calculated alternator power and efficiency.

ALTERNATOR MODIFICATIONS

In order to improve the overall performance, two types of alternator were designed. These design work were focused on to increase consumption power and enhance its efficiency from 80 to above 90%. The design specifications are summarized in Table 3.

The type I has modified alternator, and no change has been done about the engine. The type II has modified alternator and opposite piston configuration to minimize mechanical vibration during operation. These alternators are under fabrication, and the test results will appear in late 2000.

Table 3 Alternator design specifications.

	TYPE I	TYPE II
Moving parts	Magnet	Ibid.
Piston Diameter Stroke	1 piston ϕ 50 20mm	2 opposite pistons ϕ 50 12mm
Capacity	650 W	750 W
Target eff.	90 %	92%

CONCLUSION

A ground test model of solar receiver system for future space experiments has been fabricated and the performance test has been carried out. As a result, it was found that the receiver system was able to provide enough temperature for enabling engine operations under the simulated in-orbit cycle of solar heat.

In addition, the performance evaluation of NALSEM500 Stirling alternator has been performed to technically verify our design. Significant results and future research tasks are summarized below:

1. Maximum indicated power and efficiency were 585W and 32%, respectively, at a design frequency of 30Hz.
2. Alternator output power and system efficiency were 427W and 20%, respectively, at 30Hz. The alternator efficiency reaches 80%.
3. It was found that the alternator should be modified. Because the engine output power exceeded the alternator capacity.

In order to improve the overall performance, two types of alternator has been designed. These alternators are under fabrication, and the test achievements will appear in late 2000.

ACKNOWLEDGMENT

The study on the receiver system had been carried out , worked jointly with "Basic Research on High-Efficiency Cavity Receiver for Solar Thermal Energy Utilization in Space" as a part of "Ground Research Announcement for Space Utilization," promoted by NASDA and Japan Space Forum.

REFERENCES

1. Naito, H., et al., Ground-Engineering Study on Solar HP/TES Receiver for Future ISS-JEM Experiment Program, Proc. 34th IECEC, 1999, 1999-01-2587.
2. Eguchi, K., et al., Design and Analysis for Space Heat Pipe Stirling Alternator of NALSEM 500, Proc. 19th ISTS, 1994, pp.493-499.
3. Nakamura, Y., et al., Linear Alternator Design for NALSEM Stirling Engines, Proc. 6th ISEC, Rotterdam, Netherlands, May 1993.
4. Hoshino, T., et al., Basic Research on Solar Stirling Power Technology for Future Space Applications, Proc. 34th IECEC, 1999, 1999-01-2681.
5. Naito, H., et al., An Experimental Study of a Solar Receiver for JEM Experiment Program, Proc. 35th IECEC, 2000, AIAA-2000-2996.

NOMENCLATURE

P_{CH}:
Initial charge pressure of working gas.

Te:
Expansion space temperature.

W_i:
Indicated power

W_G:
Alternator power output

W_{G0}:
Alternator power output including coil loss

W_P:
Power piston power

W_{DM}:
Displacer motor driving-power

HI:
Electrical heater power input

η_i:
Indicated efficiency, expressed by W_i/HI

η_s:
System efficiency. (W_G-W_{DM})/HI

AIAA-2000-2844

RESULTS FROM THE MICROMINIATURE THERMIONIC CONVERTER DEMONSTRATION TESTING PROGRAM

Donald B. King*, Kevin R. Zavadil†, Judith A. Ruffner‡
Sandia National Laboratories
Albuquerque, NM 87123

James R. Luke
New Mexico Engineering Research Institute
Albuquerque, NM 87106

Research is in progress to develop microminiature thermionic converters (MTCs) using semiconductor integrated circuit (IC) fabrication methods. The use of IC techniques allows the fabrication of MTCs with an emitter to collector spacing of several microns or less and with emitter and collector materials that will have work functions ranging from 1 eV to 2 eV. Theory predicts that the small emitter to collector spacing and highly emissive low work function electrodes should allow the conversion of heat energy to relatively large electrical current densities (up to tens of Amps/cm^2) at relatively high conversion efficiencies (15-25%). Tests of prototype MTCs have demonstrated energy conversion at several emitter temperatures. The power generated is less than expected because of less than optimal emission characteristics encountered with the low work function materials used for the first MTC prototypes. Efforts to improve electrode emission characteristics are in progress.

INTRODUCTION

The Defense Threat Reduction Agency's (DTRA) Advanced Thermionics Program (ATP) and the Department of Energy (DOE NE) seek to advance the state of the art of thermionic power conversion in the United States. One direction of this program is to enable revolutionary advances in thermionic converter performance by dramatically reducing the gap size of the converter and incorporating low work function materials into the converter electrodes. Sandia National Laboratories (SNL) has a program to develop the microminiature thermionic converter (MTC), a new generation of thermionic converter. The MTC is a planar, two-electrode thermionic diode. Several technological developments have enabled the development of the MTC. Commercial electronic fabrication techniques allow the MTC to be fabricated as one unit with micron-sized electrode spacing (a micron-sized gap between the electrodes plays a key role in obtaining higher energy conversion efficiencies). Also, many MTCs can be fabricated as one large conversion unit using the electronic fabrication techniques to scale the output current and voltage. This scale of fabrication may allow various MTC units to be fabricated with power outputs ranging from milliwatts to hundreds of watts and at low cost. Emitter materials with low work functions that are compatible with the electronic fabrication techniques have been developed. Work is also in progress to produce materials with low but variable work functions. The use of variable work function electrodes will allow the MTCs to function efficiently for the different heat sources and their temperature regimes; these low work functions allow the MTC to operate at modest temperatures (800-1300 K).

The MTC program has several parallel development paths. Thermal modeling of the diode structure is performed to develop structure types necessary to operate the MTC at required electrode temperatures with a minimum heat loss from the diode. Electrical modeling of the diode is performed to select the optimal diode operating conditions in terms of conversion efficiency of heat to electricity and power output. Material studies are necessary to identify the appropriate materials with which to fabricate the MTC diode structure and fabricate low work function electrode materials. The materials used in the MTC must maintain diode integrity at high operational temperatures, and the low work function materials must have appropriate emission current densities at desired operational temperatures. Low work function materials with high emission properties have been identified in the open literature but have not been optimized for use

*Advanced Nuclear Concepts Department
†Interface Reliability Department
‡Electronic and Optical Materials Department

in the MTC. Research in the materials and deposition processes must continue to enable fabrication of low work function electrodes with the proper emission characteristics for the desired MTC structures. The low work function electrodes are an integral part of the MTC structure. The MTC structure will be made of ceramic and insulator materials (such as quartz and sapphire), metals (such as tungsten, nickel, platinum and molybdenum), and semiconductors (such as silicon and silicon carbide). The metal/ceramic/semiconductor interfaces must be studied and characterized at high temperatures to determine the reliability of the MTC devices during high-temperature operation. Adhesion between materials and stresses due to thermal expansion mismatches must be carefully controlled. The susceptibility of MTCs to shock and vibration as well as the effects of combined radiation and temperature must also be studied and measured.

All of the material types must demonstrate long lifetime and reliable operation at the selected diode operational temperatures. Various fabrication techniques have also been studied to identify the most feasible method to construct the MTC. Finally, tests must be performed to confirm the low work function material emission characteristics, the compatibility of diode materials at high temperatures, and the performance of the diode itself.

THERMIONIC PROCESSES IN MTCs

Thermionic emission depends on emission of electrons from a hot surface. Valence electrons at room temperature within a metal are free to move within the atomic lattice, but very few can escape from the metal surface. To first order, the electrons are prevented from escaping by the electrostatic image force between the electron and the metal surface. The heat of the emitting surface gives the electrons sufficient energy to overcome the electrostatic image force. The energy required to leave the metal surface is referred to as the material work function, ϕ. The maximum rate at which electrons leave the metal surface, J (in A/cm^2), is given by the Richardson-Dushmann equation[1]:

$$J = AT^2 \exp(-[\psi - \mu_E]/kT_E), \qquad (1)$$

where A (in A/cm^2/K^2) is the Richardson-Dushmann (RD) constant, T_E (in K) is the emitter temperature, k (in eV/K) is the Boltzmann constant, μ_E (in eV) is the emitter Fermi energy level, and ψ (in eV) is the maximum motive experienced by an electron in transit. The Richardson-Dushmann constant is calculated from physical constants, and its theoretical value is 120 A/cm^2/K^2. Figure 1 illustrates the energy levels that an emitted electron must overcome as required by equation (1) in a two-electrode thermionic converter. Electrons leave the emitter surface and travel to the collector surface. Operating the collector at lower temperatures minimizes electron emission from the collector to emitter. The maximum motive depends on the sum of the emitter work function (ϕ_E) and the electron induced space charge, ΔV_{sc} or the sum of the diode operating voltage (V_d), the collector work function (ϕ_c), and ΔV_{sc}.

Large emission current densities can be achieved by coating the emitter surface with a low work function and high RD constant material and operating the emitter at as high a temperature as possible. Operating temperatures are limited because high temperature operation may cause any material to evaporate rapidly and limit emitter lifetime. Materials with low evaporation rates usually have high work functions. The measured RD constant rarely equals 120 A/cm^2/K^2 because of surface effects such as reflection and patches. The reflection and patch effects can drastically reduce the RD constant from 120 A/cm^2/K^2 to values as low as 0.001 A/cm^2/K^2, depending on the emitting material. Measured RD constants are often referred to as apparent or effective RD constants. Electron reflection at an emitting or absorbing surface affects electrons in the emitter (internal electrons) that leave the surface and enter the vacuum as well as external electrons (from a second electrode, for example) that are striking the surface. Materials exhibiting the patch effect have surfaces of irregular or nonuniform work function. This nonuniformity creates an electric field above the surface that causes the surface to exhibit an effective work function value that is the area weighted average of the individual patches at low diode voltage biases.

Choosing the correct electrode coating is half the battle in designing thermionic converters. Once the electrons are successfully emitted, they must traverse the interelectrode gap to the collector to contribute to the net current. Electrons emitted from the emitter produce a space charge in the interelectrode gap (IEG). For large currents, the buildup of charge will act to repel further emission of electrons and limit the efficiency of the converter. Two options have been used to limit space charge effects in the IEG: thermionic converters with small IEG spacing (the close-spaced vacuum converter) and thermionic converters filled with ionized gas.

Thermionic converters using a gas in the IEG are designed to operate with ionized species of the gas. Cesium vapor is the gas most commonly used. Cesium has a dual role in thermionic converters: (1) space charge neutralization and (2) electrode work function

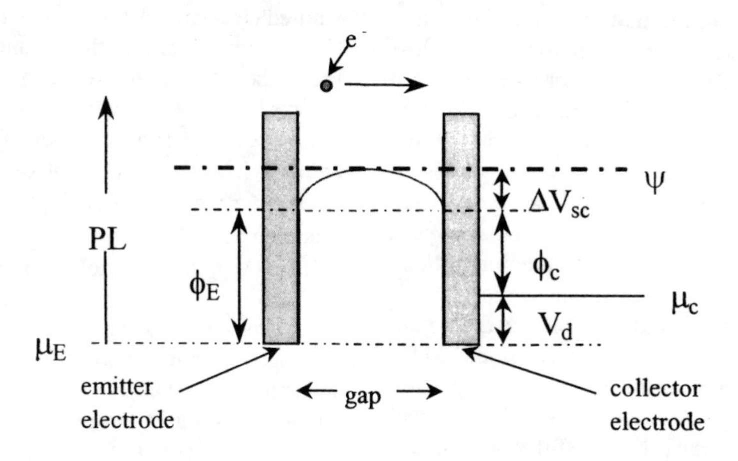

Figure 1. Motive (potential energy, PE) Experienced by Electrons During Transition.

modification. In the latter case, cesium atoms adsorb onto the emitter and collector surfaces. The adsorption of the atoms onto the electrode surfaces results in a decrease of the emitter and collector work functions, allowing greater electron emission from the hot emitter. Space charge neutralization occurs via two mechanisms: (1) surface ionization and (2) volumetric ionization. Surface ionization occurs when a cesium atom comes into contact with the emitter. Volumetric ionization occurs when an emitted electron collides with a Cs atom in the IEG. The work function and space charge minimization increases the converter power output. However, at the cesium pressures necessary to substantially affect the electrode work functions, an excessive amount of collisions (more than that needed for ionization) can occur between the emitted electrons and cesium atoms, resulting in a loss in conversion efficiency. Therefore, the cesium vapor pressure must be controlled so that the work function reduction and space charge reduction effects outweigh the electron-cesium collision effect. An example of an operational thermionic converter is found on the Russian TOPAZ-II space reactor. These converters operate at emitter temperatures of 1700 K and collector temperatures of 600 K with a cesium pressure in the IEG of just under one torr. Typical current densities achieved are < 4 A/cm^2 at output voltages of approximately 0.5 V. These converters operate at an efficiency of approximately 6%. The control of cesium pressure in the IEG is critical to operating these thermionic converters at their optimum efficiency. MTCs offer a simpler solution to space charge control for thermionic energy conversion. A gas need not be introduced into the IEG to reduce the space charge effects resulting from the current flow from the emitter to the collector. The small IEG size itself reduces the density of electrons in the gap (and their resulting current-limiting space charge). Historically, the close-spaced converter has been difficult to manufacture for large-scale operation due to the close tolerances (several microns or even submicron interelectrode gap size) needed for efficient operation. Large-scale production and operation of these close-spaced converters is now possible using IC fabrication techniques. Gap spacings as low as 0.25 to 1 microns can now be produced and maintained over relatively large emission areas.

MODELING

The MTC diode electrical characteristics were calculated using a computer code (MTCP) developed at Sandia that simulates the electrical performance of a vacuum diode. This code determines a range of possible current and voltage operational regimes for given diode operational temperatures and geometry specifications. Figures 2 and 3 illustrate the potential unit diode current density output and power output versus the voltage range through which a diode is typically operated as calculated by MTCP. Figure 2 shows the diode current density as a function of the diode operating voltage as well as the diode gap size. The ideal diode represents a fictional diode with a gap size approaching zero. As the gap size increases, the diode current decreases due to the space charge effect of electrons transiting the gap from one electrode to the other (electrons already in the gap tend to retard further electron emission). The diodes must be fabricated with as small a gap size as possible. The calculations were performed with the emitter and collector electrodes at 1200 K and 700 K, respectively. The emitter and collector work functions used in the calculation were set to 1 eV, and the apparent RD constant was set to 10 A/cm^2/K^2. The work function values and apparent RD constant chosen represent goals to achieve for material development in the MTC program. Figure 3 illustrates

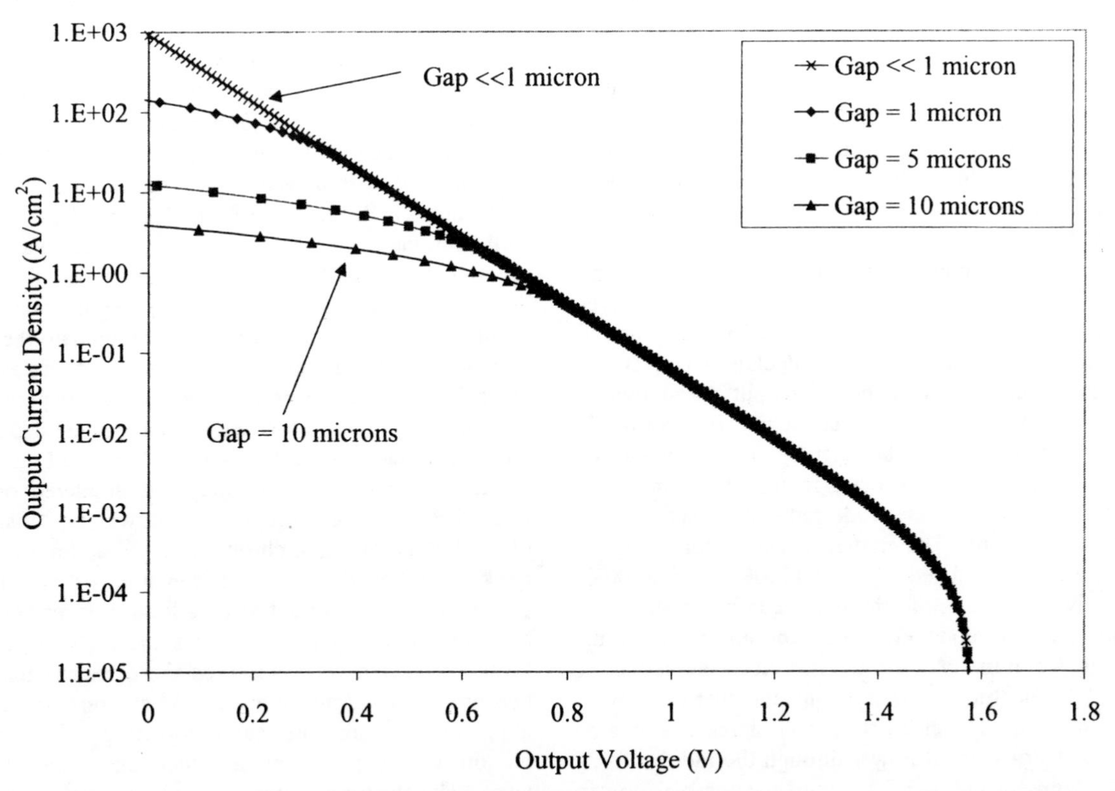

Figure 2. Calculated Current-Voltage Characteristics.

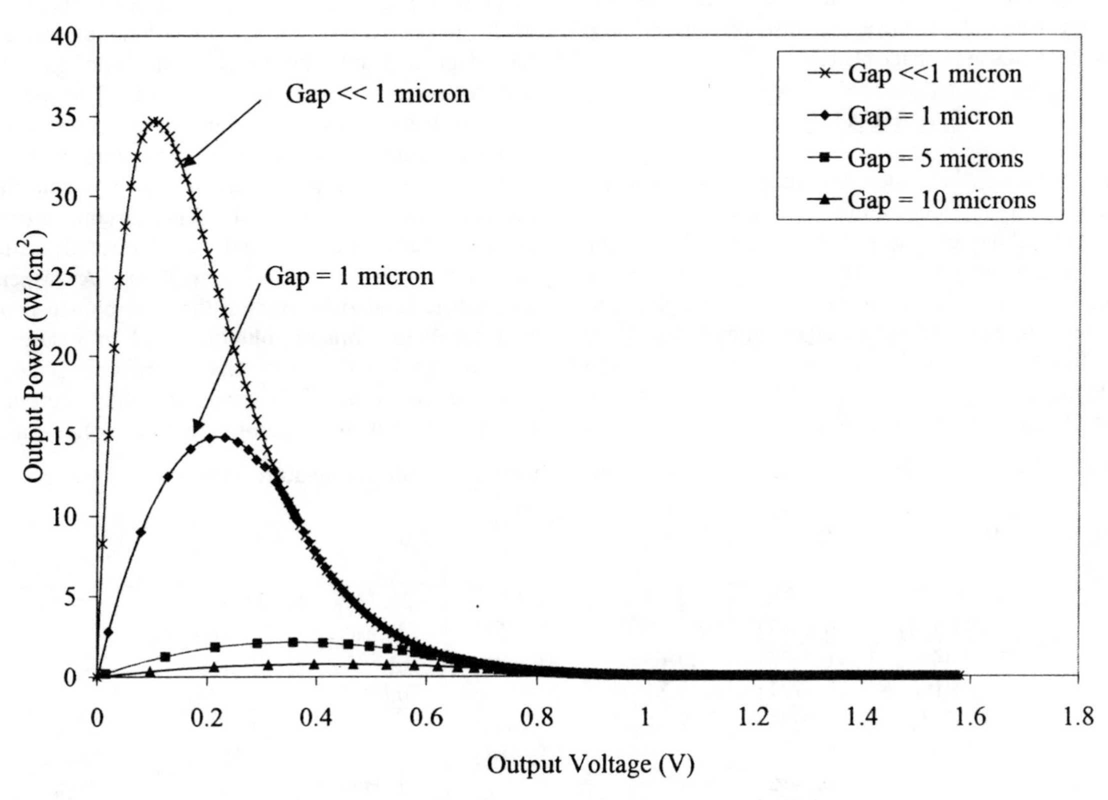

Figure 3. Calculated Power Output versus Diode Voltage.

the diode power density changes with the diode operating voltage as well as the diode gap size.

The heat load required by the MTC diode to operate at desired electrode temperatures must be calculated to determine the MTC conversion efficiencies. The conversion efficiency is defined as the electrical power output from the diode divided by the total heat input to the diode. The values for the input heat flux through the diode were obtained from thermal analyses using PATRAN, a commercial, finite element thermal solver that calculates radiative, convective, and conductive heat transfer modes. The calculated conversion efficiencies and electrical power output are shown in Table 1. The electrical power output (in Watts) is shown in parentheses. The calculations were made for a circular diode with an outer radius of 5 mm. The emitter and collector electrode radii were varied from 0.625 to 4.95 mm. The emitter and collector electrode temperatures modeled were 1200K and 700K, respectively. Increasing the emitter radius resulted in an increase of diode efficiency for emitter radii up through 4.5 mm. As the emitter electrode radius is increased, electrical power output and thermal power input increase. The electrical power increase is greater than the thermal input power through the radius of 4.5 mm. Above a radius of 4.5 mm, the thermal power input is greater than the electrical power increase. Increasing the gap size dramatically reduces the efficiency and output power because of current drop due to the increase in space charge.

FABRICATION

All elements of the diode are made with standard chemical vapor deposition (CVD) techniques, sputtering techniques, and etch techniques used by the semiconductor industry. The CVD and sputtering techniques allow us to reliably, reproducibly, and accurately grow extremely thin layers of metals and low work function materials for the electrodes. The fabrication procedure occurs roughly as illustrated in Figure 4 and described below. The prototype is fabricated from two insulating substrates as shown in Figure 4a. The substrates may have different thicknesses and may be different materials. For the prototype MTC diode, the emitter substrate is a circular sapphire substrate with a 25 mm diameter and a 0.5 mm thickness. The collector is made of quartz with a thickness of 1 mm and a cross sectional area of approximately 12 mm by 7 mm. The gap depth will be formed in the collector substrate and will form a trough running the length of the shorter cross sectional dimension of the collector. The gap depth is shown in Figure 4b. The gap depth must accommodate the total thickness of the emitter and collector electrodes as well as any thermal expansion at operational temperatures. The emitter and collector metal electrodes are deposited onto the surfaces of each wafer, as shown in Figure 4c. A 2.5 micron thick metal bilayer is sputtered on the front side of the sapphire and quartz wafers, consisting of an 0.5 micron thick chromium layer and a 2 micron thick tungsten layer. The function of the lower chromium layer is to act as an adhesion layer between the sapphire and quartz substrates and the tungsten top layer. Tungsten was chosen as the electrode material because it is a refractory metal and can operate at high temperatures with minimal out-gassing. A positive photoresist layer is then deposited onto the tungsten layer. The electrode image is then transferred through a mask onto the phototresist. The chromium/tungsten layer is etched with HCl (~10% in H_2O) and H_2O_2 at 80 °C, respectively. The low work function material is then deposited onto the metal electrodes of the emitter and collector as shown in Figure 4d. The low work function material used is a barium, strontium, calcium oxide material commonly used in vacuum tubes. The material must first be deposited as a carbonate due to the deposition method used. The barium, strontium, calcium carbonate material is deposited using a cataphoresis or electroplating method. A suspension solution is created by mixing the tricarbonate material with an acetone/binder solution in a 1 to 3 ratio. The ratio of the $BaCO_3$:$SrCO_3$:$CaCO_3$ mixture is 56:40:4. The electrodes are immersed into the solution and biased. Under the appropriate bias conditions, the

Table 1. Estimated Conversion Efficiency and Power Production of a Microminiature Thermionic Converter

	Emitter Radius (mm)						
Gap Size (μm)	0.625	1.25	2.5	3.75	4.0	4.5	4.95
1	5.80% (0.0524)	11.98% (0.2097)	18.32% (0.8382)	20.61% (1.8865)	20.73% (2.1465)	21.00% (2.7166)	--
5	2.44% (0.0198)	6.05% (0.0791)	11.65% (0.316)	14.46% (0.7113)	14.63% (0.8093)	15.01% (1.0243)	14.75% (1.2394)
10	1.14% (0.0088)	3.03% (0.0354)	6.54% (0.1413)	8.64% (0.3181)	8.77% (0.3619)	9.07% (0.4581)	8.87% (0.5543)
50	0.10% (7.36E-4)	0.29% (2.9E-3)	0.71% (0.0118)	1.02% (0.0265)	1.04% (0.0302)	1.09% (0.0382)	1.05% (0.0462)

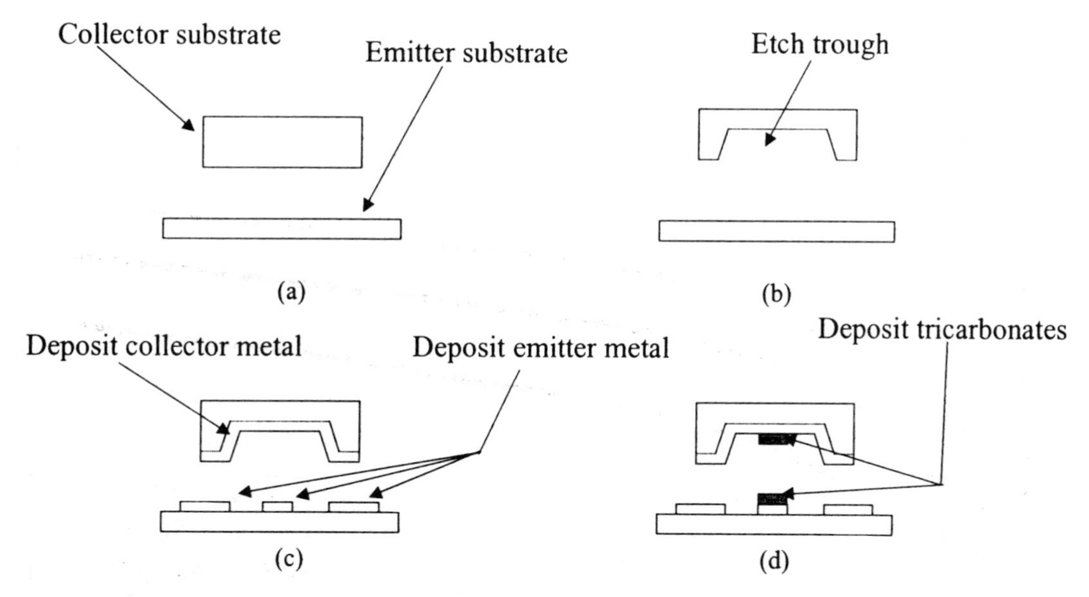

FIGURE 4. MTC Fabrication Sequence.

carbonate mixture deposits onto the electrodes.

TESTING

Prior to testing, the low work function oxide coating on the emitter and collector surfaces must be heated to condition and activate the low work function carbonate coating. The coating is delivered as a mixture of barium, calcium, and strontium carbonate ($BaCO_3$, $CaCO_3$, and $SrCO_3$) with some water content. During heating, carbon dioxide and water vapor must be driven from the carbonate mixture to form a mixture of barium, calcium, and strontium oxide. The heat conditioning profile was a bake-out temperature of 770K for at least two hours followed by an activation bake at approximately 1250K for five minutes. The diode is also electrically biased during the last thermal step. The first temperature treatment converts the carbonate to an oxide. The second temperature treatment produces free barium in the oxide layer; the free barium is essential to lowering the oxide work function.

Three prototype MTC diodes with gap sizes of 15 to 20 microns have been tested. The MTC diodes must be tested in the power consumption mode as well as the power conversion mode. The power consumption mode test is required to calculate the work function and the effective RD constant of the low work function electrode coatings; the power consumption mode test is also referred to as the Schottky mode test. The low work function material property characterization measurements are accomplished by sweeping a positive and negative voltage across the diode terminals and measuring the resulting output current at three or more operating teperatures. The voltage sweeps can range up to 200 Volts. This emission test is performed in a prototype diode structure.

Figure 5 illustrates a typical current emission profile caused by the voltage sweep during the Schottky mode test. For this test, the collector was grounded while the emitter electrode voltage was swept from 0V to -200V. For vacuum diode behavior, current emission occurs while the emitter is negatively biased. From the Schottky mode test data, the calculated emitter apparent RD constant and work function were $2.3x10^{-3}$ $A/cm^2/K^2$ and 1.26 eV, respectively. A similar test is performed on the collector. The calculated collector apparent RD constant and work function were $8.x10^{-5}$ $A/cm^2/K^2$, 1.05 eV, respectively. These ranges of values for the apparent RD constant and work function are typical of the published values for this type of oxide coating. Figure 5 also indicates that the MTC emission surface may indeed be patchy. In the absence of space charge, the ideal Schottky plot should be linear from high voltage measurements to very low voltage measurements. No space charge is predicted for these operating conditions, and the observed current-voltage curves are consistent with the effects produced by patchy surfaces. Linearity is not achieved until approximately 65 Volts indicating that the emission surface work function changes from a high value to the low value reported above as the diode voltage bias is increased. The apparent RD constant is also an indicator of the possible low work function fraction of the surface.

Figure 6 illustrates the current-voltage characteristic of the MTC for several emitter temperatures (550 through

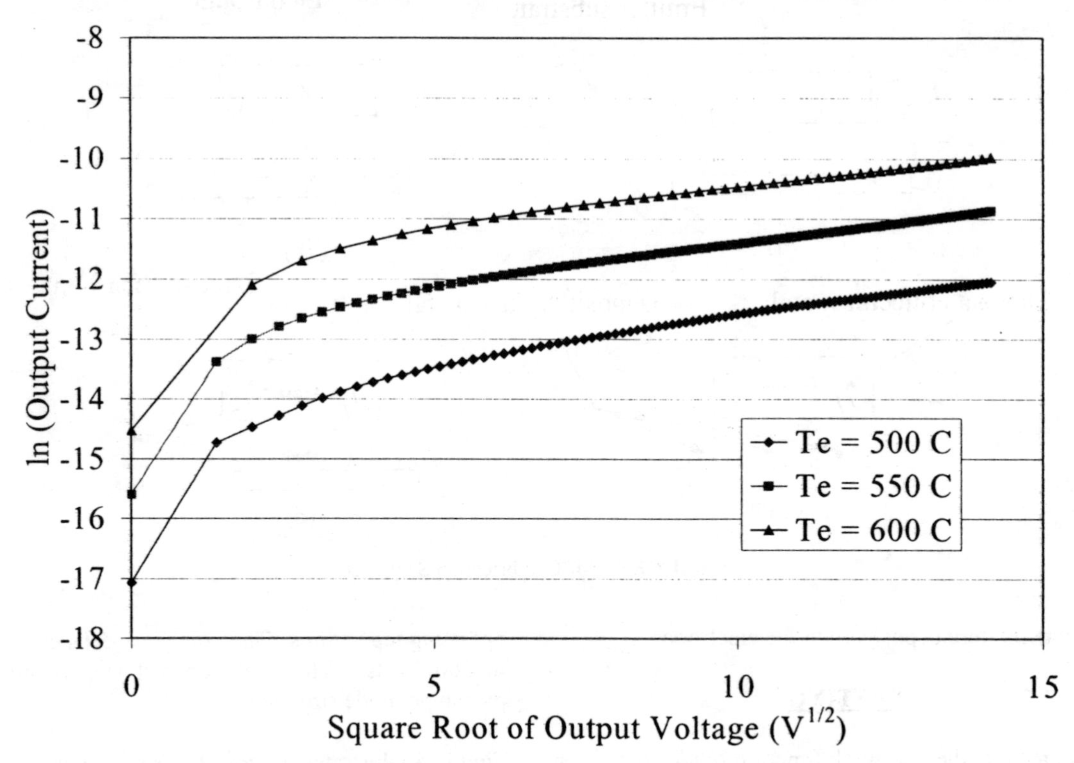

Figure 5. Schottky Emission Test Current Voltage Characteristic

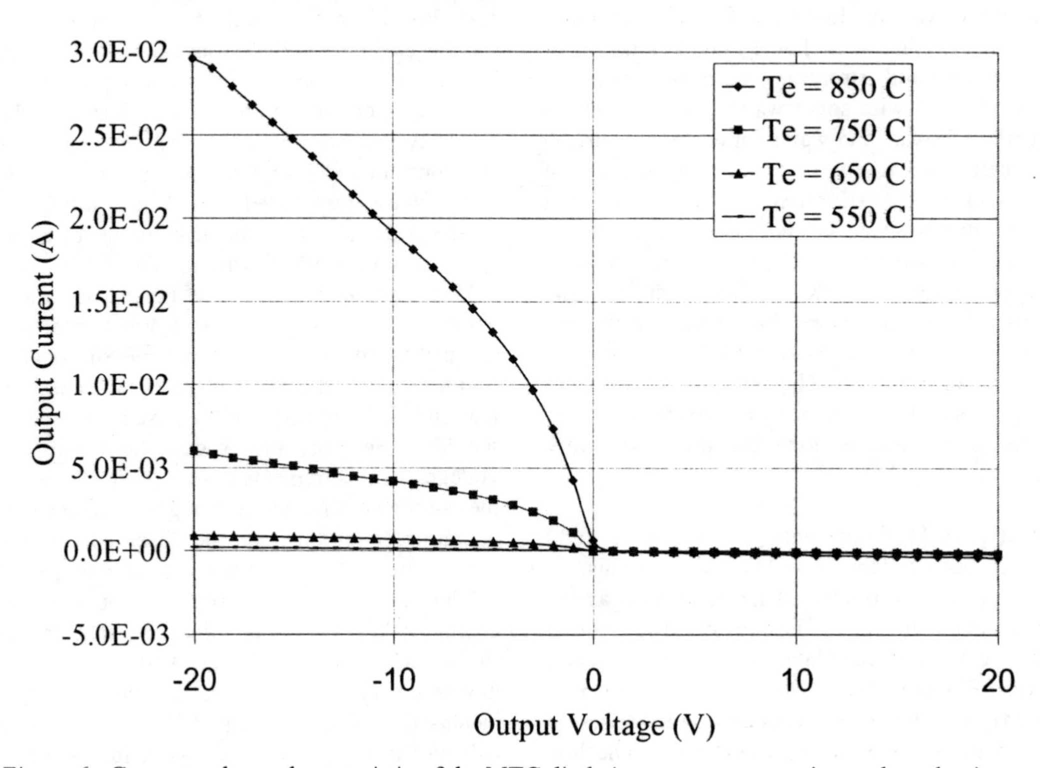

Figure 6. Current-voltage characteristic of the MTC diode in power consumption and production modes.

850 °C). The corresponding collector temperatures are 270, 309, 355, and 396 °C. The MTC produces power in quadrant 1 of Figure 6 and is a consumer of power in quadrants 2 and 4. Relatively large current densities (up to 0.3 A/cm^2) for these oxide materials are achieved at an emitter temperature of 850°C in the power consumption mode at –20 Volts bias. The reverse bias current in quadrant 4 is small because the collector temperature is low.

The MTC current production as a function of emitter temperature and diode voltage is pictured in Figure 7. Electrical power production increases by approximately three orders of magnitude over the operating emitter temperature range. The cut-off voltage also increases as the emitter temperature is increased, starting from 0.45 Volts at T_E = 600 °C and reaching 0.92 Volts at T_E = 900°C. The increase in cutoff voltage occurs because raising the emitter temperature increases the kinetic energy of the emitter electrons. Modest improvements in surface uniformity could increase current production by three to four orders of magnitude. Figure 8 illustrates the MTC power production corresponding to the current-voltage characteristics of Figure 7.

For space power applications, a thermionic diode that can operate at high heat rejection temperatures is advantageous because the energy conversion system heat rejection radiator mass and size decrease significantly as its operating temperature is increased. Figure 9 presents the power produced by operating the MTC with an emitter temperature of 900 °C. The collector temperature was varied from 700 to 850 °C. The peak power (0.2 mW) produced at a collector temperature of 700 °C was reduced by approximately 75 % (to 0.05 mW) when the collector temperature was increased to 850 °C. The cutoff voltage also decreases substantially as the collector temperature is increased.

The MTC was also used to power a Motorola 1N4004 diode. The 1N4004 was connected in series with the converter through the electrical penetrations housed in the vacuum chamber flange. Figure 10 illustrates the MTC current-voltage characteristics at various emitter temperatures with the 1N4004 current-voltage characteristic superimposed. As shown in the diagram, the MTC was able to power the 1N4004 at emitter temperatures of 600 C and above.

FUTURE WORK

A new low work function material must be developed that is easier to activate and physically and parametrically quantify. The low work function material activation sequence must be less dependent on the vacuum pressure and its constituents. The material surface must be more uniform both structurally and chemically. This material should also remain inert upon re-exposure to the atmospheric environment. Ideally, the material should also be compositionally unaffected by exposure to photoresists, wet etch solutions, and other mask materials. This capability would allow the MTC designer to pattern the diode metal and low work function material in a multi-step procedure. Currently, the $(Ba,Sr,Ca)CO_3$ cannot be re-patterned once it is deposited onto the electrode.

Highly emissive cathodes have been developed by the electron source community; apparent Richardson-Dushmann constants of up to 10 $A/cm^2/K^2$ have been realized with work functions of approximately 1.18 eV[2]. The electron source community has settled on porous tungsten films that are impregnated with a $BaCO_3$, $CaCO_3$ and Al_2O_3 mixture. Scandia (Sc_2O_3) can also be incorporated into the tungsten structure as part of the impregnate or as a top, capping layer. Metals (such as rhenium) can also form the capping layer. The tungsten structure acts as a support, an internal electron conductor, and supplies large interfacial area for barium reduction. When heated, the tungsten matrix and impurities from the metal substrate reduce the barium mixture to form free barium. Barium and oxygen are then transported to the surface and diffuse through the capping layer. Once on the surface, the barium and oxygen can be transported across the surface. If the surface temperature is too high, barium will evaporate from the surface. The low work function and large apparent RD constant are achieved when an oxygen-barium chain forms a surface dipole.

The cathode system described above is a macroscopic structure and cannot be inserted into the MTC. This design cannot be used to produce thin films, although the concepts can be incorporated into a thin film structure. The thin cathode films can be deposited using rf sputtering. Deposition is conducted in a Unifilm PVD300 system with both substrate translation and rotation to ensure film thickness uniformity. Shadow masks will be used to define electrode and cathode patterns on the substrate. The emitter is a composition of BaSrCaO or $BaCaAlO_x$, Sc_2O_3 and tungsten depending on the desired film composition. The films are deposited sequentially in a modulated structure with relative thicknesses and periods dependent on desired properties. The small gap sizes desired in a microminiature thermionic converter require thin films with a thickness uniformity and surfaces with a roughness that is a small fraction of the actual gap size. A commercial sputtering system, like the PVD300 system, is capable of better than 98% thickness uniformity over 4 inch distances. The MTC will require a deposition scheme that can be integrated

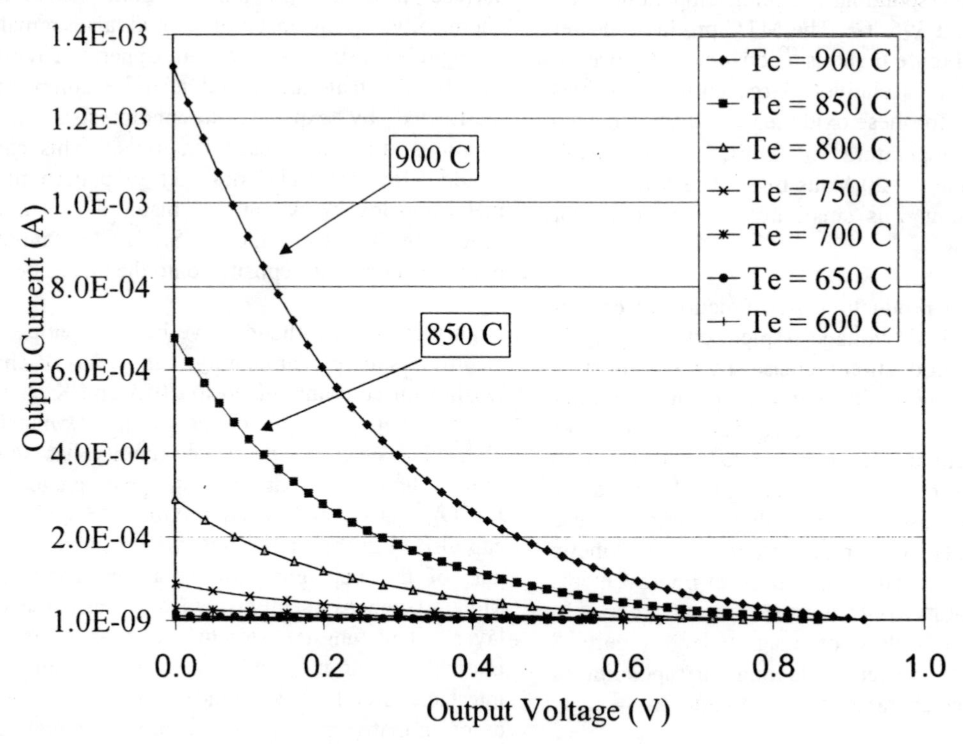

Figure 7. Current-voltage characteristic of the MTC diode in power production mode.

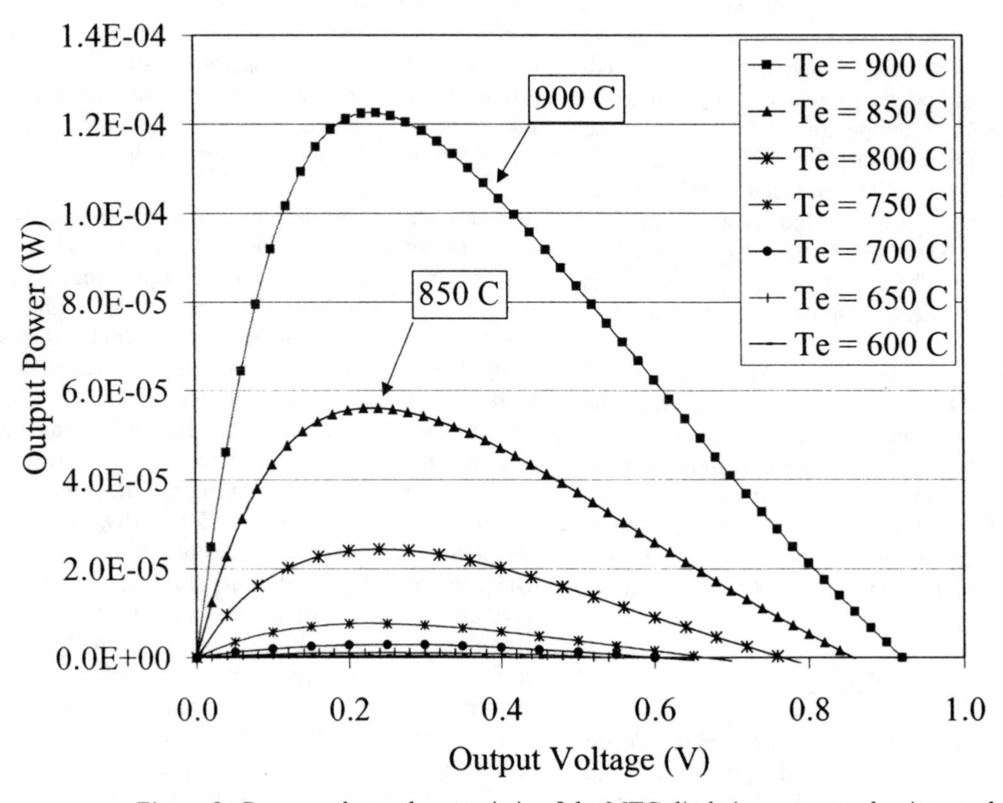

Figure 8. Power-voltage characteristic of the MTC diode in power production mode.

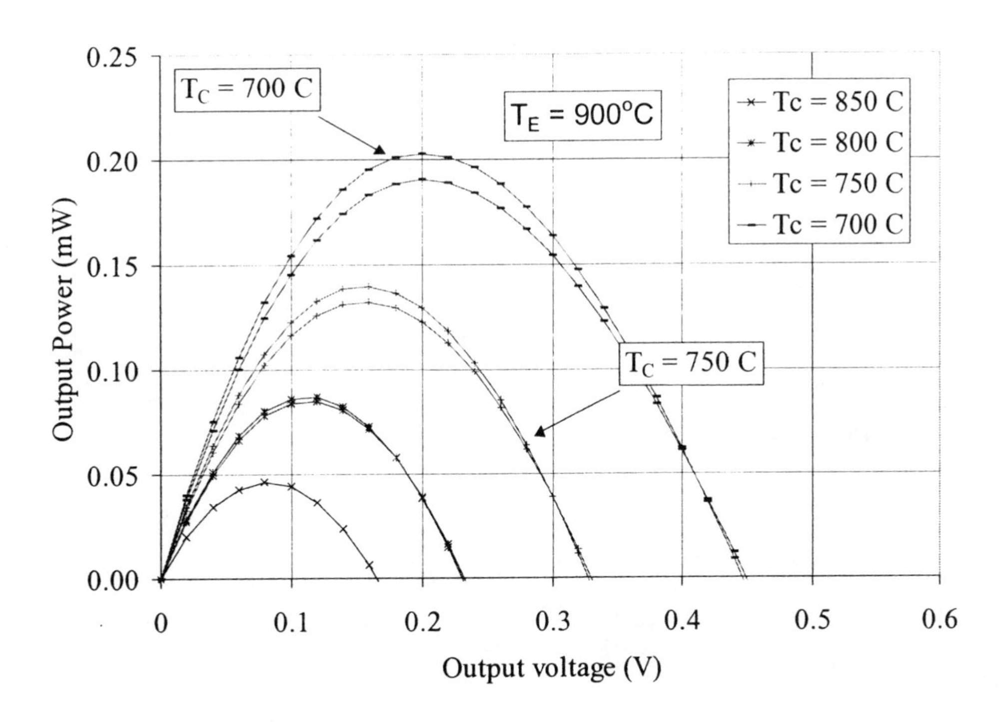

Figure 9. Power-voltage characteristic of the MTC diode in power production mode as a function of collector temperature.

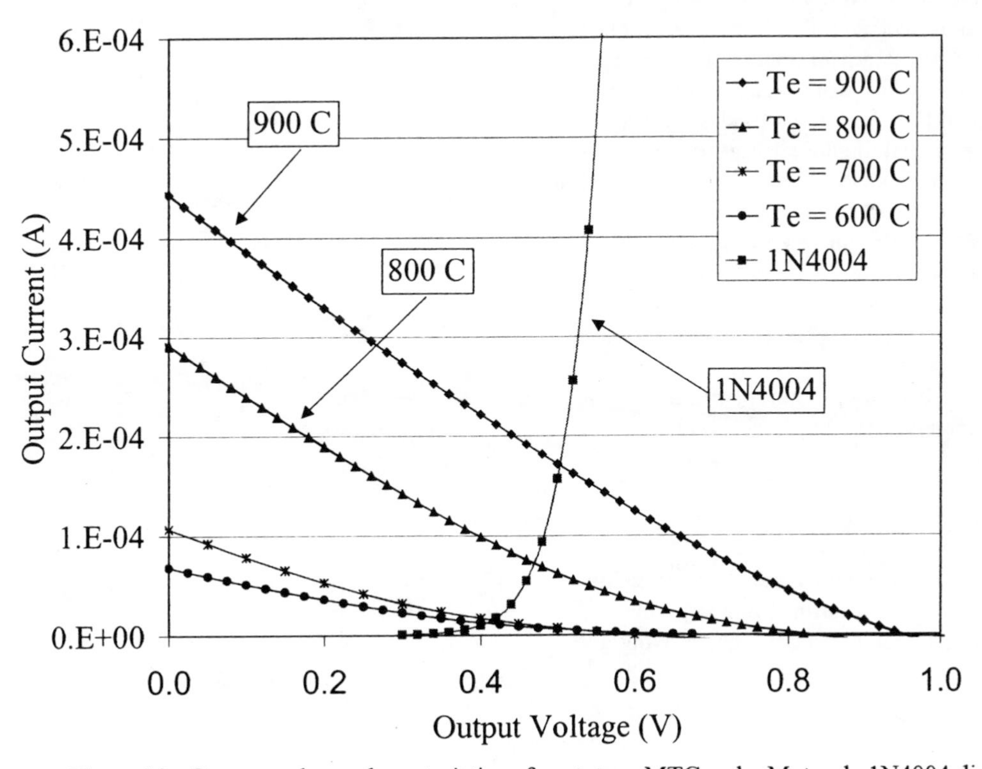

Figure 10. Current voltage characteristics of prototype MTC and a Motorola 1N4004 diode.

into a more complex fabrication process. The desired films are multi-component and sputtering is only limited by an ability to fabricate the respective materials into targets.

New generations of MTC diodes are currently being designed. One series will increase output voltage by depositing four MTC diodes in series on a single substrate. The second series will increase the MTC diode emission area to increase current output. These new diodes will also have thermocouples integrated into the diode structure. Guard rings will also be placed around the emitter structures to increase electrical isolation between electrodes. For the second design series, the collector structure will be physically placed farther away from the emitter structure and its foot print cross sectional area will be reduced. These modifications will increase electrical and thermal isolation between emitter and collector. Also, the new collector design should reduce thermal stresses in the emitter wafer and allow more testing with a backside integrated heater.

CONCLUSION

SNL is investigating the feasibility of fabricating MTCs using IC fabrication techniques. These techniques allow MTCs to be fabricated with micron-sized spacing between electrodes and low work function electrodes. Modeling has shown that thermionic diodes with these features have the potential to convert heat to electricity with relatively high efficiencies. SNL has fabricated prototype MTC diodes and has used those diodes to generate electrical power for emitter temperatures ranging from 500 °C to 950 °C. Electrical power production has been less than anticipated due to patch effects associated with the low work function oxide materials. SNL[3] is developing a new system of materials as a micro-dispenser cathode with a goal of achieving work functions of 1.0 to 1.5 eV and apparent RD constants of 10 $A/cm^2/K^2$.

ACKNOWLEDGEMENTS

SNL is a multiprogram laboratory operated by Sandia Corporation, A Lockheed Martin Company, for the United States Department of Energy under Contract DE-AC04-94AL85000.

REFERENCES

[1]Hatsopoulos, G. N., and Gyftopoulos, E. P., Thermionic Energy Conversion Volume 1: Processes and Devices, Cambridge, MIT Press, 1973, pp. 90-121.

[2]R.E. Thomas, J.W. Gibson, G.A. Haas and R.H. Abrams, *IEEE Trans. Electron Dev.* 37(3), 1990, 201-208.

[3]K. R. Zavadil, D. B. King, J. A. Ruffner, Low Work Function Thermionic Emission Materials, SNL Report SAND99-2982, November 1999.

TESTING OF A CLOSE SPACED THERMIONIC CONVERTER

J. R. Luke*
New Mexico Engineering Research Institute
University of New Mexico
Albuquerque, NM

F. J. Wyant
Sandia National Laboratories
Albuquerque, NM

S. A. Eremin
SIA Lutch
Podolsk, Russia

A close-spaced thermionic converter was recently tested at the New Mexico Engineering Research Institute. The converter operates in the Knudsen mode with an interelectrode gap of 5 μm at operating conditions (1400K emitter temperature and 800K collector temperature). Direct measurements of such a small gap in an operating thermionic converter have not been reported previously. The gap was measured by directing a laser beam through the interelectrode gap, which produced a diffraction pattern whose width is a function of the gap width. The interelectrode gap was measured for emitter temperatures ranging from 500K to 1400K. The measured value at 500K was 250 μm, and the interelectrode gap decreased with increasing emitter temperature up to 5 μm at 1400K. In this test of the prototype close-spaced converter, the emitter and collector became electrically shorted whenever the gap was less than 10 μm. This short made it impossible to obtain I-V performance data from the converter. Post test disassembly revealed that the short was due to a foreign particle embedded in the collector surface. A new converter is under construction, and will be tested in the near future.

INTRODUCTION

A new type of close-spaced thermionic converter was designed and built at SIA Lutch in Podolsk, Russia and tested at the New Mexico Engineering Research Institute (NMERI). The Lutch close spaced thermionic converter was designed to have a very small gap (5-6 μm) at operating temperature and operate in the Knudsen mode, which should result in greater efficiency than a wide-gap ignited mode converter. The major challenge in constructing a close spaced converter is ensuring that the electrodes remain flat and parallel at operating temperature, in spite of the large heat flux from the emitter to the collector. Part of the evaluation of this converter was to measure the gap and compare the measured value with the design value. Measuring the gap is also challenging, because the gap is extraordinarily small, about 1/10 the diameter of a human hair, or 10 wavelengths of light. The converter and gap measurement system are described in this paper.

* AIAA member

EXPERIMENTAL SETUP

A schematic of the close spaced converter is shown in Fig. 1. The emitter and collector (labeled 1 and 2 in Fig. 1) were single crystal tungsten and molybdenum alloys, respectively, for high strength at temperature. The swallowtail shape of the electrodes helps maintain uniform temperature across the electrode faces. The electrodes were carefully polished so that deviation from flatness was no greater than 0.3 μm over their 25 mm diameter. Three ceramic spacers (labeled 8 in Fig. 1) were incorporated into the collector. The collector was mounted on a flexible membrane (#6 in Fig. 1), and helium pressure was used to bring the collector spacers into contact with the emitter. The spacers provide electrical isolation from the emitter, and maintain the interelectrode gap at the designed value. The thermal expansion of the collector metal and spacers was carefully arranged to ensure that the gap would be 5 μm at the design conditions of 1400K emitter temperature and 800K collector temperature. The gap is 350 μm at room temperature, and shrinks to 5 μm at 1400K. Therefore the gap was a strong function of temperature. The converter was equipped with a sapphire window (#3 in Fig. 1) to provide optical access to the interelectrode gap. This sapphire ring also served as the insulator between the emitter and collector.

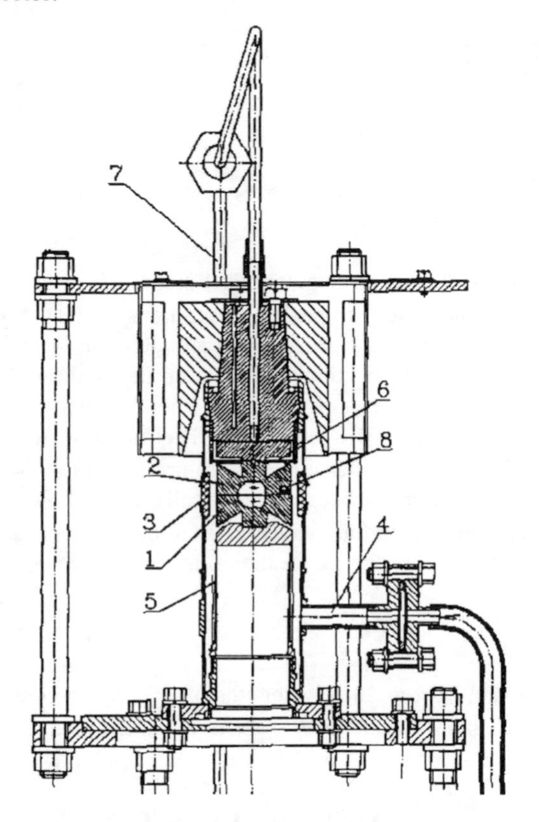

Fig. 1. Cross section of the close spaced converter.

Cs entered the converter through tube 4 in Fig. 1, from a Cs reservoir inside the bell jar. Helium pressure was applied to the membrane through tube 7 in Fig. 1. The He system was equipped with a capacitance manometer for accurate pressure measurement.

Emitter thermocouples were attached to the inside of the emitter support tube (#5 in Fig. 1) by spot welding. A tantalum filament electron bombardment heater was inserted in this tube. Two optical pyrometers, one electronic and the other a visual disappearing-filament type, were used to measure the actual electrode temperatures. Both the emitter and collector were provided with a black-body hole. The emitter thermocouples consistently read 100 to 150 degrees higher than the pyrometer due to their location near the emitter heater.

The objectives of the test were to measure the interelectrode gap as function of temperature and He pressure, and to obtain I-V data. After considering several gap measurement techniques including capacitance and optical imaging, a diffraction technique was selected. Because the emitter and collector are flat and parallel, they constitute a single slit. By directing a laser beam through the interelectrode gap, a Fraunhofer diffraction pattern can be projected onto a screen. The width of this pattern is a function of the gap width. Further details of the gap measurement system are given in the next section. The pattern on the screen was photographed, and the image was analyzed using image processing software to compute the gap width. A photograph of the experimental setup showing the laser, screen, and camera is shown in Fig. 2.

Fig. 2. The close spaced converter test stand, laser, optical pyrometer, screen, and camera.

GAP MEASUREMENTS

Diffraction occurs as a result of the wave nature of light. Light is diffracted by all objects, and the smaller the object (closer to the wavelength of light), the greater the angle through which the light is diffracted. Because the emitter and collector are flat and parallel, they constitute a single slit of width b. In the far field approximation, where plane waves are incident on a slit whose width is small and the distance from the slit is relatively large, Fraunhofer diffraction results in a pattern of light and dark interference fringes that spread out perpendicular to the slit (Fig. 3). The distribution of intensity as a function of the angle θ is given by the sinc function[1]:

$$I(\theta) = I(0)\,\mathrm{sinc}^2(\beta) \tag{1}$$

where

$$\mathrm{sinc}(\beta) = \frac{\sin(\beta)}{\beta} \tag{2}$$

and

$$\beta = \frac{\pi b}{\lambda}\sin\theta \ . \tag{3}$$

In Equation (3), b is the width of the slit and λ is the wavelength of the light used. The sinc function, plotted in Fig. 4, has minima where $\beta = \pm\pi, \pm 2\pi, \pm 3\pi$, etc. Note the order of magnitude difference in intensity between the central maximum and the subsidiary maxima. The width of the slit can be obtained experimentally by projecting the diffraction pattern onto a screen and measuring the distance between the first minima in the pattern, as shown in Fig. 3. The slit width b can be computed from

$$\pi = \frac{\pi b}{\lambda}\sin\theta \approx \frac{\pi b}{\lambda}\frac{x}{2D} \tag{4}$$

in the limit of small θ.

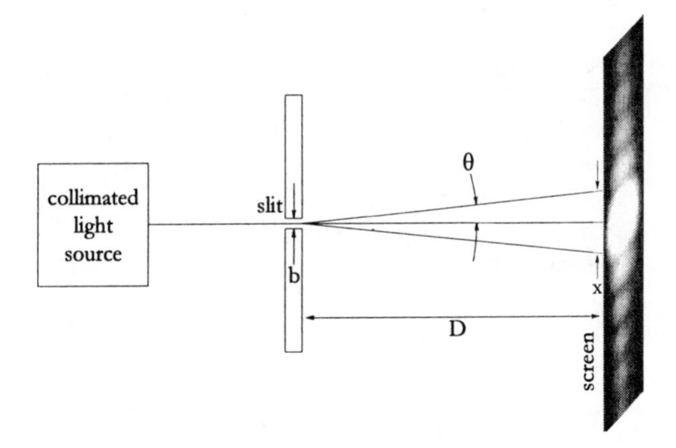

Fig. 3. Fraunhofer diffraction from a single slit.

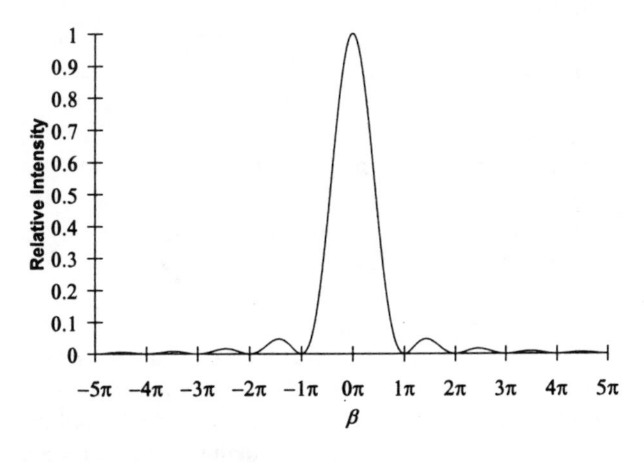

Fig. 4. The sinc function.

The diffraction pattern was photographed with a CCD camera. The signal from the camera was collected and analyzed by a computer using a frame grabber card and image processing software. Fig. 5 shows the diffraction pattern produced using a He-Ne laser and a 350 μm slit, used for calibration. The slit in this case was the gap between the jaws of a micrometer.

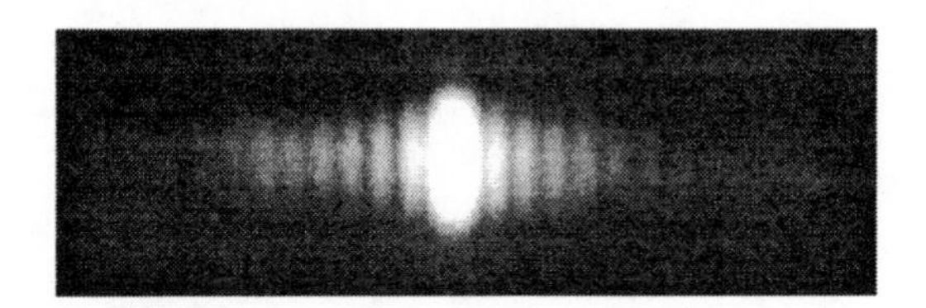

Fig. 5. Diffraction pattern from a 350 μm slit.

Fig. 6 shows the measured intensity profile of the diffraction pattern in Fig. 5, as produced by the image processing software. The y-axis in the figure is the relative intensity of the image, where 1 is black and 255 is white. Note the saturation of the CCD by the central maximum, and the lack of a completely dark background. At least seven of the side peaks can be distinguished. Fig. 7 shows the sinc function, which is the theoretical intensity distribution, with the amplitude adjusted to correspond with Fig. 6. Although there is some blooming due to saturation of the CCD, the locations of the minima between the side peaks can be clearly determined. The experimental results agree well with the theoretical result.

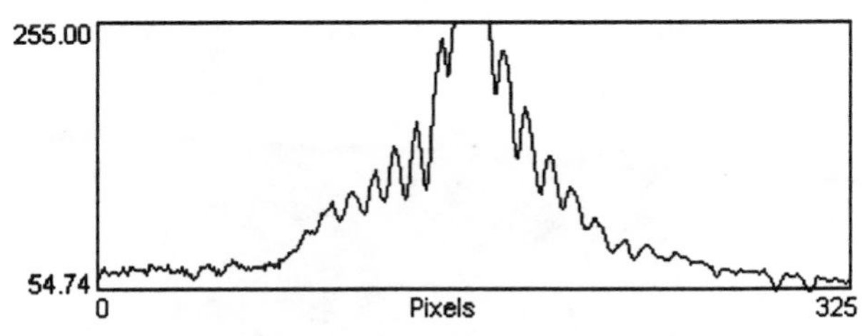

Fig. 6. Measured diffraction pattern – note saturation of central maximum.

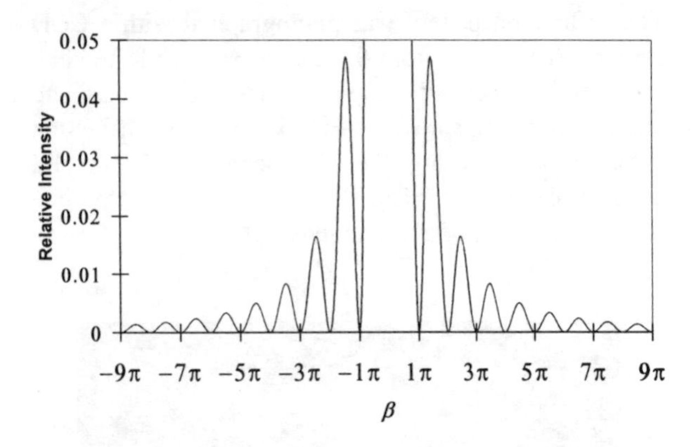

Fig. 7. Sinc function, with amplitude adjusted to correspond with Fig. 4 above.

The glass bell jar on the vacuum chamber and the sapphire ring in the device were not optically perfect, and they introduced distortions into the diffraction pattern. The position of the laser was adjusted to find the clearest patterns, but slight movements of the laser could make the difference between a good pattern and an unreadable one. Figs. 8 and 9 show two diffraction patterns taken at the same time and under the same conditions, at slightly different locations relative to the center line of the converter. Compare Figs. 8 and 9 with Fig. 5, which was taken with the same laser, but with no glass or sapphire to distort the beam. Both Figs. 8 and 9 are considered "good" diffraction patterns for this experiment. In both, the central peak and the first minima are distinguishable. In Fig. 8 some of the side peaks can also be seen. The diffraction patterns in Figs. 8 and 9 give different gap measurements, 194 μm and 183 μm respectively, a difference of 5.8 %.

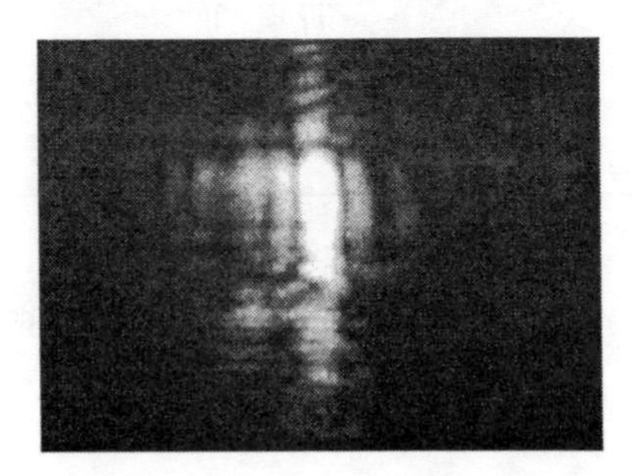

Fig. 8. Diffraction pattern through the sapphire ring and bell jar.

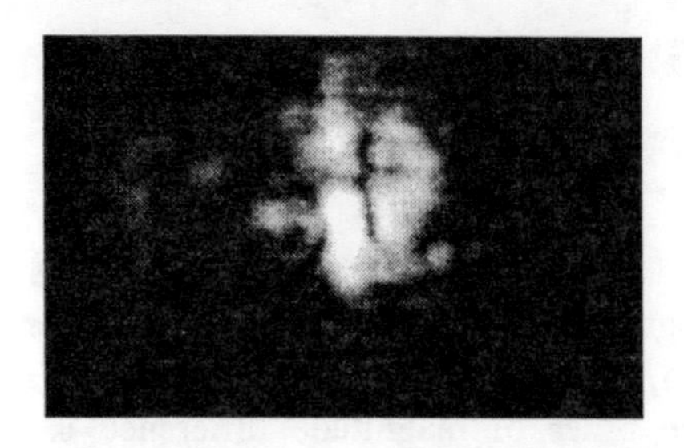

Fig. 9. Diffraction pattern under same conditions as Fig. 8, different position relative to center line of device.

At the operating temperature of the device, the electrodes glow visibly red as well as emitting infrared. The black and white CCD camera that was used at the beginning of the test was sensitive to infrared radiation, making the red He-Ne laser diffraction pattern hard to distinguish from the background. Also at the operating temperature, the gap was the smallest and the diffraction pattern was the widest, therefore it was very dim. All of the above effects combined to make it very difficult to measure the small gap at operating temperature. For these reasons, a green He-Ne laser was substituted, and a green filter and a color CCD camera were used to obtain higher quality diffraction patterns at the small gap sizes. The laser diffraction gap measurement system worked very well in the range of gap sizes from 250 μm to 5 μm. It was even possible to obtain gap measurements while a Cs discharge was taking place between the electrodes during an I-V sweep.

EXPERIMENTAL RESULTS

The close spaced converter was constructed with a 350 μm gap at room temperature. Fig. 10 shows the measured gap width as a function of emitter temperature. The measurements taken at temperatures less than 200°C with the laser diffraction system are not reliable, because the separation of the maxima in the diffraction pattern is smaller than the distortions caused by the glass bell jar and sapphire ring. As the gap decreases with increasing emitter temperature, the diffraction pattern becomes wider and more easily interpreted. When the He pressure was greater than zero, the gap width increased, particularly for emitter temperatures greater than 800°C. The collector temperature was maintained at the same value in all the runs. Adding He lowered the thermal resistance between the collector and its radiator and required the use of the collector heater to maintain the collector temperature. Heating from the outside expanded the outer shell of the device and increased the gap at intermediate temperatures. At 1400K, the gap remained at approximately 5 μm independent of He pressure. The minimum gap is determined by the thickness of the ceramic spacers and not by the He pressure.

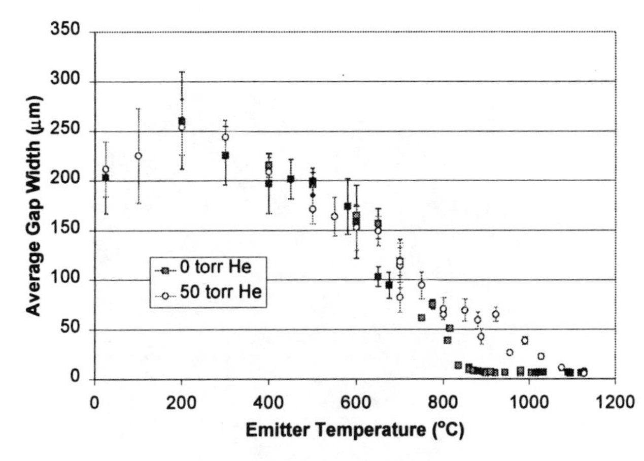

Fig. 10. The gap width as a function of emitter temperature.

When the emitter temperature reached 1250K at P_{He} = 200 torr, or 1400K at P_{He} = 20 torr, the emitter and collector shorted, apparently by touching. Gap measurements were made at different locations relative to the device center line to determine whether the electrodes were parallel to each other. Some degree of tilt was indicated, but it could not be concluded definitely whether the electrodes were tilted enough to be touching at one side. The shorting conditions were reproducible. The smallest the interelectrode gap could be without shorting was about 10 μm. No combination of collector temperature, He pressure, time at temperature, or rate of change of temperature, could prevent the electrodes from shorting at less than 10 μm. It was not possible to obtain I-V data at the design condition of 1400K because the electrodes were always shorted at this condition.

I-V CHARACTERISTICS

Since it was not possible to obtain I-V characteristics at the design conditions due to the emitter and collector shorting, dynamic I-V characteristics were obtained at a low temperature. Fig. 11 shows the I-V characteristics obtained with the device. The curve at 800°C and 2 torr was obtained first. Although the close spaced converter was not intended to operate in the ignited mode, this curve shows ignited mode operation. The reason for this is that the operating conditions were far off the design conditions, and the gap was approximately 40 μm.

The emitter temperature was subsequently increased to 950°C (second curve in Fig. 11). Increasing T_E should have shifted the I-V curve up and to the right, but it did not. It was postulated at this point that the Cs pressure in the device was abnormally low. To test this hypothesis, the Cs reservoir temperature was increased to the value corresponding to 2.5 torr, and the next curve shown in Fig. 11 was obtained. Increasing Cs pressure should shift the I-V curve up and to the right, but it did not, confirming that the Cs pressure in the device was abnormally low. Low Cs pressure could result from an empty Cs reservoir, or a cold spot somewhere in the Cs system. Although the Cs system was well designed, well insulated, and well instrumented, and none of the thermocouples on the Cs system indicated a cold spot, the Cs system was tested for cold spots by increasing the power to the Cs system and pipeline heaters. Increasing the temperature of the Cs system had no effect on the apparent Cs pressure inside the device.

On the supposition that the Cs reservoir was empty because the transfer of Cs from the ampoule to the reservoir had not been complete, the Cs transfer process was repeated. The last curve in Fig. 11, at 900°C and 2 torr, was taken after the second Cs transfer. This curve shows no emission current, indicating that there was no Cs in either the ampoule or the reservoir. Therefore Cs must have been lost from the system due to a leak.

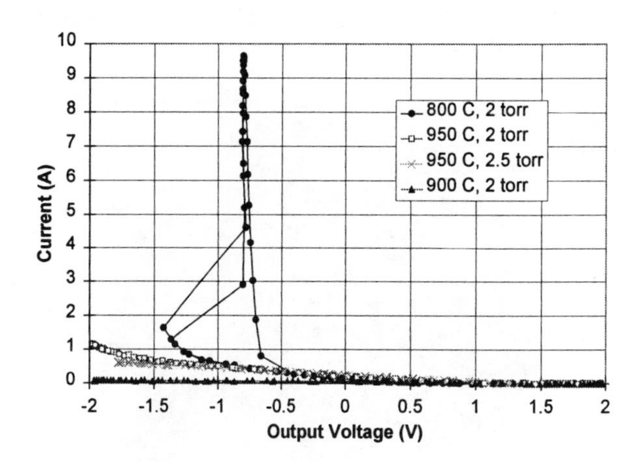

Fig. 11. Dynamic I-V characteristics.

On the day after the last I-V curve shown in Fig. 11 was obtained, the emitter heater failed. Inspection of the emitter heater filament after the failure revealed that it was twisted and melted and several strands had fused together. An attempt to leak test the device after the heater failure revealed an enormous leak between the Cs system in the device and the surroundings. The leak was found by pressurizing the Cs system with flowing argon, and listening for the hiss of argon with a stethoscope. The leak was in the emitter support tube (item 5 in Fig. 1), at the top, where the heater was located and where the emitter thermocouples had been spot welded. This failure of the device terminated the test.

POST TEST DISASSEMBLY AND INSPECTION

The converter was returned to Lutch for inspection. Before disassembly, the interelectrode gap was measured using the same optical instrument that had been used to determine the gap at room temperature prior to delivery to NMERI. The measured values ranged from 320 to 405 μm. The range of measured gap values and their average were the same as measured before the converter was delivered for testing. This result indicates that the membrane was not damaged during transportation or testing. Subsequent analysis of the membrane found that its structure and elasticity were unchanged.

The bottom flange and the Cs inlet tube were removed, and the device was taken apart to inspect the components by cutting the welds on a lathe. The emitter tube was partially melted in two places near the top, in the vicinity of the thermocouples and the heater. A hole was visible in one of the melted areas. Inside the emitter tube, the Ta foil that had been used to attach the emitter thermocouples had melted.

Fig. 12. The emitter surface after testing.

On the emitter surface, shown in Fig. 12, the three contact areas that had been scratched by the collector spacers can be seen, as well as another small spot that was discolored. The collector surface, shown in Fig. 13, had a corresponding spot. A small particle was found in that spot on the collector surface. Fig. 14 is a magnified image of that spot. This particle caused the electrodes to short when the gap was less than 10 μm.

Fig. 13. The collector surface after testing.

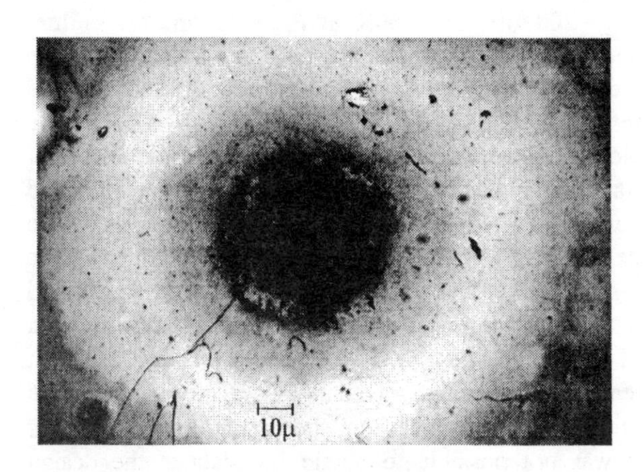

Fig. 14. The particle on the collector surface.

SUMMARY AND CONCLUSIONS

A close spaced thermionic converter was designed and built, and tested. A system was designed to measure the interelectrode gap in the range of 250 to 5 μm. The gap measurement system worked very well, and was able to determine the gap width with 20% accuracy. The close spaced converter was operated through its entire design temperature and He pressure ranges. The design goal of 5 μm gap at operating conditions was achieved. No I-V performance data could be obtained with this converter due to the 10 μm particle that shorted the electrodes. The device failed due to a Cs leak and emitter heater failure. The tests verified the design calculations and thermal modeling, and provided design information which has been used to construct a second close spaced converter. This second device will be tested in Russia in June of 2000.

REFERENCE

1 E. Hecht and A. Zajac, Optics, Addison-Wesley, New York, 1974, pp 336-340.

AIAA-2000-2847

THE RESULTS OF INVESTIGATIONS OF HIGH TEMPERATURE HIGH VOLTAGE THERMION DIODE

V.V. ONUFRYEV

RF, 107005, Moscow, Second Baumanskaya Street, house 5,
State Technical University by N..E. Bauman
Science and Research Institute of Power Engineering Industry
Phone (095)263-63-91, Fax (095)267-98-93
E-mail: irina@mx.bmstu.ru.
E-mail: niiem@power.bmstu.ru

ABSTRACT

Now the question of the current conversion for the concordance of the parameters of the power source and its consumers (ERT) are being spared the great attention in consequence of the elaboration of the powerful nuclear reactor on the basis of TEC for the provided with electrical energy the Space Apparatus. The questions concerned with the elaboration of the high temperature high voltage switches for the convert and concordance systems of the electrical parameters of powerful TEC and electric thrusters are connected with the problem of the securing of the inter electrode gap strength to the reverse electrical breakdown by the high temperatures of electrode on the level 1000...10000 V. The experimental breakdown characteristics and voltage-current characteristics (VCC) on the metal-ceramic model of the high temperature high voltage diode have been obtained. The next parameters of diode were optimized: cathode and anode temperature, operating pressure into inter electrode gap, the length of inter electrode gap.
As the result of the research the dependencies of the thermo-physics parameters of the high temperature high voltage diode have been obtained. The satisfactory coincidence between the experimental and theoretical data has been obtained.

INTRODUCTION

The regime of the reverse arc breakdown (the ignition of the self-maintaining arc discharge) was investigated in cesium vapor in the main experimentally. The results of this research works showed that the value of the breakdown voltage depends on the vapor pressure and its temperature.

In process of the experimental and theoretical investigation of this problem the main tasks were defined:

- the analysis of the reverse electrical breakdown in vapor inter electrode gaps (IEG);
- the basis of the physical conditions of the realization of the high-voltage weak-current discharge into IEG of thermion diode;
- the elaboration of the calculate method of the characteristics of diode in reverse current regime and the definition of the criterion of the reverse breakdown;
- the elaboration and experimental investigation of the diode model.

Now the carried out investigations of the reverse breakdown can divide on three groups:

- the breakdown in thermion diodes;
- the breakdown in long IEG;
- the breakdown in specific IEG.

The breakdown in thermion diodes (the ignition of non-self-maintaining discharge) was studied well [1-3]. It was stipulated for the processes of excited atoms ionization. The value of breakdown voltage is not more the atom ionization potential: $U_b < \varphi_{ion}$, when the particles have the collision motion: $l_{ia} < \Delta_{ieg}$ and arrives up to 1,5...2,3 V in dependence on the vapor pressure p_{cs} and IEG length Δ_{ieg}.

With the decrease of the pressure in IEG leading to the dis-collision motion of particles the value of the breakdown voltage of IEG increases because the energy expenditure of ionization increases too.

When the ratio $Kn = l_{ia} / \Delta_{ieg} >> 1$ there are the superfluous ion charge near the cathode and the distribution of the current and voltage into IEG corresponding to correlation $j_d = U_d^{3/2}$ as in vacuum diode. In this case the main ionization factor is the electron-atom collisions. The calculate results shows the possibility of the realization of high-voltage and weak-current discharge into IEG (the value of the breakdown voltage is up to $U_b = 100\ \varphi_{ion}$ [4, 5], that fact was confirmed experimentally in Cs diodes [6]. Notice that the process of the ignition of the non-self-maintaining arc discharge for the constant vapor pressure p_{cs} takes place at two different values of the discharge voltage U_b because of the dependence of breakdown voltage $U_b(p_{cs})$ has the C-form.

The regime of the reverse breakdown (the ignition of the self-maintaining arc discharge at "cold" electrode) was investigated into Cs IEG mostly experimentally [7, 8]. This problem was investigated into long IEG (Δ_{ieg} = 6...12 mm, the cathode temperature was T_k = 773...1073 K), where the ratio Kn < 1 (the collision motion of particles). The results of the probe measuring showed the discharge voltage U_b is concentrated near the cathode, that is the cathode layer properties define the conductivity of IEG. The electron beam forming into cathode layer and causal of the ionization in discharge column is observed. In dependence on the emissive ability of the cathode the reverse arc discharge can exist in two forms:

- weak-current form - when the current density is j_d = 0,001...0,01 A/sm^2 and the value of the discharge voltage is δU_d= (5...8)φ_{ion} which is concentrated on the cathode layer;
- high-current form - when the current density is j_d = 1...10 A/sm^2 and voltage drop is δU_d = 5...7 V.

For high-current form of the discharge at vapor pressure p_{cs} < 6,65 Pa the anode spot (the region of exited atoms and ions) was observed [7-9], which was discovered in the gas discharges. This anode spot turns into the spherical clot near the anode at the increase of the discharge voltage.

The increase of the IEG length - Δ_{ieg} (in metal-ceramic devices) permitted to decrease the influence of the surface effects and to increase the insulator resistance: in result - to increase the value of the breakdown voltage up to 30...60 V when the vapor pressure p_{cs} is 30...60 Pa and the IEG length is Δ_{ieg} = 12...30 mm [10, 11]. The minimum value of the breakdown voltage U_b = 5...7 V was observed at pressure p_{cs} = 2...3 Pa. Noted that the insulator resistance influences on the value of the breakdown voltage essentially.

Two mechanisms of the breakdown process were offered:

- the breakdown develops by the action of the volume ionization processes caused by the thermo-electrons (when the pressure p_{cs} < 10 Pa and the influence of the surface effects of the insulator is slightly), as in thermion diode;
- the breakdown develops near the insulator surface in consequence of the micro-discharges (when the vapor pressure p_{cs} > 10 Pa).

The investigations of the breakdown in IEG with the specific geometry (the holes, splits) showed the high dielectric properties of the vapor: the value of the breakdown voltage was up to 700 V in the pressure range p_{cs} = 20...133 Pa, it depended on the pressure essentially. The critical cathode temperature is discovered [12, 13].

The experimental investigation results of the breakdown in high-temperature Cs vapor permitted to reveal the main factors and causes of its rise: it is necessary to exclude: 1) the thermo-emission processes, 2) the surface insulator conductivity into IEG.

The increase of the value of the breakdown voltage in high temperature Cs vapor IEG is conditioned by:

- the change of the motion character of the particles,
- the accumulation of ions near the cathode,
- the decrease of the ionization of exited atoms.

Thus, the main contribution into the securing of the non- conducting condition of IEG carries in the ion cathode layer.

INVESTIGATIONS

The author of this paper carried out the investigations of the reverse breakdown in cesium vapor and the discharge of the disclose state of the high-voltage thermion diode (HVTD), [14 - 19]. HVTD has the following pairs of the electrodes (cathode - first), fig. 1:

- stainless steel - stainless steel;
- stainless steel - tantalum;

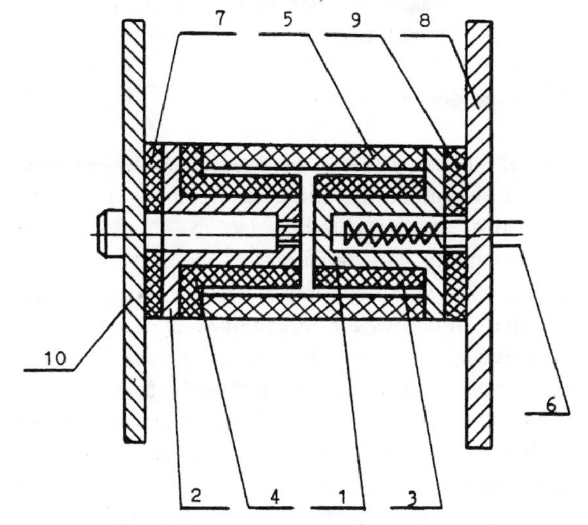

Fig.1. The scheme of the high-voltage thermion diode.
1 – cathode, 2 – anode, 3,4 – insulators, 5 – dielectric body, 6 – incandescence spiral, 7,9 – insulators, 8,10 – disk.

- molybdenum - stainless steel;
- molybdenum - molybdenum.

The range of the thermo-physics parameters was:
- operating pressure of cesium vapor p_{cs} = 1...30 Pa;
- operating pressure of Ba vapor p_{Ba} = 0,01...0,6 Pa;
operating vapor temperature $T_{cs} = T_a$ = 580...1050 K.

For the control of the heat condition of the high-temperature model it was supplied the heat sources.
Regime of investigation of reverse arc breakdown in thermion diode.

The curve form the dependence of the breakdown voltage U_b of IEG on the vapor pressure and its temperature $U_b(p_{cs}, T_a)$ are showed on fig. 2 - 3.

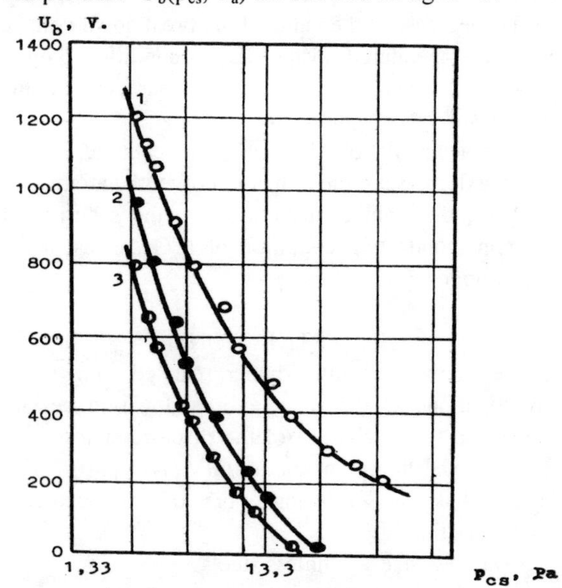

Fig.2 The electric strength of IEG of HVTD to the reverse arc breakdown.
1 – T_a = 559 K, 2 – T_a = 709 K, 3 – T_a = 745 K,
the electrodes –molybdenum, Δ_{ieg} = 3,6...8 mm.

The influence of the emissive negative electrode properties (anode of diode is "cathode") on the breakdown voltage value U_b was investigated experimentally. In the filmy regime of the cathode surface (the work function $\varphi_a = \varphi_{cs}$ = 1,89 eV) the emissive properties are conditioned by Cs film and the value of the "cathode" current density (the sum ion current and thermo-electron current) isn't changed when the differential materials of the "cathode" are used (at the constant "cathode" temperature). In this case the value of the breakdown voltage is equal for differential "cathode" materials.
As the cathode temperature increases and the pressure decreases (p_{cs} = 1...10 Pa), when the Cs film on the cathode surface is the spot film regime the emissive properties of the "cathode" material begins to influence on the current density and heat losses in the discharge: the difference in the value of the breakdown voltage is observed [14-18].

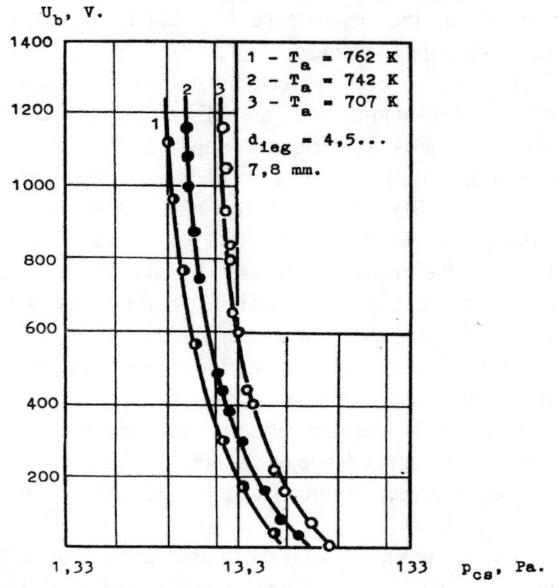

Fig. 3. The electric strength of IEG of HVTD to the reverse arc breakdown.
The anode – tantalum, cathode – stainless steel.
Anode of HVTD in regime of reverse breakdown has negative potential (cathode).

The negative electrode temperature T_a (the "cold" electrode) influences on the value of the breakdown voltage essentially when it is more than 720...770 K: the thermo-emission current density j_r increases up to two orders and the value of the breakdown voltage decreases up to 70...100 V. The further increase of the "cathode" temperature T_a (more 800 K) leads to the fact that the "cathode" electron beam becomes the main source of the ions in the discharge (the source of heat losses) and breakdown in IEG is observed at low values of voltage U_b = 5...7 V (when p_{cs} > 13 Pa). When the value of pressure p_{cs} < 5 Pa the breakdown begins at the voltage up to U_b = 100...300 V.

The range of the vapor temperature in which its influence on the value of the breakdown voltage absents practically was founded T_{cs} = 580...620 K.

The temperature of the positive electrode don't influence on the value of the breakdown voltage.

The experimental investigations of the reverse breakdown in Ba vapor was carried out on the metal-ceramic model at the following conditions: the length of IEG is 2,5 mm, the vapor temperature ($T_{Ba} = T_a$) is 850...900 K, the pressure 0,06... 0,261 Pa. The value of the breakdown voltage was U_b = 2000...2500 V.

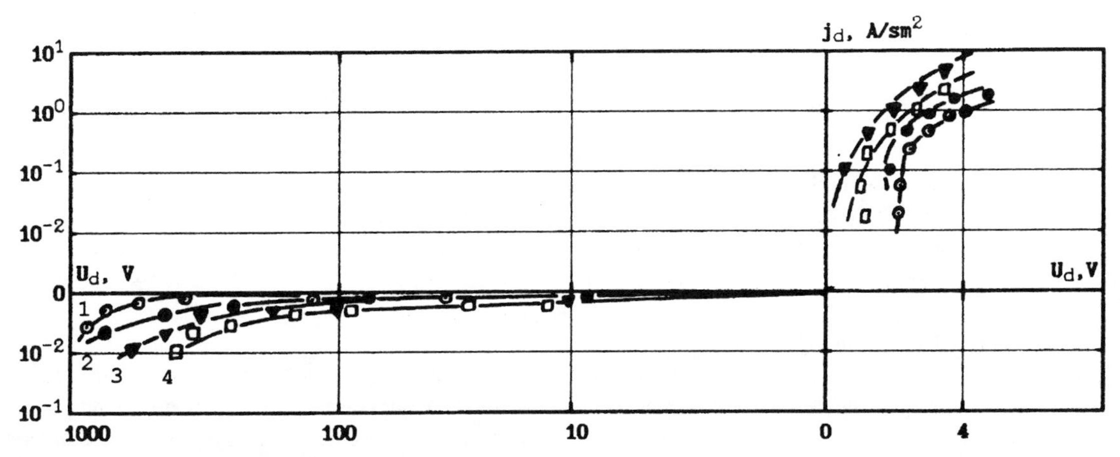

Fig.4 The voltage-current characteristics of HVTD. 1 $T_a = 678$ K, $p_{cs} = 0{,}06$ torr, 2 – $T_a = 697$ K, $p_{cs} = 0{,}06$ torr,
3 – $T_a = 690$ K, $p_{cs} = 0{,}1$ torr, 4 – $T_a = 604$ K, $p_{cs} = 0{,}12$ torr. Electrodes – stainless steel.

This experimental results about the main role of the thermo-physics parameters showed that the cathode temperature T_k and the length of IEG Δ_{ieg} don't influence on U_b. This fact permits to optimize parameters T_k and Δ_{ieg} of HVTD for the increasing its efficiency.

The experimental results is corresponding to theoretical calculations.

The investigation of disclosed state of HVTD (regime of the non-self-maintaining arc discharge).
In disclosed state there is the non self maintaining discharge with thermion cathode into IEG. The value of the voltage drop in arc discharge is up to 1,5...3 V, when the value of the current density is over 2...6 A/cm^2. The experimental voltage-current characteristics (VAC) are showed on fig. 4.

Experimental VAC shows the high efficiency of high temperature high voltage thermion diode: the ratio of direct voltage drop to reverse voltage drop is smaller 0,002, that efficiency is more than 99 %.

At the same time, the result obtained makes it possible to substantiate the physical principle underlying the construction of a new type of device - a plasma thyristor. In contrast to the existing types, the part of control electrode in the plasma thyristor is played by the ion layer of glow discharge, or to be more exact, by the region of exited atoms - the cathode glow space. A specific feature of this control electrode is the fact that its conductivity can be controlled by non-electric method (the supply of energy to the region of exited atoms can be effected by injecting a radiation of a beam of "hot" atoms and the withdrawal of energy, by injecting a beam of "cold" atoms).

The equivalent circuit of IEG in the glow-discharge regime is a series of "p" and "n" layers, fig. 5, similar to those in the semiconductor thyristor.

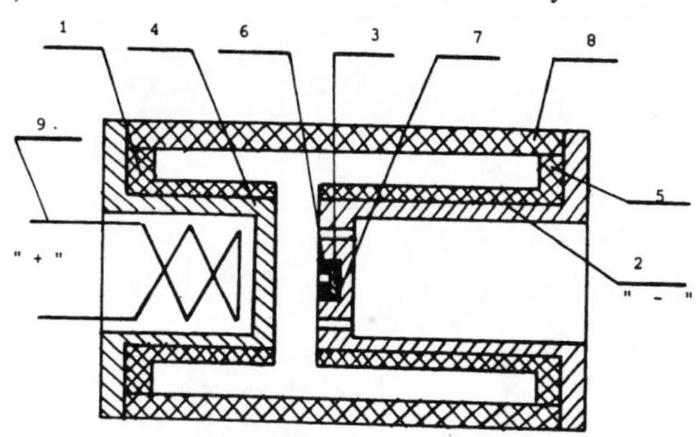

Fig. 5. The scheme of plasma thyristor.
1 – anode, 2 – cathode, 3 – control electrode, 4 – 8 insulators, 9 – incandescence spiral.

The ion layer is a "p" region whose stability is governed by the energy balance in the cathode glow space. The barrier height of the "p" region, which determines the conductivity of the latter, is a function of its thermo-physical properties: pressure and temperature. Injecting into ion layer atoms with the temperature of $T^*_{a0} > T_{a0}$ will case the ion layer to breakdown at a voltage of $U_{ign} < U_b$ and will thus initiate an arc discharge [20].

By acting on the region of exited atoms (the cathode glow space) in the ion layer one can change the discharge conditions (effect glow-to-arc transition) without changing the polarity of the electrodes and thus operate the device in the on-off switching mode. The thyristor-type control of glow discharge materially extends its potentialities.

The experimental control characteristics of the plasma thyristor are showed on fig. 6 [21].

As a result can say next:

1. The high-temperature HVTD with the cesium vapor can be used in the high temperature current conversion system (HTCCS) for powerful ERT successfully.

2. The using HVTD with the cesium vapor permits to design the current conversion system on the operative voltage up to 1...10 kV and current 100...1000 A at the operating temperature 600...700 K.

3. The using HVTD in HTCCS of current conversion permit to decrease the mass-dimension characteristics of the space nuclear power supply with TEC up to 5 - 10 times.

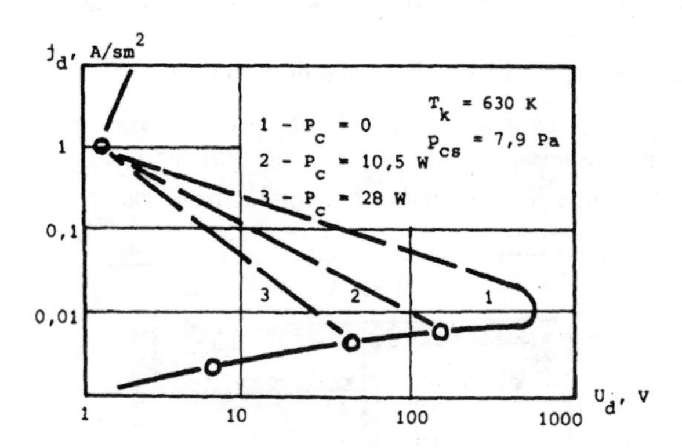

VCC of plasma thyristor.

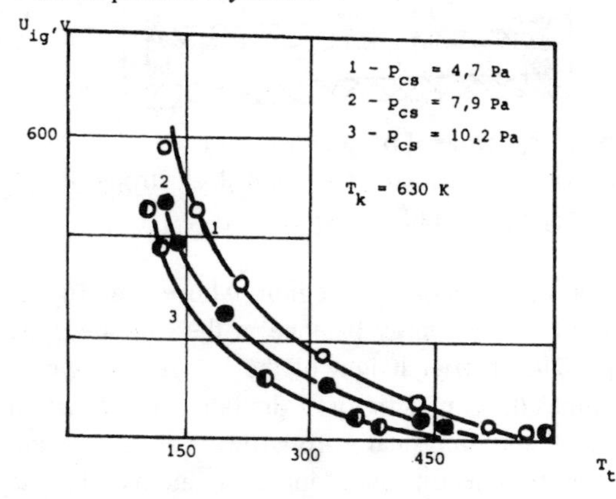

Fig. 6. The control characteristics of plasma thyristor. T_t - the difference between the cathode and control electrode temperatures.

OPTIMIZATION OF THE PARAMETERS OF HIGH-TEMPERATURE CURRENT CONVERSION SYSTEMS FOR POWERFUL EPT.

The operation of the high-temperature current conversion systems being the thermo-physics design in the structure of the space nuclear power supply with TEC depends on the many factors: the operating switches temperature, the value of the electrical conversion power, the control current method, the efficiency of current control, the values of the operating current and voltage of EPT etc. It influences on the characteristics of the HTCCS in essential order:

- mass-dimension;
- electric;
- exploited;
- economic;
- ecological.

The state parameters optimization of HTCCS permits to increase its electrical power and specific mass power, current and voltage conversion efficiency largely. The optimization of the whole HTCCS divides on two parts:

- optimization of the parameters of the controllable thermion switches (GPS) - the entrance stage of current converter;
- optimization of the parameters of the high-voltage thermion diode (HVTD) - the exit stage of the current converter.

The parameters optimization of GPS and HVTD is founded on the purpose function - its maximum power efficiency with taken into consideration the following factors: the operating temperature T_w, the losses of the electrical power in switch δP_l, the expenditures of the current control electrical power δP_{cont}, specific mass electrical power P_g.

For number of the optimizing parameters of thermo-physics state of the switch (diode) are following: the operating vapor pressure p_{cs}, the electrode temperature T_k, T_a. Yet a certain optimizing parameter is the geometrical - the value of the inter electrode gap (IEG) length - Δ_{ieg}. The optimization of the first three permits to increase the electrical characteristics of GPS or HVTD, namely - the operating voltage U_w, operating current I_w, operating power P_w, passing to EPT. The optimization of the geometrical factor permits to arrive the maximum electric efficiency at the expense of the decrease of the power losses in switch.

The problem of the switch optimization is multi-parametrical because on its solution will influence the purpose function, external and internal factors. In this paper the authors shows one of the version of the approach for that. As the entrance parameters of optimization are using the operating parameters: operating current and voltage, temperature, the parameters of the current control circus, the value of power losses; as the purpose function is the specific mass electrical power of the switch or diode.

Using the experimental data of mass electrical power of the semiconductor current converters may assume that the HTCCS are profitable at the value of the specific mass electrical power $P_g > 1000$ W/kg (it surpasses this parameter of the semiconductor converter in two - three times). Therefore the problem of the parameters HTCSS optimization is reduced to the definition of the range of the operating parameters - p_{cs}, T_a, T_k, where the value of P_g is realized at the observance of the operating voltage value (this is voltage of TEC supply and operating EPT voltage) [22].

On the foundation of this approach the analytic dependence between the switch's electrical characteristics and its thermo-physics parameters has been obtained:

$$p_{cs}^{opt} = A\,k\,T_{cs}(1+\gamma_i)^a\, j_b^{\beta}\, U_b^{\ \nu}\, F(\chi_a^*)^l$$

where A - constant, k - Boltsman constant, T_{cs} - vapor temperature in IEG, γ_i - the secondary emission coefficient, j_b - the breakdown current density, U_b - the breakdown voltage, $F(\chi^*a)$ - the function depending on vapor heat conductivity, l, a, β, ν - coefficients.

Calculations executing on this formula are showed the range of the possible cesium vapor pressure and its boundary fig. 7. Results of the experimental investigations is correlated with the calculations.

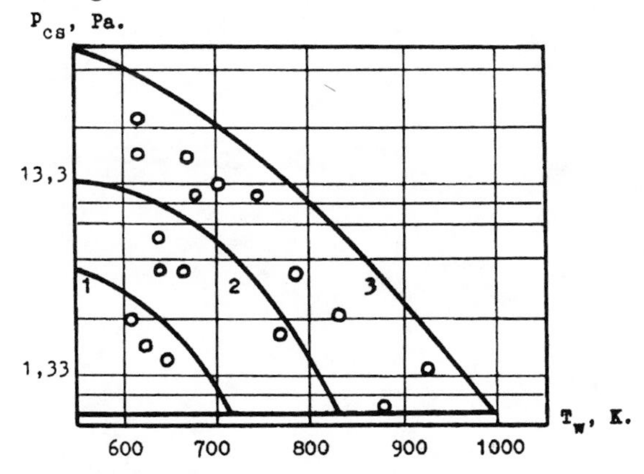

The range of optimum vapor pressure into IEG of HVTD.
1 – U_b = 1500 V, 2 – U_b = 1000 V, 3 – U_b = 500 V.
o - experimental data.

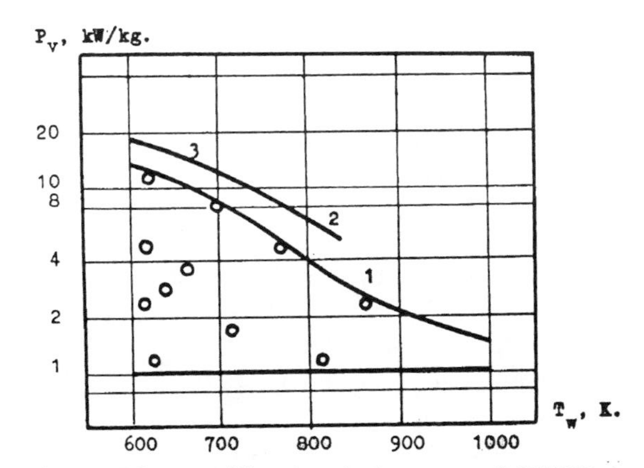

Fig. 7. The specific electrical power of HVTD in dependence on its operating temperature – T_w.
1 – U_b = 500 V, 2 – U_b = 1000 V, 3 – U_b = 1500 V.
o - experimental data.

On the foundation of this co-ordination of the electric strength characteristics of the switch with its parameters of the disclose state of IEG (where the current density of the non-self-maintaining arc discharge is the function of the cathode temperature and vapor pressure) may carry out the optimization of the p_{cs}^{opt} and T_k^{opt} in order to achievement the maximum of the arc discharge current density - j_d^{max}. As result may get the value of the maximum specific electrical power of the switch:

$$P_w^{max} = j_d^{max}(p_{cs}^{opt}(U_b, T_{cs}^{opt}), T_k^{opt})\, U_w$$

where T^{opt}_k - optimum cathode temperature.

The calculate results and our experimental data are showed on fig. 7.

The investigations showed that the using of cesium filling switches are most efficiently at the its operating temperature not more 700...730 K, because the increase of this parameter lead to reduction of the operating voltage and gets worse the electric strength of IEG.

CONCLUSION

Our research works showed that the cesium filling switches with thermion non-self-maintaining arc discharge are most perspective elements of the current conversion system for the space power propulsion plants because it surpass the all well-known types on the level of the operating temperature and specific mass electrical power.

REFERENCES

1. Ushakov B.A., Nikitin V.D., Emelyanov I.A. The basis of the thermion energy conversion. M.: Atomizdat, 1974.- 288 p. (On Russian).

2. Physical basis of the thermion energy conversion. / Editor Stakhanov I.P.- M.: Atomizdat, 1973.- (On Russian).

3. The Thermion converters and low-temperature plasma. / Editors Moyjes B.Ya., Pikus G.E.- M.: Nauka, 1973.- 480 p.- (On Russian).

4. Bogdanov A.A., Martsynovsky A.M. //Journ. Techn. Phys.- 1984.- V. 54.- N. 12.- PP. 2386-2389.- (On Russian).

5. Babanin V.I., Ender A.Ya. //Journ. Techn. Phys.- 1981.- V. 51.- N 11.- PP. 2260-2270.- (On Russian).

6. Antonov A.A., Kaplan V.B., Martsynovsky A.M., Novikov A.B. // VII Conf. "Phys. low-temperature plasma". (Thesis of report).- (Tashkent, 1987).- 1987.- PP. 71-72.- (On Russian).

7. Kiryushenko A.I., Lebedev M.A. //Journ. Techn. Phys.- 1971.- V. 41.- N 6.- PP. 1170-1173.- (On Russian).

8. Guskov Yu.K., Lebedev M.A., Stakhanov I.P. // Journ. Techn. Phys.- 1964.- V.34.- N 8.- PP. 1451-1461.- (On Russian).

9. Lebedev M.A. //Journ. Techn. Phys.- 1964.- V. 34.- N 9.- PP> 1705.- (On Russian).

10. Belousenko A.P., Vasilchenko A.V. // Thermophysics of High Temp.- 1984.- V. 22.- N 6.- PP. 1031-1033.- (On Russian).

11. Vasilchenko A.V., Nikolaev Yu.V. // Thermophysics of High Temp.- 1984.- V. 22.- N 8.- PP. 1227-1229.- (On Russian).

12. Lebedev M.A., Guskov Yu.K. //Journ. Techn. Phys.- 1963. - V. 33.- N. 12.- PP. 1462-1463.- (On Russian).

13. Guskov Yu.K., Lebedev S.Ya., Rodionova V.G. //Journ. Techn. Phys.- 1965.- V.35.- N 1.- PP. 156-157.- (On Russian).

14. V.V. Onufryev "To the question of the using the thermion vapor filling diode as high-voltage high-temperature diode" Jubilee Conf. of Kvasnikov A.V. 100- anniversary. June. 1992.- M.: MAI, 1992.- PP. 65.- (on Russian).

15. V.V. Onufryev "Influence of the atom energy balance in the cathode layer on the self-maintaining arc discharge ignition", //Second Intersociety Conf. "Nuclear Power Engineering in Space".-(Report).- (Sukhumi, October 1991).- 1992.- PP. 278-289.- (On Russian).

16. S.D. Grishin, V.V. Onufryev, P.Yu. Pekshev "The plasma switches on the base of the non-self-maintaining arc discharge of the low pressure for the electrical current conversion", YII Conf. "Physics of Gas Discharge". June 1994. (on Russian).

17. V.V. Onufryev, S. D. Grishin "Influence of the thermo-physics parameters on the value of the breakdown voltage of the glow discharge to self-maintaining arc discharge in Cs and Ba vapors", //Conf. "Physics and Technique of Plasma".- (Minsk, Sept. 1994).- 1994.- PP. 96-99.- (On Russian).

18. Onufryev V.V. Patent of Russian Federation N 2111605. Date 20.05.1998.- H 01 j 45/00.- (On Russian).

19. Onufryev V. V., Grishin S. D. // Thermo-physics of High Temp.- 1996.- V. 34.- N 3.- PP. 482-485.- (On Russian).

20. Onufryev V.V. Patent of Russian Federation N 2144716, H 01 j 17/40, from 27 Oct. 1993. Date 20 January 2000. (On Russian).

21. Onufryev V.V The plasma thyristor with the metal vapor //VIII Conf. "Phys. Gas Discharge".- (Ryazan, June 1996).- 1996. - PP. 100-102.- (On Russian).

22. Onufryev V.V., Grishin S.D., Savichev V.V. Optimization of the parameters of the high-temperature plasma current conversion systems of the powerful ERT. // 24-th IEPC.- (Moscow, Sept. 1995).- IEPS-131.- PP. 199-200.

AIAA-2000-2848

THE BASIS OF THE CHOICE OF OPTIMUM TEMPERATURE REGIME OF THE CURRENT CONVERTER AND DIMENSIONS OF THE THERMION SWITCHES

V.V. Onufriyev, Marakhtanov M.K., Muboyadzyan S.A., Senyavsky V.V.
RF, 107005, Moscow, Second Baumanskaya st., house 5,
State Technical University by N.E. Bauman
Science and Research Institute of Power Engineering Industry
Phone (095)263-63-91, Fax (095)267-98-93,
E-mail: irina@mx.bmstu.ru
E-mail: niiem@power.bmstu.ru

ABSTRACT

In present the question of the current conversion in order to concordance of the space power supply and the powerful ERT of the space vehicle is paid the great attention [1, 2] with elaboration of the perspective power-propulsion systems (PPS), which converse the electrical power of the value up $10^5 \ldots 10^6$ Watts.
The operating characteristics and resource of the Space Nuclear Power Systems based on the Thermion Energy Conversion (TEC) is depending on the parameters of the current conversion (CC) systems and the boarding cable network. In this work the results of theoretical investigation of the specific mass-energy characteristics of the current converter, wires and its temperature regime, the choice of the optimum voltage of the converter and specific power of the switches are presented.

INTRODUCTION

TEC is essentially the source of the DC of the low level voltage [1 - 3], but ERT demands the operating voltage level up to 1000 -10000 V in dependence on the parameters of the ballistic trajectory of Space Vehicle (SV), because the value of the specific impulse of ERT is proportional to the value of the acceleration voltage (operating voltage) of ERT through the electrostatics method of the acceleration the operating matter. As a consequence for the purpose of the concordance of the parameters of TEC and ERT the current converter (CC) is used on SV.
In this connection the choice of the parameters of the CC plays one of the main roles in optimization of the mass of the power-propulsion system of SV. The choice of the optimum temperature regime of the elements of CC is connected with the special purpose - optimization its mass-energy characteristics (the definite of the minimum specific mass of CC).

THE CHOICE OF THE OPTIMUM OPERATING TEMPERATURE AND VOLTAGE OF CURRENT CONVERTER.

One of the part of the CC mass is the mass of heat cooling system (HCS), which depends on the operating temperature, physical properties of its materials, the area value of the heat cooling radiator. The increase of the operating temperature of CC leads to the decrease of the mass of the HCS, but the mass of the electrical switches and transformer increase too.
As a result there is the optimum value of the operating temperature of CC, such that the value of the electrical power losses N_{los} is minimum, and the area value of the heat cooling radiator is minimum too, that secures the minimum specific mass of CC. In this case the value of the operating voltage of CC is optimum voltage.
The solution of this problem is based on the analysis of the electrical power losses into CC. The value of the relative electrical power losses connected with the switch's efficiency, is the function of the operating voltage U_W and depends on the its constructive and its physical characteristics, caused of the voltage drop in the inter electrode gap (IG) - δU_d and leakage current - I_{rd}. The electrical power of the switch - N_W is the fixed parameter.
The formula of the optimum value of the operating voltage has been obtained:

$$U_{wopt} = \left(N_w \delta U_d / I_{rd}\right)^{1/2}$$

The value of the electrical power losses of CC in optimum regime is:

$$N_{los} = 2\left(N_w \delta U_d I_{rd}\right)^{1/2}$$

On the fig. 1 there are the curves of the optimum operating voltage of CC. This calculate result shows that the value of the optimum operating voltage increases with the increase of its electrical power in proportion the degree (1/2).

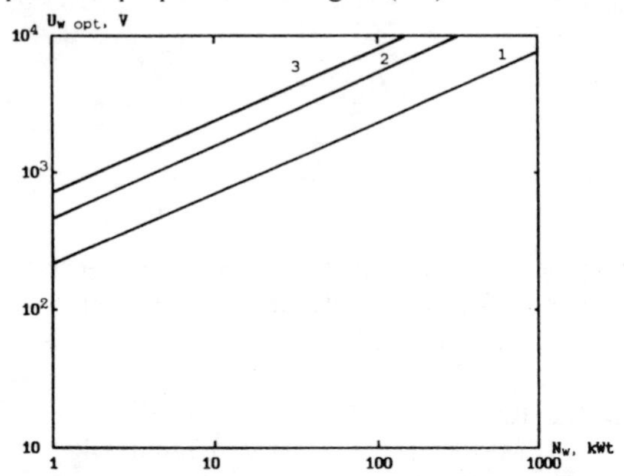

Fig. 1. The dependence of the optimum operating voltage of the high temperature high voltage diode on its electrical power.
1 - $\delta U_d = 2{,}5$ V, $I_{rd} = 0{,}05$ A, 2 - $\delta U_d = 5$ V, $I_{rd} = 0{,}05$ A, 3 - $\delta U_d = 5$ v, $I_{rd} = 0{,}01$ A.

The operating temperature of CC depends on the switch's type using in its structure.
For the semiconductor switches the dependence of its operating voltage from the temperature can present as:

$$U_w = \frac{U_{300}}{T_w^k}$$

where U_{300} - operating voltage at $T_W = 300$ K,
k - coefficient.

In this case the value of the electrical power losses is:

$$N_{los} = \delta U_d \frac{N_w T_w^k}{U_{300}} + I_{rd} \frac{U_{300}}{T_w^k}$$

In the general case the voltage drop in disclose state of switch depends on its operating temperature nonlinearly $\delta U_d = f(T_W)$, this is correct for the leakage current I_{rd}. In this connection in the design the analytic and empirical formulas of the voltage drop and leakage current can be used. In this case using the constant values of this parameters. In result the formula of the optimum operating voltage of the switches and its operating temperature has been obtained:

$$U_{wopt.sw} = U_w T_w^k$$

Notice, that with the increase of the operating temperature of CC the number of the switches will increase too. In result the number of the switches can be found from the correlation:

$$k_u = U_{wopt} / U_{wopt.sw}$$

On the base this expression the dependencies of the specific mass of the current converter has been obtained. The calculate results of the specific mass - operating temperature of the current converter are showed on fig. 2. The analogous results of the specific mass - operating temperature has been obtained for the current converter with plasma switches, fig. 2.

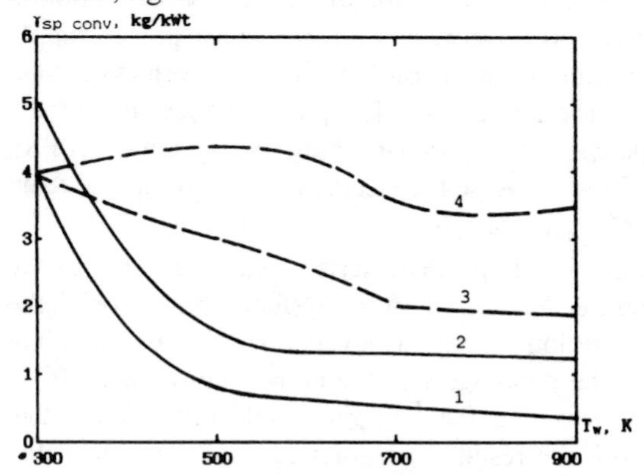

Fig. 2. The dependence of the specific mass of the current converter from its operating temperature.
1, 2 – plasma CC, 3., 4 – semiconductor CC, $\gamma_1 = 0{,}1$ kg/kW, $\gamma_2 = 0{,}2$ kg/kW, specific mass of the radiator – 30 kg/m^2. $U_{opt} = 1000$ V.

The form of the curves shows, that for semiconductor switches takes place the optimum value of operating temperature (700...900 K), when converter's specific mass is minimum. For the plasma switches the specific mass of CC decreases with the increase its operating temperature.
The dependence of the value operating temperature of current converter with plasma switches from its electrical power has been obtained too, fig. 3.

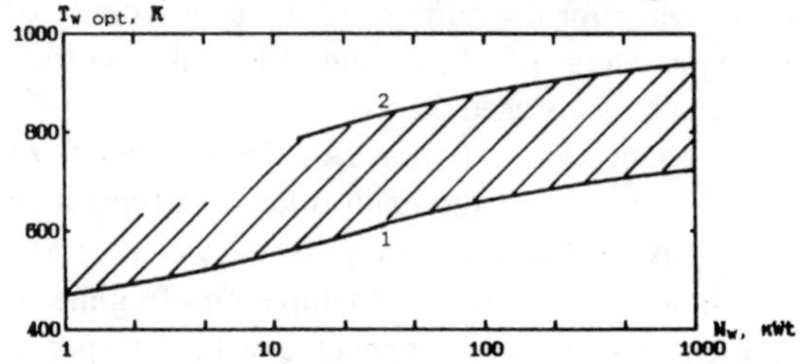

Fig. 3. The dependence of optimum operating temperature of the high temperature high voltage thermion diode from its electrical power.
1 – Cs filling, 2 – Ba – filling.

The value of operating temperature is the function of the pressure of operating matter into IG and electrical power of CC:

$$T_w = Bp_{cs}\left(N_w\right)^{1/2}$$

where B - is constant, depending on the construction and physical properties of elements of the converter.

The analysis and numerical appraisals showed that for every level of the electrical power of the switch there are the optimum values of the operating temperature and voltage, when the energy losses are minimum. The value of the optimum operating voltage is co-ordinate with the value of the optimum operating temperature to the condition of the absence of the reverse arc breakdown into IG.

The optimum value of the operating temperature of the thermion high voltage diode, fig. 3, is between 500...730 K for Cs filling and between 800...950 K for Ba filling.

THE CHOICE OF THE OPTIMUM TEMPERATURE REGIME OF THE WIRES

The presence of the high temperature space power plant and current conversion in structure SPNS is conditioned the using of the high temperature current wires.

The question of the wire's optimum temperature is connected with the optimization of the its mass with the additional mass of the energy source, which is necessary for the compensation of the electrical power losses into wires.

The calculation of the mass-temperature characteristics based on the method of the definition of the optimum current density into wires of SPNS [4]. The second part of our work is the investigation of the role of the high temperature current wires in the Space Power Nuclear System (SPNS).

The presence of the high temperature power supply and current converter in SPNS demands the high temperature wires too.

notice, that the increase of the SPNS's mass takes place in proportion to its electrical power, and the decrease of the cross-section of the wire accompanied the lowering its mass, leads to the increase of the electrical resistance and power losses, which in result leads to increase of the SPNS's mass.

The wire's mass can be defined from the correlation:

$$M_w = \gamma_w L \frac{I_w}{j} = \gamma_w LS$$

where γ_W - is the density of the wire's material, L - the length of the wire, S - its cross-section, I_W - the operating current of wire, j - current density.

The additional mass of SPNS necessary for the compensation of the electrical power losses into wires is defined as:

$$\Delta M_{sp} = \gamma_{sp} I_w j \rho_0 \left(1 + \alpha \Delta t_w\right)$$

where ΔN_{sp} - the value of power losses into wires, γ_{sp} - the specific mass of SPNS, ρ_0 - the specific electrical resistance of the wire's material, α - the temperature coefficient of the electrical resistance, $\Delta t_w = T - T_0$ - the average value of the wire's heating, T - the wire's operating temperature, T_0 - its beginning temperature, σ - Stephan-Bolzman constant , ε_W - the degree of the black.

As a result the sum of this mass (wire and additional of SPNS) is defined:

$$M_w + \Delta M_{sp} = LI_w \left[j\rho_0 \left(1 + \alpha \Delta t_w\right) + \frac{\gamma_w}{j} \right]$$

On the investigation of this correlation the equation of the optimum value of the wire's temperature has been obtained, when the sum of the mass wires and the additional mass of the Space Power Source for the compensation of the electrical energy losses in the high temperature wires is minimum:

$$T^8 + 0{,}8\left(\frac{1}{\alpha} - T_0\right)T^7 - \left(0{,}6\frac{\gamma_w \beta}{\gamma_{sp}\sigma\varepsilon_w} + 1{,}2T_0^4 + 0{,}8\frac{T_0}{\alpha}\right)T^4 -$$

$$0{,}8\left(T_0^5 - \frac{\gamma_w \beta}{\gamma_{sp}\sigma\varepsilon_w}\frac{1}{\alpha} + \frac{\gamma_w \beta}{\gamma_{sp}\sigma\varepsilon_w}T_0\right)T^3 - 0{,}2\frac{\gamma_w \beta}{\gamma_{sp}\sigma\varepsilon_w}T_0^4 + 0{,}2T_0^8 = 0$$

where $\beta = S / P$, P - the perimeter of the cross-section.

This expression shows, that the value of the optimum wire's temperature depends on the specific mass of SPNS, physical properties of wire's materials, its geometry and beginning temperature. And the optimum temperature is independent on the value electrical power of SPNS. The dependencies of the wire's optimum temperature and its specific mass are showed on fig. 4.

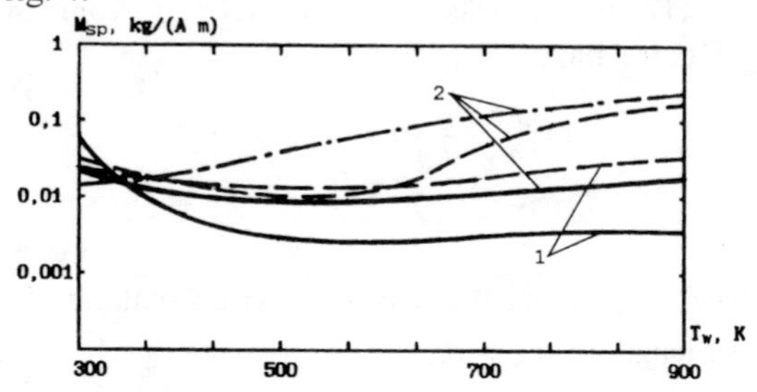

Fig. 4. The dependence of the specific mass of current wire from its operating temperature.
Materials: 1 – Cu, 2 – Mo. Calculate parameters:
$\gamma_{sp} = 100$ kg/kW, $\varepsilon_w = 0{,}85$.
$\beta_1 = 0{,}01$, $\beta_2 = 0{,}1$, $\beta_3 = 0{,}001$.

The numerical results showed that for the optimum wire's temperature its specific mass (on the unit of the length and unit of the current) is minimum at operating temperature 500...800 K.
When between the SPNS and ERT there is the intermediate unit (current converter) the value of the electrical power losses increases, that leads to the additional growth of SPNS's mass for the compensation this power losses.

The additional mass of SPNS is defined:

$$\Delta M_{sp1} = \frac{\gamma_{sp} \Delta N_{sp}}{\eta_{cc} \eta_w}$$

where η_w - the energy efficiency of wires, ΔN_{sp} - the value of the additional power losses into intermediate unit. The value of wire's energy efficiency is defined:

$$\eta_w = 1 - \frac{LT_w^2}{U_w} \sqrt{\frac{\varepsilon_w \sigma \rho_0 (1 + \alpha \Delta t_w)}{\beta}}$$

where U_W - the operating voltage.
Notice that this method is universal and independent from the number intermediate units and wire circuits. The value of the wire's efficiency can be defined for each part of the wire's circuit of SV. Thus the value of the optimum operating temperature and mass of the wire has been obtained.

THE CHOICE OF THE WIRE'S MATERIALS

When the β, γ_{sp}, ε_w, is equal for differential materials of the wire, its optimum temperatures ratio is defined as the ratio of its material's densities in order 0,25. In this case the efficiency of using wire's material is defined from next correlation:

$$\Psi_{1-2} = \frac{M_{1w}}{M_{2w}} = \frac{\dfrac{\gamma_{1w}}{j_{1opt}} + j_{1opt} \rho_{01} (1 + \alpha \Delta t_{1w}) \gamma_{sp}}{\dfrac{\gamma_{2w}}{j_{2opt}} + j_{2opt} \rho_{02} (1 + \alpha \Delta t_{2w}) \gamma_{sp}}$$

which transforms into:

$$\Psi_{1-2} = \left(\frac{\gamma_{1w}}{\gamma_{2w}} \right)^{1/2} \left(\frac{\rho_{01} \left[1 + \alpha \left\{ \left(\dfrac{\gamma_{1w} \beta}{\gamma_{sp} \varepsilon_w \sigma} \right)^{1/4} - T_0 \right\} \right]}{\rho_{02} \left[1 + \alpha \left\{ \left(\dfrac{\gamma_{2w} \beta}{\gamma_{sp} \varepsilon_w \sigma} \right)^{1/4} - T_0 \right\} \right]} \right)^{1/2}$$

The parameter of relative specific mass of the wires is the function of its physical and construction parameters and operating temperature. The calculate results of the characteristics of differential materials are showed in tab. 1.

Material of wire	γ_w, kg/m^3	$T_{w.opt}$, K	M_{sp}, kg/(A m)	$M_{sp}/M_{sp.Li}$
Li	530	464	0,001656	1
Al	2680	615	0,00208	1,265
Be	1800	510	0,0021	1,33
Cu	8920	835	0,00322	1,89
Co	8900	1050	0,00384	2,32
Mo	10200	825	0,00643	3,88
Nb	8570	863	0,00859	5,2

For light space power plants ($\gamma_{sp} = 5$ kg/kW, $\beta = 0{,}02$ and $\varepsilon_w = 0{,}85$) notice that in the range from 700 to 800 K Al has the best characteristics, then goes Be, Cu, Ag. In the range from 800 to 2000 K Li has the best mass-temperature characteristics, then goes Co, Mo, Nb. Notice that mass - temperature parameter of Li wire is smaller above 4 times then this parameter of Mo wire. This fact permits to construct the wires of SV from the casing of high temperature metal filling liquid Li. This wires have the minimum mass.

Analysis shows that increase of the beginning temperature of the wire leads to increase of its specific mass and demand to use the high temperature materials for wires (which have the greater density), that fact in turn leads for further increase of specific mass of wires.

Recent remark demands to realize the next alternative versions of the wire's construction:

- low temperature wires of the light metals (Al, Be) with thermos insulation from the high temperature units of SV,
- high temperature metal wires,
- high temperature wires as the casing with the liquid metal (Nb, Mo, steel and liquid Li).

Recent version is best suited to using in EPS because it permits to decrease the mass up to 50...75 % by comparison with the high temperature metal wires and up to 40...60 % by comparison with the low temperature metal wires.

THE CHOICE OF THE DIMENSION OF THE THERMION SWITCHES

The current converter has not only one switch, because its operating voltage may be higher, then the value of the switch's operating voltage. In result the current converter has some parallel circuits with switches in everyone.

Let us consider the current converter including the "q" parallel circuits with "p" switches in everyone. In this case the electrical power of single switch is defined by correlation:

$$N_{sw} = \frac{N_{cc}}{pq}$$

The value of operating current of single switch is defined:

$$I_{sw} = \frac{N_{cc}^{1/2}}{\left(\delta U_d / I_{rd}\right)^{1/2} pq}$$

For the thermion switch (diode) the value of its electrical power is defined by correlation:

$$N_{sw} = AT_k^2 \exp\left(-\frac{e\varphi_k}{kT_k}\right) S_k \left(\frac{\delta U_d}{I_{rd}} N_{cc}\right)^{1/2}$$

where: φ_k - effective work function of the cathode, T_k - cathode temperature, S_k - its surface area. Notice, that φ_k is the function of the ratio (T_k / T_{cs}), and we can approximate it as a polynomial of degree 2 in the (T_k / T_{cs}), A - Richardson-Deshman constant.

For the relatively simple approximation of the cathode work function φ_k

$$\varphi_k = a\left(\frac{T_k}{T_{cs}}\right)^2 + b\left(\frac{T_k}{T_{cs}}\right) + c$$

the value of the optimum cathode temperature is defined by expression:

$$T_k^{opt} = \frac{k}{ea} T_{cs}^2 + \frac{1}{ea} T_{cs} \sqrt{k^2 T_{cs}^2 + cae^2}$$

The solution (specific electrical power) depends on the characteristics of the cathode surface material, but don't depend on the current converter's parameters: its operating voltage, power. Thus the value of the optimum specific electrical power of the single switch in current converter is defined as the function of two parameters: its operating temperature and electrical power of the current converter. The calculate results are showed on fig. 5.

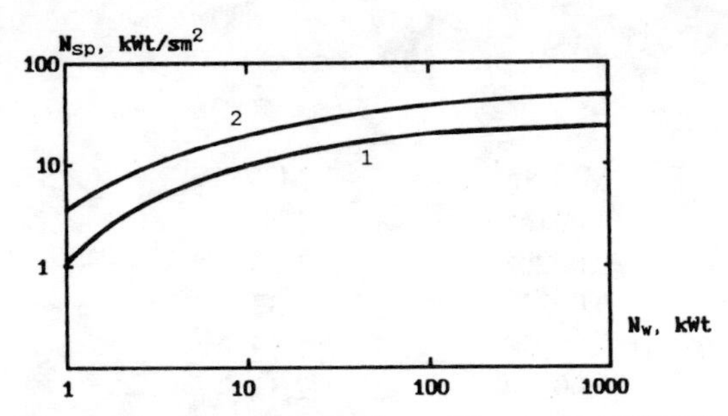

Fig. 5. The dependence of specific power of HVTD from the electrical power of the current converter.
1 – Cs filling, 2 – Ba – filling.

CONCLUSION

1. The influence of the high temperature on the mass characteristics of the current converter has been defined.
2. The optimum value of the switch's operating voltage and voltage of current converter in dependence on its electrical power has been obtained. Notice that the voltage depends on the electrical power in order 0,5.
3. The range of the optimum temperature regime of switches and current converse has been obtained. Its influence on the mass characteristics are showed. The range of optimum operating temperature of plasma switches current converter is function of its operating pressure into IG and electrical power.
4. The optimum temperature regime of current wires has been defined for differential its materials. The value of optimum operating wire's temperature is the function of its physical properties and geometry.
5. The value of specific optimum electrical power of the switches in dependence on the electrical power of current converter has been obtained.

REFERENCES

1. Ushakov B.A., Nikitin V.D., Emelyanov I.A. The basic of the thermionic energy conversion. M.: Atomizdat, 1974.- 288 p. - (On Russian).
2. Physical basic of the thermionic energy conversion / Editor Stakhanov I.P.- M.: Atomizdat, 1973.- (On Russian).
3. V.P. Ageev, P.I. Bystrov, A.A. Maslennikov, V.V. Senyavsky, V.D. Yuditsky //The Electronuclear Tug In The Program Of Space Developing/ Problems of avia and space technique.- 1994.- N 2.- pp. 36 - 39. (On Russian).
4. Troitsky S.R. The method of the defenition of the optimum wire's current density of powerful Space power plant.// High-capacity Space Thermoemissive Nuclear Power Plants and Electric Rocket Engines. - 1995.- N 3-4.- pp. 178 - 186. (On Russian).

ETHANOL VEHICLE COLD START IMPROVEMENT WHEN USING A HYDROGEN SUPPLEMENTED E85 FUEL

Gregory W. Davis, Edward T. Heil, Ray Rust
Kettering University
Flint, Mi

Ethanol Vehicles present a significant cold starting problem. Below 11 degrees Celsius, ethanol will not form a rich enough fuel vapor-air mixture to support combustion. The addition of 15% gasoline helps to alleviate this problem, by allowing the vehicle to start primarily on the vaporized gasoline compounds. This approach does not completely eliminate the problem while substantially worsening cold start emissions. Supplementing E85 (85% ethanol, 15% gasoline) with hydrogen improves cold starting significantly. An effort was made to determine the minimum amount of hydrogen which could be used to supplement E85 and which would suffice in providing acceptable cold start performance. This was accomplished by experimental injection of pure hydrogen into the intake stream of a dedicated E85 fuelled vehicle. The minimum acceptable level of hydrogen was found to be 8% by volume at cold start conditions of –4 °C (25 °F). Initial work on an E85 hydrogen reformation technologies is also reported.

INTRODUCTION

The United States has a need to develop alternative fuel technologies in order to both reduce its dependence upon foreign supplies of petroleum and to reduce air pollution. Further, in order to meet the ultra-low emissions levels (ULEV) II standards proposed for 2004 in California the use of additional strategies beyond the use of three-way catalytic after treatment systems on conventional gasoline engines will be required. Alternative fuels have been shown to reduce emissions levels and are promising alternatives to gasoline. Unfortunately, compromises have to be made when using these fuels. Technology must be developed which allows for reasonable trade-offs in cost and convenience when using alternative fuels.

Some interest in using alcohol as a fuel for internal combustion engines has been present since the early 1800's.[1] Recently, however, the U.S. government has actively encouraged the use of ethanol-blended gasoline. The reasons for this include reducing dependence on imports of foreign oil, increasing market opportunity for agricultural crops, and providing environmental benefits. Current production of ethanol is mostly from the fermentation of sugar and grain crops, but other renewable biomass resources can also be used.

Despite all of these advantages, alcohol fueled vehicles still exhibit a major problem when starting in cold temperatures. The addition of 15% gasoline helps to alleviate this problem, by allowing the vehicle to start primarily on the vaporized gasoline compounds. This does not, however, completely eliminate the problem. Various approaches are being investigated to improve cold start for ethanol. One of the promising methods is to use hydrogen-supplemented ethanol during cold start. Injecting hydrogen into the intake charge aids cold startability since hydrogen does not need to be vaporized and it has an extremely fast burn rate, making it an excellent choice as a cold starting supplemental fuel. Further, hydrogen is unique among alternative fuels in its ability to burn over a wide range of air-fuel mixtures thus aiding combustion, which helps reduce emissions.

Since hydrogen is not readily available as a fuel, it most likely would need to be made, on-board, using the ethanol fuel. Thus an effort was made to determine the minimum level of hydrogen supplementation that would provide acceptable cold start performance. Results from experiments performed on a dedicated E85 fuelled vehicle with hydrogen as supplemental fuel are reported. A significant improvement in cold start was experienced.

COLD START ISSUES

Alcohol fuels exhibit numerous differences in their motor fuel characteristics when compared with petroleum-based fuels. These differences are shown in Table 1.

	Gasoline	Ethanol	Methanol
Heating Value (Volume) [mJ/liter]	32	21.3	15.9
Vapor Pressure @ 100 °F [kPa]	62-90	17	32
Heat of Vaporization [kJ/kg]	400	900	1110
Research Octane Number	91-100	111	112
Stoichiometric Air/Fuel Ratio by mass	14.7	9	6.4
Vapor Flammability Limits by Volume [%]	.6-8	3.5-15	5.5-26

Table 1. Properties of gasoline and alcohol fuels [redrawn from reference 1]

As shown in Table 1, one of the major differences between the fuels is the vapor pressure and heat of vaporization. The low volatility and high latent heat of vaporization of alcohol fuels results in a low vapor pressure at lower temperatures. This severely reduces the cold start performance of an alcohol-fueled vehicle as compared to a gasoline vehicle.[2] Combustion starts if the vapor phase mixture of fuel and air is within flammability range for the fuel. For ethanol, the vapor flammability limit by volume is 3.5-15% (Table 1). At temperatures below 10° C, the vapor/air ratio is below the lean limit and will not properly combust. On a gasoline engine this problem is usually solved by over-fueling during start up to ensure enough fuel evaporation to initiate combustion.[3]

The cold start problem is compounded by the fact that the low stoichiometric air-fuel ratio for ethanol means that there is less air available per unit mass of fuel in the fresh charge to supply energy for vaporization. Also, unlike gasoline, which can be blended to achieve desired phase equilibrium behavior (winter blends are more volatile than summer blends), ethanol is a pure substance with a known fixed phase equilibrium behavior.

Improving Ethanol Cold Start

The cold start problem can be improved by using blended ethanol and gasoline (E85 - 85% Ethanol and 15% Gasoline) but it cannot be completely eliminated. If no other cold start strategies are implemented, the engine must be severely over-fueled during cold starts to provide a rich enough mixture of gasoline vapor and air for combustion. Essentially, the vehicle is cold started on the 15% gasoline that is in the blended fuel because at cold temperatures the ethanol will not evaporate. This results in high hydrocarbon emissions during cold starts.

The high latent heat of vaporization of E85 as compared to gasoline takes energy from spark for vaporization leaving less energy for ignition, which might cause a misfire. Supplying larger ignition energy can help with this problem. In fact, a high-energy ignition system is used on this vehicle. Unfortunately, due to the rich mixture requirements of E85, fuel wetting of the spark plugs may occur during cold starts. This problem can lead to ignition shorting due to the high conductivity of E85 compared to gasoline, further increasing the chance of misfire.[4]

The use of hydrogen to supplement the fuel can greatly improve the cold start of alcohol fuels. Hydrogen does not need to be vaporized, and the higher flame speed and low minimum ignition energy helps in rapid initiation of stable and self-sustaining flame.[5] Also it enables extremely lean operation, making it an excellent choice as a cold starting supplemental fuel. Further, hydrogen is a unique among alternative fuels in its ability to burn over a wide range of mixtures in air, which helps to minimize emissions. It has been demonstrated that mixtures of hydrogen and gasoline will burn at an air to (mixed) fuel ratio of 30, resulting in very low NOx levels and increased efficiency.[6]

COLD START TESTING WITH HYDROGEN INJECTION

All testing was done on a dedicated E85 fueled vehicle, and all parameters of the vehicle were kept constant throughout testing. The vehicle used is a 1999 Chevrolet Silverado that was converted to E85 for the 1999 Ethanol Vehicle Competition.[7] Because of the differences between gasoline and ethanol fuel, some of the Original Equipment Manufactured (OEM) powertrain control module (PCM) engine calibrations were modified (Table 2). Fueling parameters that were modified in the vehicle included changing the value for fuel stoichiometry and increasing the fuel delivered during crank prime pulses. The stoichiometric fuel/air ratio was changed from the stock value of 0.0683 to a value of 0.1 by mass. This represents a change of 1.5 from the OEM setting. This allows the PCM to properly calculate fueling for the E85 mixture.

The prime pulses just prior to and during cranking were also increased. Prime pulses help to ensure a rich fuel air mixture to promote initial combustion. The initial prime pulse and the pulses occurring at 90 and 135 crankshaft degrees after cranking begins were increased to account for the difference in the fuel stoichiometric ratios. This was necessary because the fuel injected during priming is not determined by the mass airflow and fuel/air ratio. The prime pulse tables in the PCM were increased by a factor of 1.5. Despite this increase, the difference in vapor pressure of E85 and gasoline was still not taken into consideration. In order to have the same vapor/air ratio needed for proper combustion during cranking, the prime pulses would have to be increased further.

Displacement (liters)	4.8
Cylinder Bore (mm)	96
Crankshaft Stroke (mm)	83
Connecting Rod Length (mm)	159.4
Compression Ratio	9.45:1

Table 2 – 4.8L Engine Parameters

Test Procedure

Because of the lack of availability of an environmental chamber, testing was conducted outdoors in cold weather. Temperature control was not possible and so there were slight variations in the temperatures between different tests. The lowest temperature reached was –4 ° C (25° F), which is warmer than the initial desired goal of –17.8° C.

Coolant temperature inside the engine block was used as the control temperature for the experiments. This was continuously monitored. During the test, data was acquired with the use of General Motors Tech 2 (TECH2) device that communicates and displays values read by the PCM. Snapshots were recorded for each test and were exported to a PC for analysis. For consistency in analyses, additional parameters were measured and recorded. These additional parameters are listed and explained below.

- Crank Start: First time the Engine Speed (RPM) recorded by the TECH2 was more than zero.
- Engine Start: The Engine Speed (RPM) reached 500 rpm and remained above that value.
- Engine Temperature, Air Temperature and Battery Voltage: as read from TECH2.

In order to estimate the effect of hydrogen addition, the time of cranking was recorded with no hydrogen and then for different percentages of hydrogen injection. During testing, hydrogen injection was initiated at the same time as engine cranking. The ignition was held in the start position with the starter turning the motor until the engine was started or was assumed to be started. If the engine stalled, a restart was required.

To eliminate the detrimental effects of changes in battery voltage at cold temperatures, the battery was continuously charged to maintain the same voltage during each test.

Hydrogen Injection System

The hydrogen was injected into the engine intake manifold. This was accomplished by the use of the exhaust gas recirculation (EGR) port and the addition of a tube fitting. This did not change the EGR operation because the EGR valve is normally closed during cold start.

Because hydrogen is highly flammable, helium was used to set the flow control and for purging the system. The desired flow rate for hydrogen was first determined using helium, then the actual level of hydrogen injection was measured during the test. Also, after each run and before shutting down the engine, helium was discharged through the passage to purge any remaining hydrogen. This ensured safe operation during the experiments.

Determination of Volumetric Hydrogen Ratio

The hydrogen ratio by volume injected into the engine was based on the calculated volume flowrate of air during engine cranking. This value could not be measured by the TECH 2 during cranking. Instead, the average engine speed was determined to be 165 RPM for this vehicle during cranking. A volumetric efficiency of 80% was assumed based on the mass air flow recorded by TECH2 during idle and the ideal flow rate possible as predicted from the engine displacement and speed. These values were then used to estimate the air flow when cranking.

The flow rate of hydrogen injected was maintained constant through the duration of each test. This means that the actual percentage of hydrogen changed as the speed and flow rate of air into the engine changed. This has an important effect on the hydrogen supplemented cold start process, which is discussed later.

COLD START TESTING RESULTS

During cold start testing the data alone does not yield a complete representation of the test unless it is combined with the driver's subjective evaluation. The driver is able to determine engine roughness and drivability better than any parameter that can be measured and reported; therefore, the following figure must be viewed along with the subjective evaluation found in the text.

Thirteen experiments were performed with varying percentage of hydrogen. Figure 1 shows the results of this testing. It can be seen that, in general, as the percentage hydrogen injected is increased, the time for cold start decreases. It can also be seen that when the percent hydrogen is high the variability of the data is low, but the opposite is true for when the percent hydrogen is low.

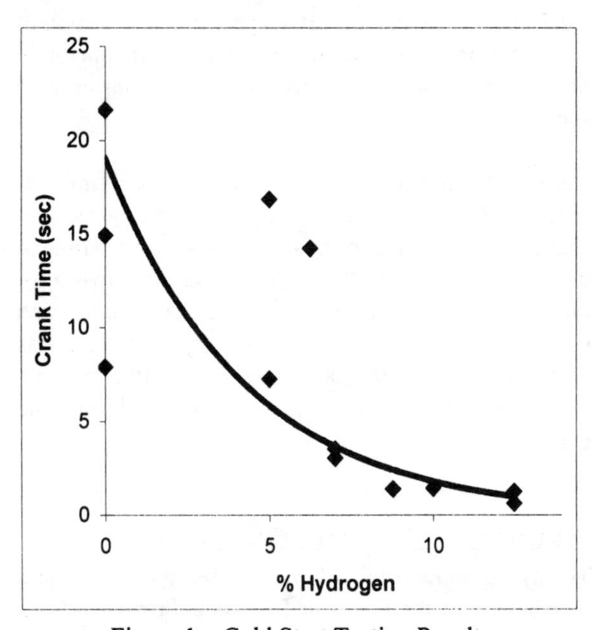

Figure 1 – Cold Start Testing Results

There are a few possible reasons for the data variations. These include: test inconsistencies, battery voltage, air temp, fuel temperature, and the hydrogen ratio variation with increase in engine speed.

Inconsistencies in the testing schedule such as the engine stalling or not starting immediately can cause the human tester to mistakenly release the key, requiring a re-start. When a re-start occurs during the test, additional fuel prime pulses are added during the start, which results in richer mixtures. Basically, when a re-start is done during a test, the engine has more fuel available to evaporate and combust; thus shortening total cranking time. This re-start, however, is unpleasant to the customer and results in poor hydrocarbon emissions. Subjectively, then, this level of hydrogen injection is unacceptable.

The battery voltage was easily controllable and the battery was fully charged throughout all the experiments so it did not vary significantly.

The air temperature affects the density of the air and thus dictates the mass flow rate. All the calculations were based on volume. A significant difference in air temperature results in a percentage of hydrogen that is not be same on mass basis even if the same volume percentage is chosen.

The temperature of the fuel will influence the amount of fuel vapor that will be produced. A higher fuel temperature will produce more fuel vapor and thus will assist the cold start.

As discussed before, the actual percentage of hydrogen seen at the combustion chamber is not constant; it varies with engine speed and flow rate. With an increase in engine speed, the actual percentage of hydrogen decreases. This behavior is very critical when the amount of hydrogen is just enough to initiate combustion. For example, 5% hydrogen (based on 165 RPM) will then go below the lean flammability limit of hydrogen, which is 4% by volume in air [reference 5], when the engine speed increases to about 200 RPM.

It was seen during testing, that when the desired level of hydrogen was set close to the lean limit of 4% hydrogen, the engine came close to starting, but it then immediately stalled or stumbled, forcing the operator to restart. This is due to the drop in hydrogen ratio as the engine speed increases, drawing in more air, while the hydrogen rate was steady.

The variation in hydrogen ratio during cranking is shown in Figure 2. The set point for this test was 7% hydrogen injection. Hydrogen injection commences immediately upon cranking and its ratio rapidly increases since the engine is just beginning to turn-over, drawing only a small amount of air. As the engine speed increased the percentage of hydrogen in the mixture decreased to below the lean flammability limit, which caused the mixture to halt combustion at about

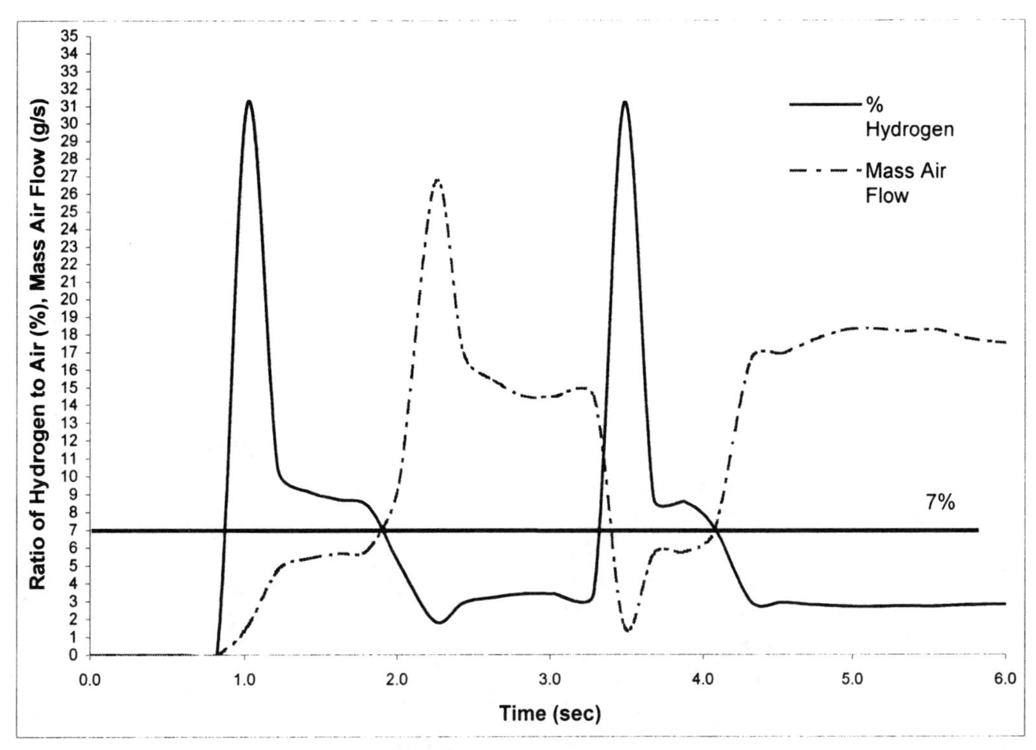

Figure 2 – Percent Hydrogen Change During Testing

3.5 seconds. During this time, the hydrogen injection continued; thus, the hydrogen ratio climbed rapidly as shown. The operator then re-started the engine. At this time, the engine started without a stall, presumably due to its warmer condition from the previous start attempt.

Based upon all of the data, including subjective evaluation, the minimum level of hydrogen supplementation for cold start improvements is about 8% by volume. This level yielded fast starts with no objectionable engine stalls or stumbles.

HYDROGEN REFORMATION TECHNIQUES

Since the distribution and storage of hydrogen most likely will remain difficult for the near future, on-board formation of hydrogen or hydrogen rich fuel offers the best short term outlook. Hydrogen can be generated by either adding energy in an endothermic reaction, or by using some of the energy of the fuel in an exothermic reaction. Endothermic reformers include thermal decomposition of the base fuel, steam reformers where steam is injected into the fuel, and exhaust gas reformation where exhaust is added to the fuel. Exothermic reformers decompose the base fuel by partially oxidizing (POX) it in a rich atmosphere. Reference 5 presents an excellent review of endothermic and some POX reformers.

Endothermic reformers have been demonstrated to produce effective lean engine operation resulting in both improved emissions, and fuel economy. One major limitation of endothermic reformers is their inability to perform well during cold starts, requiring some sort of on-board reformate storage system.

A more feasible approach for automotive applications is the use of an exothermic POX reformer. In this type of reactor the fuel is partially combusted to both provide both the energy requirements and the hydrogen rich reformate. This approach results in faster warm-up times, which are suitable for cold start operation without the need to provide on-board storage of reformate. Unfortunately, a portion of the fuel's energy is consumed during the process, resulting in a lower heating value. This means that this type of reformer is probably best suited for fuel enrichment, not operation on 100% reformate.

Early work in this area resulted in the development of a small POX reformer utilizing a catalyst to lower reaction temperatures.[8] This unit, operating at 650 °C, produced an efficiency of 80% when operating just lean of the soot formation air-fuel ratio. By extending operation of the engine further into the lean range, the energy cost required to generate the reformate was more than offset by the improvements in engine efficiency.

More recent work has demonstrated the development and use of on-board POX reformers.[9, 10, 11] These systems have been used to provide better cold starts, with reference 11 demonstrating reductions in hydrocarbon and carbon monoxide emissions of 80 and 40% using E95, respectively. Further, cold starts were achieved in less than 10 seconds at temperatures as low as –20 °C. The basic design of this under-hood reformer used a standard pulse-width fuel injector to supply fuel and a tangential air intake system to improve combustion stability. Ignition was accomplished through the use of a surface igniter of the type used for residential gas-fired heating systems. A heat exchanger was added to prevent the reformate temperature from exceeding 95 °C, in order to ensure high intake gas density and to prevent damage to the intake manifold. This system was added in parallel to the normal intake system, with reformate induction directly into the intake manifold. The system was sized in order to provide 100% reformate fueling during idle conditions; thus the main fuel system was not activated during the cold start, only the reformer. A similar system was used in reference 10 for operation on methanol with similar improvements in cold start times.

Initial Reformer Design

The previously discussed reformers were all sized to provide 100% of the engine fueling needs during cold starts. It is felt that a better approach would be to design a POX reformer which is sized to provide only enough reformate so that the minimum level of 8% hydrogen enrichment could be achieved during cranking. This system would then operate in parallel to the main intake, functioning only during cold start. The existing fuel system of the engine does not have to be modified, except to minimize cold start enrichment. It can be allowed to function normally during cold starts and, thus, the reformer can simply be turned off after the cold start. This will minimize the need for a complicated control system to handle transitions to normal operation, as is done with the 100% reformers. Finally, tis system should be smaller than the other reformers, since it does not need to provide 100% of the fuelling needs during cold start.

CONCLUSION

Cold start is a definite problem in alcohol fuels making it almost impossible to start below 10 °C without additional technology. Operating with E85, despite addition of 15% gasoline, does not solve the cold starting problem. This problem can be minimized by using hydrogen supplemented E85 during cold starts. Not only does the cranking time decrease, but the emissions are also likely to decrease since over-fueling can be minimized. Increasing the amount of hydrogen injection decreases crank time up to about 12% hydrogen by volume. Additional hydrogen doesn't provide any noticeable improvement. The minimum level of hydrogen required appears to be around 8% at cold start conditions of –4 °C (25 °F). This level of injection ensures that the hydrogen will not fall below the minimum flammability limit of 4 % for a substantial amount of time during engine startup. This information can then be used in order to size a reformer to meet cold start conditions.

REFERENCES

1 Black, F., *"An overview of the technical implications of methanol and ethanol as highway motor vehicle fuels"*. SAE 912413, 1991.

2 Markel, A. J., et. al., *"Modelling and Cold Start in Alcohol Fueled Engines"*. NREL/TP-54-24180.

3 Hodgson, J. W., et. al., *"Improving the Cold Start Performance of Alcohol Fueled Engines Using a Rich Combustor Device"*. SAE-981359, 1998.

4 Davis, G. W., et. al., *"The Effect of Multiple Spark Discharge on the Cold-Startability of an E85 Fueled Vehicle"*. SAE, 1999-01-0609, 1999.

5 Jamal, Y., et. al., *"On-Board Generation of Hydrogen-Rich Gaseous Fuels – A Review"*. Int. J. Hydrogen Energy. Vol 19, No. 7.

6 Houseman, J., et. al., *"On-Board Hydrogen Generator for a Partial Hydrogen Injection Internal Combustion Engine"*, SAE-740600, 1974.

7 Davis, G. W., et. al., *"The Development and Performance of a High Blend Ethanol Fueled Vehicle"*. FCC – 00FCC-100.

8 Houseman, J., and Cerini, D., "*On-Board Hydrogen Generator for a Partial Hydrogen Injection Internal Combustion Engine,*" SAE 740600, 1974.

9 Grieve, M. J., et al, "*Integration of a Small Onboard Reformer to a Conventional Gasoline Internal Combustion Engine System to Enable a Practical and Robust Nearly-Zero Emission Vehicle,*" 1999 Global Powertrain Congress, Stuttgart, GE, 1999.

10 Hodgson, J. W., et al, "*Improving the Cold Start Performance of Alcohol Fueled Engines Using a Rich Combustor Device,*" SAE 981359, 1998.

11 Isherwood, K. D., et al, "*Using On-board Fuel Reforming by Partial Oxidation to Improve SI Engine Cold-Start Performance and Emissions,*" SAE 980939, 1998.

ON THE RECOVERY OF LNG PHYSICAL EXERGY

G. Bisio *, L. Tagliafico - Energy and Conditioning Department - University of Genoa
Via all'Opera Pia 15 A - 16145 Genoa - ITALY

ABSTRACT

The used fraction (or thermal efficiency) of a power cycle is limited by the maximum and minimum temperatures available. The construction of LNG terminals and the need to vaporize LNG offers a thermal sink at a very much lower temperature than seawater. By using this thermal sink in a combined plant, it is possible to recover power from the vaporization of LNG. To this purpose, in this paper complex systems are considered and their pros and cons are put in evidence. A system utilizing waste energy as heat source and with a single working fluid is analyzed with sufficient details. However, the use of a single fluid is not the best solution from a thermodynamic point of view. Thus, a series of cascading cycles is also synthetically outlined. It is to be remarked that in these systems both thermal source and thermal sink are exploited as exergy sources.

NOMENCLATURE

Ex	(physical) exergy [kJ/kg]
Ex_a	actual exergy loss [kJ/kg]
Ex_{rf}	losable (or reference) exergy [kJ]
Ex_u	recovered (and actually usable) exergy [kJ]
ex	specific exergy [kJ/kg]
i	specific enthalpy [kJ/kg]
p	pressure [bar]
p_o	environment pressure [bar]
s	specific entropy [kJ/(kg K)]
v	specific volume [m^3/kg]
T	temperature [K, °C]
T_o	temperature of the environment [K, °C]
x	dryness [dimensionless]
η	used fraction (or thermal efficiency) [dimensionless]
η_o	usable fraction (or Carnot efficiency) [dimensionless]
σ_a	actual entropy production [kJ/K]
σ_{ai}	actual entropy production of the process "i" [kJ/K]
σ_{rf}	reference entropy production [kJ/K]
σ_r	relative entropy production [dimensionless]
σ_{ri}	relative entropy production of the process "i" [dimensionless]
σ_{ri}'	$= \sigma_{ai}/\sigma_{rfw}$ relative entropy production of a process "i" with reference to σ_{rfw} of the whole cycle [dimensionless]
σ_{rfi}	reference entropy production of the process "i" [kJ/K]
σ_{rfw}	reference entropy production of the whole system [kJ/K]
φ	exergy efficiency [dimensionless]
φ_i	exergy efficiency of the process "i" with reference to the parameter σ_{rfi} of the process "i" [dimensionless]
φ_i'	exergy efficiency of the process "i" with reference to the parameter σ_{rfw} of the whole cycle [dimensionless]

1. INTRODUCTION

1.1. Natural gas as a source of physical exergy

Natural gas (NG) is often found in remote locations far from developed industrial nations. For the transportation of NG from producing wells to utilization sites, two approaches are now applied with their relative pros and cons.[1-4] When possible the gas is transported by pipeline to the end user. Currently, when the gas source and the user are separated by oceans, the only viable way to transport the gas is to convert it into liquid natural gas (LNG) and convey it using insulated LNG tankers. At receiving terminals the LNG is off-loaded

* Professor, University of Genoa

into storage tanks, pumped from storage at the required pressure and vaporized for final feed to the consumers.

The natural gas output pressure required from LNG vaporizing terminals varies according to the requirements of the natural gas users. The required gas pressure at battery limit is usually as follows:

(i) steam power stations 6 bar,
(ii) combined cycle stations 25 bar,
(iii) local distribution 30 bar,
(iv) long-distance distribution 70 bar.

Since the consumption of LNG will probably increase in the future, (in addition to the use of the chemical exergy) the exploitation of the physical exergy of LNG, due to its state in liquid phase at a temperature under that of the environment, is becoming more important.

Till now, with the exception of Japan and at lower extent of Germany, most of LNG has been regassified using the thermal energy of seawater or of warm seawater effluent from a power plant, destroying in this way its physical exergy without any useful effect.

Various processes have been considered to utilize the physical exergy of liquids at the atmospheric pressure by vaporizing, at cryogenic temperatures, e.g. nitrogen, hydrogen and natural gas.[5-16]

Two general alternatives may be envisaged:

(a) direct utilization in cryogenic facilities (cold storage or various process uses);
(b) indirect utilization in the generation of electric power.

Direct utilization poses some problems:

(i) it entails coupling two plants with asynchronous characteristics;
(ii) it is essential to ensure that the operation of LNG (or liquid hydrogen) terminal is not hampered by malfunctions in the plant of the cryogenic customer;
(iii) the available exergy cannot be sufficiently exploited in relation to both its rate and its temperature level.

Indirect utilization in the generation of electric power avoids the drawbacks of direct utilization and uses the available exergy in a more efficient way.

1.2. Entropy production and exergy efficiency

The critical examination of the concept of efficiency has led Bisio to formulate some requirements for its statement based on the two laws of thermodynamics.[17] It has been demonstrated that only exergy efficiency can fulfill all the requirements; however, not all the formulations comply with the condition that efficiency should in principle vary between zero (for no useful effect) and one (maximum useful effect and no entropy production). To this purpose, it is suitable the definition of the operative domain of actual considered case. This domain is comprised within the case of no entropy production (upper limit) and the one of no useful effect (lower limit).

As a reference value, σ_{rf}, the entropy production corresponding to no useful effect is assumed; multiplied by the environment temperature T_o, it gives, for the selected operation domain, the maximum loss, a parameter called losable (or reference) exergy, Ex_{rf}. The actual entropy production, σ_a, multiplied by T_o gives the actual exergy loss, Ex_a. The ratio of the actual entropy production to the reference one is called relative entropy production, σ_r. The difference between the losable exergy and the lost one is the recovered exergy, Ex_u.

Exergy efficiency is the ratio of the recovered exergy to the losable one. One has that

$$\varphi = Ex_u/Ex_{rf} = (Ex_{rf} - Ex_a)/Ex_{rf} =$$
$$= (T_o\,\sigma_{rf} - T_o\,\sigma_a)/(T_o\,\sigma_{rf}) = (\sigma_{rf} - \sigma_a)/\sigma_{rf} = 1 - \sigma_r\,. \quad (1)$$

The recovered exergy must be actually usable in the real process; otherwise, by Ex_u, one should indicate the actually usable exergy and not the total recovered exergy.

We remark that exergy efficiency can be defined with reference both to exergies and entropy productions.

The use of the concepts of "reference entropy production" and of "losable exergy" is easily applicable in formulating the exergy efficiency for the most different processes in open and closed systems. Only in a few cases, some formal greater complication arises for exergy efficiency as to other expressions. On the other hand, however, more simple formulae lead sometimes to efficiency values of poor meaning: this applies especially in the cases of diffusers, of shock waves and in general when a large part of the inlet exergy cannot be lost in the selected operation domain.

However, in a process belonging to a cyclic or pseudocyclic series, it seems not totally meaningful to consider the relative entropy production, σ_{ri}, as the ratio of the actual entropy production of a single process, σ_{ai}, to the reference entropy production of this single process, σ_{rfi}.[18] For both conceptual and practical purposes, it seems more viable to consider the relative entropy production, σ_{ri}', as the ratio of the actual entropy production of a single process, σ_{ai}, to the reference entropy production of the whole cycle or pseudocycle, σ_{rfw}. Obviously, one has:

$$\sigma_r = \sum_i \sigma_{ri}' = 1 - \varphi\,. \quad (2)$$

2. BASIC CRITERIA FOR SUITABLE EXERGY RECOVERY

Let us consider a first cycle with maximum temperature of 1200 °C and minimum temperature of 35 °C and a second cycle with the same maximum temperature and minimum temperature of -160 °C. The usable fractions (normally called "Carnot efficiencies") would be 0.790 and 0.923 in the two cases, respectively. The difference between the two values gives the possible advantage of using the LNG vaporization as a thermal sink only in an approximate way. It is to me remarked that irreversibilities weigh in a different way in the two cases.

By using LNG as thermal sink, the fluid working in the cycle is compressed in gas phase and then the advantage of a higher difference between the extreme temperatures is notably reduced. It is always to be remembered that in a heat engine the compression work must be the lowest possible fraction of expansion work.

However, since the thermal sink at temperatures notably lower than that of the environment can be judged free in the case that LNG must be vaporized anyhow, the ratio of the exergy value of the obtained electric energy to the exergy value of the heat source is very high.

A further advantage seems attainable by the use of cascade cycles, as proposed by Wong.[19] In this way, in each cycle compression can be made in liquid phase with the corresponding pros. Wong remarks that usable fraction depends only upon the extreme temperatures and is independent of the number of cascade cycles. However, it is to be noted that these results, obtained with the reversibility hypothesis, are modified considering actual irreversibilities. Thus, only an actual analysis can allow us to reach the optimal solution.

In next sections a system using a single cycle is firstly analyzed with some details regarding the various irreversibilities.

Then, a complex cascade system, including a gas turbine, a steam cycle, a propane cycle and a methane cycle is synthetically outlined.

3. PROPOSED COMBINED SYSTEM FOR ELECTRICAL ENERGY PRODUCTION AND LNG VAPORIZATION AND SUPERHEATING USING A SINGLE CYCLE

In the sintering plants the red-hot material produced by the sintering lines has to be cooled down to ambient temperature, so as to make possible transportation on belt conveyors and storage in blast furnace stockhouses. The hot sinter, at a temperature of 600-650 °C, falls down on a continuous train of grate bottom carriages, traveling on a track set on the top of a circular channel. Large quantities of air, blown by fans into the channel, cross the hot sinter bed and cool down the sinter to the suitable temperature.

In this way, a large part of the thermal energy of the sinter is transferred to the cooling air. In most plants now, with the exception of Japanese plants, this energy is lost because the air is discharged straight into the atmosphere.

In the proposed system of Fig. 1, the thermal energy of the air crossing the sinter is utilized, taking that air to the heat exchanger E.

Nitrogen is a fluid suitable to operate in a closed cycle of the kind above-mentioned owing to its thermodynamic properties; on the contrary, air presents explosion hazards in the case of leaks of natural gas in the heat exchanger.

The layout of the global system is shown in Fig. 1. In Fig. 2 the nitrogen cycle, in which the waste energy of the sinter refrigeration is the heat source and the LNG "vaporization" and superheating is the heat sink, is represented in a T vs s diagram; it must be noted that both heat source and sink are exergy sources. We write "vaporization" since, owing to the natural gas pressure, during the heating there is always the presence of only one phase (supercritical pressure).

Let us consider the nitrogen cycle shown in Fig. 2. The values of the functions p (pressure), T (temperature), i (specific enthalpy), s (specific entropy) and ex (specific exergy) corresponding to the state of the environment and to the states of the cycle vertices (1-8) are reported in Table 1, utilizing the data of Vassermann et al.[20] In addition, this Table shows the same values for the state α that is characterized by the minimum value of exergy during the process 7-8.

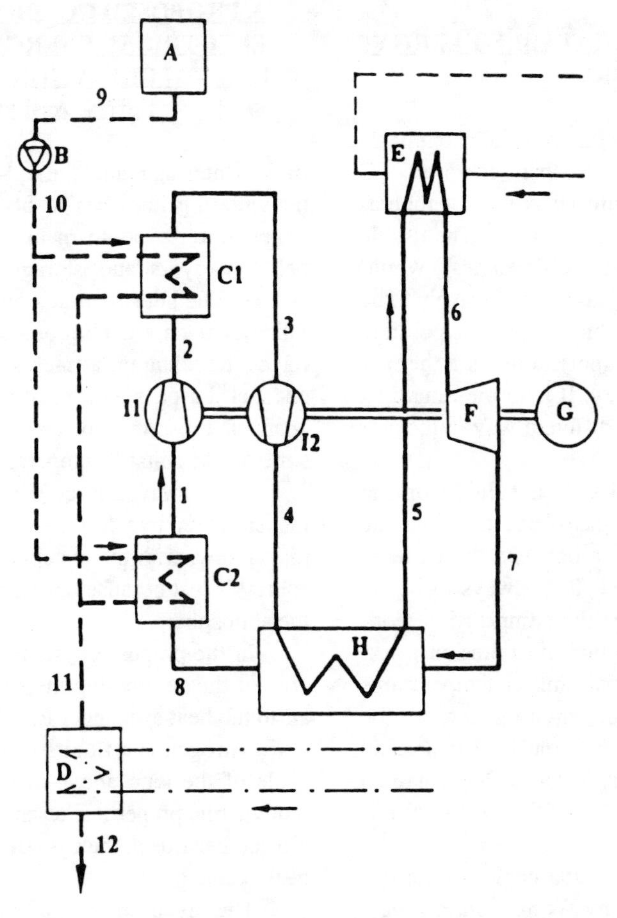

Fig. 1. Layout of a combined system utilizing a closed-cycle gas turbine with waste energy heating and LNG refrigerating: (A) LNG tank; (B) LNG pump; (C1) and (C2) LNG "vaporizers"; (D) LNG superheater by using waste energy; (E) nitrogen heater; (F) nitrogen turbine; (G) electrical generator; (H) regenerator; (I1) and (I2) nitrogen compressors.

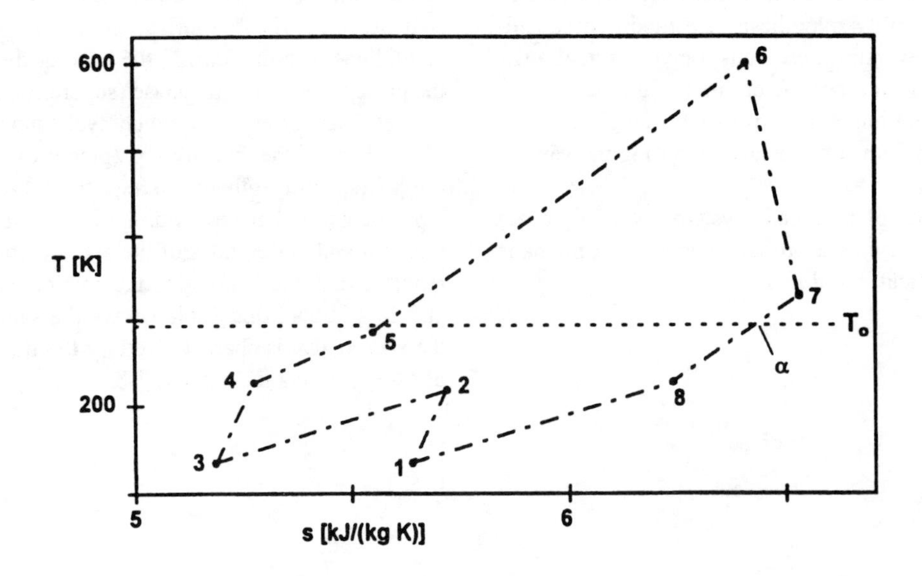

Fig. 2. T vs s diagram for the nitrogen cycle.

Table 1 - Nitrogen Cycle.					
	Environment (o)	state 1	state 2	state 3	state 4
p - bar	1	3	12.5	12.3	51.3
T - K (°C)	290.15 (17)	133 (- 140.15)	211.20 (-61.95)	133 (- 140.15)	211.60 (-61.55)
i - kJ/kg	301.0	134.7	213.7	123.8	197.2
s - kJ/(kg K)	6.810	5.657	5.710	5.182	5.231
ex - kJ/kg	0	168.2	231.9	295.2	354.5
	state 5	state 6	state 7	state 8	state α
p - bar	50	48.7	3.3	3.1	3.2
T - K (°C)	281.24 (8.09)	593.15 (320)	311.24 (38.09)	231.92 (-41.23)	290.15 (17)
i - kJ/kg	280.0	619.7	322.4	239.6	300.5
s - kJ/(kg K)	5.578	6.401	6.527	6.238	6.463
ex - kJ/kg	336.5	437.4	103.5	104.6	100.2

Table 2 - Methane Processes.					
	Environment (o')	state 9	state 10	state 11	state 12
p - bar	1	1.01325 (x_9=0)	100	99	15
T - K (°C)	290.15 (17)	111.67 (-161.48)	195 (- 78.15)	400 (126.85)	269.65 (-3.50)
i - kJ/kg	1178.7	287.2	625	1385.7	1118.1
s - kJ/(kg K)	11.557	4.946	7.05	9.820	9.951
ex - kJ/kg	0	1026.7	756.9	711	405.4

Considering the natural gas as pure methane, the values of Table 2 are obtained for its processes (where x is the dryness fraction) by using the data got from Stewart et al.[21]

By using the concepts dealt with in subsection 1.2., the values of Table 3 are obtained. The maximum value of the entropy production takes place in the turbine (process 6-7). That is put in evidence by the corresponding parameter σ_{ri}' which is the highest, even if the corresponding parameter σ_{ri} is the lowest. In addition, the parameter σ_{ri}' of the regenerator H (processes 4-5 and 7-8) is low, even if in this device there is a small increase of the exergy of one stream and a much larger decrease of the exergy of the other stream.

Table 3 - Actual (σ_{ai}) and relative (σ_{ri} and σ_{ri}) entropy productions and exergy efficiencies (φ_i and φ_i') of the single processes and relative entropy production (σ_r) and exergy efficiency (φ) of the whole system.

Processes	σ_{ai}^{o}	σ_{ri}	φ_i	σ_{ri}'	φ_i'
1-2	10.87	0.2098	0.7902	0.0572	0.9428
2-3	8.20	0.1678	0.8322	0.0432	0.9568
3-4	10.03	0.2083	0.7917	0.0528	0.9472
4-5 and 7-8	10.86	0.7981	0.2019	0.0572	0.9428
5-6	19.24	0.2249	0.7751	0.1012	0.8988
6-7	27.35	0.1274	0.8726	0.1439	0.8561
8-1	16.17	0.2834	0.7166	0.0851	0.9149
σ_r = 0.5406			φ = 0.4594		

We remark that the plant analyzed in this section is included in the set of plants characterized by the fact that there are heat exchangers in which in both streams, or at least in one stream, the extreme temperatures are one above and the other under the environment temperature.[22-23] In the first case, in both streams the exergy firstly decreases and after increases attaining a minimum near the environment temperature (not exactly at the environment temperature owing to the effect of friction irreversibilities which, however, are not very great). In the second case, the exergy has a minimum in one stream and is always decreasing in the other stream. As a consequence, one can have cases in which there is a global decrease of exergy in both streams, and other cases in which there is a small increase in one stream. In these cases any classical definition of exergy efficiency fails. However, also in this case the general statement of Reference 17 can be used, if it is applied correctly.[22] We remark that the state α is the state characterized by the minimum of exergy in the process 7-8.

We note that the exergy efficiency of the nitrogen cycle here studied is good, but not particularly high in comparison with those typical of the best combined plants.

4. ANALYSIS CRITERIA FOR A CASCADE COMBINED SYSTEM IN LNG POWER RECOVERY

The exergy potential of the LNG is used as a low temperature source in a cascade of thermodynamic cycles for power conversion of the thermal energy available from the flue gases of some topping gas turbine cycle. These flue gases could be cooled from the gas turbine outlet temperature, around 800-900 K, down to the temperature of around 400 K, the minimum temperature allowing the discharge chimney to work properly without pollution or corrosion (due to vapor condensation) problems.

The block diagram of the plant layout is sketched in Fig. 3, while the corresponding T-s diagrams for the different cycles are reported in Figs. 4-5.

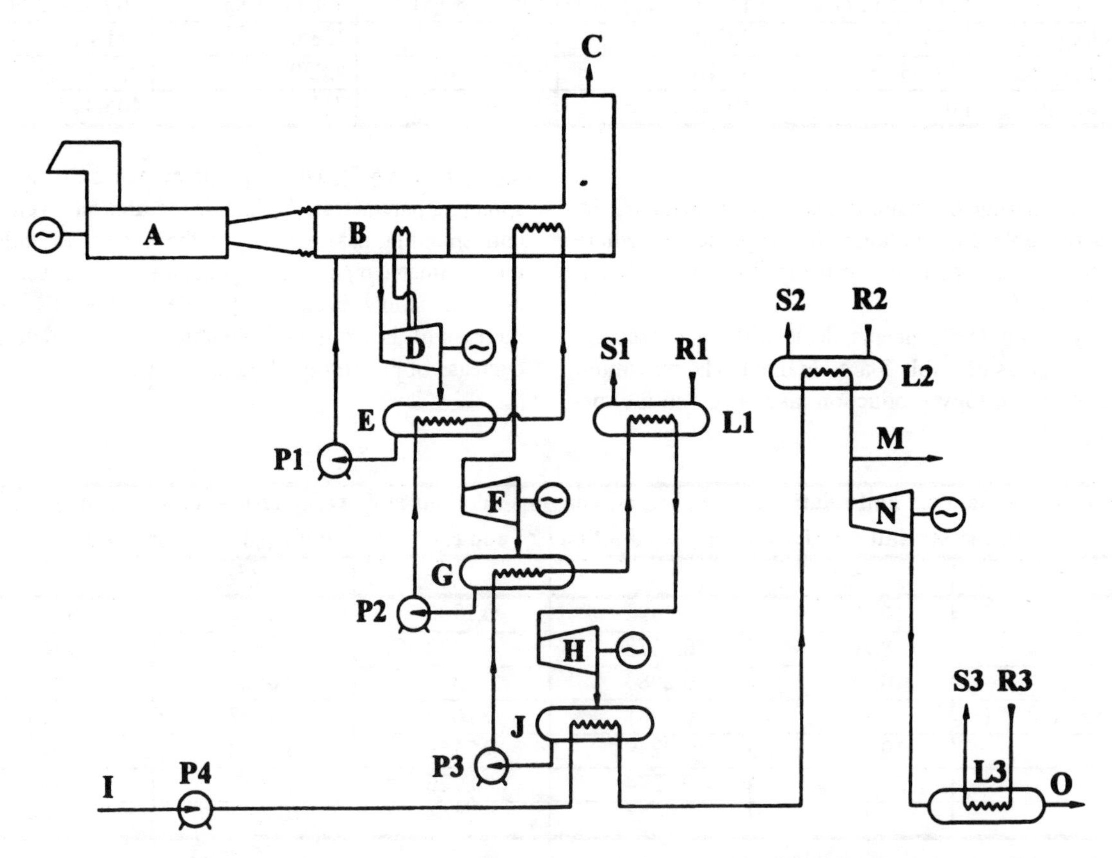

Fig. 3. LNG power recovery using a complex cycle: (A) gas turbine; (B) boiler; (C) gas exhaust; (D) steam turbine; (E) steam-propane heat exchanger; (F) propane turbine; (G) propane-methane heat exchanger; (H) methane turbine; (I) LNG inlet; (J) methane-LNG heat exchanger; (L1) seawater-methane heat exchanger; (L2), (L3) seawater-NG heat exchangers; (M) HP NG gas export; (N) NG turbine; (O) LP NG export; (P1) water circulation pump; (P2) propane circulation pump; (P3) methane circulation pump; (P4) LNG pump; (R1), (R2), (R3) seawater inlets; (S1), (S2), (S3) seawater outlets.

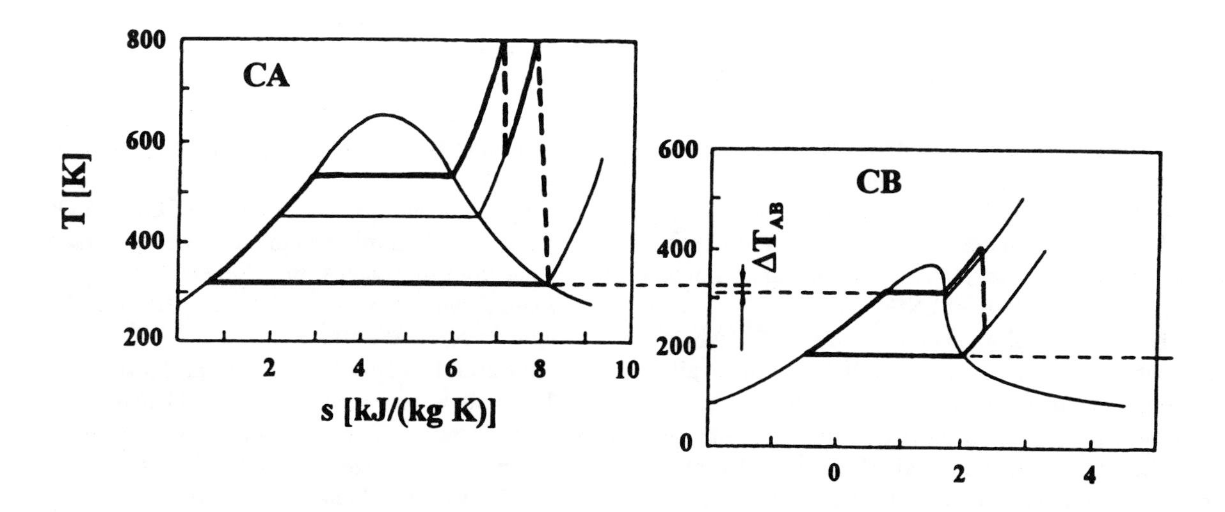

Fig. 4. Steam cycle (CA) and propane cycle (CB).

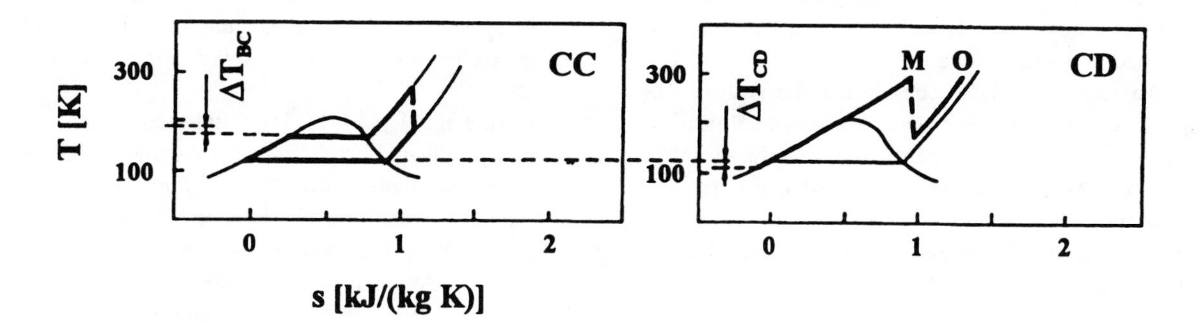

Fig. 5. Methane cycle (CC) and open LNG cycle (CD).

The flue gases are used in their higher temperature range to vaporize and superheat the steam in cycle (CA), and to superheat the propane gas in cycle (CB). The two cycles (CA) and (CB) are coupled in order to have their best match in the steam condenser/propane evaporator heat exchanger (E). The heating water (or other low-temperature source) is used to "vaporize" the methane for the user lines (M, O), and also to superheat the gas in the methane cycle (cycle CC). The LNG (methane) line is studied as a further open cycle (cycle (CD)) which can produce some exergy (useful electric power) by means of the super critical gas turbine (N).

The propane cycle (CB) is coupled to the methane cycle (CC) with reference to the condenser/vaporization heat exchanger (G), followed by the methane superheater (L1).

The methane cycle (CC) is coupled to the LNG cycle (CD) with reference to the condenser/vaporization heat exchanger (J).

It is worthwhile to note that the cycle (CD) could be performed in any way, even without any auxiliary plant development, while the other cycles (CA), (CB) and (CC) are completely new, with respect to the LNG plant, and must be designed "ex novo".

An alternative solution is that of eliminating the methane cycle (CC) and then to "vaporize" the LNG by means of the propane cycle.

Only an optimization analysis can indicate the best solution in actual plants.

5. CONCLUSIONS

It is very probable that the NG consumption increases at least at short-medium date. Thus, the employment of the NG physical exergy, owing to its pressure at pipeline outlet and moreover owing to its liquid phase at a temperature remarkably lower than that of the environment, is becoming more and more advisable.

Several processes have been realized or proposed to employ the physical exergy of liquids at atmospheric pressure by the vaporization at cryogenic temperatures of N_2, H_2 and particularly of LNG.

The indirect utilization of the above exergy avoids the drawbacks of direct utilization and uses the available exergy in a more efficient way.

In this paper, a closed-cycle nitrogen turbine, in which the sinter refrigeration air (in iron works) at a temperature of about 350 °C is the heat source and the natural gas vaporization and superheating is the heat sink, is examined.

The exergy efficiency of the nitrogen cycle is about 0.46; this value is high in relation to the temperatures of the heat source and justifies the validity of the process, obviously if the heat source and sink can be located in the same area.

An exergy analysis considering the relative entropy production of every phase of the whole nitrogen cycle is developed in this paper; this analysis visualize the relative "weight" of the entropy production rates in the various phases of the cycle.

A more complex system, recently proposed by an author, is also synthetically described.

ACKNOWLEDGMENT

The authors wish to thank Mr. Roberto Martini for his valuable help in the edition of the Figures of this paper.

REFERENCES

1. Szilas A.P., 1975, *Production and Transport of Oil and Gas*, transl. by B. Balkay, Elsevier Scientific Publishing Co., Amsterdam.
2. Harder E.L., 1982, *Fundamentals of Energy Production*, J. Wiley & Sons, New York.
3. Geist J.M., 1983, "The role of LNG in energy supply", *International Journal of Refrigeration*, **6**, 283-297.
4. Pellò P.M., Vitale G., 1984, "Centrali di spinta per sistemi di trasporto di gas naturale", *L'Elettrotecnica*, **LXX**, n° 2, 115-140.
5. Griepentrog H., Weber D., 1978, "Gasturbinen im geschlossenen Prozeß für neue Anwendungsgebiete", *Kerntechnik*, **20**, 537-543.
6. Oshima K., Ishizaki Y., Kameyama S., Akiyama M., Okuda M., 1978, "The utilization of LH_2 and LNG cold for generation of electric power by a cryogenic type Stirling engine", *Cryogenics*, **18**, 617-620.
7. Sanga Y., Shirasaki Y., Shiozawa H., Hiro-Oka T., 1982, "Control system of power generating plant from LNG cold", in *Advances in Cryogenic Engineering*, R.W. Fast, Ed., **27**, 979-989, Plenum Press, New York.
8. Shiozawa H., Hiro-Oka T., Aoki I., Inoue T., 1982, "Power generation using cold potential of LNG in multicomponent fluid Rankine cycle", in *Advances in Cryogenic Engineering*, R. W. Fast, Ed., **27**, 971-978, Plenum Press, New York.
9. Timmerhaus K.D., 1983, "Low-temperature technology utilization in the solution of energy problems", *Int. J. Refrig.*, **6**, 274-282.
10. Weber D., 1989, "Strömungsmachinen für hohe und tiefe Temperaturen bei der Kraft-Wärme-Kopplung von Stromerzeugung und Flussigerdgasverdampfung", *Forsch. Ing.-Wes.*, **55**, 155-158.
11. Kashiwagi T., 1991, "The importance of thermally activated heat pumps for solving global environmental problems", *Proceedings of the Absorption Heat Pump Conference '91*, Tokyo, September 30 - October 2, K. Watanabe, Ed., 63-71, Japanese Association of Refrigeration, Tokyo.
12. Bisio G., 1992, "Combined plants with high and cryogenic temperatures for power production and liquid hydrogen vaporization", *Proceedings of ECOS '92*, Zaragoza, June 15-18, A. Valero, G. Tsatsaronis, Eds., 413-418, ASME, New York.
13. Bisio G., Cerullo N., 1992, "Thermodynamic analysis of combined plants utilizing helium turbine with nuclear heating and liquid hydrogen refrigerating", *Proceedings of 27th IECEC*, San Diego, August 3-7, **4**, 277-282, SAE, Warrendale, PA.
14. Bisio G., 1994, "The possible utilization of low-temperature waste energy linked with the exploitation of LNG physical exergy", *Proceedings of 29th IECEC*, Monterey, CA, August 7-12, **4**, 1537-1542, AIAA, Washington.
15. Bisio G., Pisoni C., 1995, Thermodynamic analysis of solar energy utilization combined with the exploitation of LNG physical exergy, *J. of Solar Energy Engineering*, **117**, 333-335.
16. Bisio G., Massardo A., Agazzani A., 1996, "Combined helium and combustion gas turbine plant

exploiting liquid hydrogen (LH_2) physical exergy", *Transactions of the ASME, J. of Engineering for Gas Turbines and Power*, **118**, 257-264.

17. Bisio G., 1989, "On a general statement for efficiency", *Chemical Engineering Communications*, **81**, 177-195.
18. Bisio G., Benvenuto G., 1993, "Possible utilizations of the pressure exergy of natural gas", *Proc. of 28th IECEC*, Atlanta, GE, August 8-13, **2**, 167-173, American Chemical Society, Washington, D.C.
19. Wong W., 1994, "LNG power recovery", *Proc. Instn. Mech. Engrs*, **208**, 3-12.
20. Vassermann A.A., Kazavchinskii Ya.Z., Rabinovich V.A., 1971, "*Thermophysical Properties of Air and Air Components*", translated from Russian, Israel Program for Scientific Translation, Jerusalem.
21. Stewart R.B., Jacobsen R.T., Penoncello S.G., 1986, *Thermodynamic Properties of Refrigerants*, ASHRAE, Atlanta, GA.
22. Bisio G., 1995, "On the exergy efficiency of some kinds of heat exchangers", *Proceedings of ECOS '95*, Istanbul, July 11-14, Y. A. Gögüs, A. Öztürk, G. Tsatsaronis, Eds., **1**, 184-189.
23. Bisio G., Pisoni C., 1998, Thermodynamic remarks on statements of exergy efficiency for heat exchangers, *Proceedings of 16th UIT National Heat Transfer Conference,* Siena, 17-19 June, **II**, 733-744, Edizioni ETS, Pisa.

A NEW METHOD FOR THE CALCULATION OF GASES ENTHALPY

R. Lanzafame*, M. Messina°
Università di Catania, Facoltà di Ingegneria, Istituto di Macchine – Viale A. Doria, 6 – 95125 Catania - Italy
rlanzafa@im.ing.unict.it – mmessina@im.ing.unict.it
Tel: +39 095 7382414; +39 095 7382418; Fax: +39 095 330258

ABSTRACT

In this paper, the authors have investigated new polynomial functions for the determination of Enthalpy for various gases, and for their mixtures, in ICE applications.

Several polynomial functions have been investigated to determine the best matching function respect to experimental data. The polynomial functions investigated are: simple polynomials, exponential polynomials and logarithmic polynomials (third, fourth and fifth order).

The best polynomial functions have the functional form of a V order Logarithmic Polynomial, and can be used in temperature ranges of practical interest.

Logarithmic Polynomial (L.P.) coefficients have been obtained by least squares matching with thermodynamic property data from the JANAF tables.

Values of the specific heat have been calculated as a function of temperature, and they have been compared with experimental ones, in order to evaluate the percent error.

Logarithmic Polynomials (L.P.s) have been calculated for various gases: Technical air (21% O_2), N, O, H, H_2, O_2, N_2, CO, OH, NO, CO_2 and H_2O.

In literature there are many works which present mathematical functions of Enthalpy vs Temperature. A comparison between the polynomial functions that are present in literature and V order L.P. has been effected. The new L.P.s point out a major precision respect to the other polynomial functions, and the possibility to utilise a single L.P. for a wide temperature range, according to a good accuracy with experimental data.

INTRODUCTION

It is very important to have functions able to describe the correlation between enthalpy and temperature for any gases. With these functions, the designer is free to use experimental numerical tables, possessing a fast and efficient tool to determine the values of gases enthalpy.

In literature there are many works which present mathematical functions of enthalpy Vs temperature: Langen[6], Murphy[6], Chypman and Fontana[6], Faggiani[6], Caputo[12], Spencer and Flanagan[13], and also the most recent *NASA equilibrium code*[9] and the formulas presented by Annaratone[15].

These mathematical functions that are in literature have some limits: give an insufficient approximation with experimental data, or have a good approximation only for a little temperature range.

For example, Annaratone functions[15] have a good accuracy for 273 <T< 1673 K, with a percent error less than 1%; while in the *NASA equilibrium code*[9] two different polynomials are used to cover the temperature range, one for 0 <T< 1000 K and the other for 1000 <T< 5000 K.

In this work the authors have determined, for several gases, fifth order L.P.s. L.P.s show the capacity to cover wide range of temperature 0 <T< 3500 K, with a mean percent error less than 0.1%, and for some gases less than 0.01%.

The L.P. coefficients have been calculated on the basis of specific heat experimental data (JANAF tables[2,3,4 or 5]) for every gas considered (technical air (21% O_2), N, O, H, H_2, O_2, N_2, CO, OH, NO, CO_2 and H_2O).

For the calculation of L.P.s' coefficients a program has been written, and the least squares method was used.

The program dimension is 840 byte, and working with twelve decimal numbers, it executes the calculation of the coefficients in two seconds on a Pentium Computer 166 MHz with 64 Mb of RAM memory.

The program is linked with an electronic sheet to evaluate the percent error of the enthalpy (or specific heat) calculated with the L.P. respect to the enthalpy (or specific heat) from JANAF tables. Moreover the program is able to compare these obtained values with those of mathematical functions that are known in literature.

POLYNOMIAL FUNCTIONS INVESTIGATED

The polynomial functions, specific heat at constant pressure, that have been investigated are:

- Simple polynomial functions (n-order):

$$\tilde{c}_p(T) = a_0 + a_1(T) + a_2(T)^2 + \cdots + a_n(T)^n$$

*Professor, °Ph.D.

- Exponential polynomial functions (n-order):

$$\tilde{c}_p(T) = a_0 + a_1\left(e^T\right) + a_2\left(e^T\right)^2 + \cdots + a_n\left(e^T\right)^n$$

where e is the Neperian number.

- L.P. functions (n-order):

$$\tilde{c}_p(T) = a_0 + a_1(\log T) + a_2(\log T)^2 + \cdots + a_n(\log T)^n$$

where *log* is the natural logarithm.

For each polynomial function, third, fourth and fifth order has been evaluated. The coefficients of the polynomial functions have been calculated utilizing the least squares method.

The best matching with experimental thermodynamic property has been obtained with fifth order L.P.

THE LOGARITHMIC POLYNOMIAL

Fifth order L.P. has been used to fit experimental data on specific heat at constant pressure, it is:

$$\tilde{c}_p(T) = a_0 + a_1(\log T) + a_2(\log T)^2 + a_3(\log T)^3 + \\ + a_4(\log T)^4 + a_5(\log T)^5 \tag{1}$$

where a_j ($j = 0,...,5$) are calculated with the least squares method, and *log* is the natural logarithm.

In Tab. 1 the values of the L.P. coefficients are reported for each gas considered.

COMPARISON BETWEEN DIFFERENT POLYNOMIALS

For each gas, the percent error respect to experimental data on $\tilde{c}_p$, has been calculated with:

$$\varepsilon_{rel\,\%} = 100 \cdot \frac{\tilde{c}_{p(log.\,polynomial)} - \tilde{c}_{p(JANAF)}}{\tilde{c}_{p(JANAF)}} \tag{2}$$

The results obtained with V order logarithmic polynomial, reduce considerably the percent error respect to the classic methods[6], respect to Annaratone functions[15] valid only for 273 <T< 1673 K. Moreover with a single V order L.P. is possible to cover a wide temperature range (273 <T< 3500 K), while in the *NASA Equilibrium Code*[9] are used two different polynomials to cover the same temperature range.

Fig. 1 shows the percent errors on $\tilde{c}_p$ for O_2 obtained using fifth order logarithmic polynomial, Annaratone formula and the mathematical functions from NASA Equilibrium Code. Annaratone formula gives a percent error <1% only for 350 <T< 1550 K. NASA Equilibrium Code utilizes two different polynomials: first for T< 1000 K with a percent error less than 0.5%, second for T> 1000 K with a percent error less than 0.35%. By using fifth order L.P. it is possible to cover the temperature range 273 <T< 3500 K with a percent error < 0.2%.

In Fig. 2 the comparison is shown for N_2 . For T<1000 K the mathematical function from NASA Equilibrium Code report an error less than the error from the L.P. (0.05% with 0.1%). The advantage of using the L.P. consist of the possibility to cover a wide range of temperature (273 <T< 3500 K) with a very low percent error, absolutely lower than from NASA function for T> 1000 K. Annaratone formula gives good results only for a limited temperature range: 400 <T< 1700 K with $\varepsilon_{rel\,\%}$ < 1%.

The Fig. 3 shows the comparison between several calculation methods for H_2O (for H_2O Annaratone doesn't furnish any data). NASA function gives lightly best results for T< 1000 K respect to logarithmic polynomial, while the two functions represent the same percent error for 1000 <T< 3500 K.

In Fig. 4 the comparison is effected for CO_2. Annaratone formula gives good results for the same temperature range indicated in Fig. 1. For T< 1000 K the L.P. gives the same result of NASA function, while for T> 3500 K the L.P. furnish a very little percent error ($\varepsilon_{rel\,\%}$ < 0.05).

In Fig. 5 the comparison is effected for H_2. Annaratone formula gives a percent error <1% only for 300 <T< 1750 K. For T<1000 K L.P. gives a $\varepsilon_{rel\,\%}$ <0.25, and NASA function give a $\varepsilon_{rel\,\%}$ <0.15, while for T>1000 K the two polynomials have almost the same percent error.

In Fig. 6 the comparison is effected for CO. Annaratone formula gives good results for the same temperature range indicated for N_2 . For T< 1000 K L.P. gives a $\varepsilon_{rel\,\%}$ < 0.15, and NASA function gives a $\varepsilon_{rel\,\%}$ < 0.05, while for T> 1000 K L.P. gives a $\varepsilon_{rel\,\%}$ < 0.1, and NASA function gives a $\varepsilon_{rel\,\%}$ < 0.15.

In Fig. 7 the comparison is made for NO. Annaratone formula gives a percent error < 1.5% only for 300 <T< 2100 K. NASA furnishes only one function for T> 1000 with a $\varepsilon_{rel\,\%}$ < 0.2%, while L.P. gives a $\varepsilon_{rel\,\%}$ < 0.2% for the entire temperature range (273 <T< 3500 K).

Fig. 7 shows results for OH. Annaratone formula gives a percent error very high. NASA furnishes only one function for T> 1000 with a $\varepsilon_{rel\,\%}$ < 0.2%, while L.P. gives a $\varepsilon_{rel\,\%}$ < 0.3% for 273 <T< 1000 K, and a $\varepsilon_{rel\,\%}$ < 0.1% for 1000 <T< 3500 K.

EVALUATION OF GASES ENTHALPY

By integrating equation (1), and by assuming $\tilde{h}(T) = 0$ at T_0=273.16 K (T_0 = reference temperature), for each gas taken into account, it is possible to obtain the enthalpy function:

$$\tilde{h}(T) = \int_{273.16\,K}^{T} \tilde{c}_p(T)\,dT \tag{3}$$

$$\tilde{h}(T) = a_6 + \frac{T}{1000}\begin{bmatrix} a_7 + a_8(\log T) + a_9(\log T)^2 + \\ + a_{10}(\log T)^3 + a_{11}(\log T)^4 + \\ + a_{12}(\log T)^5 \end{bmatrix} \quad (4)$$

where a_6 depends only by reference temperature adopted, and:

$$a_7 = a_0 - a_1 + 2a_2 - 6a_3 + 24a_4 - 120a_5$$

$$a_8 = a_1 - 2a_2 + 6a_3 - 24a_4 + 120a_5$$

$$a_9 = a_2 - 3a_3 + 12a_4 - 60a_5$$

$$a_{10} = a_3 - 4a_4 + 20a_5$$

$$a_{11} = a_4 - 5a_5$$

$$a_{12} = a_5$$

In Tab. 2 the values of coefficient a_j (j=6,..,11) are shown. In equation (3), if $\tilde{c}_p(T)$ is expressed in [J/(mole K)], $\tilde{h}(T)$ from equation (4) will be expressed in [kJ/mole].

To confirm the good reliability of the results obtained with L.P.s (4), a comparison with enthalpy from JANAF tables was made. The comparisons with only technical air, CO_2, O_2 and OH are reported, since the percent error for the other gases is always < 0.1%).

In Fig. 9 is shown, respect to JANAF values, the percent errors on $\tilde{h}(T)$ for water vapor by using logarithmic polynomial. The percent error assumes an absolute max value of 0.025% for 2800 <T< 3500 K, while in the entire temperature range it is less than 0.01%.

In Fig. 10 is reported, respect to JANAF values, the percent errors on $\tilde{h}(T)$ for CO_2 . The percent error is very small and ranges between –0.0015% and 0.0062%.

In Fig 11. is reported, respect to JANAF values, the percent errors on $\tilde{h}(T)$ for Nitrogen. The percent error assumes a max value of 0.2% near T=400 K, while in all the temperature range it is less than 0.05%.

In Fig 12. is reported, respect to JANAF values, the percent errors on $\tilde{h}(T)$ for H. The percent error is always less than 0.01%.

SOME EXAMPLES

Let us consider, for example, the stoichiometric oxidation of hydrocarbon C_nH_m with technical air (21% O_2):

$$C_nH_m + \left(n + \frac{m}{4}\right)O_2 + \frac{79}{21}\left(n + \frac{m}{4}\right)N_2 = \\ = nCO_2 + \frac{m}{2}H_2O + \frac{79}{21}\left(n + \frac{m}{4}\right)N_2 \quad (5)$$

Let us consider now, an hypothetical adiabatic expansion for the combustion products, between two points (p_1, T_1; p_2, T_2) and let us evaluate the variation of enthalpy for the combustion products ($\Delta h_p = h_1 - h_2$), utilizing Annaratone formulas, NASA functions and L.P.s.

Let us consider three different fuels: a) methane, b)octene, c) 40 volume percent propane and 60 volume percent butane mixture.

a) stoichiometric oxidation of methane with technical air:

$$CH_4 + 2O_2 + \frac{79}{21}\cdot 2N_2 = CO_2 + 2H_2O + \frac{158}{21}N_2 \quad (6)$$

The enthalpy of the products is:

$$h_p(T) = [h(T)]_{CO_2} + 2[h(T)]_{H_2O} + \frac{158}{21}[h(T)]_{N_2} \quad (7)$$

If we consider perfect gases ,for an hypothetical expansion from T_1=1800 K and T_2=800 K, we have:

a1) from JANAF tables

$$\Delta h_p = 401.143\frac{kJ}{kg}$$

a2) from NASA functions

$$\Delta h_p = 400.881\frac{kJ}{kg}$$

a3) from L.P.s

$$\Delta h_p = 401.107\frac{kJ}{kg}$$

a4) from Annaratone formulas

$$\Delta h_p = 401.748\frac{kJ}{kg}$$

Now it is possible to evaluate the percent error with:

$$\varepsilon_{rel\%} = 100\cdot\frac{(\Delta h_p)_{JANAF} - \Delta h_p}{(\Delta h_p)_{JANAF}} \quad (8)$$

and the results are:

1. NASA: $\varepsilon_{rel\%}$=0.065%
2. LOG POL.: $\varepsilon_{rel\%}$=0.00897%
3. Annaratone $\varepsilon_{rel\%}$=0.15%

It is important to underline that the use of L.P.s furnishes the minimum percent error.

b) stoichiometric oxidation of octene with technical air:

$$C_8H_{16} + 12O_2 + \frac{948}{21}N_2 = 8CO_2 + 8H_2O + \frac{948}{21}N_2 \quad (9)$$

The enthalpy of the products is:

$$h_p(T) = 8[h(T)]_{CO_2} + 8[h(T)]_{H_2O} + \frac{948}{21}[h(T)]_{N_2} \quad (10)$$

If we consider perfect gases ,for an hypothetical expansion from T_1=1880 K and T_2=800 K, we have:

b1) from JANAF tables

$$\Delta h_p = 2341.42\frac{kJ}{kg}$$

b2) from NASA functions

$$\Delta h_p = 2340.06\frac{kJ}{kg}$$

b3) from L.P.s

$$\Delta h_p = 2341.326 \frac{kJ}{kg}$$

b4) from Annaratone formulas

$$\Delta h_p = 2344.828 \frac{kJ}{kg}$$

the results are:

4. NASA: $\varepsilon_{rel\%}$=0.058%
5. POL. LOG: $\varepsilon_{rel\%}$=0.0040%
6. Annaratone $\varepsilon_{rel\%}$=0.145%

Also in this example, the use of L.P.s furnish the minimum percent error.

c) stoichiometric oxidation of 40 volume percent propane and 60 volume percent butane mixture

$$0.4C_3H_8 + 0.6C_4H_{10} + 11.8O_2 + 22.19N_2 = \\ = 3.6CO_2 + 4.6H_2O + 22.19N_2 \quad (11)$$

The enthalpy of the products is:

$$h_p(T) = 3.6[h(T)]_{CO_2} + 4.6[h(T)]_{H_2O} + 22.19[h(T)]_{N_2} \quad (12)$$

If we consider perfect gases ,for an hypothetical expansion from T_1=1880 K and T_2=800 K, we have:

c1) from JANAF tables

$$\Delta h_p = 1161.99 \frac{kJ}{kg}$$

c2) from NASA functions

$$\Delta h_p = 1161.25 \frac{kJ}{kg}$$

c3) from L.P.s

$$\Delta h_p = 1161.86 \frac{kJ}{kg}$$

c4) from Annaratone formulas

$$\Delta h_p = 1163.64 \frac{kJ}{kg}$$

and obtain:

7. NASA: $\varepsilon_{rel\%}$=0.0636%
8. POL. LOG: $\varepsilon_{rel\%}$=0.00112%
9. Annaratone $\varepsilon_{rel\%}$=0.142%

CONCLUSIONS

In this work many polynomial functions have been valued to determine the best matching function respect to experimental data. The polynomial functions investigated was: simple polynomials (third, fourth and fifth order), exponential polynomials (third, fourth and fifth order) and L.P.s (third, fourth and fifth order).

The best polynomial functions was a V order Logarithmic Polynomial, and experimental data from JANAF tables, has been matched by using the least square method, to find the coefficients for the new L.P.s.

For each gases (Technical air (21% O_2), N, O, H, H_2, O_2, N_2, CO, OH, NO, CO_2 and H_2O), has been calculated a L.P. to fit experimental data on specific heat at constant pressure. The L.P. gives the values of specific heat as a function of temperature, for 273 <T< 3500 K.

By integrating the specific heat logarithmic polynomial, for each gas, was determined another L.P. which furnishes the gas enthalpy as a function of temperature. The coefficients of enthalpy L.P. are a combination of the coefficients of the specific heat logarithmic polynomial.

The results carried out with L.P.s have been compared with those from other mathematical functions that are present in literature (as, for example, the NASA Equilibrium Code or Annaratone formulas). The new L.P.s pointed out a major precision respect to the others polynomial functions, and the possibility to utilise a single L.P. for a wide temperature range, according to a good accuracy with experimental data.

The comparison was effected by evaluating the percent error respect to JANAF data.

The use of L.P.s have always shown a minor percent error respect to the other mathematical function that are present in literature. Sometime the NASA polynomials furnishes a percent error lightly small, but the L.P.s presents another advantage respect to the NASA polynomials. In fact the NASA Equilibrium Code utilizes two different polynomials to cover the entire temperature range, while L.P.s utilize a single polynomial to cover the same temperature range. These results are very important in order to advantage time calculations in computer applications.

REFERENCES

[1] Cox J.D., Wagman D.D., Medvedev V.A. "CODATA Key Values for Thermodynamics", Hemisphere Publishing Corp., New York, 1984.

[2] Strehlow Roger A. *Combustion Fundamentals* - Mc Graw Hill – 1985

[3] JANAF *Thermochemical Tables* 2d ed., NSRDS-NB537, U.S. National Bureau of Standards, June 1971.

[4] Chase M.W., *JANAF Thermochemical Tables, 3rd Edition*, American Chemical Society, 1986.

[5] Chase M.W., *NIST-JANAF Thermochemical Tables, Fourth Edition*, J. Phys. Chem., 1998.

[6] Acton O. e Caputo C. *Introduzione allo studio delle macchine* UTET – 1979.

[7] Hinselrath J., *Tables of thermal properties of gases* - U.S. National Bureau of Standards, Circ. 564, New York 1955.

[8] Lutz O. *Enthalpien, Entropien und Gleichgewichtskonstanten von Verbrennungs-gasen*, "Ingenieurs Archiv", vol. XVI, 1948.

[9] Gordon S. and McBride B.J., *Computer Program for the Calculation of Complex*

Chemical Equilibrium Composition, Rocket Performance, Incident and Reflected Shock and Chapman-Jouguet Detonation, NASA publication SP-273. 1971 (NTIS number N71-37775).

[10] Heywood John B. *Internal Combustion Engine Fundamentals* – Mc Graw Hill – 1988.

[11] Messina Michele. *Misure anemometriche a filo caldo in flussi a bassi livelli di turbolenza* – Tesi di Laurea – Relatore chiar.mo Prof. G. GUJ - Univ. Roma La Sapienza – Ing. Aeronautica – A.A.1991/92.

[12] Caputo C. *Un'espressione generale dell'entalpia dei gas tecnici dei cicli ad alta temperatura , mediante calcolatore elettronico.* La Termotecnica n. 10, 1961.

[13] Spencer H. M., Flannagan G. N. *Empirical Heat Capacity equations of Gases* – J. Am. Chem. Soc. – vol. 64, 1942.

[14] Svehla R., McBride B.J., *Fortran IV Computer program for the calculation of thermodynamic and transport properties of complex chemical systems* NASA technical note TND-7056,1973 (NTIS number N73-15954).

[15] D. Annaratone *Entalpia dei gas* La Termotecnica – Settembre 1999 – pp. 113 – 116.

Species	*T* range, K	a_0	a_1	a_2	a_3	a_4	a_5
21% O_2 air	273-3500	-3686.11308	2925.733112	-908.222217	138.9132785	-10.4727361	0.311791491
CO_2	273-3500	-1412.367846	1288.467702	-452.8119749	77.54809369	-6.435215333	0.207543752
H_20	273-3500	-11780.76495	8490.521798	-2414.775747	339.3366171	-23.54276815	0.645407282
H_2	273-3500	-5839.716554	3996.088289	-1073.861728	142.1986039	-9.271100003	0.23815852
O_2	273-3500	10228.342599	-7184.923331	2010.868084	-279.694958	19.348226	-0.532569
N_2	273-3500	-7513.364197	5708.380466	-1712.173896	254.2955425	-18.69983725	0.544972495
CO	273-3500	-4919.053212	3862.933492	-1190.846974	181.1646581	-13.60313045	0.403719823
OH	273-3500	-8303.673127	5947.981594	-1680.64201	234.936416	-16.24988335	0.445242635
NO	273-3500	-465.5673652	679.4364191	-285.7975952	53.25317371	-4.613952483	0.152334163
O	273-3500	-746.389217	619.7416065	-195.9792288	30.51244169	-2.347964376	0.071623359
H	273-3500	20.79	0	0	0	0	0
N	273-3500	-2789.24708	102.982819	-627.602958	93.36531882	-6.92398505	0.204789531

Tab. I: Coefficients for $\tilde{c}_p$ (T) logarithmic polynomial ($\tilde{c}_p$ =[J/K mole]).

Species	*T* range, K	a_6	a_7	a_8	a_9	a_{10}	a_{11}
21% O_2 air	273-3500	15.21839557	-9550.530943	5864.417863	-1469.342376	187.0400529	-12.0316936
CO_2	273-3500	0.015703929	-4251.098478	2838.730632	-775.1314651	107.4398301	-7.472934091
H_20	273-3500	77.80511569	-27779.33326	15998.5683	-3754.023253	446.4158353	-26.76980456
H_2	273-3500	39.28358659	-13087.80535	7248.088792	-1626.000251	184.0461743	-10.4618926
O_2	273-3500	-86.26450136	23641.43755	-13413.09495	3114.085809	-367.7392415	22.01107094
N_2	273-3500	43.118457	-18686.0585	11172.69431	-2732.15692	339.9943414	-21.42469973
CO	273-3500	23.53911025	-12625.59011	7706.536898	-1921.801703	243.6515764	-15.62172957
OH	273-3500	53.61212551	-19465.98355	11162.31043	-2607.164416	308.8408021	-18.47609653
NO	273-3500	-10.19573324	-2165.132976	1699.565611	-510.0645959	74.75566691	-5.3756233
O	273-3500	-1.685571986	-2006.109879	1259.720662	-319.9895279	41.33676639	-2.706081173
H	273-3500	-5.678910048	20.78816654	0.001027372	-0.000301167	4.2199E-05	-2.6645E-06
N	273-3500	14.17266913	-6898.378114	4109.131033	-1003.074107	125.1570496	-7.947932704

Tab. II: Coefficients for $\tilde{h}$ (T) logarithmic polynomial ($\tilde{h}$ =[kJ/ mole]).

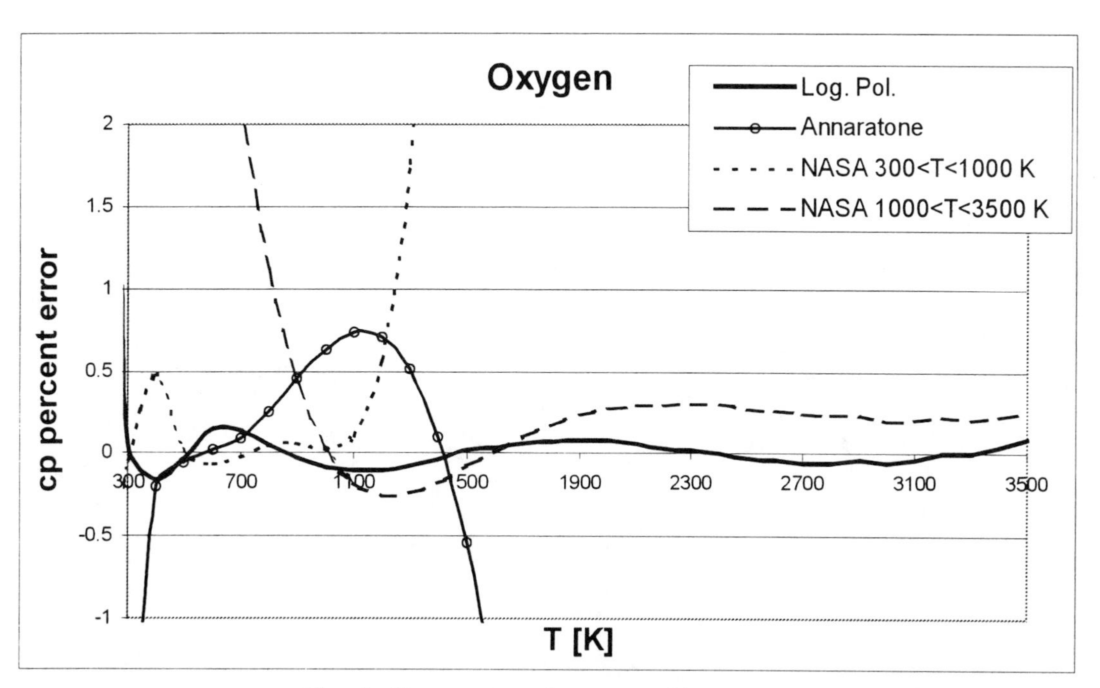

Fig. 1: Comparison between $\tilde{c}_p$ percent errors.

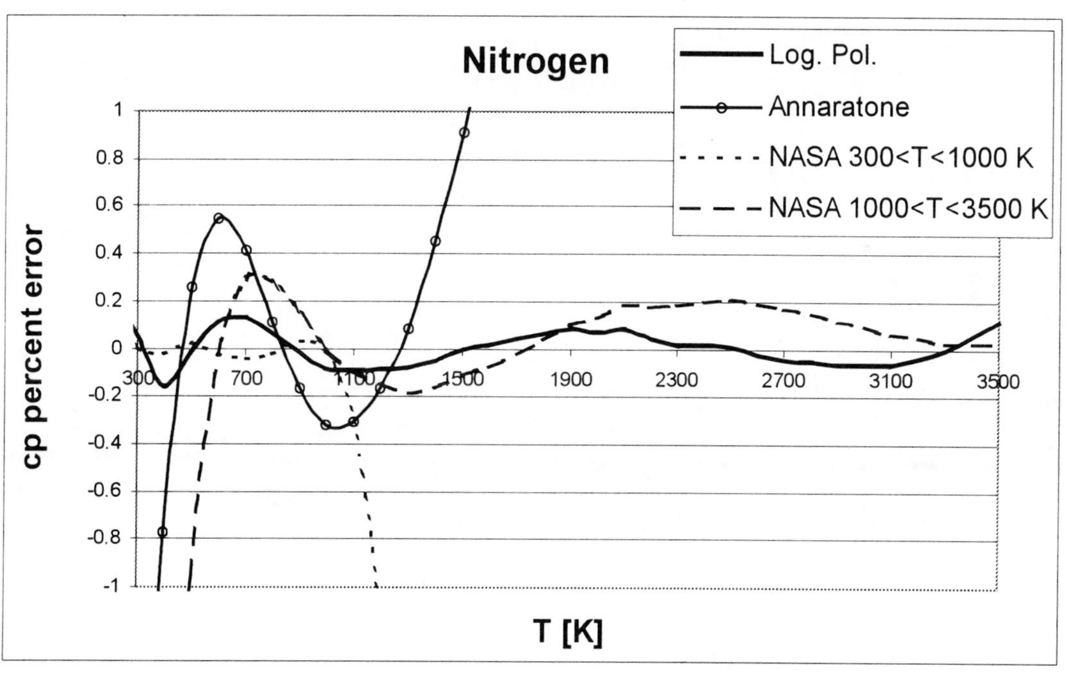

Fig. 2: Comparison between $\tilde{c}_p$ percent errors.

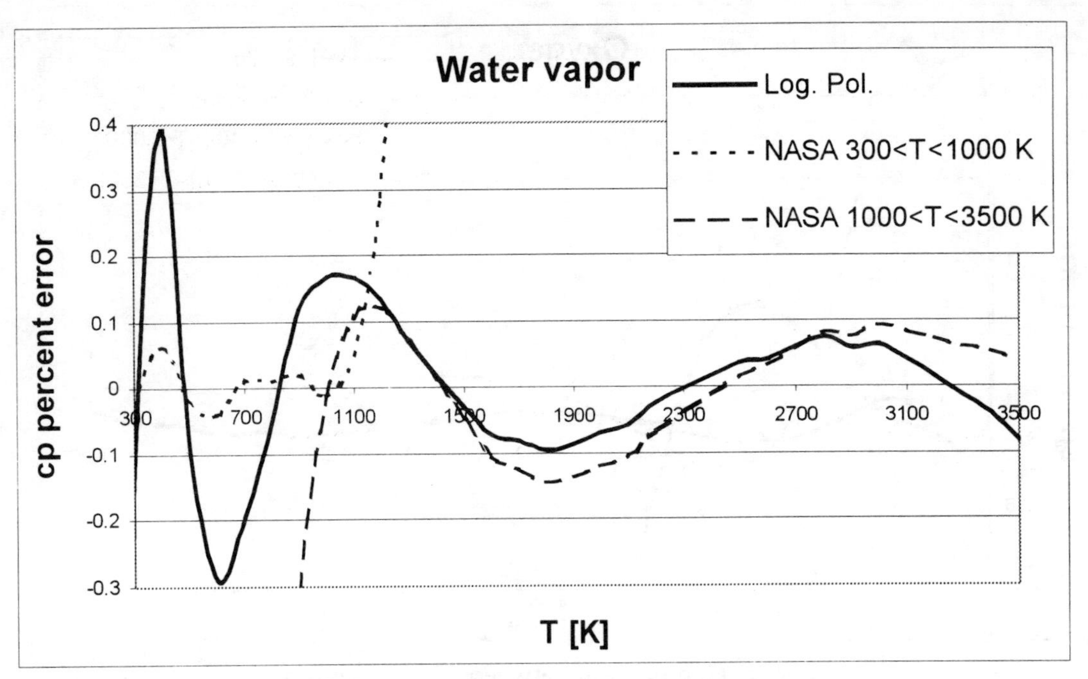

Fig. 3: Comparison between $\widetilde{c}_p$ percent errors.

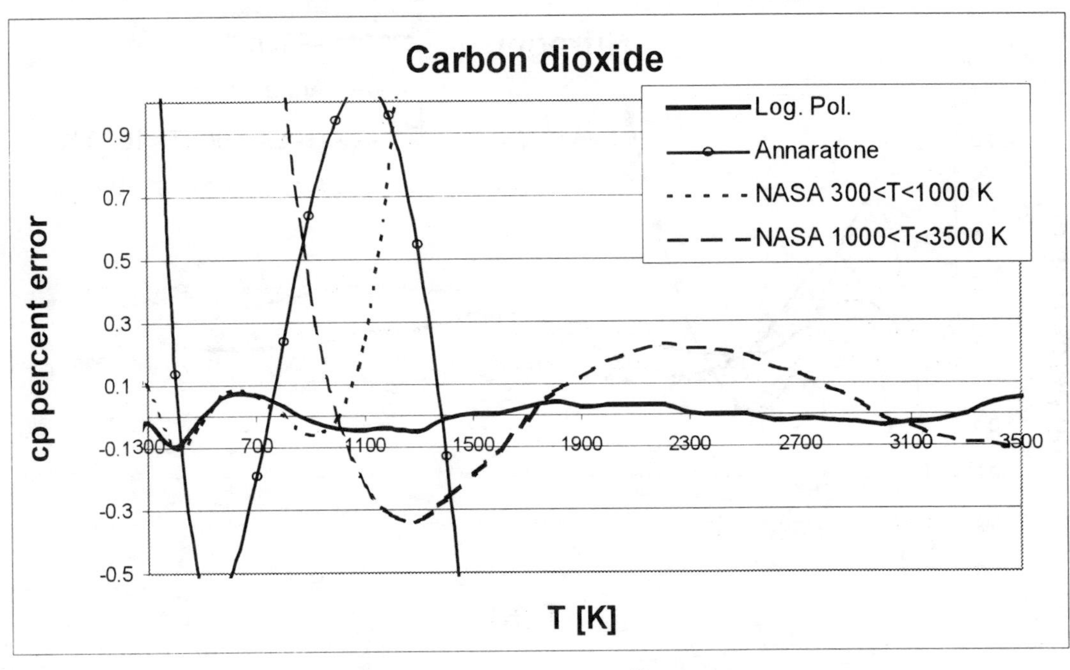

Fig. 4: Comparison between $\widetilde{c}_p$ percent errors.

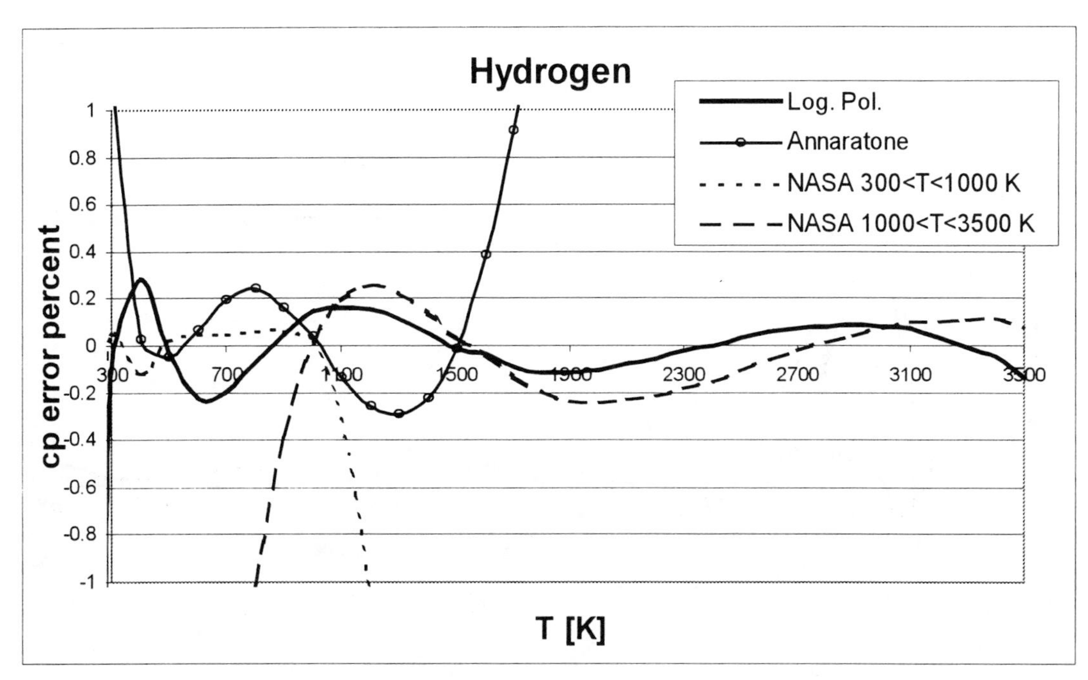

Fig. 5: Comparison between $\tilde{c}_p$ percent errors.

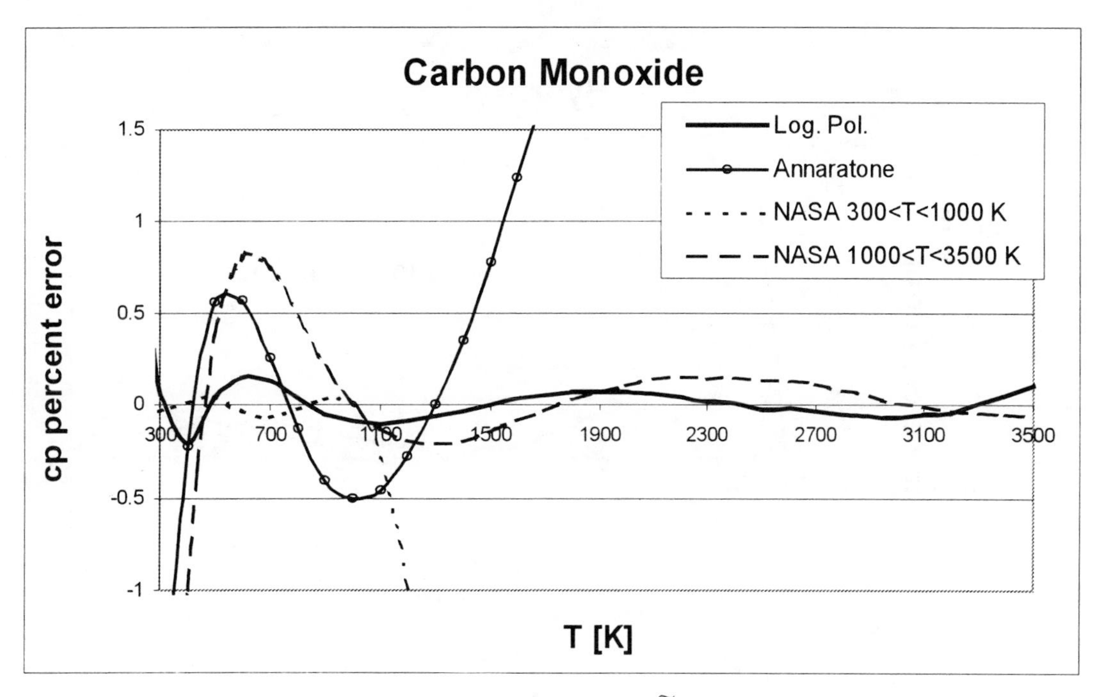

Fig. 6: Comparison between $\tilde{c}_p$ percent errors.

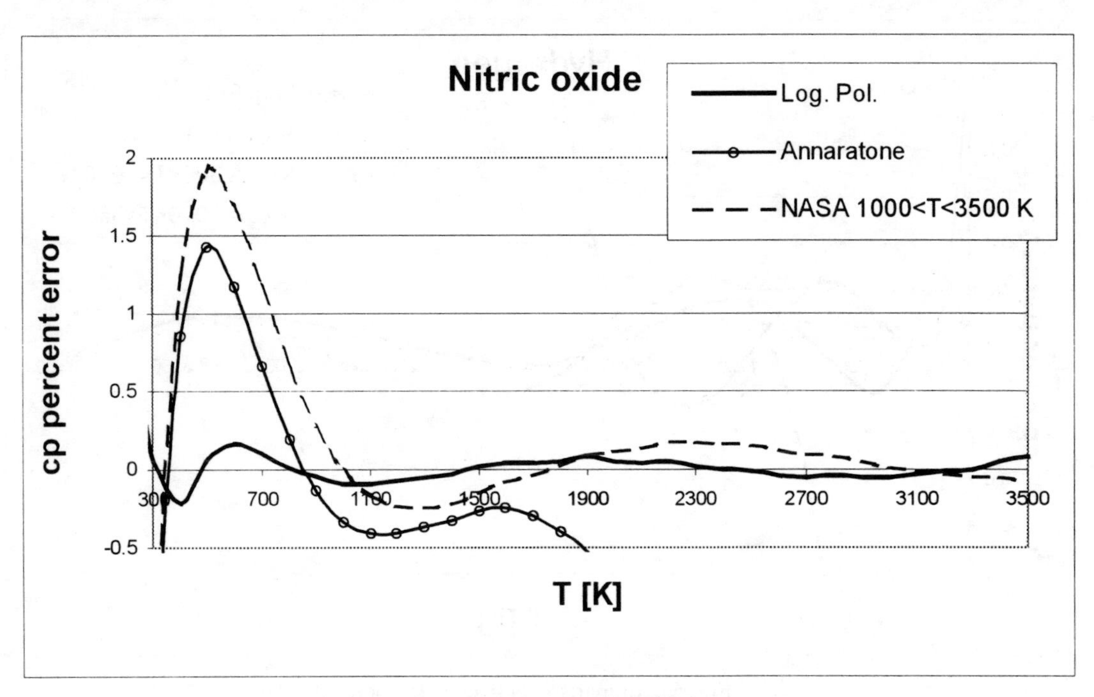

Fig. 7: Comparison between $\widetilde{c}_p$ percent errors.

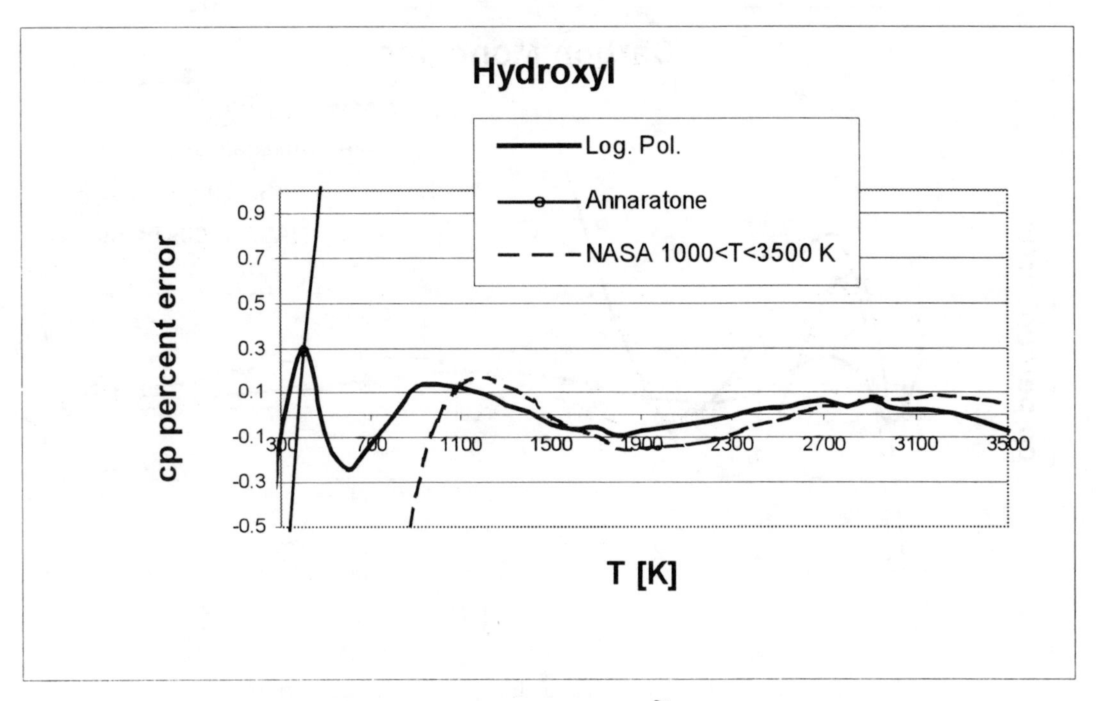

Fig. 8: Comparison between $\widetilde{c}_p$ percent errors.

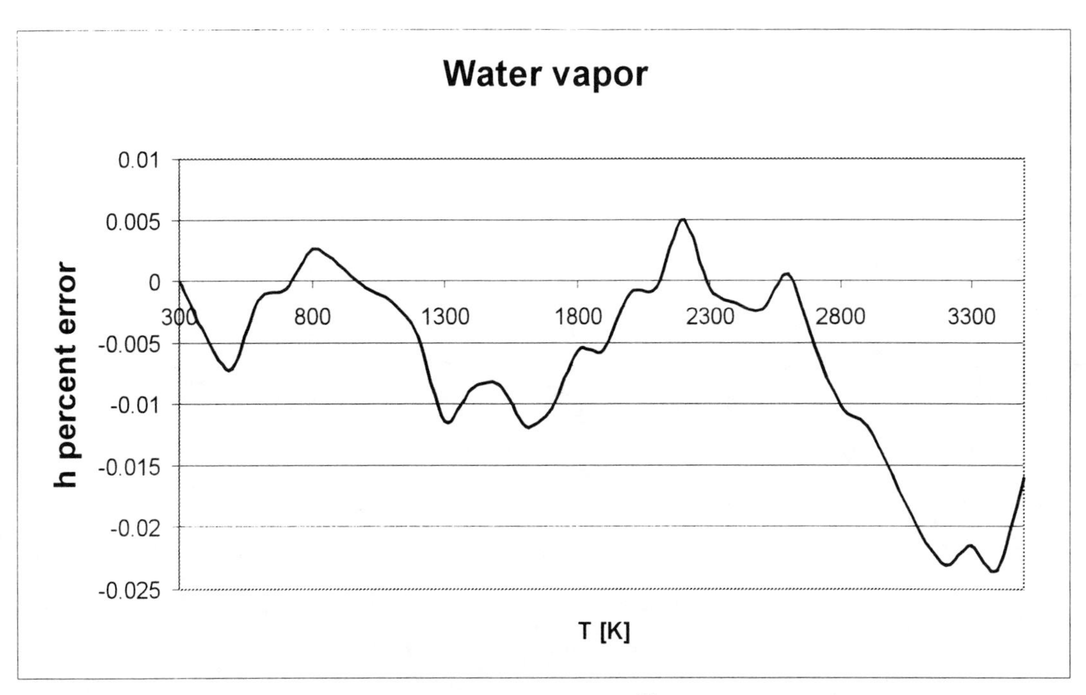

Fig. 9: Comparison between $\widetilde{h}$ percent errors.

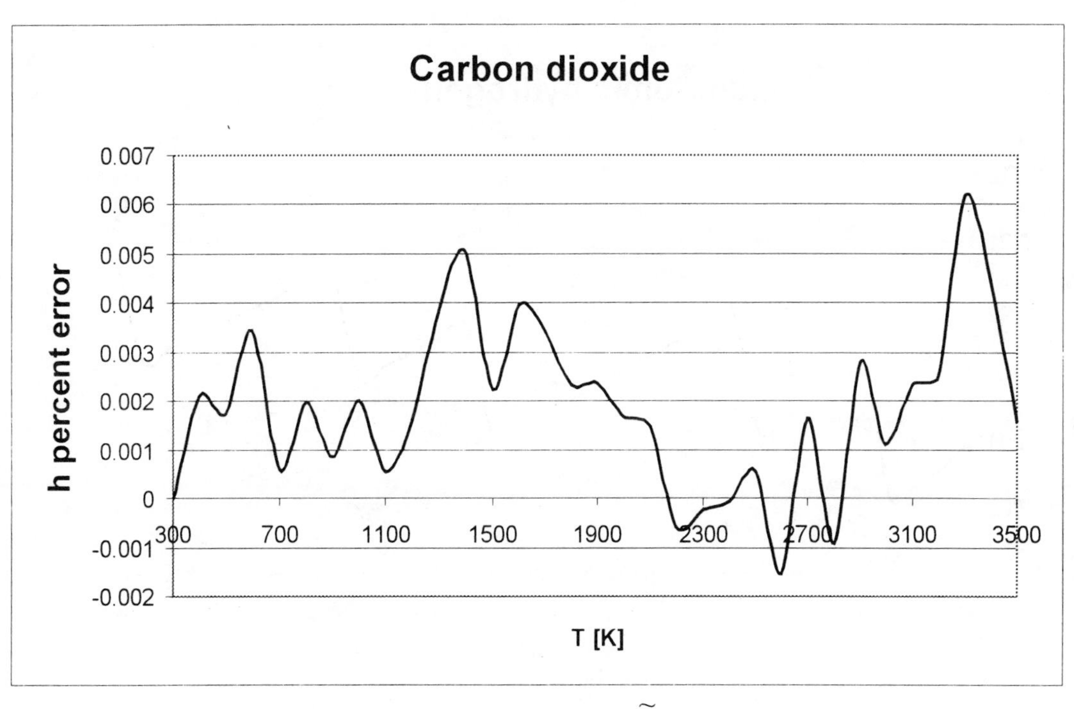

Fig. 10: Comparison between $\widetilde{h}$ percent errors.

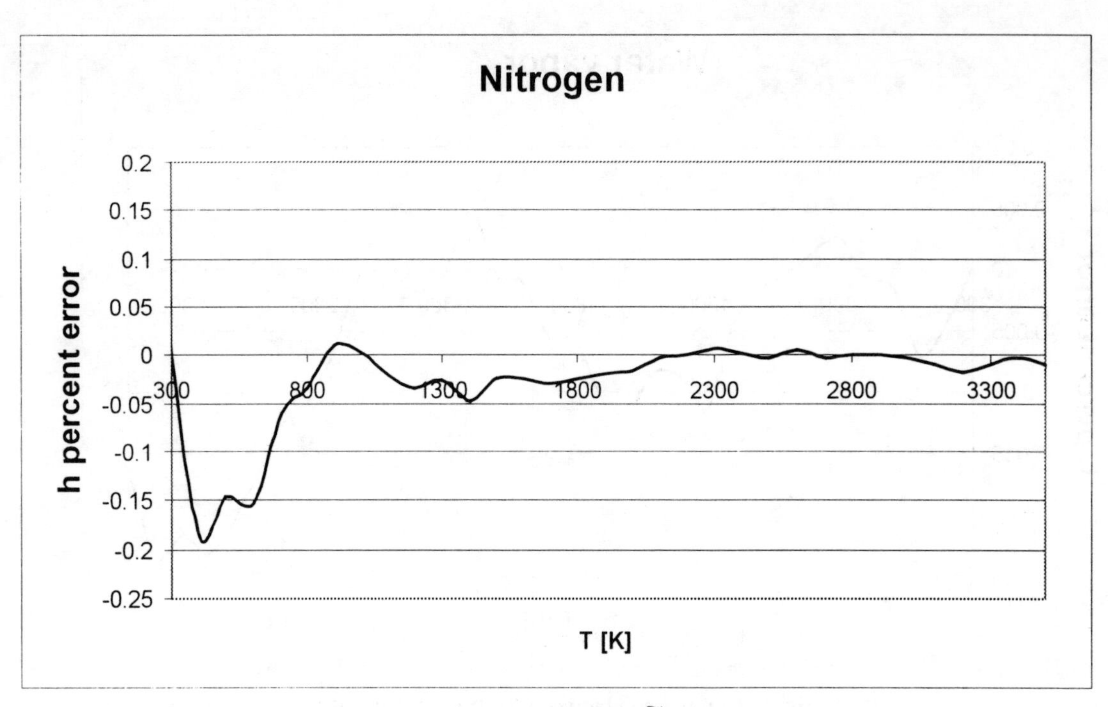

Fig. 11: Comparison between $\widetilde{h}$ percent errors.

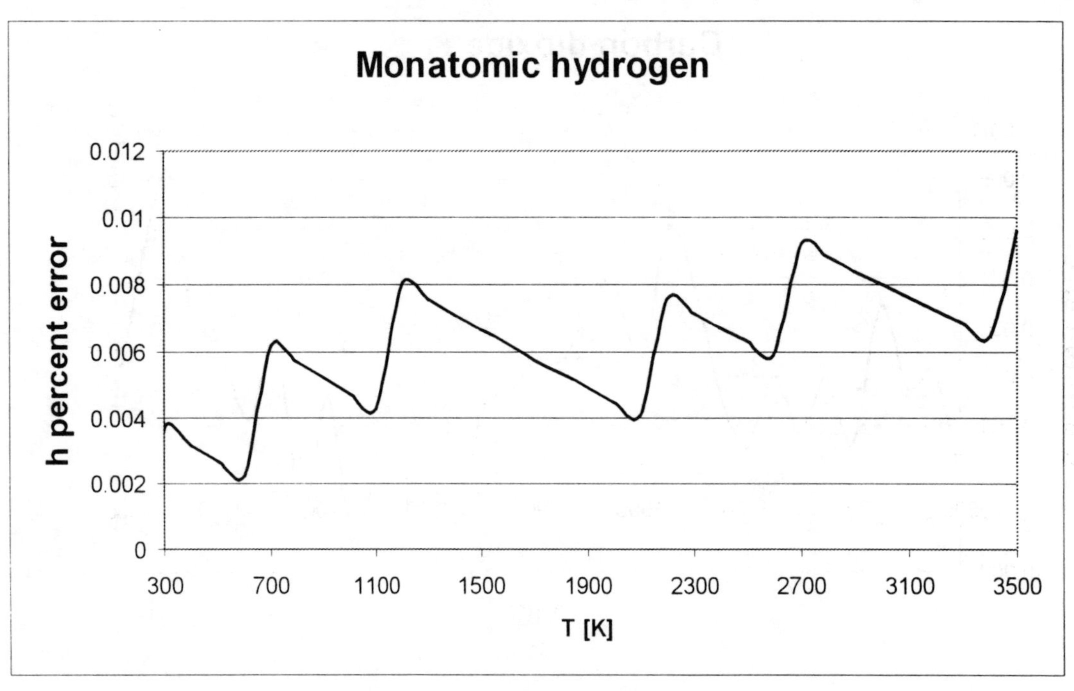

Fig. 12: Comparison between $\widetilde{h}$ percent errors.

AIAA-2000-2852

SYSTEM IDENTIFICATION AND CONTROL DESIGN OF AN ALTERNATIVE FUEL ENGINE FOR HYBRID POWER GENERATION

Jose Bustamante*, Bill Diong† and Ryan Wicker‡
The University of Texas at El Paso, El Paso, TX 79968

ABSTRACT

A throttle control system has been developed for a small V-twin engine used in a hybrid AC power generation system that incorporates two time variable power sources: a wind turbine and an internal combustion engine-generator combination. The control system utilizes a throttle actuator, a proportional integral derivative (PID) controller and software that determines desired power outputs (engine speeds) based on real-time measurements. A parallel approach to the design of the PID controller for system stability and proper performance is described. Transient response tests were performed from which suitable PID gain settings were selected by trial and error. In addition, frequency response tests were performed to develop models of the engine (system identification) at different operating points, which can then be used to analytically design the gains of the PID controller.

INTRODUCTION

Wind power is the fastest growing renewable energy technology in the world today for electric power generation. It has recently experienced annual growth rates of over 20% [1]. However, it is time variable. Thus, if wind energy is to be used for producing firm power, an additional source of power is required.

In hybrid power systems a renewable power source, like the wind, is combined with another power source that provides supplemental power as required. There are various hybrid power system strategies that combine renewable energy sources and firm power backup. A common configuration is called a wind-diesel system, which combines wind as a renewable energy source and a diesel-generator set as a source of firm power [2]. Current research in wind-diesel systems includes investigations of start and stop engine strategies, engine fuel consumption optimization, and energy storage devices. The research is aimed at maximizing the utilization of the renewable energy while minimizing the requirement for the supplemental diesel generated power. However, diesel engines have high fuel consumption at low load (many manufacturers recommend a minimum load of 40% [3]). Furthermore, use of diesel engines increases air pollution and the demand for foreign oil. If an internal combustion engine running on a renewable fuel, such as ethanol, replaces the diesel engine, overall exhaust emissions can be reduced. Furthermore, the use of ethanol can lead to energy independence and perhaps additional agricultural jobs. With this strategy, two renewable energy sources (wind and ethanol) are combined to provide the necessary power. However, as with wind-diesel systems, engine control strategies must be investigated for the system to meet the required load while minimizing fuel consumption.

Researchers with the Engines and Alternative Fuels Research Laboratory (EAFRL) at the University of Texas at El Paso (UTEP) are currently developing a hybrid power generation system incorporating two time-variable three-phase AC power sources: a wind turbine and an internal combustion engine-generator combination. These two sources are integrated using a power electronics package that synchronizes the generated power with the local utility distribution grid.

As part of a laboratory-scale demonstration, EAFRL researchers have developed a computer-controlled engine-generator combination that is installed in a wind turbine test facility. A 20 hp V-twin engine connected to a 10 hp generator provides supplemental power. A computer-controlled throttle control system responds to variations in power generated by the wind turbine simulator, maintaining the total power produced by the hybrid power system within some small tolerance of the desired power output.

---

* Graduate student, Department of Mechanical Engineering
† Assistant Professor, Department of Electrical Engineering.
‡ Assistant Professor, Department of Mechanical Engineering

The throttle control system consists of a throttle actuator, a proportional integral derivative (PID) controller, and software that determines desired power outputs (engine speeds) based on real-time measurements. This arrangement provides fast and stable engine response to load changes, while maintaining precise speed regulation. The development of the throttle control system is described in the following. In addition, design of the PID controller for system stability and proper performance is discussed.

HYBRID POWER GENERATION CONFIGURATION

SYSTEM OBJECTIVE AND DESCRIPTION

The UTEP hybrid power system provides three-phase, 480 volts, 60 Hz, utility-grade electric power. This power is to be provided by controlling and combining the power from two time-variable sources (the wind turbine and the engine-generator) using a power electronics package.

The principal components of the UTEP hybrid power system are illustrated in Figure 1. They consist of the engine-generator, the engine-generator control and interface module (EGCIM), the wind turbine control system (WTCS) and a supervisory personal computer with serial links to the EGCIM and the WTCS. Also shown in the figure are the wind turbine test bed and its separate supervisory PC.

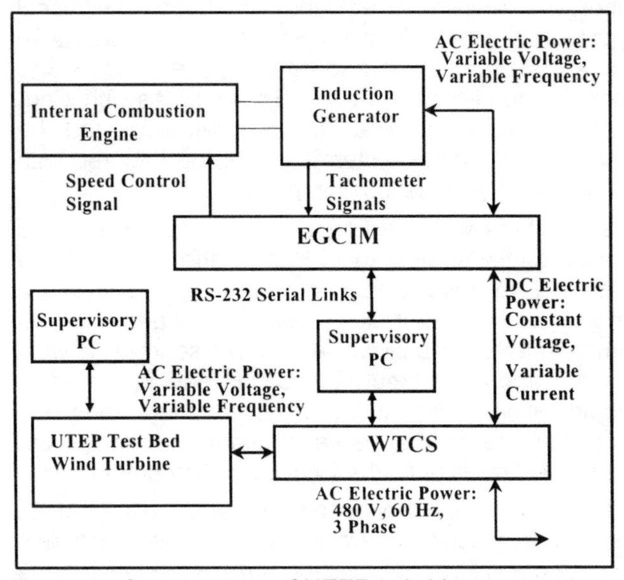

Figure 1. Components of UTEP hybrid power system.

COMPONENTS DESCRIPTION

The function of the engine-generator control and interface module (EGCIM) is to convert the variable-voltage, variable-frequency power from the 10 hp generator driven by the engine to DC voltage at a constant nominal level of 650 VDC. The EGCIM controls the power produced by the engine-generator in response to a command signal from the WTCS. After receiving the command signal, the EGCIM generates and transmits the engine speed control signal to the engine throttle control system.

The wind turbine control system (WTCS) is a four-quadrant power conversion and control system, consisting of a generator inverter, a DC bus with capacitive energy storage and a utility inverter. The DC bus isolates the "clean" utility side power from the incoming "dirty" power generated by either the wind turbine or the engine-generator. The output of the WTCS is utility-grade 3-phase, 480 volts, 60 Hz power synchronized to the grid voltage, frequency, and phase.

For the laboratory-scale power system, the power produced by the wind is simulated using a 25 hp Baldor motor controlled by a Cegelec adjustable speed drive (see Figure 2). The 25 hp motor drives a 10 hp Baldor three-phase, Y-connected, four-pole, 480 volts vector drive induction motor. The output from the 10 hp motor, acting as a generator, is connected to the WTCS, thus providing variable speed power generation. The power generated by the wind turbine test bed is controlled to follow experimentally measured wind schedules.

Figure 2. UTEP wind turbine test bed.

The engine-generator combination is shown in Figure 3. The engine selected for the experiments is an air-cooled, overhead valve Briggs & Stratton 20 hp V-twin engine. The engine is rated to provide 20 hp at 3600 rpm and 40 N-m at 2700 rpm. The engine is coupled to a 10 hp Baldor motor, similar to the one described previously, using a 1.453:1 Falk speed reducer. The speed reducer is required to match the speed requirements and characteristics of the engine-generator system. Since the speed range of the generator that can be accommodated by the EGCIM is 900 rpm to 2200 rpm, the reduction ratio allows the engine to operate between the manufacturers recommended speed range (1800-3600 rpm).

EXPERIMENTAL SETUP

DATA ACQUISITION

Data acquisition and the reference speed signal to the controller are implemented using a National Instruments Lab-PC+ data acquisition (DAQ) board and LabVIEW.

The actual engine speed signal provided from the magnetic pickup is sampled using differential inputs at a frequency of 5000 Hz, to resolve the rpm of the engine. The flywheel mounted on the shaft of the engine has one tooth that produces a high amplitude voltage signal when it passes under the magnetic pickup. This signal is used to determine the point where the engine has completed one revolution. This signal is then converted to its corresponding engine rpm by a computer program. The reference speed signal sent to the controller consists of a unipolar voltage produced by the DAQ board. The voltage range that the controller accommodates is between 0 to 5 volts. The controller is programmed so that 0 volts corresponds to 1800 rpm and 5 volts corresponds to 3600 rpm. Any desired speed within this range is set by varying the voltage signal sent to the controller.

Figure 3. Engine-generator with gearbox.

CONTROL SYSTEM

The engine-generator is controlled using a throttle control system. The principal components of the control system are shown in Figure 4 and include a throttle actuator, a PID controller, and software. Engine speed is controlled by a voltage signal sent to the PID controller from a PC. The PID controller converts this reference signal to a corresponding engine rpm and compares it to the actual engine speed which is obtained by a magnetic pickup (MPU) sensor mounted on the flywheel of the engine (as described previously). The PID controller makes the necessary adjustment to the engine speed via a DC motor that is used as a linear throttle actuator for throttle position control (shown in Figure 5). The throttle actuator is electrically connected to the PID controller and mechanically connected to the throttle valve, and replaces the mechanical governor of the engine. The throttle control system setup is based on similar control arrangements used in drive by wire systems [4,5,6,7,8].

The PID controller is a DYNA-I Digital Controller manufactured by the Barber-Colman Company. The PID controller outputs a control signal that is a linear combination of the amount of engine offspeed (proportional), the time integral of engine offspeed (integral), and the rate of change of engine offspeed (derivative). This allows the controller to maintain the engine within a desired tolerance of the set speed and obtain good transient response to speed setting changes.

A PID controller is a linear system modeled by the following equations.

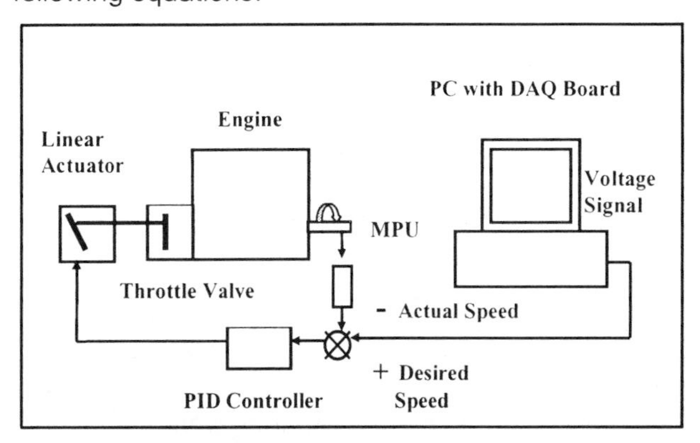

Figure 4. Schematic of throttle control system.

Figure 5. Linear throttle actuator configuration.

$$\frac{Y_C(s)}{E(s)} = K_P + \frac{K_I}{s} + K_D s \quad (1)$$

$$y_C(t) = K_P e(t) + K_I \int e(t) d\tau + K_D \frac{de(t)}{dt} \quad (2)$$

where,

$Y_c(s)$: Laplace transform of the controller output $y_c(t)$.
$E(s)$: Laplace transform of the error signal $e(t)$.
K_P, K_I, K_D: Proportional, Integral, and Derivative gains.
s: Complex frequency.

Equation (1) represents the frequency domain behavior of the PID controller, while equation (2) is the time domain description of the controller. The proportional, integral and derivative gain settings need to be selected properly to obtain a stable and properly performing system; otherwise in the worst case, the system will be unstable.

CONTROL SYSTEM DESIGN

A parallel approach using transient response and frequency response tests was adopted for the selection of the gains of the PID controller. Transient response tests were performed from which sets of acceptable PID gain settings were selected by trial and error based on overshoot/undershoot, settling time, and steady-state error requirements. Frequency response tests were performed to develop models of the engine (system identification) at different operating points. These models can then be used to design the gains of the PID controller. The final outcome is a throttle control system that allows the engine-generator combination to perform properly as part of the hybrid power generation system.

TRANSIENT RESPONSE TESTING

The trial-and-error approach used to find the set of gain settings that give the best performance consisted of varying the proportional (K_P), integral (K_I), and derivative (K_D) gains while performing a predetermined matrix of tests involving instantaneous step changes in speed and load. The test matrix comprised two speed step tests and two load step tests. Minimum and medium speeds were defined as 2000 rpm and 2600 rpm, respectively. Minimum and medium loads were defined as 13 ft-lb and 21 ft-lb, respectively. These values of the speeds and loads used in the test matrix were selected based on the requirements of driving the given generator. Step response tests were conducted using these speed and load conditions. The test matrix was devised to cover the worst transient response of the engine system based on preliminary work done on this engine. The first test conducted involved a step change in speed from minimum to medium speed while applying a minimum load to the engine. The second test consisted of a step change in speed from medium to minimum speed while the engine was subjected to a minimum load. The third test conducted involved a step change from minimum to medium load with the engine running at medium speed. Initially a fourth test involving a step change from medium to minimum load with the engine running at medium speed was going to be conducted, but due to engine related problems, the test has not yet been performed. Each of the above tests involved the use of different gain settings for the PID controller.

The test responses were then compared based on the criteria of maximum overshoot or undershoot, settling time, and steady-state error. For a speed step test, ideal performance means the engine speed follows the change in speed command with 0% overshoot or undershoot, 0 settling time, and 0 steady-state error. For a load step test, the engine speed should ideally neither droop nor rise as the load changes.

K_P	K_I	K_D
20%	3%	0%
20%	33%	0%
20%	3%	33%
40%	3%	0%
60%	3%	0%

Table 1. Gains for transient response tests.

- Tests Involving a Change in K_P

The first set of tests conducted was to determine how a change in the proportional gain affected the engine system response. The values used for K_P were 20%, 40%, and 60%. For these tests the integral and the derivative gains were kept at the minimum values allowed by the Barber-Coleman PID controller. The minimum value for K_I was 3% and for K_D was 0%. The responses shown in Figure 6 are the results of applying a step increase in speed command from minimum (2000 rpm) to medium (2600 rpm) speed with a minimum load (13 ft-lb) on the engine. One can see that no significant overshoot is present in the responses for the three different settings of K_P (20%, 40%, and 60%). One can also see that increasing K_P increases the settling time of the system. In addition, increasing K_P also causes an increase in the steady-state error and in the steady-state ripple. Steady-state ripple is the deviation from the constant value of the steady-state response.

The responses shown in Figure 7 are the results of a step decrease in speed from medium (2600 rpm) to minimum (2000 rpm) speed while applying a minimum load (13 ft-lb) to the engine. One can see that increasing K_P reduces the undershoot of the engine system. Increasing K_P also reduces the settling time and the steady-state error present in the system.

The responses shown in Figure 8 are the results of a step increase in load from minimum (13 ft-lb) to medium (21 ft-lb) load while the engine was running at medium speed (2600 rpm). This figure includes the plots of the engine speed and load (torque). One can see that increasing K_P decreases the undershoot in speed that occurs at the instant the step change in load is applied. Increasing K_P also reduces the settling time of the system. In addition, increasing K_P reduces the steady-state error of the engine system.

- Tests Involving a Change in K_I

A second set of tests was conducted to determine how a change in the integral gain affected the engine system response. The values used for K_I were 3% and 33%.

For these tests the proportional and the derivative gains were kept at the minimum values allowed by the PID controller. A K_P value of 20% was used since this was the minimum for which a response from the controller was obtained. The minimum K_D value used was 0%.

The responses shown in Figure 9 are the results of a step increase in speed from minimum (2000 rpm) to medium (2600 rpm) speed while applying a minimum load (13 ft-lb) to the engine. One can see that increasing K_I increases the overshoot. Increasing K_I also increases the settling time. In addition, a significant steady-state ripple is present when K_I is increased to 33%.

The responses shown in Figure 10 are the results of a step decrease in speed from medium (2600 rpm) to minimum (2000 rpm) speed with a minimum engine load (13 ft-lb). One can see that increasing K_I reduces the undershoot present in the system. Increasing K_I also decreases the settling time of the system and decreases the steady-state error.

The responses shown in Figure 11 are the results of a

step increase in load from minimum (13 ft-lb) to medium (21 ft-lb) load while the engine was running at medium speed (2600 rpm). Included in Figure 11 are plots of the engine speed and torque values. One can see that increasing K_I reduces the undershoot in speed caused by the step change in load. Increasing K_I also reduces the settling-time and the steady-state error.

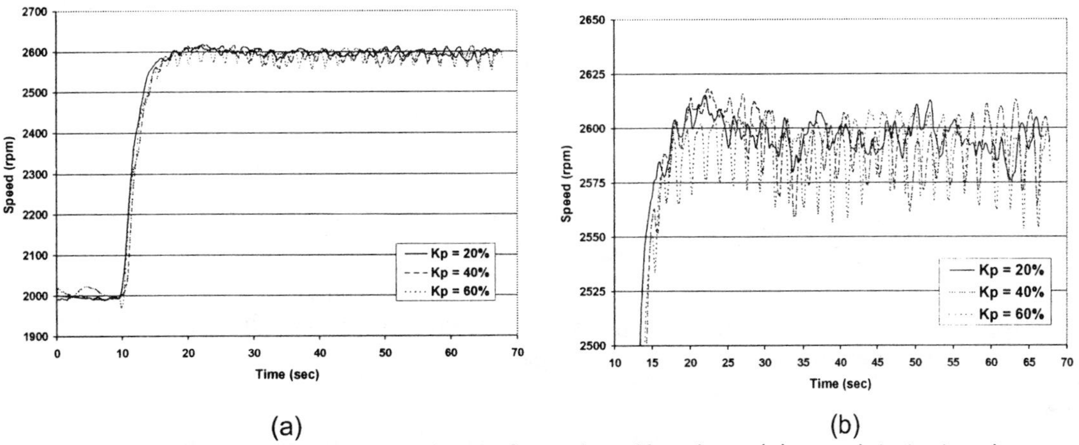

(a) (b)

Figure 6. Step increase in speed tests for various K_P values: (a) complete test region and (b) zoom-in test region.

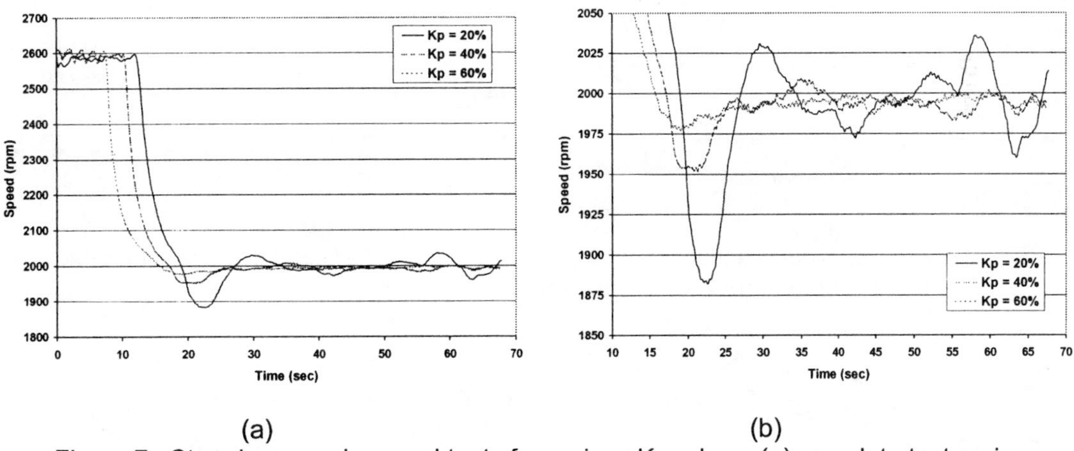

(a) (b)

Figure 7. Step decrease in speed tests for various K_P values: (a) complete test region and (b) zoom-in test region.

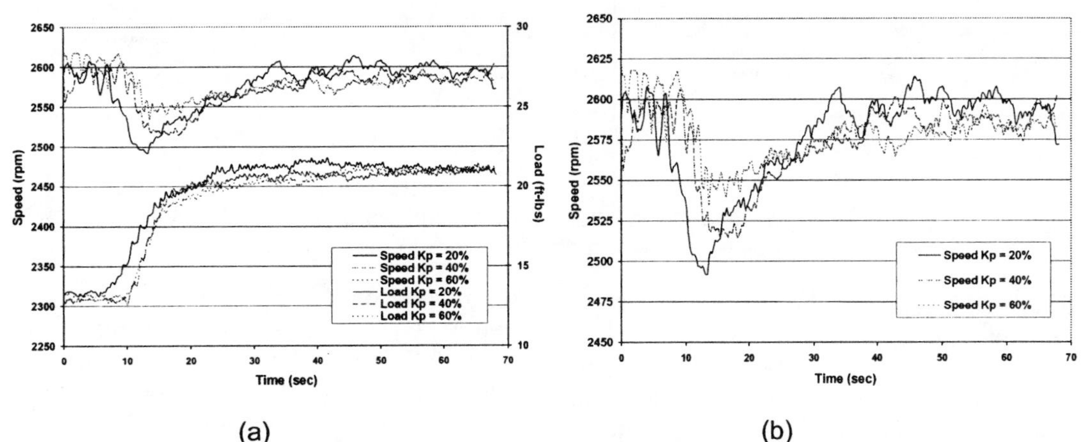

(a) (b)

Figure 8. Step increase in load tests for various K_P values: (a) complete test region and (b) zoom-in test region.

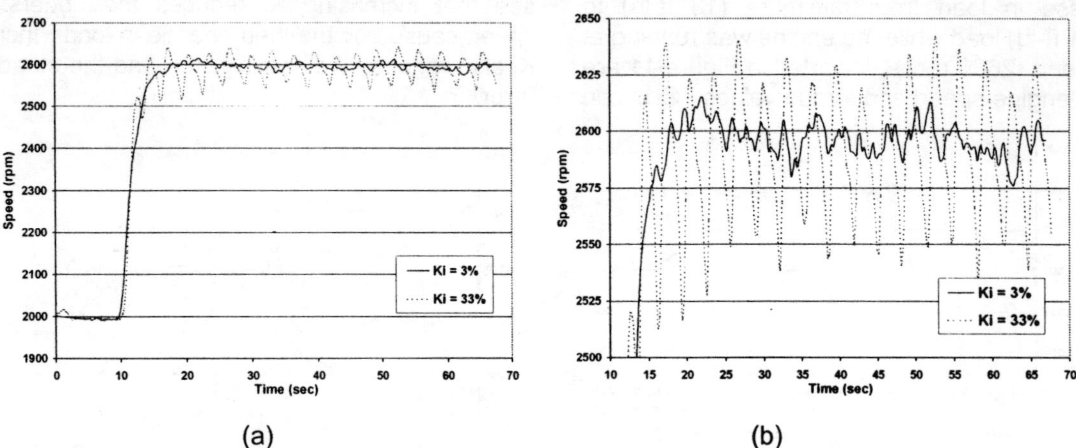

(a) (b)

Figure 9. Step increase in speed tests for K_I values of 3% and 33%: (a) complete test region and (b) zoom-in test region.

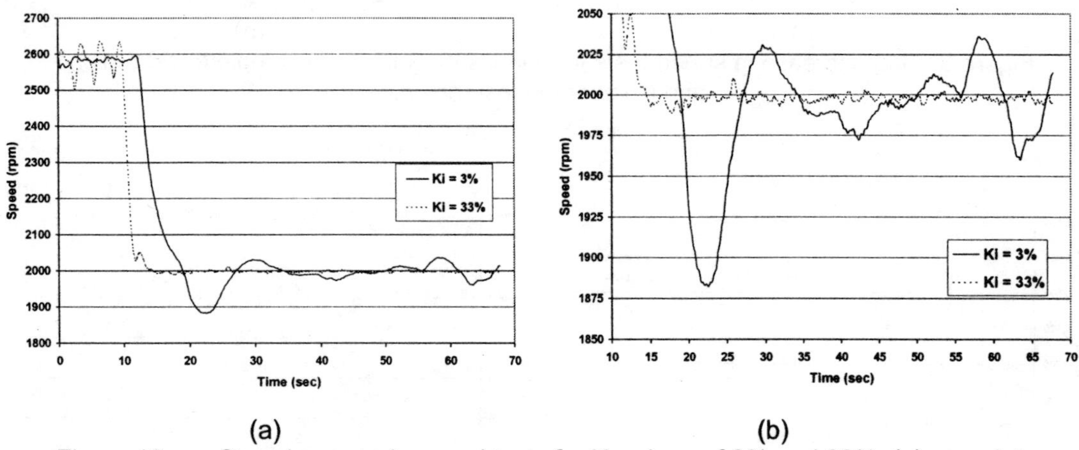

(a) (b)

Figure 10. Step decrease in speed tests for K_I values of 3% and 33%: (a) complete test region and (b) zoom-in test region.

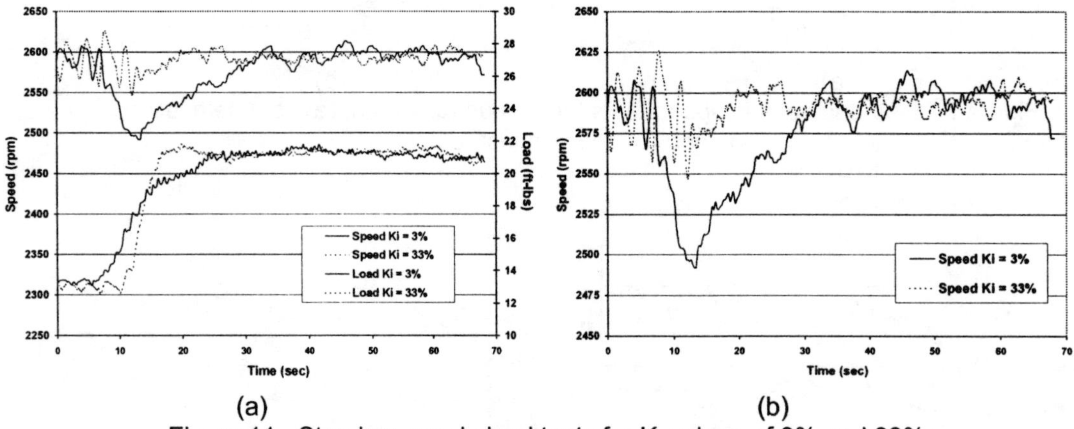

(a) (b)

Figure 11. Step increase in load tests for K_I values of 3% and 33%: (a) complete test region and (b) zoom-in test region.

- Tests Involving a Change in K_D

A third set of tests was conducted to determine how a change in the derivative gain affected the engine system. The values used for K_D were 0% and 33%. For these tests, the proportional and the integral gains were kept at the minimum values allowed by the PID controller. The minimum value for K_P was 20% and for K_I was 3%.

The responses shown in Figure 12 are the results of a step increase in speed from minimum (2000 rpm) to medium (2600 rpm) speed while applying a minimum load (13 ft-lb) to the engine. One can see that a 33% value for K_D yields a similar amount of overshoot when compared to 0% K_D. Also, 33% K_D yields a similar settling time and steady-state error than 0% K_D.

The responses shown in Figure 13 are the results of a step decrease in speed from medium (2600 rpm) to minimum (2000 rpm) speed with a minimum engine load (13 ft-lb). One can see that increasing K_D decreases the undershoot. Increasing K_D also decreases the settling time and the steady-state error of the speed response.

The responses shown in Figure 14 are the results of a step increase in load from minimum (13 ft-lb) to medium (21 ft-lb) load while the engine was running at medium speed (2600 rpm). Figure 14 includes the plots of the engine speed and torque values. One can see that a K_D value of 33% causes a slightly larger undershoot than a value of 0%. Despite this, a K_D value of 33% gives a smaller settling time when compared to a K_D value of 0%. Furthermore, a value of 33% K_D gives a smaller steady-state error than a value of 0%.

- Recommended Gain Settings

Based on the above results, three sets of gain settings are recommended. The first set of gain settings is recommended when the engine system is subjected to a step change in speed from minimum to medium speed. The values of K_P, K_I, and K_D recommended for this case are 20%, 3%, and 33%, respectively. A test that includes these gain settings is shown in Figure 15. One can see that the maximum overshoot is only 0.66%. Also, the settling time using these gain settings is close to 5 seconds after the system is subjected to the step increase in speed. In addition, the steady-state error of the system is only about 15 rpm.

The second set of gain settings is recommended when the engine system is subjected to a step change in speed from medium to minimum speed. For this case, the recommended values for K_P, K_I and K_D are 60%, 3%, and 0%, respectively. A test that includes these gain settings is shown in Figure 16.

One can see that the maximum undershoot of the system when using this set of gain settings is only about 1.15%. In addition the settling time is close to 5 seconds after the system is subjected to the step decrease in speed. Furthermore, the steady-state error of the engine system is only about 6 rpm.

The third and final set of gain settings is recommended when the engine system is subjected to a step change from minimum to medium load when the engine is running at medium speed. For this case the recommended values for K_P, K_I, and K_D are 60%, 33%, and 0%, respectively. A test that includes these gain settings is shown in Figure 17. One can see that the maximum undershoot in speed caused by the step change in load is only about 1.47% of the steady-state speed. In addition, the system response settles to within ±30 rpm of the steady-state speed at about 5 seconds after the system is subjected to the step increase in load. Finally, the steady-state error is only about 8 rpm.

FREQUENCY RESPONSE TESTING

In order to perform an analytically-based design of the controller gain settings, tests were conducted in order to obtain linear models of the engine system. Such models can be obtained from frequency response measurements of a stable closed-loop system using a Dynamic Signal Analyzer and a frequency-to-voltage converter chip. Since the small signal response of the engine system is a function of its operating load and speed, testing was performed at low, medium, and high speed and load operating points, which form a 3 by 3 test matrix. From this testing, linear models of the engine system dynamics were derived at each operating point. These models will help in designing the gains of the PID controller to achieve stability and proper performance at all operating points of the system. The final result will be an improvement of the performance of the system before incorporating it in the hybrid power generation system.

- System Identification

The test matrix used for the frequency response tests consisted of a combination of speed and load conditions. Minimum and medium speeds were defined as 2000 rpm and 2600 rpm, respectively. Minimum and medium loads were defined as 13 ft-lbs and 21 ft-lbs, respectively. For all these tests the gain settings for the PID controller were set at the minimum values allowed by the controller: 20% for K_P,

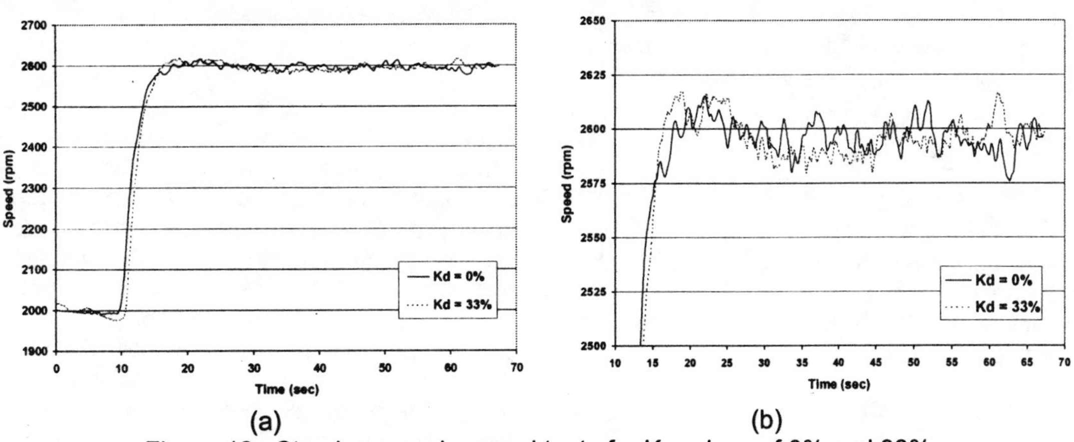

(a) (b)

Figure 12. Step increase in speed tests for K_D values of 0% and 33%: (a) complete test region and (b) zoom-in test region.

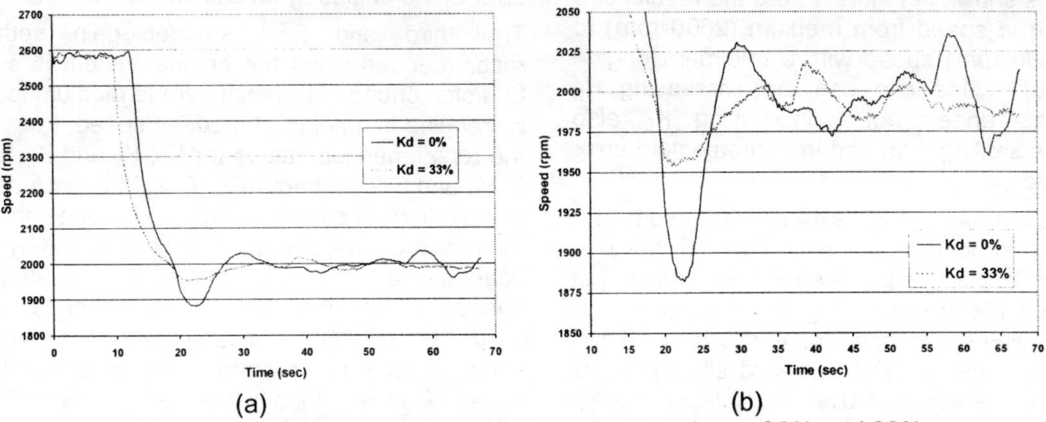

(a) (b)

Figure 13. Step decrease in speed tests for K_D values of 0% and 33%: (a) complete test region and (b) zoom-in test region.

3% for K_I, and 0% for K_D. The excitation signal used was a broadband random signal. Curve fits were then performed on the measured data.

The magnitude and phase plots obtained for the minimum speed (2000 rpm) and minimum load (13 ft-lbs) condition using a random noise signal are shown in Figure 17. The test was conducted using a Hanning window, a 300 mV peak signal level, ten averages, and 800 line resolution. It can be seen that the data are not very smooth and have significant sharp variations. These variations are probably due to nonlinear response characteristics (cross-correlation) that are being captured under the present set of measurement conditions; this issue is presently being studied so that better measurements can be obtained.

The phase data show two regions (at ~1 rad/sec and ~12 rad/sec) of variation that did not allow for an acceptable second-order curve fit to be performed simultaneously on the magnitude and phase data. This was the case for all the magnitude and phase data obtained from applying a random noise signal to the engine system at minimum speed with minimum load.

The frequency response data discussed below were similar to the frequency response data for an automotive gas turbine system obtained by Wellstead and Nuske [11]. These researchers modeled an automotive gas turbine system using a transfer function with 2 poles and 1 zero given by

$$G_A(s) = \frac{K(s+0.069)}{s^2 + .625 + 0.045} \tag{3}$$

where K is the system gain. We used this as a guide for curve fitting the engine data.

The second-order curve fit of the magnitude data shown in Figure 17(a) was performed using Matlab [The Mathworks, 1999]. This curve fit gave the following transfer function

$$G(s) = \frac{0.1(s+20)}{s^2 + 5.6s + 16} \tag{4}$$

where s is the complex frequency. Equation (4) represents a second-order frequency domain model (transfer function) of the engine dynamics at minimum speed with minimum load. A good fit of the phase data (shown in Figure 17(b)) was not obtained with this transfer function due to the significant variations present in the phase data. One can also see that the engine system at this operating point is a fairly well damped system since no resonance peak is observable from the magnitude data.

The magnitude and phase plots obtained at the medium speed with minimum load condition using a random noise signal are shown in Figure 18. The test was conducted using a Hanning window, a 300 mV peak signal level, eight averages, and 800 line resolution.

From the magnitude plot shown in Figure 18(a), it can be seen that the data shown contain significant variations. These variations are more severe in the phase data, especially in the region between 7 and 13 rad/sec. A good curve fit could not be performed simultaneously on the magnitude and phase data due to these discontinuities. This was the case for all the magnitude and phase data obtained when applying a random noise signal to the engine system at medium speed with minimum load.

A second-order curve fit of the magnitude data shown in Figure 18(a) gave the transfer function

$$G(s) = \frac{4(s+4)}{s^2 + 5s + 16} \tag{5}$$

Equation (5) represents the frequency domain model (transfer function) of the engine dynamics at medium speed with minimum load. It can be seen that the curve fit of the magnitude data gave a good approximation of the actual magnitude data. However, due to the severe variations present in the phase data, a good curve fit could not be performed on this particular set of phase data. As a final remark, the presence of a resonant peak is apparent in the magnitude data. This resonance peak indicates that the engine system at this operating condition is not as well damped as at the previous one.

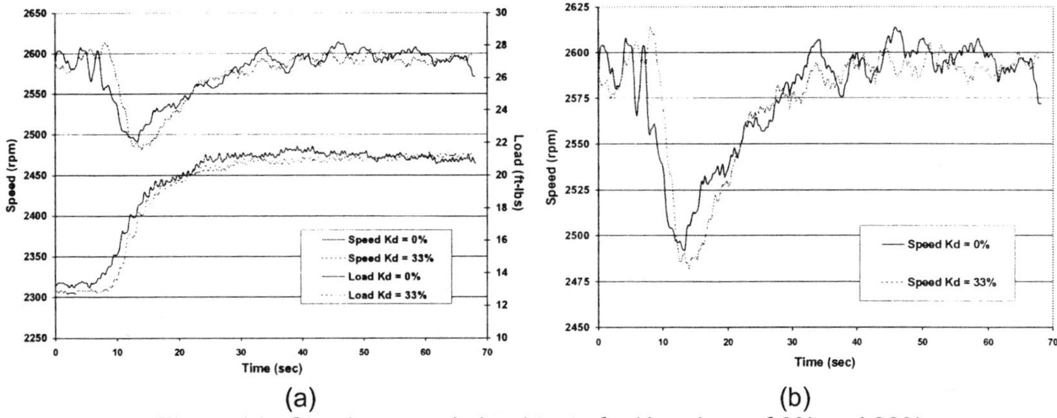

(a) (b)

Figure 14. Step increase in load tests for K_D values of 0% and 33%: (a) complete test region and (b) zoom-in test region.

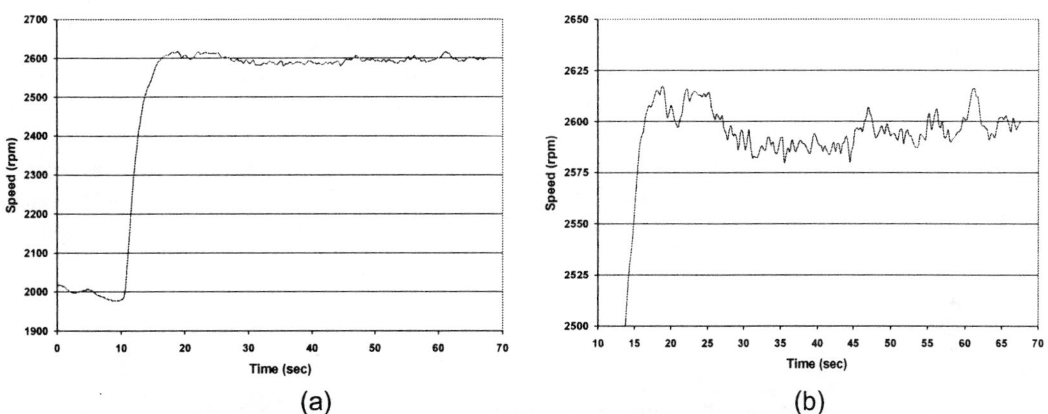

(a) (b)

Figure 15. Step increase in speed test with recommended gain settings (20% K_P, 3% K_I, and 33% K_D): (a) complete test region and (b) zoom-in test region.

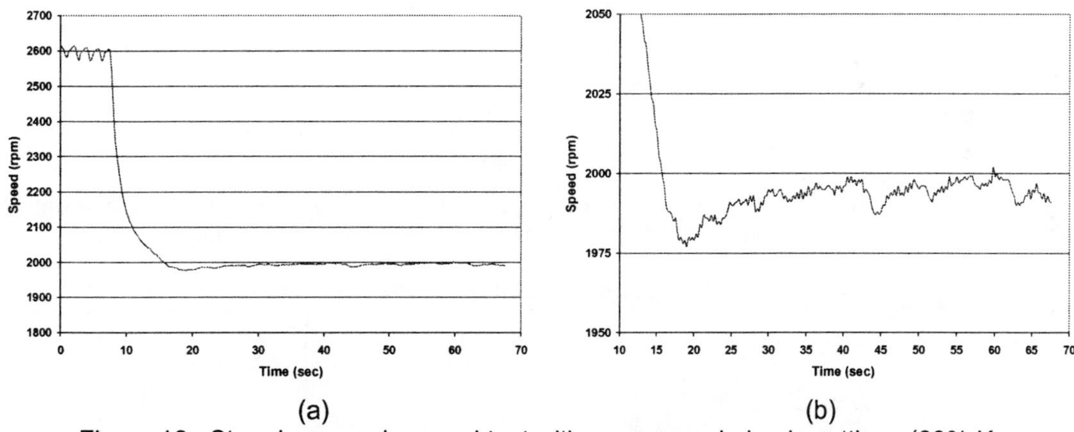

(a) (b)

Figure 16. Step decrease in speed test with recommended gain settings (60% K_P, 3% K_I, and 0% K_D): (a) complete test region and (b) zoom-in test region.

The magnitude and phase plots obtained from the medium speed with minimum load condition using a burst random signal are shown in Figure 19. This test was conducted using a uniform window, a 500 mV peak signal level, ten averages, and 400 line resolution. It can be seen that the data contain significant discontinuities similar to the ones present in the previous cases. These discontinuities are very severe in the region close to 10 rad/sec of the phase data. A curve fit of the magnitude data, shown in Figure 19(a), gave the transfer function

$$G(s) = \frac{2(s+10)}{s^2 + 3s + 25} \tag{6}$$

Equation (6) represents the frequency domain model (transfer function) of the engine dynamics at medium speed with minimum load. A good curve fit of the phase data (shown in Figure 19(b)) was not obtained using this transfer function due to significant variations present in the phase data. Finally, we can see that lightly damped dynamics are present at this operating condition since a resonant peak is prominent in the magnitude data.

Note that part of the discrepancy in the phase plots (at the higher frequencies) between the identified model and the actual data is due to system time-delay, which is mainly caused by fuel flow. As a result, if the delay is T_d seconds then the phase contribution is $-\omega \frac{180^o}{\pi} T_d$ degrees. From Figure 17(b), the time delay in the system was calculated to be about 0.25 seconds.

Transfer functions similar to the ones given by Equations (4), (5), and (6) obtained at various points throughout the operating range can be used to analytically optimize the gain settings of the PID controller. The procedure to optimize the gains of the PID controller is as follows. First, an operating point (*e.g.*, medium speed with minimum load) is selected along with the frequency domain transfer function model of the engine dynamics (*e.g.*, Equation (6)) for this operating point. Next, the PID controller is designed as part of the closed-loop system shown in Figure 20 to optimize the output transient response. In that figure, $G_P(s)$ is the transfer function of only the engine. This engine-only transfer function can be extracted from the expressions for $G(s)$ presented above by using the equation

$$G(s) = [K_P + \frac{K_I}{s} + K_D s] G_P(s) K_{fv} \tag{7}$$

where K_P is the proportional gain with a value of 20%, K_I is the integral gain with a value of 3%, K_D is the derivative gain with a value of 0%, and K_{fv} is the frequency-to-voltage converter gain of 1/60 V/Hz.

CONCLUSION

A control system for a V-twin engine has been developed. Three sets of gain settings that meet the overshoot/undershoot, settling time, and steady-state error requirements for the engine system were selected based on transient response tests involving step changes in speed and load. The first set of gain settings is recommended when the engine system is subjected to a step change in speed from minimum to medium speed. The second set of gain settings is recommended when the engine system is subjected

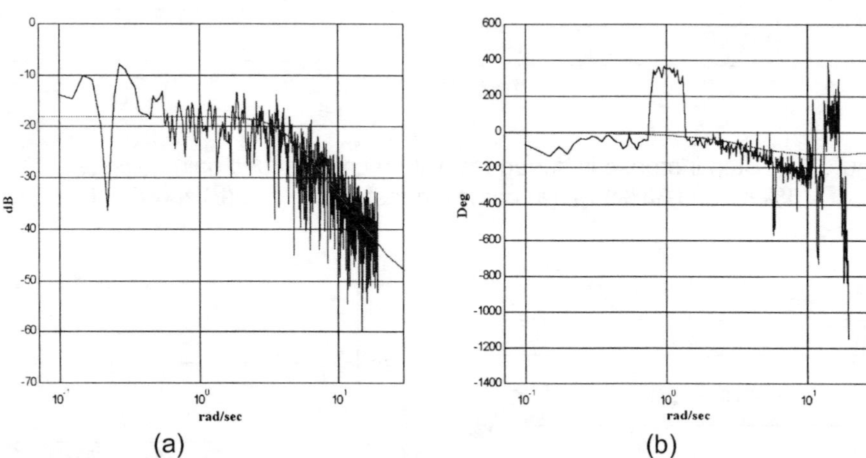

(a) (b)

Figure 17. Comparison of identified model data with actual (a) magnitude and (b) phase data at minimum speed (2000 rpm) with minimum load (13 ft-lbs) condition.

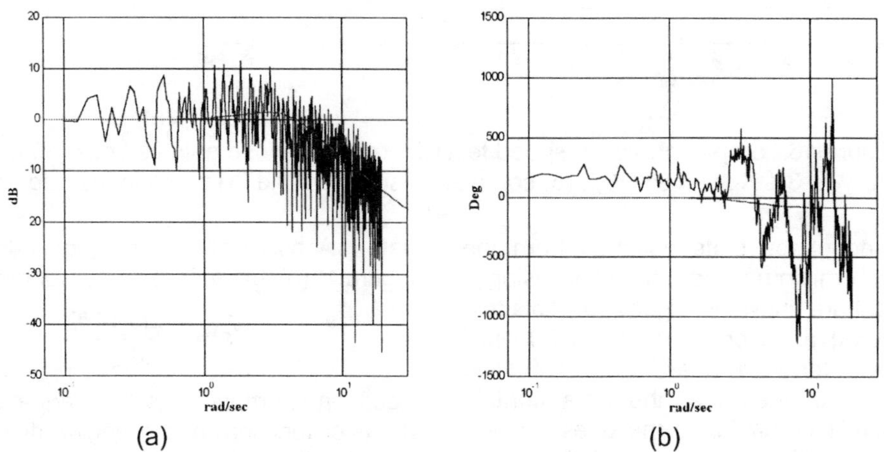

(a) (b)

Figure 18. Comparison of identified model data with actual (a) magnitude and (b) phase data at medium speed (2600 rpm) with minimum load (13 ft-lbs) condition.

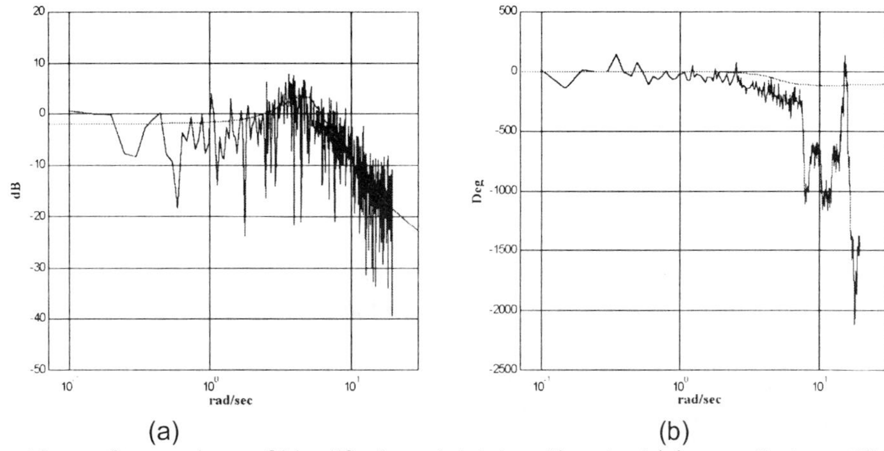

Figure 19. Comparison of identified model data with actual (a) magnitude and(b) phase data at medium speed (2600 rpm) with minimum load (13 ft-lbs) condition.

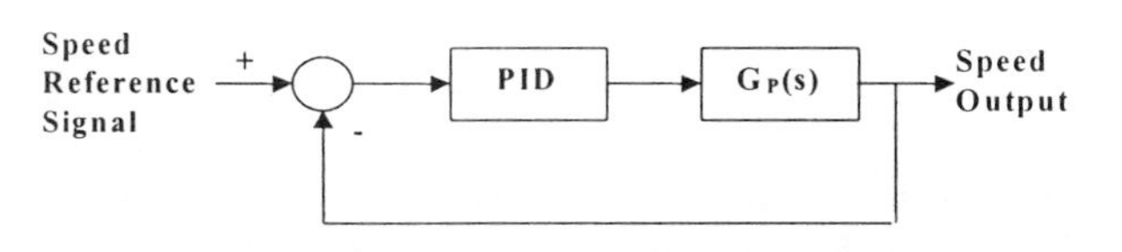

Figure 20. System for PID controller design.

to a step change in speed from medium to minimum speed. The third and final set of gain settings is recommended when the engine system is subjected to a step change from minimum to medium load when the engine is running at medium speed. But, the challenge then becomes one of designing and implementing a PID controller that can correctly detect and identify a transient condition and automatically shift from one set of settings to another.

Frequency response tests were also conducted from which transfer functions that model the engine dynamics at different operating points were obtained. From this testing, linear approximations of the engine system dynamic model were identified which will help in analytically optimizing the proportional, integral and derivative gains of the PID controller. Furthermore, these models will prove essential should the need arise for a more advanced controller than the PID in order to achieve even better performance.

Therefore, progress has been made in establishing the stability and proper performance of an engine-generator system that will be an integral part of a hybrid power generation system.

ACKNOWLEDGMENTS

The U.S. Environmental Protection Agency (EPA) has supported this research through Contract No. 524748-8501JKM with the Southwest Center for Environmental Research and Policy (SCERP - Year 8). This work has also been supported by the Texas Advanced Technology Program under Grant No. 003661-013-1997. The authors also acknowledge Briggs & Stratton and the Barber-Colman Company for their equipment donations. Individuals providing technical support include: Dr. Jamie Chapman from OEM Development Corp., Mr. Aaron Marquardt from the Barber-Colman Company and Dr. Rick Zadoks, Fernando Jasso, and Mariano Olmos from UTEP. The opinions expressed in this paper are those of the authors and do not necessarily reflect those of the sponsors or others involved in this research.

REFERENCES

1. Frank, A. L. (ed.), "Wind Power Enjoys Fastest Expansion, but Legislative Challenges Loom", The Solar Letter, vol. 7, no. 13, p. 235, June 1997.
2. Infield D.G. *et al.*, "Wind Diesel Systems – Design Assessment and Future Potential", Wind Engineering, vol. 16, no. 2, 1992.
3. Hunter, R. and Elliot, G., Wind-Diesel Systems, Cambridge University Press, p. 10, 1994.
4. Emtage, A. L. *et al.*, "The Development of an Automotive Drive-By-Wire Throttle System as a Research Tool", SAE Technical Paper No. 910081, 1991.
5. Ricardo G., G. Cotignoly, "Observer based Throttle control for drive by wire and idle control", Proc. IFAC Advances in Automotive Control, pp. 27-33, 1995.
6. W. Huber *et al.*, "Electronic Throttle Control", Automotive Engineering, vol. 99, no. 6, pp. 15-19, June 1991.
7. A. Ishida *et al.*, "A Self-Tuning Automotive Cruise Control System Using the Time Delay Controller", SAE Technical Paper No. 920159, 1992.
8. A. G. Loukianov et al., "A Robust Automotive Controller Design", Proc. IEEE Conference on Control Applications, pp. 806-811, 1997.
9. Kuo, B. C., Automatic Controls, Prentice-Hall, 1995.
10. K. Uhlen *et al.*, "Robust Control and Analysis of a Wind-Diesel Hybrid Power Plant", IEEE Transactions on Energy Conversion, vol. 9, no. 4, pp. 701-708, December 1994.
11. Wellstead, P.E., and Nuske, D.J., "Identification of an Automotive Gas Turbine", International Journal of Control, vol. 24, no. 3, pp. 297-324, 1976.

AIAA-2000-2853

2nd LAW ANALYSIS OF ON/OFF VS FREQUENCY MODULATION CONTROL OF A REFRIGERATOR

Frank Wicks
Mechanical Engineering Department
Union College, Schenectady, New York, 12308

ABSTRACT

The cooling capacity of any refrigerator or freezer should be somewhat oversized to assure that temperature can be retained over a range of conditions such as variable outside temperatures, opening of the doors, uncertainties of insulation and also for an acceptable temperature response to the introduction of new contents. Thus, a temperature control technique is required. The easiest control to implement is ON/OFF control via a thermostat. However, a more efficient control is continuous operation with temperature controlled by adjusting the compressor speed. The speed control advantage results from a lower temperature difference across the heat exchangers with continuous operation relative to ON/OFF control. This paper presents a 2nd Law analysis of a freezer that has a 12,000 (Btu/hr) capacity, but the load only requires 6,000 (Btu/hr). It is shown that speed control results in 41% electric energy savings relative to ON/OFF control.

1. Introduction

Refrigerators and freezers are typically driven by constant speed motors. Thus the most practical method for temperature control is the ON/OFF control that is performed by the thermostat as a function of temperature setting and the thermostat dead band between ON and OFF.

While this is the most practical form of control it is wasteful relative to operating all of the time while controlling temperature by varying the speed of the motor driven compressor. However, this type of operation requires a continually variable frequency power supply to the motor.

The fundamental thermodynamic reason why ON/OFF control requires more electric energy is that it results in larger temperature differences across evaporator and condenser heat exchangers and a corresponding extra power to move heat over a larger temperature difference.

The ideal coefficient of performance and thus the ideal power is defined by the Carnot cycle operating between the box and outside temperature. The extra power for an irreversible process such as a heat exchanger in a refrigerator is found by calculating the rate of entropy production and multiplying by the ambient temperature.

This paper will present a 2nd Law analysis to define the difference in extra power required for ON/OFF control versus continuous variable frequency control. It will also discuss the compressor options of piston scroll or centrifugal and solid state power electronics based frequency converters that could be used to control by modulating the motor speed.

2. System description

The analysis presents both a 2nd law analysis technique and the results for a specific refrigeration system. The specific system shown in Figure 1 will be a electric powered compression cycle commercial freezer with one ton or 12,000 Btu/hr maximum capacity with an inside air temperature of 20 F and outside air temperature of 80 F and freon to outside air heat exchange in the desuperheater and condenser and inside air to freon heat exchange in the inside evaporator. The refrigerant will be freon 12 and the compressor efficiency is 70 %. The temperature difference between the freon and the air in the evaporator and in the condenser is assumed to be 40 F, which means freon evaporates at –20 F and condenses at + 120 F.

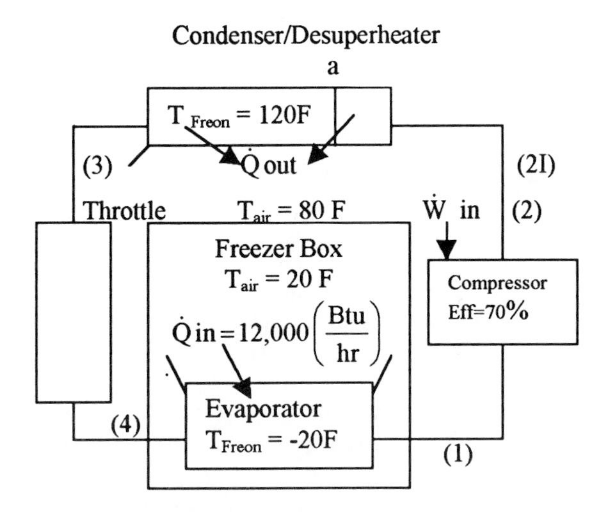

Figure 1. Schematic of Freezer and Refrigeration System
(Note: Point 2I is discharge of an ideal compressor and point 2 is the discharge of the actual compressor)

Compression cycles are typically presented on a pressure vs enthalpy diagram that superimposes the cycle over a chart of the thermal physical properties of the refrigerant. The desuperheater and the condenser operate at the high pressure and the evaporator at the low pressure. The inlet to the compressor is the low pressure saturated vapor from the evaporator and the discharge is high pressure super heated vapor.

The throttle is an adiabatic device that reduces the refrigerant pressure from the condenser pressure to the evaporator pressure. The inlet to the condenser is high pressure saturated liquid at a temperature higher than the outside air temperature. By reducing the pressure the evaporating temperature is also reduced. Thus, a portion of the refrigerant will evaporate in the throttle and the portion of the refrigerant that does not evaporate provides the thermal energy by dropping in temperature to a temperature that is lower that the freezer air temperature. Thus, a cold surface is established to cool the freezer air by absorbing heat.

3. Analysis

Next, analyze the actual cycle operating at maximum capacity by assuming freon 12 as the refrigerant and using a computer data base to obtain precise values for properties at each point as shown in Table I.

Table I, Properties
Properties at Each Point

#	T(F)	psia	v(ft^3/lbm)	h(Btu/lbm)	x(%)	s(BTU/lbmR)
1	-20	15.267	2.443	75.11	100	.17102
2I	146	172.35	.2561	93.71	-	.17102
2	190	172.35	.2891	101.68	-	.18372
a	120	172.35	.2333	88.61	100	.16241
3	120	172.35	.0138	36.01	0	.07168
4	-20	15.267	1.101	36.01	44.8	.08209

Note: 2I is discharge of an ideal compressor and 2 is the discharge of the actual compressor.

A 1st Law analysis on the evaporator now shows that a freon flow rate of 306.91 (lbm/hr) is required to provide 12,000 (Btu/hr) of cooling.

Table II
1st Law Process and Cycle
(flow rate dm/dt= 306.91 lbm/hr)

#-#	Q	=	dm/dt* (h out-h in)	+	W(Btu/hr)
1-2 (Compressor)	0	=	8,154	+	-8,154
2-a (Desuperheat)	-4,011	=	-4,011	+	0
a-3 (Condenser)	-16,143	=	-16,143	+	0
3-4 (Throttle)	0	=	0	+	0
4-1 (Evaporator)	12,000	=	12,000	+	0
Net for Cycle	-8,154	=	0	+	-8,154

Next perform the 2nd law analysis which should show that the power the actual cycle is equal to the power of the ideal cycle plus the sum of the extra powers as a result of the reversibilities as defined by equation 1.

$$\text{Wactual} = \text{Wideal} + \text{Wextra} \quad (1)$$

The Wactual is the power to the compressor and is defined and presented in Table II as 8,154 (Btu/hr).

The Wideal is defined by the Carnot cycle operating between the inside and outside air temperatures. The Carnot or ideal coefficient of performance, COP, is defined by the inside and outside air temperatures per equation 2.

$$COP\ ideal = Qcold/Wideal = Tc/(Th-Tc) = 480\ R/(540 - 480)\ R = 8.0 \quad (2)$$

For the maximum Qcold = 12,000 (Btu/hr) the corresponding ideal power requirement, Wideal=1500 (Btu/hr) = 439.6 watts.

Next determine the extra power Wextra as a sum of the irreversibilities of each of the non-ideal processes as defined by equation 3, where I(Btu/hr) is the extra power for each irreversible process. Tinf (R) is the temperature of the infinite outside heat reservoir and delta S (Btu/ R) is the entropy production for each irreversible process.

$$I = Tinf*delta\ S \quad (3)$$

It is noted that each of the defined five processes is irreversible. The desuperheater and the condenser are irreversible because heat is transferred over a temperature difference. The compressor and the throttle are both adiabatic devices, and thus there is no heat transfer, but the compressor is irreversible because it is less than 100 % efficient and a throttle also represents an irreversible process.

Accordingly, the net entropy production is calculated for each process and multiplied by the absolute temperature of the infinite environment which is 80 F or 540 R. The results are summarized in Table III.

Table III
2nd Law Process and Cycle Table

#-#	Tinf (R)	*	delta S (Btu/hrR)	=	I(Btu/hr)
1-2 (Compressor)	540	*	3.898	=	2,107
2-a (Desuperheat)	540	*	.888	=	479
a-3 (Condenser)	540	*	2.062	=	1,114
3-4 (Throttle)	540	*	3.195	=	1,726
4-1 (Evaporator)	540	*	2..273	=	1,228
Total for Cycle	540	*	12..316	=	6,654

The irreversibility represents the extra power required for each irreversible process and for the total cycle. We can now confirm equation 1.

Wactual(= 8,154) = Wideal(= 1,500) + Wextra(=6,654) (Btu/hr)

It is also observed that of the actual or total power is 8,154 Btu/hr which is equal to 2.390 kw. Also, 18.4% of the total would be required for an ideal Carnot cycle freezer and the additional 81.6% is the result of the irreversibilities as shown in Table III.

HALF CAPACITY OPERATION BY SPEED CONTROL

As a matter of good practice extra refrigeration capacity will be installed relative to the expected peak demand. The reasons are there is uncertainty over peak demand which includes how often and for how long the freezer doors are open.

The next analysis will assume that the actual demand is 6,000 (Btu/hr) rather than the 12,000 (Btu/hr) just analyzed, but the cycle still has a 12,000 (Btu/hr) capability.

One possibility for control would be On/Off which would cause the cycle to be On half the time and Off half the time. The temperature differences across the heat exchangers would be the same and thus the properties would still be defined by Table I. However, the freon flow rate would be halved, so that all values in the 1st Law process and cycle Table II and the 2nd Law process and cycle Table II would be halved.

The alternative possibility is to control by motor speed modulation, which means the temperature differences across the heat exchangers would be decreased by a factor of two and thus result in less entropy production, irreversibility and extra work. This requires redeveloping all three Tables I, II and III as Tables IV, V and VI for the speed modulated operation as presented as follows.

Table IV
Property Table at Half Load by Compressor Speed Control

#	T(F)	psia	v(ft^3/lbm)	h(Btu/lbm)	x(%)	s(BTU/lbmR)
1	0	23.85	1.609	77.27	100	.16888
2I	118	131.86	.3266	90.28	-	.16888
2	149	131.86	.3569	95.86	-	.17828
a	100	131.86	.3079	87.03	100	.16315
3	100	131.86	.0127	31.10	0	.06323
4	0	23.85	.5358	31.10	32.8	.06844

Table V
1st Law Process and Cycle Table at Half Load by Speed Control (flow rate dm/dt=129.95 lbm/hr)

#-#	Q	=	dm/dt* (h out-h in)	+	W(Btu/hr)
1-2 (Compressor)	0	=	2,415	+	-2,415
2-a (Desuperheat)	-1,147	=	-1,147	+	0
a-3 (Condenser)	-7,268	=	-7,268	+	0
3-4 (Throttle)	6,000	=	6,000	+	0
4-1 (Evaporator)					
Net for Cycle	-2,415	=	0	+	-2,415

Table VI
2nd Law Process and Cycle Table at Half Load by Speed Control

#-#	Tinf(R)	*	Delta S (Btu/hrR)	=	I(Btu/hr)
1-2 (Compressor)	540	*	1.223	=	661
2-a (Desuperheat)	540	*	.158	=	85
a-3 (Condenser)	540	*	.481	=	260
3-4 (Throttle)	540	*	.678	=	366
4-1 (Evaporator)	540	*	.543	=	293
Total for Cycle	540	*	3.083	=	1,665

The principle that the actual power is equal to the ideal power plus the sum of all the irreversibilities of the non ideal processes for operating at half maximum capability by compressor speed control rather than ON/OFF control.

Wactual(= 2,415) = Wideal (= 750) + Wextra(= 1,665) (Btu/hr)

It is also noted that the actual or total power of 2,415 (Btu/hr) is equal to .707 kw. The portion of the total for the ideal or Carnot cycle is 31% and the additional 69% is the extra power corresponding to the irreversibilities.

COMPARING RESULTS OF ON/OFF VS SPEED CONTROL

The preceding results are now summarized in Table VII which assumes that the subject freezer which has a maximum capacity of 12,000 (Btu./hr) operates for a full year at half capacity, and the cost of electricity is .12 ($/kwhr). Case A is operation by ON/OFF control and Case B is continuous operation at half capacity by compressor speed control.

Table VII
Half Power Operation by ON/OFF vs Speed Control

	Case A (ON/OFF Control)	Case B (Speed Control)
1. Operating Load	6,000 (Btu/hr)	6,000 (Btu/hr)
2. Carnot Power	750 (Btu/hr)	750 (Btu/hr)
3. Extra Power for irreversibilities	3,327 (Btu/hr)	1,665 (Btu/hr)
4. Total Power	4,077 (Btu/hr) or 1,195 watts	2,415 (Btu/hr) or 708 watts
5. Yearly Energy Cost at .12 ($/kwh)	$1,256	$744

This summary table shows a 41% electric energy saving for speed control relative to ON/OFF control.

4. Conclusions

The results show that temperature control by continuous operation and adjusting the compressor speed uses less electric energy than ON/OFF control. This was qualitatively expected because of the lower temperature difference and thus a lower pressure difference across the compressor.

The 2nd Law analysis presented in this paper provides the additional quantification of the extra power required for each irreversible process and for the entire cycle for each control mode.

Control by compressor speed would probably be most practical for positive displacement compressors of the piston/cylinder or schroll type, rather than non positive displacement centrifugal compressors that have an efficiency performance that is more sensitive to changes in speed and pressure ratios.

The variable speed motor capability is now best provided by a solid state power electronics frequency converter. Alternatively, the future may result in the conversion from AC to DC power for transmission and distribution. Thus, motors will inherently be continuously variable speed.

The preferred motor would probably be what is now known as a brushless DC motor. This is actually a three phase AC induction motor operating from a DC power supply with the switching and the potential for electric frequency and thus compressor speed adjustments by means of a solid state power electronics circuit external to the motor.

PRODUCING ELECTRIC POWER BY WET OXIDATION OF BIOMASS ETHANOL

Kiyoshi Naito, Kejing Huang[†], Akira Endo[‡], Masaru Nakaiwa[§], Takaji Akiya[¶] and Takashi Nakane[#]
National Institute of Materials and Chemical Research, Tsukuba 305-8565, Japan

ABSTRACT

A new process for producing electric power through wet oxidation of biomass ethanol is proposed in this work. The process configuration is introduced and its energy efficiency is evaluated by using simple process model. The advantages and disadvantages of the process are also indicated.

INTRODUCTION

Industrialization and enhancement of living standard worldwide cause the rise of carbon dioxide concentration in the atmosphere. This will give severe impacts on the global environment, such as the effect of global warming, if no effective measures are taken into account from now on. For protecting the global environment and reducing these ever-increasing impacts, the government of Japan has already stipulated that the release of carbon dioxide will be reduced to about 6 % of the current value by the year of 2012. The Ministry of International Trade and Industry accordingly made a plan that about 2 % of the reduction should be achieved through development of new energy and new energy saving techniques.[1] At present, the energy density for new energies is still very small except for certain techniques such as fuel cells. Moreover, these new energy technologies are usually very difficult to be used in practice because either much more energy consumption or higher investment cost must be afforded. It is, therefore, imperative to develop effective energy saving technologies to alleviate the impact on the atmosphere.

In this study, an attempt is tried to effectively use the low-concentrated biomass ethanol[2]. A system, which can generate high-temperature steam and electric power through wet oxidation[3, 4] of biomass ethanol, is proposed as a new efficient way of energy conducted by simulation, and as a result, advantages and disadvantages of the system will be pointed out.

---

[*] Visiting Researcher, Dept. of Chemical Systems.
[†] Visiting Researcher, Dept. of Chemical Systems.
[‡] Researcher, Dept. of Chemical Systems.
[§] Senior Researcher, Dept. of Chemical Systems.
[§] Senior Researcher, Dept. of Chemical Systems.
[¶] Laboratory Head, Dept. of Chemical Systems.

COMBUSTION OF BIOMASS ETHANOL THROUGH WET OXIDATION

The principle for the process design is based on the fact that liquid organic substances can be burnt with oxygen and give out large reaction heat, if they are kept on an appropriate temperature range, for instance, between 180 °C and their critical points. They can, therefore, be seen as potential thermal energy sources for producing electric power. The system configuration for producing electric power by the wet oxidation of biomass ethanol is depicted in **Figure 1**.

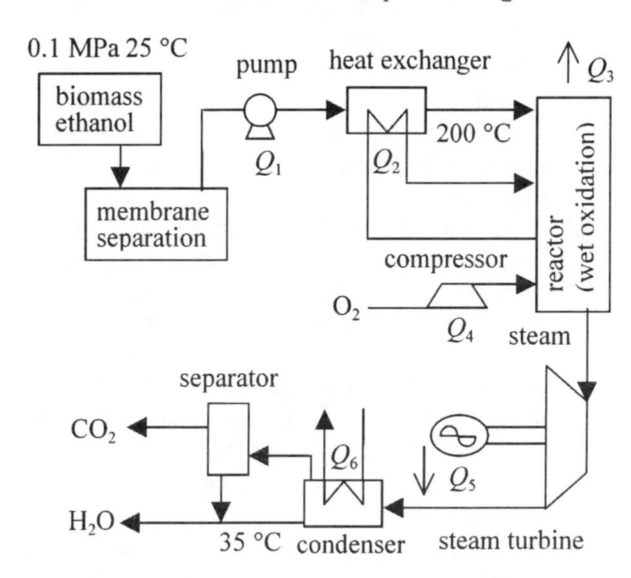

Fig.1 System for producing electric power through wet oxidation of biomass ethanol

As can be seen, the low-concentrated biomass ethanol is first condensed into a relative higher concentrated solution (>10 wt%) by the membrane separation[5, 6], because combustion of low-concentrated biomass ethanol can not release enough thermal energy for the purpose of producing electricity, as will be discussed later. The condensed biomass ethanol is preheated to about 200 °C with part of the heat generated in the reactor. The preheated biomass ethanol is then pumped into the reactor. Oxygen must be introduced into the reactor through a compressor. At an appropriate pressure, the biomass ethanol reacts with oxygen within the reactor and generates a high temperature steam with the released reaction heat. Electricity can thus be produced by a steam turbine with the high temperature steam as a power source. The unique feature of this system is no necessity of considering exhaust gas treatment in contrast to the current practice. In Fig.1, Q_1 represents the energy consumption required for pumping the biomass ethanol solution into the reactor; Q_2, the energy required for preheating the biomass ethanol from 20 to 200 °C, Q_3; the energy consumption radiated from the reactor; Q_4, the energy required for supplying oxygen into the reactor; Q_5, the energy converted into electricity, and Q_6, the energy absorbed by the cooling water. They are calculated in terms of the following assumptions.

(1) For investigating the system performance, the energy efficiency of the process will be calculated and studied under various operating conditions. Especially, the concentration of biomass ethanol solution will be subject to changes from 5 to 20 wt% and the pressure in the reactor form 5 to 20 MPa, respectively. The biomass ethanol is taken to be produced by fermentation and is usually only 5 wt% in concentration. It must be condensed into a relative higher concentration by the membrane separation. The energy for condensation of biomass ethanol solution by the membrane separation is assumed to be 2000 kJ/kg in this work.[7]

(2) The energy required for pumping the ethanol solution into the reactor Q_1 can be neglected, because it is too small compared with other items.

(3) The thermal conversion efficiency for the heat exchanger, the heat radiation of the reactor and the conversion efficiency from heat to electricity by the steam turbine, i.e., $\eta_{Q \to E}$, are assumed to be 90 %, 5 % and 80 %, respectively.

(4) The compression of oxygen is taken to follow the polytropic change. The work becomes too large, if the oxygen gas is compressed at one stage. Therefore, it is suggested to be conduced by multiple stages. For example, it is supposed to be compressed from 0.1 MPa to 5 or 10 MPa by 4 stages, and to 15 or 20 MPa by 5 stages. The compression ratio of each stage is assumed to be equal, and the inhalation gas after each stage is cooled to the same temperature of the suction gas in the first stage, the work of the compressor can be worked out by

$$-W = \frac{Nm_p RT}{m_p - 1}\left\{\gamma^{(m_p-1)/m_p} - 1\right\}\Big/\eta_{EP} \quad (1)$$

where W represents the work, N, the stages of compression, R, the ideal gas constant, T, the temperature of the suction gas, γ, the compression ratio of each stage, m_p, the polytropic index and η_{EP}, the polytropic efficiency. In this study, the temperature of the suction gas, the polytropic index and the polytropic efficiency are assumed to be 35 °C, 1.3 and 0.8, respectively.

(5) Oxygen can be supplied into the reactor by using either air (21 %) or oxygen product (99.9 %), which may be produced by the cryogenic separation. The energy for condensation of oxygen by the cryogenic separation is assumed to be 2.42 × 10^4 kJ/kmol in this work.

ENERGY FOR OXYGEN SUPPLY AND HEAT PRODUCED BY WET OXIDATION

It is necessary to send oxygen into the reactor so that ethanol solution can be burnt by the wet oxidation method. There are two ways to facilitate this purpose as mentioned above. One is to use air (21 %) directly. The other is to use oxygen product (99.9 %) produced by the cryogenic separation. **Figure 2** compares the two alternatives at different operating conditions, where the symbols ○ and ● represent the energy consumed in order to supply air and oxygen product into the reactor. Examining the case where the ethanol solution is fixed at 10 wt%, One can find that the energy consumption for using oxygen is only 20 % of that for using air directly. Calculations for different biomass ethanol concentration are also conducted and similar results are obtained. It is because for the former case a lot of work has been consumed for compressing other substances contained in the air, which have no beneficial effects on the reaction. Therefore, in this study, oxygen produced by the cryogenic separation will be adopted and supplied into the reactor.

The comparison between the heat generated by wet oxidation and the oxygen supply energy at different biomass ethanol concentration is shown in **Figure 3**. The heat generated by wet oxidation is about ten times larger than the oxygen supply energy. Moreover, the generated heat by increasing ethanol concentration is higher than the increment of oxygen supply energy introduced by the pressure enhancement. The oxygen supply energy is not very sensitive to changes in operating pressure of the reactor. For example, when concentration for ethanol is fixed at 15 wt%, the difference in process energy efficiency is less 15 % if the operating pressure has been enhanced from 5 to 20 atm.

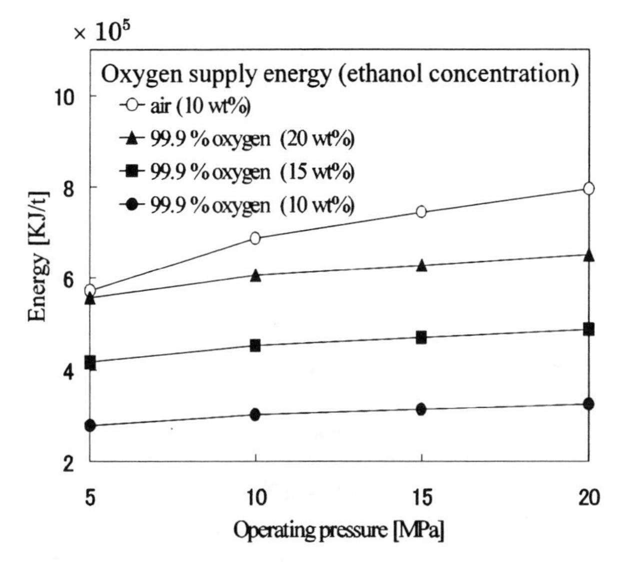

Fig.2 Energy needed for supplying air and oxygen into the reactor

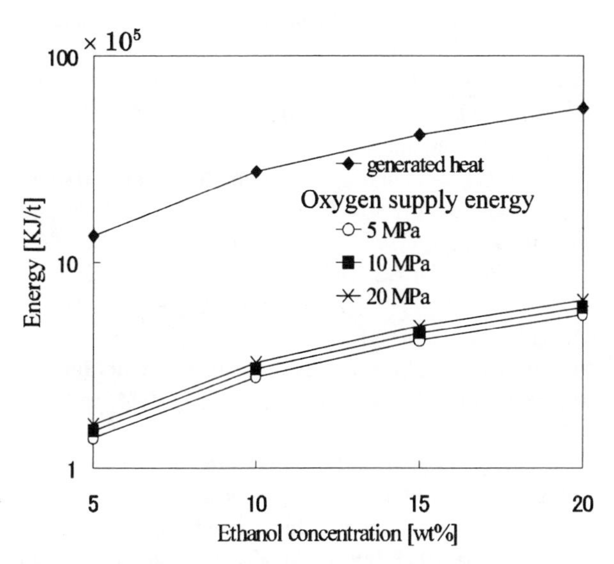

Fig.3 Comparison between heat generated by wet oxidation and oxygen supply energy

THE GENERATED ELECTRIC POWER AND OVERALL THERMAL EFFICIENCY

First, the effect of the operating condition on the system performance is investigated. The amount of generated electric power at different ethanol concentrations and pressures is shown in **Figure 4**. Because the generated heat within the reactor is very small when the concentration of biomass ethanol solution is below 10 wt%, the generation of electric power appears to be impossible. Hence, the biomass ethanol concentration must be, at least, a little bit higher than 10 wt%. The amount of generated electric power is not sensitive to the operation pressure variations, because the oxygen supply energy is not very closely dependent on the operating pressure within the reactor. For example, in the case of ethanol concentration is 15 wt%, there exist only 8 % difference in process energy consumption, when operating pressure has been enhanced from 5 to 20 MPa.

Next, the effect of the conversion efficiency from heat to electricity by the steam turbine on the overall thermal efficiency is examined in **Figure 5,** where the symbols ● and ○ represent the results whether the energy required to condense biomass ethanol is included or not. Here, the concentration of biomass ethanol solution is fixed at 15 wt% and the operating pressure within the reactor is 10 MPa. It is straightforward to understand that high electricity conversion rate is beneficial to the overall energy efficiency.

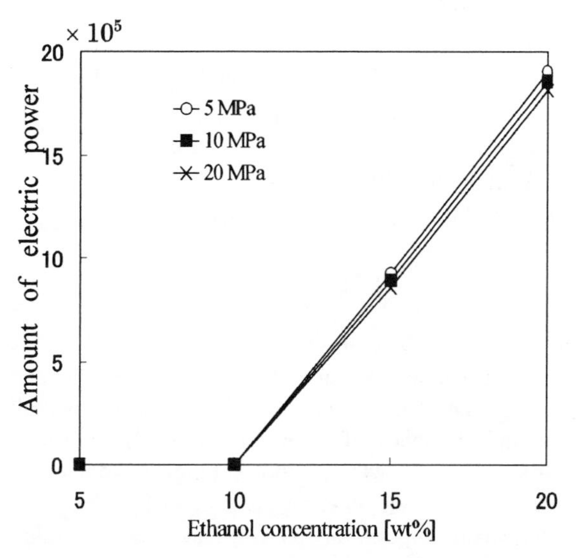

Fig.4 The generated electric power and its relations with ethanol concentrations and pressures

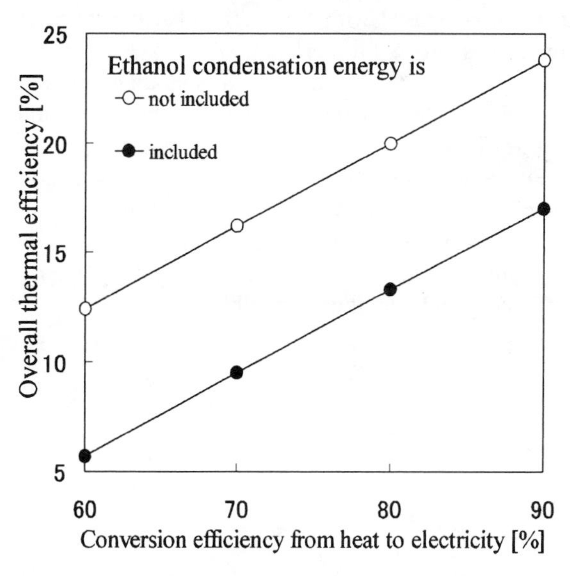

Fig.5 Conversion efficiency from heat to electricity vs. overall thermal efficiency

The summary results of producing electric power by wet oxidation at another pressure (5 or 15 MPa) are listed in **Table 1**. As can be seen, the overall thermal efficiency for this proposed system is about 20 % in any case. If the energy consumption for condensing the biomass ethanol solution is taken into consideration, the overall thermal efficiency decreases to about 13 %.

Table 1 Summary of simulation results

Ethanol solution, wt%	15	
Operating pressure, MPa	5	15
Generated heat in reactor, kJ/t	41.6×10^5	
Electric power, kJ/t ($\eta_{Q \to E}$ = 80 %)	13.5×10^5	
Oxygen supply energy, kJ/t	4.2×10^5	4.7×10^5
Overall thermal efficiency, %	20.8	19.6
Ethanol condensation energy*, kJ/t	3.0×10^5	
Overall thermal efficiency (include *), %	14.1	12.9

CONCLUSIONS

A new process for producing electric power through wet oxidation of biomass ethanol is proposed in this work. Deep insights into the process are obtained through simulation study. It is found that the overall thermal efficiency of the proposed system is around 20 %, when 15 wt% ethanol solution is adopted to react with oxygen under the pressure of 10 MPa. If the energy consumption for concentrating ethanol solution is taken into account, it will be reduced to 13 %. It is demonstrated that oxygen supply energy is drastically reduced (20 %) by pumping oxygen directly into the reactor, compared with the design of using air as an oxygen source. Furthermore, the oxygen supply energy is not very sensitive to operating pressure variations. This character permits wide and flexible applications of the system to process industry.

Future work will be focused on developing a more detailed simulation model and carrying out optimization study. Experimental work will then be prepared for the evaluation of system energy efficiency and operation feasibility.

NOMENCLATURE

m_p = polytropic index
N = stages of compression
Q = energy
R = gas constant
T = temperature of suction gas
W = work
γ = compression ratio of each stage
η_{EP} = polytropic efficiency
$\eta_{Q \to E}$ = conversion efficiency from heat to electricity in steam turbine

REFERENCES

[1]The Ministry of International Trade and Industry, White Paper on International Trade 1998, 1998, Japan.

[2]Gong, C. S., N. J. Cao, J. Du, G. T. Tsao, "Ethanol Production from Renewable Resources," *Adv. Biochem. Eng./Biotechnol.*, Vol. 65, 1999, pp. 207-241.

[3]Imamura, S-I., Y. Tonomura, M. Terada and T. Kitao, "Wet Oxidation of Organic Substrates Containing Only C, H and O as the Constituent Atoms," *Mizu Shori Gijutsu*, Vol. 20 (4), 1979, pp. 317-321, in Japanese.

[4]Boock, L. T. and M. T. Klein, "Lumping Strategy for Modeling the Oxidation of C_1-C_3 Alcohols and Acetic Acid in High-Temperature Water," *Ind. Eng. Chem. Res.*, Vol. 32, 1993, pp. 2464-2473.

[5]Sano, T., H. Yanagishita, Y. Kiyozumi, F. Mizukami and K. Haraya, "Separation of Ethanol/Water Mixture by Silicalite Membrane on Pervaporation," *J. Memb. Sci.*, Vol. 95, 1994, pp. 221-228.

[6]Yanagishita, H., D. Kitamoto, K. Haraya, T. Nakane, H. Matsuda, N. Koura and T. Sano, "Preparation of Silicalite Membrane for Pervaporation, Proc. of the 1996 International Cong. on Memb. and Memb. Processes," 1996, pp. 486-487, August 18-23, Yokohama, Japan.

[7]Nakane, T., H. Matsuda, H. Yanagishita, T. Shimazaki, I. Fujiwara, M. Nakaiwa, D-J. Lee and H. Yoshitome, "Study on the Target of Separation Factor for Development of Ethanol Preferentially Permeable Silicalite Membranes," Proc. of the 5th International Symp. on Sep. Tech.-Korea and Japan, 1999, pp. 790-793, August 19-21, Soul, Korea.

HYBRID ELECTRIC VEHICLES_YEAR 2000

Floyd A. Wyczalek *
FW Lilly Inc.
155 S. Williamsbury Road
Bloomfield Hills MI 48301-2761
248-646-9585 FAX USA
Internet:75467.2677@compuserve.com

ABSTRACT[1]

By the year 2000, the automotive manufacturers turned their attention to Hybrid Electric Vehicles (HEV). This re-direction of Electric Vehicle (EV) development effort was highlighted at the North American International Auto Show (NAIAS 2000), where, the World's major automobile manufacturers displayed twenty three hybrid electric vehicles (HEV). Thirteen HEV were configured with an internal combustion piston engine and electric traction motor, while, ten additional HEV were configured as fuel cell and battery pack hybrid vehicles (FCHV). Furthermore, the FCEV were fueled with hydrogen (H2) from H2 absorption metal hydride fuel tanks, or, from liquified hydrogen (LH2) cryogenic fuel tanks. One conclusion for HEV was that the preferred configuration includes a compression ignition (CI) engine paired with an electric motor, and, four speed manual transmission with automatic shift. A second conclusion for FCHV was that the fuel of choice is H2, and furthermore, these H2 fueled FCEV were the precursor vehicles foreshadowing deployment of a hydrogen fuel infra-structure within the first quarter of the 21st Century

INTRODUCTION

Beginning in 1990, the major automotive passenger vehicle manufacturers once again re-evaluated the potential of the battery powered electric vehicle (EV). This intensive effort to reduce the battery EV to commercial practice focused attention on the key issue of limited vehicle range, resulting from the low energy density and high mass characteristics of batteries, in comparison to the high energy density of liquid hydrocarbon (HC) fuels[1-12].

Consequently, by 1996, vehicle manufacturers turned their attention to hybrid electric vehicles (HEV). This redirection of EV effort was highlighted initially, in 1997, at the 57th Frankfurt Motor Show, the Audi Duo parallel type hybrid was released for the domestic market as a 1998 model. Also at the 1997 32nd Tokyo Motor Show, the Toyota Hybrid System (THS) Prius was released for the domestic market as a 1998 model.

This introduction of production hybrid electric vehicles triggered a new effort by most major automobile manufacturers to reduce the hybrid electric power train concept to practice. And, in November 1999, at the 33rd Tokyo Motor Show, new prototype hybrid electric passenger vehicles were introduced by: the Japanese manufacturers: Daihatsu, Honda, Mitsubishi, Nissan, Suburu, Suzuki, and Toyota, respectively. Further, in January 2000 at the NAIAS North American International Auto Show, General Motors, BMW, Daimler Chrysler, and Toyota introduced three FCHV and a conventional BMW piston engine vehicle, respectively, fueled with gaseous H2 stored as a metal hydride and/or as cryogenic LH2[13-22].

Because of this unprecedented number of HEV and FCHV, the scope of this assessment is limited to a comparative assessment of four H2 fueled vehicles: the General Motors Precept FCHV, and, the Toyota FCHV, illustrating the application of the H2 absorbing metal hydride tank, Daimler Chrysler NECAR4 FCHV illustrating the LH2 cryogenic tank, and, BMW 750-lh a conventional Internal combustion (IC) piston engine with a LH2 cryogenic fuel tank. Furthermore, the focus is now on introducing hybrid FCHV for test marketing in Europe, Japan, and in the United States[15,16,23,24].

---

[1] *Author ,President of FW Lilly Inc

Figure 1 General Motors Precept PEM fuel cell and battery pack (FCHV) hybrid electric vehicle [13]
[h:\gm_00\gmpr#b1d_gmprecept02]

Figure 3 GM Precept fuel cell H2 tank cut away shows mesh configuration of LaNi5 metal hydride alloy [13]
[h:\gm_00\gmpr#b1d_gmprecept08]

METAL ALLOY HYDROGEN GAS STORAGE

Hydrogen fueled vehicles eliminate the complex on board liquid fuel reformer system. The General Motors Precept shown in figure 1 and the Toyota FCHV shown in figure 2, respectively, illustrate the application of metal alloy absorbing hydrogen gas storage based on the well known LaNi5H6 metal hydride system.

The GM Precept 400 cell 100 kW 340 volt PEM fuel cell stack is GM developed, based on Du Pont Nafion 115/117 PEM membrane, 1 kW/liter and 1 kW/kg, and hydrogen gas absorbing LaNi5H6 alloy provides a 500 km range, see Figure 1.

Figure 3 shows the internal metal hydride alloy mesh configuration for the GM Precept hydrogen gas fuel tank. The ternary hydride alloy LaNiH6, theoretically, has twice the energy density of liquid hydrogen on a volume basis. However, in practice, the energy density is about 1.1 times liquid hydrogen, since the hydride alloy mesh is distributed in an open configuration to allow H2 access, within a fuel tank container envelope.

Toyota FCHV system configuration in figure 4, includes a similar H2 hydride alloy H2 gas fuel tank located between rear wheels of the vehicle, metal hydride alloy mass (weight) of 100 kg , tank volume of 60 liters, and stores 2.2 kg H2 gas, 88 kWh hhv energy equivalent. [hhv = higher heating value]

Toyota bipolar fuel cell stack includes a water cooled sandwich integral with each single cell, 70 kW output, mass (weight) 75 kg, and volume 65 liters. Air (O2) compressor max rate 3000 liter/min, 150 kPa max discharge pressure [13–14] [100 kPa equals 14.5037 lbf/sqin]

Figure 2 Toyota FCHV fuel cell hybrid vehicle with PEM membrane fuel cell and NiMH2 battery pack [14]
[tiger01 h:\hpias_00\ toyfucVH.jpg]

Figure 4 Toyota FCHV front drive, integral water cooled PEM fuel cell stack and NiMH2 battery pack [14]
[Tiger01 h:\phias_00\ toyfucel.jpg]

Figure 5 Daimler Chrysler model year 2000 NECAR4 PEM membrane fuel cell and battery pack FCHV
[tiger01 h:\med_hr12\ htr014b.jpg]

CRYOGENIC LIQUID HYDROGEN STORAGE

The Daimler-Chrysler NECAR4 PEM fuel cell hybrid shown in figure 5 and the BMW 750-lh conventional V12 piston engine vehicle shown in figure 6, illustrate the application of cryogenic liquid H2 fuel storage.

Figure 6 also illustrates a robotic refueling system developed by BMW and Linde to automatically refuel cryogenic H2 fuel tanks. The first robotic H2 refueling system is located at the Munich airport in Germany. The robotic system recognizes the vehicle computerized signature and automatically couples the fill arm to the cryogenic tank filler neck.

Figure 7 shows the Daimler-Chrysler fuel cell system includes two fuel cell modules, 35 kW each for a total of 70 kW. A single cell includes a PEM membrane and graphite paper sandwich. A PM permanent magnet 55 kW electric motor drives the right front wheel directly,

Figure 7 Daimler Chrysler NECAR4 fuel cell pack consists of two bipolar stacks of 160 cells each.
[tiger01 h:\med_hr12\ htr017.jpg]

and is coupled to the left front wheel through a coaxial shaft. The fuel cell system auxiliary components are mounted under the floor of the vehicle

Figure 8 shows internal structure of the BMW 750-lh cryogenic fuel tank. During refill, liquid H2 drizzles through a perforated tube running the upper tank length to re-condense and recover hydrogen gas, which otherwise would be purged from the clearance volume to the atmosphere.

Seventy layers of fiberglass and aluminum super insulation and vacuum insulate the cryogenic double walled tank, which holds 140 liters, 4.3 kg LH2, electrical energy equivalent to 171 kWh, for a 400 km range, at speeds up to 180 km/h. If internal pressure exceeds 4 bar, gaseous H2 is drawn off and safely converted to water by a catalytic reactor [13–14]
[one bar = 0.9869 atm = 100 kPa = 14.5037 psi]

Figure 6 BMW 750-lh conventional V12 piston engine vehicle and automatic liquefied H2 refueling system [13]
h:\hpias_00\bmwh_v12.jpg

Figure 8 BMW 750-lh cryogenic LH2 fuel tank cross section shows interior construction and fill neck [13]
h:\hpias_00\bmwh2flk.jpg

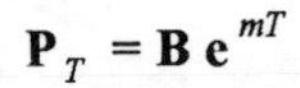

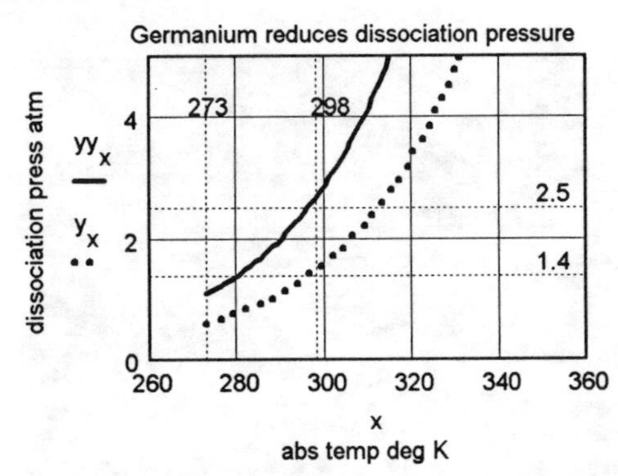

Figure 9 Low 1.4 atm dissociation pressure at 25 ^{O} C for LaNi4.8Ge0.2H5.9 hydride is ideal application [26]
[tiger01 f:\mcd0\h2hydrid.mcd]

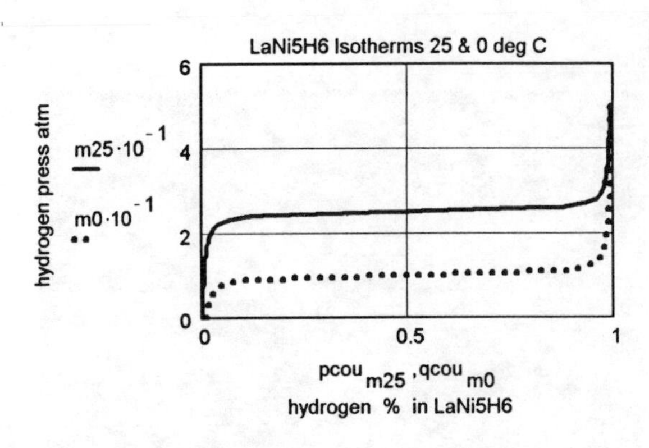

Figure 10 Pure LaNi5H6 isotherm pressures of 1.0 atm and ,2.5 atm at 0 ^{o} C and 25 ^{o} C, respectively
[Tiger01 f:\mcd0\h2hydrid.mcd]

METAL HYDRIDE H2 GAS FUEL TANK

The ternary lanthanum nickel alloy LaNi5 and hydrogen readily form a reversible and practical method for storing hydrogen at low pressures, for automotive applications, figure 9 and figure 10. Germanium alloyed with LaNi5 can tune the dissociation pressure. As an example, figure 9 compares pure LaNi5H6, (yyx) solid curve, to LaNi4.8Ge0.2H5.9, (yx) dot curve. Adding Ge0.2 reduces hydrogen storage capacity about 8% [25-26].

When hydrogen is pumped into a metal hydride tank, absorption is exothermic, while, desorption is endothermic requiring heat. The absorption pressure temperature relationship for pure LaNiH6 is expressed in [eq (1)], solid plot in figure 9:

$$\mathbf{P}_T = \mathbf{B}\,\mathbf{e}^{mT} \qquad \text{eq(1)}$$

where: $B=5.212*10^{-5}$ and $m=0.0364$
P=pressure in atm and T=temperature deg K
Atm = 101.325 kPa = 14.695 lbf/sqin

[substituting $\mathbf{10}^{0.4343}$ in place of **e** converts eq(1) to **log** base 10]

The pure hydride LaNiH6 has 2 times the energy density of liquid H2 and 12 times compressed H2 gas. However, in practice, 90%H2 LaNiH5.4 hydride is formed, and, an open mesh alloy in a fuel tank, down grades energy density further to about 1.1 times liquid H2 and about 9 times compressed H2 gas, Table I.

Table I Compare energy density mass and volume basis
Abreviations: hhv=higher heating value lhv=lower heating value

H2 Storage method	Mass basis kWh/kg	Volume basis kWh/liter
LaNiH6 hhv	1.1 kWh/kg LaNi5	3.2
Liquefied H2 hhv	39.4.	2.8
Compressed H2 100 atm	39.4 tnk not included	0.335
Gasoline lhv	2.53 tnk not included	8.82

CRYOGENIC LIQUID LH2 FUEL TANK

Figure 11 shows the BMW 750-lh vehicle auxiliary electrical power unit in the trunk, inset above the rear right tire wheel well. Cryogenic LH2 heat leak boil off gas is converted to electrical power by the fuel cell APU, and charges an auxiliary PbAcid battery. This permits the vehicle interior to be heated and/or air conditioned automatically, when parked and/or driving.

Figure 12 shows a 3 Watt heat leak rate will deplete the BMW LH2 140 liter tank in about 7 days. Figure 13 shows the fuel cell APU ideally can generate about 1 kW from the heat leak H2 gas boil off feed, and about 0.8 kW at 80% fuel cell conversion efficiency.

The PEM fuel cell APU serves to safely convert the H2 boil off gas to water and electric power. Further, a 220 V AC outlet is provided to feed excess electrical energy from the APU back into the public utility power grid, when parked in a garage [23-29].

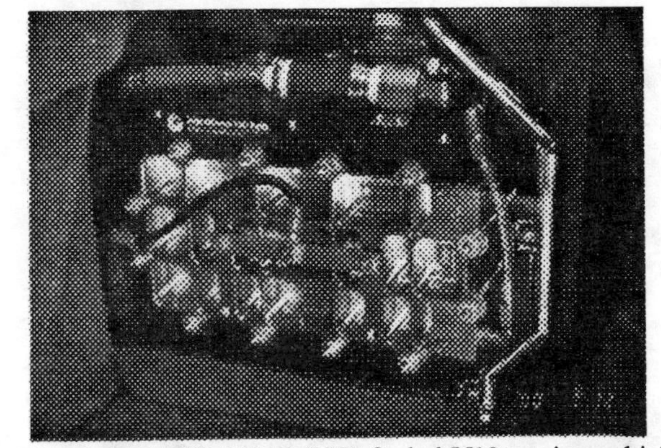

Figure 11 BMW 750-lh LH2 fueled V12 engine vehicle PEM fuel cell auxiliary power unit right trunk area [13]
[L:\hpias_00\bmwfcaux.jpg}

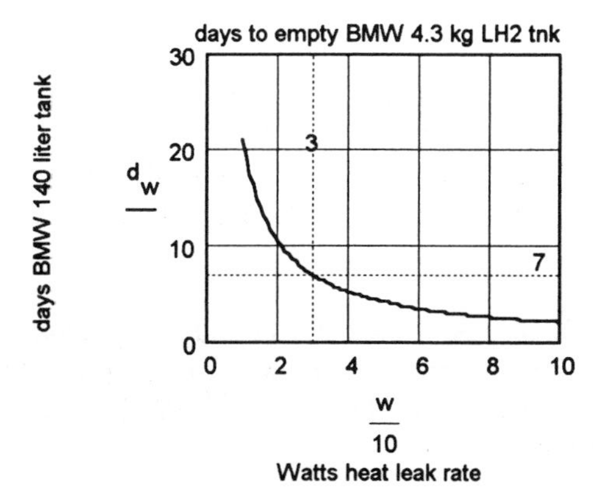

Figure 12 Heat leak rate of 3 Watts will deplete the LH2 contents of the BMW 140 liter, 4.3 kg LH2, cryogenic fuel tank in about 7 days.
[Tiger01 f:\mcd0\h2hydrid.mcd]

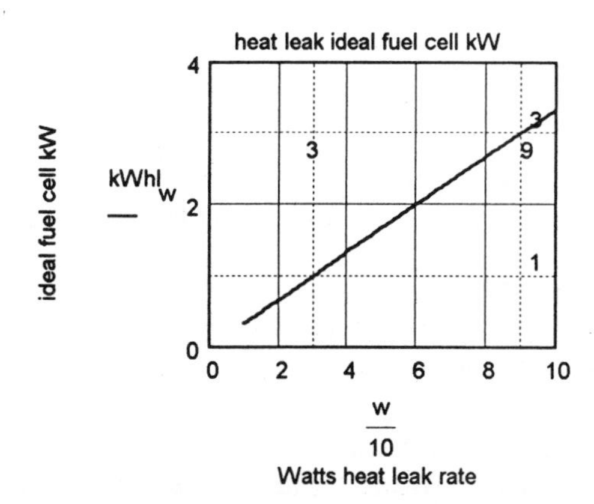

Figure 13 Heat leak rate of 3 Watts, the BMW fuel cell APU can generate about 0.8 kW from H2 boil off gas at a fuel cell conversion efficiency of 80%
[Tiger01 f:\mcd0\h2hydrid.mcd]

CONCLUSIONS

This evaluation of hybrid electric vehicles and hydrogen fuel systems led to the following conclusions.

1] The automobile manufacturers have turned to the hybrid electric power train to resolve the range limitation, and (HAC) heating and air conditioning of battery electric vehicles.

2] Two classes of hybrids are being developed: One, the piston engine and electric drive hybrid; and, two, the PEM fuel cell and battery pack electric hybrid.

3] Beginning with the year 2000, it appears that the emphasis is on hydrogen (H2) fueled propulsion systems. Furthermore, H2 is the fuel of choice for conventional piston engines and as well as the PEM fuel cell hybrid.

4] Two types of hydrogen fuel tank storage for the automotive transportation application are under development: one, Lanthanum Nickel [LaNi5] metal hydride absorption of gaseous H2, and, two, the cryogenic low temperature liquified LH2 fuel tank.

5] While the well known LaNi5H6 metal hydride absorption concept theoretically has a 2 to 1 advantage over LH2 , when reduced to practice, both are about equal on an energy volume density basis, 3.2 kWh/liter for LaNi5H6, and 2.8 kWh/ liter for cryogenic LH2.

6] However, on a mass density basis LaNi5H6 exhibits a significant weight penalty due to the mass of the metal hydride alloy. As an example, about 36 kg LaNi5 alloy are needed per kg of H2 gas absorbed.

7] The LaNi5H6 metal hydride alloy system exhibits practical isotherm dissociation H2 gas pressures ranging from 1 atm at 0 deg C to 2.5 atm at 25 deg C, and for this reason appears to be a practical H2 fuel tank storage method, in the automotive passenger vehicle application. [1 atm = 101.325 kPa = 14.695 lbf/sqin]

8] While the cryogenic LH2 system has the significant advantage of lower fuel tank mass (weight), reduction to practice results in a minimum heat leak which continuously boils off H2 gas. As an example, a 3 Watt heat leak will deplete the contents of the BMW 140 liter cryogenic LH2 fuel tank in about 7 days.

9] However, BMW has developed a unique solution by feeding the H2 gas boil off to a PEM fuel cell APU mounted in the vehicle trunk. The APU generates electric power, and, permits automatic heating and /or air conditioning the vehicle interior. As an example, a 3 Watt heat leak H2 boil off feed to the fuel cell APU delivers about 0.8 kW and converts the H2 gas to water.

10] It appears the major automotive manufacturers in Europe, Japan, and the United States are laying the foundation for deploying vehicle power train systems which utilize hydrogen as the fuel of choice, for the conventional piston engine, and, for the PEM fuel cell hybrid electric vehicle.

11] It is highly likely, that the 21st Century will witness the development and deployment of a hydrogen fueled transportation system. Which in turn will mitigate environmental pollution from our current conventional hydrocarbon based transportation system

REFERENCES

[01] General Motors (1998) ***"Driving the Next Generation EV1-II"*** GM press kit *Advanced Tech'ngy Vehicles, Spec*, NAIAS, Detroit MI, 05 Jan 1998

[02] Reisner, DE; Cole, JH; Klein, M (1996) ***"Bipolar Niclel-Metal Hydride EV Battery"***, Proceedings EVS-13 vII p37-41 Osaka Japan Oct 13-16 1996

[03] Nagano, M; Oyama, K; Sato, F; et al (1996) ***"Development of Nickel-Metal Hydride Battery for Electric Vehicles"***, Proceedings EVS-13 vI p457-464 Osaka Japan Oct 13-16 1996

[04] Wyczalek, FA (1996) ***"Passenger Electric Vehicles in Europe-1995 Frankfurt Automobile Show 56th IAA - International Automobile Ausstellung"*** IEEE AES Aerospace & Electronic Systems Magazine, v11 n02, pp11-12 Feb 1996

[05] Wyczalek, FA (1996) ***"Ultra Light Electric Vehicle Parameters"*** IEEE AES Aerospace & Electronic Systems Mag, v11 n01, pp40-44 Jan 1996

[06] Wyczalek FA (1995) ***"Ultra Light Electric Vehicles (EV)"*** Jr of Circuits, Systems, and Computers, v5 n1 pp 81-91, World Scientific Publishing Company, Singapore 1995

[07] Wyczalek, FA (1995) ***"Ultra Light Electric Vehicle Parameters"*** IPC-8 The Eighth International Pacific Conference on Automotive Engineering proceedings, v2 paper 95-31110, pp121-126, Yokohama, Japan, 04-09 Nov 1995.

[08] Wyczalek, FA (1995) ***"Ultra Light EV Global Design Characteristics"*** 28th ISATA 1995 proc, Vol-Advanced Transportation Systems paper 95ATS022 pp127-134, Stuttgart, FRGermany, 18-22 Sept 1995.

[09] Wyczalek, FA (1995) ***"Electric Vehicle Parametric Analysis"*** IECEC95 30st Intersociety Energy Conversion Eng Conf Proc 1995, paper ASME-95-360, Orlando, Lake Buena Vista,FL, 30JlyAug 1995

[10] Wyczalek, FA (1994) ***"Future Battery Electric Vehicle Technology"*** IECEC94 29th Intersociety Energy Conversion Engineering Conf Proc paper AIAA-94-3921, Monterey CA, 07-11Aug 1994

[11] Wyczalek, FA (1993) ***"Les Véhicules Batterie Eléctric-Critique/Mathématique/Évaluation"*** Proc VP/CV'93 Intnl Conf & Exhibtn Véhicules Propres, Clean Vehicles, paper 4085-C.34, LaRochelle, France, 15 Nov 1993

[12] Wyczalek, FA, Wang, TC (1992) ***"Regenerative Braking for Electric Vehicles"*** 25thISATA Zero Emission Vehicles, pp427-432 no. 920224, Firenzi, Italy, 1-5 June 1992

[13] ***AIAS 2000 North American International Auto Show*** Cobo Conv Center Detroit MI Jan 9-23 2000

[14] ***33rd Tokyo Motor Show*** (1999) Makuhari Messe, Makuhari Japan, Oct 17 Nov 3 1999

[15] Wyczalek, FA (1998) ***"Market Mature 1998 Hybrid Electric Vehicles"*** 33rd IECEC Proc paper 98-174 ANS Colorado Springs CO , 2-6 Aug 1998

[16] Wyczalek, FA (1998) ***"57th Frankfurt Motor Show Hybrid Electric Vehicles"*** IEEE Aerospace Electronics Systems Magazine ISSN-0885-8985, v13 n2 pp36-38, Feb 1998

[17] Nagasaka (1998) ***"Development of the Hybrid/Battery ECU for the Toyota Hybrid System"*** Nagasaka, Nada, Hamada, Hiramatsu, Kikuchi, Kato, SAE paper 981122, SP-1331 Technology for Electric and Hybrid Vehicles, pp19-27, SAE Intern'l Cong Expo, Detroit MI Feb 1998

[18] Toyota (1997) "***Toyota Hybrid System***" Press Kit 1997', Toyota Motor Corporation, Planning Group, International Public Affairs Division, 1 Toyota-cho, Toyota City, Aichi 471 Japan, October 1997

[19] Abe, S, et al (1997) ***"Development of Hybrid System for Mass Productive Passenger Car"*** Abe, Sasaki, Matsui, Kubo JSAE pp25-28, paper number 9739543, Oct 1997

[20] Wyczalek, FA (1996) ***"Hybrid Electric Vehicles in Europe and Japan"***, 31st IECEC'96 Proceedings, IEEE 96480 pp 1-6, 9 ref, Washington DC, Aug 1996

[21] Wyczalek, FW (1996) ***"Hybrid Electric Vehicles (EVS-13 Osaka)"*** IEEE Technical Applications Conference Record pp409-412, Northcon'/96, Washington State Conv Ctr, Seattle, 04-06 Nov 1996

[22] Wyczalek, FA (1999), *"US EPA 1998 Certification Label MPG Correlated with Consumers Union Fuel Economy Tests of 114 Vehicles"*, 34th IECEC paper 1999-01-2469, , Vancouver, 1Aug 1999

[23 Ogino, S; Kimura, Y (1996)*"Fuel Cell Powered Electric Vehicle"* Proc EVS-13 vI p671-674 Osaka Japan Oct 13-16 1996 Pt Ruthenium Anode Pt Cathode

[24] Kawahara,K; Haga, T; Suzuki, T; Asaoka, T (1996) *"Membrane Electrode Assemblies for Polymer Electrode Fuel Cells"* Proceedings EVS-13 vII p713-717 Osaka Japan Oct 13-16 1996, Du Pont Nafion 115

[25] *"Science & Technology Encyclopedia"*– 8th ed v11 Met-Nik Metal Hydrides pp 46-49 McGraw-Hill

[26] Witham, C; Bowman, RC Jr; Fultz, B. (1996) *"Gas-phase H2 Absorption and Microstructure Properties of* $LaNi_{5-x}Ge_x$ *Alloys"* Journal of Alloys and Compounds, Proc. 1996 Intnl Sym on Metal-Hydrogen Systems, Fundamentals and Applications, Aug 25-30 1996 p574-578 0925-8388 JALCEU

[27] Bosch, Robert (1993) *"Automotive Handbook"* 3rd ed pub Robert Bosch GmbH Dept for Technical Info

[28] Avallone, EA, Baumeister III, T (1997)."Marks' *Standard Handbook for Mechanical Engineers*" 10th ed McGraw-Hill 1997

[29] Millard, EB, (1953) *"Physical Chemistry for Colleges"* 7th ed p20 gas constant, p339, 563 van't Hoff equation, McGraw-Hill

TRANSPORTATION SAFETY ISSUES_ROLE OF FORENSIC MODELING

Floyd A. Wyczalek *
FW Lilly Inc.
155 S. Williamsbury Road
Bloomfield Hills MI 48301-2761
248-646-9585 FAX USA
Internet:75467.2677@compuserve.com

ABSTRACT.[1]

This paper presents three examples of forensic analysis related to airline safety. Examples of two recent fatal events are the Egyptair 990 incident and the TWA800 incident. These two examples illustrate the application of computer modeling as an aid in interpreting and understanding fatal event data. The third example illustrates the statistical analysis of the commercial airline fatality data published in the Statistical Abstract of the United States. This fatality data places the Egyptair 990 and TWA800 events into perspective. A constraint was to compare commercial aviation fatalities on a per departure basis, now in use by the NTSB National Transportation Safety Board and by the European airline industry, in contrast to the familiar, US airline industries distance traveled basis. The scope of the third example includes fatality statistics for scheduled airlines on a worldwide basis to establish an overall baseline. This part is followed by a similar analysis of the fatality data for the scheduled United States airline industry, in international and domestic passenger service. General aviation is specifically excluded since the fatality rate is almost seven times the scheduled passenger airline industry fatality rate.

INTRODUCTION

There are important differences between terrestrial passenger vehicle hazards and scheduled aircraft hazards. In automobiles, 84% of the fatalities are occupants of the vehicles and the remaining fatalities are occupants of fixed objects, or on motorbikes and bicycles, and/or pedestrians. In aircraft less than 1% of occupant fatalities involve other aircraft. Over 99% of the occupant fatalities are due to impact with a fixed object, the earth; at a remarkably steady mean impact rate worldwide of one fatal incident every two weeks, over the 29 year interval since 1970.

Furthermore, the mean fatality statistic for passenger automobiles is less than one occupant fatality per fatal vehicle incident. For US commercial aircraft fleet, 5000 scheduled-in-service aircraft, the 1970 to 1998 mean is 36 occupant fatalities per fatal incident [1,2].

Another little known factor, for the US scheduled-airlines the fatalities occur within the very small aircraft population of about 5000 aircraft-in-operation [3]. This small population of certificated aircraft make up the U.S. national domestic and major carrier services. In this case, national carriers are defined as having revenues of $75 million to $1 billion and major carriers revenue exceeds $1 billion. In contrast, the US automotive population includes 194 million vehicles-in-use, including trucks [4].

In general, we concluded that comparisons of fatalities based on a per fatal crash incident and per departure basis, improve our understanding of the hazards associated with the alternate modes of transport currently available [5]. Furthermore, for the occasional air traveler, the hazard of flight is better characterized by the *fundamental probability* parameter occupant fatalities per scheduled flight departure.

Although this study included comparisons to motor vehicles-in-use on-the-roads for personal and commercial transportation, operating within the continental United States, the scope of this report is limited. It is confined to illustrating three examples of computer modeling, relating to interpretation of contributing factors, associated with the Egyptair 990 and the TWA800 events; and, the commercial passenger airline industry fatality statistics. Consequently, terrestrial automotive vehicle results are excluded, and available in a previous report [5].

[1] *Author ,President of FW Lilly Inc

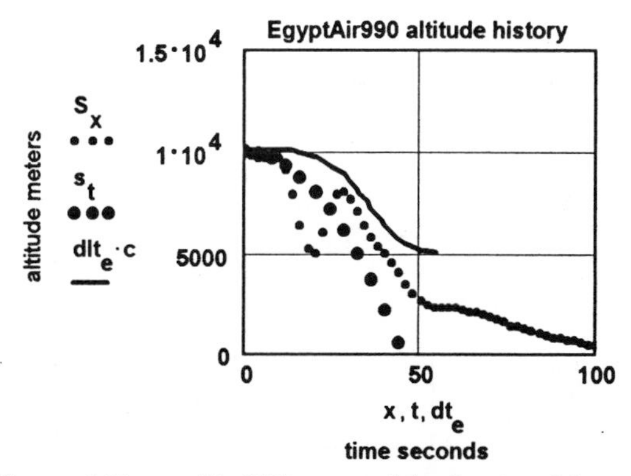

Figure 1 Egypt Air 990 event altitude_time history and nose down free fall ballistic trajectory in a vacuum
[F:\mcdx_egp\egypt_y.mcd Tiger01 wd6.2gbhd]

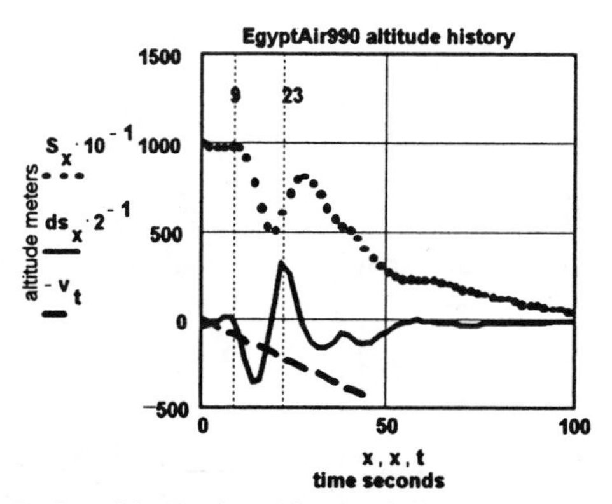

Figure 2 Radar altitude time history, inferred vertical velocity profile, and free fall velocity in a vacuum
[F:\mcdx_egp\egypt_y.mcd Tiger01 wd6.2gbhd]

EGYPTAIR FLIGHT 990 EVENT

The history of the last moments of Egyptair flight 990, on 31 October 1999, at about 0150 EST, a scheduled international flight from New York to Cairo, a Boeing 767-366-ER, crashed in the Atlantic Ocean about 60 miles South of Nantucket Island, MA [6]. Egyptair 990 was cruising at 20510 meters (33000 ft), 3 nautical miles South of the Nantucket radar. The ground track was 66 degrees and ground speed was 483 knots. The Boeing 767 started a descent from 20510 meters at 06:49:52 universal time (01:49 am EST).
[NTSB identification Docket DCA00MA006]

During the descent, there were 9 radar returns from the Nantucket radar containing altitude data. These returns show a descent from 20510 meters (33000 ft) to 10379 meters (16700 ft) in 24 seconds, an average rate of descent of 14916 m/s (24000 ft/min). The aircraft did not turn during the descent.

Later, more radar data was received from four facilities: Boston Center, New York Center, and Boston Approach Control on Nantucket, and Air Force 84th Radar Evaluation Squadron. Figure 1 shows the inferred altitude profile, (Sx) small dot symbols, based on radar altitude data, in comparison to a trajectory for free fall, (St) large dot symbols, in a vacuum.

Radar altitude data suggests Egyptair dropped faster than a free fall altitude model. However, this observation was later contradicted by (FDR) flight data recorder, pressure_altitude data, (dlte) solid plot curve, released 13 Nov 1999, which suggests another conclusion. This contradiction has not been resolved by the NTSB and released to the public domain, at this point in time.

Key events from the FDR show, initially at 20510 meters altitude and 1 second into the fatal event, the autopilot cutoff. At 9 seconds, the elevator moved to the nose down position, and the throttle pulled back to reduce power. At 23 s, the aircraft reached Mach 0.86, and the master warning sounded. At 36 s, the engine start lever went to "off". At 50 s, the FDR and transponder record stopped, and FDR data ended. Just before FDR cutoff, the aircraft moved from 40^0 nose down to 10^0, last altitude, 10193 meters, at true air speed of 574 knots.

FDR data is conclusive, Egyptair was in a 40^0 nose down dive during the 9 s to 23 s interval, ending in 10^0 during the last seconds before FDR cutoff.. This appears to be supported by the radar altitude data, (Sx) small dots, in figure 2, which compares aircraft vertical velocity profile, (dsx) solid curve, inferred from radar altitude data (Sx), to free fall velocity, (Vt) dash curve, suggesting the aircraft exceeded free fall velocity in a vacuum, in the 9 to 23 second interval. This observation is contradicted by FDR altitude data.

The aircraft then climbed back to about 14916 meters (24000 ft), before beginning a fatal plunge into the Atlantic. After the first 50 seconds at FDR failure, the radar altitude data trails off at a slower descent rate, characteristic of descending airframe breakup.

The Egyptair aircraft altitude and velocity values depicted in figure 2 have been adjusted to permit comparisons of the altitude and velocity profiles on one graph. As examples, altitude data divided by 10, velocity divided by 2. Further, velocity magnitudes cannot be considered absolute, and the velocity trajectory was evaluated only on a relative basis.

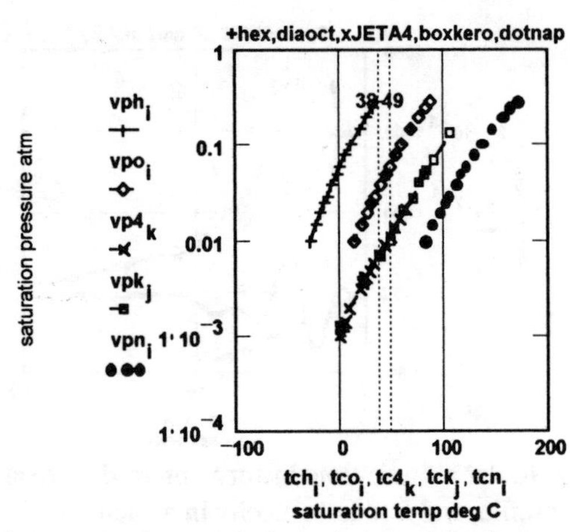

Figure 3 Reference fuels and aircraft Jet fuel saturation pressure in atm vs saturation temperature in degrees Celcius for TWA800 event
[F:\mcdx\ftnc8206.mcd pg9 Tiger01 wd6.2gbhd]

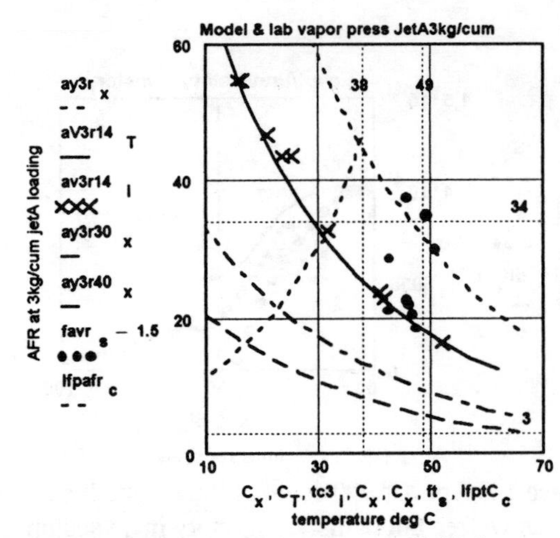

Figure 4 Correlation of laboratory and flight test data with mathematical model shows effect of altitude on TWA800 event fuel tank air fuel ratio
[F:\mcdx\ftnc8206.mcd pg16 Tiger01 wd6.2gbhd]

TWA FLIGHT 800 EVENT

The history of the last hours of Trans World Airlines, Inc., Flight 800 Boeing 747-131, N93119 fueled with low volatility kerosene type JetA in Athens, Greece, arrival New York Kennedy 17 July 1997, departure for Paris, essentially empty center fuel tank of 50 m^3 (12980 gallons) containing about 200 liters (52 gallons) of Athens refinery JetA fuel, exploding 11 minutes after takeoff at 4176 m (13700 ft) altitude, 12 km (8 miles) South of East Moriches, New York, 17 July 1997[7]
[NTSB Docket SA-516.]

The mathematical model shown by Wyczalek and Suh[11] estimates air-fuel mass ratios (AFR) as a function of fuel saturation temperature (boiling points at various pressures). As examples, the multi component JetA4 aviation fuel saturation properties are shown in figure 3 as, (vp4k) x symbol, and kerosene as, (vpkj) box. In addition, figure 3 shows properties of three single component hydrocarbons: hexane (vphi) +, octane (vpoi) diamond, naphthalene (vpni) dot.

Finally, California Institute of Technology (Caltech) (CIT), Explosion Dynamics Lab., Shepherd et al[7], also measured saturated vapor properties of TWA800 LAXJetA aviation kerosene at 3 kg/m^3 of fuel tank volume. The 3 kg/m^3 value is believed to be the fuel loading factor in the center wing tank (CWT) of flight TWA800 at the time of the explosion event.

Further, a 38° C (100° F) temperature is the JetA lean limit flash point at sea level, and, 49° C (120° F) was the center wing tank (CWT) air/fuel vapor mixture temperature at the time of the TWA800 event,. This Caltech 3 kg/m^3 JetA saturation property data was the basis for the air fuel ratio model[7]

Figure. 4 illustrates the effect of altitude on AFR at a fuel loading factor of 3 kg/m^3, as predicted by the mathematical AFR model shown by Wyczalek and Suh[11]. As examples, the model shows (ay3rx) short dash - at sea level, (aV3r14) solid curve at 4267 m (14000 ft) for the TWA800 event altitude, (ay3r30) dash-dot at 9144 m (30000 ft), and (ay3r40) dash curve at 12192 m (40000 ft).

An example of test data correlation with the AFR mathematical model, Caltech test data (av3r14)"X" symbols in figure 4 for LAXJetA3 aviation kerosene at a fuel loading factor of 3 kg/m^3, corrected to the partial pressure of air at 4267 m altitude, correlate with the curve predicted by the AFR model at 4267 m, (aV3r14) solid curve. Similar plots were made to validate correlation at seal level, 9144 m, and 12192 m.

The air fuel ratio model was validated with nine flight test AFR data points, dot symbol, obtained with Athens JetA at a 3kg/m^3 fuel loading during NTSB TWA800 flight emulation, Sagebiel et al[7]. As examples, AFR data (favr) black dots, for three center wing tank clearance space samples are shown at sea level, three points at 3048 m (10000 ft), and three points at 4267 m (14000 ft). Three sea level points are just above the model sea level AFR dotted curve. The three 4267 m points are tangent to the 4267 m model AFR solid line.

WORLDWIDE AIRLINE FATALITIES

For scheduled passenger transport operations, worldwide airline fatalities for the interval from 1970 to 1998 are summarized in the following figures. Figure 5 shows a plot of occupant fatalities per year vs fatal crashes per year[1].

This raw data exhibits an anomalous frequency distribution. As examples, fatalities per year range from about 223 to 1299 while fatal crashes per year range from about 16 to a high of 41 annually. This dispersion is due to the relatively large variation in occupant capacities of jet aircraft and/or seasonal variation in seat utilization.

Figure 6 shows a plot of cumulative occupant fatalities vs cumulative fatal events for this data. In this case, the cumulative plot shows a remarkable regularity and a linear regression line shows a Pearson's correlation coefficient of 0.999 (Pearson 1958). This linear cumulative behavior was unexpected in view of the data dispersion previously shown in figure 5. The *fundamental probability*, slope of a linear regression line, is about 30 fatalities per fatal crash

Figure 7 shows the cumulative running averages for fatal crashes and occupant fatalities with base year 1970. The running averages converge on a probable 798 fatalities=o per year and 26 fatal crashes=+ (divide by 10) annually in 1998. Figure 8 shows a plot of the ratio of occupant fatalities per fatal crash event which converges on a probability of 31 fatalities per crash, on a worldwide basis over the past 29 years.
[F:\mcd0\airwor~1.mcd pg1 Tiger01 wd6.2gbhd]

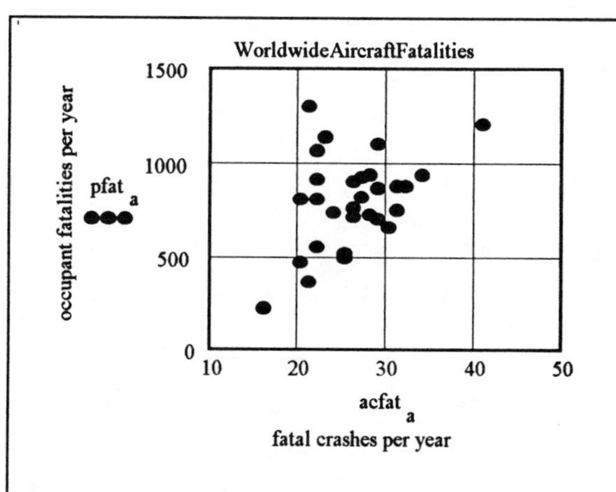

Figure 5 raw data fatalities per year and fatal crashes per year from 1970 to 1998 worldwide

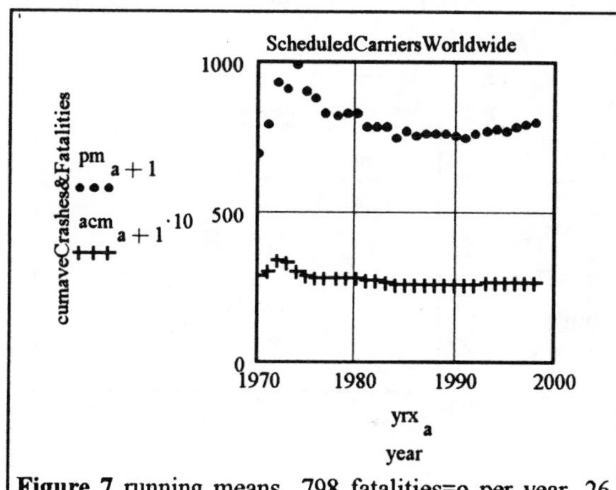

Figure 7 running means 798 fatalities=o per year, 26 fatal crashes=+ per year (divide by 10), 1970 to 1998

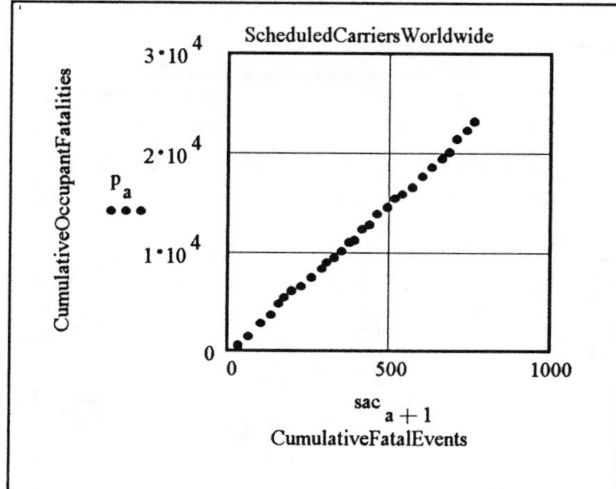

Figure 6 cumulative fatalities and events 1970 to 1998 worldwide slope=30 fatalities per crash

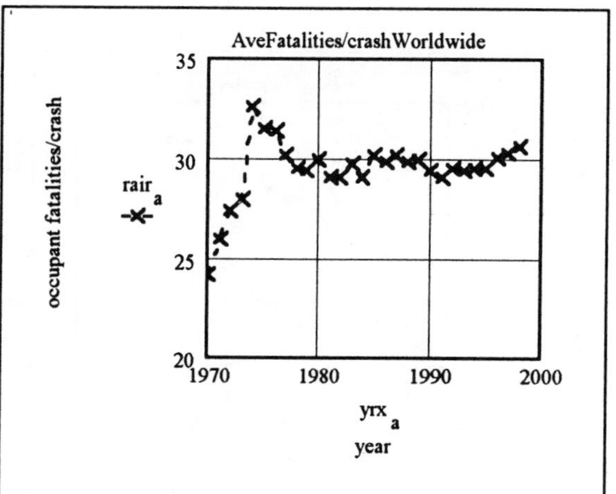

Figure 8 mean fatalities per fatal crash worldwide converges on 31 occupants per fatal event, 1970 to 1998

US DEPARTURES-PASSENGERS ENPLANED

Figure 9 shows the running mean for occupant fatalities per fatal crash converges on about 41 fatalities per fatal crash, for the US domestic and international scheduled airlines, over the 1977 to 1998 interval [2]

Figure 10 shows the number of revenue passengers enplaned annually vs the number of aircraft departures per year, for the US scheduled airline industry, for domestic and international operations. This data covers carriers certificated under Section 401 of the Federal Aviation Act. As an example, since these data are in millions, it shows 599 million revenue passengers enplaned and 8.2 million departures in 1997 [8]. The linear regression slope, revenue passengers enplaned per departure, was about 68 passengers per scheduled departure, Pearson's $r = 0.9999$.

Figure 11 shows cumulative occupant fatalities vs cumulative scheduled departures. The linear regression line exhibits a slope of about 18 fatalities per 10^6 departures, Pearson's r correlation coefficient of 0.982.

Figure 12 shows the ratio of fatalities per 10^5 scheduled departures from 1983 through 1997 [8]. The mean fatality rate, dash-dot curve, is a joint regression fit of the statistical data shown by the dot • points. In 1997 the *fundamental probability* fatality rate appears to be converging at just under 2 fatalities per 10^5 scheduled departures, among the US scheduled fixed wing jet fleet of just over 5000 aircraft [3] Pearson's r is 0.979.

For the occasional air traveler, a cumulative mean rate from 1983 to 1997, of 3 fatalities per 10^5 revenue passengers enplaned. And, for the entire period, a total of about 180 pilot fatalities. The pilot occupational rate for 1996 is 98 pilot fatalities per 10^5 pilots employed [9], since pilots are exposed to a maximum number of departures, by occupational necessity.

[F:\mcd0\airschxx.mcd pg1,4,5 Tiger01 wd6.2gbhd]

AveFatalities/crashUnitedStates

occupant fatalities/crash

rpvo c

yrm c

year

Figure 9 United States mean fatalities per crash converges on 41 occupants per fatal event, 1983 to 1998

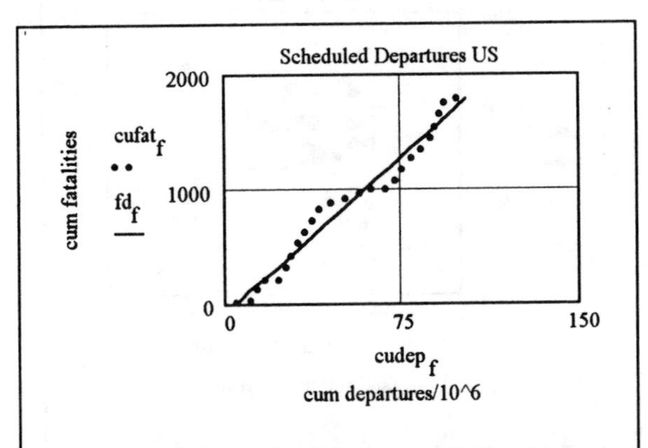

Figure 11 cumulative fatalities vs cumulative departures linear regression [solid line], slope=18 fatalities per 10^6 departures US certificated carriers 1983-1997

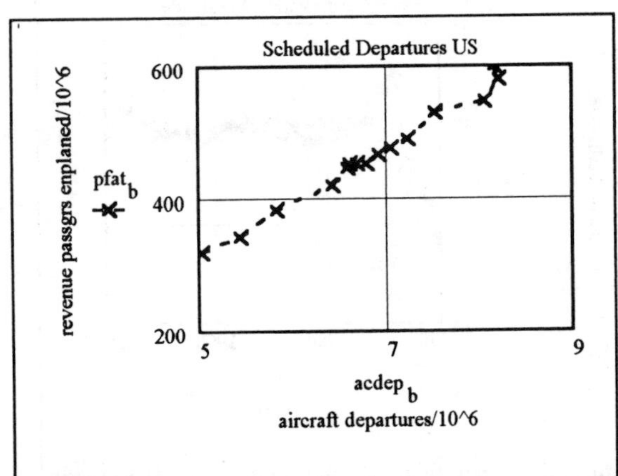

Figure 10 passengers enplaned per year vs aircraft departures per year US scheduled airlines 1983 to 1997

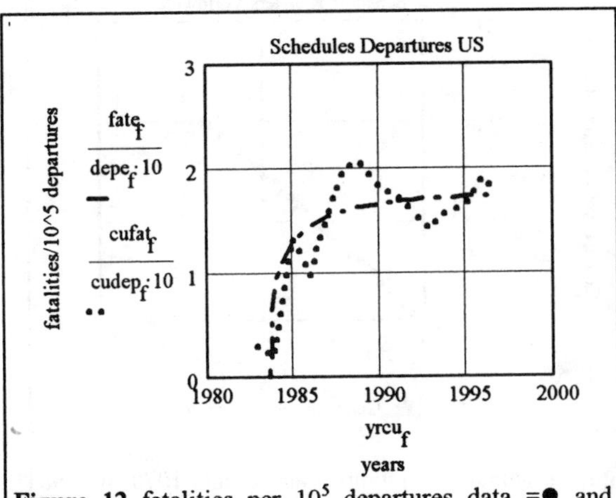

Figure 12 fatalities per 10^5 departures data =● and dash-dot joint regression, 1.8 fatalities per 10^5 departures

CONCLUSIONS

These three examples of transportation safety issues and the role of forensic mathematical modeling, resulted in the following conclusions:

1] *Egyptair flight 990 event* - Mathematical modeling applied to the radar altitude time profile date, suggested that the aircraft dropped faster than the free fall terminal velocity in a vacuum, in the Earth gravity field, during the initial 9 to 23 second event interval, after auto pilot cut off at the 20510 meter cruise altitude. However, this conclusion is contradicted by the NTSB 13 Nov 1999 FDR pressure altitude data. This conflict has not been resolved by the NTSB, nor explained in the public domain, at this point in time.

2] *Egyptair flight 990 event* - The FDR data shows that at 9 s, the elevator moved to the nose down position, and the throttle was pulled back to reduce power, at 23 s the aircraft reached Mach 0.86, and the master warning sounded. FDR data was conclusive, Egyptair was in a 40 degree nose down dive during the 9 to 23 s interval [6].

3] *Egyptair flight 990 event* - However, since the throttle was pulled back to reduce power, the initial kinetic energy of the aircraft at the cruise speed of 483 knots was a factor, in Egyptair exceeding free fall terminal velocity in a vacuum, during the 40 degree nose down 9 to 23 s interval.

4] *Egyptair flight 990 event* - At 50 seconds into Egyptair event, the FDR failed, and the radar altitude data also trailed off at an altitude descent rate somewhat slower than free fall velocity, indicating airframe breakup, and that aerodynamic free fall drag forces had become a dominant factor.

5] *TWA800 event* - Mathematical modeling illustrates the effect of altitude and temperature on the air fuel vapor ratio (AFR) within the clearance volume of the center wing tank (CWT). This model shows that the AFR test samples from the CWT of a similar Boeing 747, during a FTSB TWA800 flight simulation, carried out one year after theTWA800 event, correlate with the AFR mathematical model, at sea level and at the event altitude of 4267 meters [11].

6] *Worldwide fatal events* -, over a 29 year interval, the worldwide fatal aircraft incident rate has been remarkably constant at 26 crashes per year. Consequently, one fatal aircraft event is likely, on the average, every two weeks,

7] *United States Domestic and International events-* Within the US scheduled commercial passenger airline fleet of about 5000 fixed wing jet aircraft, in the past 16 year interval, the fatal event rate has been about 4 fatal crashes per year, or an absolute probability of 1 fatal aircraft per 1250 in service. This limited fleet-pool of aircraft contrasts with 123 million automotive vehicles-in-use on US highways [4], and including trucks, 194 million vehicles in-use-overall in 1995.

8] A more effective statistic for quantifying risk, in the case of the occasional air traveler, is based on fatalities per million scheduled departures [12]. As an example, in 1998, this probability converged at about 18 fatalities per million scheduled fixed wing jet aircraft departures.

9] Furthermore, for the occasional air traveler, the law of supermarket queues reduces risk significantly. As an example, given the top ten airport departure hubs [2], the hub you choose has only one chance in ten of encountering the fatal aircraft, since each hub does not have access to the entire 5000 US certificated jet aircraft fleet. Consequently, the probability product reduces risk to 1.8 per million [10].

10] Further, for the occasional air traveler, a cumulative mean rate from 1983 to 1997, of 3 fatalities per 10^5 revenue passengers enplaned. And, for the entire period, a total of about 180 pilot fatalities [2,8].

11] This statistical characterization of risk can be contrasted with the occupational hazard statistic of 98 pilot fatalities per 100 thousand airline pilots employed. Pilots, by virtue of occupational necessity, like truck and taxi drivers, are exposed to the maximum number of departures, by occupational necessity [9].

12] Further, these risk factors can be placed in perspective by comparisons to some of the top ten most hazardous jobs: 25 fatalities per 100 thousand truck drivers, 43 fatalities per 100 thousand taxi drivers, in addition to the 98 pilot fatalities per 100 thousand airline pilots employed [9].

13] Additional perspectives are provided for comparison in the years 1992-1994: smoking related malignancies of respiratory and intra thoracic organs 201 per 100 thousand smokers, infant mortality 840 per 100 thousand live births, abortions 37900 per 100 thousand live births [13,14,15].

DEFINITIONS

Sample fatality statistics for each year are regarded as '*a set*'. and events are fatal or not fatal. Fatal events involve at least one occupant fatality and pedestrian fatalities are excluded. Probability of occurrence is '*a posteriori*' since it is deduced from the statistical regularity of past data. *Stability of relative frequency of occurrence* of an event is determined by two methods: *one*, cumulatively summing annual statistics relative to a base year to obtain the slope of a linear regression fit; and, *two*, cumulative running averages relative to a base year. Since this paper deals with *non-finite sets*, eg. toss of a coin is a finite set, the term *fundamental probability* applies, generally, to the characterizations: means, ratios, and probabilities, shown in this paper, with the exceptions noted as *absolute probability*.

REFERENCES

1] Statistical Abstract of the United States **"Worldwide Airline Fatalities",** 1991 and 1998, 111th ed table 1075 p628 & 116th ed table 1043 p647 for scheduled air transport operations, 1999 119th ed table 1074 p663

2] Statistical Abstract of the United States, **"US Scheduled Airline Fatalities, fixed wing jet aircraft in operation"**: 1982-83 103rd ed table 1102 p635, 1984 104th ed table 1102 p635, 1986 106th ed table 1084 p619, 1989 109th ed table 1148 p613, 1990 110th ed table 1066 p622, 1991 111th ed table 1076 p629, 1992 112th ed table 1043 p619, 1993 113th ed table 1063 p637, 1995 115th ed table 1068 p657, 1996th ed table 1045 p647, 1999 119th ed table 1076 p663

3] Statistical Abstract of the United States **"US Certificated Aircraft in Operation, fixed wing turbojet"** 1982-83 103rd ed table 1099 p633, 1986 106th ed table 1081 p617, 1989 109th ed table 1044 p611, 1991 111th ed table 1072 p627, 1992 112th ed table 1038 p625, 1993 113th ed table 1058 p636

4] Statistical Abstract of the United States **"US Motor Vehicles in Use"**, 1999 119th ed table 1032 p639

5] Wyczalek, FA, **"Transportation Safety_New Math Lessons Learned",** 32nd IECEC Intersociety Energy Conversion Engig Conf 1997 proceedings AIChE paper 97137, Honolulu, Hawaii, 27Jly-01Aug 1997

6] **"DCA00MA006 NTSB Identification Docket"** Scheduled 14 CFR 129 operation of EGYPTAIR Accident occurred 31 October 1999 off Nantucket Island, MA, Aircraft: Boeing 767-366-ER, registration SUGAP, injuries 217 fatal.

7] NTSB Identification Docket **"SA_516 Fire and Explosion Group Factual Report" dated 27 Nov 1997,** Scheduled operation of Trans World Airlines, Inc. Flight 800, Boeing 747-131, N9311, Fatal Event occurred 17 July 1996. 8 miles South of East Moriches, New York, injuries 230 fatal.

8] Statistical Abstract of the United States **"US Departures and Passengers Enplaned, Passengers Enplaned per Aircraft per Year"**: 1993, 1996, 113th ed table 1056 p636, 116th ed table 1039 p645, 1999 119th ed table 1070 p661

9] Bureau of Labor Statistics, **"The Most Dangerous Jobs"**, *1996*, Bureau of Labor Statistics of the US

10] Burington, R.S., and May, D.C. Jr., 1970, **"Handbook of Probability and Statistics with Tables"** 2nd edition, Certain definitions used in statistics, Combinatorial mathematics, Elementary probability theory, Algebra of events, Conditional probabilities, McGraw-Hill Book Co

11] Wyczalek, FA, **"Fuel Tank Combustible Mixtures_New Lessons Learned",** 32nd IECEC Intersociety Energy Conversion Engineering Conference 1997 proceedings AIChE paper 97004, Honolulu, Hawaii, 27Jly-01Aug 1997

12] Birch, Stuart, **"Safety_Hull Losses All Aircraft Types per million Departures",** European AirBus Industrie 4th Flight Safety Conf 1997, SAE Aerospace Engineering magazine, p20 August 1997

13] Statistical Abstract of the United States **"US Vital Statistics, 1996"** , 116th ed section 2, pp87,94,98,103,145

14] Scientific American, 1996, **"What You Need to Know About Cancer"**, ScAm Magazine Fact Sheet, pp 128-132, v275 n3 Sept 1996

15] Warner, Kenneth, 1989, **"Smoking and Health-A 25 year Experience"**,American Journal of Public Health, v79 n2, pp141-143, Feb 1989

AIAA-2000-2857

MODIFICATIONS TO A HYDROGEN/ELECTRIC HYBRID BUS

Y. Baghzouz*, J. Fiene#, J. Van Dam#, L. Shi#, E. Wilkinson#, R. Boehm+
Center for Energy Research, University of Nevada, Las Vegas
Las Vegas, Nevada

And

T. Kell
Kell's Automotive
Las Vegas, Nevada

ABSTRACT

A program to modify a hydrogen hybrid electric bus is described. This is an electric bus that uses a hydrogen-fueled internal combustion engine to charge on-board batteries. The primary goal of the reported work is to extend the range of the bus in a cost-effective manner. Among the modifications being made are the development of a new engine, upgrading the power control system, rewiring critical elements, adapting a supercapacitor regenerative braking system, devising more effective changes to the safety hydrogen systems, and adding a high pressure gas storage system to supplement the existing hydride bed hydrogen storage. All of this work is being supplemented with the use of a dynamic simulation code to determine optimal operational strategies.

INTRODUCTION

A hydrogen-powered bus was constructed for the Atlanta Summer Olympics. This bus began as an all-electric vehicle that carried four battery packs in under-vehicle compartments. Two of the battery packs were removed and replaced by metal-hydride beds for hydrogen storage. Then the rear portion of the bus was modified so that a large, industrial V-8 engine with large electrical generator could be installed in that location. This engine was modified from its standard gasoline form to one that was specially designed for hydrogen fuel.

The system was designed that so that all of the electrical energy went through the batteries. When the battery state-of-charge reached sufficiently low levels, the driver would start the engine to accomplish a recharge. When the state-of-charge was restored to a predetermined level, the driver would then turn off the engine. More complete information on the initial vehicle design can be found in references 1 and 2 at the end of the paper.

Several problems were encountered with the bus operation. One of these was that the range was deemed to be too small. Another was that oil consumption of the engine was reported to be particularly high. Premature battery system failure was observed. This may have been related to the operation of the regenerative braking system that came with the bus. The electrical power controller was found to overheat. In addition, the flammable gas sensors that were an integral part of the hydrogen safety system of the bus did not have the appropriate hydrogen selectivity necessary.

The bus was shipped from Georgia to Nevada near the end of 1998. In 1999 a contract was awarded to the Center for Energy Research at the University of Nevada, Las Vegas, to improve the bus operation. A multifaceted effort was initiated to remedy as many of the issues as possible within budgetary constraints. This paper is a description of some of those efforts.

ELECTRICAL SYSTEM

The current electrical system of the H2 Bus consists of three main components the auxiliary power unit, the battery bank, and the electric drive system. A fourth major component that is currently under development is the supercapacitor bank. These components are connected as shown by the schematic

---

*Professor of Electrical Engineering.
#Research Assistant.
+Professor of Mechanical Engineering.

diagram in Figure 1. A short description of each component follows below.

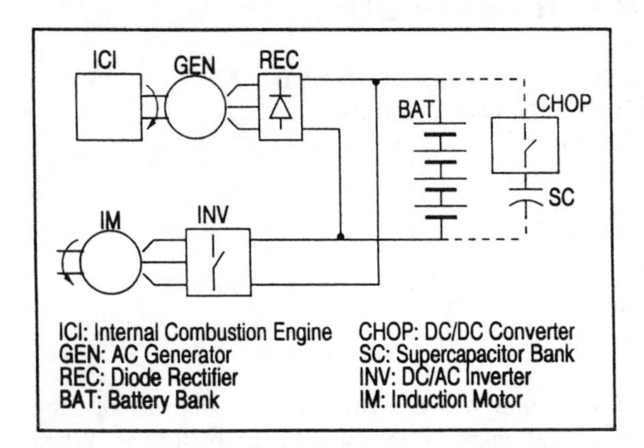

Figure 1. The electrical system layout of the bus.

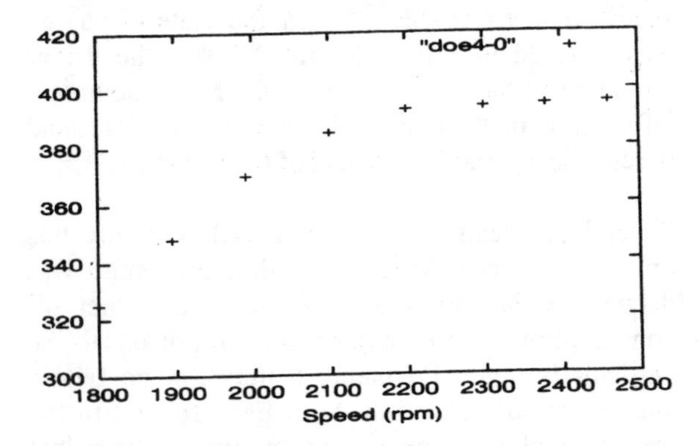

Figure 2. Open circuit voltage-frequency curve for the original generator.

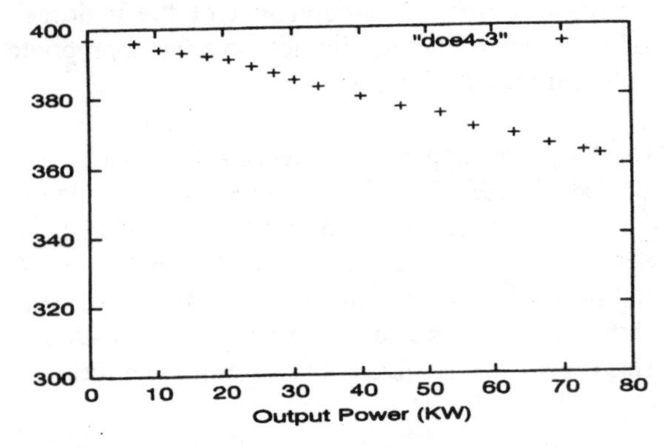

Figure 3. Original generator voltage-power curve (75 Hz/2250 rpm).

Auxiliary power unit

The auxiliary power unit consists of an engine (described in a later section of this paper) that drives a 4-pole, 277 V, 80 Hz, 70 kVA synchronous generator. The generator AC voltage is converted to DC through a diode rectifier. The generator performance characteristics were determined by conducting laboratory tests last summer. Some of the recorded data are displayed in Figures 2 and 3 which show the open-circuit DC voltage variation with rotor speed (under no-load) and with output power (under a resistive load), respectively. The generator-rectifier efficiency at full load was estimated near 83% with a voltage regulation of 8%. Plans are made to replace the generator unit with a larger and more efficient one.

Battery bank

The battery bank consists of two parallel sets of deep-cycle VRLA batteries, each containing a series connection of 28 units. Each unit is rated at 12 V, 85 Ah @ C/3. The total battery system operates at 336 V with a capacity of 170 Ah @ C/3. The equivalent maximum energy storage is 57 kWh, but only 80% of this energy (i.e., 46 kWh) can actually be used since VRLA batteries are not allowed to be discharged below 20 state-of-charge (SOC).

The original bus batteries reached their life cycle prior to the bus transfer from Georgia. Consequently, new batteries have recently been installed. Monitoring of the battery cells is accomplished by a battery management system that provides several advantages including charging of individual units to help restore them to normal state. When the bus is not in operation, the battery bank state of charge is maintained with an external battery charger which requires 480V, 100 A, 3-phase power supply. The new batteries are equipped with internal thermistors to determine the temperature of each battery. One of the current tasks related to bus improvements is considering aspects of battery thermal management that would extend battery life.

Electric drive system

The electric drive system is composed of an IGBT-based inverter that converts the DC voltage to three-phase pulse-width-modulated (PWM) AC voltages. These voltages are fed to an AC induction motor that is capable of producing 230 hp in addition to its 250 ft-lb of torque. The traction motor is wound with dual 3-phase windings and is capable of continuous operation at a maximum speed of 12,500 rpm. Both the motor and inverter electronic switches are oil cooled. Drive motor and traction battery current flow are monitored and controlled by an electronic subsystem known as the chassis vector controller (CVC). Due past thermal problems which limited the maximum power demand, the inverter has been upgraded to a more efficient and more heat-tolerant one.

Supercapacitor bank

The bus came equipped with a regenerative braking system which converts the vehicle kinetic energy back to electrical energy during braking. This feature is known to increase the driving range by up to 20%. Currently, the regenerative braking uses the battery bank to store the energy gained. Since VRLA batteries are not designed to accept large, short bursts of power, especially when near full charge, conventional regenerative braking may cause them to overcharge and result in battery damage. For this reason, the existing regenerative braking system is temporarily disabled.

A supercapacitor bank in series with a DC/DC chopper (currently at the design stage) is being developed to act as an energy buffer: It has the ability to store energy as quickly as needed, and then supply back this energy when required during acceleration.

Miscellaneous

Much of the original bus wiring was found to be very poorly documented as well as poorly installed. In many cases in violations with the NEC code were found. Much of the electrical wiring has been corrected, re-wired, and documented.

A data acquisition system (DAS) was included in the original bus system. This apparently is not designed for the expected rugged condition in normal service, its ports and connectors are not standard, and its programs run on a special computer. The entire DAS is being analyzed for further improvements.

ENGINE

The original engine

An area that is being closely examined for increase in range is the hydrogen-fueled engine. The necessary power output for the engine was established as 70 kilowatts at a speed of 2500 revolutions per minute. This was determined as a function of the operating range of the generator. Changes that had been made to the engine for the conversion to hydrogen gas included the addition of an HCI Constant Volume Injection system, the addition of exhaust gas recirculation, and modifications to the heads and pistons.

The Constant Volume Injection system is a unique sequential multi-port fuel injection method to meter the flow of hydrogen into each cylinder. The system is based upon the ideal gas relationship between pressure, volume and temperature. Not only do the pressure and volume effect the flow of gas into the engine, but changes in temperature can have adverse effects. To combat this, an electronic control system was incorporated into the injection arrangement.

An exhaust gas recirculation system was employed to reduce emissions. The system drew the exhaust from one bank (four of the eight cylinders) of the engine through four parallel coolers, into a condenser and then into a mixer at the intake manifold. This did increase the load on the engine but it was assumed that it would not drop the power output below the desired level.

The engine heads were replaced to increase the compression ratio to then improve the brake thermal efficiency. In an attempt to increase further the compression ratio and to reduce the excessive amounts of oil lost during operation, new pistons and rings were installed.

Test results, modifications and improvements

Shortly after acquisition of the bus from Georgia, extensive research was performed to determine what needed to be improved. It was decided that substantial changes needed to be made to achieve the desired performance, efficiency and emissions results. Of primary concern was increasing the operational range of the bus, which could be accomplished by improving the performance of the engine, among other aspects described in this paper.

To study what improvements could be made to the engine, it was first removed from the bus and mounted on an engine dynamometer. Extensive diagnostic tests were then performed in a controlled environment.

The use of hydrogen as a fuel for internal combustion engines has been researched greatly in recent years primarily because of its 'clean' burning properties. The greatest concern involved with emissions control is the production of NOx.

Nitrous oxide production is commonly triggered by high temperature 'hot spots' located within each combustion chamber. Coupled with high pressures within the cylinders and a relatively rich air-to-fuel ratio, levels of this harmful pollutant can easily become a problem. It was found that the rate of

NOx produced by the engine was unstable and varied between 6 and 55 parts per million as measured at the tailpipe. Reviewing the emissions data gathered via a synchronized dynamometer/exhaust analyzer in conjunction with thermistors located within each cylinder, the largest production of NOx occurred while the greatest difference between cylinder temperatures existed.

The conclusion inferred from the correlation between increased NOx production and temperature variance between cylinders was the assumption that non-uniform burns were occurring within cylinders. Mixtures of fuel and air abnormally rich were believed to be increasing the temperatures of some of the cylinders. While the NOx readings generally registered below 6 PPM, they subsequently increased to figures as high as 45 PPM at points where cylinder temperature differences reached 106°F.

In an attempt to locate the cause of this non-uniformity, a pressure transducer was installed to measure the line pressure of the hydrogen upstream of the constant-volume injector pump. Readings from this sensor revealed violent oscillations in the pressure of the fuel supply. The design of the injection system did not allow for the compensation of this anomaly; in actuality it caused large variations in the amount of fuel delivered to each cylinder. This was most likely due to the compressibility of the gaseous fuel as it passed through the volumetric metering chambers of the injector pump. The excess hydrogen introduced to the cylinders may have been the catalyst for the production of the 'hot spots' and thus the high levels of nitrous oxide.

It was then determined that further analysis of the exhaust elements would be more useful once the engine's air/fuel supply could be more accurately controlled. It was assumed that the engine would exhibit less variable emissions readings once the combustion inside each of its cylinders could be more accurately predicted and controlled.

Fuel delivery

In order to determine the possible power output of the existing engine and to evaluate its integrity, dynamometer tests were preformed while the engine was run with several fuels. A performance test was conducted on the engine in its original condition using hydrogen fuel. In order to compare the output of the engine with its gasoline fueled counterpart, a carburetor (Holley 833 CFM) was mounted to the intake manifold and the engine was run on VP-C12 (108 octane) gasoline. To study the performance of the engine as run on gaseous fuels in general, Compressed Natural Gas (CNG) was introduced through the carburetor for the next test. These two initial tests were to evaluate the performance of the engine to comparable engines for which data was already available. The engine was then returned to its original setup for hydrogen with the addition of an electronic control unit capable of adjusting the overall air/ fuel mixture. The engine was then run with a leaner mixture.

Results of these tests indicated performance derating typical of the use of gaseous fuels. Comparative (at least generally so, since different types of fuel input systems were used) maximum outputs (in kW) were found at 3000 rpm as follows: gasoline-170; natural gas-130; hydrogen (A/F ratio of 30)-75; and hydrogen (stoichiometric)-45. It was desired to have higher outputs with higher efficiency than what was demonstrated with this engine.

Several characteristics of the existing Constant Volume Injection (CVI) system gave cause for concerns related to good performance. First and foremost, the mass of hydrogen metered by the CVI was directly dependant on the line pressures upstream and downstream from the pump. Secondly, the presence of petroleum-based lubricants in the piston/ cylinder design of the CVI allowed for the contamination of the hydrogen gas before it reached the combustion chamber.

These factors made the possibility of utilizing an Electronic Fuel Injection (EFI) system an appealing one. If electronic fuel injectors could be used, little or no contaminants would be introduced to the fuel supply and the amount of fuel delivered to each cylinder could be far more accurately controlled. Furthermore, the air/fuel ratio and the injection timing could be easily optimized. The concentration of fuel within each intake of air could be controlled so that the occurrence of the aforementioned 'hot spots' might be eliminated.

Research showed that fuel injectors designed for metering hydrogen gas were not generally available. The use of injectors designed for the delivery of petroleum-based fuel raised two concerns. Because the energy density of hydrogen is much less than that of gasoline, significantly larger volumes of hydrogen must be delivered at each injection interval. And, as opposed to a petroleum-based fuel, pure hydrogen gas exhibits 'dry' properties that do not provide the lubrication needed to keep the internal mechanisms of a standard fuel injector working properly.

A testing apparatus was constructed to determine if the necessary volumes of hydrogen could be passed through a large bore fuel injector commonly used in the racing industry. Given proper upstream pressures, it was found that the fuel injectors were capable of providing the necessary mass flow rates.

Similar studies performed using electronic fuel injectors to control hydrogen flow revealed that the mechanism failed after a remarkably short duration of operation. The 'dry' hydrogen had caused the contacting metal surfaces within the injector to seize in the open position and hydrogen was allowed to flow continuously through the injector. This situation was not acceptable for many reasons, including the tremendous safety risks involved with hydrogen gas.

In an attempt to find commercially available injectors capable of operating in a hydrogen environment, injectors intended for use with Compressed Natural Gas (CNG) were then obtained. The major point of interest concerned with this type of injector was the method by which they delivered larger volumes of fuel. While the same seizure problem was likely to occur in these injectors as well, the design methodology used to obtain higher flow rates was noted and the engineering of a new injector specifically designed for hydrogen was begun. This development included considerations for the larger flow rates as developed by the CNG injectors as well as novel methods to resist the premature failure of the mechanism.

Concurrent with the aforementioned development of a new injector, the investigation for a commercially available injector continued. Custom fuel injectors further along in development were discovered that allowed us to focus our efforts on the implementation of the injection and intake systems.

Intake and exhaust systems

The new hydrogen-fueled engine was a converted internal combustion engine with high flow cylinder heads. The intake manifold, cylinder heads, and exhaust manifold were originally designed for use at variable engine speeds. Since we intend to run the engine at a constant speed (in the general range of 2500-3000 RPM), it was decided to develop manifolds and head porting with a geometry that would provide the maximum efficiency and power at the desired operational speed.

Intake and exhaust geometries are generally designed to provide the maximum efficiency of airflow over a wide bandwidth that will include all engine-operating speeds. Since the APU must only run at a single speed, the method by which air is delivered to and removed from the engine could be optimized.

Using fluid flow principles, intake runners were designed that use the inertia of the air flowing to the individual intake ports on the cylinder heads to pressurize the inlet air. When the intake valve is closed, the mass of air flowing through the tube above it compresses until the next opening of the valve. By carefully timing the valve opening, the air/fuel mixture can be introduced to the cylinder at the point of maximum upstream pressure. As will be discussed below when addressing the camshaft aspects, this has multiple benefits for our application.

We treated the exhaust headers in a similar fashion. Each exhaust runner was designed such that the mass of the air flowing through it creates a pressure drop behind each exhaust valve while the valve is closed. When the valve is opened, the exhaust gasses in each cylinder escape to pressures less than that of ambient. Engines employing similar intake and exhaust methods have yielded volumetric efficiencies greater than 120%.

Cylinder heads were then designed to accommodate the intake and exhaust runners, and to provide a higher compression ratio so that a greater thermal efficiency could be achieved.

To mate the custom fuel injectors to the new intake manifold, special mounting hardware was developed. The design consisted of four individual mounts, each of which held two injectors. These mounts were interconnected using reservoirs of ample internal volume so that the cyclic firing of injectors would not induce any substantial pressure variations.

In addition, higher precision cylinder surfaces and corresponding piston interfacing has been accomplished. This was done to minimize oil consumption and the resulting negative impacts upon exhaust emissions.

Cam

A new camshaft was designed to meet the custom needs of the hydrogen application. The variables with the most significant impact on the design of the cam profile are the range of speeds of operation and the combustion characteristics of the fuel being used. Since the engine was intended for operation at a constant speed, the valve actuation could be designed in such a way that the most efficient cycle could be obtained.

The burn-properties of hydrogen differ significantly from those of gasoline, and thus several special considerations were made in the camshaft design. The original engine exhibited frequent 'flash backs' that were presumably caused by the high flame speed and low ignition temperature of hydrogen. Unstable fuel delivery and the existence of 'hot spots' within the cylinders undoubtedly contributed to their prevalence as well. To eliminate 'flash backs' and to better control the combustion of the fuel, the valve timing was altered to fit the characteristics of hydrogen.

The amount of time that the exhaust valve remains partially open while the intake valve is being opened, otherwise known as overlap, also needed to be addressed. While overlap is commonly used on variable speed engines to control the amount of exhaust remaining in the cylinder after combustion, our methods of introducing fuel and air and extracting it once burned allowed us to more closely time the valve actuation and to use the fuel more efficiently. To prevent the escape of unburned fuel and air and reduce the possibility of premature ignition, the degree of overlap was significantly reduced.

Predicted results

Utilizing custom driver software for the electronic fuel injection system, very precise metering of hydrogen delivery can be achieved. Herein may lie the largest improvements that will be realized over the existing engine design. With the ability to control accurately the amount of fuel entering each cylinder during the cycle, the optimum control system can be developed to provide the greatest overall efficiency.

An emissions-testing apparatus has been designed with special attention to two key elements in the exhaust, namely NOx and unburned hydrogen. With predicted nitrous oxide levels falling below the detectable range of most automotive analyzers, a special, high sensitivity analyzer was required. Along with this, a flow-through hydrogen sensor was acquired for the detection of any unburned hydrogen in the exhaust stream. This will insure that any parametric changes made to the engine during development have not led to the presence of unburned hydrocarbons in the exhaust stream and will aid in the optimization of the control systems.

The concept of the hydrogen powered engine shows that there should not be any significant amount of carbon compounds in the exhaust (the only possible source being the combustion of engine oil). Nevertheless, a standard automotive analyzer will be employed to ensure that any possible exhaust emissions are within accepted levels.

The intended nominal running speed of the existing engine was determined by the efficiency and the maximum output of the generator to which it was coupled. A more efficient generator that can operate at higher speeds will be incorporated into the modifications on the bus. Since the efficiency of the engine increases proportionately with its operating speed over a specified range, the new generator will allow us to operate the engine at a more efficient speed. The APU, consisting of the engine and the generator, can now be run at its combined peak efficiency.

FUEL STORAGE SYSTEM

Existing system

The existing storage system on the bus is by means of a metal hydride system that is mounted under the bus. In this case a Lanthanum-Nickel-Aluminum (LANA) type alloy is used. The system consists of two separate beds that are cooled (usually by an external source of cool water) when hydrogen is to be added, and heated (with engine coolant in this case) when hydrogen is to be used. In its present configuration, the bed is estimated to be able to hold 15.2 kg of hydrogen. A great deal of additional information about the existing system can be found in references 1-3.

Planned system

Plans are currently underway to install additional storage on the bus. This will take advantage of some new technology being developed by DOE: high pressure composite material tanks. Current plans are to mount six 48" long x 18" diameter tanks on the top of the bus in the 2001 time frame. These tanks will allow a substantially increased amount of hydrogen fuel to be carried with the bus.

One of the design criteria is to have access to both a high-pressure-tank system and the low-pressure-hydride system by simple adjustments between the two. As new and more promising hydride bed systems are developed, it is hoped that the bus can serve as a test bed for evaluations.

SYSTEM MODELING

Introduction

To enhance the system performance, a study of the system dynamics is being carried out. In this study, the special code is being used. This software, developed by the National Renewable Energy Laboratory (NREL) and called ADVISOR (ADvanced VehIcle SimulatOR), is available from the www (see Reference 4). This is used to simulate and analyze the performance and fuel economy of this hybrid vehicle. In ADVISOR, this vehicle is handled in a series mode, which includes a fuel converter, a generator, batteries, and a motor. As the fuel converter (engine in this case) does not drive the vehicle shaft directly. Instead, it converts mechanical energy directly into electrical energy via the generator. All torque used to move the vehicle comes from the motor. The control strategy is a series power follower. The hybrid accessories are assumed to be a constant electrical power load.

Overall performance estimates and ADVISOR simulation

For the H2-fuel bus system, we know the energy capacity of hydrogen and the battery, the efficiencies of engine, generator and motor, and the electrical power load at certain speed, we can use the energy equation (1) to calculate the total load and ranges vs. speed

$$(R_b + R_p)\,\eta_m = TL \qquad (1)$$

where R_b is the rate of battery draw down, R_p is the rate of power produced from engine, η_m is the motor efficiency, TL is the total load.

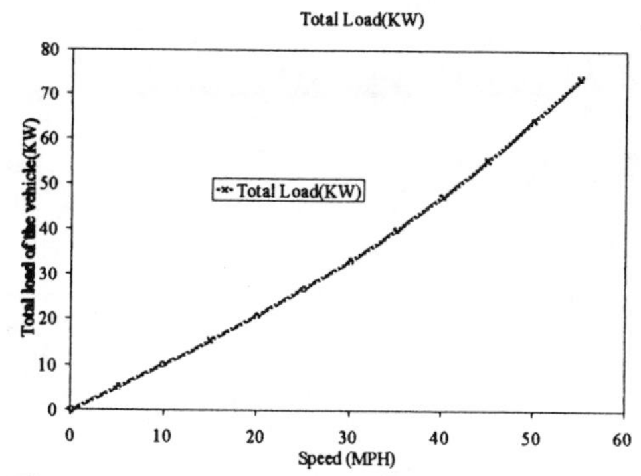

Figure 4. Electrical power load vs. speed

In the bus system, the total hydrogen available = 15.2 kg; engine efficiency = 30%; generation efficiency = 95%; the battery starts at 100% state of charge (SOC) and ends at 20% SOC; motor efficiency = 88%. The load vs. speed is found in ADVISOR 2.2.1 as shown in Figure 4. The range vs. speed is shown in Figure 5 based on the energy balance, equation (1).

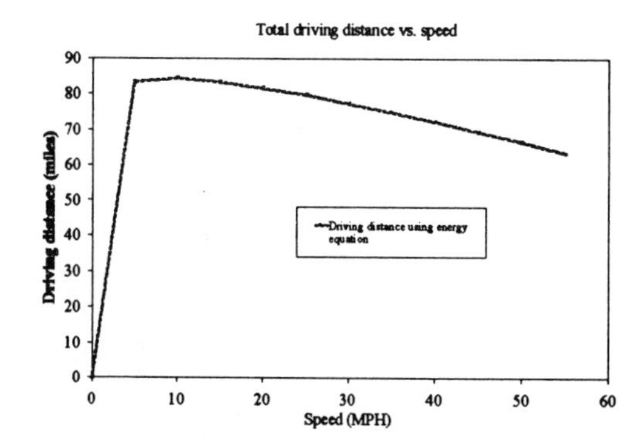

Figure 5. Maximum bus range at constant speed.

Implications of different driving cycles

On a given trip, the bus may travel at a constant speed, or it may drive with many stops and starts. In the simulation using ADVISOR, the former case can be implemented using **CYC_constant**, and the latter is implemented using **CYC_FUDS** where FUDS stands for Federal Urban Driving Schedule. In Figure 6, the simulation results for a constant speed of 55 mph are given. The bus runs solely with the electrical battery for the first 27 minutes. The engine is on when the SOC of battery is at 40%. The power from the engine and the battery is used to drive the bus at 55 mph until all the hydrogen is used up. Then the bus runs only using the battery until the SOC reaches 20% which is assumed to be the lowest SOC the battery can tolerate before requiring recharging.

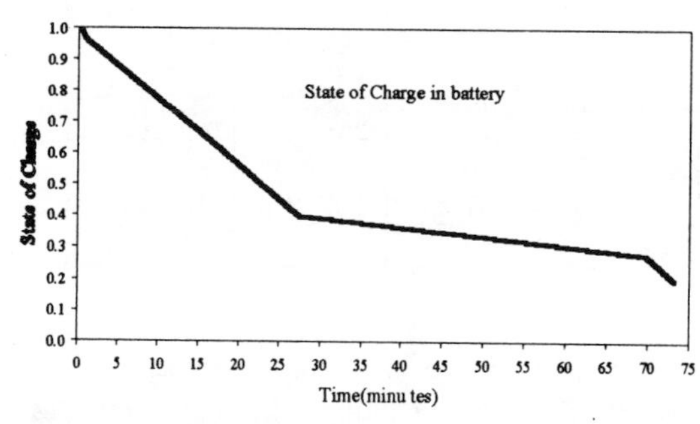

Figure 6. State of charge of battery at constant speed (55 mph).

Regenerative braking is the process by which some of the kinetic energy stored in the vehicle's translating mass is stored in the vehicle during decelerations. In most electric and hybrid electric vehicles on the road today, this is accomplished by operating the traction motor as a generator, providing braking torque to the wheels and recharging the traction batteries.

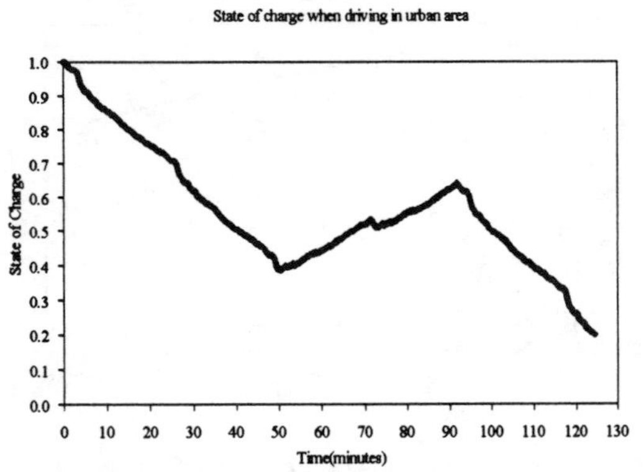

Figure 7. State of charge of battery using FUDS driving cycle.

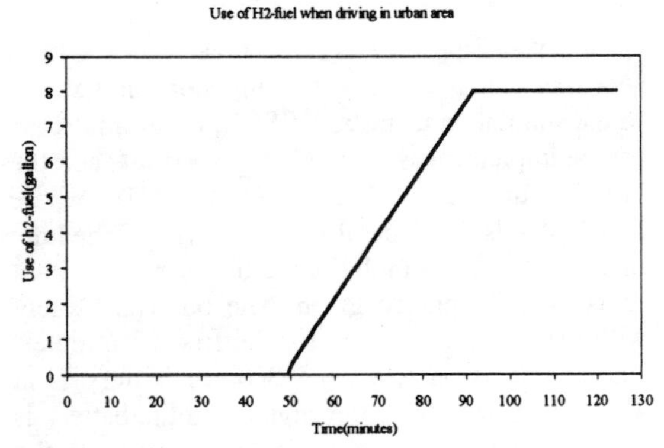

Figure 8. The temporal use of fuel during the FUDS driving cycle.

Table 1 Effects of using regenerative braking

	With reg braking	Without reg braking
Trip elapsed time	155 min	124 min
Trip mileage	48.7 mi	39.3 mi
Recharging starts at	60 min	49 min
Recharging ends at	102 min	91 min
Recharging time	42 min	42 min

In the system we are considering here, the regenerative braking will be by supercapacitors. Other than the dynamics of the two regenerative braking systems being different, the general energy considerations are similar. The energy provided by regenerative braking can then be used for propulsion or to power vehicle accessories. All scenarios were analyzed considering credit for regenerative braking or no credit.

From Table 1, it can be seen that the bus with regenerative braking has a longer range. But note that the recharging time is not affected by regenerative braking. This is a result of the fact that the engine runs the full speed after it is turned on, and it is turned off when the total fuel is used up.

CONCLUDING COMMENTS

A hybrid electric bus powered by a hydrogen fueled internal combustion engine has been described. This project is a continuation of some earlier work done by others. Extensive modifications are being made to the fuel system, engine, electrical system, and operation of the bus. All of these aspects are directed toward improving the range of the bus.

REFERENCES

1. W. D. Jacobs, L. K. Heung, T. Motyka, W. A. Summers, and J. M. Morrison, "Final Report for the H2Fuel Bus", SAE Paper 1999-01-2906, 1999.

2. W. D. Jacobs, L. K. Heung, T. Motyka, W. A. Summers, and J. M. Morrison, "Final Report for the H2Fuel Bus", Report WSRC-TR-98-00385, Westinghouse Savannah River Company, Aiken, SC, 1999.

3. L. K. Heung, "On-Board Hydrogen Storage System Using Metal Hydride," paper presented at the Hypothesis II Conference held in Grimstad, Norway, August 18-22, 1997.

4. ADVISOR Documentation, NREL website: http://www.ctts.nrel.gov/analysis/advisor_doc

ACKNOWLEDGEMENT

The support of this project by the US Department of Energy Hydrogen Energy Program through the Nevada Operations Office (Bob Golden, Project Manager) is greatly appreciated.

AIAA-2000-2858

CARBON DIOXIDE AS AN ALTERNATIVE REFRIGERANT FOR AUTOMOTIVE AIR CONDITIONING SYSTEMS

G.D. Mathur, Ph.D., P.E.
Staff Engineer, HVAC Systems
Zexel USA Corporation, Grand Prairie, TX 75080/ Decatur, IL 62521
gmathur@zexelusa.com

ABSTRACT

Thermodynamic performance of a typical automotive air conditioning system has been simulated using carbon dioxide as the working fluid. The performance of the carbon dioxide system is compared with a base case with R-134a as the refrigerant. A cooling capacity of 5.3 kW (1.5 ton) is used for this study. For the base case, evaporation and condensing temperatures of -6.7°C (20°F) and 48.9°C (120°F) are used in the investigation. For the carbon dioxide system, an evaporation temperature of -6.7°C (20°F) and a high side pressure of 130 bars (1856.5 psia) is used.

The study shows that the thermodynamic performance of the vapor compression air conditioning system with carbon dioxide as the working fluid is lower than the current R-134a systems. One major disadvantage for carbon dioxide system is extremely high operating pressures. Finally, practical design considerations and safety issues have been presented for the design of the air conditioning system with carbon dioxide as the working fluid.

INTRODUCTION

[1]Refrigerant 134a has emerged as the new refrigerant for the automotive and commercial A/C industry that has a zero ozone depleting potential (ODP) value. However, R-134a's greenhouse warming potential (GWP) is relatively high among the newly developed hydroflourocarbons (HFCs) which seems to be an obstacle for the furtherance of the use of R-134a, especially in European countries. Hence, many countries are looking for other refrigerants that do not contribute to global warming. There are many refrigerants that are currently available naturally. Examples of the so called "natural refrigerants" are: ammonia, carbon dioxide, hydrocarbons, water, helium, air etc.

A number of studies (Mathur, 1996; 1998a; 1998b; 2000; Ghodbane, 2000, etc) have been reported in the literature on hydrocarbons. Based on the thermodynamic and thermophysical properties, hydrocarbons (Propane, Isobutane, and their mixtures) are drop in replacement for R-134a. Mineral oil is compatible with hydrocarbon refrigerants. The heat transfer coefficients for boiling, condensation, and single phase (both vapor and liquid) for hydrocarbons are higher in comparison to both R-134a and R-12 (Mathur, 1998a). However, the pressure gradients for the various modes of the heat transfer process for the hydrocarbons are also higher (Mathur, 2000) in comparison to both R-134a and R-12 which will result in higher pressure drops. Due to flammability and safety issues, research in the area of hydrocarbons for mobile air conditioning has decreased. In Europe hydrcarbons are being used in stationary systems, e.g., refrigerators and freezers.

Carbon dioxide is receiving attention these days as it does not contribute to global warming. The critical temperature of CO_2 is 31.1°C, leading CO_2 systems to operate in trans-critical mode when exposed to typical ambient temperatures. Hence, the conventional condenser is replaced by a gas cooler that rejects heat under variable temperatures at high side pressure. To improve the performance of the CO_2 system, an internal heat exchanger is used. The internal heat exchanger subcools the refrigerant between the gas cooler and the expansion device and to superheat the refrigerant in the suction line of the compressor. Since carbon dioxide air conditioning systems operates at high pressures, the specific volumes are reduced (compared to R-134a). This results in compact heat exchangers.

The concept of using CO_2 air conditioning is very old. However, the use of CO_2 for mobile air conditioning system was first proposed by Lorentzen & Pettersen (1993). Ever since, CO_2 air conditioning system has been gaining popularity. This technology is mostly being developed in Europe by Germans (Koehler, et. al., 1999). A detailed historical background on CO_2 is reported by Bhatti (1997). Society of Automotive Engineers (SAE) has had 2 conferences in Phoenix/Scottsdale, AZ in the area of alternate refrigerants. During these conferences of 1998 and 1999, the vehicles with alternate refrigerants systems were tested for performance and comfort. The tested vehicles had the following air conditioning systems: CO_2, hydrocarbons, and enhanced R-134a systems. Many technical presentations were made at these two conferences for the alternate refrigerants, primarily on CO_2 systems. Some of the important presentations are as follows: Briggs (1999) on design issues for expansion valve and accumulators for CO_2; Takeuchi (1999) on the development of CO_2 scroll compressor; Yamamoyo & Komatsu (1999) and Furuya & Mathur (1999) on the performance comparison of wind tunnel testing of vehicles with CO_2 and R-134a system. The results from the studies indicate that with a CO_2 system, the performance of the vehicle air conditioning system is equal or better than R-134a for cool down or highway driving. However, the performance is not as good during idling, city driving, or at high thermal loads.

Even though a lot of testing has been conducted on the CO_2 system, none of the companies are divulging any detailed design information on these systems. Test results were presented in both SAE's symposium (1998 and 1999) on alternate refrigerant in Scottsdale. However, detailed design information was lacking. SAE has planned 3rd symposium on alternative refrigerants in Scottsdale/Phoenix, AZ in July of 2000.

Currently, limited information is available in the open literature on the performance and design of the air conditioning systems using carbon dioxide as the working fluid. In this paper, the author has presented a detailed thermodynamic analysis of a typical automotive air conditioning system operated on carbon dioxide cycle. The performance (COP) of the system has been compared with a base case with R-134a as the refrigerant. Finally, practical design considerations and safety issues have been presented for the design of the air conditioning system with carbon dioxide as the working fluid.

REFRIGERANT PROPERTY COMPARISON

Table I shows a comparison of the refrigerant property for CO_2 and R-134a at 5, 10, and 15°C. The most important difference between CO_2 and R-134a is the operating pressures. The evaporating pressure for a CO_2 system is higher by a factor 10-12 in comparison to R-134a. CO_2 has much lower surface tension in comparison to R-134a which facilitates the bubble formation in liquid flows. This will result in higher flow boiling heat transfer coefficients. The liquid viscosity of CO_2 is also much lower in comparison to R-134a. This will result in a smaller refrigerant pressure drops for CO_2 when flowing inside of a tube. The critical pressures for CO_2 and R-134a are 73.8 bars (1071 psia) and 40.6 bars (88.27 psia), respectively.

PRESENT INVESTIGATION

Table II shows the design data (ASHRAE, 1997) that was used in this investigation. The evaporator temperature was varied between -17.8 to 4.4°C (0 to 40°F) and the discharge pressure for the CO_2 system was varied between 130 to 150 bars (1885.5 to 2175.6 psia). For the base case with R-134a, a condensing temperature of 48.9°C (120°F) was used . The cooling capacity of 5.3 kW (1.5 tons) was used for this investigation. For both the systems, the vapor at the suction of the compressor is assumed to be at the saturation condition and isentropic compression is assumed. An internal heat exchanger for CO_2 was not used in this investigation. Figure 1 shows a schematic diagram for the base case with R-134a and Figure 2 shows a schematic diagram for a CO_2 system. As can be seen from the pressure-enthalpy diagram of Figure 2, a condenser is replaced by a gas cooler for the CO_2 system. In the gas cooler the heat rejection takes place at variable temperatures at high pressure.

SIMULATED RESULTS AND DISCUSSIONS

The performance was calculated using the design data from Table II. Table III shows a summary of the calculations for a given point. As can be seen from this table, the coefficient of performance for CO_2 system is 3.09 in comparison to 3.52 for R-134a. This is lower by approximately 12% than the base case. Vapor density of CO_2 at the suction condition is very low in comparison to R-134a. This is approximately one-eigth of R-134a. This is very important thermophysical property that helps in reducing the compressor displacement and reducing the size of the heat exchangers. For the same compressor capacity and RPM, the volumetric displacement of the compressor could be reduced by almost 88% (see Table III). The energy efficiency ratio (EER) determines the amount of cooling per unit input of electrical energy. As is evident from the table, EER is better for R-134a as the compression work required is lower in comparison to the CO_2 system.

Similar calculations were conducted at various operating conditions for CO_2 system and the results are plotted in Figures 3 and 4. Figure 4 shows the variation of the COP as a function of evaporation temperature at gas cooler operating pressures of 130, 140, and 150 bars, respectively. As can be seen from this Figure, the curves can be represented by straight lines. At a given gas cooler pressure, the performance increases with the increase of the evaporation temperature due to a decrease in the isentropic work by the compressor (Wark, 1993). At higher gas cooler pressures the curves come closer due successively larger isentropic compression work.

Figure 4 shows the variation of the COP as a function of the gas cooler operating pressure at evaporation temperatures of -17.8, -6.7, and 4.4°C, respectively. For a given evaporator temperature, the COP decreases with the increase of the gas cooler pressure due to increase in the compression work. In the following section practical design considerations for CO_2 have been presented.

DESIGN CONSIDERATIONS

Design of the CO_2 systems will require extensive changes to the existing R-134a design (Kobayashi, 1999; Fritz, 1999). The following are some of the major design requirements:

1. Due to higher operating pressures, the system components will have to be made with thicker walls. This will cause the weight of the system to go up and will affect fuel consumption. More fuel will be required to carry the CO_2 A/C system which will result in indirect effect of global warming.

2. The increase in the compressor power consumption will result in poor acceleration and fuel economy.

3. More installation space will be required for a CO_2 system.

4. The permeation rate of CO_2 through the elastomers will increase due to high operating pressures. Hence, the current hose materials cannot be used. Special steel wire braided hose will have to be developed.

5. The thickness of the liquid lines will have to be increased for high operation pressures.

6. New design of the charge and service ports will have to be developed. The evacuation process will have to be controlled with sensor, as a sudden drop in the high pressure will cause explosive decompression of polymer materials and dry ice will be formed that could block areas of the system. Global standards will have to be developed.

7. The current seals will have to be replaced by metallic seals (or gaskets).

8. A TXV should be designed for varying refrigerant properties. At system start-up, the fluid properties changes vary rapidly as the pressure of the system goes up.

9. New controls for the components and the system will have to be developed. Pressure regulators will have to be used at strategic point around the loop. Once the compressor shuts off, a check valve should prevent the accumulation of the CO_2 inside the evaporator. The reason is that at soak conditions, the evaporator pressure would rise very high.

10. Heat Exchangers: Evaporator performance will be better for CO_2 systems. This is due to lower surface tension and liquid viscosity. For details, see property comparison section. A serpentine evaporator (Mathur, 1998c) with heavy walls can be used for CO_2 systems. The gas cooler essentially has single-phase flow inside, and hence, results in poor heat transfer coefficients (in comparison to condensation process in a R-134a system). A current parallel flow condenser (Mathur, 1999) with heavy walls can be used as a gas cooler for the CO_2 systems. The suction line heat exchanger can be a tube-in-tube arranged in a counter-flow configuration to provide high performance. All three heat exchangers will have to be optimized for performance.

SAFETY ISSUES

A number of safety issues have to be addressed by the automobile manufacturers. The following are the major concerns:

1. Moderately high concentration of CO_2 in the passenger compartment due to a small refrigerant leakage. This is a major concern when the vehicle is being operated in recirculation mode or with low quantity of outside air.

2. Or high concentration of CO_2 due to rupture of the component. The concentration of CO_2 will instantaneously increase posing risk to the passengers in the vehicle. This will result in explosive release of energy and spraying of CO_2 snow in the cabin.

3. Flying debris due to a rupture of a component. This can pose risk to both passengers especially sitting in the front and to people on the streets.

For items # 1 and 2, a CO_2 sensor can be installed in the vehicle. If the concentration of the CO_2 goes beyond a predetermined value, one or combination of the following could be done:

a. sound an alarm

b. turn the windows open (in motorized windows)

c. set the blower in outside air mode

The effects of CO_2 on the human health (AGA Gas Handbook) are given in Table IV. As can be seen physiological effects on humans are apparent at exposure limits above 2%. Successively higher percentages of CO_2 concentration will result in shortness of breath, nausea, headaches, epileptic fits, and ultimately a stroke. Tests were conducted by SAE's subcommittee on refrigerants safety (Amin et. al, 1999) to determine the CO_2 concentration in the vehicle cabin with time. Vehicle compartment volumes from 1.5 to 4 cubic meters were tested and the refrigerant charge varied from 240 to 600 grams.

For item # 3, the system has to be designed with heavier walls to accommodate higher operating pressures. Burst disks and other safety devices should be strategically placed in the system so that the CO_2 is released from the system without causing injuries to the passengers or people on the streets.

CONCLUSIONS

Thermodynamic performance of a typical automotive air conditioning system has been simulated using carbon dioxide as the working fluid. The performance of the carbon dioxide system is compared with a base case with R-134a as the refrigerant. A cooling capacity of 5.3 kW (1.5 tons) ton is used for this study.

The study shows that the thermodynamic performance of the vapor compression air conditioning system with carbon dioxide is lower than the current R-134a systems. On major disadvantage for carbon dioxide system is extremely high operating pressures. Finally, practical design considerations and safety issues have been presented for the design of the air conditioning system with carbon dioxide as the working fluid.

REFERENCES

AGA Gas Handbook, AGA AB, Lindino, Sweden, ISBN 91-970061-1-4.

ASHRAE 1997. Fundamentals, Handbook.

Amin, J., Dienhart, B., Wertenbach, J., 1999. "Safety aspects of an A/C system with carbon dioxide as refrigerant," Automotive Alternate Refrigerant Systems Symposium, Scottsdale, AZ, June 28 - July 1.

Bhatti, M.S., 1997. "A critical look at R-744 and R-134a mobile air conditioning systems," SAE Paper # 970527.

Briggs, R., 1999. "Design issues in expanding CO_2," Automotive Alternate Refrigerant Systems Symposium, Scottsdale, AZ, June 28 - July 1.

Fritz, T., 1999. "Connecting components and inner heat exchanger for a CO_2 circuit," Automotive Alternate Refrigerant Systems Symposium, Scottsdale, AZ, June 28 - July 1.

Furuya, S. & Mathur, G.D., 1999. "A CO_2 refrigerant system for vehicle air conditioning," Automotive Alternate Refrigerant Systems Symposium, Scottsdale, AZ, June 28 - July 1.

Ghodbane, M. 2000. "On vehicle performance of a secondary loop A/C system," SAE Paper # 2000-01-1270.

Koehler, J., Lemke, N., Sonnenkalb, M., 1999. "Second year of city buses with transcritical CO_2 air conditioning units in Germany," Automotive Alternate Refrigerant Systems Symposium, Scottsdale, AZ, June 28 - July 1.

Kobayashi, N., 1999. "Concerns of CO_2 A/C system for compact vehicles," Automotive Alternate Refrigerant Systems Symposium, Scottsdale, AZ, June 28 - July 1.

Lorentzen, G., Pettersen, J., 1993. "A new efficient and environmentally benign system for car air conditioning", Int Journal of Refrig., Vol 16, No 1, pp 4-12.

Mathur, G.D. 1996. "Performance of vapor compression refrigeration system with hydrocarbons: propane, isobutane, and 50/50 mixture of propane/isobutane." International Conference on Ozone Protection Rechnologies, Washington, D.C., pp 835-844.

Mathur, G.D. 1998a. "Using natural refrigerants (hydrocarbons) in air conditioning systems." Intersociety Energy Conversion Engineering Conference, Colorado Springs, Paper # IECEC-98-049.

Mathur, G.D. 1998b. "Heat transfer coefficients for propane (R-290), isobutane (R-600a), and 50/50 mixture of propane and isobutane," ASHRAE Transaction, Vol 104, Pt 2, paper # TO-98-19-4.

Mathur, G.D. 1998c. "Performance of serpentine heat exchangers," SAE Paper # 980057.

Mathur, G.D. 1999. "Predicting and optimizing thermal and hydrodynamic performance of parallel flow condensers," SAE Paper # 1999-01-0236.

Mathur, G.D. 2000. "Hydrodynamic characteristics of propane (R-290), isobutane (R-600a), and 50/50 mixture of propane and isobutane," ASHRAE Transaction, Vol 106, Pt 2, paper # MN-00-6-3.

Takeuchi, M., 1999. "Development of CO_2 scroll compressor for automotive air conditioning systems," Automotive Alternate Refrigerant Systems Symposium, Scottsdale, AZ, June 28 - July 1.

Wark, K. 1993. Thermodynamics, McGraw Hills Book Company.

Yamamoto, K. & Komatsu, S., 1999. "Experimental evaluation of the prototype CO_2 system and the HFC-134a system in wind tunnel," Automotive Alternate Refrigerant Systems Symposium, Scottsdale, AZ, June 28 - July 1.

TABLE I: THERMOPHYSICAL PROPERTIES OF CO_2 R-134a AT 5/10/15 °C

REFRIGERANT	CO_2 (R-744)	R-134a
Saturation Pressure (MPa)	3.969/4.502/5.086	0.350/0.414/0.488
Latent Heat (kJ/kg)	214.6/196.8/176.2	194.8/190.9/186.7
Surface Tension (mN/m)	3.53/2.67/1.88	11.0/10.3/9.6
Liquid Density (kg/m3)	899.6/861.5/821.3	1277.1/1260.2/1242.8
Vapor Density (kg/m3)	114.8/135.3/161.0	17.1/20.2/23.7
Liquid Viscosity (uPa.s)	95.9/86.7/77.2	270.3/254.3/239.7
Vapor Viscosity (u Pa.s)	15.4/16.1/17.0	11.2/11.4/11.7
Liquid Specific Heat (kJ/kg K)	2.73/3.01/3.44	1.35/1.37/1.38
Vapor Specific Heat (kJ/kg K)	2.21/2.62/3.30	0.91/0.93/0.96

TABLE II: DESIGN DATA USED FOR THIS INVESTIGATION

VARIABLES	DATA
Evaporation Temperature	-17.8 to 4.4°C (0 to 40° F)
Cooling Load	5.3 kW (1.5 tons)
Condensing Temperature for Base Case (R-134a)	48.9°C (120° F)
Gas Cooler Pressure for CO_2 System	130 to 150 bars (1885.5 to 2175.6 psia)
Vapor State at Suction	Saturated Condition
Base Case	No Subcooling at Condenser Outlet
Compression	Isentropic

TABLE III. PERFORMANCE COMPARISON OF THE REFRIGERANTS

STATE /VARIABLES			R-134a	R-744
State 1	Pressure	bars	2.28	29.09
	Temperature	C	-6.7	-6.7
	Enthalpy	kJ/kg	246.76	738.79
	Latent Heat	kJ/kg	204.08	249.95
	Entropy	kJ/kg K	0.934078	3.99821
	Volume	m3/kg	0.088029	0.0126485
State 2	Pressure	bars	12.81	130
	Temperature	C	54.4	104.4
	Enthalpy	kJ/kg	282.35	796.86
	Entropy	kJ/kg K	0.934078	3.99821
	Volume	m3/kg	0.01649	0.00041827
State 3	Pressure	bars	12.81	130
	Temperature	C	48.9	26.7
	Enthalpy	kJ/kg	121.581	559.20
	Entropy	kJ/kg K	0.24211	3.230048
	Volume	m3/kg	0.0009032	0.0012006
State 4	Pressure	bars	2.28	29.09
	Temperature	C	-6.7	-6.7
	Enthalpy	kJ/kg	121.581	559.20
	Entropy	kJ/kg K	0.2593667	3.288624
	Volume	m3/kg	0.00.0345336	0.0042816
Refrigeration Effect (h1 - h4)		kJ/kg	125.18	179.59
Cooling Capacity		kW	5.275	5.275
Ref. Flowrate [Cooling Capacity/Ref Effect]		kg/hr	151.58	107.2
Isentropic Compresor Specific Work (h2 - h1)		kJ/kg	35.59	58.07
Compressor Work [Ref Flowrate * Specific Work]		kW	1.50	1.824
Compressor Capacity/Volume of (E) Refrigerant Flow [Latent Heat/Vapor Density]		kW/m3	2318.33	19761.2
Compressor Displacement of R-744 system with respect to R-134a [E,744/E,134a]			1	8.52
Energy Efficiency Ratio, EER [Cooling Capacity (Btu/hr)/Work Input W]		Btu/hr W	12.0	9.87
Condenser/Gas Cooler Specific Heat Rejection (h2 - h3)		kJ/kg	160.77	179.59
Condenser/Gas Cooler Heat Rejection [Ref Flowrate * (h2 - h3)]		kW	6.8	7.10
Coefficient of Performance, COP [Ref. Effect/ Compressor Specific Work]			3.52	3.09
COP for Reversed Carnot Cycle [Tlow/(Thigh - Tlow)]			4.8	----

TABLE IV: AFFECT OF CARBON DIOXIDE ON HUMANS

PERCENTAGE OF CO_2 IN AIR	AFFECT ON AN AVERAGE ADULT
2%	50% increase in breathing rate
3%	100% increase in breathing rate, 10 minutes short term exposure limit
5%	300% increase in breathing rate; headache and sweating may begin after 1 hour. Note this is tolerated by most persons, but is physical burdening
8%	Short term exposure limit
8-10%	Headache after 10 to 15 minutes, dizziness, buzzing in the ears, rise in blood pressure, high pulse rate, excitation, and nausea
10-18%	Cramps after a few minutes, epileptic fits, loss of consciousness, a sharp drop in the blood pressure. Note the victims will recover very quickly in fresh air.
18-20%	Symptoms similar those of stroke

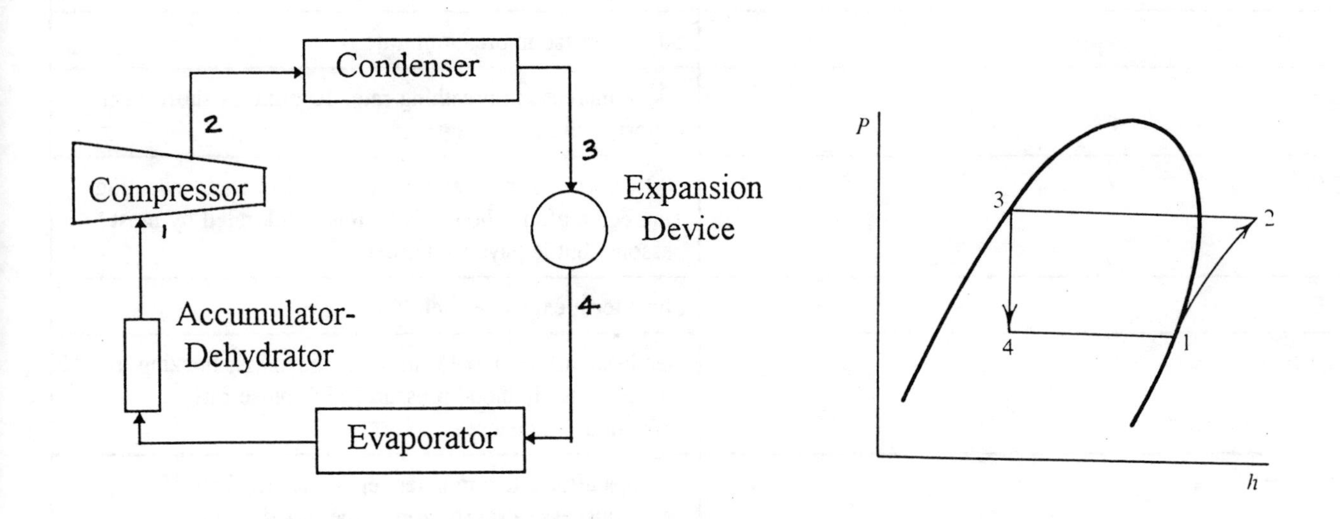

Figure 1. Schematic and Pressure-Enthalpy Diagrams for R-134a Air Conditioning Systems

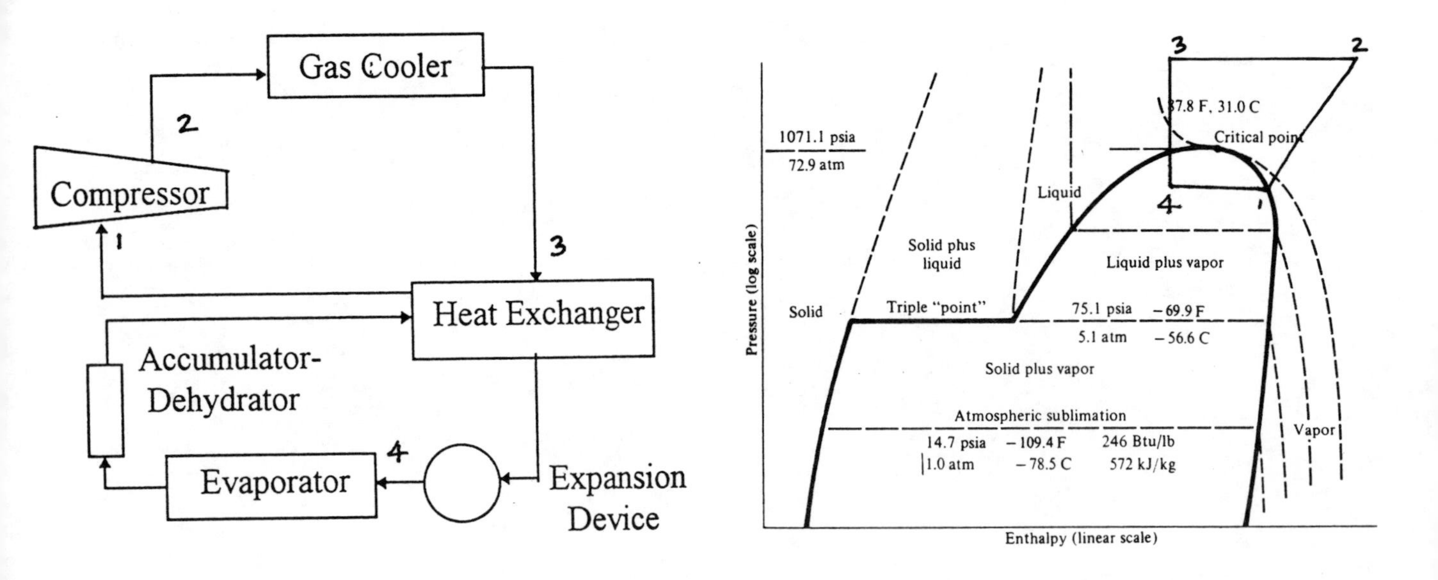

Figure 2. Schematic and Pressure-Enthalpy Diagrams for CO_2 Air Conditioning Systems

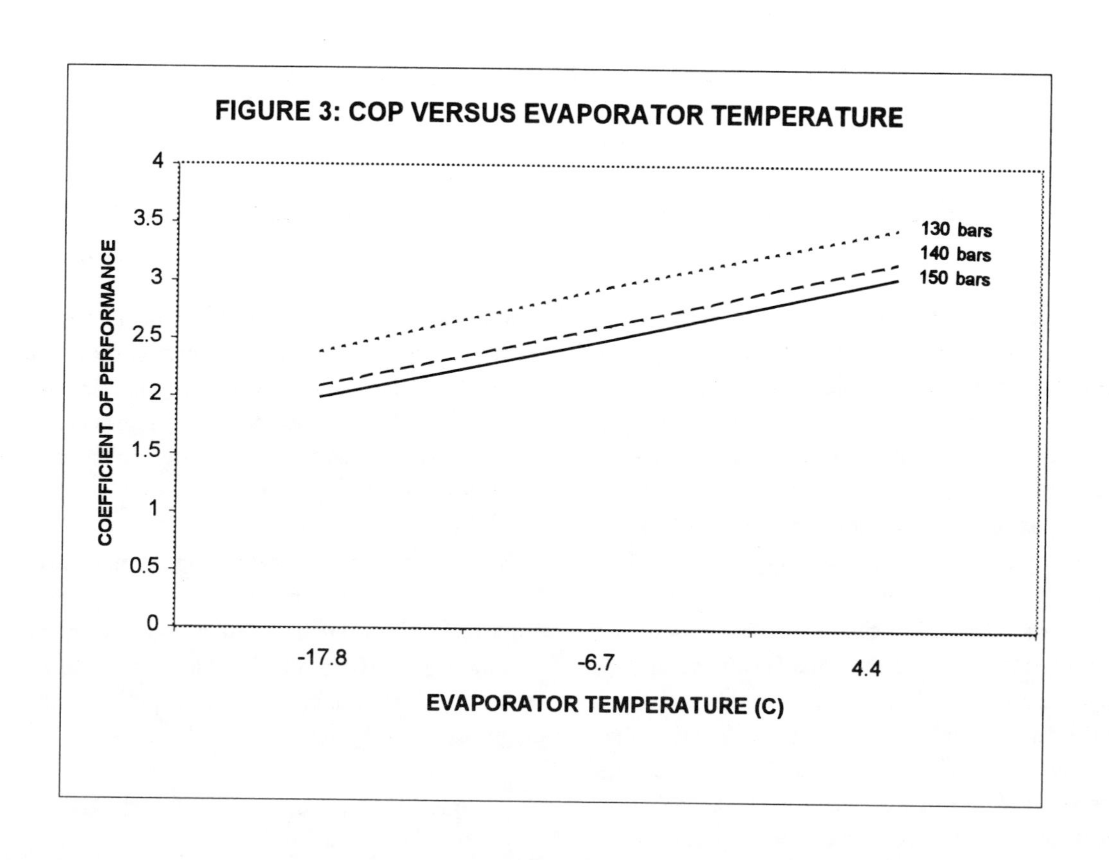
FIGURE 3: COP VERSUS EVAPORATOR TEMPERATURE
COEFFICIENT OF PERFORMANCE
4
3.5
3
2.5
2
1.5
1
0.5
0
130 bars
140 bars
150 bars
-17.8
-6.7
4.4
EVAPORATOR TEMPERATURE (C)

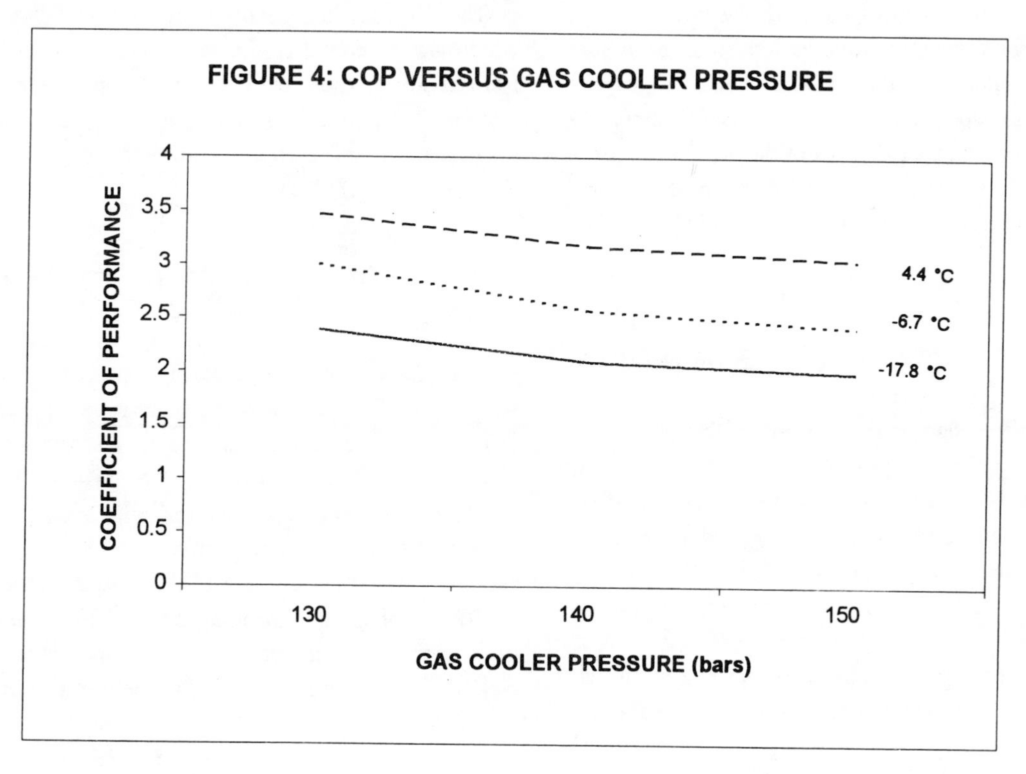
FIGURE 4: COP VERSUS GAS COOLER PRESSURE
COEFFICIENT OF PERFORMANCE
4
3.5
3
2.5
2
1.5
1
0.5
0
4.4 °C
-6.7 °C
-17.8 °C
130
140
150
GAS COOLER PRESSURE (bars)

MAXIMUM OUTPUT CONTROL OF PHOTOVOLTAIC (PV) ARRAY

Takumi Takashima*, Tadayoshi Tanaka, Masatsugu Amano, and Yuji Ando

Electrotechnical Laboratory
1-1-4 Umezono, Tsukuba, Ibaraki 305-8568, JAPAN
phone: +81-298-61-5792, fax: +81-298-61-5787, e-mail: takasima@etl.go.jp

A photovoltaic (PV) system output depends on the environmental parameters such as the solar insolation and the PV module temperature. If it is possible to predict the maximum power point under the outdoor environment and to operate at that point, the PV system can generate the maximum output every time. In this paper, a maximum power point control method that maximizes the output of a PV array is proposed. This method determines the maximum output operation point from the I-V characteristics introducing empirically the effects of the solar insolation and the module temperature. We derived two main parameters from this analysis; one is the power gain G, and another is the environmental operation parameter X. At the operation point determined by this method, G becomes larger than that of under the same environmental conditions. G becomes large with the increase of X, and the large X manly means the low solar insolation. The characteristics of PV module which will supply more power especially at large X should satisfy the following points; the fill factor of the module should be lower and the short circuit current of the module should be larger than those of currently available in the market.

INTRODUCTION

Photovoltaic (PV) systems are expected for compensating the electric power consumption and for peak-cut effect during the daytime. In these days, small PV systems for private houses are prevailing in Japan. If the system cost reduces in near future, large PV systems for commercial buildings and factories will be constructed. Since the PV system output changes according to the solar insolation and the PV module temperature, it is very important to operate the PV array at the maximum power point. However, it is difficult to calculate a maximum power point from the I-V characteristic equation of the PV module because the relations between the PV output current and the terminal voltage is not independent in the equation. Therefore, instead of the equation, the equations with the insolation and the temperature correction coefficients of the PV module were used for this analysis.

In this paper, a new method which maximizes the PV array output under the outdoor environment is proposed. This method determines the maximum operation point from the insolation and the module temperature, and this method can be applied for the fault diagnosis of the system.

MAXIMUM OUTPUT CONTROL

The PV array is generally composed of PV modules which are connected N_S in series and N_P in parallel, respectively. Then, the PV current I and the terminal voltage V at the insolation Q and the module temperature T are described as follows;

$$I = \left\{ I_S + I_{SC}\left(\frac{Q}{Q_0} - 1\right) + \alpha(T - T_S) \right\} N_P \qquad ...(1)$$

$$V = \left\{ V_S + \beta(T - T_S) - R_S\left(\frac{I}{N_P} - I_S\right) - \frac{KI}{N_P}(T - T_S) \right\} N_S \qquad ...(2)$$

Here, I_S and V_S are the output current and the terminal voltage of a PV module at the standard test conditions (STC). STC are defined as follows; the standard solar insolation (Q_0) is 1.0kW/m^2, the standard module temperature (T_S) is 25 °C, and the solar spectrum is at the air mass (AM) 1.5, respectively. Moreover, I_{SC} is the short circuit current at STC, α is the temperature coefficient of I_{SC}, β is the temperature coefficient of the open circuit voltage of the module, R_S is a series resistance of the module, and K is the curve correction factor, respectively.

These equations are based on the equations defined in the Japanese Industrial Standard (JIS) C8914. However, the application conditions of the equations defined in JIS C8914 are limited only around STC. Since we could locate the equations which Yamamoto et

---

* Senior Researcher, Energy Fundamentals Division

al. modified for wider solar flux condition than that of STC[1], we used those modified equations for our analysis.

The output power P of the PV array is calculated by Eq.(1) and Eq.(2).

$$P = IV$$
$$= I\left[V_S + \beta\left(T - T_S\right) + R_S I_S - \frac{I}{N_P}\left\{R_S + K\left(T - T_S\right)\right\}\right]N_S \quad ...(3)$$

The current I_M and the voltage V_M which maximize the output power are calculated by differentiating P with respect to I.

$$I_M = \frac{N_P}{2}\frac{V_S + \beta\left(T - T_S\right) + R_S I_S}{R_S + K\left(T - T_S\right)} \quad ...(4)$$

$$V_M = \frac{V_S + \beta\left(T - T_S\right) + R_S I_S}{2} N_S \quad ...(5)$$

Though I_M and V_M include the module temperature T, they don't include the insolation Q. Usually, the PV output current is mainly affected by the insolation, and the terminal voltage is mainly affected by the module temperature. In Eq.(3), the output power P includes the current I, and I is affected by the insolation Q as indicated in Eq.(1). Therefore, P is affected by Q. As a result, even though Q does not appear in Eq.(4) or Eq.(5), those equations include the effect of the insolation Q.

As indicated in Eq.(4) and Eq.(5), the maximum output power operation point is predicted with the module characteristic coefficients, and these module characteristic coefficients are calculated by PV makers. At that time, the predicted operation point is based on the initial conditions of the PV module. If there is any error between the predicted operation point and the real operation point, it suggests that there will be some fault in the system or the degradation of the module. Therefore, this control method can be applied for the fault diagnosis of the PV system.

The PV output power under the outdoor environment including Q and T is calculated from Eq.(1) and Eq.(2), and the maximum output power at the same condition is calculated from Eq.(4) and Eq.(5). Here, we define the power gain G, which is the ratio of the maximum output power to the power under the outdoor environment, and G is calculated as follows;

$$G = \frac{I_M V_M}{IV} = \frac{1}{1 - \left(1 - \frac{2bc}{a}\right)^2} = \frac{1}{1 - X^2} \quad ...(6)$$

Here, $a = V_S + \beta(T - T_S) + R_S I_S$, $b = R_S + K(T - T_S)$, $c = I_S + I_{SC}\left(\frac{Q}{Q_0} - 1\right) + \alpha\left(T - T_S\right)$, and $X = 1 - \frac{2bc}{a}$, respectively. X is defined as an environmental operation parameter.

From Eq.(6), the range of G and X are as follows;

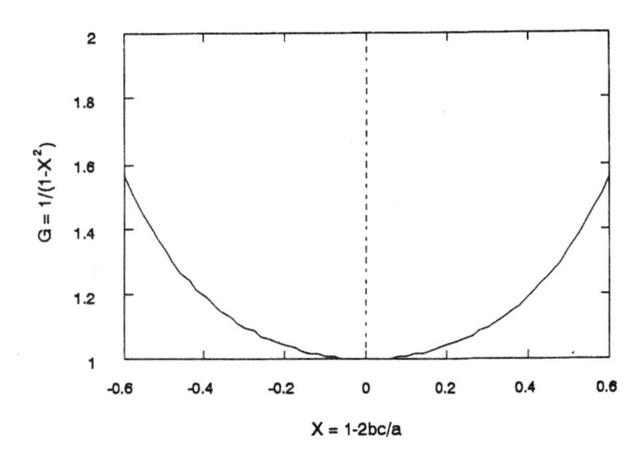

Fig.1. Relations between G and X

$$G > 0,\ -1 < X < 1 \quad ...(7)$$

Figure 1 shows the relations between G and X calculated from Eq.(6). $X = 0$ means at STC and it is clear that $G = 1$ there. As the environmental conditions are off STC, that is, as $|X|$ becomes close to 1, the value of G becomes large. From the definition of X, the closer to 1 X becomes, the smaller $2bc/a$ becomes. Since a, b, and c change according to the variations of the insolation and the module temperature, the behavior of X is also effected by those environmental parameters.

CONSIDERATIONS TO MAXIMUM OUTPUT CONDITIONS

To verify the effectiveness of the maximum power point control mentioned above, a numerical analysis was done with the parameters of some PV modules manufactured by typical PV makers in Japan. Table 1 shows those module parameters. Figure 2 shows the relations between G and X calculated by using the values of Table 1. In Fig.2, each point was calculated based on the difference parameters between the environmental conditions and STC. One parameter is the insolation ratio of the real insolation to the standard insolation, Q/Q_0, and another is the module temperature difference between the real module temperature and that of at the standard condition, $T - T_S$. Each point is on the curve of Eq.(6). When $Q/Q_0 = 1$ and $T - T_S = 0$ which means at STC, all points is at $G = 1$ and $X = 0$. X is negative value with increasing temperature difference at $Q/Q_0 = 1$,. On the other hand, when Q/Q_0 is smaller than 1, X becomes positive value and G goes large. X becomes large with the decrease of the insolation. In the range of positive X, just a small difference of X occurs when the module temperature changes.

Table.1. Characteristic Values of PV Modules

maker	maximum output P_{MAX} [W]	P_{MAX} voltage V_{MAX} [V]	P_{MAX} current I_{MAX} [A]	temperature coefficient of Isc α [A/°C]	temperature coefficient of Voc β [V/°C]	series resistance Rs [Ω]	curve correction factor K [Ω/°C]	short circuit current Isc [A]	type
A	65	21.3	3.05	0.002	-0.0858	0.49	0.0013	3.40	H-6510
B	70	17.3	4.05	0.00081	-0.072	0.50	0.0012	4.55	NT907
C	70	23.0	3.05	0.002	-0.092	0.53	0.0012	3.40	GL148N
D	62.5	20.7	3.03	0.0016	-0.096	0.50	0.0055	3.25	LA441K63S

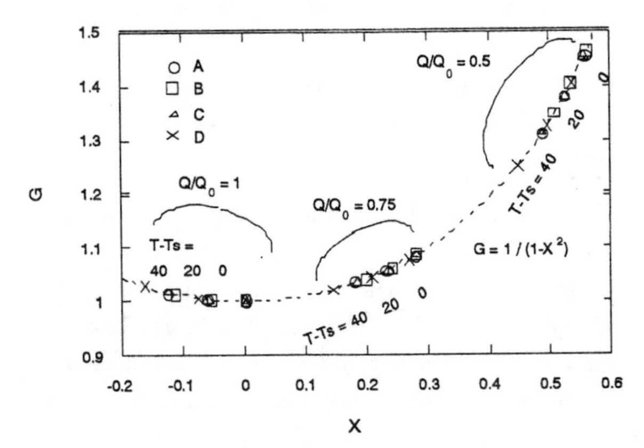

Fig.2. Relations between G and X Using Characteristic Values of PV Modules

In comparison with each module, though the difference of G is not so large in high insolation, a large difference can be seen between those modules in poor insolation. From these results, we can suppose that the characteristic values of PV modules were measured under the regulations of JIS at STC. While JIS does not regulate at other conditions of STC, we can calculate G from X even though the conditions are off STC from our analysis.

The relations between the insolation ratio G and the temperature difference T - T_S are shown in Fig.3. When $Q/Q_0 = 1$, G becomes large with the increase of $T - T_S$. Under the low insolation conditions, even though G becomes small with the increase of $T - T_S$, the value of G is larger than that of at $Q/Q_0 = 1$. While there are only small differences about G between each module in high insolation, their differences in poor insolation become large.

As shown above, G changes according to Q or T. The analysis of the PV operation point shows that the PV output will be increased when the operation point is set to the values calculated with Eq.(4) and Eq.(5) especially in poor insolation. To clarify the effect of the module temperature difference and the insolation ratio to the parameter X, their relations are shown in Fig.4. X shows the negative slope to the temperature difference regardless to the insolation, and it becomes large when the insolation becomes small.

From Eq.(6), X is described as follows;

$$X = 1 - \frac{2\left\{R_S + K\left(T - T_S\right)\right\}\left\{I_S + I_{SC}\left(\frac{Q}{Q_0} - 1\right) + \alpha\left(T - T_S\right)\right\}}{V_S + \beta\left(T - T_S\right) + R_S I_S} \quad \text{...(8)}$$

The smaller the second term of the right side of Eq.(8) becomes, the larger X becomes. Since the orders of the characteristic values of the PV module are $\beta \sim O(-10^{-2})$, $\alpha \sim O(10^{-3})$, and $K \sim O(10^{-3})$, respectively, the value of denominator of the second term of the right side of Eq.(8) becomes smaller than that of the numerator according to the increase of the temperature difference when $Q/Q_0 = 1$. Therefore, X becomes negative value. On the other hand, as the order of the term of Q/Q_0 is larger than that of others, the value of numerator becomes small when the insolation decreases. Then, X becomes small even though it is positive.

To simplify the analysis more, we assumed that the ambient and the module temperature are equal even though the insolation changes. Under this condition, Eq.(8) is described as follows;

$$X = 1 - \frac{2R_S\left\{I_S + I_{SC}\left(\frac{Q}{Q_0} - 1\right)\right\}}{V_S + R_S I_S} \cong \frac{I_{SC}}{I_S}\left(1 - \frac{Q}{Q_0}\right) \quad \text{...(9)}$$

Although the PV module with good fill factor (FF) almost shows I_{SC}/I_S is nearly equal to 1, we can find that the larger I_{SC}/I_S becomes, the larger X becomes from Eq.(9). This means that output of PV module with large I_{SC} becomes large under poor conditions even though FF of the module is small. Actually, it is reported that the maximum output of the PV modules with small FF does not decrease in spite of the change of the operation voltage in 1 MW PV demonstration plant constructed at Saijo, Shikoku, Japan[(2)]. Therefore, under poor insolation condition, a PV module with

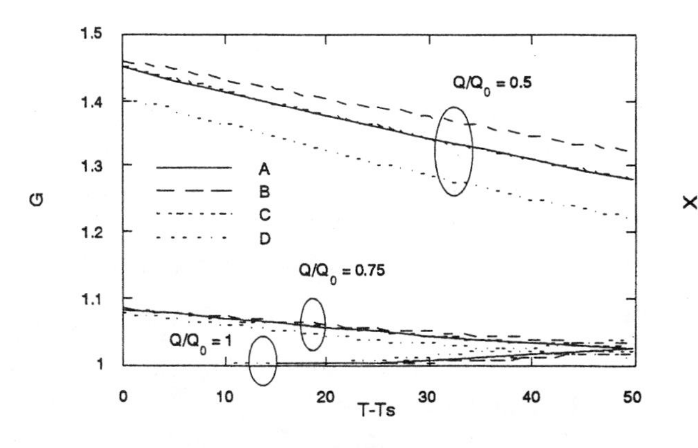

Fig.3. Relations between G and Temperature Difference, $T\text{-}T_S$

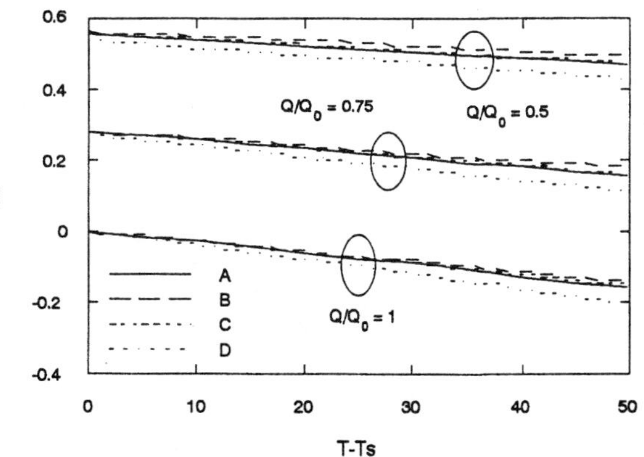

Fig.4. Relations between X and Temperature Difference, $T\text{-}T_S$

large I_{SC} is useful even though its FF is small.

CONCLUSIONS

The PV output power changes according to the environmental conditions such as the insolation change or the PV module temperature change. In this paper, the new PV control method which maximizes the PV output power was presented. From the analysis, we can obtain the following conclusions.

(1) The method which determine the maximum operation point of PV array was derived from JIS equations.

(2) To generate much power, the PV operation point should be set at the point calculated by the equations derived from this analysis especially at large X.

(3) When the PV module temperature is near the standard module temperature, the module with large I_{SC} will generate much power than that of large FF.

This control method predicts the maximum operation point from the insolation, the module temperature, and the module characteristic values. Therefore, if there is an error between the predicted point and the real operation point, there may be the system fault or the degradation of a PV array. This control method can be applied for the fault diagnosis of the PV system.

REFERENCES

(1) Yamamoto, et al., "New Method for Construction of Practical I-V Curves", Proc. of National Con. of IEEJ, No.1784, 1995.

(2) Shikoku Research Institute Inc., "Research Report of the System Design and the Operation of 1MW PV Plant", pp.80, 1990.

The Upper Limit to Solar Energy Conversion

Sean E. Wright, David S. Scott, and James B. Haddow
Institute for Integrated Energy Systems (IESVic)
University of Victoria, PO Box 3055 STN CSC
Victoria, BC, Canada, V8W 3P6
Email: strafidi@hotmail.com

Marc A. Rosen
Department of Mechanical Engineering
Ryerson Polytechnic University, 350 Victoria Street,
Toronto, Ontario, Canada, M5B 2K3
Email: mrosen@acs.ryerson.ca

ABSTRACT

Petela carried out research to determine the maximum ideal theoretical work output obtainable from BR independent of any conversion device. However, omnicolor conversion is considered by many to be the ideal theoretical process for solar energy conversion. Petela's result for blackbody radiation (BR) exergy is often thought to be of little importance because it appears to neglect fundamental theoretical issues that are specific to the conversion of thermal radiation (TR) fluxes. In this paper it is shown that Petela's BR exergy result does represent the upper limit for the production of work from solar radiation (SR) fluxes approximated as BR. This result means that reversible conversion of BR fluxes is theoretically possible. It is shown that the definition of the environment in conventional exergy analysis completely suffices for defining TR exergy. Inherent emission of TR by the conversion device is indicated to be a fundamental issue that cannot be avoided. It is explained why in contrast to non-ideal conversion inherent emission has a beneficial effect on the maximum work output for an ideal device. Various approaches used in solar energy conversion, including omnicolor conversion, are compared using second law efficiencies. In addition to being irreversible the emission of TR by these devices generally has non-zero exergy which is usually considered lost.

INTRODUCTION

Various approaches have been used to determine the maximum work obtainable from thermal radiation (TR) conversion. In the field of solar conversion it is thought that omnicolor conversion is the optimum conversion process and establishes the upper bound for solar energy conversion efficiencies. For example, Haught [1] states regarding omnicolor conversion that the "results obtained are independent of the specific form of the thermal and quantum radiation conversion device and serve as an upper bound on the efficiency with which radiant energy can be converted to useful work in any actual device." In agreement with Haught, De Vos and Pauwels [2] also state that an infinite series of optimized omnicolor collectors is "the thermodynamically optimal device for converting solar energy into work."

Petela's [3] blackbody radiation (BR) exergy result is thought by most to neglect fundamental theoretical issues that are specific to the conversion of TR fluxes. For example, Haught [1] states that "Thermodynamic treatments of the radiation field which derive the conversion efficiency from the available work content of the radiant flux neglect the limitations (reradiation, threshold absorption, etc.) inherent in the conversion process."

The viewpoint that Petela's result is not relevant to solar energy conversion centers on three main questions that arise.

First, exergy is a quantity that depends on the system and its environment, so how can an environment be defined for TR? Bejan [4] states that "there is no such thing as an "environment" of isotropic blackbody radiation (and pressure), as is assumed most visibly in the availability type derivation."

Second, how is it appropriate to assume that the conversion of BR fluxes can be reversible even though it appears that the conversion of TR fluxes is inherently irreversible? De Vos and Pauwels [2] state that "the conversion of radiation into work cannot be performed ... without entropy creation."

Third, how does the inherent emission of TR affect the maximum work obtainable? Any device that absorbs TR for conversion must also emit TR. De Vos and Pauwels [5] state that the "power flow from the solar cell is rightly considered lost." Also Landsberg [6] comments on the effect of inherent emission when he notes that Petela's efficiency is "pulled down below the Carnot efficiency because of the black-body emission from the converter which does not contribute to the useful work output."

The objectives of the present paper are to address these issues and determine the relevance of Petela's BR exergy result to solar energy conversion. Determining the correct maximum ideal work output for TR conversion allows second law efficiencies to be defined and related analysis to be performed. The results are expected to provide insight into solar energy conversion and thus may lead to practical device improvements.

BACKGROUND

Research on the ideal conversion of thermal radiation (TR) usually follows one of two general approaches. Some researchers consider the maximum work output obtainable from an enclosed blackbody radiation (BR) system. Other researchers in solar engineering have calculated the maximum conversion efficiencies of various solar converters.

Bejan [4] provides an excellent review and unifying interpretation of the different ideal efficiencies presented for the conversion of an enclosed BR system. The different efficiencies presented have different purposes. Petela's result [3] represents the exergy of the enclosed BR system:

$$\Xi_{BR} = aV\left\{T^4 - \tfrac{4}{3}T_oT^3 + \tfrac{1}{3}T_o^4\right\} \tag{1}$$

where a is a BR constant and V is the volume of the system. Petela's result represents the exergy of the BR system because the exergy does not depend on where the source radiation came from or depend on if and how the dead state TR is removed. Petela proceeds to define the maximum conversion efficiency as the ratio of the maximum work output to the energy ($U = aT^4V$) of the system:

$$\eta_P = \Xi_{BR}/U_{BR} = 1 - \tfrac{4}{3}x + \tfrac{1}{3}x^4 \tag{2}$$

where x is the ratio of the environment temperature to the temperature of the BR source. To obtain this result Petela considers an evacuated piston-cylinder device containing BR at temperature T in an environment of BR at T_o, as depicted in Figure 1. The system delivers maximum work as it settles into the dead state by an isentropic expansion process. Petela also obtains the same result for the ratio of the exergy flux of BR to its energy flux by considering the transfer of TR between two parallel blackbody (BB) surfaces.

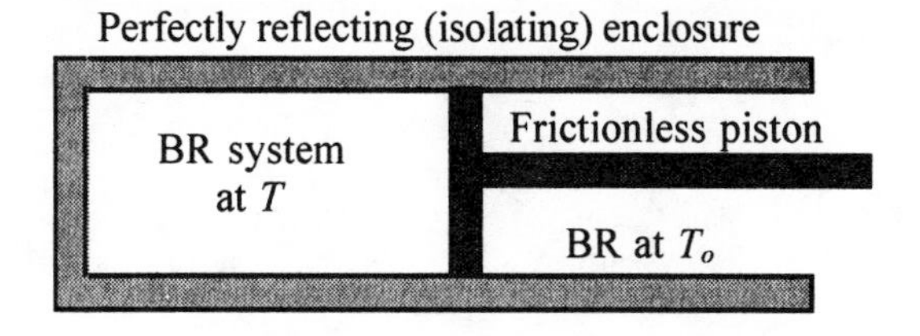

Figure 1: Piston-cylinder device.

Solar engineering researchers have calculated the maximum conversion efficiencies of both thermal and quantum devices. Figure 2 depicts a basic thermal conversion device used for efficiency calculations for isotropic BR fluxes. The thermal conductivity of the collector material is infinite to ensure that the temperature is uniform and that there is no entropy production associated with heat conduction.

The BB absorber inherently emits BR at the converter temperature T_c. The maximum work output is obtained by finding the optimum collector temperature. As collector temperature (T_c) increases the efficiency of the Carnot heat engine increases yet emission by the absorber increases as T_c^4. For solar radiation (SR) approximated as BR at a temperature $T_s = 5762$ K, the maximum energy conversion efficiency is 84.9% when $T_o = 300$ K.

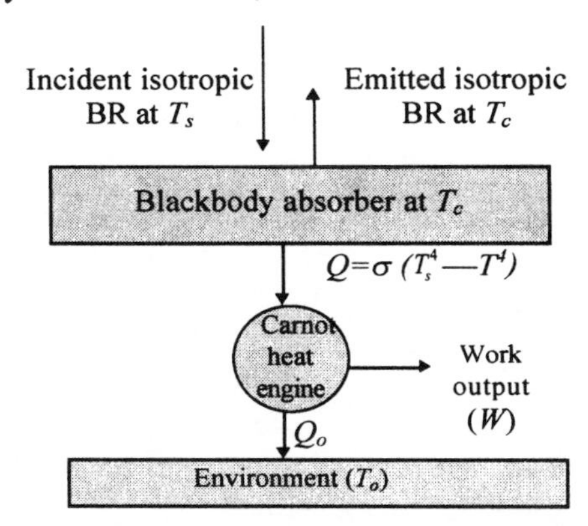

Figure 2: BR thermal conversion device.

Quantum conversion always exhibits threshold absorption, that is, only photons with a frequency above the cutoff frequency cause excitation in the absorber. Selective absorption is beneficial because it favors the absorption of relatively high-frequency solar radiation while disfavoring the emission of lower frequency TR from the converter. Selective absorption can also be achieved for thermal conversion with the use of a suitable coating material, for example see De Vos [7].

The use of multiple collectors individually optimized over different frequency ranges increases the overall conversion efficiency. For example, the quantum conversion efficiency is increased when a second collector is introduced that absorbs the radiation with frequencies below the first cutoff frequency. The maximum efficiency is achieved in the limit of an infinite number of collectors connected in tandem, each absorbing and effectively emitting at a single frequency and individually optimized at that frequency. Figure 3 illustrates an infinite tandem of omnicolor quantum conversion devices (without electroluminescence).

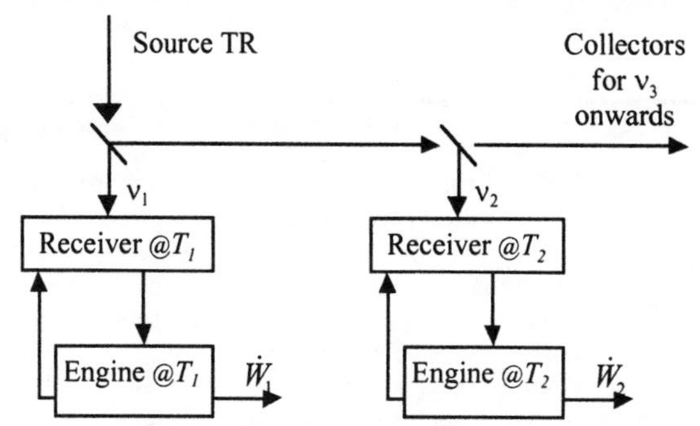

Figure 3: Infinite tandem of quantum cells.

Haught [1] shows that omnicolor quantum and thermal systems with selective absorption have the same maximum efficiency. Haught also examines hybrid (cascaded) quantum-thermal conversion to show that pure thermal and pure quantum omnicolor conversion are the two limits in a range of possible conversion processes.

Table 1 summarizes the maximum SR energy conversion efficiencies for thermal and quantum single-converter systems, and infinite-series omnicolor devices, for different levels of concentration. Lower concentration factors result in lower maximum conversion efficiencies. The entry for isotropic SR may be viewed as the case of ideal concentration (C = 43,600). Note that thermal and quantum omnicolor conversion have the same maximum ideal efficiency so there is only a single entry made for omnicolor conversion in Table 1.

Table 1: Maximum first-law efficiencies (in %) for solar radiation (SR)* conversion

Conversion Process	Direct SR C=1 Haught [1]	Concentrated, $C = 10^4$ Haught [1]	Isotropic SR*, C = 43,600	
			Bejan [8]	De Vos [7]
Single-Stage Thermal	54.0	80.0	84.9	85.4
Single-Cell Quantum	30.9	40.0	-------	40.8
Omnicolor Series	68.3	84.1	86.1	86.8

* These efficiencies are based on slightly different values for T_o/T_S

For all levels of concentration the single-cell quantum system has a lower efficiency than the corresponding single-stage thermal system. Haught [1] explains that thermal conversion with selective absorption is more efficient than the equivalent quantum conversion because quantum converters only respond to the number of photons above the threshold while a thermal converter utilizes all the energy absorbed above the cutoff.

DEFINITION OF THE ENVIRONMENT FOR THERMAL RADIATION (TR) EXERGY

The unusual nature of TR transfer within the context of exergy analysis raises the question of the appropriate definition of the environment for TR exergy. Bejan [8] states that "there is no such thing as an environment of isotropic blackbody radiation as is assumed most visibly in the availability type derivation." In Petela's [3] original derivation of TR exergy, involving a deformable reflecting enclosure, the specified environment is strictly isotropic BR in a vacuum, as depicted in Figure 1. In Petela's [3] parallel plate approach one surface is specified as a blackbody at T_o.

However, the definition of the material environment in conventional exergy analysis completely suffices for TR exergy. In Petela's [3] analysis the specification of the 'pseudo' environment of strictly BR in a vacuum has the effect of isolating the exergy of TR from other forms of exergy. To isolate TR exergy from material exergy the enclosed BR system is contained in a vacuum. But, as a result of specifying a vacuum the system has mechanical exergy in an environment with gas pressure (BR pressure is orders of magnitude less than 1 atm). To completely isolate TR exergy it is necessary to consider a BR system in a vacuum and a pseudo environment of strictly BR in a vacuum at environment conditions. Thus, the specification of a special environment of strictly BR at T_o does not cause any restrictions on the environment for TR exergy analysis. Note that the environment in conventional exergy analysis by definition is in internal equilibrium and thus any non-solid regions (such as a region of gas or a vacuum) would contain isotropic BR at T_o.

Further, the environment temperature T_o is the only relevant parameter of the environment for TR exergy. In Petela's [3] mechanical piston-cylinder approach, the final dead-state equilibrium radiation pressure is that of BR at T_o and not that of the environment gas pressure P_o. In the TR exergy calculation the BR system may be said to simultaneously achieve thermal and mechanical equilibrium with the pseudo environment, because there is only one independent intensive parameter for the BR system. For Petela's [3] piston-cylinder approach the simplest choice of parameter to specify the dead state is the pressure of BR in the pseudo environment. However, this BR pressure can be related only to the temperature (T_o) of the real environment but not to its gas pressure P_o. Therefore the only relevant parameter of the environment is its temperature T_o.

The exergy of the enclosed BR system is the theoretical maximum work output that can be obtained by bringing the BR into equilibrium with its environment. The TR has both energy and entropy whereas work output has zero entropy. During the ideal production of work from the BR system all the entropy that leaves the system must be transferred to the environment. Yet, this transfer of entropy must be accompanied by an increase of the internal energy of the environment. The environmental temperature T_o determines the magnitude of the required energy change. This temperature is defined as the slope of internal energy U versus entropy S when volume V and mole numbers N_i are fixed:

$$T_o \equiv \left(\frac{\partial U}{\partial S}\right)_{V,N_i} = \text{constant} \tag{3}$$

The extensive properties can change but the relative changes are small enough that the intensive properties such as temperature do not change.

When the entropy of the environment is increased at the rate $\dot{S}$ by BR conversion its energy must be increased at a rate of $T_o\dot{S}$. This is analogous to the production of work by the Carnot heat engine, as depicted in Figure 4. Noting that for TR conversion $\dot{E}$ and $\dot{S}$ are net quantities due to absorption of solar radiation (SR) and inherent emission.

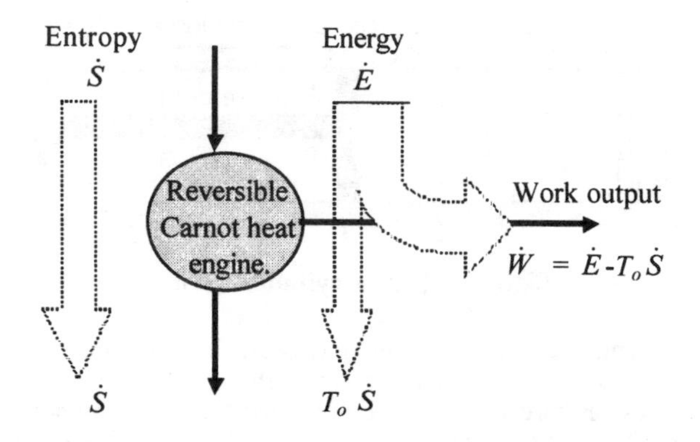

Figure 4: Reversible Carnot heat engine.

REVERSIBLE CONVERSION OF BR FLUXES

From a solar engineering perspective it appears that the conversion of BR is inherently irreversible and consequently that Petela's [3] BR exergy result for the enclosed system is not relevant. For example, Jeter [9, p. 79] feels that Petela's exergy is not representative of an actual utilization (conversion) process and states that Petela's result "is the correct formulation for the availability of cavity radiation ... however, it is not the result applicable to the collection and conversion of solar energy." Expressing a similar view, Bejan [8, pp. 495-496] states his belief that "η_p (Petela's efficiency) is no more than a convenient albeit artificial way of non-dimensionalizing the calculated work output" and that Petela's efficiency "is not a 'conversion efficiency.' "

Petela's parallel-plate approach for determining the exergy flux of BR offers no insight in resolving the question of inherent irreversibility for BR conversion. This is because the parallel-plate approach is only valid if it is assumed that reversible BR conversion is possible. In this approach this is implied because it is assumed that no part of the entropy production rate is inherent and thus the entropy production rate is directly proportional to the exergy destruction rate (Gouy-Stodola theorem applied to TR).

However, the fact that reversible conversion of an enclosed BR system is possible requires that reversible conversion of BR fluxes is theoretically possible as well. This is because the exergy radiance of BR at any point in the enclosed system is geometrically related to the internal exergy of the system. And the exergy radiance of BR must be independent of whether it is incident on a conversion device or inside an enclosure. Thus, Petela's BR exergy result for the enclosed system does represent the theoretical upper limit for the conversion of BR fluxes. After considering the geometric relationship between the internal exergy of the enclosed system and the BR exergy radiance, we will consider how reversible conversion of BR fluxes might be achieved.

Consider the thermal counterpart of Petela's mechanical piston-cylinder approach for the conversion of an enclosed BR system as depicted in Figure 5. The BR system at temperature T is in thermal contact with the environment through a reversible Carnot heat engine. The thermally conductive section of the enclosure is very small so that its changes in internal energy and entropy may be neglected. The thermally conductive section is also thin and has a high thermal conductivity such that entropy production due to heat conduction may be neglected. The remaining wall of the enclosure is perfectly insulating so that its internal energy and entropy may be neglected as well.

The work output during the process can be expressed as

$$W_{out} = \int_{T\to T_o} (1 - T_o/T)dQ \tag{4}$$

The maximum work output occurs during a reversible cooling process and it can be readily shown that Petela's result (Eqn. (2)) is obtained as expected. The BR in the cavity is isotropic and uniform so the energy and entropy radiances at any point are geometrically related to the specific (per volume) internal energy and entropy, respectively:

$$K = \frac{c}{4\pi}\frac{U}{V} = \frac{uc}{4\pi} \quad \text{and} \quad L = \frac{c}{4\pi}\frac{S}{V} = \frac{sc}{4\pi} \tag{5}$$

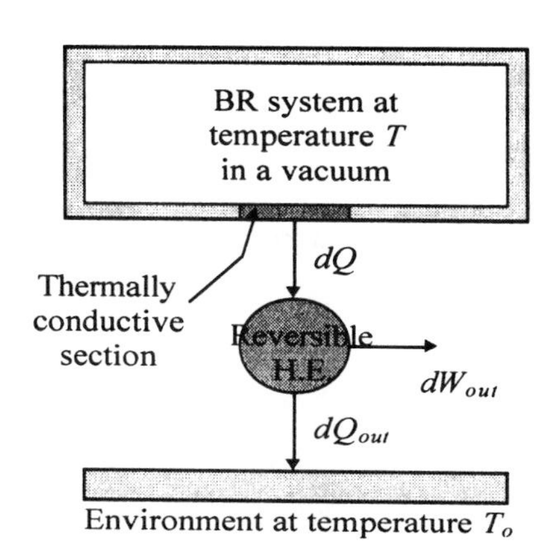

Figure 5: BR system connected to a Carnot heat engine.

Further explanation of this geometric relationship for isotropic TR is provided elsewhere (e.g., Planck [10, p. 21], Jakob [11, p. 29]). Exergy is a fundamental quantity like energy so the *exergy radiance* (N) of BR is likewise related to the *specific internal exergy* (Ξ_{BR}/V) of the enclosed BR system. Thus, using equation (1), we have

$$N_{BR} = \frac{c}{4\pi}\frac{\Xi_{BR}}{V} = \frac{\sigma}{a\pi}\frac{\Xi_{BR}}{V} = \frac{\sigma}{\pi}T^4\left[1 - \frac{4}{3}\frac{T_o}{T} + \frac{1}{3}\left(\frac{T_o}{T}\right)^4\right] \tag{6}$$

To support the validity of reversible conversion of BR fluxes consider again the thermal conversion device depicted in Figure 2. The net energy transfer is the difference between the incoming isotropic BR at T_s and the outgoing isotropic BR at T. In the limit as $T \to T_s$ one can straightforwardly show that the entropy production rate approaches zero faster than does the work production rate. This result is analogous to De Vos' [5] conclusion for quantum conversion, in which he stated that at the near open-circuit condition (the absorbed radiation equals the emitted radiation) for the photovoltaic converter, "photon transfer goes to zero in the 1st order while entropy production rate in the 2nd order."

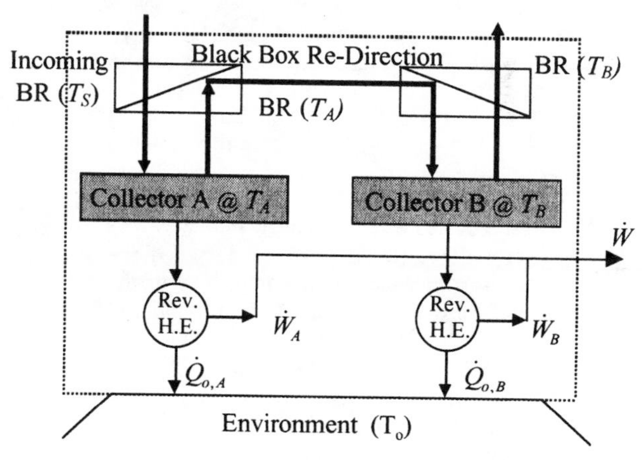

Figure 6: Two-stage thermal conversion device.

So BR conversion is reversible when there is an infinitesimal difference between the converter temperature and the BR source temperature. In fact this reversible conversion of BR fluxes is exactly what happens in the small thermally conducting region of the enclosed BR system in Figure 5 during ideal conversion. This means that the thermal conversion of a BR source flux would be reversible if there were an infinite series of absorption/emission stages with infinitesimal temperature differences between each stage. As an example of how such a multiple-stage absorption/emission device might physically operate consider the two-stage device in Figure 6. The BR energy spectrums for this two-stage thermal conversion process are depicted in Figure 7.

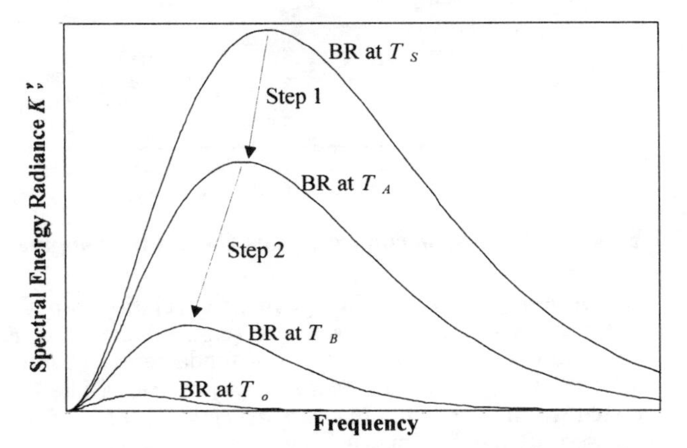

Figure 7: Energy spectrums for two-stage thermal conversion of BR.

The first-law efficiency for the two-stage process can be expressed as

$$\eta = \frac{W}{\sigma A T_S^4} = \theta_S^{-4}\left\{\left(\theta_S^4 - \theta_A^4\right)\left(1 - \frac{1}{\theta_A}\right) + \left(\theta_A^4 - \theta_B^4\right)\left(1 - \frac{1}{\theta_B}\right)\right\} \quad (7)$$

where $\theta_A = T_A/T_o$, $\theta_B = T_B/T_o$, and $\theta_S = T_S/T_o$. The maximum power, or maximum energy efficiency, is obtained when the optimum values for θ_A and θ_B are determined for a particular value of θ_S subject to the constraint:

$$1 \le \theta_B \le \theta_A \le \theta_S \quad (8)$$

For solar radiation approximated as BR with emission temperature of 5762 K and environment temperature T_o = 300 K the temperature ratio θ_S is equal to 19.21. For this condition Petela's efficiency (using Eqn. 2) is 93.1% compared to the single-stage thermal efficiency of 84.9% and the two-stage thermal efficiency of 89.7% (from Eqn. 7 when θ_A = 12.5 and θ_B = 5.9). This single-stage efficiency for isotropic BR (84.9%) is near to Petela's ideal efficiency (93.1%). However, the difference between the two-stage and the ideal efficiencies is less than half of the difference between the single-stage efficiency and the ideal efficiency.

The physical re-direction of the BR emitted from stage 1 has been represented by 'black-box' devices. One possibility for the inner workings of the black box re-direction device is that the material used has directionally dependent absorption, reflection and transmission properties. The material in the re-direction device would transmit the relatively high-frequency incoming solar radiation while reflecting the emitted BR from stage 1. Regardless, the use of the black-box re-direction device in Fig. 6 is acceptable from a theoretical perspective because simply re-orientating the TR does not necessarily change its energy or entropy.

THE EFFECT OF INHERENT EMISSION ON BR EXERGY

All material that absorbs TR must also emit TR. Thus, any TR conversion device inherently emits TR. Researchers in solar engineering usually feel that inherent emission cannot be ignored while other researchers believe that inherent emission can be ignored to determine the maximum work output for TR conversion. For example, Landsberg [12, p. 2786] gives the impression that inherent emission can be ignored by presenting the photon exergy flux as the energy flux minus the product of the environmental temperature and the entropy flux ($H - T_oJ$), with no term for inherent emission.

Further, some researchers indicate that inherent emission by the conversion device would reduce the maximum ideal work output. For example, Landsberg [6, p. 563] states that Petela's efficiency is "pulled down below the Carnot efficiency because of the black-body emission from the converter which does not contribute to the useful work output."

Analysis of Petela's parallel-plate approach gives the impression that the exergy irradiance of BR is $H - T_oJ$ with no inherent emission term. However, this is a misunderstanding that arises because the Guoy-Stodola theorem, used in Petela's approach, can only give the *net* exergy flux of TR while the inherent emission terms cancell out in a net exergy flux calculation [13].

The exergy of the enclosed BR system in the dead state is zero. Inherent emission causes the final dead state of the system to be BR at T_o. Thus, the exergy flux of BR at T_o is zero. However, the expression $H - T_oJ$ gives a non-zero negative value for the exergy of BR at T_o and therefore is not the correct expression for the exergy flux of BR.

It is true that inh. emission by a non-ideal thermal conversion device reduces the work output because the TR emitted by the converter is well above T_o. The reverse is true for ideal conversion, as in this case inherent emission has a beneficial effect. The reason for this reversal can be seen by considering the reversible black-box conversion device depicted in Figure 8. The device absorbs the incoming isotropic BR with emission temperature T and emits BR at T_o with zero exergy flux.

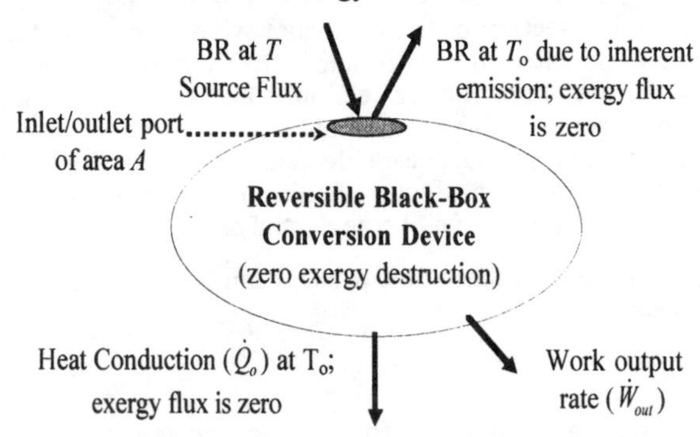

Figure 8: Black-box model for ideal BR conversion.

There is no entropy flow with work transfer so the entropy of the source radiation can leave the device by only two paths; BR at T_o and heat conduction at T_o. Note that for the Carnot heat engine there is only one path by which entroy leaves the device; by heat conduction at T_o (see Figure 4). The entropy-to-energy ratio of heat conduction at T_o is equal to $1/T_o$ while that of BR at T_o is equal to $(4/3)/T_o$.

Although BR at T_o and heat conduction at T_o both have a zero exergy flux, BR at T_o is a better means of rejecting entropy to the environment than heat conduction at T_o because the required energy flow rate to the environment is lower. As a result, inherent emission results in an additive term ($+\frac{1}{3}x^4$) in the exergy expression for ideal reversible conversion:

$$W_{out} = \sigma AT^4\left(1-\tfrac{4}{3}x+\tfrac{1}{3}x^4\right) \quad (9)$$

The reason that the Petela efficiency ($W_{out}/\sigma AT^4$) is less than the Carnot efficiency ($1 - x$) is not due to inherent emission. Rather it is due to the fact that the source flux has a high ratio of entropy to energy. That is, the entropy flux of BR at T is a factor of 4/3 higher than that of heat conduction at T with the same energy flux.

SECOND-LAW EFFICIENCIES

The resolution of Petela's BR exergy result as the true upper limit to solar-energy conversion allows second-law efficiencies to be evaluated. The second-law efficiency of a particular conversion process is simply the ratio of the maximum work output for that process to the exergy of the available SR accounting for its characteristics (e.g., isotropic or unconcentrated or attenuated by the atmosphere). Table 2 lists the second-law efficiencies corresponding to the first-law efficiencies in Table 1. Note that in Tables 1 and 2 the numerical values for the different references used are based on different values of T_S and T_o; Haught uses 6000 K and 300 K, De Vos 5762 K and 288 K, and Bejan 5762 K and 300 K.

The second-law efficiencies in Table 2 indicate that the performances of the conversion processes considered are more efficient than indicated by the first-law efficiencies. The second-law efficiencies are the true indicators of efficiency as they are relative to ideal conversion. The first-law efficiencies are unrealistic because they compare the work output to a theoretically unachievable upper limit, an upper limit established by assuming the energy flux of the source radiation is entirely convertible to work.

Table 2: Second-law efficiencies (in %) for solar radiation conversion

Conversion Process	Direct SR, C=1 Haught [1]	Concentrated, $C = 10^4$ Haught [1]	Isotropic SR, C = 43,600	
			Bejan [8]	De Vos [7]
Single-Stage Thermal	57.9	85.7	91.2	91.5
Single-Cell Quantum	33.1	42.9	-------	43.7
Omnicolor Conversion	73.2	90.1	92.5	93.0
Two-Stage Thermal	-------	-------	96.3	------

The second-law efficiencies in Table 2 for single-stage conversion are relatively high, especially for the isotropic case (91.2%). The single-stage efficiency is high because the temperature ratio T_S/T_o is large resulting in a relatively high value for the converter temperature (T_C) and a high value of the Carnot factor for thermal conversion ($1-T_o/T_C$). On the other hand, a high value for the converter temperature (T_C) also causes the energy loss due to emission to be higher and consequently the energy flow through the heat engine to be lower. However, the strong dependence of the BR energy flux on the emission temperature (proportional to T_C^4) reduces the significance of losses due to emission and allows the balance point for the maximum work output to be at a high converter temperature (T_C) and efficiency. Note that BR at half the emission temperature has only one sixteenth the energy flux.

For thermal conversion there is a direct correspondence between high second-law efficiency in Table 2 and the maximum temperatures involved in the conversion process. The efficiency of two-stage conversion is the highest (96.3%) followed by omnicolor (92.5%), and single-stage conversion (91.2%) for isotropic SR fluxes. Correspondingly, the maximum temperatures for two-stage conversion are 1770 K and 3760 K, for omnicolor conversion work is produced from a range of temperatures with a peak near 2450 K and with significant work production at 3300 K, and for single-stage conversion is 2465 K. The operating temperature for single-stage conversion decreases to 1900 K for C = 10,000 and 860 K for unconcentrated BR.

Omnicolor Conversion

The maximum efficiency (or maximum power) analysis of omnicolor conversion has been presented by many researchers as giving the upper limit to solar energy conversion [1,2]. However, from the second-law efficiency in Table 2 it is evident that omnicolor conversion is not the optimum conversion process for BR fluxes. For simplicity we will focus on explaining why omnicolor thermal, rather than omnicolor quantum, conversion is not the optimum conversion process for BR fluxes.

The non-ideal behavior of thermal omnicolor conversion is mainly due to irreversibilities in the conversion process because the exergy losses due to emission are very low compared to the exergy flux of the BR source flux. To clearly see the source of irreversibilities and the magnitude of losses due to emission we must first consider in detail the exchange of TR by a typical cell in the set.

A typical cell in the set absorbs a sliver of SR in the frequency range (ν,ν+dν). A typical cell also emits and receives TR from both cells adjacent to it, one at lower frequency and temperature, and the other at higher frequency and temperature (electroluminescence). Figure 9 depicts the exchange of TR between two adjacent cells in the set, one cell at T_1 and frequency cutoff ν_{o1}, the other cell at a higher temperature T_2 and cutoff ν_{o2}.

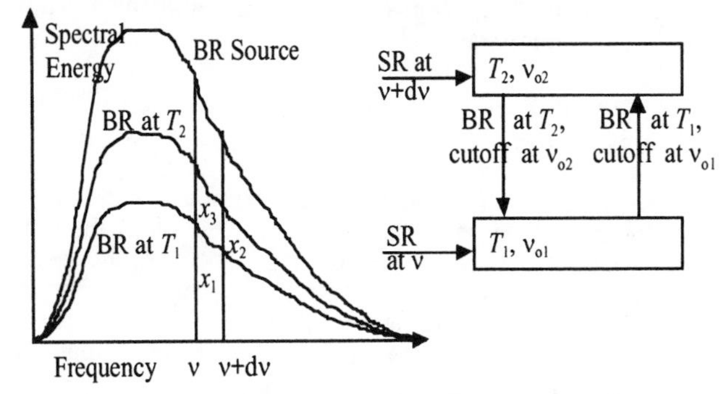

Figure 9: Energy spectrums for TR exchange between two adjacent cells in a set for omnicolor thermal conversion.

A portion of the TR emitted by cell 1 at T_1, represented by area x_1 in Figure 9, has a frequency lower than the cutoff frequency ν_{o2} and is not absorbed by any cells with $T>T_1$, and is consequently emitted by the stack. As a result, the net emission from the stack is BR with the frequency-dependent temperature $T(\nu)$. Consequently, the threshold behavior of emission from each cell appears as single-frequency emission from the stack. Figure 10 depicts the emission spectrum for omnicolor thermal conversion, SR approximated as BR at 5762 K, and a reference BR spectrum at 3800 K. Figure 10 shows that the emitted TR is not BR at T_o and thus has non-zero exergy flux but the emitted energy flux is still very low compared to the source flux.

The exchange of TR between adjacent cells in the set also results in heat transfer by TR down the stack, represented by area x_2 in Figure 9. This heat transfer results in exergy destruction and is inherent to the omnicolor configuration. However, the irreversibility due to this heat transfer is a minor part of the total irreversibility of the process. Also, note that area x_3 is the product of two differentials and thus has no significance in the limit of an infinite series of cells.

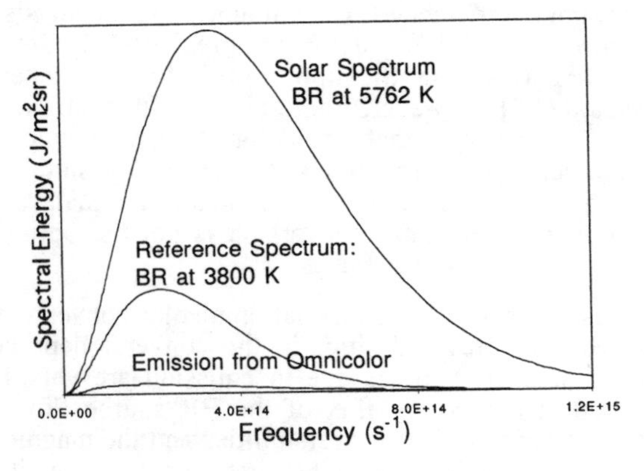

Figure 10: Source and emission energy spectrums for omnicolor thermal conversion.

The main source of irreversibility in each cell is due to the absorption of a sliver of SR with a high source temperature accompanied by emission with a relatively low cell temperature $T(\nu)$. Ideal thermal and omnicolor thermal conversion can be qualitatively compared by considering the energy flow through the device as a function of temperature. The net work produced is a product of the energy flow as a function of temperature and the Carnot factor for that temperature.

Figure 11 illustrates qualitatively the energy flow (incoming source TR minus emission) as a function of temperature for both processes, as well as the Carnot factor for the heat engine portion of the conversion device. From this plot it can be seen that, compared to ideal thermal conversion, the thermal omnicolor process inherently extracts work at much lower temperatures where the Carnot factor for the heat engine is relatively low. For ideal conversion the BR is converted in a continuous manner from T_s to T_o.

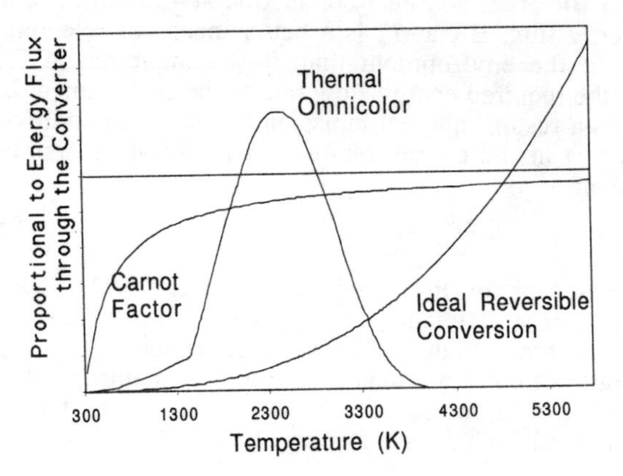

Figure 11: Energy flow vs. temperature.

The main source, and apparently only source, of non-ideal behavior in each cell for omnicolor quantum conversion appears to be due to the absorption of a sliver of SR accompanied by emission with a relatively low cell temperature T_o. For omnicolor quantum conversion there are no exergy losses because the TR emitted by the stack is BR at T_o. Also, there is no heat transfer along the stack and thus no associated exergy destruction. Thus, in Figure 9 the areas x_2 and x_3 do not exist for omnicolor quantum conversion.

Concentration

Concentration increases the operating temperature and thermal efficiency of thermal solar converters that are used in practice. Also, concentration is of benefit because it reduces the required converter surface area and size for a given solar collection area. The values in Table 2 based on Haught's [1] energy efficiency results indicate that the values of second-law efficiency for concentrated SR are substantially higher than those for the unconcentrated case. However, this difference occurs only because the Haught considers a specific case of unconcentrated SR conversion.

For Haught's unconcentrated case (typical in practice) emission by the collector occurs over a large solid angle (2π sr) while the incident SR is contained in a very small solid angle. On the other hand, for the concentrated case absorption and emission occur over the same solid angle. Thus, the energy losses due to emission are more significant for the unconcentrated case considered by Haught than for the concentrated case. Concentration is strictly a practical performance improvement measure as it cannot change the spectral distribution of the source TR and cannot increase the maximum theoretical efficiency obtainable from a source flux.

Threshold behavior improves the performance of a single-converter thermal system because it reduces the fraction of energy loss due to emission. However, Haught [1] observes that under full concentration threshold behavior has a marginal benefit. The reason for this phenomena is that under full concentration absorption and emission are over the same solid angle and thus losses due to emission are less significant and consequently the benefit of threshold behavior becomes marginal.

CONCLUSIONS

We believe that the results presented in this paper have resolved much of the controversy over identifying Petela's BR exergy result as the upper limit to the conversion of SR fluxes (approximated as BR). Petela's result for BR is thought by many to be irrelevant to the conversion of BR fluxes because it appears to neglect fundamental theoretical issues that are specific to the conversion of TR fluxes. However, these concerns are addressed by:

1) showing that reversible conversion of BR fluxes is theoretically possible,

2) showing that the conventional definition of the environment in exergy analysis completely suffices for TR exergy calculations whereas the special environment considered by Petela only serves to isolate TR exergy from other forms of exergy, and

3) explaining why inherent emission by the conversion device has a positive effect on the maximum ideal work output although it results in losses in practice,

Identifying the upper limit to BR conversion is of fundamental thermodynamic importance as it allows second-law efficiencies to be defined and provides insight by identifying sources of non-ideal behavior in known conversion processes. The requirements for ideal conversion of BR are illustrated by a black-box conversion process that has neither irreversibilities nor exergy losses.

It is explained why single-stage conversion has a relatively high second-law efficiency and that for thermal conversion there is a direct correspondence between even higher efficiencies and the maximum operating temperatures involved in the conversion process. Omnicolor thermal conversion provides a slight increase in efficiency over single-stage conversion but does not appear to be a practical option because of the impact of higher operating temperatures on material property requirements and because of the difficulty involved in producing work over a range of source temperatures. Two-stage thermal conversion offers a greater efficiency improvement than omnicolor thermal conversion and requires work production at only two sources temperatures. However, two-stage thermal conversion involves the highest operating temperatures and requires selective re-direction of the BR fluxes between stages.

In contrast to thermal conversion, omnicolor quantum conversion offers substantial improvements over the single-cell quantum system for all levels of concentration. It is yet to be determined if reversible quantum conversion of BR fluxes is theoretically possible although it appears that it may not be possible because of the inherent threshold behavior of quantum conversion. If this is the case then omnicolor quantum conversion is the optimal quantum conversion method, although it is irreversible, and then ideal reversible conversion of BR fluxes is only theoretically possible by thermal methods.

ACKNOWLEDGMENTS

This work was partially supported by the Natural Sciences and Engineering Research Council of Canada.

REFERENCES

1. A. F. Haught, 1984, "Physics Considerations of Solar Energy Conversion", *ASME Journal of Solar Energy Engineering*, Vol. 106, pp. 3-15.
2. A. De Vos and H. Pauwels, 1986, Discussion, *ASME Journal of Solar Energy Engineering*, Vol. 108, pp. 80-84.
3. R. Petela, 1964, "Exergy of Heat Radiation", *ASME Journal of Heat Transfer*, Vol. 86 , pp. 187-192.
4. A. Bejan, 1987, "Unification of Three Different Theories Concerning the Ideal Conversion of Enclosed Radiation", *ASME Journal of Solar Energy Engineering*, Vol. 109, pp. 46-51.
5. A. De Vos and H. Pauwels, 1981, "On the Thermodynamic Limit of Photovoltaic Energy Conversion", *Journal Applied Physics*, Vol. 25, pp. 119-125.
6. P. T. Landsberg, 1984, "Non-Equilibrium Concepts in Solar Energy Conversion", *Proceedings of the NATO Advanced Study Institute on Energy Transfer Processes in Condensed Matter*, June 16-30, 1983, Sicily, Italy, edited by Baldassare Di Bartolo, New York Plenum Press.
7. A. De Vos, 1992, *Endoreversible Thermodynamics of Solar Energy Conversion*, Oxford University Press, Oxford.
8. A. Bejan, 1997, *Advanced Engineering Thermodynamics*, 2nd edition, John Wiley and Sons, New York.
9. S. Jeter, 1986, Discussion, *ASME Journal of Solar Energy Engineering*, Vol. 108, pp. 78-80.
10. Max Planck, translation by Morton Mausius, 1914, *The Theory of Heat Radiation*, Dover Publications, New York.
11. M. Jakob, 1949, *Heat Transfer*, Vol. I, Wiley, New York.
12. P. T. Landsberg, 1995, "Statistical Thermodynamic Foundation for Photovoltaic and Photothermal Conversion. I. Theory", *Journal of Applied Physics*, Vol. 78 (4), pp. 2782-2792.
13. S. E. Wright, (2000, expected), "*The Exergy of Thermal Radiation and its Relevance in Solar Energy Conversion*", Ph.D. Thesis, Mechanical Engineering, University of Victoria, Victoria, Canada.

NOMENCLATURE

Table 3: Energy and entropy nomenclature

	Energy		Entropy	
	Symbol	Units	Symbol	Units
Flow Rate	$\dot{E}$	W	$\dot{S}$	W/K
Irradiance	H	W/m^2	J	W/Km^2
Radiance	K	W/m^2sr	L	W/Km^2sr

a	BR constant = $(7.61)10^{-16}$ J/m^3K^4 = $4\sigma/c$
A	Surface area of collector (m^2)
c	Speed of light = $(2.9979)10^8$ m/s
C	Concentrating factor
k	Boltzmann's constant = $(1.38)10^{-23}$ J/K
h	Planck's constant = $(6.626)10^{-34}$ J s
M	TR exergy irradiance (W/m^2)
N	TR exergy radiance (W/m^2sr)

N_i	Mol number of species i in the environment
P	Pressure (N/m^2)
q	Heat flux (W/m^2)
$đQ$	Infinitesimal heat transfer (J)
$\dot{Q}$	Heat transfer rate (W)
S	Internal entropy of the system (J/K)
s	Specific internal entropy-per unit volume (J/m^3K)
T	Material emission temperature (K)
U	Internal energy of the system (J)
u	Specific internal energy-per unit volume (J/m^3)
V	Volume (m^3)
$\dot{W}$	Work output rate (W)
x	Temperature ratio T_o/T
η	Efficiency
ν	Frequency (s^{-1})
π	Physical constant 3.14159…
θ	Temperature ratio T/T_o ($\theta = 1/x$)
σ	Stefan-Boltzmann constant =(5.67)10^{-8} W/m^2K^4
Ξ	Internal exergy of the system (J)

ABBREVIATIONS

BR: Blackbody Radiation
CV: Control Volume
SR: Solar (thermal) Radiation
TR: Thermal Radiation

SUBSCRIPTS

A	Stage A
B	Stage B
C	Converter
N	Net
o	Environment
o,A	To the environment from stage A
o,B	To the environment from stage B
out	Outgoing
S	Source

INTRODUCTION INTO APPLIED RENEWABLE THERMODYNAMICS FOR DIRECT SOLAR ENERGY CONVERSION SYSTEMS

Anatoly T. Sukhodolsky
General Physics Institute, Russian Academy of Sciences, Vavilov Street 38 Moscow Russia

ABSTRACT

The direct conversion of solar light into mechanical power within non-equilibrium phase transitions of first order had been presented at previous IECEC'1998 on the concept of the renewable thermodynamics of self-organized heat cycles to describe several new physical phenomena in liquids. This paper is to promote the principles of renewable thermodynamics created by observation of spontaneous physical processes converting light "by itself" into mechanical power for new engineering design of the actual engineering constructions powered by sunlight. Several solar renewable heat cycles are modeled by using perfect gas powered by light to find how thermodynamics of reversible cycles can be involved into concept of applied renewable thermodynamics to study available thermal and actual efficiencies of prospective constructions. The both concepts well-known reversible and proposed renewable are considered in parallel to find the fundamental restriction for direct conversion of sunlight in different new power and propulsive systems. The master equation for motion of a vector state having both mechanical and thermal degrees of freedom proposed to solve problem of efficiency. The problem to have the actual efficiency beyond thermal efficiency (Carnot theorem) has formulated as a principally available objective for new generation of direct solar energy conversion systems

INTRODUCTION

The aim of this paper is to promote renewable thermodynamics of spontaneous physical phenomena[1] into a new sphere of artificial constructions – solar heat engines. As far as problem of efficiency is a key point of following consideration let us begin, for the sake of introduction, with the problem: why in spite of maximum thermal efficiency (as it follows from reversible theory of heat[2]) we can not find around us any Carnot engine actually working. Theory of motive power of heat[2] based on the principle of Carnot to have two reservoirs: ***source*** (thermal energy by temperature) and ***sink*** of thermal energy (by temperature), which both have to be available within any heat engine. According Second Law of thermodynamics crated from this principle, a reservoir of thermal energy alone can not provide with any new energy. In framework of renewable thermodynamics of spontaneous phenomena this principle is assumed to formulate on modernized entropy of Plank[3] to treat any phenomena self-organization of renewable heat cycles in laser physics. By this suggestion, any natural thermal process between source powered by light and sink undergoes entropy production[1]

(1)

As far as such definition applied for system source&sink very far from equilibrium, it assumed to be valid for any temperature difference between given source and sink. The classical reversible thermodynamics takes place within renewable thermodynamics for a particular case with fundamental maximum of thermal efficiency given by theorem of Carnot. As an introduction into the problem of efficiency for applied renewable thermodynamics let us consider Carnot cycle with work substance as a mole of perfect gas having state function for pressure volume and temperature with gas constant

(2)

uniformly powered by black body radiation (5700K) of solar radiation having energy . If we take initial normal condition (point (**0**), in Fig.1, , it is easy to find pressure and volume for the point (**1**) of the end first adiabatic stroke of compression. In this point we can open a source of light to isothermally expand perfect gas by point (**2**) having volume and pressure

$$v_2 = v_1 \exp \frac{\Delta L}{RT_h}, \quad p_2 = R\frac{T_h}{v_2} \qquad (4)$$

If then light is shut off, the temperature of sink can be

reached by second adiabatic stroke to point (**3**)

$$v_3 = \sqrt[k-1]{\frac{p_2 v_2^{\,k}}{RT_0}} \quad p_3 = \frac{RT_0}{v_3} \qquad (5)$$

The final isothermal stoke by temperature is to close cycle to return perfect gas into initial state (**0**) (Fig.1).

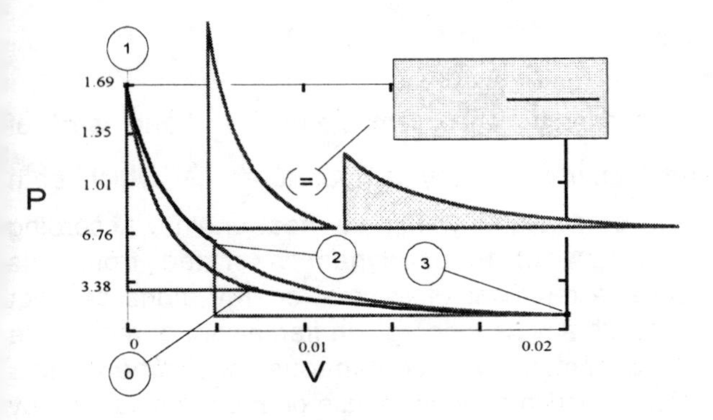

Figure 1. Renewable Carnot cycle by optical pumping.

Let us notice that we calculated only turn points of Carnot cycle powered by light (actual paths will be considered below). By non-equilibrium treatment of such Carnot cycle with Eq.1, the next problem is to find condition to have entropy production . For this purpose let us calculate work of two adiabatic strokes (**0-1**) and (**2-3**) using Eq.2-5

(6)

According (Eq.1) entropy production always takes place as spontaneous transfer of thermal energy from hot to cold. But, from Eq.6 we can conclude that work Carnot cycle with entropy production can be realized only by having an addition reservoir of mechanical energy to satisfy Eq,6. In other words, to put Carnot cycle in actual motion along with given temperature difference the reservoir of mechanical energy should provide with work . If we take the principle of Carnot to have thermal reservoirs source and since for application within renewable thermodynamics, we should add the necessity to have ***in advance*** a reservoir of mechanical energy to put Carnot cycle in actual motion. It is easy to see that the minimal mechanical necessary in advance depends only on temperature difference between source and sink and can be much more than work of cycle

(7)

Such an elementary introduction is to refresh the principle of Carnot given originally quite long ago in reversible theory of heat by Clausius[2] in new renewable thermodynamic treatment. In context of this elementary calculation, we should accept that besides source of thermal energy to make any engines, we should also consider a source of mechanical energy. As far as Carnot cycle should have very high level of start-up mechanical energy (for example mechanical energy of a fly-wheel), it has more application for speculative than actually working engines

THE ELEMENTS OF APPLIED RENEWABLE THERMODYNAMICS

This section is to introduce the problem of motive power of solar light having black-body temperature around 5700K as a motive power to pump work substance inside a heat engine in non-equilibrium condition.

Entropy of solar light and first law

Let us use term "***optical pumping***" of laser physics in next consideration to differ a way to add energy into a work substance inside a machine (for example inside a piston & cylinder system) by light dL from heat dQ. In framework of reversible thermodynamics, the equilibrium reversible entropy is to describe motive power of heat δQ by using model of a reversible engine to find maximum available heat[2]. Equation (1) is to describe actual process of entropy production by heat transfer between hot and cold that allows having derivative .

We will follows Gibbs[4] to use symbol of variation for entropy of Clausius for any reversible processes.

(8)

because, according Gibbs[4], the fundamental equation of Clausius[2] should be written always in variational form

(9)

to study all thermodynamics quantities near a point of equilibrium. While differential for Eq.1 is coincident with actual process of entropy production along time , if follow Gibbs, variation within Eq.8,9 means speculative, undertaken mathematically on paper increment of function along with any other thermodynamics quantities in Equation (9) in point of equilibrium (by fixed time). As it was shown earlier[1] any attempts to build a picture of time motion within equation (9) (by substituting operator of differential instead variation) leads to the paradox with entropy of Clausius as a unique function of time. Therefore, again, the difference of renewable thermodynamics is to use actual time derivative instead of variation of any thermodynamic function. To be more specific in promotion of Eq.1 instead of Eq.8 into practical engineering, let us use Eq.8 in meaning of a time differential for a simplest process of increasing of internal thermal energy of a solid ( ) having temperature and heat capacity up to temperature of heater by energy of light . We can find difference of internal energy as and difference in entropy of Clausius as . As a result, we can measure all quantities. But, the entropy of Clausius in differential form is a superfluous thermodynamic quantity that can not provide us with any new knowledge. In contrast, Eq.1 in renewable thermodynamics is to describe fundamental difference between energy and thermodynamic entropy along any actual motion. Suppose, we have two equal solids with equal capacity , by temperatures, say 100K and 10K, to describe the energy of a source and a sink in definition of entropy Eq.1. Physical sense of consists of followings. If we have no work cycle by heat transfer between these bodies, the final temperature during an idle process of heat transfer is . In case of Carnot cycle to conduct thermal energy we should accept that by using Eq.1. As a result of work cycle, the equilibrium temperature becomes $\sqrt{}$. Evidently, maximum energy created along Carnot cycle proportional to $\sqrt{}$. In case of non-equilibrium treatment of Carnot principle by involving Eq.1, entropy is not to duplicate energy as Clausius entropy does if instead of variation we try to build actual processes. Physical sense of is to describe a new feature of system as a whole – distribution of thermal energy between source and sink.

Let us now consider energy of solar light dL that comes from transparent cold surrounding with temperature to optically pump a work substance at temperature

While we can not build any rational variation by interaction of solar radiation having black body temperature 5700K with perfect gas of work substance by temperature we can not use fundamental equation Eq.9, because of practically full irreversibility of absorption of light by perfect gas a work substance. At the same time, entropy of Plank (Eq.1) in application for this case provide us with thermodynamic entropy of light by direct pumping any solar engines in following form

$$dS = \frac{dE}{T} - \frac{dL}{T_0} = -\frac{dL}{T_0}\,\frac{T - T_0}{T} \qquad (10)$$

to be the basic entropy in renewable thermodynamics[1].

The physical sense of negative sign of entropy by direct optical pumping any solar engines is the spontaneous flow of energy from cold transparent matter to hot work substance. To be more specific in relation to Second Law, the optical pumping any processes on Earth is accompanied by a positive sign of entropy production (Eq.1) if we involve source (sun) into consideration. But as far as source is quite far, we can consider light alone as a source of energy to have Eq.10. Let us notice, that namely, this fundamental feature of optical pumping is responsible for direct creation of mechanical energy in opto-hydraulic actuator[4]and very high capillary motive forces. The renewable thermodynamics was introduced earlier to describe the heat-mass transfer in liquids powered by light for *actual physical phenomena* working as engines. The motion by such approach is described on the concept of entropy production inevitably applicable for any natural phenomena. Again, the aim of this introduction is to find out how the concept of non-equilibrium entropy of Planck used before to treat physical phenomena of self-organization can be used to describe motive power of sunlight in theory of solar engines

The First Law of applied renewable thermodynamics is nothing more as energy balance by optical pumping when energy of light is consumed: a) to change internal energy of system $dE = c_v dT$ b) to make work $dW = pdV$ and c) to heat surrounding by spontaneous heat-transfer dQ

$$dL = c_v dT + pdV + dQ \qquad (11)$$

The goal is to find a fundamental restriction for maximum available light to be converted into work and after to renew initial volume of system. Firstly, let us clearly distinguish in any actual cycle powered by solar light the problems: (***Conversion***) -) how to convert solar energy into work by maximum efficiency and (***Renewal***) -) how to renew initial volume of system with minimum waste of energy. Let us notice that all examples of (***Non-Renewable***) processes, such as chemical reaction of explosion or combustion accompanied by mechanical energy production are beyond current approach. Evidently, as it was shown above, to renew initial condition for Carnot cycle we should spend a visible part of net energy. It take look on Eq.(11) it is easy to see the preference of adiabatic optical pumping ($dQ = 0$) to avoid the useless consumption of input energy to heat surroundings. The regime to be near adiabatic pumping realized in most of laser experiments is also available by solar pumping. By typical thermal conductivitiy, time of thermal relaxation into surrounding can be in times bigger than time necessary for local solar optical pumping.

ADIABATIC OPTICAL PUMPING.

Again, the irreversible waste of energy by dQ in First Law makes seek the problem of the maximum light to be converted into mechanical power into sphere of adiabatic processes powered by light within the law of energy conservation

$$dL = c_v dT + pdV \qquad (12)$$

We are still keeping in mind the two problems: to product as much as possible mechanical energy from light and to renew initial volume of system to start the process again. If suppose, that work substance is a mole of perfect gas within state equation (Eq.2) we can find from Eq.12 that

$$dL = c_v dT + \frac{RT}{V} dV \qquad (13)$$

and after dividing all members by T we can have following equation with all terms having dimension of entropy:

$$\frac{dL}{T} = \frac{c_v dT}{T} + \frac{c_p - c_v}{V} dV \qquad (14)$$

In case of thermodynamics of perfect gas, for the material parameters c_v, c_p assumed to be not a function of temperature, from Eq.14 follows that actual efficiency to convert an element of energy of light by adiabatic optical pumping depends on the inevitable increasing internal energy $c_v dT$ that takes visible part of energy from production of work pdV as desire output. In other words, for given temperature, energy of light divided between these two members in formula (14) along their entropy contribution. The problem is to find the better process (cycle) for a finite energy ΔL to convert as much as possible its part into work $\Delta L = \Delta E + \Delta W$. While reversible thermodynamics based on thermal efficiency of reversible Carnot cycle is beyond of large part of mechanical energy necessary for actual process, the key point of applied renewable thermodynamics is to solve the problem of efficiency by introduction of ***mechanical energy*** in scope of thermodynamics studies. The problem to unit thermodynamics and mechanics was discussed before by Truesdel[10], who also stated the problem of a huge gap between engineering origin of classical thermodynamics and its application to describe physical phenomena occur "by itself". Let us now consider the problem of internal energy.

THERMAL AND ACTUAL CONVERSION OF SOLAR LIGHT

The prime point of reversible thermodynamics is concept of the internal energy E that in most of text-books[8] is defined as work of adiabatic compression ($dQ = 0$) of a work substance. As light is also able to directly change energy in adiabatic condition, we should firstly define the basic difference between two adiabatic processes:

$$pdV = dE \quad \text{and} \quad dL = dE \qquad (15)$$

The internal energy of perfect gas given alone is state function. If so, we should expect to have 100% conversion in case we go away from a state with internal energy E_0 into more energetic (excited) state E_0 ? $E_1 = E_0 + \Delta L$ by light and, after, still being in the same adiabatic conditions we can let excited system to make work E_1 ? $E_2 = E_1 - \Delta W$. To find out the difference between two ways (Eq.15) to directly change internal energy by light and by work, let us consider the same mole of perfect gas by normal pressure p_0, V_0, T_0 in initial state of equilibrium having internal energy E_0. By fixed volume V_0, let perfect gas absorb energy of light ΔL to go into state E_1 with temperature difference

$$\Delta T = \frac{\Delta L}{c_v} \qquad (16)$$

Let us note this is an excited state with regards to initial energy E_0 along entropy definition (Eq.10). Now, we can obtain two possible ways to produce work during two adiabatic processes below called as _thermal_ and _actual_.

Thermal efficiency for adiabatic conversion of light.

The thermal efficiency of direct work production can be found if assume that $E_2 = E_0$ that can take place by $T_2 = T_0$. From equation for adiabatic process

$$V_0 (T_0 + \Delta T)^{k-1} = V_2 T_0^{k-1} \qquad (17)$$

we can find volume of gas in state E_2

$$v_2 = v_0 \left(1 + \frac{\Delta T}{T_0}\right)^{k-1} \qquad p_2 = \frac{RT_0}{v_0 \left(1 + \frac{\Delta T}{T_0}\right)^{k-1}} \qquad (18)$$

In order to find work within relation for adiabatic

expansion

$$\Delta W = \frac{p_2 v_2 - p_1 v_1}{k - 1} \tag{19}$$

we can constitute $V_1 = V_0$ and $p_1 = \frac{R(T_0 + \Delta T)}{V_0}$ into (19) that together with (18) provide us with thermal efficiency of direct conversion

$$\eta_t = \frac{(c_p - c_v)\Delta T}{\Delta T(k-1)} = 1 \tag{20}$$

The fact of 100% thermal efficiency simply follows from definition of internal energy along Eq.15, because, it is easy to see that $E_2 = E_0$ .By the equal temperatures should be equal their internal energies if specific heat is a constant. Again, as far as we do not state the problem to go back by any either reversible or renewable ways to have this process again, the thermal efficiency 100% follows from definition of internal energy of perfect gas given alone for both adiabatic processes (Eq.15).

Actual efficiency for direct conversion.

The idea to involve term *actual process of work production* follows from renewable thermodynamics of created natural phenomena to consider any real processes progressed in time within entropy production as fundamental causality of any motion. As we assumed that efficiency any renewable cycle cannot be more than by direct adiabatic conversion, let us now consider such an actual process. From the same state excited by light let system expand into point with pressure $p_2 = p_0$, in contrast to final point of thermal process $E_2 = E_0$. Making such assumption we are aware that, actually, we can not share more energy of chaotic molecular motion for mechanical energy than on way of going into this point of mechanical equilibrium if we like to renew process again without additional sources of energy. Using adiabatic equation

$$v_0 p_1^{\ k} = v_2 p_2^{\ k} \tag{21}$$

we can find volume and temperature in point of "mechanical" equilibrium as

$$v_2 = v_0 \left(\frac{p_1}{p_0}\right)^{\frac{1}{k}} \quad T_2 = p_0 v_0 \frac{\left(\frac{p_1}{p_0}\right)^{\frac{1}{k}}}{R} \tag{22}$$

to substitute into Eq.19 to find efficiency of actual direct adiabatic conversion as

$$\eta_a = 1 - \left(\frac{T_0}{T_0 + \Delta T}\right)^{\frac{k-1}{k}} \tag{23}$$

that appears to be less than efficiency of reversible cycle

$$\eta_c = 1 - \left(\frac{T_0}{T_0 + \Delta T}\right) \tag{24}$$

because (23) is obtained for heat source powered by light that is exhausted, while ΔT in (24) is a constant.

There is no problem to accept 100% efficiency of direct conversion until we told nothing about how to repeat the same process again. The main in approach of renewable thermodynamics to consider motive power of solar light is based on idea of actual efficiency. We underlined the necessity a reservoir mechanical energy to have most effective Carnot cycle actually working, but let us try to build a cycle with perfect gas when ***to renew*** means to be back within an actual processes without any additional energy sources. As far as the second actual adiabatic process at the end has difference temperatures in condition of mechanical equilibrium, to be back it is not necessary to use any additional sources of mechanical energy because spontaneous entropy production accompanied balance of temperatures is enough. Basically, this is main idea of natural renewable cycles[4,5] that is evaluated in this paper to estimate efficiency of solar light in new heat engine design. In other words, the idea to be back and to renew process by spontaneous entropy production in physical phenomena here is to estimate on classical example of a mole of perfect gas by actual adiabatic process as introduction into problem of the applied renewable thermodynamics.

Before introducing a master equation to evaluate different renewable cycles by optical pumping, the several initial points follows from previous consideration of perfect gas by optical pumping are summarized:

1. In case of renewing any cycle without mechanical sources, the efficiency of any renewable cycle by

optical pumping can not more than available by direct adiabatic conversion.

2. Work substance (perfect gas) in state of molecular-kinetic equilibrium given alone is able to produce mechanical energy only within mechanical non-equilibrium (p ? p_0).

3. The efficiency of any renewable thermodynamic cycle can be found to solve a path (cycle) of actual motion involving the law of the momentum conservation into thermodynamic studies.

MASTER EQUATION

The aim of this paper is to not review different application of reversible fundamental Eq.9 to solve the problem of efficiency within reversible thermodynamics with heat as motive power. The problem of available heat in reversible thermodynamics, which is divided between different terms in Eq.9 is solved by assumption that any arbitrary reversible cycle can be covered by a map of reversible Carnot cycles. Again, the idea of reversibility can not involve mechanical energy into consideration that along with elementary introduction into problem should share the visible part of net energy of cycle.

Work of motive forces in renewable thermodynamics

The principal difference of non-equilibrium thermodynamics of renewable cycles is to go from variational principles of reversible thermodynamics to equations of actual motion. In this case, it is not necessary to build virtual (generally, in mind) cycle of Carnot to answer question that is the share of work in any actual process of absorption of light in Eq.12. In fact, this is one equation with two unknown independent variables to describe actual motion of any system powered by light. The main in approach of applied renewable thermodynamics is to solve the ***problem of efficiency by adding the Newtonian equation*** of motion to make number of variables equals to number of equations. But, in order to keep the all equations of reversible thermodynamics valid for work substance (here, perfect gas), as an equilibrium part of a non-equilibrium renewable system we should make two new assumptions. 1) Let us add an equation for motive force F_m acting along distance dx, for example as a force to put a piston in motion 2) As far as reversible work defined only in state of equilibrium by p inside equal to pressure outside let us introduce an internal pressure equal pressure of perfect gas to keep valid standard definition of reversible work. By this assumption relation between reversible work substance and mechanical part of system can be expressed over following equation

(25)

The Eq.25 has fundamental significance to put reversible work of equilibrium perfect gas into renewable irreversible approach. The internal pressure assumed to be equal to equilibrium pressure has physical sense because the velocities of molecular motion of perfect much lager than velocity of mechanical motion of piston with mass by changing position dx. Hence, any actual motion of piston can be assumed by reversibility of work substance that put it in motion. In other words, the action of exterior for perfect gas in equilibrium is action of internal pressure on internal site of piston to satisfy definition of equilibrium work in reversible thermodynamics[8]. For perfect gas given alone in state of equilibrium, it dose not matter that process occurs outside for accepted model of internal pressure. As a result, the all results of reversible thermodynamics of perfect gas can be involved as a part into scope of applied renewable thermodynamics.

Vector of state and its motion

Basically, we should very distinguish renewable thermodynamics of physical phenomena when heat cycles are self-organized "by itself" considered earlier[1], and renewable thermodynamics of artificial heat engines as the subject of this paper. But, the common idea of both approaches is to seek maximum efficiency for two real entities: chaotic thermal and synergetic mechanical energy in actual mutual motion. Let us begin to build a master equation for a vector of state for a non-equilibrium cycliner&piston. engine

$$\mathbf{Y}(t) = \begin{matrix} x(t) \\ u(t) \\ T(t) \end{matrix} \qquad (26)$$

involving three main degree of freedom: position $x(t)$, velocity $u(t)$ of a piston and temperature $T(t)$ of a work substance to unit thermal and mechanical properties. Again, we build the model of actual motion to evaluate possibility any engineering construction to effectively generate mechanical power by solar light. In general, the master equation can be presented as equation of actual motion of state vector:

$$\overset{\langle}{\mathbf{Y}} = \begin{matrix} x'(t) \\ u'(t) \\ T'(t) \end{matrix} = \begin{matrix} t, x(t), u(t), T(t) \\ t, x(t), u(t), T(t) \\ t, x(t), u(t), T(t) \end{matrix} \quad (27)$$

that can also be written as action of a operator of motion on state vector[10]

$$\overset{\langle}{\mathbf{Y}}(t) = \mathbf{M}\mathbf{Y}(t) \quad (28)$$

Notice, the master equation (Eq.28) has the mechanical properties of motion such as inertia and second derivative of time (acceleration) added to equations of reversible thermodynamics of perfect gas to stimulate an actual path on its thermodynamic surface of two independent variable (Eq.12). While the operator of actual motion can be rather tempting for renewable thermodynamics of natural phenomena[1], we will prefer below equations written directly as matrix $\overset{\langle}{\mathbf{Y}}(t)$ for any concrete problem. By proposed approach, the problem of actual efficiency and of master equation in renewable thermodynamics becomes the problem of systems theory[11] with energy of light as ***Input*** (Sunlight energy) to find state vector of system $\mathbf{Y}(t)$ that along with actual path provide us desired ***Output***.(Mechanical energy).

OPTO-MECHANICAL ACTUATOR

This section is to find actual motion and efficiency of conversion in opto-mechanical actuator that consists of perfect gas as work substance and piston of definite mass to close gas in a transparent cylinder pumped by solar light.

Natural non-linear vibrations

Let us begin from a first possible application of master equation to solve the problem of spontaneous generation of opto-mechanical vibrations in adiabatic conditions. The equation for motion of given mass by perfect gas as work substance powered by light in system of cylinder&piston can be written as

$$\overset{\langle}{\mathbf{Y}}(t) = \begin{matrix} u(t) \\ \frac{RT(t)}{x(t)A} - p_0 \frac{A}{M} \\ (1-k)T(t)\frac{u(t)}{x(t)} \end{matrix} \quad (29)$$

Numerical solution of (29) was done by the initial non-equilibrium (excitement) of a mole of perfect gas by volume $v_0 = \frac{RT_c}{p_0}$ and pressure $p_0 = 101325Pa$, that is put into a system of cylinder of radius $r = 0.15m$ under piston of mass $M = 25kg$ and square $A = \pi r^2$ and exited up to temperature $T_h = 50($ by solar light so fast that we can use following initial state vector $Y^0_1 = \frac{v_0}{A}, Y^0_2 = 0, Y^0_3 = T_h$. Here $R = 8.31$ is gas constant $T_c = 27$ is temperature of surroundings. Solution for natural adiabatic vibrations is presented in Figure 2. From solution follows that free oscillations that can be induced by solar light in opto-mechnaical actuator has very non-linear character that is due to conversion of mechanical energy into thernal form and vice versa. It is easy to show that mechanical energy that is converted during oscillations from kinetic energy of piston to potential energy due to difference of internal end external pressure is indeed restricted by formula of actual efficiency (Eq.23) obtained for equilibrium very slow adiabatic expansion of perfect gas.

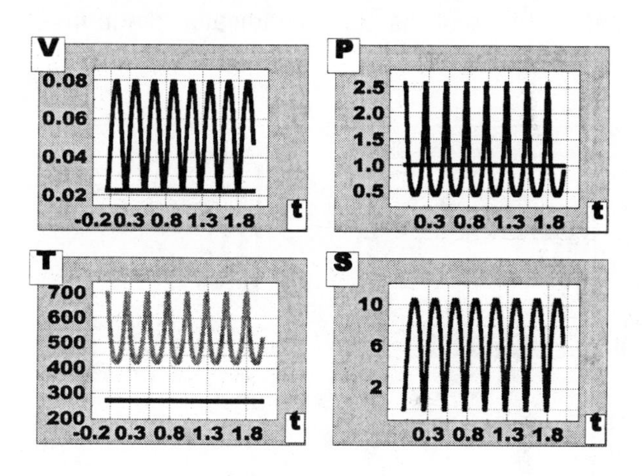

Figure 2. Non-linear free vibrations of opto-mechanical actuator

Opto-mechancial actuator in surroundings

Let us write down master equation (Eq.30) for actuator in more common form to take into account: 1) entropy production proportional Newtonian law of heat transfer 2) mechanical energy production 3) mechanical energy production along optical pumping

$$\dot{Y}(t) = \frac{-\frac{RT(t)}{x(t)A} - p_0 \frac{A}{M} - \beta u(t) \quad u(t)}{}$$

$$\frac{(k+1)\ I(t)A - RT(t)\frac{u(t)}{x(t)} - \alpha(T(t) - T_c)}{R}$$

(30)

Solutions were made for following cases.

Entropy production

The solution for natural vibrations by only non-zero thermoconductivity to surroundings; by $\alpha ? 0=3$ presented in Figure 3 . We can see that typical curve of entropy production is similar to behavior of dynamic mechanical systems with dissipation. As far as path on TS diagram is driven by mechanical motion, the thermodynamic path is actual indicator diagram of considered actuator.

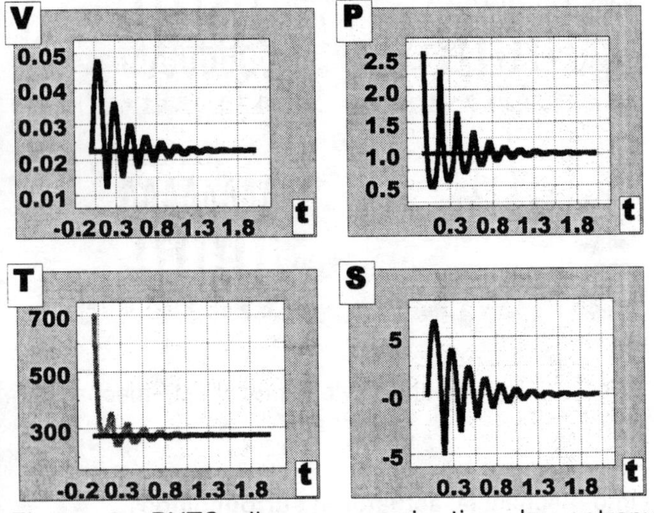

Figure 3. PVTS diagram versis time by entropy production.

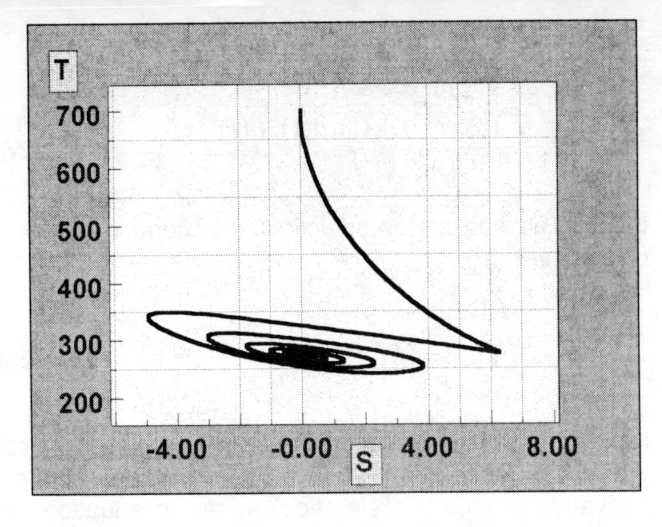

Figure 4. T-S diagram for entropy production

Mechanical energy production

In this case we can damp opto-mechanical oscillation by interaction of actuator with surroundings over mechanical degrees of freedom (constant responsible for conversion into electrical power). Figure 5 is to present results of time behavior. Basically, the solution has two entities: entropy of Clausius for perfect gas and non-equilibrium of Planck that is produced. Figure 4 is to show actual path of Clausius entropy within standard T-S diagram according Eq.14. But, It is interesting to find principal difference thermal system for previous entropy production (Fig.4) and current mechanical energy production (Fig.6). In spite of quite similar pictures of oscillation, the entropy diagrams are very different. In case of thermal energy decreased by creation of kinetic energy of piston, TS-diagram is degenerated into a curve on thermodynamic surface with all periodical oscillation closed by motion along the same path (Fig. 6).

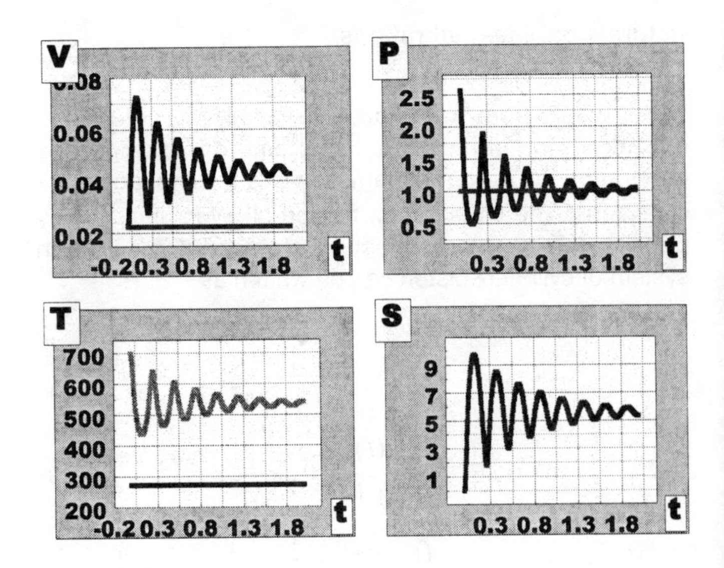

Figure 5. PVTS diagram versis time by work production.

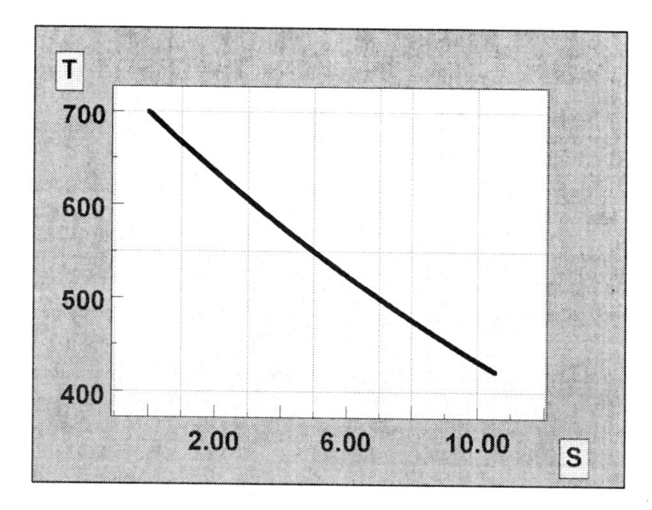

Figure 6. T-S diagram for work production

Optical pumping

The renewable thermodynamics is to convert light into electrical power by solar excitement of opto-mechanical actuators. Let us calculate behavior of all thermodynamics quantities of perfect gas the same system but when light with power 4.5KW of solar light. Advantage of master equation to consider any actual renewable cycles including any variants of optical pumping. Figure 7,8 present the particular results of calculation for direct conversion of light to accelerate a mass of a magnet in an induction coil. We take the same mechanical parameters but now, light is open during 1 sec, and after all thermal energy that is kept is involved into oscillation to covert into mechanical power that is damped by conversion into electrical current. .But only a definite part from all energy can be converted into electrical energy. We can change parameter that responsible for conversion of thermal energy find a better path. Again, from Figure 8, we can see the principal difference between initial active zone of optical pumping and a final path, when system periodically keeps the same path.

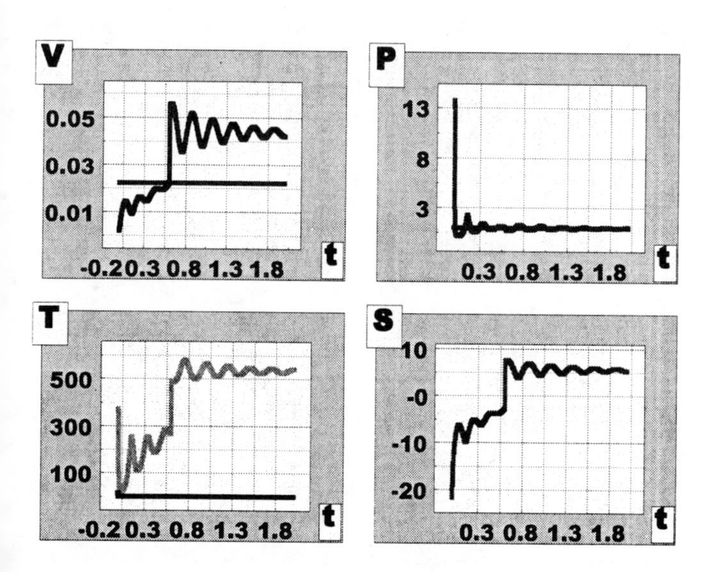

Figure 7. PVTS diagram versis time by work produced along optical pumping.

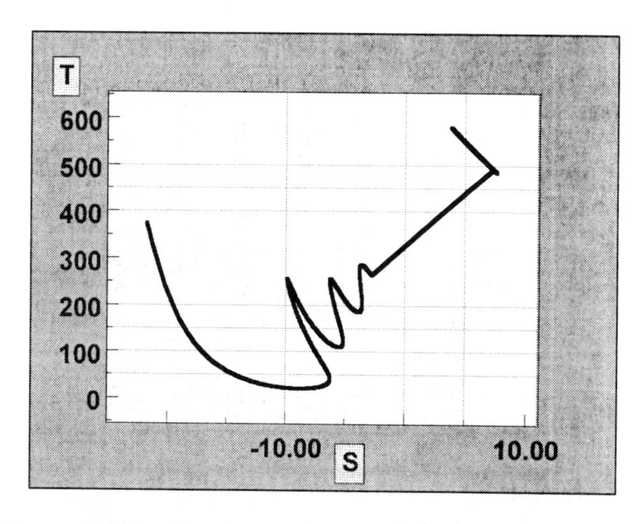

Figure.8. TS-diagram for work production during optical pumping

Let us make a summary for application of mater equation for consideration different models and direct pumping of opto-mechanical actuators.

1. Efficiency of any from calculated actual processes can not be more than is given by formula (23) of actual efficiency of direct conversion that is in accordance to initial postulate.

2. All PVTS diagram calculated for equilibrium thermodynamics quantities of perfect gas allow making conclusion that for any actual process cycles on TS (PV) diagrams never are closed

3. The problem of efficiency in applied renewable thermodynamics becomes the problem to get as much as possible direct conversion up to point of mechanical non-equilibrium. Thermal non-equilibrium as a result of any stroke can be renewed by entropy production without additional sources of energy.

RENEWABLE CARNOT CYCLE

The actual renewable cycle for perfect gas with turn points coincident with turn points Carnot cycle form introduction is considered.

Let us apply master equation for renewable Carnot cycle from introduction into problem of actual efficiency. Suppose we have a reservoir with perfect gas by normal condition $T_0 = 27$; $p_0 = 101325 Pa$ with a cylinder of volume $v_0 = 0.022 m^3 mole^{-1}$ fixed by a pass less diaphragm to insulate this volume from outside. The main figure in equilibrium Carnot cycle is temperature of source that is a constant. Let us take it, say $T_h = T_0 + \Delta T$ by $\Delta T = 10$ and try to build cycle that pass over all four turn equilibrium points of classical Carnot cycle but under solar pumping that calculated in introduction.

Adiabatic compression

Let us suppose that first stroke is conduct by a magnet mass that has a initial velocity

$$\sqrt{\quad} \tag{31}$$

as a necessary "reservoir" of mechanical energy to prepare direct conversion stroke into point 1 should be now equal to

(32)

From (Eq.31,32) follows that only by initial velocity of magnet $u_0 = 4.063\, m/sec$ we can prepare direct stroke of conversion to keep turn point of Carnot cycle (3) near given temperature difference

Optical pumping

Solar light can not have isothermal power supply, but calculation shows that in case of quite definite relation between kinetic and potential energy of mechanical system in interaction with external coil to convert thermal energy into electrical current and assuming that all energy of light is pumped during 0.02 sec we can get next turn point (2) of Carnot cycle by velocity equal zero. By numerical modeling we found there is several specific parameters in master equation to get the next point before adiabatic expansion in state of equilibrium. Again, the all energy of Carnot cycle is put during this direct conversion stroke by optical pumping because the next stroke of adiabatic expansion is only to give back kinetic energy of magnet.

Adiabatic expansion

No new energy is created from light by this stoke. Again, perfect gas gives back the kinetic energy of piston (magnet). Again, we can pick up such parameters of interaction between thermal and mechanical and electrical parts of system to have velocity of piston equal zero at the start point (2) of this stroke.

Renewable stroke

In isothermal point (3) pressure inside is less than external pressure. To renew in this case means to put system given in point (3) that has no mechanical equilibrium into contact with surrounding for entropy production. This stoke was simulated by a small mass diaphragm $m_d = 0.5 kg$ that is still to insulate work substance by entropy production with $\alpha = 10^{6}$ that is involved into dissipative vibrations to return system into initial point of Carnot cycle.

Restricted volume of this paper can not allow presenting obtained PVTS diagrams. But, as a main result, the basic most thermally effective Carnot cycle for reversible thermodynamics appeared to be not so useful because of following reasons:.

1. The classical Carnot cycle in renewable variant appears to very difficult for realization because of necessity to have the high level of initial mechanical non-equilibrium to actually work within maximum of thermal efficiency as it provides theorem of Carnot.

2. It is easy to see that our desire to close cycle for perfect gas makes disturb one of the main principles of renewable thermodynamics derived earlier form spontaneous phenomena ***to renew process in point of mechanical equilibrium***.

3. Evidently, that as soon as internal pressure during adiabatic expansion becomes equal to external pressure, it is not necessary to do something else. Actual renewable Carnot cycle should be open in this point for entropy production.

ON SECOND LAW IN RENEWABLE THERMODYNAMICS

The second Law for renewable thermodynamics of physical phenomena was formulated as the theorem of Carnot to restrict efficiency of any renewable cycle accompanied by entropy production equal zero that provide us earlier with formula[1]:

$$\eta_c = 1 - \frac{T_0}{\Delta T} \ln \frac{T_0 + \Delta T}{T_0} \tag{33}$$

Such a form of Second Law is for the restriction of maximum light available to be converted by optical pumping a work substance up to ΔT with regards to surrounding with temperature T_0 In general, formula (33) was found for self -organization of heat cycles by mass-transfer, when any element of thermal energy produce work within thermal efficiency restriction applicable for scope of this paper.

Let us note, that as far as entropy production . (Eq.1) in renewable thermodynamics involved to treat Carnot principle right from origin, there is no problem with a part of Second Law in part to direction of any spontaneous motion. Let us now point out the second part of Second Law as for the efficiency of conversion. Firstly notice again, that equilibrium entropy in each figures is equilibrium entropy evaluated as any form right members of equation (14). This was made to be more specific as for behavior of the standard reversible perfect gas in scope of renewable thermodynamics to make a "bridge" between reversible and proposed approach. Non-equilibrium entropy of Planck (that used to derive formula (33)) is applicable for any temperature difference between source and sink, therefore it can be useful for consideration of systems very far from equilibrium.

While the problem of maximum ***heat*** available for conversion in reversible thermodynamics has been solved by using the idea of equilibrium entropy of Clausius[2] and virtual Carnot cycle having maximum of ***thermal efficiency*** , the problem of maximum ***available light*** in applied renewable thermodynamics is proposed to solve by calculation of actual cycles within constraints imposed by Newtonian equations of motion. The new terms: ***motive forces*** and ***internal pressure*** as necessary elements to join thermal and mechanic features of any systems in renewable thermodynamics are proposed to join reversible and renewable concepts on a common methodological level.

But, there is a principal point in this context, if actual efficiency for motive power of solar light within renewable thermodynamics can be more than maximum of thermal efficiency of Carnot cycle ? Evidently that formula (33) and all cycles for perfect gas can not provide us the actual efficiency more than thermal of Carnot cycle (24).

At the same time, the all context of previous experimental research[4,5] allowed to state the problem of the efficiency in renewable thermodynamics by following way. The basic cycle of renewable thermodynamics is adiabatic conversion and next renewal by entropy production in state of mechanical equilibrium. To make Carnot cycle be in actual motion, the very large reservoir of mechanical energy is necessary, which, sometimes, is larger that net energy of cycle.

The renewable thermodynamics moves the problem of available efficiency for conversion of light from pure thermal processes into scope of actual processes with mechanical energy within consideration. This mechanical energy can involve not only energy of macroscopic motion of machine elements. This energy can be due to huge energy intermolecular interaction or chemical energy (or even biological energy) that can not be measured by simple temperature difference. For example, self-organization of bio-systems in Nature very far form equilibrium provides with many examples of conversion into mechanical power with actual efficiency much bigger than Carnot efficiency derived for thermal equilibrium.

The problem, to have the actual conversion of solar light with efficiency more than the thermal efficiency of Carnot cycle is still in question. But, the fact that the renewable thermodynamics units both the description of physical phenomena occur by itself and the design of heat engines allow expecting to have a form of Second Law within concept of renewable thermodynamics to describe both phenomena and engines beyond thermal efficiency of Carnot cycle.

CONCLUSION

As a result, the application of renewable thermodynamics created within the new physical phenomena to describe prospective solar engines can be summarized by following way.

1. The master equation to consider opto-mechanical systems by optical pumping is proposed to calculate motion for a state vector on basis on new concept of both motive forces and internal pressure.

2. Energy of solar light available for renewable conversion is assumed to find within the actual efficiency followed from the solution of real thermodynamic path of work substance driven by mechanical properties of system.

3. The non-linear natural vibrations of opto-mechanical actuators are found for conversion into electrical power. It is shown that maximum of actual efficiency of conversion can not be more than by direct adiabatic conversion.

4. The main difference of renewable and reversible approach was demonstrated on idea of thermal and actual efficiency of direct conversion of solar light applicable within any prospective constructions.

5. By solving master equation for actual renewable Carnot cycle, the useless of its practical realization is shown that is due to a high level of initial mechanical energy to realize actual efficiency in spite of high thermal one.

6. The aim of this paper was not to find a new speculative way to directly "utilize" solar energy. Several such attempts to grasp solar energy by adiabatic system like piston&cyliner can be found if to be more familiar with a critic review in book on advanced thermodynamics by Bejan[12].

7. The goal was to share the several practical solutions for prospective problems to calculate indicator diagrams for electrical power production. The main problem was to find a specific mathematical tool to apply for different new solar engines such as for example NESIS [13]

REFERENCES

1. Sukhodolsky A.T.*An introduction to thermodynamics of renewable cycles for direct solar energy conversion* IECEC-98-318 33rd Intersociety Engineering Conference on Energy Conversion Colorado Springs, CO, August 2-6, 1998
2. Clausius R, *On the motive power of heat, and on the laws which can be deducated from it for the theory of heat* LXXIX pp. 500-513, 1850
3. Max Plank *Treatise on thermodynamics* Dover publication 1780 Broad Way, New York 1945
4. Gibbs J.W. The scientific papers, vol 1 Thermodynamics, Longmans, Green and Co. New York, 1906
5. Sukhodolsky A.T. *Thermo-hydraulic actuator as a new way for conversion of solar energy in space* IECEC-98-390 33rd Intersociety Engineering Conference on Energy Conversion Colorado Springs, CO, August 2-6, 1998
6. Sukhodolsky A.T. *The self organization of heat cycles by capillary convection for extraction of water from a binary solution* IECEC-98-391 33rd Intersociety Engineering Conference on Energy Conversion Colorado Springs, CO, August 2-6, 1998
7. Prigogine I., *Introduction to thermodynamics of irreversible processes*, Interscience Publishers}, York, London, 1961
8. Kubo Ryogo *Thermodynamics, An advanced course with problems and solutions*, North-Hollnd Publishing Company, 1968.
9. Stive William and Raymond W.Harrigan, *Solar Energy Fundamentals and Design,* John Wiley&Sons, 1985
10. C.Truesdell, *Six Lextures on Modern Natural Philosophy* Springer-Verlag Berlin 1966
11. Stephen W.Director, Ronald A.Rohrer Introduction to systems theory, McGraw-Hill Book company, 1972
12. Adrian Bejan *'Advanced Engineering Thermodynamics'*, John Wiley and Sons, 1988, New York, Chapter on solar power, pages 466-527. (author has a pleasant duty to thank Professor Bejan for kind faxing several pages from the book)
13. A.T.Sukhodolsky, VibroSolar business description http://www.come.to/vibrosolar

AIAA-2000-2863

STUDY OF PHOTOCATALYTIC 2-PROPANOL DEHYDROGENATION FOR SOLAR THERMAL CELL

Meng Ning, Yuji Ando, Tadayoshi Tanaka and Takumi Takashima

Electrotechnical Laboratory
Tsukuba, Ibaraki JAPAN

ABSTRACT

Solar thermal cell converts low temperature solar thermal energy into electric power. It consists of chemical reactions of 2-propanol dehydrogenation and acetone hydrogenation, and a fuel cell. In this study, we studied photocatalytic 2-propanol dehydrogenation in order to apply directly sunlight to solar thermal cell. We report the results of 2-propanol dehydrogenation when noble metal (Pt, Ru, Rh and Pd) supported on TiO_2 was used as a catalyst (designated hereafter as Metal/TiO_2). A metal loading ratio was 5 wt%. Three kinds of TiO_2 (Ishihara sangyo kaisha. LTD: ST-01, ST-21 and ST-31) was used in this study. The particle size of ST-01, ST-21 and ST-31 was 7, 20 and 7 micrometer, respectively. The specific surface area of ST-01, ST-21 and ST-31 was 321, 60.9 and 275 m^2/g, respectively. We examined the activity of various combinations of noble metal and TiO_2 for the reaction and it is revealed that Pt/ST-01 has the highest activity. Compared with 2-propanol dehydrogenation with noble metal supported on activated carbon, which is not photoreaction, the characteristics of photocatalytic 2-propanol dehydrogenation with Metal/TiO_2 are summarized in small acetone retardation, large reaction rate under non-boiling condition and large reaction rate under high acetone concentration.

INTRODUCTION

Solar energy is clean and inexhaustible energy source and a lot of studies have been conducted. In Japan, it is not effective to operate high temperature solar thermal system such as the SEGS plant because the direct insolation is extremely poor. However, it is not difficult to get low temperature solar thermal energy about 100 °C.

From above point of view, we have proposed "solar thermal cell", which converts low temperature solar thermal energy about 100 °C into electric power and we have conducted fundamental studies about it.[1] It consists of chemical reactions of 2-propanol dehydrogenation (Eq. (1)) and acetone hydrogenation (Eq. (3)) and a fuel cell.

$$(CH_3)_2CHOH\,[l] \rightarrow (CH_3)_2CO\,[l] + H_2\,[g] \qquad (1)$$

2-propanol   Acetone   Hydrogen

On the negative electrode;

$$H_2\,[g] \rightarrow 2H^+ + 2e^- \qquad (2)$$

On the positive electrode;

$$(CH_3)_2CO\,[l] + 2H^+ + 2e^- \rightarrow (CH_3)_2CHOH\,[l] \qquad (3)$$

We have the below two concepts for the configuration.

Liquid-phase solar thermal cell

Figure 1 shows the concept of liquid-phase solar thermal cell. The solar thermal energy heats the negative electrode of the cell about 100 °C and the positive electrode is cooled by water. 2-propanol is decomposed into acetone and hydrogen on the negative electrode with an appropriate catalyst (Eq. (1)) and the hydrogen is separated into proton and electron there (Eq. (2)). The proton moves to the positive electrode through the electrolyte and the electron moves to the positive electrode through the outside electric circuit. The acetone is desorbed from the negative electrode and supplied to the positive electrode. The acetone, proton and electron regenerate 2-propanol on the positive electrode with an appropriate catalyst (Eq. (3)) and 2-propanol is supplied to the negative electrode again. "Liquid-phase" is named by the reason that liquid-phase 2-propanol is supplied to the negative electrode.

Gas-phase solar thermal cell

Figure 2 shows the concept of gas-phase solar thermal cell. It is composed of fuel cell and 2-propanol dehydrogenation reactor. 2-propanol dehydrogenation reactor is heated by the sun. 2-propanol dehydrogenation

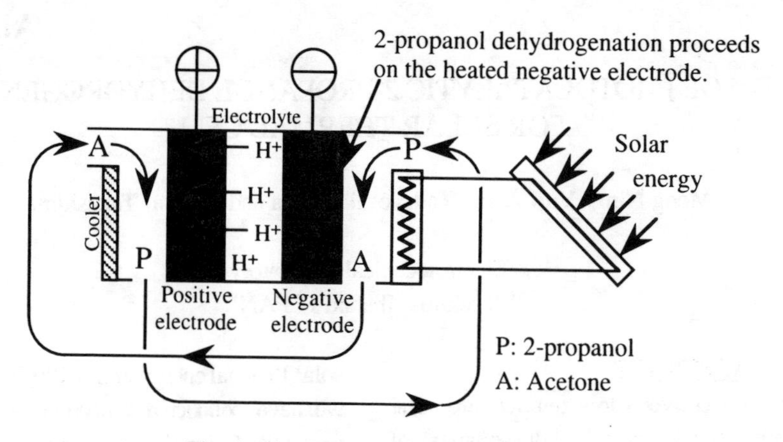

Fig. 1 Liquid-phase Solar Thermal Cell

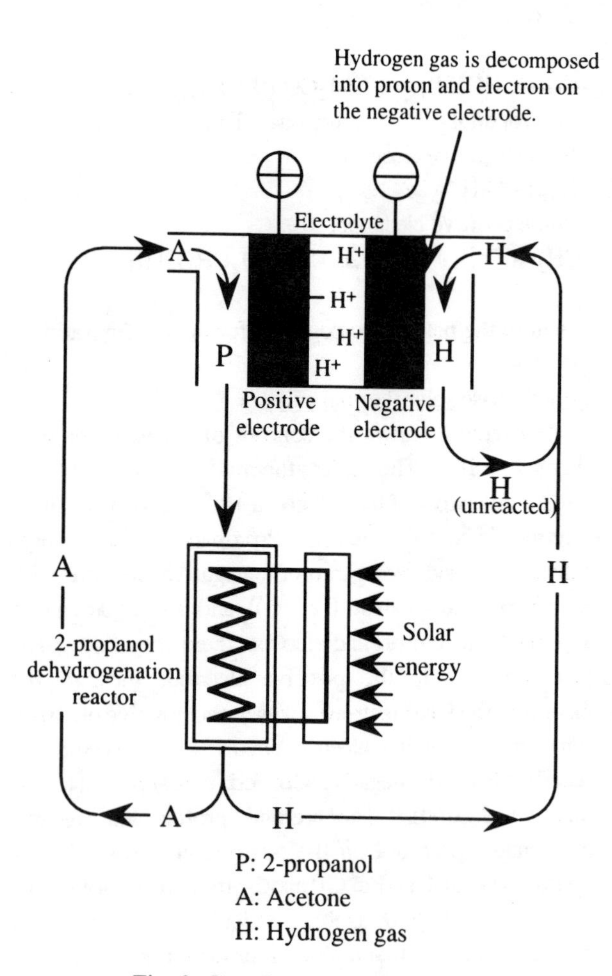

Fig. 2 Gas-phase Solar Thermal Cell

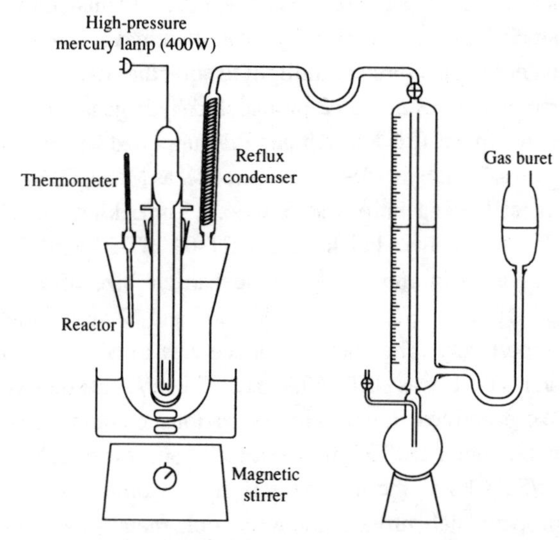

Fig. 3 Reaction apparatus for photocatalytic 2-propanol dehydrogenation

shown by Eq. (1) proceeds in the reactor and acetone and hydrogen gas are produced. Hydrogen gas is supplied to the negative electrode of the fuel cell. On the other hand, produced acetone is supplied to the positive electrode of the fuel cell. Through the process of acetone hydrogenation shown by Eq. (3), electric power is produced at the fuel cell. 2-propanol, which is reaction product, is supplied to 2-propanol dehydrogenation reactor again. "Gas-phase" is named by the reason that hydrogen gas is supplied to the negative electrode.

In this study, photocatalytic 2-propanol dehydrogenation proceeds by directly sunlight. Therefore, the gas-phase solar thermal cell is suitable in the case of using photocatalyst.

It is reported that noble metal supported on activated carbon (designated hereafter as Metal/carbon) have large reaction rate for 2-propanol dehydrogenation under boiling and refluxing condition.[2] We studied photocatalytic 2-propanol dehydrogenation with noble metal supported on TiO_2 (designated hereafter as Metal/TiO_2). The band gap of TiO_2 is 3.0 eV and it is not suitable for solar energy utilization. Therefore, we selected TiO_2 as a photocatalyst in order to study fundamental characteristics of photocatalytic 2-propanol dehydrogenation.

EXPERIMENTAL

Figure 3 shows the reaction apparatus for photocatalytic 2-propanol dehydrogenation. A prescribed amount (1000 mg) of catalyst and 300 ml of reactant (2-propanol or 2-propanol and acetone mixed solution) were placed in an inner irradiation quartz reactor equipped with 400 W high pressure mercury lamp. In order to remove oxygen in the reactor, nitrogen was passed through the reactor. The reactor was heated with an oil bath and the reaction temperature was measured by a thermometer. When the reaction temperature was lower than 2-propanol boiling point (82.4°C), 2-propanol was stirred by a magnetic stirrer in order to suspend the catalyst in the reactant. When the reaction temperature was 2-propanol boiling point, the magnetic stirrer was not used because the catalyst was sufficiently suspended by boiling. Irradiation time was 20 hour. The product hydrogen was collected in a gas burette (1000 ml) through a refluxing condenser cooled at 20 °C and the product acetone was cooled and condensed by the refluxing condenser and was returned to the reactor. Both gas and liquid-phase products were analyzed by gas chromatography (Shimadu GC-8AIF, PEG 1000 column).

As the photocatalyst, three kinds of TiO_2 (Ishihara sangyo kaisha. LTD: ST-01, ST-21 and ST-31) were used in this study. They were anatase-type TiO_2. The particle size and specific surface area of ST-01, ST-21 and ST-31 were summarized in Table 1. The surface of ST-31 is treated with an acid. Four kinds of noble metal (Pt, Ru, Rh and Pd) were supported on TiO_2 by a photochemical deposition and then they were used for photocatalytic 2-propanol dehydrogenation. A metal loading ratio was 5 wt%. The catalyst preparation method was shown in Scheme 1.

Table 1 Characteristics of TiO 2

	Particle diameter μm	Specific surface area $m^2 \cdot g^{-1}$
ST-01	7	321
ST-21	20	60.9
ST-31	7	275

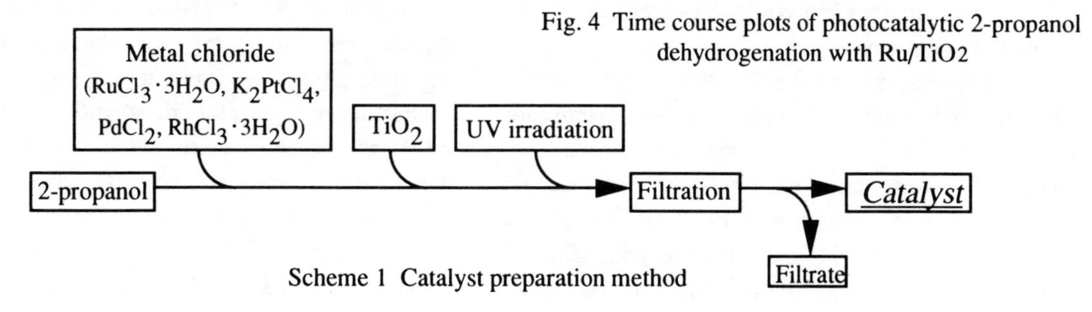

Scheme 1 Catalyst preparation method

RESULTS AND DISCUSSION

It is reported that Ru/carbon or Pt/carbon has large reaction rate for 2-propanol dehydrogenation under boiling and refluxing condition.[2] So, we studied photocatalytic 2-propanol dehydrogenation with Pt/TiO_2 or Ru/TiO_2 at first.

Figure 4 shows the results of photocatalytic 2-propanol dehydrogenation with Ru/TiO_2 under boiling and refluxing condition. The order of the activity for the

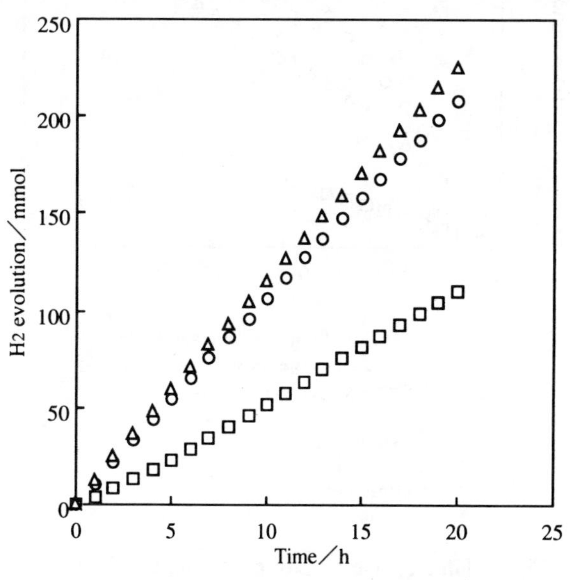

○ : TiO2 ST-01 (specific surface area 321m ^{2}/g)
△ : TiO2 ST-21 (specific surface area 60.9m ^{2}/g)
□ : TiO2 ST-31 (specific surface area 275m ^{2}/g)

Catalyst : 5 wt%-Ru/TiO 2, 1 g
Reactant : 300 ml of 2-propanol
Heating temperature : 200℃
Boiling and refluxing condition, non-stirring

Fig. 4 Time course plots of photocatalytic 2-propanol dehydrogenation with Ru/TiO2

reaction with Ru/TiO_2 was Ru/ST-21 >= Ru/ST-01 > Ru/ST-31. More than 200 mmol of hydrogen gas was produced in 20 hour with Ru/ST-21 and Ru/ST-01.

Figure 5 shows the results of photocatalytic 2-propanol dehydrogenation with Pt/TiO_2 under boiling and refluxing condition. The order of the activity for the reaction with Pt/TiO_2 was Pt/ST-01 > Pt/ST-21 > Pt/ST-31. More than 350 mmol of hydrogen gas was produced in 20 hour with Pt/ST-01.

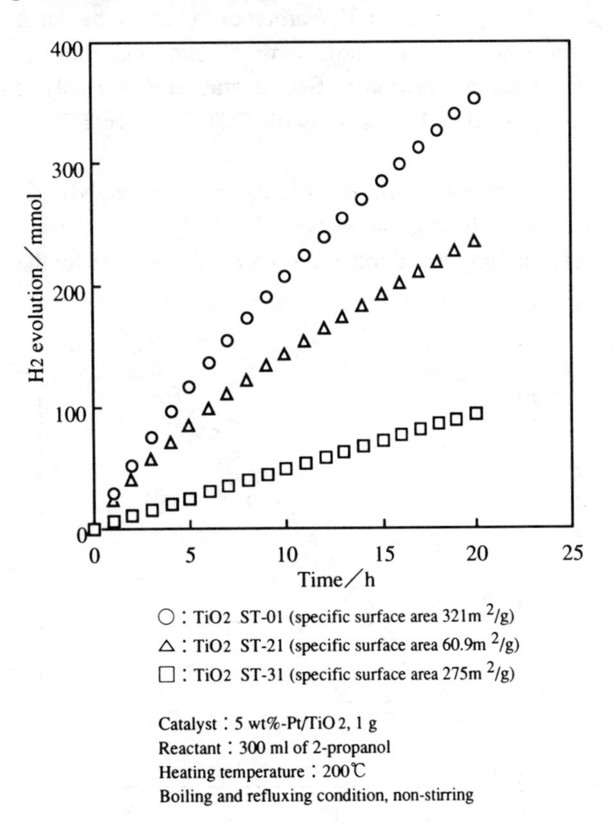

Fig. 5 Time course plots of photocatalytic 2-propanol dehydrogenation with Pt/TiO_2

In both Ru and Pt, the activity is the lowest when they were supported on ST-31. The reason is supposed that the surface of ST-31 is treated with an acid. The order of activity between Ru/ST-01 and Ru/ST-21 was different from the order between Pt/ST-01 and Pt/ST-21. The reason is not clear.

In any case, the activity of Ru/ST-01 and Pt/ST-01 was considerably high. So, photocatalytic 2-propanol dehydrogenation with Rh/ST-01 and Pd/ST-01 was also studied. Figure 6 shows the results of photocatalytic 2-propanol dehydrogenation with Pt/ST-01, Ru/ST-01, Pd/ST-01 and Rh/ST-01 under boiling and refluxing condition. In both Rh/ST-01 and Pd/ST-01, about 100 mmol of hydrogen gas was produced in 20 hour and their activity was smaller than that of Pt/ST-01 and Ru/ST-01.

It is reported that when Pd/carbon was used to 2-propanol dehydrogenation under boiling and refluxing condition, the hydrogen evolution rate was nearly zero and a large quantity of acetone was formed in the liquid-phase.[3] The reason is supposed to be spillover phenomenon of hydrogen. In contrast to the thermochemical reaction with Pd/carbon, a certain amount of hydrogen gas was evolved in this case.

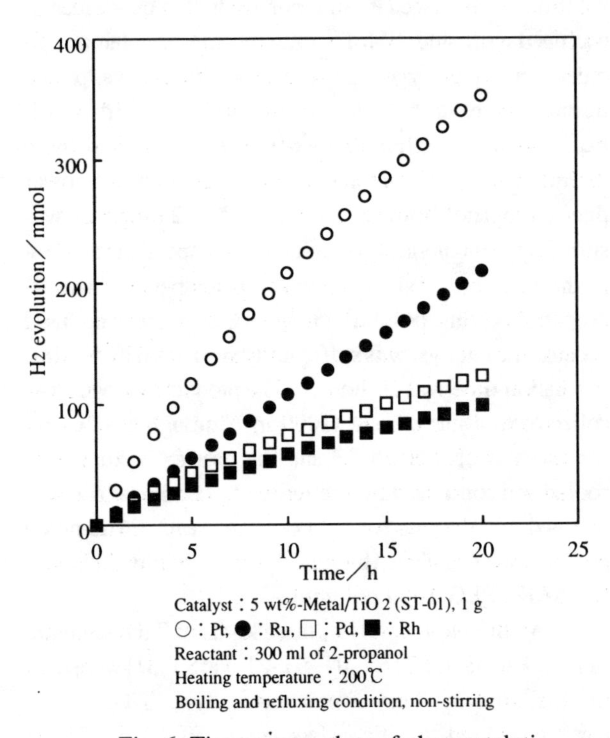

Fig. 6 Time course plots of photocatalytic 2-propanol dehydrogenation with Metal/ST-01

We compared the reaction rate of the photocatalytic 2-propanol dehydrogenation with Pt/ST-01, Pt/ST-21 and Pt/ST-31 under boiling condition or under non-boiling condition. Figure 7 shows the effect of boiling and refluxing condition on photocatalytic 2-propanol dehydrogenation with Pt/TiO_2. It is pointed out that the reaction rate of 2-propanol dehydrogenation with suspended catalyst under non-boiling condition was much smaller than that under boiling condition.[4] In contrast to the thermochemical reaction, the reaction rate of the photocatalytic 2-propanol dehydrogenation with Pt/TiO_2 under non-boiling condition was nearly equal to that under boiling condition. So, it is supposed that boiling condition is not necessarily needed for the photocatalytic

2-propanol dehydrogenation.

In above results, we don't discuss the effect of acetone, which is produced in the process of 2-propanol

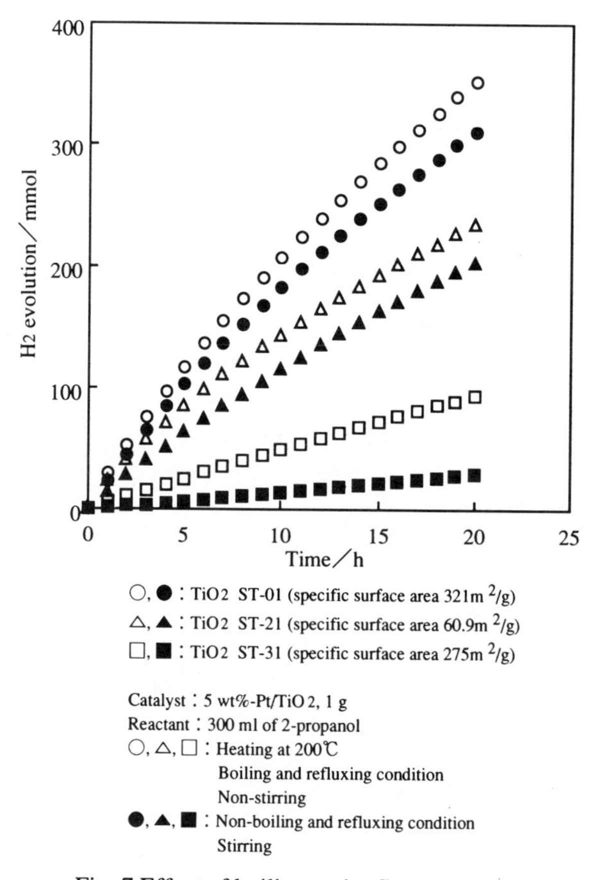

Fig. 7 Effect of boiling and refluxing condition on photocatalytic 2-propanol dehydrogenation with Pt/TiO2

dehydrogenation, on the reaction rate. When acetone is existing in 2-propanol solution, it is reported that reaction rates of 2-propanol dehydrogenation with suspended catalysts have been well described with the following rate equation.[5]

$$v = k / (1 + K [acetone]) \quad (4)$$

where v is the reaction rate, k the rate constant and K the equilibrium constant of acetone adsorption, respectively.

It is also reported that the reaction rate with Ru/carbon or Pt/carbon becomes almost zero when acetone concentration is 10 %.[6] So, we examined the effect of acetone concentration in the reactant on the 2-propanol dehydrogenation. Figure 8 shows the effect of 2-propanol concentration on photocatalytic 2-propanol dehydrogenation with Ru/ST-01. Even if acetone concentration is 50 %, the reaction rate of the photocatalytic 2-propanol dehydrogenation with Ru/ST-01 is sufficiently large and about 60 mmol of hydrogen gas was produced in 20 hour. This result is a contrast to the thermochemical reaction with Ru/carbon or Pt/carbon. It means that the equilibrium constant of acetone adsorption in the photocatalytic 2-propanol dehydrogenation with Ru/ST-01 is much smaller than that in the 2-propanol dehydrogenation with Ru/carbon or Pt/carbon. So, it is supposed that the photocatalytic 2-propanol dehydrogenation is effective especially in high acetone concentration.

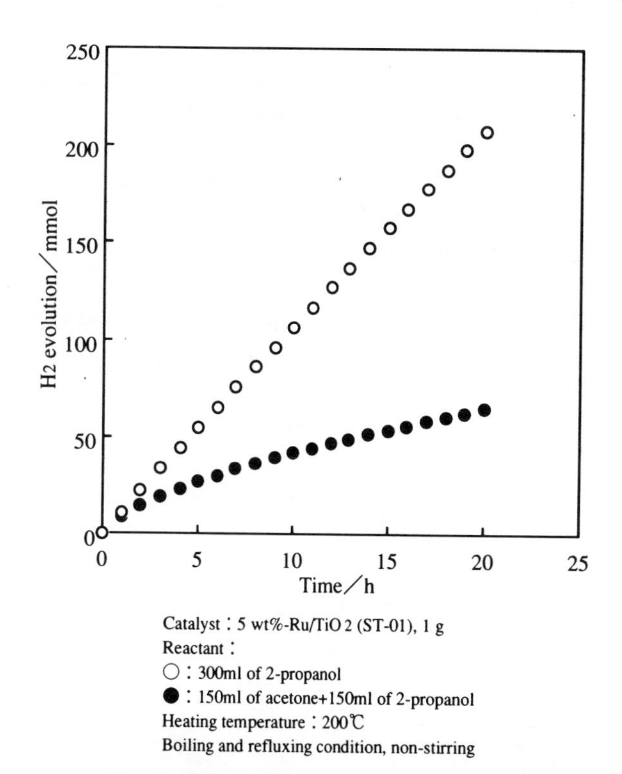

Fig. 8 Effect of 2-propanol concentration on photocatalytic 2-propanol dehydrogenation with Ru/ST-01

CONCLUSIONS

In order to apply directly sunlight to solar thermal cell, we studied photocatalytic 2-propanol dehydrogenation with Metal/TiO_2. The following results were obtained from the experiments.

(1) The activity of Ru/ST-01 and Pt/ST-01 was considerably high. More than 350 mmol of hydrogen gas was produced in 20 hour with Pt/ST-01.

(2) In contrast to 2-propanol dehydrogenation with Pd/carbon, a certain amount of hydrogen gas was evolved

in photocatalytic 2-propanol dehydrogenation with Pd/ST-01.

(3) In contrast to Ru/carbon or Pt/carbon, it is supposed that boiling condition is not necessarily needed for the photocatalytic 2-propanol dehydrogenation.

(4) In contrast to 2-propanol dehydrogenation with Metal/carbon, the photocatalytic 2-propanol dehydrogenation is characterized by small equilibrium constant of acetone adsorption.

REFERENCES

1) Y. Ando, T. Doi, T. Takashima and T. Tanaka, *32nd IECEC*, Honolulu, July, 1860 (1997).
2) Y. Saito, M. Yamashita and E. Ito, *8th World Hydrogen Energy Conference*, Honolulu, July, B1-6 (1990).
3) Y. Ando, X. Li, E. Ito, M. Yamashita and Y. Saito, *Studies in Surface Science and Catalysis*, Vol. 77, 313 (1993).
4) Y. Saito, M. Yamashita, E. Ito and N. Meng, *Int. J. Hydrogen Energy*, Vol. 19, No. 3, 223 (1994).
5) D. E. Mears and M. Boudart, *AIChE J.*, **12**, 313 (1966).
6) N. Meng, Y. Ando, M. Yamashita and Y. Saito, *J. Hydrogen Energy Systems Society of Japan*, Vol. 18, No. 2, 36 (1993).

CHF FOR UNIFORMLY HEATED VERTICAL TUBE UNDER LOW PRESSURE AND LOW FLOW CONDITIONS

W. Jaewoo Shim*
Department of Chemical Engineering,
Dankook University,
Seoul, 140-714, Korea

Jae Hyok Lim† and Gyoo Dong Jeun‡
Department of Nuclear Engineering,
Hanyang University
Seoul, 133-791, Korea

ABSTRACT

The description of critical heat flux (CHF) phenomena under low pressure ($P \leq 10$ bar) and low flow ($G \leq 300$ kg/m^{2}s) condition is further complicated by the large specific volume of vapor and the effect of buoyancy that are inherent in the condition. In this study, a total of 834 data points of CHF in uniformly heated round vertical tube for water were collected from 5 different published sources. A comparative analysis is made on available correlations and a new correlation is presented. The new CHF correlation comprise of local variables, namely, "true" mass quality, mass flux, tube diameter, and two parameters as a function of pressure only. This study reveals that by incorporating "true" mass quality in the local condition hypothesis, the prediction of CHF under these conditions can be obtained accurately, overcoming the difficulties of flow instability and buoyancy effects. The new correlation predicts the CHF data significantly better than those currently available correlations, within the root mean square errors of 7.16, by the heat balance method.

INTRODUCTION

Accurate prediction of critical heat flux (CHF) under the low pressure and flow conditions has been recognized to be particularly important in design and safe operation of nuclear power plants and heat exchanger equipment. Loss of coolant accidents (LOCA) of light water reactor and thermal safety performance analyses of power reactors or research reactors are concerns of many researchers in recent years. The prediction of CHF under these conditions is complicated by the effect of buoyancy and flow instability; the flow becomes less stable due to the large specific volume of vapor at low pressure and the effect of buoyancy becomes significant at the low flow conditions (Mishima and Nishihara[1]). In the past several empirical correlations have been investigated using local condition hypothesis, non-local condition hypothesis, or mixed local and non-local hypothesis with limited success. In recent years many investigators (Mishima and Nishihara, Weber and Johannsen[2], Chang et al.[3], Baek et al.[4], Lee et al.[5], Kim et al.[6]) have studied CHF phenomenon under these conditions, however, the basic mechanism of CHF is not yet fully understood. A survey of literature revealed that the experimental data points of interest are much scarcer than at conditions of high pressure and high flow.

The aim of the present paper is to provide a comparative analysis of currently available correlations and propose a new local correlation for up-flow of water through a uniformly heated vertical tube under the low pressure and flow conditions. For the correlation development, this study has emphasized on modifying Deng generalized local correlation[8] that predicts CHF in both subcooled boiling and saturated boiling regions. The Deng correlation relates CHF only by the local

* Associate Professor (To whom correspondences should be made: email: wjshim@dankook.ac.kr)
† Graduate Student, ‡ Professor

variables and it was developed and tested for prediction of CHF of water in uniformly heated round tube using over 10000 CHF data from many different sources. The model prediction had a standard deviation of error less than 10 % and an average error less than 2 %. The range of their data has pressures of 2 bar to 200 bar, tube diameters of 3 mm to 37 mm, Mass velocity varied from 100 kg/m^2s to 18000 kg/m^2s, and inlet qualities from –2.5 to 0.7, respectively. The importance of this correlation is that it predicts CHF in all flow patterns and boiling regimes by two parameters as a function of pressure, only. In fact, an accurate description of local conditions at the CHF seems to be very effective in simplifying the treatment of complexity of flow. However, in the range of low pressure and low flow conditions Deng's correlation over predicts CHF, for one of the sources, for example, as much as rms of 48.99 % . Thus, it seems to warrant a study to modify the Deng's model and extend the applicable range low pressure and flow conditions.

EXPERIMENTAL CHF DATA

For this study, a total of 834 data points of CHF in uniformly heated round vertical tube for water are collected from 5 different published sources. Experimental data used here are those presented by Kim et al., Thompson and Macbeth[9], Lowdermilk et al.[10], Becker et al.[11] and Mishima[12]. The data used in the analysis are pre-screen data via heat balance methods, within 7% error, and they comprise of 513 data from Kim et al., 57 data from Thompson & Macbeth, 101 data from Lowdermilk et al., 110 data from Becker et al., and 53 data from Mishima. The ranges of the collected experimental data are $1.01 \leq$ P(pressure) ≤ 10 bar, $1.3 \leq$ D(diameter) ≤ 24 mm, $65 \leq$ L(length) ≤ 3120 mm, $13 \leq$ G(mass flux) $\leq$ 300 kg/m^2s, $0 \leq \Delta H_i$ (enthalpy)≤ 653.6 kJ/kg, and $0.118 \leq q_c$ (CHF)≤ 4.006 MW/m^2, respectively. The ranges of operating conditions for the data are summarized in Table 1.

AVAILABLE CORRELATIONS

Among the correlations available, results are reported only for those providing a consistent prediction, and these are Weber and Johannsen, Chang et al., Baek et al., Lee et al., and Shah[7]. These correlations are essentially inlet condition correlations. With exception of the Shah model, predictions by the correlations are extended, using the experimental data, beyond the recommended operating ranges. The purpose of this analysis is to check the feasibility of extending the validity range of existing correlations. The correlations considered in the present analysis are Weber and Johannsen, Chang et al., Baek et al., Lee et al., Shah, and Deng correlations.

Weber and Johannsen derived their correlation using 249 experimental data. The data used for the correlation development cover the following ranges of parameters: $1 \leq P \leq 12$ bar, $10 \leq G \leq 300$ kg/m^2s, $0 \leq \Delta T_i \leq 156.4$ K, $43.4 \leq L \leq 1200$ mm, $1.3 \leq D \leq 23.9$ mm, and $4.47 \leq L/D \leq 100$. The correlation included the L/D effect on CHF that further depends on pressure and mass velocity.

Chang et al. performed experiment with round tubes with diameters of 6 mm and 8.8 mm for mass velocities below 220 kg/m^2s under atmospheric pressure, and they proposed a correlation based on the inlet conditions. They observed that under low pressure and flow conditions, CHF increases with mass velocity (G) and decreases with L/D. Further, given L/D, CHF decreases as D increases, and the effect of inlet subcooling on the CHF is negligible. The local correlation using the thermodynamic equilibrium quality was determined not to be adequate for these conditions. Chang et al. model consists of two correlations: one for very low mass velocity and another for higher mass velocity.

Baek et al. combined Chang et al. equations into one correlation by modifying the exponents. The inlet subcooling is included their correlation to reflect its effect at a higher mass velocity. The applicable ranges of the correlation are summarized as follows: $1 \leq P \leq 10$ bar, $3.05 \leq D \leq 23.9$ mm, $4.47 \leq L/D \leq 314.2$, $-231 \leq G \leq 250$ kg/m^2s, $0.03 \leq q_c \leq 7.125$ MW/m^2.

Lee et al., using the genetic programming method, developed the inlet correlation based on 414 CHF data from KAIST CHF data bank[13] for upward water flow in vertical round tubes under low pressure and low flow conditions. The genetic programming finds both the functional form and the fitting coefficients of the correlation. The ranges of the data used to develop their correlation are: $1.1 \leq P \leq 10.98$ bar, $4.57 \leq D \leq 13.1$ mm, $43 \leq L \leq 3120$ mm, $13 \leq G \leq 499$ kg/m^2s, $13 \leq \Delta H_i \leq 684$ kJ/kg.

Shah correlation consists of two correlations: one for the upstream condition correlation (UCC) and another for the local condition correlation (LCC). The 834 CHF data points used in this work can be predicted by the UCC. Shah correlation can be applied to 23 fluids (water, refrigerants, cryogen, chemicals and liquid metals). The data used includes the following: tube diameter range is from 0.315 to 37.5 mm; the tube length to diameter ratio range is from 1.3 to 940; mass velocity range is from 4 to 29051 kg/m^{2}s; reduced pressure range is from 0.0014 to 0.96; inlet quality varies from –4 to 0.85; and critical quality range is from –2.6 to 1.

Deng local correlation consists of two parameters function of pressure and 3 local condition parameters: G (mass flux), D (diameter), and X_t (true quality). The true mass flux of vapor at the tube exit (GX_t) is the most significant correlating parameter. In developing the general correlation, it is assumed that the mass velocity is sufficiently high that the effect of gravity is neglected and the flow is assumed to be stable with no significant oscillations. The effect of the pressure variation along the tube is neglected and the calculation is performed with the assumption that the pressure at each point equals to the system pressure at the tube exit. The correlation is a simple form but its accuracy shows a good agreement with experimental data. Predictions are compared with over 10000 CHF data and show a standard deviation of error less than 10 % and an average error less than 2 %. The tested conditions are the following: $2 < P < 200$ bar, $2 < D < 37$ mm, $50 < L < 6000$ mm, $100 < G < 18000$ kg/m^{2}s, $-2.5 < X_i < 0.7$ and $-0.8 < X < 1.0$. A general form of the correlation is

$$q_c = \frac{\alpha}{\sqrt{D}} \exp\left[-\gamma\sqrt{GX_t\left(1+X_t^2\right)^3}\right] \quad (1)$$

where,

$$\alpha = 1.669 - 6.544\left(\frac{P}{P_c} - 0.448\right)^2$$

$$\gamma = 0.06523 + \frac{0.1045}{\sqrt{2\pi\left(\ln(P/P_c)\right)^2}} \exp\left(-5.413\frac{\left(\ln(P/P_c)+0.4537\right)^2}{\left(\ln(P/P_c)\right)^2}\right)$$

The results from a comparison of CHF predictions from the correlations mentioned are presented in Table 2 in terms of the calculated average and rms error based on the experimental values. Table 2 shows that the accuracy of Deng correlation deteriorates significantly. It is analyzed that this result is due to insufficient understanding effect of pressure and true mass flux of vapor at the low pressure and low flow conditions. A graphic representation of the results of the comparison is presented in Figure 1 ~ Figure 6.

A NEW LOCAL CORRELATION

It is revealed that the accuracy of Deng correlation deteriorates remarkably when the correlation is applied to the low pressure and flow conditions. To relate the effect of mass flux of vapor and CHF is relatively new approach. This study has been performed in Heat Transfer Research Facility of Columbia University of recent years. But it seems to indicate that the deteriorating results are from the insufficient understanding of the effect of mass flux of vapor and pressure. The results of the systematic investigation on the effect of major variables on CHF, showed that the accuracy of prediction can be improved by use of a correction factor that can incorporate the effect of the mass flux of vapor and the effect of slip. Further α and γ of Deng correlation are inadequate in these low pressure and flow conditions. Thus values of α and γ are modified to give a more accurate prediction in the range of pressure below 10 bar and mass velocity below 300 kg/m^{2}s. In addition, the mass flux of vapor and the slip factor are incorporated in the form of an exponent l. The resulting expression of the new local correlation is

$$q_c = \frac{\alpha}{\sqrt{D}} \exp\left[-\gamma\left(GX_t\left(1+X_t^2\right)^3\right)^l\right] \quad (2)$$

where,

$$l = 0.4779 + 0.45827\, e^{-GX_t/22.80697} + 27.28264\, e^{-GX_t/2.02777} + 0.3214\, e^{-GX_t/127.68917}$$

$$\alpha = 1.9244 - 6.3608\left(\frac{P}{P_c} - 0.46029\right)^2$$

$$\gamma = 0.05805 + \frac{0.10387}{\sqrt{2\pi\left(\ln(P/P_c)\right)^2}} \exp\left(-5.40715\frac{\left(\ln(P/P_c)+0.4543\right)^2}{\left(\ln(P/P_c)\right)^2}\right)$$

The parameters of α and γ are functions of reduced pressure, and l is a dimensionless number.

Under a given constant pressure and the quality at OSV (Onset of Significant Vaporization), CHF is a continuous function of local variables, namely, diameter (D), mass flux (G), and the true steam quality (X_t). For a fixed geometry, CHF is exponentially decreasing function of true mass flux of vapor (including effect of slip). The importance of equation (2) is that CHF in all flow patterns and boiling regime is predicted only by local condition at which CHF occurs, and the two parameters of the function of pressure are sufficient for good prediction. The average and rms errors are given in Table 3 and a graphic representation is given in Figure 7. Table 3 shows the possibility that CHF under the low pressure and flow conditions can be predicted reasonably well if accurate local conditions and two parameters of function of pressure can be obtained.

For a given tube and pressure, a true steam quality (X_t) is obtained from the thermodynamic quality and the quality at OSV.

$$\frac{dX_t}{dX} = 1 + \frac{X_t - X}{X_{osv}(1 - X_t)} \quad (3)$$

where,

$$X_t = \begin{cases} 0 \quad at\ X = X_{osv} \quad if\ X_i < X_{osv} \\ 0 \quad at\ X = X_i \quad if\ X_{osv} < X_i < 0 \\ X \ at\ X_i > 0 \end{cases} \quad (4)$$

Integrating equation (12), it can be shown to be,

$$X_{osv} \ln \frac{X - X_t}{X_b} + \ln \frac{1 - X + X_{osv} - X_{osv} X_t}{1 - X_b + X_{osv}} = 0 \quad (5)$$

where,

$$X_b = \max(X_{osv}, X_i)$$

Thermodynamic quality (X) is obtained from the heat balance as,

$$X = X_i + \frac{4qL}{h_{fg} GD} \quad (6)$$

X_{osv} is obtained from Saha-Zuber correlation[14], in which the onset of significant vaporization quality was tested for flow boiling of water, freon-114, freon-22 in tubes and annuli. The data fall into two distinct regimes on St (Stanton number) vs. Pe (Peclet number) plot; vapor generation is either hydro-dynamically controlled ($Pe \geq 70000$) or thermally controlled ($Pe < 70000$).

$$X_{osv} = -\frac{q}{SGh_{fg}} \quad (7)$$

where,

$$S = \begin{cases} 0.0065 \quad if\ Pe \geq 70000 \\ 455 / Pe \quad if\ Pe < 70000 \end{cases}$$

$$Pe = \frac{GDC_{pf}}{k_f}$$

The prediction of CHF is carried out by HBM(Heat Balance condition Method)[15]. At each constant pressure, the prediction is made by initiating X_t with equation (4), calculating q_c from the correlation (2), then X_{osv} from equation (7), and X from heat balance by equation (6). Once X_{osv}, X are known, a new X_t can be calculated by a equation (3) or (5). Then by replacing the initial value of X_t with the updated X_t, the iteration continues until X_t value converges between iterations (% error < 10^{-5}). During the calculation, it is assumed that CHF occurs always at the tube exit and the calculation is performed with the assumption that the pressure at each point is equal to the exit pressure.

CONCLUSION

Most of the authors treated the experimental data as nonlocal in the low pressure and flow condition. A conclusion from this investigation has drawn that CHF depends only on local conditions. With the aim of establishing the validity of local condition for the prediction of CHF at low pressure and flow conditions, Deng correlation has been analyzed and modified by using a total of 834 data points of the ranges in $1.01 \leq P \leq 10$ bar ; $1.3 \leq D \leq 24$ mm ; $65 \leq L \leq 3120$ mm ; $13 \leq G \leq 300$ kg/m^{2}s ; $0 \leq \Delta H_i \leq 653.6$ kJ/kg ; $0.118 \leq q_c \leq 4.006$ MW/m^2. Average and rms errors have been calculated for a new local correlation and five inlet condition correlations (Weber and Johannsen, Chang et al., Baek et al., Lee et al., and Shah). Comparison of predictions of each correlation with experimental data has been shown. Among the correlations Lee et al., Shah and the new local correlations show a good agreement with experimental data. Through this study it is shown that the possibility of which CHF is reasonably predicted by local correlation under low pressure and flow conditions. Generally, it has been reported that the local condition hypothesis is not valid for short tubes such as L/D < 80. This study, however, shows the possibility of applying local condition correlation to these conditions. Accurate description of local conditions at the CHF seems to be very effective in simplifying the treatment of complexity of flow instabilities that is inherent at such low pressure and flow conditions. From the result of this study, as shown in Figure 7, we conclude that CHF decrease exponentially with "true mass flux" of vapor and CHF depends only on the local conditions.

It is important to understand what the term "local conditions" means here. The thermodynamic (equilibrium) quality, X, is generally inadequate to describe the actual physical condition at a given point in the tube. In this paper, as in Deng's model, it is assumed that CHF is a function of the actual physical conditions existing at the point of CHF which are here assumed to be determined by the values of the variables G, X_t, and P at the point of CHF. Notice, however, that the value of X_t depends on the inlet conditions when the inlet quality exceeds the OSV value (see equation 4). Thus when X is used as the correlating variable the correlation will show a dependence on L/D or, equivalently, the inlet quality X_i. This is a very important consideration since, in general, there is no obvious way to generalize a nonlocal correlation to other conditions, such as nonuniform power profiles. For the local correlation, however, there is no freedom of choice and the generalization to nonuniform profiles is immediate and unique.

Another point requires mention. In the uniform tube, CHF always occurs at the exit, but in calculating the value of X_t at the exit, it has been assumed that the pressure at any point from OSV to the exit is approximately equal to the exit pressure. In this way, flow and heat transfer has been effectively de-coupled. In most cases, the resulting error is small but this need not always be true. Without such simplifications, the development of a CHF correlation would have been extremely difficult, if not impossible. Additional investigation is necessary to extend the current work. We hope that our work can be used to develop a universal CHF model.

Table 1. Ranges of experimental data used for present calculation

Source	No. of points	P [bar]	D [mm]	L [mm]	G [kg/m^2s]	ΔH_i [kJ/kg]	Q_c [MW/m^2]
Kim et al.	513	1.04~9.51	6~12	300~1770	19.5~277	46~ 653.6	0.12~1.60
Thompson & Macbeth	57	1.03~10	3.05~24	152~2500	14.9~298.4	0~595.2	0.24~2.06
Lowdermilk et al.	101	1.01	1.3~ 4.78	65~ 991	27.1~298.4	312~ 331	0.191~ 4.006
Becker et al.	110	2.16~10	7.76~13.1	600~ 3120	102~300	78~644	0.281~ 1.991
Mishima	53	1.01	6	344	13~296	84~293	0.118~ 1.486
Total	834	1.01~10	1.3~24	65~3120	13~300	0~653.6	0.118~ 4.006

Table 2. Calculated error of correlations

Correla -tion	Average error (%)					Rms error (%)				
	Kim et al.	Thomp-son & Macbeth	Lowder -milk et al.	Becker et al.	Mishima	Kim et al.	Thomp-son & Macbeth	Lowder -milk et al.	Becker et al.	Mishima
Weber & Johann-sen	13.47	6.51	-10.99	8.43	-3.95	19.27	13.37	15.58	9.92	13.15
Chang et al.	11.72	-4.60	2.63	-2.75	-9.75	26.73	17.15	29.33	23.25	12.13
Baek et al.	12.23	-3.53	6.79	-5.38	-12.20	23.28	13.03	28.55	20.46	16.02
Lee et al.	-1.53	-6.17	-9.88	-5.78	-7.82	9.77	12.35	15.81	10.58	19.64
Shah	3.20	-3.65	-2.49	-6.10	-2.97	10.38	14.70	12.84	9.56	15.75
Deng	33.48	17.00	13.89	2.36	41.33	41.24	35.03	25.51	6.78	48.99

Table 3. Calculated error of a new correlation

Source	Average error (%)	rms error (%)
Kim et al.	2.44	6.94
Thompson & Macbeth	-2.61	5.45
Lowdermilk et al.	-2.66	9.37
Becker et al.	-3.30	5.87
Mishima	-6.05	9.52

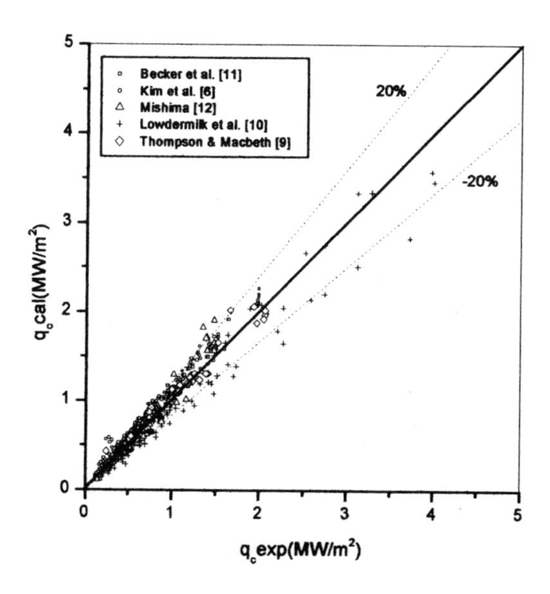

Figure 1. Calculated vs. experimental CHF of Weber and Johannsen correlation[2]

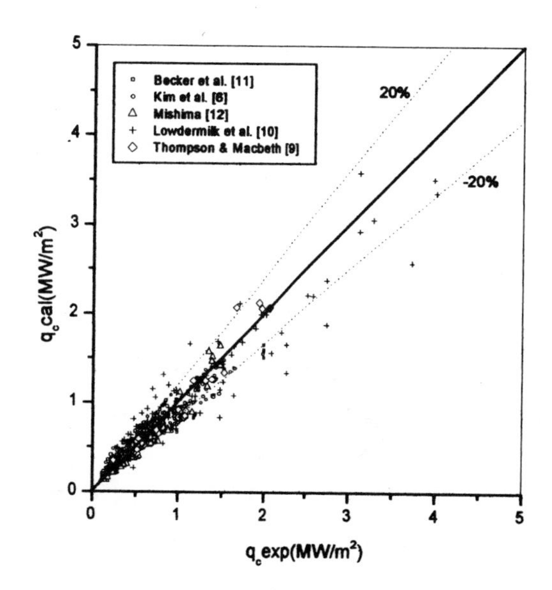

Figure 3. Calculated vs. experimental CHF of Baek et al. correlation[4]

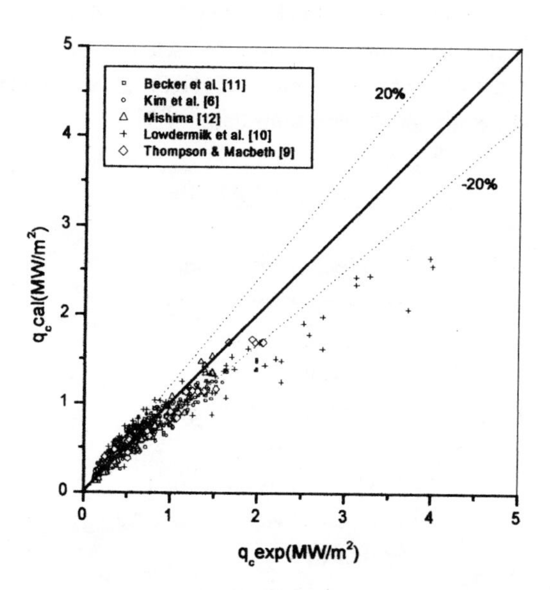

Figure 2. Calculated vs. experimental CHF of Chang et al. correlation[3]

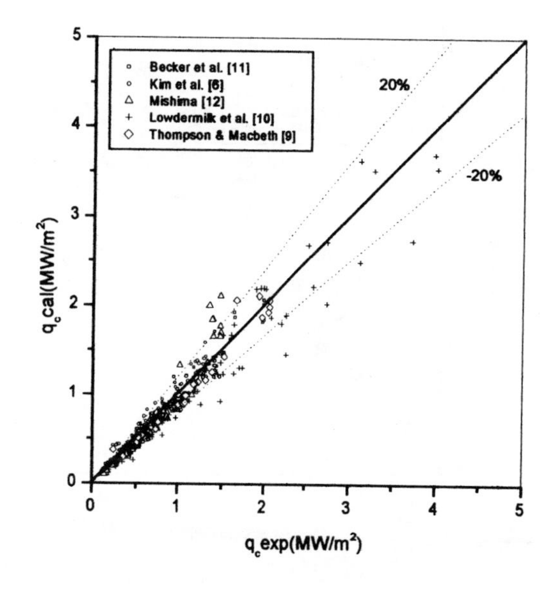

Figure 4. Calculated vs. experimental CHF of Lee et al. correlation[5]

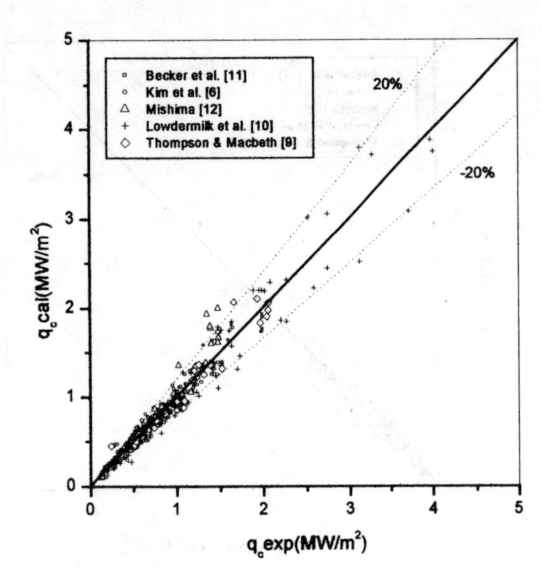

Figure 5. Calculated vs. experimental CHF of Shah correlation[7]

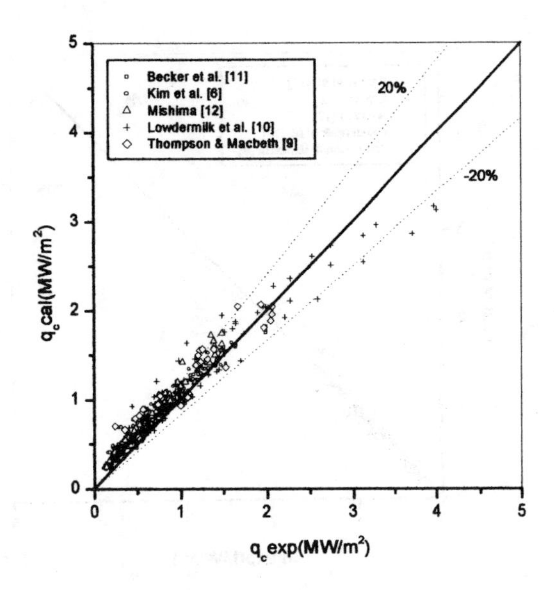

Figure 6. Calculated vs. experimental CHF of Deng correlation[8]

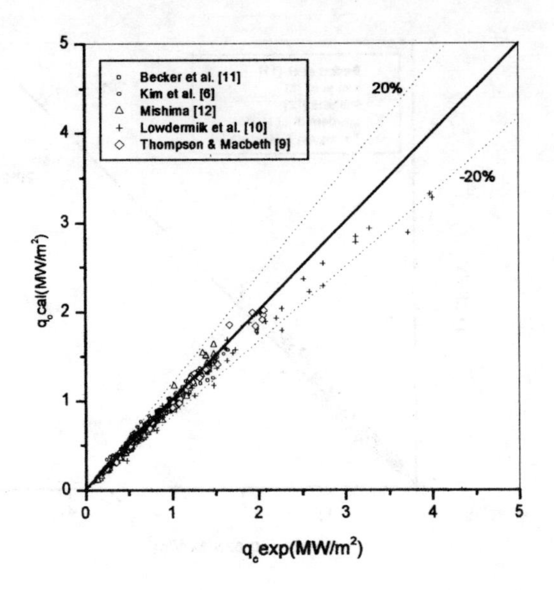

Figure 7. Calculated vs. experimental CHF of a new local correlation

ACKNOWLEDGEMENT

The authors wish to thank Korea Advanced Institute of Science and Technology (KAIST) and Heat Transfer Research Facility (HTRF) at Columbia University for sharing their data.

NOMENCLATURE

L	heated length (m or mm)
D	tube diameter (m or mm)
P	system pressure (bar)
G	mass velocity (kg/m^{2}s)
ΔH_i	inlet subcooling (kJ/kg)
ΔT_i	inlet subcooling (K)
q_c	critical heat flux (MW/m^2)
h_{fg}	latent heat of vaporization (kJ/kg)
X_i	inlet quality
X	thermodynamic quality
X_{osv}	thermodynamic quality at onset of significant vaporization (OSV)
X_t	true steam quality
S	Stanton number
Pe	Peclet number
P_c	critical pressure of water (221 bar)
α	parameter which depends on pressure (MW/m$^{1.5}$)
γ	parameter which depends on pressure ((kg/m^{2}s)$^{-1/2}$)

Subscripts

f saturated liquid

g saturated vapor

REFERENCE

[1] K. Mishima and H. Nishihara, Effect of channel geometry on critical heat flux for low pressure water, Int. J. Heat Mass Transfer Vol. 30, No. 6, pp. 1169-1182, 1987

[2] P. Weber and K. Johannsen, Study of the critical heat flux condition at convective boiling of water: temperature and power controlled experiments, Proceeding 9th Int. Heat Transfer Conference, Vol. 2, pp. 63-68, Jerusalem, 1990

[3] S. H. Chang, W. P. Baek and T. M. Bae, A study of critical heat flux for low flow of water in vertical round tubes under low pressure, Nuclear Engineering and Design Vol. 132, pp. 225-237, 1991

[4] W. P. Baek, S. K. Moon and S. H. Chang, A modified CHF correlation for low flow of water at low pressures, Two-Phase Flow Modelling and Experimentation, 1995

[5] D. G. Lee, H. G. Kim, W. P. Baek and S. H. Chang, Critical heat flux prediction using genetic programming for water flow in vertical round tubes, Int. Comm. Heat Mass Transfer, Vol. 24, No. 7, pp. 919-929, 1997

[6] H. C. Kim, W. P. Baek and S. H. Chang, Critical heat flux water in vertical round tubes at low pressure and low flow conditions, To be published at Nuclear Engineering and Design, 1999

[7] M. M. Shah, Improved general correlation for critical heat flux during upflow in uniformly heated vertical tubes, Heat and Fluid Flow, Vol. 8, No. 4, pp. 326-335, 1987

[8] Z. Deng, Prediction of critical heat flux for flow boiling in subcooled and saturated regions, Ph. D. thesis, Columbia University, 1998

[9] B. Thompson and R. V. Macbeth, Boiling water heat transfer burnout in uniformly heated round tubes: A Compilation of world data with accurate correlations, AEEW Rept. R 356, 1964

[10] W. H. Lowdermilk, C. D. Lanzo, and B. L. Siegel, Investigation of boiling burnout and flow stability for water flowing in tubes, NACA TN 4382, 1958

[11] K. M. Becker, G. Strand et al., Round tube burnout data for flow of boiling water at pressure between 30 and 200 bar, Report KTH-NEL-14, 1971

[12] K. Mishima, Boiling burnout at low flow rate and low pressure conditions, Ph. D. Thesis, Kyoto Univ., Japan, 1984

[13] S. H. Chang et al., The KAIST CHF data bank (Rev. 3), KAIST-NUSCOL-9601, 1996

[14] P. Saha and N. Zuber, Point of net vapor generation and vapor void fraction in subcooled boiling, Proceeding Of the 5th Int. Heat Transfer Conference, Tokyo, paper B4.7, pp. 175-179, 1974

[15] F. Inasaka and H. Nariai, Evaluation of subcooled critical heat flux correlations for tubes with and without internal twisted tapes, Nuclear Engineering and Design, Vol. 163, pp. 225-239, 1996

AIAA-2000-2868

DEVELOPMENT AND DESIGN OF COLD HEAT EXCHANGER OF PULSE TUBE COOLER

Wei Dong, Marco Lucentini, Vincenzo Naso
Dr. Wei Dong - dong@uniroma1.it
Dr. Marco Lucentini - luce@uniroma1.it
Prof. Vincenzo Naso Ph.D. - naso@uniroma1.it
University of Rome "La Sapienza"
Dept. of Mechanical and Aeronautical Engineering
Via Eudossiana, 18 – 00184 – Rome – ITALY
Fax: +39-06-4881759 Tel: +39-06-44585271/258

ABSTRACT

During the past fifteen years heat transfer in oscillating flows become the subject of increasing interest in the engineering community. Applications of oscillating flows include, for example, the cooling electric equipment or alternative, environmentally safe refrigeration technologies, such as pulse tubes or Stirling refrigerators. Important components of these refrigerators are their heat exchangers. In such devices the working fluid is subject to oscillatory forcing which is a key part of the process, as opposed to situations where oscillations are generated with the aim to enhance heat transfer. Heat transfer in oscillating, and often compressible flow, has not yet been completely understood, and the lack of design methodologies for heat exchangers in such flow is one reason that efficiencies of these devices are limited. A practical design method of cold heat exchanger was developed in this paper. The influence of the working gas oscillating on heat transfer was considered in this method. The design parameters related to gas oscillation can be calculated using known methods.

NOTATION

A_f	cross – sectional area for flow
c_p	specific heat
D	plate to plate spacing
d_h	hydraulic diameter
H	a space of cooler
h	heat transfer coefficient
L	length of cold heat exchanger
k	thermal conductivity
$\dot{m}$	mass flow rate
n	number of plate
P	pressure
Pr	Prandtl number
q	heat transfer rate
Re_{max}	amplitude of the instantaneous Reynolds number
Re_{ω}	dimensionless frequency
T	temperature
U_{max}	maximum velocity
X_{max}	tidal displacement of gas parcel
α	coefficient of thermal diffusivity
μ	dynamic viscosity
ν	kinetic viscosity
ρ	density
$\bar{\tau}_w$	averaged shear stress of the gas
ω	angular frequency

INTRODUCTION

Heat exchangers are devices which enhance the transfer of the heat and are vital components of every cryocooler. They exist in a wide variety of types, shapes, sizes and arrangements and are made of all kinds of materials. In order to exploit the thermoacoustic effect for heat pumping in pulse tube refrigerator (PTR), heat exchangers are attached both ends of the pulse tube. The cold heat exchanger removes heat from a cold temperature reservoir and the hot exchanger rejects the pumped heat and absorbed acoustic work to the environment at temperature. Here we analyze only the cold heat exchanger of PTR.

In general, the design methods of the heat exchanger are for steady flow and the heat is exchanged between two gas streams owing unidirectionally with steady volume flow rates. These methods are not suitable for oscillating flow. For this reason, the available design method was developed with taking into account the heat transfer correlation for oscillating flow conditions in heat exchanger. The proposed design method allows independent optimization of heat exchanger of PTR.

1. HEAT TRANSFER PROCESS OF WORKING GAS IN COLD HEAT EXCHANGER

A closer look into the cold heat exchanger, illustrated in the magnified region of Figure 1, reveals the mechanism responsible for thermoacoustic heat pumping, by considering the oscillation of a single gas parcel of the working fluid along a stack plate. The gas parcel starts the cycle at a temperature T. In the first step it is moved to left, towards the regenerator, by the acoustic wave. Because the space between the stack plate is much smaller than a thermal penetration depth in regenerator, thus it experiences adiabatic compression, which causes its temperature to rise from T to T_+. In this state the gas parcel is warmer than the regenerator stack plate, and irreversible heat transfer towards the regenerator's stack plate takes place (dQ_h in Figure 1). The resulting temperature of the gas parcel after this step is near T. On its way backing to stack plate of cold heat exchanger and leaving cold heat exchanger towards pulse tube, the pressure then goes down the gas parcel experiences adiabatic expansion and decreases down temperature to T_-. At this state the gas parcel is colder than the stack plate and irreversible heat transfer from the stack plate towards the gas parcel takes place (dQ_c). After this steps, the gas parcel has completed one cycle.

In order to maintain good thermal contact between the working fluid and the solid heat capacity across a large cross sectional area, the stacks are finely subdivided into many parallel channels, with hydraulic radius comparable to thermal penetration depth in stacks of heat exchanger.

Two characteristic length scales of the heat exchanger core, the thermal penetration depth $\delta_k = \sqrt{2\alpha / \omega}$ and the viscous penetration depth $\delta_v = \sqrt{2\nu / \omega}$ are of main importance for the design purposes. α and ν are thermal and momentum diffusivities, respectively, of the working fluid, and ω is the angular frequency. The thermal penetration depth δ_k describes the layer around the stack plate where the thermoacoustic effect occurs as the distance that heat can diffuse through the fluid. During the corresponding time interval, the thickness of the layer of fluid around the stack plates is restrained fluid movement under the influence of viscous force, thus contribute less to the thermoacoustic effect. In order to exchange heat with the working gas the heat exchange components in pulse tube refrigerator must have lateral dimensions of the order of δ_k or smaller. The ratio of these two penetration depths is called the Prandtl number of the work gas. The Prandtl number is close to unity for typical gases, so viscous and thermal penetration depths are comparable. Hence, the heat exchangers in PTR always suffer from substantial viscous effect.

2. OPTIMAL COLD HEAT EXCHANGER OF STACK OF PARALLEL PLATES

In this section, we will determine the optimal spacing for maximum heat transfer from a stack of parallel plates that are cooled by forced convection. We consider that working gas in the cold heat exchanger moves in oscillating way between the regenerator and pulse tube. To evaluate the heat transfer capability of cold heat exchanger, the knowledge of Reynolds number for oscillating flow is required.

Consider the geometry of Figure 2 in which the cold flow temperature T_- and pressure drop established by the compressor ΔP are fixed. In this analysis, the flow is assumed laminar, and the plate temperature is assumed uniform at the same level T_w. Each plate has a thickness t that is sufficiently small than D.

To determine the plate to plate spacing D is the same as determining the optimal number of plates that can fill a space of thickness H

$$n \cong \frac{H}{D} \tag{1}$$

The length of cold heat exchanger, L, should be a little larger than the tidal displacement of the gas parcel, X_{max}, since this allows gas parcel to be in contact with both the regenerator and the pulse tube. The tidal displacement defined in this way describes the peak to peak displacement of the moving gas parcel.

$$L \le X_{max} \quad \text{and} \quad X_{max} = \frac{d_h Re_{max}}{4 Re_\omega} \tag{2}$$

$$Re_{max} = \frac{d_h \dot{m}_{max}}{\mu A_f} = \frac{U_{max} d_h}{\nu}, \text{ and}$$

$$Re_\omega = \frac{\omega d_h^2}{\nu} \tag{3}$$

where d_h is hydraulic diameter, $\dot{m}$ is mass flow rate, A_f is cross – sectional area for flow, Re_ω is dimensionless frequency and Re_{max} is amplitude of the instantaneous Reynolds number.

a). *The full developed flow limit.*

Let us consider first the limit D→0, when each channel is slender enough so that the flow is full developed all along L. In the same limit, the fluid outlet temperature approaches the plate temperature T_w. The average fluid velocity in each channel is given by

$$U = \frac{D^2}{12\mu}\frac{\Delta P}{L} \tag{4}$$

While the total mass flow rate through the stack of high H is

$$\dot{m} = \rho U H = \rho H \frac{D^2}{12\mu}\frac{\Delta P}{L} \tag{5}$$

The mass flow rate is expressed per unit length in the direction perpendicular to figure 3. The total heat transfer rate removed from the entire sandwich by the $\dot{m}$ stream is

$$q_a = \dot{m}c_p(T_w - T_-) = \rho H \frac{D^2}{12\mu}\frac{\Delta P}{L}$$

(6)

In conclusion, in the limit D→0, the total cooling rate decreases as D^2. This trend is illustrated qualitatively as curve (a) in Figure 3.

b) The boundary layer flow limit.

In the opposite extreme, D→∞, the boundary layer that lines on one surface becomes "distinct". In other words, each channel looks like the entrance region to parallel plate duct. The overall pressure drop is fixed at ΔP. The overall force balance on the control volume H x L requires

$$\Delta P \cdot H = n \cdot 2 \cdot \bar{\tau}_w \cdot L$$

(7)

in which n is the number of the channels and $\bar{\tau}_w$ is the L averaged shear stress of gas

$$\bar{\tau}_w = 0.1328 R_e^{-1/2} \frac{1}{2}\rho U^2$$

(8)

Combined, equation (7) and (8) yield

$$U = \left(\frac{1}{1.328} \frac{\Delta P \cdot H}{nL^{1/2}\rho v^{1/2}} \right)$$

(9)

The total heat transfer rate from one of the L long surfaces (q_1') can be calculated by recognizing the overall Nusselt number for Pr> 0.5;

$$\frac{hL}{k} = \frac{q''}{T_w - T_-}\frac{L}{k} = 0.664 P_r^{1/3}\left(\frac{UL}{\nu}\right)^{1/2}$$

(10)

which leads to

$$q_1' = q''L = k(T_w - T_-)0.664 P_r^{1/3}\left(\frac{UL}{v}\right)^{1/2}$$

(11)

The total heat transfer rate released by the entire stack is 2n times larger than

$$q_t' = 2nq_1' = 2nk(T_w - T_-)0.644 P_r^{1/3}\left(\frac{UL}{v}\right)^{1/2}$$

(12)

In view of the n and U expression listed in eq, the total heat transfer rate becomes

$$q_t' = 1.208k(T_w - T_-)H \frac{P_r^{1/3} L^{1/3} \Delta P^{1/3}}{\rho^{1/3} v^{2/3} D^{2/3}}$$

(13)

The second conclusion we reached is that in the large D limit, the total heat transfer rate decreases as $D^{2/3}$. This second trend has been added as curve (b) to the same graph (Figure 3.)

DISCUSSION AND CONCLUSION

According to above limit conditions, maximum of actual (unknown) curve $q'(D)$ can only occur at an optimal spacing D_{opt} that is of the order as the D value obtained by intersecting the asymptotes q_a and q'. It is easy to show that the order of magnitude statement. $q_a' - q_b'$ yields the following spacing:

$$\frac{D_{opt}}{L} \cong 2.73\left(\frac{\mu\alpha}{\Delta P L^2}\right)^{1/4}$$

(14)

This estimate agrees very well with the more exact result obtained by the maximum of actual $q'(D)$ sketched in figure 3. The order of magnitude of the maximum package heat transfer rate that corresponds to $D = D_{opt}$ is obtained by combining eqs (6) and (14):

$$q_{max}' \leq 0.62\left(\frac{\rho \Delta P}{Pr}\right)^{1/2} Hc_p(T_w - T_-)$$

(15)

The brief scale analysis represented in this section can be repeated for the situation in which only one surface of the board is Joule-heated to T_w, and the other surface can be modeled as adiabatic. The only change is that 2n is replaced by n in eq. (12), so that results become

$$\frac{D_{opt}}{L} \cong 2.10\left(\frac{\mu\alpha}{\Delta P L^2}\right)^{1/4}$$

(16)

$$q_{max}' \leq 0.37\left(\frac{\rho \Delta P}{Pr}\right)^{1/2} Hc_p(T_w - T_-)$$

(17)

It is obviously that the change in the thermal boundary conditions of one plate to plate channel affects only the value of numerical coefficient in the expressions for D_{opt} and q_{max}'.

A practical design approach of cold heat exchanger was discussed in above. The influence of the working gas oscillating on heat transfer was considered this approach. The design parameters related to gas oscillation can be calculated using known methods.

REFERENCES

1. B. Lawton, G. Klingeberg, "*Transient Temperature in Engineering and Science*" Oxford University Press Inc., 1996.
2. E. J. Hoffman, "*Power Cycle and Energy Efficiency*", Academic Press, Inc., 1996.
3. R. S. Berry, V. Kazakov, "*Thermodynamic Optimization of Finite Time Process*", John Wiley & Sons, Ltd., 2000,
4. F. de Monte, G. Galli, "*An Analytical Oscillating Flow Thermal Analysis of the Heat Exchanger and Regenerator in Stirling Machines*", IEEE Proc. 1996.
5. Eric M. Smith, "*Thermal Design of Heat Exchangers*", John Wiley & Sons, Ltd., 1997,

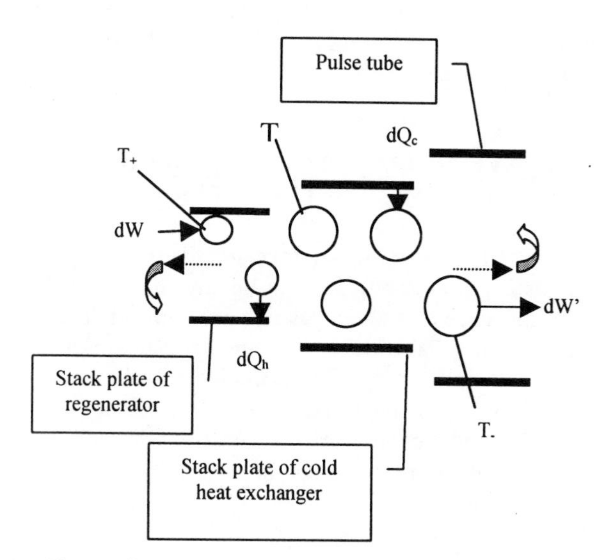

Figure 1 Illustration of the heat pumping cycle along a stack plate by considering the oscillation of one gas parcel.

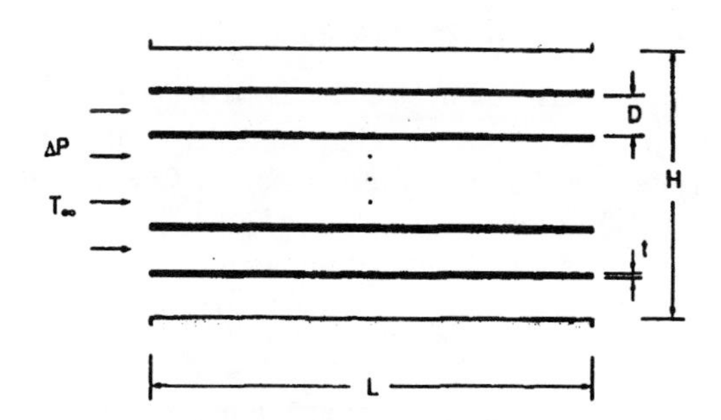

Figure 2. Stack of parallel plate cooled by fluid.

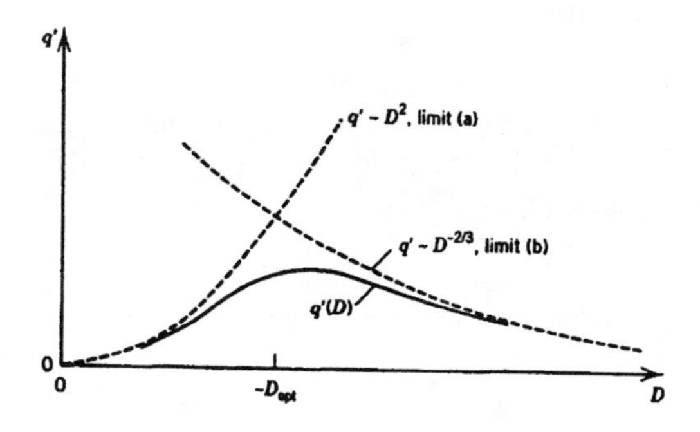

Figure 3 The optimal board to board spacing as the intersection of the fully developed flow asymtote (a) with the distict boundary layer asymptote (b)

AIAA-2000-2869

ENERGY TRANSFER PROCESSES OF WORKING GAS IN PULSE TUBE

Vincenzo Naso, Wei Dong, Marco Lucentini
Dr. Wei Dong - dong@uniroma1.it
Dr. Marco Lucentini - luce@uniroma1.it
Prof. Vincenzo Naso Ph.D. - naso@uniroma1.it
University of Rome "La Sapienza"
Dept. of Mechanical and Aeronautical Engineering
Via Eudossiana, 18 – 00184 – Rome – ITALY
Fax: +39-06-4881759 Tel. +39-06-44585271

ABSTRACT

Pulse tube cooler is a device for cooling to low temperatures. It runs on a modified Stirling cycle but have no cold moving parts and the gas flow in pulse tube can carry heat energy away from a low temperature point under certain conditions. An orifice or other device controlling the flow at the end of the cooler provides the condition for cooling to occur. Pulse tube coolers have great commercial potential; offering low cost, low vibration and high reliability and they can be designed in variety of configurations including linear, U-tube (folded 180 in the middle) and concentric. They can be designed to cool to 200 K, to as low as 2 K, or to any temperature in between. The cooling power ranges from milliwatts to kilowatts depending on cooler size and operating temperature.

In development of pulse tube cooler it is important to study the energy transfer process of working fluid in pulse tube. The enthalpy flow in the pulse tube is the gross cooling power of the cooler. Since the pulse tube section is nearly adiabatic, there are large temperature oscillations yielded by pressure oscillation. To evaluate the temperature oscillation, a mathematical model was developed. With this model, the analytical result is that the temperature ratio of working fluid is the function of pressure ratio between hot and cold heat exchanges.

INTRODUCTION

Most models are based upon ideal one dimension flows without accounting the minor effect. We consider that the analytical system consists of cooler pulse tube and hot heat exchanger. This system under consideration of minor effect is shown in Figure 1. We focus our attention to the energy transfer process with accounting the minor effect in pulse tube. To evaluate the minor effect, a mathematical model was developed. With this model, the analytical result is that the temperature ratio of working fluid is the function of pressure ratio between hot and cold heat exchanges. With this model, we can also control the cooling condition with controlling the minor effect.

1.WAVE PROPAGATION AND ENERGY TRANSFER IN PULSE TUBE

Pulse tube acts like an adiabatic thermal buffer. The gas is not in good contact with any solid. Hence the temperature of a given gas just goes up and down with pressure adiabatically. We consider that working gas in pulse tube moves in oscillating way between the cold heat exchanger and hot heat exchanger (see Figure 1). And a plane sound wave passes through in the flow direction in a gas of density ρ can be represented by

$$\rho = \rho_0(1+\zeta) \tag{1}$$

where

$$\zeta = \zeta_0 e^{-2\pi i(\alpha z - \nu t)} \tag{2}$$

In eq.2 ζ_0 is the amplitude of the wave, $\alpha = 1/\lambda$ is wave number and ν is frequency. If the wave is considered to be stationary, the its propagation in the gas corresponds to a flow of gas with a equal to sound velocity, c. If the process takes place in a hypothetical tube of unit cross section, then the mass traversing section a per second is equal to $\rho_a c_a$. Thus, as mass is conserved, $\rho_a c_a = \rho_b c_b$. Then, the rate of change of momentum per second at section a is $(\rho_a c_a)c_a$, with an analogous expression for momentum, per second at section b. Since a unit cross section has been assumed, according to Newton's second law, the difference between these quantities is equal to the pressure difference between the section a and section b. The result is then given by

$$\rho_b c_b^2 - \rho_a c_a^2 = P_a - P_b \tag{3}$$

which can be written as

$$P_a - P_b = \rho_a^2 c_a^2 \left(\frac{1}{\rho_b} - \frac{1}{\rho_a} \right) = \frac{Mc_a^2}{\widetilde{V}_a^2} \left(\widetilde{V}_b - \widetilde{V}_a \right) \quad (4)$$

then,

$$c_a^2 = \frac{\widetilde{V}_a^2}{M} \left(\frac{P_a - P_b}{\widetilde{V}_b - \widetilde{V}_a} \right) \quad (5)$$

where M molecular weight. For sound wave of infinitesimal amplitude, c_a can be replaced by c, yielding,

$$c^2 = \frac{\widetilde{V}^2}{M} \left[-\left(\frac{\partial P}{\partial \widetilde{V}} \right)_{ad} \right] \quad (6)$$

The subscript 'ad' in eq.6 indicates that the process is adiabatic, as it is too rapid to permit the exchange of heat between the regions of compression and rarefaction. Thus, temperature equilibrium is not achieved during the passage of the sound wave.

For an adiabatic process in an ideal gas, $PV^{\gamma} = constant$; then,

$$\left(\frac{\partial P}{\partial \widetilde{V}} \right)_{ad} = \frac{\gamma P}{\widetilde{V}} \quad (7)$$

and the sound velocity is given by

$$c = \sqrt{\frac{\gamma RT}{M}} \quad (8)$$

In the analysis of sound propagation presented above it was assumed that the sound wave was not absorbed by the gas. However, if so called dissipative processes are present, then sound absorption will occur that depends on the various transport coefficients.

Sound waves were considered to be of infinitesimal amplitude. The sound velocity was treated as a thermodynamic quantity, as given by eq.(8). However, in a real fluid, as the intensity of the wave becomes greater, the sound velocity increases with the pressure along an adiabatic. The fact that the speed of the sound is greater in the high pressure region of a sound wave explain, at least in part, the formation pressure wave.

Considering the sinusoidal, one dimensional wave, the three basic relations, the conservation of mass, momentum, and energy of the interacting molecules, can be expressed in the form

$\rho_a c_a = \rho_b c_b$, for the conservation of mass (9)

$\rho_a c_a^2 + P_a = \rho_b c_b^2 + P_b$, for the conservation of momentum (10)

$$\frac{1}{2}c_a^2 + h_a = \frac{1}{2}c_b^2 + h_b, \quad (11)$$

In eq.(11) the enthalpy per gram is given by $h = \widetilde{H}/M$, where M is the molecular weight. Since $H = E + PV$, eq.(11) can also be written as

$$\frac{1}{2}c_a^2 + \varepsilon_a + \frac{P_a}{\rho_a} = \frac{1}{2}c_b^2 + \varepsilon_b + \frac{P_b}{\rho_b} \quad (12)$$

with $\varepsilon = \widetilde{E}/M$. These conditions yield the fundamental relations between the thermodynamic quantities. The value of c_a is the velocity of the pressure wave into stationary gas on the low pressure side. The substitution of the expression for c_b from eq.(9) into eqs.(10) and (11), after certain amount of algebra, will lead to following equation in the form

$$\varepsilon_b - \varepsilon_a = \frac{1}{2}(P_b - P_a)\left(\frac{1}{\rho_a} - \frac{1}{\rho_b} \right) \quad (13)$$

This relation can be used to determined the temperature rise across a normal pressure wave if the relations between internal energy and temperature are known. For an ideal gas, the ratio heat capacities can be written in the form $\gamma = 1 + R/\widetilde{C}_V$. Then, the internal energy per unit mass at each side of the pressure wave can be expressed as

$$\varepsilon = \frac{1}{M}\left(\frac{\partial \widetilde{E}}{\partial T} \right)_V \qquad T = \frac{P}{(\gamma - 1)\rho} \quad (14)$$

Substitution of eq.(14) into eq.(13) leads to the relation

$$\frac{\rho_b}{\rho_a} = \frac{(\gamma - 1)P_a + (\gamma + 1)P_b}{(\gamma + 1)P_a + (\gamma - 1)P_b} \quad (15)$$

2. MINOR EFFECT IN PULES TUBE

The physical configure of pulse tube is illustrated in Figure 1 (a). The working gas in pulse tube moves in oscillating way between the cold and hot heat exchanger. The cross sectional area of the channel change in section a and section b (see Figure 1 (b)). In oscillation flow at high Reynold number, additional pressure drops are associated with the transitions between channels. These effects are known as minor effect.

For oscillation flow, the pressure drop arising from minor losses is characterized by the dimensionless minor loss coefficient K,

$$\Delta P = K \frac{1}{2} \rho u^2 \quad (16)$$

$\Delta P_a = K_a \frac{1}{2} \rho_a u_a^2$ and $\Delta P_b = K_b \frac{1}{2} \rho_b u_b^2$

The coefficient K is illustrated in Figure 2 and only the function of geometry.

3 DISCUSSION AND CONCLUSION

The oscillation of the temperature, which is named thermal oscillation, is associated with the oscillation of the pressure. When the working gas leaves cold heat exchanger, it is in expansion and it is in compression as the working gas enters into hot exchanger. Consider minor effect, The pressure in section a and b is given:

$$P_a' = P_a - \Delta P_a \text{ ,and } P_b' = P_b + \Delta P_b \qquad (17)$$

Substitute eq (17) to eq (15), the density ratio given by eq.(15) approaches

$$\frac{\rho_b}{\rho_a} = \frac{(\gamma - 1)P_a' + (\gamma + 1)P_b'}{(\gamma + 1)P_a' + (\gamma - 1)P_b'} \qquad (18)$$

For an ideal gas, the temperature ratio is give by

$$\frac{T_b}{T_a} = \left(\frac{P_b'}{P_a'}\right)\left(\frac{(\gamma - 1)P_a' + (\gamma + 1)P_b'}{(\gamma + 1)P_a' + (\gamma - 1)P_b'}\right) \qquad (19)$$

and

$$\tau = 1 + \left(\frac{(\gamma + 1)(\beta^2 - 1)}{(\gamma + 1) + (\gamma - 1)\beta}\right) \qquad (20)$$

where $\tau = T_b / T_a$ and $\beta = P_b' / P_a'$

It is obviously in equation (20) that τ is the function of β and when β is greater than one, then τ is also greater than one. This mean that if pressure ratio is greater than one, the gas flow in pulse tube can carry heat energy away from a low temperature point. So, we can control the thermal oscillation with controlling the value of K.

REFERENCES

1. G. Walker, "Cryocoolers", Plenum, New York, 1983.
2. J. M. Lee and P. Kittel,, "Higher Order Pulse Tube Modeling", 9th Inter.Cryocooler Conf., 1996.
3. GREG Swift, "Thermocaustics: A unifying perspective for some engine and refrigerator" Fourth draft, 1999.

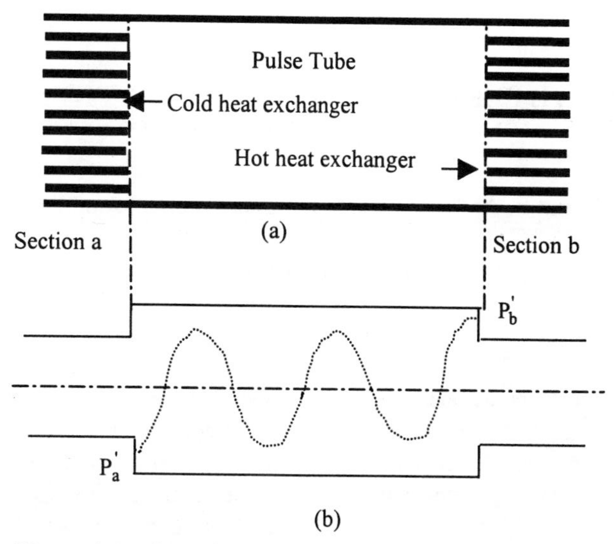

Figure 1.Analytical system
(a). Physical model of system
(b). Equivalent channel of the system

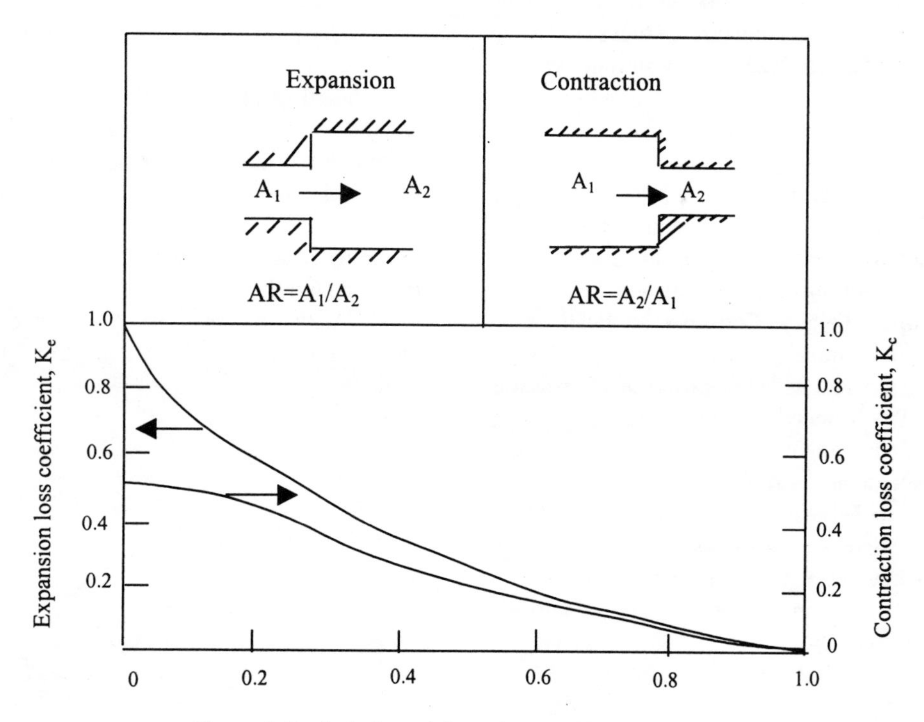

Figure 2. Typical chart giving minor coefficient K.
Reproduced from Fox and McDonald.

AIAA-2000-2870

ELECTROCHEMICAL PERFORMANCE OF A MULTI-TUBULAR FUEL CELL AND ELECTROLYZER ARRAY

Michael C. Kimble, Everett B. Anderson, Karen D. Jayne, Alan S.Woodman, and Hartmut H. Legner
Physical Sciences Inc.
Andover, MA 01810

Abstract

Miniaturization of tubular membrane and electrode assemblies (MEAs) allows compact fuel cells and electrolyzers to be developed with power density projections of 6 kW/liter and 11 kW/kg. These tubular MEAs have been assembled into multi-tube arrays and connected in either serial or parallel arrangements. Pressure testing and leak testing of these arrays has shown no signs of leakage indicating the viability of assembling these tubes into electrochemical stacks. The tubular MEAs in these arrays have been studied by assessing the electrochemical performance of each tubular MEA in the stack. These investigations have indicated no significant performance issues between the interior and exterior tubes in the array. The operation of these tubular MEA arrays in the electrolysis mode supports the use of these miniaturized tubes for a reversible fuel cell and electrolyzer energy storage system.

Background

The application of miniaturized proton exchange membrane fuel cells to low power applications has been receiving attention lately, namely, to increase the energy density over that provided by today's state of the art lithium-ion battery. Consumer applications including cellular phones, radios, laptop computers, and portable hand tools all have low power levels ranging from 1 to 50 W. Numerous studies have been conducted indicating that the batteries in these systems are limiting the device performance. Typical energy densities for a rechargeable battery range from 100 to 150 W-hr/kg whereas energy density requirements of 500 W-hr/kg have been projected to meet the capabilities that electronic designers have put into portable electronic devices.

Beyond consumer electronics, there are many applications where a low power fuel cell (e.g., 30 to 100 W) can improve the energy density such as for small unmanned aerial vehicles, astronaut extra-vehicular mobility units, and lunar and planetary rovers. Typically, most fuel cells that have been designed for these applications have simply shrunk the large kilowatt class of planar stacks into a smaller stack, keeping the same planar design. Consequently, energy densities of these systems at power levels lower than 100 W are often less than what is attainable with a battery. To overcome this limitation, Physical Sciences Inc. (PSI) has been developing a miniaturized PEM fuel cell that is designed specifically for power levels on the order of 1 to 500 W. Our approach is based upon using a micro-tubular membrane and electrode assembly that results in a higher stack power density and, consequently, a higher energy density when the fuel storage and ancillaries are included.

Technical Approach

The miniaturized tubular membrane and electrode assembly consists of an ion-exchange membrane separator that has an anode on the interior of the tube and a cathode on the exterior. Figure 1 shows a schematic of a single tubular membrane and electrode assembly (MEA). A conical metallic diffuser cone is inserted slightly into the interior of the MEA making contact with the interior electrode. A crimp ring is mounted on the outer electrode surface that may then be

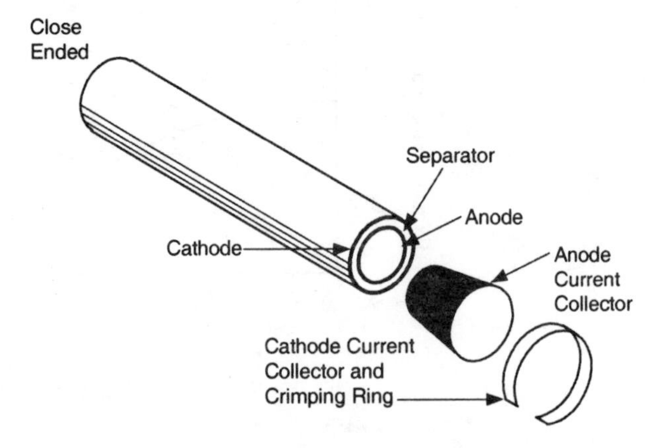

Figure 1. Schematic of a single tubular MEA.

inserted into a manifold as shown in Figure 2. In this arrangement, hydrogen gas may be introduced into the interior of the tube while an oxidant such as oxygen or air is introduced across the outer surface of the MEA. Multiple tubular MEAs may be placed on the order of 10 to 20 mils from one another resulting in an array of tubular MEAs. Figure 3 shows a photograph of a 14-tube assembly.

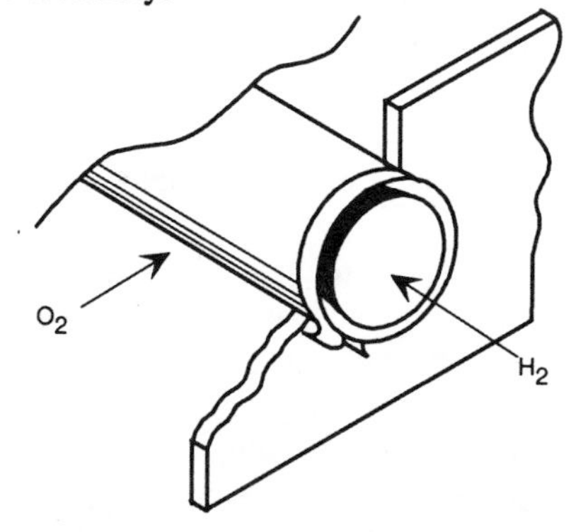

Figure 2. Schematic of a tubular MEA mounted into a manifold.

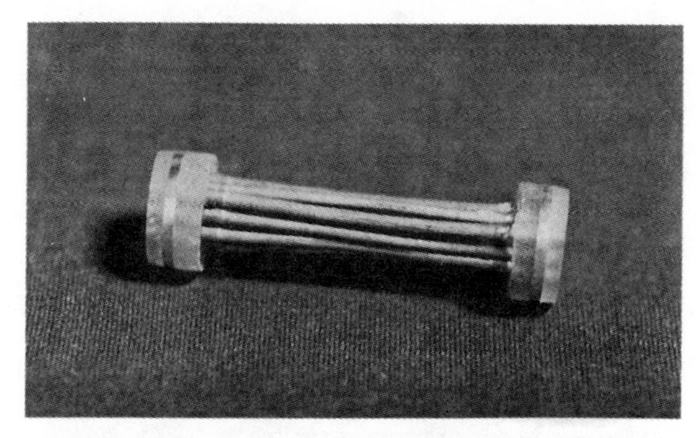

Figure 3. Photograph of a 14-tube array containing miniaturized tubular MEAs.

The advantages of the tubular MEA have been described before in previous publications.[1-3] Reducing the tube diameter towards 25 mil ID results in a maximum in the volumetric and gravimetric power density of the single tubular MEA. This effect is shown in Figures 4 and 5, respectively. The minimum and maximum performance curves correspond to the expected PEM fuel cell polarization performance window overlaid onto the tubular design parameters such as the inner diameter and electrode thicknesses. It should be noted that the power density numbers shown in Figure 4 account for a 10 mil thick oxidant gap around the cathode since this region is required to flow oxygen or air across each tube.

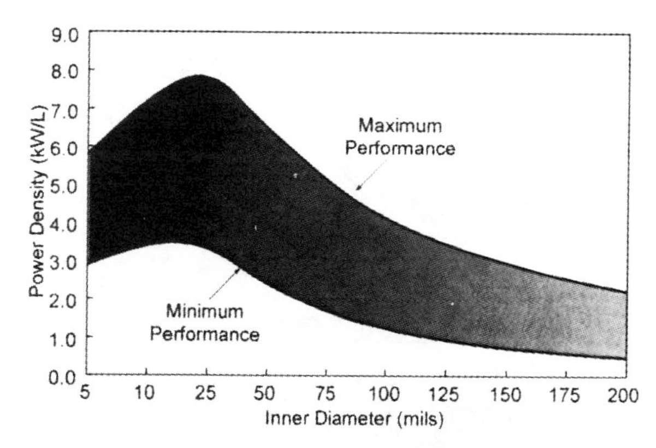

Figure 4. Volumetric power density projections for a single tubular MEA.

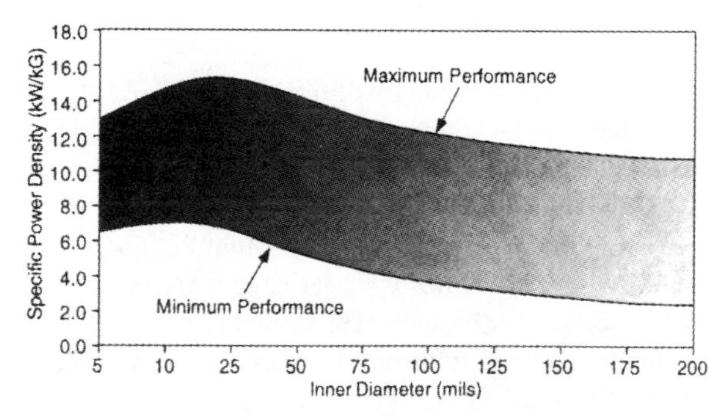

Figure 5. Gravimetric power density projections for a single tubular MEA.

The design approach for our tubular fuel cell technology exploits the fact that a low power fuel cell (e.g., 100 W) operating at 12 V produces 8.3 Amps. This current is orders of magnitude lower than that attained in large fuel cell stacks such as those that power automobiles. Due to the high currents in these automotive stacks, bipolar plates are required to transfer the current between cells. However, for low current applications, the bipolar plate may be removed using an edge-tab current collector to transfer electrons.

The traditional planar bipolar plates have an embedded flow-field that serves to introduce reactants across the electrodes and to remove products. With such an approach, there is excessive overhead in that the ratio of the volume that produces power to the volume required to flow reactants in the system is low. This is shown pictorially in Figure 6 where the shaded region represents the volume where the electrical energy is generated and the dashed region represents the volume necessary to produce the electrical energy.

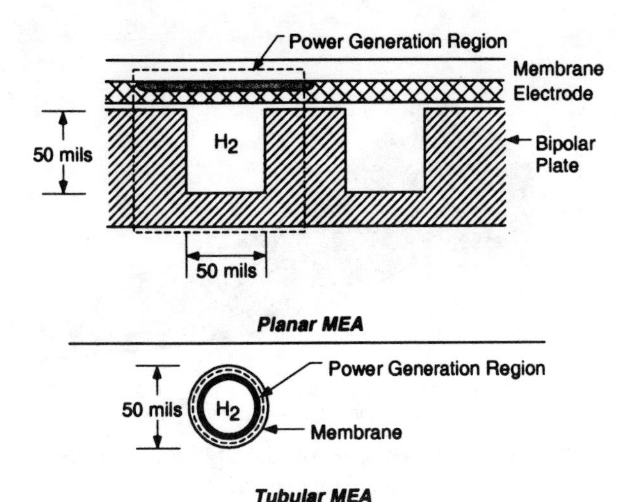

Figure 6. Pictorial schematic of the power generating regions in a fuel cell.

One of the advantages of the tubular MEAs is that the tubes may be connected in a common manifold in a combination of serial and parallel arrangements. This allows multiple tubes to be connected in parallel to build up current in a tube bundle. Multiple tube bundles may be connected in series to build up voltage in the stack. Additionally, the tubular fuel cell technology allows addressable arrays to be developed so that current and voltage may be readily switched. For instance, some electronic devices such as a radio or cellular phone have two distinct power requirements depending on whether the device is in the transmit or receive mode. Multiple electrical connections may be made to each tube in the fuel cell array where the electronic device or controller selects which tubes should be connected and in what arrangements, serial orparallel. As an example, connecting five of our micro-tubular MEAs in series will deliver 2.2 V at 0.45 Amps. If higher amperage is required at the same voltage, say for transmitting information, three more tubes may be connected in parallel by the controller to give 2.2 V at 2.45 Amps. This is shown in Figure 7. When the information is done being transmitted, the controller reverts back to the serial connection of five tubes. One of the main reasons we design the fuel cell in this manner is to help minimize the size of the fuel storage system, the major driver for mass and volume in the system. Thus, for the example discussed here and shown in Figure 7, adding three extra tubular MEAs adds a slight weight penalty to the fuel cell stack, but helps reduce the fuel storage system since the tubes may be designed to operate at high voltage efficiencies. The end result of this current collection approach is that the hydrogen gas is reacted under more efficient conditions thus minimizing the quantity of fuel stored in the system.

Fuel Cell Performance

In order to test the tubular MEAs, a new test apparatus was developed that allows single and multiple tube arrays to be electrochemically assessed. This apparatus, shown in Figure 8, consists of a chamber where either a single tube or an array of tubes may be mounted. Oxygen or air is introduced into a water chamber that is heated to generate the desired partial pressure of water. This humidified oxidant stream then travels through a slot flowing over the tubular MEAs.

Hydrogen gas is introduced into another water chamber that is heated to set up the desired water partial pressure. This gas stream is then introduced through the tubular MEAs using modified fittings to accommodate the tubular MEA array. The fuel gas may be closed at the exit to simulate dead-ended fuel delivery designs. An instrumented test system allows up to 14 tubes to be individually evaluated electrochemically.

A single tubular MEA with an outer diameter of 70 mils was investigated for its electrochemical performance operating in the fuel cell mode. Variants to the catalyst and electrode structure have been improving the performance of this single tubular MEA as shown in Figure 9. In this test, saturated oxygen and hydrogen and 70 °C at a cell temperature of 70 °C were used at approximately 15 to 20 psig. As shown, continuous improvements in the catalyst and electrode structures have been increasing the polarization

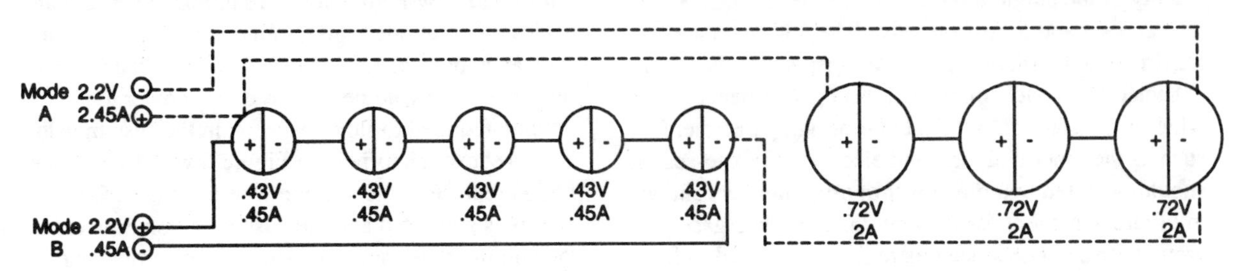

Figure 7. Addressable array voltage-current switching for the miniaturized tubular fuel cell.

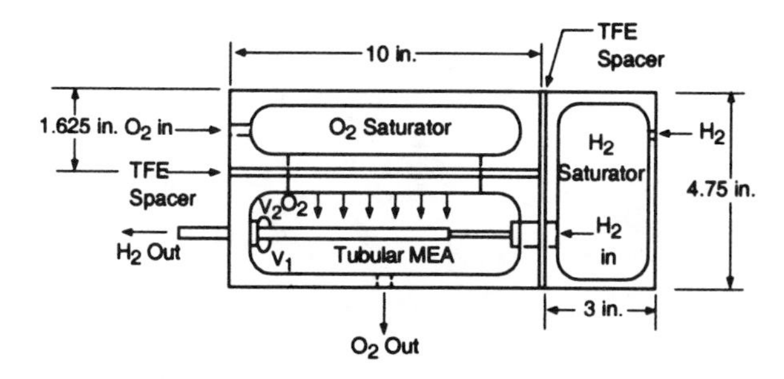

Figure 8. Schematic of the tubular MEA test fixture.

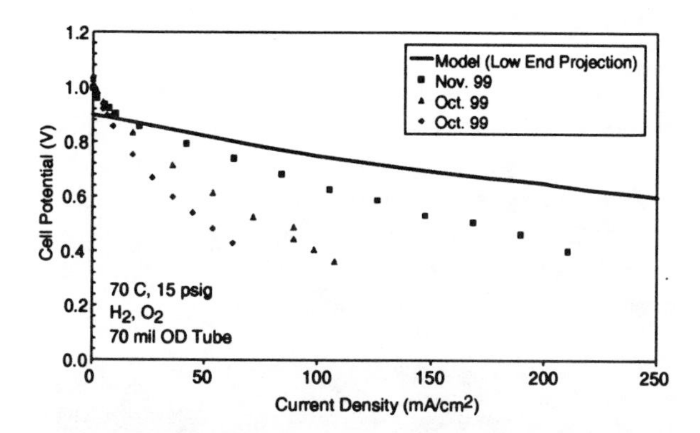

Figure 9. Single tube MEA polarization behavior operating in the fuel cell mode.

behavior toward our low-end goal defined by the model projection in Figure 9. A corresponding power density assessment is shown in Figure 10 for this data showing that the single tubular MEA is performing at 1.2 kW/liter. It should be noted that the volume used in this volumetric power density assessment includes the current collectors and a 10 mil thick oxidant region around the cathode.

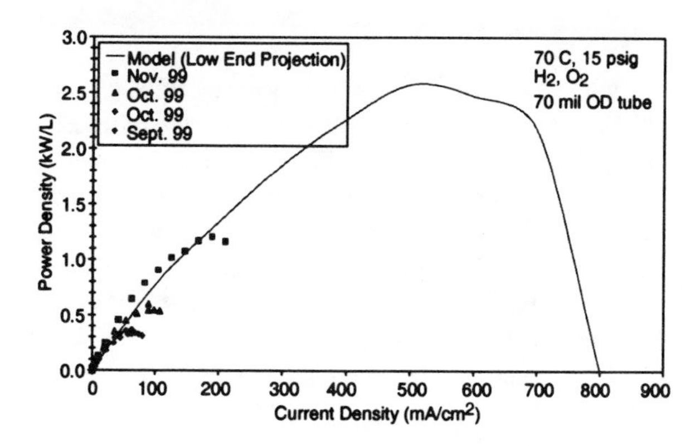

Figure 10. Power density of a single tubular MEA operating in the fuel cell mode.

Electrolysis Performance

One of the features being developed with the tubular MEA technology is the application of the tubes to both the fuel cell and electrolyzer modes. In this manner, the tubular MEAs may be operated in the electrolyzer mode to oxidize water into oxygen and hydrogen gas that are stored until later use as the oxidant and fuel in the fuel cell. Thus, we have investigated the use of these tubes in the electrolyzer mode.

Water was introduced into the chamber shown before in Figure 8 that holds the tubular MEA. A 5 cm long tubular MEA with an outer diameter of 110 mils was used for the test. Figure 11 shows the electrolyzer data for three methods of current collection. In the first method, the anode and cathode current collectors are located on opposite ends of the 5 cm long tube, termed the down-tube connection method. As shown in Figure 11, little current is extracted from the tube resulting in high polarization losses due to a low conductivity in the electrodes. In the second current collection method, the anode and cathode current collectors are located on the same end of the 5 cm long tube, termed the one-end connection. In this mode, a higher current is obtained with lower polarization losses. This result indicates that the low conductivity of the tubular MEA is allowing only a small region around the end-connection to be electrochemically active. This is supported by the third current collection method that jumpers both ends of the tubular MEA together so that current is extracted from each end. As shown in Figure 11, the polarization behavior is improved indicating that more of the tubular MEA is being used electrochemically.

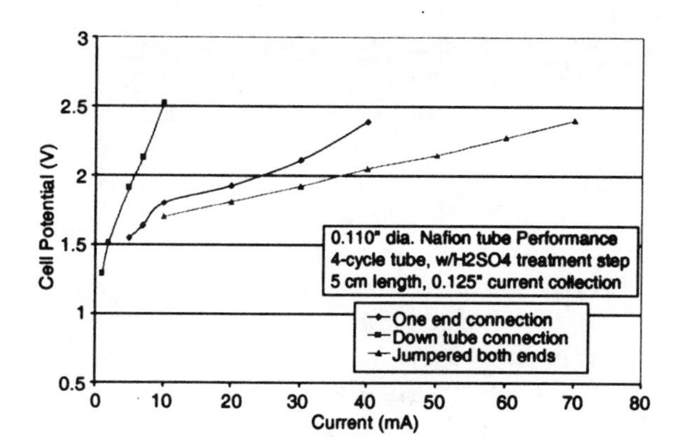

Figure 11. Electrolysis behavior on a tubular MEA at 70 °C.

The electrode conductivity issues discussed above are due to the manufacturing process used in making the tubular MEAs. Recent work has improved the electrode conductivity to about 4,600 S/cm.

Evaluation of Tubular MEA Arrays

An assessment of the performance of an array of tubular MEAs was conducted by fabricating a 14-tube array assembly, similar to that shown earlier in Figure 3. Our design goal is to place these tubular MEAs approximately 20 mils from each other (thus, each tube has an approximate 10 mil oxidant thickness about the cathode). In order to assess the flow characteristics of this approach, namely, is there sufficient flow to all outer surfaces in the tubular array, we modeled the oxidant flow over the tubular array structure.

The modeling of tubular micro fuel cells depends upon the flow surrounding the cells and the electrochemical reactions on the internal and external electrodes. The critical link between the fluid dynamics and the electrochemistry is provided by the gas exchange at the surface of the cylindrical fuel cell tubes. In essence, the tube flow is one that occurs over an effectively porous surface. This coupled flow problem is rather complex. As a first approximation to the external flow problem, the flow over the tubes may be considered as flow over a solid surface.

Flow past a single cylindrical tube can be rather complex depending upon the flow Reynolds number. Laminar wakes, unsteady phenomena and turbulence appear as the Reynolds number increases. These flow features become much more complicated when flow from neighboring tubes can interfere with the natural flow and provide additional nonlinear behavior. Fortunately, the Reynolds number based on tube diameter for our simulations is in the so-called low to moderate Reynolds number regime with values from about 1 to 10. This implies that laminar wakes may play a part in the computational flow assessment.

A general-purpose computer code *COMPACT-2D Version 4.0* was used for the analysis of flow past a 14-cylinder array as shown in Figure 12. The flow simulations necessitated very fine grid point distribution around and within the cylinder array. The flow progressed from a slot on the left side of the chamber and progressed to the right over the tube assemblage. When the spacing between the tubes was very narrow (e.g., 5 mils), the pressure drop through the array was so large that most of the flow went around the array. Such behavior is not desirable since it does not utilize the gas flow properly. As the spacing between the cylinders was increased to 20 mils, the flow behaved better and as shown in Figure 13, the entire tube array is exposed to a fairly uniform flow for the conditions of the simulation shown.

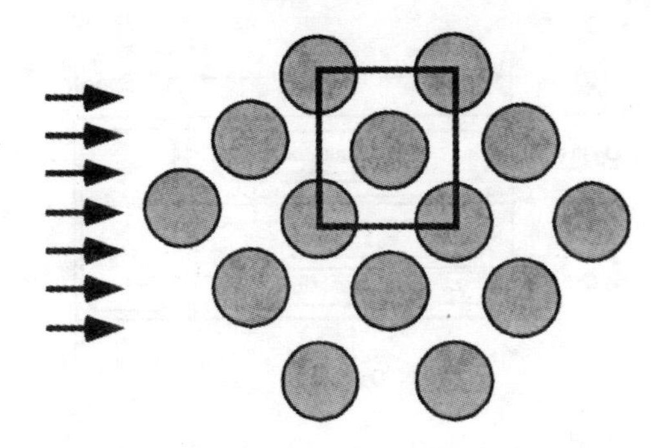

Figure 12. Staggered array pattern for modeling the flow over a 14-tube MEA array.

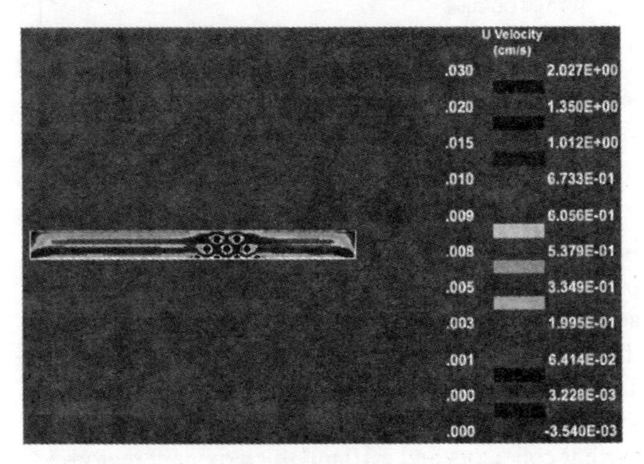

Figure 13. Velocity contours over the 14-tube MEA array.

Future flow modeling will couple the flow behavior to the electrochemistry and treat the tubes as porous. This will complicate the local flow over the tubes, but it is not anticipated to change the basic flow over the array very much. A feature that may complicate matters more is the presence of a reaction by-product: liquid water. Such two-phase effects may impact the operation of these micro fuel cells significantly and will be evaluated in subsequent modeling efforts.

An electrochemical assessment of the 14-tube array was conducted by measuring the polarization performance of each tube operating in the fuel cell mode. The oxidant, air, was introduced over the tube array at room temperature and pressure as shown in Figure 14.

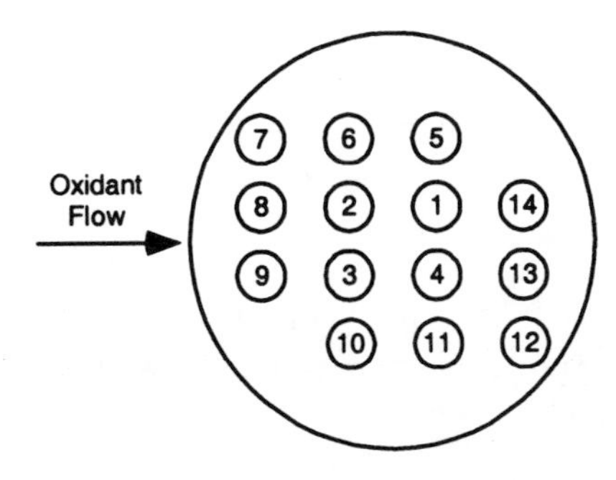

Figure 14. Numbering convention for the 14-tube array.

Applying a load to the MEAs, the tube potentials vary between 0.4 and 0.6 V as shown in Figure 15 except for tube number 14 which exhibited low performance most likely due to a defect in the electrode structure. An assessment of the potential profiles of each tube indicates no significant trends in the tube location and the resulting potential. This data supports our modeling efforts in obtaining sufficient flow to all tube surfaces.

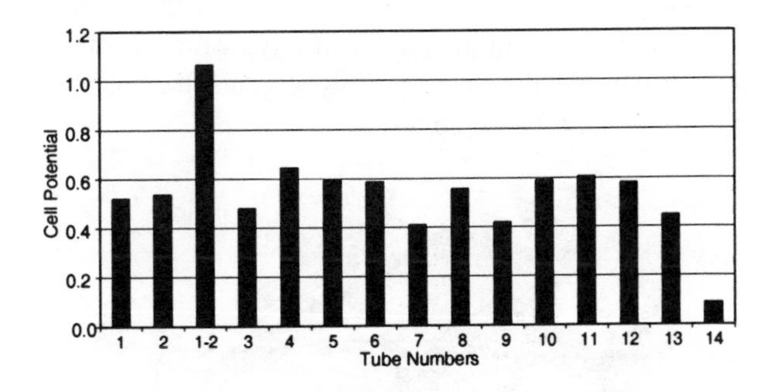

Figure 15. Tube potentials under load in the 14-tube array.

Summary

Design studies on a miniaturized tubular fuel cell and electrolyzer shows that high volumetric (3.5 to 8 kW/liter) and gravimetric (7 to 15 kW/kg) power densities are attainable at the tube level by reducing the size of the tubular MEAs to approximately 25 mils ID and by placing the tubular MEAs approximately 20 mils apart. Packaging these tubes in an array along with a housing would reduce these power density projections slightly, however, they would still be competitive in providing well over 1 kW/liter and 1 kW/kg stack power densities. Electrochemical performance testing of these tubular MEAs is showing approximately 1.2 kW/liter performance at the tube level operating in the fuel cell mode. Electrolysis testing has also been conducted in support of generating the reactant gases from water as a part of a regenerable energy storage system.

Mathematical modeling and experimental studies have been conducted on 14-tube MEA arrays indicating sufficient oxidant flow to all external tube surfaces. Electrochemical performance testing of these arrays show no signs of flow starvation to the external cathode surfaces. Present work is focusing on improving the catalyst and electrode structures to increase the polarization performance to our targeted level.

Acknowledgment

This material is based upon work supported by the Propulsion Directorate of the United States Air Force Research Laboratory, Wright-Patterson Air Force Base under Contract F33615-99-C-2902.

Contact

Correspondence should be sent to Dr. Michael C. Kimble, Physical Sciences Inc., 20 New England Business Center, Andover, MA 01810-1077; e-mail: kimble@psicorp.com; Telephone: (978) 689-0003; Fax: (978) 689-3232.

References

1. M.C. Kimble, E.B. Anderson, A.S. Woodman, and K.D. Jayne, "Regenerative Micro-Fuel Cells and Electrolyzers," Proceedings of the 34th Intersociety Energy Conversion Engineering Conference, Vancouver, British Columbia, August 1999

2. M.C. Kimble, E.B. Anderson, K.D. Jayne, and A.S. Woodman, "Micro-Tubular PEM Fuel Cells," Presentation at The Electrochemical Society Fall Meeting, Hawaii, October 1999.

3. M.C. Kimble, E.B. Anderson, A.S. Woodman, and K.D. Jayne, "Micro-Fuel Cells and Electrolyzers for UAVs," SAE Aerospace Power Systems Conference, #1999-01-1371, Pg. 75, Mesa, AZ, April 1999.

AIAA-2000-2877

Dynamic Test Results Of A Proposed Gphs Support System And Housing For Application In An Advanced Radioisotope Power System

S. Davis, A. Reynoso, R. Shaffer, G. Mintz, F. Norris
BWXT of Ohio
1 Mound Road
Miamisburg, OH 45343
Phone: (937) 865-4436; Fax: (937) 865-3485
E-mail: Davisb@doe-md.gov

V. Kumar, R. Carpenter, A. Schock
Orbital Sciences Corporation
20301 Century Blvd.
Germantown, MD 20874
Phone: (301) 428-6173; Fax: (301) 353-8619
E-mail: kumar@fsd.com

Abstract

This paper presents a detailed description of Orbital Science Corporation's (OSC) design of a new support system for the General Purpose Heat Source (GPHS) using the aeroshell's existing indentations for interlocking the heat source modules. The housing and support system were recently fabricated and tested to the Jet Propulsion Laboratory (JPL) draft Advanced Radioisotope Power System (ARPS) test requirements at the the Department of Energy's (DoE) Mound facility in Miamisburg, Ohio. The paper presents the test environments and results. This paper also illustrates some of the latest ARPS system design concepts using this specific support system and housing for future deep space missions.

Introduction

In previously launched radioisotope thermoelectric power systems, large numbers of stacked GPHS modules were used to produce kilowatts of thermal energy. Four corner studs and load spreader plates supported these large GPHS stacks on their ends. Heat loss through these end supports was not significant considering the number of modules used and the length of those stacks. Current ARPS designs are smaller power sources compared to the current generation of Radioisotope Thermoelectric Generators (RTG) and use only 2-4 GPHS modules per generator. With a limited number of modules, one criteria that should be considered in the design is the minimization of thermal loss through the end supports to increase the thermal flux, and at the same time provide rigid locking against separation and rotation of the shorter stack. Any new supporting concept should not require any changes in the safety-qualified GPHS modules. Taking all these points into consideration, OSC conceived a new design for housing and supporting the GPHS stack. The support system was initially envisioned for use with the Alkali Metal Thermal to Electric Conversion (AMTEC[1]) power system and later with a Stirling[2] generator power system.

Design Description

GPHS module's aeroshell contains two indentations on each side of its two broadfaces. These indentations had been designed to mate with graphite lock members, which served to hold the RTG's stacked heat source modules together under transverse launch loads. The OSC heat source support assembly design utilizes these indentations to support the GPHS modules off the generator's housing wall by means of four support sub-assemblies. The heat source support mechanism must be strong enough to resist compressive and shear loads during launch while minimizing conductive losses between the hot heat source and the cool housing.

Figure 1 shows a cutaway view of a typical heat source assembly in the aluminum prototype generator housing, with four support assemblies.

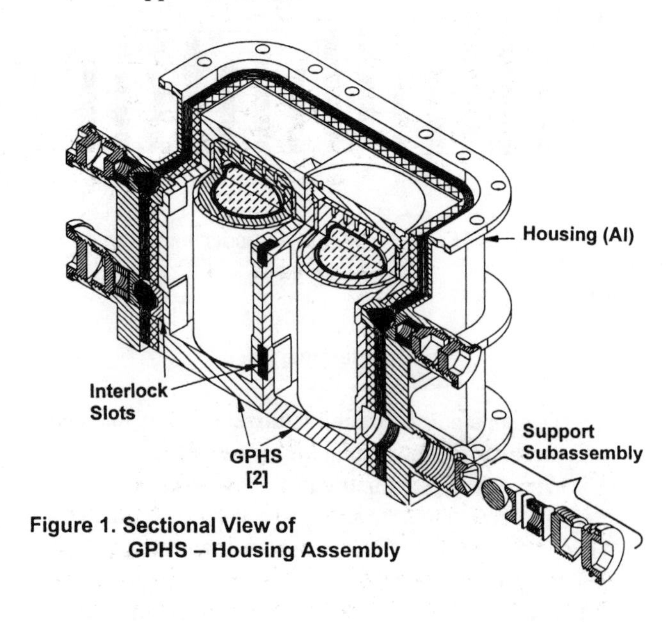

Figure 1. Sectional View of GPHS – Housing Assembly

Figure 2 shows an exploded view of components of one such support assembly, which includes the elements for sealing the heat source from the outside environment. The subassembly contains an 11.43 mm (0.45 inch) diameter zirconia thermal insulation sphere in contact with hot and cold graphite lock members. The lock member element's inner surfaces, where the zirconia ball makes contact, are pyramidal shaped to minimize heat losses. The pyramidal shaped inner surface provides four-point contact between lock member and ball and yields lower heat losses than would a circular ring contact provided by a lock member with conical seats. The use of spherical balls enables them to provide heat source support independent of any changes in their orientation caused by launch vibration.

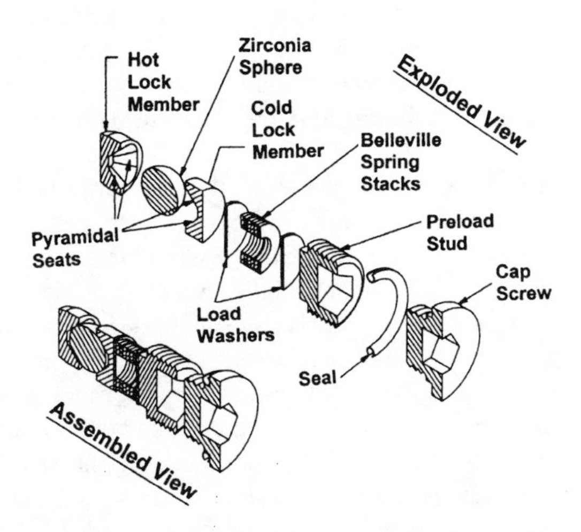

Figure 2. Exploded View of GPHS Support Assembly

During fueling of the heat source assembly, the support subassemblies are to be partially withdrawn to permit insertion of two GPHS modules. After final assembly, the cold lock member and zirconia balls are pre-loaded by nested belleville springs and threaded preload stud. The compressive force exerted by the spring stack serves to keep the hot lock member seated in the aeroshell indentations under transverse and axial launch loads. A seal ring and cap screw seal the support port.

The compressive and shear loads on the zirconia support balls computed under the assumed quasi-static 40-G axial and transverse launch loads, with two GPHS modules, which weigh 1.45 kg (3.2 lb.) each, produces an inertial load of 1136 Newton (255 lb_f) on the heat source supports. This would result in a compressive load of 568 N (128 lb_f) on each of two opposite zirconia balls or a shear load of 284 N (64 lb_f) on each of four balls.

Test Activities

The focus of the initial test was the mechanical integrity of the heat source support system under the launch load environment at ambient temperature. For this test, the OSC prototype generator housing was modified with design input from BWXT of Ohio (Mound) Power Systems Technologies (PST) personnel to facilitate ease of fabrication and attachment to support fixturing. The support system was also simplified to reduce cost and assembly time by eliminating thermal insulation and sealing components. The GPHS modules were simulated with a POCO graphite block with two stainless steel tube inserts, keeping the overall geometry and mass unaltered. The generator housing and load-path components were fabricated by Mound PST personnel in the PST machining facility. Mound personnel also fabricated the two GPHS module simulants.

Figure 3 shows a sectional view of the simplified heat source support assembly, fabricated and tested at Mound. The aluminum housing accommodates two simulated GPHS graphite modules and receives four support sub- assemblies. Each support assembly comprises one zirconia sphere, two lock members with pyramidal inner surfaces to seat the sphere, spring retainer, series-parallel belleville spring stacks, load washer and preload stud.

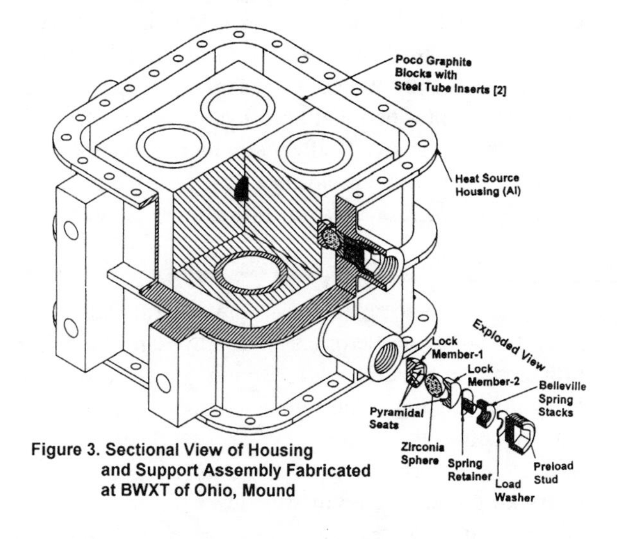

Figure 3. Sectional View of Housing and Support Assembly Fabricated at BWXT of Ohio, Mound

The tests were performed utilizing Mound's PST vibration test facility, comprised primarily of a Ling Dynamic Systems V980 shaker system and a m+p VCP9000 test controller. The test environments were derived from Cassini Flight Acceptance test requirements and Jet Propulsion Laboratory draft specification document JPL D-16118, <u>OP/SP-ARPS</u>

Environmental Design and Test Requirements. The Sine and Random vibration tests utilized averaged control (averaging the four accelerometer channels on the mounting plate) adjacent to the housing attachment point. To limit response of any of the data channels, real-time force notching was implemented during the performance of the Random tests as outlined in draft document JPL D-16118, OP/SP-ARPS Environmental Design and Test Requirements. The Sine-Transient and Pyroshock tests utilized single-point control adjacent to the housing attachment point. For further definition of the tests performed, reference Mound procedure OPA990025, OSC Generator Vibration Test, which outlines the test requirements.

The GPHS module simulants were inserted into the generator housing and secured in-place with the spring support assembly components. The modules were secured in-place with aluminum corner shims to center the modules in the generator housing so the load could be applied equally with the pre-load studs. To set the force loading on the load-path spring supports the pre-load studs were screwed into the threaded channels from a no-load position until the spring washers were compressed 0.813 mm (0.032 inch) to obtain the prescribed preload. This yielded a pre-load of approximately 568 N ($128 lb_f$) applied to each pre-load stud. The pre-load value was derived from expected vibration and shock test loads. An initial set of measurements were taken with depth micrometers to obtain the total displaced position (depth) of the pre-load studs so that later measurements could be made during testing activities to ascertain if any spring compression/pre-load had been lost. Corner blocks and shims were removed after all four supports were installed to the same preload. A sketch of the generator housing showing test axes orientation is shown in Figure 4.

The GPHS module simulants were instrumented with three miniature accelerometers (one in each of the three orthogonal axes). The total weight of the assembled generator housing, with simulated modules and load-path components, was
4.07 kg (8.96 lb.). The generator housing assembly was then transported to the vibration test cell, mated to a set of force transducers and vibration plate, and then attached to the vibration system slip-table, configured for X axis testing

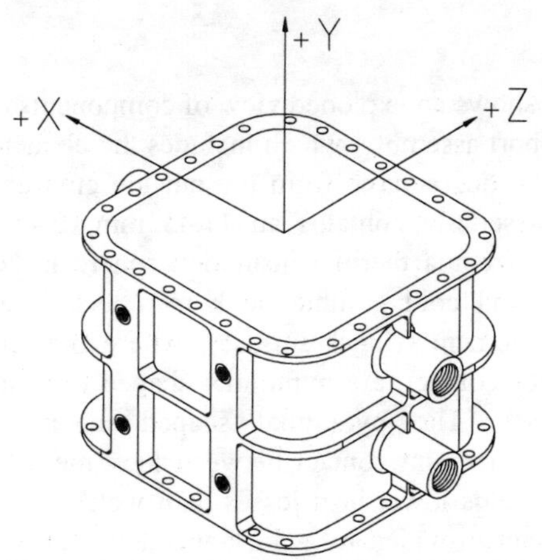

Figure 4. Generator Housing Axes Designation

Prior to the performance of the dynamic tests, an instrumentation and equipment verification was conducted to ensure that the instrumentation's measuring accuracy conformed to the operating tolerances as outlined in the test control document.

A series of vibration and shock tests, performed in each of the housing's three orthogonal axes, consisted of a pre-axis 1/2 Gpeak Sine Sweep test, 7.23 Grms Random test, four Flight Acceptance level Sine-Transient tests, a post-axis 1/2 Gpeak Sine Sweep test, and a 500 G SRS Pyroshock test (three pulses at full level). The Pyroshock tests were performed after all of the other vibration tests were completed for all three axes. In addition, a 10 Gpeak Protoflight Sine Sweep test was performed for the X axis, only, because housing components exhibited a resonance below 100 Hertz. The reference Protoflight Sine Test, Random Test, Sine-Transient Test, and Pyroshock Test profiles are depicted in Figures 5 through 8, below:

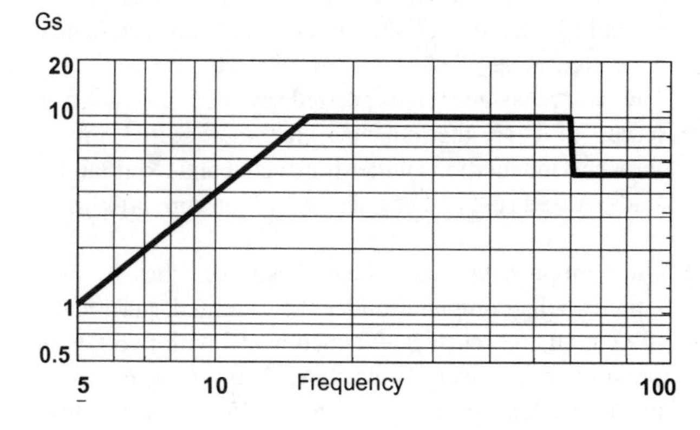

Figure 5: SINE VIBRATION PROTOFLIGHT TEST

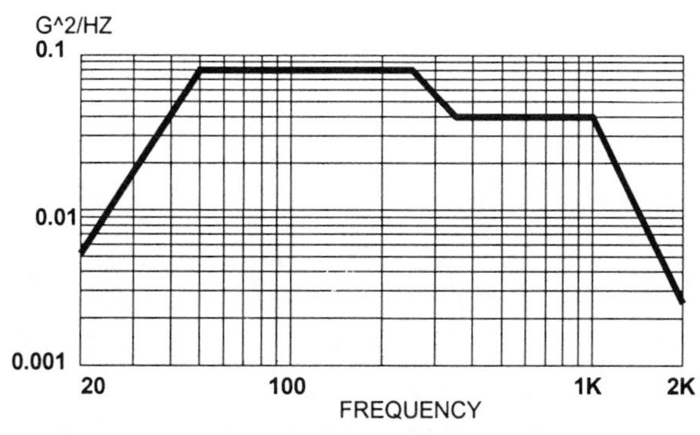

Figure 6: RANDOM VIBRATION PROTOFLIGHT TEST

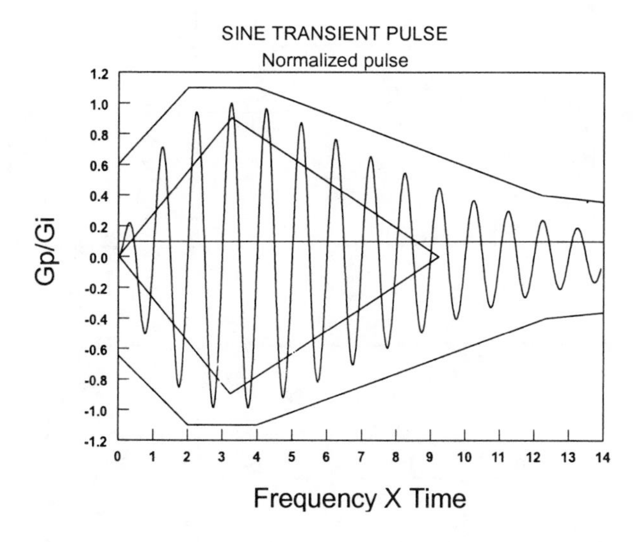

Figure 7: NORMALIZED SINE TRANSIENT PULSE

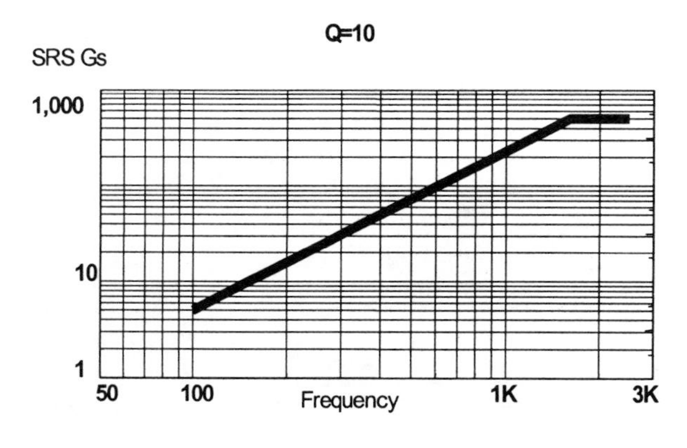

Figure 8: SRS PyroSHOCK TEST

The X, Y, and Z axes vibration tests were performed per operating procedure, OPA990025. After completion of the test series for each axis, one of the Y axis housing end-caps was removed and a visual inspection of the inner housing was performed to ascertain the condition of the simulated heat sources and load-path components. Also, a set of measurements were taken with depth micrometers to compare the depth position of the load-path pre-load studs with initial measurements made during component assembly and to ascertain if any spring compression/pre-load had been lost. The accelerometer and force transducer channel assignments and the resultant values for the response channels for the vibration tests are summarized below:

The accelerometer and force transducer control/data channel assignments are as follows:

Channel 1	Accelerometer	Control (Test axis)	Vibration plate
Channel 2	Accelerometer	Control (Test axis)	Vibration plate
Channel 3	Accelerometer	Control (Test axis)	Vibration plate
Channel 4	Accelerometer	Control (Test axis)	Vibration plate
Channel 5	Accelerometer	X axis	Housing sidewall
Channel 6	Accelerometer	Y axis	Housing sidewall
Channel 7	Accelerometer	Z axis	Housing sidewall
Channel 8	Accelerometer	X axis	Module face
Channel 9	Accelerometer	Y axis	Module face
Channel 10	Accelerometer	Z axis	Module face
Channel 11	Force Transducer	Test axis	Vibration plate
Channel 12	Force Transducer	Test axis	Vibration plate
Channel 13	Force Transducer	Test axis	Vibration plate
Channel 14	Force Transducer	Test axis	Vibration plate
Channel 15	Sum of Forces	Test axis	Vibration plate

X AXIS SINE TEST (Pre Axis)						
	Chan 5	Chan 6	Chan 7	Chan 8	Chan 9	Chan 10
Highest Response Amplitude	4.89 Gp	3.13 Gp	4.55 Gp	6.83 Gp	0.62 Gp	0.56 Gp
Response Frequency	251.95 Hz	269.50 Hz	1465.8 Hz	251.95 Hz	210.2 Hz	389.3 Hz

X AXIS RANDOM TEST						
	Chan 5	Chan 6	Chan 7	Chan 8	Chan 9	Chan 10
Highest Response Amplitude	0.91 G^2/Hz	nil	0.38 G^2/Hz	2.0 G^2/Hz	nil	nil
Response Frequency	224 Hz	nil	1436 Hz	224 Hz	nil	nil
Response Grms	10.82 Grms	3.31 Grms	6.96 Grms	12.67 Grms	0.46 Grms	1.13 Grms

X AXIS SINE-TRANSIENT TESTS							
Frequency/ Amplitude	Control Gp	Chan 5 Gp	Chan 6 Gp	Chan 7 Gp	Chan 8 Gp	Chan 9 Gp	Chan 10 Gp
15.75 Hz. 3.4 Gp	3.37	3.45	~0.05	0.20	3.58	0.24	~0.10
31.5 Hz 3.4 Gp	3.37	3.55	~0.10	0.15	3.76	0.11	0.31
50.0 Hz 3.4 Gp	3.37	4.15	~0.10	0.15	4.63	~0.10	~0.20
79.73 Hz 3.4 Gp	3.43	4.56	~0.10	0.15	5.12	~0.05	~0.10

X AXIS SINE TEST (Post Axis)						
	Chan 5	Chan 6	Chan 7	Chan 8	Chan 9	Chan 10
Highest Response Amplitude	5.47 Gp	1.36 Gp	4.51 Gp	8.14 Gp	0.58 Gp	.65 Gp
Response Frequency	240.47 Hz	1481 Hz	1488.7 Hz	240.47 Hz	206.9 Hz	216.8 Hz

Y AXIS SINE TEST (Pre Axis)						
	Chan 5	Chan 6	Chan 7	Chan 8	Chan 9	Chan 10
Highest Response Amplitude	1.44 Gp	14.3 Gp	6.1 Gp	~0.43 Gp	5.50 Gp	0.62 Gp
Response Frequency	1488.7 Hz	873 Hz	1488.7 Hz	400 Hz	186.6 Hz	200 Hz

Y AXIS RANDOM TEST						
	Chan 5	Chan 6	Chan 7	Chan 8	Chan 9	Chan 10
Highest Response Amplitude	nil	5.04 G^2/Hz	0.60 G^2/Hz	nil	0.82 G^2/Hz	nil
Response Frequency	nil	664 Hz	1428 Hz	nil	96 Hz	nil
Response Grms	2.97 Grms	21.84 Grms	7.53 Grms	1.03 Grms	7.69 Grms	1.40 Grms

Y AXIS SINE-TRANSIENT TESTS							
Frequency/ Amplitude	Control Gp	Chan 5 Gp	Chan 6 Gp	Chan 7 Gp	Chan 8 Gp	Chan 9 Gp	Chan 10 Gp
15.75 Hz. 3.4 Gp	3.35	~0.05	3.51	0.23	~0.10	4.50	~0.10
31.5 Hz 3.4 Gp	3.58	~0.05	3.69	0.21	~0.10	5.46	~0.05
50.0 Hz 3.4 Gp	3.50	~0.10	3.66	0.20	~0.25	5.43	~0.10
79.73 Hz 3.4 Gp	3.40	~0.20	4.38	~0.30	~0.80	6.42	~0.20

Y AXIS SINE TEST (Post Axis)						
	Chan 5	Chan 6	Chan 7	Chan 8	Chan 9	Chan 10
Highest Response Amplitude	1.63 Gp	19.7 Gp	6.80 Gp	~0.41 Gp	3.77 Gp	0.79 Gp
Response Frequency	1488.7 Hz	868.8 Hz	1488.7 Hz	280 Hz	157.3 Hz	220.2 Hz

Z AXIS SINE TEST (Pre Axis)						
	Chan 5	Chan 6	Chan 7	Chan 8	Chan 9	Chan 10
Highest Response Amplitude	7.01 Gp	4.83 Gp	28.91 Gp	0.94 Gp	0.95 Gp	4.96 Gp
Response Frequency	1450.7 Hz	1458.2 Hz	1450.7 Hz	253.3 Hz	189.5 Hz	197.5 Hz

Z AXIS RANDOM TEST						
	Chan 5	Chan 6	Chan 7	Chan 8	Chan 9	Chan 10
Highest Response Amplitude	0.13 G^2/Hz	0.90 G^2/Hz	1.06 G^2/Hz	nil	nil	0.45 G^2/Hz
Response Frequency	1512 Hz	1088 Hz	1504 Hz	nil	nil	104 Hz
Response Grms	6.77 Grms	6.24 Grms	21.70 Grms	1.05 Grms	0.85 Grms	6.47 Grms

Z AXIS SINE-TRANSIENT TESTS							
Frequency/ Amplitude	Control Gp	Chan 5 Gp	Chan 6 Gp	Chan 7 Gp	Chan 8 Gp	Chan 9 Gp	Chan 10 Gp
15.75 Hz. 3.4 Gp	3.45	~0.05	~0.05	3.40	~0.10	~0.50	3.87
25.0 Hz 3.4 Gp	3.51	~0.10	~0.05	3.53	~0.10	~0.50	3.92
63.0 Hz 3.4 Gp	3.41	~0.20	~0.20	3.50	~0.50	~0.50	4.42
100 Hz 3.4 Gp	3.52	~0.60	1.69	3.54	0.82	1.23	6.47

Z AXIS SINE TEST (Post Axis)						
	Chan 5	Chan 6	Chan 7	Chan 8	Chan 9	Chan 10
Highest Response Amplitude	8.86 Gp	4.31 Gp	34.66 Gp	0.92 Gp	1.72 Gp	3.24 Gp
Response Frequency	1435.7 Hz	1473.4 Hz	1435.7 Hz	250.6 Hz	190.5 Hz	177.2 Hz

The successful completion of the post-axis Sine Sweep test concluded the vibration portion of the testing activities.

A set of measurements were taken with depth micrometers to compare the depth position of the load-path pre-load studs with initial measurements made during component assembly and to ascertain if any spring tension/pre-load had been lost. The measurements taken agreed with the initial measurements taken during assembly. Also, a Y axis housing end-cap was removed and a visual inspection of the inner housing was performed to ascertain the condition of the simulated heat sources and load-path components. The modules and load-path components appeared to be in proper alignment. A very light coating of graphite fines was on the inner walls of the housing (mainly in the corners of the inner-housing) with slightly heavier deposits directly below the module/load-path graphite locking members. No other anomalous conditions were noted. After the inspection, the end-cap was re-secured to the housing.

The Z, X, and Y axes 500 G SRS (Shock Response Spectrum) Pyroshock test was performed next. Three pulses were performed at the full-test level. Each of the full-level pulses was performed within control tolerances. Again, after each test axis one of the Y axis housing end-caps was removed and a visual inspection of the inner housing was performed to ascertain the condition of the simulated heat sources and load-path components. The resultant values for the control (channel 1) and response channels for each Shock test are summarized as follows:

Z AXIS PYROSHOCK TEST							
	Control Gp	Chan 5 Gp	Chan 6 Gp	Chan 7 Gp	Chan 8 Gp	Chan 9 Gp	Chan 10 Gp
Pulse #1 502.5 GSRS	131.2	771.9	385.1	1289.3	35.23	34.4	19.1
Pulse #2 503.5 GSRS	131.4	775.5	385.9	1289.9	34.5	34.4	19.1
Pulse #3 504.5 GSRS	131.3	777.4	381.6	1290.5	34.8	34.4	19.1

X AXIS PYROSHOCK TEST							
	Control Gp	Chan 5 Gp	Chan 6 Gp	Chan 7 Gp	Chan 8 Gp	Chan 9 Gp	Chan 10 Gp
Pulse #1 510.1 GSRS	134.8	No Data	52.7	182.1	17.8	8.0	6.2
Pulse #2 510.0 GSRS	134.7	No Data	54.5	186.1	18.3	7.6	6.4
Pulse #3 511.0 GSRS	134.8	No Data	54.4	187.4	18.1	7.9	6.7

Y AXIS PYROSHOCK TEST							
	Control Gp	Chan 5 Gp	Chan 6 Gp	Chan 7 Gp	Chan 8 Gp	Chan 9 Gp	Chan 10 Gp
Pulse #1 510.9 GSRS	134.6	112.7	565.2	337.3	9.0	12.4	7.9
Pulse #2 509.6 GSRS	135.2	114.5	566.7	338.2	8.8	11.9	7.4
Pulse #3 509.1 GSRS	135.3	118.9	566.9	336.0	9.1	12.3	7.7

The generator housing was removed from the attachment fixturing for disassembly of the generator housing. Both end caps were removed and the interior of the housing examined. Small deposits of graphite fines were noted in the corners of the -Y axis housing end-cap. Also, as seen previously, a very light coating of graphite fines were on the inner walls of the housing (mainly in the corners of the inner-housing) with slightly heavier deposits directly below the module/load-path graphite locking members. No chipping, cracking, or abrasions of the module graphite were noted.

Some difficulty was encountered while attempting to insert the module corner shims (for removal of the spring loads) between the modules and the housing corners and it was noted that the modules were slightly shifted to the -Z direction. Prior to removal of the pre-load studs, a set of depth measurements was taken with depth micrometers for comparison with the initial measurements made during component assembly. No change in position of the pre-load studs was noted as the measurements taken agreed with previous depth measurements.

Both sets of load path components were removed from the generator housing and were examined for wear/abrasion. The fine weave pierced fabric (FWPF) graphite lock-members/sphere-seats, which interfaced with the zirconium spheres, were burnished (indicating wear) at their edges where they seated into the module lock-member depressions. The lock-members also had burnished spots where the zirconium spheres seated against the lock-members. The FWPF graphite sphere-seats also had burnished wear spots at the sphere interface locations. In addition, the sphere-seats each had a wear line burnished around their circumference where they had rubbed against the wall of the load-component channels of the housing. The rubbing of the sphere-seat against the channel wall resulted in the depositing of a thin layer of graphite on the channel wall. The zirconium spheres exhibited indications that they were rotating while seated between the lock-member and sphere-seat as there were swirled graphite deposits encompassing the spheres. The load washers and pre-load studs showed slight signs of wear where they interfaced. The springs were checked for indications of wear (such as cracking and flattening) but no such indications were found.

The module corner shims were removed from the housing and the housing was lifted away from the simulated modules. The POCO graphite modules were inspected for indications of wear (chipping, cracking, and abrasion) however the only wear indications found were where the load-path lock-members seated against the module lock-member depressions. The bottom surface of each lock-member depression was burnished around its circumference due to wear. Other than the above mentioned burnishing of the graphite parts, no damage or stress cracking of the modules or load path components was noted.

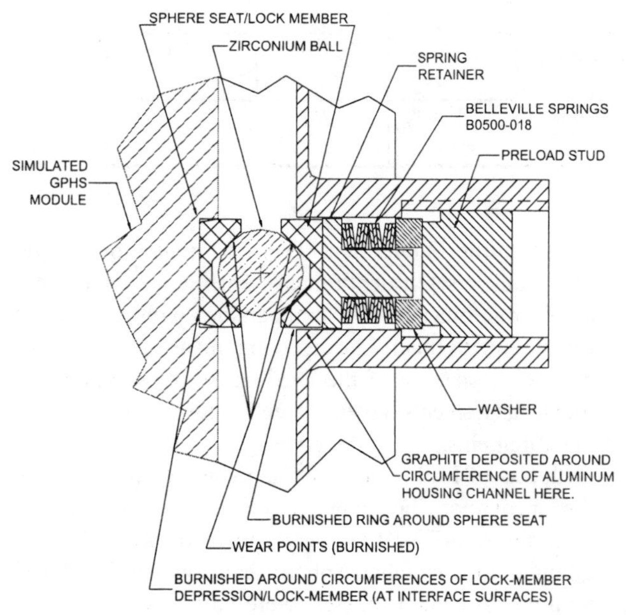

Figure 9. Load-Path Component Post-Test Condition

Figure 9, above, shows the various component wear points identified during disassembly of the generator housing.

Future Applications

Figure 10 shows an earlier design of OSC's AMTEC generator (with 'Chimney' cells), which was conceptualized for use on the six year Europa mission, using the tested heat source support system and housing. This compact generator is anticipated to produce an estimated electrical power output of 106 watts with 21.8% generator efficiency. The generator is expected to weigh about 9kg (19.8 lb.). The Europa mission power requirements are currently estimated to be 325 watts at End Of Mission (EOM) and it is anticipated that a four-generator system will meet this power requirement.

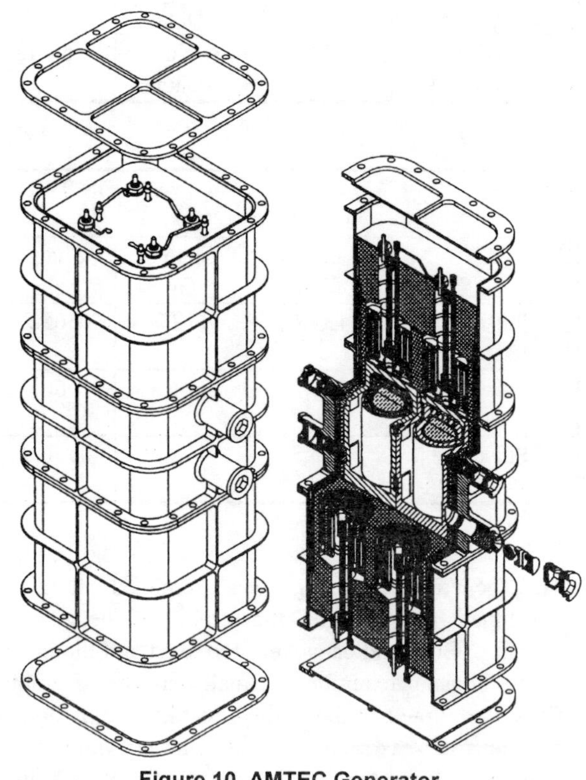

Figure 10. AMTEC Generator

An alternate design, utilizing a Stirling generator with a similar heat source support system, is shown in Figure 11. The heat source housing and two Stirling converters, depicted in Figure 11, are independently mounted onto the radiator panel and then to the spacecraft. Estimated electrical power output of this generator configuration at the end of the six year Europa mission is around 121 watts with a theorized 26% generator system efficiency. Three of these generators will meet the current power requirement of 325 watts at EOM with a comfortable margin. This Stirling generator weighs about 16.1kg (35.4 lb.). In this configuration, the heat source housing not only supports the GPHS but also provides a rigid mechanical coupling through its strengthening ribs to cancel out the vibration produced by the opposing Stirling converters.

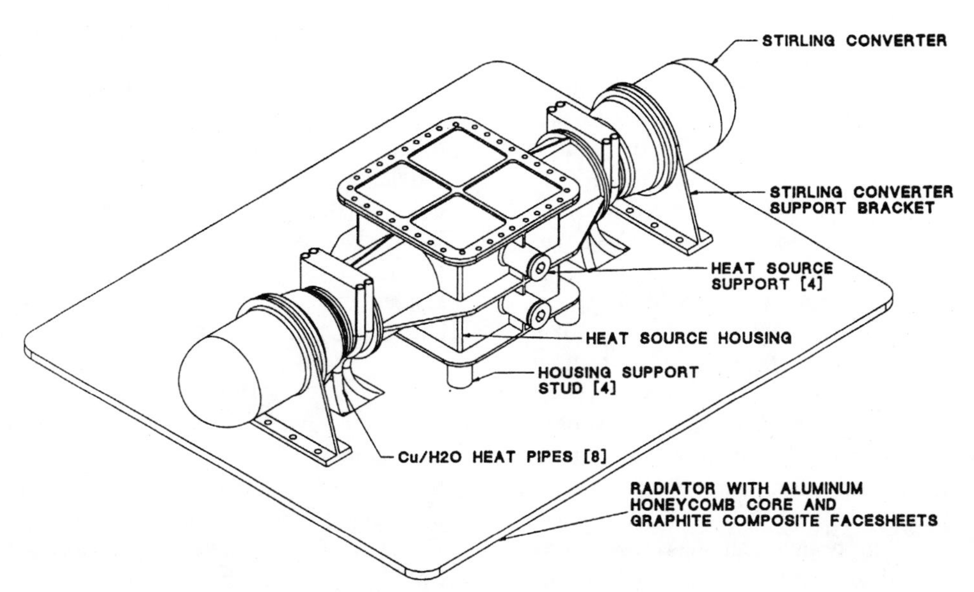

Figure 11. Radioisotope Stirling Generator

References

1. A. Schock, V.Kumar, H. Noravian, C. Or "Recommended OSC Design and Analysis of AMTEC Power System for Outer-Planet Mission," in Space Technology and Applications International Forum 1999, edited by M.S. El-Genk, AIP Conference Proceedings 458, American Institute of Physics, New York, 1999b, pp. 1534 - 1553

2. C. Or, V. Kumar, R. Carpenter, and A. Schock "Self-Supporting Radioisotope Generators With STC55W Stirling Converters," in Space Technology and Applications International Forum 2000, edited by M.S. El-Genk, AIP Conference Proceedings 504, American Institute of Physics, New York, pp. 1242-1251

ADVANCED RADIOISOTOPE POWER SYSTEMS REQUIREMENTS FOR POTENTIAL DEEP SPACE MISSIONS

Jack F. Mondt
Jet Propulsion Laboratory
4800 Oak Grove Drive
Pasadena, CA 91109-8099
jack.mondt@jpl.nasa.gov

ABSTRACT

This paper describes future mission power requirements for missions being investigated to accomplish NASA's Solar System Exploration (SSE) theme. Many potential missions to the outer planets (Jupiter and beyond) require advanced radioisotope power systems (RPS). Technology development for Advanced RPSs is being carried out as apart of NASA's Solar System Exploration Technology Program. The power levels for future planetary deep space science missions are broken down into the following four classes. Milliwatt-class (40 to 100milliwatt) that could provide both thermal and electric power for Mars weather stations. One to two-watt class for small in situ surface science laboratories and for aerobot atmospheric science laboratories for bodies in the solar system. Ten to twenty-watt class for micro satellites in orbit, surface science stations and aerobots. One hundred to three hundred watt-class to power orbiter science spacecraft, to power drills to obtain core samples of outer planet bodies, for powering subsurface hydrobots and cryobots on accessible bodies and for data handling and communicating data from small orbiters, surface laboratories, aerobots and hydrobots back to Earth. Advanced RPSs in the one hundred-watt class are also favored over solar power for obtaining comet samples on extended-duration comet missions.

INTRODUCTION

A brief description of each mission concept and a summary of the corresponding potential RPS technology needs are presented and discussed. The NASA Solar System Exploration, (SSE), theme is divided as follows: 1) Exploring Organic Rich Environments, 2) To Build Planet... 3) Bring Mars to Earth... 4) Robotic Outposts, and 5) Exploration of the far Outer Solar System the Kuiper Belt and Beyond. Studies of mission concepts to satisfy the science objectives of the SSE theme are being carried out at JPL to assess feasibility and to investigate the potential benefits of advanced technology. Science objectives and priorities for the studies were obtained from the SSE Roadmap by Elachi, et al 1996 and from the SSES Campaign Strategy Working Groups. Results of these studies are reflected in this paper. The desired power levels, efficiency, mass and voltage projected for missions that are enabled or strongly enhanced by an advanced RPS are described. These results should be considered preliminary since further studies including additional missions are currently under way.

OBJECTIVES

The objectives of Exploring Organic-Rich Environment are; 1) Inventory the organic materials found in diverse planetary environments across the outer solar system, 2) Compare these materials with those on which Earthly life is based, 3) Explore these environments to search for insights into the nature and variation of prebiotic chemistry. The present/planned missions for exploring organic-rich environments are Galileo New Millennium, Cassini/Huygens, Europa Orbiter and Pluto/Kuiper Express. The Near-Future missions are Europa Lander, Titan Organic Explorer and Neptune Orbiter with Triton Flybys. Example Farther-Future missions are Triton Lander, Europa and Titan Sample Return and Kuiper Belt Explorer.

The objectives of To Build a Planet... are; 1) analyze the building blocks of which planets are made, 2) observe the dynamics processes involved in forming planets and planetary systems and 3) study the diverse outcomes of planet formation. The Present/Planned missions to begin to meet these objectives are NEAR, Lunar Prospector, Stardust, Deep Impact and CONTOUR. The Near-Future missions are Comet Nucleus Sample Return, Saturn Ring Observer, Venus Sample Return. Examples of Farther-Future missions are Asteroid Sample Return, Lunar Giant Basin Sample Return,

Mars's Seismic Network, Multiplanet Probes and Mercury in situ Explorer.

The objectives of Bring Mars to Earth... are; 1) determine the biological history and potential of Mars, 2) understand the evolutionary history and resource inventory of Mars, 3) establish a continuous robotic presence and prepare for human exploration, and 4) engage the public in the challenge and excitement of Mars exploration. The Present/Planned missions are Mars Global Surveyor, Mars Rover 2003 or Mars Orbiter 2003, Mars Sample Return, Mars Micro-missions and Mars Telecom Satellites. The Near-Future missions are subsurface sampling, in-situ analysis of sites of biological interest, coordinated exploration networks, global telecom/navigation networks. Examples of Far-Future missions are self-sustaining interactive networks, seamless Earth-Mars Internet and Human-assisted laboratories.

Robotic Outpost is a new paradigm of exploration. The objectives of the paradigm are exploration of surface and subsurface via long-life robots, surface dynamics and tides subsurface search for prebiotic/biotic signature and interactive studies among spaceborne, surface and subsurface sensors. Europa may one day become a target for a comprehensive "Mars-like exploration program".

Exploration of the outer solar system, i.e. (the Kuiper Belt and InterStellar Probe), is emerging as an important step toward a complete understanding of the evolution of our solar system and the development of life. Studying these types of missions will focus far-term mission and technology planning. The required technology capabilities for Interstellar Exploration missions will greatly enhance near-term solar system missions, i.e. solar sail, optical communications, autonomy, radioisotope and reactor power systems, structures, propulsion, etc.

The Near-Future and Farther-Future missions that are potential users of radioisotope power sources are listed and described below.

Solar System Exploration Mission Set that Defines RPS Requirements

Near-Future Missions:

- Comet Nucleus Sample Return
- Mars Weather/Seismic Stations
- Europa Lander, including Europa subsurface concepts
- Titan Organic Explorer
- Neptune Orbiter with Triton Flyby
- Saturn Ring Observer
- Mars Robotic Outpost and Future Sample Returns

Farther-Future Mission Examples:

- Advanced Outposts
- Robotic Outposts
- Kuiper Belt Explorer
- InterStellar Probe
- Europa Ocean Science Station
- Sample all Accessible Bodies

Comet Nucleus Sample Return

A Comet Nucleus Sample Return (CNSR) mission would obtain 200 gram of the comet nucleus taken from multiple sampling sites - using a subsurface sampling apparatus such as a drill or a tethered penetrator. Challenging science goals for the mission include deep drilling (10 to 100m's) and obtaining samples from multiple sites. A mother ship would return the samples to Earth. A wide range of mission profiles including variations of the relative intelligence of the mother ship and the surface elements have been suggested.

Because of the large propulsive energy (Delta V) requirements associated with first rendezvousing with a comet and then returning to Earth, advanced solar-powered propulsion technology is enabling for all comets of interest to the science community. (Nuclear reactor powered propulsion would also be applicable but is not currently being considered for NASA science missions.) The most likely form of this would involve advances relative to the current state-of-the-art of solar electric propulsion (SEP). CSNR mission requires an advanced SEP with a power level of ~10 kilowatts and a mass of 300 kilograms or less including the power system. Improvement in the solar array specific mass is required to meet this requirement.

Techniques for approach, landing, anchoring, sample collection, and sample preservation were also identified as enabling for this mission. Many of these are well along the technology development path because of the development activities that were started for the cancelled ST-4 (Champollion) mission. An advanced 50 to 100 watt, 20% efficient, 10 kilogram advanced radioisotope power system for deep drilling to obtain subsurface samples would enhance this mission. In the far term, a solar sail and a 100-

watt, 8 kilogram advanced radioisotope power system would offer the capability of accomplishing the mission with a smaller launch vehicle and/or a shorter flight time.

Mars Weather/Seismic Stations

Mars Weather/Seismic Stations mission will establish a global network of 24 stations on the surface of Mars that will record the planet's weather and seismic activities for ten Mars years (18.8 Earth years). This long-term continuous presence will define the nature of the variability in the Martian climate system including the behavior of the polar reservoir of carbon dioxide. It will also provide weather-monitoring infrastructure for future Mars missions and of great scientific benefit to any simultaneous observations from orbiters, landers or Earth. The internal structure of Mars can also be studied by seismological measurements. At multiple sites, a minimum of three, the seismological analysis will be able to determine precisely the location of the seismic source. Passive seismic sounding will be used to determine depth profile of seismic velocities.

A low power, 40milliwatt, long-life, 20 year, advanced radioisotope power system is enabling for this mission. The power system will be used to collect atmospheric pressure, temperature, opacity and seismic data and charge ultra-capacitors to store energy for communicating this data to a Mars's orbiter for communications back to Earth.

Europa Lander

A Europa lander will conduct chemical analyses of near-surface ice and organics and will study the interior structure of Europa. In the most ambitious concepts, a "cryobot" would melt or burrow through the ice to explore the postulated underlying ocean. The trajectory being considered would insert into Jupiter orbit and use a series of satellite flybys lasting approximately 1 year to remove energy from the orbit prior to a descent to the surface. During the one-year Jupiter/satellite flybys a 150 to 200 watt ARPS enables the spacecraft to obtain science data and to operate the communications system in the high radiation environment. Regardless of the main propulsion system used to reach Jupiter, a significant portion of the launch mass would be allocated to transporting a chemical propulsion system to Jupiter for these operations.

Technology advances are needed on a broad front to enable a landed mission on Europa. The mission is very demanding energetically, calling for a combination of lightweight, radiation-tolerant systems and improvements in the performance and hardware mass of chemical propulsion. Many of the lander concepts examined would benefit from availability of small radioisotope based power systems (1 to 20 watts). Navigation to the landing site is also a significant challenge, but perhaps the most critical area is for development of systems to perform the desired science. This includes, in most concepts, systems to acquire samples of ice from a meter or so below the surface, to concentrate the samples, and to perform a broad range of organic chemical analyses. In the long term, it also may include a radioisotope power system to power a "cryobot" through the ice with a small RPS to power "hydrobot" systems for ocean exploration.

Titan Organic Explorer

A Titan Explorer would primarily study the distribution and composition of organics on the Saturnian moon, as well as look at the dynamics of the global winds. Aerocapture at Titan, avoiding a Saturn orbit insertion, is currently the most attractive trajectory option. A 150 to 200 watt advanced RPS that is 18% efficient and has mass of 20 kg or less for the Orbiter is enabling.

A variety of mission profiles have been proposed for Titan based on a variety of models of surface and atmospheric states. Cassini data will shed light on the validity of these models, but in the mean time, because of the importance of Titan as a potential host for prebiotic chemistry, it makes sense to take the early steps toward a quick follow-on to Cassini. This includes work on organic chemistry analysis systems (some overlap with work needed for Europa) and on delivery systems including aerocapture, balloon systems, and landers. For long life (>3 months) lander concepts, small radioisotope power sources, 1 watt to 20 watt, will also be enabling.

Neptune Orbiter with Triton Flybys

This mission would use a full complement of remote sensing instruments to characterize both the planet and its largest moon. To accomplish this with affordable launch vehicles and acceptable mission duration we need a very low mass spacecraft (as envisioned in the current work on "system-on-a-chip" technology), advanced solar powered propulsion systems (using SEP or solar sail in one or more close in orbits of the sun to accelerate the spacecraft for a quick trip to Neptune), and aerocapture into Neptune orbit.

Return of a high volume of science information from the distance of Neptune requires a 150 to 250 watt, 15 to 20 kg, and 18% efficient long life radioisotope power system. Low Mass, small antenna optical communication system is also required for the orbiter spacecraft. The study emphasized use of optical communication along with advanced techniques for selection, editing, and compression of the data. This mission would be enhanced with a 20 kWe, 500-kg nuclear electric propulsion system, (nuclear reactor powered propulsion would also be applicable but is not currently being considered for NASA science missions), for propulsion to Neptune, retro-propulsion into a Neptune/Triton orbit and for power communicating with Earth.

Saturn Ring Observer

The Saturn Ring Observer science objectives are; 1) to understand the ring process and evolution as a model for the origin of planetary systems, 2) to determine ring particle physical properties, dynamics and spatial distribution, and 3) to determine the composition of the ring particles. A 200-kilogram class spacecraft launched on a Delta or Atlas class launch vehicle would accomplish the mission. The mission would be an 8-10 year flight to Saturn with ballute or aeroshell aerocapture at Saturn to get into a low-inclination Saturn orbit with periodic plane changes to avoid ring-plane crossing.

Advanced solar electric propulsion or solar sail technology and a 150 to 250 watt, 18% efficient, 15 to 20 kg advanced radioisotope power system are enabling for this mission. This mission would be strongly enhanced with a 20 kWe, 500-kg nuclear electric propulsion system for propulsion to Saturn, retro-propulsion into a Saturn orbit, maneuvering at Saturn and for power to obtain scientific data at Saturn and high-speed communication with Earth.

Mars Robotic Outposts and Future Sample Returns

Mars Robotic Outpost and Future Sample Returns will provide a continuous robotic presence on Mars and prepare for future human exploration. These outposts will determine resource inventory on Mars and help scientists and the public understand the evolutionary history of Mars. The continuous presence on Mars with frequent reports and new information will engage the public in the challenges and excitement of Mars's exploration.

Mars polar terrain science laboratory would be strongly enhanced with a 50 to 100 watt 20% efficient; 8-kg advanced RPS with a different radioisotope as the heat source for four-year missions and continuous day and night surface and subsurface investigations. Central labs for analyses and coordination of all activities on Mars would be strongly enhanced with a long life, 15 year, 150 to 250 watt advanced RPS. Large distributed robots, wide area 3D exploration and sample acquisition and handling systems would be strongly enhanced with a 150 to 200 watt advanced RPS, especially in the Polar regions of Mars.

RPS Technology Requirements

NASA JPL has further studied and refined the potential missions to the Solar System as outlined in the Solar System Roadmap by Elachi et al 1996. These further mission studies have more specifically defined the technology requirements. The entries in Table 1 are quantitative; giving a value or range of values derived from the particular mission concept as shown.
The technology requirements for advanced radioisotope power source based on the mission studies listed above are updated as shown in Table 1 from the requirements listed in Table 1 by Mondt and Nesmith, 2000.

The criticality level for needing an advanced radioisotope power system for each mission shown in the table is defined as follows:

(1) Enabling - provides for achieving the science objectives of the mission with an affordable launch vehicle.

(2) Strongly enhancing - provides substantial increase in payload or reduction in cost or risk.

(3) Enhancing - provides increase in payload or reduction in cost or risk.

TABLE 1. Advanced RPS Top Level Requirements for Near-Future Missions

The first priority requirement for all radioisotope power systems is that the RPS meet all nuclear safety requirements.

Power (We)	Mass (kg)	Lifetime (Yr.)	Efficiency (%)	Voltage (Vdc)	Potential Missions
0.040 to 0.100	0.25	20	4 - 5	5	(1) Mars Weather/ Seismic Stations
1 to 2	0.5	5-10	5 - 10	5	(1) Europa Lander Surface Laboratory
10 to 20	2	15-20	15	5	(2) Surface In-situ Laboratories Aerobots or Aero-rover
50 to 100	7 to 10	4 to 5	18 to 20	28	(2or3) Rover and Sample Retriever
100 to 200	8 to 20	10	18 to 21	28	(1) Europa Lander
150 to 250	10 to 25	15	18 to 21	28	(2) Titan Explorer
150 to 250	10 to 25	15	18 to 21	28	(1) Neptune Orbiter
150 to 250	10 to 25	15	18 to 21	28	(1) Saturn Ring Observer
100 to 200	8 to 20	15 to 30	18 to 21	28	(1)InterStellarProbe

CONCLUSIONS

Near-Future and Farther-Future missions Exploring Organic-Rich Environments, To Build a Planet, Bring Mars to Earth, Robotic Outposts, Sampling all Accessible Bodies and Interstellar Exploration requires continuing NASA's Deep Space Systems advanced radioisotope power systems (ARPS) Technology Program. Accomplishing these deep space missions requires new, advanced, innovative, dual-use, smaller radioisotope power systems.

Using the most optimistic solar-based power system causes the launch masses of these missions to grow beyond the capability of affordable launch vehicles. Advanced radioisotope power systems are lower mass and easier to land and use than solar power systems for extended-duration comet sample acquisition activities. Innovative dual-use smaller radioisotope power systems and heat sources are needed for the future surface, subsurface, orbiting and airborne science laboratories. Solar power and other power systems alternatives will continue to be considered and evaluated for these missions.

ACKNOWLEDGMENTS

The work described in this paper was performed at the Jet Propulsion Laboratory, California Institute of Technology, under contract with the National Aeronautics and Space Administration. The JPL Mission and System Architecture Section under the guidance of the JPL Solar System Exploration Office developed the mission concepts described in this paper. The author acknowledges and thanks Robert Gershman and Doug Stetson for their work in developing these mission concepts and the preliminary requirements for the advanced radioisotope power systems technology.

REFERENCES

1. Elachi, C, et al, 1996, "Mission to the Solar System Exploration and Discovery: A Mission and Technology Roadmap", September 27, 1996, Version B.
2. Mondt, J., and B. Nesmith (2000) "Future Planetary Missions Requiring Radioisotope Power Systems", Space Technology and Applications International Forum, January 2000, Albuquerque, NM, American Institute of Physics CD ROM ISBN1-56396-920-3.

PROTOTYPE SOLAR PANEL DEVELOPMENT AND TESTING FOR A MERCURY ORBITER SPACECRAFT

Carl J. Ercol, Jason E. Jenkins, George Dakermanji, Andrew G. Santo
The Johns Hopkins University
Applied Physics Laboratory
Laurel, MD 20723-6099

Lee S. Mason
NASA John H. Glenn Research Center at Lewis Field
21000 Brookpark Road, MSAAC-2
Cleveland, OH 44135

ABSTRACT

A Mercury orbiting spacecraft imposes many design challenges in the area of spacecraft thermal control and electrical power generation. The Discovery Mission MESSENGER (MErcury Surface, Space, ENvironment, GEochemistry, and Ranging), being designed and built by The Johns Hopkins University Applied Physics Laboratory (APL), will orbit and survey the planet Mercury for one year. In order to reduce cost and schedule risk while increasing the probability for mission success, the MESSENGER solar arrays will be constructed from conventional "off the shelf" materials and technologies.

INTRODUCTION

This paper describes the high temperature and high intensity illumination testing which were used to thermally evaluate a series of prototype engineering solar panel concepts. Since all of the materials, construction techniques and technologies chosen for the solar panel construction are typically used for fairly benign thermal environments, qualification or proof of concept testing was necessary to assess the risks associated with such a thermally demanding mission. The solar array concept for the MESSENGER Mission consists of two double sided rotatable wings, allowing for tailored temperature control during power generation as solar distance decreases. One side of each array is fully packed with 5.5 mil single junction gallium arsenide (GaAs/Ge) cells. The opposite side is packed with a 70%/30% mixture of Optical Solar Reflectors (OSRs) and the same type of GaAs/Ge cells. Each side of each solar panel is designed to maximize power generation and minimize operating temperature for a given solar distance range. From launch until the spacecraft reaches about 0.60 Astronomical Units (AU), where the solar constant is about 2.8 greater than at Earth, the fully packed side of each array is responsible for power generation. Once inside of 0.60 AU, the array is flipped to the 30% cell side which takes advantage of the higher solar constant while maintaining the solar panel operating temperature below 150°C.

The solar panel structure uses sandwich construction comprised of high thermal-conductivity graphite-epoxy (Gr/Ep) face-skins and an aluminum-honeycomb core. The primary thermal environment that drives the MESSENGER prototype solar panel design and the dual-sided configuration is the high solar intensity condition experienced when at planet perihelion, where the solar constant is eleven times that at Earth. The mirrored side of the array safeguards against solar panel temperatures exceeding 250°C in the event of a direct Sun pointing anomaly when at Mercury perihelion. And, the dual sided design minimizes the solar panel overall size and mass by using the fully packed face at solar distances associated with the beginning of the mission when temperatures are very benign and the solar constant is low.

A large portion of the proof of concept thermal vacuum testing was accomplished using a custom designed high temperature infrared oven to test solar panel specimens between +300°C and –105°C. The oven has allowed for accurate and repeatable component testing while proving to be very reliable and cost effective. The more expensive and complicated high intensity solar illumination tests were done only after thorough infrared temperature cycle testing and high temperature soaks verified solar panel materials and construction. The high-illumination testing was done at NASA's John H. Glenn Research Center at Lewis Field in the Tank 6 high intensity vacuum chamber. These tests were used to verify the prototype solar panel thermal design and to demonstrate ability to predict the expected test panel temperature at a high intensity solar simulated environment. The paper describes the details and results of the prototype tests conducted along with plans for the flight solar panel development and test.

BACKGROUND

Mercury has never been explored by a remote orbiting spacecraft. The only man-made space probe to visit Mercury was Mariner 10. Built and launched in the United States, Mariner 10, a 3-axis stabilized solar powered spacecraft, has provided the only images and scientific exploration of Mercury. Using three flybys, Mariner 10 was able to map only about 45% of the planet surface during a one-year period between 1974–1975. The dark side flybys were all at planet aphelion and never near the sub-solar point. Mariner 10 was designed and tested to withstand a solar-only 5 Sun (one Sun is equal to the solar flux at 1 AU or 1365 W/m^2) environment, ignoring the intense omni-directional heat radiated from Mercury's surface on the Sun-lit side.

Due to severe mass restrictions and extremely harsh thermal environments, a Mercury orbiting spacecraft poses many engineering and operational challenges. MESSENGER is a 3-axis stabilized solar powered spacecraft using a high-performance all chemical propulsion system fully integrated into an all graphite-epoxy structure. The power system will utilize two low mass dual-sided solar array wings that can be rotated and flipped as necessary to control solar panel temperatures as the spacecraft approaches Mercury perihelion. The mission design uses a ballistic trajectory with multiple Venus and Mercury gravity assists. The MESSENGER spacecraft will eventually orbit Mercury for one Earth year (or four Mercurian years) and return a wealth of scientific data and complete planet coverage, something not accomplished by Mariner 10.

The MESSENGER mission was recently selected by NASA to be the eighth program in the highly successful series of the Discovery missions. Mission cost and launch vehicle choices are very constrained under NASA Discovery guidelines. The largest acceptable launch vehicle for a NASA Discovery mission is a Boeing Delta II 7925H-9.5 (Maximum Launch Mass=1066 kg). Driven by the 2700 meter per second mission Δ-velocity (ΔV) requirements, over one half of the launch mass is allocated to propellant. It quickly becomes apparent that due to the high ΔV nature of this mission, the spacecraft mass allocated to useful payload is limited. Spacecraft mass must be used with great discretion since the main purpose of the mission is to get maximum science return. Figure 1 illustrates the MESSENGER spacecraft. As shown, the dual-sided solar panel wings extend beyond the protective umbra created by the thermal shade, making the solar arrays the only critical component exposed to direct high intensity solar illumination. Designing the spacecraft for minimum mass will require the attention of all spacecraft sub-systems, and the seemingly vulnerable solar array is no exception. A mass saving benefit from the dual-sided approach is that the overall solar panel area is small because each face is optimized for power at the worst case solar distance and temperature. Each face of the dual sided solar array is designed to produce power at a cell operating temperature of 150°C or lower. As illustrated by Figure 2, the array is flipped from the fully packed side to the OSR side and back as the solar distance varies between approximately 0.45 and 0.60 AU. Figure 3 illustrates the variation in respective solar array face rotation angle as a function of solar distance as the array temperature is held at or below 150°C. The rotation angle is zero when the Sun line is perpendicular to the active face of the solar array.

Why Use a Dual-Sided Solar Array Concept?

The MESSENGER mission first baselined a single sided fully packed solar array using single junction GaAs/Ge cells bonded to graphite composite substrates. As illustrated by Figure 4, the fully packed solar array has to be rotated in excess of 75° to maintain the steady state solar array temperatures below the desired operating point of 150°C while at Mercury perihelion (11-Sun illumination condition). A major deficiency with this design is that a direct Sun pointing anomaly at the perihelion solar distance could easily cause the array to approach 400°C. Solar cell and Gr/Ep adhesives are only rated to maximum temperatures between 200 and 260°C.

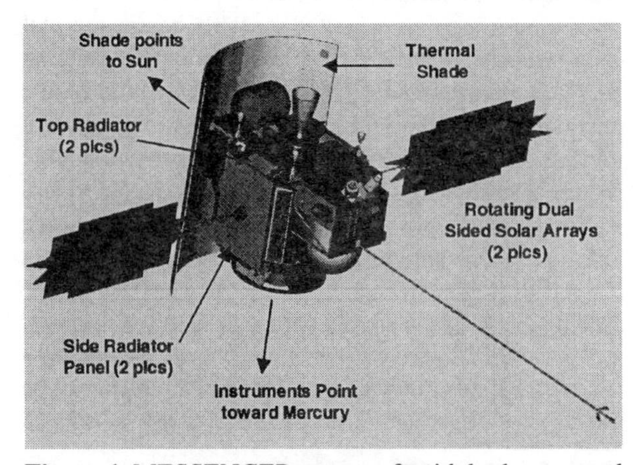

Figure 1. MESSENGER spacecraft with body mounted multi-layer insulation (MLI) removed.

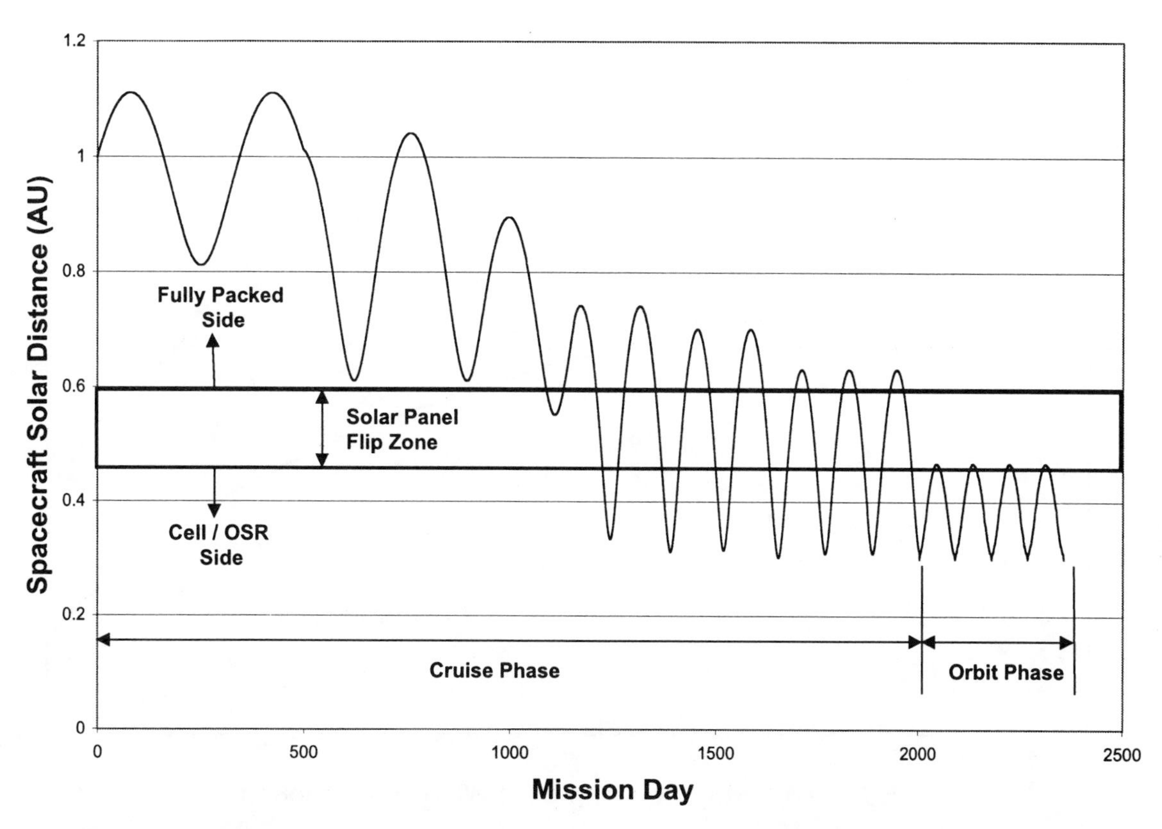

Figure 2. The MESSENGER mission profile. Cruise phase is five years and orbit phase is one year.

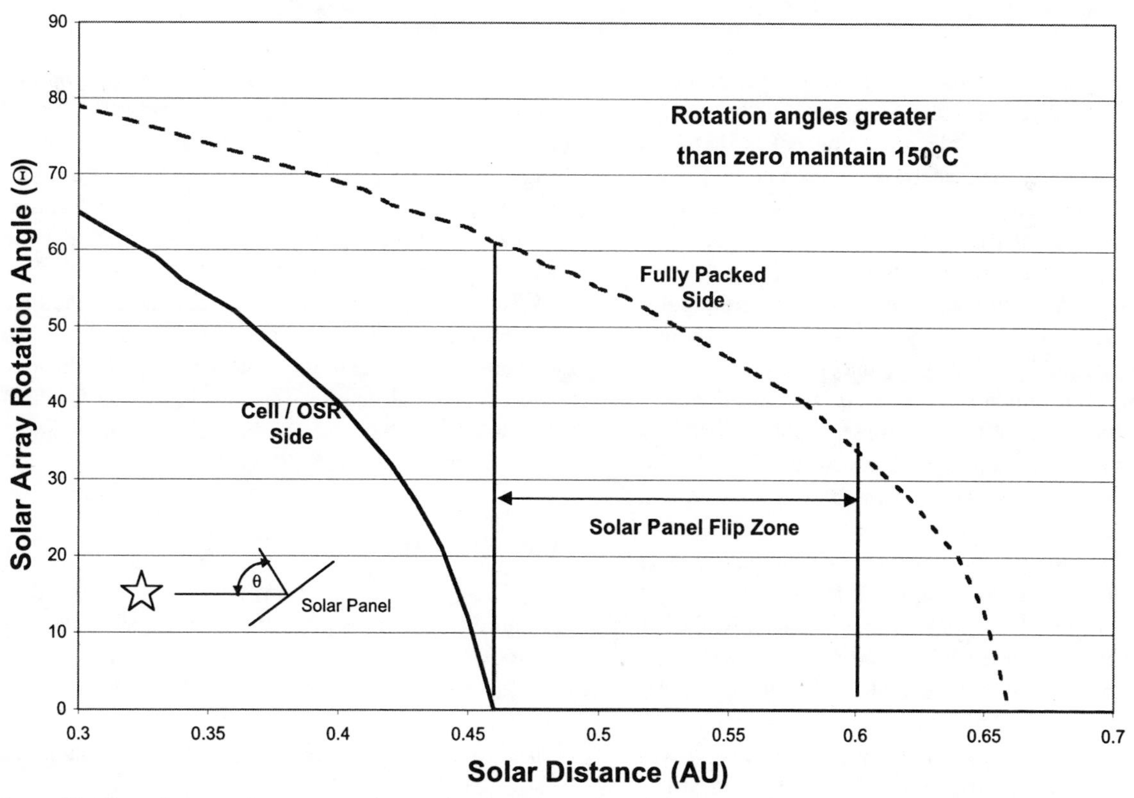

Figure 3. The solar panels are rotated to maintain the maximum temperatures at or below 150°C. The dual sided solar arrays have a wide flip zone to reduce operational constraints and minimize risks.

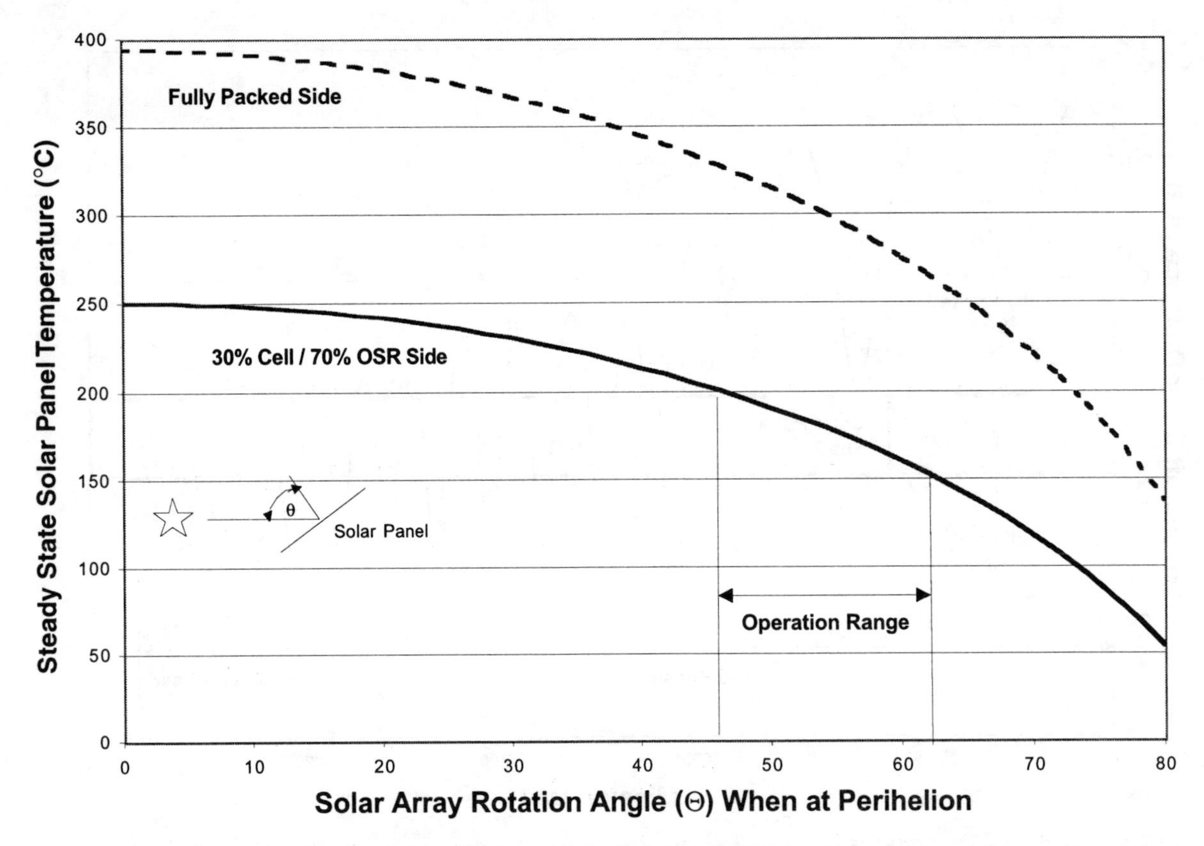

Figure 4. The OSR array side will maintain the steady state solar panel temperature below 260°C if Sun pointed when at perihelion. Note that the fully packed side could reach 400°C under the same condition.

Trying to qualify this technology to temperatures in excess of 400°C was not a feasible option. Another critical deficiency with the single sided concept is that it is thermally unstable. The analysis represented by Figure 4 illustrates that controlling the solar array temperature to 150°C while at perihelion could be very difficult because of the sensitivity between tilt angle and temperature. Also, at solar panel aspect angles greater than 65°, solar cell reflectivity begins to dominate power calculations and introduce large uncertainty.

Once the single sided concept was abandoned, a trade study was done to determine the effects of adding mirrors with the cells in order to reduce the overall solar panel temperature. Power system analysis showed that the total surface area necessary to do the complete mission using only a single solar array face that combined OSRs with solar cells would be almost a factor of four larger than the original fully packed design. The OSR design also has to incorporate thicker face skins to thermally connect the solar cells to the OSRs. Although the single solar array face that utilized OSRs with solar cells was not feasible from a spacecraft mass perspective, it was from a thermal perspective. The next trade study combined a fully packed solar array face with an OSR and solar cell face. This design would allow the MESSENGER power system to select the appropriate face depending on solar distance and solar panel temperature. Since each of the solar array faces will be designed to produce power over a defined solar distance range, the dual-sided approach allows for the panel area optimization because the worst case power condition will always dictate this parameter and set the maximum face size. This in conjunction with the deficiencies described for the single sided fully packed solar array made the dual-sided concept extremely attractive. More detailed power and thermal analysis were performed, and the dual-sided solar array became the baseline for the winning Discovery proposal, MESSENGER.

Prototype Solar Panel Construction

The prototype solar arrays were constructed of conventional materials that gave the highest probability of success for steady state temperature excursions between 200 and 260°C. The flight solar array concept calls for solar cells and OSR mirrors to be bonded to high conductivity graphite epoxy face sheets comprised of K13C2U cloth and RS3 and RS4 resin systems using conventional, NuSil Technology CV-2568, silicone adhesive, illustrated by Figure 5. The solar cells and cover

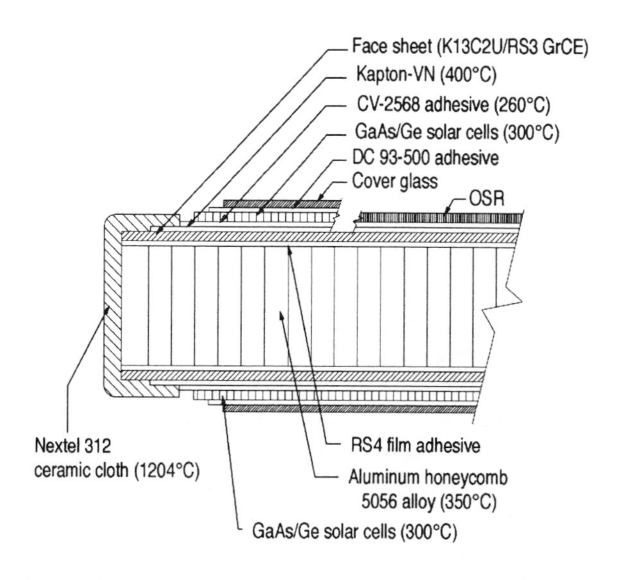

Figure 5. Solar panel cross-section showing maximum vendor recommended steady state temperatures. Solar panel materials and construction techniques were tested to temperatures in excess of 250°C.

glasses, Kapton™-VN, and aluminum honeycomb core as components are rated to temperatures at or above 300°C. The adhesives used to bond these different solar array components together creating the solar array assembly are the materials that have the lowest manufacturer's rated temperature. For example, Dow Corning rates DC-93500, which is used to bond cover glasses to solar cells, to only 200°C. Manufacturers of the resin systems RS3 and RS4, used to bond the individual face sheet plys together and to the honeycomb core, are only rated to 232°C. So it was decided to purchase and test multiple configurations of populated and unpopulated Gr/Ep sandwich panels that all used the same cyanide-ester resin systems and solar cell adhesives. Six different test panels were purchased and are listed in Table 1. Figures 6 and 7 depict the mirrored and non-mirrored sides of the prototype dual-sided solar array panel that was tested and is identified as Panel 4 in Table 1.

TECSTAR, Inc. did the solar cell and OSR bonding and the electrical wiring for all the test panels. The solar cells were TECSTAR 5.5 mil thick GaAs/Ge "standard" production cells cut to the required size. The solar cell

Table 1. Six prototype solar panels were purchased for testing under MESSENGER thermal environments. The prototype panels were fabricated to prove that conventional solar panel construction techniques and materials could be used in the high-temperature and vacuum environment an orbiting spacecraft would experience at Mercury.

Generic Panel Identifier	Configuration Description	Construction Details
Panel 0	Unpopulated Test Panel #1	A Kapton-insulated substrate with inserts. Face sheet composition is M55J/RS3 2.5mil UDPP with RS4 sandwich film adhesive and 3.1 lb/ft^3 hexcell core.
Panel 0A	Unpopulated Test Panel #2	A Kapton-insulated substrate without inserts (UDPP K13 C2U/ RS-3, 2.5 mil, with 4.5 lb/ft^3 hexcell core.) Panel 0A is populated on one side with individually wired heater elements which are arranged as the cells and OSRs of panels 3 & 4 (below). The other side of the panels is populated by Silver Teflon to emulate OSRs.
Panel 1	Single Sided Populated Panel #1	A substrate identical to Panel 0 populated with 12 older technology GaAs/Ge "reject" cells. Cells are mechanically sound, though electrically poor. Cell size is larger than intended for MESSENGER flight.
Panel 2	Single Sided Populated Panel #2	A substrate identical to Panel 0 populated with 36 high-temperature GaAs/Ge cells and representative OSRs. The cells are larger than what will be defined for the flight panels. OSRs were placed on the panel to qualify materials and methods, and as such are not positioned to be of thermal use.
Panel 3	Dual Sided Configuration Panel #1	A substrate identical to Panel 0A is populated on a single side with 9 high-temperature GaAs/Ge cells from the Panel 2 production lot which are cut down to 4×2-cm. OSRs are cut and arranged with the cells to be representative of the flight thermo-optical design. Each string and individual corner cells of strings A and C are wired for electrical monitoring. Thermocouples are bonded between two cells each of circuits A and C and between two OSRs each of the exterior OSR strings.
Panel 4	Dual Sided Configuration Panel #2	A substrate identical to Panel 0A populated on the front side identically to Panel 3. The backside is populated with 2 by 4 cm cells cut from the Panel 2 production lot. The backside panel is also instrumented. With two flight-candidate platinum thermistors.

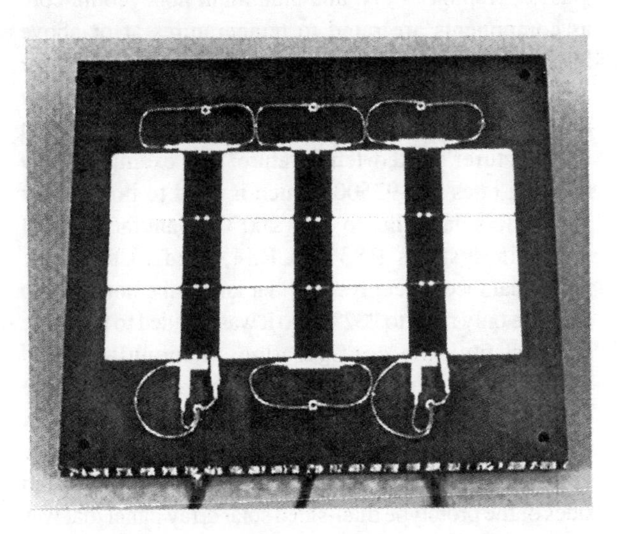

Figure 6. The prototype 30% solar cell and 70% OSR face of the dual sided array.

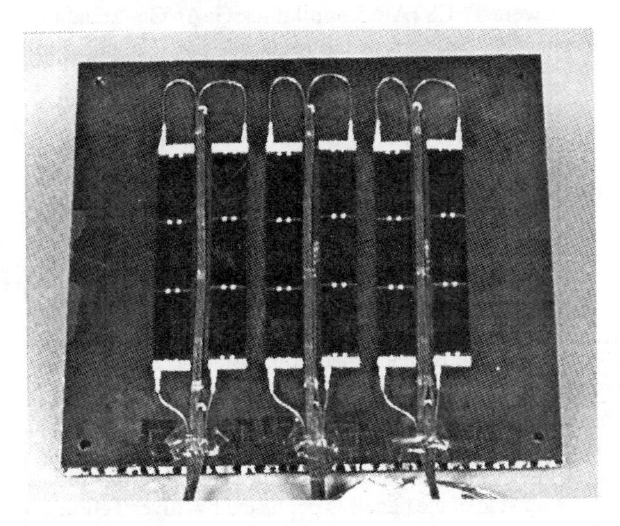

Figure 7. The prototype fully populated face of the dual sided array.

interconnects end termination were all parallel gap welded and high temperature Kapton was co-cured with the GrEp to form the electrical barrier between the solar cells and the graphite panel. Ceria doped cover glass, three tenth of a millimeter thick, coated with dual anti-reflective coating was bonded to the cell using DC 93500 adhesive. NuSil 2568 adhesive was used to bond the cells and OSR to the kapton insulation on the panels.

High-Temperature Test Program Development

Development of a test program and testing philosophy was essential to verify that conventional solar array materials and manufacturing processes could be used for high temperature applications. Thermal vacuum testing of the samples would have to represent the minimum and maximum temperature excursions experienced during the expected 65-minute eclipse as well as the nominal solar panel operating temperatures when illuminated. Also, because the solar arrays represent a single point failure, the worst case steady state survival temperature would also be tested to show design robustness in the event of a Sun pointing anomaly when at Mercury perihelion.

Each prototype test panel would first be baked out for 24 hours at 80°C. This bake-out helps to remove trapped air pockets in the solar cell adhesives that could rupture at the high temperatures and damage the cells. The test program combines solar panel acceptance testing that is typical for an Earth orbiting spacecraft with the higher temperature cycle and soak testing required for a Mercury orbiting spacecraft. Electrical performance tests would be done after each set of cycle tests and after each of the high temperature soaks to verify electrical functionality and document any temperature incurred damage.

Six conventional workmanship cycles would be performed between +120°C and –105°C, establishing a test baseline. High temperature cycling would follow the workmanship cycles and would have maximum plateaus at +150°C, +180°C and +200°C and the same lower temperature of –105°C. Also, high temperature soaks representing steady state survivability in the event of a Sun pointing anomaly at perihelion had to be demonstrated. Each of the soak temperatures, 230°C, 240°C and 250°C, would be held for a minimum of one hour to verify material and construction robustness, removing any solar array recovery time constraint if an anomaly were to occur. At the completion of each level of cycle testing and after each of the high temperature soaks the test panel would be removed from the test chamber and visually inspected for damage such as substrate delamination or cracked cells or OSRs. After the visual inspection, the test panel would be taken to a laboratory for electrical performance testing if necessary. Once the inspections and electrical test is complete the prototype solar panel would be re-installed for further temperature cycle testing.

The first high temperature tests were done using a large (9 m^3) liquid nitrogen (LN_2) cooled vacuum chamber and infrared (IR) heat lamps. A dummy black aluminum test panel, which represented a typical prototype solar panel, was located inside the chamber and heated at high intensity with the IR lamps. Thermocouple measurements made on the aluminum panel showed temperature gradients in excess of 70°C when the maximum dummy panel temperature was 310°C with the chamber walls at –180°C. A few failed attempts were made to increase the flux uniformity by repositioning the IR lamps and dummy test panel. It quickly became evident that this test setup method would impose a very high risk to the sample being tested. Also, because the test samples had to be removed for inspection after each set of thermal cycles, the test setup proved to be ineffective from both cost and schedule standpoints and therefore was abandoned.

A smaller and simpler test apparatus had to be designed. The new test apparatus had to deliver uniform sample temperatures between −180 and 300°C, it had to be time and cost effective, it had to be reliable, and it had to be simple to operate. The new apparatus would be designed to operate inside the large 9-m^3 chamber so that vacuum, liquid nitrogen, thermocouple and electrical interface resources could be easily utilized. The large chamber would not have to be temperature controlled, eliminating the long transitions from vacuum to ambient conditions. The small (0.125 m^3) test apparatus, known as the "E-Box", is shown in Figure 8. The E-Box is thermally isolated from the large chamber, the east chamber, illustrated by Figure 9, using stainless steel chains and high-temperature multi-layer insulation allowing for independent temperature control. Thirteen 1000-watt resistive OMEGALUX™ strip heaters combined with an LN_2 cooling loop plumbed from an internal east chamber LN_2 feed will provide the necessary temperature control capability for the E-Box. The E-Box has two 2600 cm^2 (400 in^2) compartments that are identically heated and cooled, allowing for multiple sample testing.

Figure 9. The E-box in test configuration installed in the East chamber. Vacuum and high temperature kapton MLI insulate the E-Box from the room temperature walls of the East chamber.

Figure 8. The E-box prior to large chamber installation. The E-box is liquid nitrogen cooled, and heated using 13 one kilowatt strip heaters.

Infrared Testing using the E-Box

Pathfinder infrared thermal vacuum testing in the E-Box started on April 23, 1998 using two unpopulated GrEp test panels. One test panel, identified as Panel0, was constructed from M55J fibers formed into two 0.25mm substrates using RS3 and co-cured with aluminum honeycomb as a sandwich using RS4. The second test panel, identified as Panel0A, was constructed from K13C2U fibers formed into two 1.0 mm face sheets using RS3 and co-cured with aluminum honeycomb as a sandwich using RS4. The pathfinder testing was used to verify the survivability of the RS3 and RS4 resin systems at extended high temperature before the expensive populated panels were subject to the same sequence of tests. The test procedure, which includes chamber breaks, inspections and E-Box power levels, was also refined to minimize the overall test time.

The low conductivity test panel, Panel0, shown in Figure 10, was extensively thermocoupled to measure the temperature gradients induced by heating and cooling. Since the Panel0 face sheets have very poor thermal conduction characteristics, temperature gradients measured between any two thermocouples illustrated quantitatively the heating uniformity. With an average temperature of 250°C for each test panel, the maximum differential measured between the warmest and coolest thermocouple on Panel0 was less than 5°C. The pathfinder thermal vacuum testing with the unpopulated test panels verified that the cyanide-ester resin systems could be taken to temperatures above the manufacturer's specification for extended time without damage or delamination of the sandwich construction. The pathfinder testing also verified that the heating and cooling of the test samples was very uniform and low risk while the turnaround time between cycle testing, hardware inspections and reinstallation of the sample in the E-Box was small.

Upon the satisfactory completion of the pathfinder testing, solar panel prototypes populated with mechanically-sound, electrically-rejected, but functional, GaAs/Ge cells and OSRs were then tested using the same thermal profile and inspection procedures established during the pathfinder tests. Table 2 lists a summary of the results of all the panel level infrared testing performed at APL. Figure 11 depicts the as-run thermal vacuum test data for Panel3 and Panel4, representing the prototype dual sided solar panel configuration.

Figure 10. The E-Box being used to test Panel0 and Panel0A. Panel0 is visible in the right chamber with Panel0A (not seen) installed in the left chamber.

Simulated High-Intensity Solar Illumination Testing

Once the dual sided prototype engineering panels were successfully temperature cycled and soaked using the E-Box, Panel3 was taken to the John H. Glenn Research Center (GRC) at Lewis Field Tank 6 thermal vacuum test facility. The Tank 6 facility is a 18.5-meter long by 7.7 meter diameter horizontal vacuum chamber with a liquid nitrogen cryogenic-wall and a solar simulator. The solar simulator uses nine 30 kW Xenon arc lamps and was designed to provide no less than 1 Sun on a 4.6 meter diameter target, 17.4 meters from the source with a subtended angle of less than 1°. In order to achieve the high intensity solar flux levels required to simulate Mercury perihelion, the targets were located at a distance of approximately 5.8 meters from the source. The vacuum system can provide ambient pressures as low as 10^{-6} torr. Tests and analysis done at APL prior to GRC solar testing indicated that heat transfer concerns are mitigated at pressures below about 150 millitorr and that 10 Suns with a room-temperature sink is equivalent to 11 Suns with a liquid nitrogen or deep space sink. Therefore all testing was conducted without high-vacuum pumps and liquid nitrogen cooling, resulting in a nominal tank pressure of 25 millitorr and an average ambient tank temperature of 22°C.

The solar array panel was installed, aligned and electrically checked on Monday, February 22, 1999. The vacuum system was started at about 10:00 am on February 23. The tank pressure was 28 millitorr at 12:49 pm when the first solar simulator lamp was started. Video and infrared cameras were mounted at window ports near the simulator window to observe the test article. A total of eight lamps were started sequentially in the following order: 5, 4, 6, 7, 2, 8, 9, 1. The lamp startup order was determined by change coupled device (CCD) imaging on a Teflon sheet prior to test panel insertion and was selected to minimize the initial flux variation on the test panel. Lamp number 3, which was planned to be used after lamp 6, failed to start and required the addition of lamp 1 as an alternative. After each lamp was started and adjusted to its operating level, solar cell I-V curves were acquired. Panel temperatures were allowed to stabilize with each lamp operation.

Table 2. MESSENGER thermal/vacuum testing at APL.

Test Report	Date	Test Item	Test Type	Test case Summary
97-032	12/18-22, 1997	Solar Panel Test Plate (Aluminum)	Engineering Development	1 Cycle (-120°C <> +310°C)
98-013	4/23–5/14, 1998	Panel0 and Panel0A Panel1 and Panel2	Thermal/Vacuum Performance	Bakeout 85°C 6 Cycles (-105°C <> +115°C) 8 Cycles (-105°C <> +150°C) 4 Cycles (-105°C <> +180°C) 4 Cycles (-105°C <> +200°C) 4 Cycles (+23°C <> 220,230, 240, 250°C)
98-023	5/22–26, 1998	Panel1	Electrical Performance T/V (Illumination)	7 Cases (+50°C <> 200°C)
98-033	8/5–24, 1998	Panel2	Long Duration T/V Test	1 Case (+23°C <> +180°C)
99-004	1/27–2/5, 1999	Panel3 and Panel4	T/V Performance Test	Bakeout 85°C 6 Cycles (-105°C <>115°C) 4 Cycles (-105°C <>150°C) 4 Cycles (-105°C <> 180°C) 4 Cycles (-105°C <> 200°C) 2 Cycles (+23°C <>230, 250°C) 6 Cycles (-105°C <>115°C)
99-008	2/8–9, 1999	Panel2	T/V Performance	1 Cycle (-100°C <> 260°C)
99-010	2/10-12, 1999	Panel3 and Panel4	Electrical Performance T/V (Electrical Continuity at High Temperature)	Bakeout +100°C Three Hour Soak @ 250°C
99-017	3/1-4, 1999	Panel2	T/V Performance	Bakeout 100°C 1 Cycle (+100°C <> 280°C) 2 Cycles (+23°C <> 300,*320°C)

*Test Panel Failed

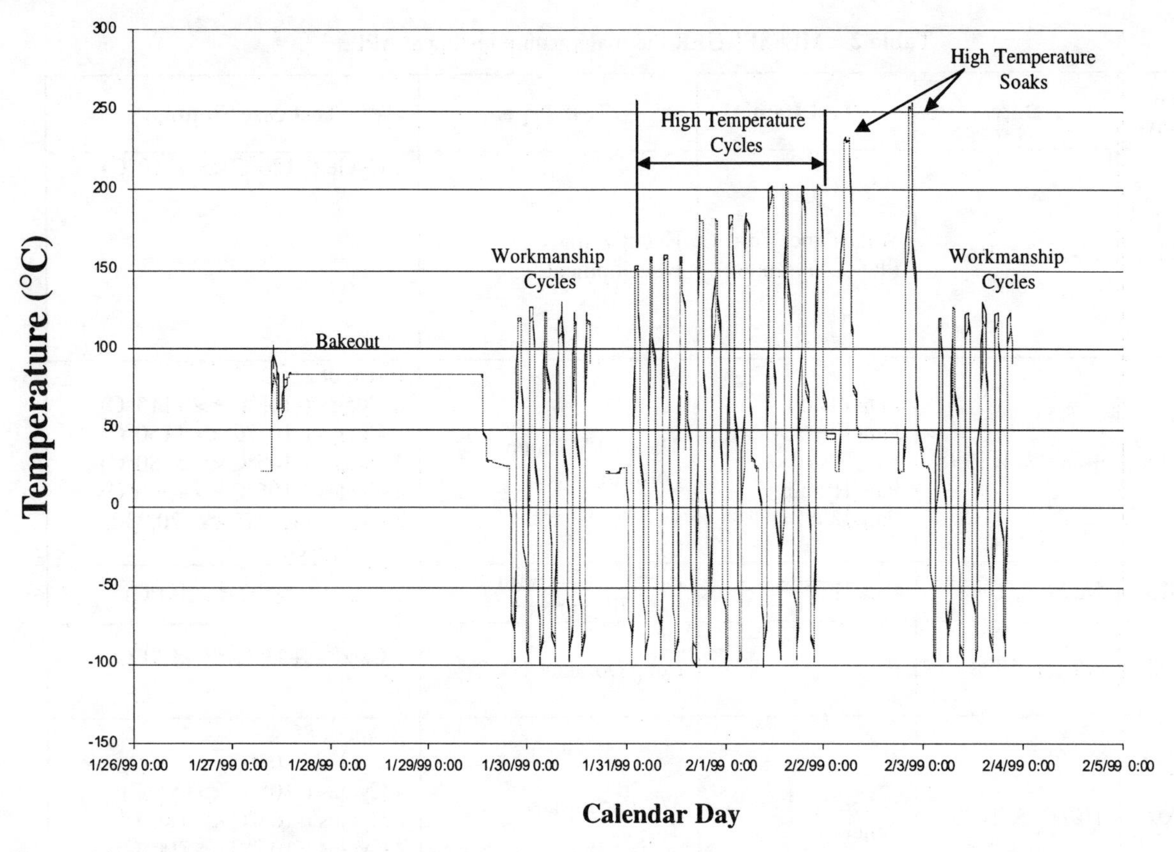

Figure 11. The actual test profile as run for Panel3 and Panel4, the dual sided prototype solar panels.

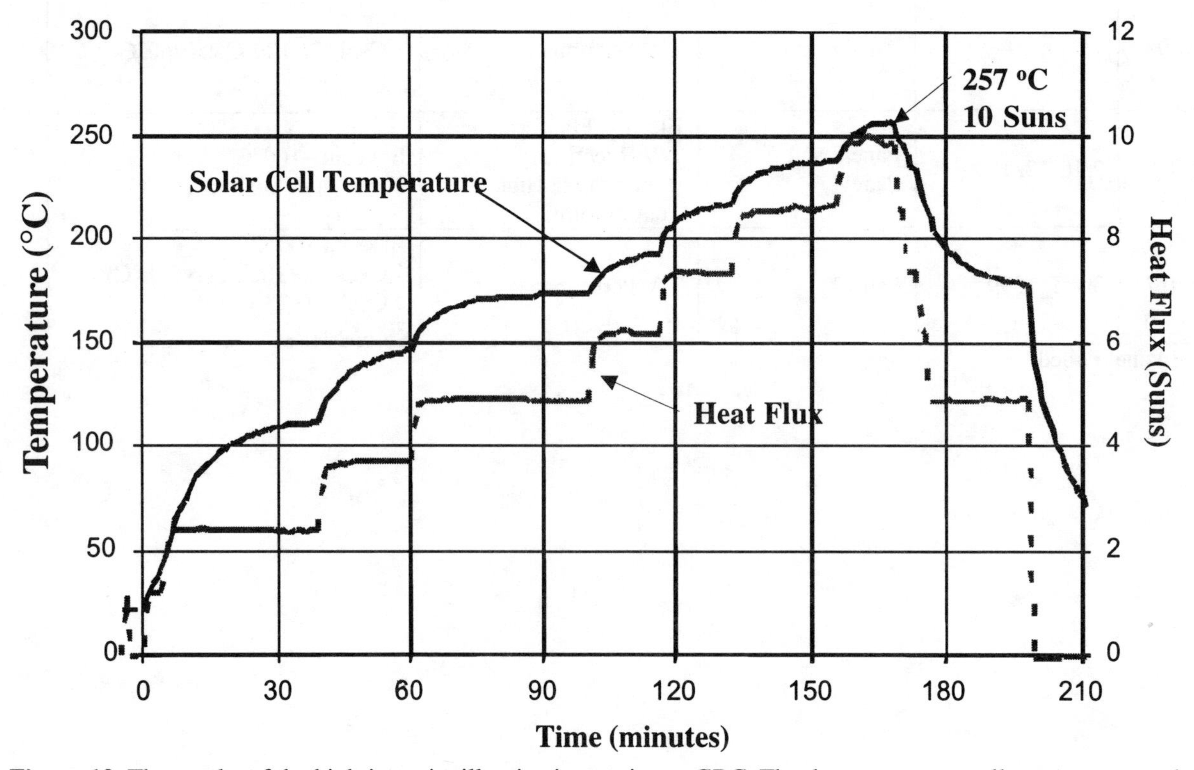

Figure 12. The results of the high-intensity illumination testing at GRC. The data represents cell temperature and simulated solar flux.

Figure 12 shows the maximum measured panel temperature and estimated local solar flux at that thermocouple location over the 200-minute illumination period. With four lamps operating, the maximum-recorded temperature was 174°C with a localized flux of 4.9 Suns. The eight-lamp configuration resulted in a maximum panel temperature of 257°C with a localized flux of 10.0 Suns. Both test points provided excellent agreement with pre-test analyses. No visible or electrical abnormalities were observed during the test. Post test removal and inspection of the panel showed no visible effects of the testing. Darkening of some non-optical adhesives occurred but without effect to adhesion. Pre-test and post test I-V characterizations performed at GRC's Solar Cell Measurements Lab with a Spectrolab X-25 were essentially identical.

CONCLUSIONS

The APL prototype test program met the goals of thermally evaluating the materials and processes used for conventional solar array construction. The MESSENGER prototype dual sided solar array has proven to be thermally robust and temperature tolerant. Results of the thermal testing presented verifies that the conventional adhesives typically used for the construction of cyanide-ester based GrEp solar panel substrates and solar cell bonding processes are not adversely effected by temperatures below 300°C. As illustrated, multiple solar panel samples using two different pitch based fibers passed all required cycle and dwell testing, with one sample, Panel2, being tested to high temperature failure (>310°C). The thermal modeling which was used to establish the maximum temperature limits during E-Box testing (250°C) was confirmed during the solar illumination testing at GRC (257°C). And the comprehensive prototype test program has verified that the actual maximum panel temperature is within the maximum survivable limits of the chosen materials and construction techniques by over 40°C. The infrared thermal vacuum testing with the E-Box provided inexpensive and conclusive test data regarding the integrity of the materials and construction techniques chosen for the engineering panels, avoiding the more complicated and expensive solar simulation testing until design concepts were well defined.

The prototype tests to date were intended to demonstrate that the basic material and processes used for a "standard" near Earth type solar array can also be used for a solar powered spacecraft in Mercury orbit. Tests to verify the effects of eclipse induced thermal shock, solar cell reverse biasing when shadowed during high intensity illumination, and adhesive darkening during high-temperature/high-intensity UV illumination were not conducted, but are presently being evaluated. Efforts are currently in progress to evaluate the outgassing properties of the candidate adhesives at high temperature. Also under consideration is the use of IR and UV reflective coatings, applied to solar cell cover glasses and tested at elevated temperature and with high intensity UV illumination. Although the prototype solar panel testing to date has been weighed heavily toward thermal design and panel survivability, future testing will focus on refining the flight panel electrical design, assessing long term exposure effects on solar array materials and coatings and determining solar array performance under the MESSENGER mission environments.

REFERENCES

1. MESSENGER Discovery Concept Study, The Johns Hopkins University Applied Physics Laboratory, March 1999.
2. Powe, J. S., E. L. Ralph, G. Wolf, and T. M. Trumble, On-orbit performance of CRRES/Hesp experiment, Proc 22nd IEEE Photovoltaic Specialists Conference, 2, 1409–1415, Las Vegas, NV, 1991.
3. Ercol, C. J. and A. G. Santo, Determination of optimum thermal phase angles at Mercury perihelion for an orbiting spacecraft, *Proc. SAE International Conference on Environmental Systems (ICES),* Paper 972534, July 1999.
4. Mason, L. S., High flux, solar vacuum testing of a solar array panel and heat shield, NASA John H. Glenn Research Center at Lewis Field, test report, March 1999.

ACRONYM LIST

APL	The Johns Hopkins University Applied Physics Laboratory
CCD	Charge coupled device
GRC	Glenn Research Center
IR	Infrared radiation
LN_2	Liquid nitrogen
MESSENGER	MErcury Surface, Space, ENvironment, GEochemistry, and Ranging
MLI	Multi-layer insulation
OSR	Optical Solar Reflector
UV	Ultra-Violet

ACKNOWLEDGMENTS

The authors wish to acknowledge the support of Harold Fox, Bill Wilkinson, Dennis Miller, Anthony Scarpati, and Bill Hamilton of APL and Steve Geng, Wayne Wong, Phillip Jenkins, Dave Scheiman, Dave Brinker, Mike Piszczor, Larry Schultz, Wayne Condo, Terry Jansen, Randy Mele, Bob Braun of GRC who contributed greatly to the success of the testing described in this paper.

AIAA-2000-2882

FLIGHT TEST OF A TECHNOLOGY TRANSPARENT LIGHT CONCENTRATING PANEL ON SMEX/WIRE

Theodore G. Stern
Composite Optics, Incorporated
San Diego, CA

John Lyons
NASA Goddard Space Flight Center
Greenbelt, MD

Abstract

A flight experiment has demonstrated a modular solar concentrator that can be used as a direct substitute replacement for planar photovoltaic panels in spacecraft solar arrays. The Light Concentrating Panel (LCP) uses an orthogrid arrangement of composite mirror strips to form an array of rectangular mirror troughs that reflect light onto standard, high-efficiency solar cells at a concentration ratio of approximately 3:1. The panel area, mass, thickness, and pointing tolerance has been shown to be similar to a planar array using the same cells. Concentration reduces the panel's cell area by 2/3, which significantly reduces the cost of the panel. An opportunity for a flight experiment module arose on NASA's Small Explorer / Wide-Field Infrared Explorer (SMEX/WIRE) spacecraft, which uses modular solar panel modules integrated into a solar panel frame structure. The design and analysis that supported implementation of the LCP as a flight experiment module is described. Easy integration into the existing SMEX•LITE wing demonstrated the benefits of technology transparency. Flight data shows the stability of the LCP module after nearly one year in Low Earth Orbit.

Background

Solar concentrating photovoltaic arrays have been developed over the last twenty years to a state of flight readiness. The main impetus for their development has been to decrease the cost of prime power generation, in terms of $/Watt, by replacing expensive high efficiency space-quality solar cells with less expensive optics. The optics effectively replace solar cell area by collecting insolation (sunlight) and directing it to the now smaller cell areas. Another potential advantage of solar concentrators is the ability to more effectively shield the solar cells against hazardous environments, while maintaining the low mass needed for space solar arrays.

Although several flight experiments have flown both reflective (mirror) and refractive (lens) concentrating panels, the first concentrating arrays to fly as prime power for space missions have only appeared in the last three years. The first flight of a concentrator array for prime power occurred with Deep Space 1 using the SCARLET modular fresnel lens technology.[1] More recently, the LCP technology flew as part of the prime power system for the SMEX mission described in this paper.[2] A third concentrator configuration using large reflector panels has also flown on a large communications spacecraft.[3]

A perception of increased risk to mission success has been a factor preventing even more widespread use of concentrator technology. The primary areas of concern are associated with degradation of optics, thermal control of the solar cell under increased and possibly uneven illumination, and the need to more accurately point the arrays towards the sun. Optical degradation is a particularly difficult risk to manage because the sources of contamination products and their environmental interactions are difficult to predict, model, and test on the ground.

LCP Design

Technology Transparency Objective

A major objective of the Light Concentrating Panel (LCP) development was to retire the perceived risks of a solar concentrator panel by using the concept of "technology transparency." In this approach, the product being developed is modeled on an existing design with the goal of achieving the same or better performance at lower cost, without significant changes in form, fit, or function.

For a solar panel, we identified those features and characteristics that typically are key inputs to the system engineering of a solar array. The characteristics of a planar panel that we wanted to emulate include specific mass (Watts/kilogram), specific area density (Watts/m^2), packaged panel efficiency (Watts/m^3 which reduces to panel thickness for an equivalent area density), power profile over time (orbital environment degradation), and panel pointing tolerance. In addition, we wanted to achieve a similar manufacturability using existing cell and panel technology, and provide familiar interfaces to the array hold-down and deployment mechanisms.

Panel and Element Design

The LCP design that addressed these technology transparency objectives is shown in Figure 1. It is a panel of similar thickness to large planar panels (about 2.5cm), comprising an array of concentrator elements. Each element consists of a four-sided mirror trough with each wall having a high solar reflectance, and a solar cell positioned at the bottom aperture. The mirror trough and cell are mounted upon a base plate consisting of a sheet of material having high thermal conductivity. The solar cells interconnect to a flex-circuit wiring layer which is integral to the base plate using conventional interconnect derived from electronic packaging technology.

Although the mirror trough walls and base plate can be considered individually as components of each element, in the construction of multi-element panel, these components span multiple elements. Using longer egg-crated (interleaved) mirror strips and a large multi-element base plate allows fewer piece-parts and provides an integral structural design which helps the panel overall performance, by minimizing the need for added structure in larger panels. Both mirrors and base plate are fabricated from thin sheets of graphite fiber reinforced composite (GFRC) material for achieving high stiffness and lightweight in its structural function.

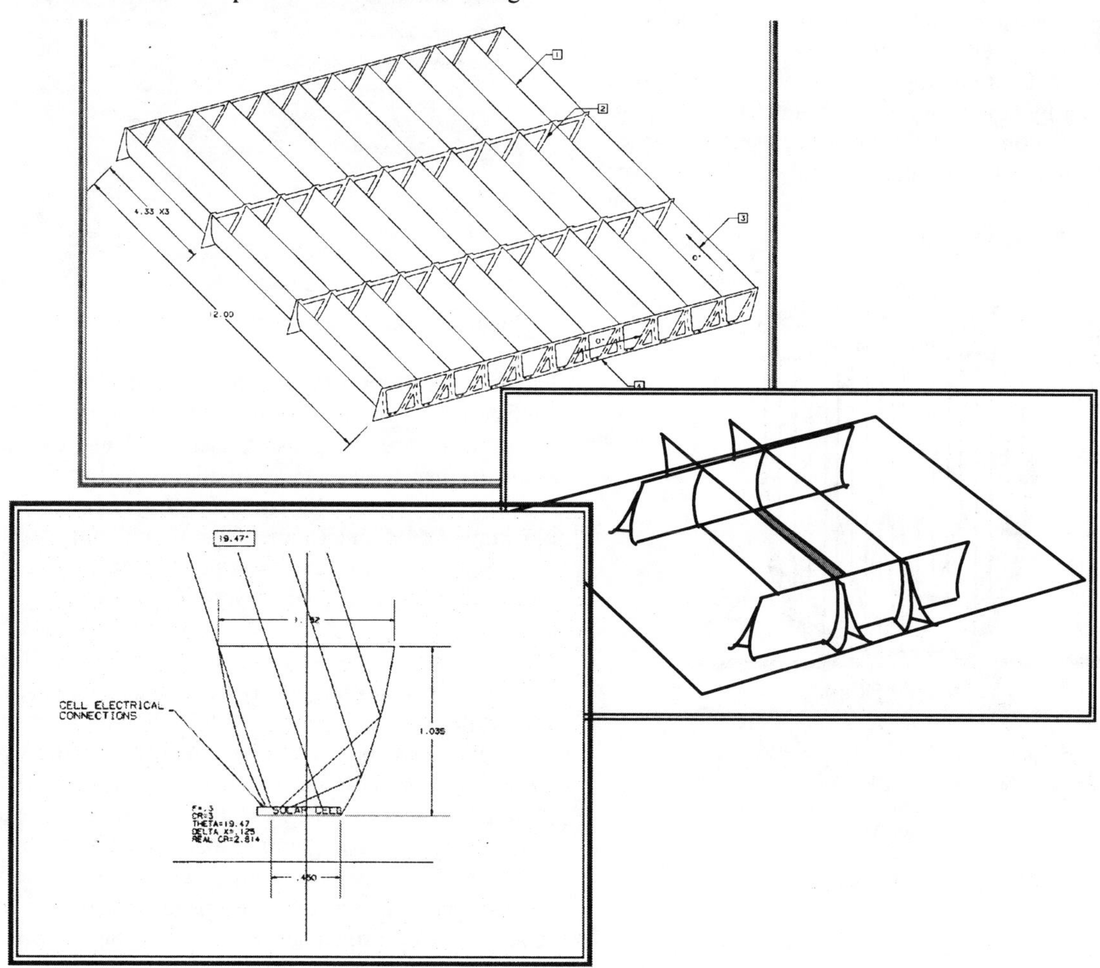

Figure 1. Conceptual Design of the Light Concentrating Panel

The theory of operation of the LCP is basic and analogous to planar arrays. Within each concentrator element, incoming sunlight is directed to the solar cells both by direct illumination, and by reflection off the trough mirror walls (see Figure 2). For most of the mirror, a single reflection suffices to place the ray of sunlight onto the solar cell, whenever the element is pointed toward the sun within the required acceptance angle. The acceptance angle is defined by the maximum angle that the panel can be pointed off of normal to the sun before reflected rays begin to miss, or "walk off," the solar cell. The ray-tracing example in Figure 1 illustrates this, showing the incoming sunlight arriving at near the maximum acceptance angle for that design, about 20 degrees off normal; further off-pointing would cause reflected rays to walk off the left side of the solar cell.

Reflected light incurs a small power loss due to mirror spectral absorptance and non-specularity (scatter). Only in the small corners of the trough are two or more reflections often incurred prior to reaching the cell. Light reaching the cell assembly passes through a transparent coverglass and is converted to electricity by the photovoltaic cell. Thermal control of the cell is accomplished by radiating directly off the cell front surface, and by conduction through the cell to the base plate for radiation across the entire back surface. Since the best mirrors are poor emitters, the mirrors themselves do not contribute significantly to the thermal control of the element.

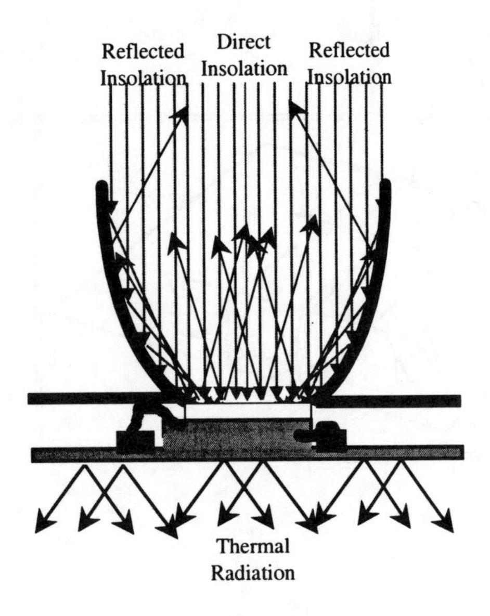

Figure 2. Basic LCP Element Operation

Elements can be treated electrically as individual solar cells, and are connected in series parallel relationships to achieve appropriate panel voltage and current. The main difference in interconnecting elements, as opposed to solar cells in a conventional planar array, is that there is a significant spacing between the cells. This provides a benefit compared to planar arrays that must pack the solar cells as tightly as possible to maximize light collection. The LCP elements hide these inter-cell areas underneath useful mirror area making assembly and repair of individual cells considerably easier. The assembly approach uses surface mounted solar cells and thermosonic wire bonds to create a cell array that is similar in technology to today's large flex-circuit assemblies, which enables implementation of automated bonding and assembly equipment used in those kinds of applications. The cell spacing that enables this is shown in the photograph of Figure 3, depicting a cell array prior to application of the mirror.

Figure 3. Cell Array Prior to Application of Optics

Panels comprising arrays of these elements in a fixed configuration can be accordion folded and deployed in the same manner as planar panels, and use the same kinds of panel and array auxiliary hardware as do conventional planar panels, including array harness, blocking diodes, inter-panel hinges and linkages and tie-downs. As with a planar panel blank "keep-out" areas can be implemented for interface hardware.

We have shown[2] the performance of the generic design of the LCP concentrator to be similar, in terms of Watts/kilogram and Watts/m^2, to a rigid planar panel using the same type of solar cell. This analysis has shown that, at a baseline concentration ratio of 3:1, the pointing tolerance of the LCP in the most sensitive axis is +/-20 degrees. The pointing tolerance was confirmed using a representative ground test model. Thermal analysis has shown that the panel reaches a peak temperature on orbit that is about 20C higher than an equivalent planar rigid panel. Table 1 summarizes the results of these prior studies.

Table 1. Comparison of concentrator and planar performance and cost

Panel Type	Structure Specific Mass (kg/m^2)	Photovoltaics Specific Mass (kg/m^2)	Cell Temp (degC)	Area Specific Power (W/m^2)	Mass Specific Power (W/kg)	Typical Panel Costs ($/Watt) In Quantity
Planar GaAs	1.25	1.48	50	228	84	$450
LCP GaAs	1.80	0.53	70	224	95	$250
LCP MBG	1.80	0.53	70	279	120	$280

SMEX/WIRE Flight Test Panel

A flight model of the solar concentrator was fabricated, tested, and flown as part of the SMEX/WIRE primary power system. SMEX/WIRE was an astronomy mission that used advanced composite designs for the bus structure and the solar panels. One of the advanced technologies demonstrated on WIRE was the SMEX•LITE solar panel, which uses modular solar panel modules, each ~(21cm x 44cm), mounted onto a composite window-frame. In the approach, solar array substrates of a standard physical and electrical configuration are pre-populated with high efficiency cells and mounted onto a composite frame designed for a specific mission. For this mission, two wings with nine modules each were used. NASA allocated one module on each wing for advanced technology, allowing an LCP module to substitute for one of the standard planar modules in one wing. The LCP flight panel was designed based on SMEX•LITE module requirements, which are summarized in Table 2. Compliance to requirements was demonstrated for each item through analysis, inspection or test.

The LCP module electrical design, which connects individual SAC elements in series to meet voltage requirements, and provided an appropriate voltage tap point, is shown in Figure 4. The Printed Wiring Board (PWB), or flex circuit implementation, used to connect the various mounted solar cell assemblies according to the schematic of the electrical design, is shown in Figure 5, and includes the wiring traces providing the required voltage tap, and feed-through of the flex-circuit to the panel backside. A sketch of the solar cell configuration within an element is shown in Figure 6.

Table 2. SMEX•LITE requirements used in the design of the LCP panel for WIRE

Item	Requirement	Verification
Physical	Dimensions of 43.61cm X 20.93 cm	Inspection
Power Output	35 Volts Vmp and 17W at maximum operating temperature	Test / Analysis
Voltage Tap	Located so that Voc of the upper string <26.3V @ -80°C	Analysis
Bypass Diode Protection	Bypass diodes to be provided for each cell assembly	Inspection
Blocking Diode Protection	Redundant blocking diodes for both taps	Inspection
Isolation	100MegOhms between cell circuit and structure	Test
Contamination / Outgassing	<1% Total Mass Loss; <0.1% Volatile Condensable Material	Analysis / Heritage
Mass	<0.27kg not including substrate	Design / Test
Environmental Durability	Accidental Damage / Malfunction, Panel Outgassing, Vibration/ Acoustic, Thermal cycling, Ultraviolet Radiation, Charged Particle Radiation, Atomic Oxygen, Micrometeoroid / Space Debris, Shadowing	Heritage / Analysis (Thermal cycling & vibration durability verified by test)

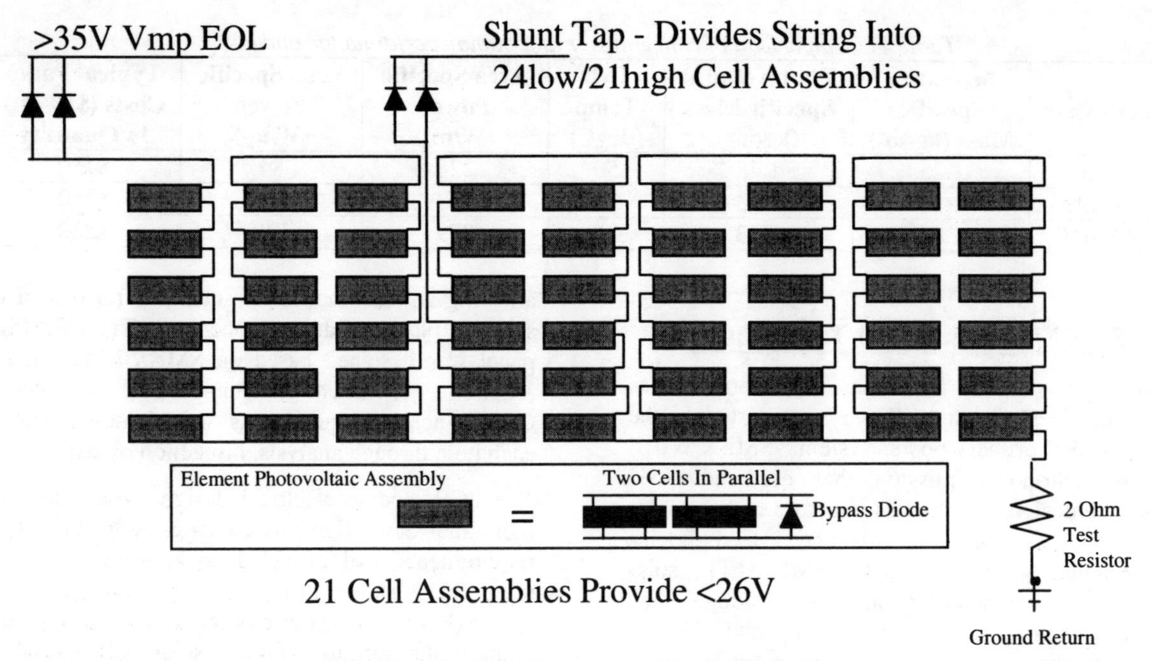

Figure 4. Solar Array Concentrator electrical schematic

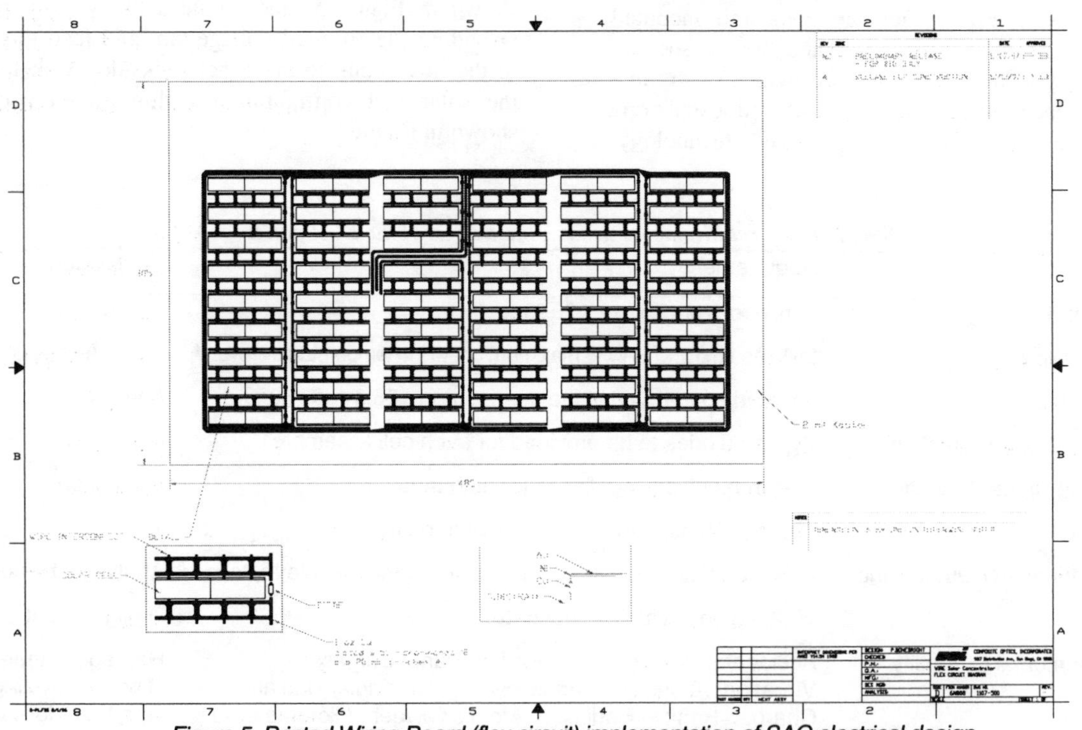

Figure 5. Printed Wiring Board (flex circuit) implementation of SAC electrical design

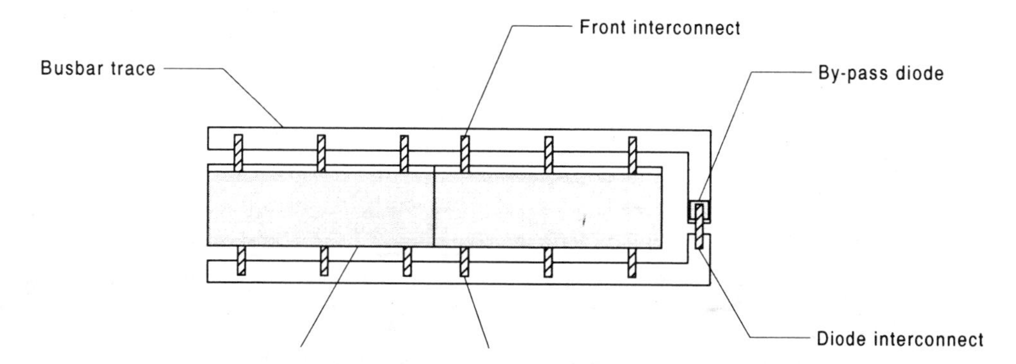

Figure 6. Solar Cell Assembly design.

SMEX/WIRE LCP Module Design Analysis

Structural, Thermal and Electrical/Photovoltaic analysis was performed on the specific LCP module design to assure that it would meet requirements. The LCP module's structural durability was analyzed by creating a Finite Element Model of the structural components of the design, adding in the non-structural mass, constraining the model to represent the launch and operating conditions and determining the stresses imposed on the various structural elements. The results of the maximum principal stresses, shear stresses and buckling loads indicated a minimum margin of safety of greater than 2 for all components of the design, indicating a robust structural response.

Two thermal models were generated with applicability to this program to determine the operating temperatures of the solar cell and other components in space. A detailed or micro-model of the LCP element, was used to determine the detailed temperatures of the components in a steady state and orbital varying thermal environment. The micro-model assumed the SAC was not influenced by the supporting array structure, and so could use the repeatability of the design to allow increased model resolution. A second, less detailed or macro-model was developed specifically for this effort and used to determine the effect of the connection between the LCP module and the supporting solar array frame. Both models were used as inputs to the electrical analysis, and also reviewed to determine if maximum allowable temperatures for the components might be exceeded and to determine if differential temperature induced stress from coefficients of thermal expansion could potentially be a problem.

The micro thermal model nodal model is shown in Figure 7. The model was run for the maximum case environment with worst case end-of-life material thermal properties listed in Table 3, in order to evaluate the worst-case temperature conditions. The result of the micro-thermal model under these conditions is shown in Figure 8. The peak solar cell temperature is 103C, and the temperature gradients throughout the concentrator element, i.e. from cell to radiator to mirror, are relatively small, in the range of 3C. Since the chosen orbit for SMEX/WIRE is sun-synchronous, we expected the earth load on the concentrator to be small, and therefore the peak temperature to more resemble that seen in this model in the time near eclipse entrance, or about 85C.

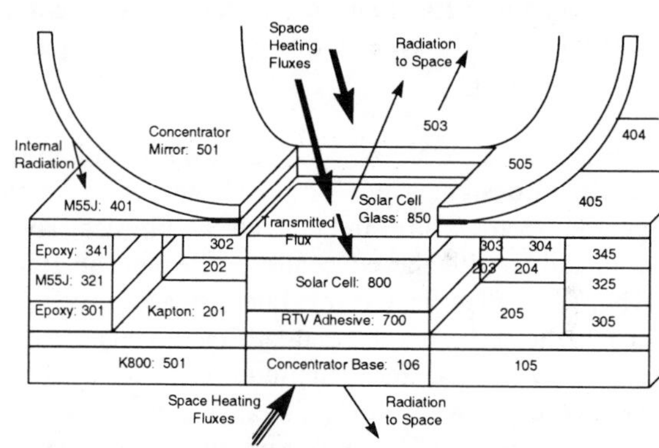

Figure 7. Micro-thermal model nodal arrangement.

Table 3. Micro thermal model worst case, end-of-life material properties and environments.

Property/Environment	Value
Solar Cell Absorptance	0.85
Solar Cell Efficiency	0.15
Mirror Absorptance	0.15
Backside Absorptance	0.39
Solar Flux	1393 W/m2
Earth IR	236 W/m2
Earth Albedo	42% of Solar

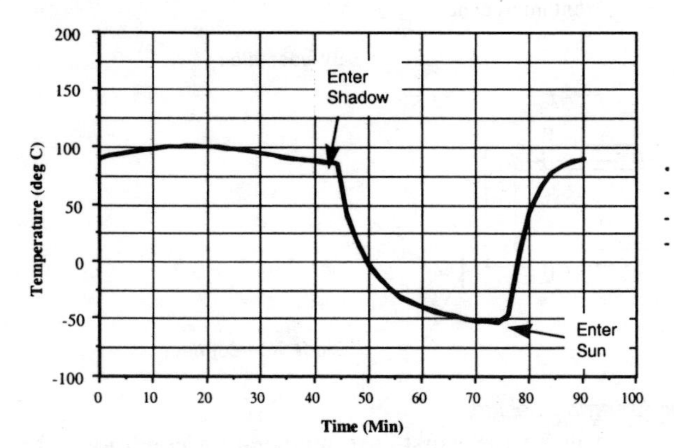

Figure 8. Results of the micro-thermal model.

Using the micro-thermal model results as a baseline, we then determined the effects of mounting the concentrator on the SMEX•LITE frame using the macro thermal model. The analysis was used to determine temperature gradients from one element to the next based on blockage of radiating elements and the conduction of waste heat into the frame structure. The magnitude of this thermal gradient was 5C and results in an estimation of peak solar cell temperature increasing to 108C. This peak solar cell temperature was then used in the electrical analysis for sizing the solar cell strings.

The electrical analysis was based on a preliminary electrical design, which arranged the cells in series parallel relationships and created the physical layout compatible with the concentrator element apertures. The solar cell string electrical design, as shown in the schematic in Figure 4, used sufficient cells in series so as to exceed the minimum end-of-life (EOL) voltage of 35 volts. In addition, a low temperature analysis at -80°C was used to determine the appropriate placement of the voltage tap within the solar cell string. The schematic was implemented in a flex circuit configuration that accounts for the placement of the cells within the substrate assembly, the potential approaches for bussing the current and the amount of metallization required to maintain a low series resistance loss.

The electrical analysis was divided up into a beginning-of-life (BOL) analysis to determine the voltage and current of the design at peak temperature, and an end-of-life (EOL) analysis to evaluate the effects of contamination, space radiation, micrometeoroids and thermal cycling that can degrade the output of the solar cell strings. The BOL analysis is summarized in the power chain analysis shown in Table 4. The EOL degradation factors are broken up into their effects on voltage and current, separately and summarized in Table 5. Finally, an analysis of the low temperature operation of the solar cells has shown that the tap voltage at -80°C is less than 26 volts, less than the maximum tap voltage requirement of 26.3V.

Table 4. Beginning of life analysis of LCP module electrical outputs.

Item	Efficiency	W/m^2
Insolation	—	1350
Optical Efficiency	0.95	1283
18%GaAs @100°C	0.15	192
Wiring Efficiency	0.99	190
Mismatch Loss	0.99	188
Net Efficiency	0.14	188
Power for $0.091m^2$	=	17.2 W

Table 5. End of life analysis of LCP module outputs considering environmental degradation factors

Item	Vmp	Imp	Pmp
BOL Outputs	38.7V	0.444A	17.2W
Optical Degradation		0.95	
Coverslip Darkening		0.985	
Radiation Degradation	0.99	0.99	
Series Resistance Increase (From Thermal Cycling)	0.98		
EOL Outputs	37.5V	0.411A	15.4W
Output After Resistor	36.7V	0.411A	15.1W

SMEX/WIRE Power System Design

WIRE employs a direct energy transfer (DET) power system that maintains a bus voltage between 22 and 34 volts. A partial linear sequential shunt regulates the solar array output. The solar array and battery are sized to support a load of 182 watts at end of life. The mission design life is 4 months in a full-sun, sun-synchronous orbit with a perigee altitude of 470 km and an apogee altitude of 540 km.[4]

Solar Array Module Electrical Configuration

The WIRE solar array consists of two wings with 9 solar cell module modules on each wing. The baseline SMEX•Lite solar array modules each consist of one string of 50 GaAs/Ge solar cells in series. The cell size is 4 x 4 cm. The module direction alternates to minimize stray magnetic fields. Figure 9 is a diagram of the WIRE solar array electrical configuration. The LCP module is wired in parallel with the baseline module modules. Each module is tapped to allow shunting of current, and contains parallel-redundant blocking diodes.

The electrical configuration of the LCP module, shown in Figure 9, presents the same electrical interfaces to the power system as for the planar modules. Current from the LCP module is measured by taking the voltage across a 2Ω resistor in series with the solar cell strings on the return side of the module. The LCP current thus measured is at the bus voltage less the voltage drop across the resistor. The solar array is sized for a voltage at peak power of 34 volts at end of life at the maximum predicted array temperature. In sunlight the bus voltage is approximately 31 volts, depending on the voltage-temperature limit selected for the battery. When the LCP module is not shunted, the measured current will be slightly on the short-circuit-current side of the knee of the module current-voltage curve. When the LCP is shunted, it will operate closer to short-circuit current. The 2Ω resistor is on the return side of the module so that total module current can be measured whether or not the module is shunted. There are two temperature sensors on each WIRE solar array wing. One temperature sensor is on the back side of the LCP.

Telemetry System Design

For communications, WIRE uses an S-band transponder that interfaces directly with the spacecraft computer system (SCS). The SCS has an 80386 processor with a 387 coprocessor that host the command and data handling (C&DH) software. The spacecraft controls all data flow using a MIL-STD-1553 data bus.[5] Power system telemetry, including solar array currents and temperatures, is conditioned in the Spacecraft Power Electronics (SPE) unit.

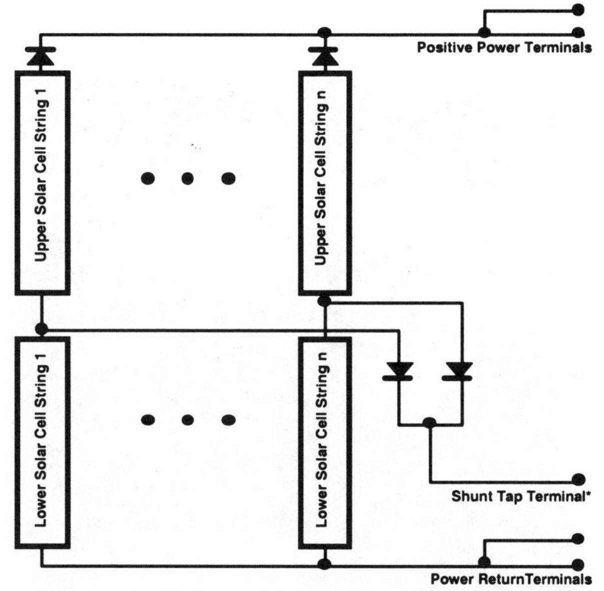

Figure 9. WIRE Solar Array Electrical Configuration

SMEX/WIRE Concentrator Module Integration & Verification

The as-built configuration of the LCP module is shown in Figure 10. Figure 11 shows how the LCP module integrated into the SMEX•LITE panel. It is noteworthy that no special modifications to the modular panel were needed to incorporate the LCP module – this was a natural fallout of the technology transparency of the design.

Figure 10. LCP Implemented as a SMEX•LITE Module

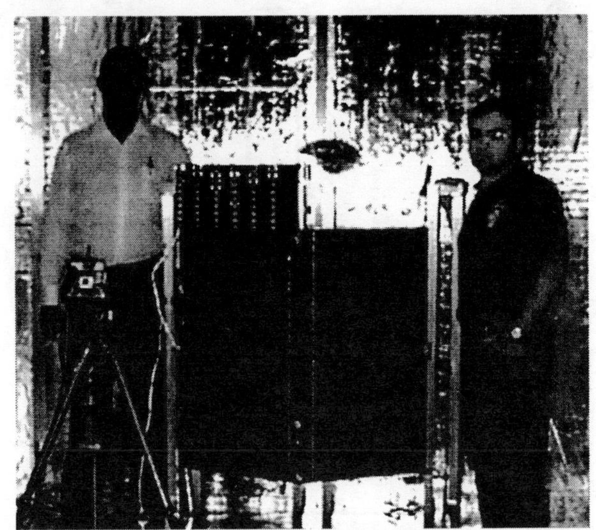

Figure 11. LCP Module Integrated into SMEX•LITE Wing

After integration of the LCP module into the wing, NASA performed ground testing including solar simulator illumination of the panel. The results of the solar simulation test are shown in Figure 12. Because the LCP module was an integral part of the array circuit, it could not be tested separately. To determine its performance, the panel was tested twice – once fully illuminated and a second time with the concentrator module covered with black cloth. The results show the extra current provided by the LCP as expected. The small bump in voltage seen in the lower right hand corner of the I-V curve is indicative of additional cells placed in series to accommodate the higher on-orbit operating temperatures, which cause greater voltage degradation.

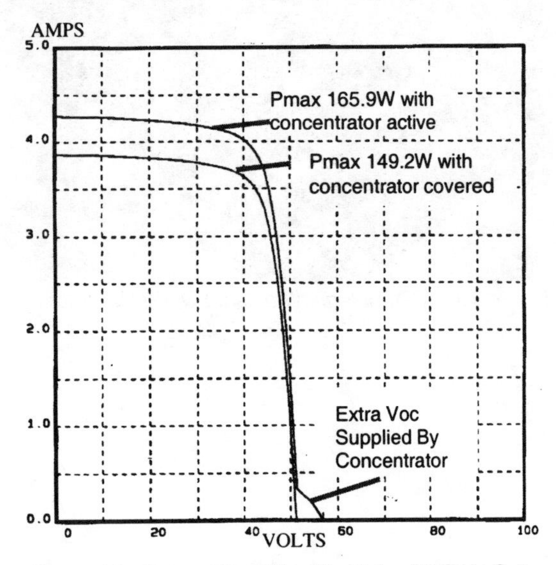

Figure 12. Ground Test Results Using NASA's Solar Simulator

Evaluation and Interpretation of Telemetry

SMEX/WIRE was launched in March of 1999. Data received for the first six months of operation of the concentrator module has shown the expected level of performance. These data (Figure 13) show a small degradation of current that is expected from initial mirror contamination from spacecraft and solar panel outgassing products. Temperature data obtained from the LCP module indicates a backside temperature of 75C, which is consistent with the thermal analysis considering the low earth load in sun synchronous orbit.

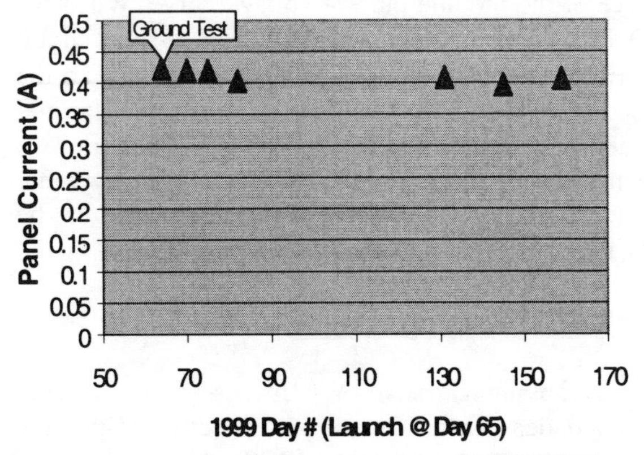

Figure 13. Space Flight Data from SMEX/WIRE Experiment

Conclusions

A concentrator panel has been designed and demonstrated that has equivalent properties to a planar panel, and is technology transparent in application. Since the design provides a plug-in replacement for a rigid solar panel, a flight experiment was easily implemented on the SMEX•LITE solar array, which uses modular rigid panels. Design, analysis and ground testing of a concentrator module for SMEX•LITE showed equivalent performance to the existing planar modules using the same type of cell. The flight testing of the LCP design on SMEX/WIRE has verified the predicted performance and shown the ability to maintain performance in the space environment.

References

1 Eskenazi, Michael I.; Murphy, David M.; O'Neill, Mark J.; Piszczor, Michael F.; **Present and Near-Term SCARLET Technology**, AIAA Aerospace Sciences Meeting & Exhibit, 36th, Reno, NV, Jan. 12-15, 1998.

2 Stern, T., **Technology Transparent Light Concentrating Panel for Solar Arrays**, 18th AIAA International Communications Satellite System Conference, Oakland, CA, April 10-14, 2000

3 Stribling, R., **Hughes HS702 Concentrator Solar Array Overview**, Space Power Workshop, Torrance, CA, April 10-13, 2000

4 Everett, D. F., and Sparr, L., **Wide-Field Infrared Explorer Spacecraft System Design**, Proceedings of the IEEE Aerospace Applications Conference, 1996, p. 150.

5 Everett, D. F., and Sparr, L., **Wide-Field Infrared Explorer Spacecraft System Design**, Proceedings of the IEEE Aerospace Applications Conference, 1996, pp. 152-153.

AIAA-2000-2885

SOLAR POWER SATELLITE: POWER LOSS THROUGH PINHOLES INTO PLASMA

Henry Oman *
Consulting Engineer
Seattle, Washington

Abstract

A solar power satellite in geosynchronous orbit can deliver power through a microwave beam to an Earth-surface receiving station for distribution by electric-power utilities. This non-polluting power is available 24 hours a day. Spacecraft launch cost is predicted to drop from $10,000 a pound to $1000 a pound, and the energy-conversion efficiency of the new multi-junction solar cells is approaching 40 percent. These developments can make the solar power satellite an economically competitive source of power. This satellite would be assembled in low-Earth orbit and then boosted to geosynchronous orbit with ion-propulsion rockets. Our tests and analyses show that a substantial leakage current will flow through pinholes in the solar array whenever the high-voltage positively charged portions of the array are in a plasma environment. For example, in passing through the peak plasma zone in the Van Allen belts at 500 km altitude the power loss can be 7.72 percent of the array's output. However, the propulsion engines will need only around one-fourth of the array's output. In geosynchronous orbit the electron density is only 100 electrons per cubic centimeter, and the leakage current will be insignificant during normal operation. However, the ion-propulsion engines, when fired to correct the spacecraft attitude, will release a plasma that could carry away 56 kA of current from a 40-kV solar array. Capturing the electrons from the plasma with a 20-volt shield before they drift into the solar array's high-voltage zone is one solution to this problem.

Introduction

The world's annual consumption of electric energy is expected to nearly double to 22 trillion kilowatt hours in 20 years. Producing the needed power generation with fuel-burning power plants would significantly increase greenhouse-effect environment warming. Constructing more nuclear power plants involves overcoming popular opposition. A typical nuclear power plant delivers 1000 megawatts, so 90 new plants would be needed each year for supplying this growing load. Earth-surface solar arrays can produce power without polluting the air when sunlight is available. However, alternative power sources would be needed for times when clouds cover the sun and during evenings when the electric utilities must supply peak loads. A solar power satellite in geosynchronous orbit can generate power 24 hours a day and transmit it in a microwave beam to an Earth-surface receiving station. At this station the microwave power received by the antennas is rectified and converted into 60-Hz power for distribution by electric utilities. No air pollution would result.

Past studies showed that the solar power satellite would be economically impractical because of the $10,000-per-pound cost of hoisting the components into low-Earth orbit for assembly. Now re-usable launch vehicles are being developed. Their cost goal is under $1000 per pound of components raised into orbit. The assembled satellite can then be propelled from low Earth-orbit into geosynchronous orbit with ion rockets powered by its solar-array. Also, the new multi-junction gallium-arsenide solar cells promise 35% conversion efficiency. With these new cells the solar array area per gigawatt can be cut to one-half of our previous estimates, which assumed use of silicon solar cells. Preliminary designs of solar power satellites being developed in Russia and America were reviewed at the 34th IECEC in 1999.

A practical solar power satellite must be large, like 100 km^2 in area, because it needs a large antenna which can form a microwave beam that is sharp enough to be captured within a reasonable-size Earth-surface receiving station, like 60 to 100 sq km in area. For example, an antenna 1 km in diameter can deliver from 5 to 10 gigawatts (GW) of power. A typical nuclear power plant generates only around one GW.

Portions of the solar array will need to be at over one kilovolt above structure in order to avoid overwhelming conductor weight for carrying power to the microwave transmitter. In our previous studies we discovered that significant power loss would occur when the satellite's solar array is in operation in the presence of plasma. Natural plasma is present in the Van Allen belts at low altitudes. Plasma will also generated by the firing of ion-propulsion rockets during the climb up to geosynchronous orbit, and during subsequent attitude corrections. In this paper we review the results of our analysis and testing which quantify the power losses caused by current leakage into the plasma.

*Associate Fellow AIAA

Space Environment

Assembling a solar power satellite in geosynchronous orbit would require hoisting to low-Earth orbit the huge quantities of propellants that would be needed to lift all of the satellite's parts to geosynchronous orbit. A less-costly alternative is assembling the satellite in low-Earth orbit, and propelling it into geosynchronous orbit with electric thrusters that receive power from the same solar-power plant that later makes power for delivery to Earth. During this transit the electrically propelled satellite will operate in varying radiation and plasma environments. During subsequent altitude-corrections in geosynchronous orbit the ion thrusters will contribute to the plasma environment. .

The environment for the solar power satellite is summarized in Figure 1, where the altitude is plotted on the horizontal scale in Earth radii, nautical miles, and kilometers. The scale is logarithmic, except that the center of the Earth has been brought back from minus infinity to the center of the circle.

The Earth's magnetic field does not directly affect the high-voltage solar array of the satellite, but it controls other phenomena that do affect it. These phenomena vary by orders of magnitude in intensity as a result of solar-activity induced changes in the Earth's magnetic field and particle arrival rates. For example, the F_1 and F_2 ionosphere layers, which occupy the altitudes between 100 nautical miles (NM) and synchronous orbit, are affected by not only the magnetic fields, but also the time of day and the season of the year. During the day the solar ultraviolet ionizes the oxygen and nitrogen of the air, producing over 10^6 electrons per cm^3 (Figure 2). At night the recombination of electrons with ionized oxygen produces the air glow. At synchronous orbit the natural electron count falls to 100 per cm^3, and 60 cubic kilometers must be swept to collect one coulomb of charge.

Solar cell degradation is the most severe in the Van Allen belts that contain trapped electrons and protons. Up to 10^6 protons with greater than 4 MeV of energy can bombard every square cm of solar cells each second.

Pinhole Current Collection

Pinholes in insulation are paths through which current can escape from high-voltage solar arrays. K. L. Kennerud measured the current flowing from plasma through pinholes[1]. He varied pinhole size, insulation type, area of the electrode surrounding the insulation, shape of pinhole, and type of insulation adhesive. He also varied the plasma density, voltage level and polarity, length of plasma exposure, and background pressure. His pertinent results are plotted in Figures 3 and 4. At low voltages the leakage of current from plasma into a positive electrode depends strongly on the hole size, and is independent of voltage. At high voltages (1000 to 20,000 volts) the current was roughly equal for all three hole sizes, and did not vary significantly with voltage.

The observed current saturation at high voltage was a surprise. However, it was confirmed with other tests. For example, tests in the same chamber showed that the current did not saturate on a sphere-shaped electrode. At high voltage the measured current flowing into the smallest pinhole was several orders of magnitude greater than that flowing into a sphere having the same area as the hole. This pinhole-saturation phenomenon was not affected by the plasma source or chamber size. Additional tests showed that this saturation is repeatable and depends primarily upon plasma density and area of insulation surrounding the pinhole. The tests and results are described in Kennerud's report [1].

Solar Array Power Loss Through Plasma Leakage

The space between 400 km altitude and the orbit of geosynchronous satellites contains neutral atoms, free electrons, positive ions, and high-energy particles. The high-energy particles, although damaging to solar cells and optical surfaces, are not numerous enough to carry significant current. The free electrons, generated each morning when ultraviolet photons ionize neutral atoms, have energies of around one to two electron volts. This energy is dissipated in reactions with neutral atoms and ions, thus increasing the temperature of the medium to the region of 500° K to 2000° K. The temperature of an electron is related to its energy by Boltzmann's constant, 8.6171 X 10^5 eV per $^\circ$K.

An electrically neutral gas containing free electrons and ions in equal numbers is called a plasma. A positively charged spherical electrode, say one cm in diameter, when inserted into the plasma will collect electrons. The volume in which electrons are influenced by the electrode, called a sheath, is much larger than the sphere. Some electrons will orbit around the electrode and escape back out of the sheath. Current collection is then said to be orbit-limited.

The solar-power-satellite's high-voltage solar array looks more like a sheet electrode than a spherical probe. For example, let us assume that 10 km^2 of solar power satellite array is deployed to supply 1500-volt power for the electric propulsion thrusters that raise the satellite from a low 500-km orbit to geosynchronous orbit. K. L. Kennerud analyzed this array by using fundamental equations that had been developed by I. Langmuir [2].

Kennerud's technique converted the planar array into a sphere having the same area. Then he calculated the radius of the electron sheath surrounding the array. His experiments with small positively-charged solar-cell panels correlated well with his predictions. With a negatively charged panel which collected ions, his experimental measurements did not correlate well with theoretical predictions, perhaps because the ion sheath extended to the chamber walls.

Using Langmuir's equations, we determined that at 500 km altitude the electron sheath extends to a few meters above the plane of the solar cells. This applies to the range of electron concentrations, electron temperatures, and array voltages of interest. The calculation of leakage current then simplifies into analyzing the rate at which electrons drift into an electron sheath which has essentially the same area as the solar array. This electron current (j_r) is simply:

$$j_r = \frac{N_e \, E_e}{3.7 \times 10^{11}} \quad \text{in amperes per cm}^2$$

where N_e = electron density in electrons per cm^3
E_e = electron energy in eV

The calculated leakage currents from a 1500-volt array for several altitudes are shown in Figure 5. These calculations were based on the electron densities in Figure 2 and the electron temperatures calculated by W. R. Doherty[3].

Solar Array Extracts Electrons from Plasma

A negatively charged solar array would attract ions rather than electrons. However, ions are less mobile than electrons, and the ion current would be much smaller than the electron current which would be observed with a positively charged solar array. Thus the positively charged solar array is the worst case.

A flow of electrons from the plasma to the solar power satellite must equal the flow of electrons out of the satellite. Otherwise the satellite will become negatively charged with respect to the plasma, and will cease attracting electrons. This flow of electrons is provided during orbit transfer by the electron emitters that are installed to neutralize the ions emitted by the thrusters. After the satellite reaches geosynchronous orbit the thrusters are used occasionally for attitude control. There the natural electron density is only about 100 per cm^3, so with thrusters turned off the power lost through plasma leakage, even at 44 kV, would be trivial.

The source of charge-exchange plasma is downstream of an electron thruster where the beam ions interact with neutral atoms that are escaping from the thrusters (Figure 6). The charge-exchange plasma becomes a conducting path for the electrons going from the thruster to the solar array. This was first discovered during tests of the ATS-6 spacecraft in a large vacuum chamber [4]. The spacecraft was biased at +15 volts relative to the thruster, and substantial electron flow into the spacecraft was observed.

The ions in the charge-exchange plasma will be propelled by electric fields away from the ion beam and toward the back of the thruster. The positive ions in this plasma, not being readily absorbed by the solar array, constitute a minor part of the leakage current. The electrons, on the other hand, can funnel through holes in any solar-array insulating surface, and rob significant power from the solar array.

H. R. Kaufman and C. C. Isaacson derived an equation that gives the charge-exchange ion-generation rate[5]. We entered into their model the following possible values for a solar power satellite:

Beam current:	64 kA per corner
Propellant utilization factor	0.892
Atomic weight of argon propellant	39.94
Argon charge-exchange cross-section	2.5×10^{-19} m^2
Beam radius	16.97 meters

The values for the solar power satellite were derived as follows. Each argon thruster has an 80-ampere beam current, so 800 thrusters at each corner would produce 64 kA of beam current. The argon propellant has an atomic weight of 39.94 atomic-mass units, and a charge-exchange cross section of 2.5×10^{-19} m^2 at 10^6 meters per second. The effective beam radius was assumed to be that of a circle having the same area as the 800 thrusters, each 120 cm in diameter. Our calculations are summarized in the Proceedings of the conference, "Effect of Ionosphere on Space and Terrestrial Systems" [6].

Kaufman's equation gives a charge-exchange rate of 3.52×10^{23} ions per second, which equals 56.4 kA. However, if we assume that each of the neutral atoms is ionized only once, then the limiting rate of generation of ions for the charge-exchange plasma should correspond to the rate of release of neutral ions by the thruster. The limiting charge-exchange current would be only 7.75 kA.

The difference between the 56.4 kA predicted with Kakufman's equation and the 7.75 kA of neutral atoms coming out of the thrusters might be explained as follows: Within the current-flow path, extending from the thruster area to the solar array, the heavy argon ions might be considered as standing still relative to the much lighter electrons. There is always a one-to-one ratio of ions to electrons in this current-flow path, thus preserving the neutral character of the plasma. However, the electrons flow at high velocity from the neutralizers to the solar array, and each electron doesn't stay long in the plasma. A similar phenomenan is observed in thermionic heat-to-electric-power converters, where relatively few cesium ions neutralize the space charge and promote a copious flow of electrons to the anode.

Limiting Power Loss Through Plasma

Kaufman showed that using a cone shield around the output of an ion thruster wouldn't significantly reduce the electron current. The cone does move the apparent source of plasma downstream, and further from the solar array. However, the cone also increases the density of neutral atoms, and the net result is an increase in the leakage current arriving at the solar array.

Insulating the solar array would eliminate the leakage-current loss, were it not for the pinholes. Kennerud showed that electrons will funnel into a pinhole from the plasma sheath surrounding the solar array. If the electron current is great enough, the pinholes will enlarge as the surrounding insulation sublimes away.

A more promising method is to collect the electrons with an anode before they can get into the solar array. If the charge-exchange plasma generation is limited by the supply of neutral atoms released by the thrusters, then the collecting plate will carry only 7.75 kA, which at 20 volts represents only 155 kW or 0.3 percent of the power supplied to the thrusters. The anode could be a thin sheet of metal that is connected to the thruster structure though a 20-volt power supply.

If nothing else works, the ion thrusters could be spaced away from the solar array. We have not yet calculated how far this spacing must be.

Conclusions

Generation of power at high voltage, around 40 kV, is advantageous on a solar power satellite where power has to be carried as far as 10 km to rf amplifiers. However, leakage current through plasma can constitute a significant power loss from a high-voltage solar array. For example, at 300 km altitude a 2-kV solar array can barely generate enough power to feed plasma losses. The satellite could be assembled in a 477-km orbit, and transferred to geosynchronous orbit with 1.8-kV thrusters. At this voltage the losses in the power busses would be significant, but still only about one-fourth of the total solar array would be needed for powering the thrusters.

In geosynchronous orbit, where the solar power satellite generates power and sends it to Earth in a microwave beam, the electron density is only 100 per cm^3. Power lost in leakage current from a 40-kV solar array would be trivial, except when ion-propulsion engines are operated for attitude and orbit corrections. Accelerated ions colliding with drifting propellant atoms generate positive ions that form a conduit for carrying electrons to pinholes in the solar array. An electron current, as high as 56 kA, could represent a substantial power loss from a 40-kV solar array. One solution could be to collect the electron current in a 20-volt shield. Another is to mount the accelerators on long structures to keep positive ions away from the solar array.

References

1. Kennerud, K. L., "High Voltage Solar Array Space Environment Model," Boeing document D2-121333-1 (1970)."
2. Langmuir, I., and Bldogett, K.. B., "Currents Limited by Space Charge Between Concentric Spheres," Physical Review 23, page 49 (1924).
3. Doherty, W. R., and Wilkinson, M. C., "High Voltage Solar Array Space Environment Model," Boeing document D2-121333-1 (1970).
4. Worlock, R. and associates, AIAA Paper No. 74-1101 (1973).
5. Kaufman, H. R., and Isaacson, C.C., "The Interactions of Solar arrays with Electric Thrusters," AIAA Paper No. 76-105.
6. Oman, Henry, "Solar Power Satellite and its Interactions with Plasma and the Ionosphere," in Conference Proceedings, "Effect of the Ionosphere on Space and Terrestrial Systems," John M. Goodman, Editor, U.S. Government Printing Office Stock Number 008-051-00069-1, pages 362-370.

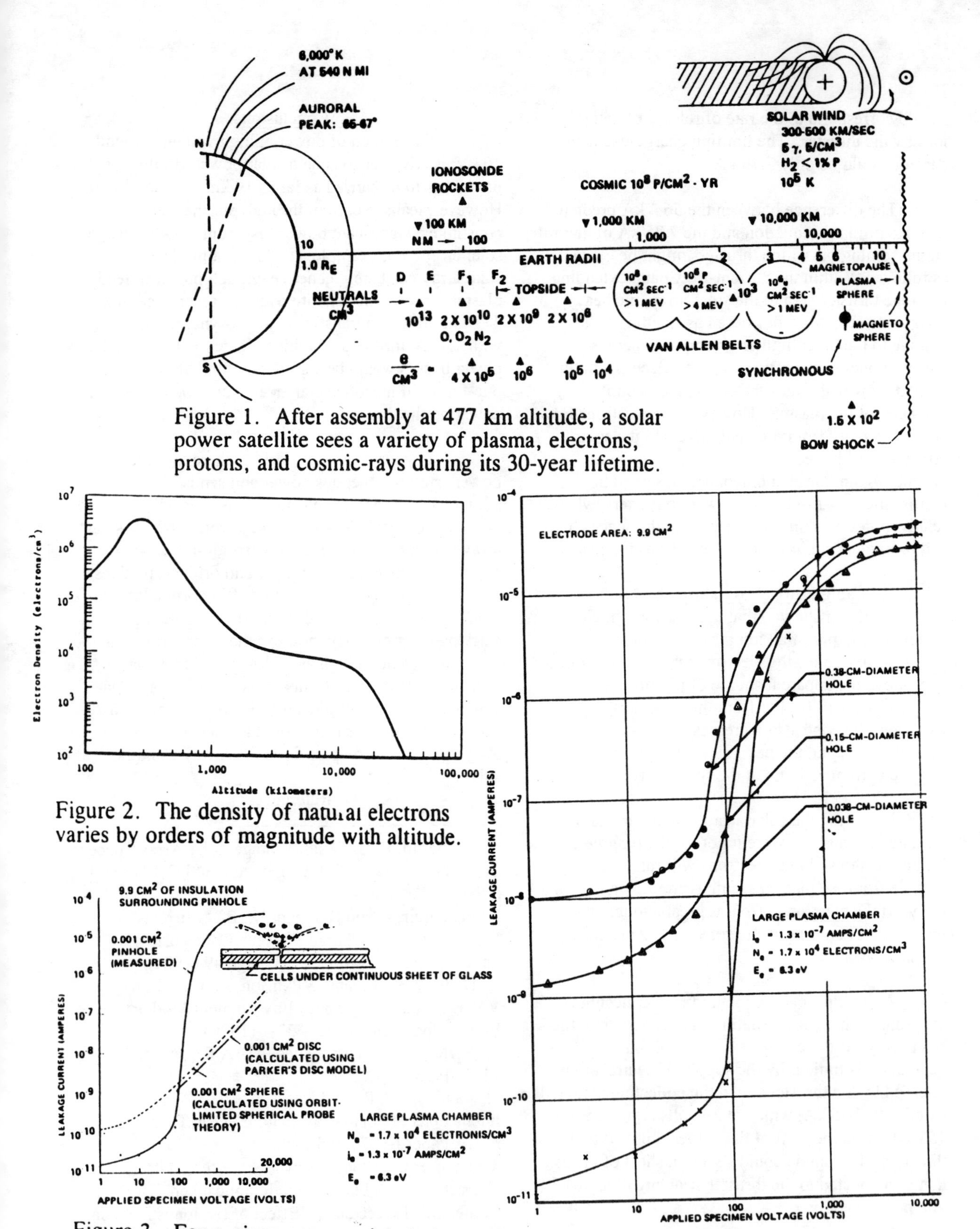

Figure 1. After assembly at 477 km altitude, a solar power satellite sees a variety of plasma, electrons, protons, and cosmic-rays during its 30-year lifetime.

Figure 2. The density of natural electrons varies by orders of magnitude with altitude.

Figure 3. For a given area, a pinhole is a better than a sphere for collecting electrons from a plasma, if the voltage difference is greater than 100 volts.

Figure 4. At above 200 volts the hole size in 125-micrometer-thick kapton no longer affects the leakage current.

Array Altitude, Km	Electron Density, N_e Electrons/cm^3	Electron Temperature ^oK	Leakage Current nA/cm^2	Leakage Current Amperes per 1500 V String*	Power Loss, Percent of Generated
500	6×10^5	3,000	824.5	0.8494	7.72
700	2×10^5	3,000	274.8	0.2831	2.57
1,000	7×10^4	3,000	96.19	0.0990	0.90
2,000	2×10^4	3,200	28.38	0.0292	0.265
5,000	1×10^4	4,400	16.64	0.0171	0.156
10,000	8×10^3	5,400	14.75	0.0152	0.138
20,000	2×10^3	9,000	4.76	0.0049	0.044
30,000	1×10^2	13,600	0.29	0.0003	0

* The string is 0.404 m by 255 m, with an area of 133.02 m^2

Figure 5. At low altitude, a 1500-volt, 133-square-meter solar array can lose 7.72 percent of its power because electrons from the natural plasma sheath will flow through pinholes in the cell covering. At geosynchronous altitude this loss is insignificant if the ion thrusters are not operating.

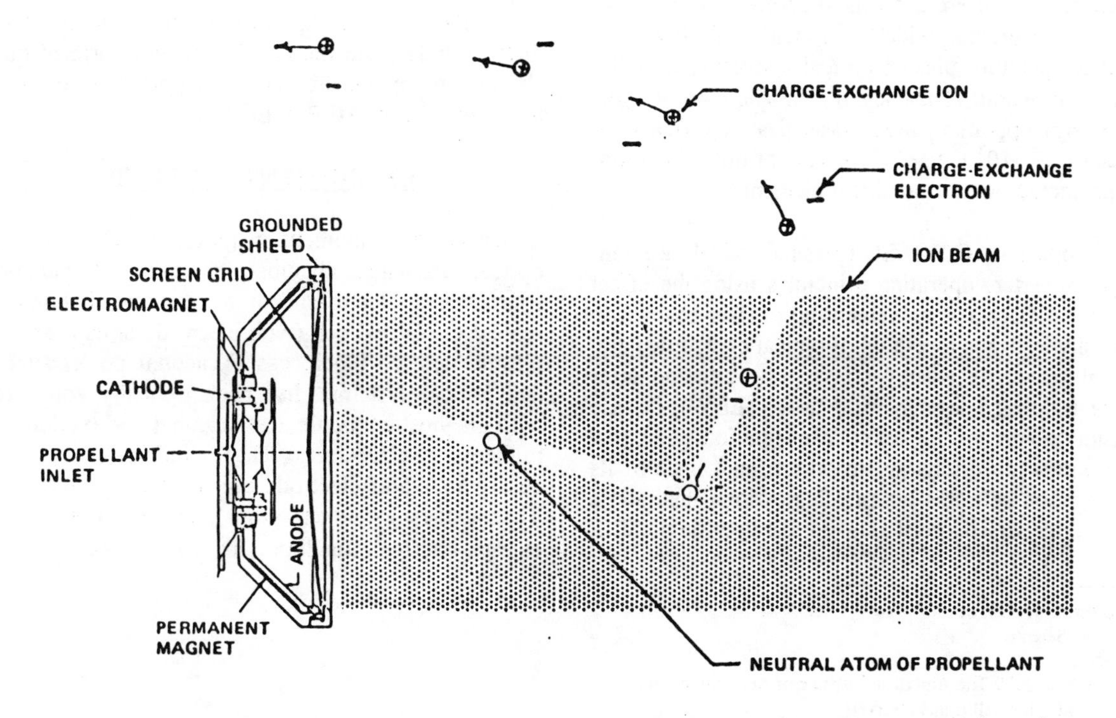

Figure 6. Drifting charge-exchange ions are generated when high-speed ions in a thruster's output-stream ionize neutral atoms of propellant. An emitter supplies neutralizing electrons that follow the ion beam. Some of these electrons are attracted to the positive ends of the solar-cell strings.

AIAA 2000-2888

EXPERIMENTAL RESEARCH OF RADIATIVE GASDYNAMIC AND THERMOPHYSICAL PROCESSES IN PLASMA OPTICAL CONVERTER OF LASER RADIATION INTO AC ELECTRICAL CURRENT. I.

Yuri S. Protasov*, Yuri Yu. Protasov**, Victor.I. Suslov**, Victor D. Telekh***
State Laboratory for Photon Energetics, Bauman Moscow State Technical University, Moscow, Russia

ABSTRACT

New method of cw laser radiation direct energy conversion into energy of AC electrical current using plasma optical cell with quasistationary optical discharge is described. The results of experimental research of intercell radiative gasdynamic and thermophysical processes determined the efficiency for all stages of energy conversion in optical discharge, supported by cw CO_2 laser radiation are presented.

INTRODUCTION

Plasma optical photon energy converter (POPEC) (as analog of thermoionic converter with plasma light heating) – is an advanced method of high efficiency conversion of radiation – laser (coherent) and thermal wideband into DC – AC electric power. POPEC module is high power density plasma-optical cell, which absorbs an intensive (E^{sp}_{rad} >10^5 W/cm^2) radiation by inverse bremsstrahlung with the plasma electrodes (and active photoprocesses) forming optical discharge of different forms supported by intensive coherent or thermal wideband radiation. POPEC – as high temperature plasma optical converter of radiation with overall efficiency η_{el} > 50%, close – cycle long term operation, high waste heat rejection temperature T_r ~10^3 K used a typerow of optical components, including new nonlinear elements[1, 2].

The proposed AC POLEC (plasma optical laser energy converter) operating principles using the effect of optical discharge movement up to laser beam with intensity less then threshold of optical breakdown are: in collimated or weakly converging radiation laser beam is formed laser burning wave (LBW), i.e. the plasma moves up the beam from ignition point (cw low power arc discharge) to an artificial limit of its propagation here this plasma slag extinguished and a new optical discharge starts from the ignition point. If the plasma passes by electrodes connected in a special way then AC power/current is generated.

Experimental and theoretical analysis of main electrothermophysical and radiative gasdynamic phases (stage I – gas breakdown and plasma optical discharge ignition; stage II – supporting of LBW with regulated dynamics and macrostructure by weak focusing cw laser beam) can be presented as following:

– analysis of thermodynamic functions, optical and transport properties of cw optical discharge plasma in conditions of POLEC cell;

– development of physical model for numerical simulation of spatial distributions of plasma parameters and discharge dynamics in quasistationary and critical regimes;

– experimental research of POLEC output characteristics and energy balance of converter cell with different active medium.

We report here on experimental and theoretical efforts to understand the physics of new form of quasistationary optical discharge supported by cw CO_2 laser radiation in AC POLEC cell.

EXPERIMENTAL SET UP

Experimental equipment includes POLEC cell, test optical discharge chamber, modules of radiation sources and electro-optical diagnostics. Laboratory scale AC POLEC cell (with electrode units chamber, plasma ignition block, gas – vacuum power supply and optical systems), has been optically connected with diagnostic module, test chamber and modules of radiators and receivers. The cylindrical and planar configuration of electrode unit (hollow porous W – Th cathodes with inert gases), POLEC optical discharge active media (Ar, He, Xe, p_Σ ~

* Professor, Sc. Dir. of Lab
** Senior Scientist
*** Docent

$(0.1-50)\cdot 10^5$ Pa) – permit operation of the converter cell in different regimes with wide band thermal or laser radiation.

Optical system (cw CO_2 laser, $E_L \sim 0.5 - 3\cdot 10^3$ W) provides an intensity $I_{0L} \sim 10^4 - 5\ 10^7$ W/cm^2 in different zones of POLEC discharge cell by focusing system. Module of optical diagnostics (including absorption and emissive spectroscopy, interferometry, probe technique and calorimetry) has been used for measurement of plasma velocity, temperature, absorption coefficients, density fields and it temporal evolution – for quantitative parametric characterization of dynamics and macrostructure of optical quasistationary discharge. Test module has been equipped by differential optical gas protection system and metrology system of cell' solid state optical elements and channels of optical system – for it' degradation control and measurements of optical characteristics (absorption/reflection coefficients, spectral radiation functions etc.) in vacuum condition and inert gas mixtures[3, 4].

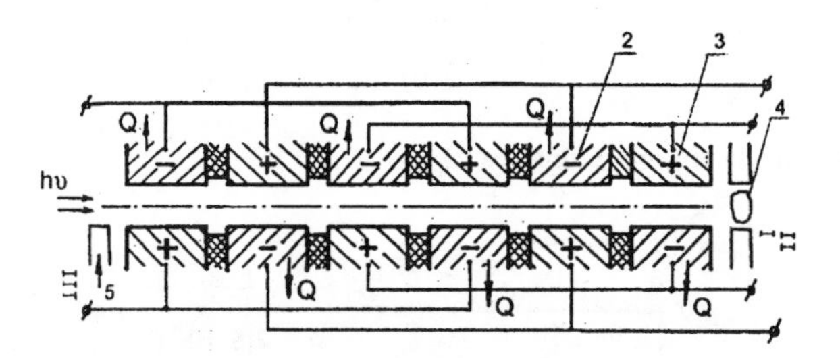

Fig. 1. POLEC electrode unit: 1 – laser radiation, 2 – cathode, 3 – anode, 4 – ignition arc plasma, 5 – system of discharge switching.

RESULTS

For AC current generation cw laser radiation is focusing in the POLEC discharge chamber, with an active gas at heightened pressure (for example, $p_{Xe} > 2\ 10^5$ Pa), and in zone z_1 (fig. 1) the low-current stationary electric arc between group of tungsten electrodes is ignited. The power density of laser radiation in section z_1 is more than threshold power density (for optical discharge maintenance), and in this case from section z_1 of axis the laser burning wave (LBW) will be spread; as plasma of the optical discharge intensively absorbs laser radiation, the distribution of power density behind the optical discharge will change. The maintenance of quasistationary optical discharge is threshold process, plasma of the discharge is not transparent for laser radiation, therefore, the standing of back boundary is determined by section, in which one the power density reaches threshold value. As approaching section z_2, in which one absorbed laser power already has not enough for maintenance of the self-supported optical discharge, the part by the absorbed optical discharge of power of laser radiation diminishes and the power density in section z_1 (in field of arc ignition) increases and will exceed threshold of optical discharge generation (new optical discharge in section z_1). The further movement of the first discharge to section z_2 will give decreasing of its size and quenching. In this moment from section z_1 starts to be spread the following discharge and process is iterated in time. Thus, usage of phenomenon of slow combustion of plasma (LBW) in weak focusing laser beam at availability of permanently existing ignition plasma allows to generate for AC POLEC periodic regime of optical discharge burning/movement.

For correct sizing of electrodes, their positional relationship, and disposition of POLEC cell electrode group concerning to caustic of using optics – the experimental research of dynamics of cw optical discharge in weak focusing laser beam has been carried out. The qualitative analysis shows, that the discharge velocity has nonuniform character (fig. 2, 3). After passage of caustic area – when the boundary front reaches section, in which density of laser power decreases below than threshold level, the velocity of it diminishes up to zero. The back boundary at this time prolongs movement, that leads to diminution of the sizes of the discharge and its quenching (thus the radial size of the discharge is commensurable with axial). Such character of movement is typical of all used active gases and mixtures in the mentioned above pressure range and power density of laser radiation.

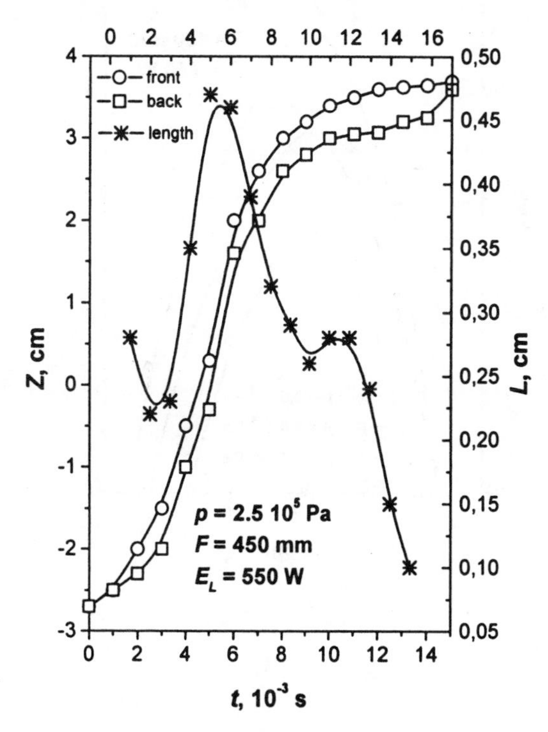

Fig. 2. Temporal evolution of front and back discharge boundaries $z(t)$, and discharge length $L(t)$.

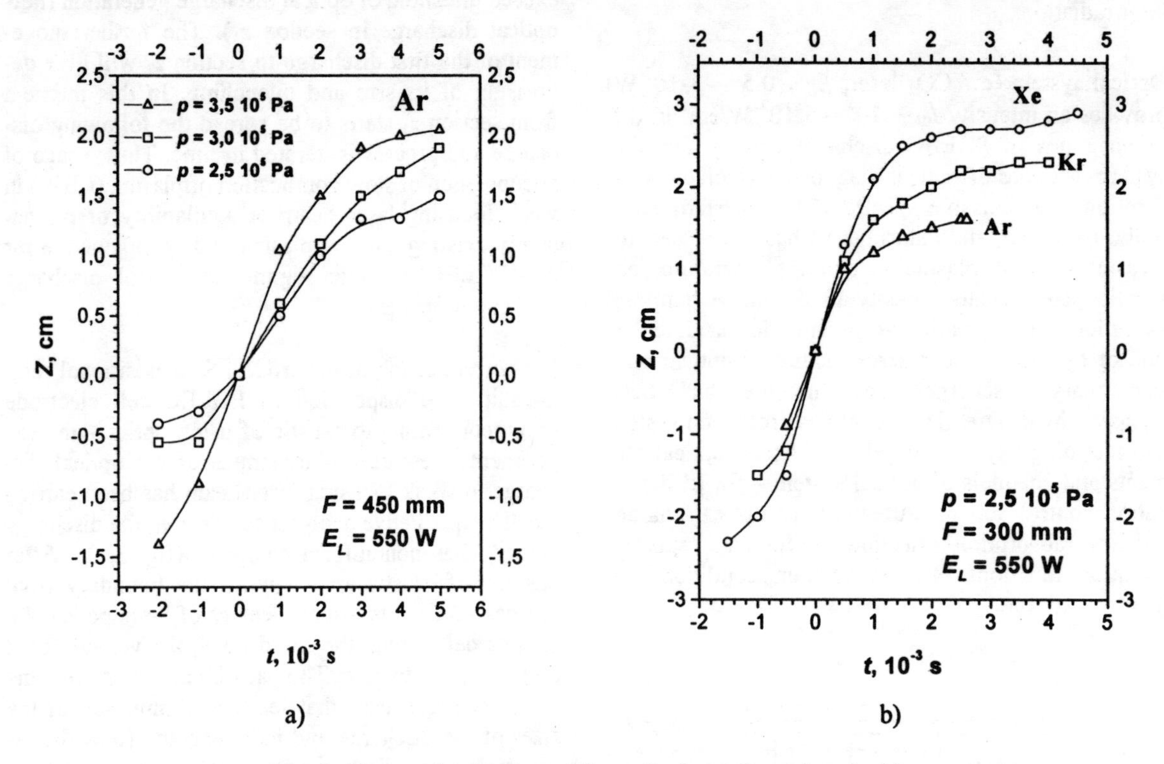

Fig. 3. Temporal evolution of front discharge boundary (a, b)

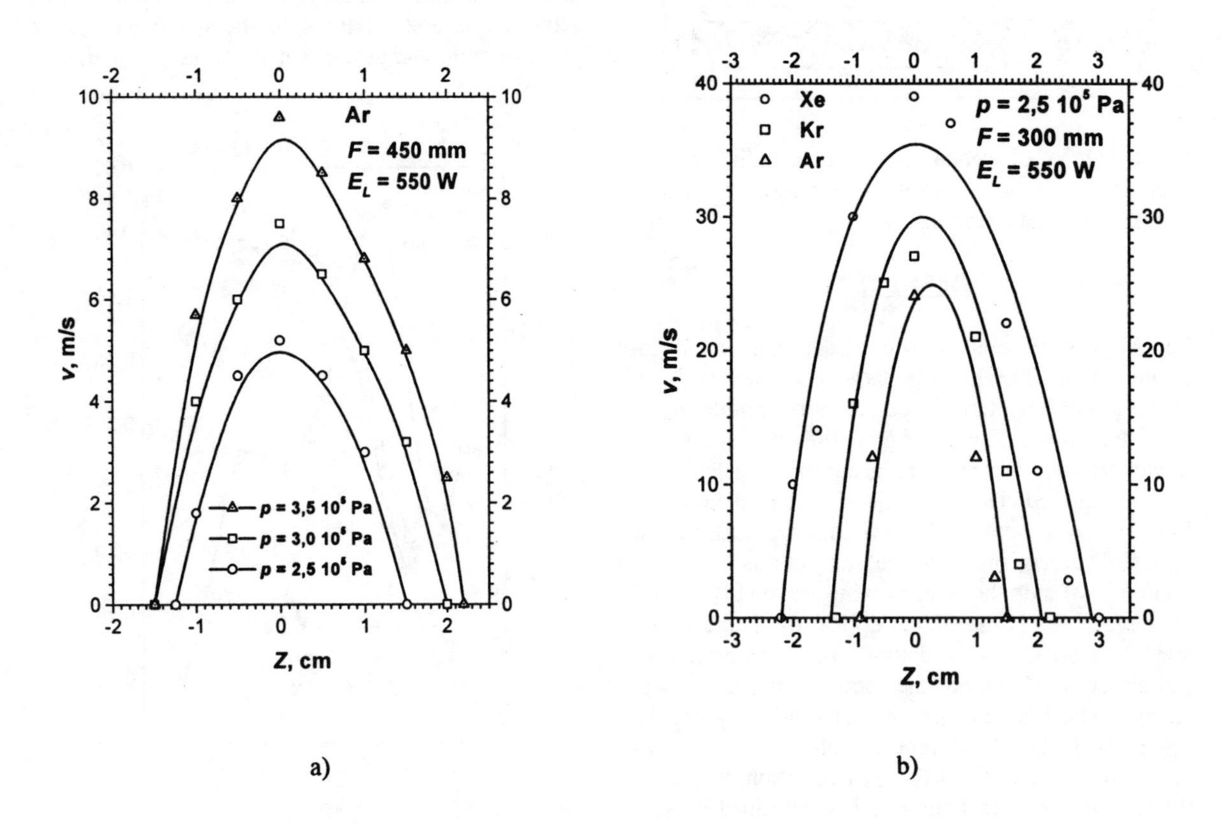

Fig. 4. Spatial velocity distribution of discharge boundary front (a, b)

In an initial stage, when optical discharge goes on concurrent beam, it movement has accelerated character. At transition in area of divergent beam the discharge rate diminishes up to zero, and, since $\Delta\tau \sim 10$ μs, the discharge starts "to hover" (localized) in space/cell gap.

The existence of the fixed discharge for long time, commensurable with period of one process of a motion of the optical discharge from ignition point up to a point of quenching – negatively influences for output characteristics of AC POLEC (current and voltage disruptures of additional laser radiation absorption). The forced quenching of the optical discharge in a zone of "hovering" can be realized by using radiative plasmodynamic methods of LBW generation[4].

Experimental study of influence of pressure both composition of plasmaforming medium and focal conditions of optical discharge dynamics (fig. 4, 5) shows, that with pressure grows in all used gases at change of focal distance the magnification of area of distribution of the discharge both in concurrent, and in divergent laser beam is watched (in a divergent beam the threshold of optical discharge formation is reduced). Experimental data allow to determine limiting values of power density, at which one will be derivated and starts to be spread the optical discharge from ignition arc at a fixed current by last. At pressure growth the power density, at which optical discharge is ignited, is reduced (fig. 5).

The essential influence of chemical composition of cell active medium – plasmaforming gas for optical discharge dynamics in cell geometry is illustrated by fig. 6, 7. Other relevant parameter of the discharge motion in weak focusing laser beam is the velocity of propagation of the discharge in the field of caustic of focalizing optics. Theoretically process of distribution of the optical discharge is described by system of equations of radiative gasdynamics with boundary conditions determined by experimentally[2]. The analysis of experimental data shows, that at passage of middle of caustic the rate of propagation of the discharge reaches the maximum (fig. 6, 7), the analysis of dependence of discharge velocity in a caustic of focalizing optics for different gases from pressure shows, that the velocity is determined, first of all, by the relation between gas pressure and its specific heat capacity (by electron temperature of optical discharge plasmain active zone). The axial size of the discharge as well as its velocity has maximum as a function of pressure (fig. 6).

The conversion efficiency (η_e, η_T) of laser radiation in POLEC cell is determined by optical properties of POLEC materials (spectral coefficients of reflection and absorption) and optical discharge plasma in converter cell chamber. The results of experimental and theoretical determination of absorption coefficients for xenon vapor – plasma are presented in table 1, and the results of numerical simulation of thermodynamic functions, chemical composition and transport coefficients for active medium in discharge zone (T ~ 5 – 25 kK) are illustrated by fig. 7, 8. This thermo-optical characterization of POLEC active medium in a wide spectral, density and temperature ranges have been used for generation of electronic data base

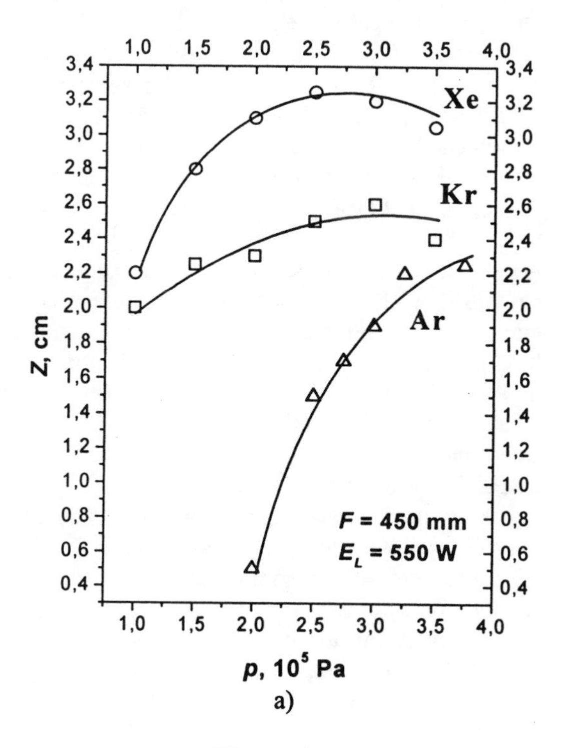

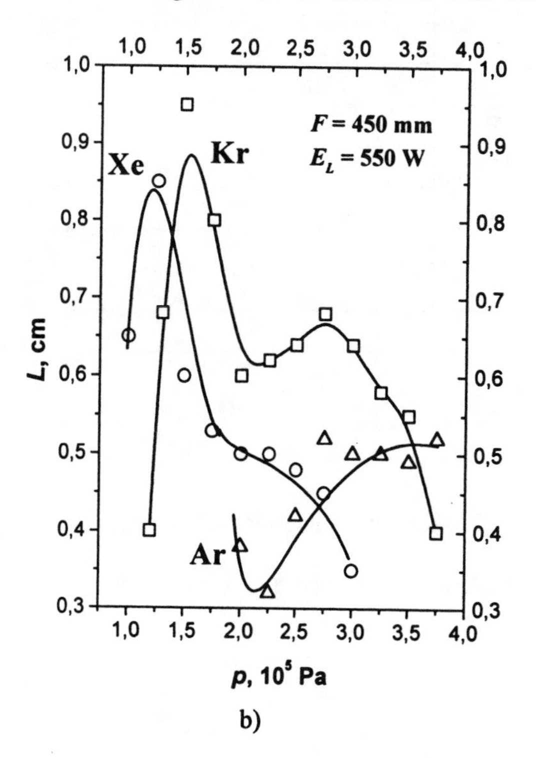

Fig. 5. Sizes of active discharge zone as a function of pressure (a, b).

"TOT-99" (Thermodynamic, Optic, Transport); it formalism and structure are descried in[3].

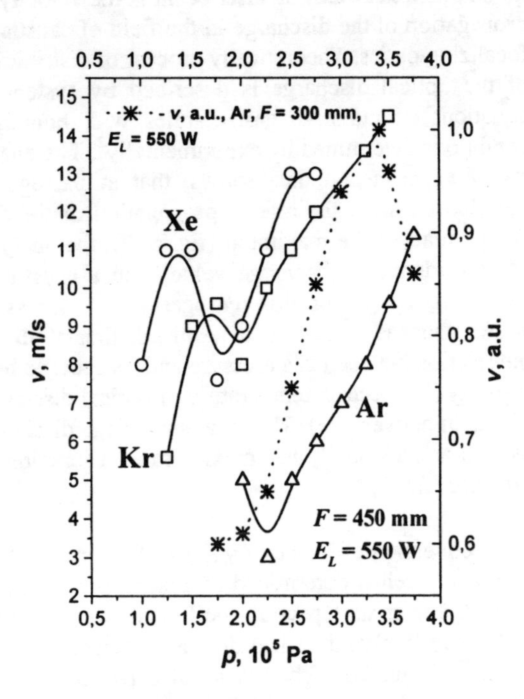

Fig. 6. Discharge velocity in caustics of focusing system.

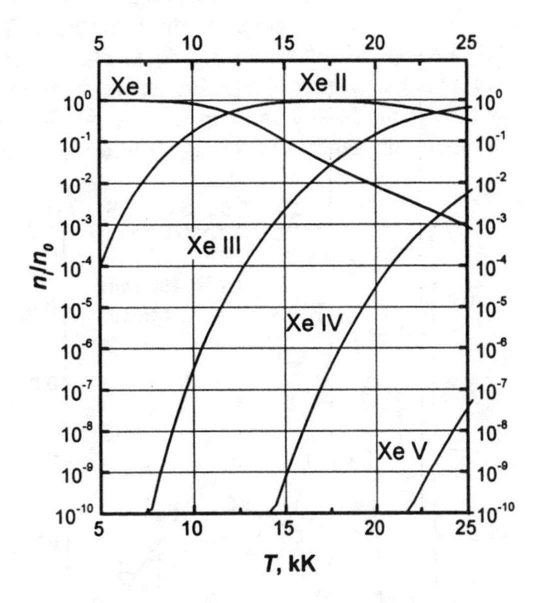

Fig. 7. Chemical composition of Xe plasma.

Table 1 – Coefficient of absorption κ'_ν, cm^{-1}, Xe, $n_0 = 10^{18}$ cm^{-3}

T, kK	λ, μm				
	10.6	1.03	0.693	0.4416	0.241
5.8	$1.9\ 10^{-5}$	$6.9\ 10^{-7}$	$6.0\ 10^{-7}$	$1.7\ 10^{-6}$	$6.3\ 10^{-6}$
11.6	3.1	$3.7\ 10^{-2}$	$2.6\ 10^{-2}$	$2.2\ 10^{-2}$	$6.2\ 10^{-2}$
17.4	17	$1.2\ 10^{-1}$	$7.6\ 10^{-1}$	$4.8\ 10^{-2}$	$1.4\ 10^{-1}$
23.2	35	$2.4\ 10^{-1}$	$1.1\ 10^{-1}$	$5.5\ 10^{-2}$	$1.3\ 10^{-1}$
29.0	62	0.4	0.2	$8.0\ 10^{-2}$	$9.0\ 10^{-2}$

One of the main parameter of the converter is the frequency of generated alternating current. The process of distribution of optical discharge in cell conditionally can be separated into two phases: first – direct motion (when $v_0 > 0$) and, second, when the discharge "hovers" in a neighborhood of a point withextreme power densityof radiation. It has been determined experimentally, that with pressure increasing the time of "hovering" is decreasing (active gas pressure growth connected with the growth of an absorption coefficient and, therefore, to decreasing of the discharge length). Thus the axial and radial sizes of active discharge zone become commensurable, that gives in magnification losses of energy from the discharge for account of thermoconductivity through end' surfaces (therefore back boundary moves faster than forward at high pressures).

So, general description of optical discharge macrostructure, analyses of current voltage characteristics and radiative gasdynamic intergap and transport processes in a wide range of pressure of active gas mixture and lasing parameters – have been used for characterization of spatial dynamics of optical discharge zones in laser beam caustic; local plasma velocity maximums, period of plasma axis oscillations and movement are necessary (and have been used) for numerical simulation of main phases of energy conversion in AC POLEC.

As a result: – optimal pressure level, chemical/ionic composition of active medium – are determined for maximum absorption efficiency of laser radiation in a long plasma gap of quasistationary optical discharge in POLEC cell at intensities of laser radiation less then threshold of an optical breakdown of active inert gases and mixtures; – the criterion of LBW generation is determined for providing plasma of quasistationary optical discharge macrostability. The conditions of optimal laser energy input into AC POLEC elementary cell (geometry and sizes of electrode' gap, arc ignition discharge unit and their mutual disposition relatively caustic of focusing optics) are characterized. On qualitative level it has been shown, that AC frequency amplitudes, phases quantity, voltage level of AC POLEC cell – can be regulated by variation of plasma – gas pressure level, geometry of

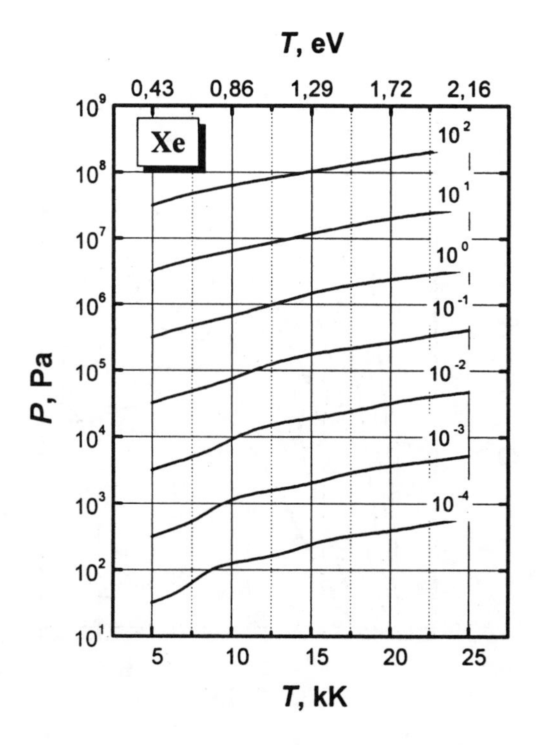

Fig 8a

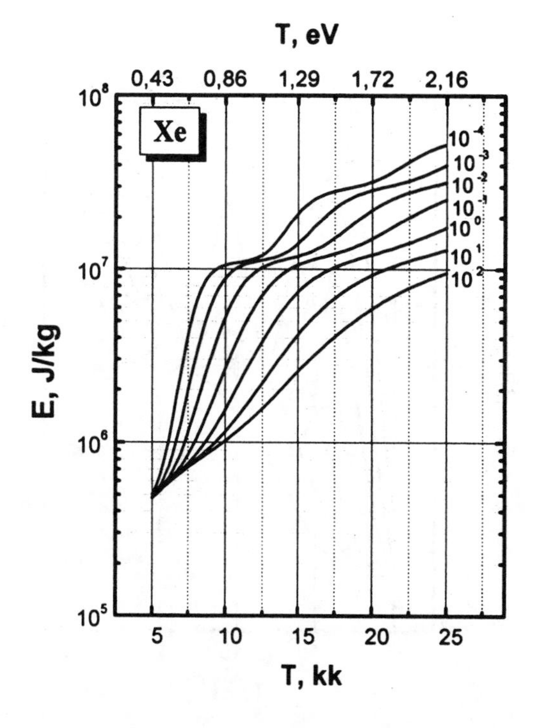

Fig. 8b

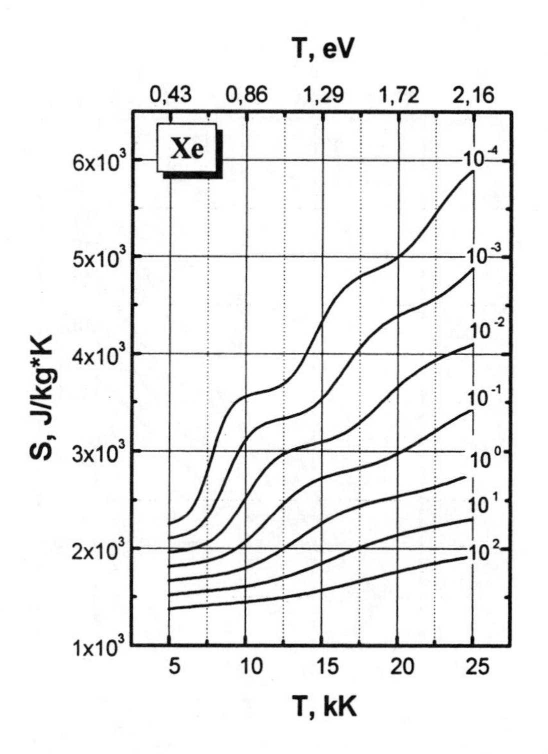

Fig. 8c

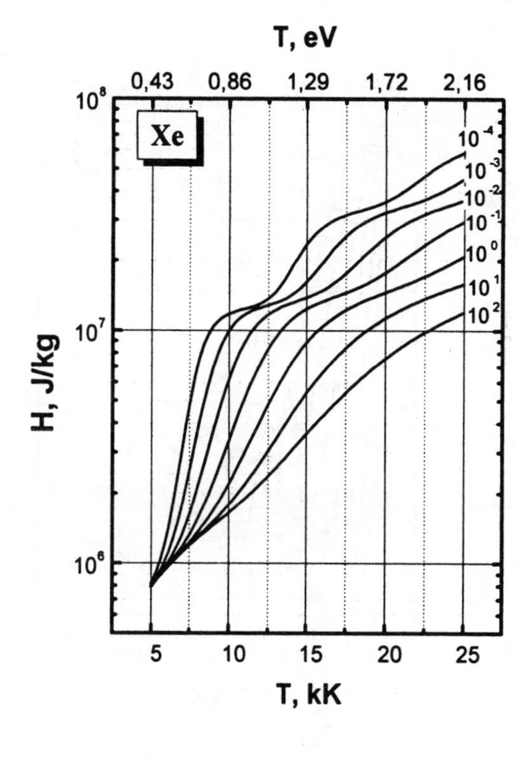

Fig. 8d

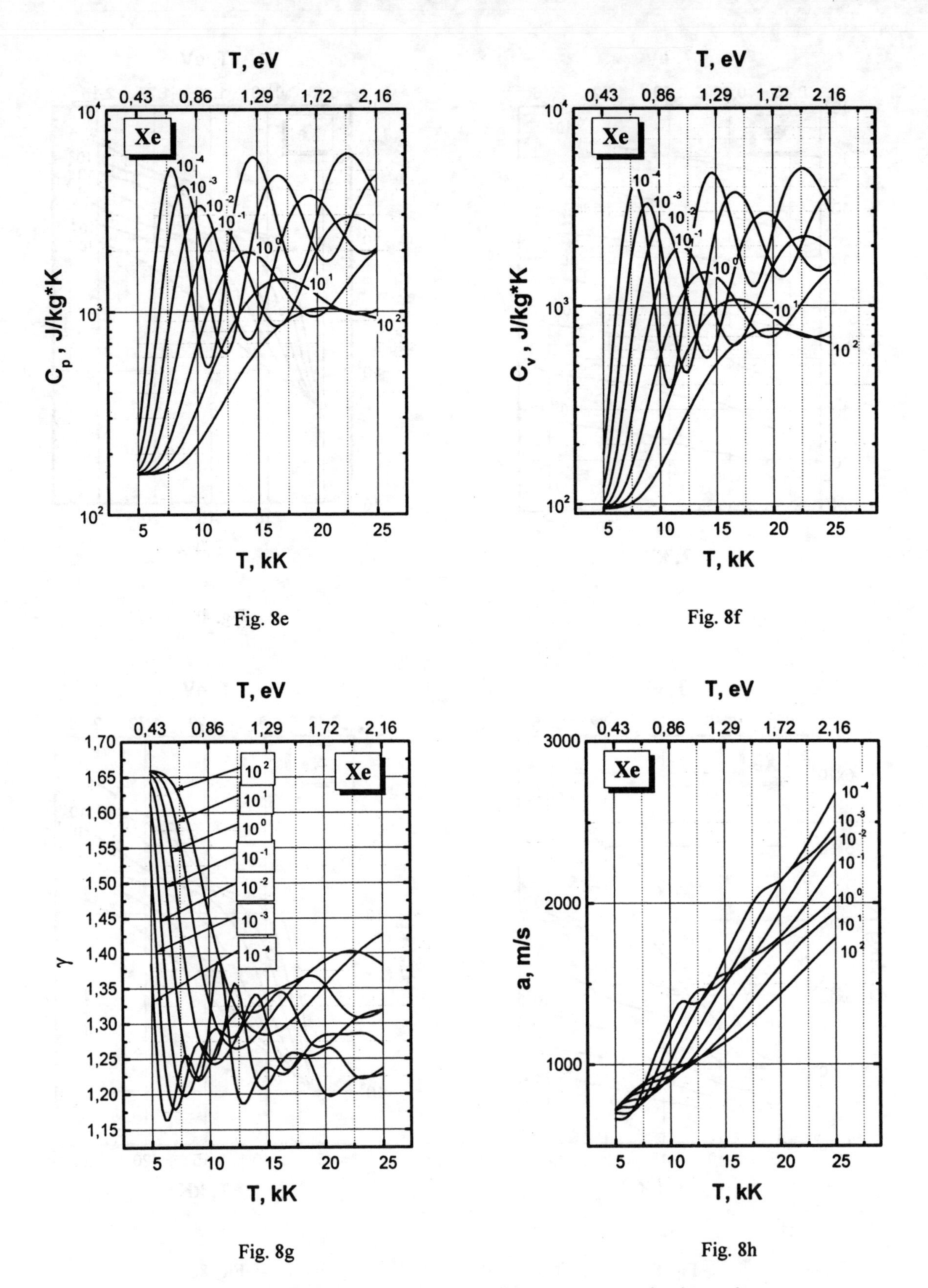

Fig. 8. Thermodynamic functions of Xe POLEC active medium: a – pressure; b – internal energy; c – entropy; d – enthalpy; e, f – specific thermocapacity; g – adiabatic coefficient; h – sound velocity

optical discharge zone and amount of electrode couples in gap of converter cell.

REFERENCES

1. Rasor, N.S., (1973) Thermoionic, Plasma and Photoemission Energy Conversion. In: Proc. of Laser-Energy Conversion Symposium (K.W. Billman, Ed.), pp. 51-63, NASA TM X-62, New York.

2. *The Radiative Plasmodynamics, V.1*, Yu. S. Protasov, ed. (Energoatomizdat, Moscow, 1991).

3. *Thermodynamic, optical and transport properties of active media of plasma and photon power generation stations*, Yu.S. Protasov ed. (BMSTU PH, Moscow, Russia, 1999)

4. Protasov Yu. S. and Chuvashev S.N. *The physical electronics of gas discharge devices. V.II. The plasma electronics*, (Vyshay shcola, Moscow, Russia, 1995)

5. Protasov Yu. S., et. al., "The similarity law for optical spectra", JETP Letters, **49, 3,** 23-27 (1989)

COMPARATIVE STUDIES OF ADVANCED MHD TOPPING POWER GENERATION STSTEMS

Naoyuki Kayukawa*

Hokkaido University, Kita-ku, Kita 13, Nishi 8
Sapporo, 060-8628 Japan

ABSTRACT

Efficiencies of four different types of MHD topping combined systems were compared where the topping units were the MHD generator and a thermochemical coal converter. The bottoming system was either combined gas and steam turbine units, a steam injection–type gas turbine, a steam turbine unit or a steam turbine coupled with a biomass unit. We also considered an IGCC system with conventional gasification schematics as a reference case. We showed that the system combined with the steam turbine and biomass units exhibited the highest efficiency of over 60% with presumably attainable MHD unit efficiencies although it is workable only under sunlight and its power scale may be limited by the pond area of the biomass plant. The present results emphasize capabilities of an advanced power generation system with extremely low environmental impacts and high efficiencie among those of so far proposed fossile fuel–fired power plants on the basis of the MHD and the state-of-the-art heat recovery technologies.

Keywords: High Efficiency Combined System, MHD Generator, Thermochemical Converter, Gas Turbine, Steam Turbine, Boimass

INTRODUCTION

The open cycle MHD generator is a heat engine that is workable under highest temperatures at present and ,therefore, has high thermodynamical efficiency capabilities. However, the generator unit efficiency itself is at most 30% because the working gas must be electrically conductive. This means that a combined system incorporated with an efficient heat recovery schematics is the key issues for realization of an MHD topping high efficiency power system. It has been thought so far that the most matured heat recovery scheme is the use of a recuperative type heat exchanger to preheat the MHD combustion air up to about 2000 K. Although a ~60% efficiency may be attainable in the recuperative–type heat recovery coupled with an advanced Rankine cycle [1], the recuperative air heater is not compatible with direct coal combustion MHD cycles due to coal slag behavior [2]. The coal slag also results in the degradation of the power output performance and enhanced damages of channel materials under high temperature, electrically stressed and alkaline seeded environment.

As an alternative to the direct coal–fired MHD–steam cycle, a concept of coal synthetic gas – fired MHD cycle with heat recovery by thermochemical coal conversion (TCC) was proposed by Batenin et al. [3] and extended studies were made by Bystrova et al. [4] on an MHD topping combined system with TCC and steam injection – type gas turbine units. Recently, Batenin et al. proposed a further advanced concetp that combined the MHD–steam injection-type gas turbine system with a biotechnological cycle [5] where the economical advantage of the bio–intracycle was emphasized, while less discussions was made on the efficiency performance. The TCC heat recovery concept and fuel upgrading reactions can apply to extremely high temperature exhausts like the MHD power generators. The present author proposed an extended MHD–TCC–gas turbine system coupled with a steam cycle [6]. The last author carried out a comparative studies of major so far proposed concepts of MHD topping combined systems driven by the TCC and a conventional pre-gasification schematics [7]. It was shown that when the MHD ehaust heat could be regenerated as chemical energy in the TCC unit until the exit temperature reduced to the solid ash temperature, the MHD–TCC–gas turbine–steam turbine system had the best efficiency performance. It had a capability to achieve a higher efficiency than the refernce IGCC system even on the basis of presently available MHD and heat recovery technologies.

This report is an extension of the work presented in Refs. [6, 7] in view of the simplification of the system, more realistic TCC design and of the environmental aspect. We considered four types of bottoming systems combined with the MHD–TCC topping system, namely combined gas turbine and steam turbine units (case I), a steam injection–type gas turbine (case II), a steam turbine only (case III) and combined steam turbine and biomass unit (case IV).

---

* Professor, Center for Advanced Research of Energy Technology

The last two systems are new combinations. The TCC design was modified with taking a more realistic concept for the ash rejection into account. Also carried out a comparison of the MHD topping system efficiencies with that of the coal gasification gas turbine–steam turbine combined cycle (IGCC, case V) operated under the conventional gasification heat supply processes.

ASSUMPTIONS

In order to calculate the performance of different systems under minimum ambiguity and to compare those upon common basis, we introduced following assumptions: (1) Adiabatic overall reactions were assumed in the gasifyer and combustors. The heat of reactions was estimated by the standard formation enthalpies. The syngas components are CO and H_2 with a volume ratio $CO/H_2=\alpha$. An equilibrium calculation shows that $\alpha \simeq 2$ and it is almost independent of the pressure. (2) Coal was represented by X_c kmol of C. The mass and heat of the coal slag or ash were ignored. (3) As for the biomass product we assumed chrollela which is presented by the carbohydrate CH_2O with the standard formation enthalpy -158.45×10^3 kJ/kmol. (4) The pebble bed–type gasifyer concept [8] was employed for the TCC design with the exit temperature being higher than the ash melting temperature, namely T_C = 1500 K. (4) The same efficiency of the corresponding power generation units was assumed in each case. (5) The inlet and outlet temperatures of the heat recovery units were identical in all cases. (6) The pressures each in the MHD combustor, gas turbine combustor and the steam turbine were 1 MP_a, 3 MP_a and 21 MP_a, respectively. A multi-stage superheated steam turbine unit was supposed. The fuel and the oxidant compression processes were isothermal. (6) The oxidant for the IGCC case gasifyer and the MHD combustor in cases I, II and III was supplied from liquid oxygen state at 100 K and 3 MP_a. The work for liq.O_2 production was assumed as q_{LO_2} = 27.3 MJ/kmol. (7) An amine–type CO_2 absorption was assumed in cases I ~ IV and the energy consumption was neglected. (8) In case IV, the CO_2 separated from the synthetic gas mixture was pressurized and fed to the biomass unit. The oxygen enriched air from the biomass unit was purified in the oxygen plant. No liquefaction was assumed and the oxygen purification work was estimated by $q_{LO_2}/5$ in this case. (9) Any units for NO_x and SO_x reduction were not considered. The desulfurization may be achieved automatically in the MHD topping cases I ~ IV due to alkaline seed and sulfur reaction taking place in the steam generation and heat recovery units.

The applicability of the assumptions listed above have been discussed in detail in the Ref. [7].

DESCRIPTION OF THE SYSTEMS

The configurations of the cases I, II, III and IV systems are presented in Figs. 1, 2, 3 and 4, respectively. In the MHD–TCC topping cases, the thermochemical coal converter (TCC) is arranged next to the MHD diffuser where coal is gasified upon mixing with the MHD exhaust gas and additional superheated steam. The high temperature MHD exhaust heat is recovered primaliry as the chemical energy of the synthetic fuel.

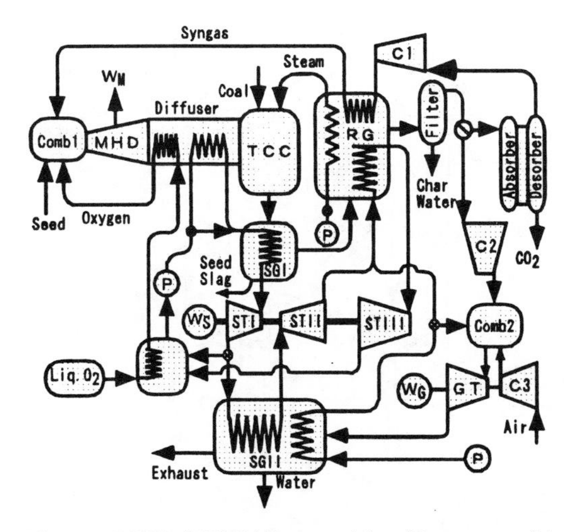

Fig.1. MHD/TCC/Gas turbine/Steam turbine combined system (case I).

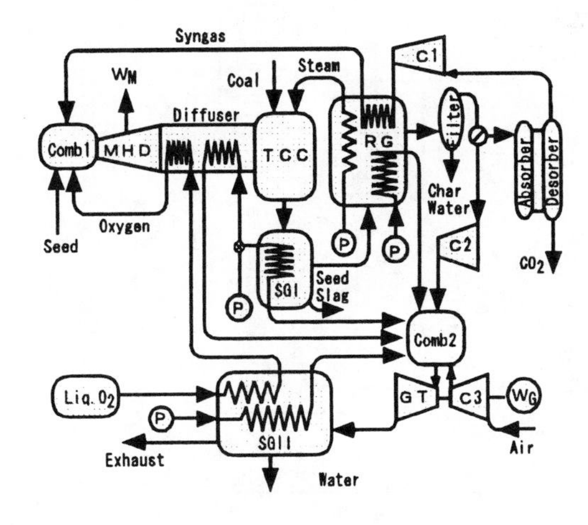

Fig.2. MHD/TCC/Steam injection-type gas turbine combined system (case II)

The TCC is an alimina–pebble bed type gasifyer proposed and developed by Yoshikawa et al. [8]. Sufficiently long time can be achieved for the reaction of coal and oxidant mixture passing through layered pebble balls. The melted slag is ejected forcibly by the syngas mixture and separated from the gaseous phase in the steam generator I (SGI) where the heat of syngas is recovered by superheated steam in the temperature range 1500 K to 1100 K. The SGI is a new unit added to the MHD–TCC–GT–ST system in Ref[7]. Due to this modification, the influence of the Rankine cycle on the overall performance is enhanced and the efficiency dependence on the syngas fraction χ will also be changed as given below in the present paper. The steam generation should also be performed at the MHD diffuser boiler in order to supply constant heat to the TCC even when the amount of the syngas to the MHD combustor is changed in the cases I and II.

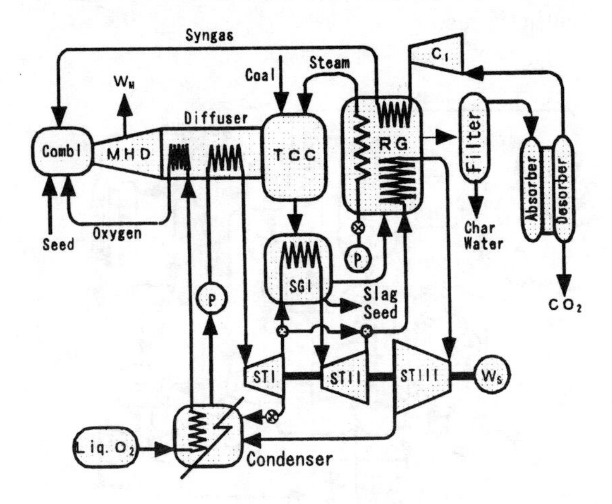

Fig.3. MHD/TCC/Steam turbine combined system (case III)

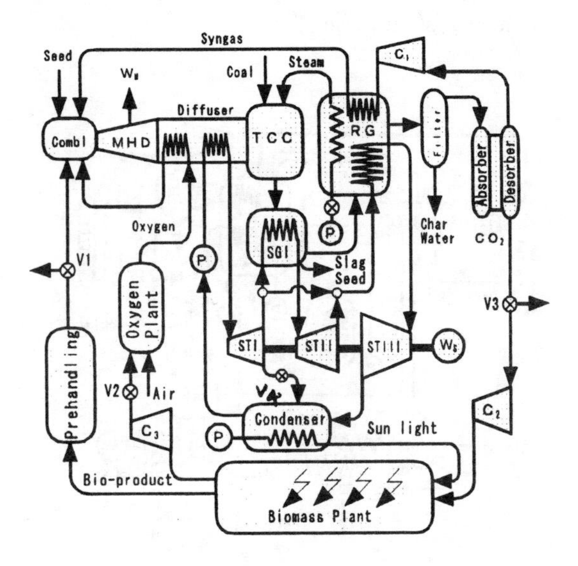

Fig.4. MHD/TCC/Steam turbine/Biomass combined system (case IV)

The syngas is then sent to the regenerator (RG) of which inlet temperature is 1100 K and, therefore, a metal tube type design can be employed. Here, the heat from 1100 K to 355 K is recovered by steam generatoin each for the TCC and the steam turbine in cases I, III and IV and for the gas turbine in the case II. In the RG, the heat is further recovered by the syngas fuel sent back to the MHD combustor at about $T_f = 1000$ K. After the RG the solid and liquid components are removed from the syngas at the filter. In cases I and II, a fraction χ of the mixture is separated at the distributor (D) to the CO_2 separator and the fraction $1-\chi$ with CO_2 is sent to the gas turbine combustor with pressure 3 MP_a. In cases III and IV, all part of the syngas is sent to the CO_2 separator. The CO_2 free syngas, CO and H_2 mixture, is pressurized up to 1 MP_a, preheated at RG and fed back to the MHD combustor. The oxidant is pure oxygen preheated up to $T_{O_2} = 800$ K at the MHD diffuser. The syngas containing CO_2 distributed to the bottoming unit in cases I and II burns with air at the gas turbine combustor. The combustion gas expands through the gas turbine unit, goes down to the steam generator II (SGII) and finally to the stack at 400 K. The gas turbine inlet temperature is assumed to be $T_G = 1500$ K.

In case II, the molar amount of injected steam from the diffuser boiler, the SGI and the SGII are determined so as to meet the assumed gas turbine temperature T_G under the constraint that the steam temperature is higher than the saturation temperature at the given pressure. In case I, III and IV the steam to the steam turbine is assumed to be 850 K and 21 MP_a. Also, in order to meet the gas turbine inlet temperature condition, an appropriate amount of saturated steam is injected from SGII to the combustor. Therefore, the steam molar amount supplied from the topping and bottoming steam generation units is different each other. This may be achieved by valves and return path from the first stage turbine to the condenser as indicated in the figures 1, 3 and 4. The following calculations show that the maximum steam is supplied from the diffuser boiler and the least is from the SGII. In all cases, we assumed $\eta_G = 0.35$ as the gas turbine net axial efficiency and $\eta_S = 0.458$ for the steam turbine unit. The essential difference between the systems II and III is the biomass unit. Because the system IV may be workable only under the sunlight, it should can be switched to the system III or be operated in combined mode by the supply of additional O_2 during the time when the sunlight is insufficient.

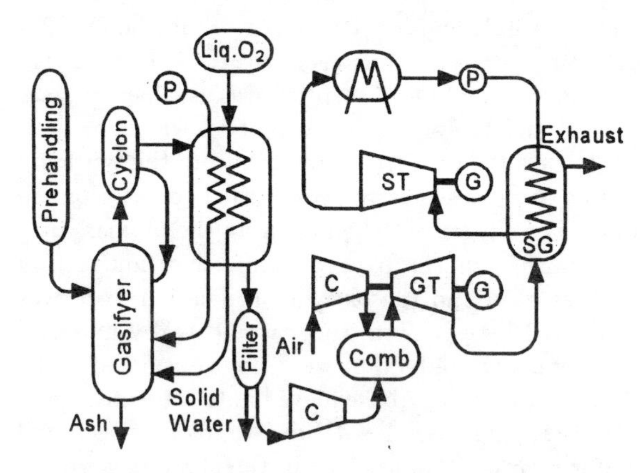

Fig.5. Coal gasification gas turbine–steam turbine combined system (case V, IGCC)

Fig. 5 shows a system of the integrated gasification combined cycle (IGCC) with a topping gas turbine and a bottoming steam turbine units. Taking the limit of the turbine inlet temperature into account, we assumed no fuel preheat. The heat of the gas turbine exhaust is recovered by steam generation until the exhaust gas temperature reduces to the same value 400 K at the stack.

MASS AND ENERGY BALANCE

Assuming X_c kmole C input into the TCC and X_b kmol of the biomass product to the combustor in the case IV, we calculate the mass and energy balance in major components.

TCC AND MHD COMBUSTOR

Major gasification reactions at temperature above ~1000 K may be described by

$$C + CO_2 = 2CO + H^0_{fCO_2} - 2H^0_{fCO} \quad (1)$$

$$C + H_2O = CO + H_2 + H^0_{fH_2O} - H^0_{fCO} \quad (2)$$

where $H^0_{fCO_2}, \cdots$ are the standard formation enthalpy of $CO_2, \cdots$. In these reactions the oxidants are CO_2 and H_2O which are supplied from the MHD exhaust gas. We assume that mole numbers of C consumed in the reaction (1) and (2) are x and y, respectively and that the mole ratio of the syngas components is $CO/H_2=\alpha$. Then, we obtain $x + y = X_c$ and $2x + y = \alpha y$ from which CO and H_2 mole numbers can be calculated as

$$N_{CO} = \frac{2\alpha X_c}{\alpha + 1}, \qquad N_{H_2} = \frac{2X_c}{\alpha + 1} \quad (3)$$

The absorbed heat in the gasification process is given by

$$Q_a = -N_{CO}H^0_{fCO} - N_{H_2}H^0_{fH_2O} + \frac{\alpha - 1}{\alpha + 1}H^0_{fCO_2} \quad (4)$$

In case IV, we assumed X_b kmol input of CH_2O into the MHD combustor. The combustion reaction of the carbohydrate may be written by

$$CH_2O + O_2 = CO_2 + H_2O \quad (5)$$

Since $\chi\%$ of the syngas is introduced again to the MHD combustor in the cases I~IV where $\chi = 1$ in the cases III and IV, then the TCC recieves CO_2 from the MHD diffuser by the amount $\chi N_{CO}+X_b$ and that of H_2O is $\chi N_{H_2}+X_b$. Therefore, CO_2 mole number contained in the syngas at the TCC exit is given by

$$N_{CO_2} = \frac{2\alpha\chi - \alpha + 1}{\alpha + 1}X_c + X_b \quad (6)$$

This requires a condition for the syngas partitioning in cases I and II ($X_b = 0$) as

$$\chi \geq \frac{\alpha - 1}{2\alpha} \simeq 0.25 \quad (7)$$

Under this condition CO_2 oxidant is sufficient in the MHD exhaust. Note that this condition is always met in cases III and IV where we can assume χ is unity. Because H_2O consumed and H_2 produced in gasification reaction (2) is equal and $1 - \chi\%$ of the syngas products should be sent to the bottoming unit in cases I and II, an additional steam input to the TCC is required to balance the hydrogen in the topping loop. The minimum amount of steam to be introduced to the TCC is $N_w = N_{H_2}(1 - \chi)$ in the cases I and II.

The combustion heat released in the MHD combustor is given by

$$\begin{aligned} Q_M &= N_{CO}(H^0_{fCO} - H^0_{fCO_2}) - N_{H_2}H^0_{fH_2O} \\ &+ X_b\left(H^0_{fCH_2O} - H_{fCO_2} - H_{fH_2O}\right) \end{aligned} \quad (8)$$

Here, the heat of CO and H_2 combustion is equal to $\chi(X_cQ_0 - Q_a)$ where $Q_a < 0$ and Q_0 is the combustion heat of 1 kmole of C. This means that the combustion heat of the syngas produced thermochemically with MHD exhaust gases is higher than the combustion heat of the primary fuel by the amount of absorbed heat. The energy amplification is an important consequence of the thermochemical heat recovery schematics.

The oxygen consumption in the MHD combustor amounts to $N_{O_2} = N_{CO}/2 + N_{H_2}/2 + X_b$. The corresponding liq.$O_2$ production work in cases I, II and III is $W_{LO_2} = q_{LO_2}N_{O_2}$ where $X_b = 0$. On the otherhand, in case IV the oxygen can be provided by the biomass unit if we assume that all amount of CO_2 separated from the syngas mixture at the CO_2 separation unit can be absorbed and converted photosynthetically to O_2. The following reaction can apply under absorption of the solar energy by 477×10^3 kJ/kmol for 1 kmole of CH_2O.

$$CO_2 + 4H_2O = CH_2O + 3H_2O + O_2 \quad (9)$$

In this case the exhaust from the biomass unit may be highly O_2–enriched air, so that we assume no liquefaction but O_2 purification work by $q_{LO_2}/5$ times the oxygen mole number, X_c+X_b. The latent heat of liq. O_2 is assumed to be supplied at the condenser of the steam turbine system in the cases I and III while it is supplied at the bottoming steam generator in the case II. The gaseous oxygen is preheated at the the diffuser boiler. The oxygen heat introduced to the combustor is evaluated by

$$Q_{O_2} = N_{O_2}\left(\xi q^{298}_{LO_2} + \int_{298}^{T_{O_2}} C_{pO_2}dT\right) \quad (10)$$

Here, $\xi = 1$ for cases I~III and $\xi = 0$ for the case IV, respectively. Also, the following amount of heat is introduced to the combustor with the preheated syngas fuel at T_f.

$$Q_f = \chi \int_{298}^{T_f} \sum_i N_i C_{pi} dT, \quad j = CO, H_2 \quad (11)$$

where $\chi = 1$ in cases III and IV. Then, the total heat input to the MHD combustor is given by

$$Q_T = Q_M + Q_f + Q_{O_2} \quad (12)$$

We evaluate the MHD generator output work by $W_M = Q_T\eta_M$. The fuel compression work is $W_{c1} = \chi(N_{CO}+N_{H_2})w_{c10} = 2\chi X_c w_{c10}$ where w_{c10} is the molar isothermal compression work for assumed pressure ratio of 10 at $T = 300$ K.

In the above calculation, we assumed a constant C input to the gasifyer even when χ is changed so that a constant gasification energy is required. Therefore, in order to keep the heat input from the MHD diffuser to the TCC constant in cases I and II, it is needed to supply an amount of heat Q_{S1} to steam loop for the steam turbine (case I) or to the gas turbine (case II). With given TCC exit temperature T_C, the heat balance at the TCC can be written as follows.

$$\begin{aligned} Q_T(1-\eta_M) - Q_{S1} &= Q_{O_2} - \xi N_{O_2} q_{LO_2}^{298} - Q_a \\ &\quad + \sum_j N_i q_i(T_C) \end{aligned} \qquad (13)$$

where j=CO, H_2 and CO_2 and $q_j(T_C)$ is the molar enthalpy of the component j at T_C. In Eq.(12), Q_T, Q_{O_2} and N_j are dependent upon χ while Q_a, N_{CO} and N_{H_2} are constant. Therefore, when χ is low and η_M is high, there may occure the case where the gasification energy supplied by the MHD exhaust gas becomes insufficient. The lower limit of χ can be obtained from the condition $Q_{S1} \geq 0$ from Eq.(13). The condition can be evaluated numerically for an MHD unit efficiency $\eta_M = 0.3$ and $T_C = 1500$ K, $T_f = 1000$ K and $\alpha = 2$ as

$$\chi_{min} \simeq 0.575 \qquad (14)$$

The lower limit χ_{min} decreases with the decrease in η_M and increase with the increase in T_C. The condition (14) is a sufficient condition of a steady operation of the TCC in the case I and case II systems because if it can be met, the condition for χ derived from CO_2 balance in Eq.(7) is always satisfied.

SGI AND RG

In order to allow a metal tube–type syngas regenerator (RG), we assume the exit temperature of the steam generator I (SGI) as $T_{SGI} = 1100$ K. Then, the heat recovered by steam at SGI is given by

$$Q_{S2} = \sum_i N_i \{q_i(T_C) - q_i(T_{SGI})\} \qquad (15)$$

where i = CO, H_2 and CO_2 and the molar numbers are given in Eqs.(3) and (6), respectively. The heat balance at RG can be written as

$$\sum_i N_i q_i(T_{SGI}) = Q_{S3} + Q_f + Q_w - \sum_i N_i q_i^{355} \qquad (16)$$

where Q_{S3} is the heat transfered to steam lines from RG, Q_w is the heat of steam supplied to TCC and $\sum_i N_i q_i^{355}$ is the heat of the component i at the assumed RG exit temperature 355 K.

BOTTOMING UNITS

The molar numbers of the syngas mixture distributed to the bittoming in the cases I and II are as follows:

$$\left.\begin{aligned} N_{CO}^g &= \frac{2\alpha(1-\chi)X_c}{\alpha+1} \\ N_{H_2}^g &= \frac{2(1-\chi)X_c}{\alpha+1} \\ N_{CO_2}^g &= \frac{(2\alpha\chi-\alpha+1)(1-\chi)X_c}{\alpha+1} \end{aligned}\right\} \qquad (17)$$

The work consumed at the fuel compressor in the gas turbine loop is then given by $W_{c2} = w_{c30}\sum_j N_j^g$ where j=CO, H_2 and CO_2 and w_{c30} is the molar isothermal compression work for the pressure ratio 30.

The combustion heat at the gas turbine combustor is estimated by

$$Q_G = N_{CO}^g(H_{fCO}^0 - H_{fCO_2}^0) - N_{H_2}^g H_{fH_2O}^0 \qquad (18)$$

The air compression work is given by $W_{c3} = w_{c30}(1-\chi)X_c/x_{O_2}$, where x_{O_2} is the O_2 mole fraction in air.

Case I

In case I system, some amount of steam, Q_S^g should be introduced from the gas turbine heat recovery boiler (SGII) to the combustor in order to meet the assumed turbine inlet temperature $T_G = 1500$ K. The excess heat of the turbine exhaust should be transfered to the steam turbine as Q_{S4}. Then, the heat balance at SGII can be written as

$$\begin{aligned} (Q_G &+ N_S^g q_S^g)(1-\eta_G) - W_{c3} \\ &= Q_S^g + Q_{S4} + \sum_j N_j^g q_j^{400} \end{aligned} \qquad (19)$$

Here, j=CO_2, H_2O, N_2 and S, where $N_{CO_2}^g$ is given by $N_{CO}^g + N_{CO_2}^g$ from Eq.(17), $N_{H_2O}^g$= $N_{H_2}^g$ and $N_{N_2}^g$ =$(1-\chi)(1-x_{O_2})X_c/x_{O_2}$. q_S^g is the injected steam molar enthalpy evaluated with the saturation temperature T_S^g at 3 MP_a. The jth component molar enthalpy q_j^{400} should be evaluated with the assumed stack temperature 400 K. The mole number N_S^g can be obtained by the energy balance in the combustor by the following formula.

$$N_S^g = \frac{\sum_j Q_j - Q_G}{q_{H_2O}^g - q_S^g} \qquad (20)$$

where, j=CO_2, H_2O, N_2 and $Q_{CO_2} = (N_{CO}^g + N_{CO_2}^g)q_{CO_2}^g$, $Q_{H_2O} = N_{H_2}^g q_{H_2O}^g$ and $Q_{N_2} = N_{N_2}^g q_{N_2}^g$, And the molar enthalpies q_j^g and q_S^g should be evaluated with the temperature T_G and T_S^g, respectively.

The total gas turbine work W_{GT} is the sum of the net axial work W_G =$(Q_G+Q_S^g)\eta_G$ and the air compression work W_{c3}, as given by

$$W_{GT} = (Q_G + Q_S^i)\eta_G + W_{c3} \qquad (21)$$

The heat of steam Q_{S4} can be calculated from Eq.(19) with the injected mole number N_S^g given by

Eq.(20). Then, the work output from the steam turbine is given by

$$W_{ST} = \eta_S \sum_{k=1}^{4} Q_{Sk} \quad (22)$$

Case II

In case II, the superheated steam generated at the topping units, at the diffuser boiler, the SGI, the RG and at the bottoming SGII should be injected to the gas turbine combustor. The heat and mass balance at the combustor and at SGII can be written by the following equations.

$$Q_G + \sum_{k=1}^{4} N_{Sk} q_{Sk}(T_{Sk}) = \sum_{k=1}^{4} N_{Sk} q_{Sk} + \sum_j Q_j^g \quad (23)$$

where the righthand side should be estimated with the turbine inlet temperature T_G. The energy balance for the steam generator II can be written as

$$(1 \quad - \quad \eta_G)\left(Q_G + \sum_{k=1}^{4} N_{Sk} q_{Sk} + \sum_j N_j^g q_j\right)$$
$$-W_{c3} = \sum_{k=1}^{4} N_{Sk} q_{Sk}^{400} + \sum_j N_j^g q_j^{400} \quad (24)$$

where $\sum_j N_j^g q_j = \sum_j Q_j^g(T_G)$. The heat Q_{S4} and the mole number N_{S4} can be determined from the above two equations (??) and (24). Then, the gas turbine work is given by the following formula.

$$W_G = \left(Q_G + \sum_{k=1}^{4} N_{Sk} q_{Sk}(T_G)\right)\eta_G \quad (25)$$

Case III and IV

The case IV system is workable as the case III system during the time when the sunlight is insufficient and therefore, the performance can be evaluated by the formulae given above. When the biomass unit is active, we suppose that CO_2 delivered at the CO_2 separator could all be supplied to the biomass unit with the pressure 0.2 MP_a. Low temperature water ($\sim$ 300 K) for the photosyntetic reaction (9) can be supplied through the condenser in the steam turbine system. If we assume that X_b kmole carbohydrate is fed to the MHD combustor, the total carbohydrate production is $X_b + X_c$ kmol, and that is equal to the oxygen production. The net biomass production is X_c kmol which may be used for the commercial food services. The total heat released at the MHD combustor is given by Eq.(8) Assuming that the heat Q_{S1}, Q_{S2} and Q_{S3} is transfered to steam each at the diffuser boiler, the steam generator (SGI) and at the regenerator (RG), then we obtain the total heat of steam from the heat balance consideration applied to the TCC, SGI and RG as follows.

$$Q_S \quad = \quad \sum_{i=1}^{3} Q_{Si} = (Q_M + Q_b)(1 - \eta_M) - \{Q_{O_2}$$
$$-\xi(X_c + X_b) q_{LO_2}^{298}\} \eta_M - Q_f - Q_a$$
$$-\sum_i N_i q_i^{355} \quad (26)$$

Here, i=CO, H_2, CO_2 and H_2O, and $N_{CO} = 2\alpha X_c/(\alpha+1)$, $N_{H_2} = 2X_c/(\alpha+1)$, $N_{CO_2} = X_c + X_b$ and $N_{H_2O} = X_b$. Q_b is the combustion heat of the carbohydrate given by $Q_b = X_b\,(H_{fCH_2O} - H_{fCO_2} - H_{fH_2O})$. As the latent heat of liquid oxygen is supplied from the cooling water through the condenser in case III and no O_2 liquefaction is assumed in case IV, we should read $X_b = 0$ and $\xi = 1$ for case III and $\xi = 0$ for the case IV. The steam mole numbers N_{S1}, N_{S2} and N_{S3} are calculated at given steam temperature 850 K and at the pressure 21 MP_a.

IGCC

As for the gasification reactions in separate unit, we add following reaction to provide gasification energy to Eqs.(1) and (2).

$$C + O_2 = CO_2 - H_{fCO_2}^0 \quad (27)$$

Assuming mole numbers of C consumed in the reactions (27),(1) and (2) as x, y and z, respectively, we obtain

$$\left.\begin{aligned} x + y + z &= X_c \\ 2y + z &= \alpha z \end{aligned}\right\} \quad (28)$$

The heat balance between the gasifyer and regenerator (RG in Fig.5) may be written as follows with the assumption of no heat transfer to the steam and the RG exit temperature of 355 K.

$$\begin{aligned} -xH_{fCO_2}^0 &+ y(H_{fCO_2}^0 - 2H_{fCO}^0) \\ &+ z(H_{fH_2O}^0 - H_{fCO}^0) = (2y+z)q_{CO}^{355} \\ &+ zq_{H_2}^{355} + (x-y)q_{CO_2}^{355} \end{aligned} \quad (29)$$

Solving Eqs.(28) and (29) for x, y and z the mole numbers of syngas components can be obtained as follows.

$$\left.\begin{aligned} N_{CO} &= 1.386\alpha X_c/(\alpha + 0.863) \\ N_{H_2} &= 1.386 X_c/(\alpha + 0.863) \\ N_{CO_2} &= (0.863 - 0.386\alpha)X_c/(\alpha + 0.863) \end{aligned}\right\} \quad (30)$$

The combustion heat is given by the same formula as Eq.(18). The turbine exhaust gas components are then given by $N_{CO_2}^g = N_{CO} + N_{CO_2}$, $N_{H_2O}^g = N_{H_2}$ and $N_{N_2}^g = (N_{CO} + N_{H_2})(1 - x_{O_2})/2x_{O_2}$. The oxidant air compression work is $W_{c3} = (N_{CO} + N_{H_2})w_{c30}/2x_{O_2}$ The gross turbine output is given by $W_{GT} = Q_G\eta_G + W_{c3}$ where the second term gives the net work output.

With the molar number $N_j^g = N_{CO_2}^g$, $N_{H_2O}^g$ and $N_{N_2}^g$ given above, the heat to the steam turbine is evaluated by

$$Q_S = Q_G(1 - \eta_G) - W_{c3} - \sum_j N_j^g q_j^{400} \quad (31)$$

The steam turbine work is given by $W_S = Q_S\eta_S$.

Taking the power consumptions for liq.O_2 production, compression work and pump work into consideration, we may evaluate the total net work and the system efficiency by the following formulae.

$$W_{\rm tot} = \sum_{\rm K} W_{\rm K} - \sum_{l} W_{cl} - \sum_{m} W_{{\rm O}_2 m} - \sum_{n} W_{pn} \quad (32)$$

and

$$\eta = \frac{W_{\rm tot}}{Q_0 + \beta Q_{\rm b}} \quad (33)$$

Here, $\beta = 0$ for the casse I, II and III while $\beta = 1$ for the case IV. The indices should be read as follows: K=M, G and S for the case I, K=M and G for the case II, K=M and S for the cases III and IV and K=G and S for the case V, respectively. Further, in the cases I and II, $l = 1 \sim 2$, $m = 1$ and $n = 1 \sim 4$ and in the case III, $l = m$ and $n = 2$. In the case IV, $l = 1 \sim 2$ (W_{c3} for biomass O_2 compression work is considered in W_{O_2}), $m = 1$ and $n = 1 \sim 3$. Finally, for the case V, $l = 1$, $m = 1$ and $n = 2$.

RESULTS AND DISCUSSION

In Fig. 6, we compare the system efficiency dependence upon the MHD unit efficiency each of the case I, II, III, IV and V with fixed gas turbine and steam turbine efficiencies at $\eta_G = 0.35$ and $\eta_S = 0.458$, respectively. The syngas composition was fixed at $\alpha = 2$. The syngas fraction $\chi = 0.85$ in cases I and II.

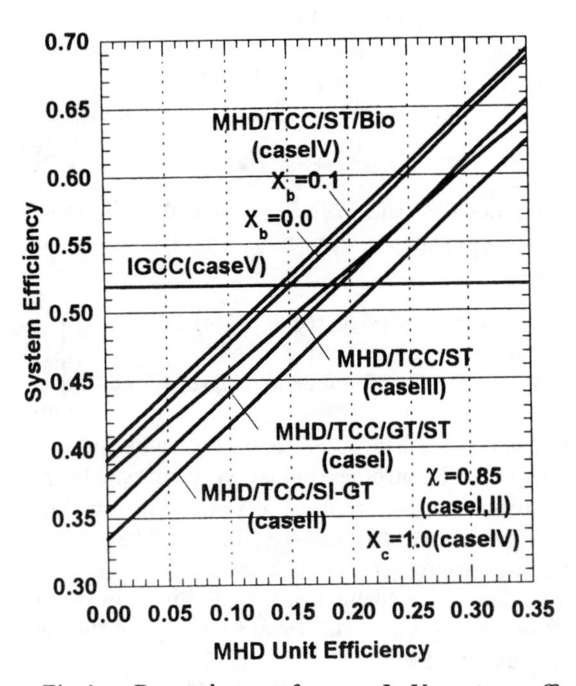

Fig.6. Dependence of cases I~V system efficiencies upon the MHD unit efficiency. $\alpha = 2$, $\chi = 0.85$ (I, II), $\eta_G = 0.35$, $\eta_S = 0.458$ and $X_c = 1$

We see that the highest efficiency can be achieved in the case IV, MHD–TCC–steam turbine–biomass system. The efficiency is comparable or higher than that of the IGCC even for the MHD unit efficiency of 15% that is the design value of the past HPDE experimental facility [9]. The clean fuel-fired MHD unit efficiency of 25~ 27 % is presumably attainable with the present MHD technology in reference to the desigh value of the U-500, i.e., $\eta_M = 0.24$ [10]. In this case, the efficiency of the case IV reaches 61~62 %. It is to be noted that these performance could be attainable with a piping-type heat regenerator design concepts. In Fig.7 the effect of the biomass fuel input to the MHD combustor is illustrated.

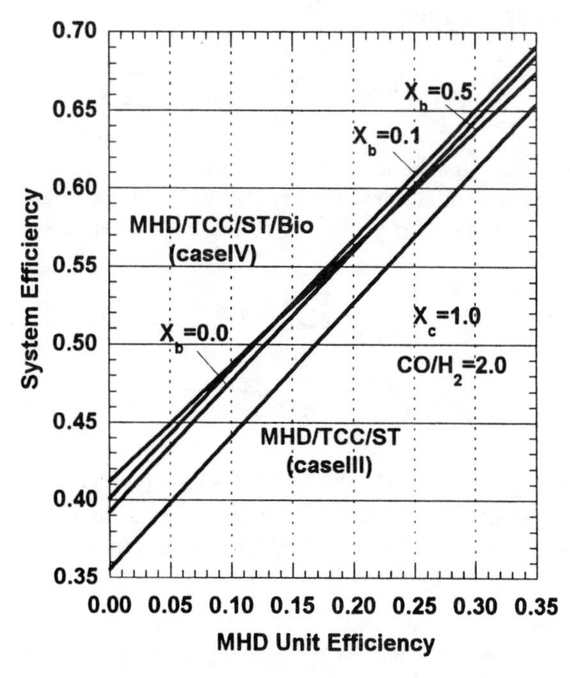

Fig.7. Effect of biomass combustion on the efficiency performance of MHD–TCC–ST–Bio system. $\alpha = 2$, $X_c = 1$, $\eta_G = 0.35$ and $\eta_S = 0.458$.

The line of the case IV with $X_b = 0.0$ and that of the case III differ in the oxygen supply schematics. The difference arises mostly from the liquefaction work for the case III. The case IV lines indicate a tendency that a higher efficiency is attained for larger X_b input in low η_M range and for smaller X_b in high η_M range. This is attributed mostly to the increase in the primary heat input (the denominator in Eq.(33)) and increase in CO_2 and O_2 compression work with increase in X_b and vice versa. However, the effect of the biomass input is not appreciable at any case. Under controlled CO_2 concentration and temperature the dry chrollela production is $\sim$ 100 gr/m^2/day. That is 107 gr/m^2 of O_2 production with a moderate sunlight radiation flux of $\sim$ 300 W/m^2. This requires the minimum specific area of the biomass pond of an order 10^5 m^2/MW$_e$ [5]. Therefore, the power plant scale of the MHD–TCC–ST–Bio system may be 100 MW$_e$ at most and it should be operated as the MHD–TCC–ST system under insufficient sunlight.

In Fig. 6, it is shown that the performance of the MHD–TCC–steam injection type gas turbine is the lowest. If it is combined with the biomass system, the efficiency will be increased by about the same values as the difference between the case IV and the case III systems. In comparison of the case I with the case III which is essensially the case I under the condition $\chi = 1$, we see that both performances are almost identical for the MHD unit efficiency of 23~27 %. However, the case III performance may have less capabilitiy than the case I in view of further increases in the total efficiency becuase a further developement in the steam turbine performance might not largely be expectable. However, in regards to the rapid progress in gas turbine technologies, it is certain to say that a much more higher efficiency could be achievable in the case I system because a combination of the MHD cycle with the gas turbine cycle is possible only in this system.

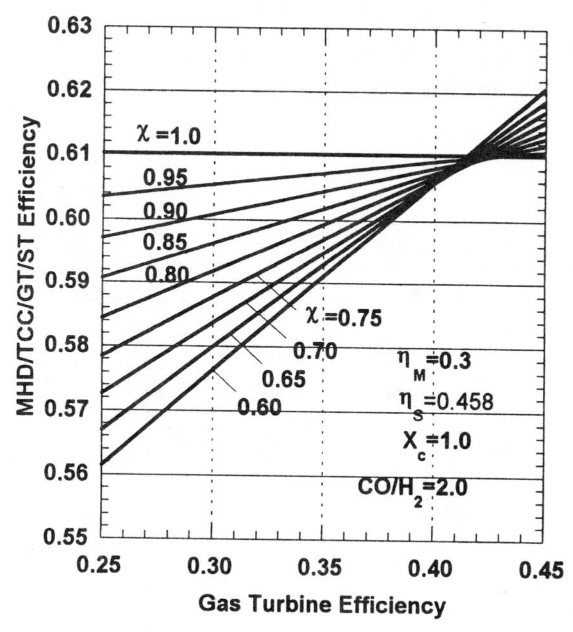

Fig.8. Efficiency dependence of case I system upon the gas turbine efficiency and the syngas fraction to MHD unit.

Therefore, it is worthwhile to see the case I efficiency dependence upon the gas turbine performance. The calculated results are presented in Fig.8 where we can see that the integrated performance of the MHD–TCC–GT–ST system is strongly dependent upon the fuel partition except the case where $\eta_G = 0.41 \sim 0.43$. When η_G is low, the total efficiency is more dependent upon the topping MHD efficiency ($\chi \sim 1$) and vice versa. It is independent of η_G when $\chi = 1$ where the case I system is identical to the MHD–TCC–ST system. This result shows that the efficiency will overcome that of the case III and reaches approximately the same lebel of the case IV in the range of $\eta_G \geq 0.42$, and that if the case I system could be combined with the biomass unit, an efficiency over 65 % could be attainable. Finally, it is to be emphasized that in view of the environmental impact and fuel saving, the further advancement of the coal–based power plant performance is achievable on the basis of the thermochemical conversion which may be applicable only to high temperature heat engines like an MHD generator.

CONCLUSIONS

The efficiency performance of four types of advanced MHD topping systems with the thermochemical heat recovery concept was estimated and compared on common basis. The topping was the MHD–TCC units and the bottoming system was either the combination of a gas turbine and a steam turbine, a steam injection type gas turbine, a steam turbine only or a combined steam turbine and biomass complex. Revisions of the past works were made for the TCC and the gas turbine inlet temperature from more realistic design point of view. The MHD–TCC topping system performances were also compared with that of the conventional integrated coal gasification combined cycle. Several idealized assumptions were introduced for comparison purposes. Howevere, the unit efficiencies of the MHD generator, the gas turbine and the steam turbine and the temperature of heat exchangers and steam generators were those being attainable with the state-of-the-art technologies. The major conclusions can be summarized as follow:

1. In regarad to the efficiency, the best system is the combination of MHD–TCC–steam turbine–biomass units. Under the present assumptions, it exhibited the efficiency of 62% with 30% MHD unit efficiency.
2. Even with the so–far demonstrated or designed scale of MHD generators, this system performance is comparable or better than the IGCC counterpart.
3. The power scale of the MHD/Biomass system may be 100 MW_e and it should be operated in combination with the MHD–TCC–steam turbine system.
4. From the viewpoint of an advanced base load plant, the MHD–TCC–steam turbine system is the best candidate.
5. The MHD–TCC–gas turbine–steam turbine system has a capability of further advancement together with the developement in the gas turbine technology.

In the present study no consideration was made on NO_x and SO_x treatments. However, it is well confirmed that the MHD generator performance and generator channel durability are relatively insensitive to contaminations and these environmental impacts could be minimized well below the standards level [11] in MHD topping systems presented in this report.

Nomenclatures

C_p: specific heat [J/kmol·K]
H_f^0: standard formation enthalpy [J/kmol]
N: mole number [kmol]
Q: heat [J]
T: temperature [K]
W: work output [J]
X: mole number of C [kmol]
p: pressure [P_a]
q: molar enthalpy [J/kmol]
w: molar work consumption [J/kmol]
α: CO/H_2 ratio
χ: syngas fraction to MHD unit
η: efficiency

Subscrips

C: TCC exit
M: MHD unit
G: gas turbine
S: steam, steam turbine
SGI: steam generatorI exit
T: topping, total
b: biomas
c: C, compressor
f: fuel
LO_2: liquid O_2

References

[1] Seikel, G.R., Coal- fired open- cycle MHD plants, The Science and Tech. of Coal and Coal Utilization, (1984), Plenum Press, New York.

[2] Shioda, S., Recent progress in research of closed cycle MHD power generation, Trans. of IEE Japan B, Vol.111 (1991), pp.237–241. (in Japanese).

[3] Batenin, V.M. and Kovbasiuk, V.I., MHD Power Generation with the Conversion of Fuel, Proc. 10th Int. Conf. on MHD Elect. Power Gen., (1989), pp. III.59–60, Tiruchirappalli, India.

[4] Bystrova, O.V., et. al., Evaluation of Characteristics of MHD Power Plant with Thermochemical Conversion of Fuel, Proc. 11th Int. Conf. on MHD Elect. Power Gen., (1992), Vol.1, pp.114–121, Beijing, China.

[5] Batenin, V.M., et al., MHD Power Plants with Intracycle Thermochemical Regeneration in Power–Biotechnological Cycle, Proc. MHD and High Temperature Technology 1999, Vol.1, (1999), pp. 243–252, Beijing, China.

[6] Kayukawa, N., Aoki, Y. and Ohtake, K., MHD/Gas Turbine/Steamturbine Triple Combined Cycle with Thermochemical Heat recovery, The AIAA 28th Plasmadynamics and Lasers Conf., (1997), AIAA Paper No. AIAA 97–2370, Atlanta.

[7] Kayukawa, N., Comparison of MHD Topping Combined Power Generation Systems, Energy Conversion and Management, (2000), (in press).

[8] Yoshikawa, K., High Efficiency Energy Extraction from Coal/ Wastes using High Temperature Air, Proc. Int. Conf. on MHD and High Temp. Technol., (1999), pp.603–630, Beijing, China.

[9] Felderman, E.J., Whitehead, G.L. and Christensen, L.S., HPDE Performance in the Faraday Mode, Proc. 20th Symposium on Engi. Aspects of MHD, (1982), Paper 4.5, Irvine, Calf.

[10] Kirillin, V.A., Sheindlin, A.E., Morozov, G.N. and Bryuskin, A.S., The Present Status and Future Introduction to the Power Industry of Magnetohydrodynamic Plants, Thermal Engineering, Vol.**33**, No.2, (1986), pp. 59~67.

[11] Morrison, G.F., Coal - Fired MHD, IEA Coal Research, (1988), p.24, IEA, London

AIAA-2000-2892

NEW THERMODYNAMIC FUNCTION, FUGERGY

A. Naimpally, Professor of Chem. Engg.
California State University
Long Beach, CA 90840

ABSTRACT

A new thermodynamic function, fugergy, g, has been defined to facilitate computation of the flow availability function, B. The definition of fugergy is dBT = RTo dln g where R is the gas constant and To is the external temperature. A formula for the computation of fugergy is also presented; tabulated values of the fugergy are obtained and presented as a function of: the reduced pressure, the reduced temperature and the reduced external temperature. Formulas for fugergy for the special case of a gas obeying the virial equation as well as VanderWaal's equation are presented in the paper. Finally, the method to compute the changes in the flow availability using the fugergy tables is shown. This is illustrated by means of a numerical example.

1.0 Introduction

It is proposed to define a new thermodynamic function, fugergy, which can be used for the computation of the flow availability function, exergy. There are two types of availability that are used for performing an energy systems analysis: The chemical availability for which the thermodynamic functions—chemical potential—would be used; and, the flow availability or Exergy. Both of these types of availability have been described by Kotas (1), Dibenedetti (2) and others. The thermodynamic function fugergy would have a relationship to exergy that is similar to the relationship between fugacity f and the Gibbs Free Energy, G, where G = H - TS. The word "fugergy" is chosen from a combination of two words: "fug"acity and ex"ergy". Although several authors have defined different types of flow availability or exergy, the most commonly defined one is the flow availability, B ((3), (4)) where

$$B=H-T_0S$$

The utility of the function fugergy would be as follows:

1) The tables of fugergy would be useful for computations of flow availability analysis and it would not be necessary to obtain detailed values of the enthalpy and the entropy of the substance under consideration. The only data for the substance that would be required would be the critical properties of the substance—the critical temperature, Tc, the critical pressure, Pc, and the critical volume, Vc.
2) The generalized tables of fugergy would include a parameter, Tor, the reduced external temperature defined as:

$$T_{or} = T_0 / T_c$$

Thus, the computations for the flow availability analysis can be performed by using any chosen value of the reference external temperature, To. Although the temperature To = 25°C has been generally chosen to be the reference temperature, one is not constrained to use this value in performing the flow availability calculations. Depending upon the conditions external to the plant or the system under consideration, one can choose the appropriate value of To.

3) The thermodynamic function fugergy, g, should be useful in quickly performing computation of flow availability when doing calculations with different trial values in optimizing parameters in a given plant or system.

2.0 Definition of Fugergy

For the case of an ideal gas, the value of dB, an infinitesimal change in the value of exergy, at constant temperature can be computed as follows:

$$dB_T = d(H-T_0S)/\text{Constant T. Ideal gas.}$$

$$= (dH-T_0dS) \Big/ \text{Constant T. Ideal gas.} \qquad (1)$$

The right-hand side of the equation can be rewritten as:

$$dH - T_o dS = dH_T \Big|_{I.G.} - T_o - T_o (dS) \Big|_{T, I.G.} \quad (2)$$

But $dH \Big|_{Const\ T,\ I.G.} = - C_p dT$ for an ideal gas (3)

Therefore,

$$(dH - T_o dS) \Big|_{\substack{Constant\ T \\ Ideal\ gas.}} = C_p dT \Big|_{\substack{Constant\ T \\ Ideal\ gas.}} - T_o dS \Big|_{\substack{Constant\ T \\ Ideal\ gas.}}$$

$$= O - T_o dS \Big|_{\substack{Const\ T. \\ Ideal\ gas.}} \quad (4)$$

One can write $dS = \left(\frac{\partial S}{\partial T}\right)_P dT + \left(\frac{\partial S}{\partial P}\right)_T dP$

At constant temperature, since $dT = 0$:

$$dS_T = 0 + \left(\frac{\partial S}{\partial P}\right)_T dP \quad (6)$$

$\therefore ds = -\left(\frac{\partial V}{\partial T}\right)_P dP$ using the standard Maxwell's equation in Thermodynamics. (7)

Therefore:

$$dB_T = 0 - T_o \left(-\frac{\partial V}{\partial T}\right)_P dP \quad (8)$$

$$= T_o \left(\frac{\partial V}{\partial T}\right)_P dP/ \quad \text{for ideal gas.} \quad (8)$$

For an ideal gas:

$$\left(\frac{\partial V}{\partial T}\right)_{P_{I.G.}} = \left(\frac{\partial}{\partial T}\left(\frac{RT}{P}\right)\right)\Big|_P \quad (9)$$

$$= R/P \quad (10)$$

$$dB_T \Big|_{Ideal\ gas} = T_o \frac{R}{P} dP \quad (11)$$

$$= R\ To\ d(\ln P) \quad (11)$$

At this point it would be worthwhile to recapitulate the relationship between fugacity and the Gibbs Free Energy:

$$dG \Big|_{\substack{Const\ T. \\ I.G.}} = RT\ d(\ln P) \quad (12)$$

The fugacity of any substance is defined as:

$$dG_T = RT d\ln f. \quad (13)$$

with the boundary condition

$$\lim_{P \to o} \left(\frac{f}{P}\right) = 1 \quad (14)$$

Analogous to this, one can define dB for any pure substance at constant temperature, dB_T by the following equation.

$$dB_T = RTo\ d(\ln g) \quad (15)$$

Where g is defined as the fugergy of the pure substance.

Furthermore, since all substances tend to behave as ideal gasses as the pressure $P \to o$, it would be necessary for the value of

$$dB \Big|_{Pure\ Substance} \longrightarrow dB_T \Big|_{I.G.} \quad \text{as } P \longrightarrow 0$$

viz, it would be necessary for the fugergy g to be of a form so that:

$$\lim_{P \to o} \left(\frac{g}{P}\right) = 1. \quad (16)$$

For the case of mixtures, a similar function, partial fugergy can be defined:

$$d\ \bar{G}_{iT} = RT_o d\ \ln \bar{g}_i$$

$$\lim \frac{\bar{g}_i}{X_i P} \to 1$$

as $P \to o$
Xi constant.

3.0 Derivation of a formula relating Fugergy to the compressibility factor and the fugacity:

Starting from the definition of fugergy (equations (15) and (16)) one can write the value of dB_T as (from equation (1)):

$$dB_T = R\,To\,d(\ln g) = dH_T - To\,dS_T \qquad (17)$$

From the definition of Gibbs' free energy, G, one can write:

$$G = H - TS \qquad (18)$$

or,

$$dG_T = dH_T - T\,d\,S_T \qquad (19)$$

From equations (17) and (19):

$$RTo\,d\ln g = dG_T + (T-To)\,dS_T \qquad (20)$$

From the definition of fugacity:

$dG_T = RTd\ln f$ where f is the fugacity of the pure substance (21)

Equation (21) becomes:

$$RTo\,d(\ln g) = RTd(\ln f) + (T-To)\,dS_T \qquad (22)$$

Now:

$$dS = \left(\frac{\partial S}{\partial T}\right)_P dT + \left(\frac{\partial S}{\partial P}\right)_T dP \qquad (5)$$

At constant temperature:

$$dS_T = 0 + \left(\frac{\partial S}{\partial P}\right)_T dP = \left(\frac{\partial S}{\partial P}\right)_T dP$$

From the standard Maxwell equation, (S) = one can write equation (5) as:

$$dS_T = -\left(\frac{\partial V}{\partial T}\right)_P dP \qquad (7)$$

Therefore:

$$RT_o d\ln g = RTd\ln f - (T-T_o)\left(\frac{\partial V}{\partial T}\right)_P dP \qquad (23)$$

One can now use the second part of the definition of fugergy viz.

$$\lim_{P \to o} \frac{g}{P} = 1 \qquad (16)$$

as well as the definition of fugacity viz.

$$\lim_{P \to o} (f) = 1 \qquad (14)$$

as follows:

Adding and subtracting RTo d lnP on the left-hand side of equation (23), as well as adding and subtracting RT d(lnP) on the right-hand side of the same equation, one obtains:

$$RT_o\,d\ln\left(\frac{g}{P}\right) + RT_o d\ln P = RTd\ln\left(\frac{f}{P}\right) + RT\,d\ln P - (T-T_o)\left(\frac{\partial V}{\partial T}\right)_P dP \qquad (24)$$

or,

$$RT_o\,d\ln\left(\frac{g}{P}\right) + RT_o\,dnP = RT\,d\ln(f) + R(T-T_o)d\ln P - (T-T_o)\left(\frac{\partial V}{\partial T}\right)_P dP \qquad (25)$$

It is now desired to put the last term on the right-hand side of equation (25) in terms of the compressibility factor, Z:

Since $P\underline{V} = ZRT$ (26)

Therefore, $\left(\frac{\partial(PV)}{\partial T}\right)_P = \left(\frac{\partial(ZRT)}{\partial T}\right)_P$ (27)

or, $P\left(\left(\frac{\partial V}{\partial T}\right)_P\right) = R\left(T\left(\frac{\partial Z}{\partial T}\right) + Z\right)$

$$\left(\frac{\partial V}{\partial T}\right)_P = \frac{R}{P}\left(T\left(\frac{\partial Z}{\partial T}\right)_P + Z\right) \qquad (28)$$

Substituting equation (28) into equation (25), one obtains:

$$RT_o\,d\ln\left(\frac{g}{P}\right) + RT_o d(\ln P) = RTd\ln\left(\frac{f}{P}\right) + RTd(\ln P) - (T-T_o)\frac{R}{P}\left(T\left(\frac{\partial Z}{\partial T}\right)_P + Z\right)dP. \qquad (29)$$

Cancelling the common factor R from the terms and dividing by To, the above equation becomes:

$$d \ln \left(\frac{g}{P}\right) = \frac{T}{T_o} d\ln \left(\frac{f}{P}\right) + \frac{(T-T_o)}{T_o} \left(1-T\left(\frac{\partial Z}{\partial T}\right)_P -Z\right) \cdot d\ln P. \quad (30)$$

Equation (30) can be integrated between the lower limit of P = o to the upper limit of P. At the lower limit the value of (f/P) is equal to 1 (from the definition of fugacity) and the value of (g/P) is also equal to 1 (from the equation (16)). Therefore, upon integration at a constant temperature, equation (30) becomes:

$$\int_1^{g/P} d\ln\left(\frac{g}{P}\right) = \frac{T}{T_o}\int_1^{f/P} d \ln \left(\frac{f}{P}\right) + \frac{(T-T_o)}{T_o}\int_0^P \left(1-T\left(\frac{\partial Z}{\partial T}\right)_P -Z\right) d\ln P. \quad (31)$$

Therefore, equation (31) becomes:

$$\ln \left(\frac{g}{P}\right) = \frac{T}{T_o} \ln \left(\frac{f}{P}\right) + \frac{(T-T_o)}{T_o}\int_{P=0}^{P}\left(1-Z - T \left(\frac{\partial Z}{\partial T}\right)_P\right) d\ln P. \quad (32)$$

The fugergy, g, is thus related to the fugacity and the compressibility factor Z, by means of equation (32).

4.0 Computation of fugergy: Tabulated values of fugergy

The formula for fugergy is:

$$\text{Ln} \left(\frac{g}{P}\right) = \left(\frac{T}{T_o}\right) \text{Ln} \left(\frac{f}{P}\right) + \frac{(T-T_o)}{T_o}\int_0^P \frac{\left(1-Z-T \left(\frac{\partial Z}{\partial T}\right)_P\right)}{P} dP. \quad (32)$$

In order to simplify this formula and obtain a generalized formula that can be used for computation purposes, the following simplifications are carried out.

1) Divide and multiply the first term on the right-hand side by the constant Tc, the critical temperature of the gas.

Thus the term $\left(\frac{T}{T_o}\right) \ln \left(\frac{f}{P}\right)$ becomes:

$$\frac{(T/T_c)}{T_o/T_c} \ln \left(\frac{f}{P}\right) = \frac{T_r}{T_{or}} \ln \left(\frac{f}{P}\right) \quad (33)$$

Here:

Tr = T/Tc = Reduced temperature of the gas.

Tor = To/Tc = Reduced external temperature of the gas.

2) Divide and multiply the second term by Tc, the critical temperature of the gas. Also, multiply and divide the term inside the integral sign by P_c, the critical pressure of the gas:

The second term now becomes:

$$\frac{(T-T_o)/T_c}{T_o/T_c}\int_0^{P/P_c} \left(1-Z-T_r \left(\frac{\partial Z}{\partial T}\right)_P\right) d(P/P_c).$$

$$= \frac{(T_r-T_{or})}{T_{or}}\int_0^{P_r} \left(1-Z-T_r \left(\frac{\partial Z}{\partial T_r}\right)_{P_r}\right) dPr \quad (34)$$

3) The term $T \left(\frac{\partial Z}{\partial T}\right)_P$ can be simplified further by multiplying and dividing it by Tc, the critical temperature of the gas. The term thus becomes:

$$(T/T_c)\left(\frac{\partial Z}{\partial (T/T_c)}\right)_P = T_r \left(\frac{\partial Z}{\partial T_r}\right)_{P_r} \quad (35)$$

The formula for the fugergy thus becomes:

$$\text{Ln}\left(\frac{g}{P}\right) = \left(\frac{T_r}{T_{or}}\right) \ln \left(\frac{f}{P}\right) + \frac{(T_r-T_{or})}{T_{or}}\int_0^{P_r} \frac{\left(1-Z-T_r \left(\frac{\partial Z}{\partial T_r}\right)_{P_r}\right)}{P_r} dP_r \quad (36)$$

According to the law of corresponding states, the value of the compressibility factor, Z, should be a function of the reduced pressure and reduced temperature of the pure substance. However, as is well known, it has been found necessary to incorporate a third parameter. Two approaches have been used to date: The acentric factor method adopted by Prausnitz and other researchers from the University of California, and the critical compressibility factor, Zc, adopted by Hougen and Watson from the University of Wisconsin. In this paper, the approach of Hougen and Watson (5) shall be adopted from the viewpoint of convenience of final results.

The value of Zc for most substances has been found to vary from 0.23 to 0.31 with a large number having values of Zc from 0.26 to 0.28. Therefore, numerical computation of values of (g) has been performed using equation (36) above with the data for Z versus (Pr, Tr) at Zc = 0.27 obtained from Reid and Prausnitz and Sherwood (6). Numerical computation has been done using the trapezoidal rule for numerical integration. Furthermore, tabulated values of the fugacity coefficient, (f), have

been used in arriving at the final values of (g),

which can be named as the "fugergy coefficient") as a function of:

a. Reduced pressure, Pr (= P/Pc).
b. Reduced temperature, Tr (= T/Tc).
c. Reduced external temperature, Tor (= To/Tc).

All of the values in Table 1 are for the case of substances with Zc = 0.27.

In addition to Table 1, values of the fugergy coefficient, (g), for 2 other values of Zc:

Zc = 0.23 and Zc = 0.29. These are presented in tables 2A and 2B respectively, in a manner similar to table 1 with (g), being given as a

function of (Pr, Tr, Tor).

In a similar manner, a formula for $\bar{G}i$ can be shown to be

$$\ln \frac{\bar{g}_i}{x_i P} = \frac{T_{r_n}}{T_{or_n}} \ln \frac{\bar{f}_i}{x_i P} + \frac{T_{r_n} - T_{or_n}}{T_{or_n}} \int_0^{P_{r_n}} \left(1 - \bar{Z}_i - T_{r_n} \left(\frac{\partial \bar{Z}_i}{\partial T_{r_n}}\right)_{P_{r_n}}\right) d \ln P_{r_n}.$$

5.0 <u>Values of fugergy for the case of a gas represented by the vivial equation.</u>

The virial equation of state is as follows:

$$Z = 1 + \frac{B_i}{V} + \frac{C_i}{V^2} + \frac{D_i}{V^3} + \ldots$$

with B_i, C_i, D_i,....... being functions of Temperature, T. The same virial equation can be put in the form of an expression consisting of terms that are given as powers of P:

$$Z = 1 + B_i^1 P + C_i^1 P^2 \; D_i^1 P^3 + \ldots.$$

Where Van Ness (7) has given equations relating B_i^1, C_i^1, D_i^1, ... to B_i, C_i, D_i...

$$B_i^1 = B_i/RT$$

$$C_i^1 = (C_i - B_i^2)/RT^2$$

&

$$D_i^1 = (D_i - 3B_iC_i + 2B_i^3)/RT^3$$

Now, $\ln \left(\frac{g}{P}\right) = \frac{T}{T_o} \ln \left(\frac{f}{P}\right) + \frac{(T-T_o)}{T_o} \int_0^P \frac{\left(1 - Z - T\left(\frac{\partial Z}{\partial T}\right)_P\right) dP}{P}$

Let $Z = 1 + B_i^1 P + C_i^1 P^2 + D_i^1 P^3 +$

$$\therefore \frac{T}{T_o} \ln \left(\frac{f}{P}\right) = \int_0^P \frac{(Z-1)dP}{P}$$

$$= \left\{ \int_0^P \frac{(B_i^1 P + C_i^1 P^2 + D_i^1 P^3)\, dP}{P} \right\} \frac{T}{T_o}$$

$$= \left\{ \int_0^P B_i^1 + C_i^1 P + \ldots\, dP \right\} T/T_o$$

$$= \left\{ B_i^1 P + \frac{C_i^1 P^2}{2} + \frac{D_i^1 P^3}{3} + \ldots \right\} \frac{T}{T_o} \quad (37)$$

Second term on RHS of equation at the top of the page.

$$= \frac{(T_r - T_{or})}{T_{or}} \int_0^P \frac{\left(1 - Z - T\left(\frac{\partial Z}{\partial T}\right)_P\right) dP}{P}$$

$$= \left(\frac{T_r - T_{or}}{T_{or}}\right) \int_0^P \left[\left(-B_i^1 P - \frac{C_i^1 P^2}{2} - \frac{D_i^1 P^3}{3}\right) - T\left(+P\frac{dB_i^1}{dT} \ldots\right)\right] dP$$

$$= \frac{(T_r - T_{or})}{T_{or}} \left\{ (-B_i^1 P - C_i^1 \frac{P^2}{2} - D_i^1 \frac{P^3}{3} \right\}$$

$$-T \left(\left(\frac{dB_i^1}{dT}\right)(P) + \left(\frac{dC_i^1}{dT}\right)\left(\frac{P^2}{2}\right) + \ldots\right) \quad (38)$$

Ln (g/p) = Sum of equations (37) and (38).

5.1 <u>Special Case: Values of Fugergy for a High Pressure Gas Which Obeys the VanderWaal's Equation:</u>

The VanderWaal's equation is:

$(p + a\ )(V-b) = RT.$

where, p = Pressure of the gas;
T = Temperature of the gas;
V = Specific Volume of the gas;
a,b: Values of Vanderwaal's constants for the gas
R: Value of the gas constant.

The equation can be expanded as:

$$p + \frac{RT}{V-b} - \frac{a}{V^2}$$

Multiplying by V:

$$pV + RTV\left(\frac{1}{V-b}\right) - \frac{a}{V}$$

$$pV = RT\,(V-b)^{-1}V - \frac{a}{V}$$

$$pV = RT\,(1-\frac{b}{V})^{-1} - \frac{a}{V}$$

The term $(1-b)^{-1}$ can be expanded as:

$$(1-b)^{-1} = 1 + b + b^2 + b^3 + ...$$

Therefore, the value of pV becomes:

$$pV = RT + \frac{RTb-a}{V} + \frac{RTb^2}{V^2} + \frac{RTb^3}{V^3} + ...$$

The value of Z therefore becomes:

$$Z = 1 + \frac{b - \frac{a}{RT}}{V} + \frac{b^2}{V^2} + \frac{b^3}{V^3}$$

The requirement that the values of $\left(\frac{\partial P}{\partial V}\right)_T$ and $\left(\frac{\partial^2 P}{\partial V^2}\right)_T$ be equal to zero at the critical point gives the values of the constants a and b in terms of the critical constants:

$$B + \frac{V_c}{3} = b$$

and $a + \frac{27RbT_c}{8} = \frac{9RT_cV_c}{8}$

The above discussion is a standard one for the VanderWaal's model and is given in books like Faires (8) and Van Ness and Smith (7).

The value of the second term on the right-hand side can be obtained from the virial expression for the value of Z:

As seen before

$$Z = 1 + \frac{b-(a/RT)}{V} + \frac{b^2}{V^2} + \frac{b^3}{V^3} + \frac{b^4}{V^4} + ...$$

This is in the form of the general expression:

$$Z = 1 + \frac{B_1}{V} + \frac{C_1}{V^2} + \frac{D_1}{V^3} + \frac{E_1}{V^4} + ...$$

with B_1, C_1, D_1, E_1,...being functions of temperature T.

$$B_1 = b - \frac{a}{RT}$$

$$C_1 = b^2$$

$$D_1 = b^3$$

$$E_1 = b^4$$

The same virial equation can be put in the form of an expression consisting of terms that are given as power of P.

$$Z = 1 + B_i^1 P + C_i^1 P^2 + D_i^1 P^3 + ...$$

where Van Ness and Smith (7) and Daubert (9) have given equations relating

B_i^1, C_i^1, D_i^1.. to B_1 C_1 D_1

$$B_i^1 = B_i$$

$$C_i^1 = (C_i - B_i^2)/RT)^2$$

$$D_i^1 = (D_i - 3B_iC_i + 2B_i^3)/(RT)^3$$

For the case of the VanderWaal's equation, the values of B_i^1, C_i^1, D_i^1... bcome:

$$B_1' = (bRT - a)/R^2T^2$$

$$C_1' = \left(b^2 - \left(b - \frac{a}{RT}\right)^2\right)/(RT)^2$$

$$D_1' = \left(b^3 - 3\frac{(b-a)}{RT}b^2 - \frac{2(b-a)^3}{RT}\right)/R^3T^3$$

The above values of B_i^1, C_i^1, D_i^1 can be

substituted into the equation for ln (g) derived in the earlier subsection (Sec. 5.0).

6.0 Computational Methods for Exergy by Using Values of Fugergy.

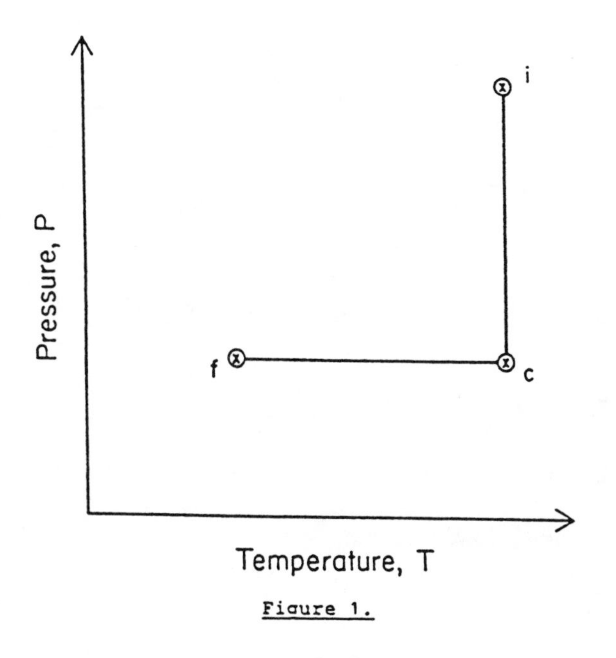

Figure 1.

To compute the difference between the values of exergy between points f and i: Let point c be at the intersection of a line power from point i parallel to the T-axis, and a line drawn from point f parallel to the P-axis.

$$B_i - B_f = B_i - B_c + B_c - B_f$$

$$= \Delta B_{ic} + \Delta B_{cf}$$

Now,

$$\Delta B_{ic} = \Delta B_{ic_T} = \int_c^i dB_T = \int_c^i RT_o \, d\ln (g)$$

$$\therefore \Delta B_{ic} = RT_o \ln (gi/gc).$$

Also,

$$\Delta B_{cf} = \Delta B_{cf,\ \text{const}\ P}$$

Writing $dB = \left(\frac{\partial B}{\partial T}\right)_P dT + \left(\frac{\partial B}{\partial P}\right)_T dP$

Since dP = o along the line cf

Therefore, $dB_{cf} = \left(\frac{\partial B}{\partial T}\right)_P dT$, $\Delta B_{cf} = \int dB_{cf} = \int_f^c \left(\frac{\partial B}{\partial T}\right)_P dT$

From the definition of B,

$$B = H - T_oS.$$

$$\therefore \left(\frac{\partial B}{\partial T}\right)_P = \left(\frac{\partial H}{\partial T}\right)_P - T_o\left(\frac{\partial S}{\partial T}\right)_P$$

$$= \left[C_p - \frac{T_o C_p}{T}\right] \quad \text{since} \left(\frac{\partial H}{\partial T}\right)_P = C_p$$

$$\& \left(\frac{\partial S}{\partial T}\right)_P = \frac{C_p}{T}$$

The values of specific heat are generally given as a power series:

$$Cp = \alpha + \beta T + \gamma T^{-2} + ...$$

$$\therefore \Delta B_{cf} \cong \int_{T_f}^{T_c} (\alpha + \beta T + \gamma T^{-2}) \frac{(T-T_o)}{T} dT$$

$$= (\alpha - \beta T_o)(T_c - T_f) + \beta (T_c - T_f^2) - \gamma \left(\frac{1}{T_c} - \frac{1}{T_f}\right) + \frac{\gamma T_o}{2}(T_c^{-2} - T_f^{-2})$$

Thus:

$$\Delta B_{if} \cong RT_o \ln \frac{(g_i)}{g_c} = (\alpha - \beta T_o)(T_c - T_f) + \beta (T_c^2 - T_f^2) - \gamma\left(\frac{1}{T_c} - \frac{1}{T_f}\right) + \frac{\gamma}{2} T_o (T_c^{-2} - T_f^{-2})$$

7.0 Example: Use of the Fugergy Function in Computations Concerning Exergy Analysis:

Problem: Find the maximum obtainable work from steam at 100 lb/in^2 and 1000°F = 1460°R when it is brought down to a final state of 40 lb/in and 500°F = 960°R in a steady-state flow process (Lee and Seers (10)). (Fig. 2)

Answer: Method #1: Use of Property Tables.

The values of the property of steam at the initial and final states are:

H_i = 1530.8 Btu/lbm; S_i = 1.9193 Btu/lbmR

H_f = 1284.8 Btu/lbm; S_f = 1.8140 Btu/lbmR
The value of the maximum useful work, per lbm of steam, is therefore:

$W_{MAX} = B_i - B_f$

$B_i = H_i - T_oS_i$ = 1530.8 - 520°R.x 1.19193. = 532.8 Btu/lbm.

$B_f = H_f - T_oS_f$ = 1284.8 - 520x 1.8140 = 341.6 BTu/lbm.

The maximum useful work per pound mass of steam is therefore:

$W_{MAX} = B_i - B_f$ = 532.8 - 341.6 = 191.2 Btu/lbm.

<u>Method</u> #2 <u>Use of Fugergy Tables:</u>

From critical tables, for water,

Tc = 374.15°C = (374+273)°K = 647.15°.K.
Pc = 28.4 atm.
dc = 0.323 gm/c.c. (dc = density at the critical point).
or Vc = 3.095 cm 3/gm.
= 55.17 cm^3/gm - mole.

R = 82.057

$\therefore Z_c = P_cV_c$ = (218.4) (55.17)

= 0.226.

$\therefore \Delta B_{ic} = B_i - B_c$

$= RT_o \ln (g_i/g_c)$

Point i:

P = 100 psi

$\therefore P_{r_i} = P/P_c$ = 0.031

T = 1000F = 810K

T_{ri} = 810/647.15

= 1.25

T_o = 60 F = 288.15K

$\therefore T_{or}$ = (288.15/647.15)

= 0.445

Point c:

P_{rc} = (40/218.4)

= 0.18

T_{rc} = 1.25

T_{or} = 0.445

Interpolating values in the tables of fugergy:

$\left(\frac{f}{p}\right)_i \approx 1.03$ & $\left(\frac{f}{p}\right)_c \cong 1.01$ (Table 2A)

$\therefore \Delta B_{ic}$ = RT_o Ln (1.03X100)/(1.01X40)

= 53.72 Btu/lbm

$\therefore \Delta B_{cf}$ = $C_p (T_c - T_f)$

Assuming C_p = (5/2) R (Ideal gas)

$\therefore \Delta B_{cf}$ = (5/2) (1460 - 960)R

= 138 Btu/lbm

$\therefore \Delta B$ total = (138 + 53.72)

= 191.72

$\therefore$ Error due to use of fugergy tables = (191.72 - 191.20)/(191.20) X 100%.

= 0.27%

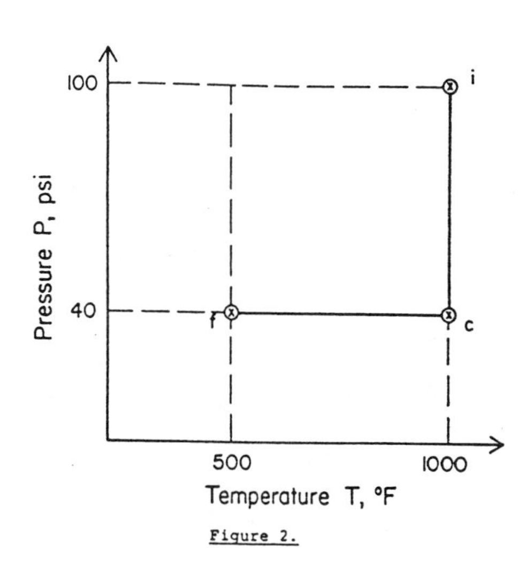

Figure 2.

8.0 Conclusion:

From the above treatment it is seen that the thermodynamic function fugergy can be used for computational work used in availability analysis of flow systems. Tabulated values of fugergy, which are presented in this paper, can be used for purpose of these analyses. In addition, the closed form equation that is given for the case of high pressure gases can be used for the case of problems that only involve the gaseous phase: An approximate solution can be thus obtained which can be further refined by using tabulated values of properties of the substance. While the illustrative example presented in the paper was concerned with a system ((H_20) that has been studied in detail (and tabulated values of properties are obtainable), the function fugergy would especially be of use for those substances where detailed tabulated values of properties of the substance are not available. Functions such as fugacity, etc., have been put forward in the past to meet specific needs and the function fugergy has been put forward to meet the needs of analysis based on first and second laws.

REFERENCES

1. Kotas, T. J., "Exergy Method of Thermal and Chemical Plant Analysis," Chemical Engineering Research and Design, Vol. 64, No. 3, pp. 212-229 (1986).

2. Moran, Michael J., "Availability Analysis Applied to Efficient Energy Use," Prentice-Hall, Inc., 1982.

3. Dibenedetti, P. G., "The Thermodynamic Fundamental of Exergy," Chemical Engineering Education, Summer 1984, pp. 116-121.

4. Keenan, J. M., Mechanical Engineering, 54, p. 199 (1932).

5. Chin, C. H., and Newton, C. L. "Second Law Analysis in Cryogenic Process," Energy—The International Journal, 5, p. 899 (1979).

6. Hougen, D. A., Watson, K. M., and Ragatz, R. A., "Chemical Process Principles," Part II — Thermodynamics. John Wiley and Sons, Inc., New York, 2nd Edition, 1959, pp. 569-575.

7. Reid, R. C., Prausnitz, J. M., and Sherwood, T. K., "The Properties of Gases and Liquids," McGraw Hill, New York, 1977.

8. Smith, J. M., and Van Ness, H. C., "Introduction to Chemical Engineering Thermodynamics, McGraw Hill, New York, 3rd Edition, 1975.

9. Faires, V. M., and Simmons, C. M., "Thermodynamics," MacMilan Publishing Co., Inc., Sixth Edition, 1978, p. 275.

10. Daubert, T. E., "Chemical Engineering Thermodynamics," McGraw Hill, New York, 1985, pp. 22-24.

11. Lee, J. F., and Sears, F. W., "Thermodynamics," Addison-Wesley, Reading, Mass., 2nd Edition, 1962, p. 199.

NOMENCLATURE

a: Constant in the VanderWaal's equation.

b: Constant in the VanderWaal's equation.

B_1, C_1, D_1 Coefficients in the virial equation given in powers of specific volume.

$B_:^1, C_:^1, D_:^1$ Coefficients in the virial equation given in powers of pressure.

B: Available energy of the substance.

Bi: Available energy of the substance at state i.

B_f Available energy of the substance at state f.

C_p: Specific heat of the substance.

d_c Density of the substance at the critical point.

dB_T: Differential change in the available energy at constant temperature for a substance.

Differential change in the available energy at constant temperature for an ideal gas.

dG_T: Differential change in the Gibbs Free Energy at constant temperature for a substance.

Differential change in the Gibbs Free Energy at constant temperature for an ideal gas.

dH_T: Differential change in the Enthalpy of the substance at constant temperature.

Differential change in the Enthalpy of an ideal gas at constant temperature.

dP: Differential change in the pressure.

dS_T: Differential change in the entropy of the substance at constant temperature.

Differential change in the entropy of ideal gas at constant temperature.

dT: Differential change in the temperature.

f: Fugacity of the substance.

g: Fugergy of the substance.

G: Gibbs Free Energy of the substance.

H: Enthalpy of the substance.

H_i: Enthalpy of the substance at state i.

H_f: Enthalpy of the substance at state f.

P: Pressure

Pc: Pressure at the critical point.

R: Universal Gas Constant

S: Entropy of the substance.

T: Temperature

T_c: Temperature at the critical point.

To: External temperature.

T_r: Reduced temperature (=T/Tc)

T_{OR}: Reduced external temperature (=To/Tc).

V: Specific volume of the substance.

Vc: Specific volume of the substance at the critical point.

Z: Compressibility factor.

Zc: Compressibility factor at the critical point.

TABLE 1.

Values of as a function of (Pr, Tr, Tor) for Zc = 0.27

Table 1(a): At Tor = 0.1

At Tor = 0.1

Tr	Pr					
	0.2	0.5	1.0	1.25	1.5	2.0
0.5	4.354×10^{-10}	1.276×10^{-9}	2.118×10^{-10}	1.918×10^{-10}	1.927×10^{-10}	1.852×10^{-10}
1.0	1.028×10^{-1}	1.755×10^{-3}	1.670×10^{-7}	2.724×10^{-9}	4.175×10^{-10}	5.071×10^{-11}
1.5	1.689×10^{-1}	4.657×10^{-2}	7.701×10^{-3}	2.127×10^{-3}	6.530×10^{-4}	7.581×10^{-5}
2.0	8.121×10^{-1}	4.502×10^{-1}	2.385×10^{-1}	1.518×10^{-1}	8.698×10^{-2}	2.828×10^{-2}

Table (1b): At Tor - 10

At Tor = 10

Tr	Pr					
	0.2	0.5	1.0	1.25	1.5	2.0
0.5	2.482×10^{-1}	9.377×10^{-2}	4.820×10^{-2}	3.989×10^{-2}	3.358×10^{-2}	2.645×10^{-2}
1.0	1.182	1.569	3.040	3.797	3.928	3.863
1.5	1.099	1.161	1.245	1.318	1.386	1.522
2.0	1.000	1.018	1.039	1.053	1.074	1.114

Table (1c) Tor = 50

At Tor = 50

Pr / Tr	0.2	0.5	1.0	1.25	1.5	2.0
0.5	2.921×10^{-1}	1.085×10^{-1}	5.631×10^{-2}	4.579×10^{-2}	3.915×10^{-2}	3.078×10^{-2}
1.0	1.206	1.658	2.479	4.502	4.729	2.730
1.5	1.115	1.192	1.298	1.388	1.475	1.648
2.0	1.002	1.025	1.051	1.070	1.096	1.147

TABLE 2

Table (2A) as a function of (Pr, Tr, Tor) for Zc = 0.23

Table (2A-a) Tor = 0.1

Pr / Tr	0.2	0.5	1.0	1.25	1.5	1.7
0.5	292.95	4482.78	53530.12	288370.1	652130.1	1164.73
1.0	2.504	2.64	.02151	.003115	.257	3.84 x 10
1.5	1.058	1.073	2.936×10^{-2}	2.226	.432	.635
1.7	.908	.9085	.4338	.457	.690	.712

Table 2A(b) Tor = 1.0

Pr / Tr	0.2	0.5	1.0	1.25	1.5	1.7
0.5	.492	.350	.256	.208	.188	.1745
1.0	.930	.829	.650	.541	.465	.420
1.5	1.002	1.003	.8816	.9338	.970	.984
1.7	.996	.996	.964	.9663	.984	.985

Table (2A-c): Tor = 10

Pr / Tr	0.2	0.5	1.0	1.25	1.5	1.7
0.5	.2595	.1358	.0753	5.05×10^{-2}	4.16×10^{-2}	3.625×10^{-2}
1.0	.9123	.9074	1.9134	.9061	1.052	1.0279
1.5	.96	.996	1.239	1.341	1.0523	1.028
1.7	1.005	1.005	1.044	1.042	1.019	1.018

Table (2A-d): Tor = 25

Pr / Tr	0.2	0.5	1.0	1.25	1.5	1.7
0.5	.249	.125	.0694	4.595	3.463	3.264
1.0	.9067	.9015	1.033	.9378	1.156	1.81
1.5	.9962	.9953	1.267	.864	1.058	1.031
1.7	1.005	1.005	1.049	1.047	1.022	1.01

Table (2Ae): Tor = 50

Pr / Tr	0.2	0.5	1.0	1,25	1.5	1.7
0.5	.245	.129	6.756×10^{-2}	4.453×10^{-2}	1.159	1.832
1.5	.996	.995	1.277	.8623	1.06	1.032
1.7	1.0058	1.0058	1.052	1.0485	1.0227	1.021

Table 2A-f: Tor = 60

Pr / Tr	0.2	0.5	1.0	1.25	1.5	1.7
0.5	.2446	.1244	6.725	4.430	3.618	3.134
1.0	.905	.899	.9433	.9504	1.160	1.836
1.5	.996	.995	1.278	.8621	1.060	1.032
1.7	.006	1.0058	1.052	1.0437	1.0228	1.0208

Table 2B: as a function of (Pr, Tr, Tor) for Zc = 0.29

Table (2B-a): Tor = 0.1

Pr / Tr	0.2	0.5	0.8	1.0	1.15
0.5	1.444×10^{-6}	4.87×10^{-6}	8.134×10^{-7}	7.383×10^{-7}	1.114×10^{-6}
1.0	1.021×10^{-1}	2.346×10^{-3}	1.799×10^{-5}	1.941×10^{-8}	1.505×10^{-10}
1.5	6.130×10^{-1}	1.540×10^{-1}	3.175×10^{-2}	1.027×10^{-2}	4.873×10^{-3}
2.0	7.877×10^{-1}	4.978×10^{-1}	3.224×10^{-1}	2.541×10^{-11}	2.038×10^{-1}

Table (2B-b): Tor = 1.0

Pr / Tr	0.2	0.5	0.8	1.0	1.15
0.5	9.079×10^{-2}	4.094×10^{-2}	2.302×10^{-2}	1.901×10^{-2}	1.769×10^{-2}
1.0	9.530×10^{-1}	8.560×10^{-1}	7.560×10^{-1}	6.770×10^{-1}	6.110×10^{-1}
1.5	9.723×10^{-1}	9.074×10^{-1}	8.397×10^{-1}	7.918×10^{-1}	7.662×10^{-1}
2.0	9.847×10^{-1}	9.557×10^{-1}	9.306×10^{-1}	9.172×10^{-1}	9.040×10^{-1}

Table (2B-c) Tor = 10

Pr / Tr	0.2	0.5	0.8	1.0	1.15
0.5	2.741×10^{-1}	1.011×10^{-1}	6.471×10^{-2}	5.248×10^{-2}	4.655×10^{-2}
1.0	1.191	1.544	2.192	3.844	5.583
1.5	1.018	1.083	1.165	1.223	1.270
2.0	1.007	1.020	1.035	1.043	1.049

Table (2B-d) Tor = 25

Pr / Tr	0.2	0.5	0.8	1.0	1.15
0.5	2.950×10^{-1}	1.073×10^{-1}	6.870×10^{-2}	5.616×10^{-2}	4.979×10^{-2}
1.0	1.209	1.606	2.353	4.316	6.471
1.5	1.021	1.096	1.191	1.259	1.314
2.0	1.008	1.025	1.042	1.052	1.060

Table (2B-e) Tor = 50

Pr / Tr	0.2	0.5	0.8	1.0	1.15
0.5	3.024×10^{-1}	1.095×10^{-1}	5.744×10^{-2}	5.744×10^{-2}	5.073×10^{-2}
1.0	1.215	1.627	2.410	4.486	6.797
1.5	1.022	1.101	1.200	1.271	1.329
2.0	1.009	1.026	1.044	1.055	1.063

Table (2B-f): Tor = 60

Pr / Tr	0.2	0.5	0.8	1.0	1.15
0.5	3.036×10^{-1}	1.009×10^{-1}	7.055×10^{-2}	5.766×10^{-2}	5.091×10^{-2}
1.0	1.216	1.631	2.419	4.515	6.853
1.5	1.022	1.101	1.201	1.273	1.331
2.0	1.009	1.026	1.045	1.055	1.064

Table (2B-e) Tor = 50

Pr / Tr	0.2	0.5	0.8	1.0	1.15
0.5	3.024×10^{-1}	1.095×10^{-1}	5.744×10^{-2}	5.744×10^{-2}	5.073×10^{-2}
1.0	1.215	1.627	2.410	4.486	6.797
1.5	1.022	1.101	1.200	1.271	1.329
2.0	1.009	1.026	1.044	1.055	1.063

Table (2B-f): Tor = 60

Pr / Tr	0.2	0.5	0.8	1.0	1.15
0.5	3.036×10^{-1}	1.099×10^{-1}	7.055×10^{-2}	5.766×10^{-2}	5.091×10^{-2}
1.0	1.216	1.631	2.419	4.515	6.853
1.5	1.022	1.101	1.201	1.273	1.331
2.0	1.009	1.026	1.045	1.055	1.064

AIAA-2000-2893

STUDIED OF RECYCLED FUEL OIL FOR DIESEL ENGINE EXTRACTED FROM WASTE PLASTICS DISPOSALS
(Case of the Cracked PE Oil)

*Shinji MORIYA Nihon University
Hiroichi WATANABE Nihon University
Rikio YAGINUMA Nihon University
Tomonori MATSUMOTO Nihon University
Morihisa NAKAJIMA Saitama kiki Co.Ltd
Masataka TSUKADA Ecology System Co.Ltd
Naotsugu ISSHIKI Tokyo Institute Technology

ABSTRACT

In developed countries there are so many discarded plastic disposals from various kinds industry and cities. One of authors company (Ecology System) has developed a new recycle system and apparatus to liquefy these plastic disposals into fuel oils. These discarded plastic are melted and evaporated first, and their vapors are condensed and extracted after passing through
various metal catalyzers for decomposition.
The recycled oils are different one by one according to the kind of raw plastic, and generally it has high heating value with rather big density, and their total weight are about
60_80 % of raw plastics.
The recycled oils are fed to Diesel engines of several factories as a 30_40% blended fuels with gas oil.
Recycle of plastics by this method above will be very good contribution to energy and
environments of the earth.
In this paper is showed used of the Cracked PE (Polyethylene) blending oil for engine performance.

_INTRODUCTION

In advanced countries, there are produced so

*ASSOCIATE PROFESSOR
DEPARTMENT OF MECHANICAL ENGINEERING
COLLEGE OF ENGINEERING

many waste plastics from industry and cities,
and they have generated the problem of how to recycle them.
In the past, those waste plastics were recycled in several way. For example, they have been mechanically transformed into construction material or low quality resin.
There were also converted into pelletized fuel or boilers. But in reality, many of those pellet plants generate too much heat and unfavorable emissions accompanied with high soot and HC.
Here, one of the author company (Ecology System) has developed new plastic recycle system which can recover 80~90% fuel oil from waste plastics by pyrolysis and liquefaction process.
The recovered oil is kind of new valuable alternative fuel, and it can be used not only
In boilers, but for Diesel engines by blending.
The above recycling process, quality of recovered oil, and example of engine utilization are described in this paper.

DESCRIPTIN OF RECYCLE PROCESS

In Fig.1 the general thermatic flow lines of process of this waste plastic recycling system are shown.
As seen there, plastics of various kinds are
Fed into the main pyrolysis oven, in which plastics are melted and evaporated at about 350~400_., where heavy HC molecules are decomposed into ligher molecules contacting metal catalyzers.
Then vapor is condensed in the condensers cooled by air and or water, and it becomes the recovered oil.
The recovered oil can be used as fuel for Diesel engines and boilers.
By this system not only waste plastics but waste tires, waste industrial or edible oils are able to be converted into useful fuel oils.
This system can be constructed in both batch type or continuous type, but by the reason of better maintenance ability, batch type is used more than the latter.
In Fig.2 one illustration figure of batch type plant is shown and Fig.3 a photograph of one plant of this type is shown.

CHARACTERISTICS & FEATURES OF THIS PLANT

By this recycle plant, waste plastics of PE(Polypropylene), PP(Polypropylene), PS(Polysyrene), acrylite polyamide and FRP can be recycled into recovered fuel except chloride plastics as vinyl chloride which generates chloride acid.
From 100% solid plastics fed to system, 70~90% liquid fuel can be recovered with 3~6% non condensable gas and 7~14% solid residue.
When other materials as wood,paper,and metal etc. are mixed in the waste, the percentage of effective recovery decreases and solid residual increases respectively.

Sometimes water is contained in the waste, it is vaporized in early stage safely in batch type process. But in continuous type process, other materials than plastic should be kept as small as possible.

This plant is quite safe at work, because all parts are kept at positive pressure lower than 100mmaq at all times resulting no air leak into system.

Also it contains safety valves. All parts are made of stainless steel and they resist corrosion of the process.

The metal catalyzers are made of various kind of metals including Ni and Cu, and or ceramics and zeolite in shape of punched plate and wire mesh type.

_UALITY OF RECOVERED OIL

Recovered oils from this plant are a little different from each others depending on raw plastic materials. On Table 1, the main characteristic qualities of recovered oil are shown. Table 2, the properties of the distilled oil are shown.

Fig.4 shown the measured relation between Viscosity and Temperature. By this figure following can be said that, recovered oil is low viscosity.

Fig.5 shows a section of ignition test equipment inside the heating oven, where ,ignition temperature is measured by thermometers placed in the oven space.

Ignition time is determined by measuring time duration between dropping of fuel from micro syringe until ignition of the happens.

On Fig.6, natural ignition temperature measured by drop method is shown, being compared with gas oil.

By looking at the table, it can be seen that the recovered oils from this plant have a wide gravity range from 0.783 to 0.9, and high calorific values near or a little above 44 MJ/kg, proving to be good fuel.

Their ignition points by drop method are similar to gas oil at except PS, and their viscosity is generally low.

So recovered oils from waste plastics may be said to be very good fuel oil similar gas oil, except they have a little residues and gamms.

Fig.7 shows a ignition temperature in PE30%Removed oil case.

QUALITY OF REMOVED OIL

The recovered oil for waste plastic is the very low viscosity. So, for a change the higher viscosity is the cracking the waste plastic oil. The cracking oil is the Removed oil. Here is the PE removed oil.

Fig.8 shows the relation between Dynamic viscosity and Temperature. By this figure following can be said, that removed ratio oil is a lager, viscosity is the highest.

EXAMPLE OF ENGINE USAGE

Some of the recovered oils are used in Diesel engine by blending at gas oil.

Main specification of Diesel engine used for the experiment of various blended fuels are shown in Table 3.

Fig.9 shows the measuring and analyzing system of the engine.

Fig.10 shows experimental test rig.

Fig.11 shows the relation between specific Fuel Consumption and Load in the case of PE30%Removed oil and gas oil. By this figure following can be said, that the PE30%Removed oil is a almost similar the
Gas oil.

Fig.12 shows the relation between Nitorogen Oxid (NOx) and Load. By this figure the following can be said that, when the low load, PE30%Removed Oil is low NOx better Gas Oil.

Fig.13 shows relation between Smoke and Load. By this figure following can be said that, PE30%Removed Oil is the low smoke the almost all area Load better Gas Oil.

Fig.14 shows P-_ Diagram at Load 3/5, 3000rpm. By this figure the following can be said, PE30%Removed Oil is the low combustion pressure better Gas Oil.

Fig.15 shows Heat Release Rate at Load 3/5, 3000rpm. By this figure the following can be said, PE30%Removed Oil is the low heat release rate better Gas Oil.

CONCLUSION

As described above, one of authors' company(Ecology System) has developed new recycle plant for waste plastics by pyrolysis liquification process, using metal catalyzer, and it can recover 80~90% oil from raw plastics.

The recovered oil has high heating value of around 45MJ/kg and it is possible to be used in Diesel engine.

One example of the engine experiments is shown here. It can be said PE30%Removed oil is used successful, and they decrease NOx, Smoke better Gas Oil.

Previously waste plastics were very difficult to recycle and gave much difficulty in combustion in incinerators and boilers.

The method of this paper gives quite a new way for waste plastics recycling, because waste plastics are converted into useful alternative fuel of engines and boilers with less influence on the environment.

CONTACT

TAMURACHO,KORIYAMASHI,
FUKUSHIMAKEN, JAPAN
DEPARTMENT OF MECHANICAL ENGINEERING
COLLEGE OF ENGINEERING
NIHON UNIVERSITY
TEL: +81-24-956-8766
FAX: +81-24-956-8860
E-mail: moriya@mech.ce.nihon-u.ac.jp

REFERENCE

(1) S.Moriya,N.Isshiki,Proc.Int.Conf.onSmall Engines and Their Fuels,1993, Chain Mai
(2) N.Isshiki,S.Moriya,etc.Proc.Int.Conf.on Small Engines and Their Fuels,1995, UK
(3) S.Moriya,N.Isshiki,etc. Proc.Int.Conf.on IECEC 96, Washington,DC

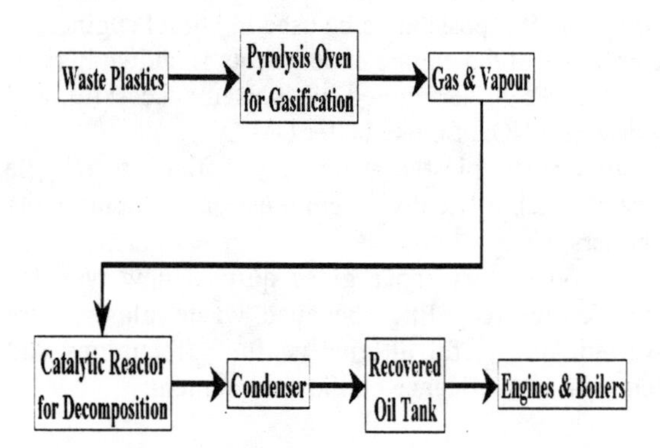

Fig.1 Thermatic Flow Line of the Process

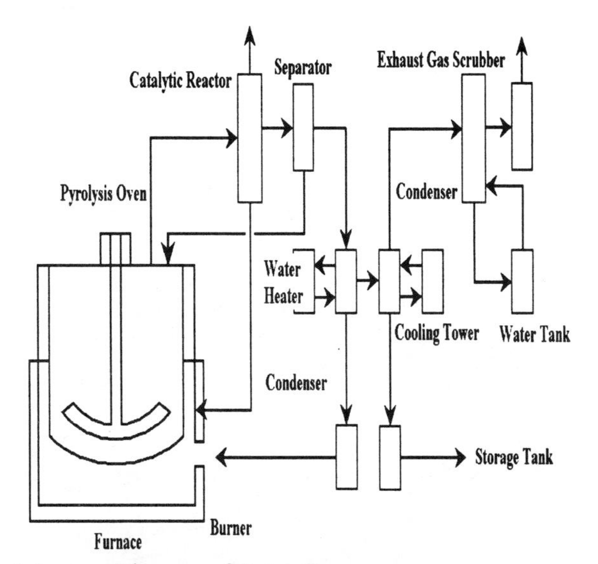

Fig.2 System Diagram of Batch Type Process

Fig.3 Photograph of one Plant

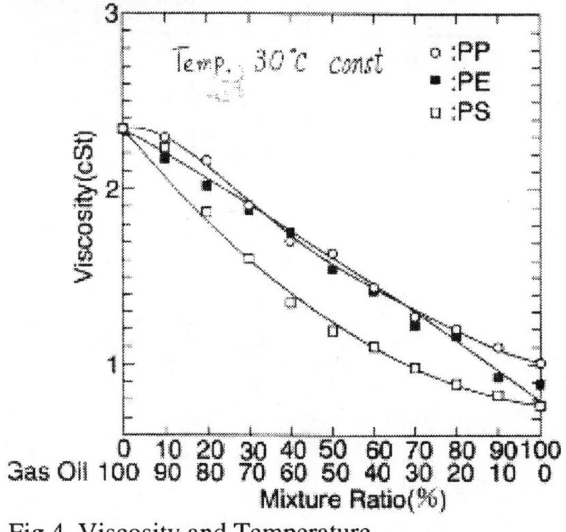

Fig.4 Viscosity and Temperature (Pure Fuel)

Table 1 Characteristics of Recovered Oil and Gas oil

_	_	—	—	—	Gas oil
Specific Gravity	(15/4_)	0.783	0.804	0.9	0.82
Flash Point	(_)	21	-	25	_45
Calorific Value	(MJ/kg)	45	44.1	41.8	45.7
Dynamic Viscosity	(cSt)	1.52	1.2	1.84	2.74
Moisture	(wt_)	0.02	0.06	0.04	-
Ash		<0.01	<0.01	<0.01	-
Carbon		86.08	87.2	90.73	-
Hydrogen		13.66	12.6	9.06	-
Nitorogen		0.14	0.16	0.08	-
Salfur		<0.01	0.02	<0.01	-
Oxygen		0.12	0.02	0.13	-
_/_	_	6.3	6.92	10.01	-

Table 2 The Properties of the Distilled Oil

_	_	—	—	_30 %	_50 %	Gas Oil
Specific Gravity	(15/4_)	0.783	0.804	0.808	0.84	0.82
Initial boiling Point	(_)	34	52	120	124	120
50% Received Temperature	(_)	214	190	252	251	268
90% Received Temperature	(_)	320	311	335	328	332
Mean Molecular Weight	_	137	193	207	327	-
Calorific Value	(MJ/kg)	45	44.1	45.2	43.9	45.7
Dynamic Viscosity (30_)	(cSt)	1.52	1.2	2.0	2.3	2.74

Fig.5 A Section of Ignition Test Equipment (Inside Heater)

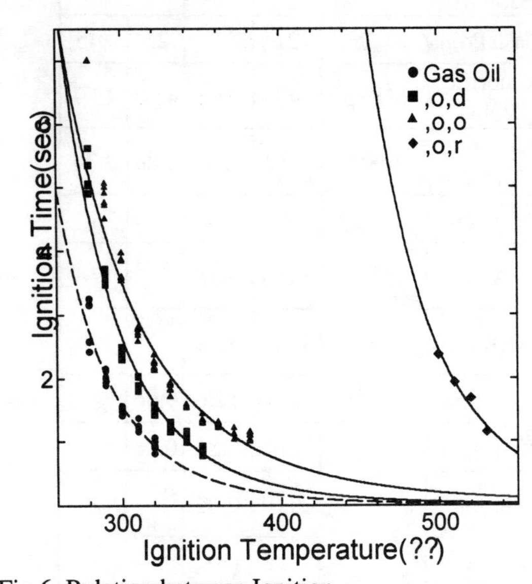

Fig.6 Relation between Ignition Temperature and Ignition Time

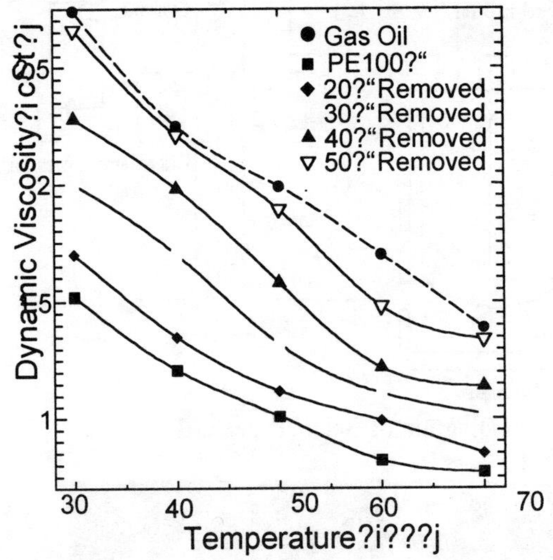

Fig.8 Dynamic viscosity and Temperature
_ ___PE Removed Oil__

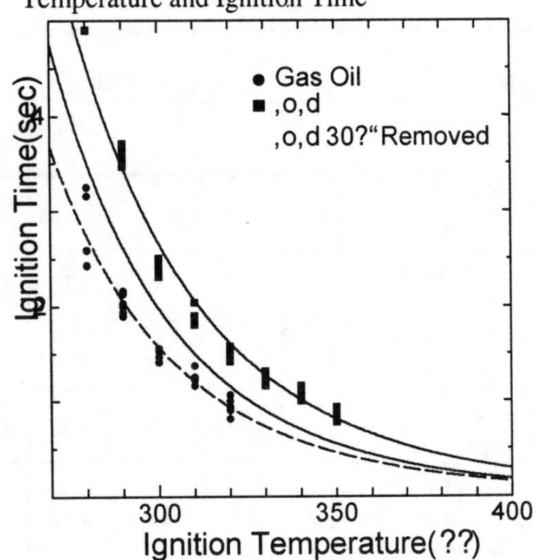

Fig.7 Ignition Temperature and Ignition Time (PE30%Removed oil)

Table 3 Specification of Diesel Engine

Engine Name		-	L60
Type		-	Air-Cooled 4-Cycle Diesel Engine
Combustion Type		-	Direct Injection
Number of Cylinders		-	1
Cylinder Bore _ Stroke		mm	75 _ 62
Stroke Volume		mm^3	273
Compression Ratio	_	-	20
Power	Maximum	PS	5.5
		kw	4
	Continuous Declared	PS	6
		kw	4.4
Injection of Fuel Pressure		MPa	19.6
Type of Injection Valve	_	-	Multi Pole

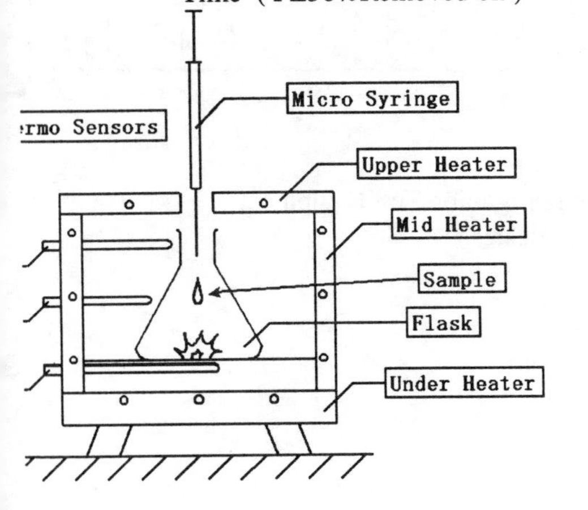

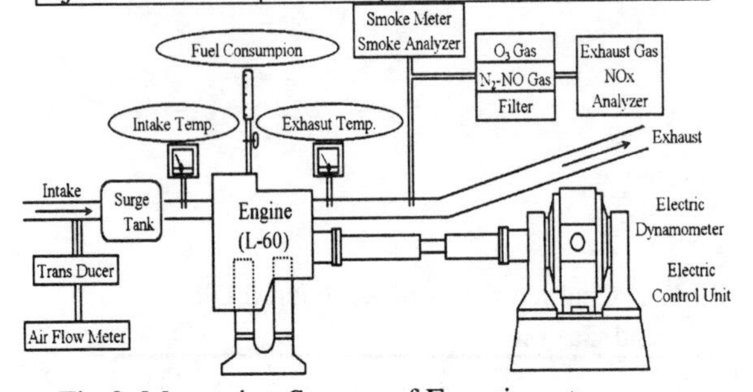

Fig.9 Measuring System of Experiments

Fig.10 Experimental Test Rig

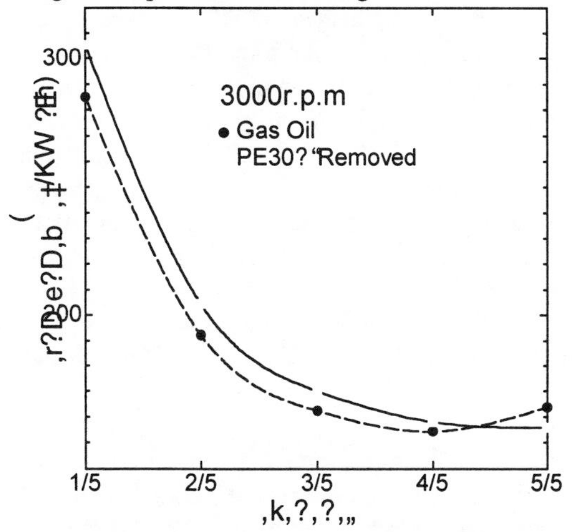

Fig.11 Specific Fuel Consumption and Load (PE30%Removed oil)

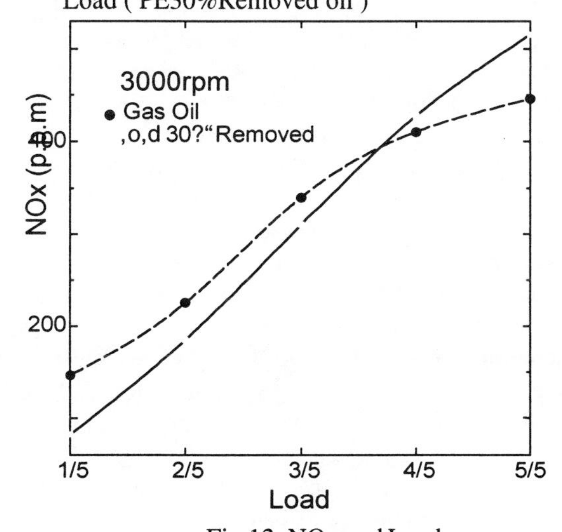

Fig.12 NOx and Load

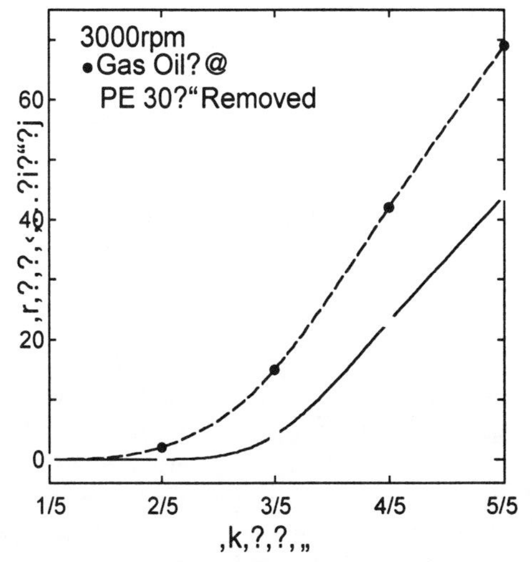

Fig.13 Smoke and Load.

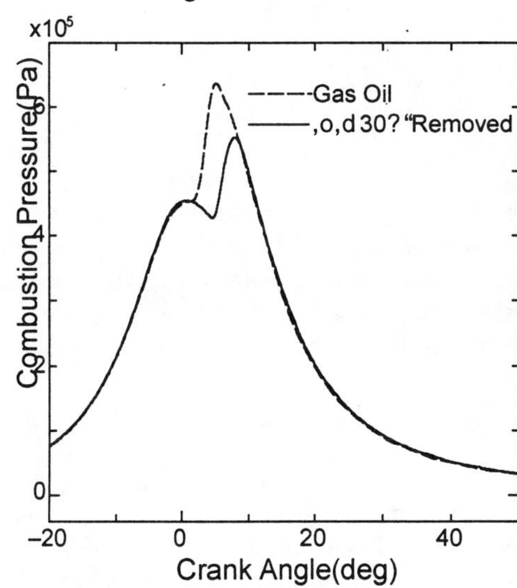

Fig.14 P-_Diagram(Load 3/5, 3000rpm)

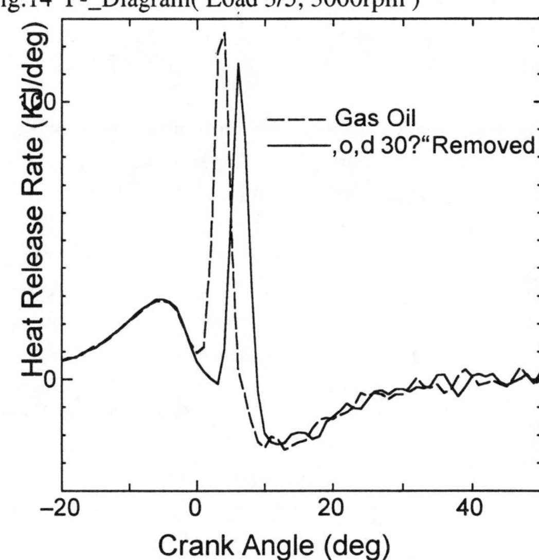

Fig.15 Heat Release Rate (Load 3/5, 3000rpm)

32.3% EFFICIENT TRIPLE JUNCTION $GaInP_2$/GaAs/Ge CONCENTRATOR SOLAR CELLS

D. Lillington, H. Cotal, and J. Ermer
Spectrolab Inc.
12500 Gladstone Ave., Sylmar, CA 91342

D. Friedman, T. Moriarty, A. Duda
National Renewable Energy Laboratory
1617 Cole Blvd., Golden CO 80401

Abstract

This paper describes progress toward achieving high efficiency, multijunction solar cells for cost effective application in terrestrial PV concentrator systems. Small area triple junction $GaInP_2$/GaAs/Ge solar cells have been fabricated with an efficiency of over 32% when measured by NREL under an AM1.5D spectrum at 47 suns concentration. Small changes to the device design can achieve similar efficiencies at concentration ratios of about 500 suns, resulting in cell costs of $0.5 to $0.6/W today, at production volumes of approximately 50 MW/year. This makes them highly cost effective in existing concentrator systems, compared to flat plate technologies. The future development of new 1 eV materials for space cells, in conjunction with further reduction in Ge wafer costs, promises to achieve solar cells of >40% efficiency that cost $0.4/W or less at these concentration ratios.

Introduction

The U.S. PV industry is a world leader in research, development, and manufacturing technology and holds a major share of the world market for PV systems. While the global opportunities for PV are immense, the challenge of emerging international competition has forced U.S. industry to re-evaluate its overall strategy and to define a road map to ensure future industry leadership [1]. High on the list of priorities for competitiveness are cost reduction and efficiency improvement to achieve end-user system costs of $3/W or less. Concentrators have for many years been recognized as a means to substantially reduce PV system costs, since the cost of optics and a tracker are generally less than the cost of the solar cell itself. However, PV sales have, historically been dominated by small power output products, such as calculators, road-side flashing lights and emergency telephones, and other small remote telecommunications applications. While concentrators cannot compete for this type of market, recently, PV sales have begun to include larger installations, e.g. 1-4 kW systems for off-grid homes, similar size systems for rural electrification, and even grid-connected systems up to 300 kW that will be installed in Arizona in the year 2000. Indeed a very positive event took place in Arizona this year in which The Corporation Commission mandated that, under the Solar and Environmentally Friendly Portfolio Standard, electricity providers must derive 1.1% of their total product from renewable resources by 2007, with 50% of the mix being from solar energy. Furthermore, 0.4% of the total product must be derived from renewable sources by January 1, 2001. As this trend toward larger systems continues, concentrators will have an opportunity to enter the market at a substantial growth rate.

Recent performance improvements, and cost reductions achieved in the manufacture of $GaInP_2$/GaAs/Ge latticed matched space solar cells [2] has prompted Spectrolab to pursue the application of this product to terrestrial PV concentrator systems. AM1.5D efficiencies of over 30% can now readily be achieved on triple junction cells using device designs and high yielding production processes similar to those used for space cells. Because of this synergy, and the resulting "dual use" of the substantial capital investments that have already been made, it is possible to achieve very low $/W cell costs at relatively modest manufacturing rates. This makes III-V cells extremely attractive for use in medium to high X concentrator systems to achieve substantial near-term reductions in PV generation costs. In addition, the leveraging of semiconductor manufacturing capacity afforded by the use of concentration allows very rapid ramp up to achieve manufacturing capacity in excess of 150 MW per year using existing Spectrolab manufacturing infrastructure, should the demand arise.

Solar Cell Design And Performance

Terrestrial triple junction $GaInP_2$/GaAs/Ge solar cells were fabricated using space solar cell device structures modified for optimum performance under the AM1.5D terrestrial spectrum. The device cross section is shown in Figure 1, and has been extensively described elsewhere [2]. The principal difference in the device design compared to a space cell, were a thicker, more

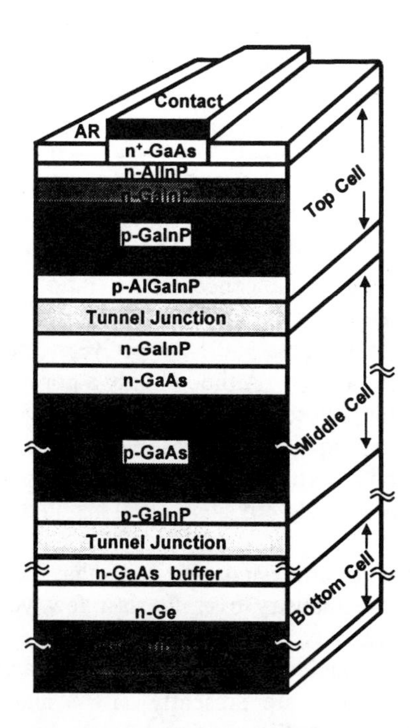

Figure 1. Cross-Section of Triple Junction Cell

heavily doped emitter to reduce distributed sheet resistance losses at high concentration, and a thicker $GaInP_2$ top cell to achieve current matching under an AM1.5D spectrum. The solar cell material was grown at Spectrolab in a production MOVPE reactor, and processed at NREL using mask sets specifically designed for concentrator cells. The area of each cell was 0.0139 cm^2. Solar cells measurements were performed by NREL. The cells were first measured at one sun under an X-25 simulator to determine the one-sun short circuit current and were then measured as a function of concentration under a High Intensity Pulsed Solar Simulator (HIPSS).

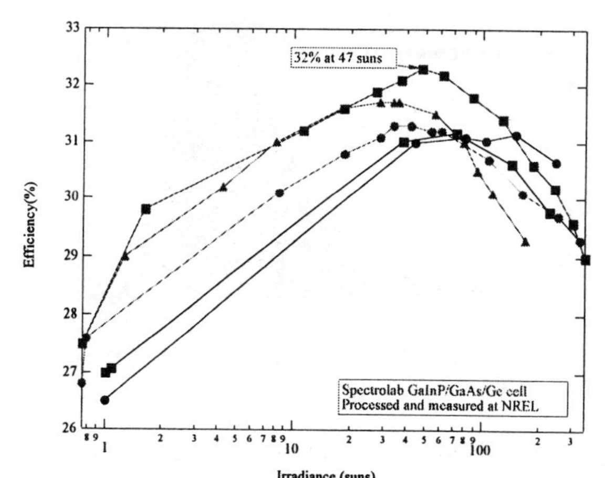

Figure 2. Efficiency vs. Concentration Ratio for TJ Cells (AM1.5D, 25 °C)

Figure 2 shows the measured efficiency vs. concentration ratio on a number of cells. The IV curve of the best cell and a typical cell spectral response are also shown in Figures 3 and 4 respectively. Many of the cells had an efficiency >30% at low to medium concentration (up to about 50 suns). The best cell had an efficiency of 32.3% at 47 suns, and 29% at 350 suns (AM1.5D, 25 °C).

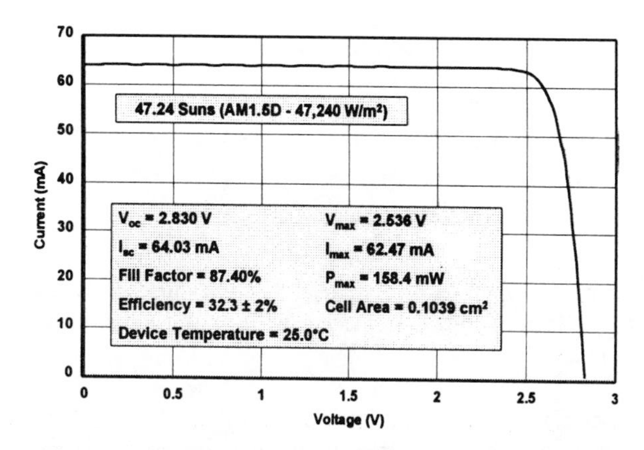

Figure 3. IV Curve of 32.3% Efficient Triple Junction Cell. (measurements performed by NREL)

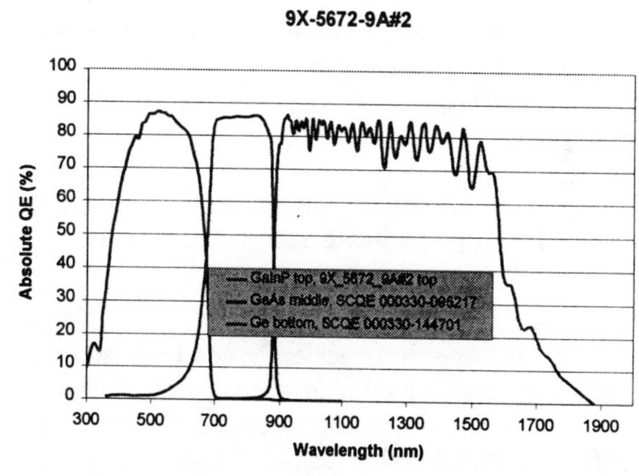

Figure 4. Typical Triple Junction Cell Quantum Efficiency (measurements performed by NREL)

The roll-off in performance of all the cells at higher concentration ratios was attributable to a reduction in open circuit voltage (V_{oc}) as intensity increased beyond about 50 suns. This was somewhat surprising since V_{oc} normally increases linearly with the logarithm of short circuit current and concentration ratio. A series of diagnostic tests revealed a "less than ideal" back contact that caused an "opposing" voltage to be generated at illumination intensities. On very similar devices, removal of the back contact and subsequent replacement with a low resistance ohmic contact, eliminated this roll-off, and resulted in the expected relationship between Voc and concentration ratio as shown in Figure 5.

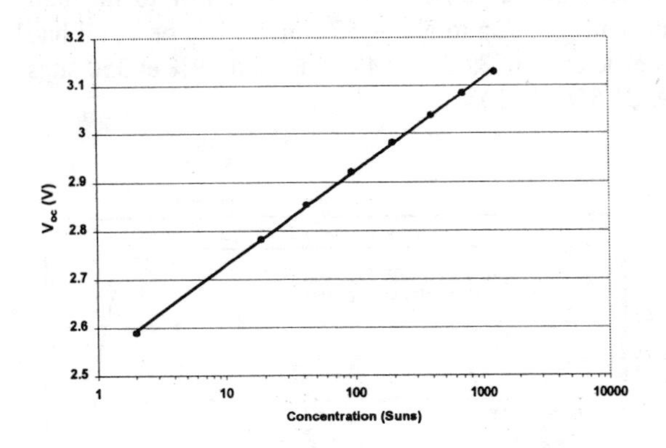

Figure 5. Open Circuit Voltage vs Concentration Ratio after Back Contact Replaced with Improved Ohmic Contact

Since no sign of a tunnel junction limiting characteristic appeared in the I-V curves, tunnel junction performance was believed to be nominal, even at the current densities resulting from concentration levels of 500 suns or more. Future devices to be fabricated will incorporate a more robust, low contact resistance back metallization, that is expected to achieve the same high efficiency but at concentration ratios greater than 500 suns. Results on these devices, together with absolute efficiency data will be the subject of a future publication.

Multijunction Solar Cell Manufacturing Maturity And Performance Improvements Road Map

Although multijunction solar cells were proposed over two decades ago [3] to achieve very high solar cell efficiencies, their adoption for space power systems has only been relatively recent. Reasons for the lack of progress included both a dearth of high quality source materials to obtain semiconductor layers with long diffusion length, in addition to the absence of large MOVPE growth systems for cost-effective manufacturing of the appropriate, lattice matched, solar cell materials. However the driving need for large power systems for direct broadcast satellites in the 10 kW range, and the future need for power systems approaching 25 kW, coupled with a government need for higher power systems, precipitated major R&D, manufacturing technology, and capital equipment investments that were made in the mid 1990's, to bring this technology to the marketplace. Technical progress was rapid and today, multijunction cells are the baseline technology for providing primary power for virtually all large space power systems. Spectrolab, alone, has manufactured over 460 kW of dual junction and over 30 kW of triple junction cells for space satellites. Over 85 kW of this product is now operating flawlessly in space. This represents almost 17 million cm^2 of yielded production, equivalent to about 175 MW of terrestrial solar cells at 500 suns concentration ratio! Global industry capacity for terrestrial multijunction cells almost certainly exceeds 500 MW per year (measured at these concentrations). Although only a portion could be allocated for terrestrial use, it is an important point to note that the equipment investments have already been made, and a portion of these assets are ready to be put to work for cost effective terrestrial manufacturing.

The increases in conversion efficiency that have been achieved by the industry over the last few years are equally impressive. Minimum average AM0 production efficiencies of radiation hard dual and triple junction space cells are presently 21.6% and 24.7% respectively as shown Figures 6 and 7.

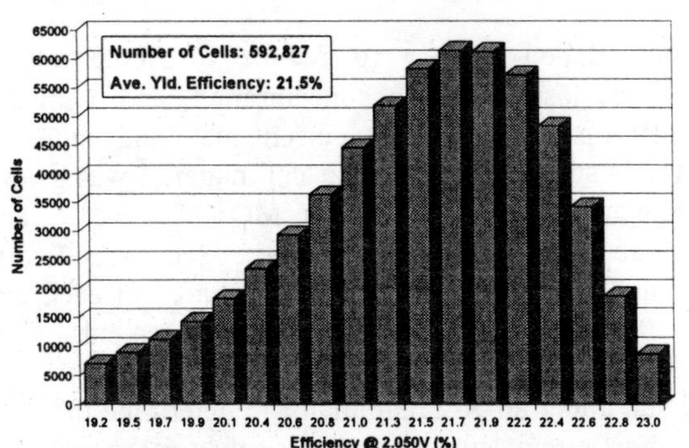

Figure 6. Efficiency Distribution for >465 kW of Planar Dual Junction Space Solar Cells

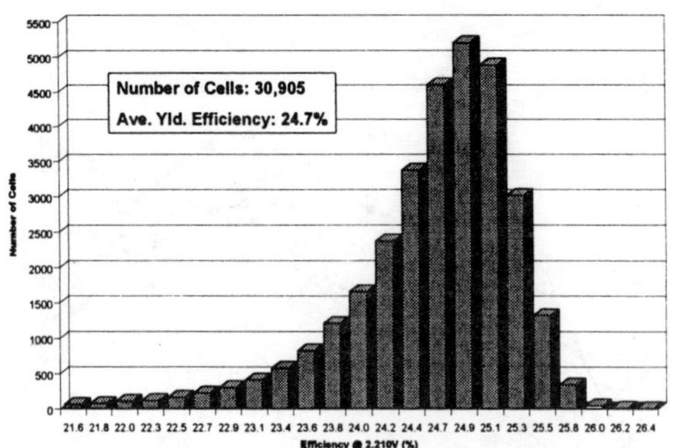

Figure 7. Efficiency Distribution for >30 kW of Planar Triple Junction Space Solar Cells

Improved radiation hard triple junction space cells of 26% efficiency will be available for widespread satellite application by mid-year 2000. This upward trend in conversion efficiency is likely to continue, spurred by government and industry investments to achieve space solar cell efficiencies in the mid-30% range using new 1 eV materials in 4-junction solar cells.

The substantial leverage of space solar cell improvements on the performance of terrestrial solar cells is summarized in Figure 8.

and 4-junction cells utilizing new 1 eV materials, as high as 47% and 52% respectively when illuminated under an AM1.5D terrestrial spectrum at 500 suns concentration. No other solar cell technology realistically offers the same opportunity to achieve such high conversion efficiency at relatively modest technical risk, and affordable cost.

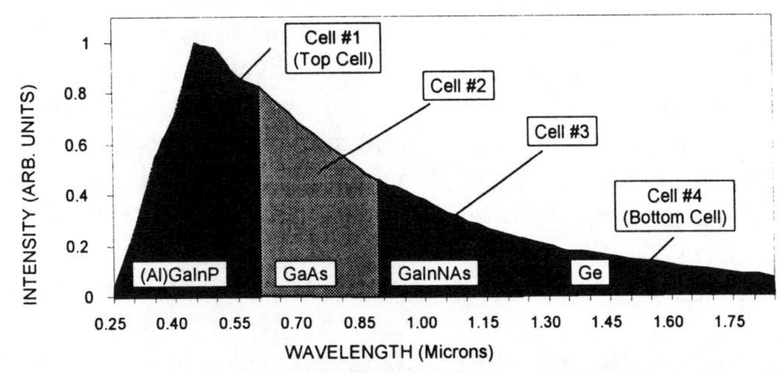

	Minimum Average Production Efficiency Contribution (AM0, 28 C)			
	2J $GaInP_2/GaAs$	3J $GaInP_2/GaAs/Ge$	3J AlGaInP/GaAs/GaInAs	4J AlGaInP/GaAs/GaInAs/Ge
Cell 1	13.7%	13.7%	17.2%	17.2%
Cell 2	9.9%	9.9%	10.4%	10.4%
Cell 3		2.4%	6.1%	6.1%
Cell 4				1.5%
Min Av Production, 1 sun AM0	**22.0%**	**26.0%**	**30.0%**	**35.2%**
Idealized Efficiency (AM0) (1)	31.0%	35.0%	38.0%	41.0%
Min Av Production AM1.5D, 500 suns	**26.0%**	**30.0%**	**35.0%**	**40.0%**
Idealized Efficiency (500 suns, AM1.5D) (4)	36.0%	42.0%	47.0%	52.0%

[4] Friedman et al, 2nd WCPEC, Vienna, 1998

Figure 8. Estimated Minimum Average Production Efficiencies by Product Type

Because of the differences in the illuminating spectrum, and the elimination of the requirement for radiation hardness, the conversion efficiency of a given cell product type under a concentrated AM1.5D spectrum is generally higher than under the 1-sun AM0 spectrum. This is true once the cell layers have been optimized for that particular spectrum. By way of example, the state-of-the-art, minimum average production efficiency for planar (1 sun) triple junction space solar cells is approximately 26% (available mid-year 2000). The same product, once optimized for the terrestrial spectrum, can achieve a minimum average production efficiency of at least 30% at 300 to 500 suns concentration. This prediction is based on relatively conservative modeling using realistic cell loss mechanisms, and production distributions that allow high manufacturing yields, and is also supported by the empirical data reported here. It is expected that new 1 eV materials currently being developed for space will allow minimum average efficiencies of over 40% (500 suns AM1.5D) within the next two to three years. Indeed, Friedman *et al* [4] and Kurtz *et al* [5] have predicted idealized efficiencies for improved 3-junction

Key Challenges For Cost Effectiveness And Growth Of Market Share

The key issues to be addressed before industry-wide acceptance of III-V concentrator systems are those of cost, and reliability. Present industry perceptions are that multijunction solar cells are too costly for use in terrestrial systems and that the optics and trackers in which they operate, are complex and unreliable. Furthermore, there is a belief that their application is limited to all but the largest stand-alone utility systems, that are in themselves the most cost-competitive with existing energy sources.

However, the widespread adoption of III-V solar cells for commercial and government space applications as described earlier, has brought about large reductions in raw materials costs, and increases in manufacturing yields. Given these cost reductions, and their proven reliability in space power systems, it is probable that III-V solar cells, when used in medium to high concentration ratio systems, offer one of the best opportunities for energy generation cost reductions over the next few years, when compared to other PV technologies.

Cost Reduction

Target system costs of \$3/W [1] will require "bare cell" costs to be on the order of \$0.5 to \$1 per watt depending on the particular concentrator module and system design. These are very aggressive targets, but ones that can likely be achieved by systematically addressing major cost drivers such as the cost of raw materials (the Ge wafer presently comprises over one third of the overall cell price), direct labor, and performance improvements. Since the \$/watt cost at the cell level reduces approximately linearly with concentration ratio, (because cell cost (in \$/sq. cm) is largely independent of concentration ratio), there are substantial advantages in operating cells at medium to high concentration ratios as shown in the cost projections of Figure 9.

concentration ratios, and that 40% efficient 4-junction cells will ultimately cost no more than a 30% triple junction cell, when manufactured in high volume. These are reasonable first order cost assumptions, that have certainly been validated by a highly competitive space industry over the last several years. Furthermore a manufacturing rate of about 50 MW/year was assumed corresponding to a throughput of about 150,000 wafers (of 100 mm diameter) per year. Under these assumptions, terrestrial concentrator cells can be cost effectively manufactured today, to provide the PV industry with cells in the range of \$0.5/W to \$0.6/W at 500 suns concentration, and about \$0.5/W if a lower price Ge wafer is available to an acceptable specification. The development of solar cells of >40% efficiency will further reduce manufacturing costs to between \$0.3/W and \$0.4/W within the next few years, at relatively modest risk. These costs are highly attractive with respect to achieving the DoE cost goals, when put into the concentrator module manufacturers

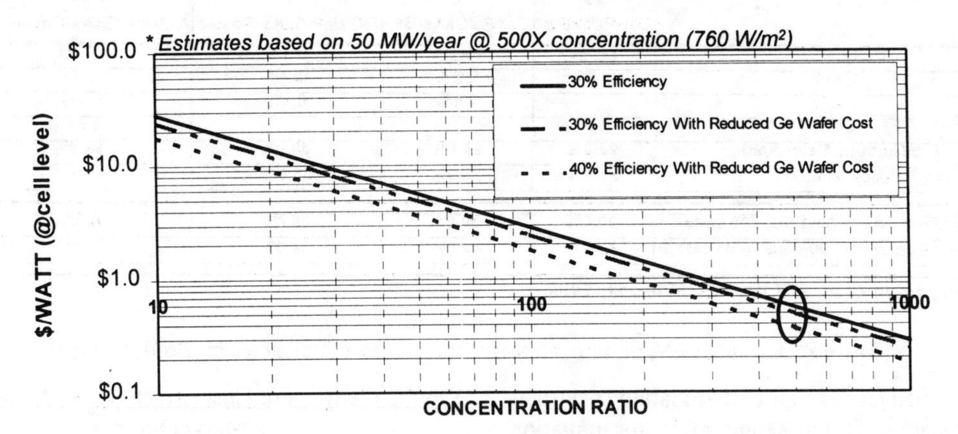

Figure 9. Estimated Cell Cost in \$/W vs. Concentration Ratio for Different Product Types

Several different cost vs. concentration ratio scenarios are included in this figure, namely:

1. 30% efficient triple junction cells (these are considered to be immediately available in large quantities in year 2000)
2. 30% triple junction cells manufactured using lower price wafers procured to a "terrestrial specification" (it is assumed that concentrator cells will be more forgiving of wafer specification compared to 1 sun space cells because of lesser influence of defects on performance at higher current densities)
3. 40% efficient, 4-junction cells that are assumed to be available within the next 2 to 3 years.

For each of the scenarios shown it was assumed that efficiency remained constant over a wide range of

cost models for linear, point focus fresnel, and dish-based concentrator systems.

Reliability

Both dual and triple junction solar cells have been extensively qualified for space application, and have undergone the rigors of long periods of illumination, temperature-humidity testing, deep thermal cycling, and shock and vibration testing without degradation. However terrestrial environments are extremely challenging, and the effects of long term exposure to concentrated sunlight (i.e. 500 suns or more), and the adequacy of existing cell packaging technology, are not well understood at this time, and should not be underestimated. Each of these issues must be proactively addressed through controlled laboratory, and extensive field tests before multijunction cells can be widely implemented in terrestrial systems. This is an area where both government laboratories and industry must each play active roles.

Conclusions

Triple junction concentrator cells, based on very mature space manufacturing technology, have been fabricated with an efficiency of >32% at 47 suns, AM1.5D. Low risk improvements will allow high volume, industry-wide availability of cost effective, 30% efficiency cells in the near term at a cost of between $0.5 to $0.6/W, at production volumes of about 50 MW/year. Longer term, the opportunity exists for 500X concentrator cells costing <$0.4/W through the introduction of improved 4-junction solar cells based on new 1 eV materials, currently being developed for space solar cells.

A significant immediate challenge is to achieve market acceptance of this new technology. Once this is established, through laboratory-based reliability testing, and multiple field demonstrations, demand for multijunction cells and concentrator systems could grow very rapidly due to the near term cost reductions afforded by this new technology. The III-V solar cell industry is ready to meet this demand, since most of the equipment capitalization has already been made.

Acknowledgements

The authors also wish to thank N. Karam, T. Cavicchi, R. King and D. Krut of Spectrolab, and S. Kurtz, J. Olson, and K. Emery of NREL for their contribution to this work.

References

[1] U.S. Photovoltaics Industry PV Technology Roadmap Workshop, Chicago, Illinois, (June 22-25, 1999)
[2] N H. Karam et al, “Development and Characterization of High Efficiency GaInP/GaAs/Ge Dual- and Triple Junction Solar Cells”, IEEE Trans ED (1999) 2116 – 2125
[3] M. F. Lamorte et al, “Computer Modeling of a Two-Junction Monolithic Cascade Solar Cell”, IEEE Trans ED, (1980), Vol ED-27, No. 1, 231-249
[4] D. J. Friedman et al, “1 eV GaInNAs Solar Cells for Ultrahigh-Efficiency Multijunction Devices”, 2nd World PVSEC (1998) Vol 3, 3-7
[5] S. R. Kurtz, D. Myers, and J. M. Olson, Proc. 26 th IEEE PVSC (1997), 875

AIAA–2000–2897

PERFORMANCE CONTRACTING OF A PARABOLIC TROUGH SYSTEM AT THE FEDERAL CORRECTION INSTITUTION–PHOENIX

Andy Walker and Jeff Dominick
National Renewable Energy Laboratory
Golden, Colorado

ABSTRACT*

Although renewable energy technologies benefit from their "green" image, they also suffer from a perception of being expensive, hard to maintain, and unpredictable in their performance. The National Renewable Energy Laboratory (NREL) formed a partnership with a solar hot-water heating manufacturer and a Federal Correctional Institution (FCI) in Phoenix, Arizona, to demonstrate the first federal renewable energy savings performance contract (ESPC) as a solution to the problems mentioned above. A 17,040 sq. ft. solar water heating system was installed at the Phoenix FCI. The system delivered 1,161,803 kWh of heat and generated revenue of $70,025.18 during the period from March 1999 to January 2000. Under the terms of this first federal renewable ESPC, FCI Phoenix obtains all the benefits of solar water heating while transferring the first-cost and performance risks to the Energy Service Company (ESCO—in this case Industrial Solar Technology, or IST). This addressed all of the perceived risks of renewable energy. The methods used by the ESCO to address the first-cost and performance risks are the subject of this paper.

BACKGROUND

The Phoenix FCI is a prison located 25 miles north of Phoenix. The prison consists of 16 buildings, including five housing buildings, a cafeteria, laundry and services building, and other administration and support buildings. The facility houses approximately 1,500 inmates. Electric water heaters in each building represent a significant energy requirement (about 15% of the total electricity use) and a major cost for the facility.

In 1991, FCI Phoenix was identified as a promising solar thermal application because the prison used electricity as its only energy source ($0.0694/kWh average [APS Tariff Number E-32, 1991]), and was located in an area with a high incidence of direct beam solar radiation (5.2 kWh/m^2/day average tracking around one horizontal east-west axis [NREL, 1994]).

Energy Savings Performance Contracting

Rather than purchase a solar hot-water heating system outright, FCI Phoenix chose to enter a renewable ESPC with an ESCO, in this case IST. ESPCs were first authorized for federal facilities by the Energy Policy Act of 1992 [42 USC 8287]. ESPCs have been used extensively to retrofit more energy efficient technologies such as electronically ballasted florescent lighting into federal facilities, but this is the first federal use of an ESPC for renewable energy technology.

Under the terms of an ESPC, a federal site such as FCI Phoenix pays no up-front costs for system design, installation, or hardware. The ESCO installs and operates the system and is paid out of savings realized in the site's utility bill. Because renewable technologies generate energy (in this case preheated water), the ESCO charges the facility for the energy delivered, much as a utility company charges for conventional energy. The ESCO charges 10% less for the preheated water than the prison would pay the utility to electrically heat the water, so the prison is guaranteed savings. At the end of the 18-year contract term, the prison will own the solar hot-water heating system and will no longer pay for the energy provided. At that time, they may contract out the operation and maintenance of the system, or may choose to perform those procedures in-house.

In this pay-for-energy arrangement, the ESCO was responsible for performing and funding the design, fabrication, and installation of the solar water-heating system. No prison or NREL funding was used for these functions. In performing these actions, the ESCO assumed all the associated risks of accurately measuring the existing site's hot-water load and solar resource and then predicting those values for the future contract term. They also had to make predictions about future utility rates because the value of their preheated water was a function of the site's prevailing cost of electricity. Lastly, they had to make predictions about the efficiency of their solar collectors and the reliability of their equipment to continually produce preheated water.

The ESCO could only self-fund a portion of the $649,000 project cost, and they could only directly control the system efficiency and reliability risks. As in most ESPCs, the ESCO chose to obtain third-party financing to pay for the balance of the project. Interest rates are determined by market rates, the perceived risk of the project, the stability of the company implementing the project, the facility receiving the project, and other factors. For this project, the ESCO got a 10-year loan for $418,000 at 11.5% interest, which was 5.75% higher than a Treasury Bill of comparable term. The uncontrolled risks of hot-water load (a function of prison population, potable water supply temperature, storage temperature, delivery temperature, use profiles, etc), weather, and utility rates contributed to the high interest rate the ESCO obtained for this project.

SYSTEM DESIGN

In an effort to determine hot-water load, electrical current to each water heater was measured by the ESCO and NREL for two weeks in July and October 1994, when the prison population was approximately 1,250 inmates. The peak power consumption of all heaters over a 15-minute averaging period (the utility billing increment) was 228 kW, and the average daily water-heating energy consumption was 3,408 kWh.

Other parameters affecting the load were either measured or estimated. Potable water supply temperatures were known to fluctuate between 80°F and 90°F. The existing electrically heated hot water system was assumed to have an efficiency of 90%, a storage temperature of approximately 130°F, and a supply temperature of 115°F. Based on these measurements and assumptions, a solar trough hot-water system was designed to deliver 1,521,000 kWh of preheated water to the existing prison hot-water system annually.

System Description

The system is installed on 1.2 acres of land adjacent to the perimeter fence on the west side of the facility. The system consists of 120 parabolic trough solar collector modules totaling 1,584 m^2 (17,040 ft^2) of net aperture area. There are two identical unpressurized water storage tanks with a total volume of 87 m^3 (23,000 gallons) located inside a steel shed. The shed also houses pumps, piping, controls, Btu-meters that measure energy delivery to the prison, and other ancillary equipment. A schematic diagram of the system is presented in Figure 1.

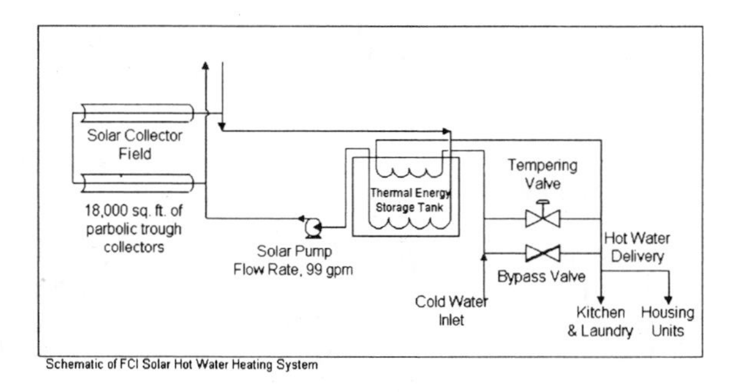

Figure 1. Schematic diagram of solar water heating

A propylene-glycol solution circulates through the collector field and through a heat-exchanger coil submerged near the bottom of each hot-water storage tank. Upon demand, potable water circulates through two identical heat-exchanger coils submerged near the top of each storage tank. Each of the two solar-side heat-exchanger coils consists of 18 coils of ¾-in copper pipe 60 feet long, for a heat exchange area of 21.2 m^2 (228 ft^2) per tank. On the load side, each heat exchanger consists of 20 coils with a heat exchange area of 30.8 m^2 (332 ft^2). A tempering valve mixes the solar preheated water with cold water to achieve the desired delivery temperature of 55°C (132°F). For safety and protection of the plastic piping, a shutoff of the solar and cold-water bypass occurs at 60°C (140°F). The prison's electric water heaters and their tempering valves remain in service, and can add heat as necessary to supply the desired temperature to each fixture.

First Year Performance

The system delivered 1,161,803 kWh (3,964 million Btu) of solar heat from March 1, 1999, to February 29, 2000. The sale of this energy provided $70,025 in revenue to the ESCO. Valued at the average $/kWh cost from the utility bill for each month, the prison would have paid $77,805.74 in utility cost during the same period. This resulted in net savings of $7,780.56. Monthly solar energy delivery and revenue are listed in Table 1. The system delivered a minimum of 80,315 kWh (274 million Btu) in January 2000 to a maximum of 130,979 kWh (447 Million Btu) in October 1999. Figure 2 shows the integrated total of hourly heat from the solar field to storage and from storage to the load. Experiments undertaken to understand the flow of hot water and energy consumption throughout the institution resulted in the Navajo housing unit (one of five) being disconnected from the solar hot-water supply from August 31 until September 23, 1999. These experiments reduced the solar hot water use by up to 15% during those months.

Table 1. Monthly Energy Delivery and Payments to IST

Month	Delivered Solar Energy (kWh/month)	Monthly Payment from FCI to IST ($/month)
March 1999	106,732	6835.54
April 1999	82,411	5939.41
May 1999	104,486	5736.68
June 1999	97,192	5381.61
July 1999	84,144	5870.45
August 1999	92,849	5478.58
September 1999	94,946	5988.06
October 1999	130,979	7785.01
November 1999	102,998	6609.50
December 1999	95,953	5128.45
January 2000	80,315	4713.34
February 2000	88,798	4558.55
Annual Total	**1,161,803**	**70,025.18**

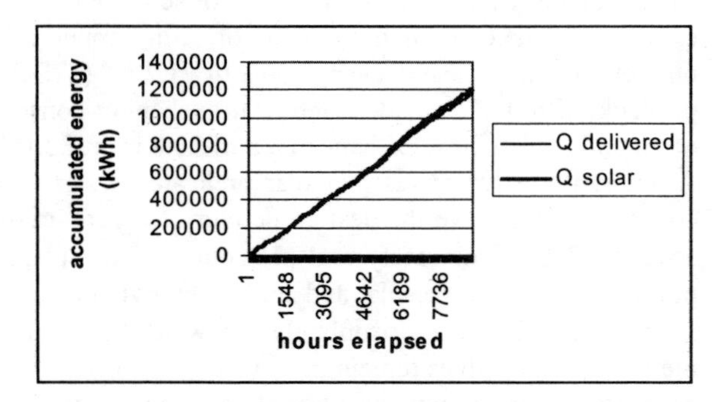

Figure 2. Energy from solar field to storage (Q solar) and from storage to load (Q delivered) integrated total.

Despite the obvious success, the first year's energy delivery was slightly below the ESCO's design estimates in a time of rising prison population. Because of the contract's risk allocation, this condition did not "cost" the prison (other than unrealized additional savings), but it was worrisome to both the ESCO and NREL. This shortfall was discovered early in the year, and a quick review of the data showed that the solar hot-water system's efficiency and availability were not at fault, and the solar resource were within predicted ranges. This left uncertainties in the hot water load as the most likely cause of the shortfall. The aforementioned flow and hot-water energy consumption measurements were made to investigate this theory.

Prison Hot Water Load Study Methods

The hot-water heating systems serving the kitchen and laundry (Building C) and four of the five housing units were inspected. In general, each system was comprised of a hot water tank or tanks with electrical heating elements, plus a large storage tank downstream of the heated tanks. A recirculation pump and piping interconnected the heated tanks and the storage tank. Downstream of the storage tank, a tempering valve allowed mixing the cold potable supply water with the output of the storage tank to ensure a 115°F supply to the end-use points.

Upon checking out the heaters, it was evident that the temperature markings on the thermostats were different from the temperature reading of the thermometer on each heater. Typically, the thermostat marking at which the heater would activate was higher than the actual water temperature by 10–20°F. In addition, the thermostat settings were generally higher than the incoming solar water temperature. The setting on the thermostats appeared to be an attempt to overcome deficiencies in the heating system.

In three out of the four housing units visited, the tempering valve was not working. Hot water from the storage tank was being tempered in an uncontrolled fashion. Thus, the only way to obtain water at a satisfactory temperature was to increase the temperature of the water leaving the storage tank. Such actions reduced the amount of solar energy that could be supplied, increased the amount of electricity used, and resulted in unnecessary run time on the heating elements.

The Navajo Unit was the only one of the four units visited where the tempering valve and all other components of the hot-water heating system were fully functional. Maintenance personnel told the ESCO that the tempering valve in the fifth housing unit was also working. Facility personnel are aware of the equipment needs, but there is no funding to purchase spare parts or undertake the needed repairs. Navajo was expected to give a true picture of energy-use patterns in the housing units under "design conditions." Navajo and Building C were therefore selected as the sites for the installation of the measurement equipment.

Solar preheated water and tempering water flowrates into the building hot-water system were measured for more than 90 days, along with tank inlet and storage temperatures, tempering valve outlet temperature, and electric heater power. For 23 of those days, the solar preheated water was cut off to Navajo and replaced with potable supply water to demonstrate the pre-solar operation of the electrically heated hot-water system.

Prison Hot Water Load Study Results

As seen in Figure 3, the Navajo tempering, which we had thought was good, was actually tempering the water very erratically, often to less than 115°F. During pre-solar operation, it became apparent that the heating and storage system was undersized and that the tempering valve continued to reduce the hot-water supply temperature even when the temperatures fell below 105°F (see Figure 4). Water-heater thermostat setpoints had been raised to over 140°F in an attempt to overcome this, decreasing the overall hot-water system efficiency.

Efficiency calculations for the pre-solar condition were computed by dividing the energy in the hot water delivered by the measured energy supplied to the water heaters. The calculations showed that the Navajo system was 70% efficient, and Building C was 50% efficient.

CONCLUSIONS

Several conclusions can be made from this data. First, the hot water demand now exceeds the pre-solar system's capacity, and the prison relies on the solar system power and capacity to meet its loads. This unanticipated and uncompensated attribute of the ESCO's solar preheat system has allowed the prison to avoid the costs of system expansion to meet the rising prison population. However, the solar preheat system has been underutilized because its flowrates are reduced and diluted by the malfunctioning tempering valves. Unnecessary electricity is used to boost the preheated water temperatures in an attempt to overcome the tempering valve problem.

The resulting decrease in hot-water system efficiency is likely a long-standing condition and probably introduced significant error into the hot-water load estimates made during system design. This would have over-predicted hot water consumption and would have oversized the solar preheat system. Fortunately, this error was partially offset by the 20% increase in the prison population since 1995.

FUTURE PLANS

It is clear that further improvements can and should be made to increase savings to the prison and increase revenue to the ESCO. The tempering valves should be repaired and maintained to accurately temper the hot water. When this is done, the electric water-heater setpoints can be reduced to below the solar preheat temperature, further reducing the prison electric bill by improving system efficiency.

Although these steps will bring performance closer to predictions, the solar preheat system remains oversized for a prison population of 1,250. The prison and the ESCO may want to address this issue and other risk factors that are beyond the ESCO's control in a good-faith negotiation to reexamine and potentially share these risks. Future renewable projects will benefit from understanding the risks better and identifying some unanticipated benefits of renewable energy systems.

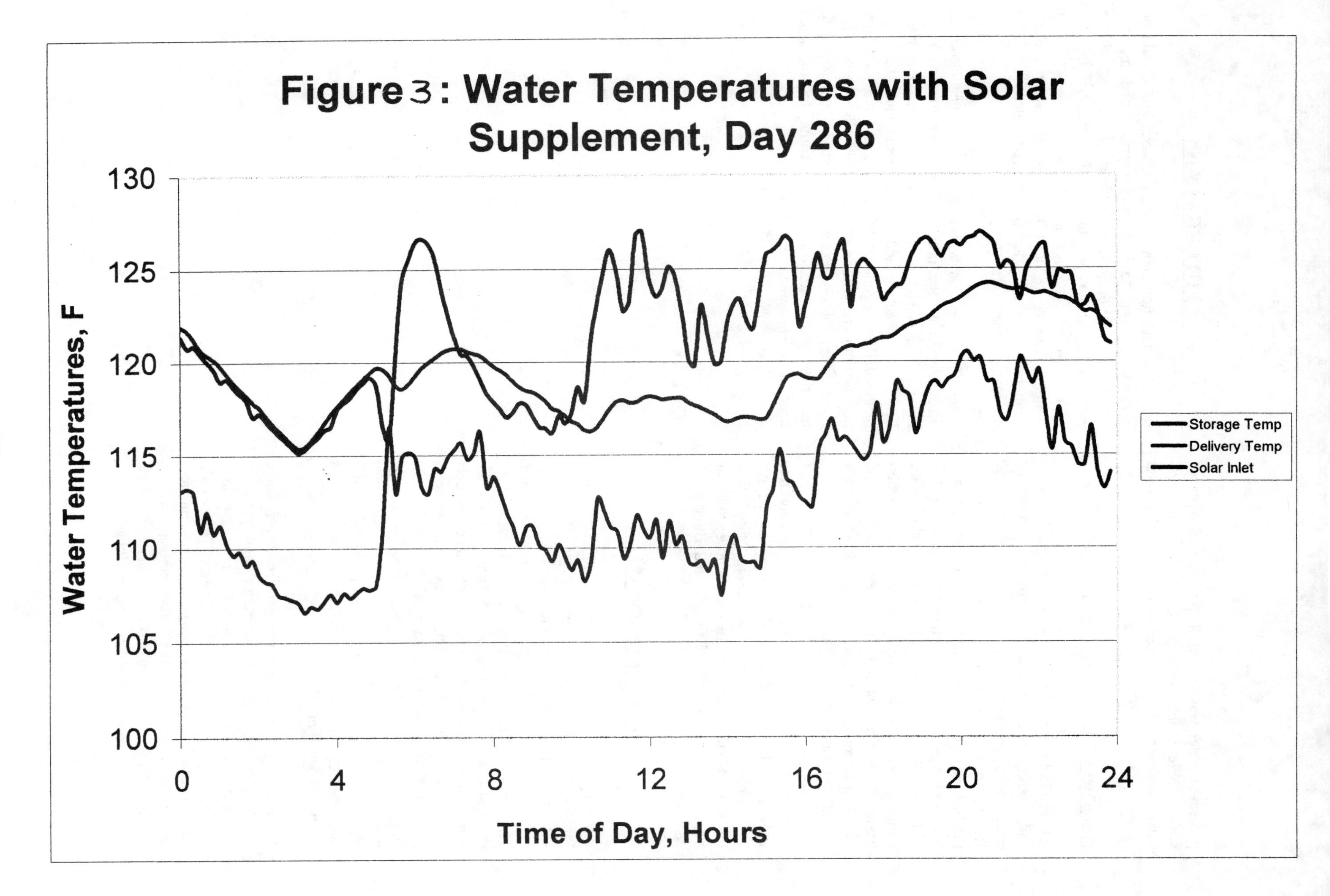

Figure 3: Water Temperatures with Solar Supplement, Day 286

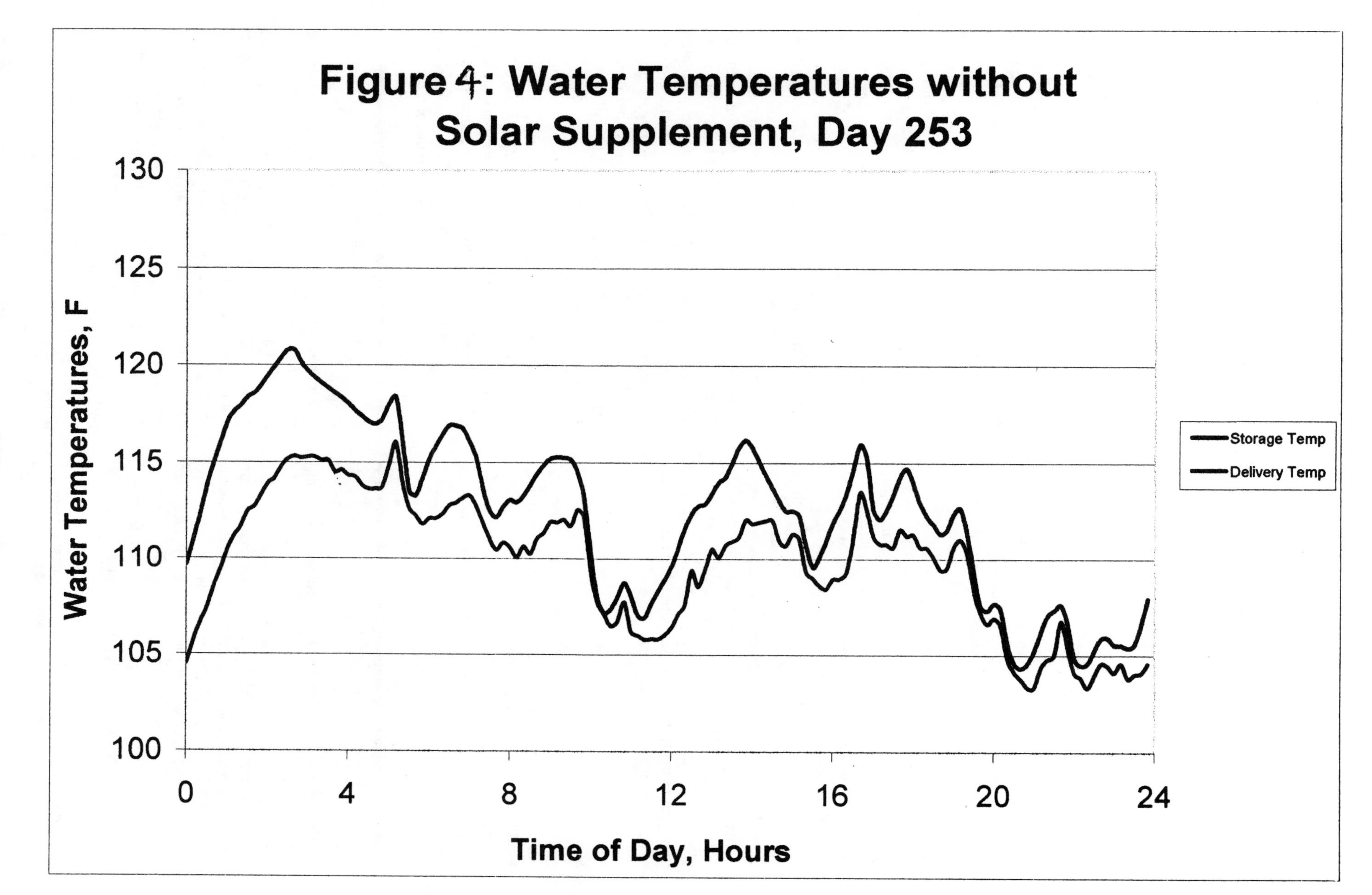

Figure 4: Water Temperatures without Solar Supplement, Day 253

AIAA-2000-2898

MULTI-CRITERIA DECISION-MAKING PROCESS FOR BUILDINGS

Dr. J Douglas Balcomb and Adrianne Curtner
National Renewable Energy Laboratory,
Golden, Colorado

ABSTRACT

The paper focuses on a process designed to facilitate two key decision points early in the building design process that are critical to a building's ultimate sustainability. As vital decisions are made during the building's design, the process and accompanying tools assist the design team in prioritizing their goals, setting performance targets, and evaluating design options to ensure that the most important issues affecting building sustainability are considered. Both the methods used and the tools required to carry out the methods are described. The process has been conceived to make the most efficient use of both the time and resources of the design team, suggesting that it will actually receive widespread use. The process is being tested within the context of the International Energy Agency Task 23, *Optimization of Solar Energy Use in Large Buildings* and has also been comprehensively tested by the United States Federal Energy Management Program.

INTRODUCTION

Practicalities of building design require that the design team make very efficient use of its resources and time. This paper recommends a process and describes two computer tools that will aid the design team in ensuring a sustainable design at vital junctures in the design process.

There are two key points early in the building design process in which the ultimate sustainability of the final building is determined. The first is during pre-design, when the most important energy-efficient strategies should be selected. The second is during preliminary design, when a choice is made between two or more competing design schemes that have been proposed. The Multi-Criteria Decision-Making (MCDM) process is designed to guide design teams through these stages in a way that makes sustainable building design easy and inexpensive. It facilitates the communication of team priorities, the setting of performance goals, and the evaluation of proposed building designs within the context of the conventional building design process to ensure ultimate building sustainability. The team can avoid the expense of corrections later in the process and be assured that the design will be the most sustainable possible by investing a few hours of time at critical points in the design process.

Early identification of effective energy efficiency measures and building energy use targets is key to sustainable building design. During pre-design, an energy-performance simulation tool can be used to select appropriate strategies. The MCDM process calls for the definition of a reference-case building that identifies how a typical building of the type, size, and location being considered uses energy. The potential for savings is identified by simulating an alternate low-energy case building that incorporates all potential energy-efficiency strategies. An initial selection of strategies is made based on simulating each strategy individually and ranking their effectiveness. Realistic performance targets are set based on these initial results. These performance targets ensure that all proposed designs will achieve a desired level of energy efficiency.

During preliminary design, the team reaches a point where two or more design options have been proposed. Each design incorporates the strategies selected earlier. A decision must be made: which option should be pursued? However, to achieve a design for a building that will be sustainable, the team must consider a multitude of criteria other than energy. A new tool, MCDM-23, automates many of the tasks involved in making an informed selection weighing a broad range of criteria. Inputs to MCDM-23 include both quantitative results for each scheme, many of which can be calculated using ENERGY-10, and qualitative results, which can be developed by expert judgement and discussion within the design team. The MCDM-23 program uses a weighting scheme to determine the overall building score based on the results of the qualitative and quantitative evaluations of the proposed building designs. The team develops the weights by expressing their priorities for six criteria (enumerated in Table 2), both quantitative and qualitative, used by the MCDM-23 program. The end products of the program are worksheets and associated diagrams that

---

quantify how various design schemes stack up according to the team's criteria.

Advantages of the MCDM-23 program are that it provides an organized framework for decision making and a way to document how the decisions were made. The latter is particularly important in public buildings. One useful product of the procedure is a star diagram, such as the graphic shown on the following page, which identifies how the selected design scheme compares to that of a typical building.

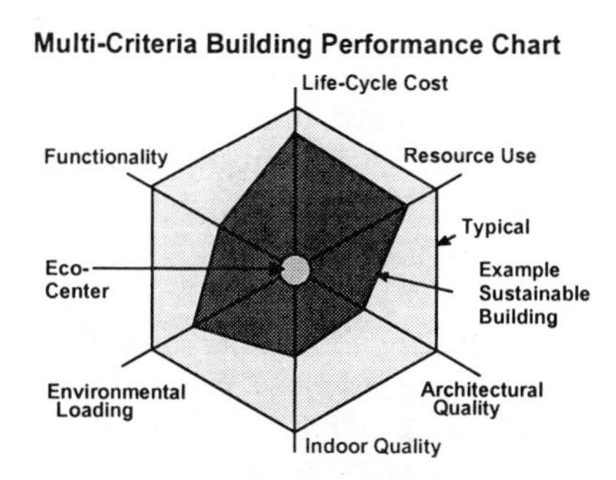

Figure 1: The star diagram is a compact way of displaying the whole picture in one graphic that can be easily interpreted by anyone—architect, engineer, client, energy analyst, building official, reporter, or layman. A smaller footprint is better on this diagram. MCDM-23 automates making this plot based on information derived from the worksheet.

THE DESIGN PROCESS

Energy efficiency has historically and rightfully been considered the most important building attribute required to achieve a sustainable building. Energy efficiency calls for a whole-building design approach, selecting and integrating those strategies that are most effective in each particular situation. However, to achieve a design for a building that will be sustainable, the design team must consider a multitude of other criteria. Some of these criteria are inherently qualitative in nature and others are amenable to quantitative evaluation.

Given enough time and resources, a design team could sift through the multitude of issues involved in making informed decisions and make the dozens of building simulation calculations required to provide input. Typically, however, resources have not been allocated for any of this work, and only rarely is the design budget increased to accommodate these steps. The only solution to this problem of limited time and resources then is for the design team to make very, very efficient use of them.

The twelve-country International Energy Agency Solar Heating and Cooling Task 23, which concerns Optimization of Solar Energy Use in Large Buildings, has taken on the challenge of recommending a design process and associated design tools for helping design teams to make the most effective use of their time and talent. Their efforts have resulted in the MCDM process, which identifies two critical points in the design process where a small effort can produce big results in terms of building sustainability. Significantly, these points occur very early, when decisions must be made based on very limited information.

For our purposes, the building design process can usefully be divided into four main phases, as shown in Table 1.

Table 1. Four Major Steps in the Building Design Process

Working Name	Other Names Used
Pre-design	Programming Strategic Planning Pre-Project Investigation of Basics
Preliminary Design	Schematic Design Preliminary Studies Project
Design Development	Preparation of Realization Definitive Proposal
Construction Documents	Building Documents Realization

Chances are that any one practitioner will be used to different terms. We needn't belabor the precise definition of these steps, only acknowledge the importance of early phases, especially the first two. Usually the steps are defined contractually between the design team and the client, making it expensive and difficult to undo decisions made in a previous step during a later step. This puts a high premium on making correct decisions early and on making decisions efficiently based on information available at the time.

THE RIGHT INFORMATION AT THE RIGHT TIME

The first critical juncture in the design process is the transition from pre-design to preliminary design. The most efficient use of the designer's talents will be made if they can proceed with the design with many decisions that affect energy efficiency and sustainability having already been made. This sounds impossible, but in fact it is quite simple, given the right tool. Most building attributes required to achieve energy efficiency and sustainability have little or nothing to do with the details of building geometry. Those issues that are affected by geometry can be dealt with through straightforward means. Many critical decisions can truly be made before design starts.

An energy analyst makes some preliminary calculations, then presents the results to the designers, who can proceed confidently into preliminary design with better information. Similarly, the MCDM process facilitates the discussion of many other aspects that affect sustainability. With a clear understanding of the priorities of the design team and other stakeholders, the designers can address all important issues at the start. Not only dos the MCDM process set performance goals and energy-use targets, it helps to identify effective design strategies through which the goals may be realized, leaving the designers relieved to be able to focus on other issues. It may be necessary to re-affirm some decisions later, but the work will be significantly reduced.

The second critical juncture occurs toward the end of preliminary design. During the preliminary design phase, the designers typically develop two or more design options, which we will call schemes. Because the designers were informed by the results of the pre-design energy and sustainability evaluations, many key decisions will have already been made. As a result, each scheme will likely perform quite well on both accounts. Ideally, each scheme will meet the performance targets set earlier. However, the selection of energy design schemes is also dependent on factors other than energy performance and sustainability. Thus, another type of tool, in addition to the energy-analysis tool used in pre-design, may be used to facilitate the decision process at this point. The tool required at this juncture is one that facilitates multi-criteria decision-making, and the comparison of dissimilar criteria.

The MCDM process is very efficient. Decisions are made early enough and quickly enough that backtracking is avoided. All that is left to the end of the design process is to confirm that performance requirements have been met.

THE TOOLS

There are two tools required to facilitate this process.

1. The first tool is an hour-by-hour energy-performance simulation program that has been programmed to automate the several steps required. The tool being used in Task 23 is ENERGY-10, a design tool developed at the National Renewable Energy Laboratory (NREL) and widely used in the United States. This tool, while it does not incorporate all the features desired, does enough to serve as a useful prototype. The evaluations done by ENERGY-10 are sufficiently comprehensive to capture the subtle interactions between the various energy-efficient strategies to produce accurate results.

2. The second tool automates the several steps involved in using the MCDM procedure being developed within Task 23. The end products of the program are worksheets and associated star diagrams that quantify how two or more design schemes stack up according to six main criteria that have been selected by the Task 23 group. The tool is a computer program called MCDM-23 and was developed at the National Renewable Energy Laboratory specifically for Task 23. It can easily be modified as the requirements of Task 23 evolve.

The MCDM-23 program does not reduce the building design process to a prescriptive procedure. Rather, it provides a framework within which to carry out the several tasks inherent in a partly qualitative decision-making process. It facilitates rather than dictates. The eight key steps are: (1) determine weights and sub-weights for qualitative and quantitative criteria by expressing team priorities, (2) evaluate a reference building and enter scores, (3) consider two or more design schemes, (4) calculate scores for quantitative criteria, (5) determine scores for qualitative criteria, (6) enter scores for each scheme, (7) print a worksheet and star diagram for each scheme, and (8) select the winning scheme. ENERGY-10 has been described extensively and need not be discussed further here.[1,2] This paper describes the MCDM-23 tool. Before beginning, however, it is important to introduce an important concept, the reference-case building.

THE REFERENCE BUILDING

The fact that there is no design to evaluate during the pre-design phase actually makes life easier, not more difficult. A very useful procedure is to identify a simple rectangular building geometry that has the principal attributes of the building being designed – (1) it is in the same location (weather characteristics), (2) it is of the same size (floor area), and (3) it fits the same building-use category (occupancy characteristics). There may be other constraints defined at the beginning by the site or other pre-conditions, such as the number of stories, the building orientation, or the choice of heating, ventilating, and air-conditioning (HVAC) equipment. If not pre-determined, these can be defaulted.

This rectangular building, which we will call a reference building, can be evaluated quite easily. The geometry is simple. Unknown characteristics can be defaulted to typical construction practice in the locality or to conform to prescriptive code regulations. The resulting building serves a very useful purpose: it is an initial benchmark. Subsequently, the design team may choose to redefine the benchmark building (often called a base-case); however, it is better at the beginning to keep the reference building as simple as possible. The reference building serves purposes beyond pre-design.

DECISION MAKING IN THE PRE-DESIGN PHASE

Decision-making in the Pre-Design Phase involves articulating design team priorities and selecting building energy performance goals. The MCDM process spans the pre-design and preliminary design portions of a building project.

Building Energy Use Performance Targets

A few key simulation results are most informative for setting performance goals during pre-design. These calculations need to be made quickly. The critical steps are:

1. Evaluate the reference building. An hour-by-hour simulation is done based on a typical reference year of weather data for the locality. Occupancy characteristics and energy load profiles are matched as closely as possible to building use.

2. Create an alternate building, which we will call a low-energy case. This is done by globally modifying the reference building description to affect the application of a set of energy-efficient strategies. The basic building geometry is not changed. The strategies might include: daylighting with associated dimming of artificial lights, energy-efficient lights, improved insulation throughout, improved windows, reduced infiltration, passive solar heating, shading windows, adding thermal mass, higher efficiency HVAC, relocation of ducts inside the thermal envelope, improved HVAC controls, and using an economizer cycle. It is better to be inclusive rather than exclusive and to apply each strategy aggressively.

3. Evaluate the low-energy case building. Although the identification of strategies and the degree to which they are applied might have been arbitrary, the aggregate result of applying all of them will result in a useful second benchmark. The result may appear as shown in Figure 2. This shows the potential for improvement, not taking into account improvements that may result from changes to the building geometry.

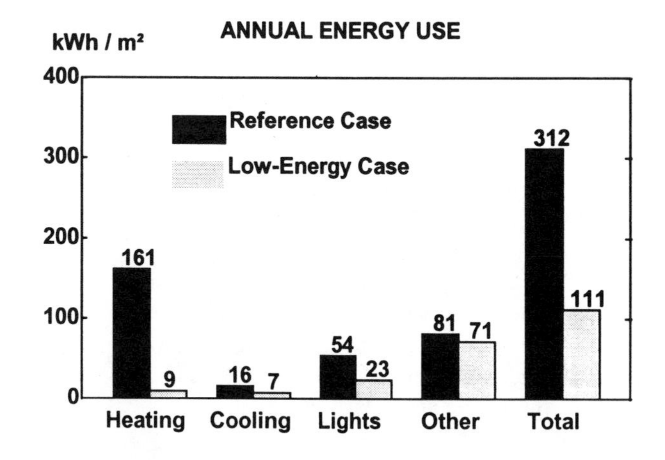

Figure 2: Simulation results for an office building in Cambridge, England. Twelve energy efficiency strategies were applied to the reference case to create the low-energy case building. Both are 18 x 55 m (1000 m^2) one-story with east-west major axes.

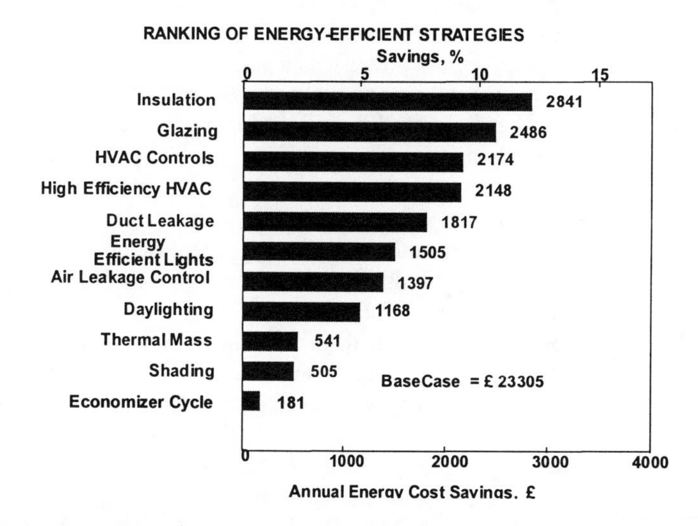

Figure 3: Ranking of strategies for the building shown in Figure 2. Each strategy was applied individually, simulated, and the results were saved. The graph shows the results sorted according to energy savings. Note that many strategies are important. On the basis of this ranking, one might choose the top eight strategies.

4. Add each strategy individually to the reference building and evaluate. Repeat for all strategies. Rank the strategies by some criteria, such as energy savings, reduction in operating cost, or reduction in life-cycle cost. The result may appear as shown in Figure 3. This serves to identify the most effective strategies and facilitates an initial screening. Because of

interactions between the strategies, the sum of all the individual savings will likely be greater than the total realized in step #3.

5. Re-create the low-energy case building applying only the most effective strategies. Evaluate this building. The result won't be very different from the result in step #3, because only the least-effective strategies were discarded.

6. Based on the result of step #5, select one or more performance targets for the project. This could be stated in terms of either energy or operating cost. For example, the target for the Cambridge office might be set at 120 kWh/m^2. This is both aggressive and clearly achievable.

   This process, which can be carried out in less than a half-hour, serves several purposes. The computed reference case building energy use clearly identifies how a typical building of this type and size operates. The potential for savings is identified by the results of the low-energy case. An indication of which strategies will be most profitably pursued is made, and realistic performance targets are set. Additionally, defining a reference case and corresponding low-energy case building sets a standard and scale by which actual concepts may be judged. The importance of this will become more evident later in this paper.

Because a simple rectangular geometry was used, some may suspect that the strategies identified by this procedure will be different than those that would be identified using the final building geometry. Experience shows that this is not the case. Strategies that were marginal may change, but the selection of the major strategies is very unlikely to differ.

A key advantage of this procedure is that it makes very efficient use of the time of both the designers and the energy analysts. The calculations can be done very quickly – perhaps 10 or even 100 times more quickly than if they were done on a complex geometry because of the inherent difficulty of describing and continually modifying the building description. In reality, the chances that a comprehensive differential evaluation will be carried out are very small unless it is done at the very beginning. It is very time-consuming and expensive to do it later.

Daylighting performance, of course, depends critically on the details of building geometry. The purpose of the daylighting calculation based on the rectangular geometry is not to simulate the final design but to identify the potential for daylighting to save energy. This can be done with a simple geometry in which daylighting is done by windows and skylights, preferably with a geometry that is thin enough to provide adequate side-lighting of lower levels.

If daylighting is one of the strategies selected in pre-design, then the burden rests on the designers to develop building designs that will achieve roughly the same degree of dimming (or greater). Preferably, this will be accomplished using sophisticated daylighting strategies such as light shelves, roof monitors, and clerestories that will achieve well-balanced lighting throughout the building, helping the building to score well in all categories.

Other Building Performance Issues

The MCDM process facilitates not only the determination of an energy performance goal, but also of other goals relating to building sustainability. Design team members meet to determine the group priorities as defined within the framework of the MCDM-23 program. The group priorities are expressed as weights within this program.

In MCDM-23, six major criteria (enumerated in Table 2) are identified. The selection of criteria was accomplished within the International Energy Agency (IEA) Task-23 group, which includes a mix of 25 highly knowledgeable and experienced designers, engineers, and analysts. These criteria can be important for all building projects, however, the most important criteria may vary from project to project. The MCDM-23 tool therefore offers the possibility to change or exclude some of the criteria. The criteria and sub-criteria default weights are shown in Table 2. The default weights indicate relative priorities. For example, the defaults for the main criteria weights are all set to be equal, indicating that each of these criteria are considered equally. This is merely a default and will preferably be changed by individual design teams.

In Table 2, there are two different types of weights listed. The first are the coefficients for the scores given to the six main criteria. The second weights, called sub-weights, are the coefficients that multiply the raw building scores determined from evaluating the designs. Multiplying the raw building design scores by their sub-weights and adding all weighted sub-scores within a criteria category gives the overall score for that category. Each of the overall category scores are then multiplied by their respective weights and are aggregated to form the overall building score.

The design team determines the main criteria weights by prioritizing the six main criteria used by the

MCDM-23 program. Team members indicate the relative importance of criteria by evaluating pairs of criteria. This is achieved by answering such questions as: "On a scale of 1 to 10, how do life cycle cost and resource use compare to each other in importance? (Here, 1 means that resource use is totally important, 5 means that resource use and life cycle cost are of equal importance, and 10 means that life cycle cost is totally important.) This process is repeated to determine weights for the sub-criteria for architectural quality, indoor quality, and functionality. For the quantitative criteria: life-cycle cost, resource use, and environmental loading, the sub-weights can be calculated. In the case of life-cycle cost, the program incorporates the life-cycle cost equations; the user enters the relevant financial parameters (e.g. discount rate, building lifetime, mortgage interest, etc.) and the program calculates the three weights (coefficients).

The MCDM-23 program then uses a statistical method called the Analytical Hierarchy Process to assign weights to each criteria based on the team's answers to the criteria-pairing questions. By determining these weights prior to beginning the building design, the team arrives at mutual conclusions about which goals should receive priority in the design process. This consensus about design priorities is a great help to designers.

The design team will probably not agree on these criteria and weights. However, the weights can all be changed and the names of the sub-criteria under architectural quality and functionality can all be changed to better reflect the team's perspectives. With this flexibility, the choice of criteria should be acceptable to most teams. Gaining acceptance at the beginning is very important.

DECISION MAKING DURING THE PRELIMINARY DESIGN PHASE

Typically, the team reaches a point in the preliminary phase of the design process where two or more design options have been proposed and a decision must be made. Incidentally, a similar situation is reached in a design competition, where a jury must choose between alternative design proposals. The purpose of MCDM-23 is to facilitate making the best decision.

The five steps in using MCDM-23 during the preliminary design phase are as follows:

1. The design team determines their preferences for the relative importance of the six main criteria as described above. (This actually occurs in the pre-design phase, immediately before the design work commences, but is an inherent part of using the MCDM-23 program.)

2. The energy analyst enters performance values of the reference case into the MCDM-23 program. The energy,

Table 2. Selection Criteria

Criteria	Default Main Weight	Sub-Criteria	Default Sub-Weight
Life cycle cost	1/6	Construction cost	.68
		Annual operation cost	19.4
		Annual maintenance cost	19.4
Resource use	1/6	Annual electricity, kWh/m^2	3
		Annual fuels, kWh/m^2 (of heat equivalent)	1
		Annual water, kg/m^2	0.15
		Construction materials, kg/m^2	0.03
		Land, m^2/m^2	300
Environmental loading	1/6	CO_2-emissions from construction, kg/m^2	1
		SO_2-emissions from construction, kg/m^2	90
		NO_x emissions from construction, kg/m^2	45
		Annual CO_2 emissions from operation, kg/m^2	30
		Annual SO_2 emissions from operation, kg/m^2	3000
		Annual NO_x emissions from operation, kg/m^2	1500
Architectural quality	1/6	Identity	0.25
		Scale/proportion	0.25
		Integrity/coherence	0.25
		Integration in urban context	0.25
Indoor quality	1/6	Air quality	0.35
		Lighting quality	0.25
		Thermal quality	0.20
		Acoustic quality	0.20
Functionality	1/6	Functionality	0.45
		Flexibility	0.15
		Maintainability	0.25
		Public relations value	0.15

operating cost, and environmental loading associated with operation have already been calculated using ENERGY-10. The added construction cost is zero by definition. Scores for the qualitative criteria are usually all 5s (indicating typical performance).

3. The energy analyst calculates the performance of each of the schemes being proposed and enters these numbers into the MCDM-23 program.

4. The design team determines the scores for the qualitative criteria using the 0-to-10 scale, and the results are entered into the MCDM-23 program.

5. Worksheets and star diagrams are printed, and the team studies these to make their recommendations to the client.

The MCDM-23 program results rely on three key concepts. One is that building energy performance is determined relative to the performance of the reference case building, which has already been defined. This provides a scale for quantifying performance. The second concept is that the relative score is linear*. If the performance metric of a particular criteria is 20% better than the corresponding score of the reference case building, then the relative score is 20% better.

The third concept is that the criteria weights are used as scaling factors to relate the scores on one criterion to the scores on all other criteria. This makes it possible to aggregate all the scores and the weights into an overall measure of goodness. The weights define acceptable trade-offs between criteria; thus, they are related to the scales on which the attributes are defined. Hence, if criterion A has a weight that is twice that of criterion B, this should be interpreted that the decision makers value 10 points on the scale for criterion A the same as 20 points on criterion B and would be willing to trade one for the other. Put another way, the decision makers should be indifferent to a trade between 1 unit of A and 2 units of B.

Two scoring scales are used in the MCDM-23 process. A scale of 0 to 10 is used for those criteria for which it is conventional to think in terms of "bigger is better". In this case, 0 is the minimum acceptable, 10 is the maximum achievable, and 5 is typical. These criteria are architectural quality, indoor quality, and functionality.

A scale of 0 to 2 is used for those criteria for which "smaller is better". For this scale, 0 is the maximum achievable, 2 is the minimum acceptable, and 1 is typical. These criteria are life-cycle cost, resource use, and environmental loading. Any confusion or ambiguity introduced by the use of two scales is more than offset by the convenience introduced by using an appropriate scale for each criteria. The two scales can be readily mapped one to the other. The mapping is a simple linear transformation.

MCDM-23 automates the following simple score calculation procedure. For the three smaller-is-better criteria, the relative score is determined by dividing the numeric score of the scheme being evaluated by the corresponding score of the reference building. For example, if life-cycle cost of the scheme being evaluated is $60,000, and the life-cycle cost of the reference building is $100,000, the relative score is 0.6.

For the three bigger-is-better criteria, the scores are first mapped to the smaller-is-better scale and then the process described above is repeated. For example, the architectural quality of the scheme being evaluated is 8 on the 0-to-10 scale and the architectural quality of the reference building score is 5 (by definition). The 8 converts to a 0.4. The 5 converts to 1. The relative score is therefore 0.4. Scores are displayed on their appropriate axes, making the different scales apparent and avoiding confusion.

MCDM-23 is inherently different than a rating tool such as the GBC tool, LEED, or BREEAM in that it is designed as an aid for decision making prior to final design rather than a means of scoring a completed building. An additional major advantage of the MCDM-23 tool is that it provides a compact and readily understandable means for documenting how decisions were made. This is particularly important in public buildings.

The MCDM-23 program automates calculations, enters results into worksheets, and plots star diagrams for each scheme. The design team then makes their selection based on all information available to them, including the MCDM-23 results.

All intermediate results are displayed on the worksheets and can easily be verified using a calculator. The star diagram is simply a graphic representation of the performance of a scheme

* It would not be at all difficult to modify the linearity assumption. An advantage would be that the increasing difficulty and cost of achieving greater and greater performance gains could be more accurately accounted. This would require some minor re-programming in MCDM-23. Appropriate nonlinear algorithms would need to be defined and agreed upon.

compared to the reference. Users can quickly compare the performance of each scheme in terms of the main criteria by visual inspection of its star diagram. At this point the star diagrams can be compared side-by-side. If the schemes do not exhibit evident differences, then they are probably not significantly different.

By its nature, preliminary design is an iterative process, cycling through a series of steps until the design meets all criteria. Each cycle typically involves design, evaluation, review, and revision. Both MCDM-23 and ENERGY-10 lend themselves to the evaluation step in each cycle. Again, emphasis must be placed on using the tool quickly so that the evaluations do not hold up the designers unreasonably.

Another use of the tools during preliminary design is performing sensitivity analyses. Because these are computer tools, they can be run repetitively incrementing a single parameter over a range of values. The parameter can then be set to achieve the best performance. In this context, "best" could be the overall score, taking into account all effects captured in the evaluation. Historically, most such optimizations have been done on the basis of minimizing life-cycle cost. Having a tool such as MCDM-23 available can broaden the nature of the optimization, quite possibly leading to significantly different choices.

TESTING THE PROCESS AND TOOLS

Several of the 12 country groups participating in IEA Task 23 are testing the process described in this paper. Most are using the tools described herein and are getting good results. This is an ongoing activity that will be evaluated to refine the process and tune up the tools. One result will be recommendations of the task regarding tools for trade-off analysis.

A group in the United States Federal Energy Management Program used the process and tools in a rigorous manner during the design of a new weather station for Caribou, Maine, which has a severely cold climate. The design team was particularly impressed that the process led them to discuss all of the six criteria and to agree on priorities. This was something that had not happened previously in their experience. Three design schemes were proposed and evaluated. The MCDM-23 tool provided valuable information at the right time. Two schemes met the previously agreed-upon energy-consumption targets. Ultimately, the best ranking scheme was selected.

CONCLUSIONS

Widespread application of well-known energy-efficiency measures and attention to other factors that affect building sustainability will require streamlining these considerations during the design process. Two vital points in the process have been identified and tools developed that make it practical for a design team to actually carry out the required steps within the time and budget constraints normally imposed.

The stakes are enormous. Energy consumption, the single most important factor affecting sustainability, can typically be reduced by 50%, compared to conventional contemporary construction. This can usually by achieved without increasing the construction cost because of reductions in the installed capacity of heating, ventilating, and air-conditioning equipment, which typically accounts for 15% to 20% of the initial cost. These reductions often amount to 40% of the HVAC cost, paying for the added cost of all the other improvements that make the HVAC down sizing possible.

The MCDM process described in this paper ensures that the most important issues that will affect building sustainability are considered equitably as vital decisions are being made during design, where they are most effective and cause no extra expenditure.

ACKNOWLEDGEMENTS

We wish to thank the 25 members of IEA Task 23 team, who have all contributed in various ways to the development of this process. Anne Grete Hestnes is operating agent of Task 23. The MCDM-23 tool was programmed in Visual Basic by Jun Tanimoto while visiting NREL on a post-doctoral assignment. Funding for the development of both ENERGY-10 and MCDM-23 is from the United States Department of Energy, Drury Crawley, Program Manager.

REFERENCES

1. Balcomb, J. Douglas (1997), ENERGY-10, A Design Tool Computer Program, proc. Building Simulation '97, International Building Performance Simulation Association, September 8-10, 1997, Prague, Czech Republic

2. Balcomb, J. Douglas (1998), ENERGY-10, Designer Friendly Simulation for Smaller Buildings, Building Performance, Issue 1, Spring 1998, Journal of the Building Environmental Performance Analysis Club, UK.

TECHNOLOGY COOPERATION AGREEMENT PILOT PROJECT (TCAPP)

Authors:
Ron Benioff
Collin Green
Colin Haller
Pat Keegan
David Kline
Debra Lew
Jeannie Renne

AAIA 2000-2899
Technology Cooperation Agreement Pilot Project
National Renewable Energy Laboratory (NREL)
Golden, CO

35th Intersociety Energy Conversion Engineering Conference
24-28 July 2000
Las Vegas, Nevada

ABSTRACT

The Technology Cooperation Agreement Pilot Project (TCAPP) is helping developing countries design and implement actions to attract investment in clean energy technologies that will meet their economic development goals, while mitigating greenhouse gas emissions. TCAPP is currently facilitating voluntary partnerships between the governments of Brazil, China, Egypt, Kazakhstan, Korea, Mexico, and the Philippines, the private sector, and the donor community on a common set of actions that will advance implementation of clean energy technologies. TCAPP is also assisting 14 countries in the Southern African Development Community with a regional technology cooperation needs assessment that was recently initiated by the Climate Technology Initiative (CTI). In the above countries, TCAPP has made strong progress in promoting the development of clean energy business investment partnerships. TCAPP has also become the leading model for implementation of technology transfer under the U.N. Framework Convention on Climate Change.*

TCAPP GOALS and APPROACH

TCAPP was launched by three U.S. government agencies—the U.S. Agency for International Development (USAID), the U.S. Environmental Protection Agency (USEPA), and the U.S. Department of Energy (USDOE)—in late 1997. TCAPP provides a model for implementing technology transfer as described in Article 4.5 (see inset below) of the United Nations Framework Convention on Climate Change (FCCC)*.

> "The developed country partners … shall take all practicable steps to promote, facilitate and finance, as appropriate, the transfer of, or access to, environmentally sound technologies and know-how to other Parties, particularly developing country Parties, to enable them to implement the provisions of the Convention."
>
> —Article 4.5 of the United Nations Framework Convention on Climate Change*

TCAPP Goals

TCAPP is designed to achieve the following major goals:

- Foster private investment in clean energy technologies that meet development needs and reduce greenhouse gas (GHG) emissions.
- Engage host country and international donor support for actions to build sustainable markets for clean energy technologies.
- Establish a model for international technology transfer under the FCCC*.

Principles of Technology Cooperation

The TCAPP approach reflects three key principles of climate change technology cooperation:

1) Technology cooperation must be host-country driven. Climate change technology cooperation priorities should be selected based on the potential benefits to the country as well as the potential GHG emission reductions. Countries should also build on existing programs and institutional roles, as well as previous climate change studies, national communications, or action plans.

2) Large-scale technology transfer can best be achieved through a sustained, coordinated, and strategic set of actions to harness private sector action in sustainable markets for clean energy technology. The most important role for government is to enable private sector activity, since commercial markets are the primary vehicle for technology transfer.

3) Successful technology cooperation requires collaboration at many different levels. This includes host country government agencies, businesses, nongovernmental organizations (NGOs), and technical experts working together to select priorities and design and implement actions. Country technical experts, international businesses and investors and international donors to secure the international investment and technical support necessary for effective implementation of technology cooperation actions.

TCAPP Program Elements

As shown in Figure 1, TCAPP includes four major program elements:

1) A country-driven process for selecting priorities, preparing market development strategies, and defining and implementing technology cooperation actions
2) International coordination and technical support to facilitate the work of the country teams and their collaboration with international businesses and donors.
3) Business participation in the development of strategies and the design and implementation of investment actions
4) Donor participation in development and implementation of strategies and technology cooperation actions.

Figure 1. Major Program Elements

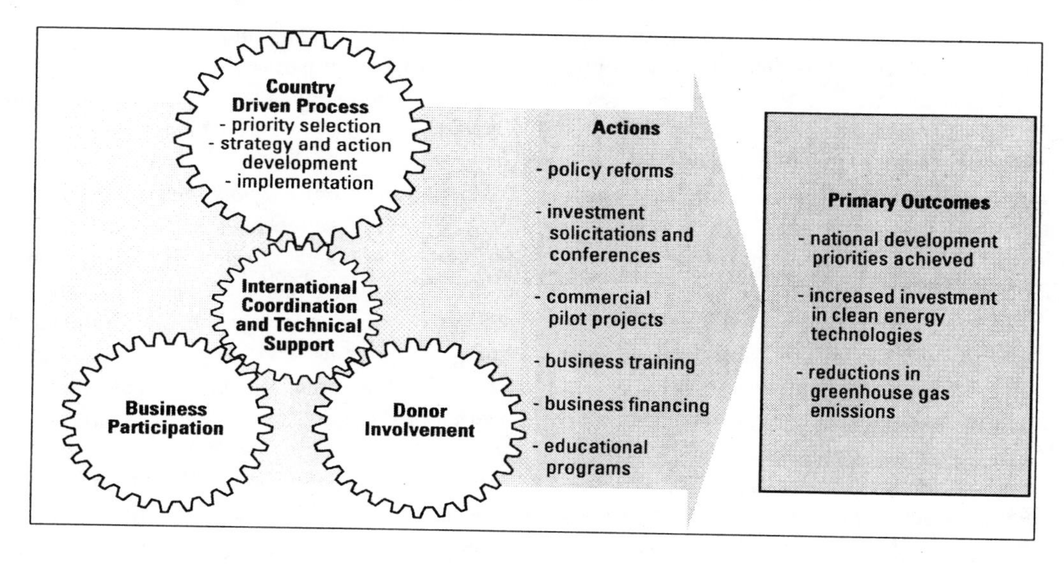

Figure 2. The TCAPP Country Driven Process

Priority Selection Phase	Strategy and Action Development Phase	Implementation Phase
1. Country Teams Formed 2. Establish Prioritization Process 3. Compile Information on Technologies and Barriers 4. Select Technology Cooperation Priority Areas 5. Prepare Technology Cooperation Framework	6. Prepare Market Strategy and Select Actions 7. Design Actions to Address Legal and Institutional Barriers 8. Design Investment Actions	9. Implement Technology Cooperation Actions 10. Evaluate Lessons Learned

Country Driven Process—Perhaps the single most important feature of TCAPP is that it is host-country directed. Countries structure their approach, select technology cooperation priorities, develop strategies to promote long term sustainable markets for these technologies, and define and manage implementation of actions to best meet national development priorities. The work of the country teams follows three basic phases of activities (see Figure 2).

Business Participation—TCAPP's market orientation could not be achieved without the involvement of host country and international companies, who provide input on the priorities, market barriers, strategies, and design of investment actions and participate in the activities themselves. The Business Council for Sustainable Energy (BCSE), a nonprofit association of clean-energy technology companies that facilitates the participation of more than 300 companies in the TCAPP business network.

Donor Involvement—Bilateral and multilateral donor agencies engage in the TCAPP process by helping to refine overall technology strategies, by identifying country actions that they can support, and by integrating the country TCAPP activities with other donor programs. International donor agencies have also provided input on the design of TCAPP through various forums, including the special TCAPP meeting for international donors in October 1998. At this meeting more than 30 representatives of donor agencies discussed priorities with senior officials from each of the five countries participating in TCAPP at that time. TCAPP is also collaborating with international donor agencies through the Climate Technology Initiative (CTI), a developed-country-supported partnership. CTI recently initiated the Climate Technology Implementation Program (CTIP), which follows the basic TCAPP approach.

International Coordination and Technical Support—TCAPP is a proactive intervention into the marketplace by a coordinated group of public and private organizations in the host country and from the international community. The U.S. National Renewable Energy Laboratory (NREL) leads and coordinates the work of the country teams, businesses, investors, international donors, and technical experts.

Actions and Outcomes—TCAPP programs are intended to further national development priorities while increasing investment in clean energy technologies and reducing greenhouse gas emissions. A variety of actions may be taken to help achieve those outcomes:

- Implementing policy reforms to remove legal or institutional barriers to clean energy development
- Issuing investment solicitations to recruit company participation and convening investment conferences to help companies find partners and financing
- Running pilot or demonstration projects and programs
- Training to build local business capacity
- Assisting clean energy businesses in securing financing for business growth and technology implementation
- Implementing educational programs to encourage host country businesses to increase their investments in clean energy technologies.

TCAPP COUNTRY ACTIVITIES

The countries participating in TCAPP have made excellent progress in defining priority areas and developing and implementing technology cooperation actions. Highlights for each country are summarized below and presented in Table 1 at the end of this summary.

Brazil

The Ministry of Mines and Energy leads a very active and effective TCAPP program that is coordinated by a Brazilian TCAPP interagency committee. The Brazilian team has established five technology cooperation priorities: energy efficiency in truck and bus transportation, direct use (including cogeneration) of natural gas, industrial energy efficiency, rural renewable energy, and fuel cells. Each priority area is connected to a national program that is supporting the design and implementation of specific investment actions. The Brazilian team has made significant progress with design and implementation of investment actions for these priorities. Among other activities, the Brazilian team recently held a cogeneration investment conference in May, 2000 and is organizing a transportation investment conference for September, 2000. The Brazilian team has also prepared an investment solicitation to attract companies into renewable energy market development and is developing fuel cell projects.

China

The TCAPP effort in China is unique due to the high-level agreement that the Minister of the State Development and Planning Commission recently signed with the Administrator of the U.S. Environmental Protection Agency to formalize this climate change technology cooperation work. Results of this technology cooperation work will be provided as a contribution to the implementation of technology transfer under the FCCC.* Under this agreement the Chinese TCAPP team will focus on the following technology priorities: high efficiency electric motors, grid-connected wind electric power, efficiency improvements in coal-fired industrial boilers, cleaner coal technologies for power generation, and two additional priorities to be selected this year.

China has just begun to implement technology cooperation actions for wind and motor technologies. For wind power, these actions include developing competitive solicitations with regional power bureaus, wind turbine testing and certification, and resource assessment. For efficient motors the actions include a financing workshop, and motors standards, labeling, and testing and certification. Lead points of contact for boilers and coal technologies have also been established in both the United States and China.

Egypt

The Government of Egypt has identified technology transfer as one of its highest priorities under the FCCC.* The Egypt TCAPP activity, which has just begun, is attracting the participation of all key agencies in the Egyptian government with the Egyptian Environmental Affairs Agency in the lead role. At meetings in Cairo in September, 1999, agency representatives selected an initial set of priority technology areas: industrial energy efficiency and enhanced natural gas use, lighting efficiency and renewable-powered lighting, renewable energy applications in rural areas, and small-scale cogeneration applications. These preliminary priorities were approved by the Egyptian Ministerial Committee on Climate Change in November and specific investment actions in these areas are currently under development.

Kazakhstan

Kazakhstan was one of the first countries to join TCAPP and has been a leader in developing this process. The Kazakhstan TCAPP team identified four technology priorities: power plant carbon efficiency program (fuel-switching, combined-cycle gas, and improved heat rate); energy-saving and district-heating improvements; wind power; and small hydro. Work has been delayed due to changes in the government and on their team, but Kazakhstan has formed a new TCAPP team, led by the Ministry of Energy, Industry and Trade, and its affiliate KEGOC, with broad-based participation from other key agencies. KEGOC is a joint stock company set up to develop improved electric utility systems. A new working group has selected promising investment projects for development from the following sectors: small hydropower, combined heat and power, and gas utilization from oil refineries. These projects are currently being refined and presented to international investment partners to support their further development..

Mexico

The TCAPP Mexico team is led by the National Commission for Energy Conservation (CONAE), the leading energy-efficiency implementing agency under the Secretary of Energy. CONAE led the prioritization process which resulted in the selection of three technology priority areas that build on current CONAE programs: efficient lighting in public buildings, solar water heating for residential and commercial uses, and improved steam generation and distribution systems. CONAE, in collaboration with other agencies in Mexico, is now implementing investment actions for these three areas. These investment actions include: an energy service company (ESCO) pilot program for industrial and municipal facilities (a solicitation was recently sent to over 2000 facilities in Mexico to identify pilot projects), the design of financing mechanisms for the purchase of solar water heaters and promotion of business partnerships for the manufacture of solar water heaters, and joint work on steam generation and distribution projects.

Philippines

The Philippines TCAPP effort has included a very strong consultative process with high level support from the Philippines government. The Philippines Department of Energy (P-DOE) leads the TCAPP work in the country in close consultation with a broad group of other agencies and NGOs. The Philippines team selected priorities in mid-1998 including energy efficiency, renewable energy, and cross cutting initiatives. The Office of newly elected President Estrada became engaged in the TCAPP effort and together with the P-DOE determined that future TCAPP work should focus on increased use of renewable energy in rural areas to address their primary goal of poverty alleviation. Based on feedback from businesses in the Philippines and other key stakeholders, it was clear that policy reforms were needed to get the market moving. Therefore the Philippines team developed a series of "fast track" policy recommendations (such as, increasing tax incentives, revising policy guidance, and providing other types of government support for use of renewable energy technologies). In March, 2000, the Philippines Secretary of Energy approved the fast track recommendations. In addition to assisting P-DOE implement the reforms, the TCAPP team is also implementing specific investment actions, including advancing a pilot program for solar powered agricultural water pumping, supporting the development of wind-diesel hybrid projects, and development of hydropower retrofit projects together with rural electric cooperatives..

Republic of Korea

The Republic of Korea joined TCAPP in January 1999. The Ministry of Commerce, Industry and Energy (MOCIE) of the Republic of Korea chairs a TCAPP steering committee and has designated the Korea Energy Management Corporation (KEMCO) to lead the implementation of technical TCAPP activities for Korea. A scoping meeting in March 1999 resulted in the development of criteria for prioritization and in the selection of three priority technologies: energy management, methane recovery from organic waste, and waste energy recovery using heat pumps. The Korean TCAPP team together with the U.S. team facilitated joint meetings between U.S. ESCOs and Korean facilities and ESCOs in January, 2000. These meetings identified opportunities to help retrofit the Hyundai auto plant in Korea and to pursue a methane recovery project. These projects are currently under development.

Southern African Development Community

The 14 countries in the Southern African Development Community (SADC) are initiating a regional program to identify the clean energy technologies that have the greatest potential across the SADC region for meeting sustainable development needs while reducing greenhouse gas emissions through accelerated private investment. The SADC Environment and Land Management Sector (SADC ELMS), is the lead agency for managing the project's approval process within the SADC system. SADC ELMS is working with teams of government officials, energy business representatives, and experts from each of the 14 countries to carry out the project. This initiative is in response to a request for such a regional needs assessment by ministers and other senior SADC officials attending the CTI/Industry Joint Seminar on Technology Diffusion on March 17-18, 1999, at Victoria Falls, Zimbabwe. This project applies the TCAPP approach to this regional effort conducted in collaboration with the CTI. The participating countries have established general priority areas and are currently completing national consultations to define regional investment actions.

Future Directions

During the next year, the TCAPP effort will focus on four primary activities. The nature of the activities will be determined by host country teams:

- *Successful implementation of technology cooperation actions in each country/region.* TCAPP needs to continue to demonstrate to developing countries how climate change technology cooperation will accelerate investment in their technology priorities and meet their development needs.
- *Increased private sector participation.* While more than 300 international energy companies are participating in TCAPP through the BCSE's network, the level and depth of the participation

of international and host country businesses and investors needs to be increased.

- *Enhanced participation of bilateral and multilateral donor organizations.* NREL will work toward expanding the participation of donor organizations to help countries secure donor support for key actions and will continue to encourage donor organizations to consider applying the TCAPP approach to other developing countries, as is occurring through the Climate Technology Initiative.
- *Evaluating TCAPP as a model for technology transfer under the FCCC.* The TCAPP developing country, business, and donor partners, together with NREL and other technical experts, will evaluate the effectiveness of the TCAPP approach as a potential model for implementing technology transfer under the FCCC.*

Table 1. Summary of Country TCAPP Activities

	Lead Agency	Technology Priorities	Selected Actions under Development/ Implementation
Brazil	Ministry of Mines and Energy (MME)	• Energy efficiency in diesel truck cargo transportation • Direct use of natural gas • Electrical energy effciiency • Fuel cells • Renewable energy in rural electrification	• Efficient truck and bus transportation conference and trade show • Solicitation for Expanded Renewable Energy Field Test Program • Identification of engineers and energy auditors for industrial plants • Development of a cooperative R&D strategy for ethanol based fuel-cells • Cogeneration investment strategy workshop
China	State Development Planning Commission (SDPC)	• Higher efficiency power generation technology, specifically circulating fluidized bed combustion • High efficiency electric motors • Advanced industrial boilers • Wind power generation	China has just begun Phase II of its TCAPP work. During the next twelve months China will, in collaboration with the United States: • Complete in-depth market and technology analysis for wind and motors and propose technology cooperation strategies and actions • Hold interagency meeting to select at least six actions for implementation in 2000 • Prepare detailed action proposals for the selected actions • Initiate implementation of the six actions by March 2000 • Lead points of contact for boilers and coal technologies will also be established in both the U.S. and China
Egypt	Egyptian Environmental Affairs Agency (EEAA)	• Industrial energy efficiency, natural gas use* • Lighting efficiency and renewable lighting* • Renewable energy in rural areas* • Small scale cogeneration applications*	* Priorities are preliminary and subject to revision by the National Climate Change Committee
Kazakhstan	Ministry of Energy, Industry and Trade and the Kazakh Electricity Grid Operating Company (KEGOC)	• Power Plant Carbon Efficiency Program (fuel switching, combined-cycle gas, and improved heat rate) • Energy-saving and district-heating improvements • Wind power • Small hydro	Investment projects in the following sectors have been proposed by the TCAPP working group and are under development • Small hydropower • Small combined heat and power systems • Utilization of gas from oil refineries
Mexico	National Commission for Energy Conservation (CONAE)	• Nationwide expansion of efficient lighting in public buildings • Solar water heating pilot program • Nationwide expansion of steam generation and distribution systems	• Develop potential energy efficiency lighting projects • ESCO workshop • Expand Solar Water Heating Program • Support "Solar Millennium 2000" event
Philippines	Department of Energy (P-DOE)	• Renewable energy for rural development (photovoltaics, wind energy) • Energy efficiency and demand-side management (energy-efficient boilers, appliances, and equipment) • Cross-cutting technology support (renewable energy and energy efficiency technology center/policy development)	Implementing "fast track" policy reforms • Revise new and renewable energy (NRE) project accreditation requirements • Exercise full P-DOE powers under the mini-hydro law • Use the host community fund for NRE development • Revise Executive Order 462 • Issue new policy statement on NRE Developing additional investment actions
Republic of Korea	Korea Energy Management Corporation (KEMCO)	• Energy management • Methane recovery from organic waste • Waste energy recovery using heat pumps	Technology cooperation actions under development
Southern African Development Community	SADC Environment and Land Management Sector (SADC ELMS)	Technology cooperation priorities currently being established	

* - United Nations Framework Convention on Climate Change, UNFCCC Secretariat, Bonn, Germany.

AIAA-2000-2901

HEAT RECOVERY FROM AIRCRAFT ENGINES[1]

S. Pasini
U. Ghezzi
R. Andriani
L. Degli Antoni Ferri
Energetic Department
Polytechnic of Milan
20133 Milan, Italy

ABSTRACT

The paper deals with the problem of heat recovery from aircraft engines, with specific reference to jet engines and turbo propeller engines. In actual propulsion systems, the trend for cycle maximum temperatures is for very high values, with a consequent increasing of discharge temperature, due account also given to operative modalities of the discharge nozzle, often working in off design conditions. Thus, the heat discharged strongly influences the performances of the system, affecting them even if already optimized with reference to other cycle parameters. In this way, heat recovery appears an obliged way for further increments of performances. In the space sector, however, the recovery of the discharged heat appears strongly problematic because of engine configuration and specific limitations, as weight, overall dimensions, maximum reliability, etc. Nevertheless, this possibility shows great interest.

The present paper evaluates first the possible modalities to utilize the recovered heat, emphasizing how the regenerative processes should be considered among the most attractive ones. Then, the different types of jet engines are analyzed to evaluate the possible obtainable recoveries. The analysis shows how the recovery not only influences positively efficiency and therefore specific fuel consumption, but allows also to amplify the importance of other operative parameters. At last, the practical possibility of heat recovery is evaluated, trying to define the configuration of suitable heat exchangers and to identify the most appropriate location within the propulsive system.

INTRODUCTION

In the field of air breathing propulsion two main areas of interest can be considered. The first one deals with high flight Mach number engines, operating in the hypersonic fields and able to reach orbital speed. The studies of such engines are growing very fast, due to the possibility offered by this kind of engine to be used as launchers for orbital missions. The other main research field is related to the possibility to develop engines rated at very high level of thrust with high efficiency and low specific fuel consumption. The necessity of reducing fuel consumption is not only related to the need of lowering the rising costs of fuel, but also to the possibility of increasing the maximum payload. Both effects greatly affect the global costs of transportation and the economy of airlines companies. The reduction of fuel consumption is achieved by improving the global efficiency of the engine, the latter depending directly upon thermal and propulsion efficiency. In the last decades this aim has been achieved with the introduction of the turbofan engines, allowing a reduction in specific fuel consumption of about 50%. The separate flows turbofan engines, especially at high by pass ratio, reduce fuel consumption by increasing the propulsion efficiency. From the point of view of the thermal cycle, the development of new materials and cooling techniques has allowed the realization of turbine blades able to withstand high temperature levels, thus increasing the maximum temperature and consequently the maximum pressure of the cycle. These processes are under way, and, for what concerns the improvement of propulsion efficiency, new turbofans at very high by pass ratio are going to be built.

These new engines can reach such values of BPR to be considered as a connecting bridge between turbofan and turboprop engines. The results obtained working in this direction have been very sensitive and it is now easy to foresee that future improvement limits can only be narrower. For this reason it becomes necessary to consider other aspects of the propulsion system if a further improvement of the global efficiency is asked for. An analysis from the energetic point of view of the thermal cycle shows that a great amount of heat is wasted in the exhaust, and indicates how a recovery, total or more realistically partial, of this heat can lead to a greater efficiency of the system. This consideration is not new, and the heat recovery from the exhaust is a well consolidated practice in industrial gas turbine power plants for ground application. The thermal cycle in this case is called regenerative and regeneration the practice of heat recovering. A classical regenerative cycle of a gas turbine engine for industrial purpose is shown in Fig. 1.

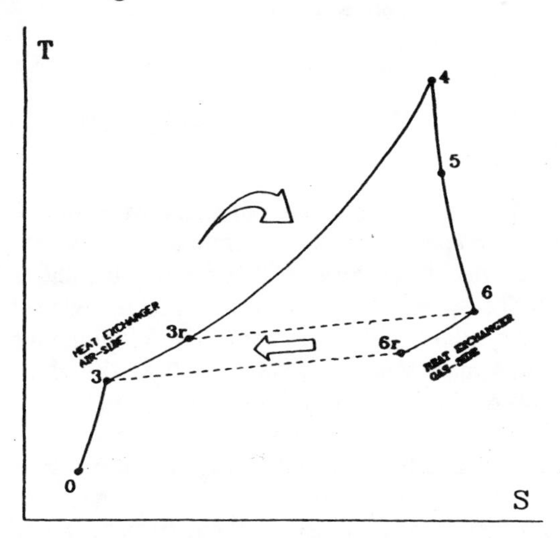

Fig. 1. Thermodynamic diagram of a turboprop engine with regeneration.

The main difficulty of the extension of this practice to propulsion systems has always been represented by the size and weight of the heat exchanger. However, the improved manufacturing possibilities and the need of new strategies for increasing engine performances induce to take into due account this possibility. From a theoretical point of view there are several methods to exchange heat, but the most suitable for propulsive applications seems to be that shown in Figs. 2a and 2b.

Comparison of Figs. 1 and 2 outlines that in the case of a turbojet engine heat is exchanged (unchanged other cycle characteristics) at a higher temperature level, and this is due to the lower expansion ratio of the turbine. Moreover, the heat exchange influences the enthalpy drop available in the exhaust nozzle, hence the discharge velocity.

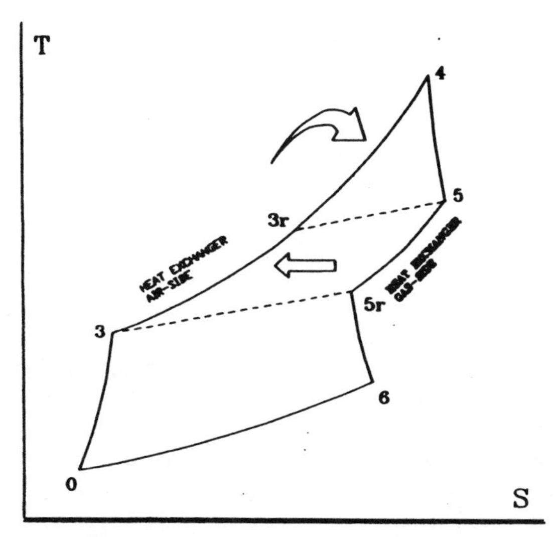

Fig. 2a. Thermodynamic diagram of a turbojet engine with regeneration.

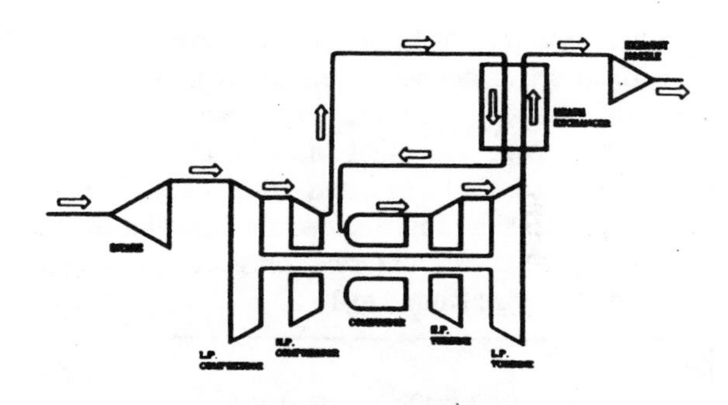
Fig. 2b. Scheme of a turbojet with regeneration.

For this reason regeneration influences not only the thermal efficiency, reducing heat in the exhaust, but also the propulsion efficiency, lowering the residual kinetic energy of the jet.

Since the global efficiency and the specific fuel consumption are a function of both thermal and propulsion efficiency, from these first simple considerations it becomes evident how regeneration can have influence on the performances of the engine. However, the reduction of the enthalpy drop available in the nozzle, and consequently of the jet velocity, reduces the specific thrust of the engine.

To establish whether regeneration can have some advantages, a numerical thermodynamic code was developed to allow to simulate the thermal cycle and the main specific performances of a jet engine with regeneration for different operative conditions. Thus, the possibility to recover heat from the exhaust has been evaluated for a simple turbojet, a turbofan and a turboprop engine in design conditions.

It has also been analyzed, in a two separate flows turbofan engine, the different way to transfer heat between the two flows. Finally, some features of the off-design behavior have been pointed out for a regenerative turbojet engine.

TURBOJET

The possibility to introduce the regeneration is first considered for a turbojet engine. In this case the heat exchanger is placed between the last turbine stage and the exhaust nozzle, as already shown in Fig. 2.

The difference with respect to the conventional jet engine is the presence of the heat exchanger, represented by the transformations 3-3r and 5-5r. Along the 3-3r transformation (air side), the air from the compressor is heated by the gas from the turbine flowing through the heat exchanger along the 5-5r transformation (gas side). The efficiency R of the regeneration is the ratio between the actual heat transferred from hot gas to cold air and the theoretical amount that could be exchanged in an ideal counter flow heat exchanger of infinite surface, that is:

$$R = \frac{h_{3r} - h_3}{h_5 - h_3}$$

As a first example, the behavior of thermal efficiency for a fixed set of operative conditions is now taken into consideration, using the regeneration efficiency R as a parameter. The flight Mach number was 0.8, the altitude 10000 m, and the turbine inlet temperature 1600 K. Fig. 3 shows the behavior of the thermal efficiency at different pressure ratios and with R in the range from 0 to 1.

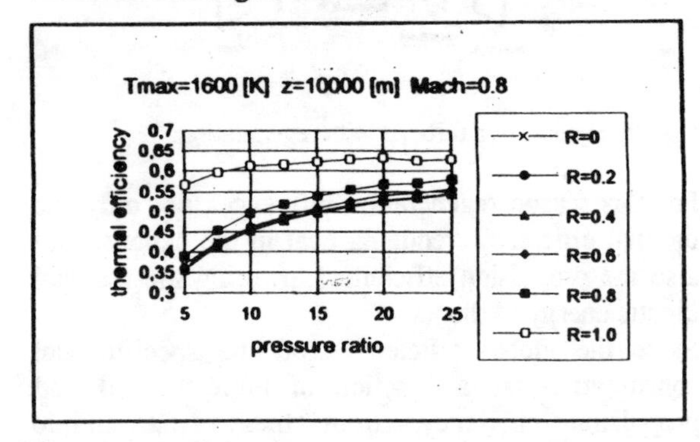

Fig. 3. Thermal efficiency of a turbojet engine with regeneration.

The heat exchanger is responsible for a pressure loss both on air and gas side, depending mainly upon its size; the high mass flow rate of a turbojet and the small size needed can come out with a pressure drop too great to be tolerated if a high degree of regeneration (at least 0.7) is not achieved. For the calculation, a pressure drop of 5 % was assumed for both sides of the heat exchanger. With reference then to the specific fuel consumption (*TSFC*), it can be said that it depends on both thermal and propulsion efficiency, being expressed as:

$$\eta_p = \frac{2v_0}{v_0 - v_j}$$

where v_0 is the flight speed and v_j the exhaust gas velocity.

TURBOFAN

The possibility to apply regeneration in a turbofan engine is now considered. This seems interesting since in a turbofan, especially at high by-pass ratio, the very large fraction of the thrust is provided by the cold flow, while the gas generator provides the required power. Thus, the enthalpy level before the exhaust nozzle of the gas generator can be lowered without loosing a great amount of thrust, while increasing the efficiency.

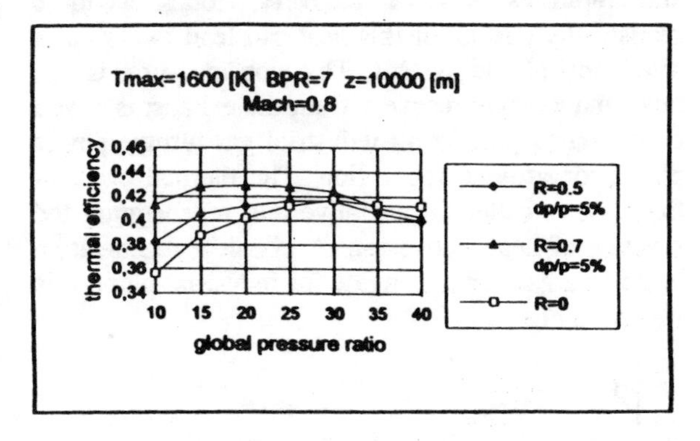

Fig. 4. Thermal efficiency of a turbofan engine with regeneration.

Fig. 4 shows the thermal efficiency of a turbofan engine compared with a regenerative one. Two cases are reported, respectively with R equal to 0.5 and 0.7. In both cases, a pressure drop of 5 % on the two sides of the heat exchanger is assumed. For this simulation, the maximum temperature was 1600 K, the altitude 10000 m, the flight Mach number 0.8 and BPR fixed at 7.

The thermal efficiency in the case with heat recovery is higher than in the conventional one of about 4 % when $R = 0.5$, while it rises to about 10 %, if $R = 0.7$. These effects are sensible at low pressure ratios. In fact, for $PR = 20$, the value of the thermal efficiency for which $R = 0.5$ and the conventional one practically coincide; for pressure ratios greater than 30, the curves cross each other and the conventional case seems better.

This is due to the fact that above a certain value of pressure ratio the heat cannot be exchanged any more, while the pressure drop in the heat exchanger still remains. For this reason this section of the plot shows no practical interest.

TURBOPROP

The natural development of this study leads to the introduction of the regeneration in a turboprop engine. In fact, this kind of engine seems to be, at the state of the art, the most indicated to utilize an heat exchanger, both for thermodynamic and constructive reasons.

In fact, the mass flow rate through the gas generator is smaller if compared to that of a turbojet, thus implying a smaller and lighter heat exchanger: an important characteristic since these technological aspects can be determinant for a possible introduction of regeneration in aircraft power plants. Fig. 5 shows, in a T-S plane, the thermodynamic cycle of a turboprop engine with regeneration.

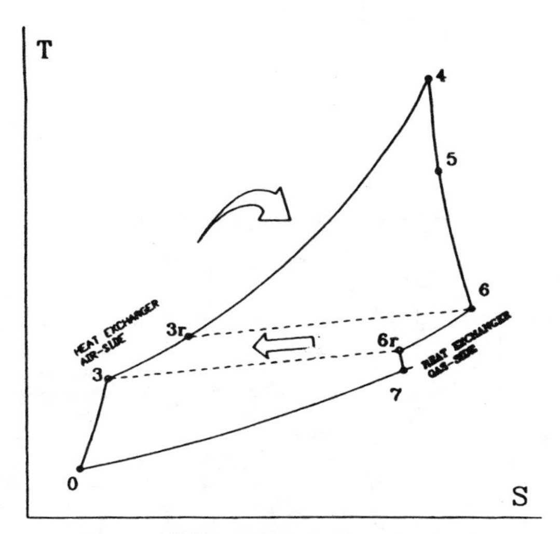

Fig. 5. Thermodynamic diagram of a turboprop engine with regeneration.

As before, the behavior of a turboprop engine with regeneration was numerically simulated, determining the main engine characteristics, such as power, fuel consumption and cycle efficiency.

Attention was focused onto one set of parameters simulating both engine and operative condition, such as turbine inlet temperature (1500 K), altitude (7000 m) and flight Mach number (0.65). Another characteristic parameter of a turboprop engine is the alpha parameter, defined as the ratio of the ideal enthalpy drop available for the power turbine to the total ideal enthalpy drop, as shown in Fig. 6.

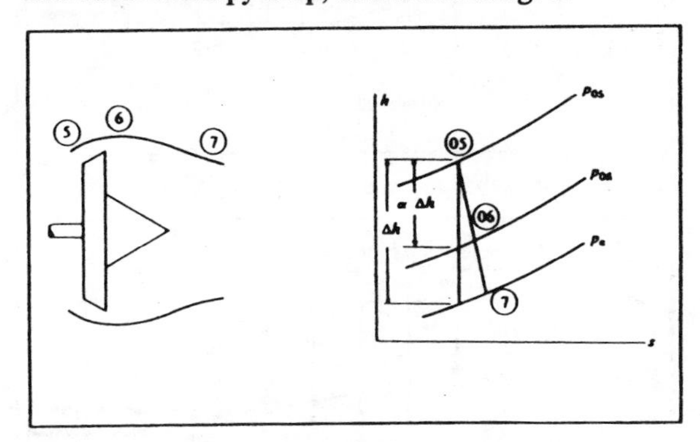

Fig. 6. Enthalpy drop distribution between turbine and exhaust nozzle.

The remaining fraction of enthalpy drop accelerates the gas in the exhaust nozzle. For the simulation the value of alpha was 0.88, a value which, for these operative condition, minimizes the fuel consumption.

Also in this case, the plots are parameterized with regeneration effectiveness, according to the same scheme of the previous examples. As already seen for the turboprop case, the introduction of regeneration causes a pressure drop across the heat exchanger. In the calculation this fact has been taken into account by introducing a pressure drop of 4 % on the air side, and 6 % on the gas side, deducing these values from similar cases.

In Fig. 7 is reported the behavior of the shaft power at different pressure ratios.

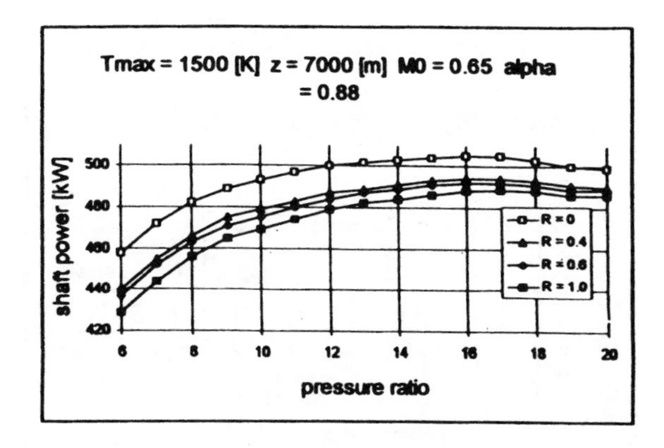

Fig. 7. Shaft power in a turboprop engine with regeneration, effectiveness as a parameter.

HEAT EXCHANGE

In the previous analysis, for both jet engine and turboprop, heat is subtracted by the hot gas before entering the exhaust nozzle. This practice seems to ensure good results, al least according to numerical simulation; it is not however the only thinkable configuration of a regenerative set up for propulsion systems. In fact, the heat can be exchanged between the two flows both during and after the expansion in the exhaust nozzle.

In the first case, heat is subtracted from gas while this is expanding in the exhaust nozzle; in the second case, heat is extracted from gas after this is expanded in the nozzle, nearly at external pressure, before being expelled into the atmosphere.

The different ways have consequences primarily on the amount of heat transferred and consequently on the behavior and performances of the engine.

Here are reported, using the same analysis as previously done, the plots of the main engine thermodynamic characteristics computed for three different methods of heat exchange: the conventional one (heat subtracted before expansion in the nozzle); during expansion (heat transferred during the expansion in the nozzle), and after expansion (heat transferred after the expansion in the nozzle and before expelling it into atmosphere). Figs. 8 through 12 show the engine characteristics computed in the three specified ways, assumed $R = 0.4$.

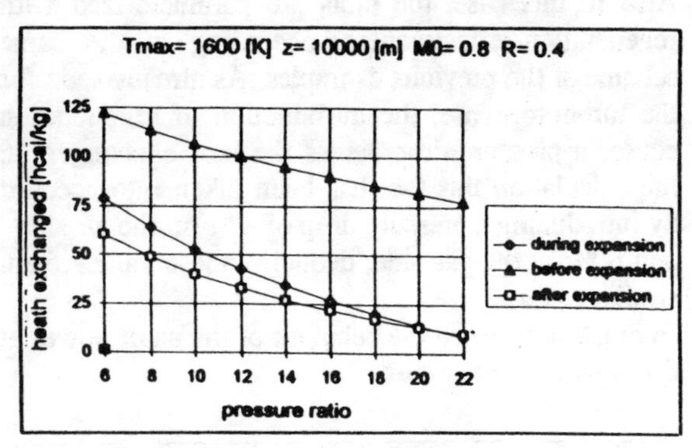

Fig. 8. Heat exchanged as a function of pressure ratio.

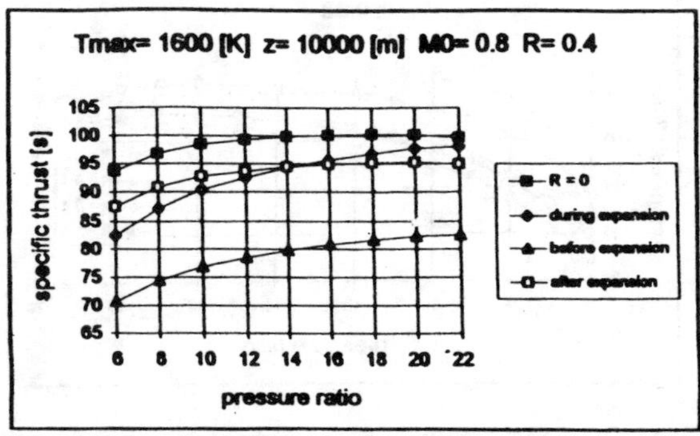

Fig. 9. Specific thrust as a function of pressure ratio, with effectiveness of regeneration equal to 0.4.

It can be observed that the conventional case allows to obtain the minimum level of fuel consumption. In fact, with reference to the behavior of TSFC, the points representing this case appear constantly below the other lines, especially at high values of pressure ratio. As far as TSFC and heat exchange are then concerned for the case with heat transferred before expansion, better results are obtained if compared to the others. Then the thermal efficiency, evaluated when heat is transferred during the expansion, is always greater, while the propulsion efficiency shows a two-face behavior: while at low pressure ratios heat subtracted during expansion seems convenient, at higher values of pressure ratio the heat exchange becomes more convenient if performed after the expansion.

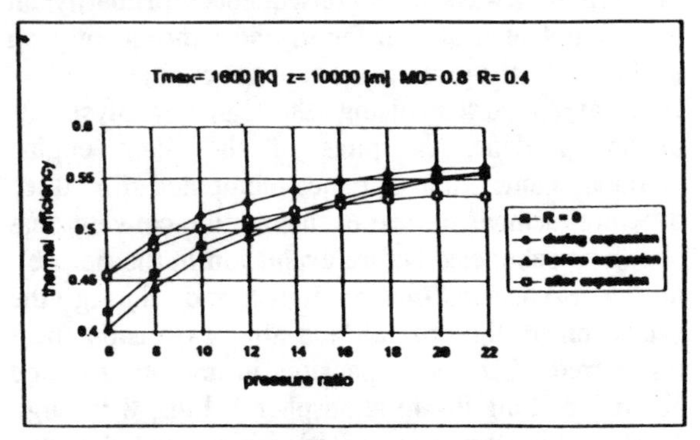

Fig. 10. Thermal efficiency as a function of pressure ratio, with regeneration effectiveness as a parameter.

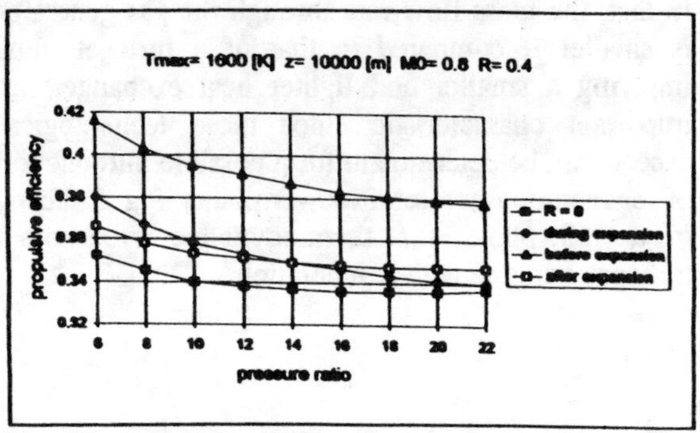

Fig. 11. Propulsive efficiency of a regenerative turbojet engine, for three methods of heat subtraction.

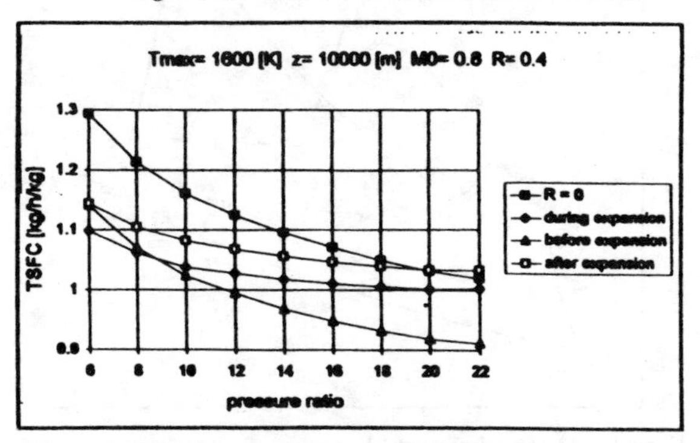

Fig. 12. TSFC as a function of pressure ratio for three methods of heat subtraction.

The analysis has then been repeated with $R = 0.8$ (this value assumed since representing a sort of limit for actual possibility of heat exchangers). The results of these simulations are reported in Figs. 13 and 14.

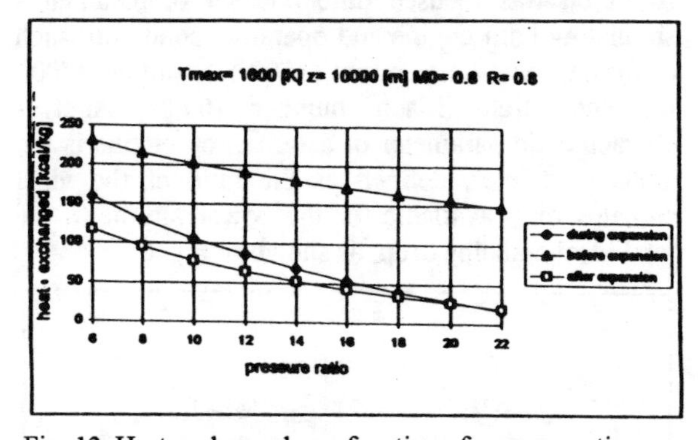

Fig. 13. Heat exchanged as a function of pressure ratio.

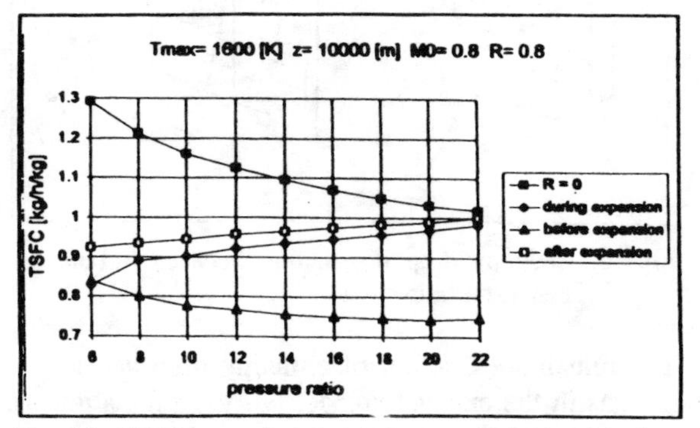

Fig. 14. TSFC as a function of pressure ratio, for three methods of heat subtraction.

OFF-DESIGN

It is interesting to study how the presence of the heat exchanger can influence the behavior of the engine in off-design conditions. Generally, the off-design conditions of a gas turbine engine can be synthesized through the knowledge of the operating line of the compressor (relationship of compressor pressure ratio versus equivalent mass flow rate).

In fact, once fixed the aircraft operative conditions and the main engine parameters, as flight Mach number, altitude, turbine inlet temperature, and nozzle behavior, an univocal relationship can be drawn between compressor pressure ratio and engine performances. From the pressure ratio it becomes so possible to evaluate the thermal cycle of the engine and from this the specific performances, such as specific thrust, thermal, propulsion and global efficiencies, and specific fuel consumption. By means of a numerical thermodynamic code the off-design behavior of a regenerative turbojet engine has been simulated.

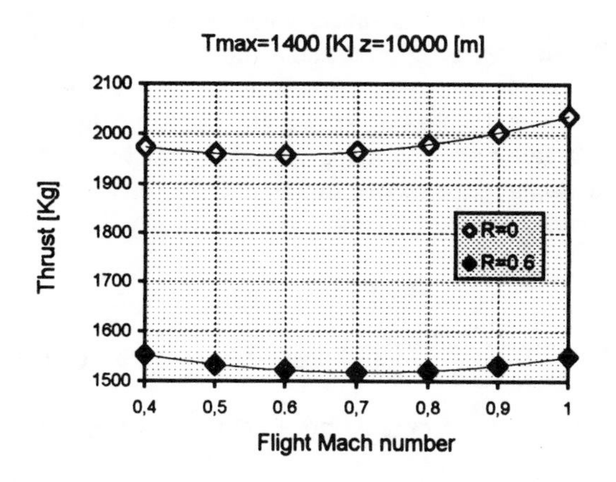

Fig. 15. Relationship of Thrust versus Flight Mach number

Fig. 15 shows the behavior of the thrust of the simple turbojet engine and the turbojet with R = 0.6 for different values of flight Mach number. Fig. 16 shows, for the same operative condition, the behavior of the specific fuel consumption, while Fig. 17 reports the ratio of the thrust at different flight Mach numbers to the thrust at the design point, the flight Mach number for both engines being 0.8.

It is interesting to see if, for both cases (conventional turbojet and regenerative one), the introduction of the heat exchanger modifies the behavior of the compressor. The points of the operative line of the compressor represent the different conditions at which the compressor works during its off-design conditions. The design point of the engine is represented by one point of the line in the compressor map. The line shows the relationship between the pressure ratio of the compressor and the corrected mass flow rate through it, as it is shown in Fig. 18.

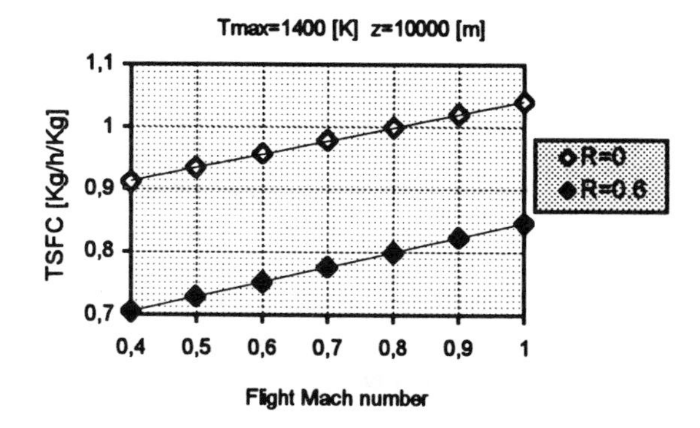

Fig. 16. Relationship of TSFC versus Flight Mach number.

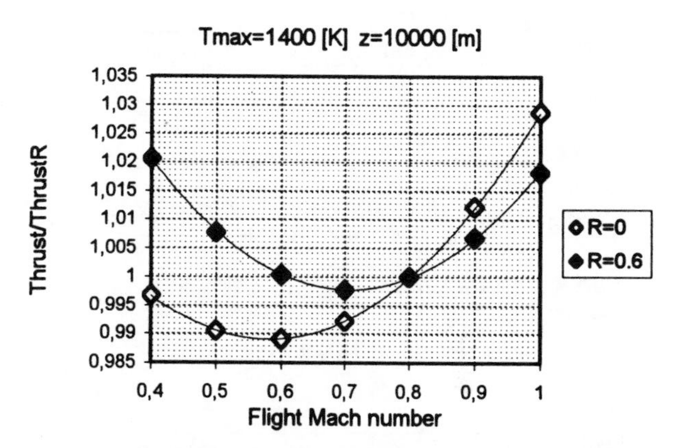

Fig. 17. Relationship of ratio Thrust to Thrust at design condition versus Flight Mach number.

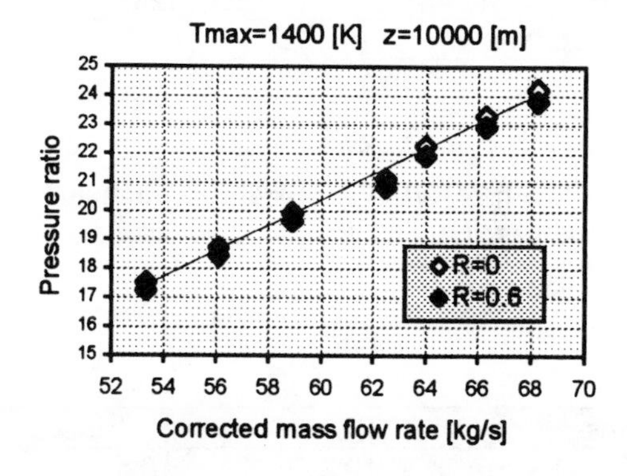

Fig. 18. Operative lines of simple and regenerative jet engines at different Flight Mach numbers.

HEAT EXCHANGER DESIGN CASE

A case is shown to demonstrate the real possibility to design a heat exchanger to be used as a regenerator in a typical gas cycle. The regenerator is located after the turbine and before the nozzle and is interested axially by the hot gas. The air, coming from the compressor, enters perpendicularly the regenerator, splitting exactly into two flows and realizing therefore a perfect cross flow compact heat exchanger, both fluids unmixed, made on both sides by same plain plate-fin surfaces.

To achieve a better configuration of the regenerator, only a part of the hot gas mass flow rate is employed (two cases are given with a fraction equal to 50 and 75 %), while the whole air from the compressor is used.
Operating conditions are specified in Tab. 1.

Tab. 1

Air flow rate	20 kg/s
Gas flow rate	10 and 15 kg/s
Air entering pressure	8 bar
Air entering temperature	600 K
Gas entering pressure	3 bar
Gas entering temperature	1100 K
Regenerator O.D.	0.8 m
Regenerator I.D.	0.4 m
Regenerator length	0.9 and 1.2 m

Surface characteristics, identical for both sides, are specified in Tab. 2.

Tab. 2

Plate spacing	6.35 mm
Hydraulic diameter	2.87 mm
Fin thickness	0.15 mm
Transfer area/volume between plates	1288 m^2/m^3
Fin area/total area	0.773

The results of calculations for the first case (100 % air mass flow rate, 50 % gas mass flow rate) and for the second case (100 % air, 75 % gas) are given in Tab. 3.

Tab. 3

Results	1st case	2nd case
Efficiency	0.8	0.63
Air exit temperature	792 K	845 K
Gas exit temperature	730 K	785 K
Heat exchanged	3.600 Mcal/h	4.600 Mcal/h
Pressure drop, air side	2.8 %	2 %
Pressure drop, gas side	2.8 %	16.4 %

These results show that if more heat is recovered without changing accordingly the overall dimensions of the heat exchanger, the efficiency decreases and the pressure drop may increase up to unacceptable values.
The two cases presented are anyway quite meaningful since the results obtained, even if still on a theoretical ground, show a certain real feasibility for the design.

SUMMARY AND CONCLUSIONS

The analysis done shows that heat recovery in jet engines can be a very interesting practice not only under the energetic point of view but also more generally in relation to the possibility of improving the engine characteristics (lower pressure ratio and consequently lower levels of NOx).
The main problem related to this practice is the introduction of a heat exchanger suitable to meet the requirements of an aircraft engine.
A short analysis has been done and the results have put in evidence that it is possible to build a heat exchanger suitable to be mounted on aircraft power plants, both for its size and weight, and also for the amount of heat exchangeable.
This study is a preliminary step on the evaluation about the possibility of introduction of regeneration in a jet engine.
A deeper theoretical study, involving the off design behavior of the engine and different kinds of jet engines, seems anyway necessary before starting an experimental phase. However, the results are encouraging and the actual possibility of building a heat exchanger suitable for aircraft engines appears less remote respect only few years ago.

REFERENCES

[1] U. Ghezzi "Motori per Aeromobili", CLUP, Milano, 1990.

[2] European Commission Research Directorate's Fifth Framework Program.

[3] R. Andriani, U. Ghezzi "Off-Design Analysis of a Jet Engine With Heat Recovery", 38th AIAA Aerospace Sciences Meeting and Exhibit, Jan 10-13, 2000 Reno, NV. Paper 00-0743.

[4] R. Andriani, U. Ghezzi, L. Ferri "Jet Engines with Heat Addition During Expansion: a Performance Analysis", 37th AIAA Aerospace Sciences Meeting and Exhibit, Jan 11-14, 1999 Reno, NV. Paper 99-0744.

[5] R. Andriani, U. Ghezzi "Heat Recovery in Reciprocating Engine", 35th AIAA/ASME/SAE Joint Propulsion Conference and Exhibit, July 20-24, 1999, Los Angeles, CA. Paper 99-2664.

[6] R. Andriani, L. Ferri, U. Ghezzi, F. Gamma "Heat Recovery in Turbofan Engines: a Performance Analysis" 33rd AIAA/ASME/SAE Joint Propulsion Conference and Exhibit, July 7-9, 1997, Seattle, WA. Paper 97-3158.

[7] R. Andriani, U. Ghezzi, F. Gamma "Regeneration in propulsion", 32nd AIAA/ASME/SAE Joint Propulsion Conference and Exhibit, July 1-3, 1996, Lake Buena Vista, FL. Paper 96-2792.

[8] R. Andriani, U. Ghezzi, F. Gamma "Analysis of the subdivision of the enthalpy drop in a turboprop engine with regeneration", 31st AIAA/ASME/SAE Joint Propulsion Conference and Exhibit, July 10-12, 1995, San Diego, CA. Paper 95-2753.

[9] G. Pellischek, B. Kumpf "Compact heat exchanger technology for aero engines", 10° International Symposium on Air Breathing Engine (X ISABE), Nottingham, September 1-6, 1991.

[10] R. Andriani, U. Ghezzi "Influence of heat addition at variable pressure on engine performances", 48th IAF International Astronautical Congress, October 6-10, 1997, Turin - Italy. Paper: IAF-97-S.5.02.

[11] Ascher H. Shapiro "The Dynamics and Thermodynamics of Compressible Fluid Flow - Vol.1", The Ronald Press Company, New York, 1953.

[12] Gordon C. Oates "Aerothermodynamics of Gas Turbine and Rocket Propulsion", AIAA, Education Series.

[13] "Jet Propulsion Engines", Editor: O. E. Lancaster, Princeton, New Jersey Princeton University Press, 1959.

[14] I. H.Ismail, F. S. Bhinder "Simulation of Aircraft Gas Turbine Engines" Journal of Engineering for Gas Turbines and Power, January 1991, Vol. 113.

[15] R. Andriani, L. Ferri Degli Antoni, F. Gamma, U. Ghezzi "Thermodynamic analysis of regeneration in jet propulsion", Third Italian/British Workshop on Heat-Engine, Firenze, 8-10 Ottobre 1992.

[16] Arthur H. Lefebvre "Gas Turbine Combustion", Hemisphere Publishing Corporation, 1983.

[17] A. M. Mellor, Design of Modern Turbine Combustors, Academic Press Limited, 1990.

[18] Philip G. Hill, Carl R. Peterson "Mechanics and Thermodynamics of Propulsion", Addison-Wesley Publishing Company, 1970.

[19] U. Ghezzi, S. Pasini, L. Ferri Degli Antoni, R. Andriani, F. Gamma "Performance and configuration analysis of jet-engine off-design behavior", 11° International Symposium on Air Breathing Engine (XI ISABE), Tokyo, September 19-24, 1993.

[20] R. Andriani, F. Gamma, U. Ghezzi "Effetto della rigenerazione sul comportamento fuori progetto di un turbogetto semplice", XII Congresso Nazionale AIDAA, Como, 20-23 luglio 1993.

[21] Richard T. C. Harman "Gas Turbine Engineering", The MacMillan Press LTD, 1981.

[22] William W. Bathie "Fundamentals of Gas Turbines", John Wiley & Sons, 1984.

AIAA-2000-2902

THERMAL PERFORMANCE OF LANDSAT-7 ETM+ INSTRUMENT DURING FIRST YEAR IN FLIGHT

Michael K. Choi*
NASA Goddard Space Flight Center
Greenbelt, MD 20771

ABSTRACT

Landsat-7 was successfully launched into orbit on April 15, 1999. After devoting three months to the bakeout and cool-down of the radiative cooler, and on-orbit checkout, the Enhanced Thematic Mapper Plus (ETM+) began the normal imaging phase of the mission in mid-July 1999. This paper presents the thermal performance of the ETM+ from mid-July 1999 to mid-May 2000. The flight temperatures are compared to the yellow temperature limits, and worst cold case and worst hot case flight temperature predictions in the 15-orbit mission design profile. The flight temperature predictions were generated by a thermal model, which was correlated to the observatory thermal balance test data. The yellow temperature limits were derived from the flight temperature predictions, plus some margins. The yellow limits work well in flight, so that only several minor changes to them were needed. Overall, the flight temperatures and flight temperature predictions have good agreement. Based on the ETM+ thermal vacuum qualification test, new limits on the imaging time are proposed to increase the average duty cycle, and to resolve the problems experienced by the Mission Operation Team.

INTRODUCTION

ETM+ is the instrument on the NASA Landsat-7 spacecraft, which was successfully launched into orbit on April 15, 1999. The spacecraft performs wide-area multi-spectral imaging of the Earth's land mass from a sun-synchronous near polar orbit (altitude 705 km, 98.2° inclination). Figure 1 shows the ETM+ on the spacecraft. It consists of two units: scanner and Auxiliary Electronics Module (AEM). The scanner is an advanced version of the Thematic Mapper (TM) flown on Landsat-4 and -5. The AEM is a new component and has no flight heritage. Both units are conductively isolated from the spacecraft. During the nominal imaging phase of the Landsat-7 mission, the scanner mapper aperture always points at the Earth. Figure 2 shows the scanner.

Figure 1. ETM+ on Landsat-7 Spacecraft.

The major power dissipation, 159 W measured in 1998, was in the Main Electronics Module (MEM), which consists of two power supplies (P/S) and twenty-eight printed wiring boards (PWBs). Heat dissipated by the MEM power supplies and PWBs is conducted to a white-paint radiator, which has a thermal louver. In flight, the radiator/louver is on the anti-sun side of the spacecraft, and the louver base-plate radiates heat to space. The initial louver set points were 15°C fully closed, and 25°C fully open.

---

*Associate Fellow

Prior to the ETM+ thermal vacuum test, the louver set points were reduced to 7°C fully closed, and 17°C fully open. The purpose of the set point change was to decrease the temperature of the MEM power supplies. Except for the MEM thermal louver, the opening of the scanner mapper aperture sunshade, and radiative cooler aperture, the ETM+ scanner is insulated with 7 m^2 of multi-layer insulation (MLI) thermal blankets. The remainder of the power dissipation, 65 W, is in the electronics components inside the scan cavity. Albedo and Earth infrared radiation enter the mapper aperture sunshield opening.

Except for a .0716 m^2 white paint radiator, the AEM is insulated with 2 m^2 MLI blankets. The AEM has a total power dissipation of 113 W, measured in 1998, when the instrument operates. Heat dissipated by the electronics is conducted to the white-paint radiator. In flight, the radiator is on the anti-sun side and it radiates heat to space.

The Full Aperture Calibrator (FAC) is a new component and has no flight heritage. When deployed, the FAC is in front of the external opening of the sunshield, and when stowed, it is in front of the FAC stow cover. Figure 3 shows the FAC and FAC motors.

FLIGHT TEMPERATURE PREDICTIONS FOR IMAGING

Modifications to the scanner and AEM thermal models were performed after the observatory thermal balance test in 1998 to give good agreement between the temperature predictions and the test results. The correlated ETM+ thermal model was used to obtain flight temperature predictions. Table 1 presents the worst cold case and worst hot case flight temperature predictions generated before launch.[1]

Figure 2. ETM+ Scanner.

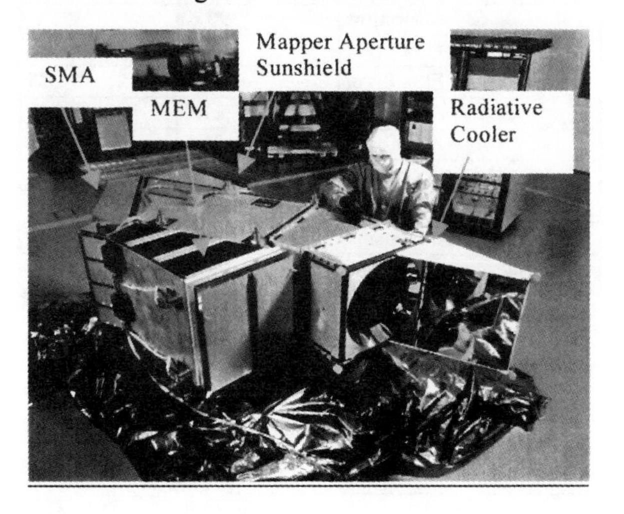

Figure 3. FAC and FAC Motors.

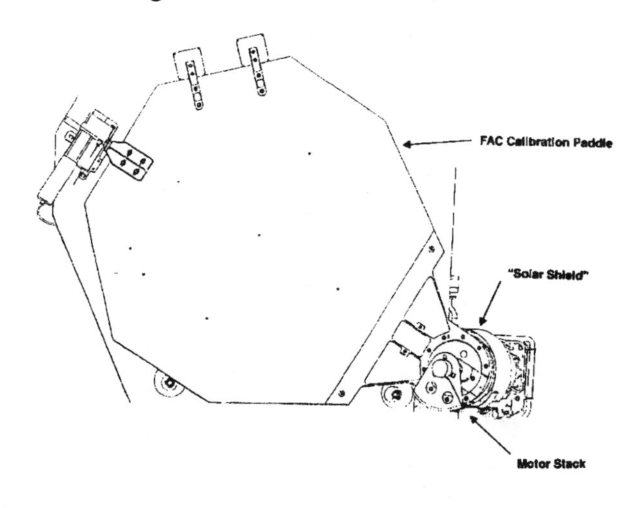

YELLOW TEMPERATURE LIMITS IN FLIGHT

The flight temperature predictions in the nominal 15-orbit mission profile, plus margins, were used as the yellow limits for most of the ETM+ components. Table 2 presents the lower and upper yellow temperature limits generated before launch.[1]

FLIGHT TEMPERTURES

After Landsat-7 was successfully launched into orbit, the first three months were devoted to the bakeout of the ETM+ scanner radiative cooler, cooler cool-down and on-orbit checkout. In mid-July 1999, the ETM+ began the normal imaging phase of the Landsat-7 mission. The flight temperatures[2] are presented in Figures 4 through 17. Note that day "0" is July 15, 1999 in these figures. The minimum temperatures are in the standby mode, and the maximum temperatures are in the imaging mode.

MEM Power Supply Heat Sink

Figure 4 presents the flight temperatures of the MEM heat sink for the power supplies. The thermal louver on the MEM radiator has new temperature set points. The flight temperatures of the MEM power supply heat sink are 13.4°C minimum and 15.9°C maximum. The flight temperature predictions were 13°C minimum and 15.9°C maximum for P/S 1, and 13°C minimum and 16.4°C maximum for P/S 2. Therefore, the flight temperatures and flight temperature predictions have excellent agreement. The louver prevents the MEM P/S heat sink temperatures from falling below 13.4°C in the nominal imaging mode in flight. The ETM+ instrument thermal vacuum test and observatory thermal vacuum test verified that the standby heater on the MEM radiator turns on only when the MEM heat sink temperature reaches 12.6°C.

Therefore, the MEM standby heater does not turn on in either the imaging mode or standby mode in the imaging phase.

Table 1. ETM+ Imaging Temperature Predictions (°C Except when Noted in K).

Component	Worst Cold Case	Worst Hot Case
MEM Heat Sink (PS #1)	13	15.9
Band 4 Post Amp	14.3	21.8
FAC Primary Motor	-1.7	23.8
AEM Heat Sink	-1.7 /5.3	14.0
Low Chan Amb PreAmp	4.7	28.4
Sunshield	21.6	27.2
Cold PreAmp (Band 7)	-8.5	1.5
Radiator Fin (+Y)	-15.4	-11.3
Baffle (Heater)	25.9	27.0
Baffle Tube	20.6	21.7
Baffle Support	19.6	20.8
MEM Heat Sink (PS #2)	13	16.4
Pan Band Post Amp	14.2	21.5
FAC Redundant Motor	7.6	22.4
High Chan Amb PreAmp	4.7	28.5
MUX 1 Electronics (Active)	3.8 /25.8	26.0
MUX 1 P/S (Active)	3.8 /25.8	25.0
Scan Ang Monitor	23.7	23.3
Cooler Amb Stage	-12.4	-9.2
Cooler Door	-65.0	-35.2
Primary Mirror Mask	11.8	14.5
Secondary Mirror Mask	18	21.5
Primary Mirror	11.8	14.5
Secondary Mirror	18	21.5
SLC Temp	14.1	24.7
SLC 1 Electronics	14.3	24.9
Cal Shutter Hub	7.2	13.9
Cal Lamp Drive	15.4	23.0
Cal Lamp Housing	8.6	14.3
CFPA Control	91.4 K	91.4 K
SiFPA	8.2	16.8
CFPA Monitor	91.4 K	91.4 K
SMA +X Flex Pivot	23.7	23.2
SMA -X Flex Pivot	23.7	23.2
SMA +Z Housing	23.1	25.0
SMA -Z Housing	23.1	25.0
SMA Electronics	24.6	27.8
SMA Torquer	23.7	23.3
Telescope Housing	14.9	18.8
Telescope Baseplate	7.6	11.5

Table 2. Yellow Temperature Limits in Flight (°C Except when Noted in K).

Component	Lower Yellow	Upper Yellow
MEM Heat Sink (PS #1)	12	21
Band 4 Post Amp	12	25
FAC Primary Motor	-4	30
AEM Heat Sink	0	20
Low Chan Amb PreAmp	5	30
Sunshield	16	27
Cold PreAmp (Band 7)	-10	4
Radiator Fin (+Y)	-19	-6
Baffle (Heater)	25	30
Baffle Tube	19	25
Baffle Support	18	25
MEM Heat Sink (PS #2)	12	21
Pan Band Post Amp	12	25
FAC Redundant Motor	-4	30
High Chan Amb PreAmp	5	30
MUX 1 Electronics	0	42
MUX 2 Electronics	0	42
MUX 1 Power Supply	0	42
MUX 2 Power Supply	0	42
Scan Ang Monitor	22	25.5
Cooler Amb Stage	-16	-5
Cooler Door	-65	-25
Primary Mirror Mask	10	17
Seconadry Mirror Mask	15	25
Primary Mirror	10	17
Secondary Mirror	15	25
SLC Temp	10	29
SLC 1 Electronics	10	29
Cal Shutter Hub	5	19
Cal Shutter Flag	5	19
Cal Lamp Drive	15	26
Cal Lamp Housing	5	19
CFPA Control	90 K	93 K
SiFPA	10**	20
CFPA Monitor	90 K	93 K
SMA +X Flex Pivot	21.5	25
SMA -X Flex Pivot	21.5	25
SMA +Z Housing	21.5	25
SMA -Z Housing	21.5	25
SMA Electronics	21.5	33
SMA Torquer	21.5	25
Telescope Housing	10	24
Telescope Baseplate	2	17

** 10°C is the telemetry saturation temperature.

Figure 4. Flight Temperatures of MEM Heat Sink.

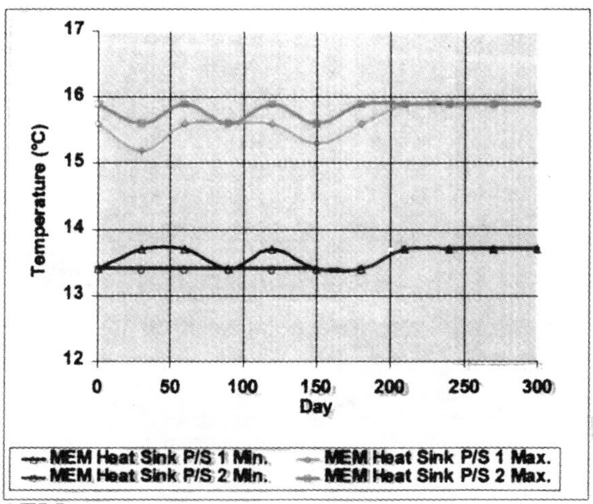

Figure 5. Flight Temperatures of AEM.

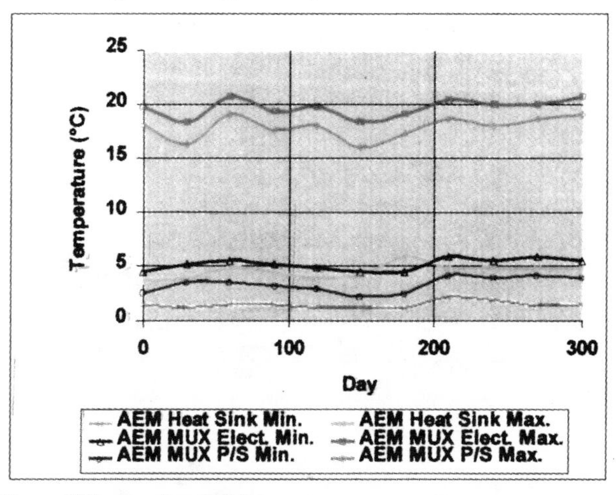

AEM

Figure 5 presents the flight temperatures of the AEM. The AEM heat sink temperature is 1.5°C minimum and 6°C maximum. It is within the yellow temperature limits of 0°C to 20°C. The flight temperature predictions were 0°C minimum in the worst cold case, and 14°C maximum in the worst hot case.

The AEM MUX 1 electronics flight temperature is 4°C minimum and 21°C maximum. It is within the yellow temperature limits of 0°C to 42°C. The flight temperature predictions were 3.8°C minimum in the worst cold case, and 26°C maximum in the worst hot case.

The AEM MUX 1 P/S flight temperature is 2.5°C minimum and 19°C maximum. It is within the yellow temperature limits of 0°C to 42°C. The flight temperature predictions were 3.8°C minimum in the worst cold case, and 25°C maximum in the worst hot case.

The explanations for the differences between the flight temperatures and flight temperature predictions are as follows. The worst cold case temperature predictions were based on the cold limiting interface temperature measured during the observatory cold thermal balance test. The worst hot case temperature predictions were based on the hot limiting interface temperature measured during the observatory hot thermal balance test. However, the interface temperature in flight is in between. Although the AEM is thermally isolated from the spacecraft by six titanium washers, the interface temperature still has an effect on the AEM temperatures. Also, the worst hot case maximum temperature prediction was based on the end of life absorptance of the white paint on the radiator. Despite the AEM is on the anti-sun side of the spacecraft, there is albedo incident on the radiator.

Scan Mirror Assembly

The quality of the ETM+ science data is dependent on the temperature of the Scan Mirror Assembly (SMA), including the mirror itself and the flexible pivots. Active control heaters on the SMA bulkhead are designed to maintain the SMA +Z Housing and –Z Housing at 24°C±0.5°C in the standby and imaging modes.

Figures 6 and 7 present the flight temperatures of the SMA. From Figure 6, the SMA +Z and –Z Housing temperatures are as low as 22°C. It occurs during the standby mode. The SMA standby heaters have insufficient heater capacities. Due to the Landsat-4 and 5 heritage design, the capacity of the standby heaters on the SMA bulkhead remains unchanged, which is 62.8 W at 28 V. However, the ETM+ has more radiative heat loss from the scan cavity to space than the Landsat-4 and 5 Thematic Mapper (TM) due to the following reasons. As mentioned earlier, the FAC is a new component on the ETM+, which has no Landsat-4 and 5 heritage. First, heat is conducted from the scanner bulkhead to the FAC motor stack and calibration paddle and is then radiated to space. Secondly, there is also a parasitic heat loss by conduction from the scanner mainframe to the FAC stow cover, despite that they are thermally isolated from each other. Thirdly, the mapper aperture of the ETM+ is slightly larger than that of the TM. Therefore, the view factor from the scan cavity to space is larger.

From Figure 6, the temperatures of the SMA +Z and –Z Housing are 23.2°C to 25.1°C in the imaging mode. This explains why the quality of the science data has not been affected. The temperatures in the imaging mode are warmer than the standby mode because the SMA Electronics box, which is mounted to the SMA bulkhead, operates in the imaging mode only.

The minimum flight temperature of the SMA is about 1°C colder than flight temperature predictions.

Figure 6. Flight Temperatures of SMA.

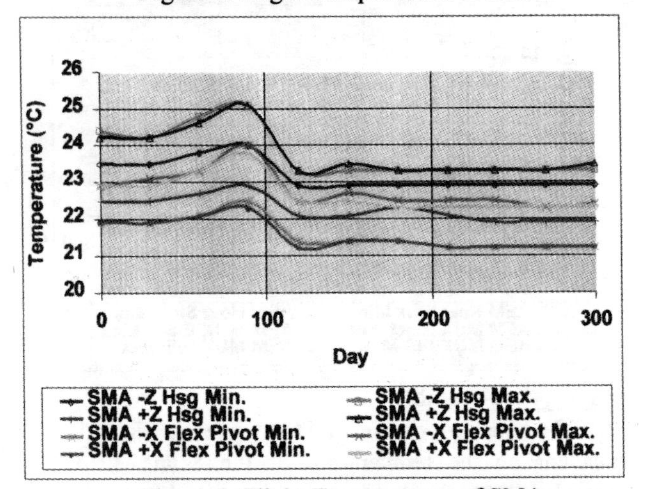

Figure 7. Flight Temperatures of SMA.

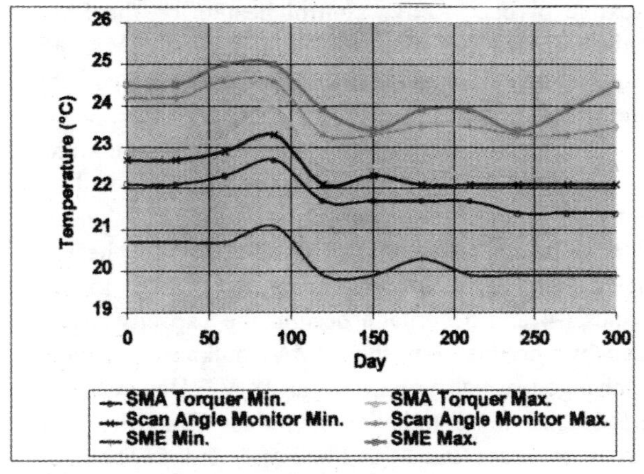

Primary and Secondary Mirrors

Figure 8 presents the flight temperatures of the primary and secondary mirrors, and the masks of these mirrors. The temperatures of the primary mirror and primary mirror mask are 11.5°C minimum and 13.2°C maximum. They are well within the yellow limits of 10°C to 17°C. The flight temperature predictions were 11.8°C minimum in the worst cold case, and 14.5°C maximum in the worst hot case. The flight temperatures and temperature predictions have good agreement.

The flight temperatures of the secondary mirror and secondary mirror mask are 15.5°C minimum and 19.5°C maximum. They are within the yellow limits of 14°C to 25°C. The flight temperature predictions were 18°C minimum in the worst cold case, and 21.5°C in the worst hot case. The flight temperatures and flight temperature predictions have good agreement.

The worst hot case temperature predictions are slightly warmer than the flight temperatures because the end of life solar absorptance was used for kapton, which is the outer cover of the MLI blankets on the ETM+ scanner.

Figure 8. Flight Temperatures of Primary and Secondary Mirrors.

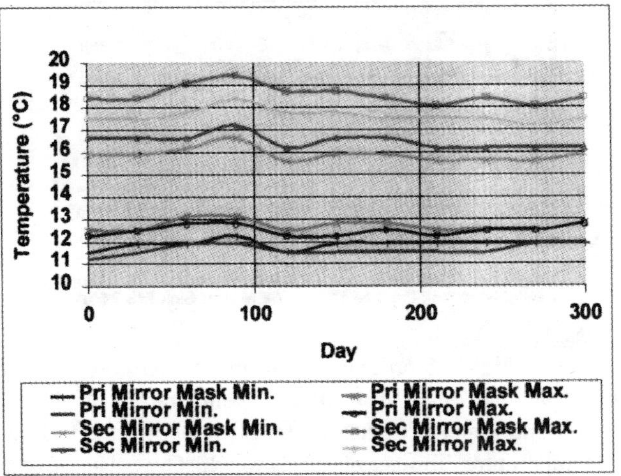

Post Amplifiers and Pre Amplifiers

Figure 9 presents the flight temperatures of the post amplifiers and ambient pre amplifiers. The temperature of the Band 4 Post Amp is 15.9°C minimum and 23.5°C maximum. It is within the yellow limits of 12°C to 25°C. The flight temperature predictions were 14.3°C minimum in the worst cold case, and 21.8°C maximum in the worst hot case. The flight temperatures and flight temperature predictions have good agreement.

The flight temperature of the Pan Band Post Amp is 15.3°C minimum and 29°C maximum. It is within the lower yellow limit of 12°C, but exceeds the upper yellow limit of 25°C. Note that the upper red limit is 55°C. The flight temperature predictions were 14.2°C minimum in the worst cold case, and 21.5°C in the worst hot case. The worst hot case temperature prediction is 7.5°C colder than the maximum flight temperature. An explanation for the difference between the maximum flight temperature and worst hot case temperature prediction is as follows. During the observatory hot thermal balance test, the maximum temperature of the Pan Ban Post Amp was 15.9°C only because the MLI on the MEM viewed the vacuum chamber walls, which were at about −180°C. In flight, the MLI is warmer due to the albedo and Earth IR radiation. On the other hand, during the ETM+ instrument thermal vacuum test, the vacuum chamber walls were at 0°C, and the maximum temperature of the Pan Band Post Amp was 30.9°C. The upper yellow limit has been increased to 31°C.

The temperatures of the High Channel PreAmp and Low Channel PreAmp are 5.9°C minimum and 25.4°C maximum. They are well within the yellow limits of 5°C to 30°C. The flight temperature predictions were 4.7°C minimum in the worst cold case, and 28.5°C maximum in the worst hot case. The flight temperatures and temperature predictions have good agreement. The worst hot case flight temperature predictions are slightly warmer than the flight temperatures because the end of life solar absorptance was used for kapton, which is the outer cover of the MLI blankets on the ETM+ scanner.

Band 7 Cold Pre Amplifier and Cooler Ambient Stage

Figure 10 presents the flight temperatures of the Band 7 Cold PreAmp and Cooler Ambient Stage. The flight temperature of the Band 7 Cold PreAmp is –6.4°C minimum and –0.4°C maximum. It is well within the yellow limits of -10°C to 4°C. The flight temperature predictions were –8.5°C minimum in the worst cold case, and 1.5°C maximum in the worst hot case. The flight temperatures and flight temperature predictions have good agreement.

The flight temperature of the Cooler Ambient Stage is –13.7°C minimum and –11.2°C maximum. It is well within the yellow limits of -16°C to -5°C. The flight temperature predictions were –12.4°C minimum in the worst cold case, and –9.2°C maximum in the worst hot case. The flight temperatures and temperature predictions have good agreement.

Silicon Focal Plane Array and Scan Line Corrector

Figure 11 presents the flight temperatures of the Silicon Focal Plane Array (SiFPA), Scan Line Corrector (SLC) and SLC Electronics. The flight temperature of the SiFPA is 10°C minimum and 14.7°C maximum. It is well within the yellow limits of 8°C to 20°C. The flight temperature predictions were 8.2°C minimum in the worst cold case, and 16.8°C maximum in the worst hot case. The flight temperatures and temperature predictions have good agreement.

The flight temperature of the SLC is 14.6°C minimum and 23.3°C maximum. It is well within the yellow limits of 10°C to 29°C. The flight temperature predictions were 14.1°C minimum in the worst cold case, and 24.7°C maximum in the worst hot case. The flight temperatures and temperature predictions have good agreement.

The flight temperature of the SLC Electronics is 14°C minimum and 22.9°C maximum. It is well within the yellow limits of 10°C to 29°C. The flight temperature predictions were 14.3°C minimum in the worst cold case, and 24.9°C. The flight temperatures and temperature predictions have good agreement.

Figure 9. Flight Temperatures of Post Amps and PreAmps.

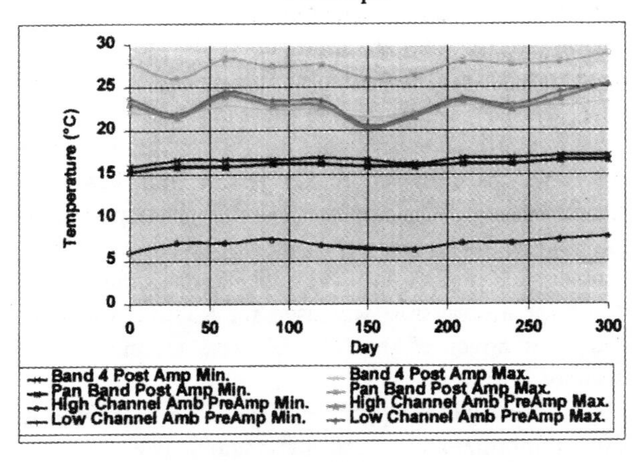

Figure 10. Flight Temperatures of Band 7 Cold PreAmp and Cooler Ambient Stage.

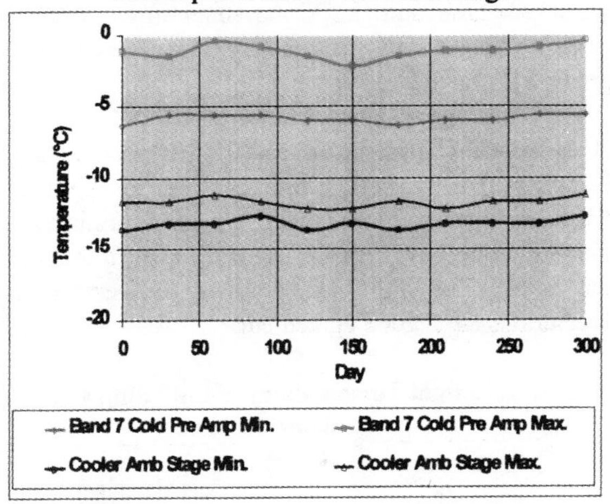

Figure 11. Flight Temperatures of SiFPA and SCL.

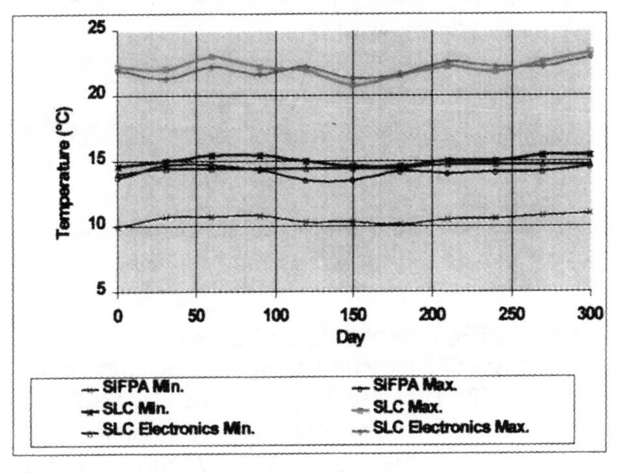

Cal Lamp and Cal Shutter

Figure 12 presents the flight temperatures of the Cal Lamp Drive, Cal Lamp Housing, and Cal Shutter Hub. The flight temperature of the Cal Lamp Drive is 14.4°C minimum and 20°C maximum. The minimum temperature exceeds the lower yellow limit of 15°C, but the maximum temperature is within the upper yellow limit of 26°C. The flight temperature predictions were 15.4°C minimum in the worst cold case, and 23°C maximum in the worst hot case. The minimum flight temperature is 1°C colder than the lower yellow limit. The lower yellow limit has been decreased to 13°C. The maximum flight temperature is 3°C colder than the worst hot case flight temperature prediction. An explanation is the end of life solar absorptance was used for Kapton, which is the outer cover of the MLI blankets on the ETM+ scanner.

The flight temperature of the Cal Shutter Hub is 10°C minimum and 12.2°C maximum. It is well within the yellow limits of 5°C to 19°C. The flight temperature predictions were 7.2°C minimum in the worst cold case, and 13.9°C maximum in the worst hot case. The flight temperatures and temperature predictions have good agreement.

The flight temperature of the Cal Lamp Housing is 9.7°C minimum and 12.8°C maximum. It is well within the yellow limits of 5°C to 19°C. The flight temperature predictions were 8.6°C minimum in the worst cold case, and 14.4°C maximum in the worst hot case. The flight temperatures and temperature predictions have good agreement.

Figure 12. Flight Temperatures of Cal Lamp and Cal Shutter.

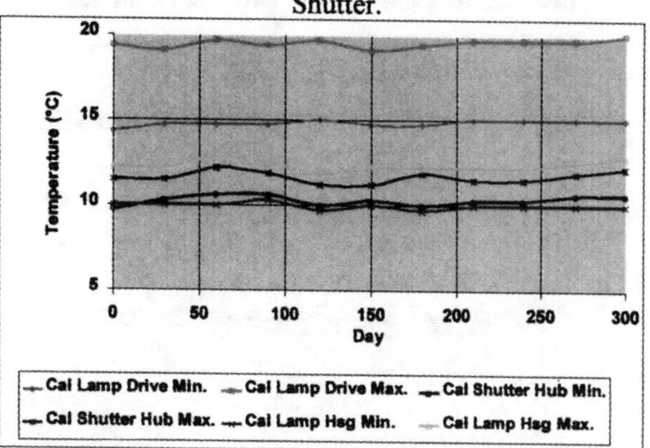

Baffle, Baffle Tube and Baffle Support

Figure 13 presents the flight temperatures of the Baffle, Baffle Tube and Baffle Support. The Baffle has active control heaters. The flight temperature of the Baffle is 27.3°C minimum and 28.3°C maximum. It is well within the yellow limits of 25°C to 30°C. The flight temperature predictions were 25.9°C minimum in the worst cold case, and 27.0°C maximum in the worst hot case. The flight temperatures and temperature predictions have good agreement.

The flight temperature of the Baffle Tube is 21.9°C minimum and 22.2°C maximum. It is well within the yellow limits of 19°C to 25°C. The flight temperature predictions were 20.6°C minimum in the worst cold case, and 21.7°C maximum in the worst hot case. The flight temperatures and temperature predictions have good agreement.

The flight temperature of the Baffle Support is 22.9°C minimum and 23.5°C maximum. It is well within the yellow limits of 18°C to 25°C. The flight temperature predictions were 19.6°C minimum in the worst cold case, and 20.8°C maximum in the worst hot case. The flight temperatures and temperature predictions have good agreement.

Figure 13. Flight Temperatures of Baffle, Baffle Tube and Baffle Support.

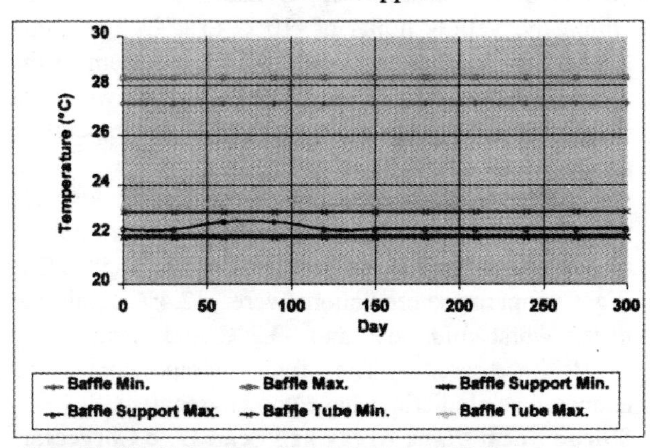

Sunshield, Telescope Baseplate and Telescope Housing

The mapper aperture sunshield prevents direct sunlight from entering the scan cavity. It is painted black on the interior and insulated with MLI on the exterior. Albedo and Earth infrared radiation enter the opening of the sunshield. Figure 14 presents the flight temperatures of the Sunshield, Telescope Baseplate and Telescope Housing. The flight temperature of the Sunshield is 21.9°C minimum and 26.1°C maximum. It is well within the yellow limits of 16°C to 27°C. The flight temperature predictions were 21.6°C minimum in the worst cold case, and 27.2°C maximum in the worst hot case. The flight temperatures and temperature predictions have good agreement.

The flight temperature of the Telescope Baseplate is 9.4°C minimum and 10.6°C maximum. It is well within the yellow limits of 2°C to 17°C. The flight temperature predictions were 7.6°C minimum in the worst cold case, and 11.5°C maximum in the worst hot case. The flight temperatures and temperature predictions have good agreement.

The flight temperature of the Telescope Housing is 15°C minimum and 16.9°C maximum. It is well within the yellow limits of 10°C to 24°C. The flight temperature predictions were 14.9°C minimum in the worst cold case, and 18.8°C maximum in the worst hot case. The flight temperatures and temperature predictions have good agreement.

Figure 14. Flight Temperatures of Sunshield, Telescope Baseplate and Telescope Housing.

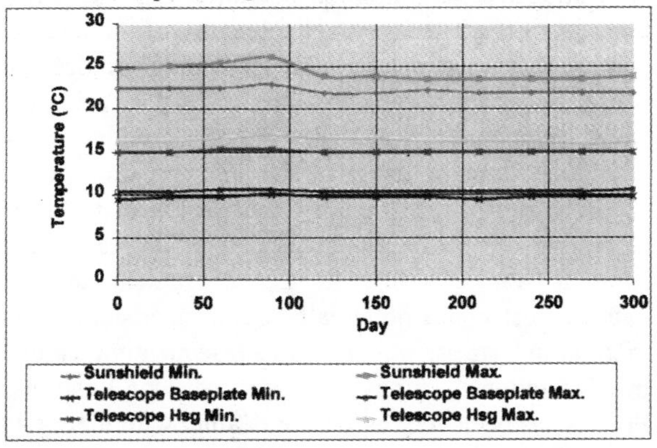

Full Aperture Calibration Motors

As mentioned earlier, the FAC is a new component and has no flight heritage. Figure 15 presents the flight temperatures of the FAC primary motor and redundant motor. The flight temperature of the FAC primary motor is 12.2°C minimum and 18.1°C maximum. It is well within the yellow limits of -4°C to 30°C. The flight temperature predictions were –1.7°C minimum in the worst cold case, and 23.8°C maximum in the worst hot case.

The flight temperature of the FAC redundant motor is 13.7°C minimum and 16.9°C maximum. It is well within the yellow limits of -4°C to 30°C. The flight temperature predictions were 7.6°C minimum in the worst cold case, and 22.4°C maximum in the worst hot case.

An explanation for the difference between the minimum flight temperature and the worst cold case temperature prediction is that the ETM+ cold limiting interface temperature in the observatory cold thermal balance test was used in the worst cold case thermal analysis. Similarly, the hot limiting interface temperature in the observatory hot thermal balance test was used in the worst hot case thermal analysis. The primary motor is mounted to the optics bulkhead, which interfaces with the spacecraft. The redundant motor is mounted on the top of the primary motor, and is thermally coupled to the FAC paddle. Both motors are external to the mapper aperture sunshield. As a result of observatory thermal balance test, a MLI blanket, with black kapton as the outer cover, was added to the "solar shield" of the motor stack at the launch site in March 1999. The MLI minimizes the heat radiation from the motor stack to space to meet the temperature requirement in the sun-pointing safehold cold case. Black kapton minimizes stray light into the scan cavity. The thermal model, with the MLI added, has not been correlated. The motors are warmer in the winter solstice than in the summer solstice because the solar flux, albedo and Earth infrared radiation are higher in the winter solstice.

Figure 15. Flight Temperatures of FAC Motors.

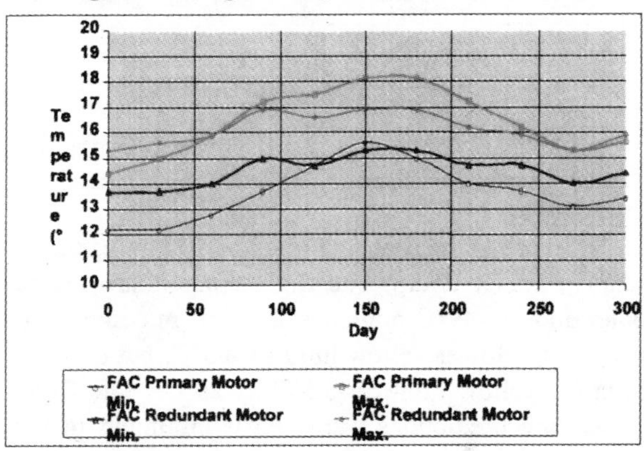

Cold Focal Plane Array

Figure 16 presents the flight temperatures of the Cold Focal Plane Array (CFPA). The flight temperature of the CFPA is 91.4 K minimum and 91.5 K maximum. It is well within the yellow limits of 90 K to 93 K. The flight temperature predictions were 91.4 K in both the worst cold case and worst hot case because the CFPA temperature is maintained at 91.4 K by active control heaters.

Figure 17 presents the CFPA flight heater current. It is 4.43 mA minimum and 5.23 mA maximum. It is significantly higher than the 2.1 mA measured during the observatory thermal vacuum test. It implies that the radiative cooler has an adequate design margin. Both the minimum and maximum currents are steady. Therefore, the parasitic heat flow to the CFPA is steady.

Figure 16. Flight Temperatures of CFPA.

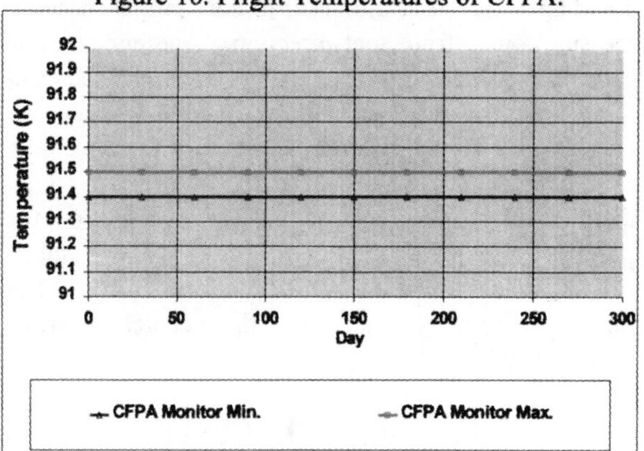

Figure 17. Flight CFPA Heater Current.

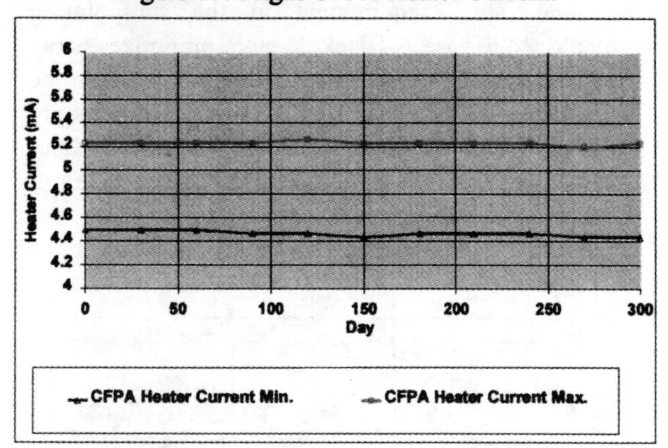

Cooler Door

Figure 18 presents the flight temperatures of the radiative cooler door. The flight temperature of the cooler door is -49°C minimum and 0°C maximum. It is within the lower yellow limit of -65°C, but exceeds the upper yellow limit of –25°C by 25°C. The flight temperature predictions were –65°C minimum in the worst cold case, and -35°C maximum in the worst hot case. The thermal coating on the Earth facing side of the cooler door is white paint, and that on the reverse side is polished aluminum. The 0°C maximum temperature in flight is significantly warmer than the maximum flight temperature prediction. An explanation for this difference is that direct solar radiation impinges on the polished aluminum side of the cooler door, when it is fully open.[3] Since the surface is specular, the solar radiation is reflected to space. It increases the cooler door temperature significantly because the ratio of solar absorptance to emittance for polished aluminum is larger than 3.0. The author showed that direct sunlight reaching the cooler door in the outgas position in 1999 because the ETM+ MEM radiator is smaller than that of the Landsat-4 and -5 Thematic Mapper.[3] Nearly 5 cm of the top of the radiator has been cut off. The 5-cm strip could have served as a blocker to the sunlight, because the sun comes from that direction. Another possible cause is that there is a difference between the Landsat-4 and -5 spacecraft bus, and the Landsat-7 spacecraft bus. So, the shielding from the sunlight for the cooler by the spacecraft bus is different. When the cooler door is fully open, solar radiation impinges on the polished aluminum side of the cooler door, and is reflected away from the cooler and into space. It has no significant impact on the cooler thermal performance.

IMAGING DUTY CYCLE

The 15-orbit mission design profile was used by the Landsat-7 Project to design and test the spacecraft bus and ETM+. The 15-orbit profile in the summer solstice was used for the worst hot case design and analysis of the different subsystems, such as thermal, power, etc. The instrument-level thermal vacuum qualification test of the ETM+ in summer of 1998 was also based on the 15-orbit profile in the summer solstice. What makes the thermal analysis, thermal vacuum testing, and mission operation difficult is that there is no flight temperature telemetry of the internal components, particularly the switching diodes, of MEM power supplies. Therefore, instead of using the flight temperature telemetry of the diodes as a maximum limit of imaging, the imaging time in the 15-orbit profile in the summer solstice is used. Operating the ETM+ at imaging duty cycles beyond what it was qualified in the instrument thermal vacuum test could potentially cause overheating and thermal overstress, which could subsequently lead to a mission failure. This is the reason why the Landsat-7 Project has been very careful in dealing with the safety of the ETM+. Operating the ETM+ at imaging duty cycles within what it was qualified in the instrument thermal vacuum test is a safe approach.

Figure 18. Flight Temperatures of Cooler Door.

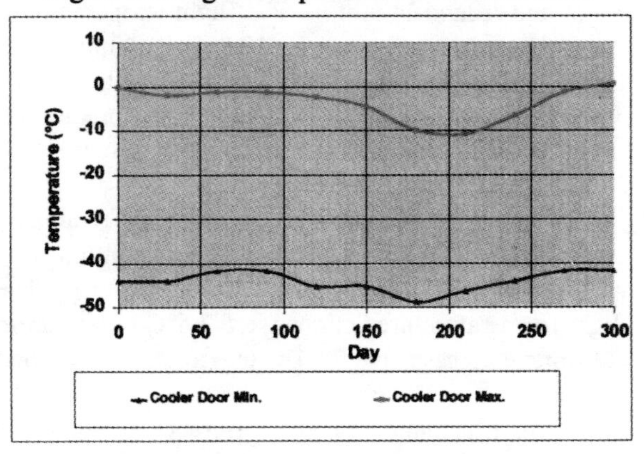

15-Orbit Mission Design Profile

Table 3 presents the 15-orbit mission design profile in the ETM+ Interface Control Document (ICD).[4] The average ETM+ duty cycle is 16.7% in the summer solstice, and is 13.5% in the winter solstice. Figure 19 shows the duty cycle versus days of the year. It is assumed to be sinusoidal.

Duty Cycle in Flight

From mid-July 1999 through April 2000, the average duty cycle in flight reported by the Landsat-7 Mission Operation team is 14%.[5] It is somewhat less than the design average duty cycle of 15%. Figure 20 presents the imaging time versus the orbit number during that period.

Table 3. ETM+ Imaging Time in 15-Orbit Design Profile (Minute).

Orbit #	Summer Solstice	Winter Solstice
1	12	8
2	7	3
3	10	6
4	11	18
5	14	13
6	24	20
7	17	13
8	10	10
9	31	29
10	19	17
11	16	15
12	17	9
13	20	15
14	22	15
15	18	9
Total	248	200
Average	16.53	13.333
Average Duty Cycle	16.7%	13.5%

Figure 19. ETM+ Design Duty Cycle vs. Days of Year.

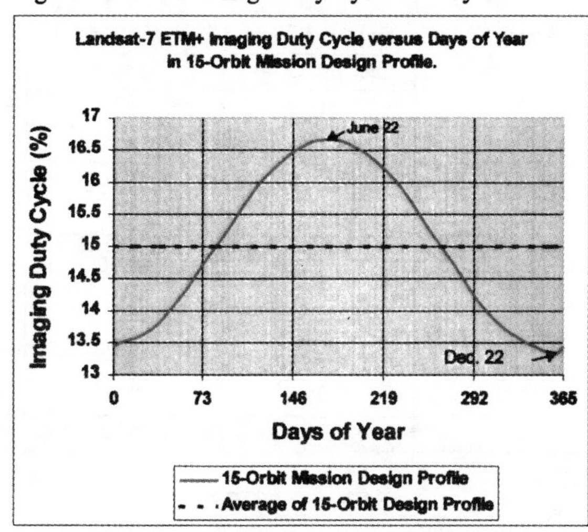

Figure 20. Imaging Time vs. Orbit Number.

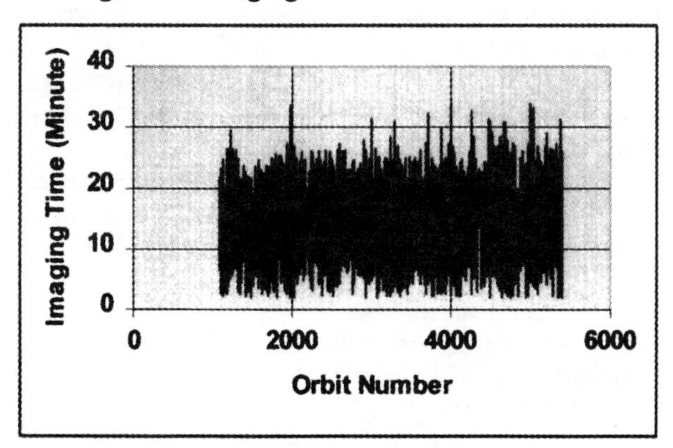

Constraints on Imaging Time

Currently, the limits of the imaging duty cycles in the ETM+ ICD[4], based on a 100-minute orbit period, are as follows:

- Short-Term: 34 minutes in any 100-minute window,
- Mid-Term (Near-Term): 52 minutes in any 200-minute window,
- Mid/Long Term (Medium-Term): 131 minutes in any 600-minute window,
- Long-Term: 230 minutes in any 1,380-minute window.

These limits are intended to prevent the MEM power supplies from overheating and thermal overstress. They were added to the ICD in June 1999 after launch. They were derived from the 15-orbit design profile in Table 3. The actual orbit period is 98.75 minutes. But, in the above limits, it was rounded off to 100 minutes.

The 15-orbit design profile in the summer solstice was used in the ETM+ instrument-level thermal vacuum qualification test in 1998. The orbit period used in the test was 99 minutes. The imaging time was 248 minutes in a 1,485 minute window, and the average imaging duty cycle in this window was 16.7%. In addition to the 15-orbit design profile, the ETM+ was also tested with 34 minutes of imaging in a 99-minute window in the instrument thermal vacuum test. This explains why the Short-Term imaging time limit in the ICD is 34 minutes, despite that the longest imaging time in the 15-orbit design profile is 31 minutes. The Mid-Term limit was derived from orbits 9 and 10, which have the longest total imaging time of any 2 consecutive orbits. It is 2 minutes higher than the total imaging time of orbits 9 and 10. An explanation is that an imaging time of 34 minutes for orbit 9 was used in deriving the Mid-Term limit. The Mid/Long Term limit was derived from orbits 9 through 14, which have the longest total imaging time of any 6 consecutive orbits. The Long-Term limit was derived from the 15 orbits. The percent duty cycle is 16.7%. But, the window is 1.2 orbits less than the 15-orbits. The Mission Operation Team reported that some of these constraints have caused interruptions during imaging.[5]

Based on the 15-orbit design profile and orbit period used in the ETM+ instrument-level thermal vacuum test, the limits of the imaging times in the ICD should be:

- Short-Term: 34 minutes in any 99-minute window,
- Mid-Term (Near-Term): 50 minutes in any 198 minute window,
- Mid/Long Term (Medium-Term): 125 minutes in any 594-minute window,

• Long-Term: 248 minutes in any 1,485-minute window.

Proposed Changes to Constraints on Imaging Time

In the 15-orbit design profile used in the ETM+ thermal-level vacuum test to qualify the instrument, orbit 9 has a 31-minute imaging, and orbit 10 has a 19-minute imaging. The Mid-Term limit was derived from these two orbits. It protects the MEM power supplies from overheating in the second orbit. Currently, the Mid-Term limit does not depend on the imaging time of the first orbit over a 2-orbit window. If the imaging time in the first orbit is less than 31 minutes, not only less thermal energy is stored in the MEM power supplies, but also the cooling time is longer. So, the total imaging time over the 198-minute window can be increased. If the imaging time in the first orbit is more than 31 minutes, not only more thermal energy is stored in the MEM power supplies, and but also the cooling time is shorter. So, the total imaging time over the 198-minute window should be decreased. Table 4 presents the proposed Mid-Term limit to increase the duty cycle and to relieve the problem of imaging interruptions experienced by the Mission Operation Team.

Also, the Mission Operation Team desires to extend the window of the Long-Term imaging limit to the daily stored command load of 40 hours. Suppose consecutive 15-orbits were run in the ETM+ thermal vacuum test, the lowest total imaging time in any 24 consecutive orbits (39.6 hours) is 384 minutes, and it is thermally acceptable to change the Long-Term limit to 384 minutes of imaging in a 39.6-hour window. The ETM+ duty cycle is 16.2% in this Long-Term limit. Therefore, the imaging limit should be 388.8 minutes in any 40-hour window. However, to maintain the ICD limit of 16.7% over 40 hours, the imaging time needs to be 400.8 minutes. After the Short-Term, Mid-Term, and Medium Term imaging limits are already satisfied, the risk of overheating the MEM power supplies by adding 12 minutes of imaging over 40 hours (24 orbits) is low. Therefore, a Long-Term imaging limit of 400.8 minutes in any 40-hour window should not be a problem thermally.

SUMMARY AND CONCLUSIONS

The Landsat-7 ETM+ instrument began the normal imaging phase of the mission in mid-July 1999. The thermal performance of the ETM+ from mid-July 1999 to mid-May 2000 is nominal. The yellow temperature limits were derived from the flight temperature predictions by a thermal model correlated to the observatory thermal balance test, plus some margins. The yellow limits work well in flight, so that only several minor changes to them were needed. Overall, the flight temperatures and flight temperature predictions have good agreement. Imaging time limits were added to the ICD after launch to protect the MEM power supplies from overheating. The Mid-Term limit was derived from orbits 9 and 10 of the 15-orbit mission design profile. Currently, the Mid-Term limit does not depend on the imaging time of the first orbit over a 2-orbit window. If the imaging time in the first orbit is less than 31 minutes, not only less thermal energy is stored in the MEM power supplies, but also the cooling time is longer. So, the total imaging time over the 198-minute window can be increased. If the imaging time in the first orbit is more than 31 minutes, not only more thermal energy is stored in the MEM power supplies, but also the cooling time is shorter. So, the total imaging time over the 198-minute window should be decreased.

Also, the Mission Operation Team desires to extend the window of the Long-Term imaging limit to the daily stored command load of 40 hours. Based on the 15-orbit mission design profile used in the ETM+ instrument-level thermal vacuum test, a new Long-Term imaging limit of 400.8 minutes in any 40-hour window should not be a problem thermally.

Table 4. Proposed Mid-Term Limit (Minutes).

First Orbit	Second Orbit	Total
34	13	47
33	15	48
32	17	49
31	19	50
30	21	51
29	23	52
28	25	53
27	27	54
26	29	55
≤25	31	≤56

REFERENCES

1. Choi, M. K., "Significance of Landsat-7 Spacecraft Level Thermal Balance and Thermal Vacuum Test for ETM+ Instrument ", SAE Paper Series 1999-01-2676, 34th IECEC, Vancouver, B.C., Aug. 1999.

2. Landsat-7 Mission Operation Center, Landsat-7 Flight Telemetry Temperature Data, July 1999-May 2000, Goddard Space Flight Center, Greenbelt, MD.

3. Choi, M. K., "Solution for Direct Solar Impingement Problem on Landsat-7 ETM+ Cooler Door During Cooler Outgas in Flight ", SAE Paper Series 1999-01-2677, 34th IECEC, Vancouver, B.C., Aug. 1999.

4. ETM+ ICD, Landsat-7 Document 430-L-0002-J, Goddard Space Flight Center, Greenbelt, MD.

5. Guit, W., Re: Notes from ETM+ Duty Cycle Meeting, e-mail to M. Choi, Mar. 16, 2000.

AIAA-2000-2904

THERMAL CONSIDERATIONS OF SPACE SOLAR POWER CONCEPTS WITH 3.5 GW RF OUTPUT

Michael K. Choi*
NASA Goddard Space Flight Center
Greenbelt, MD 20771

ABSTRACT

This paper presents the thermal challenge of the Space Solar Power (SSP) design concepts with a 3.5 GW radio-frequency (RF) output. High efficiency klystrons are thermally more favored than solid state (butterstick) to convert direct current (DC) electricity to radio-frequency (RF) energy at the transmitters in these concepts. Using klystrons, the heat dissipation is 0.72 GW. Using solid state, the heat dissipation is 2.33 GW. The heat dissipation of the klystrons is 85% at 500°C, 10% at 300°C, and 5% at 125°C. All the heat dissipation of the solid state is at 100°C. Using klystrons, the radiator area is 74,470 m^2. Using solid state, the radiator area is 2,362,160 m^2. Space constructable heat pipe radiators are assumed in the thermal analysis. Also, to make the SSP concepts feasible, the mass of the heat transport system must be minimized. Thus, the heat transport distance from the transmitters to the radiators must be minimized. It can be accomplished by dividing the radiator into a cluster of small radiators. The area of each small radiator is on the order of 1 m^2. Two concepts for accommodating a cluster of small radiators are presented. If the distance between the transmitters and radiators is 1.5 m or less, constant conductance heat pipes (CCHPs) are acceptable for heat transport. If the distance exceeds 1.5 m, loop heat pipes (LHPs) are needed.

INTRODUCTION

The concept of a Space Solar Power Station (SSPS) using a large photovoltaic solar array in geosynchronous (GEO) orbit around the Earth and transmitting this power to Earth was first proposed by Glaser in 1968.[1] The SSPS in GEO orbit is stationary with respect to a desired location on Earth. The solar array in geosynchronous orbit could receive up to 15 times as much solar radiation as that on the ground.[2]

Solar radiation is available in the GEO orbit almost all around the equinoxes when the satellite is shadowed by the Earth for up to 72 minutes per day.[3] Solar arrays covert solar energy to electricity, which is fed to microwave generators within the transmitting antenna (transmitter) between the two solar arrays. Then the transmitter transmits the RF energy to a receiving and rectifying antenna, known as "rectenna", on Earth, where the energy is reconverted into electricity.

As part of the SSP Exploratory Research and Technology (SERT) program at the National Aeronautics and Space Administration (NASA), five system design concepts were evaluated in 1999 and early 2000. They were:

- Dual RF Reflector using klystron transmitter concept.[4]
- Dual RF Reflector using solid state (butterstick) transmitter concept.[4]
- Single RF Reflector using klystron transmitter concept.[5]
- Single RF Reflector using solid state (butterstick) transmitter concept.[5]
- Integrated Symmetrical concept using klystron transmitter.[6,7]

The mission life of the SSP concepts is 40 years. Figure 1 shows the Dual RF Reflector concept. The abacus structural frame provides runs for power management and distribution (PMAD) cabling and permits a "plug and play" solar array approach for assembly and maintenance. The concentrated solar arrays always face the Sun with very little shadowing. The solar concentrator uses shifting lens to accommodate seasonal beta-tracking, and eliminates rotational joints between the cells and the abacus frame. The reflector design eliminates massive rotary joint and slip rings of the 1979 concept. The triangular truss structure provides a reasonable aspect ratio for the abacus frame. The solar concentrators convert solar energy to electricity. The DC is converted to RF energy. Two transmitters transmit a total of 3.5 GW of RF to a "rectenna" on Earth. The diameter of each transmitter is 500 m.

---

*Associate Fellow

Figure 2 shows the single RF Reflector concept. One transmitter transmits 3.5 GW of RF to a "rectenna" on Earth. The diameter of the transmitter is 500 m.

Figure 3 shows the year 1999 version of the Integrated Symmetrical concept. In this concept, 48 primary mirrors of 680-m diameter focus solar radiation to two solar arrays. The total solar energy collected is 13.38 GW. The concentration ratio is 3.8:1. The solar arrays generate 4.41 GW of electricity, which is converted to RF energy by klystrons on the transmitter. The diameter of the each solar array is 1.55 km. The diameter of the transmitter is 1 km. In the year 2000 version (see Figure 4), there are 84 primary mirrors of 500-m diameter. The diameter of each solar array is 1.1 km. The diameter of the transmitter is 500 m.

The thermal analysis in this paper is meant to show the order of magnitude type differences, and is not exact.

RADIATOR SIZING FOR WASTE HEAT FROM CONVERTING DC TO RF

Sizing of radiators in the SSP concepts using klystrons and solid state (butterstick) are presented below.

Concepts using klystrons

Using high efficiency klystrons, the efficiency of converting DC electricity to RF is 83%. Therefore, with a 3.5 GW RF output, the DC input is 4.22 GW, and the heat dissipation is 0.72 GW. A breakdown of the heat dissipation of the klystrons at the transmitters is as follows:

- 85% at 500°C.
- 10% at 300°C.
- 5% at 125°C.

Due to the three very different temperatures, three different radiators are needed for each transmitter. A space constructable heat pipe radiator[8] is selected in this analysis because it is the most developed concept, and is a proven and reliable technology. The following assumptions are used in the thermal analysis of the radiators:

- One side of the radiators is on the anti-sun side and radiates to space.
- In the GEO orbit, albedo and Earth infrared radiation are neglected.
- Emittance of the radiator, which has white paint as the coating, is 0.9.
- View factor of radiator to space is 1.0.
- The backside of the radiator is radiatively isolated from the SSP bus, which is at room temperature.

The radiators sizes required are as follows:

- 33,440 m^2 at 500°C.
- 13,030 m^2 at 300°C.
- 28,000 m^2 at 125°C.

The total size of the radiators is 74,470 m^2.

Concepts using Solid State (Butterstick)

In the solid state (butterstick) design concept, the efficiency of converting DC electricity to RF is only 60%. Therefore, with a 3.5 GW RF output, the DC input is 5.83 GW, and the heat dissipation is 2.33 GW. The temperature of the solid state electronics at the transmitter is 100°C. The radiator area required is 2,362,200 m^2, which is 6 times the total area of the two transmitters.

Klystron is more favored than solid state for converting DC to RF at the transmitters because the radiator area is 32 times smaller. If both sides of the radiator radiate heat to space, the radiator area is halved.

EFFECT OF HEAT TRANSPORT DISTANCE BETWEEN RADIATORS AND TRANSMITTERS

The mass of the heat transport system from the transmitters to the radiators in the SSP design concepts has a very large impact on the feasibility of these concepts. It is true no matter what heat rejection technology is used. The reason is that heat must be transported to an acceptable location before it can be rejected to space. The heat rejection technologies include space-constructable heat pipe radiator, rotating film radiator, rotating solid radiator, liquid droplet radiator, moving belt radiator, Curie Point radiator, and variable surface area radiator. These advanced radiator technologies were proposed in the 1970s-1980s. An assessment of these advanced radiator concepts was made by Juhasz and Peterson.[8]

LHPs have high heat transport capacities.[9,10,11] They can be used to transport the waste heat from the transmitters to the radiators. The working fluids for the LHPs in the SSP concepts using klystron transmitters are assumed to be the following in the thermal analysis:

- Potassium at 500°C.
- Thermex at 300°C.
- Methanol at 125°C.

The diameter of the LHPs is assumed to be 2.54 cm, and its wall thickness is assumed to be 0.09 cm. The LHPs for 500°C and 300°C are made of stainless steel. The LHPs at 125°C are made of aluminum. Each LHP has a single evaporator and transports 2 kW of heat from the transmitters to the radiators. Table 1 shows the distribution of the mass. The mass of the heat transport system per unit RF output per unit distance between the transmitter and radiator is 197 metric tons/GW/m. A 33% redundancy is included.

Figure 1. Dual RF Reflector Concept.

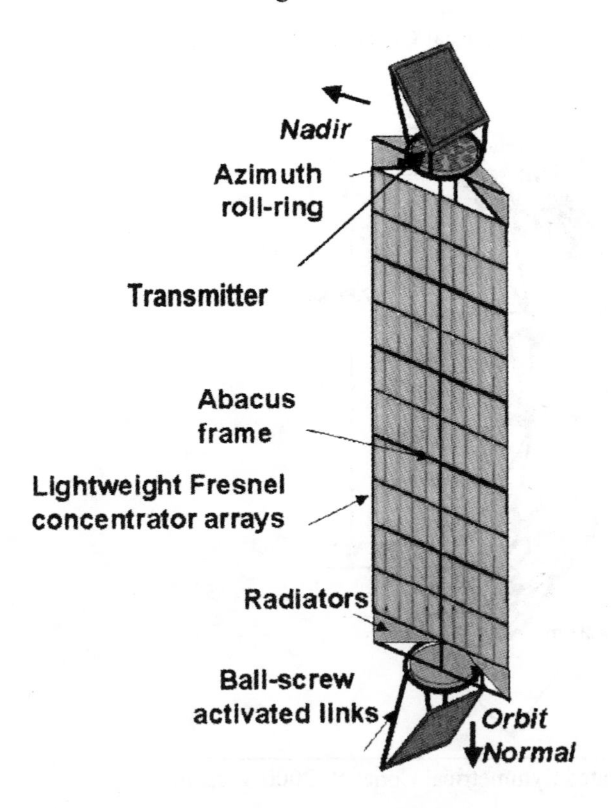

Figure 2. Single RF Reflector Concept.

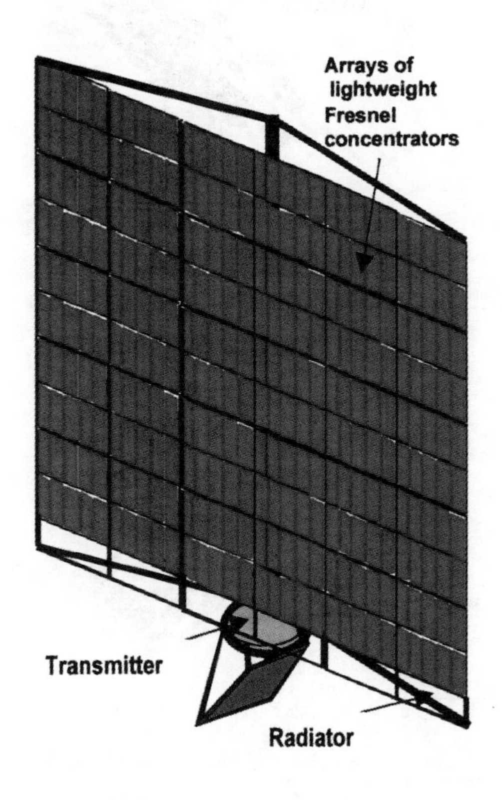

Figure 3. Integrated Symmetrical Concept (1999 Version).

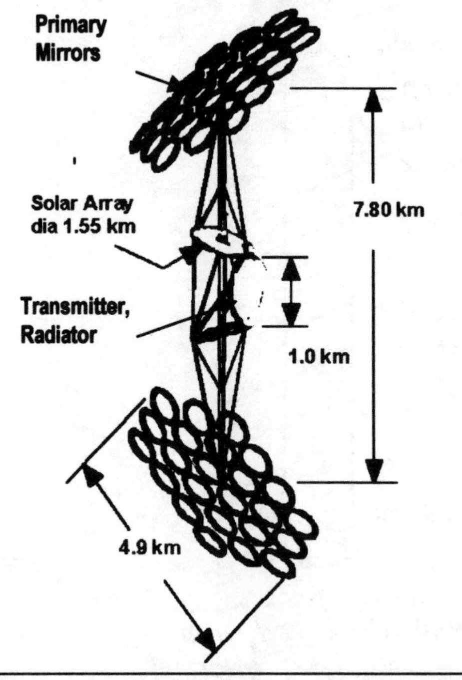

Figure 4. Integrated Symmetrical Concept (2000 Version).

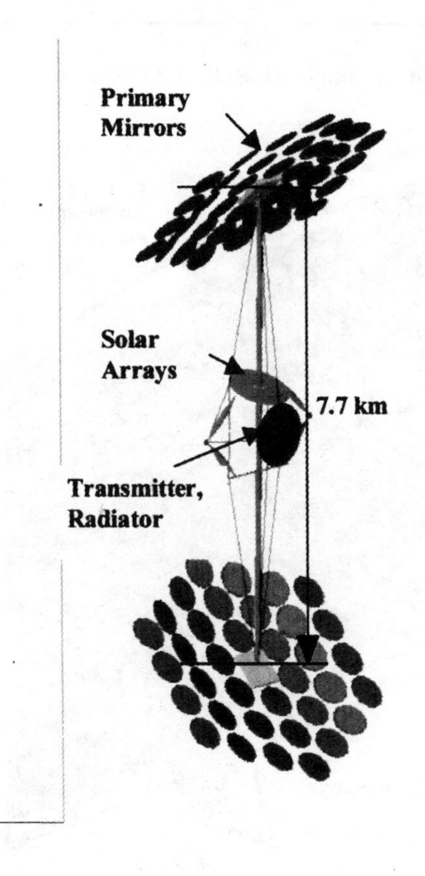

In the solid state (butterstick) concept, the working fluid for the LHPs is assumed to be ammonia, because the operating temperature of the electronics is 100°C. The diameter of the LHPs is assumed to be 2.54 cm, and its wall thickness is assumed to be 0.09 cm. The LHPs are made of aluminum. Each LHP transports 2 kW of heat from the transmitters to the radiators. Table 1 shows the distribution of the mass. The mass of the heat transport system per unit RF output per unit distance between the transmitter and radiator is 255 metric tons/GW/m. A 33% redundancy is included.

The baseline RF output of the transmitters is 3.5 GW. In the dual RF reflector concept, the radiator plane is perpendicular to the transmitter plane, and there is a large radiator for each of the two transmitters (see Figure 1). Also, the diameter of the transmitters is 500 m. To transport heat from some of the klystrons across the transmitters to the radiators, the distance is over 250 m. Suppose the average transport distance is 150 m, then the mass of the heat transport system is 103,425 metric tons. It is larger than the mass of all the other components combined in the SSP concept. In the solid state (butterstick) concept, the mass of the heat transport system is 133,875 metric tons. In the Integrated Symmetrical concept, the radiator is in the same plane as the transmitter. It has the same problem of heat transport distance as in the Dual RF Reflector concept.

To make the mass of the SSP design concepts acceptable, the distance between the transmitters and radiators must be minimized.

PROPOSED RADIATOR CONCEPTS

As mentioned earlier, several advanced radiator technologies were proposed in the 1970s-1980s. Space constructable heat pipe radiator is selected in the present thermal analysis because it is the most developed concept, and is a proven and reliable technology. Recently, a lightweight carbon-carbon composite prototype radiator was successfully fabricated and tested at NASA Glenn Research Center.[13] The design operating temperature is 700 to 800 K.

It was also mentioned earlier that for a 3.5 GW RF output, the total heat pipe radiator area required for the SSP concepts using klystrons is 74,470 m^2. In the Dual RF Reflector concept, the radiator area required for each transmitter is 37,235 m^2, which is 19% of the area of the 500-m diameter transmitter. In the Single RF Reflector concept, there is only one transmitter and therefore the radiator area is 38% of the transmitter. The front face of the transmitters transmits RF, and the heat dissipation of the klystrons is on the rear face. The klystrons are scattered on the transmitters. Therefore, locating the radiators at the rear face of the transmitters alone is insufficient to reduce the mass of the heat transport system to within budget. The radiator must be divided into a cluster of small radiators, so that the heat transport distances between the klystrons and radiators are minimized. The area of each small radiator is on the order of 1 m^2.

Figure 5 presents the first concept for the location of a cluster of small radiators at the rear surface of the transmitters. All the radiators are 1.5 m from the transmitter. There is sufficient room for the DC distribution lines and solar array support structure, and for the robots to service the transmitters. The size of each radiator is approximately 1 m x 1 m. Only one side of the radiators views space. The back of radiators faces the transmitters. The heat transport distance is only 1% of the baseline.

Figure 6 presents the second concept. The minimum distance between the transmitter and radiators is taken to be 1 m, and the maximum distance is taken to be 10 m. There is also sufficient room for the DC distribution lines and solar array support structure, and for the robots to service the transmitters. The height of the radiators is about 0.3 m, and the area is about 1 m^2. The radiators are located at cylindrical envelopes. The diameter of the cylindrical envelope closest to the transmitter is the largest. The diameter of the cylindrical envelope farthest from the transmitter is the smallest. With a special arrangement of the radiators, not only the front of the radiators has a good view to space, but also the rear can view space at least partially. The average heat transport distance is only 4% of the baseline.

In both of these radiator configurations, each radiator is integrated with a number of LHPs or CCHPs to form a modular unit. The mission life is 40 years, and the design life of heat pipe radiators is shorter. The modular heat pipe radiator units can be replaced by robots in flight when needed.

THERMAL CONSIDERATIONS OF HEAT TRANSPORT METHOD

Three methods of transporting heat from the transmitters to the radiators are compared. They are heat straps, LHPs, and CCHPs.

Heat Straps

A thermal analysis of heat straps to transport heat from the transmitters to the radiators was performed. The results are presented in Figures 7 through 9. Figure 7 shows the total cross-section area of heat straps per unit length versus the temperature gradient between the transmitter and radiator. A comparison of copper and composite material is also made in Figure 7. The thermal conductivity of copper is 3.91 W/cm-°C, and that of composite is 4.87 W/cm-°C. As the temperature gradient increases, the cross-section area decreases. For a 40°C gradient between the transmitter and radiator, the total cross section area of copper heat straps is 45,950 m^2 per meter length, and that of composite is 36,800 m^2 per meter length.

Figure 8 shows the total mass of heat straps per unit length versus the temperature gradient between the transmitter and radiator. A comparison of copper and composite material is also made in Figure 8. The density of copper is 8.9 g/cm^3, and that of composite is 1.08 g/cm^3. For a 40°C gradient between the transmitter and radiator, the total mass of copper heat straps is 412,200 metric tons per meter length, and that of composite is 39,820 metric tons per meter length. Therefore, heat straps are not acceptable to transport as much as 0.72 GW of waste heat from the transmitters to the radiators because they are too heavy.

Figure 9 shows the effect of the radiator temperature on the radiator area required for the 500°C klystrons, which contribute 85% of the total heat dissipation. The radiator area increases when the radiator temperature decreases, that is, when the temperature gradient of the heat straps increases. If the temperature gradient between the transmitter and radiator is neglected, the radiator area required for the 500°C klystrons is 33,440 m^2. If the temperature gradient is 50°C, the radiator area increases to 43,695 m^2. The mass of the radiators increases when the radiator area increases. This further makes heat straps unacceptable.

Figure 5. First Concept of Cluster of Small Radiators.

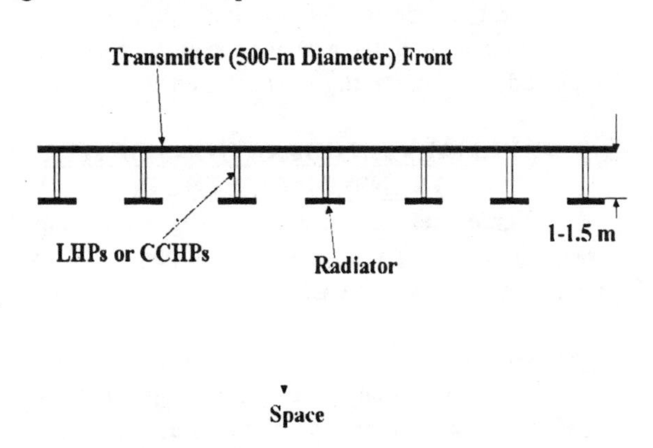

Figure 6. Second Concept of Cluster of Small Radiators.

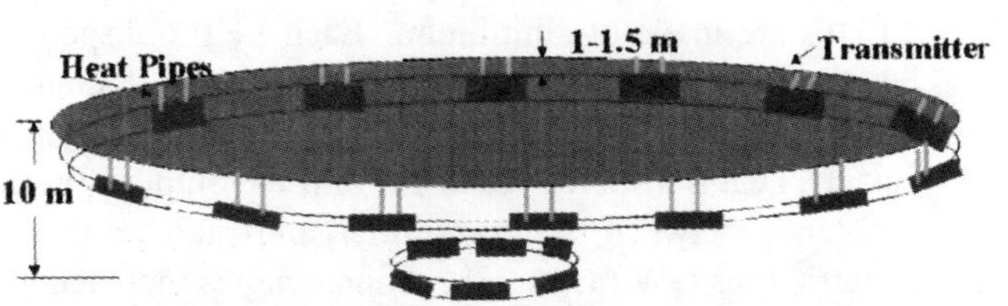

Figure 7. Cross Section Area of Heat Straps.

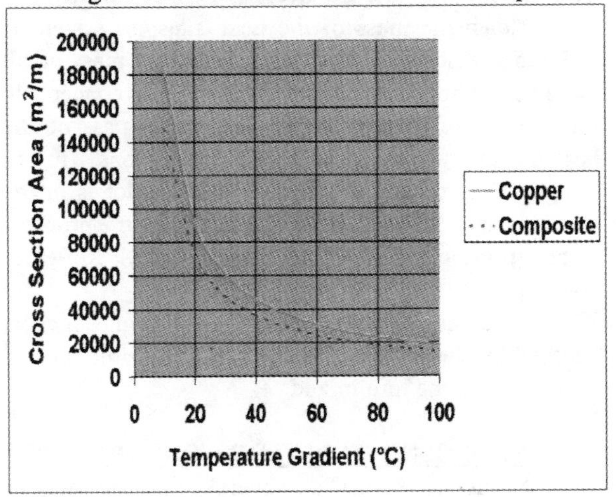

Figure 8. Mass of Heat Straps.

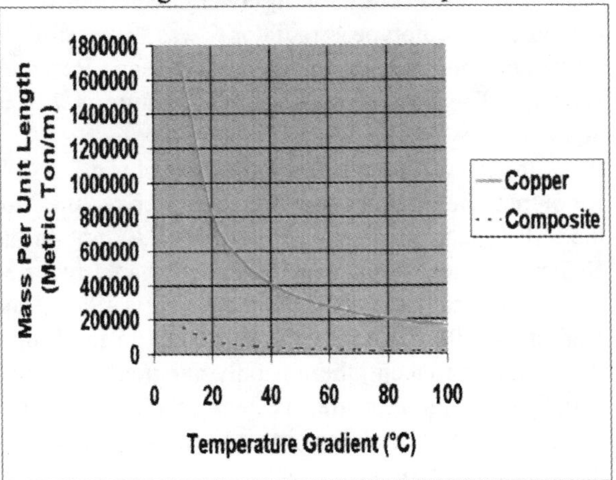

Figure 9. Radiator Area vs. Radiator Temperature for 500°C Klystrons.

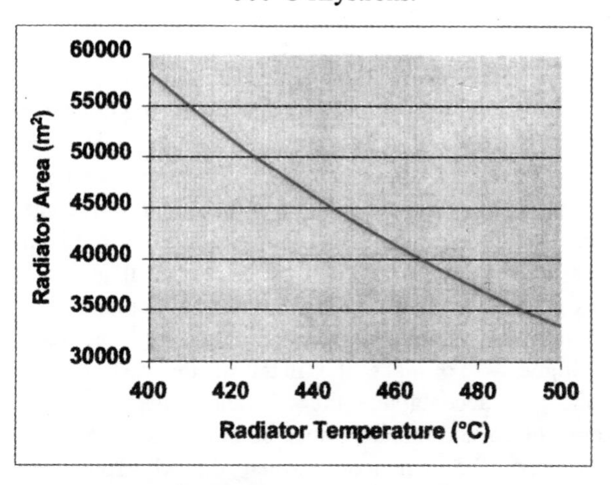

LHPs versus CCHPs

To reduce the mass of the thermal control system to an acceptable level, LHPs or CCHPs are needed. Basically, there is no temperature gradient within the LHPs or CCHPs. There will be small temperature gradients at the interfaces due to contact resistances. Table 2 below presents the mass of the thermal control system of the transmitters using LHPs in the dual RF reflector concept using klystrons. For a 3.5 GW RF output from the transmitters, the mass of the LHPs is 688 metric tons per meter distance between the transmitter and radiators. If the average distance is 5 m, then the mass of the LHPs is 3,440 metric tons. A 33% redundancy is included. The mass of the radiator and LHP condensers is 355 metric tons, and that of the evaporator is about 500 metric tons. It is assumed that the radiators are made of carbon-carbon composite. A 1.0 view factor to space is assumed for all the radiators.

Table 2 also presents the mass of the thermal control system of the transmitters using CCHPs in the dual RF reflector concept using klystrons. For a 3.5 GW RF output from the transmitters, if the distance between the transmitter and radiators is 1 m, the mass of the CCHPs is 1,423 metric tons per meter distance. The total mass of CCHPs is 1,423 metric tons. If the heat transport distance increases, the heat transport capacity of the CCHPs decreases, assuming the outer diameter of the CCHPs to be 2.54 cm (1 inch). The product of the heat transport capacity and transport distance is taken to be 1.37 kW-m. If the distance between the transmitter and radiators is 5 m, the mass of the LHPs is 7,115 metric tons per meter distance, and the total mass is 35,575 metric tons. A 33% redundancy is included.

For CCHPs to be acceptable, the distance between the transmitters and radiators is limited to about 1 m to 1.5 m. The mass of the radiator integrated with heat pipes is 355 metric tons. There is no separate mass for the evaporators.

The differences between LHPs and CCHPs are as follows. The wicks of LHPs are located at the evaporators, but the wicks of CCHPs are located along the entire length of the heat pipes. The heat transport capacity of LHPs is not dependent on the heat transport distance, but that of CCHPs is strongly dependent on the heat transport distance. During startup, based on the current technology, the entire length of the LHPs needs to be heated by heaters, for example, to melt the working fluid from solid to liquid. It would require about 30 GJ of thermal energy, or 0.8 MW of heater power for ten hours, in the system shown in Table 1. The SSP needs to provide this heater power. The heater power is needed at startup only. The heaters can be kapton film heaters bonded to the exterior of the LHPs. Heating at startup is not needed for CCHPs because there are wicks along the entire CCHPs.

Currently, the maximum operating temperature in the heat pipe technology is under 200°C (water). For LHPs to operate in the 300°C to 500°C range, liquid metal is required. This high temperature LHP technology has not been developed, but could potentially be developed for SSP by 2020.

Liquid metal CCHP technology is an existing technology. Liquid metal CCHPs were made and operated successfully as early as 1973. Recently, NASA Glenn Research Center has successfully fabricated and tested carbon-carbon composite prototype liquid potassium heat pipe radiators.[13] Liquid metal LHP technology is not an existing technology, mainly because there is no demand for it. Using the LHP technology and liquid metal heat pipe technology, it is feasible to develop liquid metal LHP technology.

Figures 10 and 11 present the mass and number of CCHPs, respectively, versus the distance between the transmitter and radiators. The product of heat transport capacity and transport distance for the CCHPs, which have a 2.54 cm (1 inch) outer diameter, is taken to be 1.37 kW-m.

The CCHPs or LHPs and radiator form an integral unit. The radiator temperature is uniform. Although the radiators are small, while compared to a single radiator of 30,000 m^2 or larger, they are still on the order of 1 m^2. Using LHPs or CCHPs, the temperature gradient between the transmitters and radiators is much less than that of heat straps, and its effect on the size of the radiators is smaller.

Figure 10. Mass of CCHPs to Tansport 0.72 GW of Waste Heat vs. Distance Between Transmitter and Radiator (Redundancy Not Included).

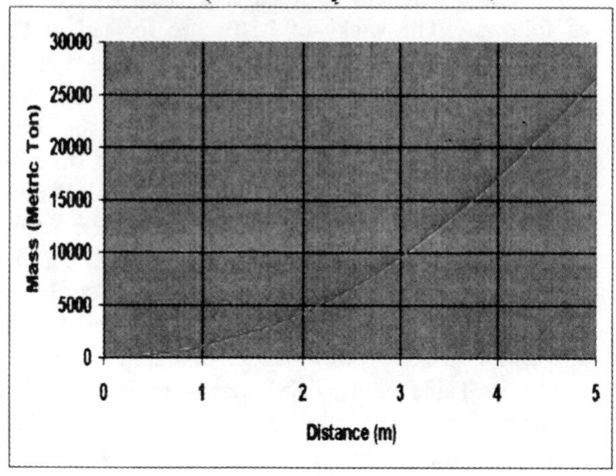

Figure 11. Number of CCHPs to Tansport 0.72 GW of Waste Heat vs. Distance Between Transmitter and Radiator (Redundancy Not Included).

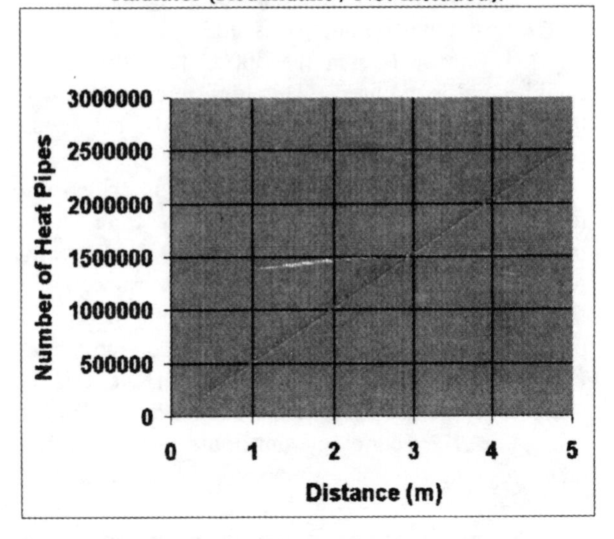

SUMMARY AND CONCLUSIONS

The thermal control system required for the solid state (butterstick) transmitter concept is much larger and heavier than that required for the klystrons concept. The reasons are as follows. First, in the klystrons concept, the efficiency of converting DC electricity to RF is 83%. In the solid state concept, the efficiency of converting DC to RF is only 60%. Therefore, the heat dissipation in the solid state concept is more than three times that in the klystron concept. Second, the klystrons operate at much higher temperatures. A breakdown of the heat dissipation at the transmitters is as follows: 85% at 500°C, 10% at 300°C, and 5% at 125°C. The solid state electronics operate at about 100°C. Since space is the heat sink for the waste heat, it is more effective to reject heat at higher temperatures.

In the baseline Dual or Single RF Reflector concept, and Integrated Symmetrical concept, there is a large radiator for the transmitter. Also, the diameter of the transmitter is 500 m. To transport heat from some of the klystrons across the transmitters to the radiators, the distance is over 250 m. If the average transport distance is 150 m, then the mass of the heat transport system is 103,200 metric tons. It is much larger than the mass of all the other SSP components combined. If solid state electronics are used, the mass is higher. To make the mass of the SSP design concepts acceptable, the distance between the transmitters and radiators must be minimized. If LHPs are used, the distance between the klystrons and radiators must be on the order of several meters. Space constructable heat pipe radiator is selected as the baseline because it is the most developed concept, and is a proven and reliable technology. The radiator area required for the klystrons concept is 74,470 m^2. Its mass is 355 metric tons. In the Dual RF Reflector design concept, the radiator area required for each transmitter is 37,235 m^2, which is 19% of that of the transmitter. In the Single RF Reflector concept and Integrated Symmetrical concept, there is only one transmitter, and the radiator area is 38% of that of the transmitter. The radiator area required for the solid state concept is 2,362,160 m^2. It is twelve times the area of that of a transmitter. Its mass is 9,921 metric tons, which is 30% to 40% of the total mass budget for the SSP. For these reasons, the SSP concepts using klystron transmitter are favored thermally. The concepts of using solid state transmitter are not practical.

The klystrons are scattered on the transmitters. Therefore, locating the radiators at the rear face of the transmitters alone is insufficient to reduce the mass of the heat transport system to within budget. The radiator must be divided into a cluster of small radiators, so that the heat transport distances between the klystrons and radiators are minimized. The area of each small radiator is on the order of 1 m^2. The envelope of the radiators is about 1.5 m from the transmitter. Another configuration of accommodating a cluster of small radiators is to locate them at cylindrical envelopes behind the transmitter. The diameter of the cylindrical envelope closest to the transmitter is the largest, and that of the cylindrical envelope farthest from the transmitter is the smallest. With a special arrangement of the radiators, not only the front of the radiators has a good view to space, but also the rear can view space at least partially. The minimum distance between the transmitter and radiators is 1 m, and the maximum distance is 10 m. The height of the radiators is about 0.3 m, and the area is about 1 m^2. If the distance between the transmitters and radiators is 1.5 m or less, CCHPs are acceptable for heat transport. If the distance exceeds 1.5 m, LHPs are needed.

REFERENCES

1. Glaser, P. E., "Power from the Sun: Its Future," Science, Vol. 162, Nov. 1968, pp.857-886.
2. Williams, J. R., "Geosynchronous Satellite Solar Power," Astronautics and Aeronautics, Dec. 1975, pp. 46-49.
3. Glaser, P. E., "Evolution of the Satellite Solar Power Station (SSPS) Concept," J. of Spacecraft, Vol. 13, Sep. 1976, pp. 573-576.
4. Carrington, C., et al., Dual RF Reflector Abacus Concept, Marshall Space Flight Center, Huntsville, AL, Oct. 1999.
5. Carrington, C., et al., Single RF Reflector Abacus Concept, Marshall Space Flight Center, Huntsville, AL, Nov. 1999.
6. Perkinson, D., et al., Integrated Symmetrical Concept, Marshall Space Flight Center, Huntsville, AL, Nov. 1999.
7. Perkinson, D., et al., Integrated Symmetrical Concept, Marshall Space Flight Center, Huntsville, AL, Feb. 2000.
8. Juhasz A. J. and Peterson, G. P., Review of Advanced Radiator Technologies for Spacecraft Power Systems and Space Thermal Control, NASA Technical Memorandum 4555, June 1994.
9. Ku, J., Operating Characteristics of LHPs, SAE Technical Paper Series 1999-01-2007.
10. Kaya, T. and Ku, J., A Parametric Study of Performance Characteristics of LHPs, SAE Technical Paper Series 1999-01-2006.
11. Kaya, T. and Ku, J., Ground Testing of LHPs for Spacecraft Thermal Control, AIAA Paper 99-3447.
12. Dunn, P. and Reay, D. A., Heat Pipes, Third Ed., Pergamon Press, 1982.
13. Juhasz, A. J., Design Considerations for Lightweight Space Radiators Based on Fabrication and Test Experience with a Carbon-Carbon Composite Prototype Heat Pipe, NASA/TP-1998-207427, Oct. 1998.

Table 1. Thermal Control Systems in SSP Concepts using Klystrons versus Solid State.

	Klystrons	Solid State (Butterstick)
RF Output from Transmitters (GW)	3.5	3.5
Efficiency of Converting DC to RF (%)	83	60
DC Input (GW)	4.2169	5.8333
Waste Heat (GW)	0.7169	2.3333
Ratio of Waste Heat to RF Output (GW/GW)	0.2048	0.6667
Distribution of Waste Heat	85% at 500°C; 10% at 300°C; 5% at 125°C	100% at ~100°C
Size of Radiator (m^2)	74,470	2,362,160
Mass of Radiator (Metric Tons) @ 4.2 kg/ m^2	197.2	9,921
Mass of Radiator Per Unit RF Output (Metric Tons/GW)	164.3	2,835
Number of LHPs @ 2 kW per Loop with 33% Redundancy	169,440	1,551,800
Material of Piping	Stainless Steel 300°C & 500°C; Aluminum 125°C	Aluminum
Working Fluid	Potassium 500°C; Thermex 300°C; Methanol 125°C	Ammonia
LHP Technology	Liquid Metal Not Current	Current
Distance Between Transmitters and Radiators (m)	100	100
Mass of LHPs @ 2 kW per Loop with 33% Redundancy (Metric Tons)	68,840	89,300
Mass of LHPs Per Unit Distance Between Transmitters and Radiators (Metric Tons/m)	688	893
Mass of LHPs Per Unit RF Output (Metric Tons/GW)	19,657	25,510
Mass of LHPs Per RF Output Per Unit Distance Between Transmitters and Radiators (Metric Tons/GW/m)	197	255
Mass of LHP Evaporator Per RF Output Per Unit Distance Between Transmitters and Radiators (Metric Tons/GW/m)	~500	~280

Table 2. Mass of LHPs versus CCHPs for Thermal Control of Transmitters using Klystrons for Concept of Cluster of Small Radiators.

	LHP	CCHP
RF Output from Transmitters (GW)	3.5	3.5
Efficiency of Converting DC to RF (%)	83	83
Waste Heat (GW)	0.7169	0.7169
Ratio of Waste Heat to RF Output (GW/GW)	0.2048	0.2048
Distribution of Waste Heat	85% at 500°C; 10% at 300°C; 5% at 125°C	85% at 500°C; 10% at 300°C; 5% at 125°C
Total Area of Cluster of Small Radiators (m^2)	84,580 (37890@500°C; 14630@300°C; 32060@125°C)	84,580 (37,890@500°C; 14630@300°C; 32060@125°C)
Mass of Radiator with Embedded Heat Pipes (Metric Tons) @ 4.2 kg/m^2	355 (159@500°C; 61@300°C; 135@125°C)	355 (159@500°C; 61@300°C; 135@125°C)
Area of Radiator Per Unit RF Output (m^2/GW)	24,170 (10828@500°C; 4182@300°C; 9160@125°C)	24,170 (10828@500°C; 4182@300°C; 9160@125°C)
Mass of Radiator Per Unit RF Output (Metric Tons/GW)	101.5 (45.5@500°C; 17.5@300°C; 38.5@125°C)	101.5 (45.5@500°C; 17.5@300°C; 38.5@125°C)
Number of Heat Pipes with 33% Redundancy	476,740 (304680@500°C; 35845@300°C; 17925@125°C)	681,050 (435260@500°C; 51210@300°C; 25600@125°C)
Material of Piping @2.54 cm Outer Diameter	Stainless Steel 300°C & 500°C; Aluminum 125°C	Stainless Steel 300°C & 500°C; Aluminum 125°C
Working Fluid	Potassium 500°C; Thermex 300°C; Methanol 125°C	Potassium 500°C; Thermex 300°C; Methanol 125°C
Heat Pipe Technology	Liquid Metal Not Current	Current, including Liquid Metal
Average Distance Between Klystrons and Radiators (m)	1**	1***
Mass of Heat Pipes with 33% Redundancy (Metric Tons)	688 (453@500°C; 52@300°C; 13@125°C)	1,423 (940@500°C; 111@300°C; 19@125°C)
Mass of per Unit Distance Between Transmitters and Radiators with 33% Redundancy (Metric Tons/m)	688 (453@500°C; 52@300°C; 13@125°C)	1,423 (940@500°C; 111@300°C; 19@125°C)
Mass of Heat Pipes Per Unit RF Output with 33% Redundancy (Metric Tons/GW)	197 (130@500°C; 14@300°C; 4@125°C)	407 (269@500°C; 32@300°C; 5@125°C)
Mass of Heat Pipes Per Unit RF Output Per Unit Distance Between Transmitters and Radiators with 33% Redundancy (Metric Tons/GW/m)	197 (130@500°C; 14@300°C; 4@125°C)	407 (269@500°C; 32@300°C; 5@125°C)
Mass of Evaporator (Metric Tons)	~500 (432@500°C; 54@300°C; 14@125°C)	Not Applicable

** Heat transport capacity not dependent on distance between transmitter and radiators.

*** Product of heat transport capacity and transport distance is 1.37 kW-m.

AIAA-2000-2905

THERMAL DESIGN TO MEET STRINGENT TEMPERATURE GRADIENT/STABILITY REQUIREMENTS OF SWIFT BAT DETECTORS

Michael K. Choi*
NASA Goddard Space Flight Center
Greenbelt, MD 20771

ABSTRACT

The Burst Alert Telescope (BAT) is an instrument on the National Aeronautics and Space Administration (NASA) SWIFT spacecraft. It is designed to detect gamma ray burst over a broad region of the sky and quickly align the telescopes on the spacecraft to the gamma ray source. The thermal requirements for the BAT detector arrays are very stringent. The maximum allowable temperature gradient of the 256 cadmium zinc telluride (CZT) detectors is 1°C. Also, the maximum allowable rate of temperature change of the ASICs of the 256 Detector Modules (DMs) is 1°C on any time scale. The total power dissipation of the DMs and Block Command & Data Handling (BCDH) is 180 W. This paper presents a thermal design that uses constant conductance heat pipes (CCHPs) to minimize the temperature gradient of the DMs, and loop heat pipes (LHPs) to transport the waste heat to the radiator. The LHPs vary the effective thermal conductance from the DMs to the radiator to minimize heater power to meet the heater power budget, and to improve the temperature stability. The DMs are cold biased, and active heater control is used to meet the temperature gradient and stability requirements.

INTRODUCTION

The BAT is one of the three telescopes on the SWIFT spacecraft. The SWIFT mission is part of the NASA Medium-Size Explorer (MIDEX) Program, and is managed by Goddard Space Flight Center (GSFC). It is scheduled to launch in 2003. The SWIFT mission is a first of its kind of multi-wavelength transient observatory for gamma ray burst astronomy. Its mission life is 3 years. The altitude of SWIFT is 600 km, and the inclination is 22°.

BAT is being developed at GSFC. There are 256 DMs in the Detector Array Plane (DAP) of the BAT instrument. Each DM consists of an 8 x 16 sub-array of CZT detectors, an ASIC with fanin, an ASIC controller, an ADC (with 4-channel mux), a biasing contact to the CZT, a transceiver to the BCDH, and the miscellaneous support electronic. The power dissipation of each DM is estimated to be 0.5779 W. Therefore, the total power dissipation of the 256 DMs is 148 W. Each DM block holds 16 DMs. The Detector Array Plate, 1.3 m x 1 m, holds all the 16 Blocks. It also provides the mounting surface and the positional stability for the Blocks. The DAP is enclosed by the graded-Z shields on the sides and coded mask at the top. Figure 1 shows the BAT design. The BCDH PWBs are mounted to the bottom facesheet of the aluminum honeycomb of the Detector Array Plate. The power dissipation of the BCDH is 32 W. Aluminum inserts conduct heat from the PWBs to the top facesheet. The current total power dissipation of the DMs and BCDH is 180 W. Figure 2 shows the CZT detector and ASIC on the DM printed wiring board (PWB).

Figure 1. BAT Configuration.

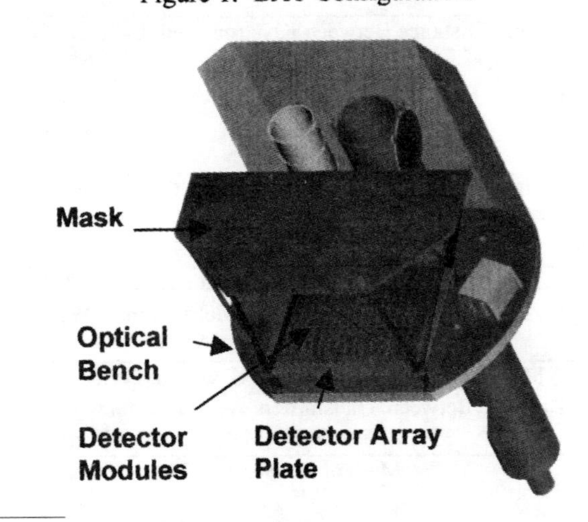

THERMAL REQUIREMENTS

The thermal requirements for the BAT detector arrays are very stringent. The maximum allowable temperature gradient of the CZT detectors of the 256 DM PWBs is 1°C. Also, the maximum allowable rate of temperature change of the ASICs of the 256 DM PWBs is 1°C on any time scale throughout the 3-year

*Associate Fellow

mission. The CZT detector works in the 0°C to 20°C range. But, a much larger radiator is needed for a 0°C CZT detector temperature than for a 20°C CZT detector temperature. Presently, a 20°C CZT detector is specified.

Figure 2. BAT DM PWB.

THERMAL DESIGN

Uncertainty Margin

Traditionally, a 5°C uncertainty margin is required for flight temperature predictions. If active heater control is used and there is sufficient heater power margin, the 5°C margin can be waived. For BAT, the flight temperature prediction of the CZT detectors must meet the 20°C±0.5°C temperature requirement, and a 5°C margin is not applicable. If active heater control is used, the heater capacity must have sufficient margin (at least 25%).

Approach

The thermal design concept of BAT to meet the stringent temperature gradient and stability requirements is as follows. The DMs are cold biased, and active heater control is used. Adequate thermal conductances between the PWBs and the DM Blocks are ensured. The DM PWBs have adequate copper planes to conduct heat from the ASICs to the Block structure. The mounting feet of each DM Block are bolted to the Detector Array Plate. Thermal gaskets are used to increase the interface thermal conductance between the Blocks and Detector Array Plate. The Blocks are made of an aluminum alloy. The Detector Array Plate is an aluminum honeycomb core with a 0.1016 cm (.04 inch) aluminum facesheet on the top and bottom. Eight CCHPs are embedded in the honeycomb core of the Detector Array. They are spaced 9.14 cm (3.6 inch) apart. The CCHPs are made of aluminum, and have a 1.27 cm (0.5 in) diameter. They interface with the top facesheet, and mounting feet of the DM Blocks are bolted to the heat pipe flanges. The working fluid is ammonia. The CCHPs have redundancy. They minimize the temperature gradient of the DMs. Presently, the temperature gradient from the DM PWB to the CCHPs is calculated to be 6°C.

Kapton heaters and electronic proportional controllers control the temperatures of the CZT detectors precisely to within ±0.5°C of the temperature requirement, and maintain the temperature stability of the ASICs. The tolerance of the controller is ±0.25°C. Temperature feedback loop is from the thermistors on the Block structure to the heater controllers. Presently, the YSI 44910 thermistor[1] is used to feed the temperature of the DM structure to the heater controller. According to YSI, the tolerance of the temperature is ±0.1°C at 20°C, and increases to ±0.2°C at –40°C. The set point of the controllers is required to be adjustable in flight such that the detector temperature can be adjusted to achieve maximum science. Each of the 16 Blocks has two redundant heater zones. Each zone has a miniature electronic proportional controller, a thermistor, and a number of kapton heaters. The heaters and electronic controllers are bonded to the interior of the Block structures using a thermally conductive epoxy.

The DAP is radiatively isolated from the coded mask at the top and graded-Z shields surrounding the BAT. Radiative isolation is achieved by adding a 0.0508 mm (2-mil) kapton, aluminized on both sides, to the interior of the coded mask and graded-Z shields, and the exterior of the DAP. The Detector Array Plate is thermally isolated from the optical bench by five titanium flexures. The exterior of the coded mask is insulated with multi-layer insulation (MLI) blankets, which have 5 inner layers. The MLI is thin enough to meet the gamma-ray attenuation requirement. The exterior of the graded-Z shields is insulated with MLI blankets, which have 18 inner layers. One possibility is to integrate the graded-Z shields into the MLI.

The location and orientation of the BAT radiator are chosen so that there is no direct solar radiation incident on the radiator for all sun angles, 45° to 180°, in the SWIFT mission. Albedo flux and Earth emitted IR radiation are the only environmental heat fluxes that affect the temperature gradient and temperature stability of the BAT detector arrays. They vary from day to night, and from summer solstice to winter solstice. Also, degradation of the thermal coating on the radiator has an impact on the temperature gradient and stability. When a coating degrades, its solar absorptance increases, and the albedo flux absorbed increases.

Radiator Thermal Coating

Albedo flux is maximum at orbit noon and it vanishes in the Earth's eclipse. Also, it is higher in the winter solstice than in the summer solstice. It is a major factor that causes the temperature instability of the BAT DM PWBs. One way to minimize the effect of albedo flux is to select a thermal coating that has a very low solar absorptance at the beginning of life (BOL), and a small degradation over a 3-year mission life for the radiator. AZ-Tek's AZW/LA-1I white paint is likely the best candidate thermal coating for the BAT radiator, based on its absorptance and emittance. The Thermal Coatings Committee at GSFC[2] recommends the following thermo-optical properties for this coating in the SWIFT mission, when it is not exposed to direct solar radiation. The solar absorptance is 0.08 and hemispherical emittance is 0.91 at BOL. The solar absorptance is 0.12 and hemispherical emittance is 0.90 at the end of life (EOL). Its high emittance provides a high radiation coupling to space. The flight data of the AZW/LA-1I white paint shows that initially the solar absorptance was 0.088, and after 48 days in space flight, it increased to 0.095, and after 197 days, it increased to 0.103.[3] They show that the degradation of this paint in space is very small. Preliminary results of an ongoing ultraviolet exposure test at GSFC show that the paint is stable. Also, the AZW/LA-1I white paint will be flying on the GSFC calorimeter on the EO-1 spacecraft, which is scheduled to launch later this year. EO-1 is an Earth orbiting spacecraft, and has an altitude of 705 km. More testing of this paint will be performed at GSFC.

Figure 3 shows the orbital transient combined albedo flux and Earth emitted IR radiation absorbed by the radiator in the hot case for different sun angles. The absorbed flux is highest at a 90° sun angle. For this reason, the worst hot case thermal analysis is performed at a sun angle of 90°.

Figure 3. Combined Albedo and Earth IR Absorbed by BAT Radiator at EOL in Winter Solstice.

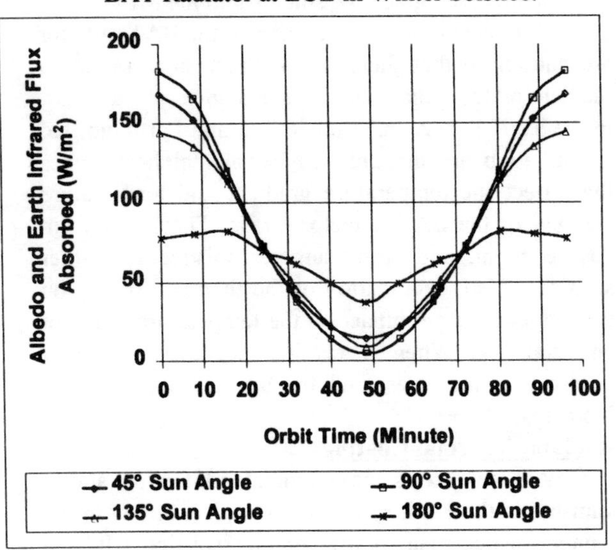

Figure 4 shows the orbital transient combined albedo flux and Earth emitted infrared (IR) radiation absorbed by the radiator in the cold case for different sun angles. The absorbed flux is lowest at a sun angle of 180°. For this reason, the worst cold case thermal analysis is performed at a sun angle of 180°.

Figure 4. Combined Albedo and Earth IR Absorbed by BAT Radiator at BOL in Summer Solstice.

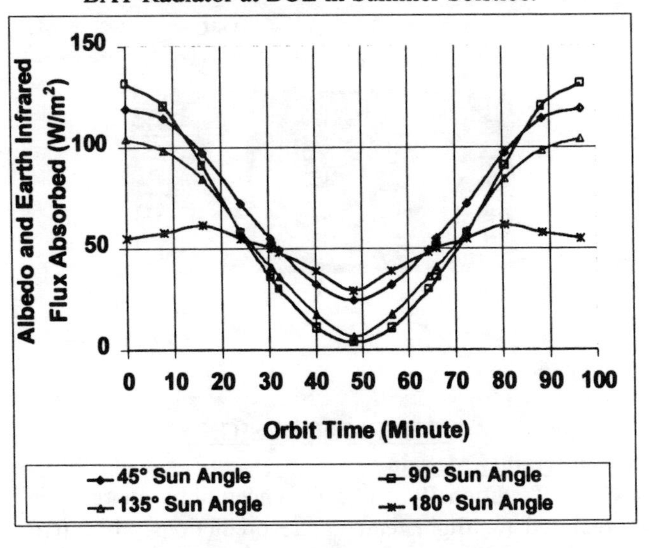

THERMAL ENERGY BALANCE

Operating Mode

For a given radiator area, the thermal energy balance of the radiator at thermal equilibrium is

$$Q_r = Q_{DAP} + Q_{Albedo} + Q_{EIR} + Q_{SC} + Q_{op\text{-}htr} \quad (1)$$

where Q_r = heat radiation from radiator to space,
Q_{DAP} = power dissipation in DAP,
Q_{Albedo} = Albedo flux absorbed,
Q_{EIR} = Earth emitted IR radiation absorbed,
Q_{SC} = heat radiation from spacecraft, particularly solar arrays, to radiator,
$Q_{op\text{-}htr}$ = heater power.

Q_{DAP} is constant (currently 180 W), and there is no duty cycle. But, Q_{Albedo}, Q_{EIR}, and Q_{SC} vary from orbit noon to Earth's eclipse, from summer solstice to winter solstice, from BOL to EOL, and from one sun angle to another. If the conduction coupling from the DAP to the radiator is constant, in order to maintain the CZT detector and ASIC temperatures constant, the radiator temperature needs to be constant, and therefore Q_r needs to be constant. As a result, $Q_{op\text{-}htr}$ needs to be varied. The value of $Q_{op\text{-}htr}$ is largest in the eclipse at BOL, and is smallest at orbit noon at EOL.

Safehold Mode

In the safehold mode, the power dissipation of the DM PWBs and BCDH PWBs is zero. The thermal energy balance of the radiator at thermal equilibrium in the safehold mode is:

$$Q_r = Q_{Albedo} + Q_{EIR} + Q_{SC} + Q_{surv\text{-}htr} \quad (2)$$

where Q_r = heat radiation from radiator to space,
Q_{Albedo} = Albedo flux absorbed,
Q_{EIR} = Earth emitted IR radiation absorbed,
Q_{SC} = heat radiation from spacecraft, particularly solar arrays, to radiator,
$Q_{surv-htr}$ = survival heater power.

Presently, the cold survival temperature limit of the BAT PWBs and BCDH PWBs is –20°C, which is 40°C colder than the operating mode temperature requirement. If the conduction coupling from the DAP to the radiator is constant, in order to maintain the BAT temperature above –20°C, the value of $Q_{surv-htr}$ in the worst cold case is very high.

USE LOOP HEAT PIPES TO MEET THERMAL REQUIREMENTS

If the conduction couplings from the DAP to the radiator can be varied, there is no need to maintain Q_r constant in Equation (1), and therefore Q_{op-htr} can be minimized in the operating mode. Also, there is no need to maintain Q_r constant in Equation (2), and therefore $Q_{surv-htr}$ can be minimized in the safehold mode. Variable conduction couplings can be obtained by using variable conductance heat pipes (VCHPs) or LHPs.

A ROM cost estimate by a heat pipe vendor shows that the cost of a VCHP system is about the same as that of a LHP system, not including the non-recurring engineering (NRE) cost. The advantages of LHPs are as follows. LHPs do not need to be horizontal to overcome the problem of gravity in ground testing. The heat transport capacity of LHPs is much higher. The diameter of LHP condensers is much smaller, and therefore they are more flexible, and can be made in a serpentine form easily. The number of LHPs needed is one-half that of VCHPs. Therefore, the number of reservoirs and active heater control for the reservoirs is one-half that of VCHPs, and the heater power for the reservoirs is significantly reduced. LHPs can have flexible lines so that the radiator panels can be swung open like a door to gain access to the DAP. The LHP technology is a reliable and proven technology.[4,5,6]

Each LHP has a compensation chamber, which is actively controlled by kapton heaters and an electronic heater controller. There are redundant heater circuits. The peak heater power is 15 W per compensation chamber. The temperature feedback to the heater controller is from a thermistor on the exterior of the compensation chamber. The set point of the controller is required to be adjustable in flight. The compensation chamber needs to be cold biased for active heater control. When the CZT detector temperature is below 20°C, the LHPs reduce the effective conductance to the radiator. When the detector temperature is above 20°C, the LHPs increase the effective conductance to the radiator. The effective conductance from the CZT detectors to the radiator is varied to maintain the LHP evaporator temperature as constant as possible. Variation of the effective conductance is accomplished by controlling the flow rate of the working fluid to modulate the heat transfer to the radiator. The heaters on the DAP act as trim heaters.

In the safehold mode, the power to the operating mode heaters is cut off. The active heater control loop is switched to the survival heater circuit. The cold survival temperature limit of the DM PWBs and BCDH PWBs is –20°C. The LHP compensation chamber heater controller set point needs to about 5°C warmer than –20°C to ensure that the BAT is safe, and the survival heater power for the DM PWBs and BCDH PWBs is minimized. It must be ensured that the LHPs can be turned off.

OPTIMIZATION OF LHP THERMAL SYSTEM

To optimize the LHP thermal system for BAT, the following factors are considered:

- Working fluid for the LHPs.
- Number of LHPs, including redundancy.
- Location of the LHPs.
- Thermal interface between the LHPs and the CCHPs embedded in the DAP.
- Cost.

Working Fluid

Presently, the power dissipation of the BAT DM PWBs and BCDH PWBs is 180 W. In the worst cold operating case, the heater power prediction to maintain the CZT detectors at 20°C is approximately 27 W. Therefore, the total waste heat to be transported by the LHPs to the radiator is about 207 W.

A working fluid for the BAT LHPs is selected to maximize the heat transport capacity and to minimize the risk of freezing. For a given heat load, which is presently 207 W, the heat transport capacity dictates the number of LHPs required and the maximum allowable heat transport distance. The minimum flight temperature prediction of the BAT radiator in the worst cold safehold case is -90°C, which is 12°C colder than the freezing point of ammonia. On the other hand, propylene has a freezing point of -185°C. The disadvantage of propylene is that its heat capacity is significantly lower than that of ammonia.

To use ammonia as the working fluid for the BAT LHPs, survival heaters and heater controllers or thermostats need to be added to the radiator and LHP lines that have no contact with the radiator. To minimize the risk of freezing, the kapton film heaters must be spread out over the entire radiator. To maintain the radiator at –70°C in the worst cold safehold case, the survival heater power needed is 52 W peak or 35 W orbital average. The peak survival heater power for the radiator alone exceeds the survival heater power budget for BAT. Additionally, survival heater power is needed to maintain the detector modules at –20°C, and the LHP reservoirs at about –15°C. Because the survival heater budget for BAT is only 50 W, it is not possible to use ammonia as the working fluid.

Another problem of using ammonia is related to the operating temperature range of the heater adhesive. The minimum temperature recommended by the heater manufacturers for the adhesive on the heater is –32°C without aluminum backing, and –54°C with aluminum backing. Even if the survival heater power budget can be increased to allow the use of ammonia, a –70°C radiator temperature is significantly colder than the minimum heater adhesive temperatures recommended by the heater manufacturers. One way to prevent the heaters from debonding is to use kapton heaters with aluminum backing and maintain the radiator above –54°C. However, it requires an additional 50 W peak survival heater power, and it is of course unacceptable. An alternative is to add aluminum plates to press the heaters against the radiator with a good contact pressure. The aluminum plates need to be bolted to the radiator. While feasible, this approach adds mass. Also, the radiator is only 0.159 cm (.0625 inch) thick, the heaters must be over the entire radiator.

Number of Loop Heat Pipes

For a given heat load and a given heat transport distance, the working fluid for the LHPs determines the number of LHPs required. To minimize the risk of a mission failure, the minimum number of LHPs is two, regardless what the working fluid is.

The BAT radiator is approximately 1.19 m wide and 0.9 m tall. If the LHP condensers are in a serpentine form across the radiator width five times, the total heat transport distance is about 5 m. For example, if two LHPs are used, each must be able to transport a minimum of 207 W over a 5-m distance. Nominally, each of the two LHPs transports 103.5 W. But, if one of them fails, the remainder needs to transport 207 W. A margin for the heat transport capacity is also needed. If one LHP, using propylene as the working fluid, is capable of transporting 207 W, plus a margin, over a 5-m distance, there is no need to consider ammonia. If propylene increases the number of LHPs by a factor of two or more, ammonia could be considered, provided the survival heater power budget for BAT can be increased and the risk of heater debonding can be minimized.

Thermal analysis by Dr. Jentung Ku[7] showed that if double condensers are used, a LHP with propylene as the working fluid achieves a heat transport capacity of approximately 300 W. Therefore, two redundant LHPs, using propylene as the working fluid, are adopted.

Location of LHPs

The compensation chambers of the LHPs need to be cold biased for active heater control. If they are outside the graded-Z shields/MLI and have a good view factor to space, they can radiate heat to space. If they are inside the graded-Z shields/MLI, they may be thermally coupled to radiators outside the graded-Z shields/MLI by heat pipes or heat straps. The disadvantages are extra mass and structural support for the radiators and heat pipes or heat straps. It is thermally less efficient because the interface area between the compensation chamber and heat pipe or heat strap is small, and the compensation chamber wall is stainless steel, which is not a good conductor. The second disadvantage of having the LHPs inside the graded-Z shields is that cutouts on the graded-Z shields/MLI are needed for penetrations of the LHP liquid lines, vapor lines, and electrical harness for the LHP heaters and heater controllers. These cutouts are potential leaks of the graded-Z shields to prevent gamma rays from entering BAT from the sides. The third disadvantage of having the LHPs inside the graded-Z shields is that if the LHPs are on the same side of BAT, not all the CCHPs have direct contact with the LHP evaporators, due to the room already used by the base of the struts. These are the reasons why the LHPs are located outside the graded-Z shields/MLI. To accommodate the LHPs outside the graded-Z shields/MLI, the Detector Array Plate needs to be extended 8.89 cm (3.5 inch) on the –Y or +Y side, or both.

With the LHPs outside the graded-Z shields, it is easy to gain access to these components for visual inspection, leak checks, adjustments or repairs, if needed. An access door is not need.

Thermal interface between LHPs and Embedded CCHPs

A trade study on having two LHPs on one side of BAT versus one LHP on each side has been performed. To accommodate two LHPs by an 8.89 cm (3.5 inch) extension of the Detector Array Plate on the –Y side, the length of each evaporator is limited to 36.12 cm (14.22 inch) in the "Flat" DM design, which is the current baseline. It is shown in Figure 5. The diameter of the evaporator is 2.54 cm (1 inch). The heat flow of about 200 W is from the +Y side of the DAP to the –Y side, and then into the LHPs across the interfaces between the LHP evaporators and embedded CCHPs. There are two header CCHPs, one on each side of the LHPs. If one LHP fails, the header CCHPs and remaining LHP can transfer all the heat from the embedded CCHPs to the LHP evaporator. The header CCHPs have a square cross section. Its size is maximized to 1.905 cm (0.75 inch) to maximize the thermal interface area with the LHP evaporators and embedded CCHPs. Figure 6 shows the LHPs and embedded CCHPs. A thermal analysis by Dr. Jentung Ku[6] showed that with a 36.12-cm (14.22-inch) evaporator length, there is sufficient heat transport capacity and pressure for the LHPs,

If there is an 8.89-cm (3.5-inch) extension of the Detector Array Plate on each of the +Y and –Y sides to accommodate one LHP on each side, the length of each evaporator can be maximized to 45.72 cm (18 inch). Figure 7 presents the layout. The 45.72-cm length is the maximum acceptable in the current technology. Adding one extension will add two header CCHPs. It increases the thermal interface area between the header CCHPs and embedded CCHPs, and reduces the temperature gradient between the embedded CCHPs and LHPs substantially. The temperature gradient is strongly dependent on the contact resistance at the interfaces. It directly affects the radiator sizing, and can be validated only in the BAT thermal balance test. Having one LHP on each of the +Y and –Y sides provides more design margins in the radiator size. It does not have any significant mechanical impact. It increases the cost by $5,000 and increases the mass by about 1.4 kg. Having one LHP on each side is thus preferred. To minimize the temperature gradient of the DMs, embedded CCHP #4 and #5, which have thermal contacts with both the LHP evaporators, will have thermal contact with only half of the LHP evaporator and one of the two header CCHPs on the -Y side.

Cost

The cost of the LHPs can be reduced by reducing the number of flight units of LHPs. Note that the cost of NRE of the LHPs and Engineering Test Unit testing is independent of the number of LHPs used for flight. ROM cost estimates by two LHP vendors revealed that the cost of a LHP flight unit is small compared to the NRE cost. Reducing the number of LHP flight units from two to one reduces the total cost by about 10%. Therefore, two redundant LHPs are adopted.

Thermal Balance Test

In the BAT instrument thermal balance test and SWIFT observatory thermal balance test, the embedded CCHPs need to be in a horizontal plane to overcome the problem of gravity. A tilt of up to 0.3 inch is acceptable. Having both LHPs on the –Y side allows the embedded CCHPs to tilt more than 0.3 inch, with the –Y side higher than the +Y side, so that the CCHPs are in the reflux mode. Having one LHP on each of the –Y and +Y sides, if the tilt cannot be maintained within 0.3 inch, the thermal system is tested in the "one LHP failure" mode. To adjust the tilt to less than 0.3 inch, an inclinometer can be placed in the vacuum chamber.

HEATER POWER REQUIREMENT

To minimize the survival heater power to within the survival heater power budget, the 8.89-cm extensions of the Detector Array Plate are insulated with MLI blankets. When the LHP compensation chambers reach –15°C in the safehold mode, they shut off the effective conductance from the evaporators to the condensers on the radiator completely. A 15 W heater power is needed for each of the two compensation chambers. It is also needed for the BAT due to radiative heat leaks from the MLI blankets, and parasitic heat conduction through the LHP lines to the radiator. Heater power is required to startup the LHPs. After launch or after the safehold mode, before the BAT instrument is turned on, the startup heaters are powered on for a short time. They are located at the evaporators. The startup heater power is estimated to be 75 W per LHP. The startup heaters for the evaporators of the LHPs need to be commanded on or off.

RADIATOR SIZING

The radiator is made of 0.15875 cm (0.0625 in) thick aluminum alloy. By detailed analysis, with the LHP condenser legs spaced 12.7 cm (5 inch) apart on the radiator, the fin efficiency is better than 96%. The area of the BAT radiator is sized to satisfy the thermal requirements of the CZT detectors in the worst hot operating case first, with a 25% heater power margin. The parameters stacked in the worst hot case are EOL thermo-optical properties, winter solstice environment, and 90° sun angle. After the radiator has been sized, the active heater capacity is sized to meet the thermal requirements in the worst cold operating case with a 25% heater power margin. With this heater capacity, the heater power margin in the worst hot case is much larger than 25%. The GSFC design values of solar constant, albedo, and Earth emitted IR radiation are used in the thermal analysis. Table 1 presents the values. The radiator size is 1.194 m x 0.902 m. A 20°C temperature gradient from the CZT detector to the radiator is assumed. It includes a gradient of 6°C from the CZT detector to the embedded CCHPs, a gradient of 9°C from the embedded CCHPs through the LHPs to the radiator, and a 5°C margin.

Figure 5. Two LHPs on One Side of BAT.

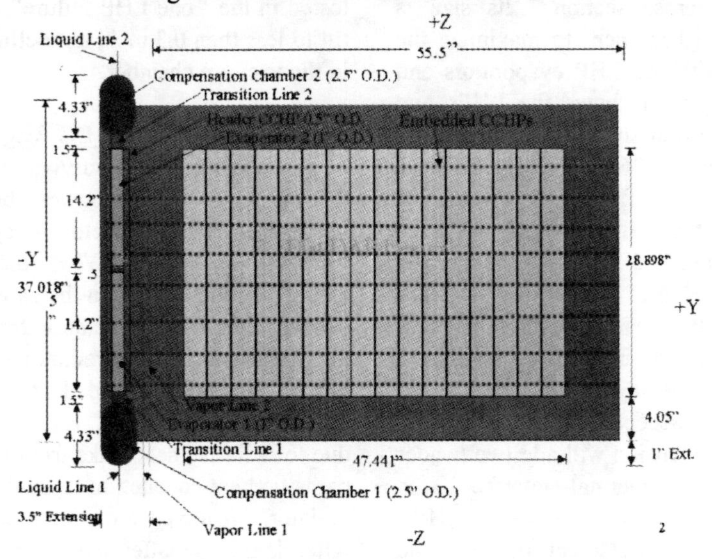

Figure 6. LHP and Embedded CCHP Configuration.

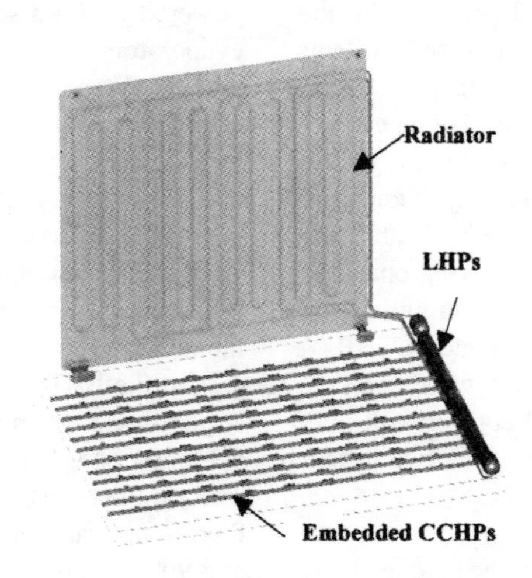

Figure 7. One LHP on Each Side of BAT.

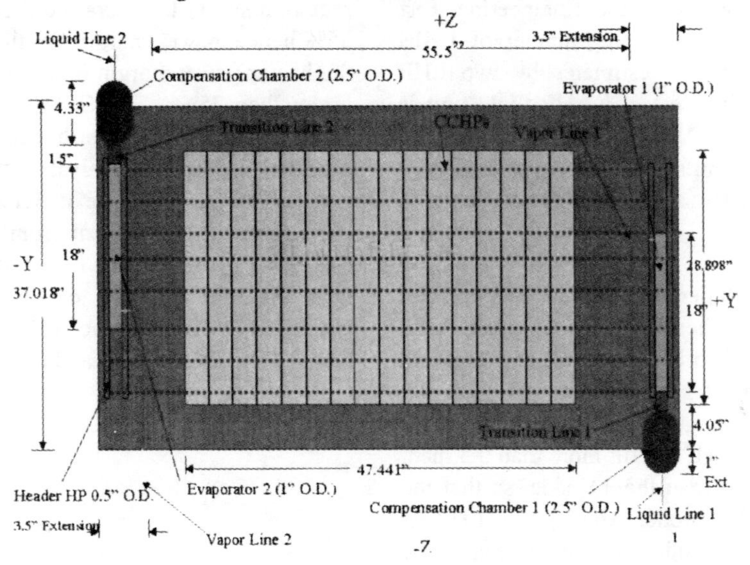

Table 1. GSFC Design Values of Environmental Constants.

	Minimum	Maximum
Solar Constant (W/m^2)	1287	1419
Albedo	0.25	0.35
Earth Emitted IR Radiation (W/m^2)	208	265

FLIGHT TEMPERATURE AND HEATER POWER PREDICTIONS

Figure 8 presents the transient flight temperature prediction of the nominal CZT detector temperature in the worst hot case in the science mode. The thermal design meets the temperature stability requirement of 1°C rate of change on any time scale. Also, the 1°C maximum temperature gradient requirement of the 256 detectors is satisfied by the embedded CCHPs and active control heaters. The heater power needed is 45 W peak, including 30 W for the LHP compensation chambers, and the orbital average heater power is 30 W.

Figure 9 presents the transient flight temperature predictions of the nominal CZT detector temperature in the worst cold case in the science mode. The thermal design meets the temperature stability requirement of 1°C rate of change on any time scale. Also, the 1°C maximum temperature gradient requirement of the 256 detectors is satisfied by the embedded CCHPs and active control heaters. The heater power needed is 50 W peak, including 30 W for the LHP reservoirs, and the orbital average heater power is 34 W. The heater power is higher than the worst hot case because the radiative heat leak through the BAT MLI blankets is larger. If LHPs are not used to vary the effective conductance from the DMs to the radiator, the peak heater power required in the worst cold case is 144 W, and the orbital average is 97 W. It exceeds the current peak heater power budget of 60 W significantly.

In this thermal design concept, there are two stages of active thermal control, one at the LHPs, and the other at the DMs. Therefore, the flight temperature predictions of the CZT detectors are very stable, in both the worst hot and worst cold cases.

In the safehold mode, the survival heater power needed is only 44 W peak, including 30 W for the LHP reservoirs, and 14 W due to radiative heat leaks through the BAT MLI blankets and the parasitic heat conduction from the embedded CCHP to LHP interface to the radiator. It is within the 50 W survival heater power budget. The orbital average survival heater power is 33 W. If LHPs are not used to vary the effective conductance from the DMs to the radiator, the peak survival heater power is 186 W, and the orbital average is 125 W. The peak survival heater power is more than three times the budget.

Figure 8. Flight Temperature Predictions of CZT in Worst Hot Case.

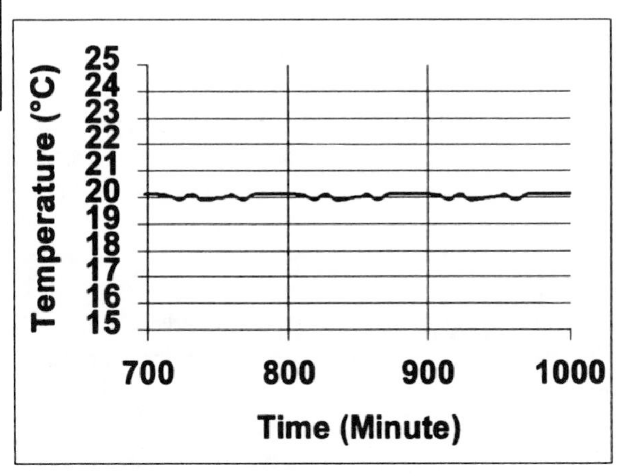

Figure 9. Flight Temperature Predictions of CZT in Worst Cold Case.

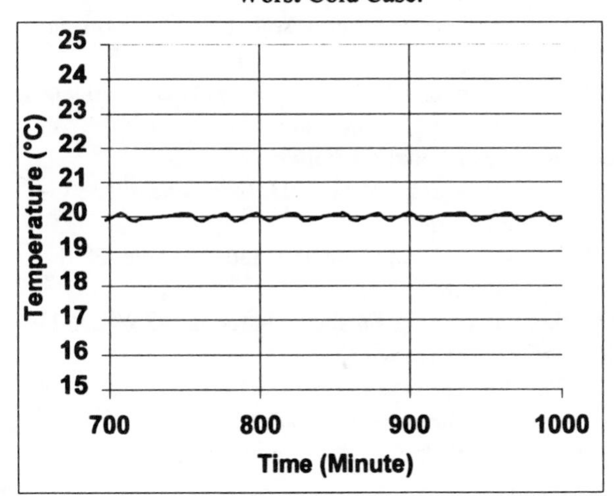

SUMMARY AND CONCLUSIONS

The maximum allowable temperature gradient of the 256 CZT detectors is 1°C. Also, the maximum allowable rate of temperature change of the ASICs of the 256 DMs is 1°C on any time scale. The CZT detector temperature is specified as 20°C. The total power dissipation of the DMs and BCDH is 180 W. The results of thermal analysis show that the thermal requirements and heater power budget are satisfied by a thermal design concept that includes the following:

- The DMs are cold biased, and active heater control is used.
- Two redundant heater circuits per DM Block.
- Each heater circuit has an electronic proportional controller with a ±0.25°C accuracy.
- Eight 1.27-cm diameter CCHPs, using ammonia as the working fluid, are embedded in the detector array

plate to minimize the temperature gradient of the DMs.

•Two LHPs, with propylene as the working fluid, transport the waste heat from the embedded CCHPs to the radiator.

•The LHPs are located outside the graded-Z shields/MLI.

•One LHP on the +Y side and one on the –Y side maximizes the thermal interface conductance, and minimizes the temperature gradient from the embedded CCHPs to the LHP evaporators.

•The diameter of the LHP evaporator is 2.54 cm, and the evaporator length is 45.72 cm.

•Each LHP evaporator has two redundant header CCHPs, which have a 1.905-cm (0.75-inch) outer diameter.

•The evaporator, compensation chamber, and transition line of each LHP, and the header heat pipes are mounted to the top of an 8.89-cm extension of the detector array plate.

•Each LHP has double condensers.

•The LHP condensers are in a serpentine form on the radiator.

•Each LHP compensation chamber has two redundant heater circuits, and each heater circuit includes an electronic proportional controller.

•The radiator is a 1.194-m x 0.092-m, and 0.15875-cm thick aluminum plate.

•The radiator is located on the anti-sun side of the SWIFT spacecraft.

•The thermal coating on the radiator is AZW-LA-1I white paint or equivalent.

The heater controllers for DMs and the LHP compensation chambers have adjustable set-points in flight. Each of the LHP evaporators has a 75 W startup heater.

REFERENCES

1. YSI Precision Temperature Group, "Resistance Data for YSI Thermistors," YSI, Dayton, Ohio.
2. Kauder, L., Re: AZ-Tek White Paint, e-mail to M. Choi, Feb. 20, 2000.
3. AZ-Tek, NASA OPM Reflectometer flight data of AZW/LA-1I white paint taken in 1997.
4. Ku, J., Operating Characteristics of LHPs, SAE Technical Paper Series 1999-01-2007.
5. Kaya, T. and Ku, J., A Parametric Study of Performance Characteristics of LHPs, SAE Technical Paper Series 1999-01-2006.
6. Kaya, T. and Ku, J., Ground Testing of LHPs for Spacecraft Thermal Control, AIAA Paper 99-3447.
7. Ku, J., Re: SWIFT thermal design meeting, e-mail to M. Choi, May 2, 2000.

AIAA-2000-2906

THERMAL CONSIDERATIONS OF SWIFT XRT RADIATOR AT-35°C OR COLDER IN LOW EARTH ORBIT

Michael K. Choi
NASA Goddard Space Flight Center
Greenbelt, MD 20771

ABSTRACT

The X-Ray Telescope (XRT) is an instrument on the National Aeronautics and Space Administration (NASA) SWIFT spacecraft. The thermoelectric cooler (TEC) for the charge coupled device (CCD) of the XRT requires a radiator temperature of –35°C or colder, and a goal of -55°C to minimize the damage by radiation. The waste heat rejected from the TEC to the radiator is in the 8 W to 20 W range. In the Phase A baseline design, the XRT radiator is mounted to the rear end of the XRT telescope tube and is very close to the bottom closeout of the spacecraft bus. The bottom closeout is multi-layer insulation (MLI) blankets. At sun angles between 90° and 180°, there is direct solar impingement on the bottom closeout. When the rolls ±5°, the XRT radiator is exposed to direct solar radiation. The radiator also has a view factor to the solar arrays. The results of thermal analysis showed that the flight temperature prediction of the radiator exceeds the temperature requirement of –35°C substantially at sun angles from 110° to 180°. A new location on the anti-sun side of the spacecraft is proposed for the radiator. It requires a heat pipe to couple the TEC and the radiator thermally. The results of thermal analysis show that the flight temperature prediction of the proposed radiator meets the temperature requirement at all sun angles.

INTRODUCTION

XRT is one of the three telescopes on the SWIFT spacecraft. The SWIFT mission is part of the NASA Medium-Size Explorer (MIDEX) Program, and is managed by Goddard Space Flight Center (GSFC). It is scheduled to launch in 2003. The SWIFT mission is a first of its kind of multi-wavelength transient observatory for gamma ray burst astronomy. Its mission life is 3 years. SWIFT has a low Earth orbit. The altitude is 600 km, and the inclination is 22°. The XRT instrument is developed jointly by Penn State University and the University of Leicester. The CCD of the XRT is cooled by a TEC, which rejects heat to a radiator. The waste heat is in the 8 W to 20 W range. In the Phase A baseline design, the XRT radiator is mounted to the rear end of the telescope at the bottom of the spacecraft. Both sides of the radiator radiate heat to space. The total area of the two-side radiator is about 6,394 cm^2. Since the radiator is close to the bottom closeout of the spacecraft, it has a good view factor to it. The radiator is made of aluminum, and its mass is 4 kg. Figure 1 shows the Phase A baseline location of the XRT radiator on the SWIFT spacecraft.

THERMAL REQUIREMENT

Presently, the temperature requirement of the XRT CCD is –105°C, and the goal is –115°C to provide sufficient margins to minimize the risk of radiation damage to the CCD. It translates to a temperature requirement of –35°C or colder, and a goal of –55°C for the TEC radiator.

Figure 1. XRT Radiator Location in Phase A.

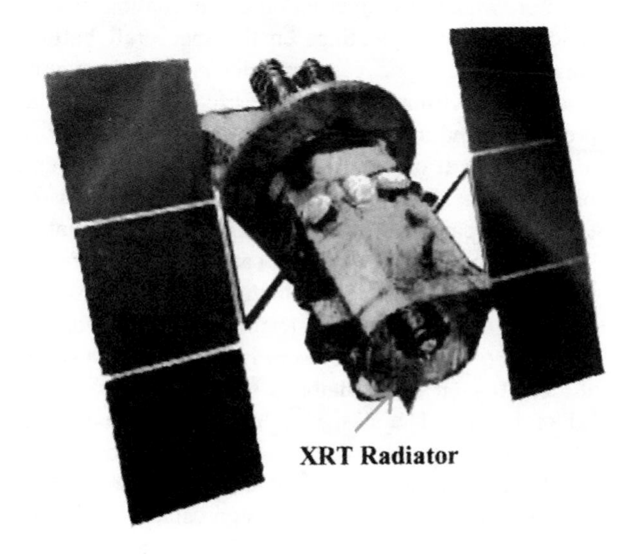

[*] Associate Fellow

THERMAL PROBLEM OF XRT RADIATOR IN PHASE A

The spacecraft bottom closeout is MLI blankets. The absorptance and emittance of the spacecraft bottom closeout of the spacecraft bus are 0.39 and 0.62, respectively, in the spacecraft thermal model. An instrument and spacecraft integrated thermal model was developed by the author in the summer of 1999, and a thermal analysis of the XRT radiator was performed. Based on the initial results of thermal analysis, the concerns on the XRT radiator location were that the flight temperature predictions of the XRT radiator are too warm at large sun angles.

The sun angle is the angle between the solar vector and the optical axis of SWIFT. Presently, the sun angle of the SWIFT mission is 45° minimum and 180° maximum. When the sun angle is 180°, the solar vector is perpendicular to the spacecraft bottom closeout. When the sun angle is between 90° and 180°, there is direct solar radiation incident on the spacecraft bottom closeout. Some of the solar flux is reflected from the spacecraft bottom closeout to the XRT radiator. Direct solar radiation absorbed by the spacecraft bottom closeout increases its temperature, and therefore increases the heat radiation to the XRT radiator. Also, the SWIFT spacecraft has a ±5° roll about the X-axis, which is the optical axis, exposing the XRT radiator to direct solar radiation. In addition to the variation of sun angle, the thermal environment varies from sunlight to eclipse. Therefore, the temperature of the XRT radiator varies transiently.

A more detailed thermal analysis was performed in Phase B, and the thermal concerns on the XRT radiator location remain. Table 1 presents the worst hot case flight temperature predictions of the XRT radiator, which has silver teflon as the coating, versus candidate thermal coatings on the spacecraft bottom closeout. The worst case parameters include winter solstice environment, end-of-life thermo-optical properties, and a sun angle of 180°. The temperature is maximum at orbit noon and is minimum at the end of the eclipse. Table 2 presents the results of using Z-93P white paint as the coating on the XRT radiator. Table 3 presents the worst hot case flight temperature predictions of the spacecraft bottom closeout versus its thermal coating. The flight temperature predictions of the XRT radiator are too warm in the sunlight, even if the coating on the radiator is silver teflon or Z-93P white paint. The design values of solar constant, albedo and Earth emitted infrared (IR) radiation have a significant impact on the flight temperature predictions the XRT radiator. The GSFC design values[1] in Table 4 are used in the thermal analysis. Tables 5 and 6 compare the GSFC albedo and Earth IR values to those used at other NASA centers as reported by F. A. Costello in 1995[2], and to those recommended by F. A. Costello based on his analysis of the Earth Radiation Balance Experiment (ERBE) flight data.[2]

Figure 2 shows the effect of the heat rejected from the TEC to the XRT radiator when the sun angle is 180°. Silver teflon is the thermal coating on both the XRT radiator and spacecraft bottom closeout. The effect is not very significant because the heat rejection from the TEC is much less than the environmental heat flux absorbed by the radiator and the heat radiation from the spacecraft bottom closeout to the radiator.

Figure 3 presents the hot case flight temperature predictions of the XRT radiator versus the sun angle. The temperature is maximum at orbit noon and is minimum at the end of the eclipse. The radiator temperature is too warm at large sun angles.

Table 1. Hot Case Flight Temperature Prediction of Phase A XRT Radiator with Silver Teflon As Coating at 180° Sun Angle.

α/ε of Spacecraft Bottom	Orbit Noon (°C)	End of Eclipse (°C)
.12/.026 (VDA)	-19	-40
.20/.21 ($Ag\text{-}Al_2O_3$ Overcoating)	-12	-37
.20/.76 (Silver Teflon)	-13	-38
.39/.62	-1	-33
.87/.77 (Black Kapton)	+29	-26

Table 2. Hot Case Flight Temperature Prediction of Phase A XRT Radiator with Z-93P White Paint at 180° Sun Angle.

α/ε of Spacecraft Bottom	Orbit Noon (°C)	End of Eclipse (°C)
.12/.026 (VDA)	-15	-44
.20/.21 ($Ag\text{-}Al_2O_3$ Overcoating)	-4	-37
.20/.76 (Silver Teflon)	-5	-37
.39/.62	+6	-35
.87/.77 (Black Kapton)	+31	-29

Table 3. Hot Case Flight Temperature Prediction of Spacecraft Bottom Closeout at 180° Sun Angle with Silver Teflon or Z-93P as XRT Radiator Coating.

α/ε of Spacecraft Bottom	Orbit Noon (°C)	End of Eclipse (°C)
0.12/0.026 (VDA)	316	-61
.20/.21 ($Ag\text{-}Al_2O_3$ Overcoating)	139	-61
.20/.76 (Silver Teflon)	33	-61
.39/.62*	95	-61
.87/.77 (Black Kapton)	146	-61

Table 4. GSFC Design Values of Environmental Constants.

	Minimum	Maximum
Solar Constant (W/m^2)	1287	1419
Albedo	0.25	0.35
Earth Emitted Infrared Radiation (W/m^2)	208	265

Table 5. Design Values of Albedo (Reported by F. A. Costello, 1995[2]).

	Averaging Time (Hr)	Set	Min.	Max.
NASA-MSFC	1.5	Max/Min	0.15	0.39
NASA-MSFC	1.5	1-99%	0.18	0.34
NASA-JSC	4.0	Max/Min	0.20	0.31
NASA-GSFC	Long	Max/Min	0.25	0.35
F. A. Costello	Long	3-σ	0.27	0.39
F. A. Costello	Long	Max/Min	0.30	0.37

Table 6. Design Values of Earth Emitted Infrared Radiation (W/m^2) (Reported by F. A. Costello, 1995[2]).

	Averaging Time (Hr)	Set	Min.	Max.
NASA-MSFC	1.5	Max/Min	198	276
NASA-MSFC	1.5	1-99%	202	257
NASA-JSC	4.0	Max/Min	232	280
NASA-GSFC	Long	Max/Min	208	265
F. A. Costello	Long	3-σ	197	241
F. A. Costello	Long	Max/Min	210	232

Figure 2. Effect of Heat Rejected by TEC to XRT Radiator in Phase A Location.

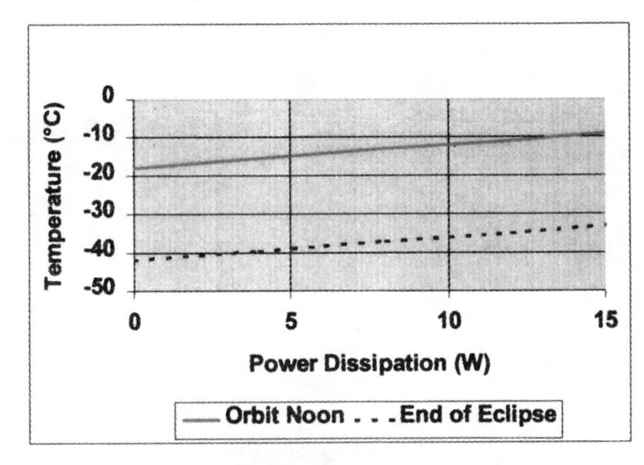

EFFECT OF SOLAR ARRAYS ON BASELINE XRT RADIATOR

In the Phase A thermal analysis of the XRT radiator, the solar arrays were neglected. Also, the heat radiation from the spacecraft interior through the spacecraft bottom closeout MLI was neglected.

The SWIFT spacecraft has two solar array wings. Each wing has a solar array drive and three articulating panels. The sun angle of the SWIFT mission is 45° to 180°. Figure 4, which is in the Thermal Synthesizer System (TSS) format, shows the position of the solar arrays when the sun angle is 45° and the baseline location of the XRT radiator in the thermal model. The following assumptions are made in the thermal analysis. The heat rejection from the XRT TEC to the radiator is 8 W. Both sides of the baseline radiator radiate heat to space. The total area of the two-side radiator is 6,394 cm^2. The mass of the radiator is 4 kg. The coating on both the radiator and spacecraft bottom closeout MLI is 0.254 mm (10-mil) thick silver teflon. The temperature of the spacecraft interior is assumed to be 30°C, and the MLI effective emittance is assumed to be 0.03.

Figure 3. Hot Case Flight Temperature Prediction of XRT Radiator (Silver Teflon) Versus Sun Angle with Solar Arrays Neglected.

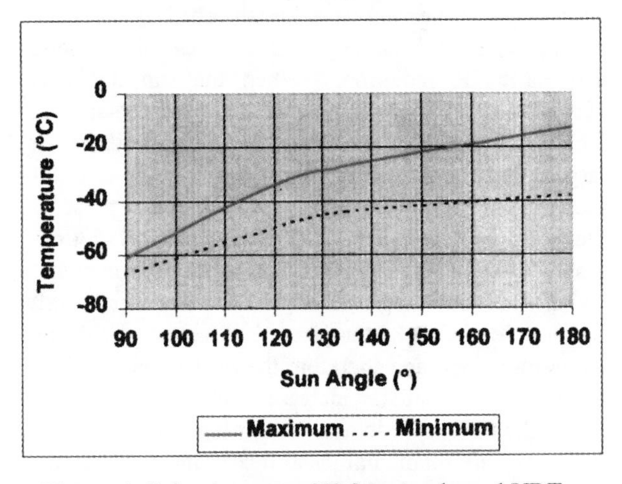

Figure 4. Solar Arrays at 45° Sun Angle and XRT Radiator in Phase A.

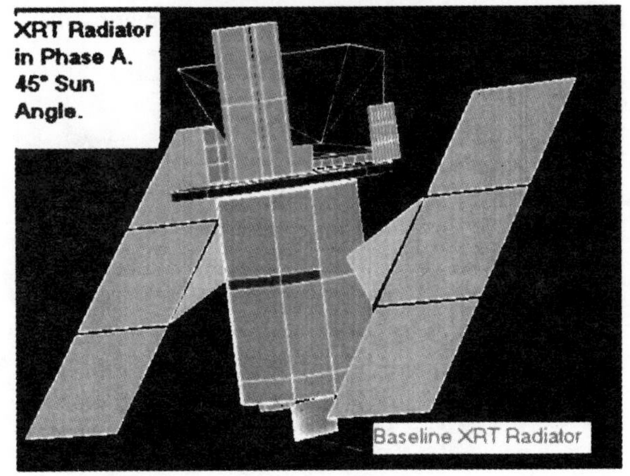

Figures 5 through 10 present the results of the thermal analysis in the hot case. Winter solstice environment and end-of-life (EOL) thermo-optical properties are used in the hot case thermal analysis. The EOL absorptance and emittance of 0.254 mm (10-mil) silver teflon are assumed to be 0.25 and 0.827, respectively. Figure 5 presents the flight temperature prediction of the XRT radiator versus the sun angle. The thermal effect of the solar arrays and heat radiation through the spacecraft bottom closeout MLI on the XRT radiator temperature can also be seen in Figure 5. The flight temperature prediction of the radiator at orbit noon exceeds the temperature requirement of –35°C at sun angles between 110° and 180°. To maintain the XRT radiator temperature below –35°C, a time constraint on the sun angles is required in flight.

Figure 6 presents the radiation coupling between the XRT radiator and solar arrays versus the sun angle. The radiation coupling is largest when the sun angle is 45° and 135°, and smallest when the sun angle is 180°. Figure 7 presents the peak heat radiation, which occurs at orbit noon, from the solar arrays to the XRT radiator versus the sun angle. Figure 8 presents the orbital average environmental heat flux absorbed by the XRT radiator versus the sun angle. It includes solar, albedo and Earth IR radiation. When the sun angle is between 45° and 90°, there is no direct solar flux incident on the radiator. When the sun angle is 90° or larger, the direct solar flux incident on the radiator increases as the sun angle increases. Figure 9 presents the peak heat radiation from the spacecraft bottom closeout MLI to the XRT radiator versus the sun angle. When the sun angle exceeds 90°, the spacecraft bottom closeout MLI temperature increases as the sun angle increases, and therefore the heat radiation from the MLI to the radiator increases. When the sun angle is less than 90°, albedo and Earth infrared radiation are the only environmental heat fluxes incident on the bottom closeout. Figure 10 presents the flight temperature prediction of the solar arrays.

Figure 5. Hot Case Flight Temperature Predictions of XRT Radiator in Phase A.

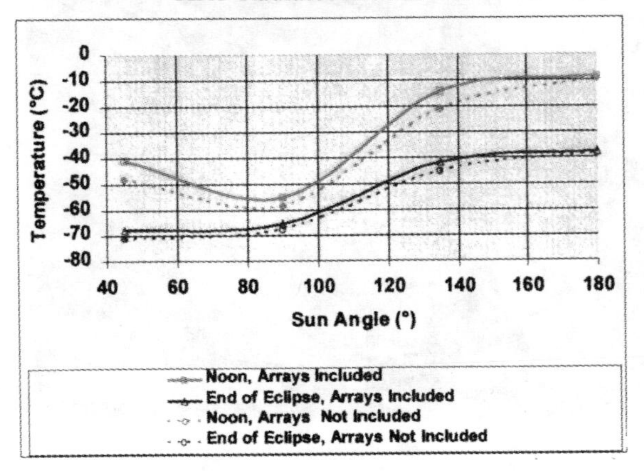

Figure 6. Radiation Couplings between Solar Arrays and XRT Radiator in Phase A.

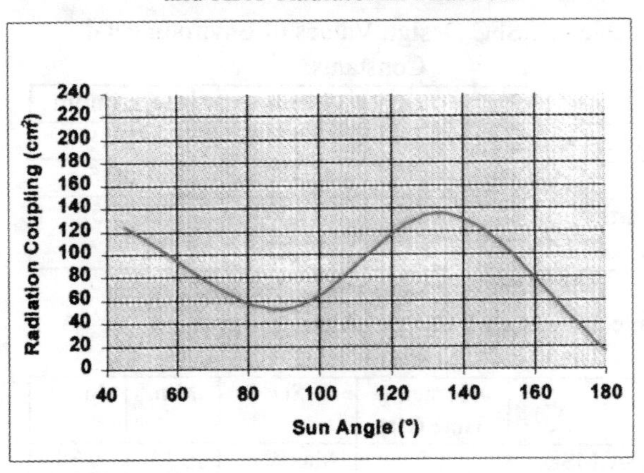

Figure 7. Peak Radiation from Solar Arrays to XRT Radiator in Phase A.

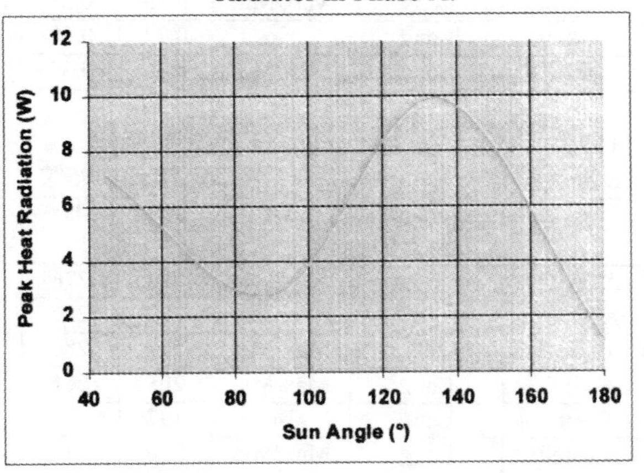

Figure 8. Orbital Average Environmental Heat Flux Absorbed by XRT Radiator in Phase A Location.

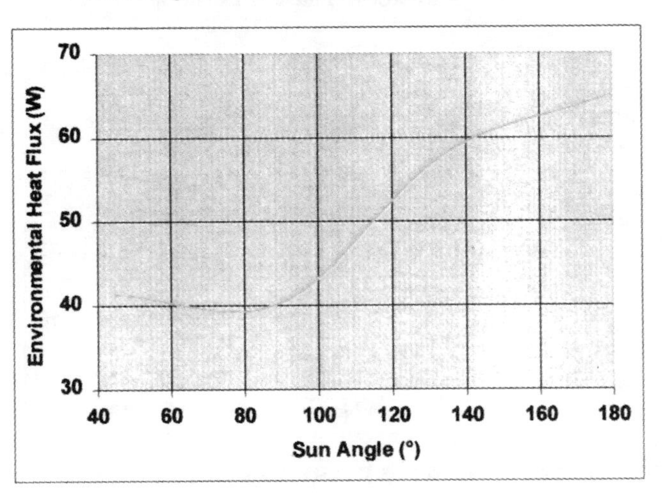

Figure 9. Peak Heat Radiation From Spacecraft Bottom Closeout To XRT Radiator in Phase A.

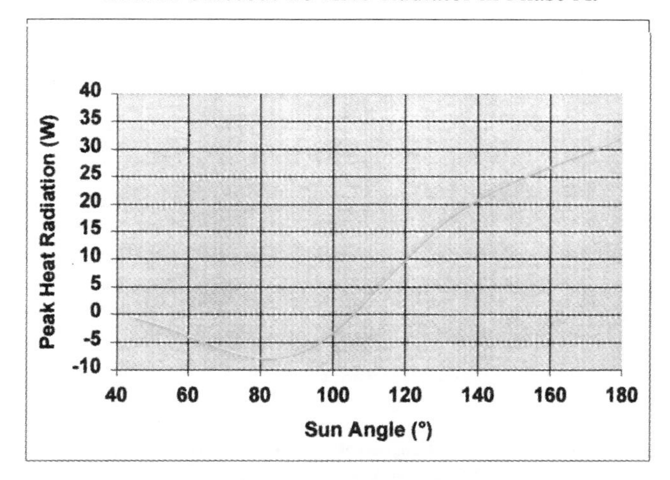

Figure 10. Flight Temperature Prediction of Solar Arrays in Worst Hot Case.

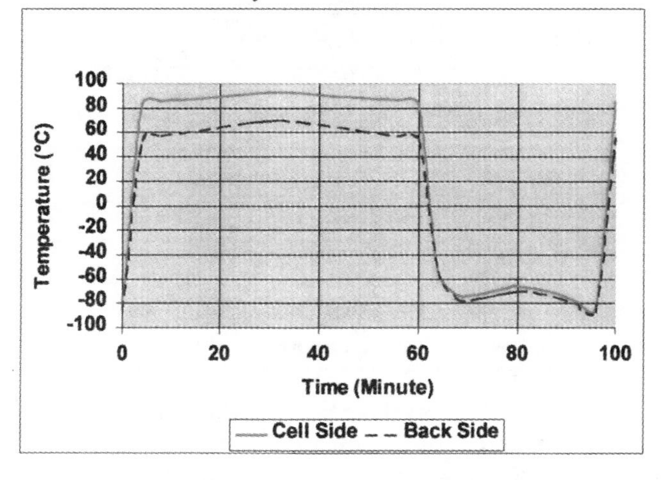

PROPOSED XRT RADIATOR LOCATION

Presently in the Phase B study, the thermal concept of relocating the XRT radiator to the anti-sun of the spacecraft bus is proposed. In this design concept, a constant conductance heat heat pipe (CCHP) transports the waste heat from the TEC to the radiator. Also, separate CCHPs minimize the temperature gradient on the radiator. The results of the preliminary thermal analysis show that the flight temperature predictions of the radiator in the new location are much colder than that of the Phase A baseline location at high sun angles. Figure 11 shows the new radiator location and the heat pipe that thermally couples the TEC and radiator.

EFFECT OF SOLAR ARRAYS ON PROPOSED XRT RADIATOR LOCATION

Figure 12 shows the position of the solar arrays when the sun angle is 45° and the proposed location of the XRT radiator on the anti-sun side of the spacecraft in the thermal model. The coating on the radiator is also 0.254 mm (10-mil) thick silver teflon. The radiator is assumed to be isothermal. CCHPs on the radiator minimize the temperature gradient. The radiator is mounted to the spacecraft bus by titanium flexures and G-10 washers, so that it is thermally isolated from the spacecraft. Presently, the conduction coupling between the radiator and spacecraft is assumed to be 0.05 W/°C. The spacecraft interface temperature is assumed to be 20°C in the hot case. The backside of the radiator is insulated with MLI blankets. The heat pipe from the TEC to the radiator is approximately 2 m long. By maximizing the thermal interface conductance between the heat pipe and TEC, and between the heat pipe and radiator, the temperature gradient between the TEC and radiator is very small.

Figure 11. Proposed XRT Radiator Location.

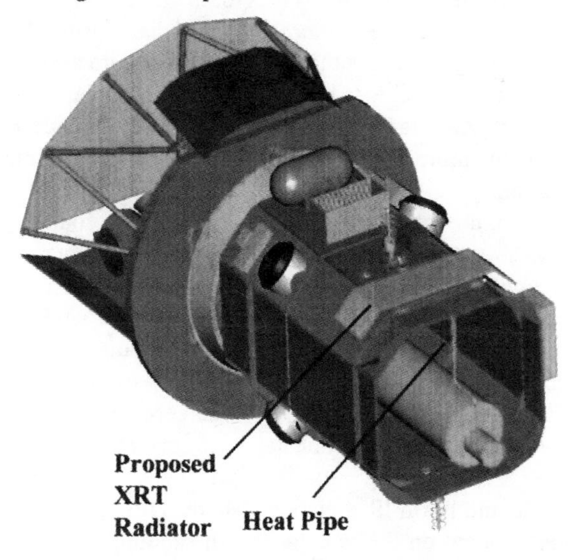

Figure 12. Solar Arrays at 45° Sun Angle and Proposed XRT Radiator.

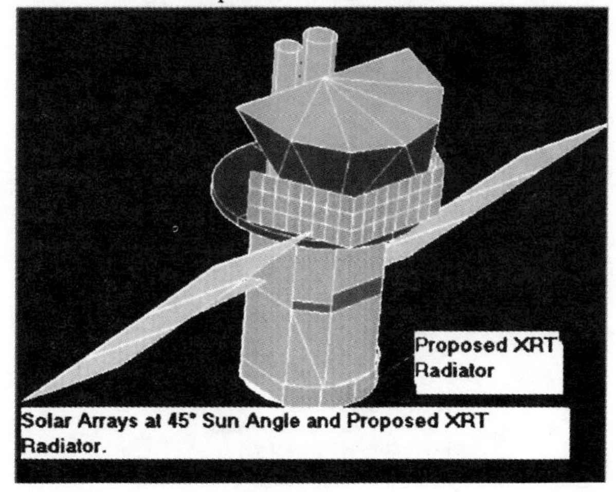

Figures 13 through 16 present the results of the thermal analysis in the hot case. In the current thermal analysis, the radiator forms a 132° sector around the spacecraft, and the height is 33 cm. The radiator area is 6,185 cm^2. The mass of the radiator is 4.4 kg. Figure 13 presents the flight temperature prediction of the XRT radiator versus the sun angle. The thermal effect of the solar arrays can also be seen in Figure 13. The flight temperature prediction of the radiator at orbit noon satisfies the temperature requirement of –35°C at all sun angles. A colder radiator temperature can be achieved by increasing the radiator area. For example, increasing the radiator height from 33 cm to 45 cm reduces the temperature at orbit noon from –35°C to –37°C at a 90° sun angle. Also, the temperature at orbit noon can be decreased by increasing the thermal mass of the radiator. The mass of the radiator includes the CCHPs for isothermalization of the radiator, and the heat pipe for transport heat from the TEC to the radiator. For example, increasing the mass of the radiator from 4.4 kg to 8.8 kg decreases the radiator temperature at orbit noon from –35°C to –42°C for a 90° sun angle. The proposed XRT radiator location on the anti-sun side is thermally more favorable than the Phase A baseline location.

Figure 14 presents the radiation coupling between the XRT radiator and solar arrays versus the sun angle. The radiation coupling is largest when the sun angle is 180°, and smallest when the sun angle is 90°. Figure 15 presents the peak heat radiation, which occurs at noon, from the solar arrays to the XRT radiator versus the sun angle. Figure 16 presents the orbital average environmental heat flux absorbed by the XRT radiator versus the sun angle. It includes albedo and Earth IR radiation. There is no direct solar flux incident on the radiator for all sun angles. Also, there is no heat radiation between the radiator the spacecraft bottom closeout.

Figure 13. Hot Case Flight Temperature Prediction of Proposed XRT Radiator versus Sun Angle.

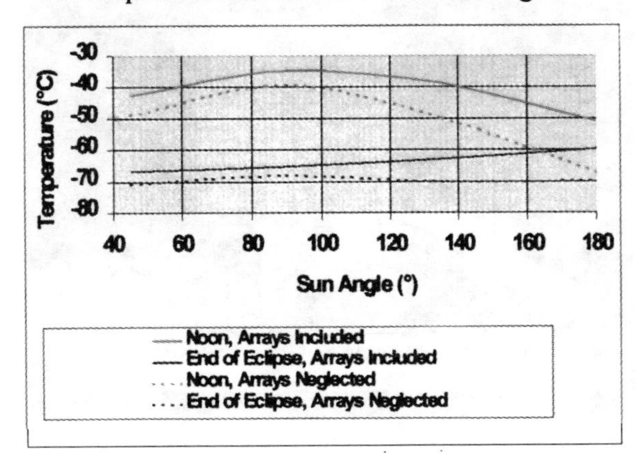

Figure 14. Radiation Couplings between Solar Arrays and Proposed XRT Radiator.

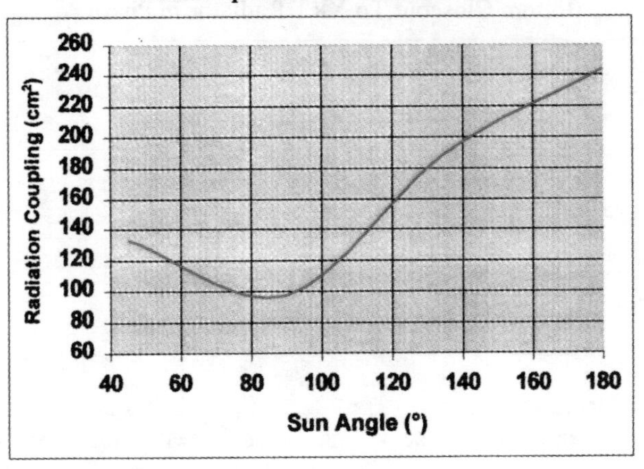

Figure 15. Peak Radiation from Solar Arrays to Proposed XRT Radiator.

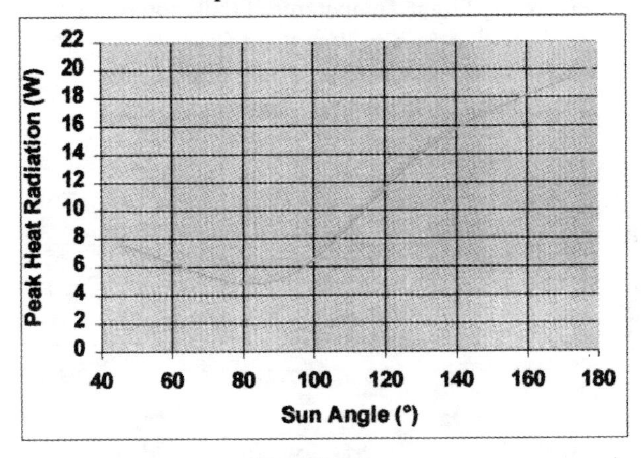

Figure 16. Orbital Average Environmental Heat Flux Absorbed by Proposed XRT Radiator.

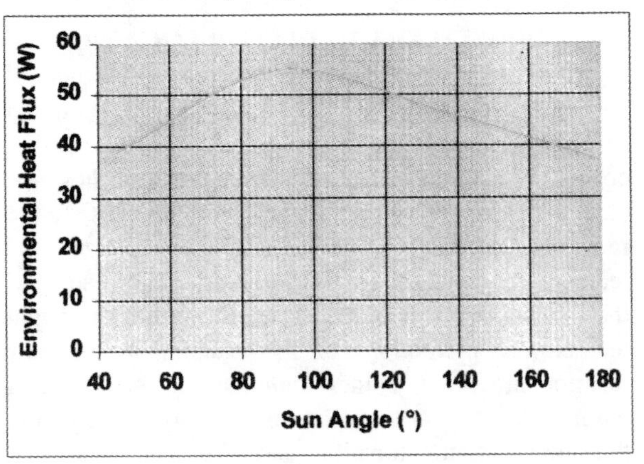

OPTIMIZATION OF PROPOSED XRT RADIATOR

Albedo flux and Earth emitted IR radiation are the only environment heat fluxes that affect the temperature of the XRT radiator at the proposed location on the anti-sun side of the spacecraft. These environmental heat fluxes vary from day to night, from summer solstice to winter solstice, and from one sun angle to another. Also, degradation of the thermal coating on the radiator has an impact on the radiator temperature.

The thermal energy balance on the XRT radiator is

$$\varepsilon A F(t)\sigma\theta^4(t) + mc_p(\theta)\, d\theta(t)/dt = Q_{TEC} + Q_{Albedo}(A,t) + Q_{EIR}(A,t) + Q_{SC}(\theta) + Q_{SA}(t) \quad (1)$$

where ε = hemispherical emittance of thermal coating on radiator,

A = radiator area,

F(t) = view factor of radiator to space, which is a function of time because position of solar arrays changes with sun angle,

σ = Stefan-Boltzmann constant,

$\theta(t)$ = radiator temperature which is a function of time,

m = mass of radiator and CCHPs,

$c_p(\theta)$ = specific heat, which is a function of temperature,

Q_{TEC} = power dissipation of TEC,

$Q_{Albedo}(A,t)$ = Albedo flux absorbed, which is a function of radiator area and time,

$Q_{EIR}(A,t)$ = Earth emitted IR radiation absorbed, which is a function of radiator area and time,

$Q_{sc}(\theta)$ = heat conduction from spacecraft mounting interface to radiator, which is a function of the radiator temperature, and

$Q_{SA}(t)$ = heat radiation from solar arrays to radiator, which is a function of time because position of solar arrays changes with sun angle.

From the above equation, for a given Q_{TEC}, the radiator temperature is dependent on A and m. Therefore, they must be optimized. Of course, m needs to be within the mass budget.

Degradation of the thermal coating on the radiator has a significant impact on the radiator temperature. When a coating degrades, its solar absorptance increases, and the albedo flux absorbed increases. One way to minimize the effect of albedo flux is to select a thermal coating that has a very low solar absorptance at the beginning of life (BOL), and a small degradation over a 3-year mission life for the radiator. AZ-Tek's AZW/LA-1I white paint is likely the best candidate thermal coating for the XRT radiator, based on its thermo-optical properties. The Thermal Coatings Committee at GSFC recommends the following thermo-optical properties for this coating in the SWIFT mission, when it is not exposed to direct solar radiation. The solar absorptance is 0.08 and hemispherical emittance is 0.91 at BOL. The solar absorptance is 0.12 and hemispherical emittance is 0.90 at the end of life (EOL).[3] Its high emittance provides a high radiation coupling to space. It also has a low degradation in the solar absorptance. The flight data of the AZW/LA-1I white paint shows that initially the solar absorptance was 0.088, and after 48 days in space flight, it increased to 0.095, and after 197 days, it increased to 0.103.[4] It shows that the degradation of this paint in space is very small. Preliminary results of an ongoing ultraviolet exposure test at GSFC show that the paint is stable. Also, the AZW/LA-1I white paint will be flying on the GSFC calorimeter on the EO-1 spacecraft, which is scheduled to launch later this year. EO-1 is an Earth orbiting spacecraft, and has an altitude of 705 km. More testing of this paint will be performed at GSFC, including measurement of the emittance at temperatures ranging from room temperature to 40 K (-233°C).

The XRT radiator and CCHPs are made of aluminum. Figure 17 presents the relationship between the specific heat of pure aluminum and temperature in the –94°C to 27°C range.[5] When the temperature decreases, the specific heat of aluminum decreases. At –70°C, the specific heat is 89% of that at room temperature. Because the thermal capacitance of the XRT radiator is the product of its mass and specific heat, and its temperature is significantly colder than room temperature, the data in Figure 17 is included in the thermal mathematical model. Note that the XRT radiator and CCHPs are made of an aluminum alloy, and not pure aluminum. The specific heat is adjusted in Figure 17 accordingly.

Figure 18 presents the worst hot case flight temperature predictions of the proposed XRT radiator versus radiator area. The effect of the mass of the radiator is also shown. The mass includes the CCHPs. The parameters stacked in the worst hot case are 90° sun angle, winter solstice environment, and EOL thermo-optical properties. The mass selected must be within the mass budget. For a radiator mass of 8.8 kg to 13.2 kg, the radiator area should be approximately 6,185 cm^2. If the mass is 8.8 kg and the area is 6,185 cm^2, the radiator temperature at orbit noon is –46°C. If the mass is 13.2 kg and the area is 6,185 cm^2, the radiator temperature at orbit noon is –49°C.

Figure 19 presents the hot case flight temperature prediction of the XRT radiator versus the sun angle. The thermal effect of the solar arrays is included. Note that in the cold case, the flight temperature prediction of the XRT radiator is significantly colder. It is important to select a proper working fluid for the CCHPs to ensure adequate heat transport capacity and no freezing.

Figure 17. Specific Heat of Pure Aluminum versus Temperature.

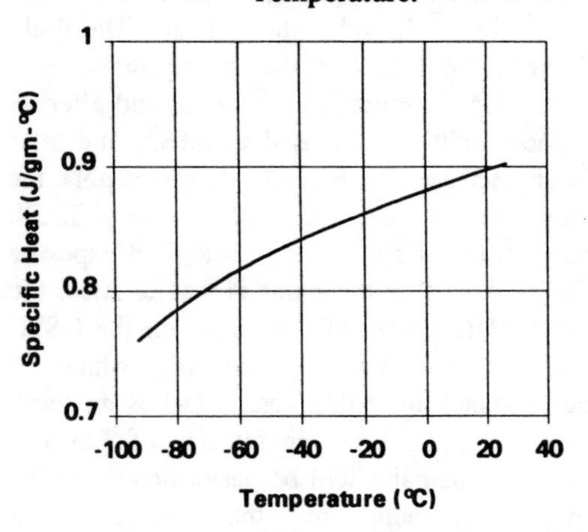

Figure 18. Hot Case Flight Temperature Prediction of Proposed XRT Radiator vs. Radiator Area at 90° Sun Angle.

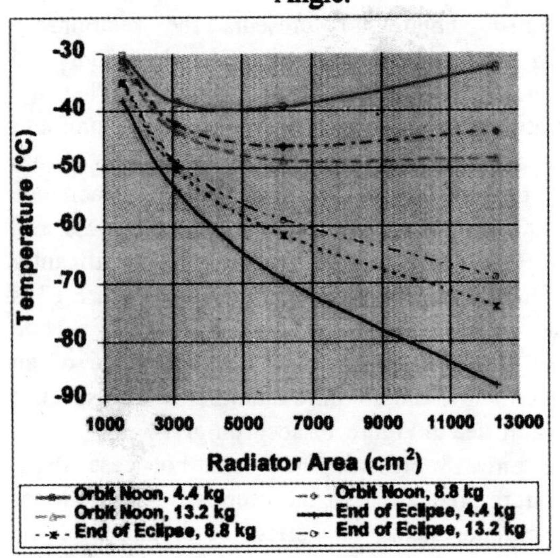

Figure 19. Hot Case Flight Temperature Prediction of Proposed XRT Radiator versus Sun Angle.

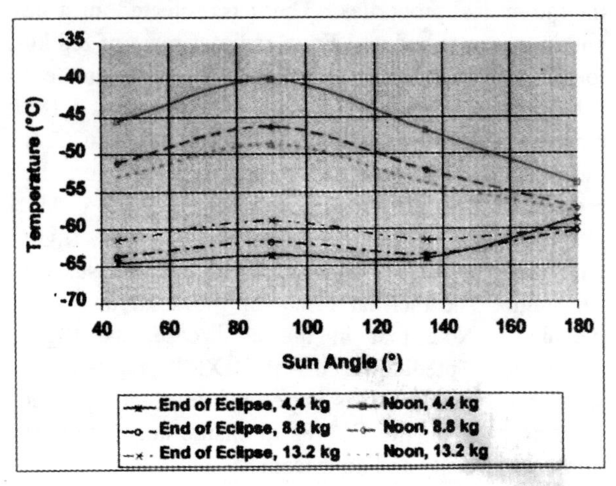

EFFECT OF TEC POWER DISSIPATION ON XRT RADIATOR OPTIMIZATION

Worst Case of 20 W

The power dissipation of the TEC was assumed to be 8 W in the thermal analysis above. But, the worst case power dissipation of the TEC could be as much as 20 W.[6] Figure 20 presents the worst hot case flight temperature predictions of the proposed XRT radiator versus the radiator area. The power dissipation of the XRT TEC is taken as 20 W. The effect of the mass of the radiator is also shown in Figure 20. The thermal effect of the solar arrays is included. From Figure 20, the optimum radiator areas for radiator masses of 4.4 kg, 8.8 kg, 13.2 kg, and 17.6 kg are 7,000 cm^2, 9,000 cm^2, 11,000 cm^2 and 13,000 cm^2, respectively. As mentioned earlier, the radiator mass includes all the CCHPs. For these optimum radiator areas, the worst hot case flight temperature predictions of the radiator at orbit noon are –31.5°C, -39.7°C, -43.8°C, and –46.2°C, respectively. Figure 21 presents the hot case flight temperature prediction of the XRT radiator versus the sun angle, using the optimum radiator areas for the four radiator masses. Note that the mass budget is 10 kg.

Sensitivity Study

Figure 22 presents the optimum radiator area versus the TEC power dissipation for the four radiator masses. The optimum area increases as the TEC power dissipation increases. Figure 23 presents the worst hot case flight temperature predictions of the XRT radiator at orbit noon versus the TEC power dissipation, when the radiator area is optimized.

SURVIVAL TEMPERATURE

In the safehold mode, the instruments are turned off, and there is no heat dissipation at the TEC. The worst cold case flight temperature prediction of the XRT radiator and CCHPs in the safehold mode is -90°C. Ammonia is not acceptable for the working fluid of the CCHPs because its freezing point is -78°C. The heat dissipation of the TEC is 8 W to 20 W, and is low. Propylene, which has a freezing point of -185°C, is possibly the best candidate working fluid. Ethane, which has a freezing point of -183°C, is another candidate.

Figure 20. Hot Case Flight Temperature Prediction of Proposed XRT Radiator vs. Radiator Area at 90° Sun Angle with 20 W TEC Power Dissipation.

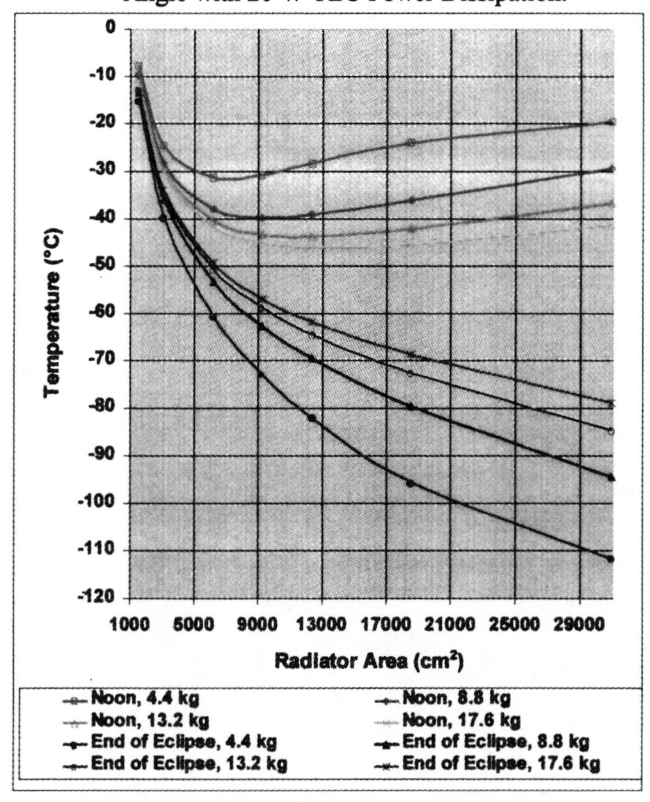

Figure 21. Hot Case Flight Temperature Prediction of Proposed XRT Radiator with Optimum Area versus Sun Angle.

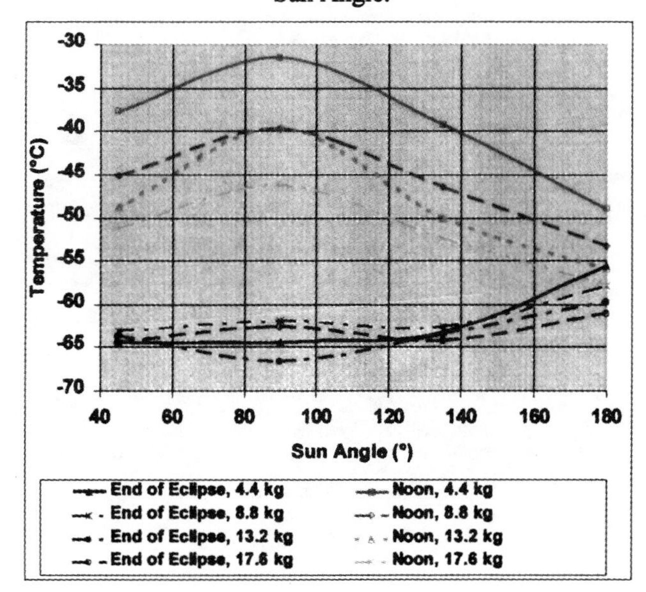

Figure 22. Optimum Area of Proposed XRT Radiator versus TEC Power Dissipation.

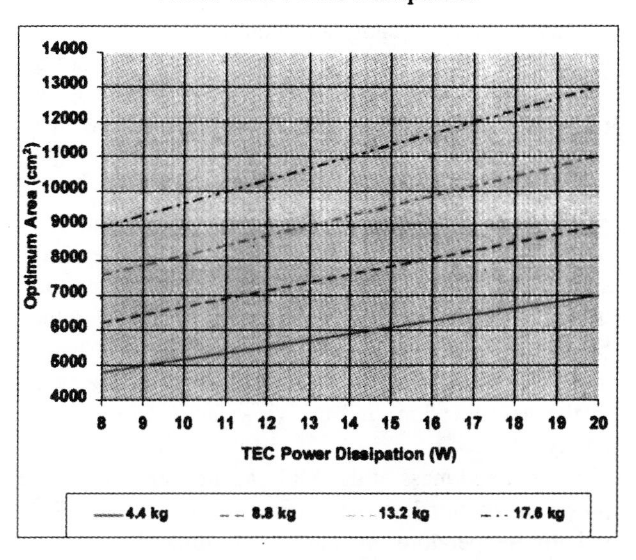

Figure 23. Hot Case Flight Temperature Prediction of Proposed XRT Radiator with Optimum Area versus TEC Power Dissipation.

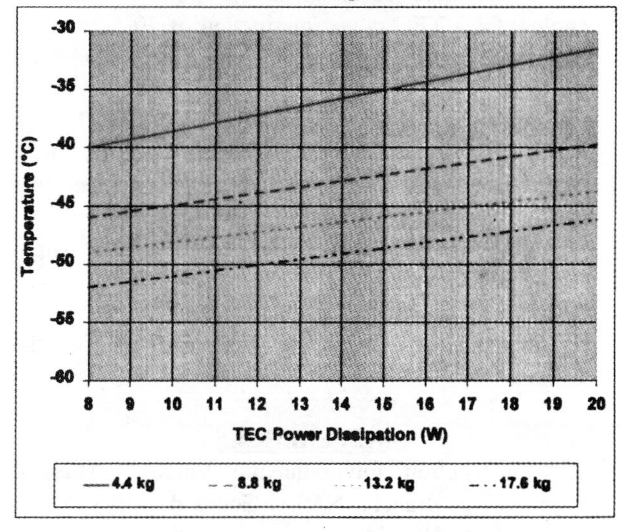

SUMMARY AND CONCLUSIONS

•The thermal effect of the solar arrays and heat radiation through the spacecraft bottom closeout MLI on the flight temperature prediction of the Phase A baseline XRT radiator at orbit noon is rather significant. It increases the hot case flight temperature prediction of the radiator by 0.5°C at a 180° sun angle and by 5.7°C at a 45° sun angle.

•If SWIFT operates at a sun angle between 110° and 180° for an extended period, the hot case flight temperature prediction of the Phase A baseline XRT radiator exceeds the temperature requirement of –35°C substantially.

•To maintain the XRT radiator temperature in the Phase A baseline location below –35°C, a time constraint on the sun angles is required in flight.

•A new location, on the anti-sun side of the spacecraft, is proposed for the XRT radiator. It requires a CCHP to transport the waste heat from the TEC to the radiator, and the radiator is thermally isolated from the spacecraft.

•The thermal effect of the solar arrays on the temperature of the proposed XRT radiator on the anti-sun side at orbit noon is significant. It increases the hot case flight temperature prediction of the radiator by 5°C at a 90° sun angle and by 17°C at a 180° sun angle. However, there is no direct solar impingement in the new location, and the view factor to space is nearly 1.0.

•The area and mass of the XRT radiator, and the TEC power dissipation have significant impacts on the XRT radiator temperature, and must be optimized.

•With a 9,000 cm^2 and 8.8 kg radiator in the proposed location on the anti-sun side, the hot case flight temperature prediction of the XRT radiator satisfies the temperature requirement of –35°C at all sun angles, for a TEC power dissipation of 20 W or less. The margins are at least 5°C. The mass includes all the CCHPs.

•Ammonia is not acceptable for the working fluid of the CCHPs because its freezing point is -78°C, which is close to the XRT radiator flight temperature prediction in the cold case in the eclipse.

•In the safehold mode, the XRT radiator flight temperature prediction is –90°C.

•Propylene, which has a freezing point of -185°C, is possibly the best candidate working fluid for the CCHPs.

REFERENCES

1. "Earth Orbit Environmental Values," Thermal Engineering Branch, NASA Goddard Space Flight Center, Greenbelt, MD.

2. Costello, F. A., Spacecraft Thermal Environment Near Earth, Guideline 34: Earth Orbit Environmental Heating, Report FAC/SWAL9-49, February 13, 1995.

3. Kauder, L., Re: AZ-Tek White Paint, e-mail to M. Choi, Feb. 20, 2000.

4. AZ-Tek, NASA OPM Reflectometer flight data of AZW/LA-11 white paint taken in 1997.

5. Johnson, V. J., ed., 1960, "A Compendium of the Properties of Materials at Low Temperature (Phase I), Part II, Properties of Solids," Wright-Patterson Air Force Base, OH.

6. Burrows, D., Re: Updated Thermal Analysis of SWIFT XRT Radiator, e-mail to M. Choi, Feb. 12, 2000.

LASER-INDUCED BREAKDOWN SPECTROSCOPY IN A METAL-SEEDED FLAME

Hansheng Zhang, Fang-Yu Yueh, Jagdish P. Singh, and Robert L. Cook

Diagnostic Instrumentation and Analysis Laboratory
Mississippi State University
205 Research Boulevard, Starkville, MS 39759-9734
Tel: (662) 325-2105 FAX: (662) 325-8465

and

Gary W. Loge

Spectrumedix Corporation
2124 Old Gatesburg Rd.
State College, PA 16803
Tel: (814) 867-8600

ABSTRACT

Laser-induced breakdown spectroscopy (LIBS) has been used to detect atomic and molecular species in various environments. It has the capability to be used as a real time monitor in a harsh and turbulent environment. Recently, LIBS has been tested at the Diagnostic Instrumentation and Analysis Laboratory (DIAL)'s combustion facility to optimize and characterize the LIBS operation in a flame. Three metals, Cu, Fe, and Ca, were injected into the combustor with a peristaltic pump. LIBS spectra were recorded at different delay times with a fixed gate width to determine the best detection gate. LIBS signal in the flame were also compared with the direct emission from the flame. It is found that the LIBS signal is stronger than the signal from the direct emission of the metal in the same flame and the LIBS method has a much better signal over background ratio. To test the calibration scheme, metal solutions were injected into the combustor with different rates. Linear calibration curves were obtained.

INTRODUCTION

Laser-induced breakdown spectroscopy (LIBS) is a laser-based diagnostic technique for measuring the concentration of various elements in a test medium. It works for solids, liquids, and gases. [1-6] The technique uses a lens to focus a pulsed laser beam on the target and generates a spark (high temperature plasma) consisting of excited neutral atoms, ions, and electrons. The optical emission from the neutral and ionized atoms is then collected to obtain an emission spectrum with the proper detection system. The intensity of the emission line is analyzed to deduce the elemental concentration in the sample. Compared to other analytical techniques, LIBS uses a very small amount of sample and no sample preparation is required. Since LIBS vaporizes and excites the sample in one step, it has the ability to perform real-time analysis.[7] LIBS has been successfully used as a real time continuous emission monitor (CEM) to detect As, Be, Cd, Cr, Pb, and Hg in the off gas of combustion facilities. [8-11] LIBS has also been applied to a harsh, turbulent, and highly luminescent coal-fired magnetohydrodynamics (MHD) gas stream.[12]

Recently, LIBS was evaluated to be used as a hydrocarbon fueled rocket engine health monitor.[13] Detection and characterization of metallic species in the plume of hydrocarbon fueled rocket engines is of special interest. Metallic species in the plume reflect wear and/or corrosion of metals in the rocket engine. Information on engine wear obtained when an engine is undergoing test is very useful. This allows the possibility of engine shutdown before catastrophic failure. It has been observed in the plume that catastrophic engine failure was usually preceded by a bright optical emission which was identified as the emission of the metal eroded from the engine parts. The traditional method to monitor engine plume during a test is atomic emission spectroscopy (AES) in the near ultraviolet and visible spectral regions.[14,15] However, in a hydrocarbon-fueled rocket engine, beside strong OH emission in the region from 300 nm to 320 nm, plume emission from atomic carbon, C_2^+, and other carbon free radical can form a very strong background and interfere with the detection of target metal emission. In addition, incomplete combustion in the plume near the engine nozzle can lead to the formation of carbon soot, which can cause strong scattering and further increase plume continuum background due to blackbody radiation from the soot particles. Another disadvantage of AES method is the location of the emission collection system which is restricted to be close to the brightest part of the plume, where usually there is a harsh, turbulent, and high temperature environment. For these reasons, AES may not work successfully to detect the presence of metal corrosion and engine wear in a hydrocarbon-fueled rocket engine.

Laser-induced breakdown spectroscopy (LIBS) provides an alternative technique that offers better sensitivity and the ability to monitor the rocket engine plume at different locations. Laser-induced breakdown spectroscopy uses a focused laser beam to produce a spark as the emission source. It doesn't relay on the plume temperature. The diagnostic target can be any part of the engine plume of interest. In the LIBS technique, a pulsed laser beam is focused to the target and generates a spark which consists of excite neutral atoms, ions, and electrons. During the plasma (spark) formation, expansion, and cooling down, the plasma temperature can reach as high as 20,000 K and then drop. At different stages, the optical emissions are dominated by different types of emission that can be identified as continuum emission due to Bremsstrahlung, line emission from ions, and line emission from atoms. The detector of the LIBS system usually uses an intensified multi-channel detector. A high voltage pulse, a time gate, controls the detector to turn on or turn off. By setting a proper time delay and width of the gate, the best signal over background ratio (S/B) can be achieved. Time-gated detection can effectively discriminate against the signal from undesirable emissions. Usually, the width of the gate is around 10 μs, any emission directly from the engine plume can be greatly reduced.

FACILITY AND EXPERIMENT

The primary research was conducted at the DIAL test stand with our LIBS system to study the feasibility of using LIBS as a hydrocarbon-fueled rocket engine health monitor.

DIAL test stand consists of a gas/kerosene-fueled combustor and 8-inch inside diameter gas stream channel with many optical ports and probe access. It can simulate various thermal processes and is ideal for instrumentation testing. The facility operation is monitored and controlled by a computer. With a heater, the air can be preheated to as high as 870 K before mixing with fuel. The fuel and air flow can be adjusted to provide actual stoichiometries between 0.8 and 1.2. The temperature of the gas stream is in the range of 800 to 2400 K at different optical ports and different combustion conditions. This experiment was set at a port 1-foot down stream from the combustor. The port used an UV grade quartz window and was purged with nitrogen to cool and reduce deposits on the window. The laser beam was focused in the flame through the window. A beam dump was mounted at the opposite port. Most of the experiments were conducted at the stoichiometry (Φ) of 1.1 and 500 lb/hr air flow.

Metal solutions were injected into the combustor with a peristaltic pump. Before the pump was attached to the facility, it was calibrated to obtain its flow rate against the pump speed. The metal solution of Fe was prepared from a plasma emission standard solution from Accu Standard Inc.. The solutions of Cu and Ca were prepared from the solid sample copper sulfate and calcium carbonate, respectively.

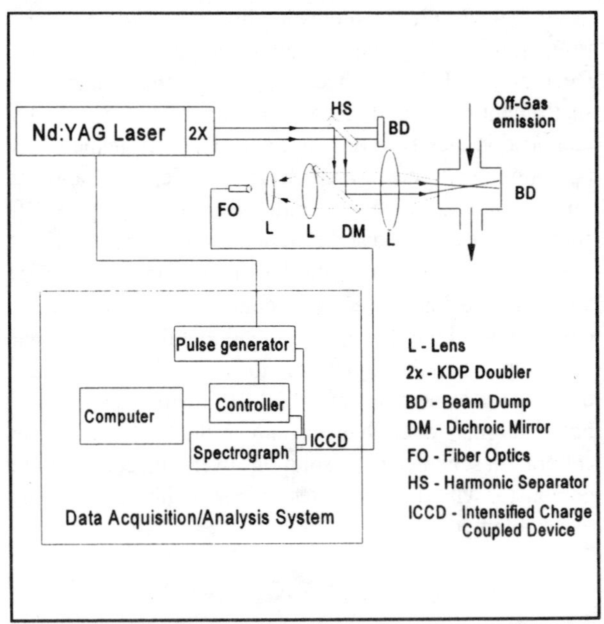

Figure 1. Schematic of the LIBS system used at the DIAL/MSU facility.

The LIBS system used in this study is described in detail in reference 16 and is shown in Figure 1. A frequency-doubled Q-switched Nd:YAG laser (Continuum Surelite III) produced 320 mJ of pulse energy at 532 nm with repetition rate of 10 Hz and the pulse width of 3 - 5 ns. The laser beam was reflected at a harmonic separator to remove its fundamental beam. The 532 nm beam was then reflected to the probe lens through a dichroic mirror which reflects 532 nm and transmits other wavelengths. An UV grade lens of 200 mm focal length was used to focus the laser beam to a spot size of 0.03 mm in the gas stream to produce the breakdown. The same lens was also used to collect and collimate emission from the induced spark. The LIBS signal was transmitted through the dichroic mirror and coupled to the fiber optic bundle with the other two lenses. The bundle was formed with 80 single fibers. The core diameter of a single fiber was 0.1 mm and the numberical aperture (NA) of the fiber bundle was 0.16. The configuration of the two ends of the bundle were round and rectangular respectively. The round end was used as an entrance to accept the LIBS signal and the rectangular end was used as an exit to couple the signal to an optical spectrograph. The spectrograph (Instruments SA Inc., Model HR460) was equipped with a 1800 or 3600 l/mm diffraction grating of dimension 75 mm × 75 mm. A 1024×256 elements ICCD detector (Princeton Instrument) with a pixel width 0.022 mm was mounted at the exit of the spectrograph to record the spectrum. The spectral region monitoring by the detector was 30 nm wide with a resolution of 0.15 nm with the 1800 l/mm grating. The detector worked in a gated mode and was synchronized to the laser Q-switch. To maximize the signal over noise, a gate pulse delay of 1 - 20 μs and width of 2 -10 μs was used in most of the work. Data acquisition and analysis were performed using a PC. Typical LIBS sampling time was 10 seconds, corresponding 100 pules accumulation per spectrum.

In this experiment we wish to optimize LIBS operation in a combustion environment, to compare LIBS signal with direct emission signal, to find the effect of different stoichiometries, and to establish a calibration curve. The combustor took two hour to heat up the input air to 870 K and to achieve a stable stoichiometry of 1.1 and 500 lb/hr air flow. (Most of the data were taken at this condition, except the data at a different stoichiometries) After the combustion is stable, metal solution was injected into the combustor and then LIBS data were taken.

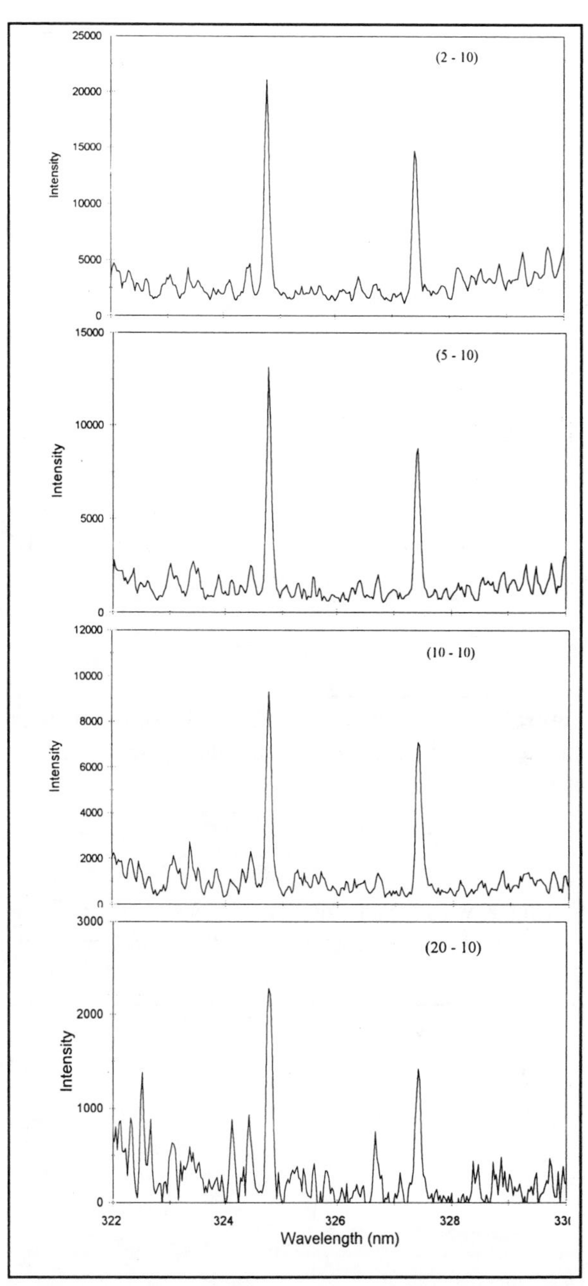

Figure 2. LIBS spectra of Cu recorded with different detector gate delay times. The notations (2,10) indicates a 2 μs delay and a 10 μs width of the gate.

RESULTS AND DISCUSSION

The best detection gate is strongly dependent on the laser energy and property of the sample. It is also related to the target element and the spectral region. To find the best experimental conditions for flame measurements, 5000 μg/ml Cu solution was inject into the combustor at a flow rate of 6.3 ml/min. LIBS spectra at the center wavelength of 325 nm were recorded at different delay times with a fixed gate width of 10 μs to determine the best detection gate. Figure 2 shows LIBS spectra recorded with four different delay times. As seen in the spectra, a 5 μs delay provides the best signal. Therefore, the gate of 5 μs delay and 10 μs width was used in the subsequent runs except when the combustion stoichiometry changed. Similar tests were performed for metals Ca and Fe. We found that the Ca 393.3 nm and 396.8 nm lines achieved the best signal at the gate of 2 μs delay and 4 μs width. Fe line 371.99 nm provided the

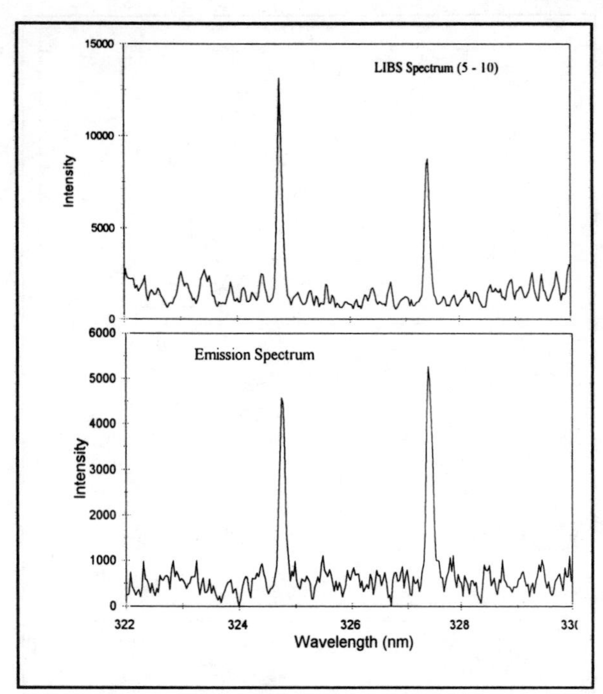

Figure 3. A comparison of Cu LIBS spectrum with directly optical emission spectrum during stoichiometry of 1.1.

best signal with the gate of 10 μs delay and 20 μs width.

In order to make a comparison of the LIBS signal with direct emission from the flame, emission spectra of the metals from the flame were recorded at the same spectral position and combustion conditions as LIBS measurement. It is found that at stoichiometry of 1.1 the LIBS signal of Cu is about three times higher than its signal from the direct emission and has a better signal over background ratio. Figure 3 shows the difference between LIBS and direct emission. In case of a stoichiometry of 0.85, due to the large amount of soot particles formed from unburned fuel, both LIBS and direct emission were enhanced (see figure 4). The absolute Cu line intensity of direct emission is a little lower comparing with the LIBS intensity. However, the Cu direct emission line sits on a very strong background, and the signal over background ratio is only 0.4. The signal over background ratio of the Cu LIBS line is about 2. The background for direct emission varies with spectral region and it is beyond our control. On the other hand, by controlling the gate, LIBS can obtain a much better signal over the background ratio. This feature allows LIBS to achieve a better detection limit.

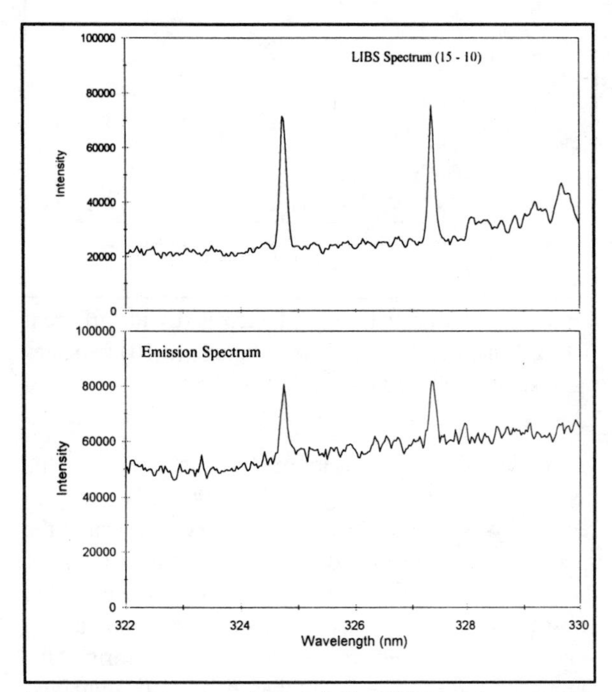

Figure 4. A comparison of Cu LIBS spectrum with directly optical emission spectrum during stoichiometry of 0.85.

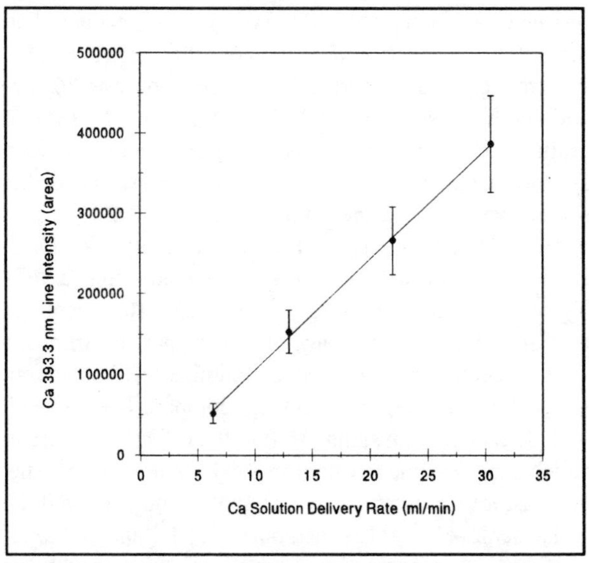

Figure 5. LIBS signal intensity of Ca versus solution delivery rate of 1000 μg/ml Ca solution into the combustor.

LIBS data at different metal concentrations were also recorded. The different metal concentrations in the flame were achieved by manually adjusting the speed of the peristaltic pump. The metal solutions of Cu, Ca, and Fe were tested. The concentrations of the metal solution used for Cu, Ca, and Fe were 5000 μg/ml, 1000 μg/ml, and 1000 μg/ml, respectively. Figure 5 shows the intensity of Ca line (393.3 nm) at different solution delivery rates. The LIBS signals are linearly dependent on the solution delivery rates (i.e., concentrations in the flame). For Cu and Fe, very similar results were obtained. A rough calculation shows the solution was diluted ~100,000 time in the gas stream. The detection limit of Ca from this data is about 100 μg/m^3.

LIBS spectra were also recorded at different combustion conditions to compare LIBS signal at various combustion condition. Figure 6 shows the LIBS spectra recorded at the combustion stoichiometries of 1.0 and 0.85 under the same laser energy and gate setting. The Intensities of Fe in both spectra are comparable. However, the spectrum recorded at stoichiometry of 0.85 (fuel rich) shows a stronger background and CN emission (will cause some interference) than that at stoichiometry of 1.0. At the fuel rich (stoichiometry of 0.85) condition, unburned fuel formed a great amount of soot particles. These particles enhanced breakdown and form a stronger spark than the spark without these particles (with the same laser energy). Thus at the same gate setting, stronger background will be observed. The net effect is that the signal over background ratio (S/B) is not particularly good.

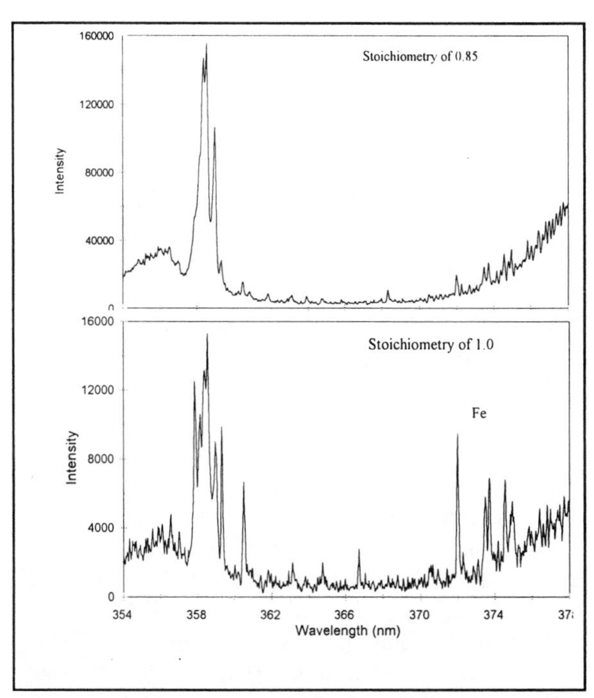

Figure 6. LIBS spectra recorded at different combustion conditions. Please notice there is ten times difference in the intensity scales.

CONCLUSION

The results of this study show that LIBS is capable of measuring metals in a hot, combustion environment. LIBS spectra show a superior signal-over-background ratio as compared to traditional emission spectroscopy. This reveals its advantage to monitor metals in a rocket engine plume containing carbon soot. This research will continue and be extended to more metals. The detection limits will be improved to satisfy practical applications.

ACKNOWLEDGMENTS

This research is supported by NASA contract NAS 13-99002. We want to express our thanks to John Etheridge and R. Arunkumar for operating the test stand.

REFERENCES

1. D.A. Cremers and L.J. Radziemski, "Laser Plasma for Chemical Analysis" in Laser Spectroscopy and its Applications , L.J.Radziemski, R.W. Solarz and J.A. Paisner eds., Chap. 5, 351-415, Dekker, New York, 1987.
2. D.K. Ottesen, J.C.F. Wang, and L.J. Radziemski, Real-time laser spark spectroscopy of particulates in combustion environments, Appl. Spectrosc. 43, 967-976 (1987).
3. D.A. Cremers, L.J. Radziemski, and T.R. Loree, Spectrochemical analysis of liquids using the laser spark, Appl. Spectrosc. 38, 721-729 (1984).
4. D. R. Anderson, C. W. McLeod, and T. A. Smith, J. Anal. Atom. Spectrosc. 9, 67 (1994).
5. K.J.Grant, G.L.Paul, and J.A.O'Neill, Quantitative elemental analysis of iron ore by laser-induced breakdown spectroscopy, Appl. Spectrosc. 45, 701-705 (1991).
6. J.P. Singh, H. Zhang, F.Y. Yueh and K.P. Carney ,"A investigation of the Effects of Atmospheric Conditions on the Quantification of Metal Hydrides Using Laser-Induced Breakdown Spectroscopy", Appl. Spectrosc. 50, 764-773, (1996).
7. L.J. Radziemski and D.A. Cremers eds. Spectrochemical Analysis Using Plasma Excitation, in: Laser Induced Plasma and Applications, Chap. 7, 295-325. Marcel Dekker, New York, NY, 1989,
8. R.L. Cook, J.P. Singh, H. Zhang, F.Y. Yueh and M. McCarthy, Advanced Analytical Instrumentation Demonstration - LIBS Measurements, SAIC's STAR Center, Idaho Falls, DIAL/MSU Trip Report 95-1, DIAL 10575 (1995).
9. J.P. Singh, H. Zhang, and F.Y. Yueh, "DOE and EPA Continuous Emission Monitoring Test at EPA National Risk Management Research Laboratory - LIBS Measurements", Research Triangle Park, Raleigh, North Carolina, DIAL/MSU Trip Report 96-2, DIAL 10575 (1996).
10. C. Cornelison, L.M.DeWitt, G.R.Hassel, G.L.Leatherman, Final Report for the Advanced Analytical Instrumentation Demonstration, Science

Applications International Corporation, Waste Management Technology Division, Idaho falls, SAIC Report number SAIC-95/1308,1995.

11. W.J.Haas Jr., N.B.French, C.H.Brown, D.B.Burns, P.M.Lemieux, S.J.Priebe, J.V.Ryan, and L.R.Waterland, "Performance testing of multi-metal continuous emissions monitors," Ames Laboratory, U.S. DOE (Iowa State University, Ames, Iowa, November, 1997).
12. H. Zhang, J.P. Singh, F.Y. Yueh and R.L. Cook, "Laser Induced Breakdown Spectra in a Coal-Fired MHD Facility, Appl. Spectrosc. 49, 1617-1623 (1995).
13. G.W. Loge, J.P. Singh, F.-Y. Yueh and H. Zhang "Hydrocarbon-Fueled Rocket Engine Health Monitoring by Laser-Induced Breakdown Spectroscopy". Final Report for Contract No. NAS 13-99002, STTR Phase 1, (October 27, 1999).
14. R.G.McCoy, W.J.Philips, W.K.McGregor, W.T.Powers and H.A.Cikanek, "Analysis of UV-VIS Spectral Radiation from SSME Plume", JANNAF Propulsion Meeting, San Diego, CA, CPIA 480, V. 205, December 1987.
15. D.G.Gardner, F.E.Bircher, G.D.Tejwani, and D.B.Van Dyke, "A Plume Diagnostic Based Engine Diagnostics System for the SSME", AIAA 90-2235, 26th Joint AIAA/SAS/ASME/ASEE propulsion Conference, Orlando, FL, July 1990.
16. J.P. Singh, F.Y. Yueh, H. Zhang, and R.L. Cook "Analytical Method Using Laser-Induced Breakdown Spectroscopy," U.S. Patent S. No. 08/705, 267 (1997).

SIMULATION OF FLYWHEEL ELECTRICAL SYSTEM FOR AEROSPACE APPLICATIONS

Long V. Truong, Frederick J. Wolff, and Narayan V. Dravid
NASA Glenn Research Center
Cleveland, Ohio

ABSTRACT

A Flywheel Energy Storage Demonstration Project was initiated at NASA Glenn Research Center as a possible replacement for the Battery Energy Storage System on the International Space Station (ISS). While the hardware fabrication work was performed at a contractor's place and at a university, the related simulation activity was begun at NASA Glenn Research Center. At the top level, the primary electrical system of ISS is simulated with one Battery Charge and Discharge Unit (BCDU) replaced by the Flywheel Energy Storage Unit (FESU) is described in another paper. The FESU consists of a Permanent Magnet Synchronous Motor/Generator, PMSM, (connected to the flywheel), the Power Electronics that connects the PMSM to the ISS dc bus, and the associated controller. While the PMSM model is still under development, the rest of the FESU model is described in this paper.

INTRODUCTION

Spacecraft electrical power systems, in general, convert solar energy into usable electrical energy. For Low Earth Orbit (LEO), a rechargeable energy storage system is necessary to supply loads during the eclipse period as well as to provide backup power. Traditionally, space qualified electrochemical storage batteries have served this purpose. However, unless there is a provision to replace such batteries often, they limit the life of the spacecraft power system itself.

As a battery replacement, an electromechanical energy storage (Flywheel) system is under development. In terrestrial systems, a flywheel is often used as a load equalizer. In space systems, flywheels have been extensively used for 'Attitude Control' purposes.

The new development intends to use the flywheel for both energy storage and attitude control purposes. It is a joint industry-government (NASA) effort. Naturally, the energy storage requirements for battery replacement are much higher compared to those for attitude control.

A comparison between flywheel and NiH_2 battery systems for the EOS-AM1 type spacecraft[1] has shown that flywheel system would be much smaller and lighter: 35% in mass reduction, 55% in volume reduction, and 6.7% in solar array area reduction. For more information on the NASA's flywheel programs, current status, business opportunities, etc. please refer to NASA Glenn Research Center's web site[2].

In support of this development, modeling and simulation activity was begun at NASA Glenn Research Center. At the top level, the ISS primary electrical system will be modeled with one BCDU-Battery unit replaced by the FESU. At the next level, the FESU itself will be modeled to account for the dynamics between the 'dc to 3 phase ac' converter and the motor model. A separate motor modeling activity is taking place, both in the circuit domain and in the Finite Element Modeling (Electromagnetic) domain[3]. This paper describes the simulation of the converter and associated control system that regulates energy transfer to/from the flywheel. A simpler representation of the motor/generator is employed as described later in the paper.

SYSTEM DESCRIPTION

Inertial energy stored in a flywheel varies as the square of its rotational speed. This permits the flywheel energy storage system a depth of discharge of 90% (possible for a battery system but at the expense of life) for a reduction in speed to only one third. The operating speed of the flywheel has been established to be between 60,000 RPM (max) and 20,000 RPM (min). The flywheel rotational inertia constant is then selected based on energy storage needed. The flywheel rotating speed simply goes up/down during charge/discharge, while observing the above mentioned limits. Mechanical and structural properties of the wheel as well as the reliability of the shaft bearings govern the upper limit. Further lowering of the lower speed limit does not yield any significant depth of discharge. Figure 1 shows the energy flow in a flywheel system for charge and discharge conditions. Figure 2 shows the top-level diagram of the simulated flywheel system. For clarity only details of phase 'A' are shown.

A 2-pole, 3-phase, Permanent Magnet Synchronous Motor/Generator is shaft-coupled to the flywheel. The voltage magnitude and frequency at the machine winding terminals are directly proportional to the speed of rotation. Thus, during motor action, for example, an inverter output with desired voltage and frequency profile is made available to cause energy transfer to the motor and, hence, to the flywheel. Naturally, the reverse process takes place when energy is transferred in the opposite direction.

For a typical synchronous motor, changing the frequency of the applied voltage changes the motor speed. This is known as open loop speed control. However, in the present case, the speed control is also based on shaft position feedback, the closed loop control. Briefly, only the winding whose axis aligns with the pole axis is energized, successively, to create torque. Hence it is necessary to know the location of the pole axis or the shaft position. It may be noted, however, that these considerations are not relevant in the present simulation, as there is no explicit shaft rotation in the simplified model of the synchronous machine.

The Power Electronics (Fig. 3) in a FESU works to convert from a dc source to ac (charge mode) and vice versa (discharge mode). While there could be many ways to bring this about, the system under simulation consists of three, single phase, dc-to-ac converters that are connected to the three motor terminals. The Y-point of the motor winding remains unconnected. The dc ends of the converters are connected, in parallel, to the dc bus. Appropriate filters are installed between the dc bus and the converter switches ('S1A' and 'S2B') to mitigate the effects of harmonics caused by switching action. The motor winding terminal is connected to the common point between the switches. Thus, any current entering or leaving a motor winding is returned via the other two windings and appropriate closed switches or the diodes in parallel (actually the body diode). A closing and opening sequence for the switches is established to cause power transfer between the dc and the ac.

As mentioned earlier, the controller (Fig. 4) operates the power electronic switches to excite the motor windings based on the shaft position. For the purposes of this simulation, the controller also limits currents in the switches and the windings to 150 A and 100 A, respectively. Reference values of currents to be limited are compared to the actual currents fed back and error signals are generated to drive the pulse width modulator that operates the switch. (Figure 5). Similarly, other desirable quantities could be controlled, i.e., rate of change of flywheel speed, rate of charge/discharge current, etc.. The controller is also capable of reversing the operation from charge to discharge based upon a command. A reference sine wave is provided to generate control error for the PWM. The frequency of this sine wave determines the frequency of the machine output voltage and, therefore, the flywheel speed.

Certain observations are made with respect to the switch action (see Fig. 5). During charge operation, only switch 'S1A' is pulse width modulated while switch 'S2B' works in ON/OFF mode. During discharge operation, only switch 'S2B' is pulse width modulated while switch 'S1A' is open all the time. However, the body diode of 'S1A' conducts if appropriate bias voltage is available.

MOTOR MODEL

An 'electrical' and not an 'electromechanical' model is used for this simulation. Thus, the model cannot simulate the motor torque equation as a mechanical quantity. However, it can simulate the electrical response by virtue of its back EMF and the winding parameters. Each phase of the motor consists of an R-L series circuit, in series with a sinusoidal voltage source that represents the back EMF (i.e. air gap voltage). The Y-point is left unconnected. Presently, the magnitude and frequency of the back EMF are held constant for a given case. Thus, we are able to simulate power exchange with the motor/generator but not energy exchange. Or, the simulation takes place at a fixed rotational speed of the machine. Motor simulation data are as follows.

3-phase, 2-pole, Permanent Magnet Synchronous Motor/Generator
Resistance per phase: 0.0145 ohm
Inductance per phase: 16.7 micro H
At maximum rotor speed of 60,000 RPM:
 Open circuit voltage, 80 volts L-L, rms
 Machine rating, 7 kVA
 Frequency, 1000 Hz
 Line current, 50 amps, rms
DC Bus voltage range: 160 to 120 volts

During normal operation the rotor speed will not go below 20,000 RPM. At that speed, the voltage and frequency will be one-third as much as that at the maximum speed. The current rating does not change.

This model assumes that the rotor is uniformly round and that saliency effects are neglected. These will be considered in the next phase of modeling.

SCOPE OF SIMULATION

As was mentioned earlier, the motor operation is simulated while at a constant speed. At the chosen speed we can simulate charge or discharge action. Due to the voltage level magnitudes on both the dc and the ac side, one should note that bucking action takes place during charging, and boosting takes place during discharging. Thus, in a given situation, the level of power being transferred will depend upon the levels of dc and ac voltages encountered. Keeping this in view, we ran five simulations, as shown in Table 1. These cases cover all the extremes of voltages on both dc and ac sides. Also, as pointed out earlier, the voltage magnitude and frequency change proportionally with RPM.

Table 1: Simulation cases based on ac and dc operating voltages.

Cases	Flywheel RPM	Back EMF, L-N, peak volts	Frequency (Hz)	DC bus voltage (V)	Simulation Results
Case 1	40,000	43	666	140	Figure 6
Case 2	20,000	22	333	120	Figure 7
Case 3	20,000	22	333	160	Figure 8
Case 4	60,000	65	1000	120	Figure 9
Case 5	60,000	65	1000	160	Figure 10

SIMULATION TOOL

Saber[4], a simulation software package developed by Analogy, Inc., is used for the simulations at hand. The tool permits complete circuit simulation as well as control simulation using logic devices.

SIMULATION RESULTS

The five cases delineated in Table 1 were simulated, and the resulting data plots are shown in Figures 6 through 10. Case 1 has voltage values midway between the extremes chosen in the other cases. The results shown are steady state operations over an arbitrary length of time. For each case, both charge and discharge operations are shown. Cases 2 through 5 represent the pairings of the high/low values of the ac and dc voltages. These are possible conditions of operation.

Figure 6 shows simulation results for Case 1. Levels of 'charge/discharge' and 'enable' commands are shown to achieve charge, disable, and discharge operations. Phase 'A' back EMF is shown which, by definition, is a sine wave of specified magnitude and frequency. Next shown is a rectified version of the phase 'A' current. Lastly, the dc current output of the FESU to the system dc bus is shown. Clearly, this current is negative during the charge mode, as it should be.

Figures 7 through 10 show simulation results for the remaining cases from Table 1. Only rectified Phase 'A' current and the dc current into the dc bus are shown for comparison. The high and low values of the ac voltages denote conditions at the operating range of the flywheel, while the high and low values of the dc voltage denote values most likely available at beginning of life and end of life of the space station. It may be noted that the motor current is seen to be limited to 100 A (mentioned) in all the cases except case 4.

As discussed earlier, bucking and boosting actions take place in charge and discharge modes, respectively, mainly due to the voltage levels available. Amount of boosting possible also depends upon the commutating inductance (the winding inductance in this case). We have verified that we could make discharge current nearly zero by arbitrarily increasing the dc voltage or decreasing the ac voltage (results not shown).

The switch model consists of a nearly ideal switch in parallel with an ideal diode (the body diode). The motor winding current waveforms do not look like rectified sine waves. In fact, one half of the waveform looks quite distorted. Snubber circuits, which were not modeled, will have influence on the shape of these waveforms.

CONCLUSION

Interaction of the power electronic dc-ac converter, connecting the dc bus to the flywheel machine is demonstrated at constant flywheel speed. Exchange and reversal of currents between the two has been shown. A simplified version of machine model was used which precludes the simulation of electromechanical behavior of the machine. Work continues to upgrade the models, leading to the demonstration of energy exchange between the dc bus and the flywheel.

ACKNOWLEDGEMENTS

The authors would like to thank Mr. Ray Beach from NASA Glenn Research Center, Cleveland, OH, and Mr. John Biess from US Flywheel Systems, Newbury, CA for their advice and support.

REFERENCES

1. Flywheel Energy Storage for Spacecraft Power Systems, Mukund R. Patel, PhD, PE, Power Systems Engineering, Yardley, Pennsylvania, 99IECEC-183
2. http://space-power.grc.nasa.gov/ppo/flywheel/
3. MAXWELL is a trademark of AnSoft Corp., www.ansoft.com
4. Saber is a trademark of Analogy, www.analogy.com

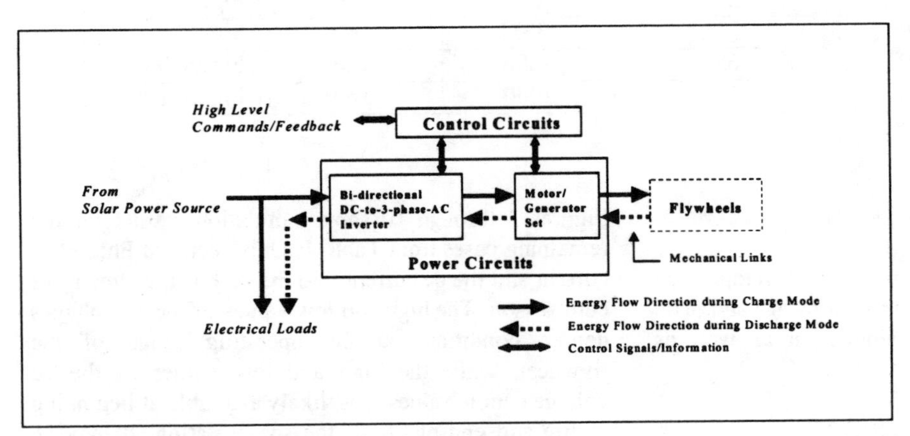

Figure 1: Energy Transfer Diagram for FESU.

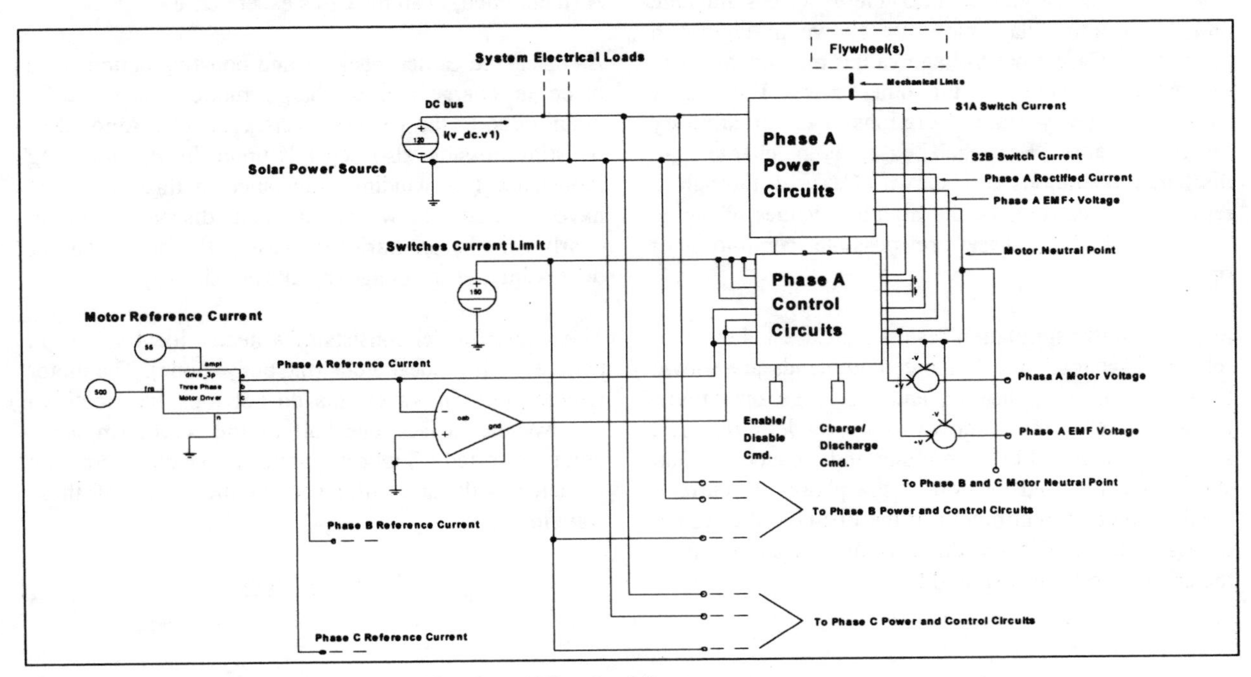

Figure 2: Top Level Diagram of the Simulated Flywheel System.

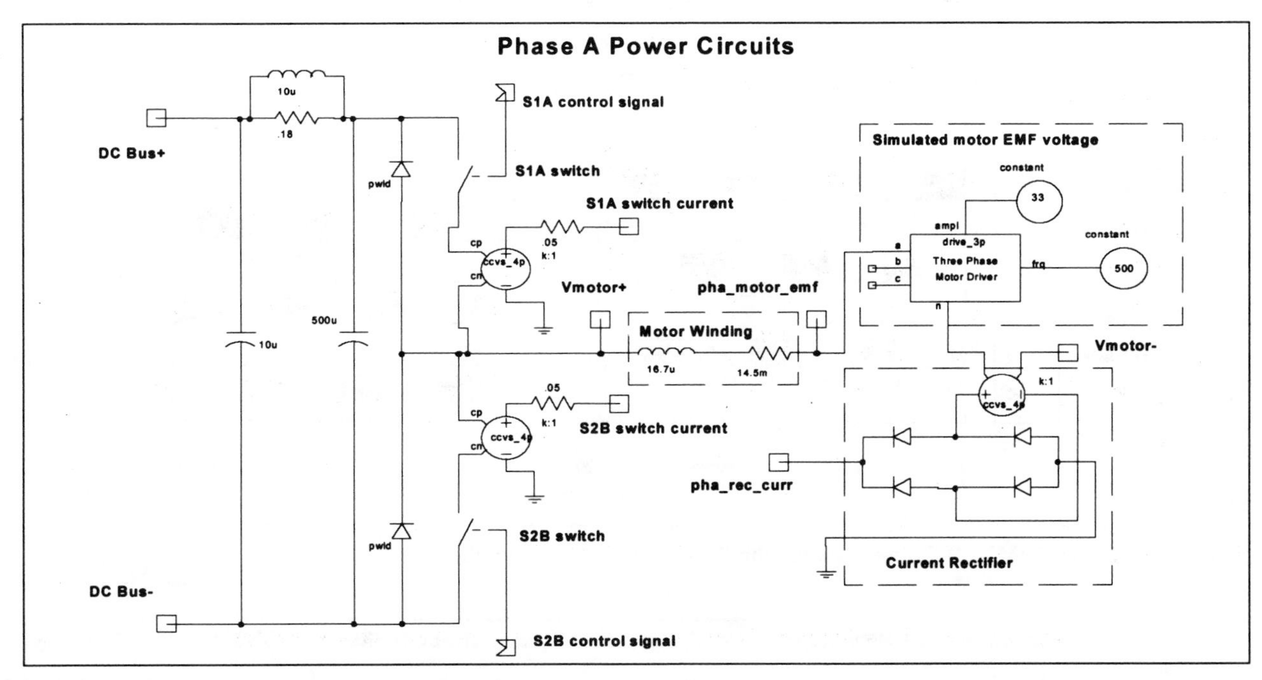

Figure 3: Details of 'Phase A' Power Circuits.

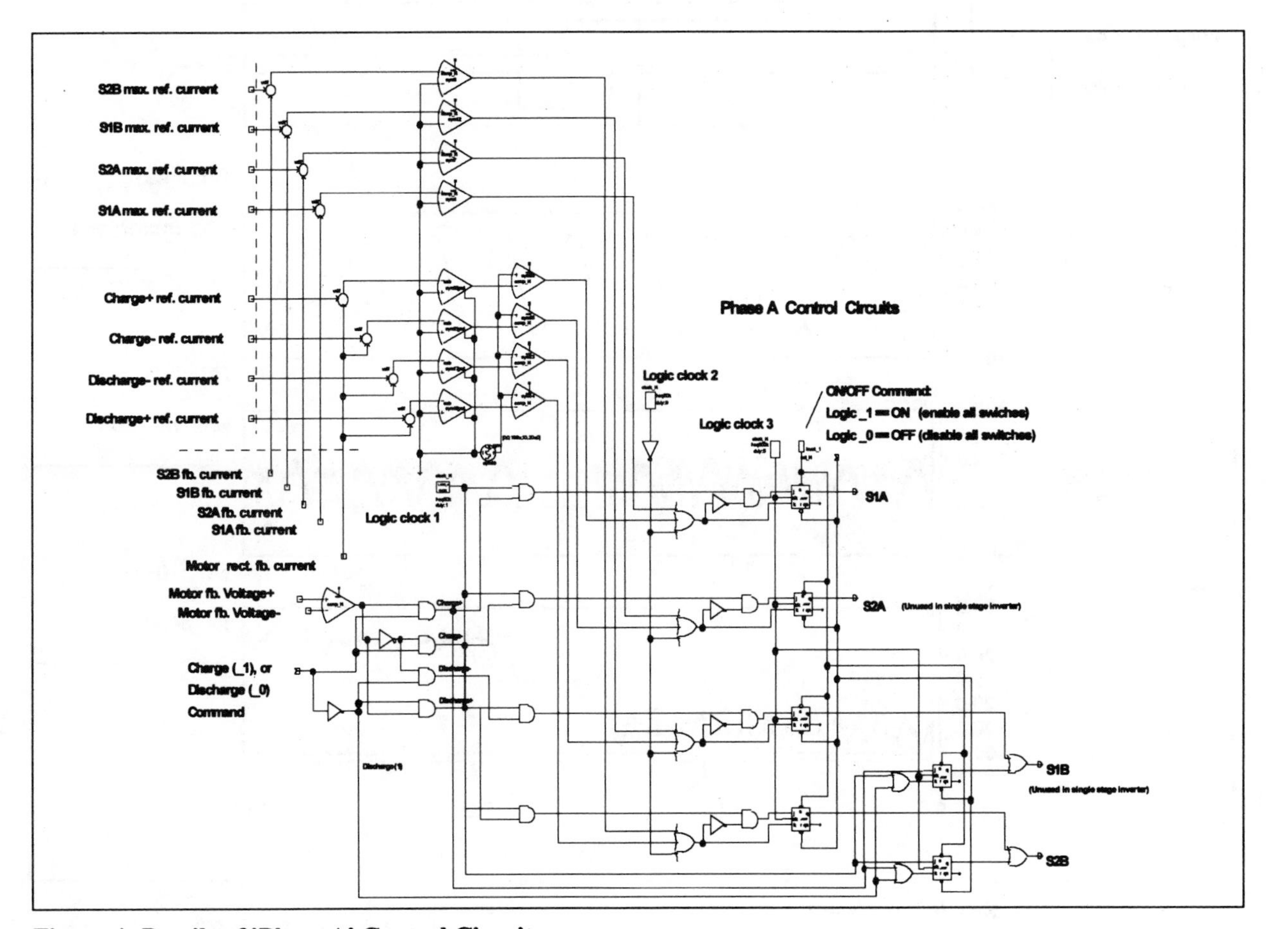

Figure 4: Details of 'Phase A' Control Circuits.

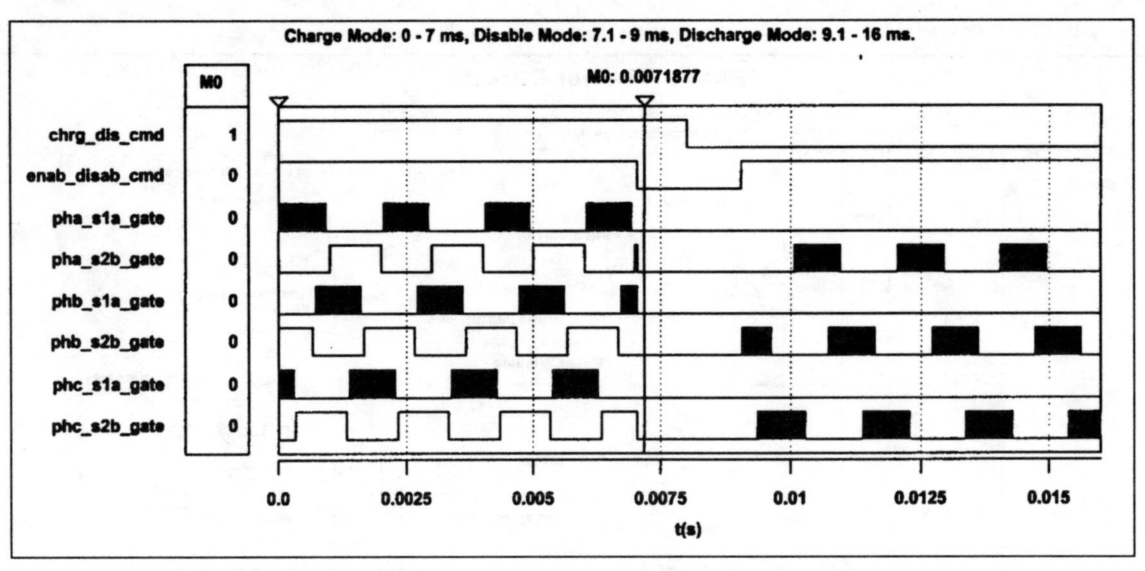

Figure 5: Typical PMW Drive Signals for the Power Electronic Switches.

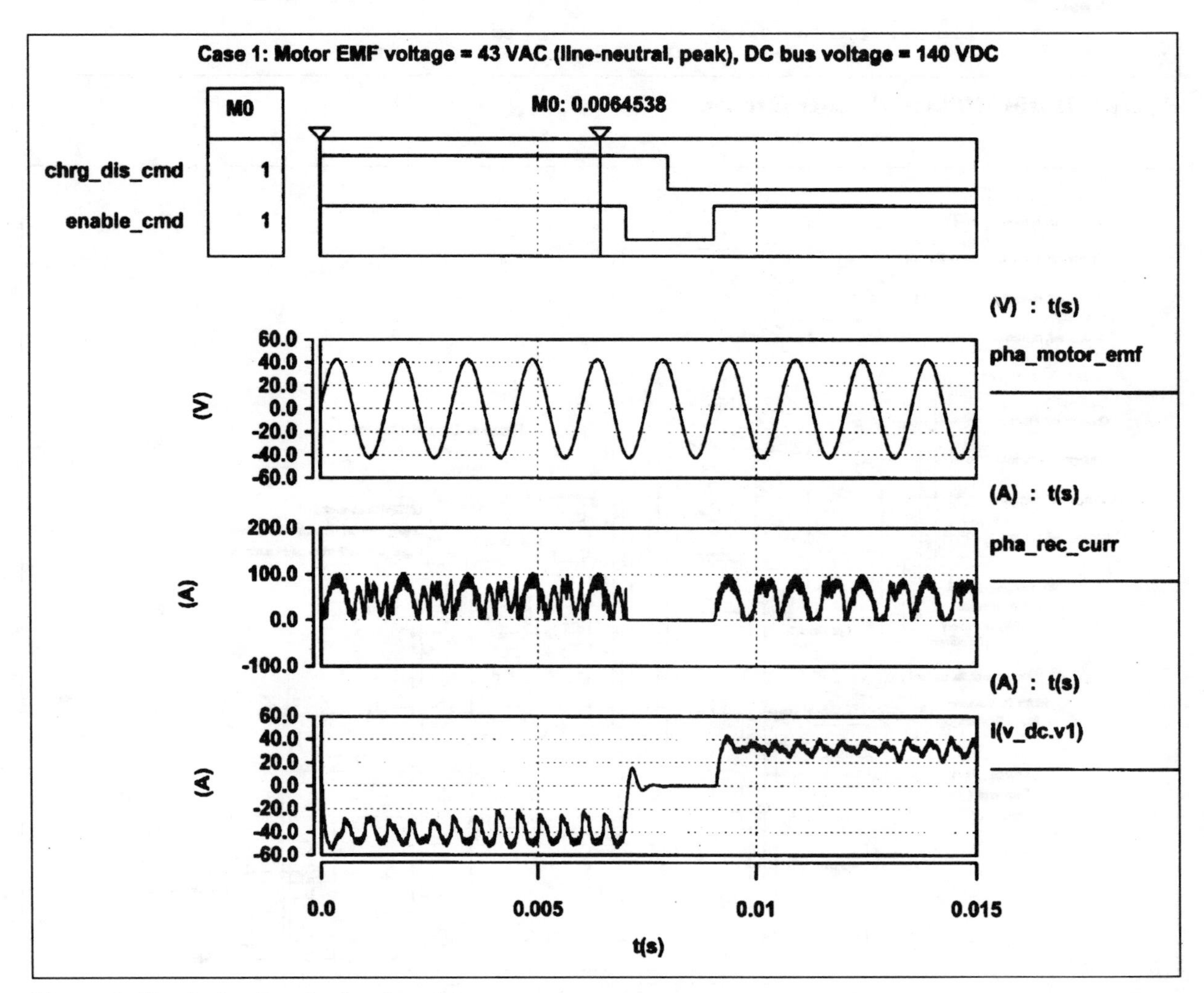

Figure 6: Simulation Results for Case 1.

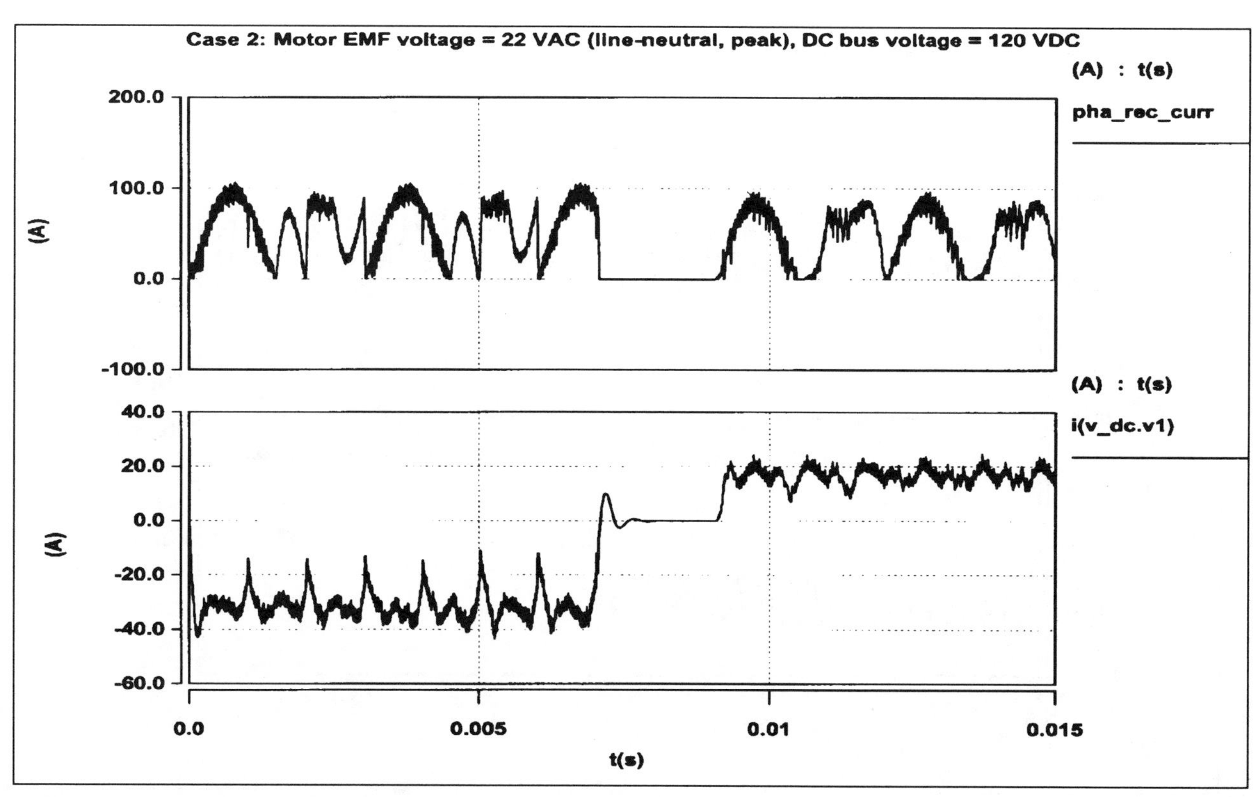

Figure 7: Simulation Results for Case 2.

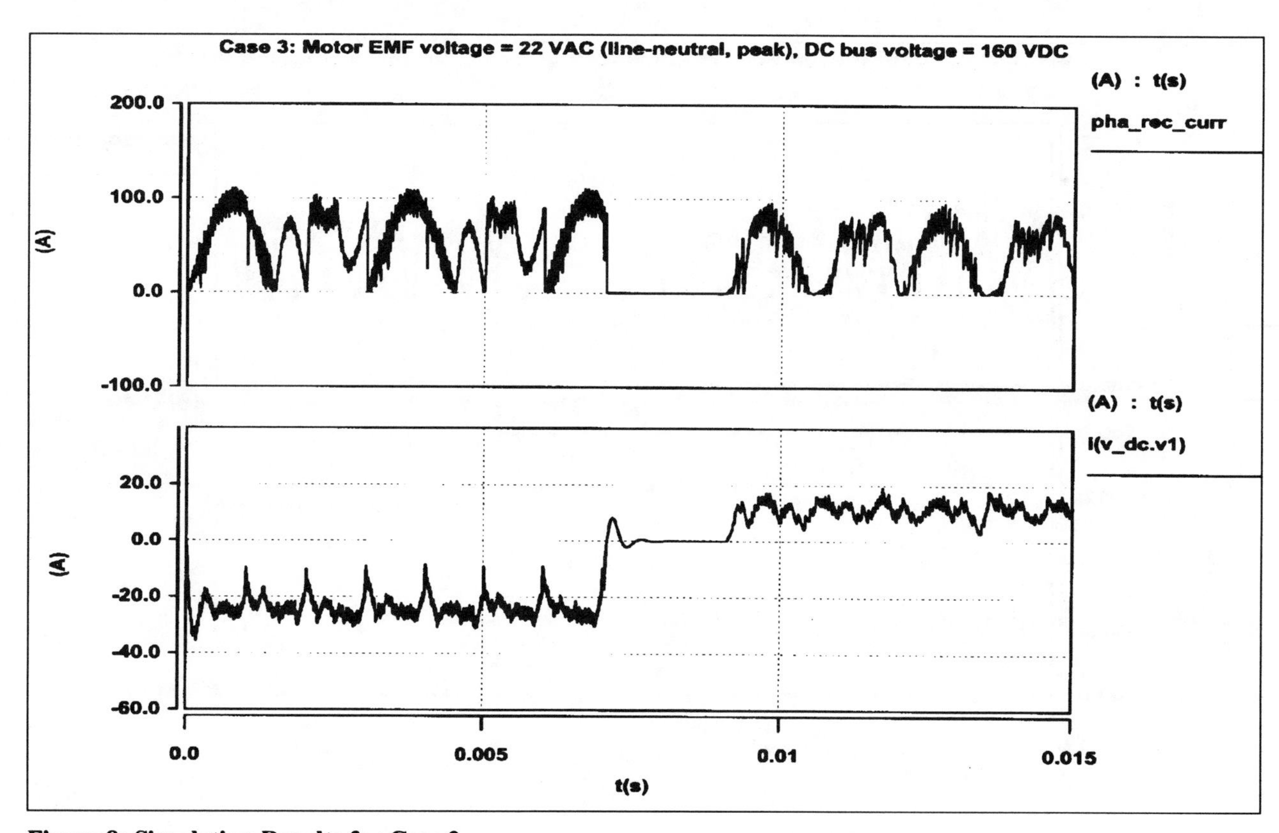

Figure 8: Simulation Results for Case 3.

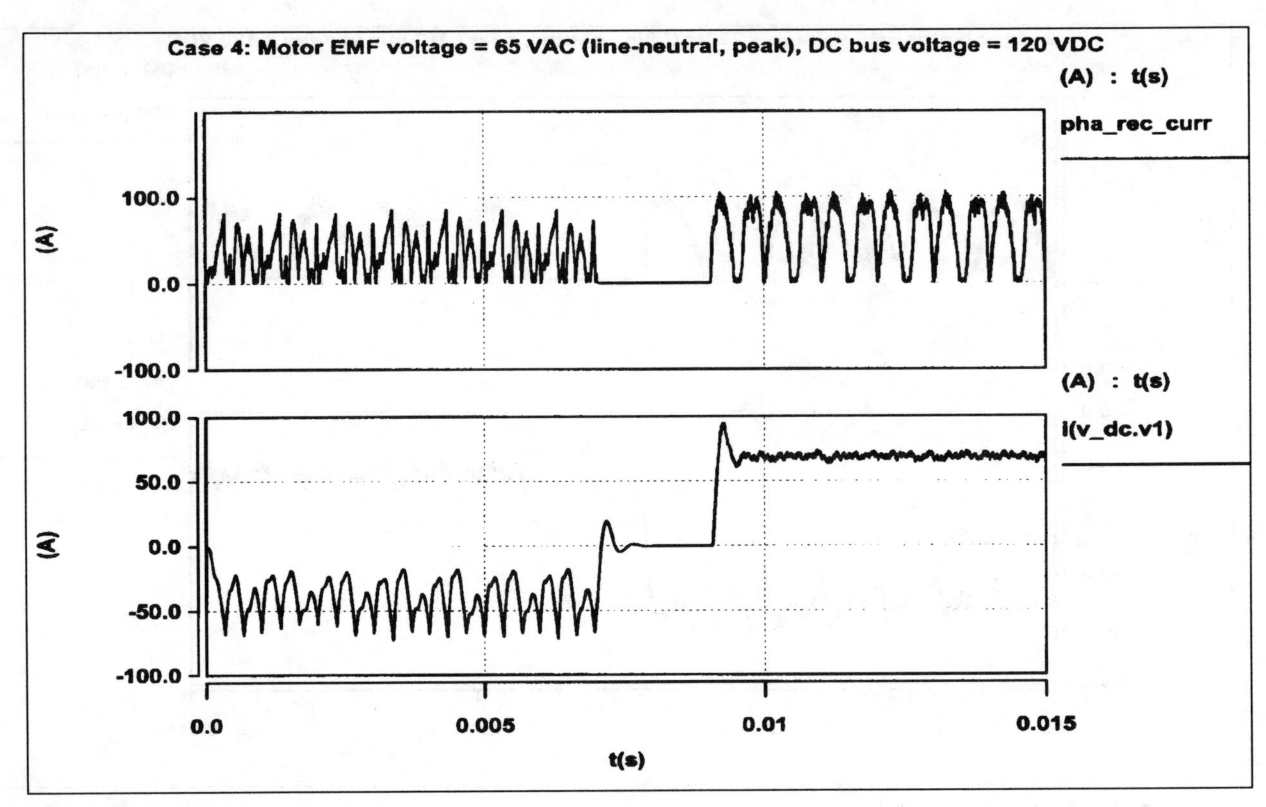

Figure 9: Simulation Results for Case 4.

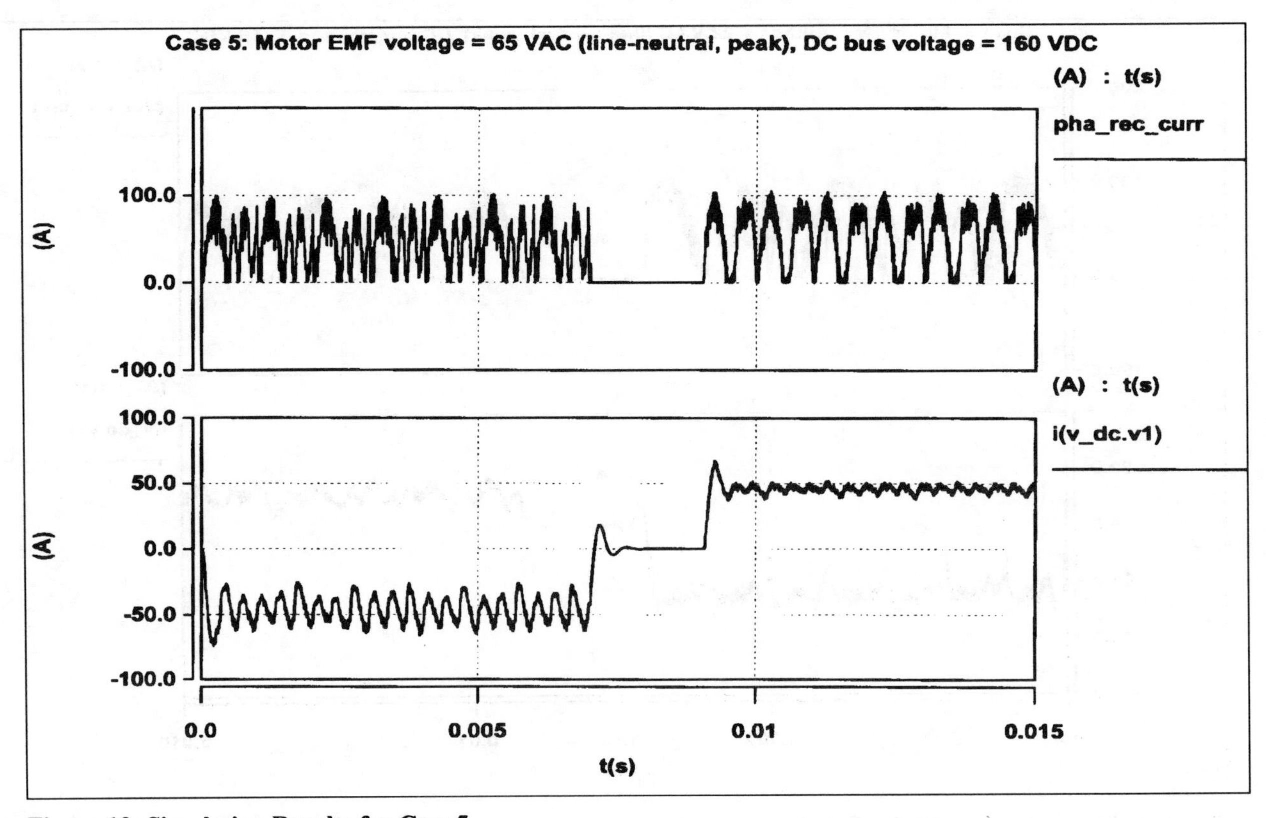

Figure 10: Simulation Results for Case 5.

ENERGY-MOMENTUM-WHEEL FOR SATELLITE POWER AND ATTITUDE CONTROL SYSTEMS

Mukund R. Patel, PhD, PE
U.S. Merchant Marine Academy
Kings Point, New York 11024

ABSTRACT

In an earlier paper by the author[1] on a 2500-watt low earth orbit satellite, such as NASA Goddard Space Flight Center's EOS-AM, the mass and volume reductions by replacing the battery with the flywheel were estimated to be 35% and 55% respectively. Further savings are possible by using the dual function flywheel that stores energy for the electrical power system and momentum for the attitude control system of the satellite. This paper analyzes the operation of such dual function flywheel, termed as the energy-momentum-wheel. As the spacecraft cannot discharge energy without discharging the momentum, the maximum depth of energy discharge is limited by the minimum momentum storage requirement on a given axis. Such mission level operating constraints are analyzed and presented in the form of circle diagrams.

1.0 INTRODUCTION

The energy storage requirement of the spacecraft power systems is traditionally met by using electrochemical batteries such as the NiCd and NiH2. The momentum storage and control requirements of the attitude control system are traditionally met by using flywheels such as momentum wheels and reaction wheels. In has been reported earlier[1] that potential savings in the electrical power system by replacing the battery with the energy storage flywheel in a 2500-watt low earth orbit satellite are 35% in mass and 55 % in volume. Further savings are possible by using dual function flywheels for storing energy for the power system and momentum for the attitude control system of the satellite[2,3,4].

Multiple energy-momentum-wheels are required for replacing the battery of the electrical power system and momentum and reaction wheels of the attitude control systems. At least one energy-momentum-wheel on each axis controls the momentum in that axis and stores the corresponding energy at the same time. Placing additional wheel on each axis would give a degree of freedom in simultaneously storing the required energy and the required momentum. One more wheel on multi-axis gimbals is needed for redundancy. Thus, a typical single fault-tolerant system would require seven energy-momentum-wheels.

2.0 POWER SYSTEM BLOCK DIAGRAM

The spacecraft power system configuration with the energy-momentum-wheel is shown in Figure-1, which represents the direct energy transfer regulated bus. Such architecture is widely used by satellite manufacturers, except that the battery is now replaced by the energy-momentum-wheel.

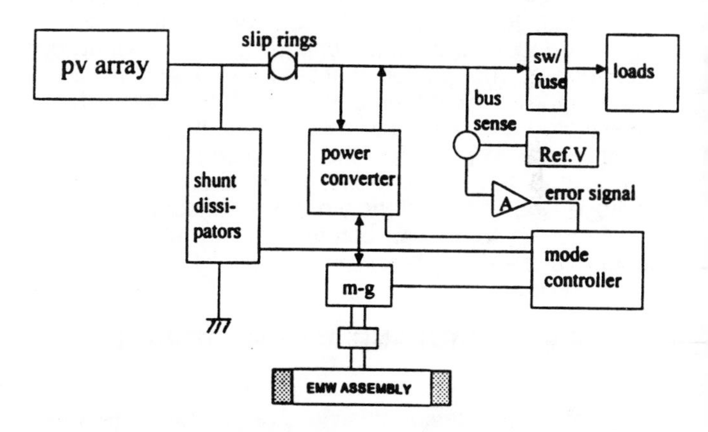

Figure-1. Direct energy transfer photovoltaic / energy-momentum-wheel architecture of the spacecraft power system

3.0 ENERGY-MOMENTUM-WHEEL ASSEMBLY

The major components of the energy-momentum-wheel assembly are the rotor rim(s), the hub, magnetic bearings, mechanical bearings, electromechanical energy converter (motor-generator machine), sensors and the power and momentum management electronics.

The rotor rim is the primary components of the energy-momentum-wheel that stores the required energy and momentum. It is made of high strength fiber-resin composite matrix, with fibers principally oriented in the hoop direction to withstand high hoop stress due to the rotational centrifugal forces, which is the greatest at the outermost tip of the wheel. With two-rim design (Figiure-2), the outer rim is made of high strength fiber composite and the inner rim is made of low cost fiber composite.

Figure-2. Two-rim design of the flywheel with high strength outer rim and low-cost inner rim.

Power and momentum management electronics, although not a part of wheel assembly, is an important part of the energy-momentum-wheel system. It is an independent box, separate from the wheel, that senses the bus voltage and bucks or boosts the bus voltage to match with the motor-generator voltage depending on the wheel speed at a given time. It provides a total interface between the motor-generator and the regulated voltage bus. It also contains the charge and discharge converters, the power management function, the momentum management function and the associated telemetry and commands. It controls the charge (motor) mode, the discharge (generator) mode and the shunt mode as required during the eclipse and the insolation periods. Additionally, it controls the torque as commanded by the attitude control system.

4.0 LAUNCH AND ASCENT POWER

A potential design issue with the energy-momentum-wheel system is whether the wheels can be spinning at the time of the launch to provide the required power during the launch and ascent. This issue must be analyzed and resolved with the launch vehicle primes. Until then, one can take a conservative approach of assuming that the wheels cannot be spinning until the satellite is in the transfer orbit and the photovoltaic array is exposed to the sun. In that case, an effective alternative for providing the launch and ascent power is the primary battery, such as the Lithium Carbon Monofluoride. The LiCFx has high specific energy and a long flight history in launch vehicles. The cells are available in several AH ratings. Once the sun is acquired in the transfer orbit, the systems is commanded to spin up the wheels to full speed (100 % state of charge). Then onward, the system maintains the energy balance over the orbit period.

5.0 ANALYSIS OF THE ENERGY-MOMENTUM WHEEL OPERATION

A flywheel having the mass moment of inertia J spinning at speed ω stores the energy

$E = \frac{1}{2} J \omega^2$, which is a scalar quantity

It also stores the momentum

$M = J \cdot \omega$, which is a vector quantity.

If two wheels are spinning on a given axis, having mass moment of inertia J_1 and J_2 and rotating at speeds ω_1 and ω_2, respectively, then the total energy storage is

$$E = \frac{1}{2} J_1 \omega_1^2 + \frac{1}{2} J_2 \omega_2^2 \quad \text{(scalar quantity)}$$

This is an equation of an ellipse in the ω_1-ω_2 plane.

The total momentum of such two wheels is

$$M = J_1\omega_1 + J_2\omega_2 \qquad \text{(vector quantity)}$$

Rearranging the above, we have

$$\omega_2 = M/J_2 - (J_1/J_2)\,\omega_1$$

which is an equation of a straight line with the

slope $-(J_1/J_2)$ and the ω_2-axis intercept of M/J_2.

Figure-3 shows a constant energy ellipse and a constant momentum line of a pair of two unequal wheels. Two wheels requiring to store a given energy and a momentum simultaneously would operate at either of the two intersection points p1 or p2.

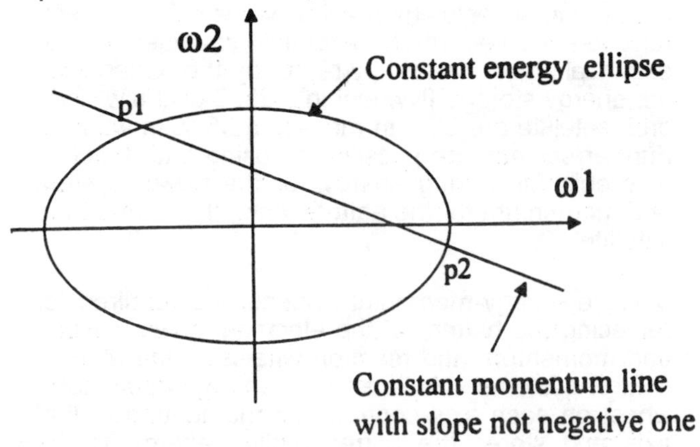

Figure –3: energy-momentum-wheel operation with two unequal wheels

Total energy stored in two equal wheels each with inertial J is

$$E = \frac{1}{2} J \cdot \omega_1^2 + \frac{1}{2} J \cdot \omega_2^2$$

Rearranging the above gives

$$\omega_1^2 + \omega_2^2 = 2E / J$$

This is an equation of a circle with radius of

$$\sqrt{(2E / J)}$$

The total Momentum of two equal wheels is

$$M = J . \omega_1 + J . \omega_2 = J (\omega_1 + \omega_2)$$

Rearranging the above leads to

$$\omega_2 = M / J - \omega_1$$

which is a straight line with slope −1 and ω_2-axis intercept of M/J.

Figure-4 shows the two possible points of operation, p1 and p2, for two equal wheels requiring to store a given level of energy while simultaneously maintain certain momentum level. The both points of operations are dynamically stable; hence the system can indeed operate at any one point which is closest to the prior operating point.

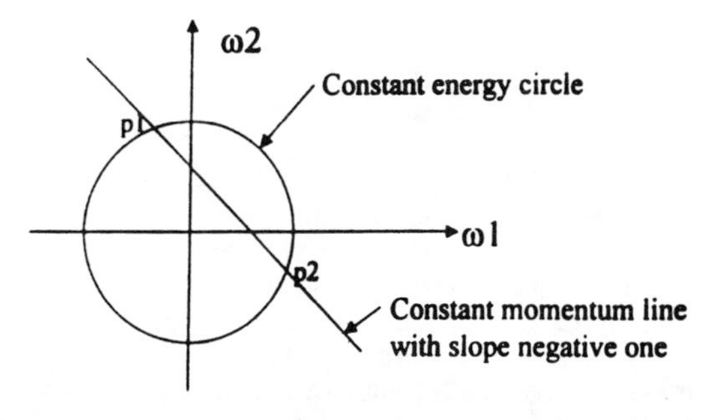

Figure-4: energy-momentum-wheel operation with two equal wheels

6.0 DEPLETING ENERGY AND MOMENTUM

As the spacecraft cannot discharge energy without discharging the momentum, the maximum depth of energy discharge is limited by the minimum momentum storage requirement on a given axis. Such an operating constraint is analyzed below.

If a pair of equal wheels that must maintain a given energy level, but change its momentum level to meet a attitude control requirement must operate at any point on the same circle, while changing the momentum line closer or farther from the origin, a shown in Figure-5. On the other hand, if the system requires to deplete the energy without depleting the momentum, the speed of the wheels must change to contract the energy circles while maintaining the momentum line as shown in Figure-6.

7.0 MISSION CONSTRAINT ON MEETING THE SIMULTANEOUS ENERGY AND MOMENTUM STORAGE REQUIREMENTS

The necessary condition for energy-momentum-wheel operation is

$$\sqrt{(2 \cdot E_{min} / J)} > M_{min}$$

or $E_{min} > \frac{1}{2} \cdot J \cdot M_{min}^2$

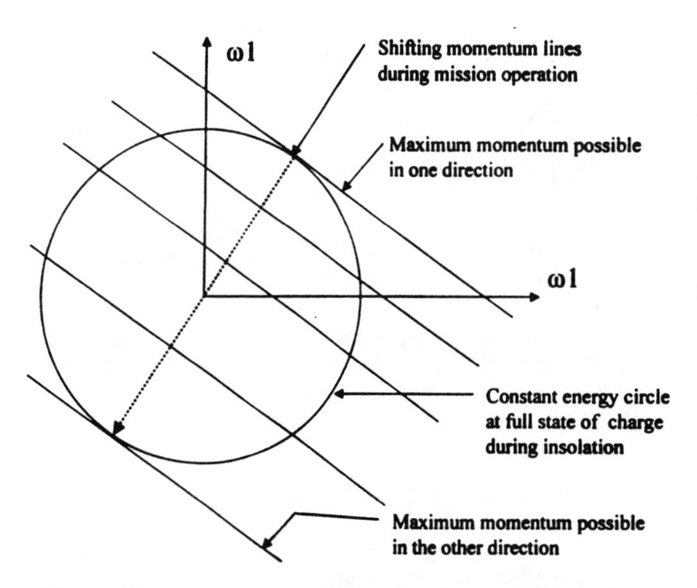

Figure 5: energy-momentum-wheel operation during sunlight

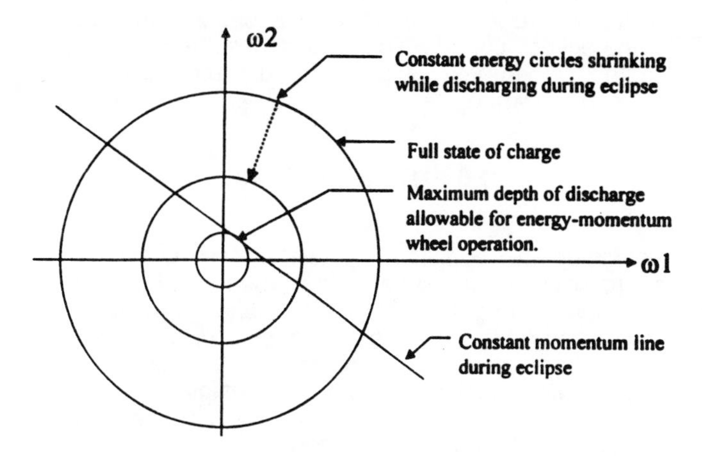

Figure-6: energy-momentum-wheel during eclipse.

In all modes of spacecraft operation, the radius of the minimum energy circle must be greater than the distance of the momentum line from the origin, so that the circle and the line have two intersecting points of operation, with the limit of the line being tangent to the circle of the minimum energy level as shown in Figure-7. Hence, the maximum permissible Depth of Discharge

$$DoD_{max} = 1 - \{ E_{min} / E_{max} \}$$

This is the upper limit on the DoD from the energy-momentum-wheel operational point of view.

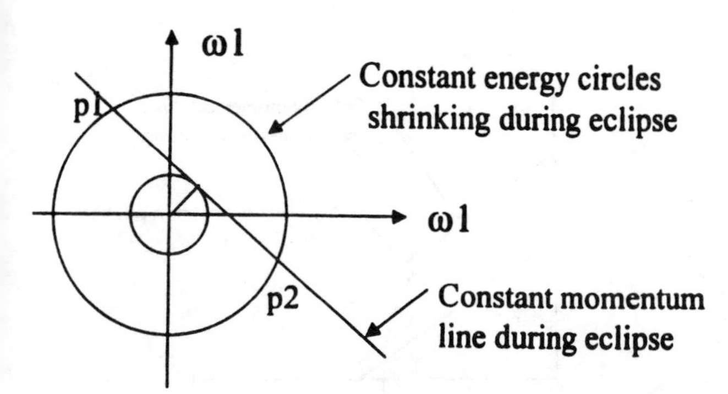

Figure-7: minimum momentum storage requirement line tangent to the minimum state of charge is the necessary condition for energy-momentum-wheel operation.

8.0 CONCLUSION

The operation of the energy-momentum-wheel for storage of the energy and the momentum in the satellite is analyzed. The analysis is developed in terms of the energy circles and the momentum lines. It is shown that in order to simultaneously meet a mission requirements of the two, it is necessary that the minimum momentum line must be tangent to the minimum energy circle at the end of the eclipse periods. This establishes the allowable depth of energy discharge of the energy-momentum-wheel.

9.0 FUTURE WORK

Designing the spacecraft power system using the energy-momentum-wheel is the next logical step in quantifying the potential benefits. Since high power low- earth-orbit satellites are expected to benefit more from this new technology, the study is underway to develop a conceptual design using the energy momentum wheel in a 2500-watt satellite power and attitude control systems.

REFERENCES

1. Patel, M. R., Flywheel energy storage for spacecraft power system, Intersociety Energy Conversion Engineering Conference, Paper Number 183, Vancouver, British Columbia, August 1-4, 1999.

2. Patel, M. R., The Flywheel energy storage as the emerging technology for spacecraft power systems, Launchspace Inc. course binder for the course 'spacecraft power system design and analysis', 1998.

3. Harris, C., Flywheels for spacecraft energy storage and attitude control, Flywheel energy storage workshop, Oak Ridge, TN, Conference Records 9510242, National Technical Information Service, Washington, DC, November 1995.

4. Wehmer, J., Spacecraft flywheel system requirements, Flywheel energy storage workshop, Oak Ridge, TN, Conference Records 9510242, National Technical Information Service, Washington, DC, November 1995.

5. Aerospace Flywheel Workshop Records, U.S. Air Force Research Laboratory and NASA Glenn Research center, Cleveland, OH, October 18-20, 1999.

CONTACT

Mukund R. Patel, PhD, PE, is a research and development engineer with over 35 years of experience in the power industry. He has served as Principal Power System Engineer at the General Electric Space Division, Fellow Engineer at the Westinghouse Research Center, Senior Staff Engineer at Lockheed Martin Astro Space, and the 3M McKnight Distinguished Visiting Professor at the University of Minnesota, Duluth. Dr. Patel earned his MSEE and PhD degrees from the Rensselaer Polytechnic Institute, Troy, NY, MSIE from the University of Pittsburgh, and ME from Gujarat University, India. He is a Fellow of the Institution of Mechanical Engineers (UK), Senior Member of the IEEE, Registered Electrical Engineer in PA, Chartered Mechanical Engineer in UK, and a member of Eta kappa Nu, Tau Beta Pi, Sigma Xi and Omega Rho. Dr. Patel has presented and published 35 papers at national and international conferences, holds several patents, and has earned NASA recognition for exceptional contribution to the UARS power system design. He is active in teaching and consulting. He can be contacted at patelmr@msn.com.

AIAA-2000-2912

NEGATIVE IMPEDANCE STABILIZING CONTROLS FOR PWM DC/DC CONVERTERS USING FEEDBACK LINEARIZATION TECHNIQUES

Ali Emadi and Mehrdad Ehsani

Department of Electrical Engineering
Texas A&M University
College Station, TX 77843-3128
Phone: (979) 845-7582
Fax: (979) 862-1976
E-mail: ehsani@ee.tamu.edu

ABSTRACT

Power electronic converters, when tightly regulated, have constant power sink characteristics at their inputs. This is a destabilizing effect, which is known as negative impedance instability, for the DC/DC converters feeding these loads in a multi-converter power electronic system. In this paper, necessary and sufficient conditions of stability for PWM DC/DC converters with constant power and conventional constant voltage loads are expressed. A nonlinear robust stabilizing controller based on the feedback linearization techniques is proposed. Furthermore, the stability of the proposed controller for the Buck converter using the second theorem of Lyapunov is verified.

INTRODUCTION

More Electric Aircraft (MEA), International Space Station (ISS), and spacecraft power systems are multi-converter power electronic systems [1], [2]. In these systems, power electronics is extensively used for generating, distributing, and utilizing of electrical energy. Furthermore, most of the loads are in the form of electric motor drives, DC/DC choppers, DC/AC inverters, and AC/DC rectifiers.

Power electronic converters, when tightly regulated, behave as constant power loads. An example is an electric motor drive which tightly regulates the speed when the rotating load has one-to-one torque-speed characteristic. Another example is a voltage regulator which tightly regulates the voltage for an electric load that has one-to-one voltage-current characteristic [3].

Constant power loads have negative impedance characteristic. This means that although, in constant power loads, the instantaneous value of impedance is positive, but the incremental impedance is always negative. In fact, the current through a constant power load decreases/increases when the voltage across it increases/decreases. This is a destabilizing effect for the system and known as negative impedance instability [3]-[7].

In this paper, feedback linearization techniques [8], [9] have been proposed to design stabilizing robust controllers for PWM DC/DC converters with constant power and constant voltage loads. For designing the stabilizing controllers, the state space averaging method is considered as a modeling tool for the power electronic converters [10], [11].

Negative impedance stabilizing controller for PWM DC/DC Buck converter has been designed and simulated under large changes in the loads and for different operations. Furthermore, the stability has been verified using the second theorem of Lyapunov for a large range of variations in the constant power and constant voltage loads. Therefore, the proposed controllers have the large-signal capabilities. Furthermore, dynamic responses have been improved.

STAND-ALONE OPERATION OF PWM DC/DC CONVERTERS

The Buck converter of figure 1 which is operating with the switching period of T and duty cycle d, in continuous conduction mode of operation, is considered. Figure 1, also, shows the conventional PI controller. The system of figure 1 with shown resistive load is stable.

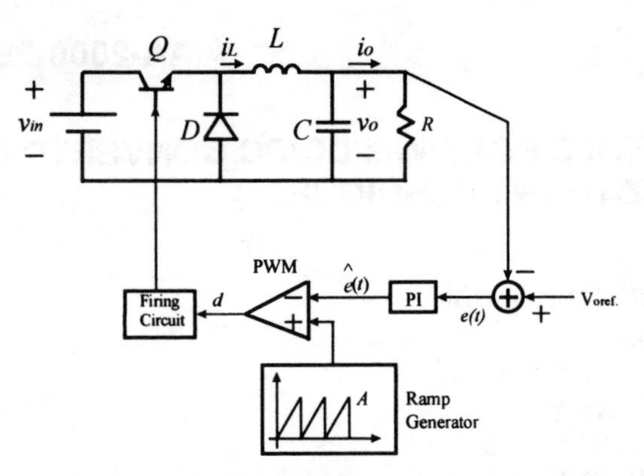

Figure 1. PWM DC/DC Buck converter with conventional *PI* controller.

However, if the load behave as a constant power sink *P*, small-signal transfer functions are given by,

$$H_1(s)=\frac{\tilde{v}_o(s)}{\tilde{d}(s)}=\frac{\frac{V_{in}}{LC}}{s^2-\left(\frac{P}{CV_o^2}\right)s+\left(\frac{1}{LC}\right)} \qquad (1)$$

$$H_2(s)=\frac{\tilde{v}_o(s)}{\tilde{v}_{in}(s)}=\frac{\frac{D}{LC}}{s^2-\left(\frac{P}{CV_o^2}\right)s+\left(\frac{1}{LC}\right)} \qquad (2)$$

The poles of transfer functions $H_1(s)$ and $H_2(s)$ have positive real parts. Therefore, the system is unstable as the effect of constant power load. Figure 2, depicts the simulation of the Buck converter with the *PI* controller designed for conventional resistive load when the converter load behave as a constant power sink. The parameters of the system are given in Table I.

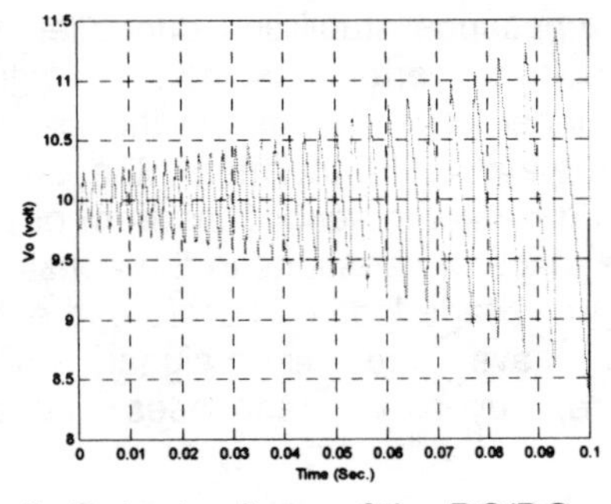

Figure 2. Output voltage of the DC/DC converter with constant power load.

In reference [6], the dynamic properties of Buck converter with constant power load are studied, and line-to-output and control-to-output transfer functions are derived for both voltage mode and current mode controls, in continuous conduction mode and discontinuous conduction mode.

Table I
Parameters of the Buck converter

V_{in}	L	C	f	R	$V_{o,Ref.}$	P
20v	1mH	10mF	10kHz	10ohms	10v	10W

STABILITY OF PWM DC/DC CONVERTERS DRIVING SEVERAL LOADS

Figure 3 depicts the interconnecting converters in the multi-converter power electronic systems. In general, there are two kinds of loads in the system. First group is constant voltage loads which require constant voltage for their operation. Second group is constant power loads sinking constant power from the source bus. The system has to provide constant power in a specified range of voltage, i.e., $V_{o,\min.} \le v_o \le V_{o,\max.}$, for them without going to instability.

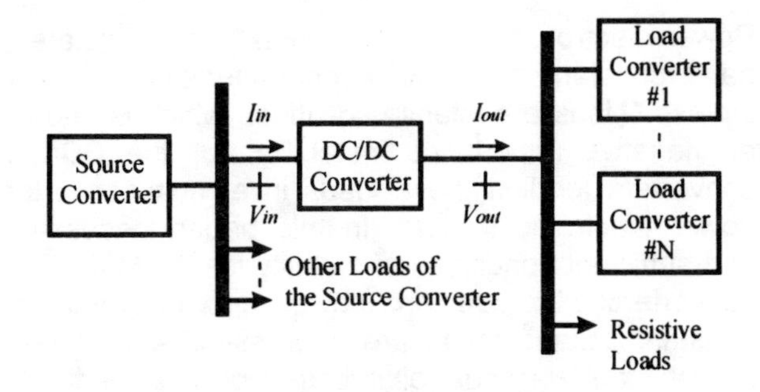

Figure 3. Interconnecting converters in the multi-converter power electronic systems.

Figure 4 shows the equivalent constant power and constant voltage loads, represented by *P* and *R*, respectively, of the DC/DC converter of figure 3. Figure 5 depicts *v-i* characteristic of the equivalent load of the DC/DC converter. If $v_o>V_o$, the slope of the *v-i* curve is positive and the equilibrium point is stable, i.e., the operation will be restored to it after a small departure from it due to a disturbance in the source or load converter. If $v_o<V_o$, the incremental impedance is negative. In this case, the equilibrium point is unstable since the operating point moves away from it after a small departure due to a disturbance. Therefore, necessary and sufficient condition for stability can be expressed as follows.

$$P_{\text{Constant Power Loads}} < P_{\text{Constant Voltage Loads}} \qquad (3)$$

In the case of equation (3), we can control the DC/DC converter using conventional controllers such as *PI* controller without going to instability. However, if equation (3) is not satisfied, the system is unstable. In our last paper [3], a non-linear stabilizing sliding-mode controller is proposed to provide constant power by the DC/DC converter in the predefined range of output voltage, i.e., $V_{o,\min.} \le v_o \le V_{o,\max.}$ [12]-[14]. However, the main disadvantage is that the output voltage is not fixed and it is varying in the above-mentioned range. Therefore, we need to have a voltage regulator to fix the voltage for the constant voltage loads.

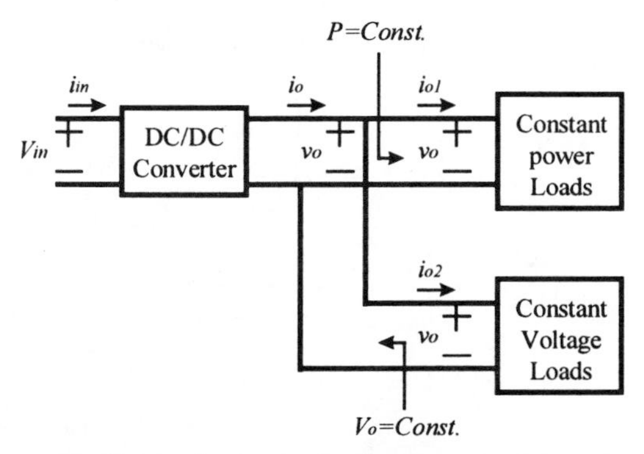

Figure 4. Equivalent constant power and constant voltage loads of the DC/DC converter.

In this new work, feedback linearization techniques have been used to design stabilizing robust controllers for PWM DC/DC converters with constant power and constant voltage loads. It means that we provide a constant voltage at the output of the DC/DC converter for constant voltage loads, at the same time, supplying constant power to constant power loads. For designing the controllers, the stability has been verified using the second theorem of Lyapunov for a large range of variations in the constant power and constant voltage loads *($R_{min} < R < \infty$, $0 < P < P_{max}$)*. Therefore, the proposed controllers have the large-signal capabilities.

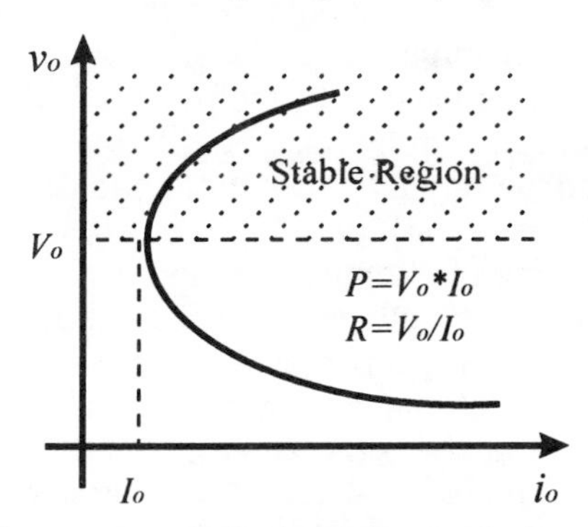

Figure 5. *v-i* characteristic of the equivalent load of the DC/DC converter of figure 4.

STABILITY CONDITION IN A DC DISTRIBUTION SYSTEM

As was explained, constant power loads have negative impedance destabilizing effect on DC/DC converters. In order to guarantee the small-signal stability, as was determined, the power of constant power loads must be less than the power of conventional constant voltage loads for a DC/DC converter. However, in a distribution system with the presence of output resistance of the source subsystem seen from the input side of the DC/DC converter, the stability margin is improved. Figure 6 depicts the equivalent circuit of a DC distribution system in a typical bus driving several constant power and constant voltage loads. *P* and *R* are representing the equivalent constant power and constant voltage loads, respectively. In figure 6, the source subsystem includes the state space averaged circuit of the DC/DC converter as well as filters and distribution system at the input side of the DC/DC converter.

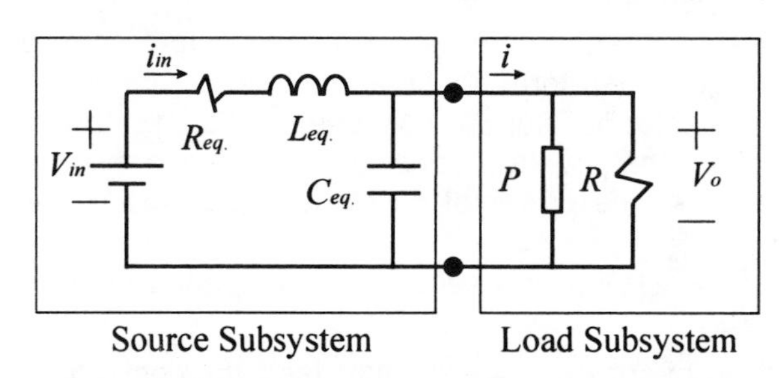

Figure 6. Equivalent circuit of a DC distribution system in a typical bus.

Voltages and currents of the circuit of figure 6 are given by,

$$\begin{cases} v_{in} = R_{eq.} i_{in} + L_{eq.} \dfrac{di_{in}}{dt} + v_o \\ i_{in} = C_{eq.} \dfrac{dv_o}{dt} + \dfrac{P}{v_o} + \dfrac{v_o}{R} \end{cases} \tag{4}$$

By linearizing the set of differential equations (4) around the operating point, the small-signal transfer function is expressed by,

$$H(s) = \frac{\tilde{v}_o(s)}{\tilde{v}_{in}(s)} = \frac{1/C_{eq.}L_{eq.}}{s^2 + \left(\dfrac{R_{eq.}}{L_{eq.}} + \left(\dfrac{1}{R} - \dfrac{P}{V_o^2}\right)\dfrac{1}{C_{eq.}}\right)s + \left(\dfrac{1+R_{eq.}}{C_{eq.}L_{eq.}}\right)} \tag{5}$$

V_o is the output nominal voltage. By considering the condition for which the poles of the system have negative real parts, necessary and sufficient condition for small-signal stability is determined by,

$$P < \frac{V_o^2}{R} + \frac{R_{eq.}C_{eq.}}{L_{eq.}} V_o^2 \quad (6)$$

This can be written as follows.

$$P_{CPL} < P_{CVL} + \frac{R_{eq.}C_{eq.}}{L_{eq.}} V_o^2 \quad (7)$$

CPL and CVL mean constant power loads and constant voltage loads, respectively. The effect of the distribution system and its resistance is shown in the second term at the right hand side of (7). This is an additional term compared to (3). It is related to the distribution system, filters, and the DC/DC converter parameters as well as the output nominal voltage. Therefore, in order to assure the small-signal stability for a converter in a distribution system, the power of constant power loads must be less than the power of constant voltage loads plus $\frac{R_{eq.}C_{eq.}}{L_{eq.}} V_o^2$. If this second term is greater than the power of constant power loads itself, the converter is stable even without any resistive load. This is an improved stability condition.

In order to stabilize the system, based on relation (7), there are four different possible solutions. First one is increasing $R_{eq.}$ which is not practical due to the power loss. The second one is decreasing $L_{eq.}$ which is not feasible also. However, in the design stage of the system, it should be considered to have minimum inductance as possible. The third method is increasing $C_{eq.}$ which is easily possible by adding a filter. In fact, adding passive filters is the solution which has been proposed to maintain the stability of the system [4], [5], [7]. The last method is increasing V_o. In most cases, there is no control over the nominal voltages of the system. However, systems with higher base voltages have better negative impedance stability than systems with low base voltages.

Another approach to stabilize the system, only if we have control over constant power loads, is manipulating the input impedances of the constant power loads to satisfy the stability condition (7) in all cases. Figure 7 shows a typical mechanical constant power load. In figure 7, the DC/AC inverter drives an electric motor and tightly regulates the speed when the rotating load has one-to-one torque-speed characteristic. The controller tightly regulate the speed; therefore, the speed is almost constant. Since the rotating load has one-to-one torque-speed characteristic, for a constant speed, torque is constant and, as a result, power which is the multiplication of speed and torque is constant. If we assume a constant efficiency for the drive system, the input power of the inverter would be constant.

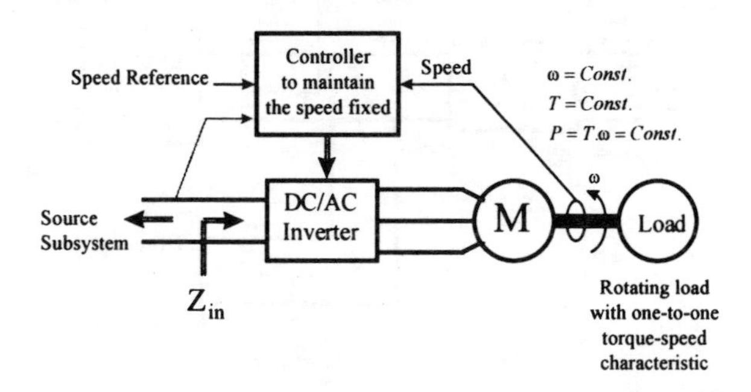

Figure 7. A DC/AC inverter which presents a constant power load characteristic to the system.

The input impedance of the constant power load is expressed as follows.

$$Z_{in} = \frac{V_o^2}{P} \quad (8)$$

By considering this input impedance, the relation (6) can be shown by,

$$\frac{1}{Z_{in}} < \frac{1}{R} + \frac{R_{eq.}C_{eq.}}{L_{eq.}} \quad (9)$$

In this method, the input impedance of constant power load is manipulated in such a way that the relation (9) is satisfied. In fact, instead of having a stable source control, changes in the load are made and the tightly regulated output is sacrificed. Based on this method, [4] and [5] propose a nonlinear stabilizing control to manipulate the input impedance of the converter/motor drive in an induction motor based electric propulsion system as well as a small distribution system consisting of a generation system, a transmission line, a DC/DC converter load, and a motor drive load. For their studies, since only constant power load is considered, the stability condition (9) becomes

$$Z_{in} > \frac{L_{eq.}}{R_{eq.}C_{eq.}} \quad (10)$$

In this paper, we suppose that there is no control over the loads. Therefore, we design robust stabilizing control for the DC/DC converters to drive constant power and constant voltage loads without imposing any restriction for the loads or the system. As a result, there is no need for adding the filters or considering any additional restriction on the design of the distribution system. We do the same method for other converters in the multi-converter distribution system to stabilize them. However, for the original source of the system, the relation (7), with parameters defined at the source bus, should be satisfied. In fact, we must have a negative impedance stable source for the system.

NEGATIVE IMPEDANCE STABILIZING CONTROL FOR PWM DC/DC CONVERTERS WITH CONSTANT POWER LOADS AND RESISTIVE LOADS

The Buck converter of figure 3 which is operating with the switching period *T* and duty cycle *d* is considered. This converter is driving several different loads which require constant voltage for their operation. Some of these loads, such as tightly regulated power electronic converters, behave as constant power sinks. Equivalent constant power and constant voltage loads of the Buck converter, as is depicted in figure 4, are represented by *P* and *R*, respectively. Upper and lower limits of *P* and *R* are given by,

$$\begin{aligned} &P_{\min.} < P < P_{\max.}(P_{CPL,\min.} < P_{CPL} < P_{CPL,\max.}) \\ &R_{\min.} < R < R_{\max.}(P_{CVL,\max.} > P_{CVL} > P_{CVL,\min.}) \end{aligned} \quad (11)$$

For designing stabilizing controllers in this paper, $P_{min.}$ and $R_{max.}$ are considered zero and infinity, respectively. Furthermore, we suppose $P_{max.}=P_{CPL,max.}=P_{CVL,max.}$. $P_{max.}$ occurs when all loads are on and behave as constant power sinks. It also occurs when all loads are on and have resistive characteristics as in constant voltage loads. $P_{min.}$ and $R_{max.}$ occur when there is no constant power load and no constant voltage load on, at the output of the DC/DC converter, respectively. They also happen when all loads are off.

Using the state space averaging method [10], [11], during continuous conduction mode of operation, the state equations of the DC/DC converter of figure 3 can be shown by,

$$\begin{cases} \dfrac{di_L}{dt} = \dfrac{1}{L}\left[dv_{in} - v_o\right] \\ \dfrac{dv_o}{dt} = \dfrac{1}{C}\left[i_L - \dfrac{v_o}{R} - \dfrac{P}{v_o}\right] \end{cases} \quad (12)$$

The small-signal transfer functions of the system of equations (12) assuming small perturbations in the state variables due to small disturbances in the input voltage and duty cycle are as follows.

$$H_1(s) = \frac{\tilde{v}_o(s)}{\tilde{d}(s)} = \frac{\dfrac{V_{in}}{LC}}{s^2 + \left(\dfrac{1}{RC} - \dfrac{P}{CV_o^2}\right)s + \left(\dfrac{1}{LC}\right)} \quad (13)$$

$$H_2(s) = \frac{\tilde{v}_o(s)}{\tilde{v}_{in}(s)} = \frac{\dfrac{D}{LC}}{s^2 + \left(\dfrac{1}{RC} - \dfrac{P}{CV_o^2}\right)s + \left(\dfrac{1}{LC}\right)} \quad (14)$$

Necessary and sufficient condition for stability is determined as follows, i.e., poles of the transfer functions have negative real parts.

$$\frac{1}{RC} - \frac{P}{CV_o^2} \rangle 0 \quad \Rightarrow \quad R \langle \frac{V_o^2}{P} \quad (15)$$

This is the same stability condition as (3). It means that if the total power of constant power loads is less than the total power of constant voltage loads, we can control the DC/DC converter using conventional controllers such as *PI* controller without going to instability. In other words, if equation (15) is satisfied, we can provide constant voltage at the output of the DC/DC converter for constant voltage loads and, at the same time, supplying constant power to the constant power loads. In the next section, using feedback linearization techniques, we design a negative impedance stabilizing controller for the DC/DC converter of figure 3 with different loads in the range of (11).

FEEDBACK LINEARIZATION TECHNIQUE

In this section, we use feedback linearization technique to control the converter [8], [9]. We look for a nonlinear feedback to cancel out the nonlinearity in the set of differential equations (12). In (12), there is no direct relation between the control input (dv_{in}) and the nonlinearity (P/v_o); therefore, the following change of variables is considered.

$$\begin{cases} x_1 = i_L - \dfrac{v_o}{R} - \dfrac{P}{v_o} \\ x_2 = v_o - V_{\text{oRef.}} \end{cases} \quad (16)$$

Where V_{oRef} is the output reference voltage for the Buck converter. Using change of variables (16), state equations (12) can be written as follows.

$$\begin{cases} \dot{x}_1 = \frac{dv_{in}}{L} - \frac{x_2 + V_{oRef.}}{L} - \frac{x_1}{RC} + \frac{Px_1}{C(x_2 + V_{oRef.})^2} \\ \dot{x}_2 = \frac{1}{C} x_1 \end{cases} \quad (17)$$

In order to cancel out the nonlinearity in (17), the following nonlinear feedback is proposed.

$$\frac{dv_{in}}{L} = k_1 x_1 + k_2 x_2 - \frac{\hat{P} x_1}{C(x_2 + V_{oRef.})^2} + \omega \quad (18)$$

$$\omega = \frac{V_{oRef.}}{L}$$

Where k_1, k_2, and $\hat{P}$ are the parameters of the controller to be designed. With the nonlinear feedback, the state equations of (17) can be shown by,

$$\begin{cases} \dot{x}_1 = (k_1 - \frac{1}{RC}) x_1 + (k_2 - \frac{1}{L}) x_2 + \frac{x_1}{C(x_2 + V_{oRef})^2}(P - \hat{P}) \\ \dot{x}_2 = \frac{1}{C} x_1 \end{cases} \quad (19)$$

If $\hat{P} = P$, the system is linear. Other parameters of the controller, i.e., k_1 and k_2, can be designed such that the resulted system has poles at appropriate places. However, we don't have any control over the loads and they may change. In fact, *P* and *R* vary according to (11). Therefore, in the next section, we design the controller to guarantee the stability of the system in the presence of loads changes.

STABILITY ANALYSIS

In order to evaluate the stability of the converter, a continuously differentiable positive definite function, $V(x)$, needs to be determined. We define $V(x)$ as follows.

$$V(x_1, x_2) = \frac{1}{2} K x_1^2 + \frac{1}{2} KC(\frac{1}{L} - k_2) x_2^2 \quad (20)$$

Where $K > 0$ and $k_2 < \frac{1}{L}$; therefore, $V(x)$ is a positive definite function. The derivative of $V(x)$ is given by,

$$\dot{V}(x_1, x_2) = K(k_1 - \frac{1}{RC}) x_1^2 + K \frac{x_1^2}{C(x_2 + V_{oRef.})^2}(P - \hat{P}) \quad (21)$$

In order to guarantee the stability of the converter, the derivative of $V(x)$ needs to be negative definite. Therefore, the parameters of the controller are chosen as follows.

$$k_1 - \frac{1}{RC} < 0 \quad \& \quad P - \hat{P} < 0 \quad \& \quad k_2 < \frac{1}{L} \quad (22)$$

Considering upper and lower limits of *P* and *R* given in (11) as well as variations in the values of the inductor and capacitor of the converter, control parameters for a robust design are given by,

$$\begin{cases} k_1 < \frac{1}{R_{max.} C_{max.}} \\ \hat{P} > P_{max.} \\ k_2 < \frac{1}{L_{max.}} \end{cases} \quad (23)$$

$L_{max.}$ and $C_{max.}$ are the maximum values that *L* and *C* can hold, respectively. Therefore, $\dot{V}(x)$ is a negative definite function and, consequently, $V(x)$ is a Lyapunov function. Hence, the closed loop system is asymptotically stable and the operating point is a stable equilibrium point.

SIMULATION RESULTS

A negative impedance stabilizing controller based on the proposed feedback linearization technique and *PI* controller has been designed and simulated. The parameters of the converter and designed robust controller are given in Table II.

Table II
Parameters of the Buck converter and stabilizing controller

V_{in}	20v
L	1mH
C	10mF
f	10kHz
$V_{oRef.}$	10v
R	20ohms, 5ohms<R<∞
P	10W, 0<P<20W
PI controller	K_p=1, K_i=100
Feedback linearization	K_1=-195, K_2=900, $\hat{P}$ =25

In order to study the converter dynamic performance under load variations, step changes in constant power and resistive loads have been investigated. Figures 8-10 depict the simulation results.

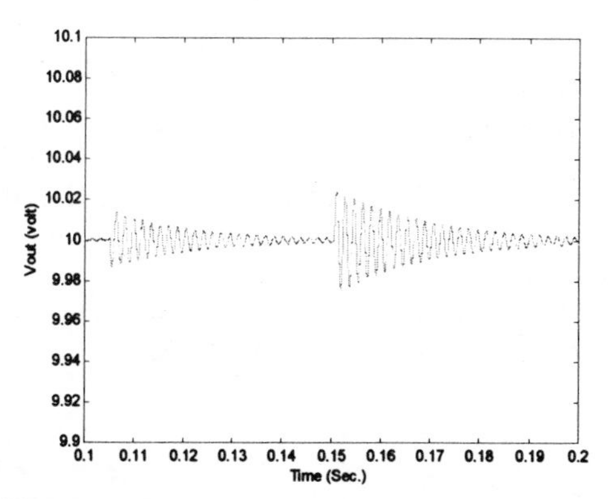

Figure 8. Dynamic response of the proposed controller for Buck converter to load step change from $P=10W$, $R=20\Omega$, to $P=10W$, $R=10\Omega$, and $P=10W$, $R=100\Omega$.

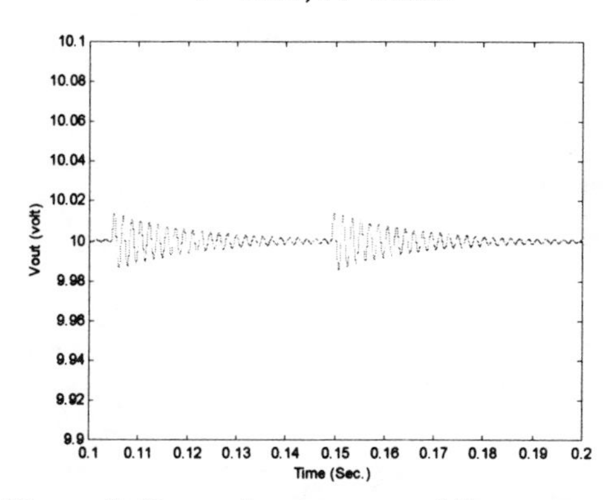

Figure 9. Dynamic response of the proposed controller for Buck converter to constant power load step change from *10W* to *5W* and *0* ($R=20\Omega$, *i.e.*, $P_{Constant\ Voltage\ Load}=5W$).

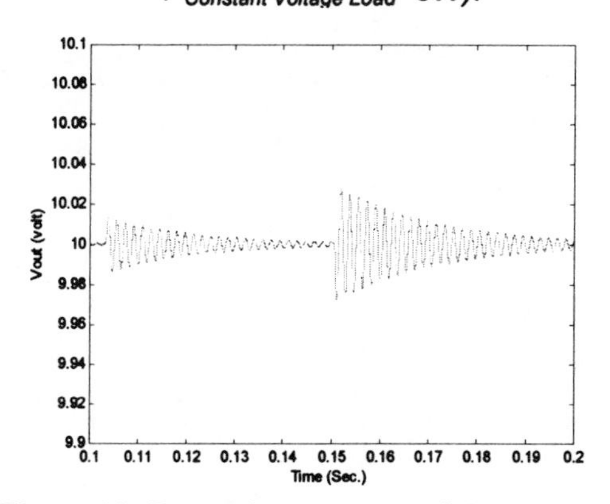

Figure 10. Dynamic response of the proposed controller for Buck converter to load step change from $P=10W$, $R=20\Omega$, to $P=10W$, $R=1000\Omega$, and $P=20W$, $R=1000\Omega$.

It must be mentioned that most of the practical constant power loads at the starting phase of their operations have positive incremental impedance characteristic. Therefore, their power is increasing until they reach the nominal power. After they hit the nominal power, the input power will be constant and they behave as constant power sinks which have negative incremental impedance characteristic. In out simulations, as is depicted in figure 11, we considered a linear *v-i* characteristic for the starting period of the constant power loads.

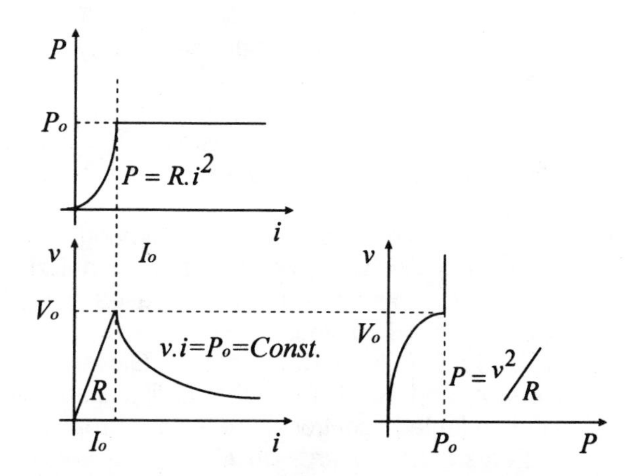

Figure 11. *v-i* characteristic of a practical constant power load.

CONCLUSIONS

In this paper, the concept of negative impedance instability of the constant power loads for PWM DC/DC converters was described. It was illustrated that if the power of constant power loads was greater than the power of constant voltage loads, the converter would be unstable. Furthermore, the necessary and sufficient conditions of stability for a DC distribution system containing DC/DC converters were explained. By applying the resultant model from the state space averaging method, a nonlinear robust stabilizing controller based on the feedback linearization techniques for Buck converter was proposed. Dynamic behavior of the controller under different operations and significant variations in loads were studied and shown to be satisfactory. To prove that the designed controller was stable, Lyapunov second theorem was used.

ACKNOWLEDGMENTS

Financial support of Advanced Vehicle Systems Research Consortium at Texas A&M University for this work is gratefully acknowledged.

REFERENCES

[1] A. Emadi and M. Ehsani, "Aircraft Power Systems: technology, state of the art, and future trends," *IEEE Aerospace and Electronic Systems Magazine*, vol. 15, no. 1, pp. 28-32, Jan. 2000.

[2] A. Emadi, J. P. Johnson, and M. Ehsani, "Stability analysis of large DC solid-state power systems for space," *IEEE Aerospace and Electronic Systems Magazine*, vol. 15, no. 2, pp. 25-30, Feb. 2000.

[3] A. Emadi, B. Fahimi, and M. Ehsani, "On the concept of negative impedance instability in the more electric aircraft power systems with constant power loads," Proc. of the *34th Intersociety Energy Conversion Engineering Conf. (IECEC'99)*, Vancouver, Canada, Aug. 1999.

[4] S. F. Glover and S. D. Sudhoff, "An experimentally validated nonlinear stabilizing control for power electronics based power systems," SAE 981255.

[5] S. D. Sudhoff, K. A. Corzine, S. F. Glover, H. J. Hegner, and H. N. Robey, "DC link stabilized field oriented control of electric propulsion systems," *IEEE Trans. on Energy Conversion*, vol. 13, no. 1, pp. 27-33, March 1998.

[6] V. Grigore, J. Hatonen, J. Kyyra, and T. Suntio, "Dynamics of a Buck converter with constant power load," *Proc. of IEEE 29th Power Electronics Specialist Conf.*, Fukuoka, Japan, May 1998, pp. 72-78.

[7] M. Belkhayat, R. Cooley, and A. Witulski, "Large signal stability criteria for distributed systems with constant power loads," *Proc. of IEEE 26th Power Electronics Specialist Conf.*, Atlanta, Georgia, June 1995, pp. 1333-1338.

[8] H. A. Khalil, *Nonlinear Systems*, Prentice-Hall, 1996.

[9] J. E. Slotine and W. Li, *Applied Nonlinear Control*, Prentice-Hall, 1991.

[10] R. D. Middlebrook and S. Cuk, "A general unified approach to modeling switching converter power stages," *Proc. of IEEE 1976 Power Electronics Specialist Conf.*, 1976, pp. 18-34.

[11] J. Mahdavi, A. Emaadi, M. D. Bellar, and M. Ehsani, "Analysis of power electronic converters using the generalized state space averaging approach," *IEEE Trans. on Circuits and Systems I: Fundamental Theory and Applications*, vol. 44, no. 8, pp. 767-770, Aug. 1997.

[12] V. I. Utkin, *Sliding Modes and their Application in Variable Structure Systems*, MIR, Moscow, Russia, 1974.

[13] H. Sira-Ramirez, "Sliding motions in bilinear switched networks," *IEEE Trans. on Circuits and Systems*, vol. CAS-34, no. 8, Aug. 1987.

[14] J. Mahdavi, A. Emadi, and H.A. Toliyat, "Application of state space averaging method to sliding mode control of PWM DC/DC converters," *Proc. of the 1997 IEEE Industry Application Conf.,* New Orleans, Louisiana, Oct. 1997, pp. 820-827.

[15] P. A. Ioannou and J. Sun, *Robust Adaptive Control*, PTR Prentice-Hall, 1996.

LITHIUM-ION BATTERIES for GEOSYNCHRONOUS SATELLITES. QUALIFICATION TEST RESULTS OF THE STENTOR BATTERY.

Jean-Pierre SEMERIE
SAFT Space and Defense BP 1039 86060 Poitiers Cedex 09 France
Phone 33 549 554 589 – Fax 33 549 554 780
e-mail: jean-pierre.semerie@saft.alcatel.fr

ABSTRACT

SAFT is engaged in development and qualification of Lithium-Ion cells and batteries for space since 1996.
This paper presents the results obtained during the qualification tests on the first battery built for the STENTOR Satellite.
The STENTOR satellite is powered by lithium-Ion only, with two 45 Volts-80 Ampere.hour battery packs, made from cylindrical shaped 40 Ampere.hours Lithium-Ion cells, and including the electronics to monitor cells voltages and balance the cell to cell capacities. Each battery pack weights less than 37 kg.
The required capacity is achieved by connecting two 40 Ah cells in parallel prior to connecting it in series. Then 11 of these group of two cells are series connected.
The battery includes an automatic by-pass systems to prevent any type of cell failures.
The two electronics are cross strapped to ensure that the two battery packs have no single point failure.
The first complete battery built in 1999, was submitted to qualification test, which comprises:
Capacity test, hot and cold temperatures.
Thermal vacuum tests.
Vibration, shock tests.
EMC tests.
The paper gives a summary of the development results:
Cell validation and safety tests.
Cell qualification tests.
Cell life tests.
Other component qualification status.
The paper then presents the battery, and the details and analysis of the qualification tests performed.

1. THE STENTOR PROGRAM

STENTOR, - *Satellite de Telecommunications pour Experimenter les Nouvelles Technologies en Orbite*- is a program which was decided by the French Government in October 1995. Its goal was to demonstrate, in orbit, the capabilities of emerging technologies, and to perform a flight qualification of some new equipment's. Among these new technologies, in the frame of energy, one can mention Ionic Propulsion and Lithium-Ion batteries.
With regard to the Lithium-Ion Battery, the goal was to demonstrate the feasibility of the concept, by designing and building a battery suitable to power a 2 kW satellite, the main driver being the schedule.
The final goal was to demonstrate the feasibility, and to have flight proven and on the shelf, a modular concept of Lithium Ion battery able to power satellites in the range of 15 to 30 kW, with no additional development or qualification costs.

2. CELL DEVELOPMENT

2.1 Development philosophy

The work done to develop a cell usable for space was already published [1] and is summarized there.
The Lithium-Ion cell was developed based on the design established for the Electric Vehicle (EV) application.
The development plan followed to obtain a cell qualified for space use was :
- Select the more advanced available electrochemistry and processes, based on the work done for EV. Major interest was to benefit from the industrial capital investments committed. During that phase, several types of cells were built and tested, including initial design of prismatic 100Ah, the cylindrical 40Ah cells "Generation 1" also called Prototype cells. Some life tests were soon initiated, and Prototype cells which cycling results were already published [2] have reach the design life of 9 years.
This selection leads to what is called "Generation 2" electrochemistry.
- Adapt the cell container design to space use, to have a leak proof design. The answer is a cell with an aluminum can and a terminal feed-through derived from a previous design used for Nickel Hydrogen cells, with Nylon seal replaced by a polypropylene one.
Demonstrate that the cell is able to sustain the mechanical environments specific to space.
This cell was developed and characterized with regard to the specific conditions of the space use.

2.2 Validation phase

During 1998, validation tests were conducted on 18 cells subjected to various test conditions, comprising functional tests, engineering and safety tests.
The main tests conducted were:
Acceptance tests.
Internal pressure measurement
Thermal cycling
Electrical performances @20°C, charge at 4V, constant current discharges(8,22,26,29 & 44A)

Electrical performances @20°C, charge at 3.8, 3.9, 3.95 & 4V, combined with constant power discharges.(33, 50 & 92W).
Electrical performances @0°C, charge at 4V, discharges at 8,22,29 &44A. charge at 3.95V, discharges at 33 & 50W.
Electrical performances @10°C, charge at 4V, discharges at 8,22,29 &44A
Electrical performances @40°C charge at 4V, discharges at 8,22,29 &44A. charge at 3.95V, discharges at 33 & 50W.
Leak test. Tests under vacuum to evidence electrolyte leaks.
Pulse discharge test. Discharges 123A for 250 ms at several Depth Of Discharges (DOD%).
Cycling @60°C: 4 capacity cycles charge at 4V, discharge at 26A.
Sine vibrations. 20g up to 100Hz .
Random vibrations: @44 gRms up to 2000Hz.
Shocks: Half sine, 300g 0.8ms
Overcharge type 1: 5 reference capacity cycles, charge 4A at 4V, discharge 29A, then 1 cycle charge at 4A until 4.2V, discharge 29A, followed again by 5 reference cycles.
Overcharge type 2: 10 consecutive cycles, charge at 4A up to 4.5V, discharge at 29A.
Overdischarge type 1. 5 reference capacity cycles, charge 4A at 4V, discharge 29A, then 1 cycle charge at 4A until 4.2V, discharge 29A until 2.4V, followed again by 5 reference cycles.
Overdischarge type 2: 10 consecutive cycles, charge at 4A up to 4.0V, discharge at 29A until reversal at – 0.8V.
Short circuit: Cell charged at 3.5V or 4V, shorted for 1 minute on 3mOms resistor,
Nail penetration test: Cell charged at 3.5V, tested with nail of diameter 3 or 20mm.
Autopsy on cells having passed mechanical tests.
The cells used for that phase had the following average characteristics:
Weight: 1166g
Capacity (4 to 2.7V, @29A, 20°C): 46.1Ah
Energy (4 to 2.7V, @29A, 20°C): 163.8 Wh
Internal resistance: 2.4 mOhm.
All the cells successfully passed the tests, and the next phase, qualification was started.

2.3 Cell Qualification

Then at the end of 1998, formal qualification program was conducted which ended at the spring of 1999.
During that qualification program, 15 cells were tested in various conditions to demonstrate the Beginning Of Life (B.O.L) performances. The tests were basically the same as those performed during the validation phase. The cell qualification test schedule is shown in Table 1. The main results are summarized in Table 2.
The average characteristics of the "G2" cells were the following:
Weight: 1127g
Capacity (charge 20A, Discharge 22A @20°C): 45.8Ah
Energy (charge 20A, Discharge 22A @20°C): 161.7Wh
Internal resistance: 2.63 mOhm.
The main characteristics of the cells which will have to be taken into account to drive the battery design are, the stored energy under a given voltage, the behavior with regard to charge and discharge control , and the reliability and safety aspects. We will consider these various points now. The cell is rated at 40 Ah, with a minimum acceptance capacity of 44 Ah at BOL and a

Table 1 .Cell Qualification Test Schedule

Cell #	1	2	3	4	5	6	7	8	9	10	11	12	13	14	15
Internal resistance measurement	1	1	1	1	1	1	1	1	1	1	1	1	1	1	1
Current leakage measurement	2	2	2	2	2	2	2	2	2	2	2	2	2	2	2
Hermeticity	3	3	3	3	3	3	3	3	3	3	3	3	3	3	3
Electrical Performances @ 0°C	4	4	4	4	4	4	4	4	4	4	4	4	4	4	4
Electrical Performances @ +10°C	5	5	5	5	5	5	5	5	5	5	5	5	5	5	5
Electrical Performances @ +40°C	6	6	6	6	6	6	6	6	6	6	6	6	6	6	6
Internal resistance measurement	7	7	7	7	7	7	7	7	7	7	7	7	7	7	7
Current leakage measurement	8	8	8	8	8	8	8	8	8	8	8	8	8	8	8
Hermeticity	9	9	9	9	9	9	9	9	9	9	9	9	9	9	9
Vibrations		10	10	10		10						10	10	10	10
Hermeticity		11	11	11		11						11	11	11	11
Current leakage measurement		12	12	12		12						12	12	12	12
Internal resistance measurement		13	13	13		13						13	13	13	13
Electrical Performances @ +10°C		14	14	14	10	14					10	14	14	14	14
short-circuit							10								
Overcharge U = 4.2V	10														
Overcharge U = 4.5V									10						
Overdischarge U = 2.4V								10							
Overdischarge U =- 0.8V										10					
Hermeticity	11	15	15	15	11	15	11	11	11	11	11	15	15	15	15
Autopsy		16		16							12				

(Numbers in the table represents the test sequence.)

minimum average voltage of 3.5 V, with the cell charged at 4.0 V and discharged until 2.7 V at 20A and 20°C.

Table 2. Qualification Summary

	Criteria	Result
Standard capacity	≥ 44 Ah	≥ 45 Ah
Internal resistance	≤ 3 m Ohm	≤ 2.75 m Ohm
Leakage current	≤ 10 mA	≤ 5 mA
Hermeticity before filling.	≤ 1. 10^{-7} atm cm^3/s	≤ 8. 10^{-8}atm cm^3/s
Diameter.	≤ 54.2 mm	OK
Length	≤ 248.5 mm	≤ 247.7 mm
Weight	≤ 1164 g	≤ 1143g
Insulation before filling	> 10^{+3} M Ohm	> 10^{+3} M Ohm
Capacity @ 0°C	≥ 28 Ah	≥ 38 Ah
Capacity @10°C	≥ 40 Ah	≥ 42 Ah
Capacity @40°C	≥ 40 Ah	≥ 48 Ah
Internal resistance (5 s, 60% DOD)	≤ 3 m Ohm	≤ 2.84 m Ohm
Leakage current	≤ 10 mA	≤ 5 mA
Hermeticity	No electrolyte leak	OK
Resonance search	Natural Frequency > 140 Hz	Natural Frequency >660 Hz
Sine & Random: voltage noise.	≤ 10 mA	< 10 mA
Mechanical integrity.	Visual & X Rays	OK
Electrical health @ 10 °C	C ≥ 40 Ah	C ≥ 41 Ah
Overcharge Ufc = 4,2 V	C ≥ 40 Ah	C ≥ 41 Ah
Overcharge Ufc = 4,5 V	No electrolyte leak	OK
Overdischarge Ufd = 2,4 V	C ≥ 40 Ah	C ≥ 43.9 Ah
Overdischarge Ufd = -0,8 V	No electrolyte leak	OK
Short-circuit	No electrolyte leak	OK
Hermeticity	No electrolyte leak	OK
Leakage current	≤ 10 mA	≤ 5 mA
Internal resistance	≤ 3 m Ohm	≤ 2.77 m Ohm
Capacity @ 10 °C	≥ 40 Ah	≥ 41 Ah
Hermeticity	Leak ≤ 2.10^{-9} g/s	OK
Autopsy	No damages	OK

2.4 Outputs for battery design

The battery has to be sized to accommodate a given loss of capacity at end of life, and life tests have to be performed to validate this hypothesis. There is no "maximum end of charge voltage" like on Nickel Cadmium or Nickel Hydrogen cells, with the possibility to "Overcharge" the cell, charging more capacity than needed, in order to obtain a "fully charged" cell. The Lithium-Ion cell acts like a capacitor: The faradic efficiency is nearly 1: All the quantity of electricity charged shall be discharged. Higher is the cell voltage, higher is the stored energy... Up to irreversible cell damages. Life time will not be affected until 4.0 V, which is the nominal charged capacity for this cell, even if by precaution, it is recommended to charge the cell for what is necessary and no more. Then above 4 V, aging of the cell increases progressively, and about 4.5 V, when all the electrodes are charged, electrolyte oxidation will occur with irreversible damages to the cell, that could leads to over-temperature, over-pressure and even fire if the highly flammable electrolyte is released and ignited. One of the consequences of this behavior, is the fact that within series connected cells, if for any reason one of the cells has a leakage current (self-discharge), different from the others in the string, its state of charge will continuously drift until severe overvoltage (in charge) or undervoltage (in discharge) of this cell. For that reason, the Lithium-Ion batteries incorporate a "balancing" capability, able to equilibrate the relative state of charge of the cells. Also one can emphasize the behavior with regard to abuse conditions, related to safety, as all "Lithium" products have a reputation of hazardous product. It has first to be pointed that at the contrary of Lithium primary cells, like Lithium-Thionyl-chloride by example, Lithium-Ion cells have no metallic lithium present inside during normal use. Lithium is only present as Ions. You have similarly Sodium Ions present in sea water with none of the risks associated to sodium! Nevertheless, Lithium-Ion cells are a system in which a given amount of energy is stored, and this is always associated with some hazard.
Figure 1 shows what happens when a cell is continuously charged at C/10 during 36 hours: 10 hours after the beginning of charge, cell voltage is about 4 V, and the charge should normally stop. Charge continuing, at T=15 hours, voltage is 4.5 V, all the Lithium Ions having been transferred from the positive to the negative, the electrolyte solvent becomes to be oxidized, with increase of temperature and pressure. This leads to the cell burst at about 10 bars , then return to stable conditions, at temperatures not exceeding 55°C, until T=36 h, when current is turned off and temperature returns to ambient. This sustained overcharge is the major hazard associated

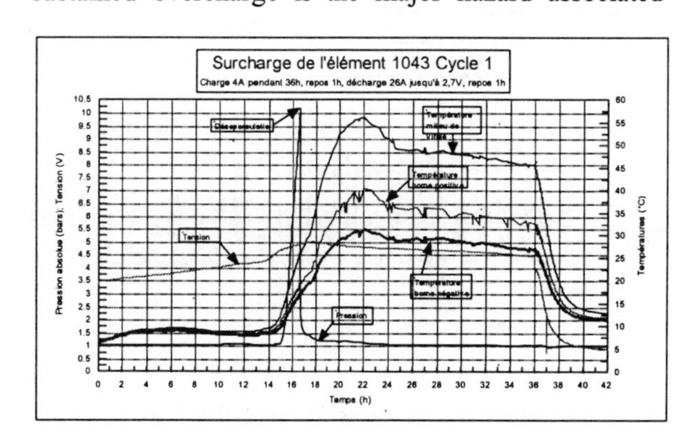

Figure 1 . Cell overcharge

with Lithium-Ion cells, as it automatically leads to cell bursting, with release of highly flammable electrolyte solvents.
Figure 2 shows the behavior of the cell under a repeated reversal, with a cell discharged at 26A down to 0 V, then forced to reversal for 1.2 hours at the same current, recharged and discharged again 5 times.

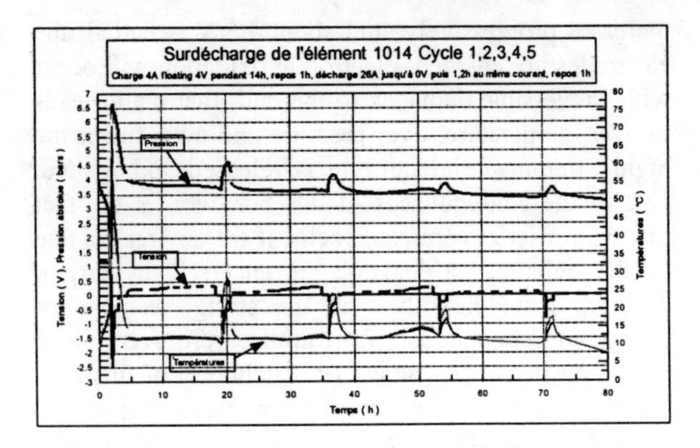

Figure 2. Cell Reversal

During reversal, the copper current collector of the negative electrode is oxidized and will deposit dendrites to the positive plate, shorting the cell. One can see that the pressure, temperature and voltage across the cell decreases continuously, showing that the cell is becoming progressively shorted and inert, acting like a single ohmic conductor. Nevertheless, in some particular cases, with higher currents, cell bursting could occur during the initial reversal, so it is recommended to avoid this condition.

2.5 Life Tests

Although STENTOR mission itself is limited to 9 years, with the two first years at full power and the 7 following ones with reduced energy need, life tests have been started since the beginning of the program, to demonstrate the ability of the cell to sustain the 15 years corresponding to a standard geosynchronous mission. This work has been reported previously[2], and is just updated here, as the 6 cells battery tested under

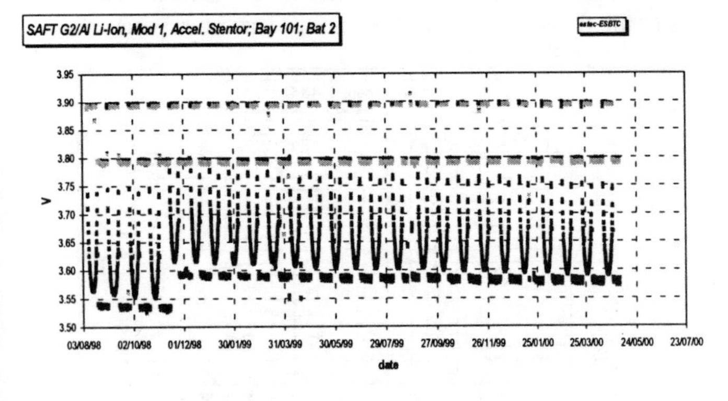

Figure 3. Semi-accelerated life test

semi accelerated cycling has achieved its 13rd year of cycling last May and is still working well.
The Figure 3 shows the end of discharge voltage of the 6 cells over the 2 years and 26 eclipse seasons of the cycling. The total capacity fade is well below the design target for STENTOR, and gives confidence to the life expectancy of the battery.

3. THE STENTOR BATTERY

3.1 Requirements

The STENTOR spacecraft power system is powered by two identical Battery Packages, connected to the power bus. Each of the two Battery Packages has his own charge and discharge regulator, and can be managed independently. The requirement for each of the battery packs is to deliver 1.1 kW during the eclipses, and also about 0.6 kW for the ionic thrusters used in the AOCS, during both equinox and solstice seasons, for duration ranging from several times 1/2 hour up to 1.5 hours, depending of the season. These values are for the first two years of mission, then for the next 7 years, a reduction of the power demand of about 30% is planned.
The battery packages are located on the north and south walls of the satellite , the thermal control being done by conduction through the wall, then by radiation to space. With regard to the avionics, it was requested that the battery has to be monitored through the same interface as the State of the Art Nickel-Hydrogen battery, which was considered as a possible backup until January 1999. So, analog individual cell voltage lines and current sensors are the same. Furthermore, no software was allowed to be resident into the battery, the On Board Computer (OBC) being in charge to manage all the operations to balance the cells into the battery. So, the battery is equipped with a digital interface MIL STD 1553B to communicate with the OBC.

3.2 Electrical Architecture

3.2.1 Cell arrangement

It is a common practice for Nickel Cadmium or Nickel Hydrogen cells to connect in parallel 10 to 100 or more plates into a single cell to achieve the required energy, but the paralleling of two different cells without protection like blocking diodes is hazardous, because cell voltage is strongly affected by temperature, and also because knowing cell voltage is sufficient to determine the cell state of charge. Any thermal unbalance between the two (or more) parallel cells could lead to unstable conditions with one cell pouring in the other one. For that reason, battery engineers have in mind that this arrangement is practically prohibited, except, as already mentioned, into a single cell, provided that constraints of maximum thermal gradient across the cell are satisfied. The situation is completely different for Lithium-Ion, where we have the unique opportunity to have a system having open circuit voltage practically independent from the temperature, and linked to the state of charge by a nearly linear relationship. So, by connecting in parallel two or more lithium ion cells, they will be maintained at the same state of charge, whatever is their temperature. This possibility is used by SAFT, and to provide the necessary energy to the STENTOR Battery Package, two Lithium-Ion cells of 40 Ah each are first connected in parallel, to create a single cell of 80 Ah called "Cell Package". Then 11 of these Cell Packages are series connected to constitute the Battery Package (Figure 4).

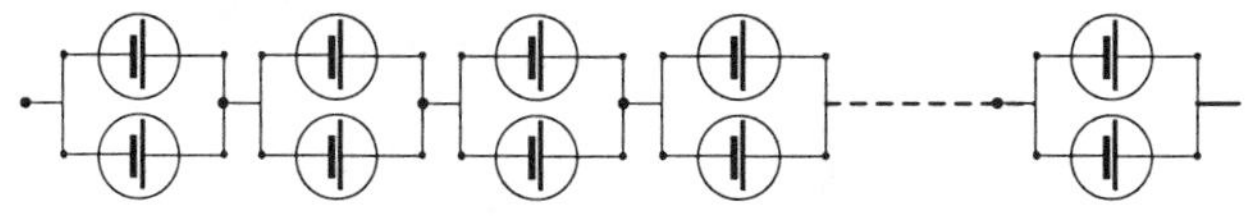

Figure 4. Cell interconnects

3.2.2 Cell Package Bypass

As the system has to be one failure tolerant, it was necessary to cope with the possibility to have an open circuit failure at the level of one of the cell packages. For that, a bypass system is implemented across each of the cell packages, to allow the battery current to pass in case of failure as it is shown in Figure 5. The Bypass

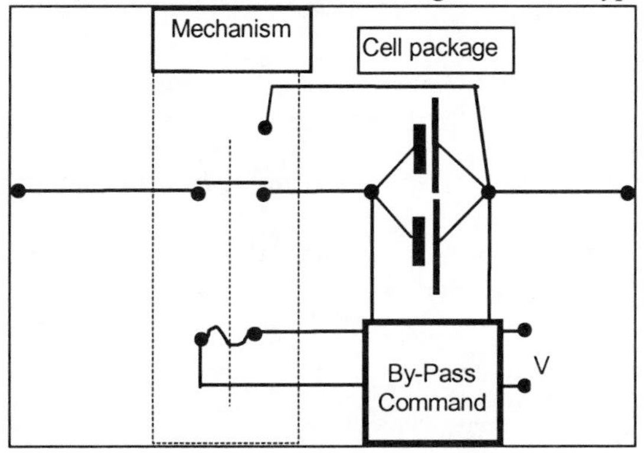

Figure 5. By-pass principle

made by N.E.A (USA) is a make before break actuator, to never interrupt the current flow into the battery. See picture in Figure 6. The bypass is activated by melting a fuse. The required energy is taken on the battery itself. The fuse is powered when the cell package is suspected to be failed, and this is detected by the

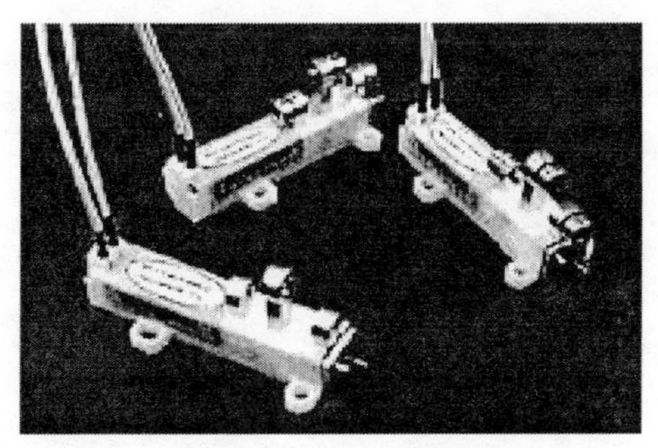

Figure 6. N.E.A Bypass

voltage across the cell package: Should the voltage raise above 4.15 V in case of overcharge or go down to negative values in case of reversal, an independent electronic circuit will trigger the current flow inside the bypass fuse. This circuit has been developed for SAFT by Alcatel ETCA, in the shape of an hybrid circuit comprising the necessary electronics to drive two bypass, corresponding to two consecutive cell packages. This system can be inhibited from the external of the battery, as the power supply of the hybrid circuit is routed through connectors to the satellite interface. So, when the system is inhibited, only the current of some mA necessary to monitor the cell package voltage is drawn out of the cells.

3.2.3 Electronics

The principle to maintain the equilibrium of state of charge of the different cell packages, to have all charged at the same time, is based on the following principle:

Near the end of charge, the voltage of each cell package is measured. There is a direct relationship between voltage and state of charge for Li-Ion cells, as explained before. Then if one cell package is detected significantly at a higher State of Charge as the other ones, it is easy to compute at how many Ah it corresponds, and then to connect it on a resistive load for a given period of time, in order to discharge the excess of Ah. The voltage measurement and the necessary telemetry and telecomand is ensured by an electronic box integrated into the battery. The electronic function is installed in a standard module

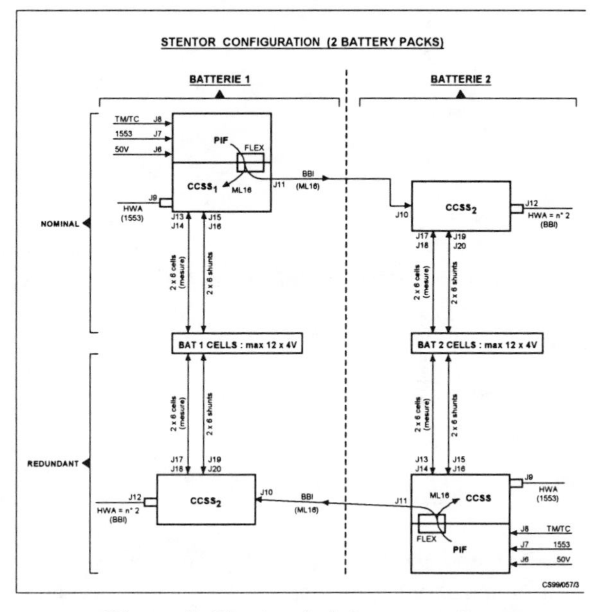

Figure 7. Electronic interconnection

able to monitor 12 cell packages in series. Two of these cards are implemented in a battery package: One nominal and one redundant. Then the information's are transmitted to the interface module of the battery package, formatted in ML16/DS16 series protocol, which is the battery internal format for data. This interface module comprises the to MIL-STD-1553-B converter, the power supply to power all the electronics. There is no redundant interface module into the battery package. The redundant one is located into the second battery package. The redundancy of electronics is ensured through the internal data bus, which carries both data and power supply from one battery package to the second one. Figure 7 shows the symmetry of the two redundant interface modules, located on the two battery packages.

3.3 Mechanical & Thermal Architecture

3.3.1 Cell Modules

The cells are inserted in machined aluminum structures able to receive 6 or 5 cells. The Figure 8 shows as the cells are inserted and thermally bonded to the structure by a polyurethane resin. These modules will be directly bolted on the satellite wall, and will be equipped with heaters, thermostats, thermistors. The resistors to

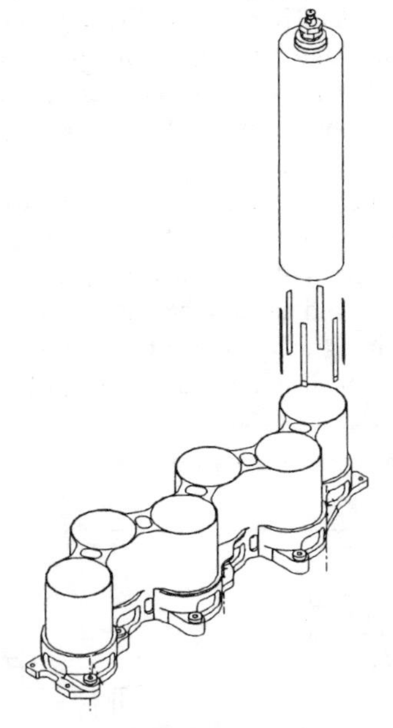

Figure 8. Cell module assembly

discharge the cells during balancing are also incorporated on this module.

3.3.2 Electronics

The unit is composed of 3 boards :

"CCSS" (cell conditioning and shunt switching) boards, nominal and redundant, and which include all electronics required for the Lithium-Ion battery management.

"PIF" (platform interface) board which include the power (50V bus) and TM/TC (1553) interfaces to the satellite on one side, and a serial bus (ML16 / DS16) on the other side to address each "CCSS" board.

3.3.3 Bypass, and connectors

The bypass are grouped by pair with the corresponding hybrid command circuit, and attached on the external walls of the electronics box. (Figure 9). All the interface connectors are also attached on the electronic box, either because they are directly in relation with the

Figure 9. Electronic command & bypass implementation

electronics, or because the electronic box provides an adequate attachment.

The complete battery (Figure 10), is made by the association of 2 modules of 5 cells, 2 modules of 6 cells and the electronic module which supports all the

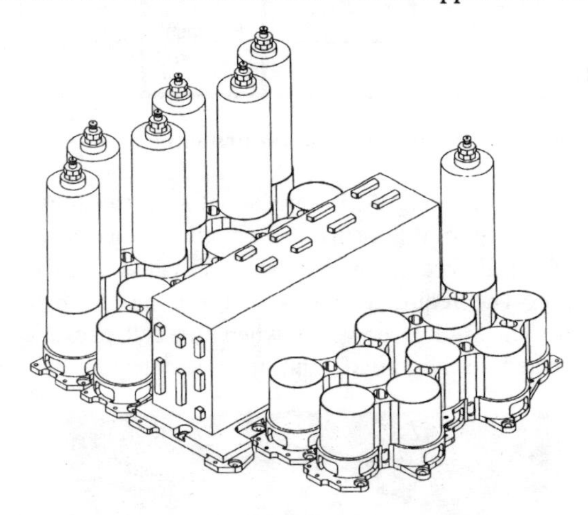

Figure 10.Stentor battery architecture

interface connectors, the bypass and their command.

4. BATTERY QUALIFICATION

4.1 Program

The qualification battery pack was built mid 1999, and the tests were performed since that time. The test schedule is as shown in Table 3. As the normal operation of the battery packs is with two items, and qualification is performed on a single Battery Pack, the nominal and redundant CCSS are linked together

Table 3. Battery qualification test schedule

Test	Description
General Physical Properties	Visual, Dimensions, Mass, Identification
Electrical Health Tests	Insulation, Continuity, Bonding, TM/TC
Initial Functional Tests	Full Performance Test Battery
Thermal Vacuum Test	Corona Test
	Thermal Balance
	Thermal Balance : Equinox
	Thermal Balance : Solstice
Reference Test Battery	Capacity Test at +20°C, Internal resistance, check TM/TC
Reference Test Battery	Capacity Test at +20°C, Internal resistance, check TM/TC
Reference Test Electronics	
Mechanical Tests per axis	Resonance search Test
	Sine vibrations with supervision electronics & battery
	Resonance search Test
	Random vibrations with supervision electronics & battery
	Resonance search Test
Reference Test electronics	
Reference Test Battery	Capacity Test at +20°C, Internal resistance, check TM/TC
Shock per axis	Shock with supervision electronics & battery
	Reference Test electronics
Reference Test Battery	Capacity Test at +20°C, Internal resistance, check TM/TC
Electromagnetic Tests	EMC with supervision electronics & battery
	ESD with supervision electronic & battery
	Magnetic Moment
Reference Test Electronics	
Final Electrical Health Tests	Insulation, Continuity, Bonding, check TM/TC
Final Functional Tests	Full Performance Test Battery

by an internal bus. This allows to test all the functionality's of the electronics with only one battery pack.
The details of vibration tests are given in Tables 4 and Table 5 for sine vibrations, Table 6 and Table 7 for random vibrations.

Sweep rate 2 octaves/minute

Frequency	Level
5 to 19 Hz	± 10 mm
19 to 70 Hz	13.5 g
70 to 100 Hz	8 g

Table 4. Sine vibrations level OX and OY

Sweep rate 2 octaves/minute

Frequency	Level
5 to 22 Hz	± 10 mm
22 to 60 Hz	15 g
60 to 100 Hz	20g

Table 5. Sine vibrations level OZ

Frequency	Level
20 to 100 Hz	+6dB/Oct
100 to 2000 Hz	0.05 g^2/Hz
Global 9.8 gRms	

Table 6. Random vibration level OX and OY.
Duration 3 minutes per axis

Frequency	Level
20 to 100 Hz	+6dB/Oct
100 to 350 Hz	0.2 g^2/Hz
350 to 2000 Hz	0.05 g^2/Hz
Global 11.8 gRms	

Table 7. Random vibration level OZ..

Duration 3 minutes per axis

4.2 Test results

The tests started end of 1999 and ended at the time of this paper. All the tests were satisfactory. The battery pack passed all the environments. These qualification tests gave the opportunity to adjust the user handbook of the battery, the software used to operate the balancing and the data acquisition, and a lot of other minor, but important problems of interface specific with this new and complex battery system.
The Qualification Battery is now integrated into the STENTOR satellite, and is used to test the communication links and interfaces.

Figure 11.The Stentor Qualification Battery Pack

4.3 Conclusion

From 1996 to 2000, SAFT developed and qualified a Lithium-Ion battery for the STENTOR satellite. This battery, now qualified, is the first demonstrator which will open the way to Lithium-Ion for Geosynchronous satellites. The STENTOR flight batteries are being built now, and will be delivered and integrated into the satellite this year. With the launch of STENTOR early 2001, a new era will be open in energy storage for space.

[1] JP.Semerie, Proceedings of the 33rd IECEC, Aug 1998, Modular Concept of Lithium-Ion Battery for Geosynchronous Orbit Satellites, IO89, Colorado Springs, CO, USA.

[2] J.P.Semerie, G.Dudley, P.Willmann, Proceedings of the 34th IECEC, Aug 1999, Evaluation and Life Testing of Lithium-Ion Batteries for Geosynchronous Satellites, 01-2595, Vancouver, BC, CAN.

Performance Characteristics of Yardney Lithium-Ion Cells For the Mars 2001 Lander Application

M.C. Smart[a], B.V. Ratnakumar[a], L. Whitcanack[a], S. Surampudi[a], L. Lowry[a]
R. Gitzendanner[b], C. Marsh[b], F. Puglia[b], J. Byers[c], and R. Marsh[d]
[a]Jet Propulsion Laboratory, California Institute of Technology, Pasadena, CA 91109
[b]Yardney Technical Products, Pawcatuck, CT
[c]Lockheed-Martin Astronautics Corporation, Denver, CO
[d]Air Force Research Laboratory, Wright-Patterson AFB, Dayton, Ohio

ABSTRACT

NASA requires lightweight rechargeable batteries for future missions to Mars and the outer planets that are capable of operating over a wide range of temperatures, with high specific energy and energy densities. Due to the attractive performance characteristics, Lithium-ion batteries have been identified as the battery chemistry of choice for a number of future applications, including Mars Rovers[1] and Landers[2,3]. The Mars 2001 Lander (Mars Surveyor Program MSP 01) will be among one of the first missions which will utilize Lithium-ion technology. This application will require two Lithium-ion batteries, each being 28 V (eight cells), 25 Ah and 9 kg (18 kg total). In addition to the requirement of being able to supply at least 90 cycles on the surface of Mars after a 1 year storage and cruise time, the battery must be capable of operation (both charge and discharge) over a wide temperature range (–20°C to +40°C), with tolerance to non-operational excursions to –30°C and 50°C. To assess the viability of lithium-ion cells for these applications, a number of performance characterization tests have been performed on state-of-art Yardney lithium-ion cells, including: assessing the room temperature cycle life, low temperature cycle life (-20°C), rate capability as a function of temperature (-30° to 40°C), pulse capability, self-discharge and storage characteristics, as well as, mission profile capability. This paper will describe the Mars 2001 Lander mission battery requirements and will contain results of the cell testing conducted to-date in support of the mission.

INTRODUCTION

NASA is planning several missions in the near future to continue the exploration of the Mars, including some missions being aimed at retrieving Martian samples back to the Earth. Various missions, such as Landers, Rovers, Mars Ascent Vehicles (MAV) and Orbiters are thus being planned and will be supported by different advanced technologies. One advanced technology in the area of power sources is the lithium ion battery, which has been selected as the baseline for the upcoming MSP 2001 Lander, which will be fabricated by Lockheed-Martin Astronautics, in collaboration with JPL. The MSP 2001 Lander was originally scheduled for launch in April 2001, however, it has recently been delayed to an expected launch date in 2003.

The goal of the Mars 2001 mission is to complete the global reconnaissance of Mars and perform surface exploration in support of a number of science objectives. In particular, an attempt will be made to 1) globally map the elemental composition of the surface, 2) acquire spatial and spectral resolution of the surface mineralogy, 3) determine the abundance of hydrogen in the subsurface, 4) study the morphology of the Martian surface, 5) provide descent imaging of the landing site, 6) study the nature of local surface geologic processes, and 7) assess the viability of future human exploration in terms of radiation-induced risks, and soil and dust characteristics. To accomplish these objectives, the MSP 2001 Lander will incorporate a number of key technologies and experimental devices, including: 1) a Mars Descent Imager (MARDI) 2) a Mars Radiation Environment Experiment (MARIE), 3) a Mars In-Situ Propellant Production Precursor (MIPP), 4) a Mars Environmental Compatibility Assessment (MECA) experiment, 5) a Robotic Arm and Robotic Arm Camera, 6) a Stereoscopic Panoramic Imager (T=PanCam), as well as, 7) a Mini-Thermal Emission Spectrometer (Mini-TES).

After evaluating a number of different cell chemistries, supplied by different vendors responding to the Lockheed-Martin BAA, Yardney Technical Products was selected as the vendor to supply the lithium-ion batteries for the MSP01 2001 Lander. The cell chemistry adopted by Yardney to meet the projected mission requirements consists of MCMB carbon anodes, $LiNiCoO_2$ cathode materials, and a low temperature electrolyte (1.0 M $LiPF_6$ EC+DMC+DEC (1:1:1)) developed at JPL.[4,5] The cell design selected by Yardney is a prismatic arrangement, which enables easy stacking of the cells to produce an eight cell battery. It must noted that the cell and battery technology development effort was made possible, in part, by the participation of the NASA-DOD consortium recently formed to establish domestic capability to manufacture lithium-ion cells and batteries in the US.[6]

POWER SUBSYSTEM FOR MSP 01 LANDER

The main power source for the MSP 2001 Lander consists of a 300 W Ga-As solar cell array. The auxiliary power source augmenting the solar array for the nighttime operations will be a Li-ion rechargeable battery. Two lithium-ion batteries will be used with the current orientation, each being 28 V, consisting of eight cells and a capacity of 25 Ah (name plate capacity). Both batteries will be contained within a single housing, and the total battery assembly should not weigh more than 18.2 Kg. Although both batteries will be operated during the course of the mission, a single battery can fulfil the needs of the entire mission, thus, one battery can be considered as being redundant. Each of the batteries will have an independent charge-control unit, with individual cell bypass features for charge control.

Li ION CELL/BATTERY REQUIREMENTS

The mission dictates that a number of performance requirements must be met by the 28 volt, 25.0 Amp-hour batteries to successfully complete the planned mission. Perhaps the most important feature of the battery is its requirement to operate (both charge and discharge) at continuous rate of C/5 over a wide range of temperatures (-20° to +40°C) once the Lander has successfully landed on the surface of Mars. The battery should be capable of providing a minimum EOL capacity of 25 Ah. The typical discharge drains will be C/5 to a maximum of 50% DOD. However, with both the batteries being connected in parallel (with a diode protection), the actual depths of discharge could be even milder than 50%. The maximum charge current is projected to be approximately 5 A (C/5). In addition to operating efficiently on the surface of Mars, the batteries should be able to withstand 50 A pulses at 0°C for short duration, during the entry, descent and landing phase (EDL). In case that the Li-ion batteries are unable to meet this criterion, a thermal battery (Li-FeS_2) is being considered as in the case of Mars Pathfinder, however, recent testing has shown that is may not be necessary. Prior to satisfying both of these requirements, the battery must survive a pre-discharge storage duration of nearly 2 years (6 months to one year pre-cruise storage) and a one year cruise period at 0° to 30°C.

Li ION CELL/BATTERY EVALUATION

In order to assess the viability of using lithium-ion technology for the Mars 2001 Lander, a test plan was formulated by Lockheed-Martin, in collaboration with JPL and Yardney, which reflects the need for data which address the various mission requirements. The test plan generally consists of determining: (i) the room cycle life performance (25°C), (ii) low temperature cycle life performance (-20°C), (iii) discharge and charge rate capability at different temperatures (-20, 0, 25, and 40°C), (iv) pulse capability at different temperatures and different state-of-charge (SOC), (v) optimum storage condition to ensure minimal loss of performance (vi) ability to perform an EDL load profile, and (vii) ability to cycle under surface temperature profile conditions. Although testing to achieve these ends was performed by all three institutions, the results of the cell testing performed at JPL only will be considered in this paper.

CELL TESTING RESULTS

CYCLE LIFE PERFORMANCE

According to the projected mission plans, the battery should be capable of providing a minimum of 90 cycles once the spacecraft has reached the surface of Mars. Due to the fluctuating temperatures on the surface of Mars during the course of a typical sol period, the battery will be required to cycle efficiently over wide temperature variations (-20 to +40°C). In addition, successful operation must be demonstrated after being subjected to an extended cruise period (~ 11 months) and an additional storage period from the date of manufacturing and time of launch. In order to assess the viability of the lithium-ion technology to meet these requirements, a combination of tests were undertaken to establish a comprehensive data base to enable predictive performance trends. One general test performed to evaluate the life characteristics involved 100 % DOD cycling of cells between a voltage range of 3.0 Vdc to 4.1 Vdc at a number of temperatures. As illustrated in Fig. 1, 20 Ahr prototype cells have been cycled successfully cycled > 800 cycles at both ambient temperatures as well as at –20°C (charged and discharged at low temperature).

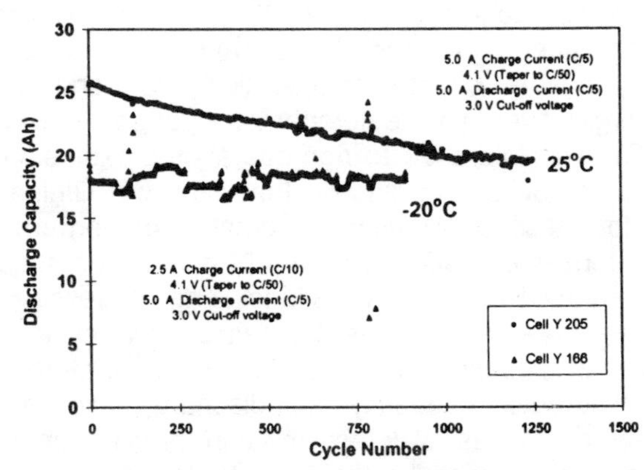

Fig.1. Cycle life performance (100% DOD) of Yardney 20 Ahr prototype lithium-ion cells at 25 and –20°C.

As illustrated by Fig 1, upon completing 750 cycles the cells cycled at room temperature delivered ~ 85% of the initial capacity, whereas, the cells cycled at –20°C were observed to deliver ~70% of the initial capacity. In addition, it is apparent that the capacity fade is much less at low temperatures compared with higher temperatures, due most likely to the increased rates of impedance build-up with increasing temperature.

CYCLING AT ALTERNATING TEMPERATURES

In addition to evaluating the cycle life performance of the cells under conditions of constant temperature, effort was focused upon determining the effect of cycling between temperature extremes for fixed number of cycles. As shown in fig. 2, cells were cycled intermittently (10 cycles) between two temperature extremes (-20°C and 40°C). As illustrated, the impact of cycling a cell intermittently at 40°C results in a dramatic decrease in amount of capacity being able to be delivered at low temperature. After completing 200 cycles under this regime, cells were observed to lose 50-75% of the initial capacity delivered at low temperature.

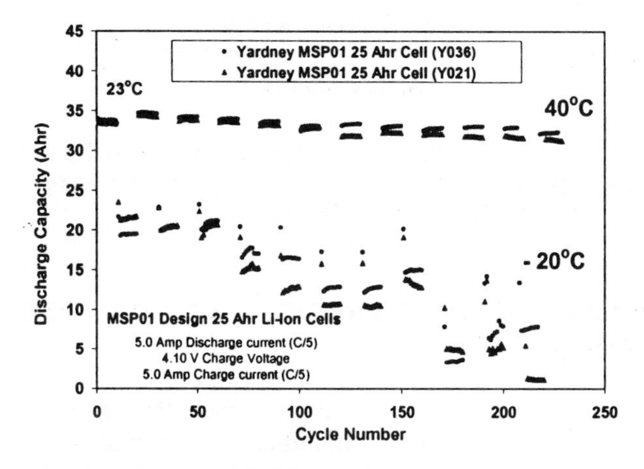

Fig. 2. Variable temperature cycling of Yardney MSP01 25 Ahr design lithium-ion cells.

This is in sharp contrast to the minimal capacity fade obtained when the cells are continually cycled at low temperature with no excursion to higher temperatures. However, it must be noted that there is little degradation of the cell performance at higher temperatures, suggesting that the observed capacity losses at low temperature are due to an increase in cell impedance, most likely due to increased passivation at the electrode surfaces resulting in resistive films preventing facile lithium ion kinetics, which is magnified at low temperatures. It must be emphasized, however, that in terms of mission requirements, this type of testing represents a worst case scenario. According to the mission profile, the cells will not experience prolonged high temperature exposure (>30°C) or prolonged low temperature exposure (>-10°C) for significant length of time, but rather will be subjected to milder conditions as discussed in the mission simulation testing section.

In an attempt to understand more fully the mechanism by which the cell impedance increases upon cycling under these conditions, a number of additional tests were performed on Yardney prototype 5 Ahr cells of similar chemistry. In these series of tests, the impact of the charge voltage upon cell degradation was investigated by either charging the cells to either 4.0 or 4.1V during the period of high temperature exposure. As illustrated in Fig. 3, the loss in performance at low temperature can be decreased by using lower charge voltages at high temperatures. One possible interpretation of these results is that oxidative decomposition of the electrolyte is accelerated at high temperatures and high charge voltage which results in the formation of increasingly resistance electrode films (both anode and cathode).

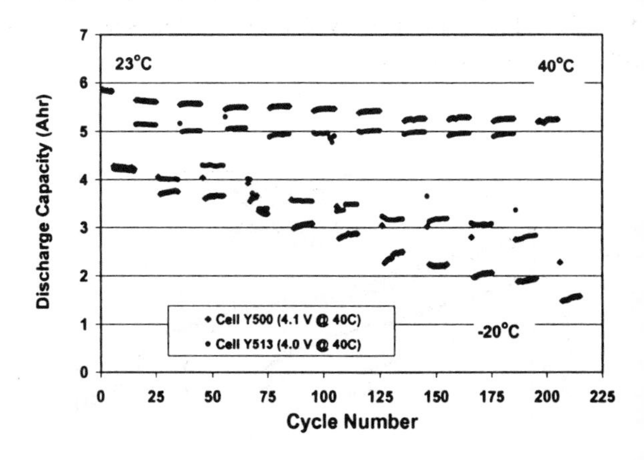

Fig. 3. Variable temperature cycling of Yardney 5 Ahr design lithium-ion cells.

This interpretation is supported, in part, by electrochemical impedance spectroscopy (EIS) measurements taken on the cells while carrying out the variable temperature testing. As shown Fig. 4, the cell which was charged at lower voltages during the high temperature cycling displayed smaller increases in cell impedance and lower film resistance.

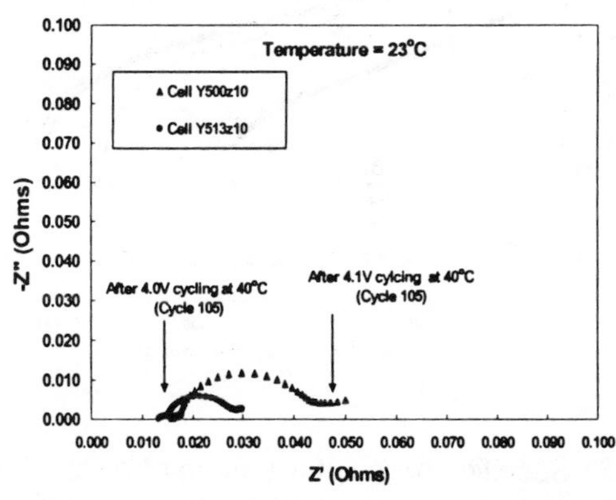

Fig. 4. EIS measurements of 5 Ahr Li-ion cells during variable temperature cycling test

DISCHARGE PERFORMANCE AT DIFFERENT TEMPERATURES

Since demonstration of efficient performance at low temperature was a major technological challenge, a large amount of emphasis was placed upon evaluating the discharge capacity over a number of different rates (C/2, C/3, C/3.3, C/5 and C/10) and temperatures (-30, -20, 0, 23, and 40°C). When Yardney MSP01 design cells were evaluated at a C/5 discharge rate (5.0 Amp discharge to 3.0 V) at different temperatures, good erformance was observed over the range of emperatures, as shown in Fig 5. At –20°C, ~ 24 Ahr of apacity was capacity was delivered (cell charged at -20°C using a C/10 charge rate to 4.1V), representing -70% of the room temperature capacity. As shown in ig. 6, the cells were also observed to deliver excellent pecific energy over a large range of temperatures, with ver 85 Wh/kg and 140 Wh/kg being delivered at –20°C nd 40°C, respectively.

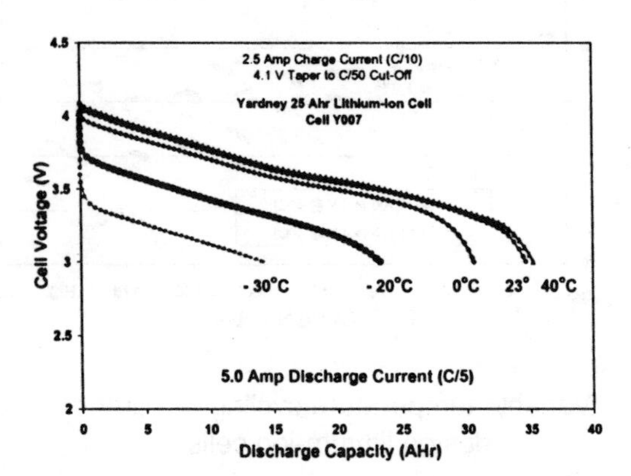

Fig. 5. Discharge capacity of a Yardney MSP01 design 25 Ahr cell at a C/5 discharge rate and at different temperatures (-30, -20, 0 , 25 and 40°C).

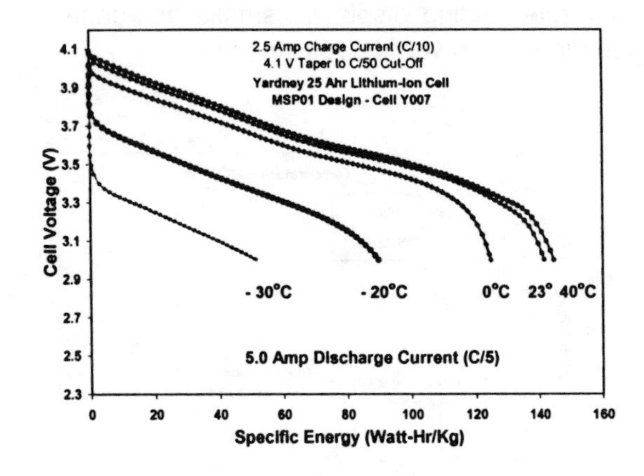

Fig. 6. Specific energy of a Yardney MSP01 design 25 Ahr cell at a C/5 discharge rate and at different temperatures (-30, -20, 0 , 25 and 40°C).

In the course of the discharge rate characterization studies, it was observed that better low temperature performance was generally obtained when the relative tests were performed sequentially from low temperatures to high temperature (-30, -20, 0, 23, and 40°C), rather than in the reverse order (40, 23, 0, -20, and –30°C). This trend underscores the sensitivity of the low temperature performance depending upon cell history and extent of exposure at high temperatures. The most dramatic illustration of this behavior was observed when cells were exposed to +50°C cycling (8 cycles). As shown in Fig .7, significant loss in low temperature capability was observed when the cells were evaluated at –20°C before and after the high temperature (+50°C) cycling.

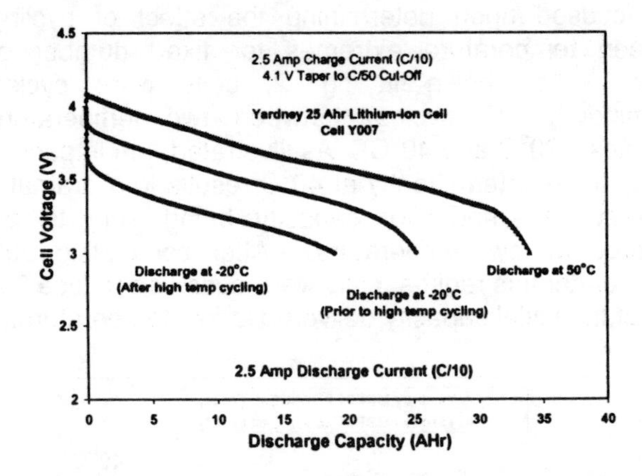

Fig. 7. Impact of high temperature cycling upon the low temperature performance of a Yardney MSP01 design 25 Ahr cell (C/5 discharge to 3.0V).

CHARGE CHARACTERISTICS AT DIFFERENT TEMPERATURES

In the same manner in which the discharge capacity as a function of temperature was evaluated, the charge characteristics were assessed at different rates and temperatures. As shown in Fig. 8, cells displayed good charge acceptance over a wide temperature range (-30° to 40°C) with ~70% and ~50% of the ambient temperature capacity realized at –20°C and –30°C, respectively.

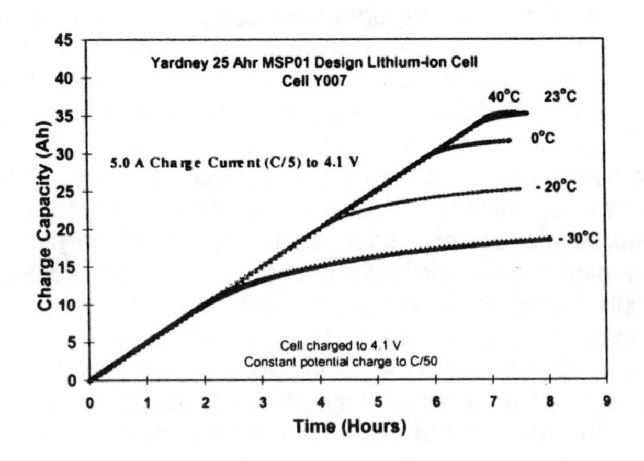

Fig. 8. Charge capacity of a Yardney MSP01 design 25 Ahr cell at a C/5 charge rate to 4.1V and a C/50 taper current cut-off as a function of temperature.

As shown in Fig. 9, essentially full capacity can be obtained in ~ 4 hours with high charge rates (C/2=12.5 Amps) at –20°C, with little variation in charge capacity as a function of constant current charge rate.

This is primarily due to the fact that the charge characterization tests were performed with a constant potential charging step with the same taper current cut-off value in common (C/50=0.500 Amps).

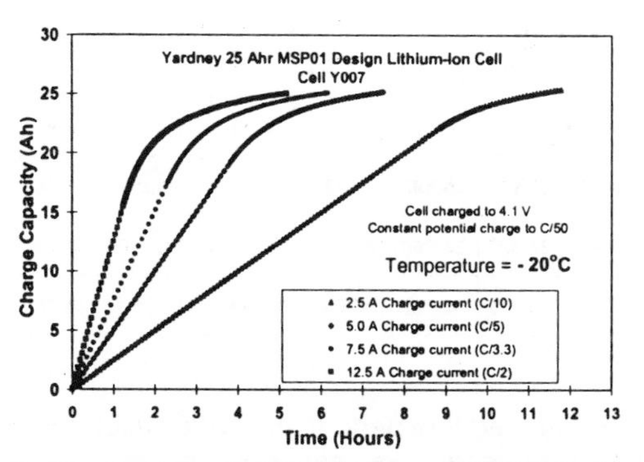

Fig. 9. Charge capacity of a Yardney MSP01 design 25 Ahr cell at –20°C at different rates (constant current charge to 4.1V with a C/50 taper current cut-off).

A general trend consistently observed throughout the charge characterization tests is that upon going to lower temperatures more significant amounts of the total charge capacity accepted by the cells occurred in the constant potential charging mode, as illustrated in Fig. 10, due to inability to sustain high charge current densities without large electrode polarization.

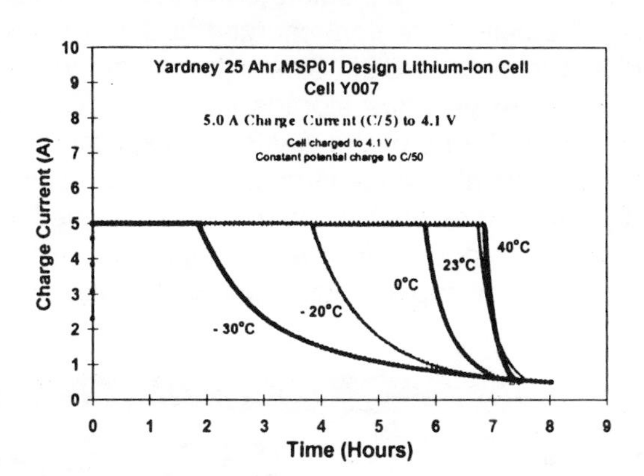

Fig. 10. Applied charge current as a function of temperature (-30, -20, 0 , 25 and 40°C) with a C/5 charge rate (5 Amps) to 4.1V and a C/50 taper current cut-off).

Although the bulk of the charge rate characterization tests were performed using a taper current cut-off value to end the constant potential charging mode, such that charge step is terminated when the current decays to less than 0.500 Amps (C/50), some characterization was performed using extended charge periods. This is especially relevant for the 2001 Lander application, since the battery will be permanently connected to the buss and will experience longer charge periods. These conditions generally lead to more capacity accepted, especially at low temperature, as illustrated in Fig. 11.

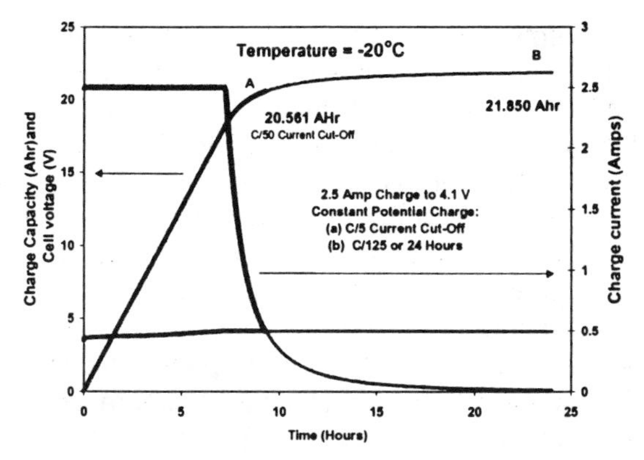

Fig. 11. Charge characteristics of a Yardney 20 Ahr prototype cell at -20°C with extended constant potential charging.

One initial concern with operating the batteries under such conditions, with very long charge periods, is that an increased rate of cell degradation was anticipated due to the length of time the cells are held at high potential (for the reasons mentioned previously). For this reason, cells were cycled (100% DOD) using especially long charge periods (24 hours) and compared with cycling results obtained when the charge period is discontinued upon reaching a taper current cut-off value of C/50 (~ 6 hour charge time with C/5 charge rate to 4.1 V). However, no significant increase in the capacity fade characteristics was observed when an extended charge period was employed, as shown in Fig. 12.

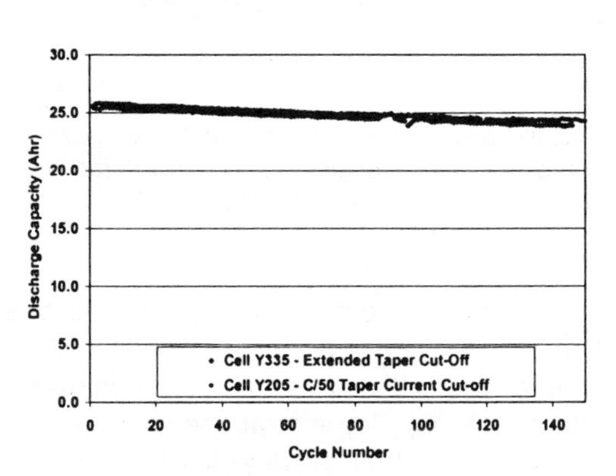

Fig. 12. Cycle life performance of Yardney 20 Ahr prototype cells using different charge regimes.

STORAGE CHARACTERISTICS

In order to assess the capability of the technology to meet the various life requirements, it was necessary to conduct a number of tests to evaluate the effect of prolonged storage upon performance. In the case of the Mars 2001 Lander, the battery must be operational after an 11-month cruise period while the spacecraft is in transit to Mars. The first set of tests were aimed at determining the effect of storage temperature and cell state of charge upon performance when the cells are stored under open circuit conditions (OCV). The cells selected for this testing were of an early generation, 20Ah capacity design. In order to represent the extremes projected for the cruise storage period, two different temperatures were selected (0 and 40°C) and two different states-of-charge (50 and 100%) were utilized. For these initial tests, the cells were: (i) first cycled (5-10 cycles) prior to storage (ii) stored at the selected temperature and state-of-charge (iii) discharged to 3.0V to determine the residual capacity and (iv) then cycled a number of times (5-10 cycles) to determine the extent of permanent capacity loss of the cells (if any) as a result of the storage period. The cells were first subjected to a two month storage period accompanied by full performance characterization before and after, followed by a longer ten month storage period. As shown in Table 1, in general minimal permanent capacity loss was observed over the range of conditions investigated, with the largest loss in capacity with the cells which were stored at high temperature (+40°C). However, if the cells are stored at low temperature (0°C) and low state-of-charge over 96% of the initial capacity is realized after one year of cell storage.

		Two Month Storage Period			Ten Month Storage Period			Total
Storage Condition	Initial Capacity	Capacity Prior To Storage (Ah)	Capacity After Storage (Ah) 5th Disch.	Rever. Capacity (%)	Capacity Prior To Storage (Ah)	Capacity After Storage (Ah) 5th Disch.	Rever. Capacity (%)	Total Reversible Capacity After 12 Month Storage
(0°C and 50 % SOC)	27.879	27.609	27.327	98.98	26.972	26.786	99.31	96.08
(40°C and 50 % SOC)	28.749	28.021	27.479	98.07	27.918	25.595	91.68	89.03
(0°C and 100 % SOC)	25.475	25.471	24.781	97.29	24.607	24.296	98.74	95.37
(40°C and 100 % SOC)	25.674	25.670	25.156	98.00	23.912	22.727	95.05	88.52

Table 1. Discharge capacity of Yardney 20 Ahr prototype cells before and after being subjected to various storage conditions (OCV storage).

In addition to determining the impact of storage conditions upon the reversible capacity at ambient temperature, the cells were also characterized at low temperature (-20°C). In general, the storage of the cells was observed to affect the low temperature capability more dramatically, and proportionately lower capacities were observed as illustrated in Table 2. In contrast to the trend observed when the cells were evaluated at room temperature, the effect of state-of-charge was seen to be more dominant than the effect of temperature upon in determining the low temperature capability. The best results were obtained with cell which was stored at 50% SOC and at 0°C, with ~66% of the room temperature capacity realized at –20°C (compared to ~ 70 % of the room temperature being delivered prior to the storage characterization tests.

		Ten Month Storage				Low Temperature Performance	
Storage Condition	Initial Capacity	Capacity Prior To Storage (Ah)	Capacity After Storage (Ah) 5th Discharge	Rever. Capacity (%)	Total Reversible Capacity (% of Initial)	1st Discharge at -20°C (5 Amps = C/5)	% of Initial Capacity
(0°C and 50 % SOC)	27.879	26.972	26.786	99.31	96.08	17.276	61.97
(40°C and 50 % SOC)	28.749	27.918	25.595	91.68	89.03	18.935	65.86
(0°C and 100 % SOC)	25.475	24.607	24.296	98.74	95.37	12.995	51.01
(40°C and 100 % SOC)	25.674	23.912	22.727	95.05	88.52	11.400	44.40

Table 2. Low temperature performance (-20°C) of Yardney 20 Ahr prototype cells before and after being subjected to various storage conditions (OCV storage).

In addition to investigating the effect of storage under OCV conditions, effort has been devoted to evaluating the viability of storing the cells connected to the buss for the duration of the storage period. This is especially relevant due to the fact that the spacecraft design is simplified if the cells are connected to the buss for the duration of the mission. In order to simulate potential cruise conditions, a number of cells (4) were stored for ~11 months connected to the buss and stored at 10°C. The cells were float charged at 3.875 V which corresponds to ~70% SOC. Similar to the methodology described for the previous storage study, all cells were characterized in terms of the reversible capacity before and after storage at various temperatures. As shown in Table 3., excellent reversible capacity was obtained after 11 months of storage under these conditions, with less that 5% permanent capacity loss observed in all cases.

	Storage After 20 Days							
Last Discharge Prior to Storage (Ahr)	1st Discharge After Storage (Ahr) 23°C	2nd Discharge After Storage (Ahr) 23°C	% of Initial Capacity (Reversible Capacity)	Permanent Capacity Loss (%)	1st Discharge After Storage (Ahr) 23°C	2nd Discharge After Storage (Ahr) 23°C	% of Initial Capacity (Reversible Capacity)	Permanent Capacity Loss (%)
33.804	26.034	33.523	99.169	0.831	25.6252	32.9636	97.515	2.485
33.962	25.959	33.534	98.738	1.262	29.059	32.266	95.006	4.994
34.153	25.445	32.788	96.005	3.995	25.639	32.999	96.622	3.378
33.727	25.922	33.460	99.210	0.790	25.478	32.917	97.599	2.401

Table 3. Discharge capacity of Yardney MSP01 design 25 Ahr cells before and after being subjected to storage (cells stored at 10°C and 70% SOC).

When the low temperature performance was assessed following the 11 month storage period, less cell to cell variation in performance was observed with cells stored at 10°C and 70% SOC on the buss compared with the group of cells stored under various OCV conditions. The consistency of the values obtained is encouraging when considering potential battery issues related to how well the cells are matching in capacity and performance characteristics throughout the mission life. Only 5-10% reduction in capacity was observed at −20°C after prolonged storage connected to the buss, as illustrated in Table 4.

	[illegible]					Low Temperature Performance After Storage Period			
	Last Discharge Prior to Storage (Ahr)	1st Discharge After Storage (Ahr) 23°C	2nd Discharge After Storage (Ahr) 23°C	% of Initial Capacity (Reversible Capacity)	Permanent Capacity Loss (%)	1st Discharge (Ahr) -20°C	% of Initial Room Temp Capacity	2nd Discharge (Ahr) -20°C	% of Initial Room Temp Capacity
[illegible]	33.804	25.62524	32.9636	97.515	2.485	22.466	66.46	19.537	57.79
[illegible]	33.962	29.059	32.266	95.006	4.994	22.099	65.07	19.437	57.23
[illegible]	34.153	25.639	32.999	96.622	3.378	22.224	65.07	19.299	56.51
[illegible]	33.727	25.478	32.917	97.599	2.401	22.397	66.41	19.647	58.25

Table 4. Low temperature performance of Yardney MSP01 design 25 Ahr cells before and after being subjected to storage (cells stored at 10°C and 70% SOC).

Overall, the results indicate that efficient storage of lithium-ion cells can be achieved while connected to the buss if proper conditions are selected. As illustrated in Fig. 13, when the discharge profiles are compared before and after storage, very little change in performance is observed, with minimal degradation of operating voltage and minimal capacity loss (~ 2.5%).

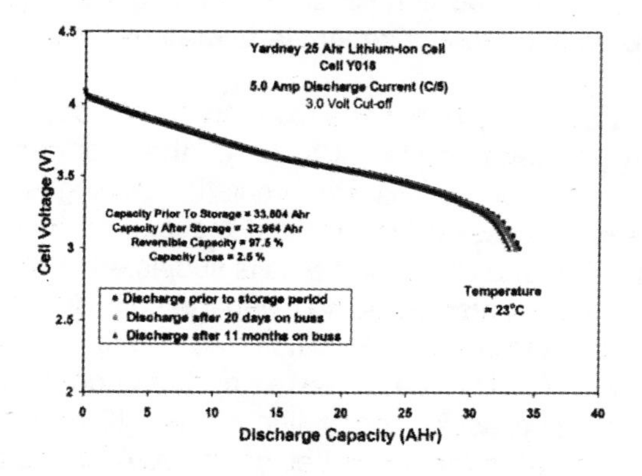

Fig. 13. Discharge capacity of a Yardney MSP01 design 20 before and after storage on the buss at 10°C (70% SOC) for 11 months.

EDL PROFILE

After completing the cruise period, the battery is expected to assist in the entry, descent, and landing process, which involves supplying power to various pyros and landing functions. The general load profile that the battery will experience during this period is shown in Fig. 14. The most demanding segment of the load profile consists of a 20 Amp discharge current onto which 30 Amp pulses are applied, thus, both contributing to produce 50 Amp loads (2C discharge rate) for short duration (100 milliseconds). In terms of mission requirements, the ability to sustain cell voltages above 3.0 V throughout the duration of this test at 0°C is the most difficult to fulfill.

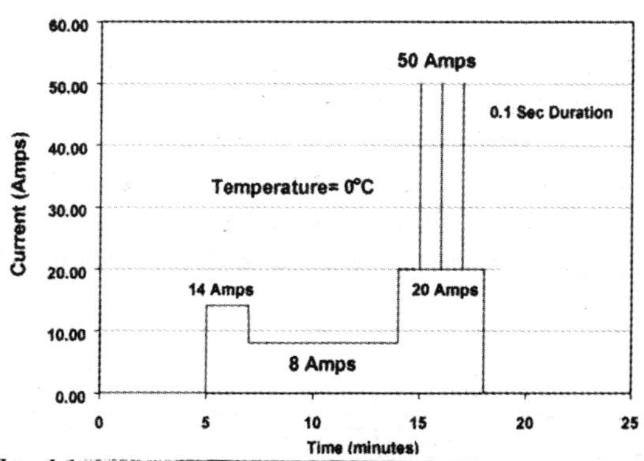

Fig. 14. Load profile of the Lander battery during the entry, descent, and landing process.

Since performance data relating to the EDL load profile is more relevant on cells which have been subjected to prolonged storage to simulate the cruise phase of the mission, the tests were performed on the group of cells previously described which were stored under OCV conditions. Due to the variation in cell performance observed after the differing storage conditions, some variation in cell polarization was expected when subjected to the high current loads. This indeed was the case, with the cells which were subjected to conditions of high state of charge displaying the greatest cell polarization and the inability to sustain a voltage greater that 3.0 V during the high current (50 Amp) pulses. In contrast, the cells stored at low state-of-charge were able to maintain much higher operating voltages throughout the duration of the load profile, never dipping lower than 3.2 V, as shown in Fig. 15. Again it should be noted that these cells were of an earlier generation, 20Ah capacity design and thus not designed to meet these pulse requirements.

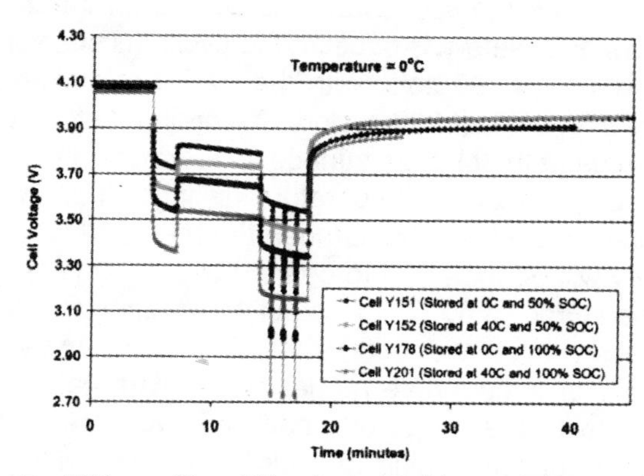

Fig. 15. EDL profile of Yardney 20 Ahr prototype cells after being subjected to ~ 1 year OCV storage.

imilar to the trends discussed earlier in relation to the w temperature capabilities, the state-of-charge during rage appears to have more influence upon the pulse apability compared to temperature of storage.

In addition to the group of cells which were tored under OCV conditions, the MSP01 design cells hich were stored on the buss for 11 months were also ubjected to the EDL profile. As shown in Fig. 16, xcellent results were obtained being capable of uccessfully meeting the mission objectives, with the perating voltages never dipping below 3.4 V throughout ie load profile.

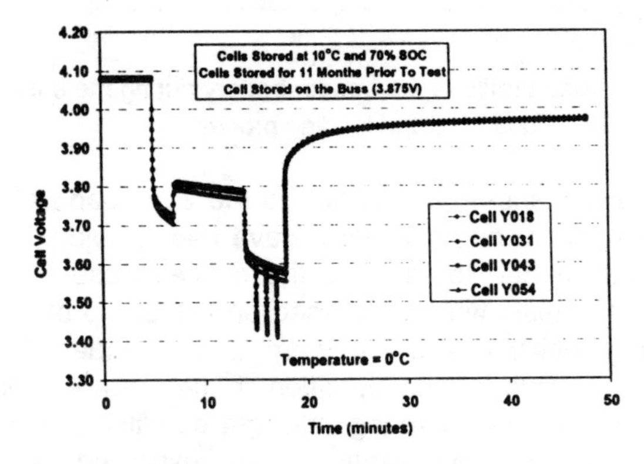

Fig. 16. EDL profile of Yardney MSP01 25 Ahr cells after being subjected to11 months storage connected to the buss.

n addition, very consistent data was obtained for the four cells studied, which were stored under identical conditions (10°C, 70% SOC = 3.875 V) prior to the pulsing test.

MISSION SIMULATION PROFILE

Once the spacecraft has landed on the surface of Mars, the battery is expected to cycle successfully for a minimum of 90 sols, with the desire of successfully completing at least 200 cycles. According to the current estimates of the Martian surface temperature profile, and the corresponding temperature swings that will be experienced within the Lander thermal enclosure, the battery will be expected to operate over a large range of temperatures (Δ 60°C). In order to simulate the battery operation over the course of the entire mission, a number of temperature ranges were investigated which correlate to the projected battery environment as the Martian season begins to change. These ranges are characterized by the widest temperature swings experienced in the beginning of the mission, and less severe, but colder temperature ranges later in the mission. Thus, continuous cycling was performed under the following conditions: (a) 20 cycles (days) over a temperature range of –20°C to 40°C, (b) 10 cycles at –20°C to 30°C, (c) 10 cycles at –20°C to 20°C, and a (d) 100 cycles at –20°C to 10°C. The electrical profile during this cycling consists of charging the cells with a constant current (C/5 rate) to 4.1V for a total charge time of 12 hrs, and a relatively mild discharge current (1 Amp or C/25 rate) for a total of 12 hrs, corresponding to 12 Ahr of capacity (~40% DOD). As shown in Fig. 17 for a typical mission simulation cycle, the beginning of the charge period occurs when the battery experiences the coldest temperatures, whereas, the beginning of the discharge period commences when the highest temperatures are experienced.

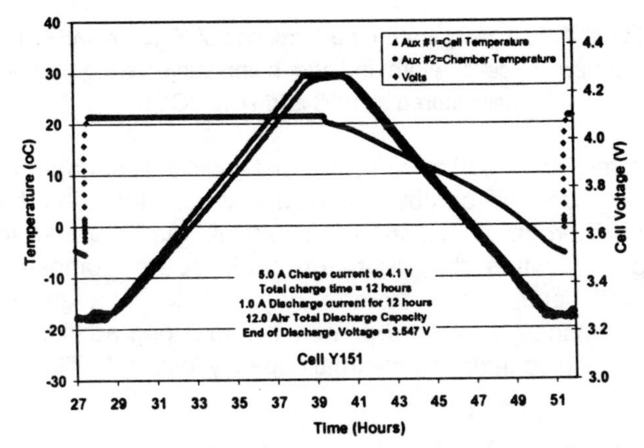

Fig. 17. Typical mission simulation cycle displaying the cell voltage response and temperature profile.

Due to the fact that a fixed amount of capacity is discharged each cycle (12 Ahr), the performance characteristics of the mission simulation cycling is most adequately expressed in terms of the end-of-discharge voltage. The end of life for the cells subjected to this test has been designated as being when the cells drop below 3.0 V upon discharge. As illustrated in Fig. 18, when prototype 20 Ahr cells were cycled under these conditions, successful completion of over 40 cycles has been observed over a number of different temperature ranges as previously described. These cells had previously been subjected to a 12 month OCV storage and EDL pulsing (described earlier) prior to the mission simulation testing. Thus, the observed cell performance is especially relevant, since the cell histories prior to the mission simulation profile testing reflect similar conditions to that expected to be experienced by the actual Lander battery. The fact that the operating cell voltages never dip below is 3.4 V, and display little capacity fade, is encouraging in terms of meeting the mission requirement previously described. Even more relevant mission simulation testing data is currently being generated on the MSP01 design cells which were previously stored on the buss at 10°C (70% SOC), which more adequately represents the actual projected storage conditions.

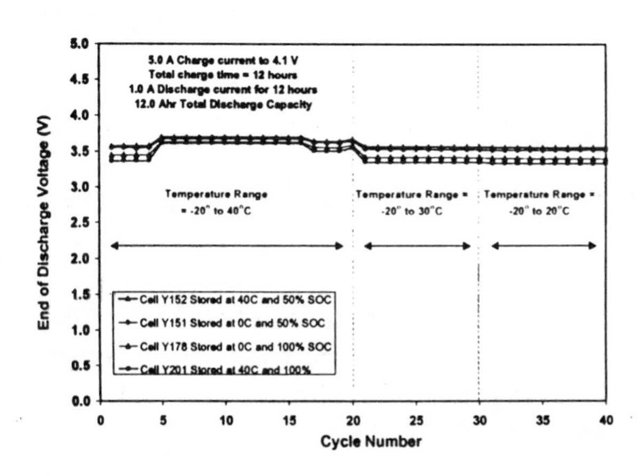

Fig. 18. Mission simulation cycling of prototype 20 Ahr lithium-ion cells.

CONCLUSIONS

A number of Yardney prototype 20 Ahr size and MSP01 design 25 Ahr size lithium-ion cells have been evaluated to determine their viability for use in the upcoming Mars Lander mission. This was accomplished by implementing a number of general and mission specific performance tests, including room temperature cycle life, low temperature cycle life (-20°C), rate capability as a function of temperature (-30° to 40°C), storage characteristics, pulse capability, as well as, mission profile cycling capability. When evaluating the cycle life performance, the technology has been demonstrated to well exceed the mission requirements of 200 cycles over a wide rage of temperatures (-20 to 40°C). Some diminishment in low temperature performance, however, was observed if the cells were cycled to extensively at higher temperatures (40°C). Good discharge and charge rate capability for the cells was demonstrated over a wide range of temperatures (-30° to 50°C), with greater than 24 Ahr being delivered at –20°C with at a C/5 rate (charge and discharge at low temperature). Thus, the performance of the cells was demonstrated to meet the low temperature requirements established for the mission. The results from a number of storage tests that were performed indicate that the least amount of cell degradation occurs when the cells are stored at low temperatures and low state-of-charge. In addition, it was demonstrated that float charging the cells at a fixed voltage (storage on the buss) is a viable method of storage and results in minimal performance degradation. Provided that the cells are stored under desirable conditions for the long cruise period, the load profile of the entry, descent, and landing phase can be effectively sustained, including 50 A pulses for short duration. When the cells were subjected to the mission simulation cycle life testing, which mimics the conditions the battery will experience on the surface of Mars, all of the cells cycled successfully. These results were obtained on cells which were previously subjected to both extended storage periods and the EDL load profile. In summary, it has been demonstrated that Yardney lithium-ion cells have met the performance requirements of the MSP01 Lander.

ACKNOWLEDGEMENT

The work described here was carried out at the Jet Propulsion Laboratory, California Institute of Technology, for the MSP 01 Lander Battery program and Code S Battery Program under contract with the National Aeronautics and Space Administration (NASA).

REFERENCES

1) B. V. Ratnakumar M. C. Smart, R. Ewell, S. Surampudi, and R. A. Marsh, Proceedings of the Intersociety Energy Conversion Engineering Conference (IECEC), Vancouver, British Columbia, Aug. 1-5, 1999.
2) M. C. Smart, B. V. Ratnakumar, L. Whitcanack, J. Byers, S. Surampudi, and R. Marsh, Proceedings of the Intersociety Energy Conversion Engineering Conference (IECEC), Vancouver, British Columbia, Aug. 1-5, 1999.
3) M. C. Smart, B. V. Ratnakumar, L. Whitcanack, S. Surampudi, J. Byers, and R. Marsh, *IEEE Aerospace and Electronic Systems Magazine*, 14: 11, 36-42 (1999).
4) M. C. Smart, B. V. Ratnakumar and S. Surampudi, *J. Electrochem. Soc.*, 146, 486, 1999.
5) M.C. Smart, C.-K. Huang, B.V. Ratnakumar, and S. Surampudi, Proceedings of the Intersociety Energy Conversion Engineering Conference (IECEC), Honolulu, Hawaii, July, 1997.
6) S. Surampudi, G. Halpert, R. A. Marsh, and R. James, NASA Battery Workshop, Huntsville, AL, Oct. 1998.

AIAA-2000-2915

Performance Characteristics of Lithium-Ion Cells for Mars Sample Return Athena Rover

B. V. Ratnakumar, M. C. Smart
R. Ewell, S.Surampudi
Jet Propulsion Laboratory, California Institute of Technology,
4800 Oak Grove Drive, Pasadena, CA 91109
and
R. Marsh
Air Force Research Laboratory
Wright-Patterson Air Force Base, Dayton, OH

ABSTRACT

Future planetary exploration missions, especially Landers and Rovers, will utilize lithium ion rechargeable batteries, due to their advantages of reduced mass and volume compared with non-lithium systems. In addition to the usual requirements of high specific energy and energy density, some applications, e.g., Mars Landers and Rovers, require the batteries to be functional over wide range of temperatures, i.e., -20 to +40°C. Several prototypes with modified chemistries commensurate with these mission needs were built by U.S. battery manufacturers, in conjunction with a NASA-DoD Interagency developmental effort and are being tested at JPL in the last couple of years. The proposed Mars Rover will have three lithium ion batteries (with one serving the purpose of redundancy) connected in parallel, each with four 6-9 Ah cells in series, to augment the solar array. The charger for these Rover batteries is being designed and built in-house. In this paper, we will present several performance characterization tests, including cycle life at different temperatures, rate capability at various charge/discharge rates and temperatures and real time and accelerated storage, which were carried out on these prototype cells in support of the Mars exploration missions.

INTRODUCTION

Mission Objectives

NASA is undertaking a detailed exploration of the planet Mars with the use of several robotic spacecraft including Orbiters, Landers, Rovers, Scouts and Microprobes. Following the recent set backs of two Mars missions, there have been some delays and even cancellations of some of the proposed missions in the immediate future, i.e., 2001 Mars Sample Return Lander and 2003 Mars Rover missions. The entire Mars Exploration architecture is currently being reviewed to formulate new, robust missions. The scientific goal of these missions, however, remains the same, i.e., to determine the geologic and climatic history of a site in the ancient highlands of Mars with conditions supposedly favorable to the preservation of evidence of possible prebiotic or biotic processes. Specific objectives of these missions would thus include 1) taking color stereo images of the Martian surface, 2) determining the elemental and mineralogical compositions of Martian rocks and soil to derive information of their geology and climate, 3) obtaining microscopic images of the rocks and soils and 4) finding and collecting the samples most likely to preserve the evidence of ancient environmental conditions and possible life and storing them for return to Earth. The early missions will cater to *in-situ* analysis of Martian samples, whereas the latter sample return missions will focus on bringing the Martian samples back to the Earth. Typical payload elements in such sample return Rovers/Landers would include Pancams for color stereo imaging, an ∝- proton X-ray spectrometer for elemental composition analyses, Mossbauer, Mine-TES and Raman spectrometers for mineralogical composition analyses, a microscopic imager for close-up imaging and a mini-corer and sample container for sample collection and storage.

Li Ion Rechargeable Batteries for Mars Rover

The main power source for the Mars (2003) Rover consists of a Ga-As solar cell array. The auxiliary power source provides power for the nighttime and peak power operations is a lithium ion rechargeable battery. The lithium ion rechargeable battery for the proposed Rover will have 16 V and 150 Wh, with the total mass and volume of the battery not exceeding 3 kg and 2 liters, respectively. Three four-cell (each of 6-7Ah) strings will be connected in parallel with a diode protection, with one string providing some redundancy. Presently, the design of the battery is compatible with either prismatic or cylindrical cells. The above-specified energy of 150 Wh is to be made available after about two years of activation, with the preferred mode of storage during flight integration and cruise being the battery off the buss and at a temperature of 0-10°C. One critical performance requirement for the battery is the need for it to operate at sub-zero temperatures at moderate rates (maximum of C/2), without any reduction in the room temperature performance. The batteries will be provided with a heater to maintain temperatures of at least -20°C (minimum). Charging of the cells will be carried out at relatively higher temperatures (0 to +30°C). The battery will be charged with an indigenously built charge control unit, which facilitates individual cell monitoring and cell balancing. A cycle life of over 200 cycles (to 80% of initial capacity) and two years of calendar life are required for the mission.

Earlier, we have briefly reported the performance characterization tests carried out on prototype lithium ion cells, fabricated by different manufacturers under a NASA/DoD lithium ion battery consortium.[1-5] Under this program, multiple manufacturers are being supported to provide the desired technological developments and their cells, either prismatic or cylindrical and with capacities ranging from 4 to 40 Ah, are being evaluated at JPL under generic performance conditions as well as those relevant to Mars Exploration programs (Landers and Rovers). In this paper, we provide the updates of such performance characterization tests on 4 - 9 Ah lithium ion cells, specifically for the applications similar to Mars Rovers. Similar tests carried out on larger cells (20 Ah) for Lander applications are being communicated in our companion paper.[6]

Li ION CELLS EVALUATION

The lithium ion cells evaluated contain proprietary electrolytes, electrode materials and designs to achieve the desired performance characteristics, i.e., low temperature performance and cycle life. It was deemed essential to keep the manufacturers anonymous in this paper to promote parallel development at each of the respective organizations. All the cells have gone through a series of tests, aimed at establishing the baseline performance data of all the cells and validating lithium ion technology for the intended missions. Accordingly, these tests consist of both generic performance tests and mission specific tests. The generic tests include cycle life at 100% DOD at three different temperatures of 25, 40 and -20°C, and rate characterization at different rates of charge and discharge (typically C, C/2, C/3, C/5 and C/10) and temperatures (40, 25, 0 and –20°C). The mission specific tests include cycling alternating high and low temperatures, and accelerated and real-time storage under conditions similar to spacecraft cruise. In addition, several miscellaneous tests have been carried out to understand the thermal characteristics, temperature-compensated voltage charging, and failure modes.

STATUS OF LITHIUM ION CELLS FOR ROVER APPLICATIONS

Cycle Life

Fig.1 displays the cycle life characteristics of various lithium ion prototype cells at 25°C. The data include cells from three different sources are omprised of four different designs, and are onfigured either in the prismatic or cylindrical hape and have initial capacities ranging from 4-9 ∖h. The cycling regime typically consists of a harge at a C/5 rate (based on the observed apacity) to a charge cut-off voltage of 4.1, followed y tapered charging at the same voltage to C/50, or or an additional three hours (usually the current imit is approached earlier than the time limit), and discharge at C/5 to 3.0 V, with a rest period of ifteen minutes between charge and discharge.

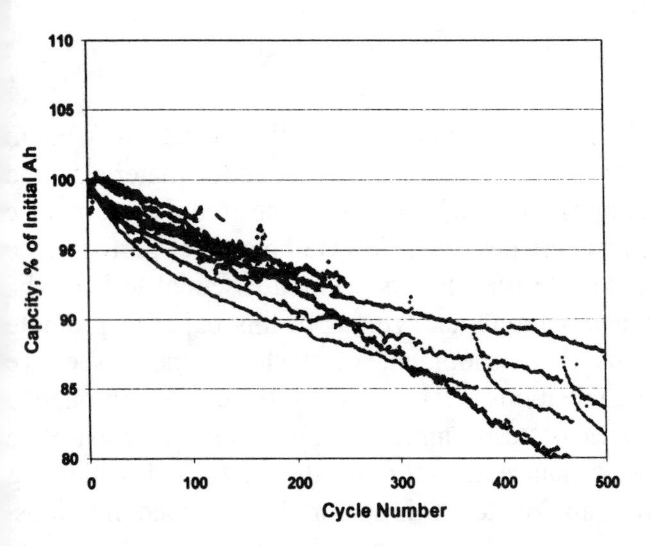

Fig. 1 Cycle life of Li ion cells (4-7 Ah) cells at 25°C for Mars Rover applications.

As may be seen from Fig. 1, the cycle life of the cells is generally acceptable from the mission point of view, with over 80% of the initial capacity being retained over 500 cycles. There is also noticeable spread in the capacity fade rate in the above cells from different manufacturers, stemming partly from the electrolytes (low temperature), the electrode materials and partly from the design parameters.

The cycle life characteristics of these cells at low temperatures (-20oC) are illustrated in Fig. 2. The cycling regime for these tests is similar to that at ambient temperature, except that the charge rates are reduced from C/5 to C/10, to avoid the possibility of lithium plating at low temperatures, parallel to lithium intercalation process.

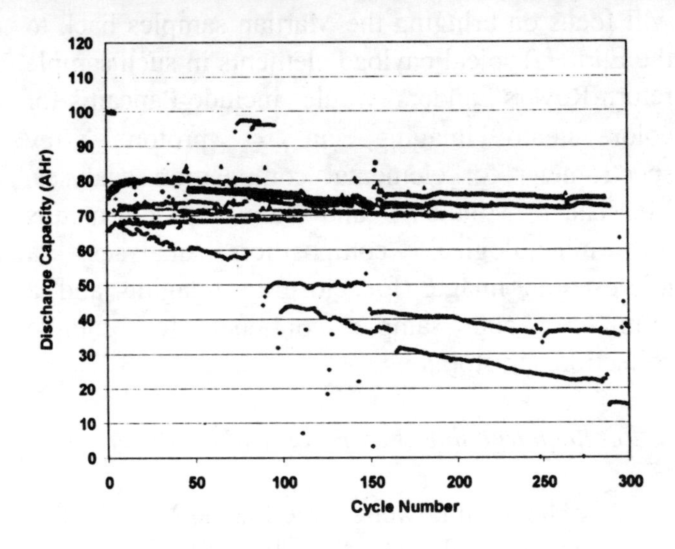

Fig. 2 Cycle life of Li ion cells (4-7 Ah) cells at –20°C for Mars Rover applications.

As may been from Fig. 2, the initial capacity at –20°C is around 70-80% of the room temperature capacity, which is quite impressive. This may be attributed to the improvements made at JPL[7] and elesewhere[8] in terms of developing low temperature electrolytes for Li ion batteries. The capacity fade rate during cycling at the low temperature is also fairly impressive, at least in most cases. However, in one set of samples, we have observed an unusually rapid capacity fade, which may be attributed to an improper cell fabrication and to the failure of the test chamber resulting in a higher cell temperatures of ~25°C.

We have seen other instances also where such occasional exposure to high temperature has become detrimental to subsequent low temperature performance. The low temperature performance is rather sensitive to any prior exposure of the cell to higher temperatures. This may be relevant to the Mars Rover missions where the battery temperature alternates between ambient and low temperatures each cycle (each day). Typically the charge temperatures are higher compared to the discharge temperature. In order to understand the capacity degradation under such conditions, We have therefore been carrying variable temperature cycling tests, where the cell is cycled at low (-20°C) and high (either ambient or 40°C) temperature for ten cycles alternately. Figure 3 shows typical performance under such variable temperature cycling.

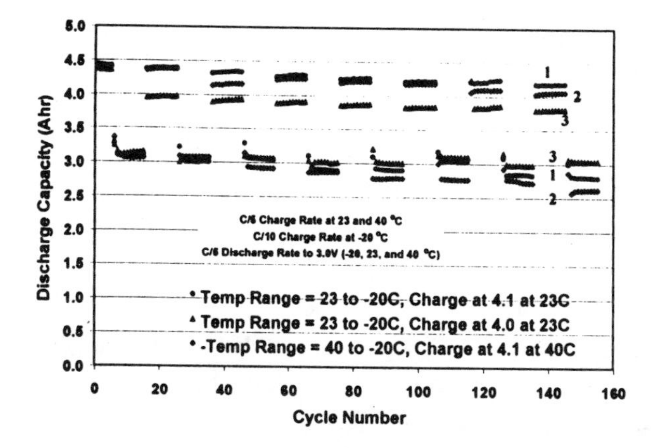

Fig. 3: Capacities of Li ion 'D' cells during cycling between –20°C and 1&3) 23°C (40°C in the case of 2) and with a charge voltage of 1& 2) 4.1 V and 3) 4.0 V.

Fig. 3 illustrates cycling of lithium ion cells under three different conditions, i.e., cell charged to 4.1 V or 4.0 V and the cell temperature alternating between –20°C and 23°C or –20°C and 40°C. The charge cut off voltage has been reduced to 4.0 to minimize the effects of electrolyte oxidation at high charge voltages, especially at high temperatures. Another variation could be to switch the cells from low to high temperature in the discharged state, which was not pursued here. As shown in the figure, the cell capacity at the high temperature (25 or 40°C) is largely unaffected, except when the charge voltage is reduced to 4.0 V. The capacity at low temperature, on the other hand, fades at relatively rapid rate. The capacity fade at low temperature decreases with a decrease in the temperature on the high end and with a decrease in the charge voltage.

Similar measurements made with another type of cells are consistent with these observations (Fig. 4). The capacity fade is slightly higher than in Fig. 3 and also the effect of lower charge voltage on the capacity fade is more prominent.

The rapid buildup of the surface films on the carbon anode and metal oxide cathode during ambient or high temperature cycling may be responsible for the subsequently poor low temperature performance. This is evident from the Electrochemical Impedance Spectroscopy measurements on these cells (Fig. 5).

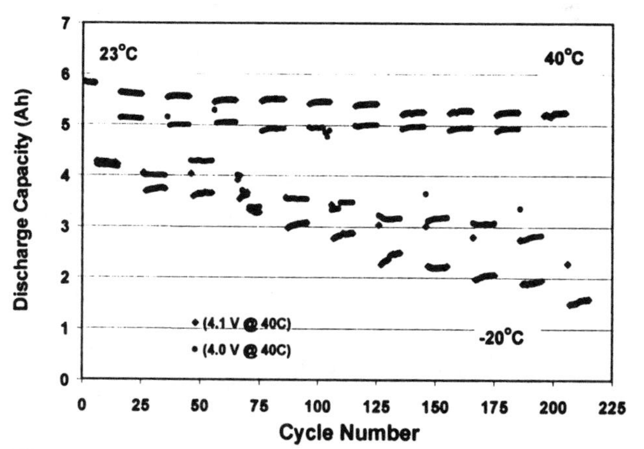

Fig. 4: Capacities of Li ion cells during cycling between -20°C and 40°C with a charge voltage of 4.1 and 4.0 V.

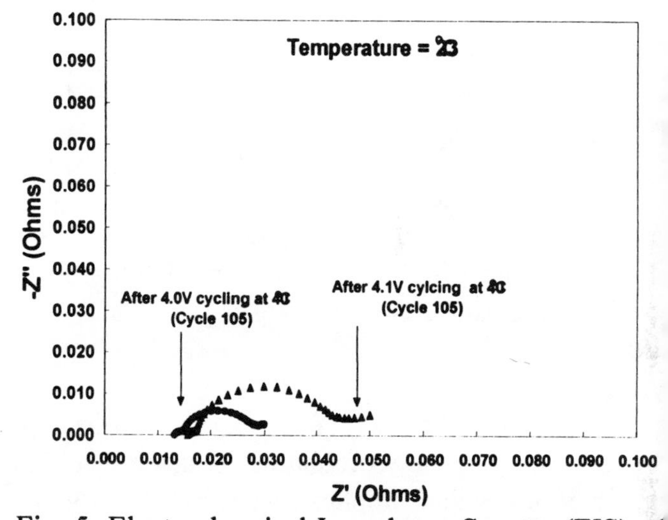

Fig. 5: Electrochemical Impedance Spectra (EIS) of Li ion cells during variable temperature cycling.

The impedance pattern of lithium ion cells is generally characterized by two relaxation loops, corresponding to each of the electrodes. Based on a comparison with the EIS response in three electrode cells, we may correlate the first relaxation (high frequency loop with the anode and the second relaxation loop with the anode. It also clear from our earlier three electrode studies on Li ion prototype cells that the surface films (SEI) on both the carbon anode as well as the metal oxide cathode build up during storage at high temperature. Likewise, there is a build up of both the relaxation loops here during the variable temperature cycling in all cases. However, the increase in the impedance is relatively small in the cell charged only to 4.0 V, compared to the cell charged to 4.1 V. It is reasonable to infer that the electrolyte-related processes corresponding to the SEI build up are potential dependent and are reduced with lower electrode potentials at cathode and higher potentials at the anode.

During the cycling, the charge to discharge capacity (C/D) ratio remains the same; in fact it gets closer to unity in the course of cycling. However, the round trip energy efficiency decreases during cycling, as illustrated by the typical data from one set of prototype cells in Fig. 6.

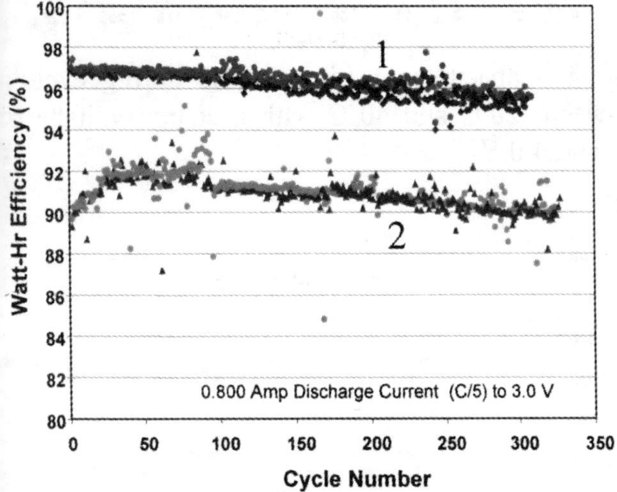

Fig. 6: Round trip energy efficiency of lithium ion cells during 100% DoD cycling at 1) 25 and 2) –20°C.

As may be seen from the figure, the Wh efficiency is typically around 97 % a for a cell of about 9 Ah, which implies that about 3 % of the energy is being lost in the form of resistive heating. At low temperatures, the corresponding numbers are much higher (about 8% going for the resistive heating) due to increased cell internal resistance at low temperatures. In both cases, i.e., cycling at ambient as well as low temperatures, the energy efficiency tends to decrease with cycling and a value slightly higher than unity.

Performance Characterization : Different Rates and Temperatures

As mentioned above, one critical aspect of the Mars Lander and Rover missions is the low temperature performance, i.e., at –20°C and at moderate rates of C/5 as mentioned above. All the types of cells, have, therefore been subjected to a series of charge and discharge capacity measurements at different rates and different temperatures. The cells were examined for their rate capability during discharge as well as charge at different temperatures. The rates included C/10, C/5, C/2 and C both for charge and discharge and the discharge tests were preceded by charge under standard conditions of C/10 charge and the charge experiments were followed by discharges at C/10. The temperature range included –30 to + 40°C, specifically –30, -20, 0, 25 and 40°C. These studies are mainly to assess the applicability of these cells for the MSR Athena Rover applications. From these tests it became clear that the cells are rather sensitive to the history, i.e., the extent of cycling and the conditions experienced prior to the test. In particular, the low temperature performance is different for the pristine cells compared to the cells cycled through characterization tests from the high end of the temperatures (25 and 40oC). This is very similar to variable temperature cycling described above. Fig. 7 shows the variation of discharge capacity of Li ion cells from one manufacturer as a function of rate and temperature.

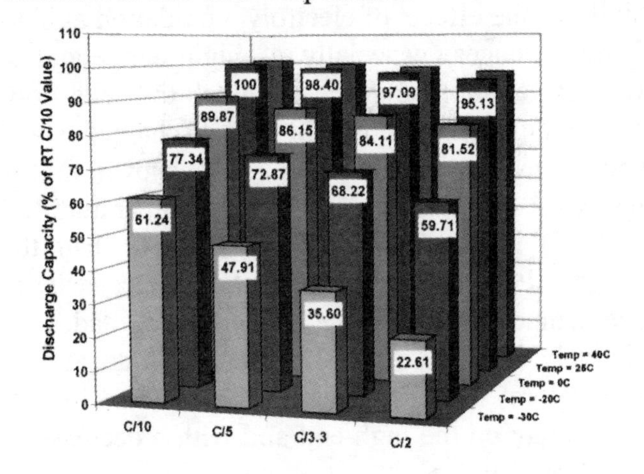

Fig. 7 : Discharge capacity as a function of rate and temperature of Li ion cell for Mars Rover.

The above data correspond to one type of lithium ion cell, which is typical of the cells containing modified cell designs, though other cells also showed capacities in the same range as a function of temperature. As shown in figure 5, Li ion cells can provide a high proportion (over 70% of RT capacity) at –20°C at moderate rates of C/5, a critical requirement for the Mars Rover missions. In order to achieve such high yields at low temperature, with the preceding charges carried out at the low temperatures, several modification have been made to the chemistries, mainly with respect to the electrolyte. Fig. 8 illustrates the typical performance of these cells at various rates at low temperature of –20°C, in the form of discharge curves.

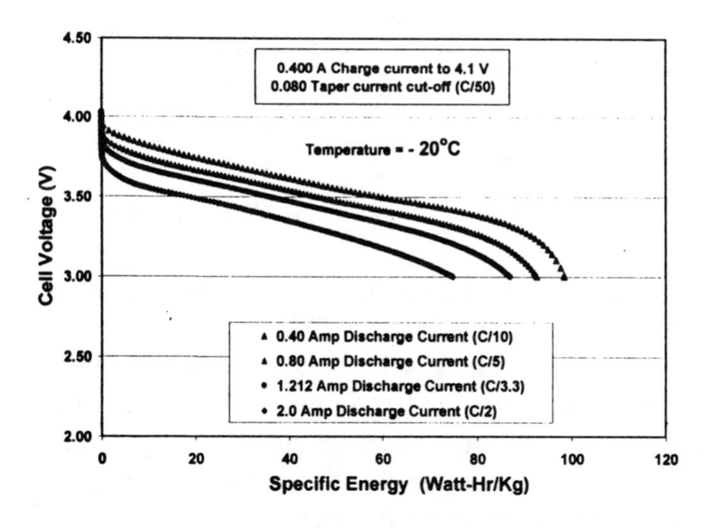

Fig. 8 : Discharge curves of lithium ion cells at discharge rates of and at a low temperature of -20°C for MSR Athena Rover applications.

The cells have shown impressive specific energies of ~ 130-140 Wh/kg at ambient temperatures. More importantly, the specific energies at low temperatures relevant to Mars Lander and Rover missions, i.e., at -20°C are equally impressive, ranging for 90 –100 Wh/kg at discharge rates. Also, the cells do function at even lower temperatures of ~-30°C, albeit at low discharge rates.

Charge characteristics

Though the batteries experience slightly warmer temperatures during charge, it was deemed necessary for the Mars mission that the cell are capable of accepting charge at low temperatures. Fig. 7 shows charge profiles of one type of prototype Li ion cell with low temperature electrolyte.

As may be seen from the figure, the cells can be charged at moderate to high rates of C/2, even at low temperature. During such rapid charging, more than 70% of the capacity is accepted in the constant current mode. It is also important to note that even at low rates of charging, and even after extended periods of charging, the charge capacity would not exceed ~ 80% of the cell (room temperature) capacity. It is probably necessary to increase the charge voltages at low temperatures, at least to the extent of compensating for the ohmic polarization, which is rather significant at low temperatures.

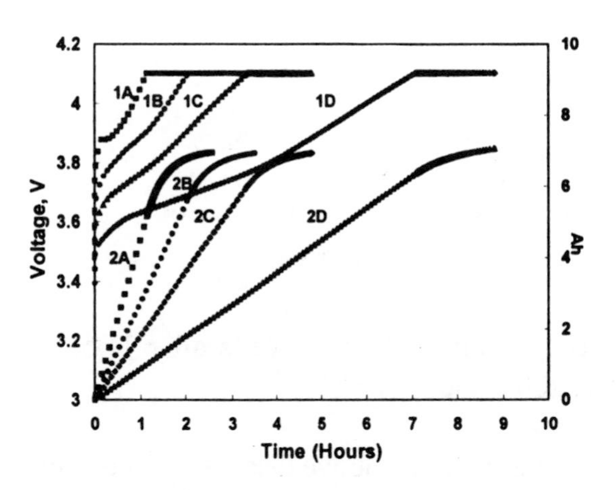

Fig. 9: Charge characteristics, i.e., 1) Voltage and 2) Capacity of Li ion cells at low temperature (-20°C) at different rates of A) C/2, B) C/3, C) C/5 and D) C/10, respectively.

EIS during cycling

We have reported in our earlier publications that the cell impedance increases during cycling. In order to monitor such changes in the cell impedance, we routinely measure the electrochemical impedance periodically during cycling (after each 100 cycles). Fig. 6 shows the typical EIS (Electrochemical Impedance Spectroscopy) plots during cycling of lithium ion cells.

As may be seen from the figure, the impedance of the Li ion cell increases appreciably during cycling. There is a small increase in the series (Ohmic) resistance but a relatively larger increase in the relaxation loop, especially the second one. From the half-cell studies on individual electrodes, we typically observe two relation loops, the high frequency loop corresponding to the surface films (SEI or solid electrolyte interface) and the low frequency loop attributable to the charge transfer process. Similar measurements in a three-electrode cell reveal that that the high frequency loop is primarily related to the processes of the phite anode and the low frequency loop responds to the processes at the cathode.

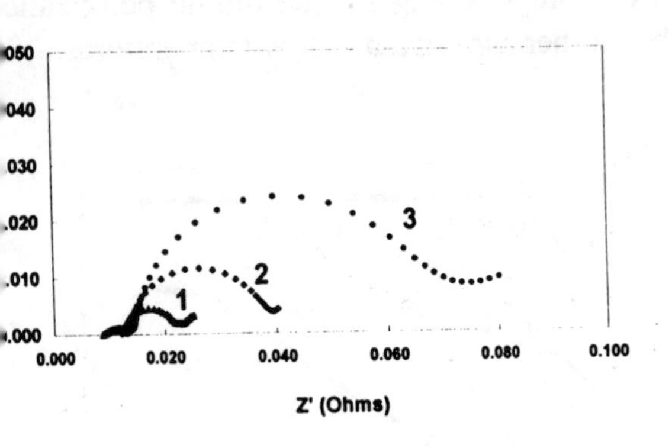

g. 10 EIS plots of Li ion cells after 1) 5, 2) 100 d 3) 200 cycles at 25°C.

Based on the above discussion, the above crease in the cell impedance upon cycling may be cribed to the cathode.

orageability of Li ion cells

The missions to Mars have typical cruise mes of 7-12 months. This combined with pre-ruise storage requires that the batteries have a alendar life of at least 2 years. Subsequent to the vo years storage, the cells need to perform well at ow temperatures, with moderate cycling. The xpected mode of storage could either be in an open ircuit condition as for the Rover or on float, as equired by the Lander. From our earlier studies on ander cells, we have observed beneficial effects of toring at low temperature, at low state of charge nd in float condition.

Similar studies have also been made on several smaller cells. Table 1 shows such storage data at two different temperatures and two different states of charge under open circuit conditions. As a result of the storage, the cells experience some permanent loss in the capacity, mainly as a result of increased cell impedance, which is a concern to the project. As shown in the Table 1, the permanent capacity is higher at high storage temperature, and high state of charge. The capacity loss ranges from a maximum of 8 % at 100% state of charge and 40°C to a minimum of 0.1 % at 50% state of charge and 0°C, respectively.

Table 1: Capacity of Li ion cells after storage for eight weeks

State of Charge	Temperature of Storage	Storage Time (Months)	Reversible Capacity (%)	Stored Capacity Loss (%)
50%	0°C	2	99.9	0.4
100%	0°C	2	97.5	7.88
50%	40°C	2	97.3	13
100%	40°C	2	92.9	13.3

We have continued storage test on these cells at 100% state of charge and 0°C conditions. The changes in the cell capacity since the cell fabrication are illustrated in Fig. 10. As may be seen from the figure, the cells have excellent storageability, which bodes well for many applications that would require long calendar life from the cells.

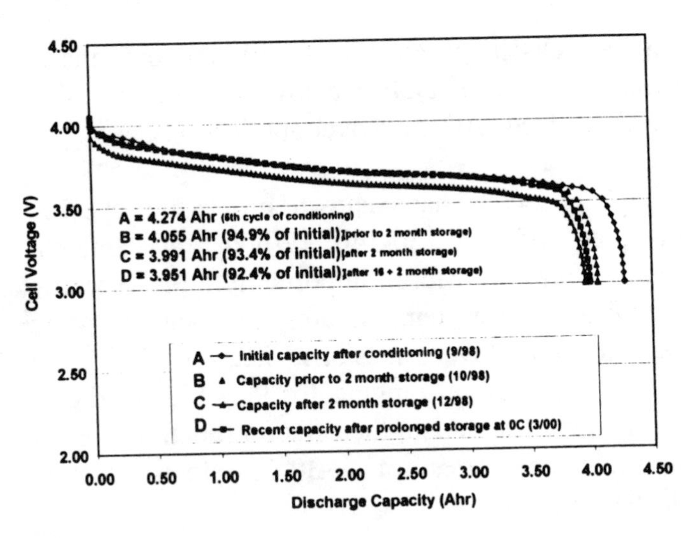

Fig. 11: Discharge curves of lithium ion cell at various stages of storage.

Charge Voltages at different temperatures

The nickel rechargeable systems, both Ni-Cd and $Ni-H_2$ undergo thermal runaway, when similar charge cutoff voltages are used across all temperatures. Since the cell thermodynamic potential decreases at high temperatures and the current efficiency for the parasitic oxygen evolution reaction increases at high temperatures, a lower charge voltage, termed as temperature compensated voltage (V_T) is used as a charge cut off to avoid any thermal runaway. In order to understand how the charge voltages need to be adjusted for lithium ion cells, we have generated coulometric titration curves, i.e., cell open circuit voltage as a function of state of charge at different temperatures .

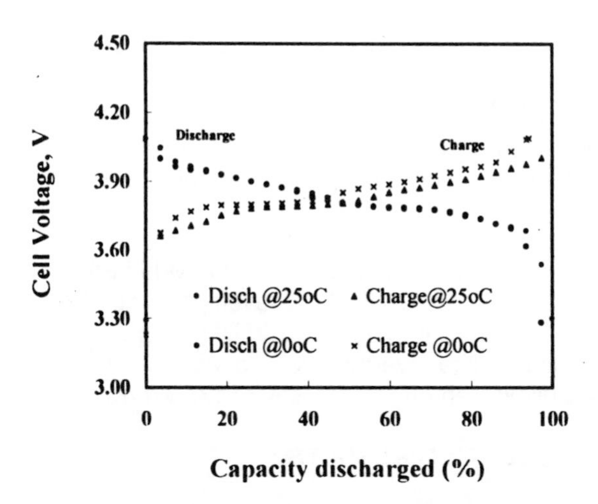

Fig. 12: OCV vs. state of charge for lithium ion cells at 25 and 0°C, during charge and discharge.

Two noteworthy points emerge from the above figure. Firstly, the OCV of a lithium ion cells does not seem to be a strong function of temperature as in the case of nickel systems. Secondly, there is no hysteresis in the charge and discharge potentials, as is typically observed with nickel systems. These, combined with the absence of a parasitic reaction as the oxygen evolution for nickel electrode, imply that the charge voltage may be compensated with temperature only based on kinetic considerations

ACKNOWLEDGEMENT

The work described here was carried out at the Jet Propulsion Laboratory, California Institute of Technology, for the Mars Exploration Program, MSP 2003 Athena Rover Sample Return Program and Code S Battery Program under contract with the National Aeronautics and Space Administration (NASA).

REFERENCES

1) B. V. Ratnakumar, M. C. Smart, S. Surampudi S. Surampudi, Proc. Annual Battery Conference, Long Beach CA, Jan. 1999.

2) M. C. Smart, B. V. Ratnakumar, S. Surampudi, J. Byers and R. A. Marsh, *Proc. IECEC,* Vancouver, Canada, Aug. 1-5, 1999, Paper # 1999-01-2638.

3) B. V. Ratnakumar, M. C. Smart, S. Surampudi, and R. A. Marsh, *Proc. IECEC,* Vancouver, Canada, Aug. 1-5, 1999, Paper # 1999-01-2639.

4) B. V. Ratnakumar, M. C. Smart, S. Surampudi, and R. A. Marsh, *NASA Battery Workshop*, Huntsville, AL, Nov. 1999.

5) M. C. Smart, B. V. Ratnakumar, S. Surampudi, J. Byers and R. A. Marsh, *NASA Battery Workshop*, Huntsville, AL, Oct. 1999.

6) M.C. Smart, B.V. Ratnakumar, L. Whitcanack, S. Surampudi , L. Lowry, R. Gitzendanner, C. Marsh, and F. Puglia, J. Byers, and R. Marsh, Proc. IECEC, Los Vegas, Aug., 2000, Paper # 2000-2914.

7) M. C. Smart, B. V. Ratnakumar and S. Surampudi, *J. Electrochem. Soc.*, 146, 486, 1999; M. C. Smart, B. V. Ratnakumar and S. Surampudi, Proc. 38th Power Sources Conf., Cherry Hill, NJ, June 1998; M. C. Smart, B. V. Ratnakumar, S. Surampudi, and S. G. Greenbaum, Proc. ECS Fall Meeting, Boston, MA, Nov. 1998; M. C. Smart, B. V. Ratnakumar and S. Surampudi, Y. Wang, X. Zhang, S. G. Greenbaum, A. Hightower, C. C. Ahn and B. Fultz, *J. Electrochem. Soc.*, 146, 3963 (1999).

8) For example, E. J. Plicta and W. K. Behl, *38th Power Sources Conf.,* 444, Cherry Hill, NJ, June 8-11, 1998; Ein-Eli, Y., *et al*, *J. Electrochem. Soc.*, **1996**, 142, L273.

AIAA-2000-2916

Cycling and Low Temperature Performance of Li Ion Cells

Haiyan Croft*[a], Bob Staniewicz[a], M.C. Smart[b], and B.V. Ratnakumar[b]
[a] **SAFT America Research & Development Center,** Cockeysville, MD 21030
[b] **Jet Propulsion Laboratory, California Institute of Technology** Pasadena, CA 91109

ABSTRACT

Lithium-ions cells, of DD and D size, are being developed under a contract with the USAF for NASA's Mars Rover missions. The cells contain spirally wound electrodes of $LiNiO_2$ positive electrode material and graphite anode in cylindrical stainless steel hardware. The electrolytes were selected based on the mission needs of good low temperature (-20°C) performance, combined with adequate stability at ambient temperature. The cells are being subjected to 100% DOD cycling at –20° C as well as at ambient temperature. In addition, the cells are being tested in 30% depth of discharge (DOD) Low Earth Orbit (LEO) regime and 60% DOD Geosynchronous Earth Orbit (GEO) regime. The cells have so far completed 5000 accelerated LEO cycles and about 500 accelerated GEO cycles. In this paper, the results of these on-going tests will be presented.

INTRODUCTION

SAFT's Li ion D cells were delivered to JPL as baseline cells. The cylindrical cell diameter is 1.32 inches, height is 2.26 inches and the average cell weight is 118 grams. The capacity of the D cells is 4 Ah using a 4.0 V charge limit. The D cells contain an electrolyte consisting of 1.0M $LiPF_6$ in 1EC:1DMC:1EMC (developed by the Army Research Lab). Rate capability and cycling were performed at different temperatures.

SAFT's Li ion DD cells (stainless steel hardware), developed under the contract with USAF for NASA's Mars Rover Missions, are 4.8 inch long and have a diameter of 1.32 inches. The case negative cells utilized a stainless steel cell case and glass to metal seals. The cell chemistry was based on a $LiNiO_2$ positive electrode, coated onto Al foil with a PVDF binder, and graphite negative coated onto copper foil with non-PVDF binder. The cells have multiple tabs on the electrodes to lower the overall ohmic component of impedance. The electrolytes in the DD cells are 1.0 M Li $LiPF_6$ in 1EC:1DEC:1DMC:2EMC (developed at JPL) and 1.0 M Li $LiPF_6$ in 15EC:25DMC:60EA + VC additive. The average cell capacity is 9.2 Ah and display an energy density of 135 Wh/Kg.

CELL TESTING RESULTS

DISCHARGE AND CYCLE LIFE PERFORMANCE AT DIFFERENT TEMPERATURES OF D-SIZE CELLS

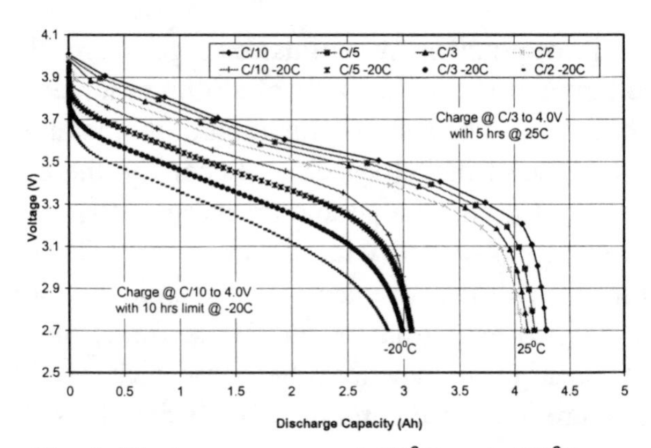

Fig. 1 Discharge curves at 25°C and –20 °C

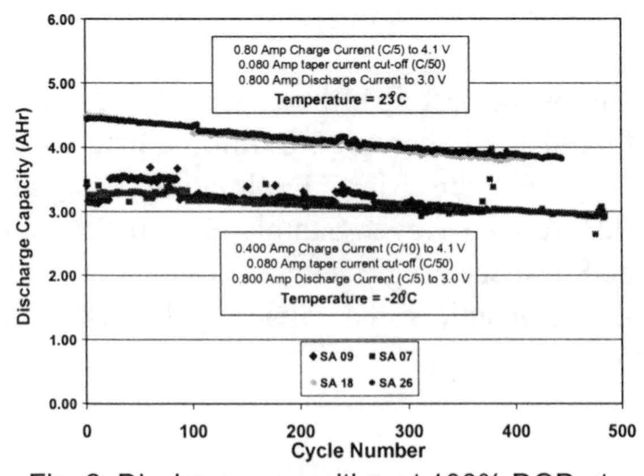

Fig. 2 Discharge capacities at 100% DOD at 25°C and –20 °C

Discharge capacities at different rates (C/10, C/5, C/3 and C/2, C and 2C) were characterized at different temperatures (-20°C, 0°C, 25°C, and 40°C). At 25°C, 2C discharge capacity is 89% compared to the capacity delivered at a C/10 rate. At C/3, the capacity at -20°C is 74% of the capacity at 25°C. Figure 1 shows the discharge curves at different rates and at different temperatures.

Fig. 2 shows the cycle life performance of D cells at 23°C and –20°C. Less than 8% capacity loss has occured in 500 cycles at -20 °C and there is about 13% loss while cycling at ambient temperature.

DD CELL CHARACTERIZATION

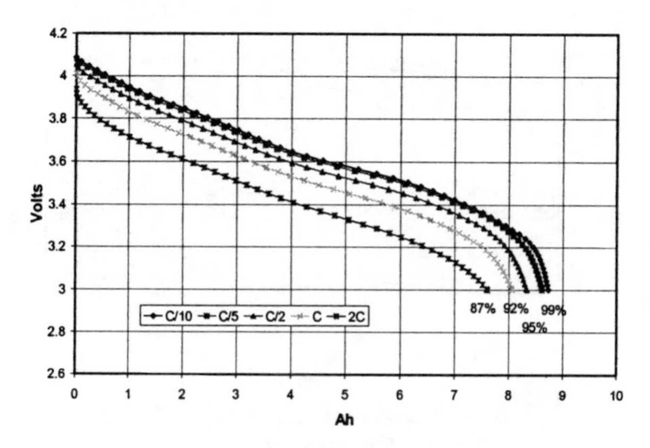

Fig. 3 DD Cell discharge Curves at 25°C

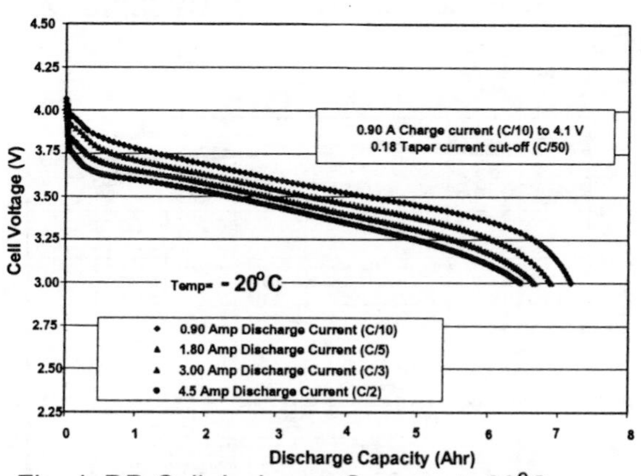

Fig. 4 DD Cell dscharge Curves at -20°C

The DD cells SAFT delivered to JPL were tested for different rates (C/10, C/5, C/3 and C/2, C and 2C) at different temperatures (-20°C, 0°C, 25°C, and 40°C). At 25°C, the discharge capacity at a 2C rate is 87% of the C/10 capacity. At C/3, the capacity at -20°C is 74% of the capacity at 25°C. Figure 3 shows the discharge curves at different rates at 25°C, whereas, Fig. 4 shows the discharge curves at different discharge rates at -20°C.

DD CELL CYCLING AT VARIOUS TEMPERATURES

Cycling at various temperatures were performed on the DD cells. Fig. 5 shows the 100% depth of discharge (DOD) cycling at 25°C. One cell has a charge limit of 4.0V, whereas, the other cell has a charge limit of 4.1V. In 600 cycles, both display about 5.4% loss of the initial capacities.

Fig. 6 shows the 100% DOD cycling tests at 23°C, -20°C and –40°C. At –40°C, 6 Ah was obtained.

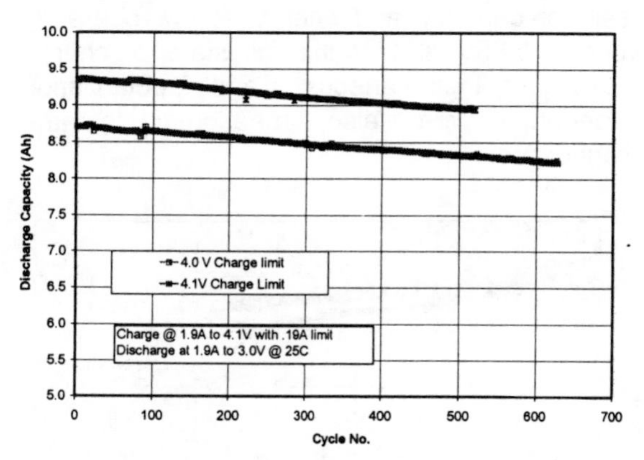

Fig. 5. DD Cells 100% DOD Cycling at 25°C

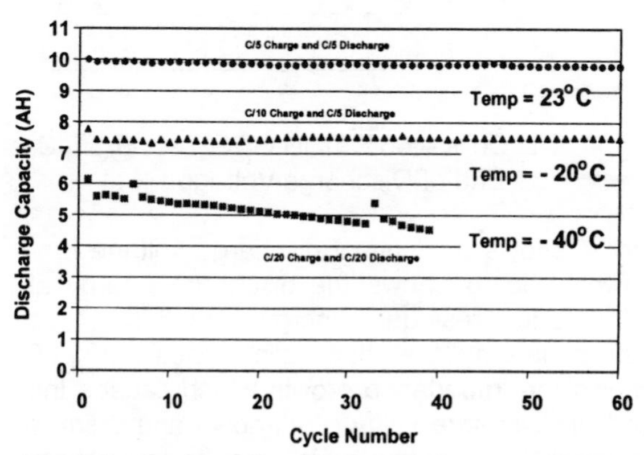

Fig. 6 Cycle Life Performance at Different Temperatures

DD cells were developed for cold temperature applications for Mars Rover mission. We tested accelerated 30% DOD Low Earth Orbit (LEO) regime and 60% DOD Geosynchronous Earth Orbit (GEO) regime cycling to demonstrate performances for planary and interplanary applications.

Accelerated 60% DOD GEO CYCLING AT 25°C

In a real time GEO test, there are forty-five eclipse cycles per season and there are two seasons per year. Thus, each year there are ninety cycles. GEO applications typically require 15 years cycle life which equates to 1,350 cycles. Accelerated 60% DOD GEO tests were started in July of 1999 at 25°C. The cycling is continuous in the same manner as SAFT's 40Ah space cells.[1, 2] During the accelerated tests, the maximum charge potential is 3.85V at 3.85A for 4.8 hours and the discharge is 60% of full state of charge capacity or 5.4Ah for 1.2 hours. Diagnostics are performed every fifty cycles for residual capacity and energy at the operating voltage 3.85V and at the full state of charge 4.0V. Instantaneous and polarization impedance are also measured in the diagnostics.

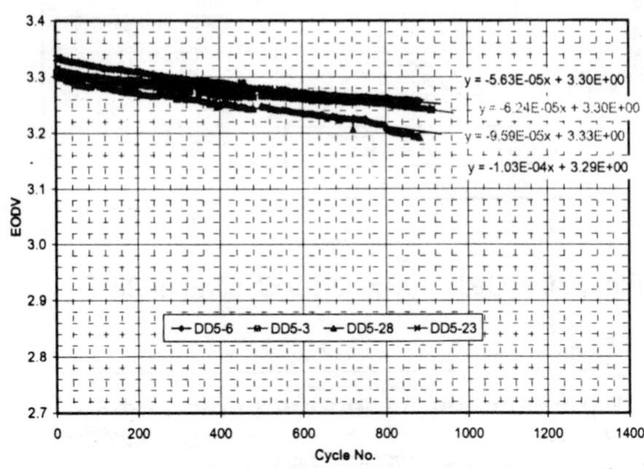

Fig. 7. DD Cells – 60% DOD GEO Test @ 25°C End of Discharge Voltage

Fig. 7 shows the end of discharge voltage run-down. Fig. 8 shows the discharge energy at 4.0V and residual energy at 3.85V, as determined by the diagnostic cycles. Fig. 9 shows the impedance growth which causes the end of discharge voltage run-down and there is material consumption which results in capacity fade. The average energy fade rate at 4.0V is .00430% per cycle. After 1,350 cycles, there would be 5.80% energy loss from cycling, which implies that End of Life (EOL) 4.0V energy would be 28.3 Wh.

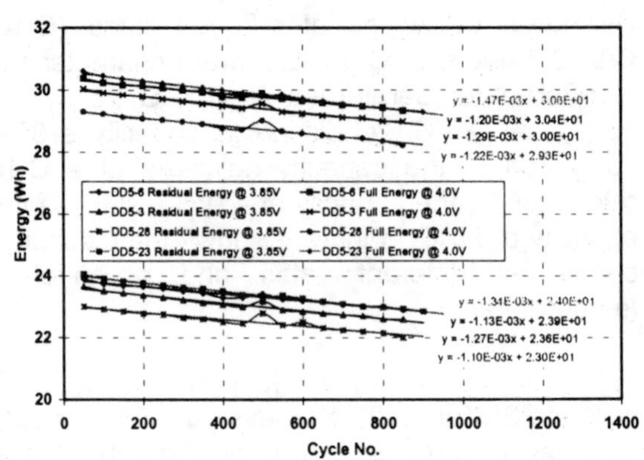

Fig. 8. DD Cells – 60% DOD GEO Test @ 25°C Energy

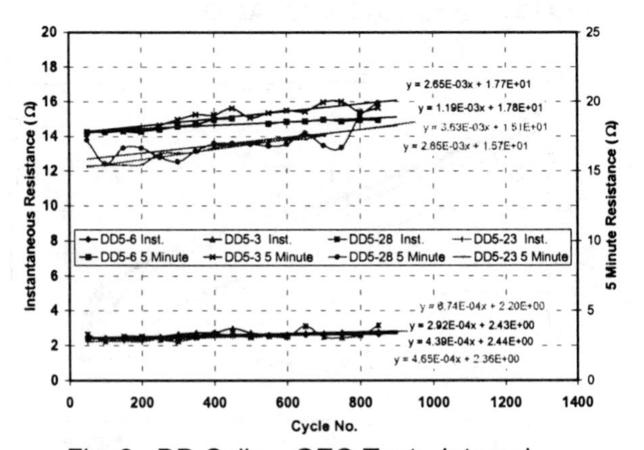

Fig. 9. DD Cells – GEO Test - Internal Resistance

Accelerated 30% DOD LEO CYCLING AT 25°C

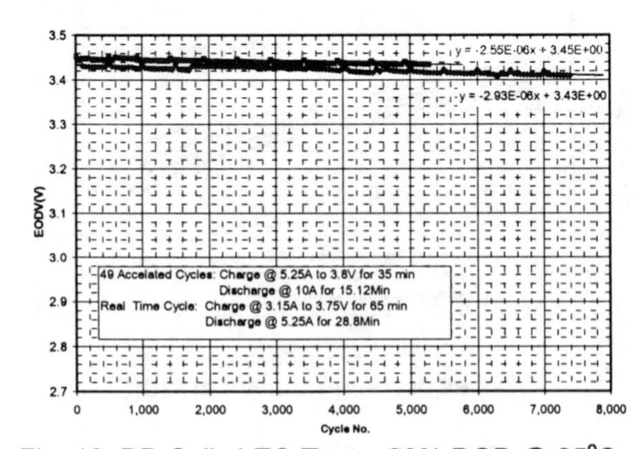

Fig. 10. DD Cells LEO Test – 30% DOD @ 25°C End of Discharge Voltage

An accelerated LEO cycle consists of a 35 minute charge at 5.25 A to 3.8 V and a 15.12 minute discharge at 10 A, which is 30% of the 4.0V full state of charge capacity. There are 48 accelerated cycles, and then an accelerated 35 minute charge at 5.25A to 3.8V and a normal LEO rate of 28.8 minute discharge at 5.25A and the 50th cycle is a normal LEO rate cycle which is a 65 minute charge to 3.75V and a 28.8 minute discharge.

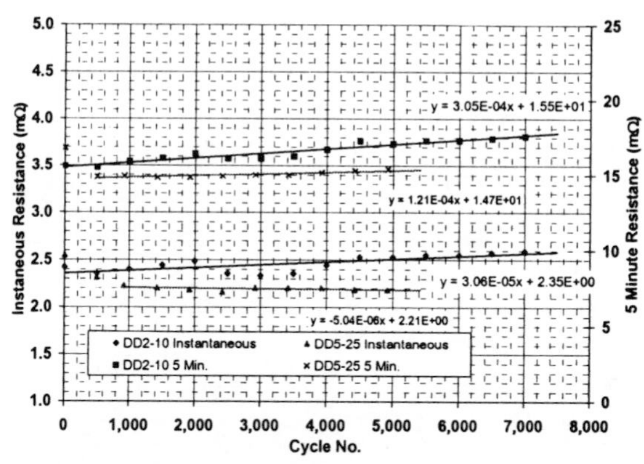

Fig. 11. Internal resistance measurements of DD-size cells undergoing LEO testing.

Fig. 10 shows the end of discharge voltage run-down of the real time cycles of two cells. Diagnostics are performed every five hundred cycles. The diagnostics include capacity measurements at both 3.8V and 4.0V, followed by dc impedance measurements. Instantaneous and polarization impedances are measured at 60% SOC, which are shown in Fig. 11. Fig. 12 shows the discharge energy at 4.0V and residual energy of the diagnostic cycles. The average energy fade rate at 4.0V is .000704% per cycle. After a typical LEO requirement of 40,000 cycles, there would be 28% energy loss from cycling, implying that EOL 4.0V energy (uncorrected for calendar life) would be 21.6 Wh (Versus a BOL of 30 Wh). The average energy fade at 3.8V is .00058% per cycle, similar to the 4.0V rate, which suggests that the energy loss is reasonably uniform over the entire energy spectrum of the cells.

Calendar life loss has been presented at the Space Power Workshop[2]. For LEO, there would be 9.65% loss in 8 years and 15.87% loss in 15 years for GEO. The results of GEO and LEO testing due to both cycling and calendar life are summarized in Table 1

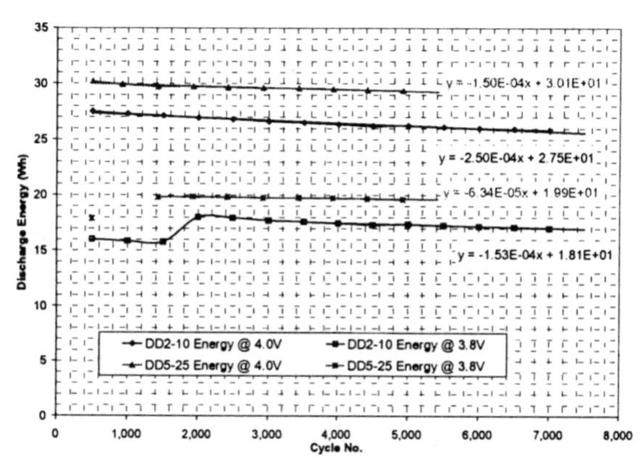

Fig. 12 DD Cells LEO 30% DOD @ 25^{0}C - Energy

Table 1 Cycle Life Extrapolation

Depth of Discharge %	Cycles Achieved	Wh Fade Rate @4V %/cycle	Typical Req. For Cycles	EOL Energy* @4V Wh 25^{0}C
30	7,500	.000704	40,000	21.6-2.9 = 18.7
60	900	.00430	1,350	28.3-4.8 = 23.3

* Corrected for calendar Life (8 years for LEO; 15 years for GEO) assuming at 25^{0}C.

CONCLUSIONS

DD cells developed for the Mars Rover Mission offer good low temperature performance and very stable performance at room temperature. The cycle life projections for either GEO or LEO profiles are excellent which suggests with high confidence that the cell chemistry is robust to losses sustained during both storage and cycling.

ACKNOWLEDGEMENT

The work described here was carried out at both SAFT R & D Center in Maryland and Jet Propulsion Laboratory, California Institute of Technology.

REFERENCES

1) R. Staniewicz, L. d'Ussel, P. Kasztenjna and W. Gollatz, Proc. Of the 1999 SAE Aerospace Power Systems, Mesa, AZ (April 6-8, 1999)

2) R. Staniewicz, Space Power Workshop, Torrance, CA (April, 2000)

AIAA-2000-2917

PROCESS DEVELOPMENT OF ELECTRODE-TAB WELDS FOR AEROSPACE, LITHIUM–ION CELLS

Ken Richardson, Brian Stein, Pinakin Shah, and Nathan Isaacs
Mine Safety Appliances Company, 38 Loveton Circle, Sparks, Maryland 21152

Martin Milden, Consultant for Batteries and Space Power
2212 S. Beverwil Dr., Los Angeles, CA 90034

ABSTRACT

This paper describes the design and process development of internal, electrode-tab connections for an aerospace, prismatic, 50-Ah, lithium–ion cell. The objective of this investigation was to formulate designs and processes that result in high-quality and uniform welds. Such welds must be sufficiently robust to withstand launch and operational environments and to support cell life in excess of 15 years. Electron Beam Welding (EBW), Laser Beam Welding (LBW), Plasma Arc Welding (PAW), Tungsten Inert Gas (TIG) welding, and Ultrasonic Welding (USW) were evaluated. Alternative current-collection schemes were also tried. In addition, the effects of process variables on weld quality were investigated. A feeder tab to comb connection system was developed for the internal cell connections. The challenges of welding thin sections of aluminum and copper were met by using an ultrasonic weld for joining the electrode tabs to an intermediate feeder tab. A pulsed, Nd:YAG (neodymium: yttrium aluminum garnet) laser was chosen to weld collector tabs to combs and combs to terminals.

INTRODUCTION

Mine Safety Appliances Company (MSA) is in the process of developing a 50-Ah, hermetically sealed, prismatic, lithium–ion cell for aerospace applications. Such a cell must be of reliable construction to withstand launch and to operate for 15 or more years in a space environment. Additionally, the cell should have low impedance, good utilization of active materials, and be producible.

Aerospace cells, commonly Ni–Cd or Ni–H_2, have used electrode tabs of nickel and steel alloys for many years. The welding processes and assembly techniques for these low-voltage electrochemical systems are well understood by cell producers. In contrast, the welding of subassemblies within lithium–ion cells presents new challenges to manufacturers.

To minimize corrosion and self-discharge within high-voltage (greater than 4 V) lithium–ion cells, current collectors and tabs are constructed from thin foils of high-purity copper and aluminum. Copper is used for the negative electrodes and aluminum for the positive electrodes. Such high-purity materials do not contain additives to facilitate welding.

Consequently, welding these materials required the development of new, highly reliable and reproducible welding processes, as well as some unique assembly techniques. Welding parameters were developed to yield high-quality and uniform welds. In addition, metallographic inspection techniques were refined for evaluation of welds using these specific materials.

INITIAL WELD DESIGN

The initial design of the comb-to-tab connection system is depicted in Figure 1. Each comb had 13 slots into which bundles of tabs were welded as shown in Detail A. The combs, in turn, were welded to pads on the bottoms of the terminals.

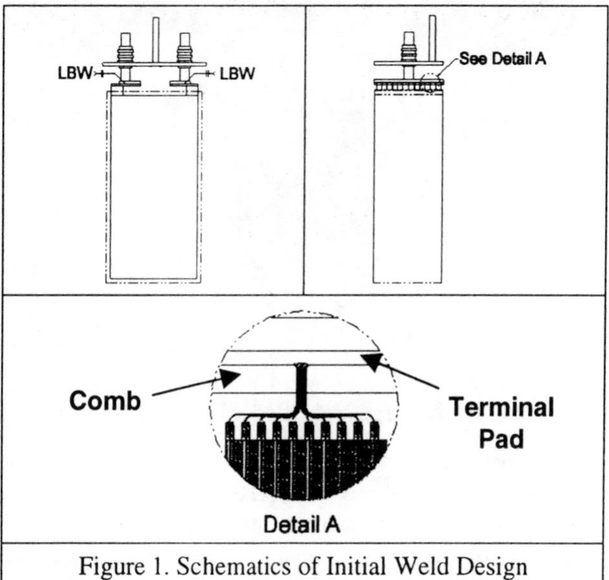

Figure 1. Schematics of Initial Weld Design

The materials used in the cell are as follows:

- Positive electrodes of Li_xCoO_2 on aluminum 1085 foil
- Negative electrodes of Li_yC_6 on UNS C11000 copper foil
- Case and header of 316L stainless steel
- Aluminum 1100 positive terminal, terminal pad, and comb
- Copper 101 negative terminal, terminal pad, and comb
- Celgard ® 2300 tri-layer separator
- 1.0 M $LiPF_6$ electrolyte dissolved in mixed carbonates.

Initial investigations focused on TIG and laser welding. Early on, it was decided that all weld joints should be autogenous rather than using filler material. This presented an interesting challenge as the high-purity alloys and the multiple foil layers that are being used were difficult to weld.

STARTING OUT

Initial trials to develop weld schedules utilized simulated comb structures consisting of two pieces of 0.032-in stock sandwiching 10 electrode tabs of either 20-μm Al stock or 18 μm Cu stock. Parts were placed in a vise to apply side pressure, and welds were made.

External visual examination of the aluminum welds revealed shiny evenly flowed surfaces. However, cross sections of the welds indicated problems beneath the surface. Figures 2 and 3 show partial or incomplete joining of the tabs to the main weld nugget and random voids throughout the solidified nugget for both LBW and TIG welds.

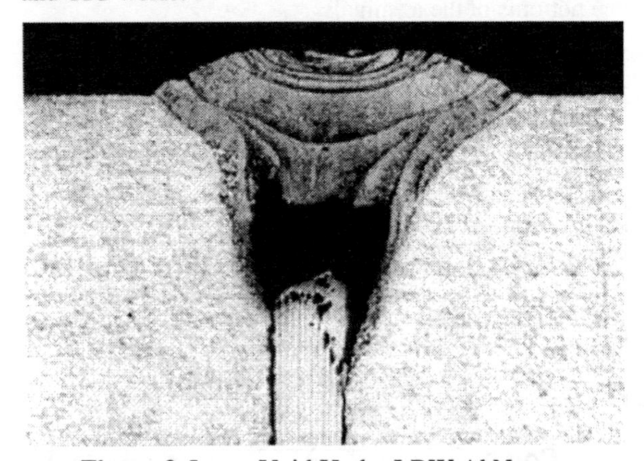

Figure 2. Large Void Under LBW Al Nugget

Cross sections of the earliest copper welds were more promising. Acceptable TIG welds were achieved for copper, but extensive heat sinking would be required to protect the separator from melting. Laser welding of copper was found to produce a nearly acceptable weld with significantly less heat flow into the parts. Because of the success with welding copper, development efforts concentrated on improving the aluminum welds.

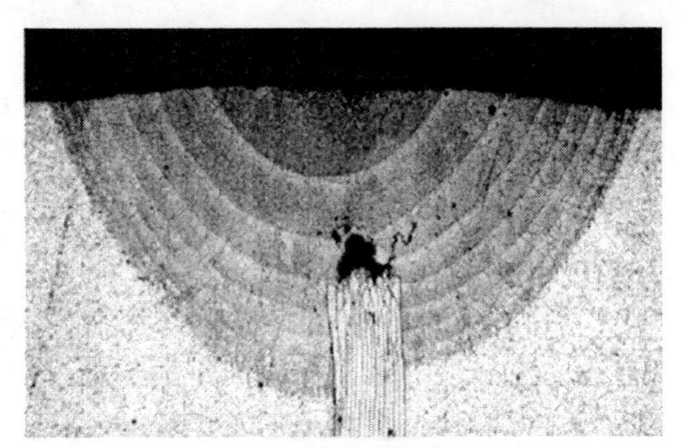

Figure 3. Large Void Under TIG Weld Al Nugget

PROBLEM SOLVING

The initial welding trials had been performed at the application labs of several LBW and TIG equipment suppliers. In all instances, aluminum welds, with excessive porosity and poor fusion, were obtained. Some of the reasons for poor results were believed to be as follows:

- Oxidized Aluminum Surfaces
- Aluminum Purity (The tab material is 99.85% Al and the comb is 99.00% Al.)
- Joint Configuration (shape, parts fit, or side pressure)
- Inadequate Gas Shielding
- Moisture Content of Metal Parts
- Welding Parameters

In addition to the equipment suppliers previously mentioned, six additional companies were consulted during the development process. These firms are all well known in the aerospace industry. The basic approach was to conduct experimental trials to eliminate or confirm the proposed causes of weld defects. These trials were done with both LBW and TIG welds. Unless otherwise stated, the weld configuration sample was a bundle of ten, 20-μm thick, Al-1085 tabs, which were sandwiched between two pieces of 0.032-in Al 1100.

Oxidized Aluminum Surfaces

An attempt was made to remove the oxide layer from Al tab bundles prior to welding. For this test, component parts were treated with Keller's Etch one-half hour prior to LBW. The result, shown in Figure 4, indicates that etching could not eliminate large voids. However, the LBW exhibited good fusion between the tabs and the weld nugget.

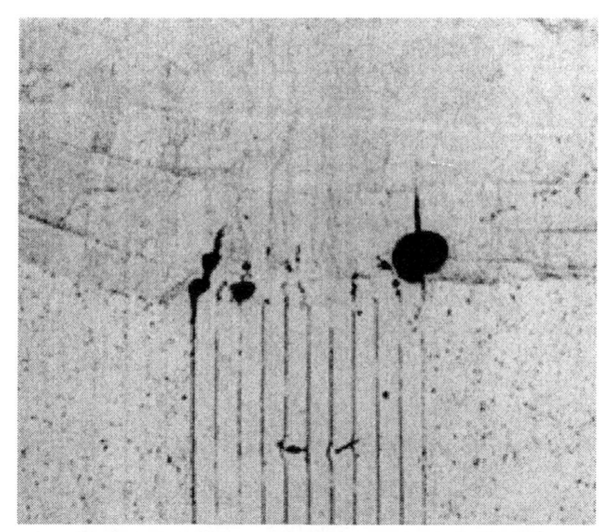

Figure 4. Effect of Keller's Etch on Al LBW

To confirm that the many oxide layers contributed to the formation of voids and porosity, a solid piece of Al 1085 was welded between two pieces of 0.032-in Al 1100. (The solid piece was 0.008-in thick, which is the same thickness as a 10-tab bundle.) The result is shown in Figure 5. This weld contains less porosity and fewer voids than the welds with multiple foil layers.

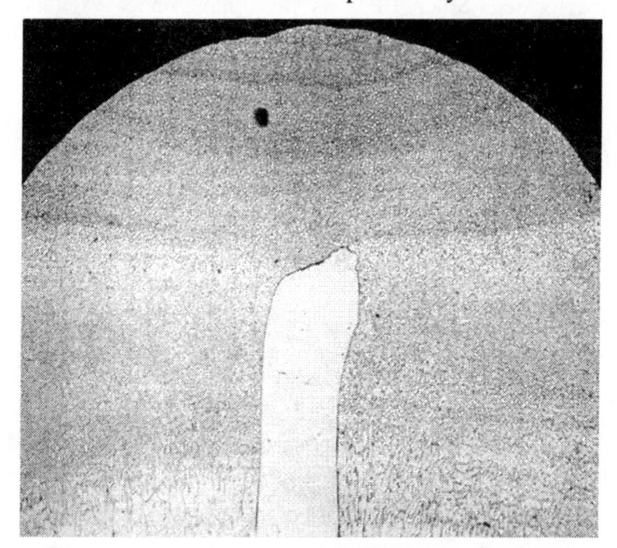

Figure 5. TIG Weld of Al 1085 Between Al 1100

Figure 6, which is an enlargement of a portion of Figure 5, shows a discontinuity around the perimeter of the 1085-Al strip. This is indicative of poor fusion between the various weld elements.

Aluminum Purity

To determine the effects of the high-purity tab material, two welds were made with 0.008-in solid tabs that were welded between pieces of .032-in, Al 1100.

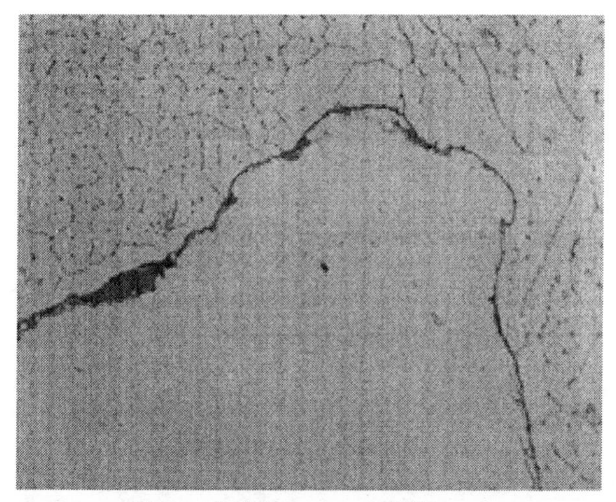

Figure 6. Enlargement of Figure 5

One 0.008-in piece was Al 1199 (99.99% pure), and a second was Al 1100 (99.00% pure). The results, shown in Figures 7 and 8, indicate porosity and voids along the edges of the tabs. However, there is excellent fusion along the top of the tab in both cases.

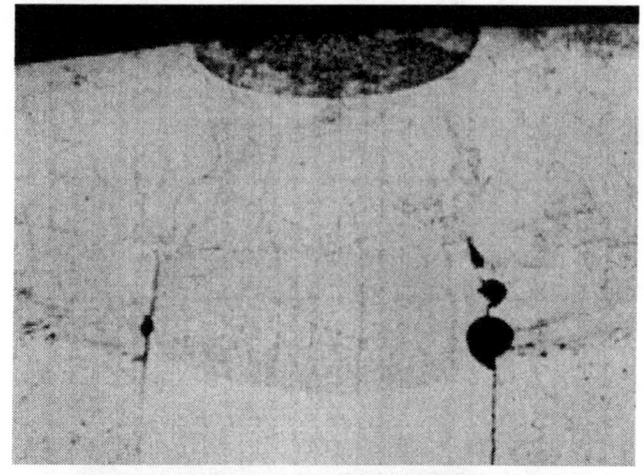

Figure 7. LBW of Al 1199 Between Al 1100

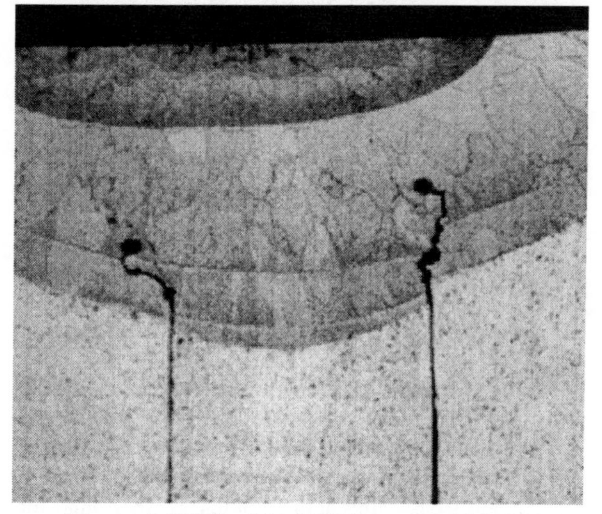

Figure 8. LBW of Al 1100 Between Al 1100

The good fusion can be attributed to the LBW being more effective than the TIG weld shown in Figure 5. In addition, it can be concluded that high-purity Al is not the primary cause of the weld problems. Porosity and voids appear to be associated with the oxidized, side surfaces of the tab.

Joint Configuration

A joint configuration experiment was conducted to test the effects of having the 10-tab bundles projecting above (Figure 9), aligned flush (Figure 10), and below (Figure 11) the surfaces of the two pieces of Al 1100.

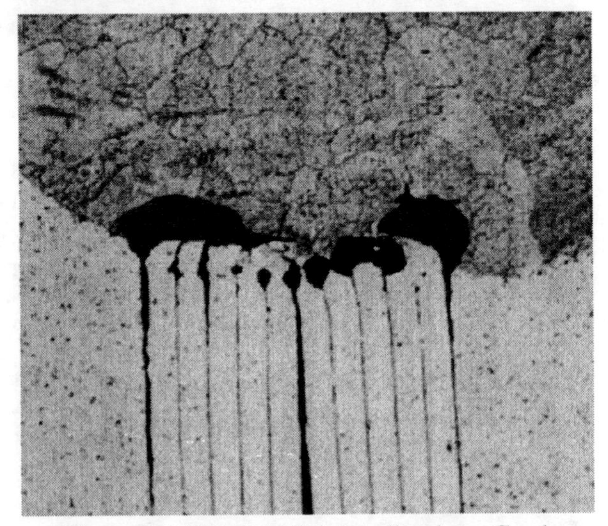

Figure 9. LBW of Al Tab Bundle Above Comb

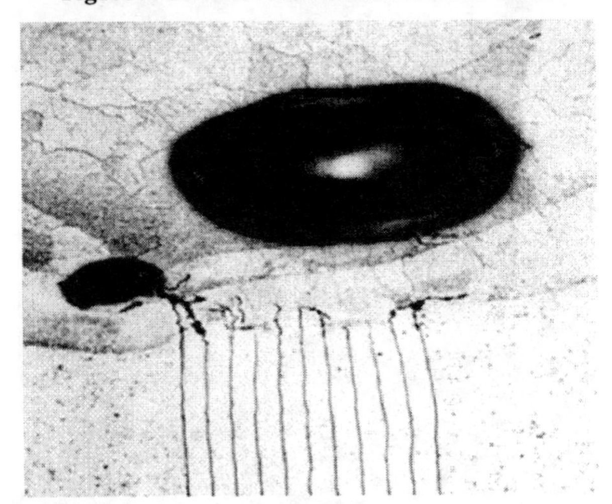

Figure 10. LBW of Al Tab Bundle Flush With Comb

Porosity and voids appear in all three of the figures shown. Considering all the welds made in this series of tests, there was no discernible difference in the porosity or voids related to the position of the tab bundles. In subsequent trials using actual combs, it was found that side pressure and precise fit-up produced better welds.

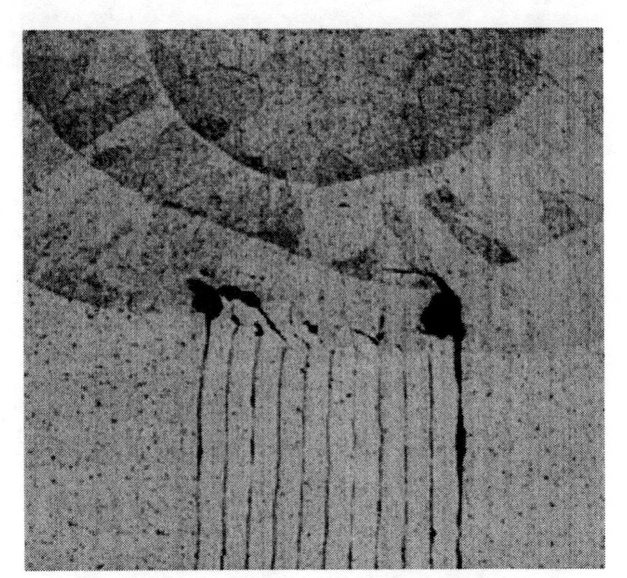

Figure 11. LBW of Al Tab Bundle Below Comb

Inadequate Gas Shielding

To study the effects of inadequate gas shielding (i.e., air entrapment), an electron beam weld trial was performed. Since EBW is performed under vacuum, this would prevent air entrapment in the weld joint.

Figure 12 is a photomicrograph from the EBW test. The same basic problem existed for this trial. There is porosity in the weld zone and no fusion between the weld elements.

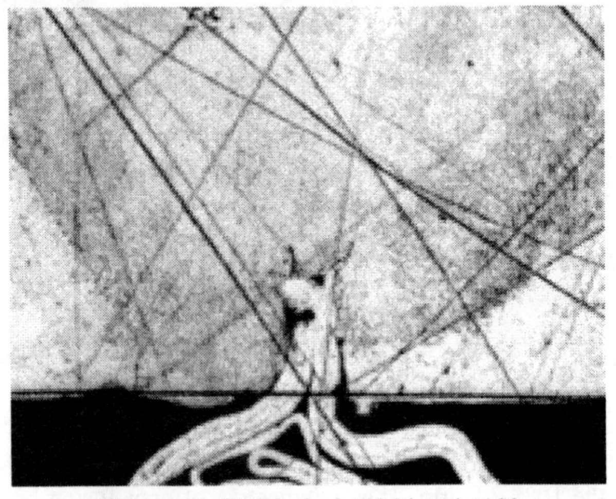

Figure 12. EBW of Al 1085 in Al 1100

Moisture Content

Efforts were made to clean and dry the parts prior to welding. Parts were cleaned in an alkaline solution, rinsed in an organic solvent, and air-dried. This was not sufficient to improve weld quality. It was later found that vacuum drying at elevated temperatures was beneficial in producing acceptable welds.

Welding Parameters

One laser supplier had LBW equipment with extensive pulse-shaping capability. Tests were performed at that firm's application lab. The best and worst results are shown in Figures 13 and 14, respectively. Voids and porosity were present in all welds.

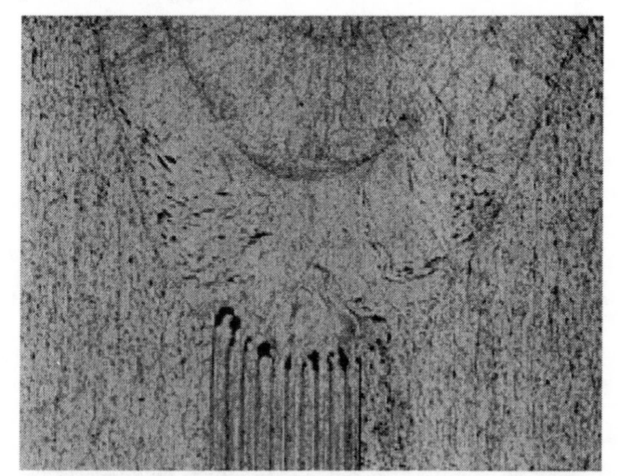

Figure 13. Best LBW Pulse-Shaping Test on Al Tabs

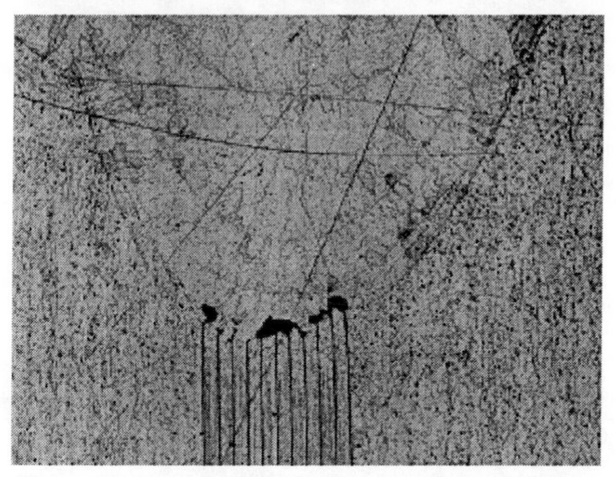

Figure 14. Worst LBW Pulse-Shaping Test on Al Tabs

Trials were also performed at another vendor's facility using a Plasma Arc Welding (PAW) technique. This weld method benefits from the inert-gas plasma that is part of the shield-gas flow. The first trials contained the best fusion to date. The best and worse photomicrographs are shown in Figures 15 and 16, respectively.

The results with Plasma were promising enough to switch from the experimental configuration of the ten-tab bundle between two pieces of Al 1100 to an actual comb of Al 1100.

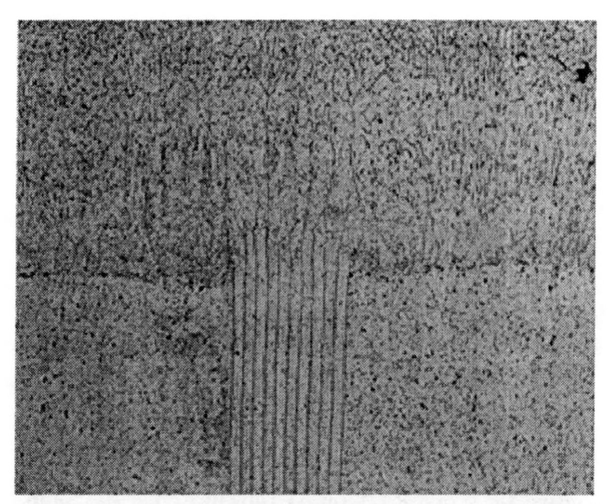

Figure 15. Best PAW on Al Tabs Between Al 1100

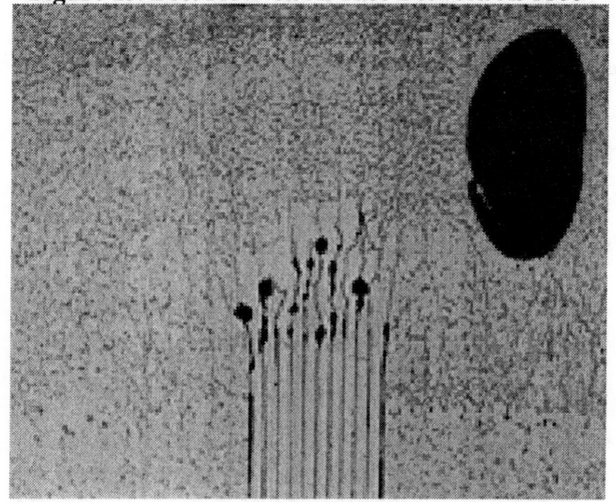

Figure 16. Worst PAW on Al Tabs Between Al 1100

It became obvious that side pressure on the comb would be required. This resulted in the modification of the comb to allow the application of side pressure. Typical welds using side pressure are shown in Figures 17 and 18.

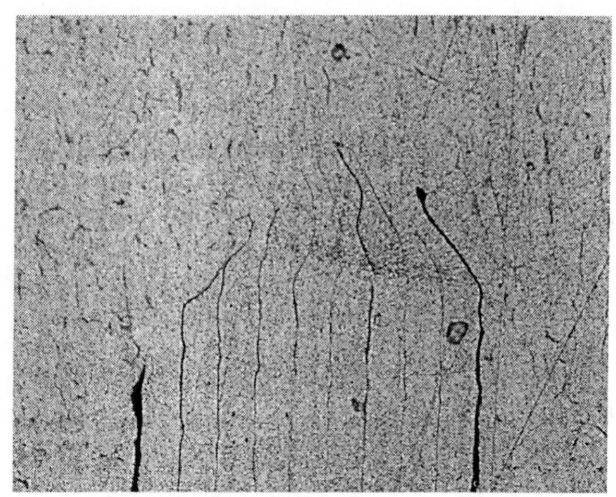

Figure 17. Best PAW on Al Tabs in Actual Al Comb

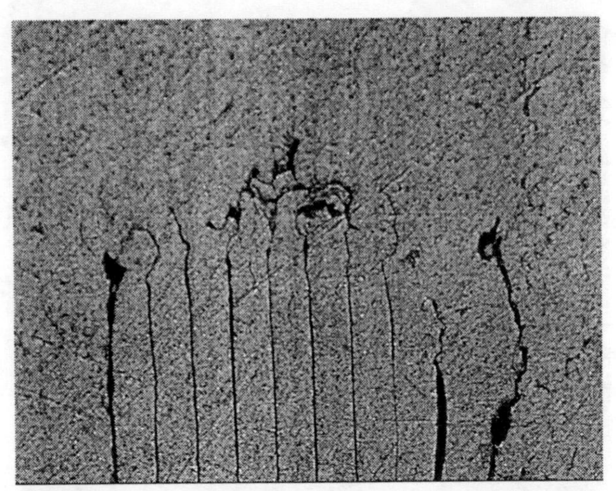

Figure 18. Worst PAW on Al Tabs in Actual Al Comb

Although the PAW welds were very promising, weld consistency could not be routinely demonstrated. Furthermore, the heat load with PAW is very high. Experiments on heat sinking the electrode tabs with water-cooled fixtures and liquid-nitrogen-cooled argon gas were only marginally acceptable. It was concluded that the PAW technique would not achieve the reliability necessary for aerospace applications.

FEEDER TAB TO COMB DESIGN

At this point in MSA's weld development, none of the welding methods tried could consistently produce an acceptable electrode tab-to-comb weld. Porosity, large voids, and poor fusion were the common problems. There were several good welds, but every attempt to duplicate them was unsuccessful. Therefore, it was decided to change the design of the tab-to-comb connections.

MSA, adapting a collection scheme used in early Ni–H_2 aerospace cells, decided to try a Feeder Tab to Comb design for its development. In this design, a bundle of tabs is ultrasonically welded to a thicker, feeder tab. This feeder tab is then welded into one of the slots in a comb. The concept is shown in Figures 19 and 20.

Although this design introduces an additional weld joint, the effect on the overall resistance of the current path is negligible. For this design, there are four welds in series between electrodes and the terminal post.

- Ultrasonic weld of tab bundle to feeder tab
- Laser weld of feeder tab to comb
- Laser weld of comb to terminal pad
- Laser weld of terminal pad to terminal post

An Amtech Ultraweld 20 was used for USW. All LBW work was performed using a Lumonics JK702H, 350 watt, pulsed, Nd:YAG laser.

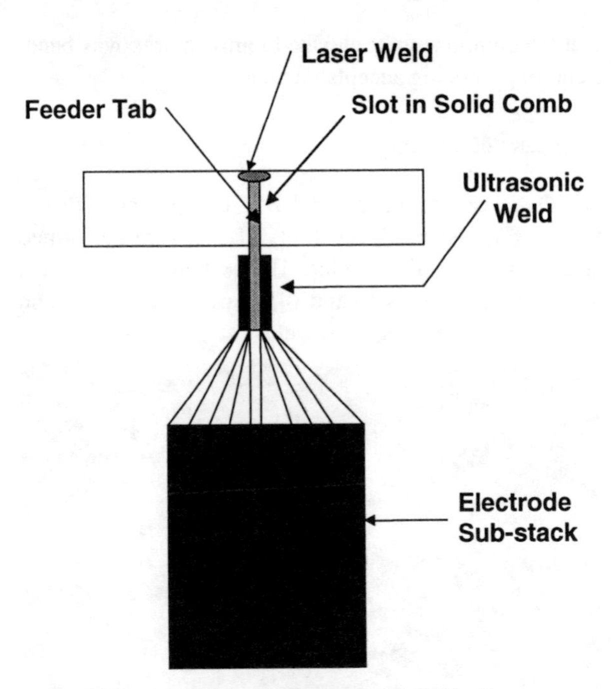

Figure 19. Feeder Tab to Comb Design

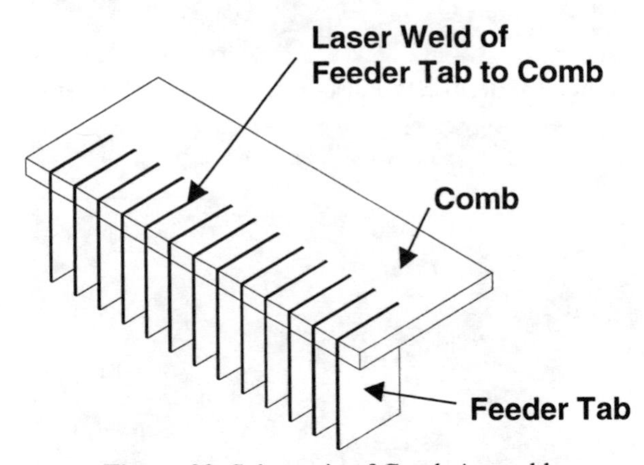

Figure 20. Schematic of Comb Assembly

Ultrasonic Welds

Ultrasonic weld development, for both copper and aluminum, was conducted at American Technology, Inc. (Amtech) in Shelton, CT. Successful welds were obtained with both materials although the copper welds were the most difficult. At present, MSA has not developed a visual method (using photomicrographs) to examine USWs. Therefore, the measures of acceptability that MSA used were the tensile strength and the electrical resistance of the individual welds. These data are shown in Tables 1 and 2.

Table 1. Tensile Strength of of Ultrasonic Tab Welds

Electrode Tab	*Copper Tensile Strength (lbs)*	*Aluminum Tensile Strength (lbs)*
1	9.3	5.4
2	8.8	5.7
3	8.2	5.0
4	9.1	5.8
5	7.5	3.6
6	7.0	4.7
7	6.0	3.7
8	8.9	3.5
9	9.1	3.0
10	6.3	4.2
Average	8.0	4.5

Table 2. Electrical Resistance of Ultrasonic Tab Welds

Electrode Tab	*Copper Electrical Resistance (mohm)*	*Aluminum Electrical Resistance (mohm)*
1	1.35	1.60
2	1.15	1.50
3	1.30	1.50
4	1.25	1.45
5	1.15	1.55
6	1.20	1.50
7	1.30	1.60
8	1.15	1.45
9	1.25	1.55
10	1.20	1.50
Average	1.23	1.52

Feeder Tab to Comb Weld

The process variables that were under consideration for this weld are as follows.

- Gas Flow Rate
- Argon–Helium Gas Mixture
- Root Gas
- Pulse Shaping, Energy, Frequency, Duration
- Feed Rate
- Multiple Passes
- Beam Location at Seams
- Orbital Welding
- Weld Direction
- Focal Point Position
- Surface Finish
- Surface Treatment (Oxide Removal)
- Feeder Tab Height
- Pressure Parts Fit
- No Weld Rails

Figures 21 and 22 show typical feeder tab to comb welds for aluminum and copper, respectively. It should be noted that some porosity is present in both aluminum and copper welds. However, fusion and penetration are excellent.

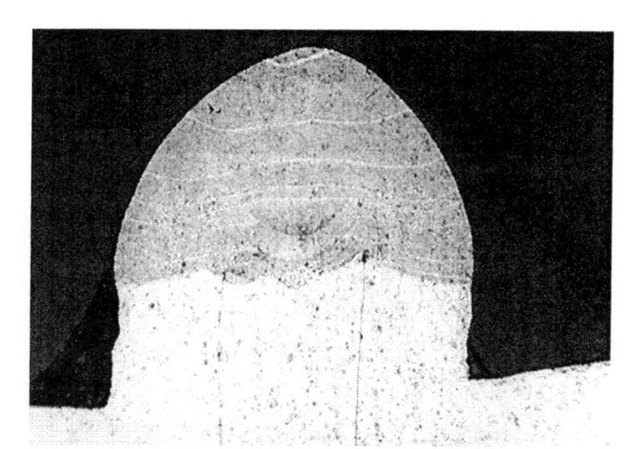

Figure 21. Al Feeder Tab to Comb Weld

Figure 22. Cu Feeder Tab to Comb Weld

LBW of Comb to Terminal Pad

Figures 23 and 24 show the comb-to-terminal pad, LBW joints for Al and Cu, respectively. In both cases fusion is excellent and porosity minimal.

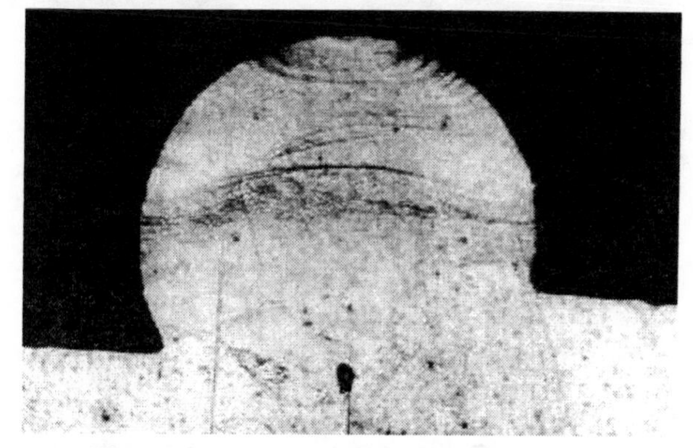

Figure 23. Al Comb to Terminal Pad LBW

Figure 24. Cu Comb to Terminal Pad LBW

LBW of Terminal Pad to Terminal Stud

Figures 25 and 26 show the terminal pads-to-terminal post LBW joints for Al and Cu, respectively. In both instances again, fusion is excellent and porosity minimal.

Figure 25. Al Terminal Pad to Terminal Stud LBW

Figure 26. Cu Terminal Pad to Terminal Stud LBW

SUMMARY

During the welding investigations, it was learned that the welding of small aluminum parts is much more difficult than the welding of small copper parts. Side pressure on the welds improved weld consistency.

The metal parts must be cleaned, vacuum dried, and maintained in a dry condition prior to welding.

Aluminum is a stable material due to its natural oxide surface film, which has a significantly higher melting point than its base metal. The weld process must overcome the negative effects of this oxide film.

An inert shield gas is especially important in the welding of aluminum. It is used to displace reactive gases from the weld zone during welding and cooling.

Copper is highly reflective to the pulsed Nd:YAG laser frequency. Because of this and the conductivity of copper, copper welds require more energy than aluminum welds.

Machined parts must have sharp edges at the surfaces to be laser welded.

It is important to have an in-house metallurgical facility for expediting weld development. As a minimum, this includes the ability to prepare and examine weld cross-sections. Expertise in the techniques for polishing and etching aluminum is also instrumental in analyzing defects in welds.

Alignment of the laser beam at the start of welding and periodic realignment is necessary to insure reproducibility.

CONCLUSIONS

Mine Safety Appliances Company has developed a low resistance, current-collection system in a prismatic, 50-Ah, Li–ion cell. This design, which connects individual electrodes to the cell terminals, should be capable of meeting demanding aerospace application requirements. Unique welding techniques have been developed for each of the joints that are used in the cell. The challenges of welding thin sections of aluminum and copper have been met by utilizing ultrasonic welding of electrode tabs to intermediate feeder tabs. All other welds within the cell are accomplished using a Nd:YAG laser.

ACKNOWLEDGEMENTS

The contributions of personnel at Amtech, Edison Welding Institute, Liburdi Pulsweld, and GSI Lumonics are acknowledged and appreciated.

AIAA-2000-2918

A MECHANICAL DEPLOYMENT STRUCTURE FOR THE POWERSPHERE CONCEPT

David A. Hinkley
Edward J. Simburger
The Aerospace Corporation
P.O. Box 92957
Los Angeles, Ca. 90009

ABSTRACT

The development of Picosatellites and Nanosatellites for low Earth orbits requires the collection of sufficient power for onboard instruments with a low-weight, low-volume spacecraft. Because the overall surface area of these satellites is small, body-mounted solar cells are incapable of providing enough power. Deployment of traditional, rigid, solar arrays necessitates larger satellite volumes and weights, and also requires extra apparatus needed for pointing. A solution to this "power choke" problem is the deployment of a thin-film solar array with omni-directional solar collection capabilities. The array would have a high collection area, low weight, and low stowage volume, and eliminates the need for a pointing mechanism. This paper considers several omni-directional deployable array geometries and details a mechanism for deploying them. The packaged size, deployed size, and expected minimum illuminated area are compared using an DARPA-Aerospace Picosatellite as the anticipated host satellite.

MOTIVATION

Solar panels are a critical satellite subsystem. They are commonly rigid planar structures populated with arrays of small silicon or GaAs cells. Three axis-stabilized satellites use motorized joints to orient the panels and project the largest area towards the sun to extract the most power. As the orbit progresses, the arrays track the sun.

The first task of a satellite once it achieves orbit is to find the sun. Using a charged battery, the guidance, navigation, and control (GN&C) computer is powered on and attempts to determine the spacecraft orientation. Once the location of the sun is known, the solar panels are correctly oriented and power flows into the batteries. Sun acquisition must be accomplished before the initial charge is depleted. These events can repeat themselves if a satellite loses attitude control and once again relies on battery charge while it attempts to regain its orientation. All told, a large amount of weight and a number of mechanical failure modes have been added to the satellite by solar arrays that need to be actively pointed at the sun. If an omnidirectional array were used instead, the motors would not be needed and the GN&C software would not be so complicated.

Satellites with no active attitude control would also benefit from an omnidirectional array. Typically, these satellites have their faces populated with 18% efficient solar cells. But if there is not enough area, then the satellite must become larger, or the power requirements must be reduced; the solar array becomes a design driving variable. The Aerospace Corporation is interested in miniature or Picosatellites.[1,2] The payloads are electronic and Microelectromechanical System (MEMS) devices being qualified for space, and, therefore, the satellites are very small. However, the power required by the payloads is more than the surface area covered in 18% cells can provide. A deployable solar array is being considered to solve this "power choke" problem.

The first DARPA-Aerospace Picosatellite (or PICOSAT) measured 4 x 3 x 1 in. and was ejected into orbit in February 2000 from the OPAL spacecraft. The PICOSAT class of satellites, typically weighing less than 1 kg, is designed to test DARPA-funded MEMS devices in space. Another mission for this size of spacecraft is called the MEMS-based PicoSatellite Inspector (MEPSI) and is directed by AFRL Rome Labs. The purpose of this mission is to inspect the host satellite in space. The PICOSAT is now serving to pathfind the techniques and technology to be used in the MEPSI. The PICOSAT, because of its small size, can achieve access to space very often as ballast on launch vehicles and host satellites. Once in space, it is ejected away from the host and begins its mission. The PICOSAT downloads the results of its payload tests at 915 MHz directly to ground stations. The first PICOSAT mission survived on primary batteries for 3 days, although properly managed, it was designed to last 21 days. The follow-on missions are incorporating more power-intensive devices and will require either larger primary batteries or a solar array solution.

Thin-film solar cell (TFSC) arrays are an attractive alternative to traditional silicon and GaAs cell arrays. They are significantly less expensive and are commonly used in commercial terrestrial electronics. The Aerospace Corporation has been actively engineering TFSC material for space application.[3] A solar array made of this material would be lighter because TFSCs are not brittle and therefore do not need a secondary rigid structure to support them. Furthermore, without this rigid structure, they can be connected together into non-planar shapes like spheres or cylinders. Those shapes can be as large as needed to supply the necessary power and they would have TFSCs all around them so that pointing towards the sun is not necessary. Without the need for pointing, the motors used to orient planer solar arrays and the flight software needed to control them would no longer be needed. This paper considers several deployable geometries and addresses the ease of packaging them into a PICOSAT.

TFSC CHARACTERISTICS

The TFSC is a sandwich of many thin layers. The active medium is an amorphous silicon junction. As shown in Figure 1, this junction is built on a polyimide sheet with aluminum deposited on one side and stainless steel on the backside. (The stainless steel is not connected to the active cell but is required to bleed off electric charge during vacuum deposition of the amorphous silicon solar cell material.) On top of the solar cell material, a conductive pattern is deposited to retrieve the charges generated by the photons. The aluminum layer on the polyimide is the other electrode. The total thickness, in its terrestrial form, is 0.002 in. (2 mils).

To use TFSCs in space, the front and back surfaces of the cells need tailored properties. As shown in Figure 2, a thin sheet of clear polyimide or other suitable polymer is adhered to the top of the cell to provide the proper thermal emmissivity so that the cell will maintain a maximum operating temperature of 80°C. The polyimide also has the benefit of providing some mechanical protection to the cell and some radiation

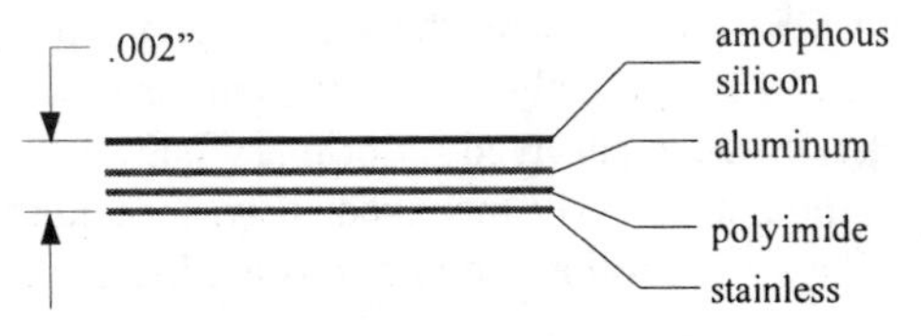

Figure 1. TFSC composition.

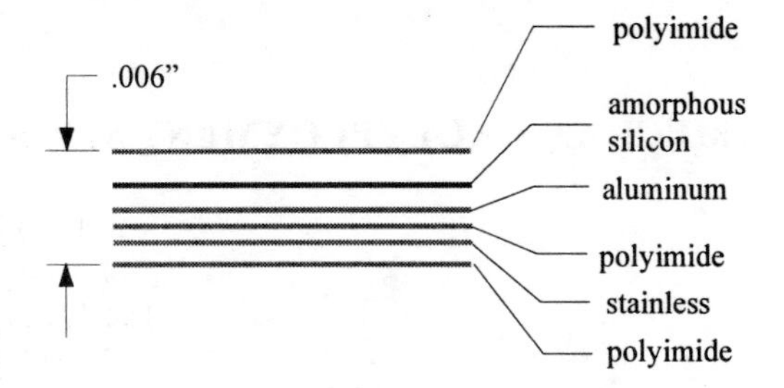

Figure 2. TFSC in spaceflight configuration.

protection. On the back of the TFSC, the stainless steel is a poor emitter of heat. The stainless steel can either be removed, leaving the underlying polyimide as a thermal emitter, or it can be covered with polyimide tape to raise its emittance. The total thickness now ranges between 4 and 6 mils depending on which method, the removal of the stainless steel or the addition of a 2-mil polyimide layer, is used on the backside.

The TFSCs presently have an efficiency of 4–5%. Each cell has a potential of 1.5 V. Alternatively, GaAs/Ge cells typically have a 1-V potential and 18% efficiency. Due to the difference in efficiency between TFSCs and GaAs, the TFSC array must have about 4 times more exposed area than a GaAs array of equal power capacity.

Unlike silicon and GaAs cells, TFSCs can be made to any size and are flexible rather than brittle. The cells are made via a roll-fed process where the rolls are 12 in. wide and hundreds of feet long. As shown in Figure 1, the polyimide center of the solar cell is its structural member. The other elements that make up the cell are deposited onto this and are very thin. The TFSCs can be bent without damage to a minimum diameter of 0.30 in. Any smaller diameter will cause delamination and ruin the cell.

The TFSC has an advantage for connecting cells into "series strings." The cells can be fabricated with monolithic interconnects that allow the series connections to be integrated into the thin-film structure itself. This contrasts with conventional silicon and GaAs cells that are wired together by soldering interconnects from the back of one cell to the front of the next until the desired voltage is achieved. This adds weight and further complicates the manufacturing process. The TFSC uses conductive ink for the positive and negative contacts at each end of a series string. This conductive ink cannot be soldered to. A solution for delivering the current generated by a series string of monolithically interconnected solar cells would be to use a conductive epoxy to provide a bond between the end of each string and the power connection bus.

DESIGNING SOLAR ARRAYS FOR THE PICOSAT

The PICOSAT 1.0 that flew in February 2000 was designed to last 21 days, limited by battery life. Had a rechargeable battery been used, the mission could have extended much longer. Therefore, with this 4 x 3 x 1 in. form factor in mind, the following TFSC arrays are depicted as a part of a PICOSAT and compared against the alternative 18% efficiency solar cells mounted onto the sides of the PICOSAT. Table 1 provides the expected power and the anticipated stowed and deployed dimensions for these arrays.

PowerSphere

The PowerSphere (Figure 3) is really a geodetic. It is a sphere made up of 32 pentagons and hexagons and is the most sun-angle insensitive solar array shape considered here. The PowerSphere can be efficiently stowed such that there is very little wasted space. One version of this design encompasses a satellite and thereby provides a uniform thermal environment for the payload.[4] Alternatively, this shape can float beside the spacecraft and provide power.

The maximum diameter of a PowerSphere that can be packed into, in this case a 4 x 3 in. area, is determined by the largest hexagon that will fit within the perimeter of the satellite in launch configuration. The PICOSAT has a perimeter shown in Figure 4 and the largest hexagon that fits is 3 in. across. The final deployed diameter is given by the formula

$$D = 2.62\, w_f, \quad (1)$$

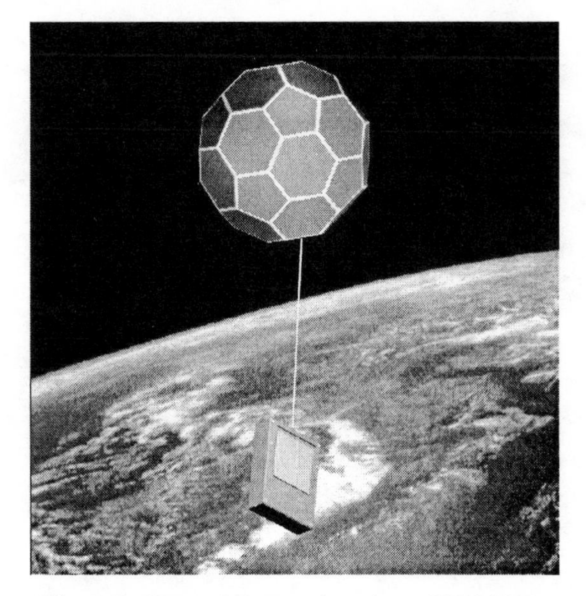

Figure 3. PowerSphere tethered to a PICOSAT.

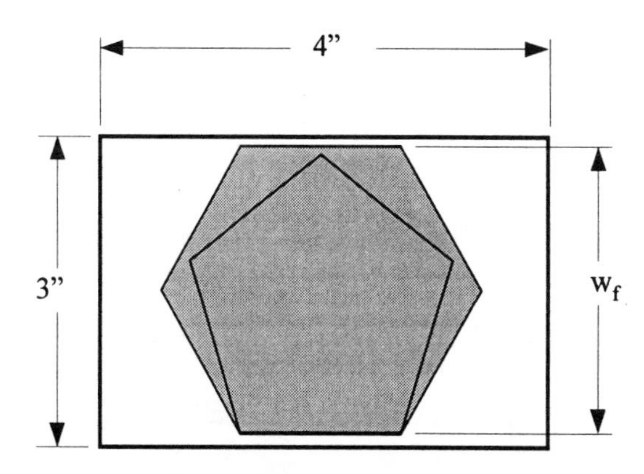

Figure 4. Plan view of stowed PowerSphere on a PICOSAT.

where w_f is the width across the hexagon flats. For the PICOSAT, the final PowerSphere diameter is 7.86 in.

The packaging height t_a is the thickness that the deployable array adds to the PICOSAT (Figure 5). However, the calculation for t_a is dependent on the stowed configuration and is computed in the last section of this paper. The result is listed in Table 1.

PartialSphere

The PartialSphere (Figure 6) is another possible TFSC array configuration. It has the benefit of simpler packaging and a larger diameter than the PowerSphere; its diameter is not limited by any PICOSAT dimension, and its round shape makes it very sun-angle insensitive. The PartialSphere connects solidly to the PICOSAT frame and has an unobstructed view of the Earth. The solar cells have a large view of space and threfore would not necessarily have the same tight thermal design constraints that the PowerSphere does.[5]

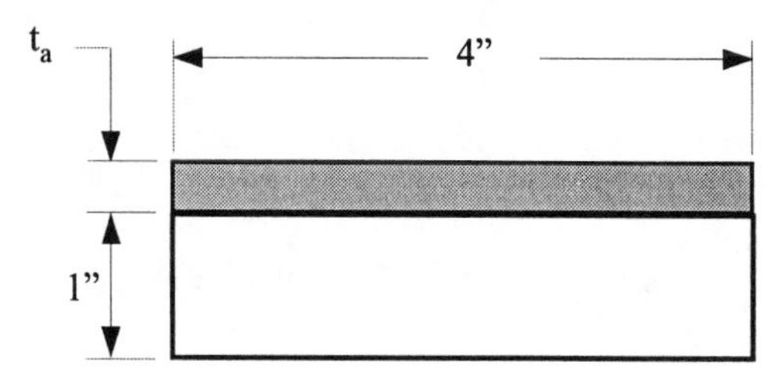

Figure 5. Side view of PICOSAT with stowed PowerSphere.

When it is stowed for launch, the PartialSphere folds flat and is rolled up as shown in Figure 7. The rolls are then tucked under a retainer that secures them for launch (Figure 8).

To maintain its shape once it is deployed and to provide the spring force for unrolling, a spring material is incorporated all along the solar cell material. The added thickness of the PartialSphere to the PICOSAT thickness without the retainer is approximately

$$t_a \cong 2[r_o + N(t_b + t_c)], \quad (2)$$

where r_o is the start radius of the roll, R is the radius of the deployed PartialSphere, t_c is the thickness of the TFSC, and t_b is the thickness of the combined polyimide and spring element. N is the approximate number of rotations that occurs when rolling up a length R of material starting with an initial radius of r_o:

Figure 6. Partial sphere thin-film array.

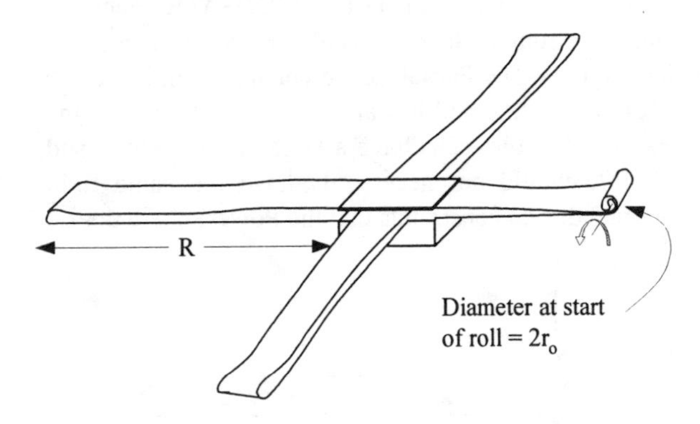

Figure 7. Folding up PartialSphere.

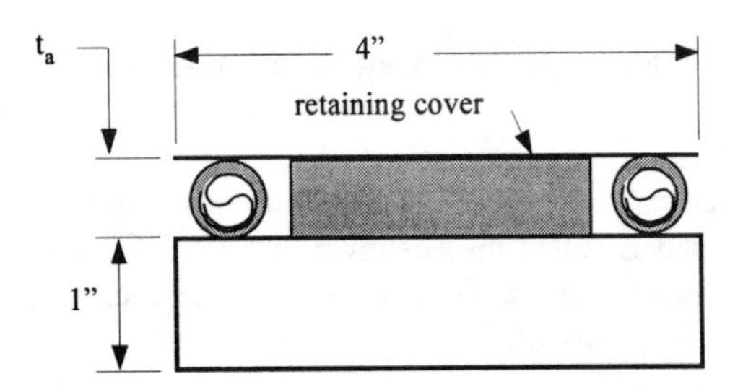

Figure 8. Stowed PartialSphere secured for launch.

$$N \cong R / (2 \pi r_o). \quad (3)$$

As mentioned before, r_o must be greater than 0.15 to prevent damage to the TFSC due to delaminatiion.

The sidelights shown on one side in Figure 6 add additional area for capturing sunlight but significantly increase the term t_a. All calculations are done in this paper do not include these sidelights.

Rectilinear

The rectilinear TFSC array (Figures 9 and 10) is the simplest configuration to stow. The power it generates will be more sensitive to its orientation to the sun due to its square shape. The legs of this array, called "z-fold strips," fold like an accordian, and deployment occurs by either some central spring that pushes the retaining top away from the body or by incorporating a spring material within the strips themselves.

Figure 9. Z-fold array beginning deployment.

Figure 10. Fully deployed Z-fold array.

A side view of the PICOSAT with a stowed rectilinear array is identical to Figure 5 but with a different thickness t_a:

$$t_a \cong (M+1)d, \quad (4)$$

where d is the hinge bend diameter, derived later, and M is the number of panels. In the deployed configuration shown in Figure 10, M = 6 since there are six individual TFSC panels per z-fold strip.

The dimensions of the TFSC panels on the z-fold strips shown in Figure 10 are determined by the plan view dimensions of the PICOSAT. In Figure 11, each of the four z-fold strips are shown. The PIOCSAT is not square, so one pair of z-fold strips is 2 in. wide, and the other is 3 in. wide. Both use 1-in. tall panels. When fully deployed, the height of each z-fold strip is approximately 6 in. and the TFSC area is either 12 in^2 and 18 in^2 depending on which side is illuminated.

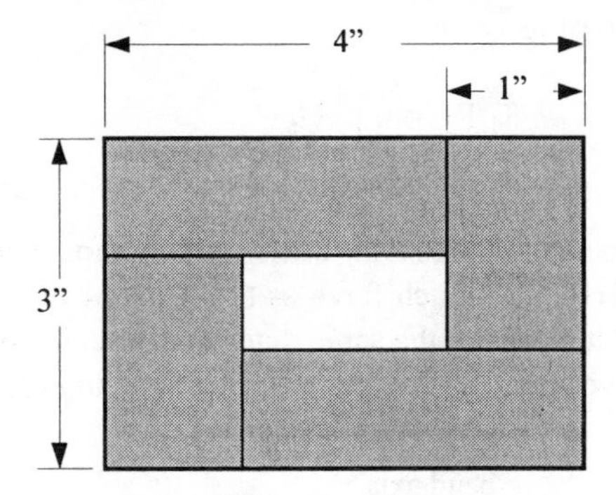

Figure 11. Plan view of stowed rectilinear array.

Table 1 summarizes the properties of each deployable array including the minimum area illuminated by the sun, total additional packaging height, and final deployed size. They are compared against high-efficiency cells placed on the exterior of the PICOSAT. Note that PICOSAT dimensions determine many of the parameters that shape the TFSC arrays considered here. The PowerSphere is completely determined. But the PartialSphere and the Rectilinear array both have a dimension R and M, respectively, that can be specified independently and thereby change the available power from these shapes.

The entries in Table 1 use the following numbers as design cases: the TFSC t_c = 0.006 in.; the combined polyimide and spring element backing for the TFSC t_b = 0.005 in.; the bend diameter d = 0.015 in., the number of panels in a rectilinear z-fold strip M = 6; the radius of a deploy PartialSphere R = 8 in. and the start radius for stowing the PartialSphere r_o = 0.20 in. From the previous section, it was shown that for an 18% cell and a 4-5% TFSC, the TFSC array would need about 4 times more area to generate the same power. All of these possible array shapes shown in Table 1 can exceed that requirement.

Table 1. Summary of Three Deployment Choices as Shown in This Section

	Minimum Deployed Area (in^2)	Packed size (in)	Deployed size (in)
Power Sphere	48	4 x 3 x 0.30	7.8 dia
Rectilinear	12	4 x 3 x 0.10	4 x 3 x 6
Partial Sphere	48	4 x 3 x 0.54	16 dia
18% cells	3	3 x 1 x 0.01	3 x 1 x .01

SMA METAL AS A HINGE

In the previous section, various TFSC arrays were shown as they might be used as the power source for a PICOSAT class satellite. In each array, the deployment occurs and the deployed shape is held by a spring element. That element, for those purposes in this paper, is a NiTi (nickel titanium) shape memory alloy (SMA) superelastic sheet metal.

SMA is a two-phase metal that through temperature or stress reverts from one phase to another. The phase transformation spreads out the stress-strain curve and allows a significant elastic strain, up to 8% before plastic strain occurs (Figure 12). This can be compared with spring steel, which yields plastically at 0.8% strain maximum. The force contained generated by that elastic strain is measured in this section.

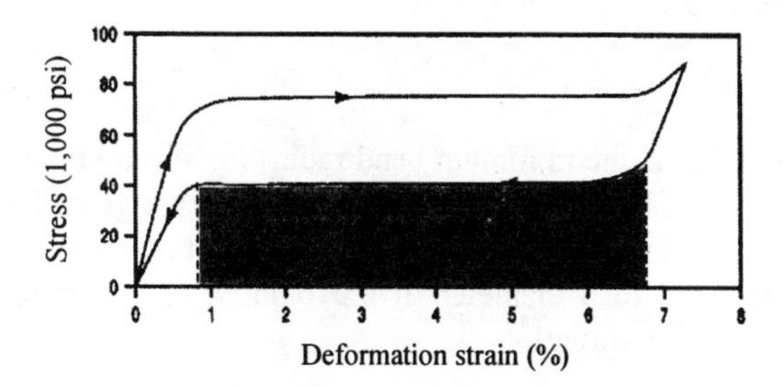

Figure 12. Stress-strain curve for NiTi superelastic metal (www.sma-inc.com).

The energy stored per unit volume in the elastic deformation of a metal is the area under the return path on the stress-strain diagram, per unit active volume.[6] For spring steel, this is

$$U^{el}_{steel} = 0.5\sigma_y^2/E$$
$$=0.5(2.87 \times 10^5)^2 / 3.0 \times 10^7$$
$$=1373 \text{ psi.} \quad (5)$$

For the case of the superelastic metal, the area under the return curve shown in Figure 12 is mainly contained in the long flat center region or

$$U^{el}_{NiTi} > (4.0 \times 10^4)(0.06)$$
$$> 2400 \text{ psi.} \quad (6)$$

This is almost double the energy in the steel. But the real benefit is the strain available. The strain determines the smallest bend that the hinge can make without a permanent set. Consider a hinge shown in Figure 13. The hinge has thickness t_S and is bending with an internal diameter d through a 180° bend.

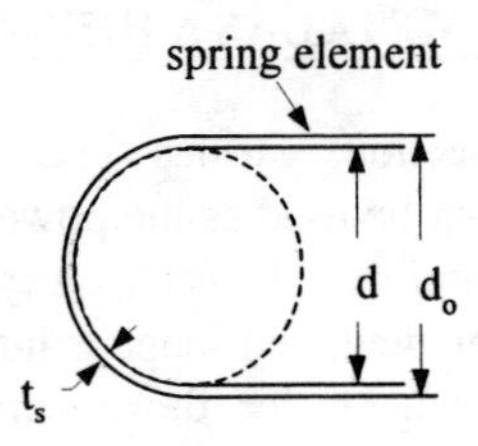

Figure 13. The minimum 180° bend.

The strain ε can be written

$$\varepsilon = \delta L/L_O$$
$$= 0.5\pi(d_O - d)/\pi d$$
$$= 0.5(d + 2t_S - d)/d$$
$$= t_S/d. \quad (7)$$

Now solve equation (7) for d:

$$d = t_S / \varepsilon. \quad (8)$$

In Table 2, the minimum bend radius for several thicknesses of NiTi, as well as for spring steel, are given for comparison. Notice that the t_S = 0.0007 in. NiTi metal can bend to a diameter of 0.010 in. without permanently deforming!

The unfolding force of the spring-hinge will determine the thickness of SMA material required. To measure the unfolding force, coupons measuring 1.5 x 0.5 in. of two different thickness SMA material were tested (Figure 14). The unfolding force was measured using a digital scale and ruler as shown in Figure 15. As the force increased, the bend diameter was reduced as listed in Table 3. The results for the t_S = 0.0007 in.

Table 2. Minimum Bend Diameter , d, of Spring Metals Before Plastic Deformation Occurs

	Steel	NiTi	NiTi	NiTi
t_s (in.)	0.001	0.002	0.0014	0.0007
ε (%)	0.8%	7%	7%	7%
d (in.)	0.125	0.029	0.021	0.010

coupon are calculated because that material was not available for testing. The ratio of the forces is proportional to the ratios of the moments of inertia of the cross sections or

$$F_1/F_2 = I_1/I_2, \quad (9)$$

where

$$I = {}^1/_{12}\, w_S t_S^{\,3}. \quad (10)$$

In equation 12, w_S is the width of the hinge, and t_s is its thickness. For $t_{s,1}$ = 0.00144 and $t_{s,2}$ = 0.0007 in., the ratio of forces is

$$F_1/F_2 = t_{s,1}^{\,3} / t_{s,2}^{\,3}$$
$$= 0.11. \quad (11)$$

Therefore, the 0.0007 in. thick coupon should require about 1/10th as much force as the 0.00144 in. coupon to hold it closed to the same diameter, or, conversely, it will have about 1/10th the force available to unfold.

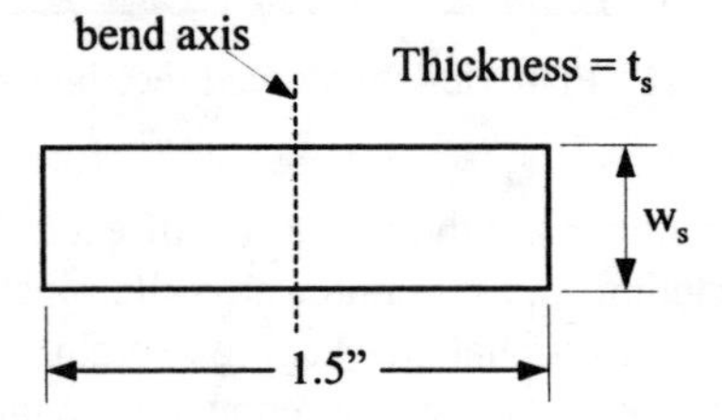

Figure 14. Test coupon size.

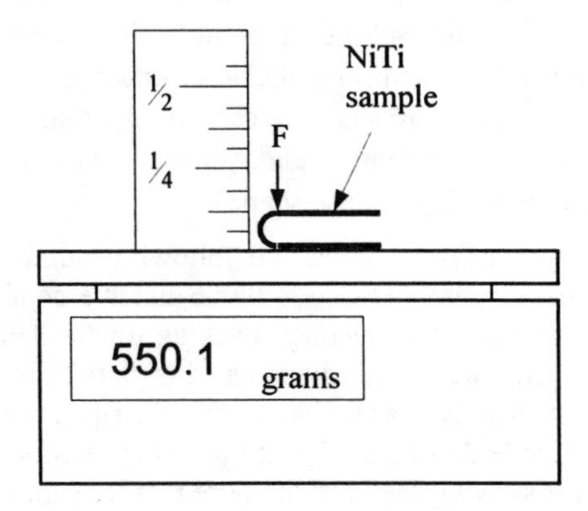

Figure 15. Force measurement setup.

Table 3. Bending Force for NiTi Samples

Bend Radius	0.00198" NiT (g)	0.00144" NiTi (g)	0.0007" NiTi (g)*
1/8"	140	60	6
1/16"	220	80	8
1/32"	550	200	20

* calculated

The results in Table 3 are for a coupon of width w_S = 0.5 in. If a wider hinge is used, these forces will scale linearly with the ratio of new width and the tested width.

POP-UP SOLAR ARRAY USING SMA

In the section "Designing Solar Arrays for PICOSATs," various pop-up array geometries configured to supply power to a PICOSAT-class satellite were suggested and compared against a common alternative of placing high-efficiency cells on the area available only on the exterior of the PICOSAT itself with no unfolding element. It was evident that pop-up arrays, even with the lower efficiency thin-film cells, would provide more power than the typical high-efficiency alternative. An obvious choice for spring-hinge material is spring steel. However, in the previous section it was shown that a better spring metal is NiTi shape memory alloy because not only does it have a significant Young's modulus, it has an incredibly large elastic strain limit, which allows the hinge to fold to very small radii without plastic strain. This significantly reduces the stowed array size. In this section, the NiTi metal is designed into the rectilinear pop-up array. This array is easy to visualize and will clearly show the limitations of any spring hinge method.

In Figure 16, a single side of a rectilinear array is shown. This z-fold strip works like an accordion and has six thin-film solar cell panels mounted to a combined polyimide and NiTi backplane of thickness t_b that performs the deployment function.

A side view of a folded strip is shown in Figure 17. The NiTi metal is used in sheet form and is sandwiched between two polyimide sheets. The hinge diameter d is greater than the theoretical minimum calculated in the previous section.

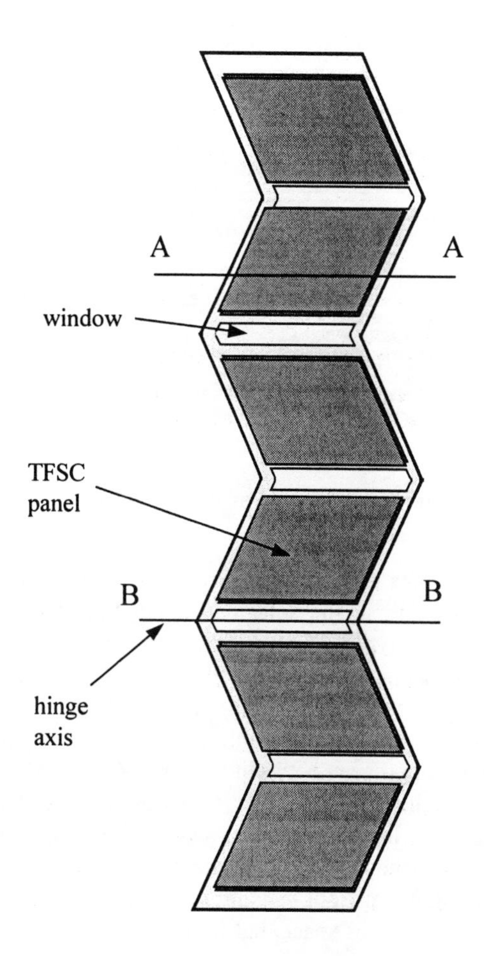

Figure 16. Z-fold strip from rectilinear array. M = 6.

From Figure 17 and eqn. (4), the amount of vertical space required to stow this array is

$$t_a = (M + 1)d = 7d. \tag{12}$$

A choice of material and thickness from Table 2 will provide a number for d. For example if t_s = 0.0007 in. NiTi is used, then

$$t_a = 7(0.010") = 0.070". \tag{13}$$

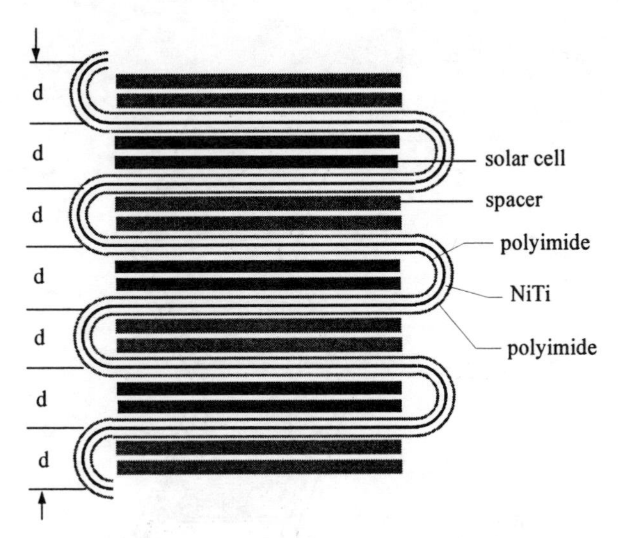

Figure 17. Side view of stowed z-fold array.

Some caveats regarding the deployed z-fold array are necessary. First, the solar cell populating the z-fold array is the typical TFSC shown in Figure 1. This material cannot bend beyond 0.30 in. diameter and therefore cannot pass over the hinges which can bend to the diameters listed in Table 2. The solar cell panels are therefore placed between hinges and are connected to each other in parallel or series as needed. Those connections have not been shown. Finally, note that the inside bends, where there are no solar cells to keep the spring metal from bending too tightly, use spacers. This is a waste of space, but they are required for this design to be viable.

The hinges are meant to be the preferred bending axes for the z-fold array. Therefore, the compliance should be significantly less there than in the middle of the solar cell. The reduced compliance is achieved in two ways. The first is the absence of the solar cell at the hinge. The second is by cutting windows in the polyimide-spring metal-polyimide structure. In Figure 18, it can be seen that the Section A-A has significantly more inertia than Section B-B and will, therefore, be stiffer, as desired.

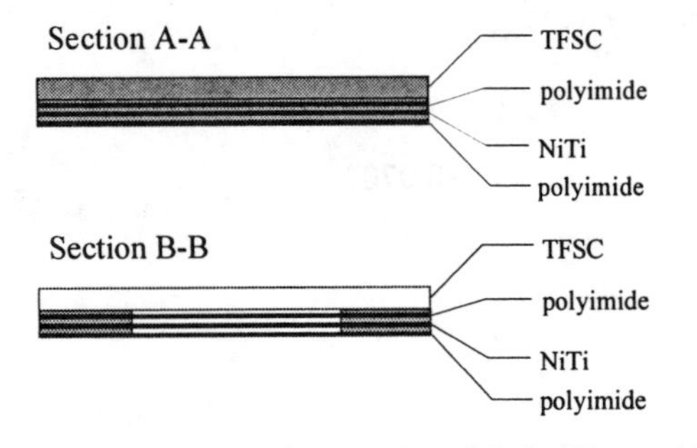

Figure 18. Cross sections A-A and B-B (Figure 16.)

A POWERSPHERE USING SMA

In the previous section, it was shown how a z-fold strip, used in a rectilinear array could be constructed with an integral NiTi metal backing to provide the spring force for deployment. The specifics of folding up the z-fold were shown and some design rules established, namely that relief is necessary at the hinges to make bending occur there and the fact that the TFSC material cannot pass over a hinge. The PowerSphere could use the same methods in its construction and operation.

The PowerSphere is not really a sphere but a geodetic shape that looks like a soccer ball, consisting of 12 pentagons and 20 hexagons. To efficiently deploy these polygons, a scheme (patent pending) whereby two hemispheres unfold and join together is used. In stage 1 of deployment, the two hemispheres separate from the PICOSAT and from each other. This is shown in Figure 19 as a cartoon. A tapered compression spring provides the force that holds the two hemispheres apart, and a power cable of length D, the desired diameter of the PowerSphere, limits the travel of the spring. The power cable serves another purpose; it electrically connects both hemispheres. Another short power cable pigtail then splices from it to the PICOSAT DC-to-DC converter. Figures 20–23 are CAD renderings of the deployment sequence. The SMA metal provides the unfolding energy.

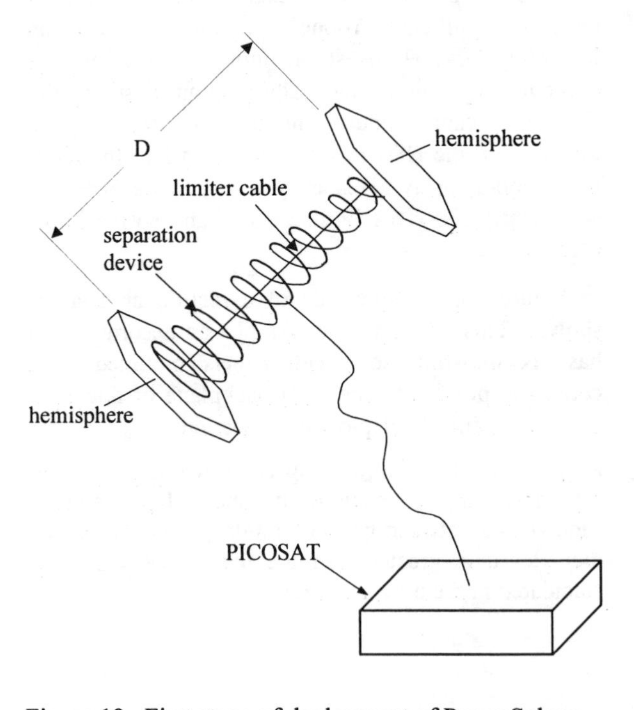

Figure 19. First stage of deployment of PowerSphere.

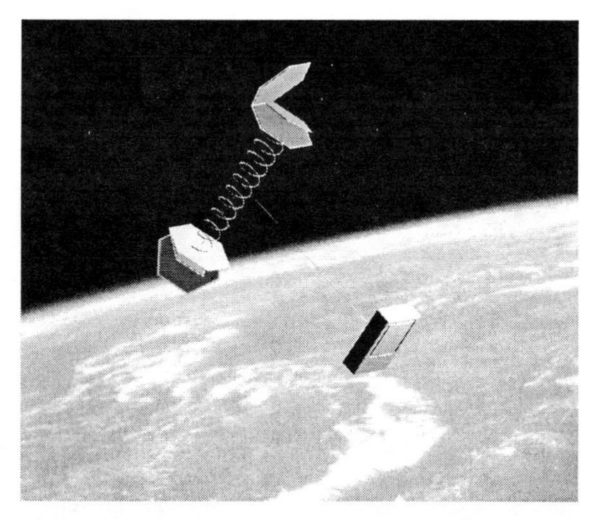

Figure 20. Second stage: unfolding begins.

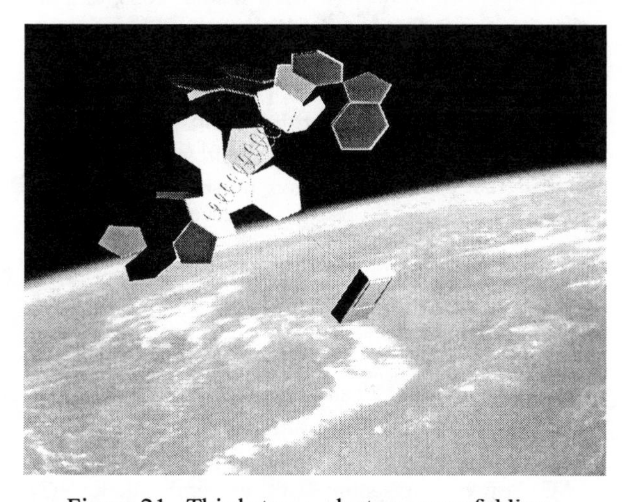

Figure 21. Third stage: clusters are unfolding.

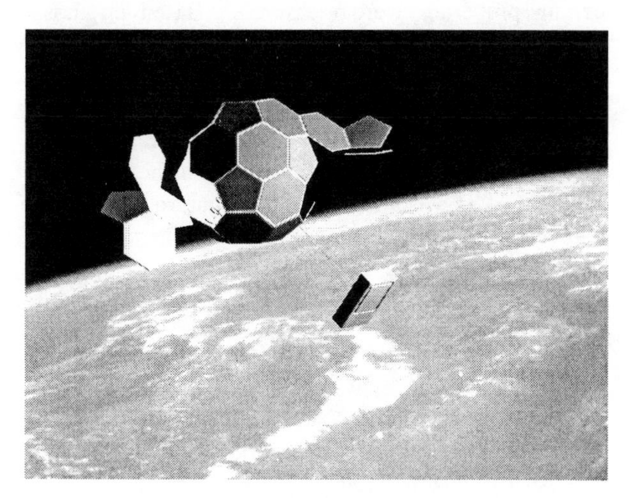

Figure 22. Fourth stage: remaining polygons are coming to their final position.

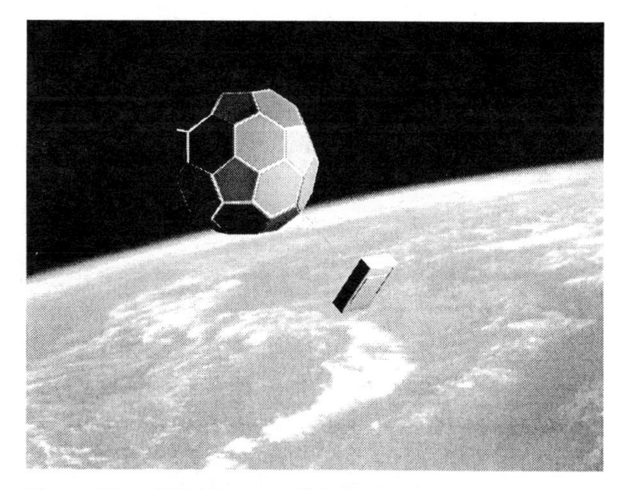

Figure 23. Final stage: all the polygons are at their correct angles.

The PowerSphere packed to fit on the PICOSAT adds an additional height of (see Figure 24)

$$t_a = 2h_a + d_s, \tag{14}$$

where h_a is the thickness of a single stowed hemisphere, and d_s is the wire diameter of the expansion spring. The term h_a is derived by adding up the thickness of the 16 polygons that stack to form just one of the two hemispheres (Figures 25 and 26). The stowed height of the hemisphere h_a depends on the hinge bend diameter d, the thickness of the TFSC backing t_b, and the thickness of the TFSCs, t_c. If d = 0.015 in., t_b = 0.005 in., and t_c. = 0.006 in., then total thickness is

$$\begin{aligned} h_a &= 3(d + 3t_b + 3t_c) - (t_b + t_c) + d \\ &= 0.148''. \end{aligned} \tag{15}$$

The total packaged height would require more than twice this thickness or

$$t_a = 2(0.148'') + d_s. \tag{16}$$

The PowerSphere packages quite well. If the t_s = 0.0007 in. SMA metal is used, then the term d will be quite small. The unfolding may be erratic because all of the hinge-springs will unspring as soon as possible rather than in a controlled sequence.

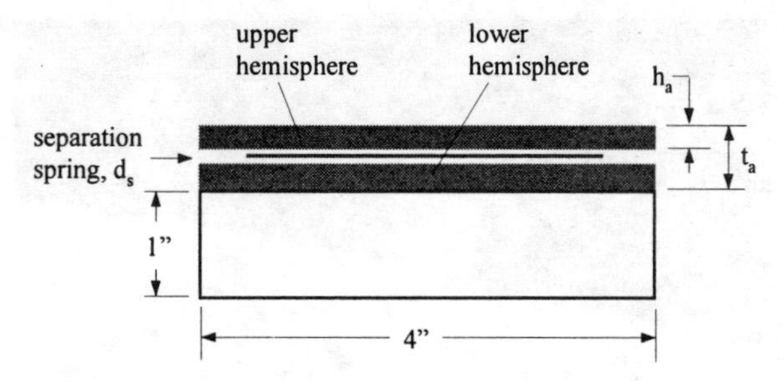

Figure 24. Side view of a PICOSAT with a stowed PowerSphere.

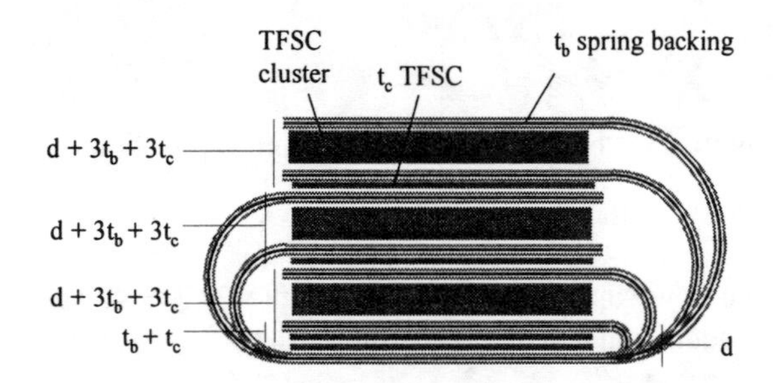

Figure 25. Detail of stowed hemisphere. See Figure 26 for detail of TFSC cluster.

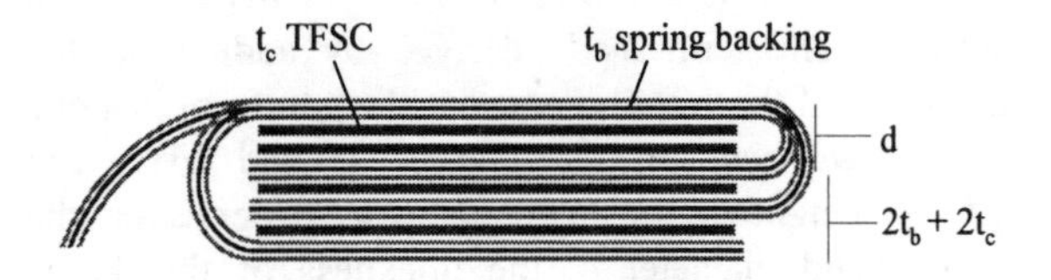

Figure 26. Detail of TFSC cluster.

The hinge sizes are variable depending on where a polygon is in the stack. In Figure 27, the pathlengths of the hinges are denoted by S_i for i = [1,6]. A typical pathlength is explicitly denoted for S_1. That length, for the same values chosen for eqn. (15) is

$$\begin{aligned} S_1 &= 0.5\,\pi\, t_a \\ &= 0.5\,(3.14)(.148") \\ &= 0.23". \end{aligned} \quad (17)$$

The other values of S_i are similarly calculated. The smallest hinge length, S_6, is

$$\begin{aligned} S_6 &= 0.5\,(3.14)\,(.015") \\ &= 0.024". \end{aligned} \quad (18)$$

These varying hinge lengths will degrade the shape of the PowerSphere unless some compensations in the geometry of the panels are made.

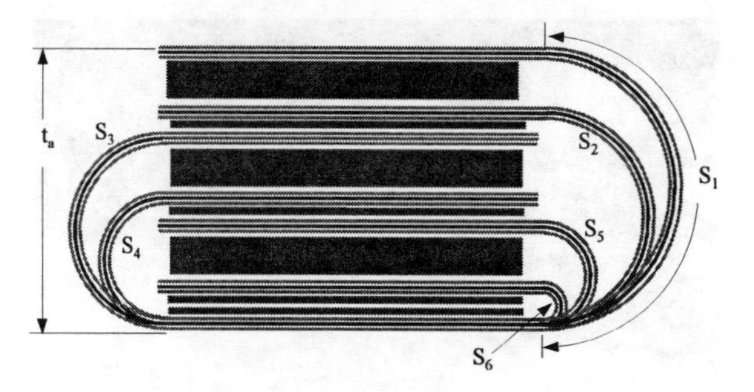

Figure 27. Hinge lengths for PowerSphere.

SUMMARY

Small satellites have a power limitation inherent in their size. They do not have enough surface area, even when populated with 18% efficient solar cells, to generate the power they need. Designers must consider deployable arrays to increase the area. Arrays that are planar in nature require pointing to the sun. This further requires motors to point them, software to control the motors, and stabilization of the platform. Satellites with no platform stablilization must use omnidirectional arrays.

Several deployable omnidirectional arrays were considered as they might be designed for use on the DARPA-Aerospace Picosatellite. Amorphous silicon thin-film solar cells were preferred over brittle silicon or GaAs because they can be used to form non-planar array shapes, can be stowed more efficiently, and are easier to deploy. A deployment mechanism for the TFSC arrays consisting of an embedded NiTi superelastic shape memory alloy was described and certain properties were measured. Based on those results, dimensions of the stowed arrays were calculated and listed for comparison. In the deployed configuration, all of the arrays had enough area to significantly surpass the power generation by body-mounted silicon or GaAs solar cells. Furthermore, the low cost of TFSCs and their rugged nature make them a good choice for small satellite power generation.

REFERENCES

1. Janson, S. W. et al., "The Concept of Nanosatellites for Revolutionary Low-Cost Space Systems," paper IAF-93-U.S.573, 44th Congress of the International Astronautics Federation, Graz, Austria, October 1993.
2. Osborn, J. V. et al., "A PicoSAT Network and MEMS Space Platform," *Proceedings of 2nd International Conference on Integrated*

MicroNanotechnology for Space Applications," 11–15 April 1999.

3. Simburger, E. J. et al., "Development of a Thin Film Amorphous Silicon Space Solar Cell for the PowerSphere Concept," *Proceedings of the 16th Space Photovoltaic Research and Technology Conference*, 31 August–2 September 1999.
4. Simburger, E. J., "PowerSphere Concept," *Proceedings of Government Microcircuit Applications Conference*, 8–11 March 1999.
5. Gilmore, David G., "Thermal Design Aspects of the PowerSphere Concept," The Aerospace Corporation, *Proceedings of The 2nd International Conference on Integrated MicroNanotechnology for Space Applications*, 11–15 April 1999.
6. Ashby and Jones, *Engineering Materials 1*, Pergamon Press, 1980.

AIAA-2000-2919

THIN-FILM PHOTOVOLTAIC SOLAR ARRAY PARAMETRIC ASSESSMENT*

David J. Hoffman, Thomas W. Kerslake, Aloysius F. Hepp
NASA Glenn Research Center, Cleveland, OH.

Mark K. Jacobs
SAIC, Schaumburg, IL.

Deva Ponnusamy
Spectrum Astro, Gilbert, AZ.

ABSTRACT

This paper summarizes a study that had the objective to develop a model and parametrically determine the circumstances for which lightweight thin-film photovoltaic solar arrays would be more beneficial, in terms of mass and cost, than arrays using high-efficiency crystalline solar cells. Previous studies considering arrays with near-term thin-film technology for Earth orbiting applications are briefly reviewed. The present study uses a parametric approach that evaluated the performance of lightweight thin-film arrays with cell efficiencies ranging from 5% to 20%. The model developed for this study is described in some detail. Similar mass and cost trends for each array option were found across eight missions of various power levels in locations ranging from Venus to Jupiter.

The results for one specific mission, a main belt asteroid tour, indicate that only moderate thin-film cell efficiency (~12%) is necessary to match the mass of arrays using crystalline cells with much greater efficiency (35% multi-junction GaAs based and 20% thin-silicon). Regarding cost, a 12% efficient thin-film array is projected to cost about half as much as a 4-junction GaAs array. While efficiency improvements beyond 12% did not significantly further improve the mass and cost benefits for thin-film arrays, higher efficiency will be needed to mitigate the spacecraft-level impacts associated with large deployed array areas. A low-temperature approach to depositing thin-film cells on lightweight, flexible plastic substrates is briefly described. The paper concludes with the observation that with the characteristics assumed for this study, ultra-lightweight arrays using efficient, thin-film cells on flexible substrates may become a leading alternative for a wide variety of space missions.

INTRODUCTION

Very lightweight and low cost photovoltaic (PV) solar arrays based on thin-film PV array technology have held much promise for future space missions. While sample thin-film cells and panels have flown in space (LIPS-III in 1987, PASP-Plus in 1994, the Mir space station in 1998) and are planned to fly (Earth Observing-1 in 2000), a complete solar array consisting of thin-film cells has yet to be built. Also, the projected array-level efficiency of thin-film PV is currently much less than that of arrays based on advanced thin-crystal silicon (Si) and multi-junction gallium arsenide (GaAs) based cells. Consequently, at the spacecraft level, the large deployed array area required for thin-film arrays offsets or even negates its lower array mass and cost benefits. Until thin-film PV efficiency improves and manufacturing methods to deposit the thin-films on lightweight substrates over large areas are refined, future space missions will most likely keep using high efficiency silicon or multi-junction PV planar and/or concentrator arrays. As thin-film PV technology for use in space improves, more applications will consider its advantages, namely low cost, low mass, improved radiation tolerance[1], and high specific power (W/kg).

Figure 1 depicts two ways to obtain very high specific power using photovoltaic arrays. Flexible planar arrays of moderate area density (1-2 kg/m^2) using either relatively heavy but very efficient multi-junction solar cells, or relatively lighter but less efficient thin silicon cells, could obtain an array-level specific power approaching 300 W/kg. To get to this level, new solar array substrates, support structures and deployment concepts may be needed in conjunction with improved cell technology.[2] Ultra-lightweight arrays (0.25-0.75 kg/m^2) using lightweight thin-film solar cells of moderate efficiency may enable the attainment of even greater array-level specific power. As the plot in figure 1 implies, ultra-lightweight thin-film arrays may be the most feasible means of approaching the very high specific power necessary to enable missions with very high power requirements, such as space solar power satellites, manned Mars or lunar missions and some solar electric propulsion concepts.[3,4]

The objective of the present assessment is to develop a model and parametrically determine the circumstances, both in terms of solar array technology and mission scenarios, for which thin-film PV solar arrays would be more beneficial than alternatives. NASA Glenn research Center's (GRC) approach to depositing thin-film cells on lightweight substrates with the aim of ultimately achieving higher efficiencies is also described.

BACKGROUND

A number of past studies have compared solar cell and array technologies for Earth orbiting missions. Ralph performed system trades for presently available and near-term crystalline and thin-film cells on rigid, flexible and concentrator arrays in Low Earth Orbit (LEO) and Geostationary Earth Orbit (GEO).[5,6] Ralph's results in reference 5 indicate that GEO arrays using high efficiency multi-junction GaAs cells have mass and cost advantages over alternatives, especially when the area penalty (increased attitude control fuel) of arrays using the less efficient thin-film cells is included. With Ralph's assumptions, thin-film cell efficiency needs to be at least 12.6% to be competitive in GEO. For LEO, Ralph concluded that while the most efficient multi-junction cell array has the lowest mass, arrays with 9% to 12.6% efficient thin-film cells have competitive area-adjusted costs.

In a similar study, Gaddy looked at the cost performance of multi-junction GaAs and advanced Si arrays on small, medium and large LEO spacecraft.[7] This study included the cost of the spacecraft support to the payload and concludes that the most efficient multi-junction arrays result in the greatest spacecraft-level mass and cost benefits.

The paper by Bell outlines a model developed by the Aerospace Corporation to "determine optimal power subsystem suites as a function of spacecraft design and total system cost."[8] Example model results are reported for a 100 satellite high-power (15kW) LEO constellation and a small, single-mission 1 kW LEO satellite. Study results for both cases favored high efficiency cell solar arrays. Because satellites in the LEO constellation were delivered to a low parking orbit and then transferred to the final 1852 km orbit, the large area of the 8%-10% efficient thin-film arrays led to significant attitude control system impacts, and ultimately higher mission costs. For the single-mission low power LEO case, the Aerospace model favored mature, low nonrecurring cost array technologies using 16% efficient Si and 21.5% efficient GaAs cells.

Each of the studies reviewed above looked at near-term thin-film cell technology on flexible, but not necessarily lightweight arrays for Earth orbiting applications. Only when the cell efficiency of a thin-film array was greater than 10% did they compare favorably with crystalline cell arrays for some of the missions studied. In the present study, the performance of ultra-lightweight thin-film arrays with assumed cell efficiencies ranging from 5% to 20% are evaluated for missions in Earth orbit and beyond.

ANALYTICAL APPROACH

One objective of this study is to estimate the improvement in cell efficiency required for thin-film arrays to be more competitive with higher-efficiency crystalline cells from a mass and area perspective. From a mass perspective, array specific mass is the figure of merit. Array specific mass can be obtained by dividing the specific area (W/m^2) by the array's area density (kg/m^2). Specific area is a function of the cell efficiency and array packing factor. Area density is a function of the cell material density and thickness and the array substrate, wiring, support structure and mechanisms. To a first order, the cell efficiency required to match the specific power of an array of a given type but using different cells (i.e. the array area density not including cells is assumed to be constant) can be estimated with the following equation,

$$\eta_{TF} = \eta_2 \left(\frac{PF_2}{PF_{TF}} \right) \left(\frac{Array + Cell_{TF}}{Array + Cell_2} \right) \qquad (1)$$

where η is the cell efficiency, PF is the array packing factor, *Array* is the area density (kg/m^2) of the array, including its wiring, substrate, support structure and mechanisms, and *Cell* is the cell area density. While the array area density is held constant in this first order approximation, in actuality, it should decrease with the use of lighter cell technology. The more detailed array model discussed later accounts for this effect.

Figure 2 shows the approximate thin-film cell efficiency required to match the specific power of a high efficiency cell array using equation 1. Cell material densities, including the coverglass, of 0.50 kg/m^2 for the Si cell, 1.0 kg/m^2 for the multi-junction GaAs cells, and 0.16 kg/m^2 for the thin-film cells are assumed in figure 2. In practice, the actual cell efficiency required to match array specific mass will also depend on the cell operating temperature and degradation of the cell efficiency from environmental effects over the mission life. Nevertheless, figure 2 can be used to discern trends. For example, the figure shows that for ultra-light arrays (area densities from 0.25 to 0.75 kg/m^2), only moderate thin-film cell efficiencies are required to match the specific power of arrays using much higher efficiency, but heavier cells. Improvements in thin-film cell efficiencies may still be necessary in order to reduce the size of thin-film arrays in order to minimize attitude control system impacts and to reduce array stowed volume and deployment complexity for missions with these concerns.

To perform the main analysis of this study, a spreadsheet model was developed that calculates the size and estimates the cost of PV arrays based on different cell and array technologies for a given set of mission requirements. Comparative metrics (e.g. W/kg, W/m^2, kg/m^2, etc.) are calculated for various array components, at the array level itself, and then at the power subsystem and spacecraft level.

Representative mission information was gathered for eight missions at various locations in the solar system with various end-of-life (EOL) power requirements. The model was applied to each mission in a parametric fashion in an effort to determine meaningful trends.

MODEL DESCRIPTION

The Array Design Assessment Model (ADAM) was developed to support evaluation of array design alternatives. ADAM includes several integrated array design modules, five databases to manage input set alternatives for running the design modules, and a user interface with input forms and model outputs. Outputs include nearly 100 items representing array performance, including PV array, other power subsystem elements, and spacecraft development. ADAM elements and estimating methodology flow are shown in figure 3.

Mission candidates in the ADAM database cover array sizes from several hundred watts to around 20 kilowatts. Size and costing relationships have not been tested for very small (<100 W) or very large (>25 kW) arrays.

PV Array Sizing

For PV array sizing, ADAM separates the array into several elements, as shown in figure 4. The model first estimates cell area requirements based on cell performance characteristics in the selected operating environment, including the effects of operating temperature, cell mismatch, interconnects, radiation, thermal cycling, contamination deposition, meteoroid and orbital debris, ultraviolet degradation, shadowing, offpointing and the array packing factor. Additional blanket layers are built up based on material selections and layer thicknesses. Many advanced features are incorporated to address scaling issues. For example, as required rigid array wing areas grow, less dense and thicker substrate core materials are used to maintain reasonable structural characteristics.

After all blanket requirements are estimated, structure and mechanical elements are added based on blanket properties and required structural characteristics. ADAM handles structural design differently for rigid and flexible arrays. For rigid arrays, a yoke is used to reduce losses from shadowing and stiffness is based on properties of the blanket panels and hinges between panels. For flexible arrays, a deployable boom is sized to support the panel and meet first fundamental frequency requirements.

For rigid panels, the model uses a sandwich structure, which includes a honeycomb core and aluminum or composite face sheets. A parametric curve, correlating mass to the substrate area has been developed based on past data, and the mass is initially estimated using this curve. The masses of other mechanical elements are computed as a fraction of the substrate mass.

The deployed fundamental frequency is one of the basic requirements of the array, and it is calculated to further validate the sizing and configuration. The natural frequency is calculated using the Jones' equation,

$$f_n = \frac{1.2769}{2\pi}\sqrt{\frac{g}{\delta_{\max}}} \quad (2)$$

where, f_n is the natural frequency in first bending mode, g is the acceleration due to gravity, and δ_{max} is the maximum deflection of array. This is a close approximation of the fundamental frequency of a uniform thin plate of arbitrary shape, having any combination of fixed, partially fixed or simply supported boundaries.

Substrate materials are selected and each layer's thickness is calculated to match the substrate mass estimated earlier. For the purpose of this calculation, the solar array is assumed to be a uniform thin plate and the total deflection under 1 g due to the bending of the substrates and the compliance of the hinge lines is calculated. Hinge stiffness is assumed to be 10^5-10^6 Nm/rad. Details like the aspect ratio of the array are chosen to achieve a fundamental frequency of about 0.5 Hz as the model default, although the user can specify other fundamental frequency values.

In the case of the flexible panel, the total mass of the blanket, cells and all other add-ons is estimated by ADAM's Blanket Design Module. Given the total mass and the aspect ratio of the array, the uniformly distributed mass on the boom is calculated. The boom used in this study is a coilable lattice boom. The diameter of the boom, which is limited to a minimum of 10cm, is chosen to provide the equivalent stiffness, necessary to achieve a user-defined fundamental frequency (typically 0.5 Hz) for the given load.[9] This approach results in boom dimensions and masses that are realistic, even though the strength of the boom in bending or buckling is not taken into account. The mass of the canister is assumed to be 1.5 times the mass of the boom. The mass of the array stowage and tensioning systems is calculated as 25% of the sum of the boom, canister, blanket and wiring masses.

For both rigid and flexible arrays, the mass of a single axis drive actuator (SADA) is accounted for and is assumed to scale linearly with the beginning-of-life (BOL) power level (1.5 kg/kW). Wiring mass for either type of array is assumed to be 1.2 kg/kW.

Array sizing accounts for energy storage to support eclipse operations or other mission requirements. ADAM includes nine PMAD and energy storage inputs to estimate other power subsystem element requirements and additional array output required for charging the storage system.

Cost Assumptions/Methodology

ADAM includes parametrics to estimate spacecraft hardware development costs in fixed year dollars (fiscal year 2000). This covers activities typically performed in Phases B/C/D. Cost estimating relationships (CERs) were developed for each ADAM Reference Mission Candidate using proven methods. For Earth orbiting missions, CERs were derived from the NASA GSFC Space Systems Quick Estimating Guide (Version 2.0, August 1997). For the other planetary missions, SAIC's Planetary Development Model was used. Heritage credits were applied to approximately 75% of each subsystem and the other 25% is assumed to be new development with available technology. Advanced technology development costs are not included. Parametrics are based on costs per kg for all spacecraft subsystems except power, and are only intended to be accurate for concepts reasonably similar to the selected Reference Mission Candidate. Spacecraft system-level assembly/integration/test costs are estimated to be 15% of the subsystem total. Cost results should be interpreted as relatively representative, not absolute values.

Power system costs are built up from several elements. Hardware costs are estimated at the component-level (e.g. cells, substrate, structures, etc.). Non-recurring costs are assumed to be 50% of the hardware costs, and assembly/integration/test labor is added at a rate of $500 per Watt. Because ADAM does not estimate advanced technology development, each array design concept is assumed to be at an equivalent technology readiness level. Savings from advanced array concepts need to offset costs to demonstrate flight readiness.

Model Inputs/Outputs

Table 1 shows a summary of ADAM's databases, inputs, and outputs. ADAM generates almost 100 output items from over 50 inputs to compare performance of different array design concepts. Four high-level inputs – Mission Type, Operating Environment, End-of-Life (EOL) Power Required, and Array Design Lifetime – interface with the ADAM databases to determine initial default values for 24 Level 1 and 30 Level 2 inputs. Level 1 inputs interface with the model databases to determine Level 2 input defaults. ADAM users can choose to operate at the high-level or modify any Level 1 or 2 input to better represent their array/mission design concept. As the ADAM databases are expanded, model capabilities are enhanced. Future versions of ADAM may incorporate more database candidates and additional/enhanced databases, inputs/outputs, and design modules.

More details describing ADAM can be found in the final review presentation for the task order contract under which the model development was performed.[10]

ANALYSIS CASES

As previously mentioned, this study assessed eight representative missions throughout the solar system: a Venus orbiter, LEO and GEO missions, a lunar lander, a Mars communication orbiter and a Mars lander, a Main Belt Asteroid Tour, and a Jupiter orbiter. Given the lightweight substrate and parametric thin-film cell efficiency assumptions used in this study, the same overall trends were found for all missions.

The results for the Main Belt Asteroid Tour (MBAT) are used to illustrate the trends from the parametric analysis. The MBAT mission was chosen because it is a relatively high power mission using solar electric propulsion (SEP). MBAT mission characteristics are as follows:

- Location 1.5 AU
- Design Life 6 years
- EOL Power Required 7.5 kW (1.5 AU)
- Spacecraft Dry Mass 560 kg
- Spacecraft Wet Mass 956 kg

The specific thin-film technology considered in the trade study is a 0.2-mil (5 micron) copper indium disulfide ($CuInS_2$, CIS_2 or CIS2) cell on 0.08-mil of molybdenum and 2 mils of a polyimide, resulting in an area density of 0.16 kg/m^2. The CIS2 cell also contains ZnO and CdS layers and is estimated to cost \$60/W. The BOL, 28 degree C, AM0 cell efficiency for CIS2 is parametrically varied from a low of 6% up to 20%. CIS2 performance metrics are compared with a presumed 35% efficient four-junction (4-j) cell based on single-crystal GaAs/Ge technology (1.1 kg/m^2 and \$400/W) and a 20% efficient single-crystal thin-Si cell (0.55 kg/m^2 and \$220/W). For reference, present state-of-the-art AM0, 1-sun efficiency is about 25% for multijunction GaAs based cells and 17% for thin Si. Both crystalline cells have a 4-mil coverglass. All cells are mounted on a 5-mil composite flexible substrate with a coilable deployment boom sized for a 0.5 Hz minimum first fundamental frequency.

RESULTS

Figures 5 through 8 show the model results in graphical form. Figure 5 plots the PV blanket and total array specific power for each array. Figure 6 depicts the total array area for each array on a relative basis, normalized to the 4-j GaAs case (50 m^2 total). Figures 7 and 8 show the array mass and cost breakdowns on a relative basis, again normalized to the 4-j GaAs case (121 kg, \$14.1M total array mass and cost).

DISCUSSION

Pertaining to specific power, figure 5 shows the arrays with 4-j GaAs and Si crystalline cells have comparable values for this key metric at both the PV blanket and total array levels. For the thin-film array, progressively higher specific power at the PV *blanket level* results as a linear function of cell efficiency. However, at the *total array level*, which includes array wiring, structures, mechanisms and a single-axis drive actuator for pointing, the increase is not linear and is much less rapid than at the blanket level. This illustrates the difficulty in attaining very high total-array-level specific power when accounting for all typical array "ancillaries".

With respect to array total deployed area, figure 6 confirms what is expected – area scales linearly with cell efficiency (assuming similar packing factors and mission cell efficiency knockdown factors). For the MBAT mission, unbalanced drag torques would not be a problem for the much larger array sizes with the lowest thin-film cell efficiencies. However, other disturbance torques and or spacecraft/array slewing to maintain SEP thrust vectors may be an issue.

Figure 7 indicates that for the assumptions underlying the present study, a moderate thin-film cell efficiency of 12% is necessary to match the total mass of arrays using crystalline cells with much greater efficiency. The array component mass breakdowns reveal the leading contributors to each array's total mass. Mechanical components, which include the array stowage and tension mechanisms and SADA contribute a significant portion to all arrays. The cell and coverglass mass dominate the crystalline PV blanket mass, while the substrate mass dominates the thin-film blanket. This highlights the point that in order to take full advantage of the mass benefits of thin-film cell technology, very lightweight substrates and support structures are necessary.

The cost breakpoint for the thin-film arrays occurs at thin-film efficiencies greater than 12% according to figure 8, resulting in an array that costs about half as much as the 4-j GaAs array. Improving the thin-film cell efficiency beyond 12% did not significantly further improve the cost benefit.

THIN-FILM CELL DEVELOPMENT AT GRC

Among the desirable attributes in any space-bound component, subsystem or system are high specific power, radiation tolerance and high reliability, without sacrificing performance. NASA GRC is currently developing space-bound technologies in thin film chalcopyrite solar cells and thin-film lithium polymer batteries.[11] The thin-film solar cell efforts at GRC are summarized below.

The key to achieving high specific power solar arrays is the development of a high-efficiency, thin-film solar cell that can be fabricated directly on a flexible, lightweight, space-qualified durable substrate. Such substrates include Kapton™ (DuPont) or other polyimides or suitable polymer films. While the results of the present study indicate that lightweight thin-film cells with moderate efficiency on lightweight substrates can compete on a mass basis, higher cell efficiencies will be required to mitigate impacts associated with large array area. Current thin-film cell fabrication approaches are limited by either (1) the ultimate efficiency that can be achieved with the device material and structure, or (2) the requirement for high-temperature deposition processes that are incompatible with all presently known flexible polymides, or other polymer substrate materials.

At GRC, a chemically based approach is enabling the development of a process that will produce high-efficiency cells at temperatures below 400 °C. Such low temperatures minimize the problems associated with the difference between the coefficients of thermal expansion of the substrate and thin-film solar cell and/or decomposition of the substrate.

Polymer substrates can be used in low temperatures processes. As such, thin-film solar cell materials can be deposited onto molybdenum-coated Kapton, or other suitable substrates, via a chemical spray process using advanced single-source precursors, or by direct electrochemical deposition. A single-source precursor containing all the required chemically-coordinated atoms such as copper, indium, sulfur and others, will enable the use of low deposition temperatures that are compatible with the substrate of choice.[12]

A combination of low-temperature electrochemical deposition and chemical bath deposition has been used to produce ZnO/CdS/$CuInSe_2$ thin-film photovoltaic solar cells on lightweight flexible plastic substrates, depicted in figure 9.[13]

CONCLUSION

Once available and space qualified, moderate to relatively high efficiency thin-film cells on lightweight flexible substrates will offer significant mass and cost benefits. This approach may even enable ultra-lightweight solar arrays to attain the very high specific mass required for future high-power missions and applications. Further, as thin-film cell efficiency improves, the packaging, deployment and attitude/control impacts of the larger array area will diminish. With these characteristics, ultra-lightweight arrays using efficient, thin-film cells on flexible substrates may become a leading alternative for a wide variety of space missions.

ACKNOWLEDGEMENTS

The authors would like to acknowledge the support of the following individuals in the development of the analytical model under task order contract NAS3-26565: Dennis Pellacio and Mike Stancati of SAIC; Ken Rachocki and Kerry Wesley of Spectrum Astro; and Eli Kawam (formerly of Spectrum Astro).

REFERENCES

1. Woodyard, J., Landis, G., "Radiation Resistance of Thin Film Solar Cells", *Solar Cells*, Vol. 31, No. 4, pp. 297-329, 1991 (also NASA Technical Memorandum TM-103715).
2. Jones, P.A., White, Stephen F., Harvey, T.J., "A High Specific Power Solar Array for Low to Mid-Power Spacecraft", Proceedings of the Space Photovoltaic Research and Technology Conference (SPRAT XII), 1992, NASA CP-3210, pp. 177-187.
3. Landis, G., Hepp, A., "Applications of Thin-Film PV for Space", Proceedings of the 26th Intersociety Energy Conversion Engineering Conference, Vol. 2, pp. 256-261, Aug. 1991.
4. Kerslake, Thomas W., Gefert, Leon P., "Solar Power System Analyses for Electric Propulsion Missions", NASA/TM-1999-209289, SAE 99-01-2449, 34th IECEC, August 1999.
5. Ralph, E.L., Woike, T.W., "Solar Cell Array System Trades – Present and Future", AIAA, 1999.
6. Ralph, E.L., "High Efficiency Solar Cell Arrays System Trade-Offs", IEEE, 1994.
7. Gaddy, Edward M., "Cost Performance of Multi-Junction, Gallium Arsenide, and Silicon Solar Cells on Spacecraft, 25th PVSC, May 1996.
8. Bell, Kevin D., Marvin, Dean C., "Power Generation and Storage Technology Selection for an Optimal Spacecraft System Design", IECEC 1999-01-2531.
9. Conley, Peter, "Space Vehicle Mechanisms: Elements of Successful Design", John Wiley and Sons Inc., 1998.
10. "Thin-Film PV Solar Array Assessment for Satellites, Final Review Report," contract number NAS3-26565, Task Order 31.
11. Raffaelle, R.P., Harris, J.D., Rybicki, G.C., Scheiman, D.A., Hepp, A.F., "A Facile Route to Thin-Film Solid State Lithium Microelectric Batteries", *J. of Power Sources*, in press, 2000.
12. Hollingsworth, J.A., Hepp, A.F., and Buhro, W.E., 'Spray CVD of Copper Indium Sulfide films: Control of Microstructure and Crystallographic Orientation", *Chemical Vapor Deposition*, Vol.3, pp. 105-108, 1999.
13. Raffaelle, R.P., et al, "Chemically Deposited Thin-Film Solar Cell Materials", presented at SPRAT XVI, Cleveland, OH., Aug 31 – Sept 2, 1999, NASA Conference Proceedings (2000).

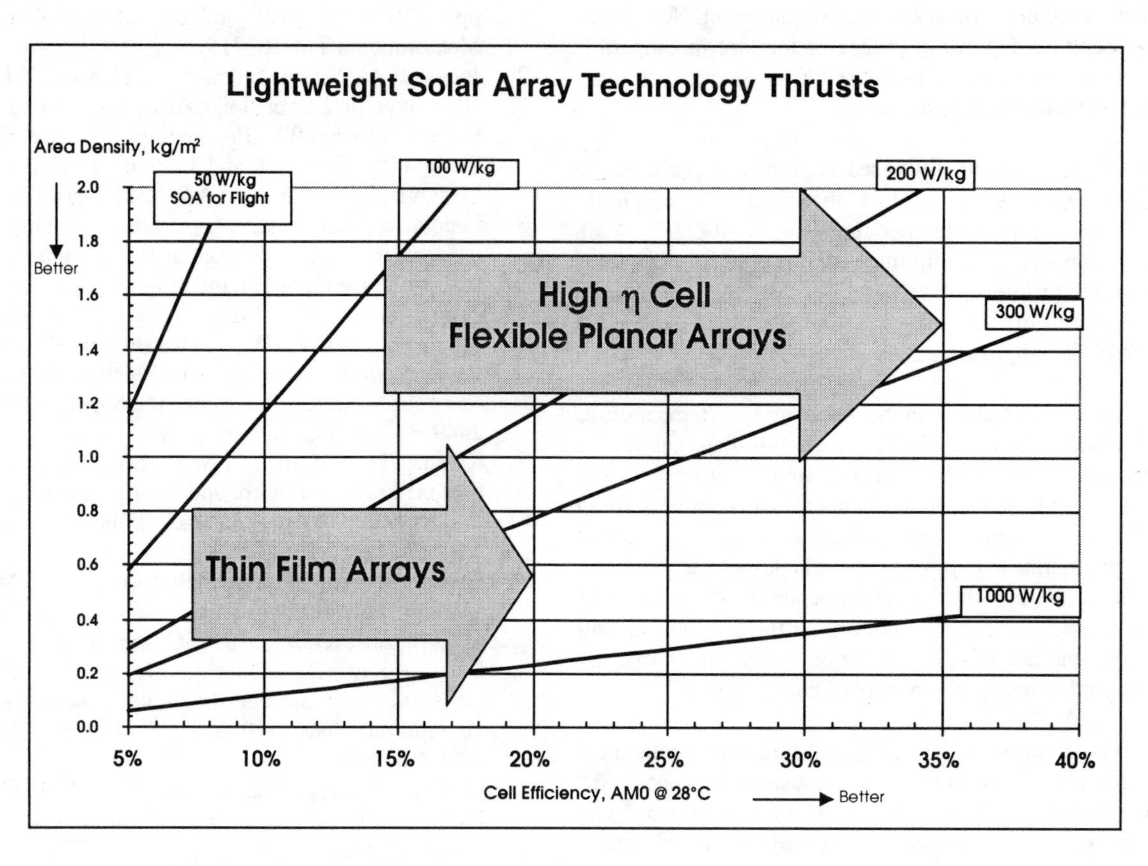

Figure 1 - Lightweight solar array technology thrusts.

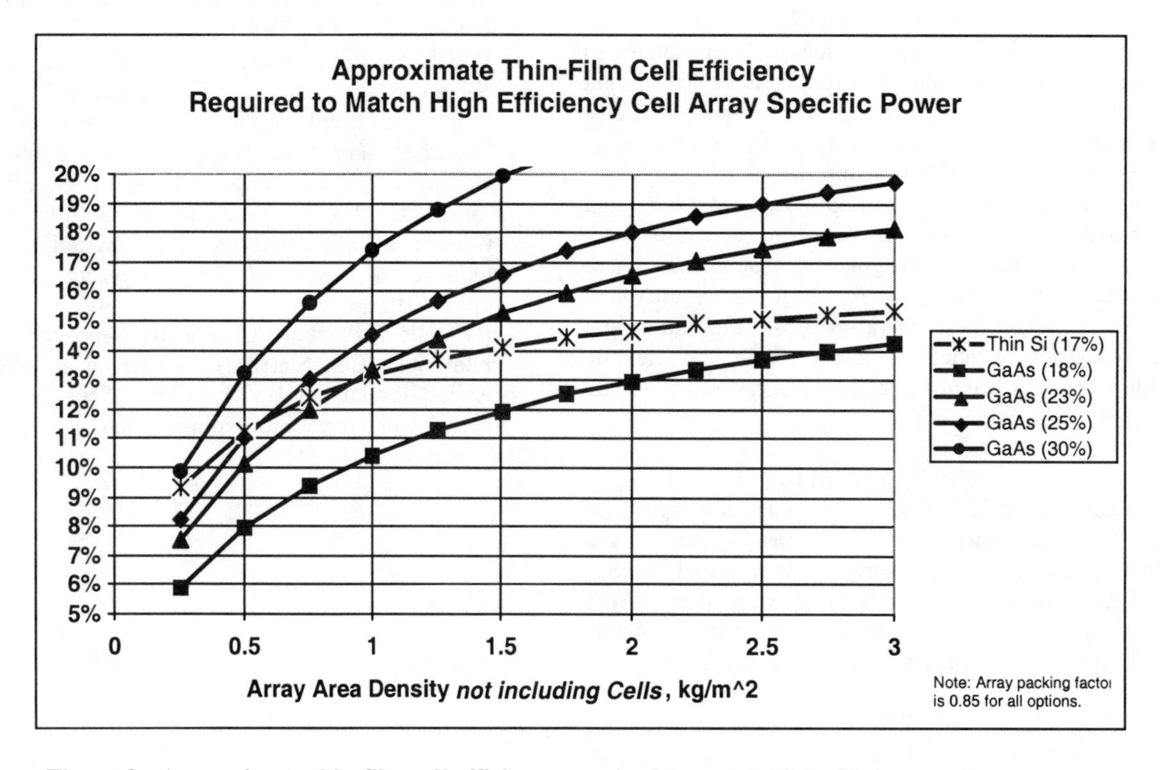

Figure 2 - Approximate thin-film cell efficiency required to match high efficiency cell array specific power.

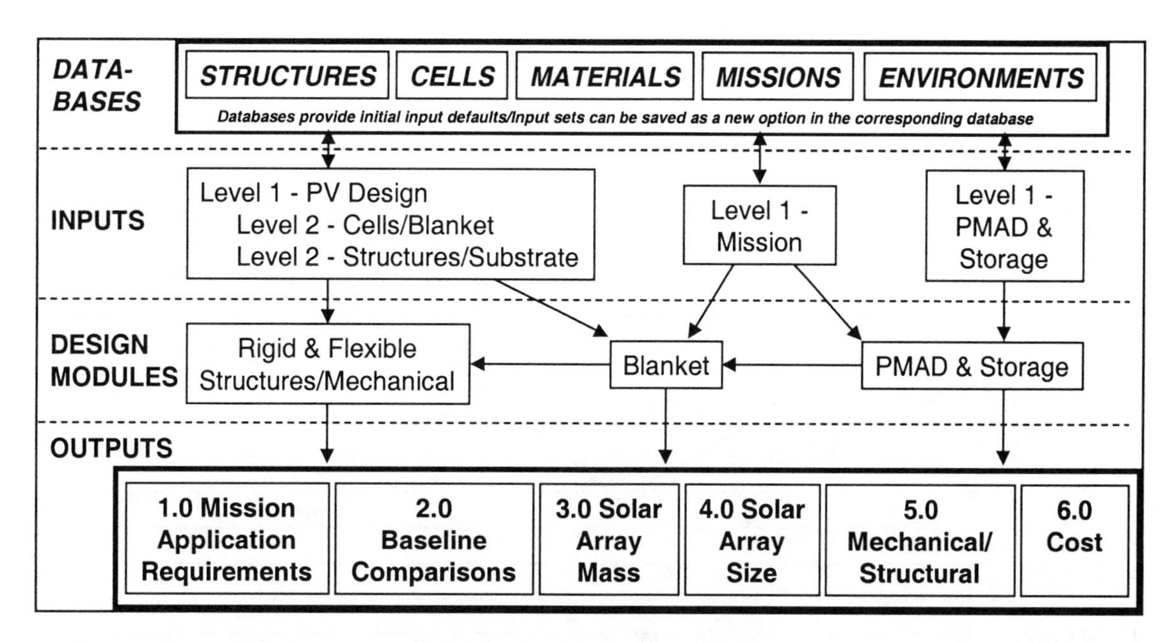

Figure 3 - ADAM Elements and Estimating Methodology Flow

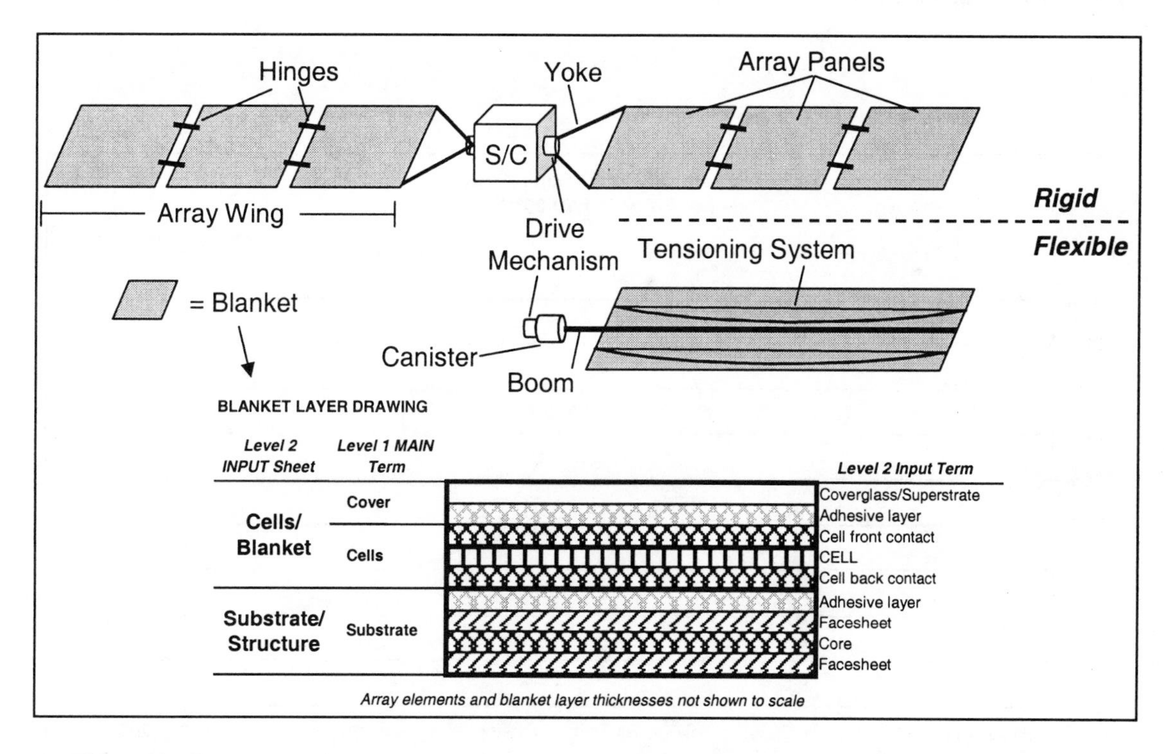

Figure 4 – Solar array hardware elements and ADAM model nomenclature.

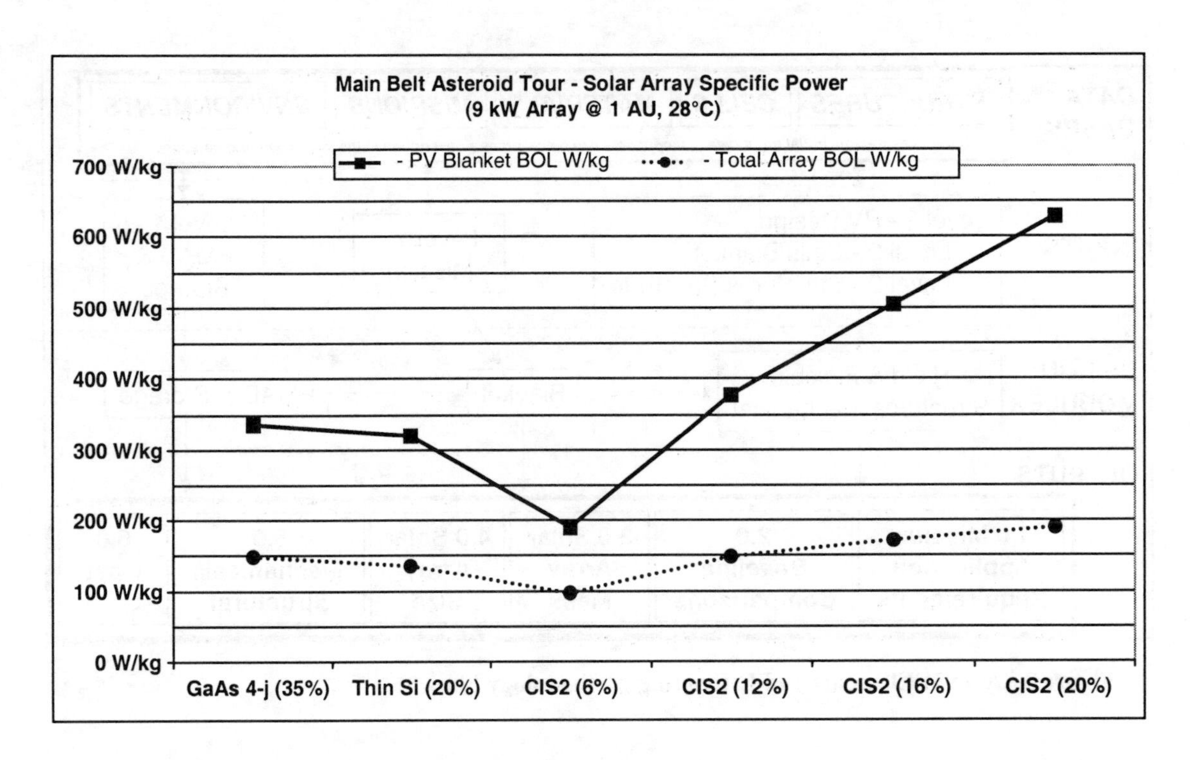

Figure 5 - PV blanket and total solar array specific power.

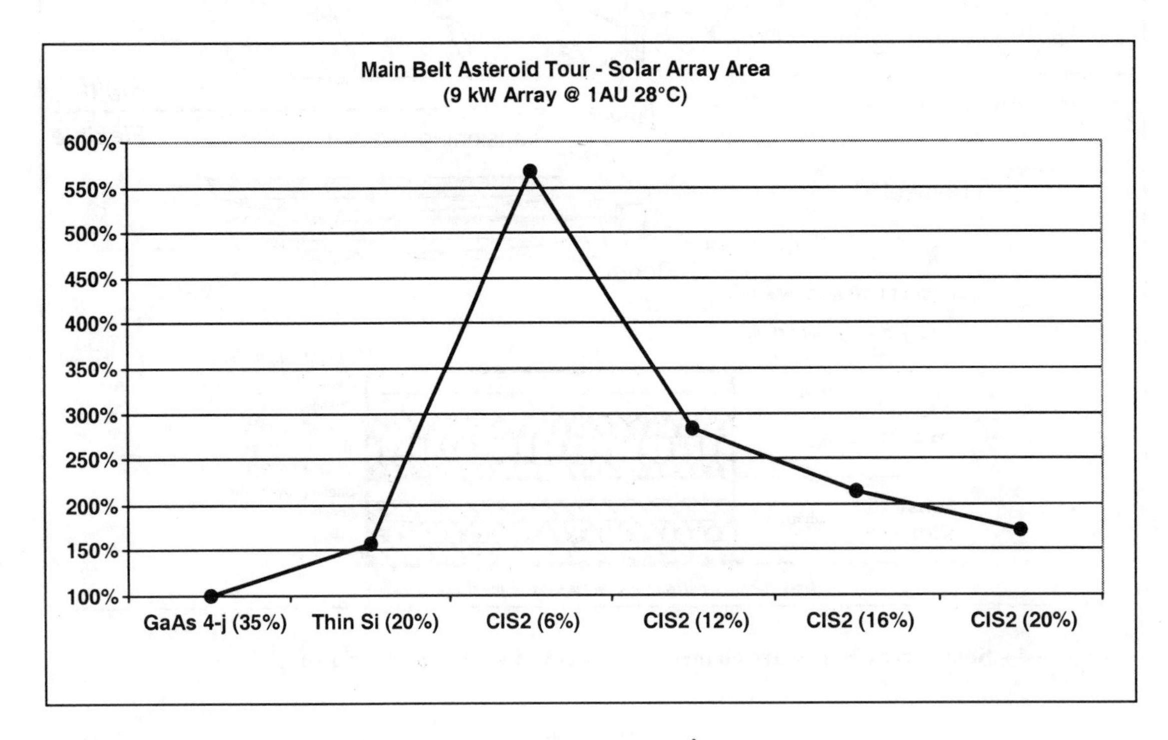

Figure 6 - Relative solar array total area (GaAs 4-j = 50 m^2 total).

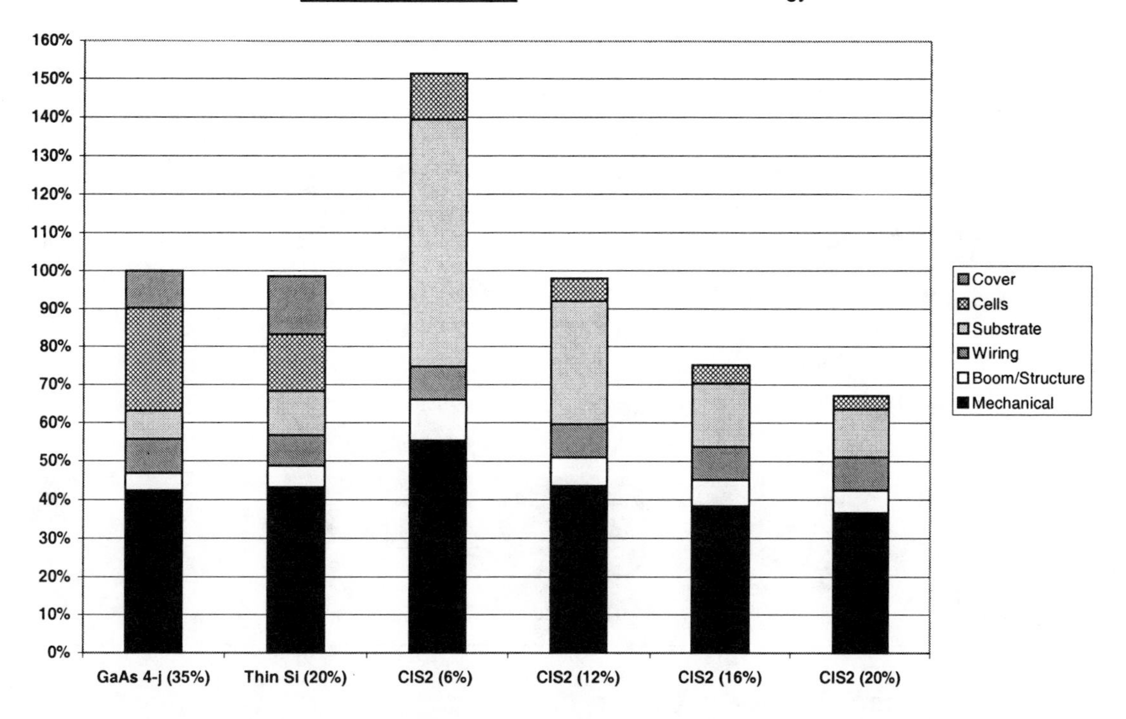

Figure 7 - Solar array mass breakdown and relative comparison (GaAs 4-j = 121 kg).

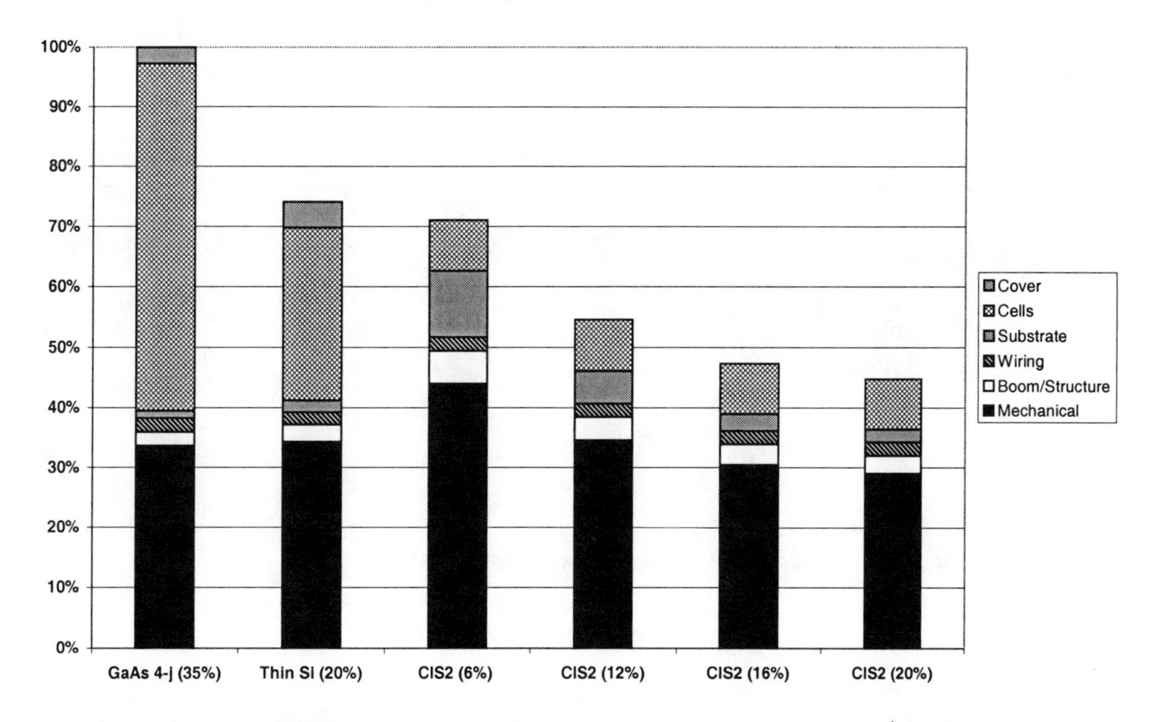

Figure 8 - Solar array cost breakdown and relative comparison (GaAs 4-j = $14.1M).

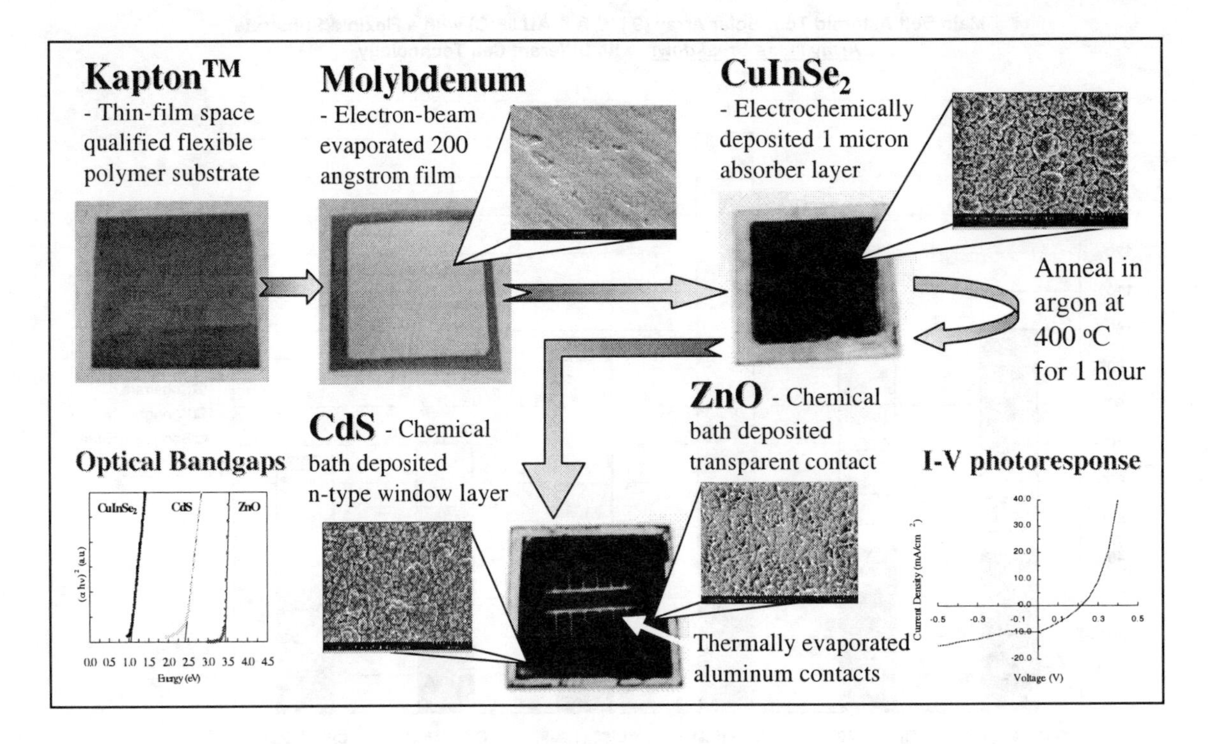

Figure 9 - NASA GRC thin-film cell approach.

Database Summary	Inputs Summary	Outputs Summary
6 Cell Candidates 1 Rigid Substrate Type 5 Flexible Substrate Types Over 50 Material Candidates 8 Mission Types - 2 Earth Orbiting - 4 Planetary - 2 Lunar/Mars Landers 8 Operating Environments (specific to Mission Types)	***4 Primary Inputs:*** - Mission Type - Operating Environment - EOL Power Required - Array Design Lifetime ***24 Level 1 Inputs:*** - 4 Mission - 11 PV Array Design - 9 PMAD/Storage ***30 Level 2 Inputs:*** - 21 Cells/Blankets - 9 Structures/Substrates	***38 Level 1 Outputs*:** - 15 Option Comparison Summary Metrics - 8 Mass Elements - 4 Size Characteristics - 3 Structural Characteristics - 8 Cost Elements ***54 Level 2 Outputs:*** - 11 Power System Items - 23 Spacecraft Elements - 20 Power System Equipment List Items

Table 1 - Summary Descriptions of ADAM Database, Inputs and Outputs

SOLAR ARRAY STRING CHARACTERISTICS IN STRANGE PLACES

Robert J. Pinkerton, Sr. Member AIAA, Motorola, Chandler, AZ

Abstract

Solar cell string characteristics are well documented when operating in the fourth quadrant of the characteristic curve under illumination (insolation). However, there is little consolidated documentation on the characteristics of strings/arrays operating under shadow or short circuit conditions. These conditions cause operation in the first or third quadrants of the device characteristic curve. This paper attempts to regress this long-standing injustice. The subject matter will be addressed in the following order:

1. Description of the device behavior in each quadrant;
2. Discussion of the conditions which cause the device to operate in quadrants I and III;
3. Discussion of the impacts to the system resulting from operation under these conditions; and
4. Discussion of the resulting design implications.

Introduction

The behavior of a semiconductor device is described by its characteristic curve. Figure 1 is the characteristic curve for a typical diode. Solar cells are photodiodes on a larger scale and therefore have the same basic characteristics as a diode. They can be both the forward and reverse biased. They have avalanche breakdowns when reverse biased.

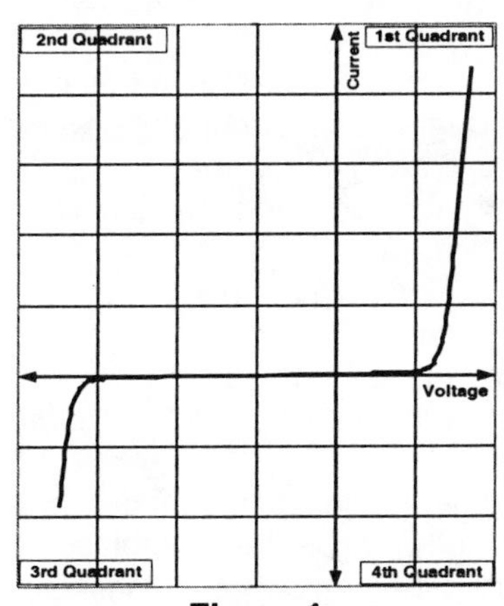

Figure 1
Typical Diode Characteristic Curve

The solar cell 'photovoltaic effect' results from putting the device under insolation. The effect on the characteristic curve is to shift the curve downward along the current axis. Figure 2 shows the effect of increasing insolation on the characteristic curve. The result is that the cell generates current and voltage. The current is proportional to the incident sunlight, while the voltage capability is almost constant from very low light levels. The photovoltaic effect only occurs in the fourth quadrant. Here it can be seen that the cell generates both current and voltage.

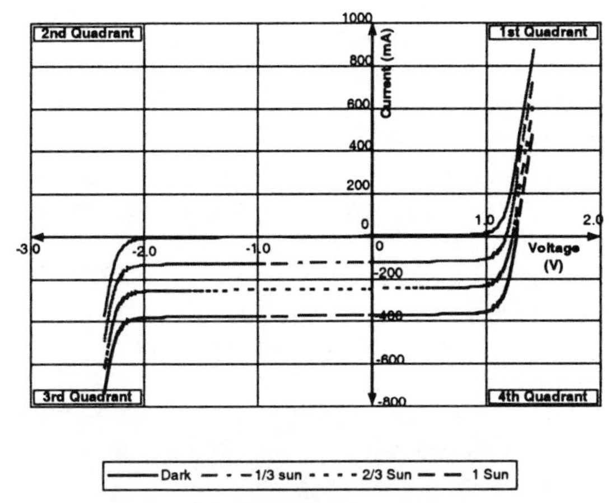

Figure 2
Illumination Effects on Characteristic Curve

When discussing most semiconductor devices, the full characteristic curve is addressed. When discussing solar cells, only the power producing fourth quadrant is typically addressed. To understand solar array behavior under unusual conditions, one must first understand basic solar array operational behavior described by these fourth quadrant characteristics. The fourth quadrant portion of the characteristic curve is called an IV curve in solar cell vernacular. To produce a solar cell IV curve, the fourth quadrant of the characteristic curve is flipped vertically about the voltage axis into the 1st quadrant. This is done so the current direction is positive, rather than negative as it is on the characteristic curve, and is done strictly for convention. Figure 3 shows a typical GaAs/Ge solar cell IV curve.

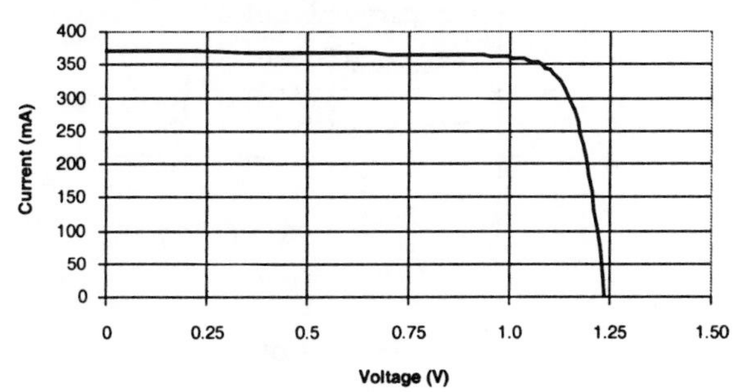

Figure 3
Solar Cell IV Curve

The standard semiconductor characteristic curve will be used when discussing the first and third quadrants. The solar cell IV curve will be used for discussions involving only the fourth quadrant behavior.

There is a fundamental difference in the method used to develop the two curves. A semiconductor is treated as a load when producing a characteristic curve. A varying voltage (magnitude and polarity) is applied to its terminals, and its current magnitude and direction are measured, or a varying current (both magnitude and direction) is applied to its terminals, and its voltage magnitude and direction are measured. The solar cell is treated as a source when producing an IV curve. A varying load is wired between the terminals of an illuminated cell and varied between 0 and infinity ohms, while simultaneously measuring the cell current and voltage. Creation of a solar cell characteristic curve requires the use of both techniques.

The standard way of describing the performance of a solar cell is by its 'three point data'. Three point data refers to the three points on the IV curve, which describe the key performance parameters, the Short Circuit Current, Maximum Power Point and the Open Circuit Voltage. The Short Circuit Current is the amount of current the device can source if its terminals are shorted. The Maximum Power Point is the point at which the device produces maximum power and is near center of the knee of the curve. The Open Circuit Voltage is the voltage the device develops when its terminals are open with no load connected.

To describe a point in an artesian plane, two values are required, so to describe three points, six values are required. A current and a voltage value describe each point. The three point data values are; short circuit current (I_{SC}), max power point voltage (V_{MP}) and current (Imp), and open circuit voltage (V_{OC}). The voltage at I_{SC} and the current at V_{OC} are, by definition, both zero, so they are not typically mentioned but are implied. These values are shown on Figure 4.

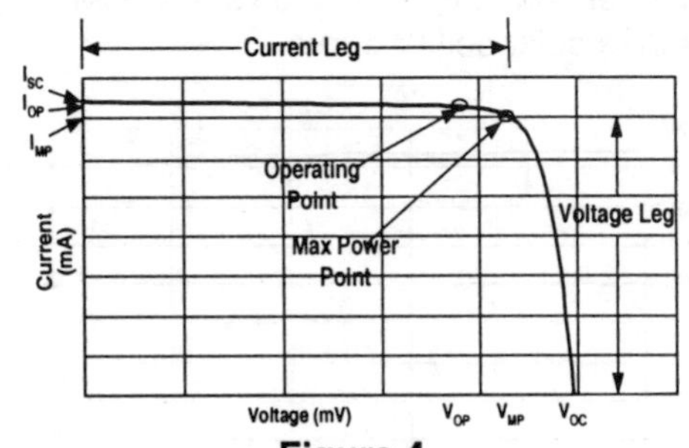

Figure 4
Solar Cell Three Point Data

The Operating Point is also of interest. The current and voltage that describe it, I_{OP} and V_{OP}, are also identified in Figure 4. In order for a string to contribute to the bus, it must produce enough voltage to reach the threshold voltage, V_T. V_T is the sum of the bus voltage, V_{Bus}, and the blocking diode voltage drop, V_{BD}, $V_T = V_{Bus} - V_{BD}$. The operating point voltage, V_{OP}, is typically the same as the threshold voltage. There are cases, however, when the string can not attain the threshold voltage. In these cases the operating point voltage falls below the threshold voltage, $V_{OP} < V_T$. This will be discussed in greater detail later in the paper.

The IV curve is described as having a current leg and a voltage leg. The current leg is the horizontal portion of the curve over which it has almost constant current. The voltage leg is the near vertical portion of the curve over which it has almost constant voltage. These terms will be used throughout the discussion.

Temperature also affects the current and voltage. As the temperature rises, the voltage decreases and the current increases. Inversely as the temperature drops, the voltage increases and the current decreases. In any case, the voltage is much more sensitive to temperature changes than is current. Figure 5 depicts the device response to temperature.

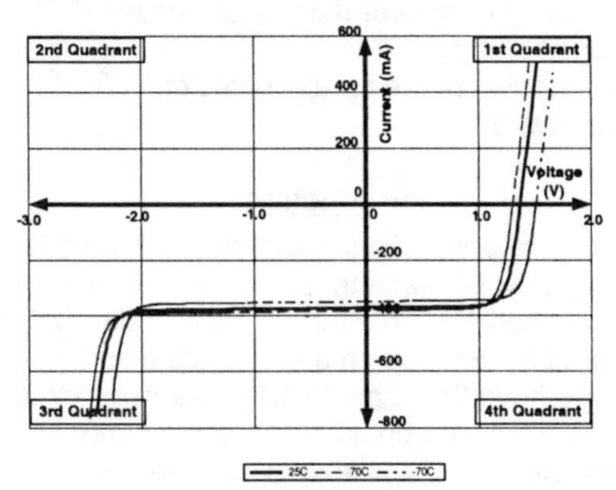

Figure 5
Temperature Effects on Characteristic Curve

Array aging also affects the operation of the system. Degradation parameters, which contribute to array aging, are:

- coverglass darkening due to ultraviolet radiation;
- coverglass adhesive darkening due to ultraviolet radiation;
- contamination of the coverglass which reduces the amount of light entering to the cell;
- micrometeoroid and orbital debris (M&OD) causing hazing of the coverglass which reduces the amount of light incident on the cell;
- micrometeoroid and orbital debris (M&OD) penetrations reducing the active area of the cell;
- increased series resistance due to broken cell interconnects, cracked cells, etc.;
- radiation damage to the junction and;

- plasma interaction, i.e. arcing.

These mechanisms primarily affect the current capability of the string and have much less effect upon the string voltage. The effect of aging is very similar to that of changing insolation.

Device Behavior

Fourth Quadrant Operation

The fourth quadrant is the operating region in which solar cells produce power. To operate in this quadrant, the cells must be illuminated. In this mode, the cells produce a positive voltage, i.e. are forward biased with a negative current flow, using the characteristic curve convention. Sunlight transforms the solar cell from a load into a source, producing a voltage.

Solar cells are near constant current devices while operating on the current leg of the IV Curve. The array operating point voltage is set by the bus voltage regulator. To maintain voltage regulation, more power is made available to the bus than is required by the load. The additional power is then dropped across a variable load controlled by the regulator. Another scheme moves the operating point on the IV curve to match the load, then a regulator steps the output voltage up or down to maintain a constant bus voltage.

First Quadrant Operation

A device operating in the first quadrant has a positive voltage (forward biased) and positive current, using the characteristic curve convention. Devices can operate in this quadrant while shadowed or illuminated. To operate in this quadrant, a parallel external source must be applied to the string, at a voltage greater than the combined string V_{OC}. Such a source could be another string(s) wired in parallel, prior to the blocking diode. A device operating in the 1st quadrant is a load.

Third Quadrant Operation

A device operating in the third quadrant has a negative voltage (reverse biased) and a negative current, using the characteristic curve convention. Devices can operate in this quadrant while shadowed or illuminated. To operate in this quadrant, a reverse voltage must be applied from a parallel external source, such as other cells in the same string. A device operating in the 3rd quadrant is a load.

Quadrant I and III Operating Conditions

Operation in these quadrants can be the result of several different operating conditions, which are summarized below. A detailed examination of these conditions is presented in the following pages.

Short Circuit

The first condition is a short circuit to chassis placed somewhere in the midst of the string. This causes the part of the string below the short to operate in the fourth quadrant and the part above the short to operate in the first or fourth quadrant, depending on the array configuration. In this case, the string has positive voltage but reversed current. Both portions of the string are forward biased. The location of the short circuit, i.e. number of cells shorted, and secondarily, the resistance of the fault will limit the voltage the string can produce.

Shadowed String

The second condition is a partially shadowed string. The shadowed portion of the string operates in the third quadrant with negative voltage (reverse biases) and forward current. The illuminated portion of the string operates in the fourth quadrant with positive voltage (forward biased) and positive current. The number of shadowed cells will limit the voltage the string can produce.

High Series Resistance

The third condition is when high series impedance is introduced into the string. This can be a result of distributed or lumped impedance at any point in the string. This causes the entire string to operate in the fourth quadrant with a positive voltage (forward biased) and a negative current. The magnitude of the series impedance will limit the current the string can produce.

System Impacts from Quadrant I and III Operation

Background

A review of a few more solar array terms is in order at this point, as they will be used throughout the remainder of the discussion. The term 'string' is used here to mean a single circuit of series connected solar cells. The string length is determined by the voltage requirements of the bus. A string must be able to produce at least the threshold voltage, V_T, under EOL (end of life) worst-case conditions. Strings are then paralleled to provide the amount of current required by the bus.

A blocking diode, which is a standard power diode, is placed in series with a solar array string(s). It is electrically positioned between the hot end of the string(s) and the hot rail of the bus. The purpose of these diodes is to block reverse current flow from the batteries into the array during eclipse. Without blocking diodes, the dark array is equivalent to strings of series connected forward biased diodes, all wired in parallel and capable of dissipating a great deal of power from the batteries. When the solar array enters the eclipse, the bus voltage, now supplied by the battery, exceeds the solar cell string output voltage. This reverse biases the 'blocking' diode, thus 'blocking' current flow back into the string. Some solar arrays are designed with one blocking diode for each string, and some are designed with one blocking diode for multiple strings. The consequences of having one blocking diode for multiple strings, versus one diode per string, will be examined during the course of this paper.

String Shorted to Chassis

A single string is shorted to chassis at some point between the first and the Nth series connected cell. The string is effectively broken into two circuits when a short is applied to the circuit, as shown in Figure 6. One circuit is between the return and the chassis, and the other is between the chassis and the bus. The short is depicted as a resistor, to imply that there is some resistance in any short.

Assumptions made for the purposes of this analyses are: 1) the resistance of the short, R_S, is very low compared to the load resistance, R_L, 2) the load voltage, V_L, is equal to the bus voltage, V_B, i.e. there is no voltage drop in the cables. The intent is to convey the concepts. These details just make the explanations cumbersome. The actual resistance of the harness, other components and the short must be taken into account when performing a complete analysis.

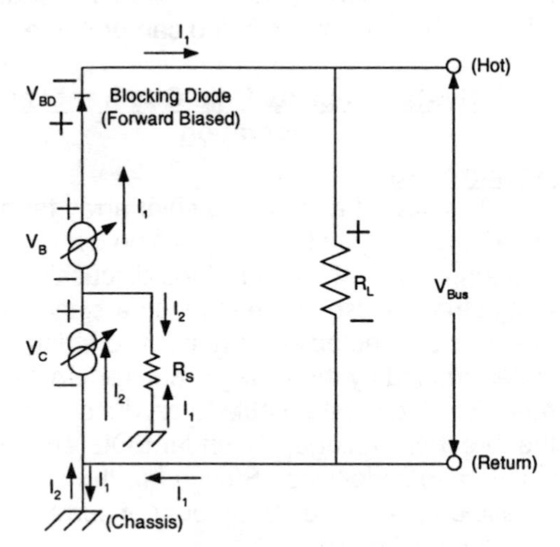

Figure 6
Shorted String
$V_B > V_T$ (4th Quadrant)

The string operates normally supplying current to the bus, in accordance with Curve A in Figure 7, prior to the shorting event. After being shorted, the shorted portion of the circuit operates at I_{SC}. Its current divides between the short and the rest of the string, as shown in Figure 6. The approximate current split is determined by the ratio of the resistance of the short, R_S, to the resistance of the load, R_L. The maximum amount of current the circuit can source to the short is I_{SC}. This remains true so long as the resistance of the short remains much less than the resistance of the load. The current into the short will decrease from I_{SC} toward zero, with increasing short resistance. Likewise the voltage contributed by the shorted portion to the rest of the string will be near zero for a short with a resistance of much less than that of the load resistance. The voltage contribution from the shorted portion of the string will approach its normal voltage contribution of about 1V/cell for GaAs, as the resistance of the short becomes much greater than that of the load.

The operational modes resulting from different string voltages will now be examined. The location of the short determines the voltage the string can produce. The closer the short is to the return, fewer cells shorted, the more voltage the string can produce. The voltage of the portion of the string above the short, V_B, will be greater than the threshold voltage ($V_B > V_T$), if the short is low enough in the string. In this case, V_B will be equal to V_T, and it will continue to provide current to the bus. The amount of current it provides, I_1, is dictated by Curve B in Figure 7. This current level is determined by finding the intersection of the operating voltage, V_T, and Curve B. The shorted portion of the string, Curve C of Figure 7, will operate with $I_2 = I_{SC}$ and $V_C = 0$. Since the short has a very low resistance, the shorted portion of the string cannot generate any voltage.

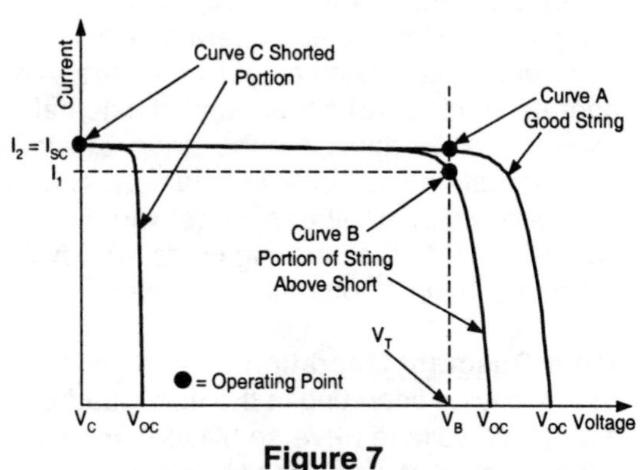

Figure 7
Shorted String IV Curves
$V_B > V_T$ (4th Quadrant)

The voltage produced by the portion of the string above the short, V_B, will be below the bus voltage if the short is higher in the string, i.e. $V_B < V_T$. This causes the blocking diode to become reverse biased, isolating the string from the bus. The circuit is shown in Figure 8. The operating point voltage for the portion of the string above the short, V_B, will go to V_{OC}. Its current will go to zero since the reverse biased blocking diode leaves no path for current flow. Figure 9 Curve B shows its operating point.

The voltage of the shorted portion of the circuit, V_C, will again be zero. Its current, I_2, will still be equal to I_{SC}. Figure 9 Curve C shows the operating point and IV curve for the shorted string. Figure 9 Curve A again depicts the string IV curve and operating point, prior to the shorting event.

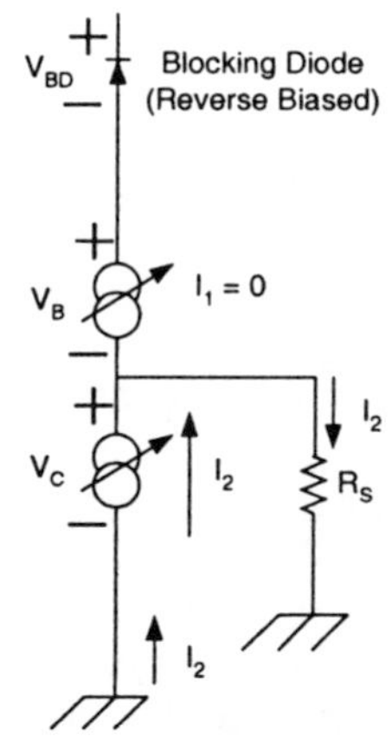

Figure 8
Shorted String
$V_B < V_T$ (4th Quadrant)

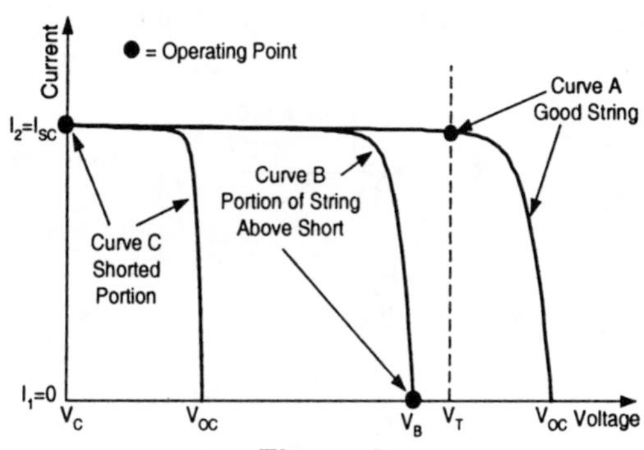

Figure 9
Shorted String IV Curves
$V_B < V_T$ (4th Quadrant)

Multiple Strings per Blocking Diode

Now the more complicated case with multiple parallel strings bused to a single blocking diode will be examined. This circuit is depicted in Figure 10. The number of strings in parallel only affects the amount of current available to the bus and/or the affected string. For convenience, three are used here, but it could be two to eight or even more, depending upon the application. Again, several values of string voltage, resulting from shorting the string in different locations, will be considered. Each case has a short higher in the string. The lowest results in $V_B > V_T$ in the 4th quadrant (Curve B Figure 7), the next lowest results in $V_B > V_T$ in the 1st Quadrant (Curve B Figure 11), and the last results in $V_B < V_T$, in the 1st quadrant (Curve B Figure 14). Again $V_C = 0$, because the short has very little resistance compared to the load. All three cases are discussed below.

When $V_B > V_T$ in the 4th quadrant, under short circuit conditions, paralleled solar array strings, bused to a single blocking diode, react the same as arrays with a single string per blocking diode. Therefore, each parallel string has the characteristics of Curve A in Figure 7. The portion of the string below the short still corresponds to Curve C in Figure 7, with $V_C = 0$ and $I_2 = I_{SC}$. I_2 still flows into the short, bypassing the load. This is true so long as the short has very low resistance compared to the load. The portion of the circuit above the short, Curve B in Figure 7, still operates in the fourth quadrant with $V_B = V_T$. I_1 is again established by locating the intersection of the operating voltage, V_T, and Curve B.

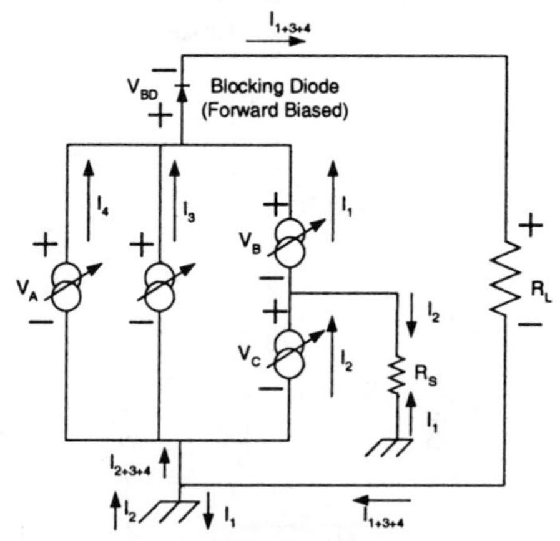

Figure 10
Shorted String with Parallel Strings
$V_B > V_T$ (4th Quadrant)

If $V_B > V_T$ in the 1st quadrant, as shown in Curve B of Figure 11, the portion of the string above the short is operating in the 1st quadrant, with current, I_1, flowing in the reverse direction into the short, as shown in Figure 12. This occurs because the voltage of the parallel strings is greater than the voltage the shorted string can generate. The parallel strings set the operating voltage and reverse the current flow through the shorted portion of the string, forcing it to operate in the 1st quadrant. Its voltage, V_B, still equals V_T, so the operating voltage remains unchanged. The shorted portion of the string still operates at $V_C = 0$ with $I_2 = I_{SC}$, as shown in Curve C of Figure 11.

The paralleled strings continue to operate normally at V_T as shown in Figure 11 Curve A. I_3 and I_4 are summed to produce the current value of Curve A. The blocking diode is still forward biased allowing the string to source current to the bus. The current available to the bus is reduced however, since the current flowing into the shorted string must be subtracted from the amount available from the paralleled strings. The current available to the bus is $I_{Load} = I_3 + I_4 - I_1$. I_1 is again established by locating the intersection of V_T with Curve B.

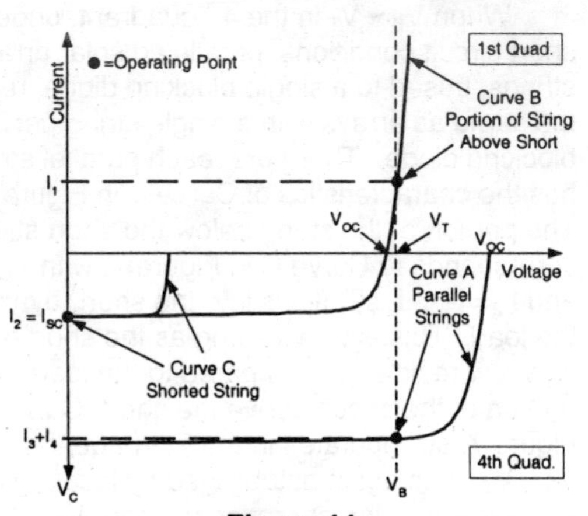

Figure 11
Shorted String Characteristic Curve
$V_B > V_T$ (1st Quadrant)

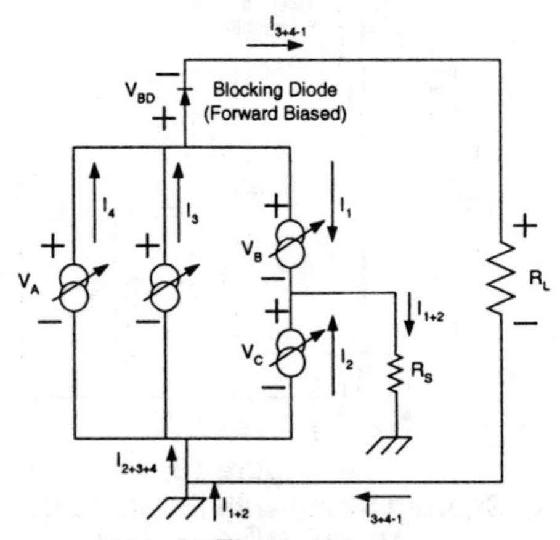

Figure 12
Shorted String with Parallel Strings
$V_B > V_T$ (1st Quadrant)

It should be noted that two major changes have now occurred in the circuit. First, the portion of the string above the short has now become a load not a source. The polarity of the affected string has not changed, but the current direction has. This means that, rather than being a current source, the string has now become a voltage drop. The shorted string went from being a source, operating in the 4th quadrant, to being a load, operating in the 1st quadrant, by reversing the direction of current flow. Second, the paralleled strings have now become a series source to the shorted string, not sources in parallel, as they were.

If $V_B < V_T$ in the 1st quadrant, the blocking diode will be reverse biased and isolate all the strings from the bus. This circuit is depicted in Figure 13. Now the shorted string sets the operating point, not the bus voltage. The operating point voltage must be above the threshold voltage, $V_{OP} > V_T$, for the strings to act as a source to the bus. Examination of Curve B in Figure 14 reveals that the shorted string can conduct the full current supplied by the parallel strings in reverse and still not drop enough voltage to reach V_T. Therefore it will set the operating point voltage.

Since these strings can no longer overcome the threshold voltage, the blocking diode becomes reverse biased, isolating them from the bus. Once again the strings are no longer in parallel, as shown in Figure 10, but are in series with the shorted string as shown in Figures 12 and 13. The result is that the healthy strings continue to act as paralleled sources, but to the shorted string, not the bus.

The portion of the circuit above the short will again attempt to go to Voc. However the paralleled string(s) will force it into reverse current flow since the parallel string(s) have greater voltage capability than their shorted partner. This current reversal will force the portion of the string above the short to operate in the first quadrant. This portion of the string is still forward biased but the current direction is reversed, sinking I_3 and I_4, again transforming it into a load, as can be seen by examining the operating point of the shorted string on Curve B of Figure 14.

The parallel strings continue to operate as current sources in the 4th quadrant.

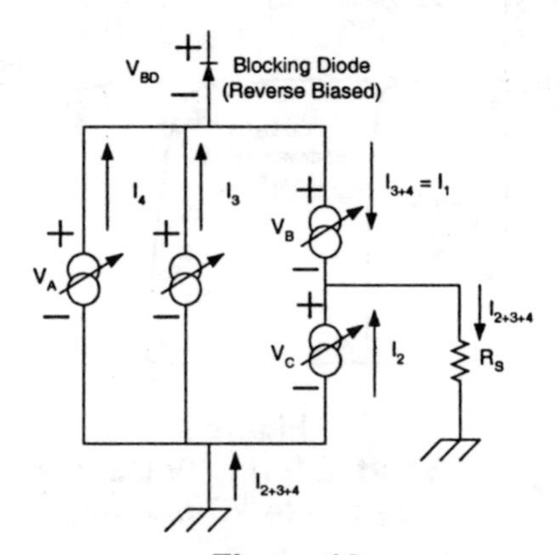

Figure 13
Shorted String with Parallel Strings
$V_B < V_T$ (1st Quadrant)

Examination of the circuit in Figure 13 reveals that the current through the unshorted portion of the shorted string must be equal to the combined current from the paralleled strings or $I_1 = I_3 + I_4$. Also we can see that if $V_C = 0V$, that V_A must be equal to V_B. Given these constraints, the operating point is determined by finding the voltage at which the currents for the parallel strings and the unshorted portion of the shorted string are equal but opposite on Figure 14, $I_1 = I_3 + I_4$.

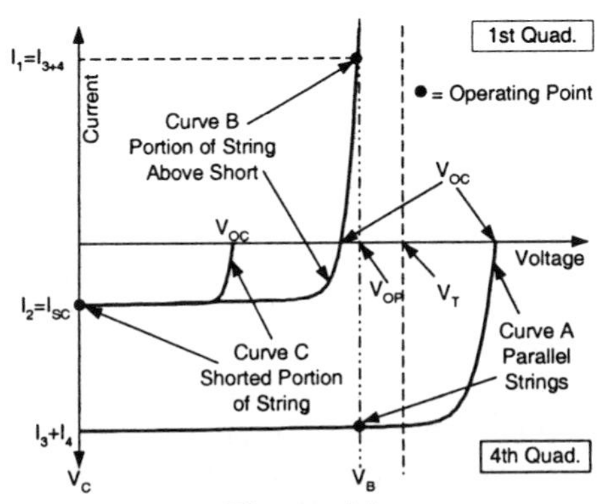

Figure 14
Characteristic Curve
Shorted String with Parallel Strings
$V_B < V_T$ (1st Quadrant)

Hot Side Short

A circuit with a string shorted to its hot side is shown in Figure 15.

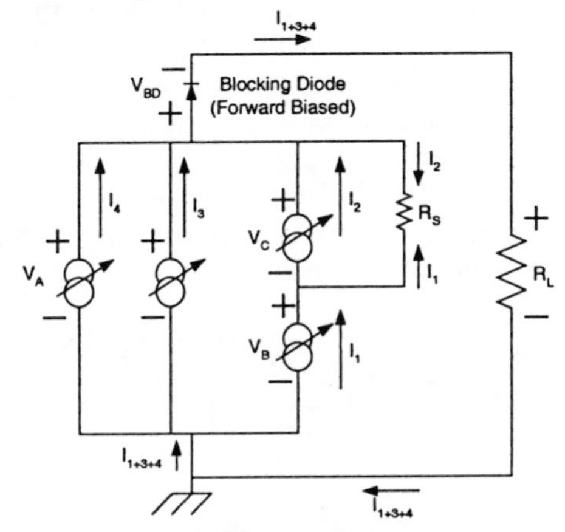

Figure 15
Circuit with Hot Side Short

This short has the same net effect on the circuit as the chassis short; however, the specifics are somewhat different. In this case, the current from the shorted portion of the circuit circulates in a very localized area of the solar panel, whereas, with the chassis fault the current must flow to the single point ground to reenter the circuit, which could be many feet away. Another difference is that the voltage from the portion of the string below the short, V_B, now determines whether the string is on or off the bus. It is the upper portion of the string, on a chassis fault, which make this determination. Whereas with the chassis fault, the closer the short was to the return, the more it contributed to the bus. With the hot side fault, the closer the short is to the hot side, the more it contributes to the bus.

The various circuit segments act the same as their chassis short counterparts. The shorted portion has a circulating current of I_{SC}, assuming a very small R_S. The non-shorted portion of the shorted string supplies current to the bus if $V_B > V_T$ in the 4th quadrant, as shown in Figure 7. It becomes a load on the parallel strings if $V_B > V_T$ in the 1st quadrant, as shown in Figure 11. It completely stops supplying current to the bus if $V_B < V_T$, as shown in Figure 14.

One other difference could be the ability to diagnose this fault. It is unlikely that a fault current sensor would detect it, as they are typically in-line between the single point ground and power return line. It would likely exhibit itself only by a low solar array current, for a given number of actively operating strings.

Return Side Short

This fault exhibits the same net effect as a hot side fault. There are subtle differences in how the return currents flow, as shown in Figure 16. Again, as with the hot side fault, the fault current would be localized to the solar cell panel so it would be hard to diagnose, and the IV curves and operating points would again be the same as shown in Figures 7, 11, and 14.

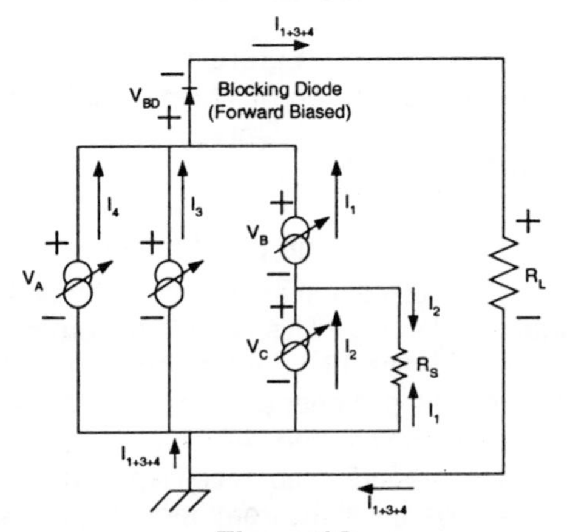

Figure 16
Circuit with Return Short

Effect of Short Location on the String

The effect of varying the location of a short along the length of a string has a similar effect to that of varying the resistance of the short itself. Moving the location of the short upward from the 1st cell reduces the string voltage and moves the voltage leg of the IV curve to the left along the voltage axis. V_T remains constant, but the operating point moves closer to the voltage leg of the curve as more cells are shorted.

Figure 17 depicts characteristic curves for a string with shorts occurring at two cell intervals between 2 and 16 cells. It assumes a 50 cell string at BOL and on-orbit temperatures. The curve keeps the same shape, but loses voltage capability with each shorted cell. The movement has little effect while the string is operating on the current leg of the curve. However, after the operating point moves past the knee of the curve, the string begins to operate on the voltage leg. Here, even small changes in voltage capability result in large changes in current output.

As can be seen from the Figure, the string only operates on the current leg with shorts up to about the fourth cell. It operates on the voltage leg for a short placed anywhere beyond cell 5. To produce enough voltage to operate at V_T, there must be enough parallel strings bused to a common blocking diode to exceed the required operating point current.

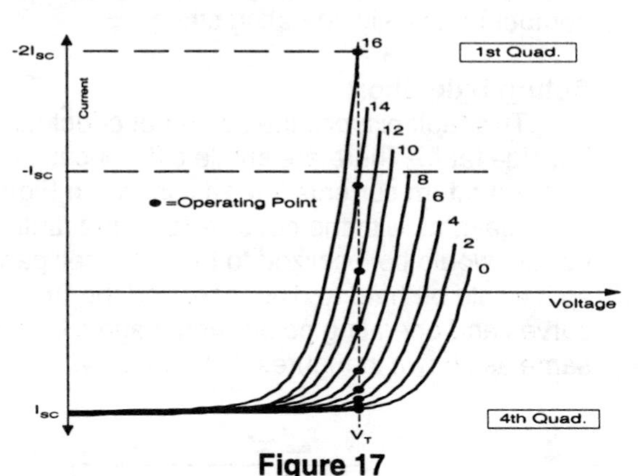

Figure 17
Effect of Location of Short on the String IV Curve

String Shadowing

Now let's examine the case of a partially shadowed string. This case has more impact on a per cell basis than does a shorted string. When a string is shorted, the string looses the voltage of the shorted cells, about 0.5V to 2.5V per cell, depending upon cell type. When a cell is shadowed, it shifts from operating in the fourth to the third quadrant, if the string has enough voltage to continue to force nominal current flow. The cell becomes reversed biased, dropping between 2 and 65V, at operating current levels, depending upon cell type. 65V is the high side for a Si cell. It must be understood that if a Si cell is operated at nominal forward current while reverse biased, the junction will be shunted, so practically speaking, bypass diodes must be used every 20 cells or so, if the operating voltage exceeds 20V. The result is an 11V drop (20 cells at 0.5V each plus a diode drop of 1V). Some GaAs cells are screened to operate a nominal current while shadowed. They drop 2 to 5V. Multijunction cells have integral bypass diodes that drop about 1V per cell. So, with a partially shadowed string, the string not only loses the voltage contribution of the shadowed cells, but also their reverse bias voltage of 1 to 11V. Therefore, a shadowed GaAs cell, which was producing ~1V, is transformed into a load dropping between 1 and 5V. This results in a 2 to 6V loss in string voltage for every shadowed cell. Shorted GaAs cells only cause approximately a ~1V loss to the string voltage for each cell below the short location.

When dealing with shorted cells, we need only concern ourselves with the voltage and IV characteristics of the unshorted portion of the shorted string. The voltage of the shorted portion, V_C, neatly goes to zero with a good hard short, so only the characteristics of the unshorted portion come into play. With a shadowed string, V_C doesn't go to zero, it becomes negative when operated at nominal currents. Therefore, with a shadowed string the sum of these two voltages and the resulting composite characteristic curve must be dealt with. Again, as was the case with a short, the unaffected portion of the string provides all the voltage. It must operate at a voltage in excess of $V_T + V_C$ to overcome not only the threshold voltage, but also the voltage dropped by the shadowed cells.

Figure 18 shows a partially shadowed, single string circuit. Writing the loop equations for the circuit we have $V_B - V_{BD} - V_{Bus} - V_C = 0$. Reordering the equation we have, $V_B = V_C + V_{Bus} + V_{BD}$. Since we have already stated that $V_T = V_{Bus} + V_{BD}$, we can say that $V_B = V_C + V_T$. Therefore, $V_B - V_C$ must be greater than V_T when operating at current I_1, if the string is to make a contribution to the bus.

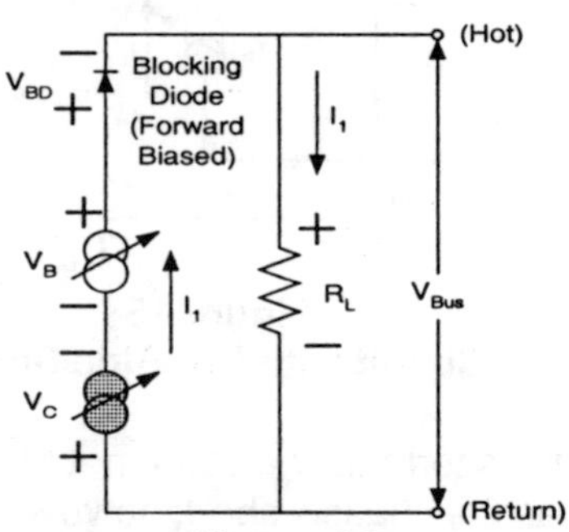

Figure 18
Partially Shadowed String
$\mathbf{V_B - V_C > V_T}$

When $V_B - V_C > V_T$, the string will operate at V_T and I_1. The string operating point is derived by finding a common current, I_1, for which $V_B - V_C = V_T$. Figure 19 depicts three curves. Curve A depicts the characteristic curve of the string prior to being shadowed. It operates in the 4th quadrant at V_T and produces I_{OP}. Curve B again shows the characteristic curve, but for the illuminated portion of the shadowed string. It operates in the 4th quadrant at V_B and produces I_1. Curve C shows the characteristic curve for the shadowed portion of the string. It operates in the 3rd quadrant at V_C and passes I_1. It is a load.

In order to operate at V_T, the illuminated portion of the string must be able to make up the voltage dropped by the shadowed portion of the string, $V_B - V_C > V_T$. This means that the operating point of the illuminated portion, Curve

B of Figure 19, must exceed V_T by at least the amount of voltage dropped by the shadowed portion of the string, V_C, Curve C on Figure 19, i.e. $V_B - V_C = V_T$.

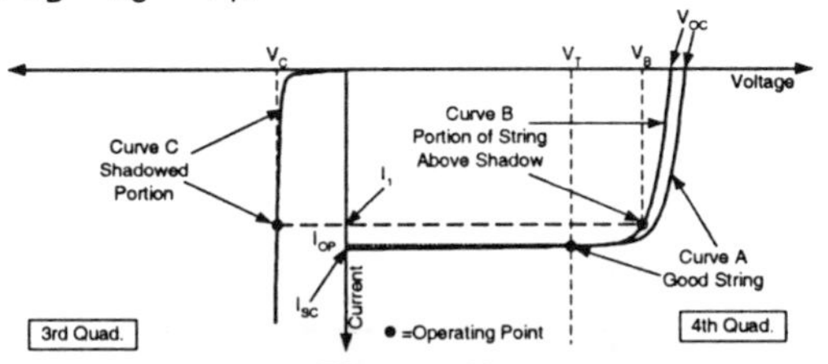

Figure 19
Characteristic Curve
Partially Shadowed String
$V_B - V_C = V_T$

If the string does not have enough voltage to overcome the shadowed cell(s), and exceeds the threshold voltage, V_T, $V_B - V_C < V_T$, the blocking diode will become reverse biased. The circuit is depicted in Figure 20.

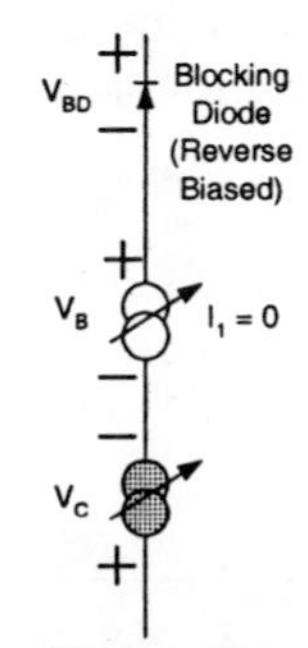

Figure 20
Partial Shadow/Single String
$V_B - V_C < V_T$

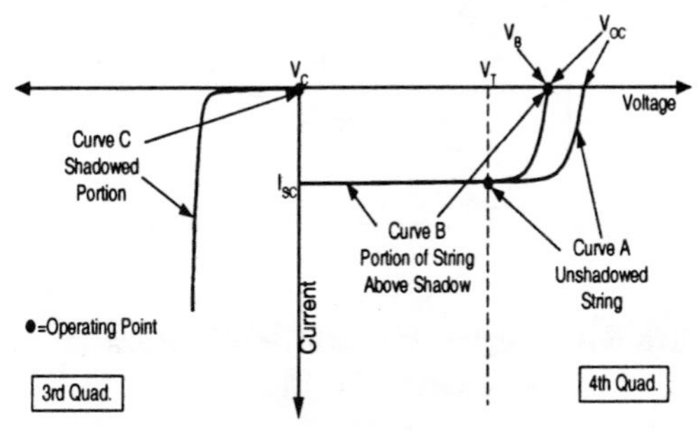

Figure 21
Characteristic Curve
Partially Shadowed String
$V_B - V_C < V_T$

The illuminated portion of the string voltage will go to V_{OC} and I_1 will go to zero, as there can be no current flow. The shadowed portion of the string will go to 0V and 0A. There is no insolation to generate power in these cells and no current from the illuminated portion to generate a voltage drop. Figure 21 depicts the characteristic curves and operating points.

Multiple Strings per Blocking Diode

The shadowed string will act the same with paralleled string(s), so long as the net string voltage, V_B-V_C, exceeds the threshold voltage, V_T. The bus voltage will establish the operating point voltage. See Figure 22 for the circuit diagram. The shadowed cells will still operate in the 3rd quadrant and drop the string voltage by V_C. The illuminated cells will continue to operate in the 4th quadrant and provide power to the bus at a decreased current level as shown in Figure 23. The illuminated portion of the shadowed string will operate at V_B. The shadowed portion will operate at V_C. The operating current is determined by finding a common current level, I_1, at which $V_B - V_C = V_T$, again as shown in Figure 23.

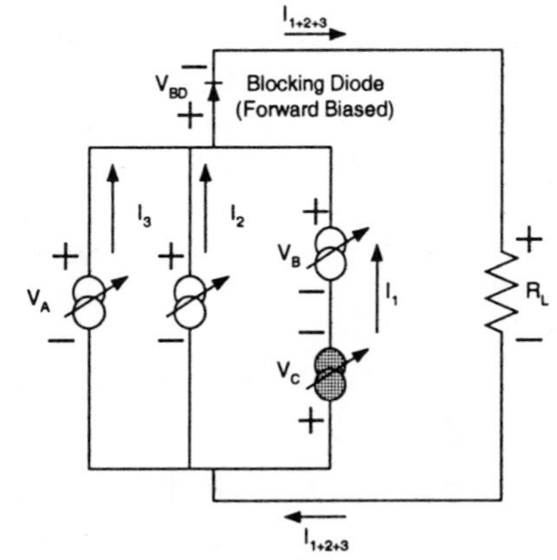

Figure 22
Partial Shadow/Parallel Strings
$V_B - V_C > V_T$

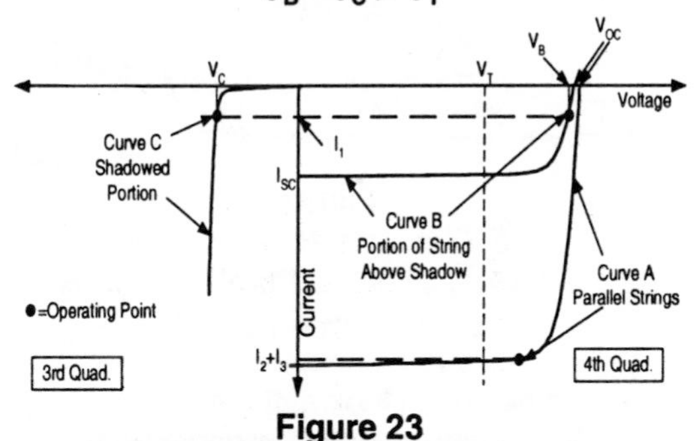

Figure 23
Characteristic Curves
Partial Shadow/Parallel Strings
$V_B - V_C = V_T$

As the number of shadowed cells increases, the affected string voltage quickly decreases, again 2 to 11V per cell at nominal current levels, as previously explained. Once the affected string voltage, $V_B - V_C$, falls below V_T, the voltage of the parallel strings will hold up the voltage of the shadowed string. There will be some reverse current flow through the affected string, forcing both portions of the shadowed string to operate as loads, as shown in Figure 24. The current reversal means that the affected string no longer operates in the 4th quadrant as a source, but is forced into the 1st quadrant as a load.

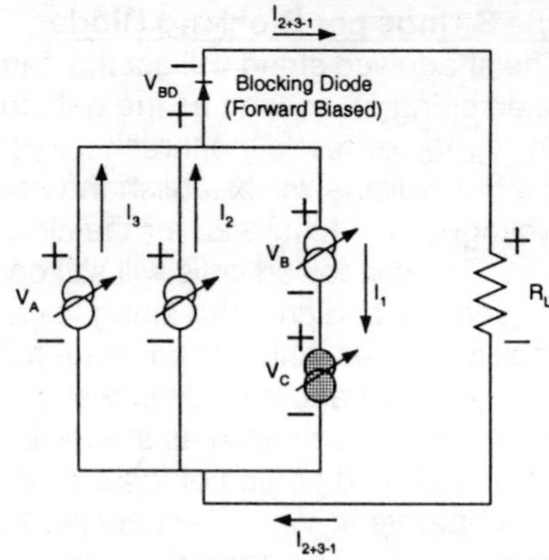

Figure 24
Partial Shadow/Parallel Strings
$V_B + V_C > V_T$ and $I_3 + I_2 > I_1$

To understand the circuit, it is best to focus on the current behavior. The parallel string combination $I_3 + I_2$ must be greater than I_1 for there to be current flow to the load. I_1 is established by finding a common operating current at which $V_B + V_C = V_T$, as shown in Figure 25.

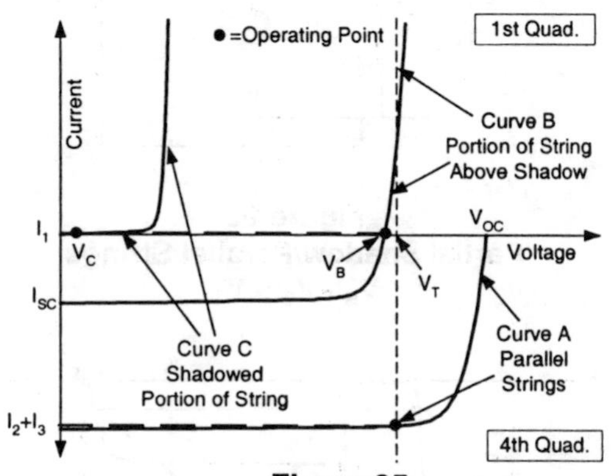

Figure 25
Characteristic Curves
Partial Shadow/Parallel Strings
$V_B + V_C = V_T$ and $I_3 + I_2 > I_1$

It is helpful to look at the case of a fully shadowed string to better understand the behavior of shadowed cells. The IV curve shifts straight up into the 1st quadrant when a string is fully shadowed, as shown in Figure 26. A review of this figure reveals that at V_T there is very little reverse current flow through the string. This is logical as solar cells are basically constant current devices in the region of V_T. Therefore, the shadowed string would be expected to conduct about the same amount of current as it does at 0V, which is about 0A, and that is exactly what it does.

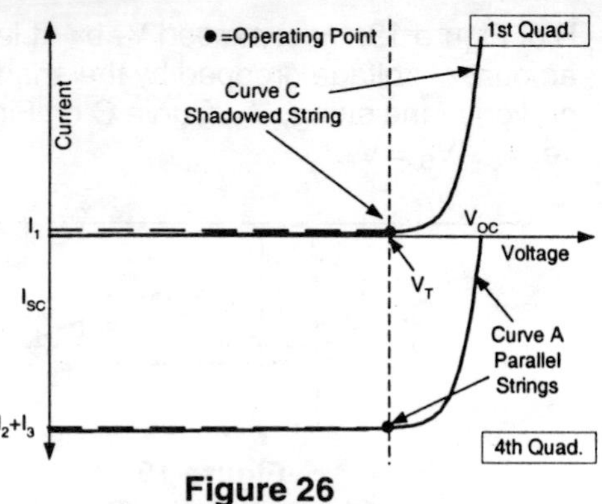

Figure 26
Characteristic Curves
Full Shadow/Parallel Strings
$V_C = V_T$ and $I_3 + I_2 > I_1$

The net result of shadowing one of a set of paralleled strings, is that some current will be lost from the shadowed string if string voltage is high enough to force nominal current through it. There will be no current to a slight negative current from the shadowed string if there is not enough voltage to force nominal current flow.

Series Impedance

Series impedance can be introduced into a string by various mechanisms. Figure 27 shows the resulting circuit.

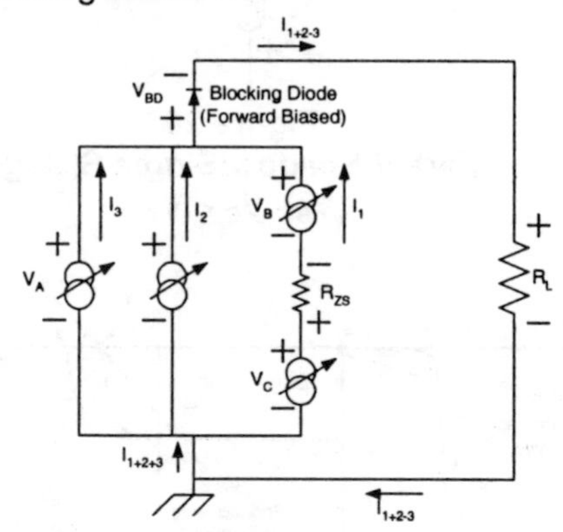

Figure 27
String with Series Impedance in Parallel with Good Strings

One cause of high series impedance is the de-bonding of cell interconnects due to thermal cycling. This increases the series impedance since the current path now has less conductor area to carry the current. Another cause can be hairline cracks in the cells. This can be caused by handling or inadequate cell to substrate bond lines, or induced by thermal cycling and/or M&OD impacts. These can be lumped or distributed impedance. The effect is the same.

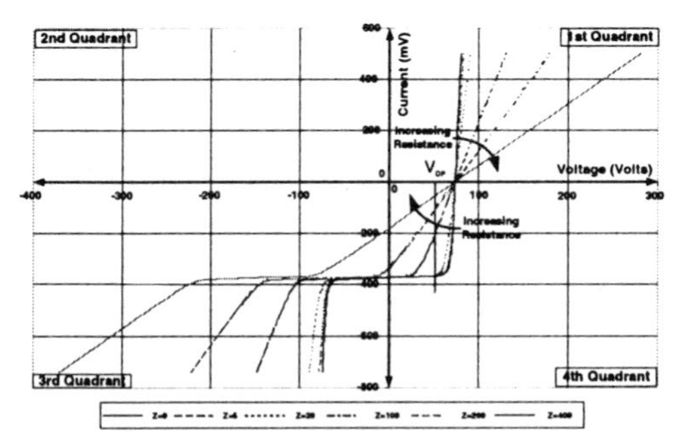

Figure 28
Characteristics of a String with Increasing Series Impedance

Figure 28 shows the characteristic curves for a string with varying amounts of induced series impedance. It clearly shows that for increased series impedance, the string produces less and less current at the operating voltage. This means that as the series resistance increases, the affected string contributes less and less current to the bus. In cases of high series impedance, the current contribution from the affected string simply goes to zero.

Paralleled strings are unaffected by the string with the high series impedance. This string does not become reverse biased or reverse its current flow to become a load as with the previously discussed anomalies. It just stops contributing to the bus.

It should be noted though, that the V_{OC} for the affected string remains unchanged. This is a key factor in how the string responds. Since V_{OC} remains unchanged, the string always has enough voltage to meet V_T. It just produces less and less current as the impedance increases. The reason the string V_{OC} remains unchanged lies in fundamental circuit analysis. There is, by definition, no current flow at V_{OC}, so the series resistance develops no voltage drop. There must be current flow through a resistance to have a voltage drop. No current, no voltage drop. Therefore V_{OC} does not change regardless of the value of R_S.

Design Implications

The failure mechanisms that have been discussed give rise to the question, "What should be done during the design phase to ensure these failure mechanisms don't cause problems later?" The following is a list of suggestions, which if implemented, will resolve these problems.

1. Ensure that there is adequate insulation between the solar cell and the substrate and that wiring has adequate strain relief and is protected from abrasion. This will reduce the probability of a shorting event.
2. Provide one blocking diode for each string. This ensures that if there is a short, only one string is affected.
3. Use a soft ground. Reference the power return to chassis through a high resistance, ~20K ohms. It should be sized to limit fault current to milliamps when full bus voltage is applied to it. This ensures that even in the event of a fault that there will be minimal power loss. It protects against all hot to chassis faults, not just solar array faults.
4. Isolate the solar array substrate from the chassis ground by a large resistance, again ~10 to 20K ohms. This solution only protects against array shorts between the hot side and chassis, while the soft ground also protects against bus faults as well.
5. Utilize a direction sensing current sensor between the chassis and the power return to monitor fault current if a short occurs.
6. Utilize current sensors on each solar array wing such that any fault locations can be isolated. Being able to identify the fault source can provide valuable data to designers so they can be designed out of the next generation of vehicles.

This is by no means a comprehensive list of solutions to solar array problems. It does not address problems such as plasma interaction, arcing or ways to prevent shorts. It also goes without saying that the changes recommended above cannot be blindly applied without careful consideration of the impacts these design solutions may have on the overall system. The suggestions do however give some ideas of design solutions to mitigate the impacts of array shorts.

INNOVATIVE FLEXIBLE LIGHTWEIGHT THIN-FILM POWER GENERATION AND STORAGE FOR SPACE APPLICATIONS[1]

Cary Clark and Jeff Summers, ITN Energy Systems, Inc., and Joseph Armstrong, Global Solar Energy, LLC, 12401 West 49th Avenue, Wheat Ridge, CO 80033

ABSTRACT

Tomorrow's spacecraft are taking advantage of miniaturized circuitry, sensors, and actuators, ***leaving the power subsystem a major portion of the spacecraft's weight, volume, and cost.*** Typically, spacecraft weight and volume reductions are paramount to improving spacecraft performance, and in some cases, a critical enabling step to the existence of a given mission. Recently, weight, volume and cost have driven challenging microsatellite designs for military, government and commercial applications equally.

However, spacecraft power systems are not easily conducive to weight, volume or cost reduction measures. Innovations in the power generation, regulation, and storage aspect of spacecraft have taken an evolutionary path with very little dramatic changes. Most solar cell and array technologies bear significant resemblance to those components flown in the early days of space exploration. Rigid, discrete crystalline solar cells are interconnected in strings to form a circuit to generate suitable voltage and current for a given mission. Because these cells are sensitive to the space environment, coverglass protection is bonded to each cell to form a protective barrier from protons, electrons, and optical (UV and IR) issues. Due to their fragile nature, these cells are bonded to a rigid substrate with electrical harnessing sufficiently robust to handle the required current from the array elements to a spacecraft bus. Thus, spacecraft designers must deal with the ensuing increase in harness weight for high-voltage arrays that reduce current demands, but have serious space environmental interaction issues. All harnessing from the arrays leads to the power storage and conditioning components within the spacecraft bus that, despite miniaturization of circuitry, consumes much of the spacecraft's volume and weight. Many of today's spacecraft batteries are not conducive to temperature extremes, and as such their temperature must be regulated. Onboard circuitry used to manage the charge state of the batteries and the power output from the solar arrays has become smaller, but the thermal management of these circuits has become a major problem with the only means for thermal control in space being radiative in nature. ***Thus, it appears that our industry is up against a major power systems barrier for future spacecraft unless a significant change in approach is taken.***

We at ITN Energy Systems and Global Solar Energy (GSE), LLC, are developing innovative power solutions for future spacecraft. Our flexible copper-indium-gallium-selenium (CIGS) photovoltaic material shows significant promise towards volume and weight reduction, using innovative stowage and deployment technologies, and has tremendous promise for cost reduction when teamed with monolithic integration to reduce touch labor (Fig. 1). Rollout or foldout array designs using flexible CIGS lend themselves to very small stowage volume, which are directly applicable to the trend towards microsatellite technologies. Specific power of the CIGS PV blanket can be over 1000 W/kg and with array specific power (including structure) 100-200 W/kg. Further development with multi-junction CIGS devices can provide array specific power of 200-300 W/kg. Other developments at ITN and GSE include an extremely long-lived solid-state flexible thin-film battery with less sensitivity to temperature that could be integrated with the solar array for localized power generation and storage. These batteries have demonstrated over 40,000 charge cycles to 100% depth of discharge, and have a virtually flat discharge behavior that holds great attraction to system designers (Fig. 2). Utilizing this concept, we have been developing a flexible integrated power pack (FIPP) approach that combines power generation, management, and storage onto a single component for space applications (Fig. 3).

[1] This work was sponsored by NASA under SBIR Phase II contract # NAS9-98059

Figure 1 - Photograph of Flexible CIGS Photovoltaics

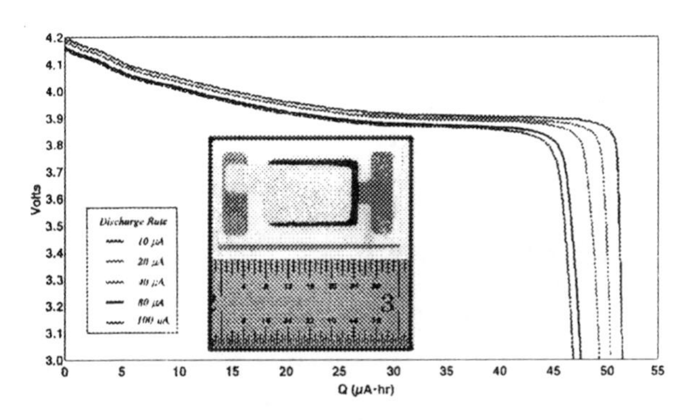

Figure 2 - Discharge Characteristics of a Flexible Solid-State Thin-Film Battery with Photo of It

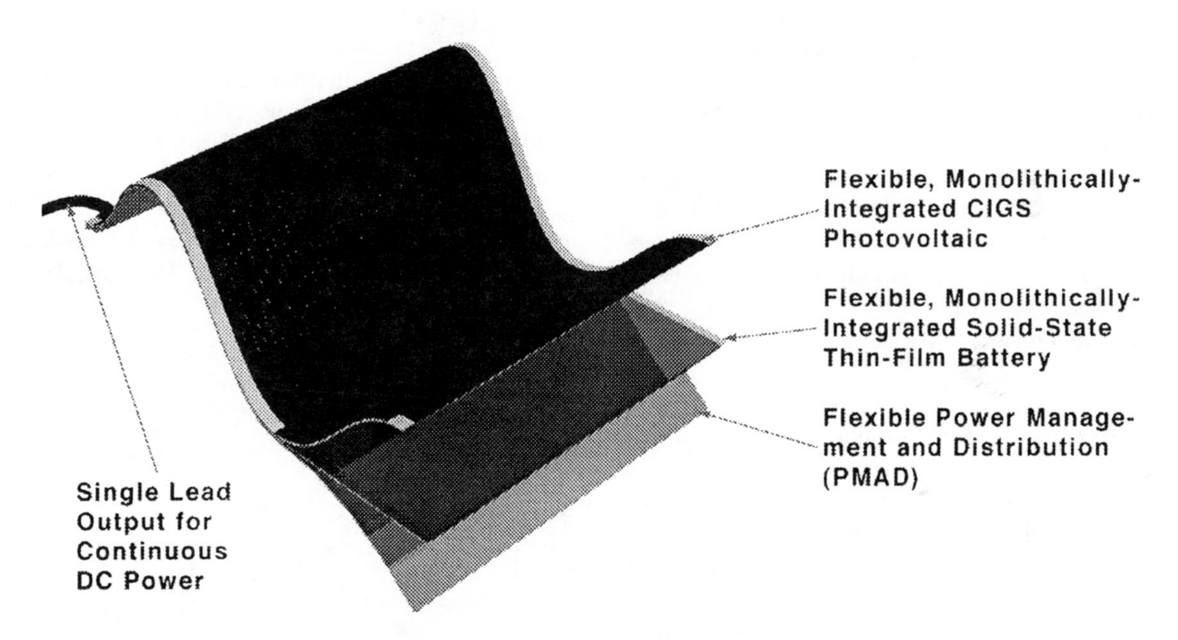

Figure 3 - FIPP under Development through Funding from NASA JSC

In this paper, we shall describe concepts utilizing flexible power generation, management, and storage for future space missions, and the state of development of these flexible technologies.

ITN'S FLEXIBLE POWER GENERATION

ITN is a founder and part owner of Global Solar Energy, LLC (GSE), a company formed specifically to manufacture flexible polycrystalline thin-film photovoltaic (PV). GSE is leading a DARPA-sponsored consortium in the development of this new photovoltaic material, comprising thin-film copper-indium-gallium-diselenide (CIGS) deposited onto a 0.05 mm-thick polyimide substrate. As part of this process, the CIGS stack is laser scribed laterally (across the web) to define individual cells, which are then connected in series by subsequent deposition of the transparent top contact. This "monolithic integration" provides a damage tolerant, highly redundant cell interconnection, which eliminates single point failures within the CIGS module. Figure 4 shows our CIGS module and a cross-section of our device for space applications.

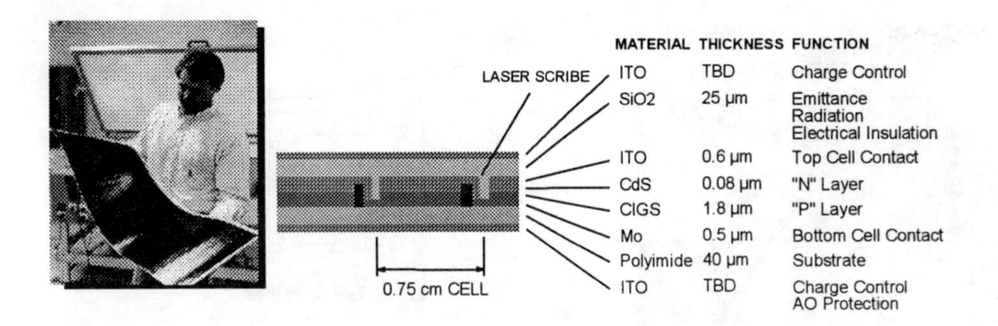

Figure 4 - Thin-Film CIGS Photovoltaics is a Multi-layered Stack of Conductors and Semiconductors with Laser-Scribed Cell Boundaries. GSE Manufactures CIGS by Roll-to-Roll Vacuum Deposition onto a Polyimide or Stainless Steel Substrate, Resulting in Flexible, Lightweight Photovoltaic Modules.

An ongoing parallel effort at GSE is the development of CIGS on a flexible 0.025-mm stainless steel substrate, which provides higher efficiencies than the polyimide substrate, due to higher processing temperatures. However, the technology for monolithic integration of these cells has not been developed to this point, requiring traditional cell-to-cell interconnections using bus bars and current collectors. The interconnects and the denser substrate result in an areal density increase of 1.6 times over the polyimide cells. These disadvantages are overcome by the higher efficiency, and therefore provide a very viable candidate for use in space.

Key features of CIGS thin-film photovoltaics include:

- Current performance: 10 percent module efficiency on flexible substrate translating to a specific power of almost 1000 W/kg for a space solar array blanket
- Unlike traditional Si and GaAs photovoltaics, thin-film CIGS is very flexible with handling characteristics similar to that of a simple 0.05 mm film of polyimide, and 0.025 mm stainless steel; it can be rolled onto a 25.4 mm diameter spool (or less), and even given a hard crease or through-puncture with little damage
- "Monolithic integration" replaces traditional cell-to-cell series interconnects, drastically reducing touch-labor costs and providing highly redundant, full-width interconnects
- Early testing by JPL has shown thin-film CIGS to be very radiation tolerant

We have demonstrated large area roll-to-roll processing of each operation in the baseline approach and have achieved small-area CIGS efficiencies of over 9.1%. On stationary substrates, 11.6% device efficiencies have been measured.

Recognizing its potential for spacecraft, the Air Force Research Lab in Albuquerque, NM recently performed AM0-spectrum testing on our thin-film CIGS. This sample, which had an AM1.5 efficiency of 10.5 percent, delivered an AM0 efficiency of 9.7 percent (Figure 5). This result is very encouraging and provides a clear path to meeting our 9.0 percent AM0 module efficiency goal.

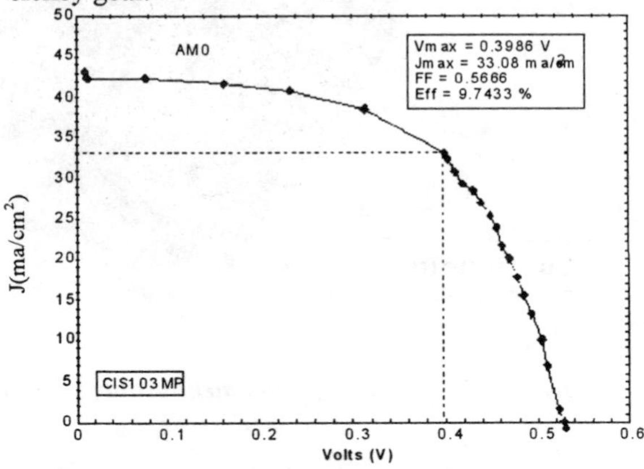

Figure 5 - AM0 Spectrum I-V Curve for GSE's Flexible, Thin-Film CIGS

Thin-Film Cigs PV Enhancements

Under ITN's High-Voltage Array SBIR through NASA Glenn, a methodology was demonstrated to connect modules in both series and parallel, as well as the ability to mount flexible module connections to draw out the power at very high voltages (up to 300vdc). GSE is developing an advanced two terminal, monolithic, tandem (multi-junction) PV cells with 15-20% anticipated AMO power conversion efficiency. Top and bottom cells of the tandem structure will be based on copper chalcogenide alloys (CIS-alloys) and will be fabricated on lightweight flexible substrates. ITN will further develop manufacturing technology to interconnect the tandem cells into modules suitable for use in space-based solar blankets yielding specific power of nearly 1500 W/kg. ***It is clear that CIGS represents a potentially revolutionary space photovoltaic material and, by virtue of its flexibility and monolithic integration, it is ideally suited for integration into spacecraft applications.***

ITN'S FLEXIBLE POWER STORAGE

"State-of-the-art" in spacecraft battery chemistries is NiH_2 which presents several technical issues, such as being heavy, occupying a large area, and very temperature sensitive. Many future satellite systems, including large constellations, are considering lithium-based batteries to avoid these issues and to capitalize on their high power density, lightweight, and the potential of leveraging this technology towards low-cost commercial products. Therefore, development of a flexible, solid state, lithium thin-film battery (TFB) has significant promise for space applications.

ITN Energy Systems is developing a ***solid-state thin-film battery*** (TFB) which has significant promise for space applications, offering high power density, light weight, and long life. This battery consists of a multi-layer stack of $LiCoO_5$ cathode, LIPON electrolyte, and Li-metal anode. Originally developed at Oak Ridge National Laboratory (ORNL) to provide backup power for microelectronics (Figure 6), this design has demonstrated high performance in small areas on ceramic substrates. ITN Energy Systems, with ORNL transitioned this technology to large-area, flexible substrates using a roll-to-roll vacuum processing technique (Figure 7).

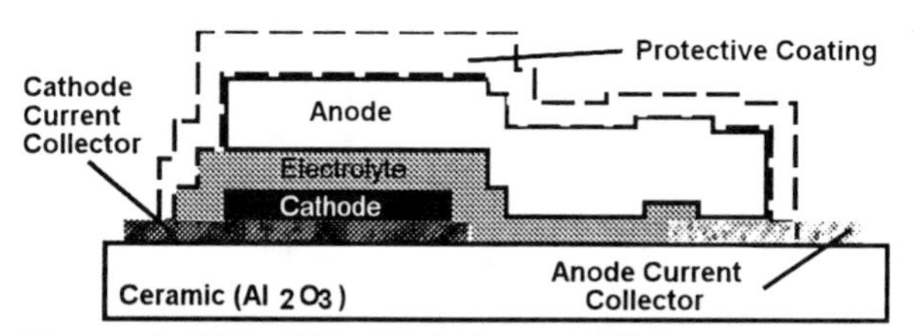

Figure 6 - Thin-Film Multi-layer Lithium Battery Design

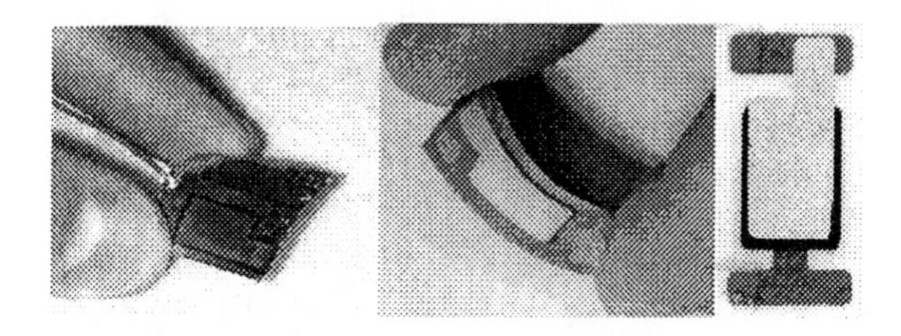

Figure 7 – TFB Deposited Onto Titanium Foil, Stainless Foil and Alumina

Key features of this battery include:

- Stable, 100 percent solid-state device
- High energy densities: up to 450 Watt-hour/liter, 200 Watt-hour/kg
- High cycle life: up to 10,000 cycles at high depth-of-discharge
- Can operate at temperatures up to 100°C
- High voltage: 4.0 V, nominal

The lithium technologies are seen to have the highest energy densities, but most lithium batteries have limited cycle-life. A key distinguishing feature of the solid state lithium battery is its very high cycle-life. Table 1 shows a comparison of the performance of thin-film lithium chemistry to other chemistries.

Table 1 Comparison of Various State-of-the-Art Battery Chemistries

Battery System	Anode	Cathode	Voltage Nominal, V	Energy Density Wh/kg	Energy Density Wh/l	Cycle Life
Baseline Li TFB	***Li***	***LiCoO2***	***4.0***	***200****	***450***	***>40,000***
Lead-acid	Pb	PbO2	2	35	70	200-250
Nickel-Cadmium	Cd	Ni oxide	1.2	50	80	400-500
Nickel-M-Hydride	(MH)	Ni oxide	1.2	50	175	400-500
Lithium ion	C	LixCoO2	3.6	150	240	>1000
Lithium-organic	Li	MnO2	3	120	265	500
Lithium-Polymer	Li	V6O13	3	200	350	500

Nickel-hydrogen	H2	Ni oxide	1.2	55	60	>10,000
High-Temperature	Na	S	2	150	170	>10000

* For a 0.025 mm Titanium Foil

By cycling between a full-charge of 4.2 V and the 3.0 V cutoff, defining 100 percent depth-of-discharge (DOD) in this device, the TFB is seen to deliver more than 40,000 cycles with less than 10 percent capacity degradation. Current capacities for a single layer, thin-film battery cell are approximately 0.15-0.20 milliAmp-hours/cm^2 (3μm cathode), yielding a specific energy of approximately 200 W-hr/kg. Deposition processes and improvements in chemistry are currently being developed to increase the capacity to 0.60-0.80 milliAmp-hours/cm^2.

A typical discharge curve for the thin-film lithium battery at room temperature and 80°C is shown in Figure 8. Similar characterization at low temperatures has not yet been completed, however, preliminary cycling data between –50 and 75°C showed no degradation in room temperature performance. This tolerance to large variations in temperature could virtually eliminate battery-driven thermal control requirements.

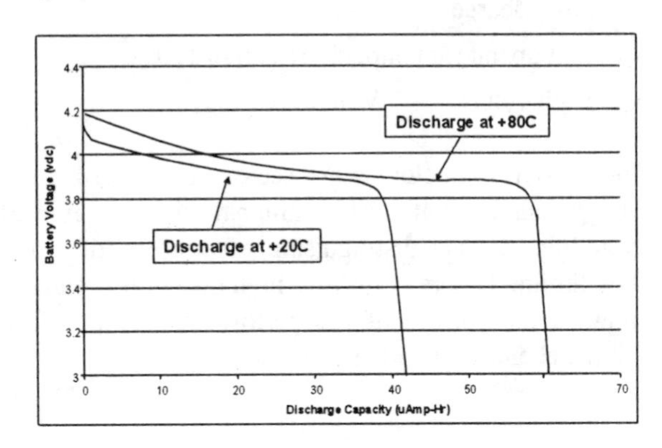

Figure 8 - Battery Discharge at 20° and 80° C

ITN'S FLEXIBLE POWER MANAGEMENT

Flex-circuits, which employ a flexible copper-on-polyimide printed circuit in lieu of the traditional rigid glass/epoxy printed circuit boards, are rapidly gaining acceptance in consumer electronics as a means of providing sophisticated electronic functionality in a product-conforming package. ITN's design for TFB charge control implements the power control electronics using surface-mount components on flex-circuitry to achieve low profile and low mass in a flexible format compatible with the TFB. Per the battery characteristics described in the previous section, the controller must provide a 4.2 V overvoltage protection to prevent overcharging and a 3.5 V undervoltage protection to prevent overdischarging.

Flex circuit spacecraft power management features include:

- Charge Controller Logic - Split into 3 sections
  - Charge Circuit
  - Discharge Circuit
  - PV/Battery Switching Circuit (CCCV)
- Supply Constant Voltage to Spacecraft in Light & Dark Conditions
- Provide TFB Over & Under - Voltage Protection
- Charges TFB While Also Providing System Power
- Limit Charge and Discharge Currents
- Small, Lightweight and Flexible Circuit

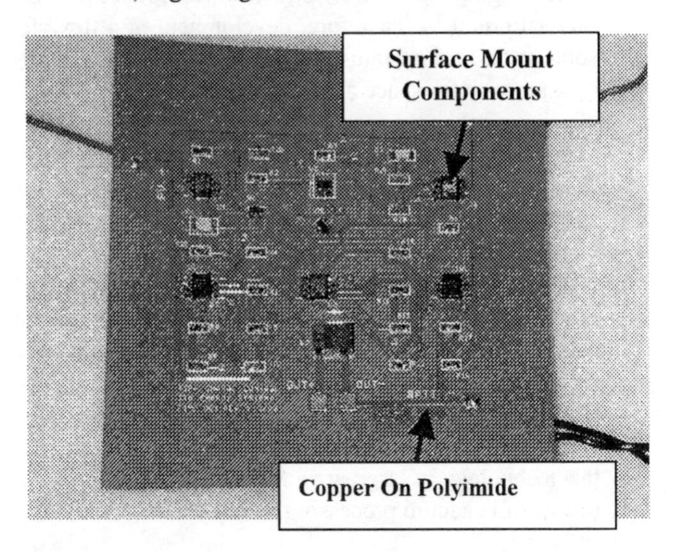

Figure 9 - ITN Has Implemented Charge Control Electronics Using Surface-Mount Components on a Flex-Circuit to Provide Low-Profile, Low Mass, and a Path to a Potential Flexible Integrated Circuits.

The Charge Controller shown in figure 9 was manufactured on a 0.127-mm polyimide substrate and is a double sided through-hole printed circuit. A top access layer, used as a solder mask and a bottom cover layer are both made of 0.025 mm polyimide, for an overall thickness of approximately 0.177 mm. This gives the controller good flexibility while still providing good solder-able and electrical functionality.

ITN'S FLEXIBLE INTEGRATED POWER PACK (FIPP)

As Johnson Space Center and other NASA centers study long-term manned space missions, one of the keys enabling technologies is development of a livable, self-sustaining habitat. However, while this habitat must represent a safe haven, it also must be practical to launch, erect, and maintain. ITN believes that *integrated systems* represent an elegant, achievable means to these ends and is currently developing a number of ***thin-film power system technologies*** previously discussed, which could be integrated with the habitat skin to create a ***Multifunctional Habitat Skin***.

GSE's product is a flexible CIGS, monolithically integrated PV module capable of significant specific power generation due to its lightweight and efficiency. Because a battery is often required for a complete PV power system, it is only logical that marrying the flexible CIGS PV with a reliable, long life flexible battery would result in a lightweight self-contained power source. Therefore, ITN is developing a revolutionary integrated skin concept comprising:

- Flexible Integrated Power Pack (FIPP), a multi-layered laminate of
  - Flexible thin-film CIGS photovoltaics,
  - Flexible thin-film Lithium battery, and
  - Flexible power management and distribution electronics
- Direct integration of FIPP with the habitat's structural skin element

Key advantages of our FIPP concept, compared to a conventional power subsystem with separate solar array, battery, and power electronics components, include the following:

- Reduced launch mass, volume, and number of launches to field a given system
- Reduced assembly procedures; assembly of habitat also deploys the power subsystem
- Reduced mass of "parasitic/low value" structural elements; increased mass fraction for science, consumables, fuel, or margin

Figure 10 illustrates the FIPP concept integrated with a habitat module; it also shows a highly integrated skin concept with additional RF and health-monitor sensing functionality. ITN is developing thin-film technologies which could ultimately enable the "Future Concept", and the current program takes the first step by focusing on integration of the power system elements.

The innovation, and the challenge, forwarded in this development is the task of combining these individual technologies (Figure 11) into a compact, highly-integrated power system while maintaining the high levels of performance which makes each technology attractive. The design for FIPP uses a unit cell as a building block for integration to large area arrays.

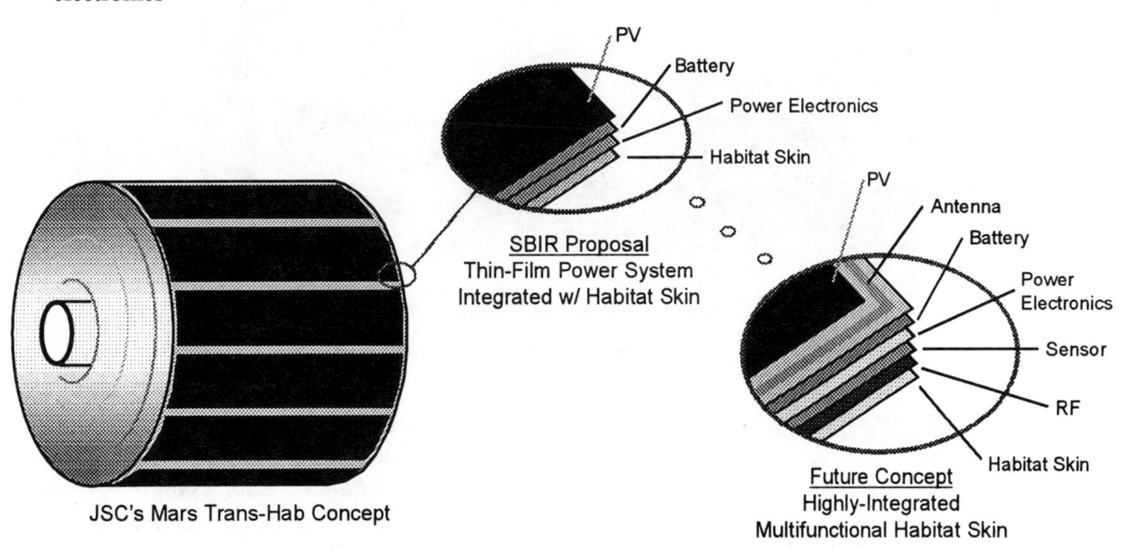

Figure 10 - ITN Envisions Habitat Systems with Significant Embedded Functionality, such as Our Flexible Integrated Power Pack (FIPP), a Fully Functional Thin-Film Power Subsystem

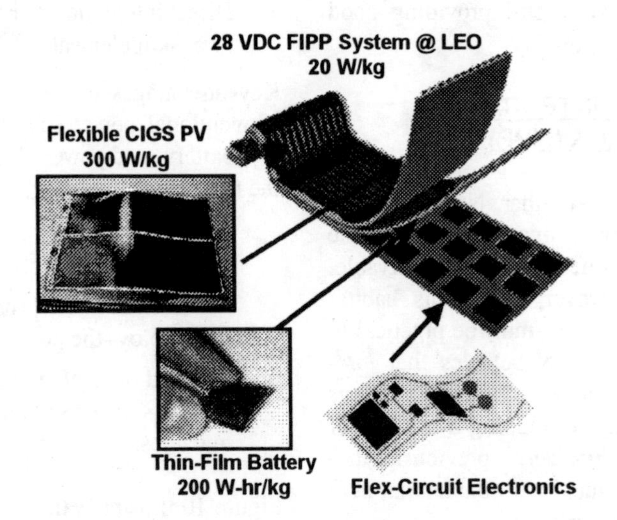

Figure 11 – FIPP Unit Cell Concept with Individual Flexible Power System technologies

A small unit cell demonstrator was built with ITN manufactured thin film and flexible components, shown in Figure 12. Three LCD volt meters shows (from left to right) that the PV has output voltage (12.94 volts), the FIPP has output volts (5.01 volts) and that batteries are being charged (8.42 volts, 4.2 volts x 2). A totalizer/hourmeter/timer with LCD readout is being used as the active load for the demonstrator and is showing the elapsed time.

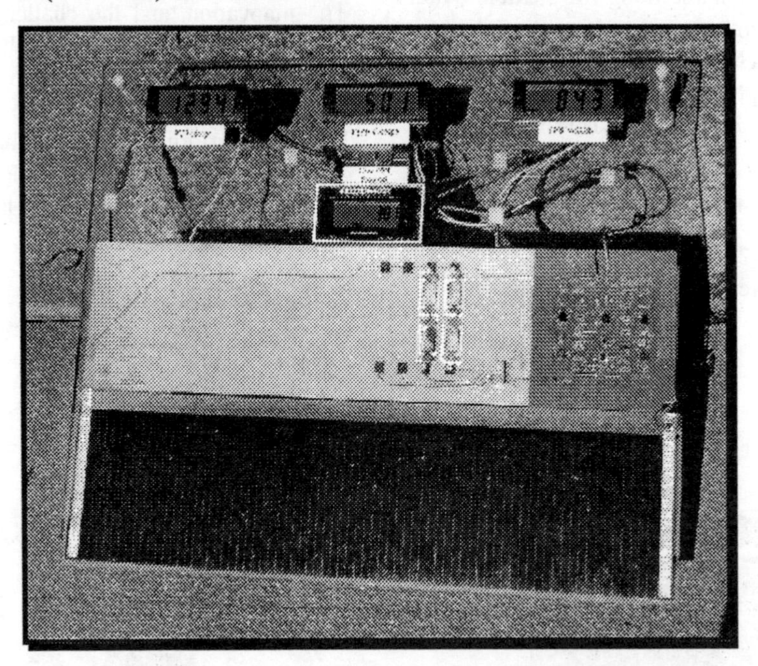

Figure 12 - Demonstrator unit Illuminated

SUMMARY

ITN's thin-film CIGS, TFB and flexible power management technologies present a stark contrast to conventional space power components i.e., rigid PV, heavy batteries, and large power electronics boxes. Each of these individual technologies provides a significant evolutionary step for its particular function. ***However, working together as an integrated system, they represent a potentially revolutionary new capability for a highly integrated, full-featured thin-film power subsystem for space.***

AIAA-2000-2926

IMPROVED SODIUM POOL TEMPERATURE CONTROL IN A SODIUM EXPOSURE TEST CELL

Michael Schuller*, Brad Fiebig†, Patricia Hudson, Alicia Williams
Center for Space Power, Texas A&M University, 3118 TAMU,
College Station, TX 77843-3118

ABSTRACT

In this paper we discuss a design change to the sodium exposure test cell (SETC) developed by JPL for evaluating alkali metal thermal to electric conversion (AMTEC) electrodes. This change, the addition of an antechamber to contain the sodium pool, was made to improve the control of the sodium pool temperature in the test, in order to improve the consistency and repeatibility of the electrode performance measurements. Experimental results and post test analysis showed that the change was very successful in controlling the location and temperature of the sodium pool. Using heater tape capable of higher temperature operation, we were able to duplicate electrochemical impedence spectroscopy results from earlier SETCs. These results indicate that when earlier SETC sodium pool temperatures were measured at 210-240 °C, the effective sodium pool temperature was actually 385-400 °C. The change in the overall length of the experimental apparatus also moved the specimens into a region in the oven with a flatter temperature profile, reducing the temperature gradient between the samples from 50-60 °C to 10-20 °C.

INTRODUCTION

Alkali metal thermal to electric conversion (AMTEC) is a high efficiency device for directly converting heat to electricity, first described by Weber in 1974.[1] AMTEC operates as a thermally regenerative electrochemical cell by expanding sodium through the pressure differential across a sodium beta" alumina solid electrolyte (BASE) membrane.[2]

While AMTEC technology is still being developed, laboratory devices have achieved efficiencies as high as 19% and system design studies indicate that efficiencies as high as 30% are achievable in the near term and 35% or more may be possible. AMTEC can provide all the advantages of a static power system (low vibration, redundancy, no wear) at efficiencies normally achieved only in dynamic systems. ARPS system designs using AMTEC have shown 27% cell and 23% system efficiencies[3], while laboratory experiments with developmental ARPS type cells have achieved 16% efficiencies.[4]

The high sodium activity gradient across the AMTEC electrolyte provides the device with high open-circuit cell voltages, approaching 1.6V, but in order to produce sufficient current densities at useful voltages for high efficiency operation, internal resistances have to be minimized. These resistances include pressure losses due to sodium flow through the device, contact and sheet resistances, and the potential-dependent resistance, which has been defined as the apparent charge transfer resistance, R_{act}. Realizing that the impedance to sodium flow and charge transfer contribute to the internal cell resistances and must be minimized, the theory of the kinetics and transport at AMTEC electrodes was developed. The morphology factor, G, and normalized exchange current density, B, were developed at this time.[5]

Considering the low pressure of the sodium on the cold side of the cell and the size of the pores encountered in thin AMTEC electrodes, the sodium transport was determined to be by free molecular flow, although other forms of transport have since been identified. Using a modified formula for pressure loss in a short tube, the Knudsen flow coefficient could be explicitly determined and from that the morphology factor, G, was derived.[6,7] The normalized exchange current density, normalized to the sodium collision rate and reaction rate at unit activity sodium, provides a measure of electrode performance which is essentially independent of electrode temperature in the range of 900-1200 K, and is a sensitive measure of the interfacial impedance.[7] The significance of these two parameters lies in their ability to be calculated using established electrochemical techniques and to provide insight into the charge transfer and mass transport characteristics of AMTEC electrodes.

The electrodes, particularly the cathode, are a key component in an AMTEC system. The ability of the electrodes to efficiently oxidize and reduce sodium and

---

*Assistant Director, Center for Space Power, AIAA Member

†Graduate student, now Research Scientist at Lynntech, College Station, TX 77840

transport both electrons and neutral sodium to and from the reaction sites is crucial to achieving high efficiency in an AMTEC system. We have been working on understanding the degradation of AMTEC electrodes and improving their performance. The sodium exposure test cell (SETC) is our key test device.

THE SODIUM EXPOSURE TEST CELL

The SETC is a simple, low cost test designed to measure the key electrochemical performance parameters of an AMTEC electrode.[7] Figure 1 shows the original layout of a SETC. Since the Advanced Radioisotope Power System program's AMTEC cell used Nb-1Zr as the structural material, the titanium liner was replaced by a Nb-1Zr liner.

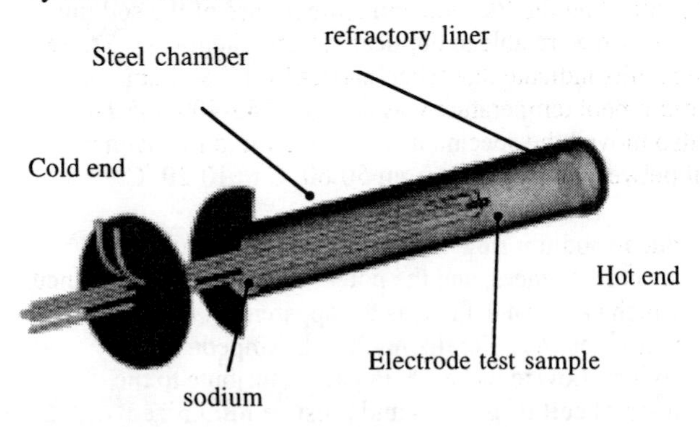

Figure 1. Cut-away of the Sodium Exposure Test Cell

The SETC consists of a single or double-walled stainless steel tube chamber evacuated and then heated at one end to AMTEC operating temperatures (900K-1200K) and kept at condenser temperatures (400K-700K) near the flange. At the flange a sodium pool provides a vapor pressure of the same order of magnitude as that found on the low-pressure side of AMTEC cells. The chamber is lined with a refractory metal to remain consistent with actual AMTEC cells and prevent volatiles from the stainless steel, such as chromium and manganese, from reacting with the test samples. The test sample, which is placed near the hot end, consists of a cylindrical β"-alumina electrolyte with four electrode bands approximately 0.25 cm wide deposited on the outer surface and separated by .25 cm spaces. Molybdenum leads are attached to each electrode band and threaded through electrically insulated feedthroughs at the flange. Thermocouples are placed at the hot end of the chamber and in the sodium pool.

Two types of measurements are made on an SETC sample, controlled potential current-voltage curves (iV curves) and Electrochemical Impedance Spectroscopy (EIS). IV curves are created by applying a potential difference between two electrode bands and measuring the resulting current flow between them. Based on these curves, a limiting current, which is used in the calculation of *G*, can be measured.

EIS is accomplished by applying a small potential, generally 5-10mV, to the electrode bands, and alternating the current, initially at high frequency, ~65kHz, reducing it in steps to low frequencies, ~0.1Hz. The resulting impedance is plotted on the complex plane producing a semi-circular curve with two real axis intercepts. From AMTEC electrochemical theory, the difference between the high and low frequency intercepts can be interpreted as the apparent charge transfer resistance, R_{act}, of the electrode, which is a resistance including both charge transfer and sodium transport effects. The high frequency intercept value is interpreted as the series resistance of the electrode, electrolyte and connecting leads. The low frequency intercept includes those factors and the impedence due to charge transfer and diffusion.

EXPERIMENTAL PROCEDURE

In order to determine the ability *B* and *G* have to predict AMTEC electrode performance, accurate measurement of the quantities needed to calculate these parameters is essential. Looking at the equations for both *B* and *G*, Eqs.[1] and [2], it can be seen that accurate measurement of both temperature and pressure at the electrode, (T_{el}, P_{el}), and of sodium pool, (T_{na}, P_{na}), is required, as well as accurate measurement of the apparent charge transfer resistance, R_{act}, and limiting current, j_{lim}.

$$B = \frac{T_{el}^{1/2}}{P_{el}P_{Na}} \frac{RT_{el}}{R_{act}F} \quad (1)$$

$$G = \frac{P_{Na}(T_{el}/T_{Na})^{1/2}F}{j_{\lim}(2\Pi MRT_{el})^{1/2}} - 1 \quad \frac{8\Pi}{3} \quad (2)$$

R is the gas constant, F is the faraday constant, and M is the molecular weight of sodium. In SETCs without a sodium pool containment extension, it has been suggested, based on post test observations of the location of the sodium pool, that the lack of containment of the sodium pool allows the sodium to wick along the walls of the chamber and acquire a temperature gradient, preventing accurate temperature readings and, therefore, accurate sodium vapor pressure calculation.

Effect of Na pool containment on B and G values in an SETC

To determine the effect of sodium pool containment on the calculation of *B* and *G* in an SETC, an extension was designed which allowed the sodium to be confined while maintaining the single vapor space of a conventional SETC. The new configuration is shown in figure 2. This configuration allowed for a more confident temperature measurement of the sodium pool and, therefore, a more accurate vapor pressure calculation.

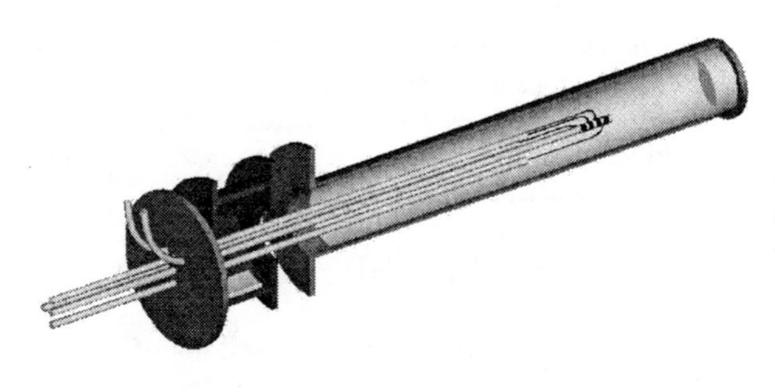

Figure 2. SETC with sodium pool extension.

In order to show the effect of the containment extension, similar electrode test samples, from SETC's with and without the containment extension, were compared. *B* values were calculated for both SETC configurations. In looking at the equation given for *B*, Eq[1], it was reasoned that, for an SETC, since the temperature and pressure at the electrode were held essentially constant, Eq[1] could be reduced to,

$$B = \frac{K}{P_{Na}^{1/2} R_{act}} \quad (3)$$

where K is a constant. In looking at calculated values of R_{act} from SETC's both with and without the sodium containment extension, it was seen that the natural log of R_{act} had a linear relationship with the natural log of P_{na},

$$\ln R_{act} = -C \ln P_{Na} + C_{arbitrary} \quad (4)$$

where the negative sign is shown to indicate a negative slope. Rearranging Eq[4] and exponentiating yields,

$$R_{act} = C_{arbitrary} P_{Na}^{-C} \quad (5)$$

Subsituting Eq[5] for *Ract* in Eq[3], rearranging and taking the natural log yields,

$$\ln B = \left(C - \frac{1}{2}\right) \ln P_{Na} + \ln K \quad (6)$$

After plotting ln(*B*) versus ln(P_{Na}) for SETC's both with and without the extension, a clear trend was noted. SETC's without the extension showed ln(*B*) had a negative slope in relation to ln(P_{Na}), while the SETC's with the extension yielded positive slope with respect to ln(P_{na}). The variation of ln(*B*) with ln(P_{Na}) is shown in figure 3. The results from six SETC's are shown, two with the sodium containment extension and four without. From this, it can be seen that the important term in Eq[6], with respect to determining which trend is correct, is to determine whether C should be greater than or less than 1/2. To do this, the equation for R_{act}, Eq[7], was broken down, or de-convoluted, into the

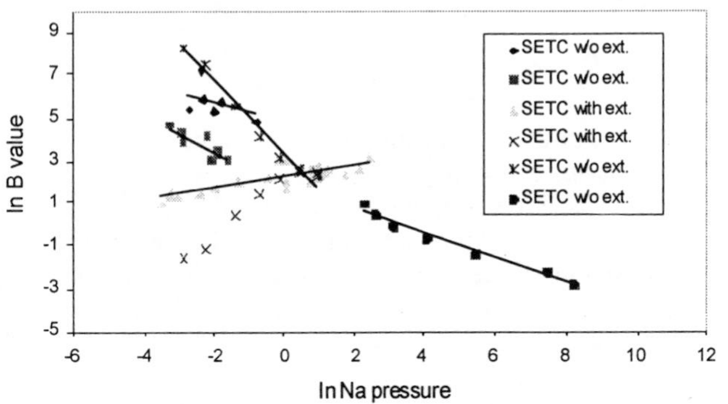

Figure 3. Plot of ln(B) vs ln(P_{Na})

contributions from both charge transfer resistance, R_{ct}, and diffusion resistance, R_D.

$$R_{act} = R_{ct} + R_D \quad (7)$$

Using fundamental electrochemical theory, an equation was derived for R_{ct} which related it to the sodium pressure. The derivation begins with the equation for the exchange current,

$$i_\circ = FAK^\circ C_{Na^+}^{(1-\alpha)} C_{Na}^{\alpha} \quad (8)$$

where, i_o is the exchange current density, *A* is the surface area of the electrode, F is the Faraday constant, K_o is the standard rate constant, *C* is the concentration of the associated subscript, and α is the transfer coefficient. Due to the ionic conductivity of the electrolyte at the electrode temperature, the concentration of the sodium ions can be considered constant, reducing the equation to,

$$i_\circ = KC_{Na}^{\alpha} \quad (9)$$

where, K is a constant. Using the ideal gas law the concentration of sodium can be expressed as,

$$C = \frac{P_{Na}}{RT_{Na}} \quad (10)$$

where R is the gas constant. Inserting Eq[10] into Eq[9], yields,

$$i_\circ = K\left(\frac{P_{Na}}{RT_{Na}}\right)^{\alpha} \quad (11)$$

The exchange current density can also be written as a function of the charge transfer resistance, Eq[12], which leads to the relation between the charge transfer resistance and the sodium pressure and temperature, Eq[13].

$$i_{\circ} = \frac{RT_{el}}{R_{ct}F} \quad (12)$$

$$R_{ct} = \frac{RT_{el}}{FK}\left(\frac{RT_{Na}}{P_{Na}}\right)^{\alpha} \quad (13)$$

Taking the natural log of Eq (13) yields,

$$\ln R_{ct} = \ln \frac{RT_{el}}{FK} + \alpha \ln RT_{Na} - \alpha \ln P_{Na} \quad (14)$$

Although there is a contribution from the sodium temperature, the magnitude of the contribution is minimal. For a temperature increase of 25%, the pressure increases by a factor of 100. Given that the transfer coefficient has been observed to be close to 1/2 in previous experiments[8], the slope of Eq[14] should approach -1/2.

Focusing on the diffusion resistance, we have,

$$R_D = \frac{k_f RT_{el}}{FP_{Na}} \quad (15)$$

where k_f is the Knudsen flow parameter, which is associated with pressure loss in the pores of the electrode. Assuming the Knudsen flow parameter is constant over the temperature range tested, which will be checked shortly, and taking the natural log yields,

$$\ln R_D = \ln \frac{k_f RT_{el}}{F} - \ln P_{Na} \quad (16)$$

The assumption that the Knudsen flow parameter was constant was justified only after a deconvolution of Nyquist plots yielded a linear relationship between $\ln(R_d)$ and $\ln(P_{na})$ and the slope approached 1, which suggests that the parameter is constant for the conditions tested. In figure 4, the natural log of the inverse of the resulting resistances from the deconvolution of the Nyquist plots are shown against the natural log of sodium pressure. The result verified that the normalized exchange current should increase as sodium pressure increased, not decrease as seen in SETC's without the sodium containment extension.

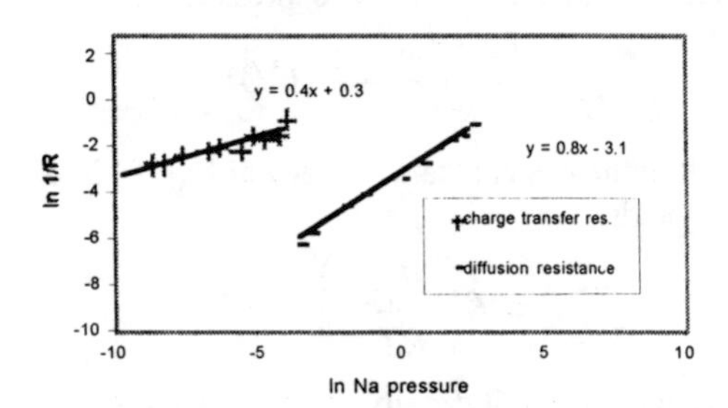

Figure 4. Plot of assumed diffusion resistance and assumed charge transfer resistance vs. sodium pressure

Taking the slopes of both the R_{ct} and R_D yielded a slope greater than or equal to 1/2, and much closer to 1 for R_D. This information was then used to determine the effect of the containment of the sodium on the relationship of B with sodium temperature and pressure. The analysis showed that the slope of the function relating $\ln(B)$ with $\ln(P_{na})$ was positive. The breakdown of the apparent charge transfer resistance into its contributions from charge transfer and diffusion revealed that the impedance data from the electrodes could be deconvoluted into a separate charge transfer resistance and diffusion resistance.

RESULTS AND DISCUSSION

The variation of ln(B) with the natural log of sodium pressure is shown in figure 3, which is repeated below. The results from six SETC's are shown, two with the sodium containment extension and four without. From this plot, it can be seen that SETC's with the extension exhibit a positive slope, while the SETC's without the extension show a negative slope.

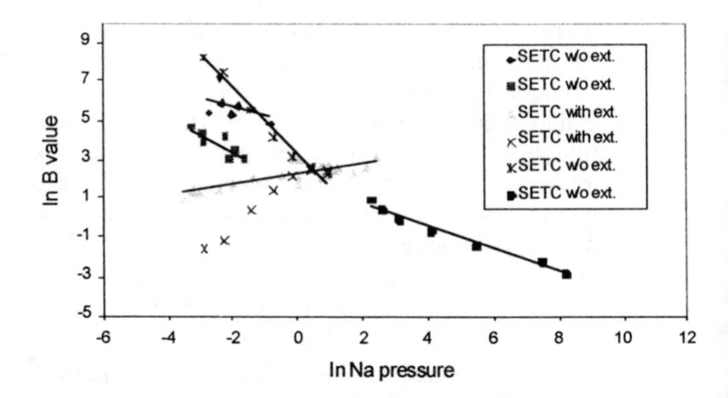

Figure 3. Plot of ln(B) vs ln(P_{Na})

Taking note of the differences in the slope of the SETC's without the sodium containment, it is possible that this is a result of the varying degrees of error in sodium pool temperature measurements, directly affecting the value of the sodium pressure. Causes of the error in measurement include differing levels of wicking due to differences in the surface condition of the SETC chamber, and the inability to accurately place thermocouples in the sodium pool.

A finite element model of the SETC has been developed at the Jet Propulsion Laboratory to model the electrical performance of the test electrodes.[9] It can be used to visualize the current profile of the electrodes. This model can also be used to determine the performance of the electrodes as it relates to parameters such as temperature, pressure, and B and G.

Figure 5 shows the 2-probe iV curve of a WRh electrode in an SETC with sodium containment extension, along with an iV curve generated by the finite element model using parameters measured at the same time the data was taken. The close agreement between the predicted curve (the darker line in the first quadrant) and the experimental curve (lighter line) supports the claim for more accurate temperature measurements for an SETC with the sodium containment extension.

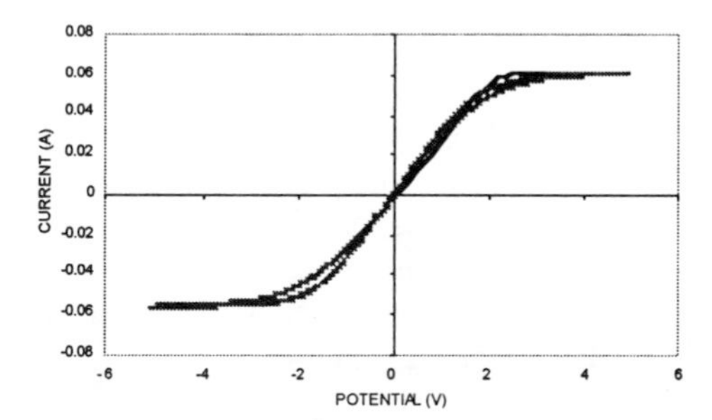

Figure 5. An iV curve for a WRh electrode in an SETC with sodium containment. An iV curve using the finite element model is also plotted.

Comparison of EIS curves

Figure 6 is an electrochemical impedence spectrum from a WRh electrode run in a SETC with a sodium pool extension. Figure 7 is an electrochemical impedence spectrum from a WRh electrode run in a SETC without a sodium pool extension. The two plots yield nearly the same value of R_{act}. The difference between the two is that figure 6 was measured using a three probe technique, which measures the resistance of a single electrode, while figure 7 was measured using a two probe technique, which measures the resistance of two electrodes simultaneously.

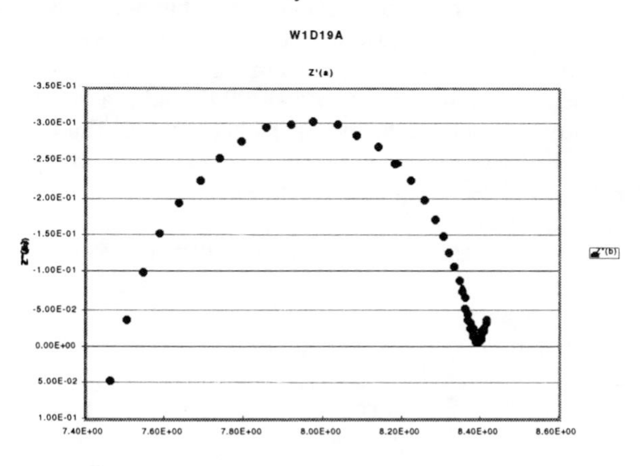

Figure 6. Electrochemical impedence spectrum, WRh electrode using sodium pool extension, T_{pool} 385 °C, T_{sample} 889-911 °C

SUMMARY AND CONCLUSIONS

The experimental and analytical results presented here show that the addition of a containment extension for the sodium pool of a SETC significantly improves the accuracy of the sodium pool temperature measurement. The extension also allows more confident control of the sodium temperature, which improves the confidence level in the electrode performance measurements made in the SETC.

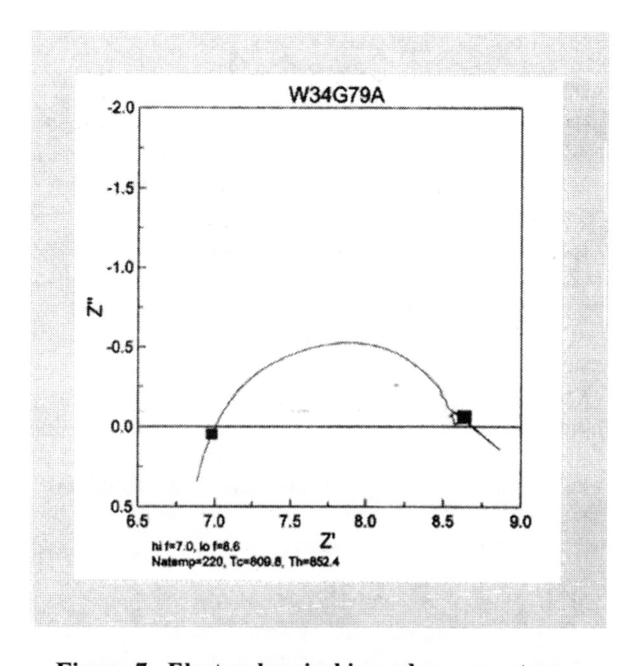

Figure 7. Electrochemical impedence spectrum, WRh electrode without sodium pool extension. T_{pool} 220 °C, T_{sample} 809-852 °C.

REFERENCES

1. Weber, "A Thermoelectric Device Based on Beta-Alumina Solid Electrolyte," Energy Conversion, 14, 1-8, 1974.
2. Cole, "Thermoelectric Energy Conversion with Solid Electrolytes," Terry Cole, Science, 221, 915 (1983).
3. Schock, et al., "Design and Performance of Radioisotope Space Power Systems Base on OSC Multi-tube AMTEC Converter Designs," 32nd IECEC, Honolulu, HI, p489, (1997).
4. Merrill, et al., "Vacuum Testing of High Efficiency Multi-BASE Tube AMTEC Cells," 32nd IECEC, Honolulu, HI, 1997.
5. Williams, et al., *J. Electrochem. Soc.*, **137**, 1709 (1990)
6. Bard, A.J., Faulkner, L.R., "Electrochemical Methods, Fundamentals and Applications," John Wiley and Sons, New York (1980).
7. Ryan, et al., "Sodium Exposure Test Cell to Determine Operating Parameters for AMTEC Electrochemical Cells," in Proceedings of the 33rd Intersociety Energy Conversion Engineering Conference, Colorado Springs, CO, ANS, 1998, paper 98-I335.
8. Williams, et al., *J. Electrochem. Soc.*, **137**, 1716 (1990)
9. Fiebig, Master's Thesis, Texas A&M University, 1999.

AIAA-2000-2927

Sensitivity of Performance and Degradation Measurements on AMTEC Electrodes in Sodium Exposure Test Cells to Experiment Conditions

Michael Schuller*, Brad Fiebig†, Patricia Hudson, Alicia Williams
Center for Space Power, Texas A&M University, 3118 TAMU,
College Station, TX 77843-3118

ABSTRACT

In this paper we discuss performance and degradation measurements made on AMTEC electrodes in sodium exposure test cells. We measured electrode temperatures, sodium pool temperatures, and apparent charge transfer resistance values, from which we derived normalized exchange current density (B) values for WRh, MoRe, Re, and Ir electrodes. Electrode temperatures ranged from 600 °C to 900 °C, while sodium pool temperatures ranged from 190 °C to 340 °C. In general, the B values for WRh, MoRe, and Re samples were in the same range, while the B values for Ir were lower. B values generally increased linearly with increasing sodium pool temperature.

INTRODUCTION

The Alkali Metal Thermal to Electric Converter (AMTEC) is a high efficiency device for directly converting heat to electricity, first described by Weber in 1974.[1] AMTEC operates as a thermally regenerative electrochemical cell by expanding sodium through the pressure differential across a sodium beta" alumina solid electrolyte (BASE) membrane.[2]

The electrodes, particularly the cathode, are a key component in an AMTEC system. The ability of the electrodes to efficiently oxidize and reduce sodium and transport both electrons and neutral sodium to and from the reaction sites is crucial to achieving high efficiency in an AMTEC system. We have been working on understanding the degradation of AMTEC electrodes and improving their performance. This paper will report results from metal electrodes run in sodium exposure test cells (SETCs) described in a companion paper[3].

The high sodium activity gradient across the AMTEC electrolyte provides the device with high open-circuit cell voltages, approaching 1.6V, but in order to produce sufficient current densities at useful voltages for high efficiency operation, internal resistances have to be minimized. These resistances include pressure losses due to sodium flow through the device, contact and sheet resistances, and the potential-dependent resistance, which has been defined as the apparent charge transfer resistance, R_{act}. Realizing that the impedance to sodium flow and charge transfer contribute to the internal cell resistances and must be minimized, the theory of the kinetics and transport at AMTEC electrodes was developed. The morphology factor, G, and normalized exchange current density, B, were developed at this time.[4]

Considering the low pressure of the sodium on the cold side of the cell and the size of the pores encountered in thin AMTEC electrodes, the sodium transport was determined to be by free molecular flow, although other forms of transport have since been identified. Using a modified formula for pressure loss in a short tube, the Knudsen flow coefficient could be explicitly determined and from that the morphology factor, G, was derived.[5,6] The normalized exchange current density, referred to the sodium collision rate and reaction rate at unit activity sodium, provides a measure of electrode performance which is essentially independent of electrode temperature in the range of 900-1200 K, and is a sensitive measure of the interfacial impedance.[6] The significance of these two parameters lies in their ability to be calculated using established electrochemical techniques and to provide insight into the charge transfer and mass transport characteristics of AMTEC electrodes.

The electrodes, particularly the cathode, are a key component in an AMTEC system. The ability of the electrodes to efficiently oxidize and reduce sodium and transport both electrons and neutral sodium to and from the reaction sites is crucial to achieving high efficiency in an AMTEC system. We have been working on understanding the degradation of AMTEC electrodes and improving their performance. The sodium exposure test cell (SETC) is a key device in our work.

---

*Assistant Director, Center for Space Power, AIAA Member
†Graduate student, now Research Scientist at Lynntech, College Station, TX 77840

Two types of measurements are made on an SETC sample, controlled potential current-voltage curves (iV curves) and Electrochemical Impedance Spectroscopy (EIS). IV curves are created by applying a potential difference between two electrode bands and measuring the resulting current flow between them. Based on these curves, a limiting current, which is used in the calculation of *G*, can be measured. This paper will not discuss our IV measurements and the G values derived from them.

EIS is accomplished by applying a small potential, generally 5-10mV, to the electrode bands, and alternating the current, initially at high frequency, ~65kHz, reducing it in steps to low frequencies, ~0.1Hz. The resulting impedance is plotted on the complex plane producing a semi-circular curve with two real axis intercepts. From AMTEC electrochemical theory, the difference between the high and low frequency intercepts can be interpreted as the apparent charge transfer resistance, R_{act}, of the electrode, which is a resistance including both charge transfer and sodium transport effects. The high frequency intercept value is interpreted as the series resistance of the electrode, electrolyte and connecting leads. This paper will focus on our results for the normalized exchange current density, *B*, and our interpretation of those results.

RESULTS

The following figures were compiled from B values calculated using electrode temperature, sodium pool temperature, and apparent charge transfer resistance values measured in 4 different SETCs. The B values have not been modified to account for the length of time the sample was at temperature and exposed to sodium. The particular SETC a sample ran in is identified by the number following the material identification in the figure title, for example WRh-14 indicates data from the tungsten-rhodium electrodes run in SETC 14.

Figures 1, 2, and 3 are for WRh electrodes run in SETCs 14, 15, and 16. SETCs 14 and 15 had very little oxygen contamination. SETC 16 had some problems with oxygen contamination, particularly at the end of its run, when a serious vacuum breach occured. The trend line superimposed on the data plot comes from a linear regression using all data points.

Figures 4 and 5 are for MoRe electrodes run in SETCs 14 and 17. SETC 14 had very low oxygen levels. SETC 17 did not have significant problems until the end of its run, when it appears to have suffered some oxidation due to a serious vacuum breach in SETC 16, which was running in parallel to SETC 17.

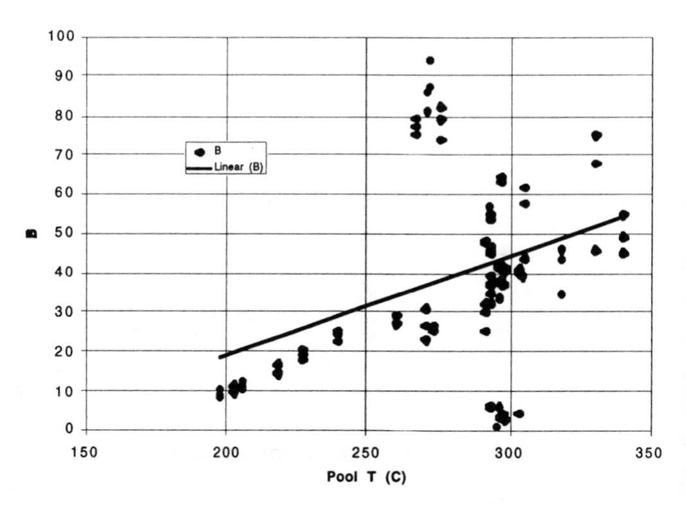

Figure 1. B vs Pool Temperature, WRh-14.

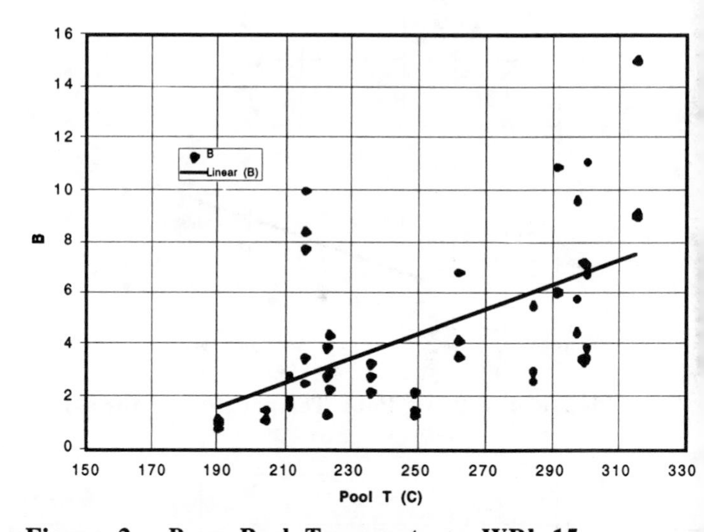

Figure 2. B vs Pool Temperature, WRh-15.

Figures 6 and 7 are for Ir electrodes run in SETCs 15 and 17. Figures 8 and 9 are for Re electrodes run in SETCs 15 and 17.

Figure 10 is a plot of B vs time for two WRh electrodes from SETC 14 between 400 and 900 hours of operation. Figure 11 is a plot of B vs time for three MoRe electrodes from SETC 14 between 400 and 900 hours of operation. This period was selected because the sodium pool temperature was roughly constant. The one zero value in Figure 11 represents a missing measurement.

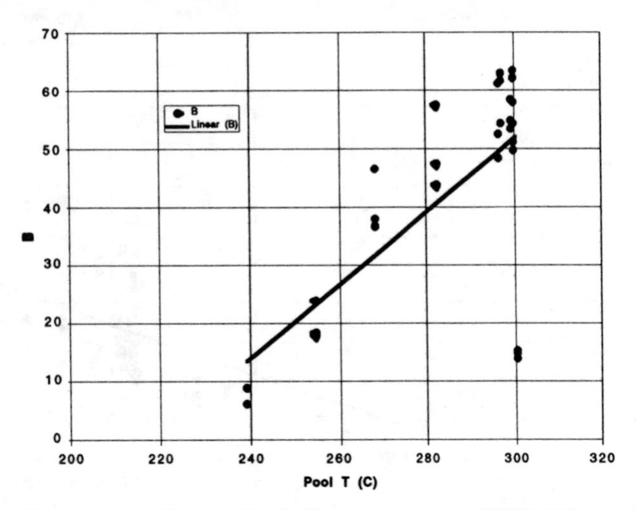

Figure 3. B vs Pool Temperature, WRh-16.

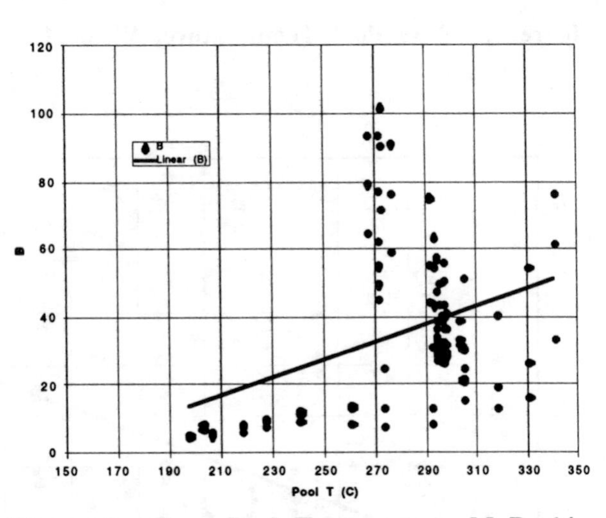

Figure 4. B vs Pool Temperature, MoRe-14.

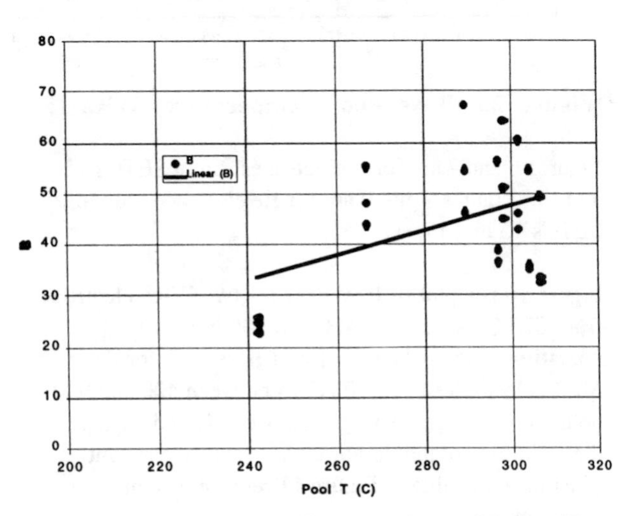

Figure 5. B vs Pool Temperature, MoRe-17.

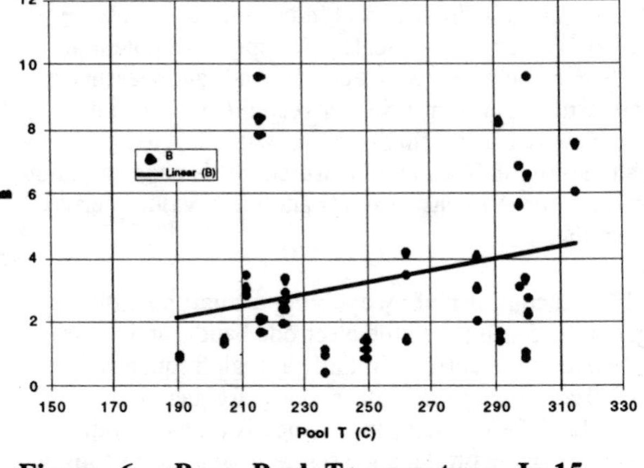

Figure 6. B vs Pool Temperature, Ir-15.

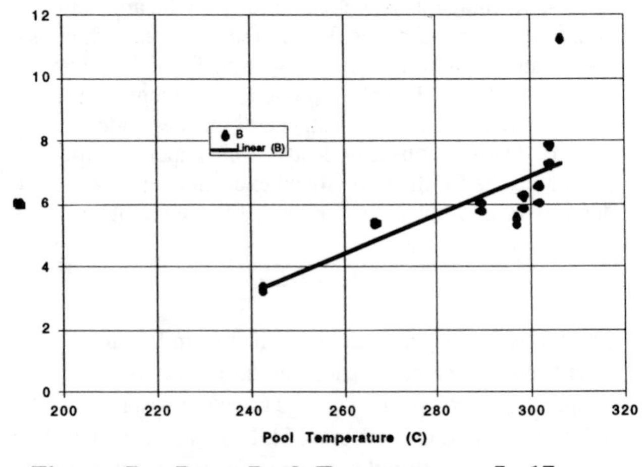

Figure 7. B vs Pool Temperature, Ir-17.

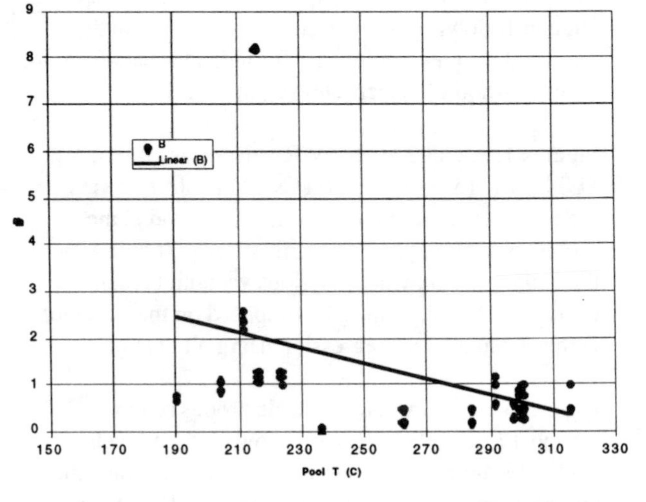

Figure 8. B vs Pool Temperature, Re-15.

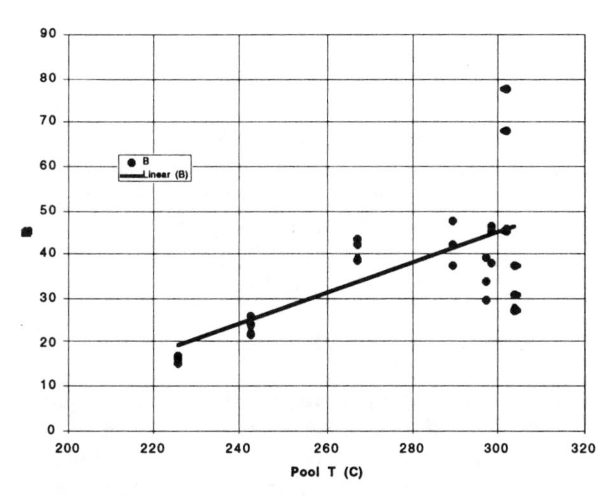

Figure 9. B vs Pool Temperature, Re-17.

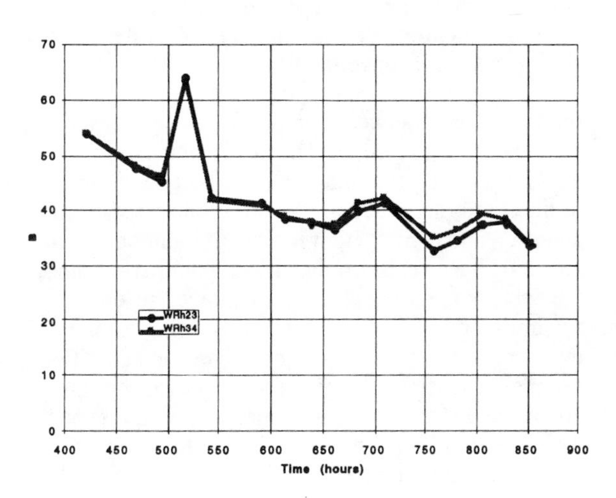

Figure 10. B vs Time, WRh electrodes.

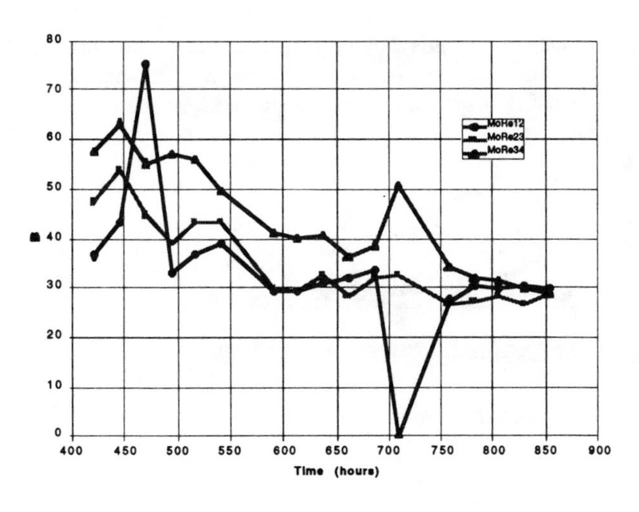

Figure 11. B vs Time, MoRe electrodes.

DISCUSSION

The values of B in Figures 1-9 show considerable variation. These variations exist between different materials, which is expected, but also between samples of the same material in different tests, which is not expected. More importantly, the value of B varies with the temperature of the sodium pool, which is contrary to the established theory for this parameter.

As seen in Figures 1-9, the value of B rises with increasing sodium pool temperature. The one exception to this trend, Figure 8, contains the results for the poorest performing electrodes tested. Given that other Re electrodes show increasing values of B with sodium pool temperature, we are confident that other factors caused the anomolous and poor performance of the electrodes in Figure 8.

To understand and explain these results, we examined a number of factors associated with the experiments. We will discuss four factors that seem to play significant roles in the results. These four factors are: 1) electrode area effects, 2) measurement or instrumentation effects, 3) interactions between samples, and 4) inconsistencies between samples.

The effects of electrode area on the calculated B value are the most significant factor in the variability of the results. The value of B is directly proportional to the electrode area, so errors in measuring the electrode area carry over into errors in B. Overlap and underlap between the electrode and the current collector complicate matters, because these mismatches cause variations in the actual area of electrode that is active during a test. In the case of overlap, the current collector can diffuse onto bare ß″ alumina, creating a de facto molybdenum electrode. In the case of underlap, the uncovered electrode area will contribute very little to the current, due to the large resistance that portion of the current that is generated in the uncovered area encounters.

Aside from uncertainties in the physical area of the electrode, the question of how much of the electrode is active, by which we mean fully involved in oxidation-reduction reactions of sodium, has a large effect on the calculated value of B. If active sites at the electrode/ ß″ interface are idle (empty) for times that are long compared to the reaction time, the result would be the same as if a smaller electrode area were used. These sites could be idle due to the low collision rate at lower sodium vapor temperatures, or due to the geometry of the sample.

The formulation of the B equation should compensate for the reduced sodium vapor pressure. It is possible, however, that the act of moving sodium through the electrode sample depletes the local sodium inventory, lowering the effective sodium pressure at the anode enough to affect the measurement of R_{ACT} and the calculation of B.

A more likely explanation for the change in B with pool temperature is that the geometry of the SETC samples causes the active area of the electrode to be less than the geometric area of the electrode, particularly at lower sodium pool temperatures, which translate to lower sodium vapor pressures over the specimens. Figure 12 shows the flow patterns for sodium ions in an SETC and in an AMTEC cell.

As shown in Figure 12, sodium ions in an SETC move along the surface of the ß″ alumina, rather than through it as in an AMTEC cell. This flow pattern would encourage sodium ions to react preferentially at sites near the inner edges of each electrode, particularly at lower vapor pressure conditions, when many sites are not constantly occupied. As the sodium vapor pressure rises, more current is pushed through the SETC, so more and more sites are constantly occupied, forcing the additional sodium ions to use more distant sites to react. Eventually, the entire area of the electrode would be in constant use, and the value of B would no longer change with temperature. None of the test setups discussed in this paper were capable of operating at sufficiently high temperatures for sufficiently long times to thoroughly test this hypothesis.

While the variable active area hypothesis explains variations in B with sodium pool temperature, it does not address the problem of variations in calculated B values between samples of the same material run in different tests. The other factors mentioned previously, however, could cause the observed variations.

Measurement or instrumentation effects are a candidate for explaining variations in performance between samples of the same material in different tests. Something as simple as a crimped or misplaced thermocouple could cause enough difference between the measured sodium pool or electrode temperature and the actual temperature to account for the difference in B values. This sensitivity is due to the exponential relationship between sodium pool temperature and sodium vapor pressure. A 10 °C difference between the measured sodium pool temperature and the actual sodium pool temperature causes a 20% to 30% change in the calculated value of B.

Since we run multiple samples in the same vacuum oven, interactions between samples are possible. If a particular specimen has or forms a material with a high vapor pressure, it can transfer material to other samples, where it might interact with the electrode or current collector, changing its properties. Post test examination, however, has not revealed evidence of significant cross contamination.

Inconsistencies between samples can also play a role in the variability of results. The key factors here are the quality of the ß″ alumina and the quality and uniformity of the sputtering process. Severe problems with adhesion and contact between the ß″ alumina and the electrode have been correlated with poor performance in the past. None of these samples showed severe separation problems, though they did show some evidence of poor electrode/electrolyte contact. The Re samples in particular, while not separated from the BASE, had lost considerable material during exposure. This material loss is most likely due to evaporation of volatile rhenium oxides.

Figures 10 and 11 show 500 hours degradation at roughly constant sodium pool temperature for WRh and MoRe electrodes. Despite some fluctuations, both materials show consistent reductions in B of 30-40%. The value of this data is reduced somewhat by the fact that the sodium pool heater failed twice during this test, causing substantial temperature swings each time.

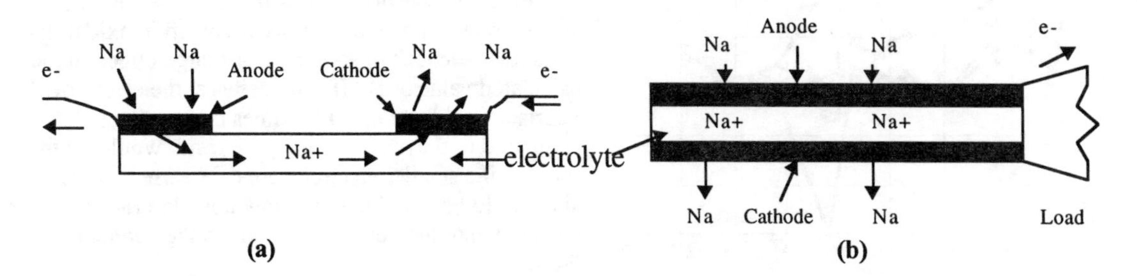

Figure 12. Schematic view showing the difference between an (a) SETC and (b) AMTEC electrode/electrolyte configuration.

SUMMARY AND CONCLUSIONS

WRh and MoRe performed very well. Re was inconsistent, performing extremely well at times, but extremely poorly at other times, very likely due to its susceptibility to oxidation. When intermixed with Mo, Re seemed to perform better, possibly because the Mo reacted with any oxygen present, thereby keeping the Re intact. Ir performed poorly, apparently because it is more mobile than we expected, based on data from the literature.

As a result of the findings presented above, we have made several changes in how we perform SETCs. First, we have reduced the number of samples in each SETC, to minimize the possible interactions between samples. Second, we have taken steps to reduce oxidation problems during bakeout and operation. Third, we are examining alternative sample geometries to reduce the "leading edge" problem. Fourth, we are examining ways to increase the pool temperature, so as to see if the calculated value of B does level off at some higher value of pool temperature. Fifth, we are redesigning our thermocouple fittings to reduce problems with misplacement and crimping.

REFERENCES

1. Weber, "A Thermoelectric Device Based on Beta-Alumina Solid Electrolyte," Energy Conversion, 14, 1-8, 1974.
2. Cole, "Thermoelectric Energy Conversion with Solid Electrolytes," Terry Cole, Science, 221, 915 (1983).
3. Schuller, et al., "Improved Sodium Pool Temperature Control in a Sodium Exposure Test Cell," Proceedings of the 35th IECEC, Las Vegas, NV, July, 2000, paper AIAA-2000-2926.
4. Williams, et al., *J. Electrochem. Soc.*, **137**, 1709 (1990)
5. Bard, A.J., Faulkner, L.R., "Electrochemical Methods, Fundamentals and Applications," John Wiley and Sons, New York (1980).
6. Ryan, et al., "Sodium Exposure Test Cell to Determine Operating Parameters for AMTEC Electrochemical Cells," in Proceedings of the 33rd Intersociety Energy Conversion Engineering Conference, Colorado Springs, CO, ANS, 1998, paper 98-I335.

AIAA-2000-2928

Post Test Examination of AMTEC Electrodes from Sodium Exposure Test Cells

Michael Schuller*, Brad Fiebig†, Patricia Hudson, Alicia Williams
Center for Space Power, Texas A&M University, 3118 TAMU,
College Station, TX 77843-3118

ABSTRACT

In this paper we discuss electron microprobe results from WRh AMTEC electrodes tested in sodium exposure test cells. We compare and contrast the morphology and elemental distribution of WRh electrodes in the as sputtered condition with those after up to 1250 hours in sodium vapor at temperatures of 800-900 °C.

INTRODUCTION

The Alkali Metal Thermal to Electric Converter (AMTEC) is a high efficiency device for directly converting heat to electricity, first described by Weber in 1974.[1] AMTEC operates as a thermally regenerative electrochemical cell by expanding sodium through the pressure differential across a sodium beta" alumina solid electrolyte (BASE) membrane.[2]

The electrodes, particularly the cathode, are a key component in an AMTEC system. The ability of the electrodes to efficiently oxidize and reduce sodium and transport both electrons and neutral sodium to and from the reaction sites is crucial to achieving high efficiency in an AMTEC system. We have been working on understanding the degradation of AMTEC electrodes and improving their performance. Physical changes in the structure of the electrodes may play a vital part in the degradation of their performance. This paper will describe and discuss observed physical changes in WRh electrodes run in sodium exposure test cells (SETCs) described in a companion paper[3].

RESULTS

Pre-test specimens

The first ten images were taken from specimens in the as sputtered state, before sodium exposure. For the most part, these were witness pieces, small beta" alumina rings sputtered at the same time as larger samples. Both top and side view pictures were taken, though this paper focuses on the top view, which more clearly shows physical changes in electrode morphology.

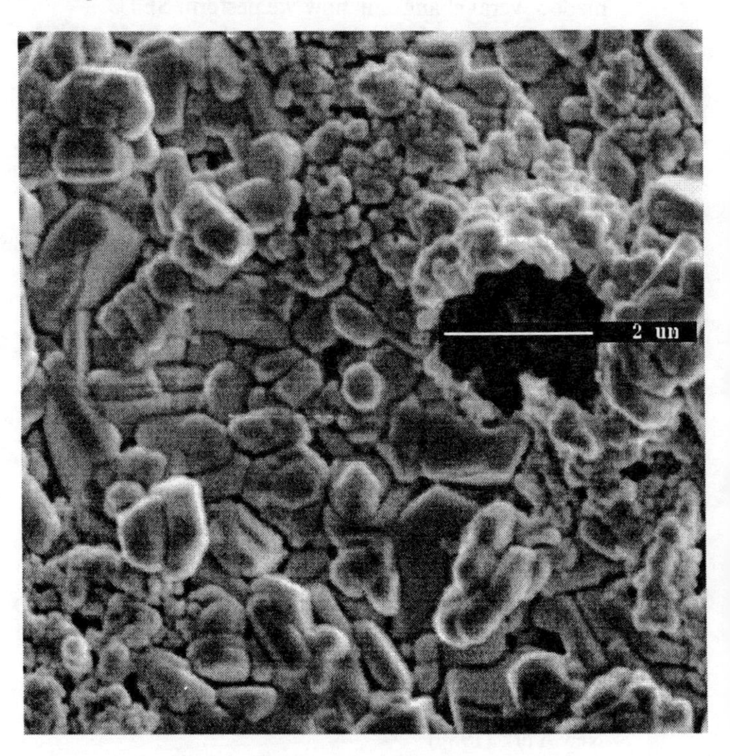

Figure 1. WRh electrode, as sputtered, top view, 5000 x SE image.

Figure 1 is a photograph of a WRh electrode, as sputtered, at 5000 x magnification, using secondary electron (SE) imaging. The boulder like nature of the WRh particles is evident. The hole in the electrode shows the beta" alumina structure beneath, and gives some sense of the thickness of the electrode, 0.7-1.2 microns in this case. This electrode was deposited using 50 watts of DC power, approximately 9 millitorr Ar pressure, and 30 minutes exposure while rotating at roughly 20 RPM.

Figures 2 and 3 are photographs of WRh electrodes, sputtered at Texas A&M and at JPL, respectively. Note the strong similarities in appearance, especially with respect to grain size and grain size distribution.

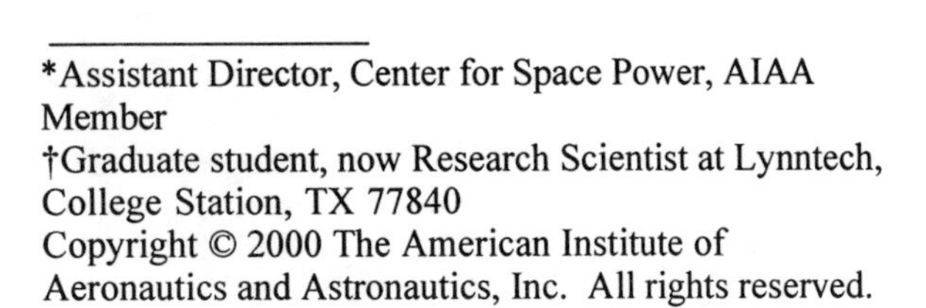
*Assistant Director, Center for Space Power, AIAA Member
†Graduate student, now Research Scientist at Lynntech, College Station, TX 77840

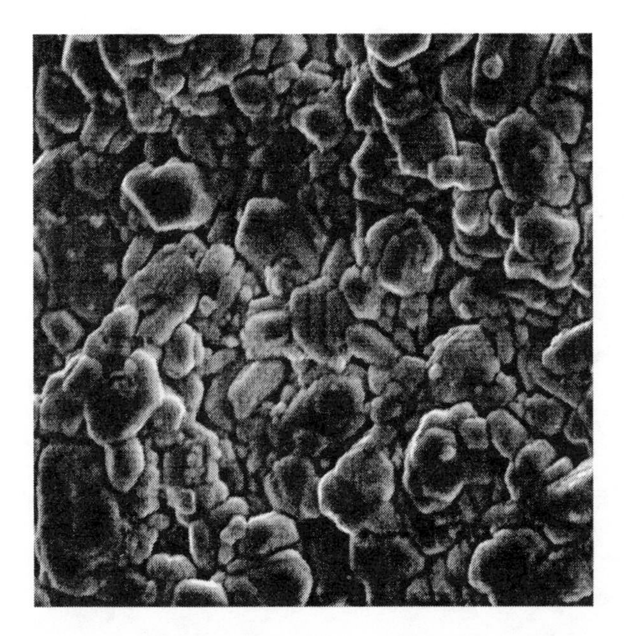

Figure 2. WRh electrode, as sputtered, 5000 x SE image.

Figure 4. WRh electrode, as sputtered, 5000 x SE image.

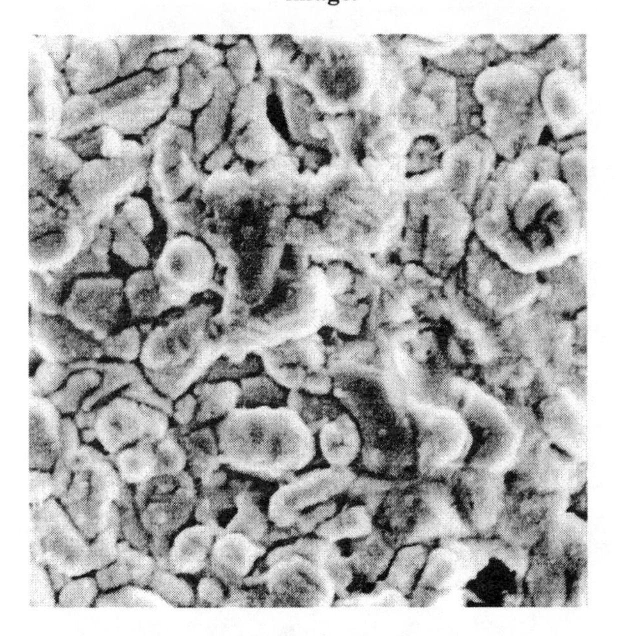

Figure 3. WRh electrode, as sputtered, 5000 x SE image. Prepared at JPL.

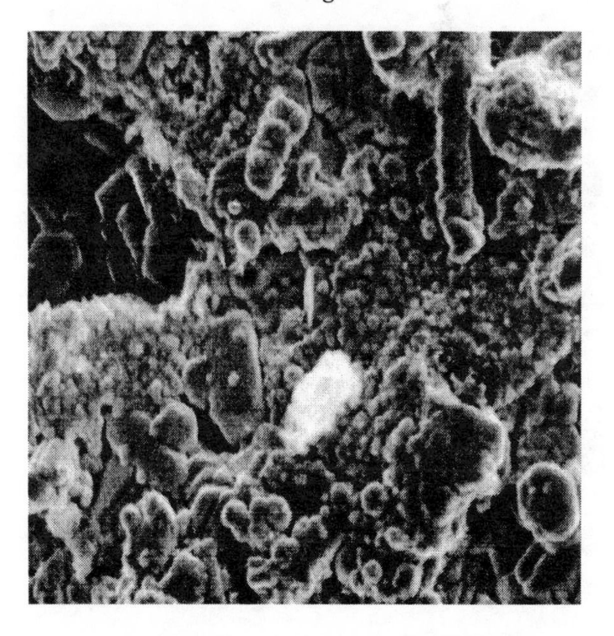

Figure 5. WRh electrode, as sputtered, 5000 x SE image.

Figures 4 and 5 are photographs of WRh electrodes sputtered for shorter times at higher power than those in Figures 1 and 2, though the time-power product was constant.

Figure 5 also has a hole in the electrode, showing the beta" alumina substrate and the thickness of the electrode, which is consistent with the depth in Figure 1. The main difference is the higher proportion of small grains in the shorter duration depositions.

Figure 6 is a photograph of an as sputtered WRh electrode at 2000 x magnification. It shows the region in the x-ray maps which follow. Specific x-ray maps which are absent from this set, but appear in later sets, were left out because the particular element, e.g. molybdenum, was not present in an energy dispersive spectrum taken of the sample. Oxygen is omitted because it tracks the aluminum distribution.

Figures 7-10 are the x-ray maps for aluminum, tungsten, rhodium, and sodium. The only item of note

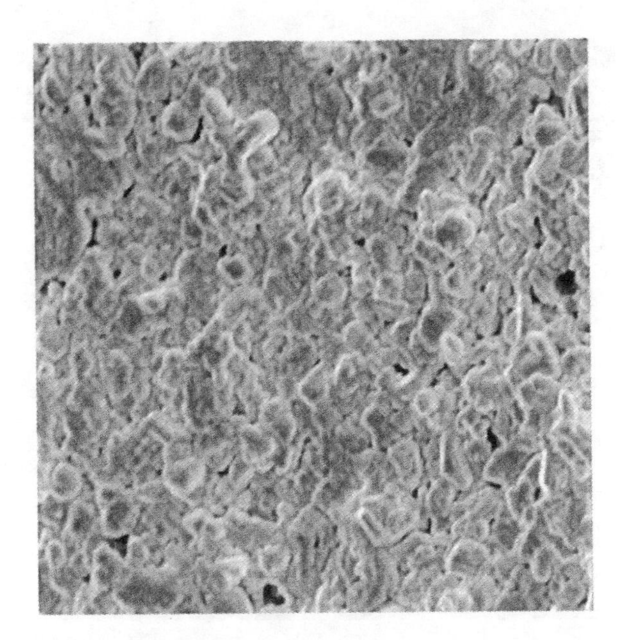

Figure 6. WRh electrode, as sputtered, 2000 x SE image. Photo is region in following x-ray maps.

Figure 7. Aluminum x-ray map.

Figure 8. Tungsten x-ray map.

Figure 9. Rhodium x-ray map.

in these maps is the somewhat segregated distribution of the tungsten compared to the rhodium.

Post-test specimens.

Table 1 summarizes the exposure conditions for the specimens reported here. The oxygen level is necessarily qualitative, based on observations of the behavior during test and the macroscopic condition of the test during disassembly.

Figure 11 is a photograph of a WRh electrode at 5000 x magnification after 360 hours of operation in a SETC at 800-940 °C. The most noticeable change is the "pancake-like" appearance of the grains, with the accompanying reduction in apparent porosity. Note, too, the very fine grains coating the larger grains. The small grains appear to be molybdenum.

Figure 12 is a photograph of the same spot as Figure 11, but at 2000 x magnification. Figures 13-17 are x-ray maps of the region in Figure 12 for aluminum, tungsten, rhodium, sodium, and molybdenum. The large bump in the lower right-hand corner is a contact point with the molybdenum current collector mesh. Its

Figures	Temperature (^{o}C)	Time (hours)	Oxygen Level
11-17	800-940	360	medium
18-19	750-850	340	low
20	800-850	460	low
21	800-850	150	high
22-29	800-850	1250	very low
30	800-850	300	low

Table 1. Specimen Exposure Conditions

Figure 10. Sodium x-ray map.

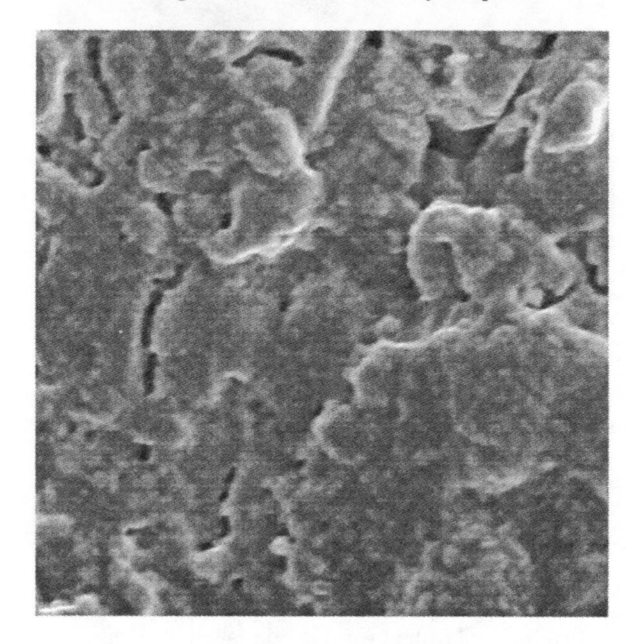

Figure 11. WRh electrode, 360 hours operation, 5000 x SE image.

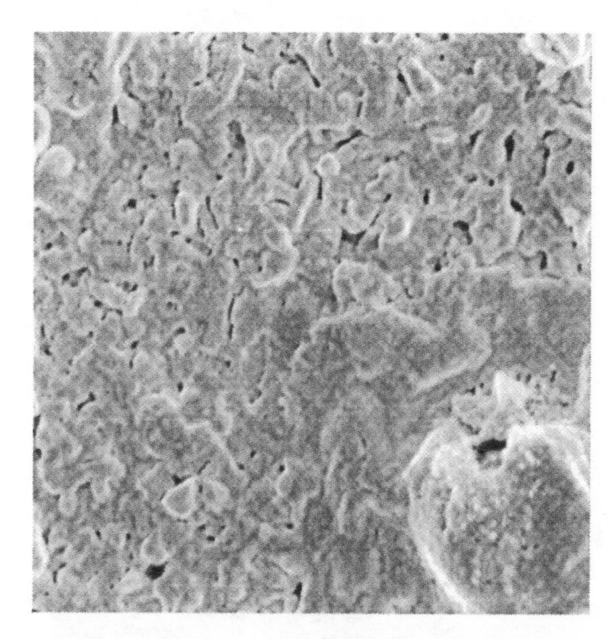

Figure 12. WRh electrode, 360 hours operation, 2000 x SE image. Image is region in following x-ray maps.

Figure 13. Aluminum x-ray map.

Figure 14. Tungsten x-ray map.

Figure 16. Sodium x-ray map.

Figure 15. Rhodium x-ray map.

Figure 17. Molybdenum x-ray map.

higher elevation is clear in the aluminum and tungsten x-ray maps, where the bulk of the mound shields the signal, causing a shadow effect. The molybdenum and rhodium clearly concentrated at the contact point, leading to depletion of rhodium in other parts of the electrode. The molybdenum, however, spread out over the rest of the electrode, too.

Based on earlier work done at JPL[4], the molybdenum would be expected to spread over an area 27 microns in diameter in 360 hours at 850°C. The actual area covered measures about 40 microns in diameter. The difference is most likely due to the formation of volatile oxides enhancing the mobility of the molybdenum. It is also possible that the mobility of molybdenum is higher on tungsten-rhodium than on molybdenum.

Figures 18-19 are photographs of WRh electrodes after 340 hours of operation in a SETC at 750-850 °C. These electrodes show far less physical change than those in the earlier figures. The reason for the difference, aside from the lower temperature of operation, is that the SETC they were in continued to pump down to higher temperature than previous ones did. The SETC containing the samples in Figures 18-19 were still being evacuated at 850 °C, while the earlier

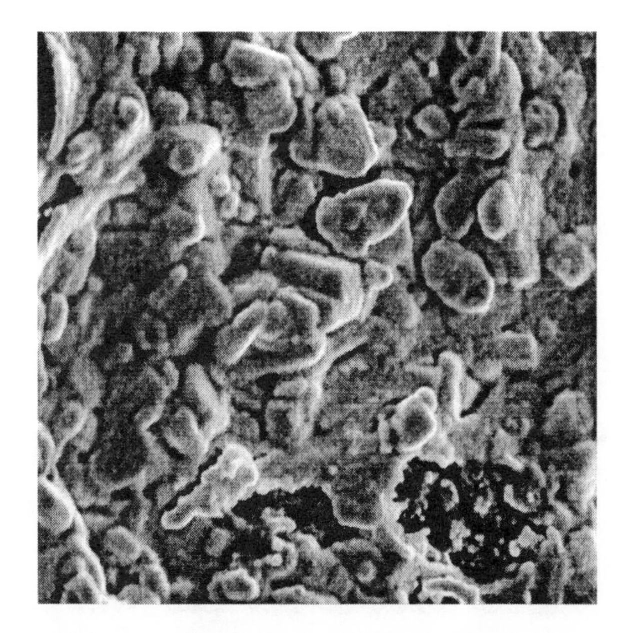

Figure 18. WRh electrode, 340 hours operation, 5000 x SE image.

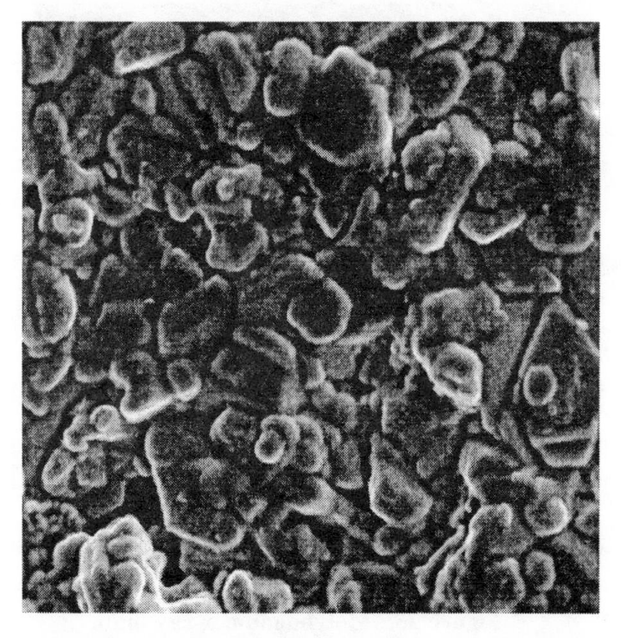

Figure 19. WRh electrode, 340 hours operation, 5000 x SE image.

samples' pump-out port plugged at 600 °C, causing a buildup of outgassed materials, especially oxygen.

Due to the lower oxygen levels the samples in Figures 18-19 encountered, they formed fewer volatile oxides, thus reducing the mobility of their materials. Figure 20 is a photograph of a WRh electrode after 460 hours of operation at 800-850 °C. This SETC was evacuated until the samples were at 850 °C. In addition, unlike most of the earlier SETCs, this one was shut down

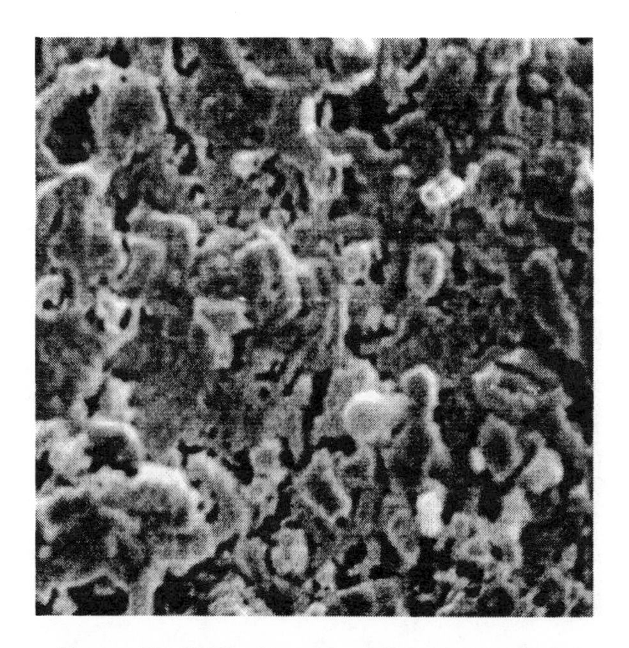

Figure 20. WRh electrode, 460 hours operation, 5000 x SE image.

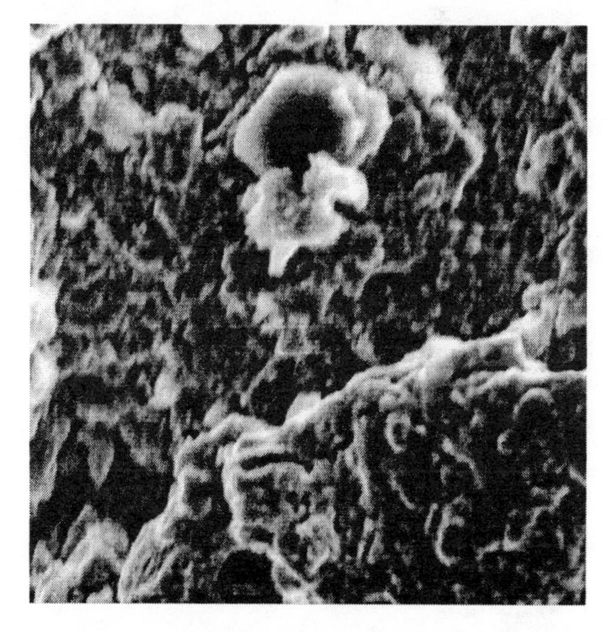

Figure 21. WRh electrode, 150 hours operation, 5000 x SE image.

while its samples were still performing at or near their initial levels.

Figure 21 is a photograph of a WRh electrode after 150 hours of operation at 800-850 °C. The SETC was shut down due to a serious vacuum breach; therefor, most of the observed changes were probably due to oxidation, rather than diffusion.

Figure 22 is a photograph of a WRh electrode operated for 1250 hours at 800-850 °C. This SETC did not

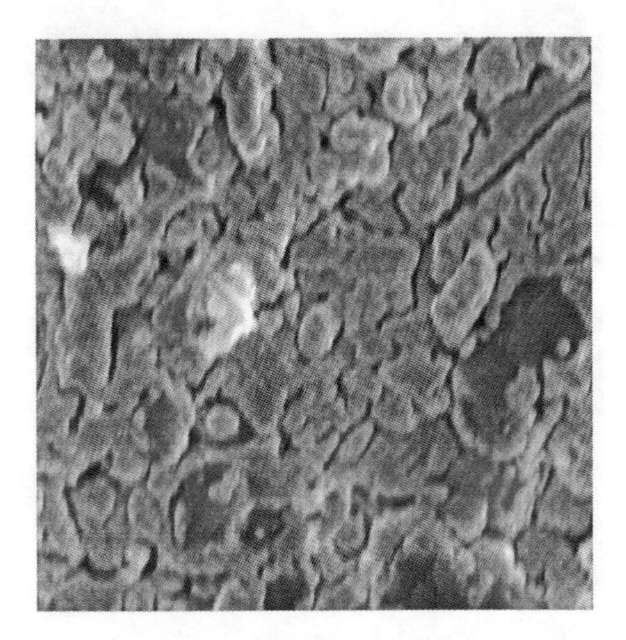

Figure 22. WRh electrode, 1250 hours exposure, 5000 x SE image.

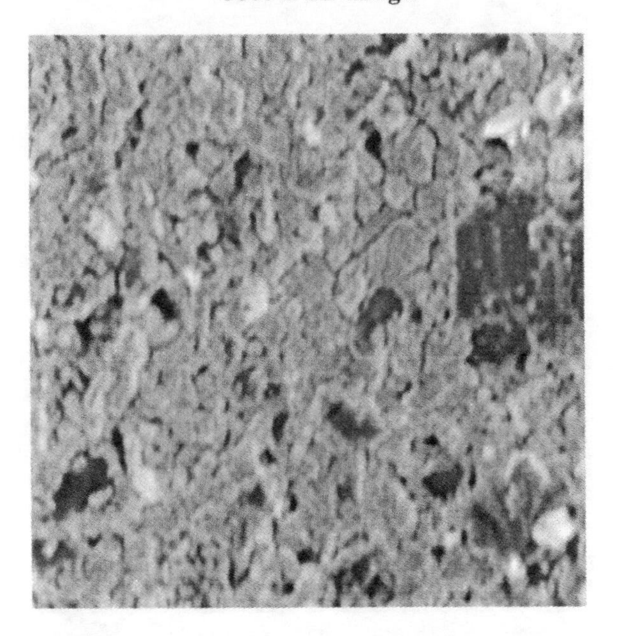

Figure 23. WRh electrode, 1250 hours exposure, 2000 x SE image. Image is region in following x-ray maps.

plug until it had been pumped out at 850 °C and it was shut down because a heater tape failed, not because of problems with the samples. The structure seen in Figure 22 is between that in Figure 19, which shows little change, and that in Figures 11 and 21, which show considerable change. Given the history of the SETC and the fact that the sodium in it was very clean upon disassembly, it is reasonable to conclude that the observed changes are due to diffusion and "normal"

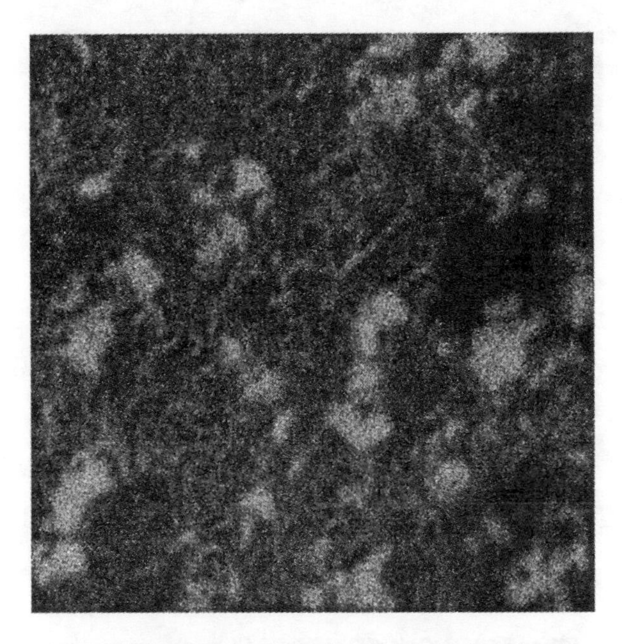

Figure 24. Aluminum x-ray map.

Figure 25. Molybdenum x-ray map.

interactions, such as would be found in an AMTEC cell.

Figure 23 is a photograph of the same spot as Figure 22, but at 2000 x magnification; it shows the region in the following x-ray maps. Figures 24-29 are x-ray maps for aluminum, molybdenum, rhodium, tungsten, chromium, and iron. The chromium and iron were originally thought to have come from evaporation and redeposition of material from the stainless steel vacuum shell containing the test samples, but, based on their pattern of distribution, we now believe the two large deposits came from steel tweezers used to manipulate

Figure 26. Rhodium x-ray map.

Figure 27. Tungsten x-ray map.

Figure 28. Chromium x-ray map.

Figure 29. Iron x-ray map.

the sample after its removal from the test. It is still possible that the dispersed chromium seen elsewhere in the x-ray map was due to evaporation and redeposition from the vacuum shell, but we have not seen chromium in any other samples, and it is unlikely that chromium came from stainless steel without manganese, which has a higher vapor pressure, also appearing.

Based on the data showing that oxygen levels in the test had a profound impact on the structure of the electrode, we modified our experimental procedure to include a 24 hour bake out period at 800-850 °C prior to sodium loading, to outgas as much oxygen as possible and remove it from the test article. Figure 30 is a photograph of a WRh electrode, operated for 300 hours at 800-850 °C in our most recent SETC, which shut down for reasons unrelated to this sample. The condition of the electrode indicates very little oxidation has taken place.

SUMMARY AND CONCLUSIONS

Small variations in the sputtering parameters, even when sputtering is performed at different facilities, have very little effect on the initial morphology of the electrodes. What variations do exist, primarily in grain

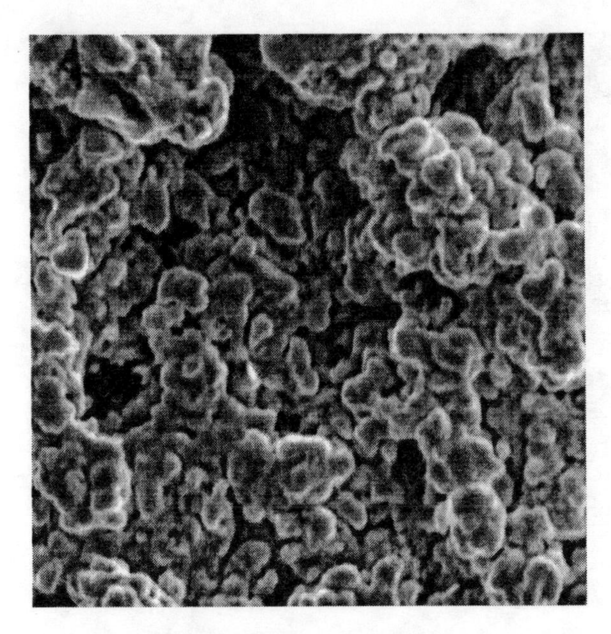

Figure 30. WRh electrode, 300 hours exposure, 5000 x SE image.

size, have little effect on performance and most likely disappear through coalescence in a very short time.

Oxygen levels inside the test apparatus have a significant effect on the morphology of the electrode. Outgassing at temperatures above the normal operating temperature of the test is necessary to reduce the oxygen concentration during testing.

The rhodium in the electrode tends to migrate to where the molybdenum current collector mesh touches the electrode, reducing the amount of rhodium present elsewhere in the electrode. At the contact points, the molybdenum diffuses and/or evaporates and redeposits onto the electrode. In regions away from contact points, molybdenum is also present, but at much lower densities.

REFERENCES

1. Weber, "A Thermoelectric Device Based on Beta-Alumina Solid Electrolyte," Energy Conversion, 14, 1-8, 1974.
2. Cole, "Thermoelectric Energy Conversion with Solid Electrolytes," Terry Cole, Science, 221, 915 (1983).
3. Schuller, et al., "Improved Sodium Pool Temperature Control in a Sodium Exposure Test Cell," Proceedings of the 35th IECEC, Las Vegas, NV, July, 2000, paper AIAA-2000-2926.
4. Ryan, et al., "Lifetimes of AMTEC Electrodes, Molybdenum, Rhodium-Tungsten, and Titanium Nitride," STAIF 2000, Albuquerque, NM, February, 2000, p. 1377

AIAA-2000-2929

THE POTENTIAL MARKET ANALYSIS OF A SMALL COGENERATION SYSTEM BASED ON STIRLING CYCLE

Wei Dong, Marco Lucentini, Vincenzo Naso
Dr. Wei Dong - dong@uniroma1.it
Dr. Marco Lucentini - luce@uniroma1.it
Prof. Vincenzo Naso Ph.D. - naso@uniroma1.it
University of Rome "La Sapienza"
Dept. of Mechanical and Aeronautical Engineering
Via Eudossiana, 18 – 00184 – Rome - ITALY

ABSTRACT

In the past decade there have been rapid improvements in small-scale combined heat and power technologies. With the deregulation of electric power, observers see little future for large, expensive central power plants that require miles of transmission wires and take decades to pay off. Future demand will be met by smaller power generation systems located closer to where the electricity is used. In the coming years, the world's electricity systems will change beyond recognition. This change is being driven by the need to address growing environmental challenge. The global policy response to climate change will probably be most important influence on the power sector in the future. Localized forms of the power production – based on cogeneration, renewable energies, fuel cells and other forms of distributed power system – have less environmental impact than conventional centralized electricity systems. In our paper, we emphases Stirling technology. The Stirling Engine (SE) is an heat machine which operates on a closed thermodynamic regenerative cycle, capable of very high efficiency and that can be used as a power system, refrigerator or heat pump. SEs may be used as power systems converting heat to work, as refrigerators for cooling and as heat pumps for elevating the temperature of heat drawn from an ambient temperature source.

1. INTRODUCTION

For years, power generation has been based on centralized system. These are large power stations, usually remote from electricity consumers, and requiring long distance, high voltage transmission network to deliver electricity to final consumers. Recent technology developments have brought a range of much smaller scale generation systems to the energy market. These include engines, turbines and micro-turbines and fuel cells. Comparison with centralized power generation these localized systems can produce electricity near to the consumers, or even in the consumer's house, office or factory. These localized power systems are more generally environmentally friendly, more flexible, more efficient and more cost effective than traditional centralized system.

In the coming years, cleaner and more localized forms of the power production will be necessary. Driving these systems into the market are the twin forces of electricity sector restructuring and technological development.

A very large number of small scale power producers provide flexibility and diversity that reinforces security of supply. In a conventional system, when a large power station fails, system security is more fragile. Finally, when power capacity is planed, with the new small scale technologies, it is now more efficient to build new plant incrementally on a small scale rather than, as in the past, constructing large plant with high capital costs in anticipation of a longer term increase in power demand. European experience over the last 20 years shows clearly that when new capacity requirements are planed, it is easy to overestimate.

2. COGENERATION SYSTEM – COMPETITIVE ENERGY SYSTEM

Cogeneration system offers local governments an opportunity to reduce the cost of providing electricity, heating and cooling to their buildings. In essence, the challenge for the electricity industry and for policymakers is to overcome the perceived challenges to existing structures of power production that smaller, cleaner cogeneration and renewable units can bring. A new way of thinking about producing electricity is needed if we are to achieve the full benefits of localized cogeneration systems.

Increasingly, energy consumers are becoming more sophisticated in their requirements for quality, reliability and cost, and want more than just the provision of the fuel and /or energy. Industrial consumers in particular realize that greater business benefits come from contracting out their energy needs. The emphasis shifts from maximizing energy sales to providing consumers with an package delivered efficiently at minimal cost to the environment. Such a package can consist of heating and cooling, lighting and automotive power at low cost. Cogeneration is an integral part of the concept of energy services and energy companies are in prime position to move towards this type of activity.

In generally, liberalization of energy market will almost certainly play a positive role. A more competitive system, at the very least, will provide new opportunities for market entrants to the power business. The liberalization of energy market and development of energy service provision point to more dynamic opportunities for cogeneration. The threat of climate change, growth in global power demand and the need to preserve security of fuel supply clearly imply that fossil fuel driven power production without utilization of the heat and at conventional efficiencies is no longer a realistic option.

With the present characterized by high fossil fuel consumption and future pointing to less fossil fuel intensive, decentralized power production, cogeneration provides a technology that uses fossil fuels with high efficiency in localized way.

3. COGENERATION TECHNOLOGY BASED ON STIRLING CYCLE

Main cogeneration cycles can be classed in following basic types according to the engines used:

- Cycle with steam turbines (Rankine cycle)
- Cycle with gas turbines (Brayton cycle)
- Cycle with internal combustion engine (Diesel or Otto cycle)
- Cycle with Stirling engine (Stirling cycle)

Here we emphases Stirling cycle. The Stirling Engine (SE) is an heat machine which operates on a closed thermodynamic regenerative cycle, capable of very high efficiency and that can be used as a power system, refrigerator or heat pump. SEs may be used as power systems converting heat to work, as refrigerators for cooling and as heat pumps for elevating the temperature of heat drawn from an ambient temperature source.

They have the following main characteristics:

- ***higher efficiency of operation***; SE has the highest theoretical efficiency permitted by thermodynamic laws. Of course the practical efficiency of SE is not so high, but the freedom of choosing the most appropriate working fluid and the opportunity of incorporating a sophisticate combustor with exhaust heat recovery allow the SE to approach its theoretical efficiency more closely than any other engine system
- ***multifuel capability***; a wide variety of heat sources can be used such as both conventional fuels (solid, liquid and gaseous fossil fuels) and new and renewable ones (biomass, waste), but also solar heat, vegetable oils, heat rejected by industrial processes, or stored in molten salts.
- ***low pollution and noise***; as heat is supplied continuously during operation and the combustion process can be easily controlled and efficient, the emission from a well designed SE system is definitively low.

4. RESEARCH AND DEVELOPMENT OF SMALL COGENERATION SYSTEM BASED ON STIRLING CYCLE

We are facing a vision of the future in which energy production installations are more numerous, smaller and cleaner than today. Energy consumers will deal with a single service supplier that can provide all their needs and charged on a single bill. Cogeration will be at the center of these developments.

To bring about the evolution of the vision portrayed here presents a major challenge to both energy and environmental policymakers. Too often, they are persuaded that there must be an economic trade-off if sustainable energy technologies are to be encouraged.

4.1. SMALL COGENERATION SYSTEM COMBINED WITH HEATING AND COOLING SUBSYSTEM

With Stirling technology, we develop a Small Cogeneration System combined with heating and cooling subsystem (named SECHP system) (see figure 1). Comparing with other cogeneration system, it has an auxiliary boiler and independent generator for dynamically supplying electrical and thermal energy (see figure 2). SECHP system is efficient, reliable, flexible and most importantly environmentally friendly. They are equipped with state-of-the-art technology to provide the best energy value and environmental controls. And while the technology is critical, understanding how it fits within the community is equally important. SECHP system takes extra steps to minimize the environmental impacts and to build a facility aesthetically in harmony with its neighbors.

4.2. ADVANTAGES OF SECHP SYSTEMS

In a typical SECHP System there are following advantages:

1. *Air emissions destruction*
2. *Power generation*
3. *Process cooling*
4. *Process heating*
5. *Low, or no electricity transmission costs and losses*

6. *Greater consumer control*

7. *Allows incremental, 'step-by-step' capacity increases*

4.3. BENEFITS OF SECHP SYSTEMS

SECHP system is one of the most effective technologies for rational use of energy, because it allows a considerable reduction of fuel consumption and produces a lower quantity of polluting emission. It can be considered an "environmental friendly" energy system. The energy advantage of cogeneration is measured in terms of the energy saving achieved when compared with the separated generation of the two forms energy. Normally, the electricity supplied to consumers comes from a plant that produces only that form of energy, while low temperature heat is generated locally in conventional boilers. Since in cogeneration system, electrical and thermal energy may be produced and exploited. The total efficiency in the case of separate production of thermal energy and electricity power is about 57%, while with the combined production it is possible to reach an efficiency of 86%.

4.4. POTENTIAL MARKET OF SECHP SYSTEM

Looking to the future of SECHP system, one can anticipate some of the developments for the coming years:

- SECHP system, possibly based on Stirling engine, is now very close to the market. It is designed for residential sector and will be ideal for counties with distributed gas networks and little or no district heating.
- For environmental reasons, renewable power generation should always have a priority and such technologies are almost always small in scale. If it is assumed these technologies, in particular those based on biomass, successfully reach the market in a major way, they will do so in strongly decentralized fashion.

The potential market of SECHP system is in following fields which frequently combine a sufficiently large heat load with a fairly constant electrical load.

1. *Hotels*
   Hotels often have a demands of heating in winter or cooling in summer for space for as much as 18 hours per day over long period. Furthermore the electricity demand is fairly constant, so SECHP units can be applied very effectively.
2. *Hospitals*
   Energy demand in hospitals tends to be continuously high throughout almost the full day and over most of the year. A high heat load is combined with a high electrical demand, and all electricity generated can be used on site. Furthermore , professional staff are often available to maintain and operate cogeneration plant. For these reasons SECHP can be very profitable installed in hospitals.
3. *Office buildings*
   A high space heating in winter is often combined with cooling requirements in summer in an office building. Because of sometimes high lighting load, electrical demand may be high and constant for 10-12 hours per day, over almost the entire year.
4. *Supermarkets*
   Large stores and supermarkets often a high demand for heating and cooling energy, while lighting levels are high for 10-12 hours per day. Most of the energy consumption is attributable to the electrical loads required to meet sale objectives bright lighting, preservation of foodstuffs stores and refrigerated displays.

The above fields are not complete and there are many other places which small SECHP system will also be feasible used.

REFERENCES

1 Kaarsberg, T.J.Bluestein, J.Romm, J. and A. Rosenfeld, 1988 "The Outlook for Small scale Combined Heat and Power in the U.S.," CADDET Energy Efficient Newsletter, June1998.

2 V. Naso: "La Macchina di Stirling", Ed. ESA, Milano 1991 (in Italian).

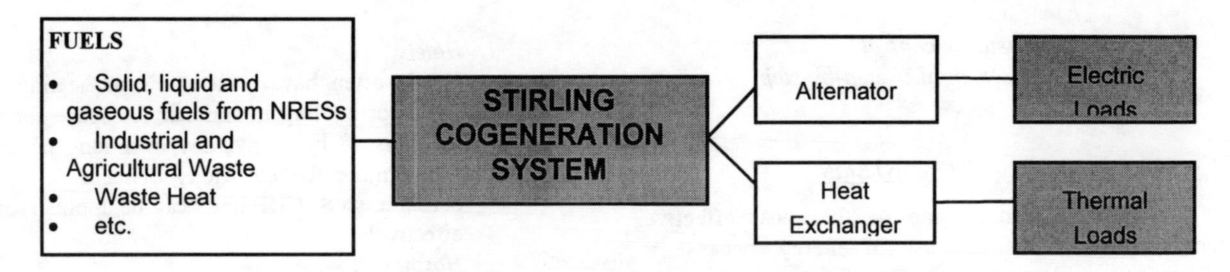

Figure 1 Stirling Cogeneration System

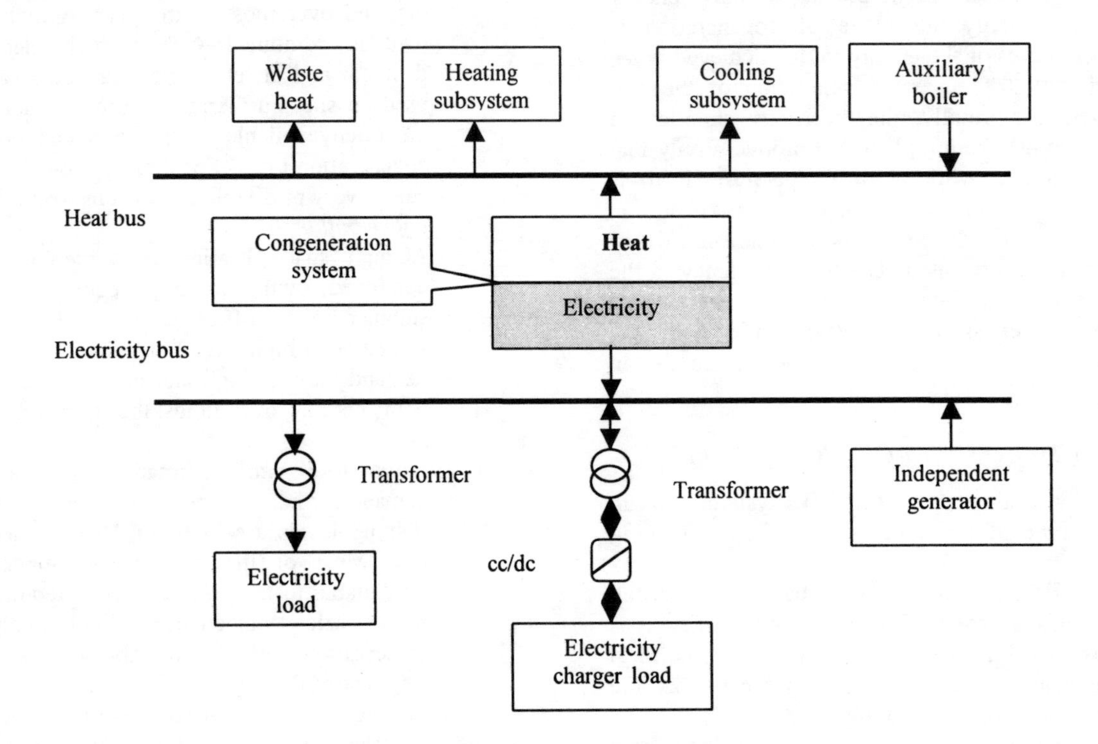

Figure 2 Energy flow of SECHP system

AUTHOR INDEX